Gallium Arsenide and Related Compounds 1993

Organizing Committee

H S Rupprecht (*Chairman*), R Diehl, (*Vice Chairman*), H J Boehnel (*Secretary*), G Meier (*Treasurer*)

Programme Committee

G Weimann (*Chairman*), G Packeiser (*Vice Chairman*), G Tränkle (*Secretary*), M R Brozel, H Dämbkes, K J Ebeling, T Foxon, V Graf, M Heiblum, K Iga, E Kapon, P Lugli, E Muñoz, D Pons, L Samuelson, J Schneider, B Schwaderer, N Yokoyama

Award Committee

J Magarshack (*Chairman*), C Hilsum, H Beneking, G Stillman, L Eastman, I Hayashi, H Yanai

International Advisory Committee

T Sugano (*Chairman*), H S Rupprecht (*Vice Chairman*), A Christou, L R Dawson, L Eastman, T Ikoma, J Magarshack, T Nakahara, D W Shaw, G E Stillman, M Uenohara, B L H Wilson

Sponsor

The Symposium was sponsored by: Deutsche Forschungsgemeinschaft, Bonn; Fraunhofer-Gesellschaft zur Förderung der angewandten Forschung e.V., Munchen; Alcatel SEL AG, Stuttgart; Daimler-Benz AG, Stuttgart; Siemens AG, München; Stiftungsfonds IBM Deutschland, Essen.

Gallium Arsenide and Related Compounds 1993

Proceedings of the Twentieth International Symposium on Gallium Arsenide and Related Compounds, Freiburg, Germany, 29 August–2 September, 1993

Edited by H S Rupprecht and G Weimann

Institute of Physics Conference Series Number 136
Institute of Physics Publishing, Bristol and Philadelphia

CODEN IPHSAC 136 1–842 (1994)

British Library Cataloguing-in-Publication Data

A catalogue record for this book is available from the British Library.

ISBN 0 7503 0295 X

Library of Congress Cataloging-in-Publication Data are available

Published by Institute of Physics Publishing, wholly owned by
The Institute of Physics, London

Institute of Physics Publishing, Techno House, Redcliffe Way, Bristol BS1 6NX, UK

US Editorial Office: Institute of Physics Publishing, The Public Ledger Building, Suite 1035, Independence Square, Philadelphia, PA 19106, USA

Printed in the UK by Galliard (Printers) Ltd, Great Yarmouth, Norfolk

GaAs Symposium Award and Heinrich Welker Gold Medal

The Gallium Arsenide Symposium Award was initiated in 1976; the recipients are selected by the GaAs Symposium Award Committee for outstanding research in the area of III–V compound semiconductors. The Award consists of $2500 and a plaque citing the recipient's contribution to the field. The Award is accompanied by the Heinrich Welker Gold Medal, established by Siemens AG, Munich, in honour of the foremost pioneer in III–V semiconductor development.

The winners of the GaAs Symposium Award and the Heinrich Welker Medal are:

Nick Holonyak	1976, for developing the first practical light-emitting diodes
Cyril Hilsum	1978, for contribution in the field of transferred electron logic devices (TELDs) and GaAs MESFETs
Hisayoshi Yanai	1980, for his work on TELDs, GaAs MESFETs and ICs, and laser diode modulation with TELDs
Gerald L Pearson	1981, for research and teaching in compound semiconductors physics and device technology
Herbert Kroemer	1982, for his work on hot electron effects, Gunn oscillators and III–V heterostructure devices
Izuo Hayashi	1984, for development and understanding of room temperature operation of DH lasers
Heinz Beneking	1985, for his contributions to III–V semiconductor technology and novel devices
Alfred Y Cho	1986, for pioneering work on molecular beam epitaxy and contributions to III–V semiconductor research
Zhores I Alferov	1987, for outstanding contributions in theory, technology and devices, especially epitaxy and laser diodes
Jerry Woodall	1988, for introducing the III–V alloy AlGaAs and fundamental contributions to III–V physics
Don Shaw	1989, for pioneering work on epitaxial crystal growth by chemical vapour deposition

Georg S Stillmann	1990, for the characterization of high purity GaAs and developing avalanche photodetectors
Lester F Eastman	1991, in recognition of his dedicated work in the field, especially on ballistic electron transport, δ-doping, buffer layer technique, and AlInAs/GaInAs heterostructures
Harry C Gatos	1992, for contribution to science and technology of GaAs and related compounds, particularly in relating growth parameters, composition and structure to electronic properties

In 1993 the GaAs Symposium Award and the Heinrich Walker Gold Medal were given to James A Turner. Jim Turner received his BSc degree in Physics from Sheffield University in 1960. He spent the whole of his working life with GaAs devices, first with Plessey Research and currently with GEC-Marconi Materials Technology Ltd. After starting his career with GaAs bipolar transistors, he began developing GaAs MESFETs, publishing the worldwide first experimental results as early as 1966. In 1968 he recognized the implications of velocity saturation in short gate length transistors, together with B L H Wilson.

Numerous publications on high frequency, low noise and high power GaAs devices followed, so did a paper on the first e-beam fabricated GaAs MESFETs in 1970. By then the GaAs MESFETs were clearly superior to the rivalling Si bipolar transistors—the work of Jim Turner and his group initiated major GaAs MESFET activities around the world. In 1974 Jim Turner fabricated the worlds first GaAs MMIC (published in 1976, together with R S Pengelly).

In the early and mid 1980's Jim Turner's group developed a manufacturable MMIC fabrication process, comparable to the multilevel metal technology for Si ICs, which is still in use today.

Jim Turner made many pioneering contributions to GaAs technology, promoting research, development and applications of GaAs MESFETs and MMICs, as well as other devices such as IMPATTs and Gunn-devices. These achievements were recognized by the Award of the Order of the British Empire (MBE) in 1981 and the Nelson Medal, a GEC award, in 1992.

Jim Turner has regularly published papers on GaAs MESFET related topics and written numerous chapters in books on GaAs device and MMIC technology. He is currently visiting Professor at the University of Cardiff, Wales.

Young Scientist Award

The International Advisory Committee of the GaAs Symposium has established a Young Scientist Award to recognize technical achievements in the field of compound semiconductors by a scientist under the age of forty. The Award consists of a cheque and a plaque citing the recipient's contributions.

The first Young Scientist Award was presented at the 1986 Symposium to Russel D Dupuis for his work in the development of organometallic vapour phase epitaxy of compound semiconductors. In 1987, the second Award was given to Naoki Yokoyama for his contributions to self-aligned gate technology for GaAs MESFETs and ICs and the resonant tunneling hot-electron transistor. The 1989 Young Scientist Award was presented to Russel Fischer for the demonstration of state of the art performance, at DC and microwave frequencies, of MESFETs, HEMTs and HBTs using (AlGa)As on Si. The 1990 Young Scientist Award was made to Yasuhiko Arakawa for his pioneering work on low-dimensional semiconductor lasers, showing the superior performance of quantum wire and quantum box devices. The next Award, in 1991, was presented to Sandip Tiwari for his work in the field of compound semiconductor devices, especially MESFETs and HBTs. The recipient of the 1992 Young Scientist Award was Umesh K Mishra for his pioneering and outstanding work on AlInAs–GaInAs HEMTs and HBTs.

The recipient of the Young Scientist Award of the 20th International GaAs Symposium, 1993 is Young-Kai Chen. He received the Award for significant advancements in the fields of high speed III–V electronic and optoelectronic devices.

Young-Kai Chen obtained his bachelor degree in Electronics Engineering from

the National Chiao Tung University, Hsinchu, Taiwan in 1976 and the MSEE degree from Syracuse University, Syracuse, NY in 1980. He received his PhD degree from Cornell University, Ithaca, NY in 1988.

From 1980 to 1985 Young-Kai Chen was working in the Electronics Laboratory of General Electric Company, Syracuse, NY on the modelling and design of Si and GaAs ICs for phase array applications. During his thesis research at Cornell University he realized the first 120 nm T-gate GaInAs/AlInAs MODFETs grown lattice-mismatched on GaAs with 100 GHz cut-off frequencies.

Since February 1988 he has been with the Solid-State Electronics Research Laboratory of AT&T Laboratories, Murray Hill, NJ, working on high frequency optoelectronic devices and ICs, non-equilibrium carrier transport and ultrafast optoelectronic processes.

In 1989 Young-Kai Chen fabricated InP/InGaAs HBTs with a 300 K cut-off frequency of 165 GHz, utilizing non equilibrium carrier transport. In 1990 he made a major advancement in high speed laser diodes, devising a colliding pulse mode locking scheme with InGaAsP lasers for very short pulses of 600 fs with 350 GHz repetition rates. He has made numerous contributions to the structuring and understanding of III–V electronic and optoelectronic devices, such as high power InP-MODFETs, InGaP/GaAs/InGaP HBTs and laser diodes.

Preface

The 20th International Symposium on Gallium Arsenide and Related Compounds was held in Freiburg i. Br., Germany, from August 29 to September 2, 1993. More than 330 participants from 22 countries came to Freiburg, making this 20th and last Symposium carrying 'Gallium Arsenide' in its name truly international. The future Symposia, beginning in 1994 in San Diego, will be on 'Compound Semiconductors'. We all hope that the future Symposia will be as successful as the past twenty conferences were, showing new frontiers for GaAs devices and circuits. The Symposium in Freiburg focused on practical applications of GaAs and III–V compounds in devices and circuits, both conventional and based on quantum effects.

The programme committee selected 138 contributions for oral and poster presentation, out of 165 submitted papers. Practically all regular papers are published in these proceedings. Nine invited papers are included, covering the present status of ultrafast GaAs transistors and integrated circuits, novel laser diodes and tunneling devices, as well as future technologies. These papers are arranged in 11 topical chapters in these proceedings, without distinction between oral contributions—both invited and regular—and posters. The final and 12th chapter contains the ten late news papers presented at the end of the conference. These voluminous proceedings would not have been possible without the dedicated work of the programme committee.

We would like to thank all members of the programme committee. Their judicious selection of the contributed papers and their thoughtful suggestions for invited speakers were indispensible for the success of the 20th GaAs Symposium. The conscientious and careful peer reviewing of the manuscripts by members of the programme committee and their colleagues is gratefully acknowledged. We also express our sincere thanks to the advisory committee and the award committees. They all helped to make this 20th International GaAs Symposium a great success.

Finally, the editors would like to thank all persons and sponsoring organizations, who made the conference a lasting scientific and social experience for the GaAs community, for their help and generosity.

Hans S Rupprecht
Günter Weimann

Contents

Chapter 6: Heterostructures and Quantum Wells

Chapter 7: Process Technologies

Chapter 8: Bulk Crystal Growth

Chapter 9: Molecular Beam Epitaxy

Chapter 12: Late News

Analog, digital and mixed-signal integrated circuits for high speed applications

M. Berroth
Fraunhofer-Institut für Angewandte Festkörperphysik, Tullastr. 72, D-79108 Freiburg
Tel. ++49–761–5159–556, Fax: ++49–761–5159–400

ABSTRACT: The presentation gives an overview of recent results upon a broad range of monolithic integrated circuits for high-speed applications based on III/V device technology.

Monolithic microwave integrated circuits (MMICs) like low noise amplifiers and down converters at 12 GHz have been successfully demonstrated with MESFET technology. For short range communication systems at 60 GHz, amplifiers and modulators with coplanar transmission lines for matching networks with excellent performance have also been developed. Highest frequency of operation is required for automotive applications, e.g. amplifiers at 76 GHz using special transistor geometries utilizing pulse doped heterostructures and 0.15 μm mushroom gates.

Digital circuits with gate delays below 10 psec are under investigation for signal processing as well as laser driver and transimpedance amplifier for 20 Gbit/s optical data transmission.

High speed and precision is required for analog-to-digital-converters as well as digital-to-analog-converters with sampling rates of more than two billions per second, using low power enhancement/depletion transistors.

1. INTRODUCTION

The GaAs world market was about 580 Mio $ 1992 . The GaAs world market in 1993 has three columns of the same weights: integrated circuits, discrete devices and captive market. However the merchant market of integrated circuits is expected to rise at much higher growth rate than the other until the year 2000 as shown in Fig. 1.

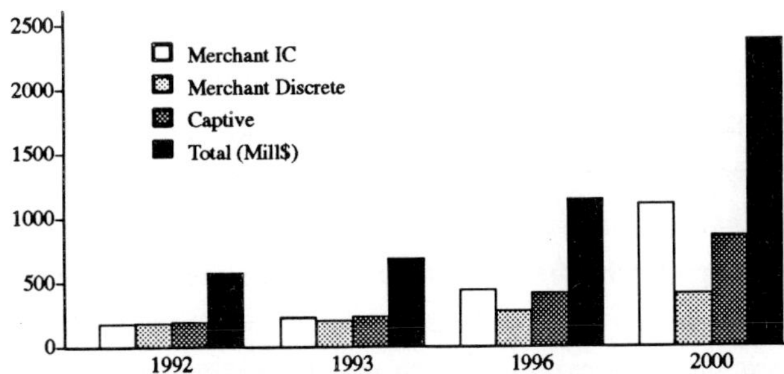

Fig. 1: GaAs Market Forecast (BIS, Dataquest, Electronics).

A survey of present and future applications of GaAs devices and circuits on the commercial market is shown in Fig. 2. All the consumer applications like direct broadcasting satellite

(DBS) receiver or mobile phone use only analog circuits. The digital applications are logic functions in supercomputers or high-end workstations. For communication systems using fibre optical links as well as high speed local area networks (LAN) for computer interconnects analog and digital as well as mixed functions circuits are necessary for serialisation and transmission of huge amount of data across coaxial cable or a fibre. Instrumentation systems are also interesting industrial applications of GaAs circuits e.g., analog-to-digital converter (ADC) at the input of sampling scopes.

	Industrial	Communication	Computers	Consumer
Analog	Instrumentation	Fibre Optic Systems DBS GPS MLS	High-Speed-LAN	DBS-Receiver Mobile phone HDTV Auto Collision Avoidance
Mixed Analog-Digital	Instrumentation	Fibre Optic Systems	High-Speed LAN	
Digital	Instrumentation	Fibre Optic Systems	High-Speed LAN Supercomputers Workstations	

Fig. 2: Present and expected applications of GaAs circuits in respect to market and circuit type.

This paper presents latest the results of the German GaAs industry and research labs and is divided into three parts. First we look at some analog integrated circuits for consumer applications like mobile communication or anti-collision radar. In the second part pure digital circuits are presented which are useful for signal and data processing. The third part describes some combinations of analog and digital functions on the same chip, which are required in data transmission systems and instrumentation including optoelectronic integrated circuits (OEIC).

2. ANALOG CIRCUITS

A break-through for low cost GaAs integrated circuits was the development of DBS converter chips like the AKD12000 from Anadigics (Wallace 1990). It replaces several hybrids in DBS receivers. Siemens has developed a similar low-noise converter (LNC) with more than 30 dB of gain with a noise figure of less than 6 dB at 10.94 - 11.75 GHz (Microwave 1992). These chips are the first monolithic integrated circuits in the microwave frequency range, which are produced in very large volume and therefore initiate lower chip costs. Even larger volumes are expected in the mobile phones, which are using GaAs devices and circuits at 0.9 and 1.8 GHz, where silicon is a strong competitor for the rf components (Pettenpaul 1993). These MMICs use MESFET technology, but HEMT MMICs for the low-cost consumer market are already under development. Especially low noise amplifiers for the DBS receivers are of great interest. At the IAF, a two stage low-noise amplifier with 0.2 μm pseudomorphic MODFETs showed a noise figure of 1.2 dB with an associated gain of 18 dB at 12 GHz, as shown in Fig. 3 (Bosch 1993). These results are corroborated on by an advanced noise modelling concept using a temperature noise model (Pospiezlski 1989, Tasker 1993).

The frequency band around 60 GHz is expected to be used for communication systems. One possible application in the V-band is the supply of the microcell stations for mobil phone systems. Also the short range communication links with data rates up to 1 Gbit/s are under investigation. ANT (Bosch Telecom) is investigating such a communication link at 60 GHz with direct digital modulation with data rates up to several Mbit/s using quadrature amplitude modulation (QAM). First components of such a system have already been fabricated. In Fig. 4 the measured gain of a 3-stage amplifier (Schlechtweg 1992a) mounted into a waveguide housing is plotted in the frequency range from 50 to 60 GHz (Lohrmann 1993). This high gain amplifier has a noise figure of about 5 dB with an associated gain of 10 dB at 55 GHz. Alcatel SEL is also developing a V-band communication link. A first travelling wave amplifier is available with 5 dB gain up to 50 GHz (Heilig 1993).

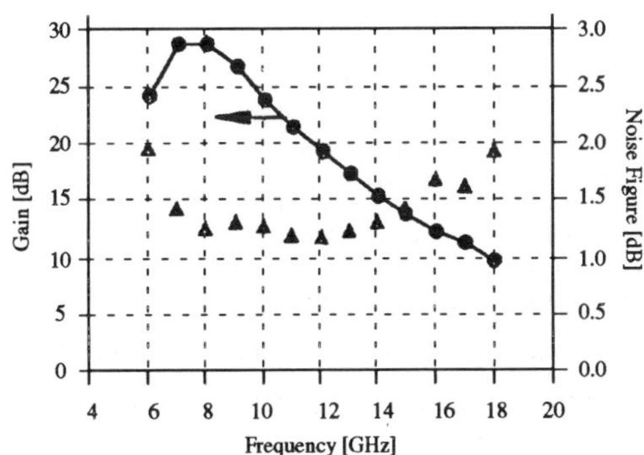

Fig. 3: Measured gain and noise figure of a 2-stage HEMT noise amplifier (Bosch 1993).

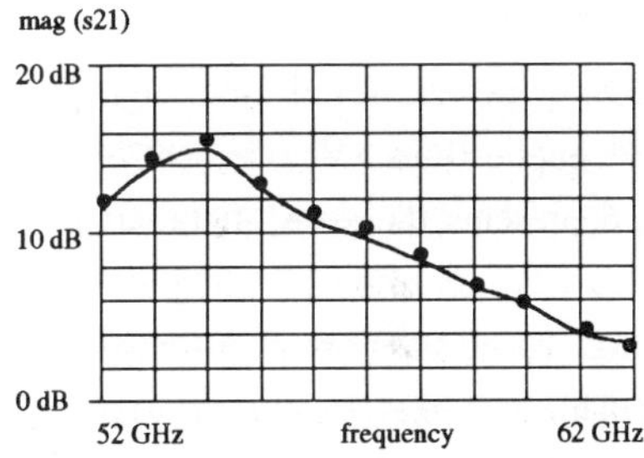

Fig. 4: Measured gain of a 3-stage V-band amplifier mounted into a waveguide housing (Schlechtweg 1992a).

The world wide assignment of the 76/77 GHz band for anticollision radars of vehicles at the WARC 1992 has initiated intensive research work at this frequency range. A few industrial companies and research labs are now able to produce amplifiers with reasonable gain at 76 GHz (Wang 1992), (Webster 1992). The IAF has developed a very high gain three stage amplifier, which also uses coplanar transmission lines for matching networks (Schlechtweg 1993b). Fig. 5 shows more than 20 dB of gain at the frequency of interest with good isolation and acceptable input and output matching. These MMICs at 76 GHz are a major challenge to the GaAs industry. Low cost production of those millimeterwave frontends in large volumes could initiate lots of other applications in the frequency range between 12 GHz, where the DBS receiver has proven commercial competitivness, and the 76 GHz auto collision avoidance radar, which could be installed into millions of cars.

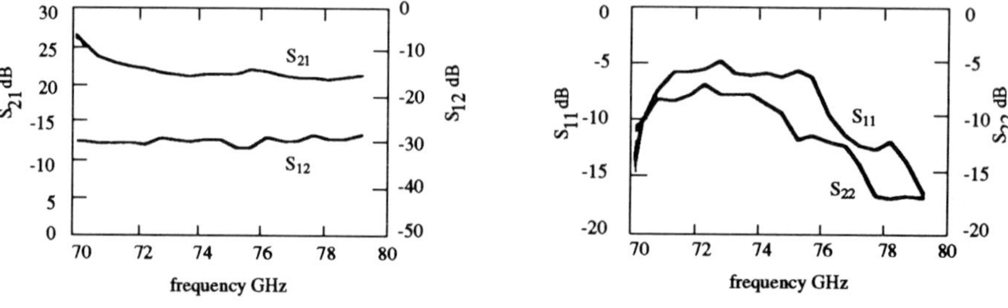

Fig. 5: High gain 70 - 80 GHz coplanar 3-stage amplifier (Schlechtweg 1993b).

3. DIGITAL CIRCUITS

For years GaAs digital circuits have competed against silicon BICMOS and bipolar ECL technology. The GaAs MESFET technology has achieved VLSI integration levels at moderate power levels for high speed applications (Vitesse 1992). Fig. 6 presents the state of the art of gate arrays in respect to integration level, speed and power consumption (Thiede 1992a), (Notomi 1991). At the IAF the reduction of power has been investigated by reducing transistor size and scaling of interconnection rules. Comparing static frequency dividers, we achieved a power dissipation of only 0.74 mW at input frequencies up to 1.8 GHz.

Vitesse	Fujitsu	IAF
350 k	45 k	6 k
2-input NOR	3-input NOR	2-input NOR
55 ps (F0 = 0)	35 ps (F0 = 1)	19 ps (F0 = 1)
0.34 mW	0.24 mW	1.35 mW
44 W (50 %)	11 W (80 %)	4 W (50 %)
0.6 µm	0.6 µm	0.3 µm
MESFET	HEMT	HEMT
4 level metallization	4 level metallization	2 level metallization

Fig. 6: Comparison of GaAs gate arrays

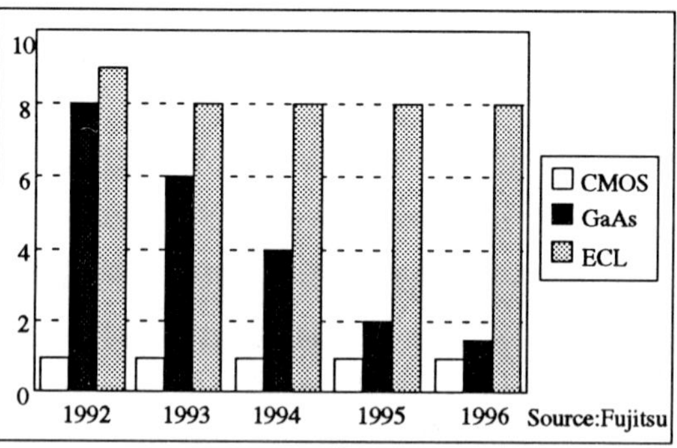

Fig. 7: Cost comparison of silicon and GaAs gate array [Fujitsu].

In regard to gate delays CMOS is no competitor, however bipolar ECL is achieving about the same clock rates, but at a higher power dissipation. As shown in Fig. 7, Fujitsu is forecasting a strong reduction of production costs of GaAs logic circuits compared with silicon bipolar.

A further increase in performance is achieved by using pseudomorphic heterostructures for digital circuits. The first digital circuit with 0.2 μm gate length has been a dynamic frequency divider developed at the IAF (Thiede 1993b). Operating at an input frequency range from 28 to 51 GHz the performance is significant improved over that achieved by any other material system or technology as shown in Fig. 8 (Jensen1987a), (Saito 1989), (Yamauchi 1989), (Thiede 1993c), (Jensen 1992b), (Klose 1992).

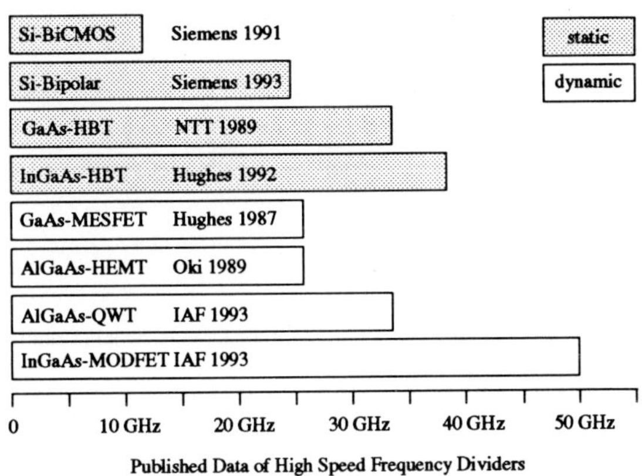

Fig. 8: State of the art of static and dynamic frequency dividers.

4. MIXED MODE APPLICATIONS

The increasing data rates in communication systems is a driving force for extremely high clock rates in logic circuits combined with large bandwidth amplifiers. Even optoelectronic (OE) components like metal-semiconductor-metal (MSM) photodetectors or lasers can be integrated with large scale integrated circuits. So GaAs is presently the only material system, which can provide real optoelectronic integrated circuits, which are required in fibre optical links. The IAF is developing single chip OE transmitter and OE receiver. In Fig. 9 the block diagram of the optical link is shown. On the transmitter side a 2:1 multiplexer with integrated laser driver has already been fabricated for data rates up to 18 Gbit/s (Wang 1992). A first monolithic integration of a laserdiode with the laser driver has successfully been demonstrated (Hornung to be published). On the receiver side the MSM-photodetector has already been integrated with a transimpedance amplifier (Hurm 1993). The -3 dB bandwidth is above 14 GHz, which is sufficient for data rates above 20 Gbit/s.

There has already been a first experiment of free space data transmission from the monolithic integrated laser driver with laser diode to the MSM-detector with monolithic integrated transimpedance amplifier. In Fig. 10, the input signal of the pulse pattern signal and the receiver signal are shown at a data rate of 7.4 Gbit/s. Further increased integration level will allow the serial transmission from highly parallel input data at 10 or even 40 Gbit/s across a fibre optical link with only two monolithic optoelectronic integrated circuits on GaAs substrate.

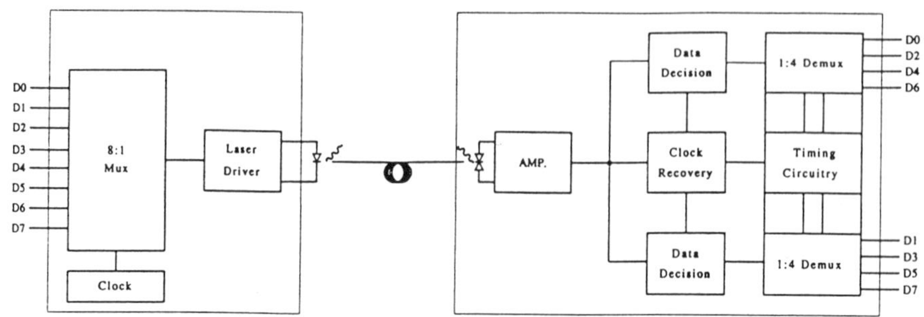

Fig. 9: Block diagram of a high speed optical link.

The increasing clock rate and signal frequencies require faster measurement systems. One of the most difficult functions in this area is a fast analog-to-digital-converter (ADC) with high resolution. The fastest ADC yet reported with 5 bit resolution can be used at clock rates up to 3.6 GHz at a power consumption of 2.5 W (Oehler 1993). This parallel converter was developed by the FhG-IIS using the 0.3 µm QW-HEMT process of the IAF. Of commercial interest, however, are the 8 - 10 bit higher resolution ADC which are under development.

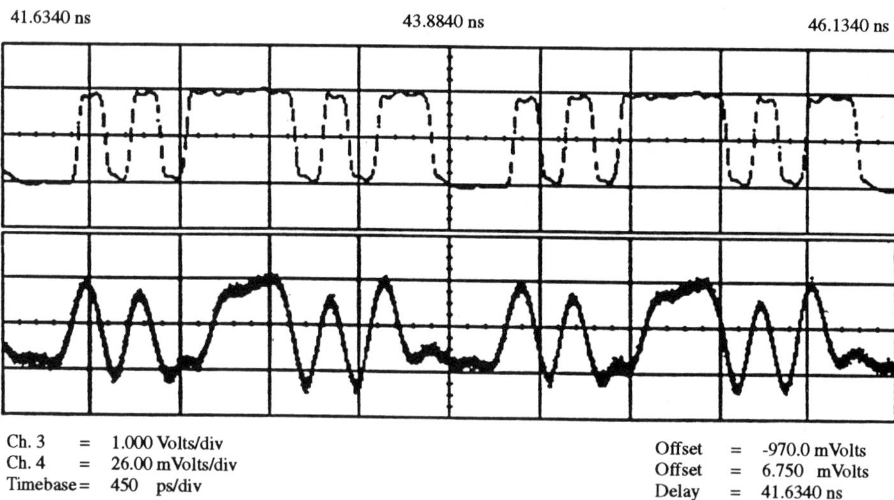

Fig. 10: Free space transmitted optical pulses using monolithic integrated optoelectronic transmitter and receiver at 7.4 Gbit/s.

However high resolution ADCs at high clock rates request not only fast devices and high levels of integration, but also good matching and reproducible devices, thus challenging technology. Triquint has already shown, that these mixed signal applications can be satisfied by GaAs technology outperforming silicon with 14 bit, 1 Gs/s digital-to-analog converter Weiss 1991).

5. CONCLUSION

Monolithic millimeterwave integrated circuits (M^3IC) meet system requirements already up to 80 GHz. Further improvement of heterostructure epitaxy and process technology will give access to frequencies above 100 GHz for M^3IC technology. The battle against silicon on the digital market is going on with a good chance for GaAs to dominate the high speed market niche. First high speed OEICs are available for data rates even above 10 Gbit/s, however more sophisticated transmitter and receiver ICs are required. A still challenging task are high speed, high resolution ADCs for measurement systems.

ACKNOWLEDGEMENT

The author is grateful to Prof. H.S. Rupprecht and all colleagues at the IAF. He is also indepted to Dr. Pettenpaul, Siemens; J. Schroth, DASA; W. Ehrlinger, ANT and E. Müller, SEL for their contributions to this presentation. He thanks J. Seibel and P.J. Tasker for critical reading of the manuscript. Part of the work was funded in the III/V-Project of the BMFT.

REFERENCES

Bosch, R., Tasker, P.J., Schlechtweg, M., Braunstein, J., and Reinert, W., "A lumped element 12 GHz LNA MMIC using InGaAs/GaAs MODFETs with optimized gate width and reactive feedback", Electronics Letters, 1993, vol. 27 no. 15, pp. 1394-1395.

Heilig, R., Hollmann, D., and Baumann, G., "A Monolithic 1-55 GHz HEMT Distributed Amplifier in Coplanar Waveguide Technology", 23. European Microwave Conference, Madrid, 1993, to be published.

Hornung, J., et al., "7.4 Gbit/s Monolitically integrated GaAs laserdiode-laserdriver structure", to be published.

Hurm, V., Ludwig, M., Rosenzweig, J., Benz, W., Berroth, M., Bosch, R., Hülsmann, A., Köhler, K., Raynor, B., and Schneider, J., "14 GHz bandwidth MSM photodiode AlGaAs/GaAs HEMT monolithic integrated optoelectronic receiver", Electronics Letters, 1993, vol. 29, no. 1, pp. 9-10.

Jensen, J.F., Salmon, L.G., Deakin, D.S., and Delaney, M.J., "26 GHz GaAs room-temperature dynamic divider circuit", IEEE-GaAs-IC-Symp., Portland, USA, 1987a, pp. 201-203.

Jensen, J.F., Hafizi, M., Stanchina, W.E., Metzger, R.A., and Rensch, D.B., "39.5 GHz static frequency divider implemented in AlInAs/GaInAs HBT technology", IEEE-GaAs-IC-Symp., Miami Beach, USA, 1992b, pp. 101-104.

Klose, H., Kerber, M., Meister, T., Ohnemus, M., Köpl, R., Weger, P., and Weng, J., "Process optimization for sub-30 ps BICMOS Technologies for Mixed ECL/CMOS Applications", IEDM, Washington, 1992, pp. 89-92.

Lohrmann, R., and Ehrlinger, W., "Packaging of Millimeterwave-MMICs for Communication Systems Applications", MIOP 93, Conference Proceedings, Sindelfingen, 1993, pp. 237-241.

Microwave Engineering Europe; "ESPRIT: a review of the progress in 1992", pp. 44-47.

Notomi, S., Watanabe, Y., Kosugi, M., Hanyn, I., Suzuki, M., Mimura, T., and Abe, M., "A 45 K-Gate HEMT Array with 35 ps DCFL and 50 ps BDCFL Gates", Journal Solid State Circuits, 1991, vol. 26, no. 11, pp. 1621-1625.

Oehler, F., Saurer, J., Hagelauer, R., Seitzer, D., Nowotny, U., Raynor, B., and Schneider, J., "3.6 Gsample/s 5-bit Analog-to-Digial Converter Using 0.3 μm AlGaAs/GaAs HEMT-Technology", IEEE-GaAs-IC-Symp. 1993, to be published.

Pettenpaul, E. Schaf, L., and Schöpf, K.J., "Enhanced GaAs Device Concepts for the New Digital Mobile Communication Systems", 23rd European Microwave Conf., Madrid, 1993.

Pospiezalski, M.W., "Modelling of Noise Parameter of MESFETs and MODFETs and their Frequency and Temperature Dependence", IEEE Trans. Microwave Theory Tech., vol. 37, pp. 1340-1350, 1989.

Saito, T., Fujishiro, H.I., Ichioka, T., Tanaka, K., Nishi, S., and Sano, Y., "0.25 μm gate inverted HEMTs for an ultra-high speed DCFL dynamic frequency divider", IEEE-GaAs-Symp., 1989, San Diego, USA, pp. 117-120.

Schlechtweg, M., Reinert, W., Tasker, P.J., Bosch, R., Braunstein, J., Hülsmann, A., and Köhler, K., "Design and characterization of high performance 60 GHz pseudomorphic MODFET LNAs in CPW-technology based on accurate S-parameter and noise models", IEEE Trans., 1992a, MTT-40, pp. 2445-2451.

Schlechtweg, M., Tasker, P.J., Reinert, W., Braunstein, J., Haydl, W., Hülsmann, A., and Köhler, K., "High Gain 70 - 80 GHz MMIC Amplifier in Coplanar Waveguide Technology", Electronics Letters, 1993b, vol. 29, no. 12, pp. 1119-1120.

Tasker, P.J., Schlechtweg, M., and Braunstein, J., "On-wafer single contact S-parameter measurements to 75 GHz: Calibration and measurement system", 23rd Europan Microwave Conf., Madrid, 1993.

Thiede, A., Berroth, M., Hurm, V., Nowotny, U., Seibel, J., Gotzeina, W., Sedler, M., Raynor, B., Köhler, K., Hofmann, P., Hülsmann, A., Kaufel, G., and Schneider, J., "16 x 16 Bit Parallel Multiplier Based on 6 K Gate Array with 0.3 μm AlGaAs/GaAs Quantum Well Transistors", Electronics Letters, 1992a, vol. 28, pp. 1005-1006.

Thiede, A., Tasker, P.J., Hülsmann, A., Köhler, K., Bronner, W., Schlechtweg, M., Berroth, M., Braunstein, J., and Nowotny, U., "28-51 GHz dynamic frequency divider based on 0.15 μm T-gate $Al_{0.2}Ga_{0.8}As/In_{0.25}Ga_{0.75}As$ MODFETs", Electronics Letters, 1993b, vol. 29, no. 10, pp. 933-934.

Thiede, A., Berroth, M., Nowotny, U., Seibel, J., Bosch, R., Köhler, K., Raynor, B., and Schneider, J., "An 18-34 GHz dynamic frequency divider based on 0.2 μm AlGaAs/GaAs/AlGaAs quantum-well transistors", Int. Solid State Circuits Conf., San Francisco, USA, 1993c, pp. 176-177.

Vitesse Product Data Book, 1992, pp. 1-8.

Wallace, P., et al., IEEE 1990 Microwave and Millimeterwave Monolithic Circuit Symposium, Digest of papers, p. 7.

Wang, H., Dow, G.S., Allen, B.R., Ton., T.-N., Tan, K.L., Chang, K.W., Chen, T.-H., Berenz, J., Lin, T.S., Liu, P.H., Streit, D.C., Bui, S.B., Raggio, J.J., and Chow, P.D., "High-performance W-band monolithic pseudomorphic InGaAs HEMT LNA's and design/analysis methodology", IEEE Trans, 1992, MTT-40, pp. 417-426.

Wang, Z.-G., Nowotny, U., Berroth, M., Bronner, W., Hofmann, P., Hülsmann, A., Köhler, K., Raynor, R., and Schneider, J., "18 Gbit/s monolithically integrated 2:1 multiplexer and laser driving using 0.3 μm gate length quantum well HEMTs", Electronics Letters, 1992, vol. 28, no. 18, pp. 1724-1725.

Webster, R.T., Slobodnik, A.J., and Roberts, G.A., "Monolithic InP HEMT V-band low-noise amplifier", IEEE Microw. & Guided Wave Lett., 1992, **2**, pp. 236-238.

Weiss, F.G., and Bowmann, T.G., "A 14-Bit, 1 Gs/s DAC for Direct Digital Synthesis Applications", IEEE-GaAs-IC-Symp. Digest, 1991, pp. 361-364.

Yamauchi, Y., Nakajima, O., Nagata, K., Ito, Hiroshi, and Ishibashi, T., "A 34.8 GHz 1/4 static frequency divider using AlGaAs/GaAs HBTs", IEEE-GaAs-IC-Symp., San Diego, USA, 1989, pp. 121-124.

Paper presented at the Int. Symp. GaAs and Related Compounds, Freiburg, 1993

High-gain, directly connected HBT amplifier

Y Ota, M Yanagihara, A Tamura and O Ishikawa

Semiconductor Research Center, Matsushita Electric Industrial Co., Ltd.
3-1-1, Yagumo-nakamachi, Moriguchi, Osaka 570, Japan

ABSTRACT: A single-chip amplifier using abrupt junction AlGaAs/GaAs HBTs has been studied to obtain high gain at a low operation voltage. The amplifier consists of two HBTs using a direct wire connection between the collector of the 1st HBT and the base of the 2nd HBT without a DC block capacitance. The current gain and power gain of the amplifier are 43 dB and 37 dB respectively at 1 GHz by applying DC bias voltage of 1.5 V to all base and collector terminals. The new MMIC amplifier will contribute to minimize the size of portable cellular telephone.

1. INTRODUCTION

Heterojunction bipolar transistors (HBTs) have been studied as high-frequency devices in many laboratories. In application to the millimeter-wave band, HBTs have the advantage of high gain (Ogawa et al 1990) in comparison with GaAs MESFETs. In the microwave band below the X band, however, MESFETs overcome the advantage of HBTs, because MESFETs have sufficient and practical gain in this frequency band (Yanagihara et al 1992) and can be produced at a lower cost than HBTs. The difference of production costs between HBTs and MESFETs becomes large especially in case of MMIC fabrication, owing to the high-cost epitaxial substrate of HBTs. Accordingly, to solve these problems has been one of the most important matters in order to apply HBTs to the large market of the microwave band.

Yamauchi and Ishibashi (1987) developed a microwave wideband HBT amplifier. The amplifier was composed of cascade connection HBTs and some resistances without level-shift diodes, and operated at around 8 - 10 V. As for the low voltage operation amplifiers, Ishikawa et al (1992) developed front-end ICs and power modules with GaAs MESFETs, which were operated with 3 to 4 battery cells. We have developed a very small two-stage MMIC amplifier which is operated at very low voltages with a simple structure.

The new amplifier consists of two AlGaAs/GaAs HBTs connected directly without an internal DC block capacitance and shows very high gain at 1.5 V which is supplied by one battery cell. An abrupt junction of a base/emitter in HBTs has a high turn-on (built-in) voltage of 1.5 V and makes it possible to form the novel construction of the amplifier. The features of the new HBT amplifier realize not only the reduction of the chip size and the production cost but also further reduction of battery cells, and consequently contribute to realize single cell systems. In this paper, we present the design, the fabrication process and the characteristics of the newly-developed direct connection amplifier.

2. DESIGN CONCEPT

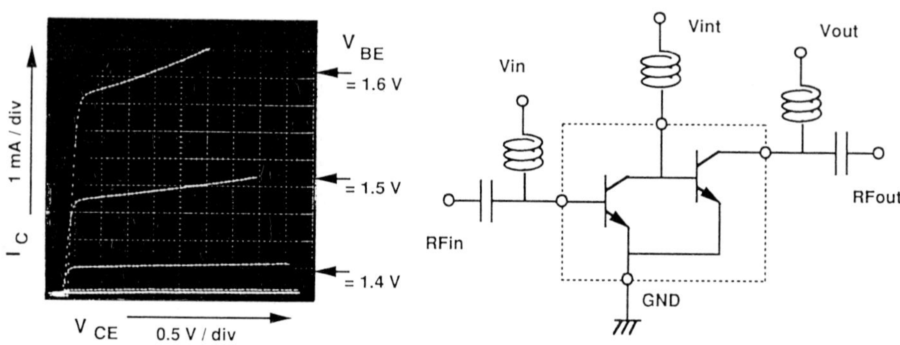

Fig. 1 I - V characteristics of the single HBT.

Fig. 2 Circuit diagram of the MMIC amplifier.

Figure 1 shows the I - V characteristics of the single HBT whose emitter area is 2 μm x 20 μm. Due to the abrupt junction, the operating point of V_{ce} = 1.5 V at V_{be} = 1.5 V is completely in the active region of the HBT, thus the gain is sufficient and the output power is practical. Accordingly, each bias voltage can be set at the same voltage of 1.5 V as the turn-on voltage of the abrupt HBT. As a result, a direct connection structure without a DC block capacitance is possible for a two-stage amplifier.

Figure 2 shows the equivalent circuit of the developed MMIC amplifier (inside of the dotted square) and a bias/input/output circuit (outside of the dotted square). An input radio frequency RF_{in} is applied to the base terminal of the 1st HBT and an output radio frequency RF_{out} is picked up at the collector terminal of the 2nd HBT. Each DC bias (an input bias voltage V_{in}, an internal bias voltage V_{int} and an output bias voltage V_{out}) is supplied through the choke-coil as shown in the figure. By repeating the direct connection of HBTs, much higher gain will be obtained.

3. FABRICATION PROCESS

Table 1 Parameters of multi-layered structure of an abrupt AlGaAs/GaAs HBT.

Layer	Dope (cm^{-3})	Thickness (Å)
n^+- GaAs	5×10^{18}	1900
Grading	1×10^{18}	300
N - AlGaAs	5×10^{17}	1500
p^+- GaAs	2×10^{19}	1000
n^- - GaAs	3×10^{16}	7000
n^+- GaAs	5×10^{18}	8000
s. i. sub	—	—

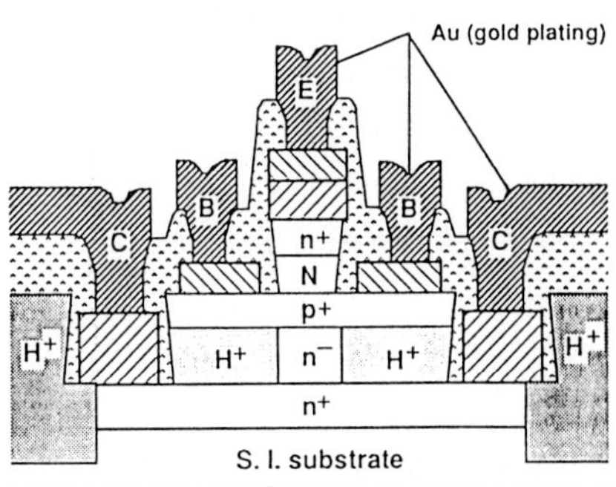

Fig. 3 Cross sectional view of the HBT fabricated by a self-alignment process.

The multi-layered structural parameters of the AlGaAs/GaAs HBT is listed in Table 1 and the cross sectional view of the HBT is shown in Fig. 3. The layers were grown by MOCVD method. The base layer consists of GaAs with a C-dope of 2×10^{19} cm^{-3} and a thickness of 1000 Å, and the emitter layer consists of $Al_{0.3}Ga_{0.7}As$ with a Si-dope of 5×10^{17} cm^{-3} and a thickness of 1500 Å. This heterojunction forms an abrupt interface and produces a high turn-on voltage of 1.5 V. With the layer parameters as listed in Table 1, the base-emitter breakdown voltage of 8 V and the base-collector breakdown voltage of 20 V are obtained. These high breakdown voltages allow practical use as a low distortion amplifier.

The HBTs were made by the one-mask multiple self-alignment process (Inada et al 1987), where an emitter electrode, base electrodes and a small buried collector were self-aligned to an emitter region as shown in Fig. 3. The processing steps are 1) forming a dummy emitter of SiO_2 or SiN on the multi layer, 2) exposing the base layer by wet etching using the dummy emitter as a mask, 3) implanting proton ions to a collector layer under an extrinsic base, 4) removing the dummy emitter and depositing an emitter electrode by pattern inversion method, 5) forming buried collector electrodes, 6) forming base electrodes using the emitter electrode as a mask, and 7) implanting proton ions for device isolation. A gold-plating metalization to reduce losses and resistances of wiring was performed with 1 µm thickness on each ohmic electrodes.

The top view photograph of the MMIC amplifier is shown in Fig. 4. The emitter area of the 1st HBT is 2 µm x 20 µm x 2 (80 μm^2) and that of the 2nd one is 2 µm x 50 µm x 2 (200 μm²). The area of the 2nd HBT was designed to show an output impedance of around 50 Ω, in order to obtain a high gain without tuning circuits.

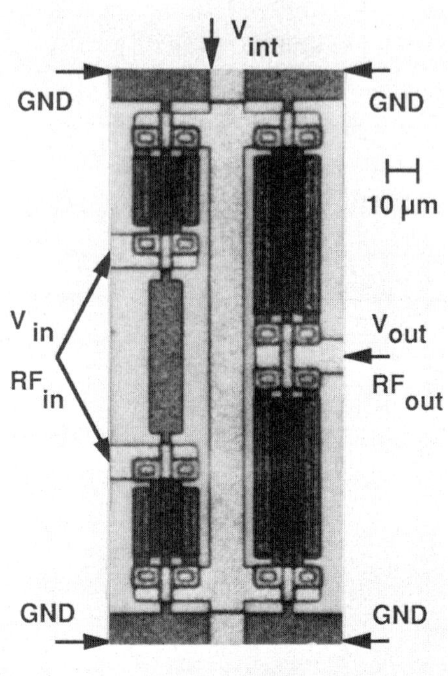

Fig. 4 Top view photograph of the amplifier.

4. RESULTS AND DISCUSSIONS

Figures 5 (a) and 5 (b) show the current gain (h_{fe}) and the power gain (the maximum stable gain MSG and the maximum available gain MAG) versus frequency, which are obtained from S-parameters measurements. The high current gain of 43 dB and power gain of 37 dB are obtained at a frequency of 1 GHz and a supply voltage of 1.5 V for each bias terminal. In individual measurements, the current gain and power gain of the 1st HBT are 25 dB and 21 dB, and those of the 2nd HBT are 25 dB and 20 dB, as shown in Figs 5 (a) and 5 (b).

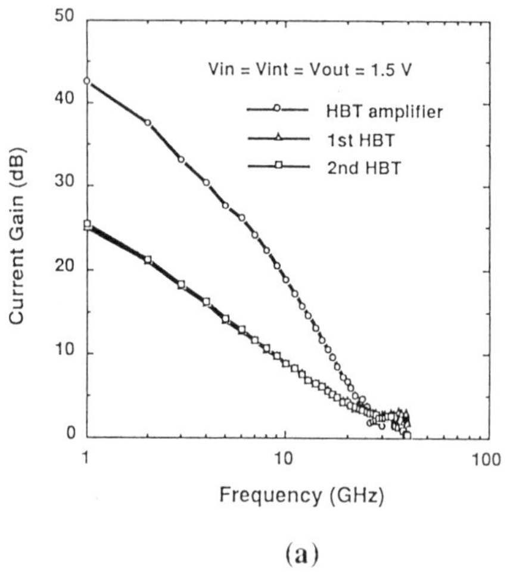

(a)

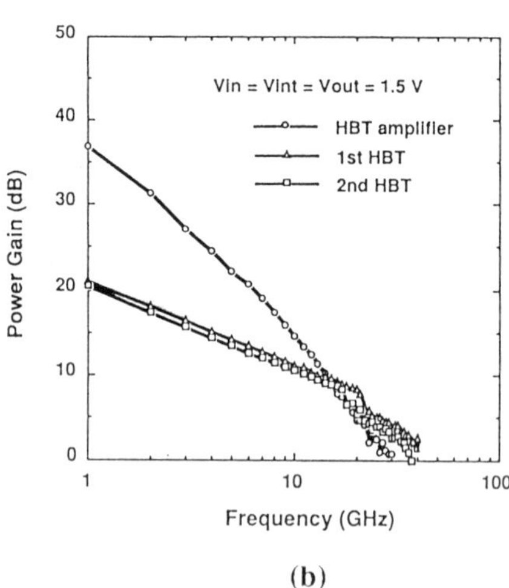

(b)

Fig. 5 Current gain (a) and power gain (b) of the amplifier (circles) versus frequency. The values of the 1st HBT (triangles) and the 2nd (squares) are obtained by individual S-parameters measurements.

The current gain of the amplifier is slightly less than the sum of that of the 1st HBT and that of the 2nd HBT. The power gain of the amplifier is shown as the maximum available gain (the stability factor K > 1), while that of the 1st and the 2nd HBT is shown as the maximum stable gain (K < 1).

Figure 6 shows the properties of P_{out} and the total current I_{total} versus the DC bias of the amplifier. The same voltage was applied to each terminal. Considering the application to actual systems, the measurements of RF properties were performed at a frequency of 1.9 GHz, which corresponds to the microwave band of the digital cordless phone in Japan (PHP) and in Europe (DECT). P_{out} and I_{total} are 2 dBm and 36 mA respectively at Pin = -30 dBm and 1.5 V in a 50 Ω impedance measurement system without tuners.

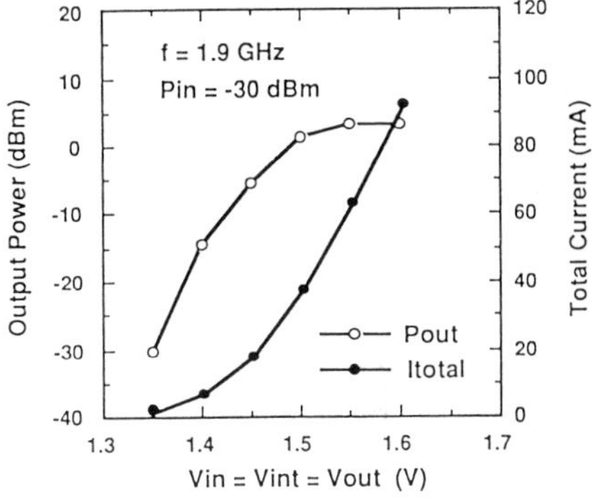

Fig. 6 Output power P_{out} and total current I_{total} dependence on DC bias. The same voltage is applied to each terminal ($V_{in} = V_{int} = V_{out}$).

The output power P_{out} dependence on the input power P_{in} of the amplifier is shown in Fig. 7. The third order intermodulation distortion IM3 is also illustrated in the figure. The measurements were carried out at a frequency of 1.9 GHz (and 1.9003 GHz for IM3 measurement). All terminals (Vin, Vint and Vout) are set at 1.5 V. The linear gain and the saturation power show excellent values of 32 dB and 10 dBm, respectively. The values of the third order intercept point IP3 is 27 dBm with I_{total} = 36 mA. The linearity figure-of-merit (IP3/Pdc) becomes 9.3, which compares well with that of the low-distortion pulse-doped MESFET by Otobe et al (1992). The low level distortion in the amplifier is suitable for a digital cellular phones using QPSK modulation.

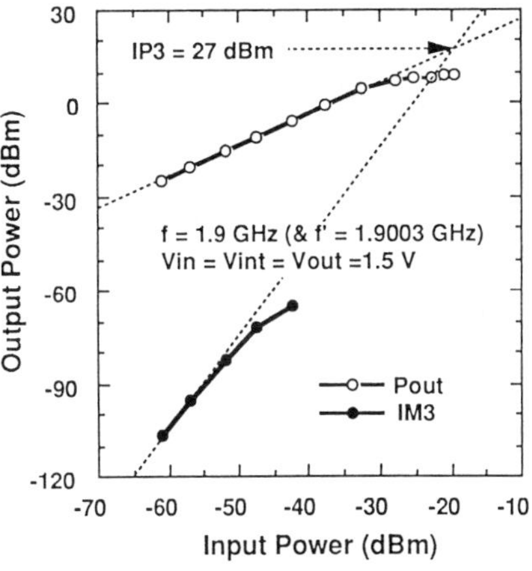

Fig. 7 Output power P_{out} and intermodulation distortion IM3 versus input power P_{in} in the amplifier. Two tone frequencies of 1.9 GHz and 1.9003 GHz are used for IM3 measurement.

5. CONCLUSION

The MMIC amplifier of directly connected HBTs has been introduced without additional processes or passive elements as matching networks. The current gain and power gain of the amplifier are 43 dB and 37 dB respectively at 1 GHz with single supply voltage of 1.5 V to all base and collector terminals. The linear gain and the saturation power are 32 dB and 10 dBm respectively at 1.9 GHz, while the third order intercept point IP3 is 27 dBm. The new MMIC amplifier can be applied to one-battery cell communication systems.

ACKNOWLEDGEMENT

The authors gratefully acknowledge Dr. T. Takemoto and T. Onuma for encouragements. Dr. M. Inada, K. Inoue and T. Uwano are thanked for helpful discussions.

REFERENCES

Inada M, Ota Y, Nakagawa A, Yanagihara M, Hirose T and Eda K 1987 IEEE Trans. Electron Devices **ED-34** pp 2405-11

Ishikawa O, Ota Y, Maeda M, Tezuka M, Sakai H, Katoh T, Itoh J, Mori Y, Sagawa M and Inada M 1992 IEEE GaAs IC Symp. (Florida) Tech. Dig. pp 131-4

Ogawa K, Hashimoto K, Uwano T and Ota Y 1990 Electron. Lett. **26** pp 2134-5

Otobe K, Kuwata N, Shiga N, Nakajima S, Matsuzaki K, Sekiguchi T and Hayashi H 1992 19th Int. Symp. GaAs and Related Compounds (Karuizawa) Inst. Phys. Conf. **129** pp 761-6

Yamauchi Y and Ishibashi T 1987 Electron. Lett. **23** pp 156-7

Yanagihara M, Ota Y, Nishii K, Ishikawa O and Tamura A 1992 Electron. Lett. **28** pp 686-7

Inst. Phys. Conf. Ser. No 136: Chapter 1
Paper presented at the Int. Symp. GaAs and Related Compounds, Freiburg, 1993

K-band dielectric resonator oscillator using a GaInP/GaAs HBT

U Güttich, H Leier*, A Marten*, K Riepe*, W Pletschen**, K H Bachem**

Deutsche Aerospace A.G., Sedanstraße 10, 7900 Ulm, Germany

* Daimler Benz A.G., Research Center Ulm, Wilhelm-Runge Str. 11, 7900 Ulm, Germany

** Fraunhofer Institut für Angewandte Festkörperforschung, Tulla-Str. 72, 7800 Freiburg, Germany

ABSTRACT: Design, fabrication and performance of the first hybrid dielectric resonator oscillator (DRO) using a GaInP/GaAs HBT as active device are described. The oscillator consists of a microstrip circuit realized on alumina substrate. The HBT has an emitter area of 1.5 µm x 20 µm and a selfaligned base contact. The vertical layer structure of the HBT incorporates a highly carbon doped base and a thin GaInP blocking layer. The fabricated oscillator exhibits an output power of +5 dBm at 21 GHz with a conversion efficiency of 13%. The phase noise has been measured to be -102.5 dBc/Hz at 100 kHz and -81.6 dBc/Hz at 10 kHz off carrier, respectively

1. INTRODUCTION

The expansion of application fields in the last years for sensor and communication systems in the microwave and millimetre-wave range results in an increasing demand for low phase noise and good output power local oscillators. They are key components in Doppler moduls, e.g. for short range proximity or motion sensors, or in receiver frontends, e.g. applied in digital satellite communications. Due to the low 1/f noise and to the excellent high frequency performance Heterojunction Bipolar Transistors (HBTs) are considered as promising candidates for the oscillator active device element. The improvements in the MOCVD growth of GaInP/GaAs have initialized a rapid increase on GaInP/GaAs HBT research in the last two years. Indeed, the GaInP/GaAs material system has several inherent advantages compared to the conventional AlGaAs/GaAs material system such as the availability of selective etchants, the large valence-band discontinuity and the reduced surface recombination of GaInP (Mondry and Kroemer 1985). Last advances in GaInP/GaAs HBT technology have made it possible to realize high performance GaInP/GaAs HBT oscillators for K- and Ka-band frequencies (Leier et al 1993).
To obtain low phase noise characteristics the use of high-Q resonators like a YIG, a metal cavity, or a dielectric resonator (DR) is a well known means. DRs have the advantage of good temperature stability (Tsironis and Pauker 1983), small size and easy mounting.

In this paper we present a GaInP/GaAs HBT dielectrically stabilized oscillator (DRO) for the K-band range with good output performance and excellent phase noise capabilities, which is well suited for the above mentioned LO applications.

2. HBT DEVICE DESCRIPTION AND FABRICATION

The HBT structure is grown by MOCVD and incorporates a 20 nm thin undoped GaInP hole barrier and a 60 nm highly carbon doped base ($p = 6.5 \times 10^{19}$ cm^{-3}). The GaAs emitter (d_e = 115 nm) is doped to $n = 7 \times 10^{17}$ cm^{-3} and the GaAs collector (d_c = 250 nm) is doped to $n = 3 \times 10^{16}$ cm^{-3}. Between the highly doped subcollector (7×10^{18} cm^{-3}) and the collector layer a 10 nm thin undoped GaInP etch stop is introduced. A GaAs cap layer with $n = 7 \times 10^{18}$ cm^{-3} is grown on top to minimize the emitter contact resistance.

Selfaligned HBT devices with an emitter area of 1.5 x 20 μm^2 are defined using a triple mesa approach. Standard optical lithography, selective wet etching and a multi-energy proton isolation are applied for device fabrication. GeNiAu and TiPtAu are the n- and p-type contact metallisations, respectively. The base metal is selfaligned to the emitter stripe leading to a 0.1 - 0.2 μm spacing between base metal and emitter.

The I-V characteristics of the fabricated HBTs indicate current gains of up to 25 with only small dependence on collector current, a breakdown voltage of BV_{CE0} = 9V and a very high Early voltage. On wafer s-parameter measurements are performed from 0.5 - 26.5 GHz in common emitter configuration.

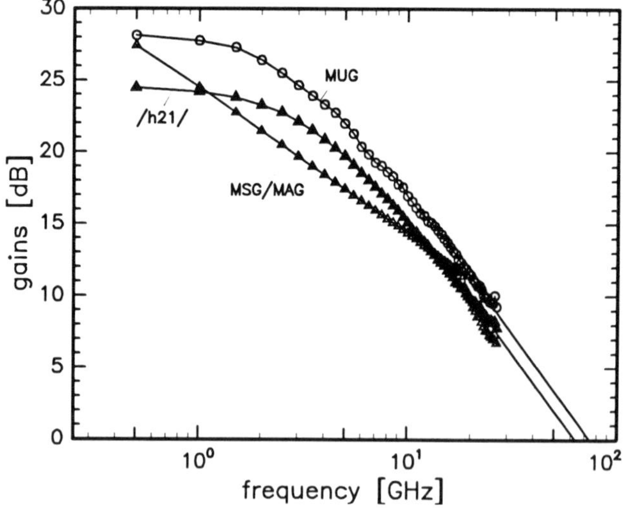

Fig. 1: Current gain |h21| and power gains (MSG/MAG and MUG) vs frequency for an selfaligned 1.5 x 20μm² GaInP/GaAs HBT

In Fig. 1 we have displayed the frequency dependence of /h21/, MSG/MAG and MUG for a 1.5 x 20 μm^2 HBT device at I_C = 20 mA and V_{CE} = 2.5 V. The cut-off frequencies f_T and f_{max} are extrapolated to 65 GHz and 74 GHz, respectively, assuming a -20 dB/decade roll off. At the oscillator frequency of 21-22 GHz MAG is determined to 9 dB

3. OSCILLATOR CONFIGURATION, DESIGN AND REALIZATION

For this work an oscillator topology with a series feedback configuration (the feedback element is the common current carrying element) and the HBT in common emitter operation as shown in Fig. 2 is chosen.

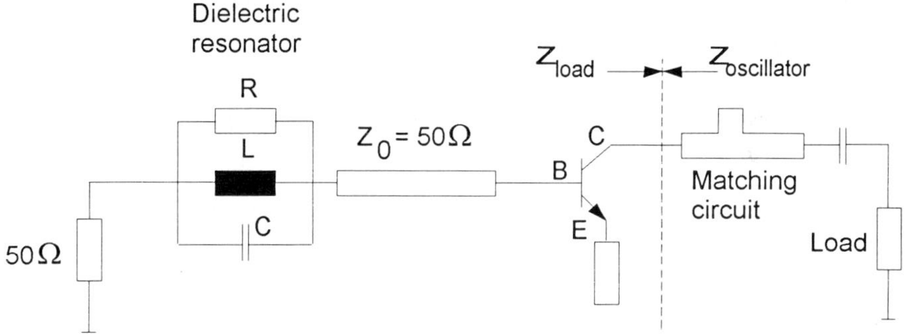

Fig. 2: Equivalent circuit of the dielectric resonator stabilized oscillator

The dielectric resonator is placed at the base side of the transistor, the RF output port is at the collector side. Due to the isolation between output port and frequency determining element, this type of oscillator exhibits a superior frequency stability and a low phase noise (Tsironis and Pauker 1983). Small signal S-parameters measured up to 40 GHz and inhouse developed linear CAD software are used for the design of the microstrip circuit. The dielectric resonator coupled to the microstrip line is modeled as a resonant circuit. The length of the stub at the HBT emitter (feedback element) and the distance between transistor base and the resonator coupling locus are optimised to obtain maximum negative resistance at the transistor collector. The simulated output impedance vs frequency at the collector (Fig. 3) shows a negative resistance of -45 Ω and a corresponding imaginary part of +10 Ω at the desired oscillation frequency of 21.2 GHz. The output matching circuit is designed to satisfy the oscillation condition and to extract the maximum output power. The oscillator output impedance Z_0 and the load impedance Z_L have to verify the empirical relation

$$Z_L = R_L + jX_L = -R_0/3 - jX_0$$

where $Z_0 = R_0 + jX_0$.

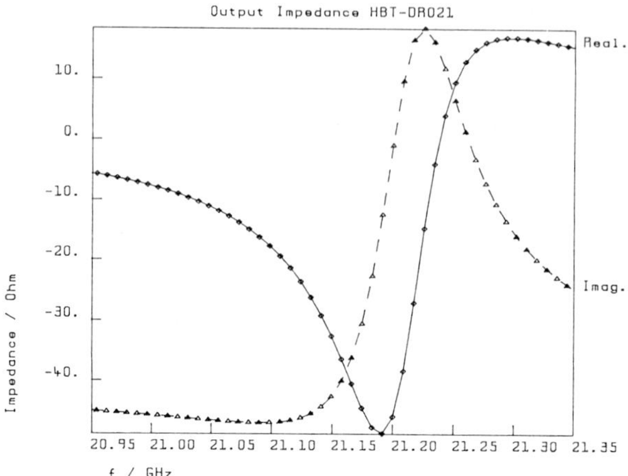

Fig. 3: Simulated output impedance at the HBT collector vs frequency

The oscillator is fabricated on a 150 µm alumina substrate. The HBT chip (0.3 mm x 0.5 mm) is inserted into a hole ultrasonically drilled into the substrate and connected by an electrically conductive adhesive to the microstrip lines. These technique avoids additional parasitic inductances by bond wires. The base termination of the transistor is formed by a microstrip line and a 50 Ω off-chip resistor, which quenches spurious oscillations. Bias networks are realized by using quarter-wavelength 75 Ω lines and 60° radial stubs. Off-chip blocking capacitors are adhered close to the substrate to avoid bias oscillations. As dielectric resonators commercially available $Ba_2Ti_9O_{20}$ - pucks with a constant of 36 and diameters of 2.8 mm to 2.9 mm (thickness 1.2 mm) are used. Fig. 4 shows the realized HBT DRO. The chip size is 6 mm x 6 mm.

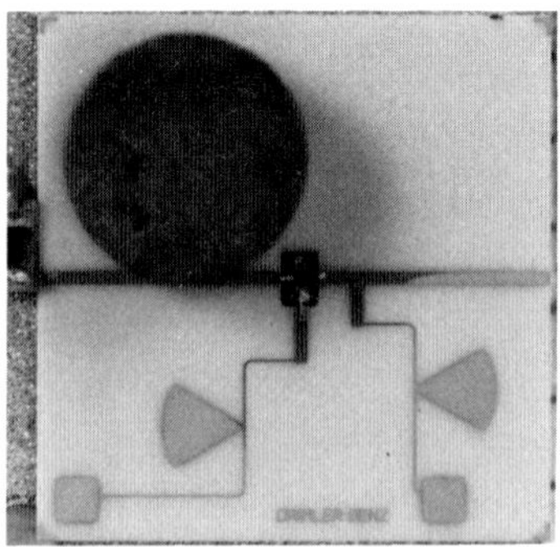

Fig. 4: Photograph of the GaInP/GaAs HBT DRO

4. OSCILLATOR PERFORMANCE

The RF performance of the GaInP/GaAs DRO has been measured in a test-fixture with a microstrip-coaxial transition, the phase noise is determined by using a HP 7000 spectrum analyzer. Dielectric resonator oscillators at 21 GHz and 22 GHz are realized using resonator pucks with slightly different geometry. The transistors are biased at $I_C = 20$ mA and $V_{CE} = 2.5$ V. An output power of +6.5 dBm with a DC to RF conversion efficiency of 13% is measured at 22 GHz. The phase noise N/C_{FM} is determined to -100 dBc/Hz at 100 kHz off carrier. At 21 GHz an output power of +5 dBm is achieved with a corresponding phase noise of -102.5 dBm at 100 kHz and -81.6 dBc/Hz at 10 kHz off carrier, respectively (Fig. 5).

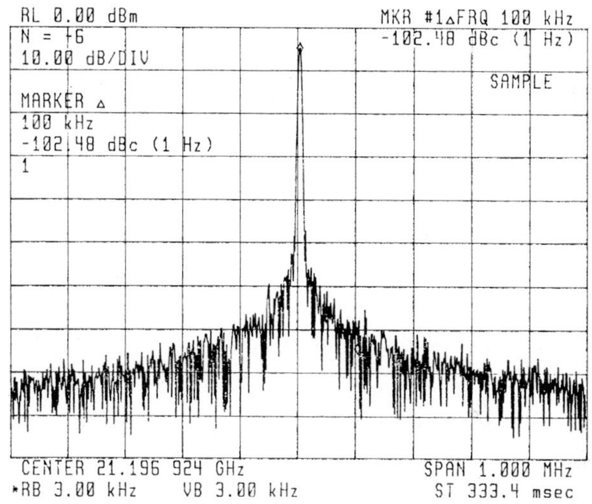

Fig. 5: Output spectrum of the 21 GHz DRO (Pout = +5 dBm, N/C_{FM} = -102,5 dBc/Hz @100 kHz and -81.6 dBc/Hz @10 kHz)

The noise behavior obtained is slightly better than recently published results on a 25 GHz AlGaAs/GaAs DRO (Ogawa et al 1990). Best reported value for a K-band MESFET DRO (23 GHz) at a modulation frequency of 10 kHz is -79 dBc/Hz (Güttich 1992); a 23 GHz Si/SiGe-HBT DRO using a comparable topology shows a phase noise of -92 dBc/Hz at 100 kHz (Güttich et al 1993).

5. CONCLUSION

State of the art topology dielectric resonator oscillators (DROs) for K-band frequencies have been fabricated using high speed selfaligned GaInP/GaAs HBT devices with an emitter area of 1.5 μm x 20 μm. At an oscillation frequency of 21 GHz an output power of +5 dBm has been measured. The excellent phase noise characteristics of -102.5 dBc/Hz at 100 GHz off carrier and -81.6 dBc at 10 kHz off carrier, respectively, make the GaInP/GaAs HBT devices well suited for local oscillator applications in sensor or digital communication systems. Monolithic integrated versions of a 24 GHz DRO and of a 35 GHz VCO are under fabrication.

6. ACKNOWLEDGEMENTS

The authors would like to thank A. Schaub for technological assistance and J. Wenger for digitizing the DRO layout. The continuous support of this work by P. Narozny and H. Dämbkes is gratefully acknowledged. TEKELEC AIRTRONIC is thanked for supplying dielectric resonator samples.

7. REFERENCES

Güttich U 1992 AEÜ 46 pp 368-70
Güttich U, Gruhle A, Luy J F 1993 Proc. 2th Conf. for Ultra High Frequency Technology (MIOP) pp 146-50
Leier H, Marten A, Bachem K H, Pletschen W, Tasker P 1993 Electron. Lett. 29 pp 868-70
Mondry M J and Kroemer H 1985 IEEE EDL-6 pp 175-7
Ogawa K, Ikeda H, Ishizaki T, Hashimoto K, Ota Y 1990 Electron. Lett. 26 pp 1514-6
Tsironis C and Pauker V 1983 IEEE MTT-31 pp 312-4

Device and circuit modeling of GaAs-based transistors

M Shur* and T Fjeldly+*

*University of Virginia, Charlottesville, VA 22903-2442, USA and +University of Trondheim, Norwegian Institute of Technology, Norway

ABSTRACT: We describe a new approach to MESFET and HFET modeling based on the universal charge control model. This approach utilizes the concept of the effective gate-to-channel capacitance which is determined by the gate-to-channel capacitance above the threshold and by the potential barrier limited charge below threshold. These models were implemented in our new circuit simulator called AIM-Spice. The model parameters are based on device physics and linked to the FET characterization techniques suitable for automatic parameter extraction.

1. INTRODUCTION

The development of accurate device models is a prerequisite for a further development of compound semiconductor technology. Accurate device models have to be based on insight into the physics of the devices. This type of insight may be obtained from numerical simulations such as self-consistent two-dimensional Monte Carlo modeling (see, for example, Hess and Kizilyalli (1986), and Jensen et al. (1991) and (1991a)). However, numerical device simulations are not directly applicable to circuit design involving tens or hundreds of devices interacting with each other and with other circuit elements, nor to device design where numerous dependencies of device characteristics on the design parameters have to be optimized, nor to device characterization where the device and process parameters must be extracted from experimental data. All these tasks require accurate analytical or semi-analytical device models. Fjeldly and Shur (1991), Shur et al (1992), and Lee et al (1993) described such models based on physical device and material parameters (rather than using look-up tables and simple interpolations of the measured device characteristics). Lee et al (1993) also presented a new integrated circuit simulator, AIM-Spice where these models were implemented. In this paper, we review GaAs MESFET and HFET models implemented in AIM-Spice.

2. UNIVERSAL FET MODEL

Our universal FET model which we applied (with minor modifications) to MOSFETs, MESFETs, and HFETs (see Lee et al (1993)) is based on the concept of the effective gate-channel capacitance:

$$c_{gc} = \frac{c_a c_b}{c_a + c_b} \tag{1}$$

where c_a and c_b are the above and below threshold gate-channel differential capacitances per unit area, respectively. (The expressions for c_a and c_b are different for MOSFETs, HFETs, and MESFETs.) The surface carrier concentration, n_s, is found by integrating eq.

(1) with respect to the intrinsic gate-to-source voltage, V_{GS}. The intrinsic channel conductance, g_{chi}, is then given by

$$g_{chi} = \frac{qn_s W \mu_n}{L} \tag{2}$$

where μ_n is the low-field carrier mobility, W is the device width, and L is the gate length. The extrinsic channel conductance is

$$g_{ch} = \frac{g_{chi}}{1 + g_{chi} R_t} \tag{3}$$

Here, $R_t = R_s + R_d$ is the sum of the source and the drain series resistances.

The output current-voltage characteristics are obtained by the interpolation between the linear and saturation region as follows

$$I_d = \frac{g_{ch} V_{ds}(1 + \lambda V_{ds})}{\left[1 + (V_{ds} / V_{sate})^m\right]^{1/m}} \tag{4}$$

where V_{ds} is the drain-to-source voltage, $V_{sate} \approx g_{ch}/I_{sat}$ is the effective extrinsic saturation voltage, I_{sat} is the drain saturation current, and m is a parameter that determines the shape of the characteristics in the knee region.

An important effect which must be taken into account for an adequate description of the FET behavior is the dependence of the threshold voltage on the drain-source voltage. This dependence can be fairly accurately described by the equation

$$V_T = V_{To} - \sigma V_{ds} \tag{5}$$

where V_{To} is the threshold voltage at zero drain-source voltage and σ is a coefficient which may depend on the gate voltage swing. As discussed by Lee et al (1993), σ also depends on the gate voltage swing:

$$\sigma = \frac{\sigma_o}{1 + \exp\left(\dfrac{V_{gto} - V_{\sigma t}}{V_\sigma}\right)} \tag{6}$$

which gives $\sigma \rightarrow \sigma_o$ for $V_{gto} < V_{\sigma t}$ and $\sigma \rightarrow 0$ for $V_{gto} > V_{\sigma t}$. Here $V_{gto} = V_{gs} - V_{To}$. The voltage V_σ determines the width of the transition between the two regimes.

The gate-source capacitance, C_{gs}, and the gate-drain capacitance, C_{gd}, are found by using an approximation similar to that used in the model by Meyer (1971):

$$C_{gs} = C_f + \tfrac{2}{3} C_{gc}\left[1 - \left(\frac{V_{sate} - V_{dse}}{2V_{sate} - V_{dse}}\right)^2\right] \tag{7}$$

$$C_{gd} = C_f + \tfrac{2}{3} C_{gc}\left[1 - \left(\frac{V_{sate}}{2V_{sate} - V_{dse}}\right)^2\right] \tag{8}$$

Here, $C_{gc} = c_{gc} LW$, V_{dse} is an effective extrinsic drain-source voltage. V_{dse} is equal to V_{ds} for $V_{ds} < V_{sate}$ and is equal to V_{sate} for $V_{ds} > V_{sate}$ and is given by the following equation:

$$V_{dse} = V_{ds}\left[1+\left(\frac{V_{ds}}{V_{sate}}\right)^{m_c}\right]^{-1/m_c} \tag{9}$$

where m_c is a constant determining the width of the transition region between the linear and saturation regimes. The capacitance C_f in eqs. (7) and (8) is the side wall and fringing capacitance which can be estimated, in terms of a metal line of length W, as

$$C_f \approx \beta_c \, \varepsilon_s W \tag{10}$$

where β_c is on the order of 0.5 (see Gelmont et al. (1991)).

This approach provides the general framework for modeling different FETs. However, many important details are different for MOSFETs, MESFETs, and HFETs. For MESFETs and HFETs with positive or slightly negative threshold voltages, the gate leakage current has to be accounted for. Also, the maximum surface carrier concentration in HFETs is limited, and at large gate voltage swings, the carriers may be induced into the wide band gap semiconductor separating the HFET channel from the gate. MESFET characteristics are strongly dependent on the doping profile in the MESFET channel.

3. MESFET MODELING

The MESFET saturation current can be expressed as

$$I_{sat} = \frac{I_{sata} I_{satb}}{I_{sata} + I_{satb}} \tag{11}$$

where I_{sata} and I_{satb} are the saturation currents above and below threshold, respectively. The saturation current above threshold is given by:

$$I_{sata} = \frac{2\beta V_{gte}^2}{\left(1+2\beta V_{gte} R_s + \sqrt{1+4\beta V_{gte} R_s}\right)\left(1+t_c V_{gte}\right)} \tag{12}$$

Here, the transconductance parameter can be written as

$$\beta = \frac{2\varepsilon_s v_s W}{d\left(V_{po} + 3V_L\right)} \tag{13}$$

where $V_L = F_s L$, F_s is the saturation electric field, d is the channel thickness, v_s is the saturation velocity, and

$$V_{gte} = \frac{V_{th}}{2}\left[1+\frac{V_{gt}}{V_{th}} + \sqrt{\delta^2 + \left(\frac{V_{gt}}{V_{th}} - 1\right)^2}\right] \tag{14}$$

We note that V_{gte} approaches asymptotically V_{th} below threshold and V_{gt} above threshold. The parameter δ determines the width of the transition region ($\sim \delta V_{th}$). Typically, $\delta = 5$ is a good choice.

The value of the transconductance compression factor, t_c, depends on the doping profile and may also depend on the properties of the substrate-channel interface and other factors. Based on numerous simulations, we have concluded that $t_c = 0$ usually gives a good fit for uniformly doped or ion-implanted devices with pinch-off voltages smaller than approximately 2.5 V. For MESFETs with pinch-off voltages of about 6 V, Statz et al. (1987) found that $t_c \approx 0.1\ \mathrm{V}^{-1}$ gave an excellent fit. The saturation current in the subthreshold regime for relatively long channel devices is given by

$$I_{satb} = \frac{qn_o\mu_n V_{th} W}{L}\exp\left(\frac{V_{gt}}{\eta V_{th}}\right) \tag{15}$$

where

$$n_o = \frac{\varepsilon_s \eta V_{th}}{qd} \tag{16}$$

The MESFET gate capacitance above threshold is given by

$$c_a = \varepsilon_s / d_d \tag{17}$$

For an arbitrary doping profile, d_d can be expressed as follows using the gradual channel approximation:

$$V_{bi} - V_{GS} + V = \frac{q}{\varepsilon_s}\int_0^{d_d} y\, N(y)\, dy \tag{18}$$

Here, V_{GS} is the intrinsic gate-source voltage, V is the channel voltage, V_{bi} is the built-in voltage, y is the distance into the semiconductor from the gate electrode and $N(y)$ is the doping density profile of the conducting channel. In eq. (16) we have assumed that the doping profile does not vary too abruptly, so that the extent of the partially depleted boundary layer separating the neutral channel from the depletion region is small compared to the characteristic length of the doping profile variation.

The threshold voltage, V_T, corresponding to the gate-drain voltage where the channel is fully depleted, is given by

$$V_T = V_{bi} - V_{po} \tag{19}$$

where the pinch-off voltage, V_{po}, can be expressed as

$$V_{po} = \frac{q}{\varepsilon_s}\int_0^{d} y\, N(y)\, dy \tag{20}$$

For a uniform doping profile, $V_{po} = qNd^2/(2\varepsilon_s)$ where d is the channel thickness.

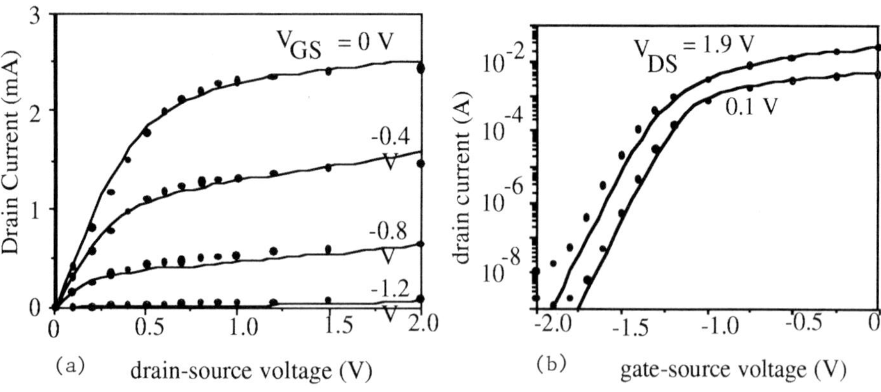

Fig. 1. Above (a) and below (b) threshold measured (symbols) and calculated (solid lines) I-V characteristics for MESFET with L= 1 μm, W = 20 μm, d = 0.12 μm, μ_n = 0.23 m^2/Vs, v_s = 1.5x10^5 m/s, m = 2.5, V_{bi} = 0.75 V, V_{To} = −1.26 V, η = 1.73, $R_s = R_d$ = 31 Ω, λ = 0.045 V^{-1}, t_c = 0, σ_o = 0.081, V_σ = 0.1 V, $V_{\sigma t}$ = 1.0 V. (From Shur et al (1992).)

This basic model (implemented in AIM-Spice) is in a good agreement with experimental data for GaAs MESFETs when no gate leakage current is important (see Fig. 1).

Further model improvements should include the incorporation of the temperature dependencies of device parameters (see Conger (1992)), a more realistic description of the leakage current, and accounting for the frequency dependence of MESFET parameters (see Conger et al (1993)).

4. HFET MODELING

In all HFET structures, a conducting channel forms in the semiconductor at the heterointerface with the wider band gap semiconductor separating the channel from the gate. The dopants in the wide gap semiconductor layer or in the channel control the device threshold voltage. Often, these donors are separated from the two-dimensional electron gas by a thin spacer layer. There are many different HFET structures (see, from example Shur and Fjeldly (1993)). However, most of these structures can be modeled using the universal FET model.

The HFET threshold voltage (at zero drain bias) is given by

$$V_{To} \approx \phi_b - V_N - \Delta E_{ceff} / q \tag{21}$$

where ϕ_b is the Schottky barrier height voltage, ΔE_{ceff} is the effective conduction band discontinuity, and $V_N = \dfrac{qN_d d_d^2}{2\varepsilon_i}$ for a uniformly doped wide band gap semiconductor where N_d is the doping concentration, d_d is the thickness of the doped layer, and ε_i is the dielectric permittivity of the wide band gap semiconductor. The electron sheet density in the channel can be expressed as

$$n_s' = 2n_o \ln\left[1 + \frac{1}{2}\exp\left(\frac{V_{gt}}{\eta V_{th}}\right)\right] \tag{22}$$

where

$$n_o = \frac{\varepsilon_i \eta V_{th}}{2q(d_i + \Delta d)} \tag{23}$$

(see Lee et al. (1993) for more details). This dependence becomes invalid at large gate-source voltages because the electron quasi-Fermi level, E_{Fn}, in the wide band gap layer may approach the bottom of the conduction band so that the carriers are induced into the conduction band minimum, and the value of the electron sheet density to be used in the expression for the intrinsic linear channel conductance can be approximated by

$$n_s = \frac{n_s'}{\left[1 + \left(n_s' / n_{max}\right)^{\gamma}\right]^{-1/\gamma}} \tag{24}$$

where γ is the interpolation parameter. A basic HFET model which does not account for the sublinear dependence of the electron sheet density on the gate-source voltage leads to the following expression for the drain saturation current:

$$I_{sat}' = \frac{g_{chi}' V_{gte}}{1 + g_{chi}' R_s + \sqrt{1 + 2 g_{chi}' R_s + \left(V_{gte}/V_L\right)^2}} \tag{25}$$

Here, $V_L = F_s L$ where F_s is the saturation field and L is the gate length, R_s is the source series resistance, $g_{chi}' = q n_s' W \mu_n / L$ is the intrinsic linear channel conductance where W is the gate width and μ_n is the low-field electron mobility. The drain saturation current is given by

$$I_{sat} = \frac{I'_{sat}}{\left[1+\left(I'_{sat}/I_{max}\right)^{\gamma}\right]^{-1/\gamma}} \tag{26}$$

where $I_{max} = q n_{max} v_s W$ is an upper limit for the channel current determined by n_{max} and the saturation velocity, v_s.

At small drain-source voltages, the HFET gate-to-channel capacitance, c_{gc} can be expressed as

$$c_{gc} = q\frac{dn_s}{dV_{gs}} \approx \frac{c'_{gc}}{\left[1+\left(n'_s/n_{max}\right)^{\gamma}\right]^{1+1/\gamma}} \tag{27}$$

where

$$c'_{gc} = \left[\frac{d_i + \Delta d}{\varepsilon_i} + \frac{\eta V_{th}}{q n_o}\exp\left(-\frac{V_{gt}}{\eta V_{th}}\right)\right]^{-1} \tag{28}$$

When n_s' becomes comparable to or larger than n_{max}, c_{gc}' will noticeably differ from c_{gc}.

As can be seen from Fig. 2, this model is in good agreement with measured data.

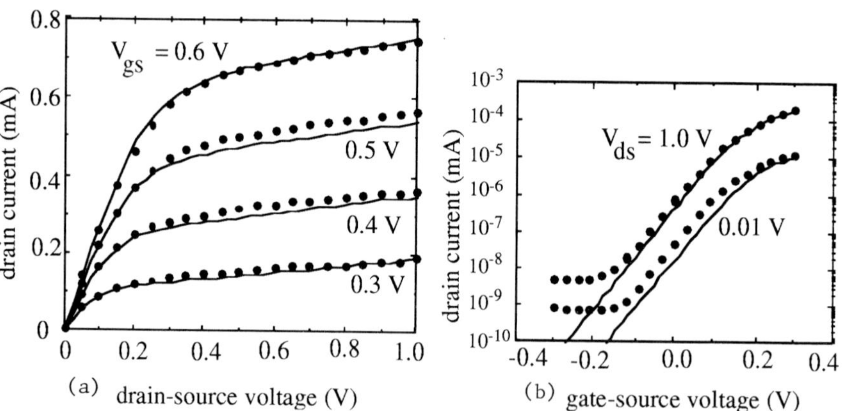

Fig. 2. Above-threshold (a) and below threshold (b) experimental (symbols) and calculated (solid lines) I–V characteristics for an HFET. L = 1 µm, W = 10 µm, ε_s = 1 .14x10^{-10} F/m, $Al_{0.3}Ga_{0.7}As$ thickness, d_i = 0.04 µm, Δd =4.5x10^{-9} m, mobility, μ_n = 0.4 m^2/Vs, saturation velocity, v_s = 1.5x10^5 m/s, threshold voltage at zero drain bias, V_{To} = 0.15 V, subthreshold ideality factor η = 1.28, source and drain series resistances, $R_s = R_d$ = 60 ohm, output conductance parameter, λ = 0.15 V^{-1}, n_{max} = 2x10^{16} m^{-2}, $m = \gamma = \delta$ = 3.0, σ_o = 0.057, V_σ = 0.1 V, $V_{\sigma t}$ = 0.3 V. (After Fjeldly and Shur (1991).)

5. LEAKAGE CURRENT

For enhancement-mode compound semiconductor FETs, the gate current can play a dominant role and may even affect the value of the drain-source current. Fig. 3 shows an equivalent circuit which accounts for the gate current. As can be seen from Fig. 4, such a model is in good agreement with measured data.

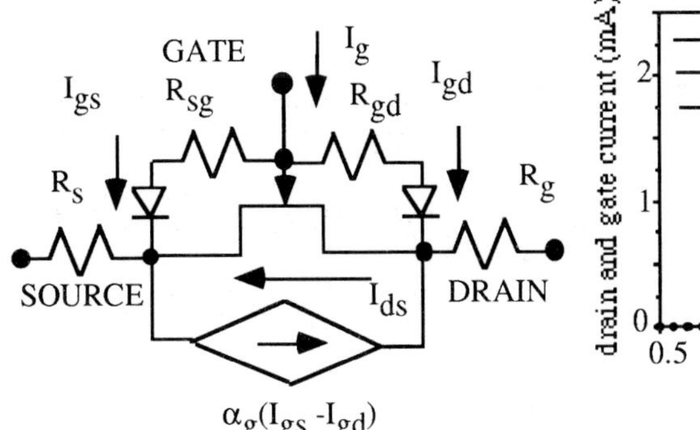

Fig. 3. FET equivalent circuits that takes into account the effect of the gate current on the channel current (after Ruden et al (1989)).

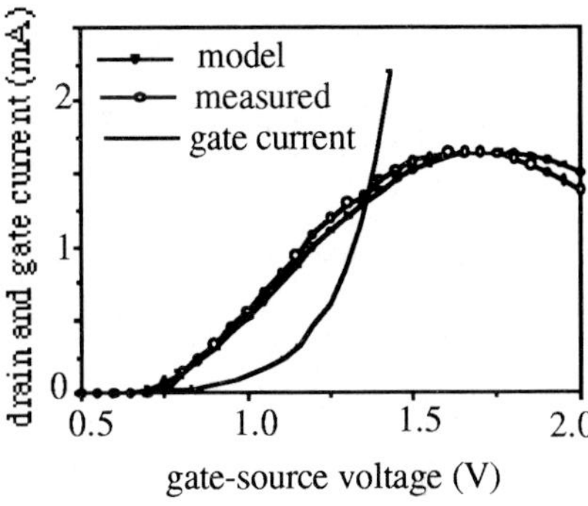

Fig. 4. Drain and gate current in a 1 μm gate length *n*-channel AlGaAs/GaAs HFET (from Ruden et al (1989))

A further improvement of such a model may be achieved by accounting for electron heating in the device channel. Berroth et al (1988) introduced effective electron temperatures at the source side and the drain side of the channel. The electron temperature at the source side of the channel is taken to be close to the lattice temperature, i.e., $T_s \approx T$, and the drain side electron temperature, T_d, is assumed to increase with the drain-source voltage to reflect the heating of the electrons in this part of the channel where the electric field is large. These two temperatures are substituted into the diode equations describing the diodes in Fig. 3.

This gate leakage model represented by the equivalent circuit in Fig. 3 is less accurate for HFETs. In these devices, the gate current is limited either by the Schottky barrier or by the conduction band discontinuity, depending on the gate bias. This may be more accurately accounted for by replacing each diode in Fig. 3 by two diodes in series having different saturation current currents and different ideality factors (see Chen et al (1988)). Also, at high gate biases, the electron heating in HFETs becomes quite substantial and the distribution of the gate current along the channel becomes very important (see Baek and Shur (1990)). Under such conditions, the gate current may sharply increase with the drain bias. The equivalent circuit showed in Fig. 3 does not adequately describe this regime of operation. Finally, in self-aligned devices, an additional gate leakage current path from the n^+ regions implanted for the source and drain ohmic contacts to gate contact may become important or even dominant (see Schuermeyer et al (1992)).

6. CONCLUSIONS

The universal FET model is adequate for the circuit simulation of GaAs-based MESFETs and HFETs. Similar models are applicable for submicron CMOS and amorphous silicon and polysilicon TFTs.

7. ACKNOWLEDGMENTS

This work has been supported by the Royal Norwegian Council for Scientific and Industrial Research, the NATO Scientific Affairs Division, Office of Naval Research, and Martin Marietta.

8. REFERENCES

Baek J H and M. Shur M 1990 IEEE Trans. Electron Dev. ED-37 1917-21

Berroth M, Shur M and Haydl W 1988 Extended Abstracts of the 20th Conference on Solid State Devices and Materials (SSDM-88) (Tokyo) pp. 255-258

Conger J 1992 Characterization, Modeling and Simulation of Compound Semiconductor Field-Effect Transistors and Integrated Circuits (University of Minnesota: Ph.D. Thesis)

Conger J, Peczalski A and Shur M 1993 J. Sol. St. Cir. to be published

Chen C H, Baier S, Arch D, and Shur M 1988 IEEE Trans. Electron Dev. ED-35 570-577

Fjeldly T A and Shur M 1991 Proceedings of the 11th European Microwave Conference Workshop Volume (Stuttgart) pp. 198-205

Gelmont B, Shur M and Mattauch R J 1991 IEEE Trans. Micr. Theory and Tech. 39 857-63

Hess K and Kizilyalli C 1986 IEDM Technical Digest (Los Angeles: IEEE) pp 556-8

Jensen G U, Lund B, Fjeldly T A and Shur M 1991 IEEE Trans. Electron Dev. ED-38 840-51

Jensen G U, Lund B, Shur M and Fjeldly T A 1991 Com. Phys. Comm. 67 1-61

Lee K, Shur M, Fjeldly T A, Ytterdal T 1993 Semiconductor Device Modeling for VLSI. (New Jersey: Prentice Hall)

Meyer J E 1971 RCA Review 32 42-63

Ruden P P, Shur M, Akinwande A I and Jenkins P 1989 IEEE Trans. Electron Dev. ED-36 453-6

Schuermeyer F L, Martinez E, Shur M, Grider D E and Nohava J 1992 Electronics Lett. 28 1024-5

Shur M, Fjeldly T A, Ytterdal Y and Lee K 1992 Intern. J. High Speed Electr. 3 201-33

Shur M and Fjeldly T A 1993 HEMT Modeling, in Semiconductor Device Modeling (London: Springer Verlag) pp. 56-73

Statz H, Newman P, Smith I W, Pucel R A and Haus H A 1987 IEEE Trans. Electron Dev. ED-34 160

Paper presented at the Int. Symp. GaAs and Related Compounds, Freiburg, 1993

Double modulation-doped strained-channel AlInAs–GaInAs–AlInAs HEMT structures operating at high drain current densities and millimeter-wave frequencies

François Gueissaz, Takatomo Enoki and Yasunobu Ishii

NTT LSI Laboratories, 3-1, Morinosato Wakamiya, Atsugi-Shi, Kanagawa-Pref., 243-01 Japan

ABSTRACT: Double modulation-doped $Al_{0.48}In_{0.52}As$ / $Ga_{0.35}In_{0.65}As$ HEMTs are shown to yield unprecedented drain current modulation characteristics associated with high maximum unilateral power and current gain cut-off frequencies (f_{max} and f_T). Fabricated 0.2 µm-gate-length HEMTs yield maximum f_{max} / f_T values of 290 / 112 GHz at 800-900 mA/mm, and 160 / 80 GHz at 1350 mA/mm. The high 2DEG density and high saturation current of the channel access regions as well as a low output conductance of the channel are the main features of the presented heterostructure.

1. INTRODUCTION

High electron mobility field-effect transistors (HEMTs) with high current density and power gain cut-off frequencies (f_{max}) are necessary in order to efficiently generate power at millimeter-wave frequencies, for example in portable communication devices. Double modulation-doped heterostructures (DMDH) have been shown to be capable of higher power density by increasing either the breakdown voltage or the maximum channel current, depending on the design (Matloubian 1991, Kwon 1993). Since the common AlInAs / GaInAs HEMT grown on an AlInAs buffer layer is a double heterostructure, one therefore simply has to introduce donor atoms in the buffer layer, close the GaInAs channel, to make it a DMDH HEMT. However, obtaining an adequate layer design for high current density devices requires further considerations such as growth temperature (Brown 1991), doping, channel thickness and composition. High electron densities are essential to achieve millimeter-wave high current density HEMTs because they result in high modulation efficiency and high saturation current densities in the access regions between the gate and the ohmic contacts. We report here on the growth of high quality $Al_{0.48}In_{0.52}As$ / $Ga_{0.35}In_{0.65}As$ / $Al_{0.48}In_{0.52}As$ DMDHs on InP by molecular beam epitaxy (MBE) at low substrate temperature, and submicron-gate-length HEMTs static and microwave characterization results showing unprecedented charge carrier modulation efficiency at millimeter-wave frequencies.

2. MATERIAL GROWTH AND CHARACTERIZATION

Introducing dopants below the channel under the regular growth conditions rapidly degrades the electric mobility of the two-dimensional electron gas (2DEG). This effect apparently originates mainly in the thermally assisted surface segregation of the dopant atoms into the channel. Therefore, reducing the growth temperature of the first AlInAs spacer layer from typically 500 °C to 350 °C or less permits mobilities to be obtained close to those of single modulation-doped heterostructures by suppressing the surface segregation of the dopants (Brown 1991). Our approach consisted in growing the entire DMDH at low temperature, about 300 to 350 °C, and reduced

V/III elements flux ratio to simplify the growth procedure. The structure utilizes planar doping, which proved to be effective even at low growth temperatures. Due to the nearly flat band condition at the top of the buffer, one has to keep the amount of doping below the channel to the strict necessary amount of charge increase in the 2DEG, that is 1.5 to $2.0 \cdot 10^{12}$ cm^{-2}. The thickness of the channel has to be reduced to less than 200 Å to avoid a double-peaked electron density distribution, which causes a loss of linearity in the charge control characteristic. Finally, the InAs mole fraction of the channel has to be increased, to prevent the parasitic carrier conduction in the barriers by increasing the conduction band discontinuity.

The layer structure is depicted in figure 1. Prior to growth, the native surface oxide is desorbed from the (100)-oriented semiinsulating InP : Fe substrates under arsenic flux in a separate chamber. The MBE machine operates (both oxide desorbtion and layer growth) with an indium-free mounting system. The entire structure is grown at 300 to 350 °C at a reduced arsenic flux reproducibly yielding mirrorlike surfaces. From several growth parameters optimization experiments, we determined that the optimum V/III elements flux ratio for growth at 500 °C has to be reduced by a factor 6 to 8 when lowering the temperature to 300 °C. The buffer consists of a single 4000 Å AlInAs layer. All AlInAs layers are lattice matched to InP and the InAs mole fraction of the GaInAs channel is set to 0.65. The first planar doping is placed near the top of the buffer layer, 50 Å below the channel, with an ionized impurities sheet density of $1.70 \cdot 10^{12}$ cm^{-2}. The channel thickness is 120 Å, yielding an almost single-peaked electron distribution without significant quantum size effect reducing its confinement properties. The second planar doping is placed on the top of a 30 Å thick AlInAs spacer, above the channel, with an ionized impurities sheet density of $5.0 \cdot 10^{12}$ cm^{-2}. A 250 Å thick AlInAs layer, in which the gate recess will subsequently be adjusted, ends the active layer sequence. Finally, 100 Å of AlInAs and 50 Å of $Ga_{0.35}In_{0.65}As$, both heavily n^+ doped, are grown successively to form the contact layer for non-alloyed ohmic contacts (Enoki 1991). The total initial electron sheet density (n_s) including electrons in the cap layer, as found from Hall measurements, is $1.2 \cdot 10^{13}$ cm^{-2} at room temperature and the associated sheet resistance 120 Ω/square. The electron mobility vs electron sheet density characteristic is given in figure 2, where the reduction of electron sheet density is obtained by carefully etching the structure's surface repeatedly. Once the three-dimensional electron gas vanishes from the barrier layer above the channel, a maximum electron mobility of 7,500 cm^2/Vs associated

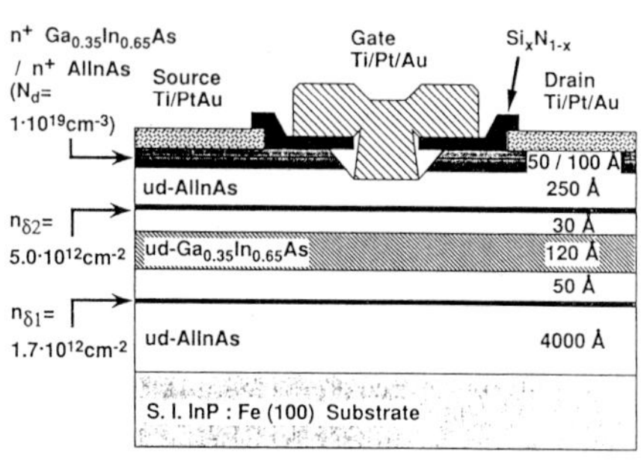

Fig. 1. Epitaxial layer and HEMT device design.

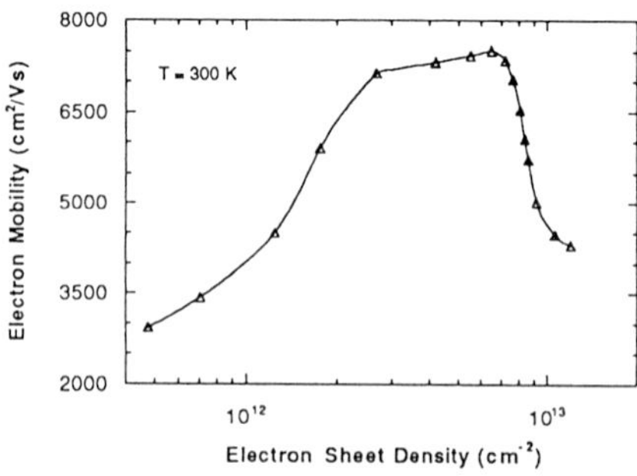

Fig. 2. Hall electron mobility vs sheet density of the DMDH at room temperature.

with an n_s of $6.5 \cdot 10^{12}$ cm^{-2} is observed at room temperature for the 2DEG alone. Self-consistent Schrödinger-Poisson energy-band calculation results are shown in figure 3 at various 2DEG sheet densities. These show that over 90 % of the electrons are confined within the 120 Å wide channel at an n_s of $6.5 \cdot 10^{12}$ cm^{-2}, where the electron mobility starts to degrade, and demonstrate the formation of an almost single-peaked 2DEG. The arrow indicates an increase of electron density in the top barrier, at an n_s of $6.0 \cdot 10^{12}$ cm^{-2}. Moreover, it is interesting to note that a saturation of mobility increase occurs when n_s reaches $2.7 \cdot 10^{12}$ cm^{-2}, where the second subband begins to be filled and intersubband scattering may add to polar optical phonon and defect related scattering. In fact, at low temperatures, the electron mobility even decreases before increasing again when the second subband is mostly filled (Störmer 1982, Gueissaz 1991).

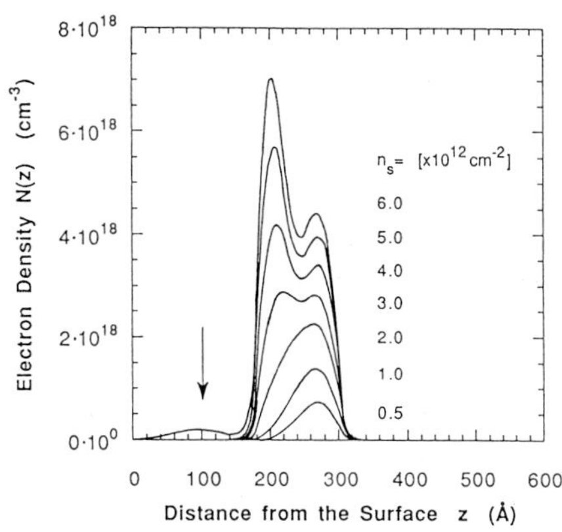

Fig. 3. Electron density distribution of the DMDH at various total sheet densities.

3. DEVICE FABRICATION AND DC CHARACTERIZATION

After defining of the active areas by mesa etching, the formation of non-alloyed ohmic contacts is done by depositing a Ti / Pt / Au metallization after cleaning the surface. Then, the entire surface is passivated with Si_xN_{1-x} by photo assisted chemical vapor deposition and 0.2 µm-gate-length recess windows are formed between the ohmic contacts (gate footprints), 0.5 µm away from the source contact, by electron-beam lithography and reactive ion etching (RIE). Optical lithography is further used to produce 0.7 µm-length gate patterns (gate top portion) aligned on the 0.2 µm windows in Si_xN_{1-x} and the structures are recessed in a citric acid : H_2O_2 : H_2O solution before depositing the Ti / Pt / Au gate metallization. A cross-section of the completed devices is shown on figure 1. The specific ohmic contact resistance is less than 0.1 Ω·mm, due to the low barrier height of Ti / n^+ GaInAs, easy tunnelling through the highly doped AlInAs / GaInAs cap and low resistivity of the undoped 250 Å thick AlInAs layer on both sides of which electrons naturally diffuse. The drain current density vs drain voltage characteristics of a 0.2 x 50 μm^2 (L_g x W_g) HEMT are shown on figure 4. The maximum saturated drain current density (I_d') at a gate-to-source bias (V_{gs}) of +0.4 V is 1450 mA/mm. Regardless of the high cap doping and channel InAs mole fraction, the drain saturation characte-

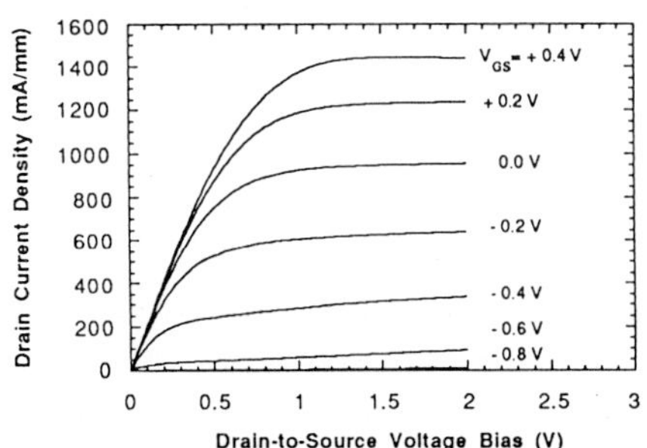

Fig. 4. Drain current density vs drain bias voltage at various gate-to-source bias voltages, starting from +0.4 V for the top curve with -0.2 V steps, for a 0.2 x 50 μm^2 (L_g x W_g) HEMT.

ristics are excellent and yield a maximum trans- to output conductance ratio (g_m'/g_d') ratio of 53, at the g_m' peak of 1620 mS/mm, with no kink effect. In the case of the DMDH, we suggest that the absence of kink effect and high output conductance generally observed in single modulation-doped heterostructures (SMDHs) originates from the potential pinning effect of the bottom-side planar doping. This prevents the accumulation of impact-ionization generated holes at the bottom interface of the channel and its modification of the vertical electrostatic potential. Moreover, inverted HEMTs, where impact-ionization generated holes can drift to the gate should also exhibit no kink-effect and have a low output conductance. Such a result was reported by Schmitz (1991).

The transconductance vs drain current density characteristic is further shown on figure 5 in comparison with state-of-the-art results (Nguyen 1992) obtained on SMDHs. The DMDH devices show an unprecedented charge carrier modulation ability for a field-effect device. In fact, the intrinsic g_m' is estimated to 2.1 S/mm with a typical total source access resistance of 0.14 Ω·mm. Moreover, the g_m' value remains higher than 1 S/mm for a gate bias swing of 0.9 V.

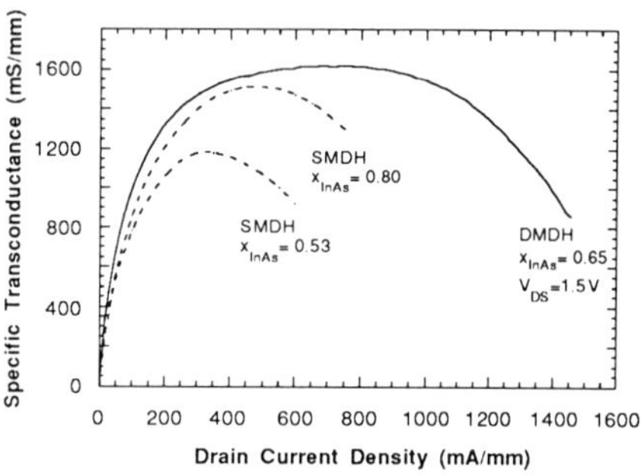

Fig. 5. Transconductance vs drain current density of a 0.2 x 50 μm^2 (L_g x W_g) DMDH-HEMT, compared with SMDH-HEMTs with L_g = 0.065 μm (Nguyen 1992).

From the gradual channel approximation of the HEMT (Delagebeaudeuf 1982), under the assumption that the saturation velocity (v_{sat}) is reached at the drain edge of the channel ($x = L_g$), we find the following expression for the ratio of n_s ($x = L_g$) to the n_s at the source edge of the channel ($x = 0$) :

$$\frac{n_s(x = L_g)}{n_s(x = 0)} = \frac{I_d'/q v_{sat}}{C''(V_{gs} - V_{th})/q} = \frac{1}{\sqrt{1 + \frac{2\varepsilon_s v_{sat}^2 L_g}{(d + \Delta d)\mu I_d'}}} \qquad (1)$$

The gate capacitance per unit area and its mean distance to the 2DEG are C" and (d + Δd), respectively. V_{th} is the threshold voltage, ε_s and μ are the channel effective dielectric constant and electron mobility, respectively. With a saturation velocity of $2.7 \cdot 10^7$ cm/s and a (d + Δd) of 140 Å for our devices, we find that the ratio of n_s given in (1) reaches 0.85 at I_d' = 800 mA/mm, where the maximum g_m' occurs, corresponding to an n_s of $2.2 \cdot 10^{12}$ cm^{-2} at the source edge of the channel. At the n_s ratio of 0.85, the gradual channel approximation is very close to the saturation velocity model and it seems that neither non-constant velocity charges nor a low 2DEG n_s could degrade the performance of the device. This holds even at 1450 mA/mm where the n_s ratio is 0.91 and the n_s at the source edge is $3.7 \cdot 10^{12}$ cm^{-2}. It follows that the current and transconductance limitations are likely to solely depend on the characteristics of the access regions outside the channel, close to the recess edges. Since the gate metallization does not overlap the edges of the recessed region, these can become the current limiting elements when the gate potential minus the Schottky barrier height reaches or exceeds the surface potential of the structure. Ultimately, the absolute maximum I_d' value of 1700 mA/mm extrapolated from the $g_m'(I_d')$ characteristic corresponds to a maximum n_s value of about $4.3 \cdot 10^{12}$ cm^{-2} at the source edge, which can possibly be fixed by the surface potential near the recess edge on the source side of the channel. Finally, it is instructive to consider the fact that the additional absolute maximum I_d' of about 700 mA/mm over that obtained from the same structure without bottom planar doping is consistent with the density $n_{\delta 1}$ of the bottom planar doping multiplied by the saturation velocity.

4. RF CHARACTERIZATION

On-wafer s-parameters measurements were performed from 1 to 40 GHz and the obtained S-matrices were then corrected for waveguide parasitics by using measurements on open and short HEMT patterns. The microwave performances of a 0.2 x 50 μm^2 device passivated with Si_xN_{1-x} are represented in figure 6 against the drain current density. The maximum current gain cut-off frequency (f_T) is 112 GHz at 810 mA/mm and the maximum Mason's unilateral power gain cut-off frequency (f_{max}) is 290 GHz. At a record drain current density of 1350 mA/mm, the f_T and f_{max} values are as high as 80 and 160 GHz, respectively. The high f_{max} figures can be explained by the low output conductance inherent to the DMDH.

Theoretically, the saturation velocity model with $v_{sat} = 2.7 \cdot 10^7$ cm/s yields a maximum f_T of about 130 GHz for this device, including parasitic capacitances and resistances. A similar discrepancy of f_T has been reported by other groups and it has been speculated (Kwon 1993) that the inverted heterointerface might have an impact on the saturation velocity, leading to lower g_m' values for a given gate capacitance. This however does not seem to be compatible with the fact that reasonably high electron mobilities can be achieved, indicating the good quality of the inverted heterointerface. We further performed two-dimensional device simulations using the drift and diffusion equations to calculate f_T from the ratio of current to total charge modulation in the structure. These calculations suggest that no increase of capacitance occurs in the DMDH HEMT under the assumption that the carriers are all confined within the channel and f_T should not be lower in the case of the DMDH structure.

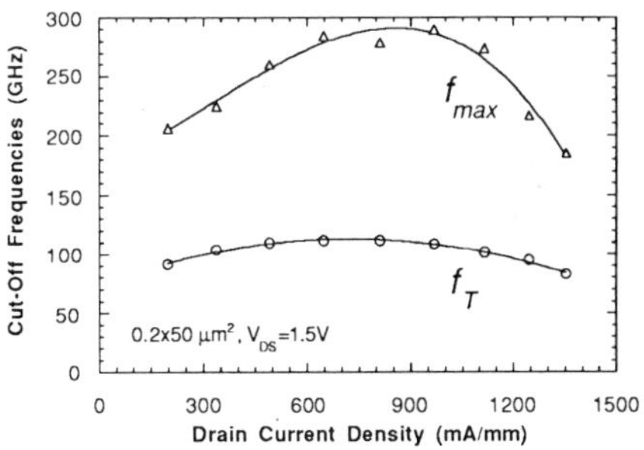

Fig. 6. Cut-off frequencies (f_{max} & f_T) vs drain current density at a drain-to-source bias voltage of +1.5 V of a 0.2 x 50 μm^2 (L_g x W_g) HEMT.

The concept of modulation efficiency (Me), also derived from the gradual channel approximation, and including the contribution of low velocity charge density appearing in the barrier layers (Q_B) in the realistic heterostructure at high n_s and equilibrium was discussed by Foisy (1988). It is expressed as :

$$Me = \frac{\delta Q_S}{\delta(Q_S + Q_L + Q_B)} = \frac{f_T}{v_{sat}/2\pi L_g} \qquad (2)$$

In this definition, Me represents the ratio of f_T to the f_T given by the saturation velocity model. Q_S and Q_L are the saturation velocity model and excess non-constant velocity channel charge densities, where Q_L becomes insignificant at high drain current values. Consequently, at high gate bias or high current, Q_B represents the degrading factor for Me and we would like to point out that the bottom planar doping significantly lowers the potential of the conduction band edge below the channel (by 0.25 to 0.3 eV in this structure). As a result, under non-equilibrium, the probability of real-space transfer of L-valley electrons into the buffer is enhanced. This possibly starts at the source side of the channel and may contribute to the Q_B term of (2). In other words, the average velocity can be reduced by such a real-space transfer and thus the f_T value.

5. CONCLUSIONS

In conclusion, we have demonstrated the growth of a double modulation-doped strained-channel heterostructure at low substrate temperature with a very high 2DEG density and a low sheet resistance. HEMT devices with 0.2 µm-gate-length fabricated on this type of material yielded unprecedented charge modulation characteristics up to drain current densities of 1450 mA/mm and f_T figures similar to their counterparts fabricated on single modulation-doped heterostructures. The f_{max} figures up to more than 1000 mA/mm are excellent as compared to any kind of field-effect device of this size. The key features of the double modulation-doped heterostructure lie in the very high 2DEG density and excellent drain saturation characteristics, both attributed to the implementation of a planar doped layer below the channel.

ACKNOWLEDGEMENTS

The first author is indebted to R. Houdré with whom this work was started at the EPFL, Switzerland. We thank T. Ishikawa from NATC for the growth of the materials, M. Tomizawa for having provided Schrödinger-Poisson 1-D and drift-diffusion 2-D simulators and K. Hirata for his support.

REFERENCES

Brown A S, Metzger R A, Henige J A, Nguyen L D, Lui M and Wilson R G 1991 Appl. Phys. Lett. 59 (27) pp 3610-3612

Delagebeaudeuf D and Linh N T 1982 IEEE Trans. Electron Dev. Vol. ED-29 (6) pp 955-960

Enoki T, Ishii Y and Tamamura T 1991 Proceedings of the Third Int. Conf. InP Rel. Mat. IPRM Cardiff

Foisy M C, Tasker P J, Hughes B and L F Eastman 1988 IEEE Trans. Electron Dev. ED-35 (7) pp 871-878

Gueissaz F and Houdré R 1991 unpublished results on low temperature electron mobility measurements in double-modulation-doped heterostructures

Kwon Y, Pavlidis D, Brock T, Ng G I, Tan K L, Velebir J R and Streit D C 1993 Fifth Intern. Conf. InP & Related Materials Paris

Matloubian M, Nguyen L D, Brown A S, Larson L E, Mendeles M A and Thompson M A 1991 IEEE MTT-S Digest pp 721-724

Nguyen L D, Brown A S, Thompson M A, Jelloian L M, Larson L E and Matloubian M 1992 IEEE Electron Dev. Lett. Vol. 13 No. 3 pp 143-145

Schmitz A E, Nguyen L D, Brown A S and Metzger R A 1991 Proceedings of the Device Research Conference IIIB-3 DRC

Störmer H L, Gossard A C and Wiegmann W 1982 Sol. State Commun. 41(10) pp 707-709

Paper presented at the Int. Symp. GaAs and Related Compounds, Freiburg, 1993

AlGaInP/GaInAs/GaAs-MODFETs with carbon doped p^+-GaAs gate structure, a novel device concept, its implementation and device properties

K.H. Bachem, W. Pletschen, K. Winkler, J. Fleissner, C. Hoffmann and P.J. Tasker

Fraunhofer-Inst. of Applied Solid State Physics, Tullastr. 72, D-79108 Freiburg, Germany

ABSTRACT: We report a fabrication scheme of a new AlGaInP/GaInAs/GaAs-MODFET device. The structure incorporates two novel features: An AlGaInP barrier which provides a larger conduction band offset than the commonly used AlGaAs and a highly carbon-doped p-type GaAs gate structure which supports a very simple fabrication scheme. Using 1 μm optical lithography we have fabricated first demonstrator devices with F_t and F_{max} of 60 and 140 GHz, respectively.

INTRODUCTION

Heterostructures incorporating the wide bandgap material $(Al_xGa_{1-x})InP_2$ have gained considerable interest for the fabrication of a variety of optoelectronic devices. These structures expand the spectral range of light emitting diodes and injection lasers into the visible region and some of these devices have promising prospects for becoming mass products in the near future. Overlooking the experience grown up in this area over the last ten years and encouraged by the successful work on $GaInP_2$/GaAs based heterojunction bipolar transistors in our group [1] and elsewhere [2] it seemed attractive to explore the potential of these heterostructures for fabrication of MODFET devices. Such an undertaking seems to be straightforward considering the many proposals on this subject published in textbooks already. However, such an approach is bearing considerable risks in fact, because the data base is quite weak. Particulary, the published band offset data scatter over a wide range, and transport properties as well as maximum available densities of two dimensional electron gases formed at interfaces adjacent to AlGaInP barrier structures have been studied only recently [3] and have not been confirmed until now. In this paper we present first results of real MODFET devices based on the heterostructure system AlGaInP/GaInAs/GaAs. Beside substituting the conventional AlGaAs barrier structure by AlGaInP we have introduced another structural feature: The Schottky gate structure has been replaced by a heavily carbon doped p^+-GaAs structure. This approach lends easy support to a self-aligned fabrication scheme and has the potential to fabricate devices with quarter micron gate structures by using optical lithography tools only.

LAYER GROWTH AND DEVICE FABRICATION

The epitaxial layer structure has been grown by low pressure metalorganic vapour phase epitaxy in a horizontal reactor cell at reduced pressure (100 mbar). Except for the heavily carbon doped p^+-GaAs layer, which was grown from trimethylgallium and trimethylarsine at a deposition temperature of 540 °C /1/, the layer stack was formed from the trimethylalkyls of aluminum, gallium and indium together with arsine and phosphine at a temperature of 650 °C. For n-type doping silane was added as dopant source. The growth rate for all individual layers was adjusted to approximately 1 μm per hour. The layer structure was grown on (100)-oriented, semi-insulating GaAs substrates (50 mm) starting with a thick (1 μm), entirely depleted GaAs buffer layer which was followed by a pseudomorphic InGaAs channel. The In content was kept relatively low (x=0.15) and thickness (20 nm) was choosen safely below the critical value. The barrier layer structure consists of three layers. The layer in the center is of 35 nm AlGaInP (Al/Ga=1/4,[Al+Ga]/In=1/1), homogeniously silicon doped over the almost entire thickness. Only, the first grown 2 nm of this layer have not been doped intentionally. This core of the barrier structure is separeted from the electron gas channel by a thin (1.8 nm) not intentionally doped AlGaAs interlayer (Al/Ga=1/5). Its purpose has been explained already, elsewhere /2/. This interlayer and the non-doped part of the core layer form the so called spacer. The top part of the barrier structure is a 5 nm thick, silicon doped GaInP layer (Ga/In=1/1). This Al free layer becomes exposed to atmosphere during the course of processing and forms the base for the ohmic contacts. It prevents oxidation of the Al-containing core layer during processing. The layer structure is completed by growing the carbon doped GaAs gate layer (6×10^{19} cm^{-3}; 250 nm), which is 100 nm thicker than the ohmic contact structure (150 nm). The design of the layer structure used for this first test of the new device concept is obviously not optimized with respect to device performance, since a layer structure that simplifies the processing to obtain a high yield was considered to be more important at this stage.

Device processing starts with deposition of the gate metal (Ti/Pt/Au=50/50/200 nm) using optical direct printing lithography and lift off technique (fig. 1a). In the next step the resultant gate metallization pattern, which is approximately 1 μm in length, is used as etch mask to form the mushroom shaped gate structure by etching off the p^+-GaAs gate layer down to the barrier layer structure whose upper GaInP layer acts as etch

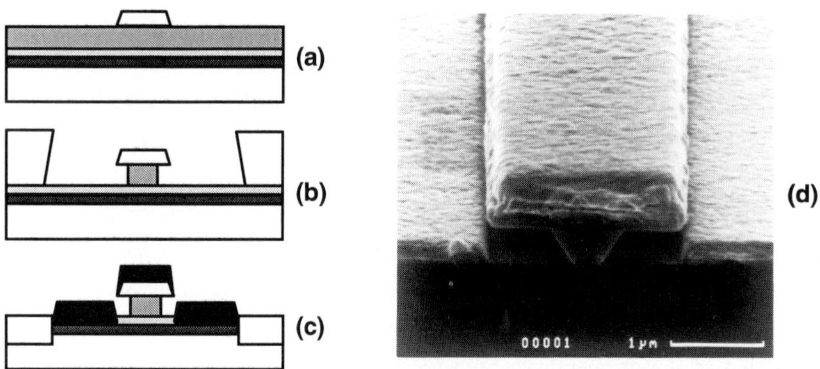

Fig. 1: Major processing steps (a, b), schematic of final device structure (c), and scanning electron micrograph of a device with an effective gatelength of 0.3 µm (d).

stop. For this process a highly selective phosphoric acid based peroxide etchant is used /3/. This process, which is also used in our HBT processing scheme /3/, works well down to 1 µm feature size, but the stiff correlation between vertical and lateral etch rate is quite restrictive in view of gate dimension control below 1 µm. A short gate footprint can be achieved (fig. 1d) by sufficient overetching. However, the large undercut produces undesired large source series resistances. Dry etching techniques offer considerable more flexibility. In this context it is worthwhile to mention that the device concept and the process scheme are well adapted to the requirements of dry etching processes. The wafer is clear from photoresist and etching takes place in large areas outside the sensitive gate areas. These conditions provide unrestricted access to the surface for inspection prior to dry etching, in situ process monitoring during etching and wafer inspection after etching. The pinch off voltage of the devices is not affected by the gate formation processes, it depends almost entirely on the details of the epitaxial layer structure except for short channel effects.

After finishing the gate structure the side wall of the pillar formed by the p^+-GaAs gate structure can be passivated by depositing a dielectric and re-etching of this dielectric. This step has been omitted for this first approach and the layer structure for the ohmic contacts has been deposited immediately after finishing the gate structure. The photoresist pattern for the ohmic contact structure is self-aligned with respect to the gate metallization structure and the details of contact processing used in this study are almost identical with the procedures for fabricating ohmic contacts on conventional AlGaAs/GaAs MODFET devices. Only the alloying temperature has been choosen 20 K higher. At the time the devices in this study were fabricated, the ohmic contact process was not optimized and the contact resistance determined from TLM measurements was relatively high (0.51 Ωmm) but meanwhile best values of 0.17 Ωmm have been realized on structures with thinner barrier layers (18 nm). Fig. 1d shows a scanning electron photomicrograph of the device after fabrication is completed. It can be seen that the gate structure has passed the preceding alloying process without any significant change of its geometry. The Ti/Pt/Au structure deposited first remains flat and did not collapse during heat treatment. Despite its fragile appearance the mechanical stability of the gate structure is excellent. Ultrasonic agitation and dry blowing with compressed nitrogen during processing did not harm the integrity of the structure. Fractures always took place near the footprint of the pillar. Shorts between the two levels of metallization have not been found. Detailed judgements on the reliability of the ohmic contacts can not be presented at the moment, but the results of DC-characterization of the devices have been reproduced 12 month after the first measurements were made. Device fabrication is completed by boron isolation implantation.

RESULTS OF DEVICE CHARACTERIZATION

The DC-characteristics of the devices have been measured on an HP-Parameter Analyser and for the microwave characterization an HP-Network Analyser was used. The results presented in this section have been obtained on FETs with gate metal stripes 1 μm long and 240 μm wide unless indicated otherwise. Due to the undercut the effective gate length given by the footprint of the p^+-GaAs gate layer is 0.3 μm. Fig. 2 shows a typical output characteristics of such devices. In general terms speaking the devices behave like conventional AlGaAs/InGaAs-MODFETs. Some shortcomings result from the non-optimized design of the epitaxial layer structure and immature processing. The relatively thick barrier layer limits the transconductions at moderate levels (200 mS/mm) and the high output conductance (≈15mS/mm) results from the very low aspect ratio and the low In content of the channel layer. The high source contact resistance (0.5 Ωmm) plus the resistance of the ungated part of the channel combine to a total source resistance of 1.2 Ωmm. This value is too

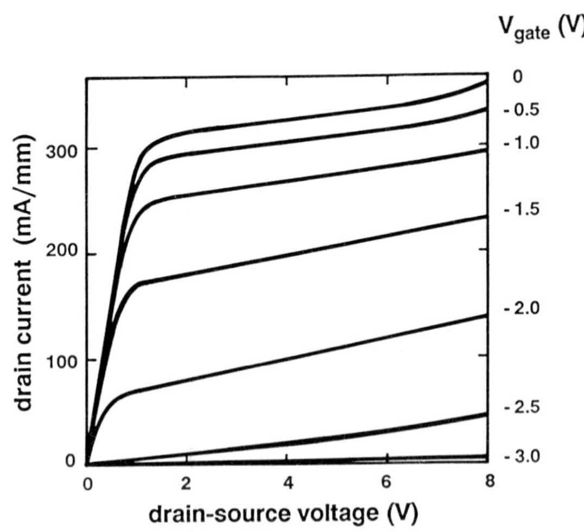

Fig. 2: Output characteristics of a FET with an effective gatelength of 0.3 μm and a gatewidth of 240 μm.

large, but both parts can be reduced by optimizing the technology for the ohmic contacts and by introducing 0.7 μm stepper lithography in combination with dry etching techniques.

Fig. 3 shows the I-V characteristics of FATFET devices (100x100 μm^2). The gate channel breakthrough occurs around 30 V and the gate leakage current is remarkable small. This feature is not unexpected because the combination of p^+-GaAs gate structure and AlGaInP barrier layer structure is likely to produce a larger tunnel barrier than a Schottky contact on a conventional AlGaAs barrier structure. However it must be admitted that final conclusions can not be drawn based on this observation because the band offset data for the heterojunctions involved are still debated in the literature. Regardless of the imperfections mentioned above the devices show already noteable RF performance (fig. 4). Typical cutoff frequencies obtained are 60 and 140 GHz for F_t and F_{max}, respectively, with best F_{max}-values of 160 GHz.

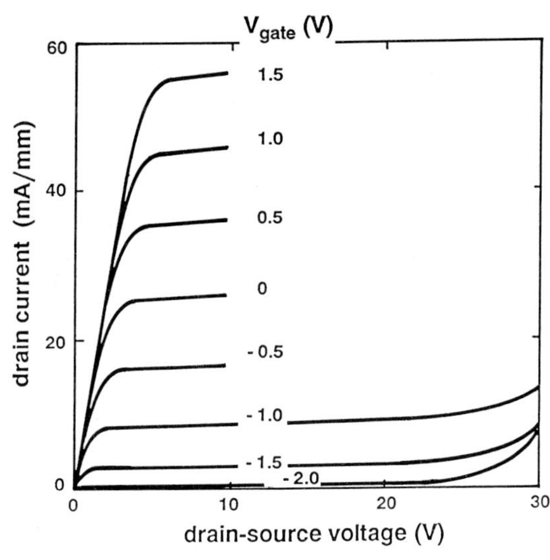

Fig. 3: I-V characteristics of a 100x100 μm^2 FATFET.

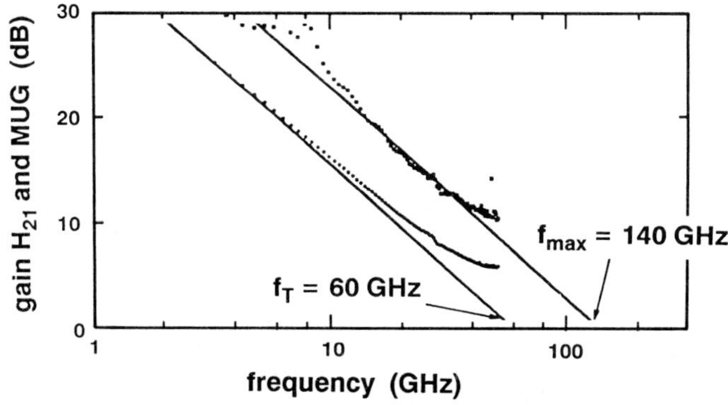

Fig. 4: RF figures of merit for the device of fig. 2.

SUMMARY

AlGaInP/GaInAs/GaAs-MODFET devices incorporating a carbon doped p^+-gate structure have been fabricated for the first time. The processing scheme is very simple and combines only process modules already in use for fabrication of III-V compound devices. The ohmic contact pattern is produced self-aligned with respect to the gate structure. Devices with gate lengths of 0.3 μm have been realized by using conventional optical lithography with 1 μm feature size and simple wet etching techniques. The devices show good RF-performance (F_t=60 Ghz and F_{max}=160 Ghz) considering that no attempts have been made to optimize the layer structure and the details of the process, yet.

Acknowledgement: We like to thank N. Herres, M. Maier and W. Rothemund for material characterization, and H.S.Rupprecht for supporting this study. This work has been sponsored by the German Ministry of Science and Technology.

REFERENCES:

[1] T. Lauterbach, W. Pletschen and K.H. Bachem, IEEE Trans. Electron Devices 39 (1992) 753

[2] T. Kobayashi, K. Taira, F. Nakamura and H. Kawai, J. Appl. Phys. 65 (1989) 4898

[3] K.H. Bachem, D. Fekete, W. Pletschen, W. Rothemund and K. Winkler, J. Cryst. Growth 124 (1992) 817

G_{ds} and f_T analysis of pseudomorphic MODFETs with gate lengths down to 0·1 μm

Jürgen Braunstein, P. J. Tasker, A. Hülsmann, K. Köhler, W. Bronner, and M. Schlechtweg

Fraunhofer Institute for Applied Solid State Physics, Tullastrasse 72, D-79108 Freiburg, Germany

Phone +49-761-5159 534, Fax +49-761-5159 400; e-mail: braunstein@iaf.fhg.de

ABSTRACT: Developing circuits for higher frequencies requires not only shorter gate lengths (l_g), but also proper scaling of the epilayer structure. A common problem with reducing only l_g is increasing output conductance (g_{ds}), thus no f_{max} improvement is achieved. To quantify the effect of scaling on MODFET performance a set of pseudomorphic MODFETs with different epilayer structures and gate lengths between 1 µm and 0.1 µm were investigated. A linear dependence of the intrinsic output resistance as a function of aspect ratio was derived. In addition no velocity overshoot was observed when analyzing $f_T l_g$. These results indicate that simple scaling rules can be applied to gate lengths even down to 0.1 µm.

1. INTRODUCTION

The crucial point in transistor performance optimization is often considered to be simply an issue of gate length reduction. This is certainly true when considering the current gain cutoff frequency (f_T) of the intrinsic device which is directly proportional to gate length for HEMTs in the saturated velocity mode. The intrinsic transistor is charged by various capacitors, so that the extrinsic performance can be substantially worse (Nguyen, 1990). Apart from a high f_T also a high power gain cutoff frequency (f_{max}) is required for MMICs. Prerequisites for high f_{max} are a high voltage gain g_m/g_{ds} and a high c_{gs} to c_{gd} ratio.

MMIC design with short gate length transistors is rather difficult as the important c_{gs} to c_{gd} and g_m to g_{ds} ratios usually become very small as the gate length decreases (tab. 1). Especially the experimentally observed behavior of the output conductance (g_{ds}) is not well analyzed and only few publications can be found. There has been an investigation by Kohn et al. deriving an experimental relationship between these 4 intrinsic equivalent circuit model parameters as follows (Kohn, 1989):

$$\frac{c_{gs}}{c_{gd}} = 1 + \beta \frac{g_m}{g_{ds}}$$

Kohn et al. found β to be on the order of 0.3. This seems to be only correct for specific gate lengths and processing conditions because we found other values for β than they did (tab. 1). β is also strongly dependent on the accuracy of the circuit model extraction. Our extraction method was described earlier (Schlechtweg, 1992). Dickmann et al. used the same formula to analyze different transistor structures also extracting a β of 0.3 (Dickmann, 1991). One objective of this paper is to provide a simpler, hopefully more process independent relationship for g_{ds}. Also the dependence of the intrinsic f_T will be analyzed.

lg/[μm]	0,16	0,18	0,2	0,3	0,5	1,0
cgs/cgd	3,35	3,83	4,55	5,54	7,19	11,5
gm/gds	15,7	16,9	21,5	25,5	33,2	44
β	0,15	0,17	0,17	0,18	0,19	0,24

Tab. 1 Intrinsic MODFET parameters and β for a gate to channel separation of 210 Å. The MODFETS were biased for peak g_m at V_{ds} = 1.5 V.

2. TRANSISTOR FABRICATION

As transistor scaling is an issue for ultimate MMIC performance this work focuses on aspect ratio variations. The aspect ratio is defined as the ratio of gate length to gate to channel separation; AR = l_g/d. Pseudomorphic MODFETs were grown by MBE on GaAs substrates with 120 Å wide $Ga_{0.75}In_{0.25}As$ channels and $Ga_{0.8}Al_{0.2}As$ barrier layers of variable thickness. δ-doping was used in all cases. The layers include etch stops for reactive ion etching allowing for precise gate to channel distance control. The gates were defined by electron beam lithography. Their length was controlled by SEM cross sections and analysis of S-parameters (Braunstein, 1993). The gate length was varied deliberately from 1 μm down to 0.1 μm resulting in aspect ratios between 47 and 4. The intermediate gate lengths are on the order of 0.5, 0.3, 0.2, 0.18 and 0.16 μm. This concept also permits comparison of transistors of the same aspect ratio with different gate lengths.

3. RESULTS AND ANALYSIS

All transistors of various gate lengths and aspect ratios were biased for peak transconductance at comparable intrinsic biases. S-parameters were measured at this bias point and intrinsic circuit model elements were extracted. Plotting of the intrinsic output resistance r_{ds} (the inverted output conductance g_{ds}) versus intrinsic gate capacitance (c_{gs} plus c_{gd}) revealed a linear dependency of r_{ds} on gate capacitance (fig. 1). Equ. (1) summarizes the results given in fig. 1. The linear fit for r_{ds} as a function of $c_{gs} + c_{gd}$ exhibits an excellent correlation of 0.98 for the wide range of aspect ratio variations. The gate capacitance, given by the transistor geometry, is directly related to the aspect ratio of the transistor. Gate capacitance in pF/mm is approximately aspect ratio times 0.11 for pseudomorphic MODFETs in this material system. g_{ds} is given by equ. (2) and r_{ds} by equ. (3). It is thus observed, validated here experimentally for l_g between 0.1 and 1 μm, that r_{ds} is a function of aspect ratio and not gate length.

$$\frac{r_{ds}}{[\Omega * \mathrm{mm}]} = \frac{c_{gate}}{[\mathrm{pF/mm}]} * 12.8 \qquad \textbf{(1)} \qquad\qquad g_{ds} = \frac{700}{AR}\ [\mathrm{mS/mm}] \qquad \textbf{(2)}$$

$$r_{ds} = 1{,}43 * AR\ [\Omega\mathrm{mm}] \qquad \textbf{(3)}$$

The f_T analysis is carried out only for comparable devices with AR < 20 or gate lengths between 0.1 μm and 0.5 μm because the transistor is in the gradual channel mode for longer gates. Once having all intrinsic circuit model elements f_T was calculated as $g_m/(2\pi(c_{gs} + c_{gd})$. l_g was calculated from the gate capacitance taking fringing capacitances into account. To check

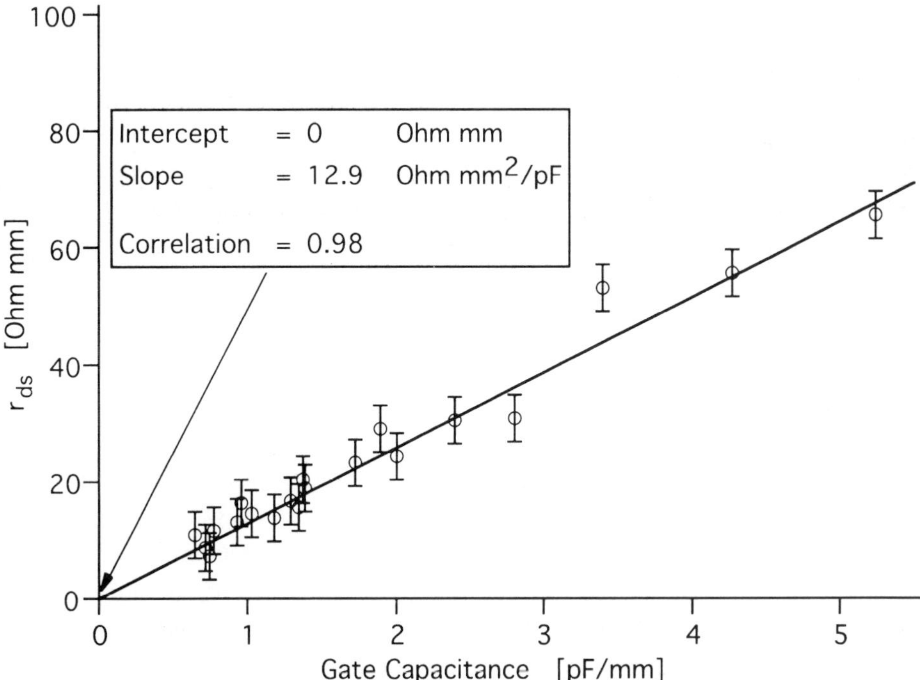

Fig. 1 Output resistance r_{ds} versus gate capacitance for various MODFETs with gate lengths between 0.1 μm and 1 μm. The MODFETS were biased for peak g_m. V_{ds} = 1.5 V.

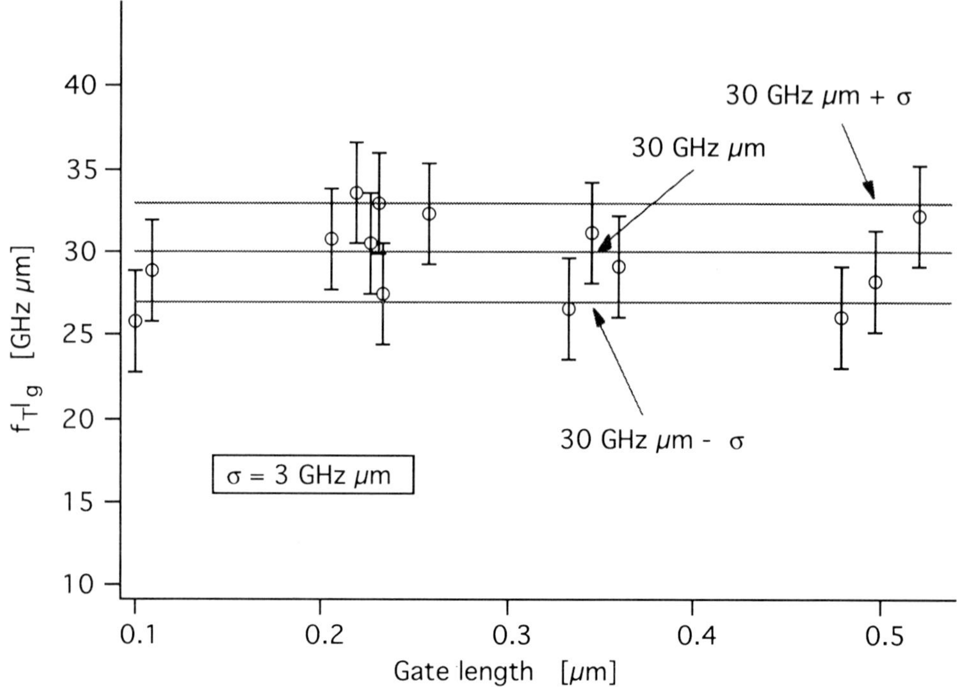

Fig. 2 $f_T l_g$ product versus gate length extracted from gate capacitance for various MODFETs with gate lengths between 0.1 μm and 0.5 μm. The MODFETS were biased for peak g_m.

how the intrinsic performance of the devices was influenced by the aspect ratio variations the $f_t l_g$ product was investigated. It directly represents the effective electron velocity (Braunstein, 1991). All transistors had the same intrinsic $f_T l_g$ product of 30 GHz*μm with a standard deviation of 3 GHz*μm. This is equivalent to an effective electron velocity of 1.9 10^7 cm/s, hence within experimental error no velocity overshoot is needed to explain the f_T values for short gate lengths because $f_T l_g$ equals $v_{eff}/2\pi$, which is a constant.

4. CONCLUSION

The reason for bad c_{gs} to c_{gd} and g_m to g_{ds} ratios normally is deficient aspect ratio scaling. For a first simple analysis c_{gd} and g_m can be treated to be independent from gate length. g_m is only a function of the gate to channel separation if the transistor is in the saturated velocity mode. c_{gd} depends on various parameters, such as technology processing conditions, surface effects, or layer design, but only weakly on l_g. The two remaining parameters c_{gs} and g_{ds} are mainly functions of the aspect ratio, therefore it has to be kept constant if c_{gs}/c_{gd} and g_m/g_{ds} should not be influenced by gate length reduction.

It was shown that the aspect ratio of gate length and gate to channel distance is the only factor describing a pseudomorphic MODFET´s output conductance even at very short gate lengths down to 0.1 μm. As a consequence, to obtain reasonable voltage gains from ultra short gate length pseudomorphic MODFETs, proper scaling of the aspect ratio is all that is required in terms of MBE layer design. In practice this may be difficult to be achieve.

Finally, gate length reduction of course is the way to achieve better current gain from the transistor. The effective carrier velocity for MODFETs with gate lengths down to 0.1 μm was calculated to be on the order of 1.9 10^7 cm/s.

5. REFERENCES

1. L. Nguyen and P. J. Tasker, "Scaling issues for ultra-high-speed HEMTs," SPIE, High-Speed Electronics and Device Scaling, 1990, p. 251.

2. E. Kohn, A. Lepore, H. Lee, and M. Levy, "Performance evaluation of GaAs based MODFETs," IEEE Cornell Conference on Advanced Concepts in High Speed Semiconductor Devices and Circuits, 1989, p. 91.

3. M. Schlechtweg, W. Reinert, P. J. Tasker, R. Bosch, J. Braunstein, A. Hülsmann, and K. Köhler, "Design and Characterization of High Performance 60 GHz Pseudomorphic MODFET LNAs in CPW-Technology Based on Accurate S-Parameter and Noise Models," T-MTT, 1992, **40**(12), p. 2445.

4. J. Dickmann, E. Kohn, S. Strähle, A. Wiersch, H. Künzel, H. Lee, and H. Nickel, "RF-Current De-Confinement in III-V HFETs," IEEE Cornell Conference on Advanced Concepts in High Speed Semiconductor Devices and Circuits, 1991, Ithaca, NY., p. 435.

5. J. Braunstein, P. J. Tasker, M. Schlechtweg, A. Hülsmann, G. Kaufel, and K. Köhler, "Relating μ-wave mapped data to physical parameters for MODFETs," 1993, **B20**, p. 37.

6. J. Braunstein, P. J. Tasker, K. Köhler, T. Schweizer, A. Hülsmann, M. Schlechtweg, and G. Kaufel, "Investigation of transport phenomena in pseudomorphic MODFETs," 18th International Symposium on GaAs and Related Compounds, 1991, Seattle, WA., p. 161.

Paper presented at the Int. Symp. GaAs and Related Compounds, Freiburg, 1993

High temperature performance of GaInP and AlInP HEMT's with WSi_x gates

Yi-Jen Chan
Department of Electrical Engineering
National Central University
Chungli, Taiwan 32054, ROC

Jenn-Ming Kuo
AT&T Bell Laboratories
Murray Hill, N.J. 07974
USA

Abstract: Both GaInP/$In_{0.2}Ga_{0.8}As$ and AlInP/$In_{0.2}Ga_{0.8}As$ HEMT's were fabricated with WSi_x gates and evaluated at high temperature operations. Based on the high stability of WSi_x gates, both devices functioned at temperature up to 300°C. GaInP HEMT's showed a g_m reduction from 176 mS/mm at 25°C to 97 mS/mm at 300°C, while V_{th} shifted from -1.44 V to -3.31 V. AlInP HEMT's also showed a g_m decrease from 135 mS/mm at 25°C to 70 mS/mm at 300°C, while V_{th} shifted from -0.4 V to -0.84 V. Both of these devices with WSi_x gates offer a potential applications to high power and high temperature operations.

I. Introduction

$Ga_xIn_{1-x}P$/GaAs (x=0.51) and $Al_yIn_{1-y}P$/GaAs (y=0.52) heterostructures, lattice matched to GaAs, have achieved a successful role for use in electronic devices, replacing AlGaAs/GaAs heterostructures on GaAs substrates. Unlike $Al_{0.3}Ga_{0.7}As$/GaAs HEMT's, GaInP/GaAs and AlInP/GaAs HEMT's demonstrated a negligible deep-trap effect at low temperatures, reported by Chan et al. 1990 and Kuo et al. 1993. No current collapse and threshold voltage (V_{th}) shift were observed for either GaInP/GaAs or AlInP/GaAs HEMT's at low temperatures, which are common problems with AlGaAs/GaAs HEMT's (Drummond et al. 1983, and Mooney et al. 1984). In addition to the HEMT's, heterojunction bipolar transistors (HBT's), built into this material system (Kawai et al. 1989, Delage et al. 1991), have also received considerable attention based on its large valence band discontinuity (ΔE_v= 0.24 eV in GaInP/GaAs system). This large ΔE_v can substantially reduce reverse current injection from the base region, consequently current gain increases.

Pseudomorphic HEMT's with $In_{0.2}Ga_{0.8}As$ channels in these heterostructure systems were also reported by Chan et al. 1992. The excess In in the conduction channels can enhance carrier transport properties. When compared to lattice-matched HEMT's, a 25% transconductance (g_m) improvement is demonstrated.

In this study we try to explore the advantage of a good Schottky performance by these heterostructures, and investigate both GaInP/InGaAs and AlInP/InGaAs HEMT's operated under high temperatures. In order to achieve high thermal stability of Schottky contacts, we used WSi_x to replace the conventional Ti/Au metal as a gate material. WSi_x materials have been applied to devices for which high temperature processes are necessary, with excellent gate performance being maintained. Pseudomorphic $In_{0.2}Ga_{0.8}As$ channel HEMT structures were also used in this study.

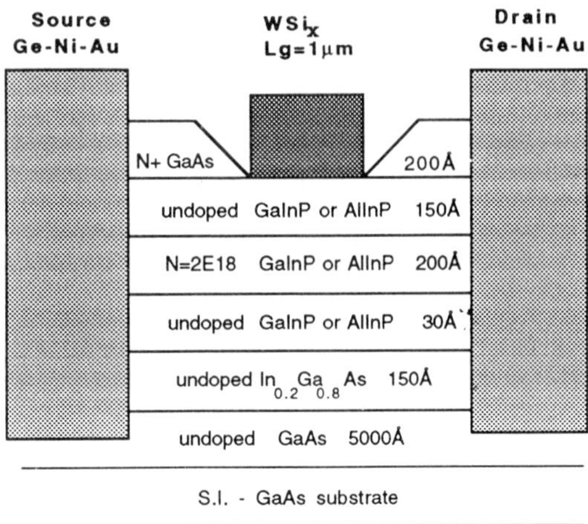

Fig. 1 Device cross-section of GaInP/ $In_{0.2}Ga_{0.8}As$ pseudomorphic HEMT's with WSi_x gates.

II. Device structure and fabrication procedure

The epitaxial layers were grown in an Intevac Gen II GSMBE system on 2-inch semi-insulating (100) GaAs substrates. Epitaxial growth was carried out using As_2, P_2 molecular beams produced by thermal decomposition of the gaseous hydrides AsH_3 and PH_3. The use of gas sources provided a means for rapidly switching between GaInP (or AlInP) and $In_{0.2}Ga_{0.8}As$ growth while maintaining precise control of the group V molecular flux. The growth rate of GaAs was 0.45 µm/hr, and the rates of both GaInP and AlInP were 0.9 µm/hr. Lattice mismatched within $\Delta a/a= 5x10^{-4}$ was easily obtained with calibration runs on the GaInP and AlInP layers.

A device cross section is shown in Fig. 1. An 5000Å thick undoped GaAs buffer was used followed by a 150 Å undoped $In_{0.2}Ga_{0.8}As$ pseudomorphic channel, and a 30 Å undoped GaInP (or AlInP) spacer. A 200 Å GaInP (or AlInP) Si-doped layer ($n= 2x10^{18}$ cm^{-3}) provided the donors, and a 150 Å undoped GaInP (or AlInP) was used to improve the Schottky contact. Finally a 200 Å n^+GaAs cap was used to reduce contact resistance.

The devices were processed with a conventional optical lithography technique. A $NH_4OH/H_2O_2/H_2O$ solution was used for etching both GaAs and $In_{0.2}Ga_{0.8}As$ layers. This etchant proved to be selective for GaInP or AlInP, attacking GaAs and InGaAs only. It therefore provides a way to deposit the gate metal on the undoped Schottky layers without any risk of uncontrolled gate recess variation. The etching of GaInP for the mesas was achieved by using a HCl/H_3PO_4 solution, and a HCl/H_2O solution for AlInP layers. Both solutions were proved not to etch GaAs and InGaAs away. Ohmic contacts were realized using Au-Ge-Ni (2000 Å) alloys by thermal evaporation followed by a 2 min. 400°C annealing. After etching the top n^+GaAs layer, WSi_x gates (1500Å) were sputtered from a $WSi_{0.3}$ target and defined by a lift-off process.

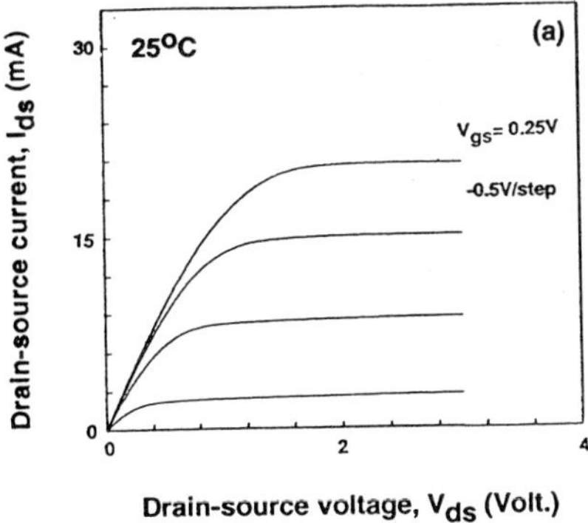

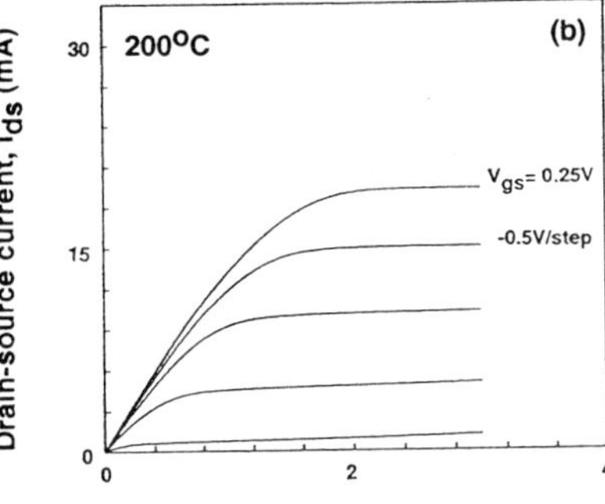

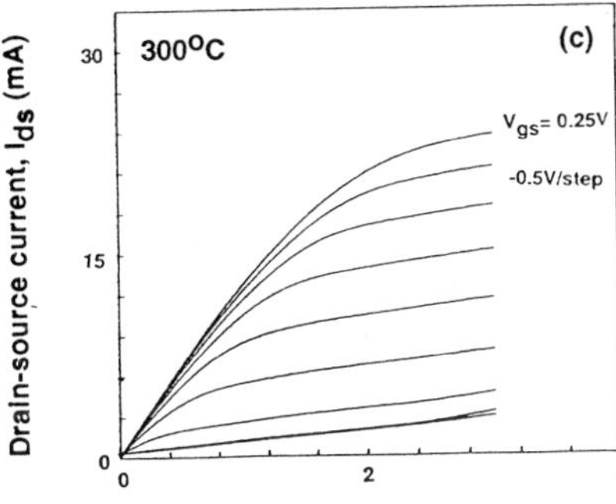

Fig. 2 DC I-V characteristics of WSi_x gate GaInP/$In_{0.2}Ga_{0.8}As$ pseudomorphic HEMT's (a) 25°C, (b) 200°C, (c) 300°C. (gate: 1μm x 75μm)

The Hall mobility for GaInP/InGaAs HEMT's was 3200 cm^2/V-sec and 15700 cm^2/V-sec at 300K and 77K respectively while the corresponding values for the sheet charge density were 4.38×10^{12} cm^{-2} and 1.64×10^{12} cm^{-2}. As to the AlInP/InGaAs HEMT's, the Hall mobility was 5390 cm^2/V-sec and 27300 cm^2/V-sec at 300K and 77K respectively while the corresponding values for the sheet charge density were 2.24×10^{12} cm^{-2} and 1.55×10^{12} cm^{-2}. The sheet charge densities increased only 5% under illumination at 77K. This indicates that the deep trap effect is not significant for either material system.

III. High temperature characteristics for the device

Fig. 2 shows the I-V characteristics of WSi_x gate GaInP/InGaAs HEMT's with a gate of 1 μm x 75 μm under different temperature conditions. Fig. 2(a) shows the I-V curves of this device operated at room temperature. A clean knee voltage together with good channel output-conductance (g_{ds}) values were 176 mS/mm and 1.2 mS/mm, respectively. As we increased the operation temperature up to 200°C (Fig. 2(b)), g_m dropped slightly to 137 mS/mm, while the channel pinch-off characteristic remained good. No significant gate leakage current was detected which indicates the high thermal stability of WSi_x gates on GaInP layers. If the temperature was further increased to 300°C (Fig. 2(c)), g_m dropped to 98 mS/mm and g_{ds} increased dramatically to 18.2 mS/mm. Although functional device characteristics can still be observed at 300°C, high gate leakage current resulted in excess channel current and soft pinch-off in the channel.

Fig. 3 shows the I-V characteristics of the WSi_x gate AlInP/InGaAs HEMT's with a gate of 1 μm x 50 μm under the same operational temperatures as presented previously. These results are similar to GaInP/InGaAs HEMT's. The peak g_m was 135 mS/mm at room temperature, and this value dropped to 79 mS/mm at 200°C and 70 mS/mm at 300°C, respectively. Although functional device characteristics can still be detected at 300°C (Fig.

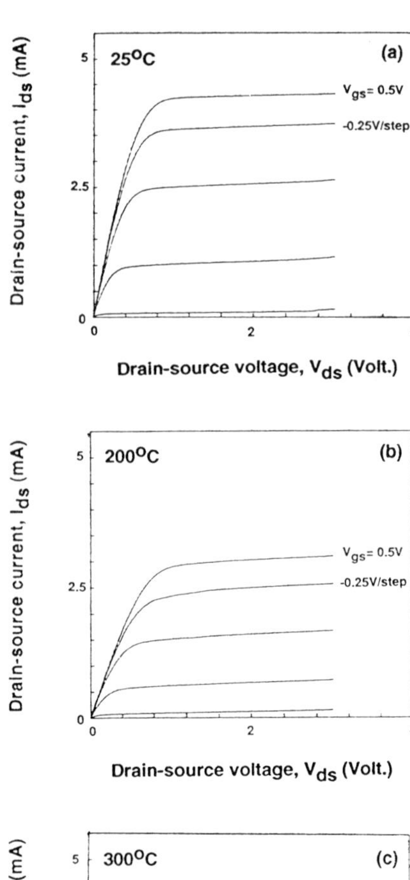

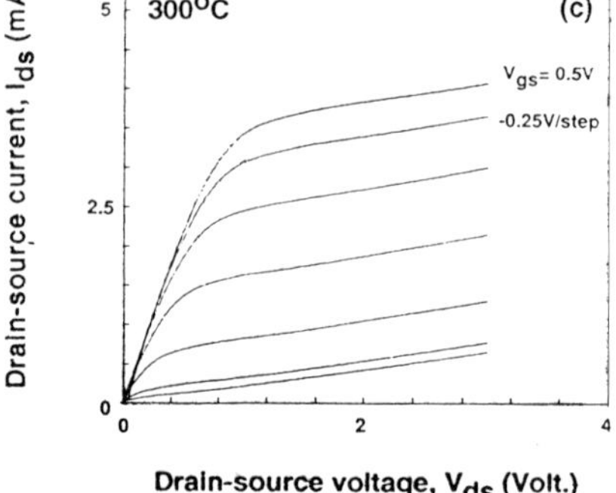

Fig. 3 DC I-V characteristics of WSi_x gate $AlInP/In_{0.2}Ga_{0.8}As$ pseudomorphic HEMT's (a) 25°C, (b) 200°C, (c) 300°C. (gate: 1μm x 50μm)

3(c)), however due to the gate leakage as for GaInP/InGaAs HEMT's, a large g_{ds} value and soft channel pinch-off were observed.

Fig. 4(a) and 4(b) summarize the temperature dependence of g_m and g_{ds} between 25°C and 300°C for GaInP/InGaAs and AlInP/InGaAs HEMT's, respectively. Both devices showed a 50% g_m reduction at 300°C as compared to the value at 25°C. However, for GaInP/InGaAs HEMT's, g_{ds} values increased dramatically for temperatures higher than 200°C. Consequently, the DC gain ratio (g_m/g_{ds}) dropped from 147 (25°C) to 60 (200°C), and only 5.4 at 300°C. This g_{ds} degradation was caused by the gate leakage current, resulting in excess current in the channel as we have discussed previously. Compared to GaInP/InGaAs HEMT's, this g_{ds} increase is not significant in AlInP/InGaAs HEMT's. g_{ds} increased from 1.4 mS/mm (25°C) to 2.2 mS/mm (200°C) and 4.9 mS/mm (300°C). This translates into a DC gain drop from 96 (25°C) to 35 (200°C) and 14 (300°C).

This performance degradation at high temperatures can also be evaluated by the threshold voltage (V_{th}) shift. Tab. 1 illustrates the V_{th} variations from 25°C to 300°C for both GaInP/InGaAs and AlInP/InGaAs HEMT's. For GaInP/InGaAs HEMT's, V_{th} shifted slightly from -1.44 V at 25°C to -1.8V at 200°C. As we further increased the measured temperature, V_{th} shifted to -2.48 V at 250°C and -3.31 V at 300°C. This large V_{th} shift between 200°C and 300°C exactly corresponds to the dramatic g_{ds} increase in Fig. 2(c). Therefore, this phenomenon together with the observed soft pinch-off at high temperatures suggests that the WSi_x gate on GaInP layers is not stable for temperatures above 200°C. The Schottky gate can not effectively modulate the channel current, which causes a g_m decrease, and a V_{th} shift to more negative values. In addition, higher gate leakage current in this case also causes the g_{ds} increase. The V_{th} shift for AlInP/InGaAs HEMT's was -0.4 V (25°C), -0.47V (200°C), and -0.84 V (300°C). Apparently, either from the I-V characteristics or V_{th} evaluations, AlInP/InGaAs HEMT's performed better under a high temperature operation. WSi_x gates are less

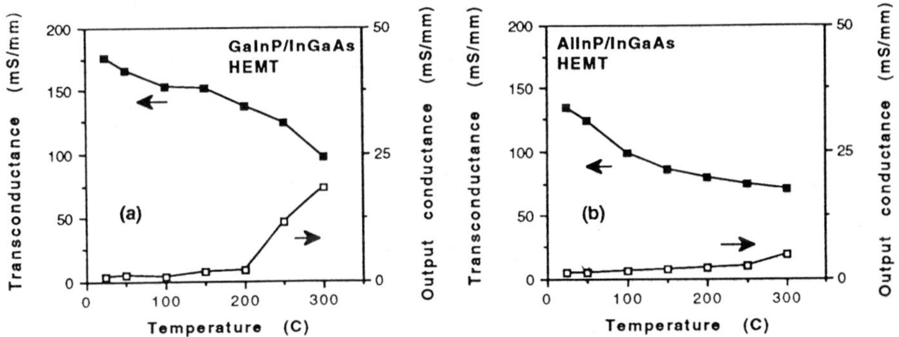

Fig. 4 Transconductance (g_m), output-conductance (g_{ds}) as a function of measurement temperatures for WSi_x gate (a)GaInP/InGaAs, and (b) AlInP/InGaAs pseudomorphic HEMT's.

Vth	25C	50C	100C	150C	200C	250C	300C
GaInP HEMT	-1.44	-1.49	-1.50	-1.52	-1.83	-2.46	-3.31
AlInP HEMT	-0.40	-0.41	-0.43	-0.45	-0.47	-0.53	-0.84

Tab. 1 Threshold voltage (Vth) variations as a function of temperature for GaInP/InGaAs and AlInP/InGaAs HEMT's.

temperature sensitive at the top of the AlInP layers which may be due to the higher bandgap in this material. The Schottky barrier heights (Ø) evaluated by C-V measurements were 1.06 eV and 0.92 eV for GaInP/InGaAs and AlInP/InGaAs HEMT's, respectively. The Schottky barrier of GaInP/InGaAs HEMT's maintain a good performance for an operating temperature up to 250°C (Ø= 0.94 eV); however, a significant degradation of the Scuottky barrier performance was observed at 300°C (Ø~ 0.7 eV). This results in an increase of gate leakage as can be seen in Fig. 2(c). After this thermal cycling test, device performance for both GaInP and AlInP HEMT's was irreversible as the temperature cooled down to room temperature.

IV. Conclusion

In summary, WSi_x gate pseudomorphic GaInP/$In_{0.2}Ga_{0.8}As$ and AlInP/ $In_{0.2}Ga_{0.8}As$ HEMT's were fabricated and evaluated for a high temperature operations. Functional device characteristics were observed at temperature as high as 300°C. However, for temperatures above 200°C, WSi_x gates became unstable in GaInP/InGaAs HEMT's and resulted in a g_m decrease, a g_{ds} increase, a V_{th} shift and a soft channel pinch-off. These performance degradations are not significant for AlInP/InGaAs HEMT's within this temperature range. As long as the devices were operated below 200°C, they both demonstrated similar characteristics compared to the results at room temperature. This suggests that WSi_x is very thermally stable, and that potentially this device can be applied to working conditions that involve high temperature operations.

Acknowledgements:

The first author (YJC) would like to thank Prof. D. Pavlidis (University of Michigan, USA) for stimulating discussions regarding GaInP/GaAs and AlInP/GaAs HEMT's. This work at NCU, Taiwan is supported by the National Science Council, ROC (NSC 82-0404-E-008-103).

REFERENCES:

Chan Y.J., Pavlidis D., Razeghi M. and Omnes F., IEEE Trans. Electron Devices, 37, p.2141, (1990).

Chan Y.J., Pavlidis D., Kuo J.M.and Huang J.H., Proceedings of Electron Device and Material Symposium, 251, Taipei, Taiwan (1992).

Chan Y.J. and Pavlidis D., accepted for publication, IEEE Trans. Electron Devices, (1993).

Delage S.L. et. al., Electronics Lett., 27, p.253 (1991).

Drummond T.J. et. al., IEEE Trans. Electron Devices, 30, p.1806, (1983).

Kawai H., Kobayashi T., Nakamura F., and Taire K., Electronics Lett., 25, p.609 (1989).

Kuo J.M., Chan Y.J., Pavlidis D., Appl. Phys. Lett., 62, p.1105, (1993).

Mooney P.M., Solomon P.M., and Theis T.N., Inst. Phys. Conf. Ser., 74, p.617, (1984).

Inst. Phys. Conf. Ser. No 136: Chapter 2
Paper presented at the Int. Symp. GaAs and Related Compounds, Freiburg, 1993

Pseudomorphic AlGaAs/InGaAs SQW-MODFETs with double sided modulation doping

J. Plauth, R. Kempter, S. Grigull, H. Heiß, M. Walther, W. Klein, G. Tränkle and G. Weimann

Walter-Schottky-Institut, Technische Universität München
Am Coulombwall, D-85747 Garching, Germany

ABSTRACT: Pseudomorphic $Al_{0.23}Ga_{0.77}As/In_xGa_{1-x}As/Al_{0.23}Ga_{0.77}As$ QW-structures with double sided modulation doping have high 2DEG-densities between $1.5 \cdot 10^{12}$ cm^{-2} and $2.8 \cdot 10^{12}$ cm^{-2} for In-contents $0 \leq x \leq 0.3$. MODFETs with T-shaped gates ($L_G = 0.19$ µm) were fabricated by e-beam lithography and a selective RIE gate recess yielding threshold voltage variations of 20 mV. The transit frequency $f_T = 93$ GHz is remarkably high for the AlGaAs/GaAs-devices and rises to 109 GHz for $x = 0.25$. At 12 GHz the latter devices exhibit a minimum noise figure $F_{min} = 0.54$ dB and an associated gain $G_{ass} = 15$ dB. Within $0.15 \leq x \leq 0.3$ DC- and RF-data remain essentially unchanged.

1. INTRODUCTION

The use of pseudomorphic $In_xGa_{1-x}As$-layers in modulation-doped FETs on GaAs-substrates has been a major step towards faster devices with less noise (Ketterson et al 1986, Nguyen et al 1988). It is well established that a high 2DEG-density is a key for high-speed performance (Nguyen et al 1988, 1992). In this work we obtain high carrier densities by homogeneous doping of both $Al_{0.23}Ga_{0.77}As$ barriers. We have investigated the influence of the In mole-fraction in the QW on device properties. As these properties critically depend on the gate-to-channel separation we have used RIE for a self-adjusted gate recess (Kaufel et al 1990).

2. DEVICE FABRICATION AND CHARACTERIZATION

Pseudomorphic $Al_{0.23}Ga_{0.77}As/In_xGa_{1-x}As$ QW-structures have been grown by MBE. The Indium mole-fraction x in the channel is 0, 15, 20, 25, 28 and 30 %. The width of the quantum wells has been kept constant at 12 nm (Fig. 1). Klein et al (1993) have shown that SQWs with well widths of 12 nm and In contents as high as 26 % are below the onset of relaxation. For the samples containing 28 or 30 % of In, however, partial relaxation of the strained layers cannot be excluded. Both barriers are homogeneously doped with $3 \cdot 10^{18}$ cm^{-3} of Si. Carrier densities measured by SdH at 4.2 K increase from $1.5\cdot10^{12}$ cm^{-2} in the GaAs-channel to $2.8\cdot10^{12}cm^{-2}$ in the $In_{0.3}Ga_{0.7}As$-channel (Table 1).

	material	doping conc. [cm^{-3}]	thickness [nm]
cap	GaAs	$5\cdot10^{18}$	60
	$Al_{0.23}Ga_{0.77}As$	-	5
upper barrier	$Al_{0.23}Ga_{0.77}As$	$3\cdot10^{18}$	25
spacer	$Al_{0.23}Ga_{0.77}As$	-	2
	GaAs (x=0:$Al_{0.23}Ga_{0.77}As$)	-	1.1
channel	$In_xGa_{1-x}As$	-	12
	GaAs (x=0:$Al_{0.23}Ga_{0.77}As$)	-	1.1
spacer	$Al_{0.23}Ga_{0.77}As$	-	3
lower barrier	$Al_{0.23}Ga_{0.77}As$	$3\cdot10^{18}$	3.3

Fig. 1: Vertical structure of MODFETs with In mole-fraction $0 \leq x \leq 0.3$

MODFETs have been fabricated from these samples with a gate-length $L_G = 0.19$ µm, a gate-width W_G ranging from 60 µm to 150 µm and a drain-to-source-distance of 2 µm. Wet chemical etching has been used for isolation of the mesas. Ohmic contacts have been formed by rapid thermal alloying of Ge/Au/Ni/Au layers. T-shaped gates have been written by electron-beam-lithography into a resist which consists of highly sensitive $P(MMA_{0.915}\text{-}MAA_{0.085})$ on top of less sensitive PMMA (M=950K). In the opening of the resist the n^+-GaAs cap-layer has been removed selectively by RIE using CCl_2F_2 at low damage conditions (13.56 MHz, $P_{RF} =$ 0.1 W/cm^2, $U_{Bias} = 90$ V). This self-adjusted gate-recess uses the AlGaAs-layer beneath the n^+-GaAs cap as an etch-stop. We estimate that approximately 8 nm of this layer have been removed after wet chemical dissolution (NH_4F:HF:H_2O) of the etch residues. For formation of the

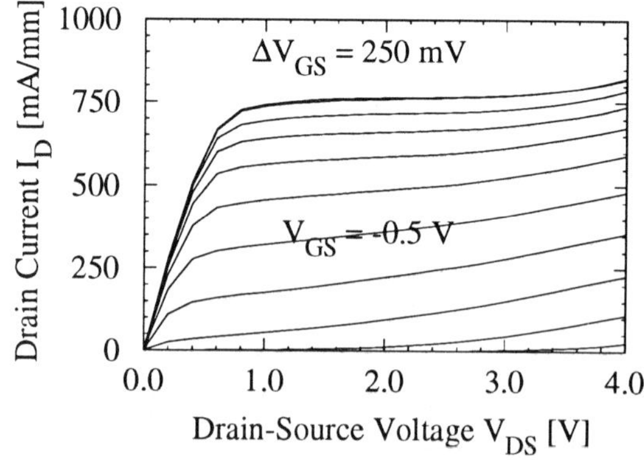

Fig. 2: Double doped MODFET with 25% In: output characteristics

Schottky-contact Ti-Pt-Au have been evaporated and lifted-off. Finally, contact pads have been reinforced by 400 nm of Au.

DC-characteristics, such as maximum extrinsic transconductance $g_{m,ext}$, maximum drain current $I_{D,max}$ and threshold voltage V_{th}, have been measured at V_{DS}=1V bias (Fig. 2). Due to the self-adjusted gate-recess by RIE we have achieved a standard deviation in threshold voltage of less than 20 mV (Fig. 3). Parasitic resistances of source (R_S) and drain (R_D) have been determined by the end-resistance method. S-parameters have been measured on-wafer within 2..26 GHz. Current gain h_{21} and maximum unilateral gain MUG have been extrapolated with a slope of -20 dB/decade for determination of their respective cut-off frequencies f_T and f_{max} (Fig. 4). The standard deviation σ of transit frequencies f_T is less than 4 GHz for all samples. By scanning electron microscopy of cleaved devices we have determined the metallurgical gate-length to be $L_G = 190 \pm 10$ nm.

3. RESULTS AND DISCUSSION

The distinct feature of the pseudomorphic devices as compared to the those with GaAs-channels is the increased peak of extrinsic transconductance $g_{m,ext}$, which is not only shifted to a higher drain current $I_{D,gmax}$ but also covers a larger range of bias (Fig. 5, Table 1). Quite

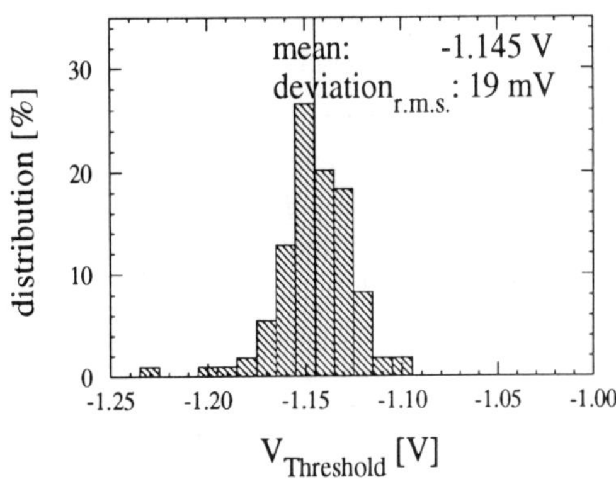

Fig. 3 MODFET with 20% In: distribution of threshold voltage V_{th} (W_G=60 .. 150 μm)

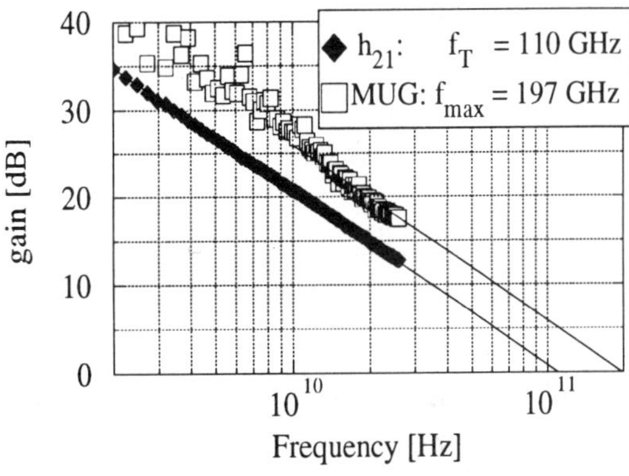

Fig. 4: MODFET with 25 % In: current gain h_{21} and maximum unilateral gain MUG (W_G=150 μm)

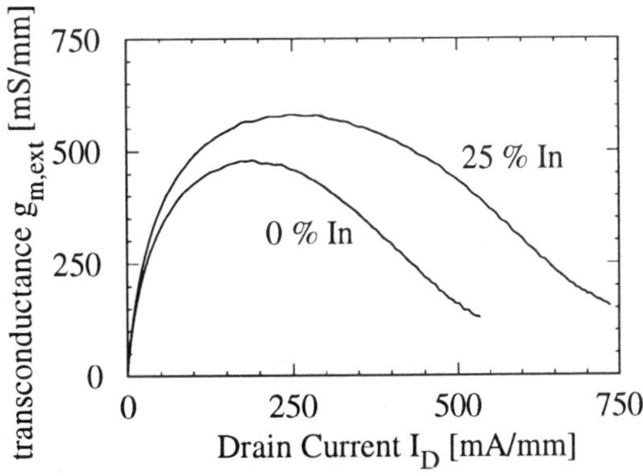

Fig. 5: Double doped MODFETs: Extrinsic transconductance $g_{m,ext}$ vs. drain current I_D (V_{DS} = 1 V)

similar is the bias-dependence of extrinsic transit-frequency f_T (Fig. 6). At V_{DS}=1V and V_{GS} tuned for maximum transit-frequency f_T (optimum bias) we find an average extrinsic f_T of 93 GHz and f_{max} = 170 GHz for the GaAs-MODFET (W_G = 150 µm). The average transit-frequency is enhanced by 17 % to f_T = 109 GHz if the channel contains 25 % of In. Noise measurements at 12 GHz yield a minimum noise figure F_{min} = 0.58 dB with an associated gain G_{ass} = 13 dB for the In-free devices. With an In-contents of 25 % these values improve to F_{min} = 0.54 dB and G_{ass} = 15 dB. The increase in In-contents from 15 to 30 % is reflected in the rise of 2DEG density N_S and, somewhat less, in the rise of maximum drain current $I_{D,max}$. The crucial figures of merit $g_{m,ext}$, $g_{m,int}$, $I_{D,gmax}$ and f_T, however, remain practically unchanged for $0.15 \leq x \leq 0.3$ (Table 1). In spite of possible partial relaxation in the devices which contain 28 or 30 % of Indium no significant degradation in electrical performance is observed.

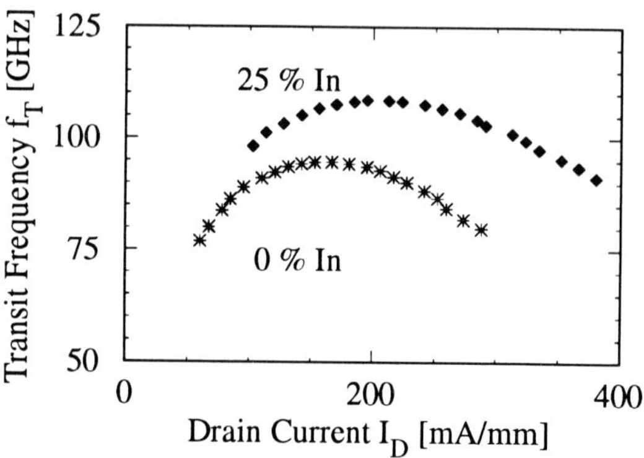

Fig. 6: Double-doped MODFETs: Extrinsic transit-frequency f_T vs. drain current I_D (V_{DS} = 0.7 V)

In mole-fraction	0 %	15 %	20 %	25 %	28 %	30 %
N_S [10^{12}cm^{-2}]	1.5	2.0	2.3	2.5	2.7	2.8
$I_{D,max}$ [mA/mm]	560	680	705	730	740	750
$R_S + R_D$ [Ωmm]	0.69	0.61	0.63	0.63	0.65	0.63
$g_{m,ex}$ [mS/mm]	490	560	560	575	570	580
$g_{m,int}$ [mS/mm]	590	670	680	690	700	710
$I_{D,gmax}$ [mA/mm]	190	255	250	250	250	270
V_{th} [V]	-1.01	-1.14	-1.13	-1.27	-1.23	-1.24
f_T [GHz]	93	102	101	109	106	103
f_{max} [GHz]	170	180	190	195	190	185
F_{min} @ 12 GHz [dB]	0.58	0.51	0.54	0.54	n. a.	0.52
G_{ass} @ 12 GHz [dB]	13	14	14	15	n. a.	15

Table 1: DC- and RF-results of double-doped MODFETs
L_G = 190 nm, W_G=150 µm except for $I_{D,max}$: W_G=60 µm
N_S from SdH-measurements at 4.2 K, $I_{D,max}$ .. f_{max} mean values at 300 K ,
noise data F_{min}, G_{ass} from selected devices

The DC-data furnish an estimate of intrinsic transit-time τ_i: From the threshold voltage V_{th} we calculate the effective gate-to-channel separation $d+\Delta d = 31$ nm. According to $v_{s,eff} = (d+\Delta d)\cdot g_{m,int}/\varepsilon_0\varepsilon_r$ (ε_r=12.46) we obtain an effective saturation velocity $v_{s,eff}$ of $1.7\cdot 10^7$ cm/s for the GaAs-MODFETs and 1.9 .. $2.0 \cdot 10^7$ cm/s for the pseudomorphic ones (Table 2). From measured L_G we calculate the intrinsic transit-times $\tau_i = L_G/v_{s,eff}$. For an estimate of extrinsic transit-time $\tau = 1/2\pi f_T$, however, parasitic delays have to be added; those which arise from the input pad-capacitance C_P and the fringing capacitance C_f' of the gate as well as the drain delay τ_d (Nguyen et al 1992). Parasitic capacitances are estimated as follows: Using the equivalent circuit of the intrinsic FET we calculate C_{GS} and C_{GD} from S-parameters at optimum bias. Their sum is plotted versus gate-width W_G. Linear extrapolation for $W_G \rightarrow 0$ yields the input pad-capacitance C_P of approximately 24 fF which is due to the 80×80 μm^2 gate-pad. The slope $\delta(C_{GS}+ C_{GD})/\delta W_G$ is interpreted as the sum of the parallel-plate-capacitor $\varepsilon_0\varepsilon_r L_G/(d+\Delta d)$ and the fringing capacitance C_f'. With these values we estimate the parasitic delays $\tau_P = C_P/g_{m,ex}W_G$ and $\tau_f = C_f'/g_{m,int}$. The drain delay τ_d is determined as the difference between the total delay and its extrapolated value at zero drain bias voltage (Nguyen et al 1992).

Table 2 summarizes intrinsic and extrinsic delays for all samples with $0 \leq x \leq 0.3$. The calculated extrinsic transit-times $\tau = \tau_i+\tau_P+\tau_f+\tau_d$ agree fairly well with the measured values $1/2\pi f_T$. An In-content of 25% in the channel enhances the intrinsic speed by 17 %. Intrinsic and extrinsic transconductances rise while intrinsic and parasitic capacitances are roughly constant. Thus the contribution of parasitic delays to extrinsic transit-time τ remains nearly unchanged (~36 %).

In mole-fraction	0%	15 %	20 %	25 %	28 %	30 %
$v_{s,eff}$ [10^7 cm/s]	1.66	1.89	1.91	1.94	1.96	1.99
$\tau_i = L_G/v_{s,eff}$	1.08 [1)]	1.01	0.99	0.95[2)]	0.97	1.01 [3)]
$\tau_P = C_P/g_{m,ex}W_G$ [ps]	0.33	0.29	0.29	0.28	0.28	0.28
$\tau_f = C_f'/g_{m,int}$ [ps]	0.17	0.16	0.19	0.13	0.09	0.11
τ_d [ps]	0.14	0.12	0.13	0.11	0.12	0.13
$\tau = \tau_i+\tau_P+\tau_f+\tau_d$ [ps]	1.72	1.58	1.60	1.47	1.46	1.53
$(\tau_P+\tau_f+\tau_d)/\tau$	37 %	36 %	38 %	35 %	34 %	34 %
measured: $1/2\pi f_T$ [ps]	1.71	1.56	1.57	1.46	1.51	1.55

Table 2: Intrinsic and parasitic delays of double-doped MODFETs with $W_G = 150$ µm, $L_G = 0.19$ µm except [1)]: $L_G = 0.18$ µm, [2)]: $L_G = 0.185$ µm, [3)]: $L_G = 0.2$ µm

4. CONCLUSION

Homogeneous doping of both $Al_{0.23}Ga_{0.77}As$-barriers on top and below the $In_{0.25}Ga_{0.75}As$-QW leads to a high 2DEG-density of $2.5{\cdot}10^{12}cm^{-2}$, as opposed to $1.5\cdot10^{12}cm^{-2}$ in the GaAs-QW. This rise in carrier density is accompanied by a significant enhancement (17 %) of effective saturation velocity and transit-frequency. However, the rise in carrier-density from 2.0 .. $2.8{\cdot}10^{12}cm^{-2}$, which is caused by variation of In-contents between 15 and 30 %, has only small influence on device performance. Fortunately, homogeneous doping of both barriers furnishes such a level of carrier density already at moderate levels of In-content. An average transit-frequency $f_T = 109$ GHz, a minimum noise figure at 12 GHz $F_{min} = 0.54$ and an associated gain $G_{ass} = 15$ dB have been achieved on devices with 25 % In. But, due to the double sided doping, even the In-free devices exhibit remarkable $f_T = 93$ GHz, $F_{min} = 0.58$ and $G_{ass} = 13$ dB (at 12 GHz).

5. ACKNOWLEDGEMENTS

We thank the SIEMENS AG for support. Measurements of RF-noise by T. Felgentreff (TU München) are gratefully acknowledged.

6. REFERENCES

Kaufel G, Olander E 1990 Mater. Res. Soc. Symp. Proc. 158, 401

Ketterson A A, Masselink W T, Gedymin J S, Klem J, Peng C, KoppW F, Morkoç H, Gleason K R 1986 IEEE Trans. Electron Devices ED-33, 564

Klein W, Böhm G, Tränkle G, Weimann G 1993 J. Crystal Growth 127, 36

Nguyen L D, Radulescu D C, Tasker P J, Schaff W J, Eastman L F 1988 IEEE Electron Dev. Lett. EDL-9, 374

Nguyen L D, Larson L E, Mishra U K 1992 Proc. IEEE 80, 494

Paper presented at the Int. Symp. GaAs and Related Compounds, Freiburg, 1993

Low leakage current InAlAs/AlAs/*n*-InAlAs structures for InAlAs/InGaAs FET applications

Hironobu Miyamoto, Tatuo Nakayama, Emi Oishi and Norihiko Samoto

Kansai Electronics Research Laboratory, NEC Corporation,

9-1, 2-choume, Seiran, Otsu, Shiga 520, Japan

ABSTRACT

We have investigated $In_{0.52}Al_{0.48}As$/AlAs/n-$In_{0.52}Al_{0.48}As$ structures to reduce the gate leakage current for InAlAs/InGaAs FET applications. The reverse leakage current of Al/Ti/$In_{0.52}Al_{0.48}As$/AlAs/n-$In_{0.52}Al_{0.48}As$ diodes exponentially decreased from 3.5 x 10^{-3} to 2 x 10^{-6} A/cm^2 at Vr=2 V as the AlAs layer thickness increased from 0 to 10nm. There was no degradation of two-dimensional electron gas characteristics in a $In_{0.52}Al_{0.48}As$/AlAs/n-$In_{0.52}Al_{0.48}As$/$In_{0.53}Ga_{0.47}As$ selectively doped structure, compared to the simple InAlAs/InGaAs structure. These results indicate that this $In_{0.52}Al_{0.48}As$/AlAs/n-$In_{0.52}Al_{0.48}As$ structure is well suited for InAlAs/InGaAs FETs.

1.INTRODUCTION

InAlAs/InGaAs field-effect transistors are of great interest for ultra-high-frequency microwave telecommunication applications. However, these transistors suffer from a large gate to source leakage current due to the insufficient Schottky barrier height(0.56-0.73 eV) of the $In_{0.52}Al_{0.48}As$. To overcome this problem , several works have sought to enhance the effective Schottky barrier height of the $In_{0.52}Al_{0.48}As$ using various layer structures. Hong et al. (1988) used a thin lattice-mismatched n-GaAs top layer on the $In_{0.52}Al_{0.48}As$ layer. Imanishi et al.(1989) decreased the AlAs mole fraction x in the $In_xAl_{1-x}As$($0.32<x<0.52$). Heedt et al.(1992) inserted a strained $In_{0.1}Al_{0.9}As$ layer into the $In_{0.52}Al_{0.48}As$ spacer of InAlAs/InGaAs FETs.

In this paper, a new structure to enhance the Schottky barrier height of the $In_{0.52}Al_{0.48}As$ through the use of $In_{0.52}Al_{0.48}As$/AlAs/n-$In_{0.52}Al_{0.48}As$ layers is reported. Since AlAs/$In_{0.52}Al_{0.48}As$ has a large Γ-band-offset, it is expected that the strained AlAs layer sufficiently enhances the Schottky barrier height. Although AlAs could not used as the top layer before because of its high reactivity, the top $In_{0.52}Al_{0.48}As$ passivation layer enables use of the AlAs layer as an electron barrier layer in this structure.

2.InAlAs/AlAs/InAlAs DIODE STRUCTURE

Figure 1 shows the energy band diagram of a metal(Al/Ti)/$In_{0.52}Al_{0.48}As$/AlAs/n-$In_{0.52}Al_{0.48}As$ diode. A strained thin AlAs layer is inserted in the $In_{0.52}Al_{0.48}As$ layer near the Schottky contact interface. A Schottky contact is formed on the top $In_{0.52}Al_{0.48}As$ layer. The top $In_{0.52}Al_{0.48}As$ layer passivates the AlAs layer. Since a strained AlAs/$In_{0.52}Al_{0.48}As$ heterostructure has a large Γ-band-offset, the AlAs layer is used as an electron barrier in this diode structure. This Γ-band-offset is estimated at 0.68 eV from the reported values of an AlAs/$In_{0.53}Ga_{0.47}As$ band-offset(1.2 eV)(Broekaert et al. 1990) and a $In_{0.52}Al_{0.48}As$/$In_{0.53}Ga_{0.47}As$ band-offset(0.52 eV)(Welch et al. 1984).

Figure 2 shows the diode structure. $In_{0.52}Al_{0.48}As$/AlAs/n-$In_{0.52}Al_{0.48}As$ structures were grown by molecular beam epitaxy (MBE) on (100) oriented semi-insulating InP substrates. Growth temperature was 500 °C. A 60-sec growth interruption was introduced at the

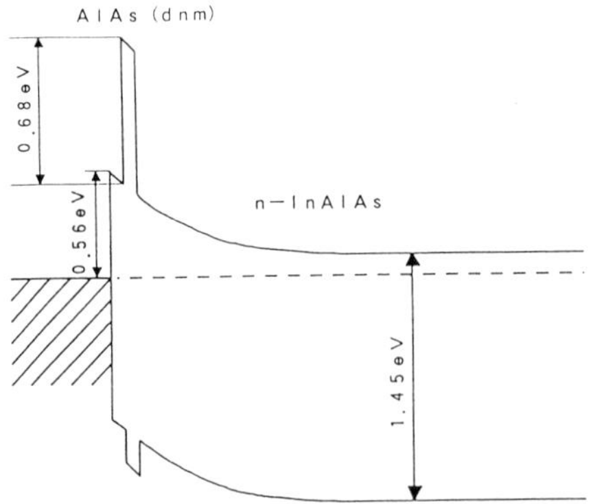

Fig. 1 Energy-band diagram of a Metal/InAlAs/AlAs/n-InAlAs diode.

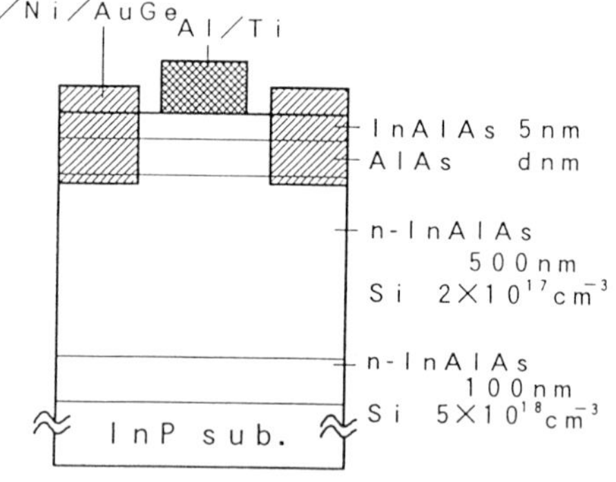

Fig. 2 Schematic diagram of the Al/Ti/InAlAs/AlAs/n-InAlAs diode.

AlAs/$In_{0.52}Al_{0.48}As$ interface to improve the interface flatness. The diode structure consists (from bottom to top) of a 100-nm heavily Si doped(5×10^{18} cm^{-3}) $In_{0.52}Al_{0.48}As$ layer, a 500-nm lightly doped(2×10^{17} cm^{-3}) layer, a thin strained AlAs layer and a 5-nm undoped $In_{0.52}Al_{0.48}As$ layer. AlAs thicknesses between 0 and 10 nm were selected to examine the effect of AlAs insertion on electrical and crystallographic characteristics.

Ohmic contacts to the diodes were formed by commonly used Au/Ni/AuGe alloying at 350 °C. Al/Ti Schottky gates were deposited on the top 5-nm $In_{0.52}Al_{0.48}As$ layer by e-beam evaporation. The Schottky diode diameter was 580 μm.

3.CURRENT-VOLTAGE CHARACTERISTIC OF $In_{0.52}Al_{0.48}As$/AlAs/n-$In_{0.52}Al_{0.48}As$ DIODES

The current voltage(I-V) characteristics of Al/Ti/$In_{0.52}Al_{0.48}As$/AlAs/n-$In_{0.52}Al_{0.48}As$ diodes were measured at room temperature. Forward current-voltage characteristics of these diodes is shown in Fig. 3. As the AlAs layer thickness *d* increases, the forward current decreases rapidly and the ideality factor *n* begins to depart from unity. The ideality factor *n* of these diodes increases from 1.05 to 1.22 with the increase of the AlAs layer thickness *d* from 0 to 3 nm. Beyond the AlAs critical layer thickness(3 nm), the ideality factor *n* approaches 2.0. Beyond the critical layer thickness, the major cause increasing the *n* value is believed to be

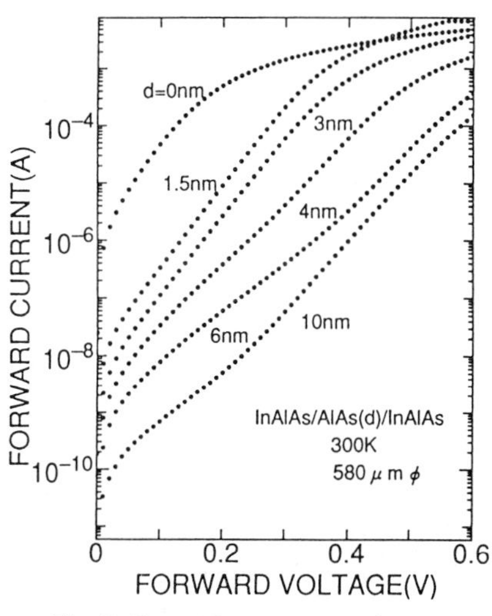

Fig. 3 Forward current vs voltage characteristics of Al/Ti/InAlAs/AlAs/n-InAlAs diodes.

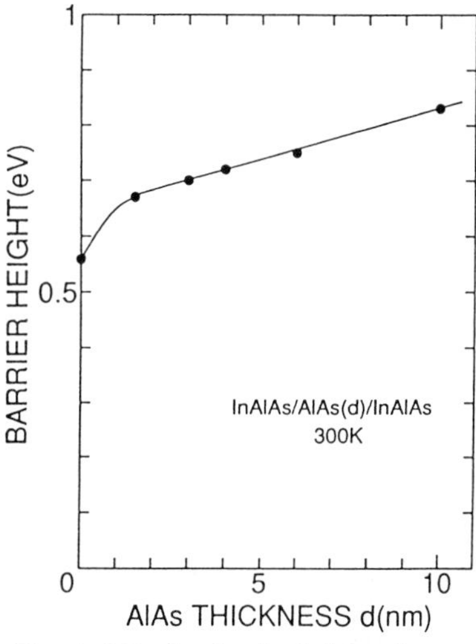

Fig. 4 Effective barrier height of Al/Ti/InAlAs/AlAs/n-InAlAs diodes.

the recombination current flow through the dislocations generated in AlAs by stress.

An effective barrier height Φb of Al/Ti/$In_{0.52}Al_{0.48}As$/AlAs/n-$In_{0.52}Al_{0.48}As$ diodes was determined from the following expression.

$$\Phi b=(kT/q)\ln(A^*T^2/Js)$$

where $A^*(=4\pi qm^*k^2/h^3)$ is the effective Richardson constant, Js is the saturation current density, T is the absolute temperature, k is Boltzmann's constant, q is the electronic charge. The value of A^* was taken to be 18 A cm^{-2} K^{-2} by assuming the effective mass $m^*=0.15m_0$(m_0 is the electron rest mass). The dependence of the effective barrier height on the AlAs layer thickness d is shown in Fig. 4. The effective barrier height is enhanced from 0.55 to 0.82 eV by increasing the inserted AlAs layer thickness d from 0 to 10 nm.

Figure 5 shows reverse current-voltage characteristics of diodes with various AlAs layer thickness. With an increased AlAs layer thickness, the reverse current decreases by a factor of 10^3. These results demonstrate that the $In_{0.52}Al_{0.48}As$/AlAs/n-$In_{0.52}Al_{0.48}As$ structure is quite effective to reduce the reverse current.

Figure 6 shows the reverse leakage current density dependence on the inserted AlAs layer thickness d at the reverse bias voltage Vr=2V. The reverse leakage current of these diodes exponentially decreases from 3.5 x 10^{-3} to 8 x 10^{-6} A/cm^2 with the increase of the AlAs layer thickness d from 0 to 4 nm, then is saturated at 2 x 10^{-6} A/cm^2. The reverse current

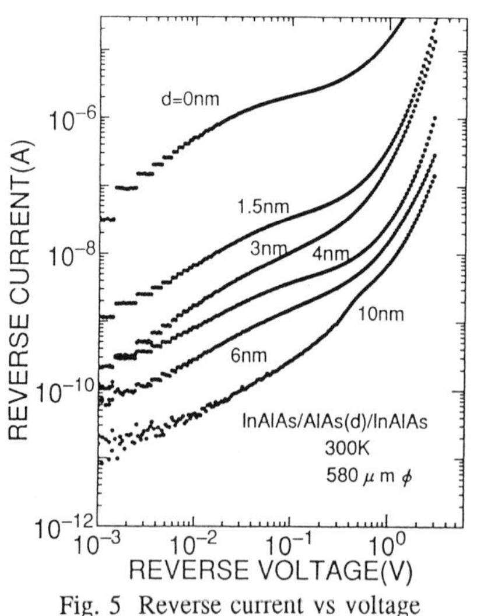

Fig. 5 Reverse current vs voltage characteristics of Al/Ti/InAlAs/AlAs/n-InAlAs diodes.

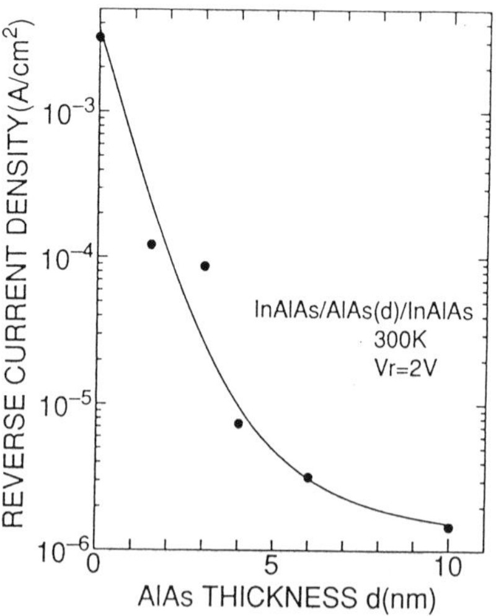

Fig. 6 Reverse current density of Al/Ti/InAlAs/AlAs/n-InAlAs diodes.

dependence on the AlAs layer thickness *d* indicates that the leakage current flows through the AlAs barrier($0<d<4$ nm) by quantum tunneling.

4.$In_{0.52}Al_{0.48}As/AlAs/n\text{-}In_{0.52}Al_{0.48}As/In_{0.53}Ga_{0.47}As$ SELECTIVELY DOPED STRUCTURES

Figure 7 shows $In_{0.52}Al_{0.48}As/AlAs/n\text{-}In_{0.52}Al_{0.48}As/In_{0.53}Ga_{0.47}As$ selectively doped structures These structures were grown by molecular beam epitaxy (MBE) on (100) oriented semi-insulating InP substrates. The hall mobility measurement on these structures was carried out and was compared with the conventional structure(d=0 nm). Figure 8 shows the dependance of hall mobility and sheet electron concentration on the thickness of the AlAs layer. For $In_{0.52}Al_{0.48}As/AlAs/n\text{-}In_{0.52}Al_{0.48}As/In_{0.53}Ga_{0.47}As$ structure, the hall mobility and sheet electron density are constant(12,000 $cm^2v^{-1}s^{-1}$ and 1.6 x 10^{12} cm^{-2} respectively) over a AlAs thickness range of $0\leq d\leq 10$ nm at room temperature. No degradation was observed even though its thickness (10 nm) is much larger than the critical layer thickness(3 nm).

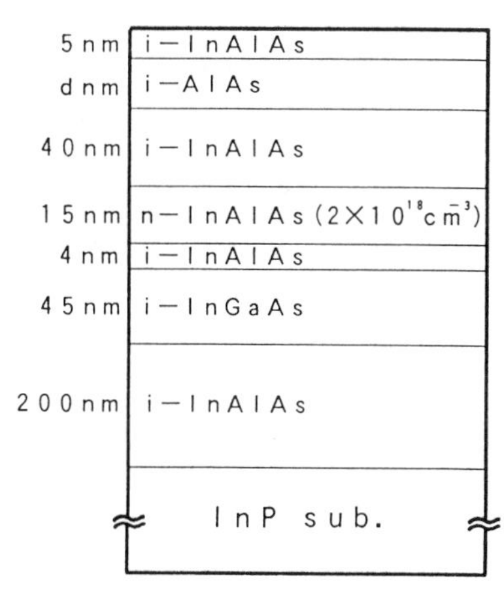

Fig. 7 Cross-sectional layer structure of the InAlAs/AlAs/n-InAlAs/InGaAs selectively doped structure.

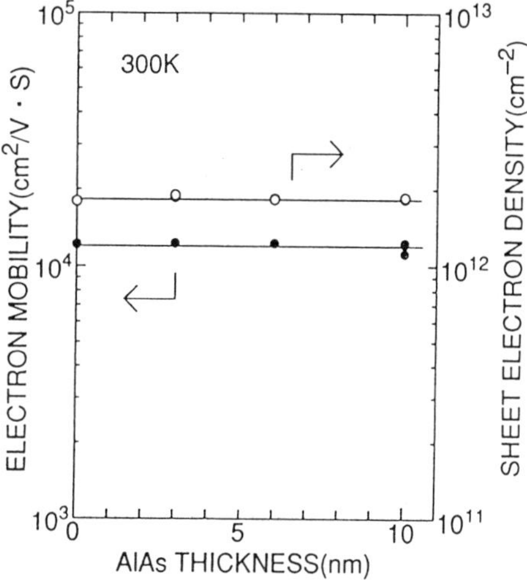

Fig. 8 Electron mobility and sheet electron density of the InAlAs/AlAs/n-InAlAs/InGaAs selectively doped structure.

5.Conclusion

We have significantly reduced the reverse leakage current of diodes by using $In_{0.52}Al_{0.48}As/AlAs/n\text{-}In_{0.52}Al_{0.48}As$ structures. The effective barrier height is enhanced by the strained AlAs insertion. There are no degradation of electron mobility and sheet electron density in a $In_{0.52}Al_{0.48}As/In_{0.53}Ga_{0.47}As$ selectively doped structure by the strained AlAs insertion. These results indicate that this $In_{0.52}Al_{0.48}As/AlAs/n\text{-}In_{0.52}Al_{0.48}As$ structure is well suited for InAlAs/InGaAs FETs.

ACKNOWLEDGEMENT

The authors would like to thank K. Onda, Y. Ando and M. Kuzuhara for discussion. They also thank Dr. Abe for his encouragement.

REFERENCES

Broekaert T P E and Fonstad C G 1990 J. Appl. Phys. 68 4310

Heedt C, Buchali F, Prost W, Fritzshe D, Nickel H and Tegude F J 1992 GaAs and Related Compounds, Karuizawa Japan(1992 Inst. Phys. Conf. Ser. No 129 941)

Hong W P and Bhattacharya P 1988 IEEE Electron Dev. Lett. 9 352

Imanishi K, Ishikawa T and Kondo K 1989 GaAs and Related Compounds, Karuizawa Japan(1992 Inst. Phys. Conf. Ser. No 106 637)

Welch D F, Wicks G W, Eastman L F 1984 J. Appl. Phys. 55 3176

Inst. Phys. Conf. Ser. No 136: Chapter 2
Paper presented at the Int. Symp. GaAs and Related Compounds, Freiburg, 1993

Characterization of the breakdown behaviour of pseudomorphic InAlAs/$In_xGa_{1-x}As$/InP HEMTs with high breakdown voltages

J. Dickmann [1], S. Schildberg [2], H. Daembkes [1], S.R. Bahl [3], J.A. del Alamo [3]
[1]Daimler-Benz Research Center Ulm, D-89081 Ulm, Germany
[2]University Rostock, Department for Insulatorphysics, D-18055 Rostock, Germnany
[3]Massachusetts Institute of Technology, Cambridge MA 02139, USA

Abstract
The experimentally based work presents a systematic analysis of the breakdown mechanisms in InAlAs/$In_xGa_{1-x}As$ HFETs in on-and off-state operation. As results of the analysis we found out that the breakdown is a two-step process of thermionic emission/tunneling and avalanche. In the off-state operation mode thermionic emission/tunneling is recognized as the dominant mechanism, while in the on-state operation mode breakdown is governed by avalanche. From burn-out experiments we found indications, which support the assumption that burn-out in the off-state operation mode is surface related, while in the on-state operation mode burn-out occurs in the channel layer.

Introduction

InAlAs/InGaAs heterostructure field effect transistors have demonstrated their potential for low noise and high speed operation. The highest cut-off frequencies and the lowest noise figure and highest associated gain at millimetre wave operation ever achieved for an HFET have been reported for this type of devices, [1-3]. However, the material system InAlAs/InGaAs/InP has not yet been exploited for power application. The low bandgap of the channel layer and the low Schottky-barrier height have presently limited the breakdown voltage and hence the output-power and efficiency performance of the devices. The characterization of breakdown phenomena has extensively been done theoretically and experimentally for GaAs MESFETs [4-7]. In all cases avalanche has been identified as the dominant mechanism. One remaining problem was to allocate the definite position where breakdown occurs and whether this position is bias dependent or not. The result of one investigation was that the localization for breakdown is always at the drain side of the gate and bias independent [7], while in other publications the result was that the localization is strongly bias dependent and can/cannot take place in the channel or the buffer layer [4,6]. Up to now, only few paper have been published on the analysis of breakdown phenomena in GaAs based heterostructure field effect transistors (HFETs) and (to our knowledge) only one on InP based HEMTs. However, the uncertainty in the breakdown analysis of MESFETs is also continued in these investigations [8-10]. A theoretical analysis of [8] came to the conclusion, that avalanche is the dominant process and the position for avalanche was localized in the wide bandgap AlGaAs. In [9] a theoretical model which includes thermionic emission/tunneling and avalanche was used to explain the breakdown behaviour in GaAs based HEMTs in the off-state mode of operation and was localized to be surface oriented. The first published experimental investigation of InP based HEMTs found indications that breakdown in the off-state mode is a two step process of thermionic emission/tunneling and

avalanche [10]. No information was given for the on-state mode of operation and the localization of breakdown.
In this paper we confirm our findings for $In_{0.52}Al_{0.48}As/In_xGa_{1-x}As/InP$ pseudomorphic HEMTs in the off-state mode of operation [10] and extend the investigation to the on-state condition. From burn-out experiments we deduce indications for the localization of breakdown in the HEMT structure.

Device layer structure

The layer structure used in this investigation is shown in Fig.1.

	InGaAs	surface depleted cap layer	
20nm	InAlAs i	barrier layer	
3nm	InAlAs $N_D = 8x10^{18} cm^{-3}$		donor layer
2nm	InAlAs i	spacer layer	
	$In_xGa_{1-x}As$ i $In_{0.53}Ga_{0.47}$ As i	L_z =40nm	hc dsub channel
	InAlAs/InGaAs i	superlattice	
	InAlAs	buffer	
	InAlAs/InGaAs i	superlattice	
	InP s.i.	substrate	

Fig. 1: Layer structure for the $In_{0.52}Al_{0.48}As/In_xGa_{1-x}As/InP$ pseudomorphic HEMTs .

More details about the fabrication process and the theoretical background of the layer structure design are given in [12, 13, 14]. The focus in the design was on high breakdown voltages and the incorporation of high indium molefractions in the InGaAs channel. An exponential doping profile was used in the InGaAs cap-layer in order to achieve a surface depleted cap layer [14], which reduces the electric field near the drain edge of the gate. The indium molefraction in the channel layer was set to x=0.53, 0.62, 0.7 and the channel thickness of the strained portion of the channel was h_c= 40nm for x=0.53, 0.62 and h_c= 15nm for x=0.7.

Experimental results

Important for a common understanding is the definition criterion for the breakdown voltage. We define the two terminal gate-source/drain breakdown voltage ($V_{BrGD/S}$) as well as the three terminal drain-source breakdown voltage (V_{BrDS}(on-state)) as the bias point for a gate leakage current of I_G=1mA/mm. For the off-state we used the criterion I_{DS}=1mA/mm. These criteria are the definition of a reversible breakdown in spite of the burn-out breakdown where the destruction of the device has happened. All data below is the mean value of at least 30 devices per wafer.

Two terminal measurements

The leakage current and breakdown behaviour of a HEMT is determined by the quality of its Schottky contact. Therefore we first analyzed the IV-characteristics of the gate-drain diode of completed HEMT devices and did not utilized test structures. In order to verify that in the following investigation all findings are based on the same diode parameters, we analyzed the diode IV-characteristic of the devices with the three different indium molefractions in the forward direction and determined the effective barrier height (Φ_B), ideality factor (n) and series resistances (R_B). For all three samples with a gate geometry of $L_G x W_G = 0.28\mu m x 30\mu m$, parameters are of the same scale ($\Phi_B = 0.27eV$, n=1.3, $R_B = 125\Omega$) and the difference was in the order of the measurement accuracy, assuring that the breakdown behaviour of the different samples is caused by the breakdown physics and not by technological differences.

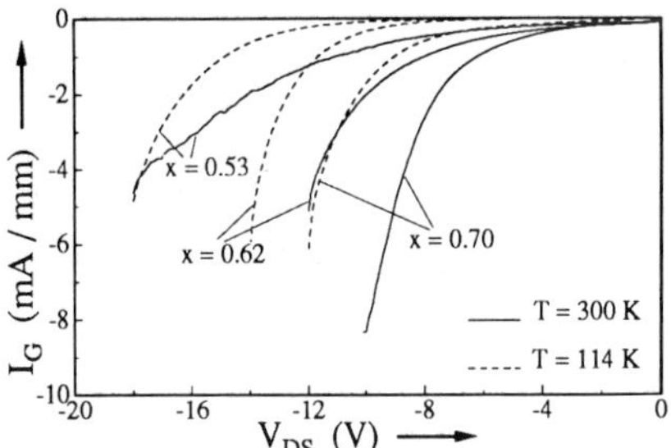

Fig.2:IV-reverse characteristic of the gate-drain diode of pseudomorphic HEMTs with an indium molefraction of x=0.53, 0.62, 0.7 in the channel at different temperatures.

In order to determine the breakdown voltage we measured the reverse IV-characteristic of the gate-drain diode of the HEMT devices at different temperatures. Since thermionic emission/tunneling and avalanche have an opposite temperature coefficient, with temperature dependent measurements it is possible to identify the breakdown mechanism which is presently active [9, 10, 15]. If thermionic emission/tunneling is the dominant process the breakdown voltage increases with reduced temperature, while for avalanche being the dominant process the breakdown voltage drops. A typical set of IV reverse characteristics as a function of temperature for the different HEMT devices is shown in Fig. . 2. From Fig. 2 the following information can be obtained. The breakdown voltage increases drastically with reduced temperature for each device, while it drops with increasing indium molefraction ($E_G\downarrow$). The improvement in detail is V_{BrGD} (x=0.53) = 11.9V,15V, V_{BrGD} (x=0.62) = 7.9V,13.9V, V_{BrGD} (x=0.7) = 6.7V, 9.31V at T = 300K, 114K, respectively. From these measurements it becomes obvious, that the breakdown behaviour for the diodes is determined by a large thermionic emission/tunneling component. A possible explanation for the drop of V_{BrGD} with increasing indium concentration is, that electrons enter the channel hot and immediately relax their energy through impact ionization [10,11]. From these results it becomes obvious that breakdown in the gate-drain diodes is a two step process with thermionic emission/tunneling as the dominant process.

Three terminal measurements

Off-state operation

We carried out three terminal off-state breakdown measurements by measuring the I_{DS}-V_{DS} characteristic at $V_{GS} \leq V_{TH}$ for different temperatures. Fig. 3 shows the IV-characteristic for the devices with x = 0.53 and x = 0.7 at room temperature and T = 114K. From these measurements we found that for low and medium drain-source voltages the HEMT shows a similar behaviour as the gate-drain diode in the two terminal measurement set-up. The breakdown voltage increases with reduced temperature and drops with increased indium molefraction. This reveals that thermionic emission/tunneling is the dominant process on this bias range (thermionic emission dominated). At higher drain bias, the behaviour changes. The IV curves cross-over, which represents a change in the gradient of the breakdown voltage as usually observed if avalanche takes place. A further fact that supports that avalanche has taken place is the ratio of I_G/I_{DS}. This ratio was determined to be well below unity for V_{DS} values above the I_{DS} cross-over point. These results clearly demonstrate that breakdown in these devices is a two step process and a function of drain bias and ambient temperature. Avalanche limits the device performance only if the thermionic field component is nearly eliminated. In order to support the above findings we confirmed these results with a newly established drain-current injection technique [10, 11]. In addition to the presented results in [10] it was found out, that the gate current approaching breakdown is weakly thermally activated at around 300K ($E_A(x = 0.53) = 0.2$eV at low V_{GD}, $E_A = 0.09$eV at $V_{GD} = 10$V), also indicating a large thermionic/tunneling component. This reveals that the breakdown mechanism is not a simple impact ionization process. The breakdown voltages were determined to be $V_{BrDS} = 5.1$V, 6.3V, 8.9V for x = 0.53, 0.62, 0.7, respectively.

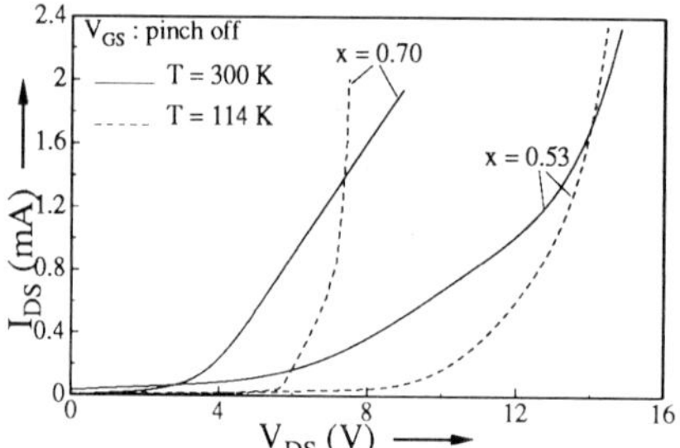

Fig.3 : I_{DS}-V_{DS} characteristic at threshold voltage (off-state) for pseudomorphic HEMTs with an indium molefraction of x=0.53, 0.7 in the InGaAs channel.

On-state operation

For the on-state operation we measured the I_{DS}-V_{DS} characteristic at $V_{GS} < \Phi_B$ for all different temperatures. Since avalanche is determined by the amount of I_{DS} that is carried in the channel layer, the measurements are carried out at V_{GS} that adjusted the same I_{DS}-V_{DS} characteristic up to the first kink (V_{DS} >1.5V) for the three devices. For all devices we observed the same behaviour. Fig. 4 shows for example the IV-characteristic and the I_G/I_{DS} ratio for a device with an indium molefraction of x = 0.53 at room temperature and T=114K. In comparison to the off-state operation, the breakdown behaviour is quite different.

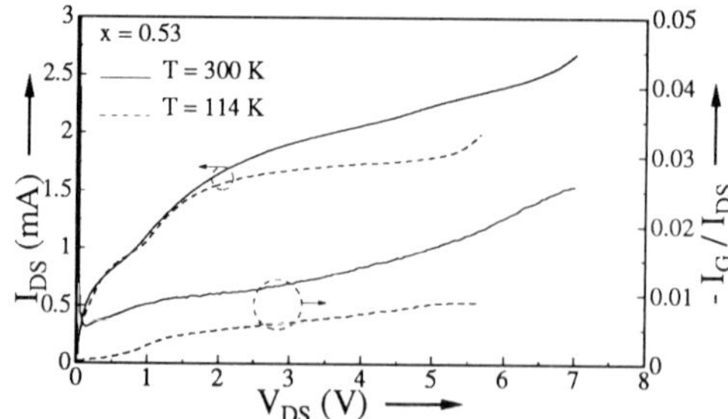

Fig.4 : I_{DS}-V_{DS} characteristic and I_G/I_{DS}-V_{DS} characteristic in on-state mode for a pseudomorphic HEMT with an indium molefraction of x = 0.53.

The drain-source current for each temperature starts to rise even at low values of V_{DS} >1.5V with an I_G/I_{DS} ratio of less than unity. At high values of V_{DS} it increases nearly exponentially and the ratio I_G/I_{DS} is far less than unity and the I_{DS} curve for T = 114K crosses that for T = 300K. These facts are clear indications for a large avalanche component that dominates the breakdown behaviour in the on-state mode of operation. In order to manifest this conclusion we now compare the I_{DS}-V_{DS} characteristic for the three different devices (x=0.53, 0.62, 0.7) at T=300K. Fig. 5 shows the measured characteristics for this case.

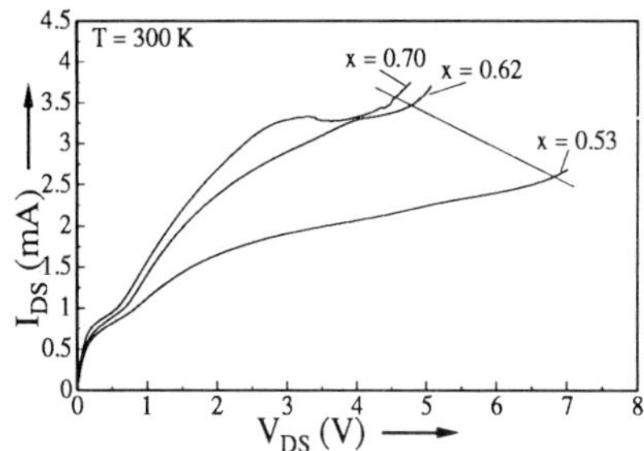

Fig.5: I_{DS}-V_{DS} characteristic at V_{GS} that adjusts the same I_{DS}-V_{DS} characteristic up to the first kink for pseudomorphic HEMTs with an indium molefraction of x = 0.53, 0.62, 0.7 in the InGaAs channel.

As indicated by the straight line the drain-source voltage for which the drastic rise in the I_{DS}-V_{DS} characteristic occurs drops with increasing indium molefraction. In other words, the drain bias point for the exponential rise in I_{DS} drops with reducing the bandgap of the channel layer. This is consistent with the theory that there exists a bandgap dependent threshold field for which avalanche takes place [5]. The breakdown voltages in this mode of operation were determined to be V_{BrDS} = 3.5V, 4.5V, 6.8V, for x = 0.53, 0.62, 0.7, respectively.

Burn-out experiment

The location of the occurrence of burn-out as a function of bias point was investigated by analyzing the SEM photographs of burned-out devices. In the experiment we drove the devices smoothly into burn-out by slowly increasing V_{DS}. This for the on-state (V_{GS}=0V) and off-state (V_{GS}=V_{TH}) mode of operation and inspected them electrically and visually (SEM). The interesting finding was, that the devices that were burned-out in the on-state mode of operation showed no sign for any deterioration, although they were electrically damaged. The devices burned-out in the off-state mode of operation showed drastically deformations in the semiconductor at the surface between gate and drain. Visually no indication for metal migration was found. These observations lead to the assumption, that burn-out appears at different places in the device as a function of bias. From these results we have evidence that off-state burn-out is surface related while on-state burn-out happens inside the layer structure, most probably in the channel layer.

Conclusion

In conclusion we have investigated the breakdown behaviour of pseudomorphic $InAlAs/In_xGa_{1-x}As/InP$ HEMTs with x=0.53, 0.62, 0.7. With our investigation we determined for the first time the breakdown mechanism in InP based HEMTs in the on-state and off-state mode of operation. Further more we were able to show, that the dominant breakdown mechanism and the localization of burn-out are depending on the bias condition and the ambient temperature used.

References

[1] P. Ho et.al., "Extremely High Gain 0.15µm Gate Length InAlAs/InGaAs/InP HEMTs", Electronics Lett., vol. 27, No.4, pp.325-326, 1991

[2] D.C. Streit, et.al., "High Performance W-Band InAlAs/InGaAs-InP HEMTs", Electronics Lett., vol. 27, No.13, pp.1149-1150, 1991

[3] L.D.Nguyen, et.al., "50-nm Self-Aligned-Gate Pseudomorphic AlInAs/InGaAs High Electron Mobility Transistors", IEEE Trans. Electron Dev., vol.ED-39, No.9, pp.2007-2014, 1992

[4] R.A. Pucel, F. Sandy, L. Holway, R. Hohlfeld, "The use of massively-parallel computers for the simulation of FETs", MM`92 Conference Proc., Brighton, GB, pp.281-287, 1992

[5] K. Ohnaka, J. Shibata, "Excess gate-leakage current of InGaAs junction fieldeffect-transistors ", Journ. of Appl. Phys., vol. 63, No.9, pp.4714-4717, 1988

[6] J.M. Ashworth, N.Arnold, "The Gate-Bias Dependency of Breakdown Location in GaAs Metal Semiconductor Field Effect Transistors (MESFETs)", Japanese Journ. of Appl. Phys., vol.30, No.12B, pp.3822-3827, 1991

[7] J.P.R. David, J.E. Stich, M.S. Stern, "Gate-Drain Avalanche Breakdown in GaAs Power MESFETs", IEEE Trans. Electron Dev., vol.ED-29, No.10, pp.1548-1552, 1982

[8] H.F. Chau, D. Pavlidis, K. Tomizawa, "Theoretical Analysis of HEMT Breakdown Dependence on Device Design Parameters", IEEE Trans. Electron Dev., vol.ED-38, No.2, pp.213-221, 1991

[9] R.J. Trew, U.K. Mishra, "Gate Breakdown in MESFETs and HEMTs", IEEE Electron Device Lett., vol.EDL-12, No.10, pp.524-526, 1991

[10] S.R. Bahl, J.A. del Alamo, J. Dickmann, S. Schildberg, "Physics of breakdown in InAlAs/InGaAs MODFETs", Tech. Digest of the 51st annual Device Research Conference, Santa Barbara, USA, June 21-23, 1993

[11] S.R. Bahl, J.A. del Alamo, "Physics of breakdown in $InAlAs/n^+$-InGaAs Heterostructure Field-Effect Transistors", 5th Intern. Conf. on Indium Phosphide and Related Materials, Paris, France, Tech. Dig. pp.243-246, 1993

[12] J. Dickmann, H. Haspeklo, A. Geyer, H. Daembkes, R. Loesch, " High Performance fully passivated InAlAs/InGaAs/InP HFET", Electronics Lett., vol.28, No.7, pp.647-649, 1992

[13] J. Dickmann, K. Riepe, H. Daembkes, H. Künzel, " AlInAs/InGaAs Pseudomorphic HEMTs: Design and Performances", 5th Intern. Conf. on Indium Phosphide and Related Materials, Paris, France, Tech. Dig. pp.461-464, 1993

[14] J. Dickmann, H. Daembkes, H. Nickel, R. Lösch, W. Schlapp, J. Böttcher, H. Künzel, "Influence of Surface Layers on the RF-Performance of AlInAs-GaInAs HFETs", IEEE Microwave And Guided Wave Letters, vol, 2, No.11, pp.472-474, 1992

[15] N. Iwata, Y. Okamoto, M. Kuzuhara, "Breakdown voltage enhancement in a GaAs MESFET with a step-doped channel under high output power operation", Inst. Phys. Conf. Ser. No.129, pp.937-938, 1992

Paper presented at the Int. Symp. GaAs and Related Compounds, Freiburg, 1993

DC characterization of $Ga_{0.51}In_{0.49}P$/GaAs inverted-structure HEMT

C. L. Huang, Y. W. Hsu, and S. S. Lu

Department of Electrical Engineering, National Taiwan University,
Taipei, Taiwan, R.O.C.

ABSTRACT: The *first* $Ga_{0.51}In_{0.49}P$/GaAs inverted-structure HEMTs have been fabricated and measured. High drain-to-source breakdown voltage ~ 20V and gate-to-drain breakdown voltage ~ 26V was achieved by using the GaInP passivation layer. Preliminary results on mobility and g_m have been obtained. By further optimization of crystal growth and device fabrication, GaInP/GaAs I-HEMT can be used as a high gain, high speed, and high breakdown (power) device.

1. INTRODUCTION

The mostly studied high electron mobility transistor (HEMT) is the "normal" metal-n AlGaAs-i GaAs structure. In the past years, the "inverted" structure HEMT (I-HEMT), in which small bandgap (GaAs) channel is on top of large bandgap (AlGaAs) supplying layer, was also studied. Several advantages of I-HEMT have been revealed. It has been shown that the transconductance of the I-HEMT is nearly independent of the doping level in the high bandgap material, while in the normal HEMT the transconductance decreases with reduced doping. Enhanced electron mobility (H. Morkoc et al 1982) and extremely high transconductance (Nicholas C. Cirillo et al 1986) were observed, which was attributed to the smaller distance between 2-D electrons and gate in the I-HEMT compared with that of normal HEMT.

Although the breakdown mechanism in a FET is still not totally understood, it is believed to be due to the avalanche breakdown in the gate-to-drain region. One way to improve the breakdown voltage was the use of a passivation layer. The passivation layer can be a low-temperature-grown GaAs layer (L.-W. Yin et al 1990) (Chang-Lee Chen et al 1991) or an undoped high bandgap AlGaAs layer (B. Kim et al 1984). Because of the high resitivity and the low avalanche rate of the passivation layer, these devices show high breakdown voltage characteristics. Thus high power transistors are obtained. Recently, there is an interest in the growth of GaInP/GaAs lattice matched material system. Because of several advantages of GaInP, such as its large direct bandgap ~ 1.92eV and little deep trap density, it has been used to fabricate optoelectronic devices, HEMTs, and HBTs. In this paper, the first I-HEMT using GaInP/GaAs material system was investigated. In our device, an undoped GaInP layer was used between the GaAs channel layer and cap layer. This undoped GaInP layer was used as a passivation layer on the gate-to-drain region and was aimed to increase the breakdown voltage of the I-HEMT.

2. DEVICE FABRICATION

Fig. 1 shows the layer structure and finished device of the I-HEMT. It consisted of a 1μm undoped GaAs buffer followed by a 500Å undoped GaInP buffer, a 200Å n^+-doped GaInP supplying layer, a 70Å undoped GaInP spacer, a 300Å undoped GaAs channel layer. A 200Å undoped GaInP layer was used as the passivation layer between gate and drain to increase the breakdown. Finally, a n^+-GaAs cap layer helped forming ohmic contacts. Note that GaInP undoped layer was etched during gate recess etching to be an I-HEMT. Wet etching technique was used. The gate length and gate width was 1.5μm and 75μm, respectively.

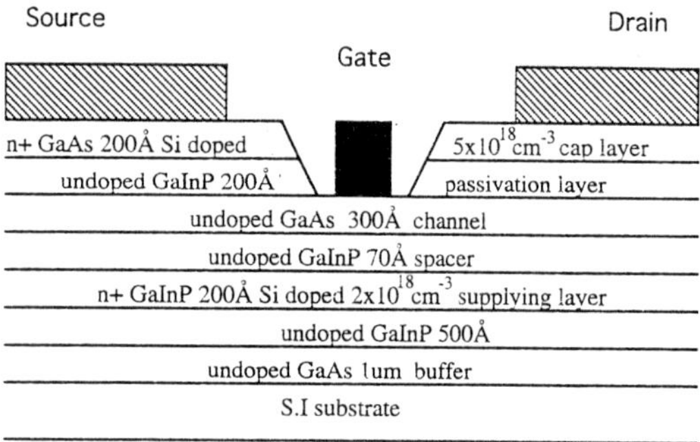

Figure 1 Schematic diagram of layer structure and the finished device of GaInP/GaAs I-HEMT

3. RESULTS AND DISCUSSIONS

A Hall measurement was done for the I-HEMT sampe whose GaAs cap layer was etched. The Hall mobility was 2400 and 11100 $cm^2/V{\cdot}s$ at 300 and 77°K, while the corresponding value for the sheet charge was $2.2X10^{12}$ cm^{-2} and $1.6x10^{12}$ cm^{-2}. The mobility was lower than the previous reports (3200 $cm^2/V{\cdot}s$, 300°K, and 17700 $cm^2/V{\cdot}s$, 77°K) (Z. P. Jiang et al 1992) of the normal GaInP/GaAs HEMT grown by the same technique (GSMBE). However, the sheet charge densities were higher than the previous results ($9.8x10^{11}$ cm^{-2} and $5.6x10^{11}$ cm^{-2}) of the normal HEMT.

The dc I-V characteristics of I-HEMT is shown in Fig. 2. It showed a very small leakage current I_{DS} in the pinch-off region even for drain bias V_{DS} up to 18.5V. A drain-to-source breakdown voltage V_{BDS} ~ 20V was obtained, which is defined as the drain voltage where drain current I_D reaches one-tenth of the saturation current I_{DSS}. This result was comparable with that (20~23V) obtained from the GaAs MISFET with undoped AlGaAs as an insulator (Bumman Kim et al 1986).

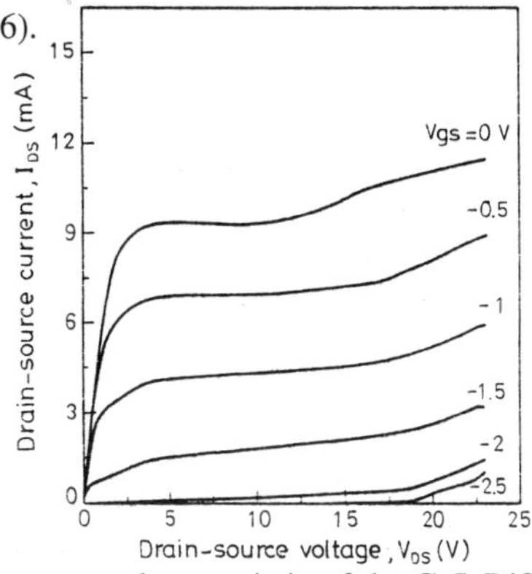

Figure 2 Common source characteristic of the GaInP/GaAs I-HEMT at 300°K

A gate-to-drain breakdown test with source opened was done as shown in Fig. 3

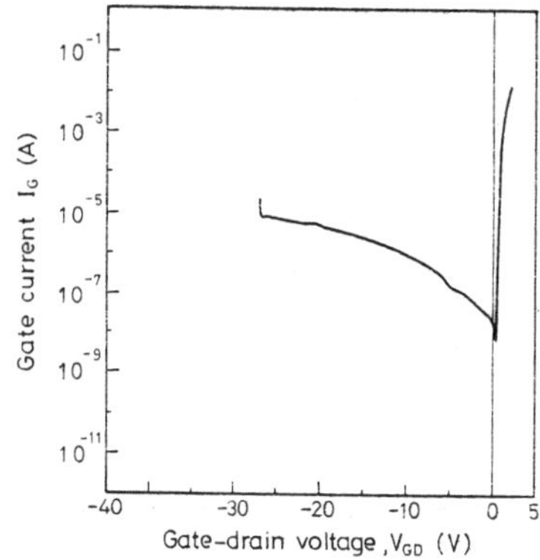

Figure 3 Gate-to-drain breakdown test of the GaInP/GaAs I-HEMT

It shows that a high gate-to-drain breakdown voltage ~ 26V was obtained from another device, where the gate-to-drain breakdown voltage was defined as the voltage at which the gate current reaches 1mA/mm. It should be noted AlGaAs/InGaAs/GaAs quantum well MISFET shows an internal breakdown voltage ~ 7V with doped InGaAs channel and a gate-to-drain breakdown voltage ~ 10 to 13V with undoped InGaAs channel (Bumman Kim et al 1989). The saturation current shown in Fig. 2 was about 130 mA/mm. The I_{DS}-V_G and g_m-V_G transfer curves are shown in Fig. 4. The maximum extrinsic transconductance g_m

was 67 mS/mm at 300°K. The threshold voltage V_{th} was -2.2V. Enhanced electric performance (higher drain current and higher g_m) was observed at low temperature.

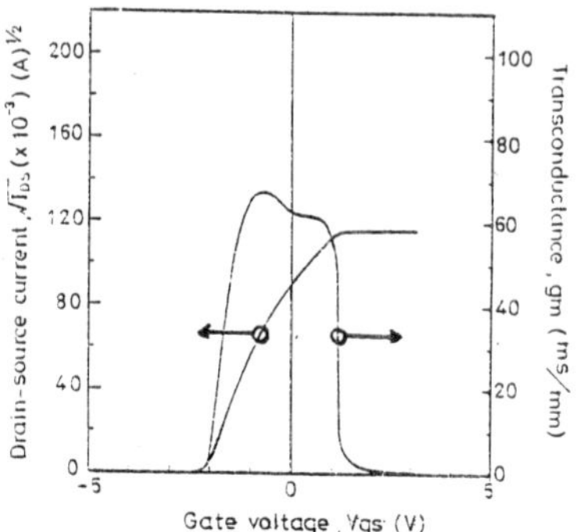

Figure 4 I_{DS}-V_G and g_m-V_G transfer curves of the GaInP/GaAs I-HEMT

For contrast experiment, the GaAs cap layer and the undoped GaInP passivation layer of gate-to-drain and gate-to-source region of the I-HEMT were both etched for comparasion with the previous results of I-HEMT. Drain current decreasing and poor breakdown characteristics were observed. This indicates that GaInP layer worked to increase the breakdown voltage.

4. CONCLUSION

The dc I-V characteristics of the GaInP/GaAs I-HEMT showed a drain-to-source breakdown voltage ~ 20V and a gate-to-drain breakdown voltage ~ 26V with a 200Å undoped GaInP insulated layer between the GaAs cap layer and channel layer. GaInP/GaAs I-HEMT showed the potential to achieve high-power transistors. Support from NSC of ROC under No. NSC82-0404-E-002-283, NSF-DMR, NSF-ECS, and IBM are acknowleged. We also thank the help from Dr. F. Williamson and Prof. M. I. Nathan.

REFERENCES

Chang-Lee Chen, Frank W. Smith, Brian J. Clifton, Leonard J. Mahoney, Michael J. Manfra, and Arthur R. Calawa, IEEE. Device Lett., vol. 12, no. 6, pp 306~308, Jun. 1991.

H. Morkoc, T. J. Drummond, and R. Fisher, J. Appl. Phys., vol. 53, pp 1030, 1982.

L.-W. Yin, Y. Hwang, J. H. Lee, Robert M. Kolbas, Robert J. Trew, and Umesh K. Mishra, IEEE. Electron Device Lett., vol. 11, no. 12, pp 561~ 563, Dec. 1990.

Nicholas C. Cirillo, JR., Michael S. Shur, and Jonathan K. Abrokwah, IEEE. Electron Device Lett., vol. EDL-7, no 2, pp 71~74, Feb. 1986.

B. Kim, H. Q. Tserng, and H. D. Shih, IEEE. Electron Device Lett., vol. EDL-5, no. 11, pp 494~495, Nov. 1984.

Bumman Kim, Hua Quen Tserng, and J. W. Lee, IEEE Electron Device Lett., vol. EDL-7, no. 11, pp 638~639, Nov. 1986.

Bumman Kim, Richard J. Matyi, Marianne Wurtele, Keith Bradshaw, M. Ali Khatibzaseh, and Hua Quen Tserng, IEEE Trans. ED, vol. 36, no. 10, pp 2236~2242, Oct. 1989.

Z. P. Jiang, P. B. Fischer, S. Y. Chou, and M. I. Nathan, J. Appl. Phys., vol 71, no. 9, pp 4632~4634

Paper presented at the Int. Symp. GaAs and Related Compounds, Freiburg, 1993

High speed $Ga_{0.5}In_{0.5}P/In_{0.15}Ga_{0.85}As$ pseudomorphic p^+-Schottky-enhanced barrier heterojunction-field effect transistors (HFETs)

J. Dickmann, M. Berg*, A. Geyer, H. Daembkes, F. Scholz+, M. Moser+

Daimler-Benz Research Center Ulm, D-89081 Ulm, Germany

* Technical University Darmstadt, Institute for High Frequency Technology, D-64283Darmstadt, Germany

+ University Stuttgart, 4 th Institute of Physics, D-70569 Stuttgart, Germany

Abstract

This paper reports about the first successful use of a $Ga_{0.5}In_{0.5}P/In_{0.15}Ga_{0.85}As$ HFET in combination with a very thin p^+- surface layer. For a device with a 1.8µm long T-gate and 30µm gate width, a maximum transconductance of g_{mmax}=200mS/mm and a maximum saturation current of I_{DSmax}=400mA/mm were measured. The drain-source burn-out breakdown voltage was determined to be in excess of V_{BrDS}>20V. The current gain and power gain cut-off frequencies were f_T=12GHz and f_{max}=42GHz, respectively.

Introduction

GaAs based HFETs are the heart of modern microwave and millimeterwave communication and sensor systems. However, their present limits in RF-power performance are in general determined by the breakdown voltage. The ways proposed to overcome this problem were either to use a double recessing [1,2] or to put a thin layer of material with a high breakdown field under the gate. This layer was an AlGaAs wide bandgap insulator layer or a GaAs layer grown at low temperature [3,4,5]. Quite recently InP, a material with a high breakdown field, low ionization coefficient and good thermal conductivity was used [6]. Further improvements in breakdown voltage can be achieved by adopting the p^+-enhanced Schottky barrier height approach by using thin highly doped surface layers as proposed in [7,8,9].

With reference to these attempts and considering its bandgap, lattice constant and crossover point from a direct to indirect semiconductor, $Ga_xIn_{1-x}P$ seems to be an attractive candidate as the barrier and electron supplying material under the gate of a GaAs based HFET. Although there is only a very small discontinuity in the conduction band between $Ga_xIn_{1-x}P$ and GaAs (ΔE_C=0.19eV-0.21eV) a first successful fabrication of a $Ga_{0.51}In_{0.49}P$/GaAs HFET with a 2DEG was reported in [10,11]. For a gate length of 1µm a maximum transconductance of g_{mmax}=163mS/mm, a maximum saturation current of I_{DSmax}=280mA/mm, and cut-off frequencies of f_T=17.8GHz and f_{max}=23.5GHz were reported.

In this paper we report about the first successful use of a combined application of $Ga_xIn_{1-x}P$ as the substitute for $Al_xGa_{1-x}As$ and thin p^+-surface layers in order to achieve a high speed device with high breakdown voltage and no need for a gate-recessing.

Device Layer Structure

The layer structure was grown by low pressure metal organic chemical vapour deposition (MOCVD) on a semi-insulating (100) GaAs substrate. The sources used were trimethylindium (TMI), trimethylgallium (TMG), arsine (AsH_3) and phosphine (PH_3). Dimethylzinc (DMZn) was used for p-doping and selenidehydrogen (H_2Se) was used for n-doping. The background residual doping levels for GaAs, $In_yGa_{1-y}As$ and $Ga_xIn_{1-x}P$ were determined to be $N_D<10^{15}cm^{-3}$, $N_D<10^{15}cm^{-3}$ and $N_D<5x10^{15}cm^{-3}$, respectively. The growth condition for the HFET layer structure was as follows. The growth temperature was set to T=750°C for the entire layer structure in order to meet the optimum growth condition for the $Ga_xIn_{1-x}P$ layer. The V/III ratios and growth rates for the GaAs, $In_yGa_{1-y}As$ and $Ga_xIn_{1-x}P$ layers were 200 / 1µm/h, 200 / 1.2 µm/h, 330 / 2.2 µm/h, respectively. Seperate investigations of the interface $Ga_xIn_{1-x}P$ grown on $In_yGa_{1-y}As$ showed a nearly perfect interface with no indication of an intermixing of arsenic and phosphor leading to an InGaAsP interface layer as observed in [12]. A detailed study will be published elsewhere. The layer structure (Fig.1) in growth direction was: 0.5µm GaAs undoped buffer layer to smoothen the growth front, a 12nm $In_{0.15}Ga_{0.85}As$ undoped channel layer followed by a 2nm $Ga_{0.5}In_{0.5}P$ spacer layer. The electron supplying layer was formed by a 20nm thick $Ga_{0.5}In_{0.5}P$ layer homogeneously doped with selenium (Se) to $N_D=3x10^{18}cm^{-3}$. On top of this layer a 10nm thick homogeneously selenium doped GaAs layer ($N_D=3x10^{18}cm^{-3}$) was grown in order to allow the formation of low ohmic contact resistances and to reduce the total access resistances. The entire layer structure was completed by the 10nm thick GaAs p^+-surface layer which was homogeneously doped with zinc (Zn) to $N_A=5x10^{18}cm^{-3}$.

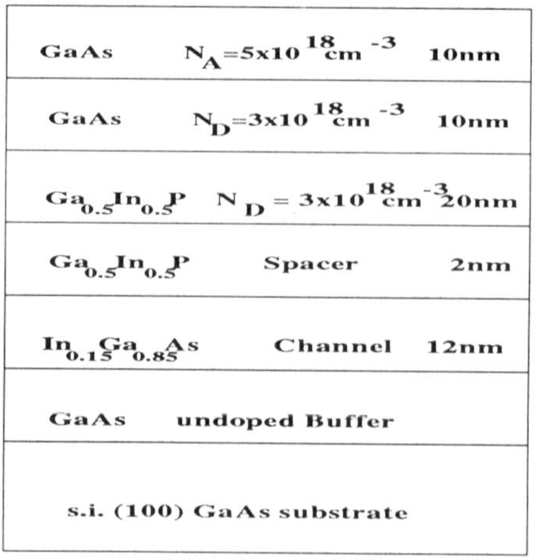

GaAs	$N_A=5x10^{18}cm^{-3}$	10nm
GaAs	$N_D=3x10^{18}cm^{-3}$	10nm
$Ga_{0.5}In_{0.5}P$	$N_D = 3x10^{18}cm^{-3}$	20nm
$Ga_{0.5}In_{0.5}P$	Spacer	2nm
$In_{0.15}Ga_{0.85}As$	Channel	12nm
GaAs	undoped Buffer	
s.i. (100) GaAs substrate		

Fig.1: Layer structure of the $Ga_{0.5}In_{0.5}P/In_{0.15}Ga_{0.85}As$ HFET

From Shubnikov de Haas (SdH) measurements a 2DEG sheet carrier concentration of $n_s = 1.25 \times 10^{12} cm^{-2}$ was determined, proving the high quality of the sample and the fact that a reasonable 2DEG concentration can be generated in such a layer structure.

Device Fabrication

In contrary to a conventional JFET process where a very thick and highly doped p^+-layer is used to form the gate, in this work a very low resistance Ti/Pt/Au T-gate is used to form the Schottky-contact on a very thin p^+-layer which in general just changes the effective barrier height of the Schottky contact.

The device fabrication starts with the fabrication of a 1.8µm long T-gate. The T-shaped cross sectional area is generated by using a five layer sequence consisting of AZ-resists and germanium intermediate layers. The top resist is exposed by conventional contact lithography and the generated pattern is then transferred into the subsequent layers via reactive ion etching as described in [13]. After the T-shaped cross sectional shape is generated, the gate Ti/Pt/Au (15nm/20nm/600nm) metallization is e-gun evaporated. The T-bar was designed to generate a 0.3µm undercut with respect to the gate foot print. Device isolation was done by mesa isolation. In order to prevent a shortening of the "p^+-n junction", a double-mesa isolation is utilized where the gate-pad and the gate electrode are located on mesa plateaus and isolated by an etch trench that works as an airbridge.

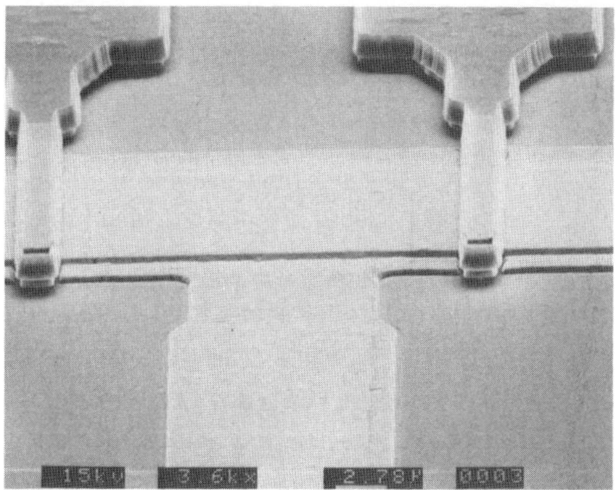

Fig.2: SEM photograph of the $Ga_{0.5}In_{0.5}P/In_{0.15}Ga_{0.85}As$ HFET

For the etching of GaAs a sulphuric based chemical solution (H_2SO_4:H_2O_2:H_2O) and for GaInP a phosphorus based (H_3PO_4:HCl) chemical solution is used. The distinguished material selective behaviour of InGaAs and GaInP is a further advantage of this material system and makes device processing more save and well controlled. The next step is the fabrication of the Ge/Ni/Au ohmic contacts by using the T-gate structure to self-align the ohmic metallization. According to the undercut of the T-bar, the drain/source-gate distance is adjusted to be 0.3µm which yields a total drain-source distance of 2.4µm. A completed device is shown in Fig.2.

Device Performance

The devices have been characterized for DC and RF-performance. Fig.3 shows a typical IV-characteristic of a device with a gate geometry of $L_G x W_G$=1.8µmx30µm.

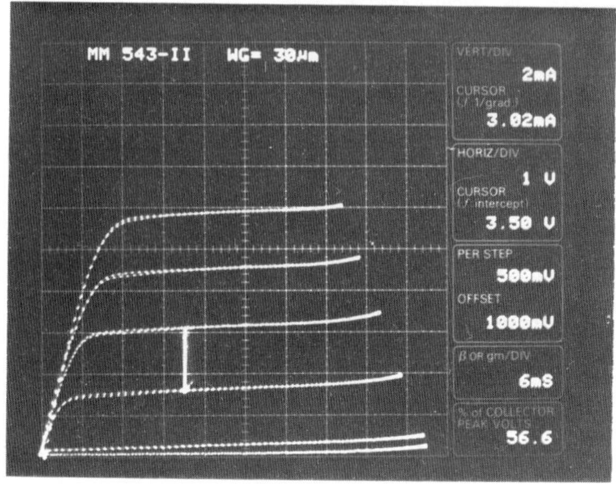

Fig.3: Typical IV-characteristic of a $Ga_{0.5}In_{0.5}P/In_{0.15}Ga_{0.85}As$ HFET. $L_G x W_G$=1.8µmx30µm , V_{GStop}=1V.

The maximum transconductance is g_{mmax}=200mS/mm and the maximum saturation current is I_{DSmax}=400mA/mm, which are quite reasonable values for this gate-length. The threshold voltage is V_{TH}=-1V. As can be seen in Fig.4, the devices behave well, displaying a sharp pinch-off, a very small output conductance (g_d<3mS/mm) and a high drain-source operation voltage in excess of V_{DS}<9V. The maximum gate-source on-voltage before the onset of a gate leakage current is V_{GS}=1V and reflects the influence of the p^+-surface layer. An outstanding feature of these devices is the very high drain-source breakdown voltage (Fig.4). The drain-source burn-out voltage is well in excess of 20V, which is possible since the p^+-surface layer concept is assumed to reduce the high electric field at the drain side of the gate. This high breakdown voltage in combination with the good saturation behaviour makes this type of device an attractive candidate for power application.

The microwave performance was measured by using on-wafer probing technique in the frequency range of f=0.5GHz to 26.5GHz. The cut-off frequencies were determined from the unit current gain and the maximum stable gain by extrapolating with a (-20)dB/decade roll off. The current gain cut-off frequency was determined to be

f_T=12GHz and the power gain cut-off frequency to be f_{max}=42GHz (Fig.5) which are good values for this gate-length. This ratio remains nearly unchanged up to drain source voltages of V_{DS}=6V. The large ratio of f_{max}/f_T=3.5 is due to the reduced feed back capacitance C_{GD}, which continuously drops with increasing drain source voltage. The evolution of C_{GD} as a function of V_{DS} reveals the advantage of the p^+-surface layer concept, which allows the depletion region to move towards the drain with V_{DS}.

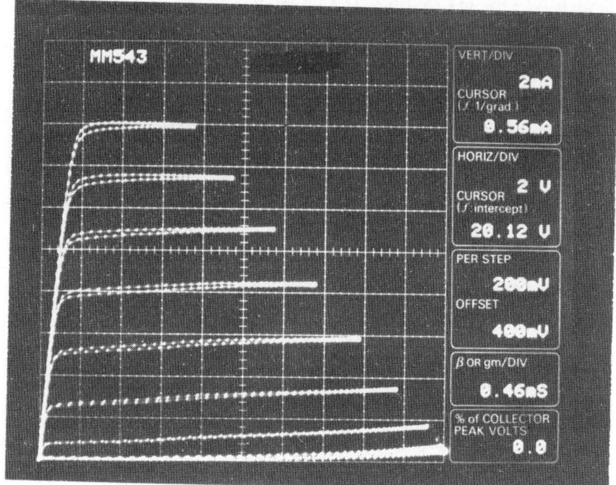

Fig.4: Typical IV-characteristic of a $Ga_{0.5}In_{0.5}P/In_{0.15}Ga_{0.85}As$ HFET. biased at high V_{DS}. L_GxW_G=1.8µmx90µm , V_{GStop}=1V.

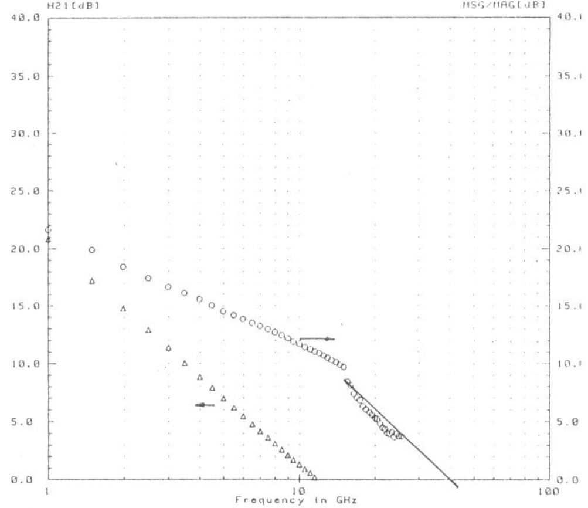

Fig.5: Current gain and power gain of a $Ga_{0.5}In_{0.5}P/In_{0.15}Ga_{0.85}As$ HFET with L_GxW_G=1.8µmx70µm. V_{DS}=5V, V_{GS}=-0.5V.

Conclusion

In conclusion we have reported for the first time a p^+-$Ga_{0.5}In_{0.5}P/In_{0.15}Ga_{0.85}As$ HFET, which showed reasonable transconductance and saturation current in combination with an excellent saturation behaviour and unprecedented breakdown voltages. This DC performance led to a high ratio of the cut-off requencies of f_T(=12GHz)/f_{max}(=42GHz) =3.5, which remained unchanged up to high drain source voltages. This device performance makes this approach attractive for power application.

Acknowledgment: We thank Joachim Hugo for his expert work in Shubnikov-de Haas characterisation.

References

[1] L.F.Lester, P.M.Smith, P.Ho, P.C.Chao, R.C.Tiberio, K.H.Duh, E.D.Wolf, "0.15 µm gate-length double recess pseudomorphic HEMT with f_{max} of 350GHz", IEEE IEDM Tech.Dig., pp.172-175, 1988

[2] H.M. Macksey, " Optimization of n^+-ledge channel structure for GaAs power FETs", IEEE Transac., vol. ED-33, pp.1818-1824, 1986

[3] T.Waho, F.Yanagawa, " A GaAs MISFET using an MBE grown CaF2 gate insulator layer", IEEE Electr. Device Letters, vol.EDL9, No.10, pp.548-549, 1988

[4] H.Hida, A.Okllamato, H.Toyoshima, K.Ohata, "A high current driveability i-AlGaAs/n-GaAs doped channel MIS-like FET (DMT)", IEEE Electron Device Letters, vol.EDL-7, No.11, pp.625-627, 1986

[5] L.W.Yin, Y.Hwang, J.H.Lee, R.M.Kolbas, R.J.Trew, U.K.Mishra," Improved Breakdown voltage in GaAs MESFETs utilizing surface layers of GaAs grown at low temperature by MBE", IEEE Electron Device Letters, vol.EDL-11, No.12, p.561-563, 1990

[6] P.Saunier, R.Nguyen, L.J.Messick, M.A.Khatibzadeh, "An InP MISFET with a power density of 0.18W/mm at 30GHz", IEEE Electron Device Letters, vol.EDL-11, No.1, p.48-49, 1990

[7] J.M.Shannon, " Control of Schottky Barrier Heigth Using Highly Doped Surface Layers", Solid-State Electronics, vol.19, pp.537-543, 1976

[8] K.Ohata, H.Hida, H.Miyamoto, " A low noise AlGaAs/GaAs FET with p^+-gate and selectively doped structure", IEEE MTT-s Techn. Digest, pp.434-437, 1987

[9] K.L.Priddy, D.R.Kitchen, J.A.Grzyb, C.W.Litton, T.S. Henderson, C.K.Peng, W.F.Kopp, H.Morkoc, "Design of enhanced Schottky-barrier AlGaAs/GaAs MODFET's using highly doped p^+ surface layers", IEEE Trans. on Electr. Devices, vol.ED-34, No.2, pp.175-179, 1987

[10] M.Razeghi, F.Omnes, Ph.Maurel, Y.J.Chan, D.Pavlidis, "$Ga_{0.51}In_{0.49}P$ / $Ga_xIn_{1-x}As$ lattice matched (x=1) and strained (x=0.85) two dimensional electron gas field effect transistor", Semicond. Sci. Techn.6, pp.103-107, 1991

[11] Y.J.Chan, D.Pavlidis, M.Razeghi, F.Omnes, "$Ga_{0.51}In_{0.49}P$/GaAs HEMT exhibiting good performance at cryogenic temperatures", IEEE Transac. on Electron Devices, vol.ED-37, No.10, pp.2141-2146, 1990

[12] K.H.Bachem,.D.Fekete, W.Pletschen, W.Rothemund, K.Winkler, "OMVPE-grown $(Al_xGa_{1-x})_{0.5}In_{0.5}P$/InGaAs MODFET structures:growth procedure and Hall properties", J. of Crystal Growth, vol.124, pp.817-823, 1992

[13] J.Dickmann, A.Geyer, H.Daembkes, H.Nickel, R.Lösch, W.Schlapp, "Fabrication of Low Resistance Submicron Gates in Pseudomorphic MODFETs Using Optical Contactlithography", J.Electrochem.Soc., vol.138, No.2, pp.491493, 1991

MOVPE growth of heavily Si-doped $In_yGa_{1-y}As$ layers and its application to nonalloyed ohmic contacts for InGaP/$In_{0.2}Ga_{0.8}As$ HEMTs

Naoki Hara, Mizuhisa Nihei, Haruyoshi Suehiro, Kazumi Kasai, and Junji Komeno

Fujitsu Laboratories LTD.,

10-1 Morinosato-Wakamiya, Atsugi 243-01, Japan.

ABSTRACT: Heavily Si-doped InGaAs layers for nonalloyed ohmic contacts of high electron mobility transistors (HEMTs) were grown by metalorganic vapor phase epitaxy. In quantitative studies of the surface roughness using atomic force microscopy, we found that lower growth temperature and thicker graded InGaAs provide good surface morphology. The surface roughness affected the specific contact resistance. We then fabricated InGaP/$In_{0.2}Ga_{0.8}As$ HEMTs with nonalloyed ohmic contacts whose characteristics were comparable to those of conventionally fabricated HEMTs.

1. Introduction

Reduction of device sizes is a prime issue for development of highly integrated high electron mobility transistors (HEMTs). Nonalloyed ohmic contacts allow source/drain and gate to be connected directly with the same metal, and their low specific contact resistance itself reduces ohmic metal size. To construct nonalloyed ohmic contacts, Woodall et al. (1981) proposed to use compositionally graded InGaAs layers.

Nonalloyed ohmic contacts with n^+-InGaAs layers grown by molecular beam epitaxy (MBE) have been fabricated for AlGaAs/GaAs HEMT LSIs (Kuroda et al. 1989), but no HEMT nonalloyed ohmic contacts using InGaAs layers grown by metalorganic vapor phase epitaxy (MOVPE) has been reported. MOVPE growth of n^+-InGaAs layers is preferable for the application of nonalloyed ohmic contacts to InGaP/InGaAs HEMTs, because only MOVPE grown InGaP/InGaAs HEMTs have shown good performance (Takikawa et al. 1991, Kuroda et al. 1992, Ochimizu et al. 1993). Capability of large area growth of InGaP/InGaAs (Ochimizu et al. 1993) as well as AlGaAs/GaAs HEMT structures (Shiina et al. 1993) is also a remarkable merit of MOVPE. However, heavily Si-doped InGaAs layers with acceptable surface morphology are difficult to grow. Good surface morphology can be obtained with low temperature growth, but the decomposition rate of disilane decreases, lowering the carrier concentration. Therefore, growth conditions need to be optimized.

In this paper surface roughness of InGaAs layers grown under various conditions was characterized quantitatively using atomic force microscopy (AFM), and relationship between surface roughness and specific contact resistance was investigated. Then InGaP/$In_{0.2}Ga_{0.8}As$ HEMTs with nonalloyed ohmic contacts were fabricated.

2. MOVPE Growth

MOVPE growth was done in an inverted horizontal atmospheric pressure reactor (Kikkawa et al. 1991). Trimethylgallium (TMGa), trimethylindium (TMIn), arsine (AsH_3), and phosphine (PH_3) were used as source materials. The dopant source was disilane (Si_2H_6). Growth temperature was 630°C except for the n^+-InGaAs layers, which were grown at 470 to 630°C. Compositionally graded InGaAs layers were grown by varying the TMIn and TMGa supplies during growth.

3. Surface Roughness and Carrier Concentration of InGaAs Layers

Surface roughness of n^+-$In_yGa_{1-y}As$/graded InGaAs structures shown in Fig.1 was investigated using AFM (Nano Scope II, Digital Instruments). The characterization of surface roughness is conventionally qualitative, e.g., as mirror-like or rough. Quantitative evaluations such as that in this study should go far toward clarifying the relationship between surface roughness and specific contact resistance, as explained in Section 4. At first effects of growth

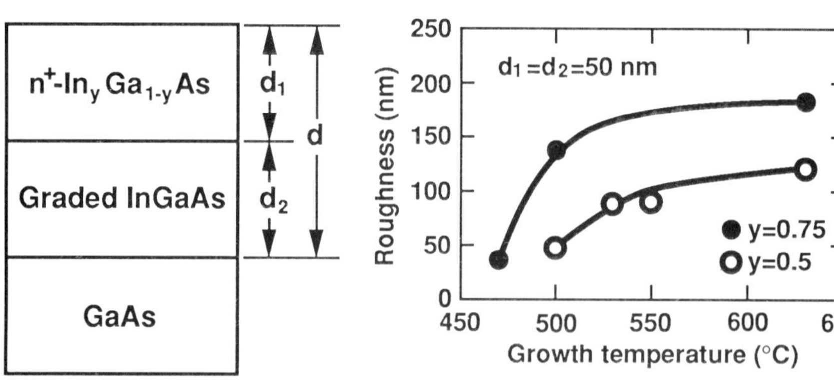

Fig. 1 n^+-$In_yGa_{1-y}As$/graded InGaAs structures.

Fig. 2 Dependence of surface roughness on growth temperature.

temperature were investigated for n^+-$In_yGa_{1-y}As$ (50 nm)/graded InGaAs (50 nm) structures. Figure 2 shows the relationship between surface roughness and the growth temperature. The surface roughness decreased with growth temperature.

Next, effects of the InGaAs thickness were investigated. It is not easy to define thickness of layers with rough surface. We used values measured with a Dektak profilometer, which neglects microscopic roughness. In one experiment, we changed the thicknesses of both the graded InGaAs and the n^+-$In_yGa_{1-y}As$ (Fig. 3). In another we fixed the thickness of the n^+-$In_yGa_{1-y}As$ at 10 nm and changed only that of the graded InGaAs. In both cases, surface roughness decreased with the thickness. We found that the surface roughness of samples with thicker graded InGaAs is small compared to the same total InGaAs thickness. It means that surface roughness is mainly determined by the graded InGaAs thickness, regardless of the n^+-$In_yGa_{1-y}As$ thickness. The effect is clearly seen in Fig. 5, which shows dependence of surface roughness on the graded InGaAs thickness. The degraded surface morphology in thinner graded InGaAs probably results from insufficient strain relaxation due to a rapid change in In composition.

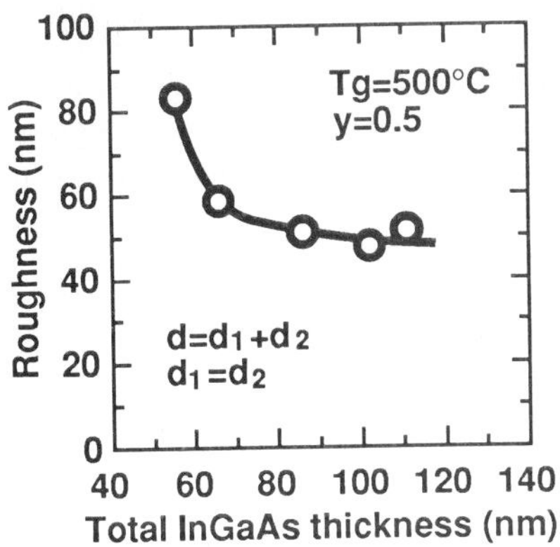

Fig. 3 Dependence of surface roughness on InGaAs thickness. The n^+-$In_yGa_{1-y}As$ and graded InGaAs thicknesses are equal.

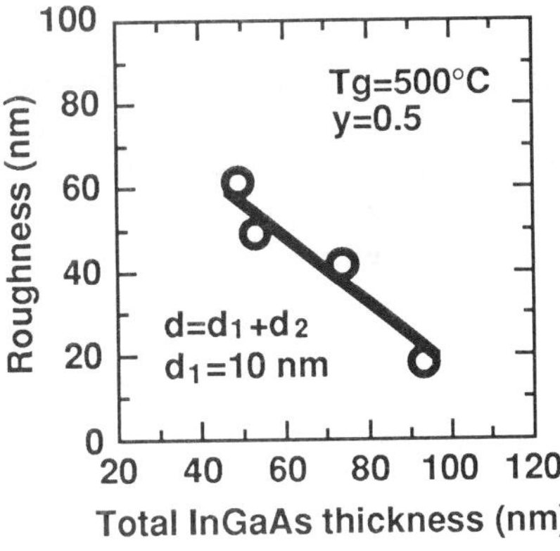

Fig. 4 Dependence of surface roughness on InGaAs thickness. The n^+-$In_yGa_{1-y}As$ thickness is 10 nm.

Good ohmic characteristics require the n^+-$In_yGa_{1-y}As$ with a high carrier concentration. Figure 6 shows the carrier concentration of 500 nm-thick n^+-$In_yGa_{1-y}As$. The disilane flow rate was 2.68×10^{-6} mol/min, which is the limit of our equipment. The carrier concentration suggests that the optimum growth temperature is about 500°C.

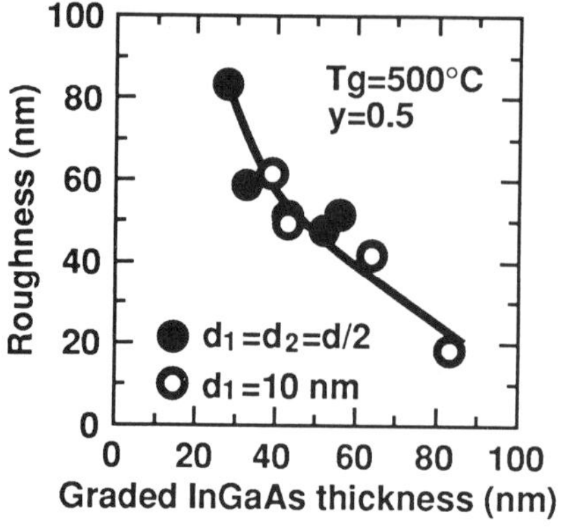

Fig. 5 Dependence of surface roughness on graded InGaAs thickness.

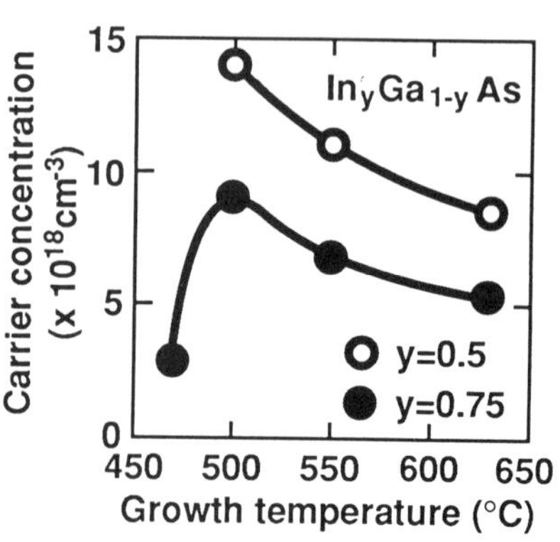

Fig. 6 Carrier concentration of n^+-$In_yGa_{1-y}As$ layers.

4. Specific Contact Resistance

Specific contact resistance was measured for samples summarized in Table I using the transmission line model (TLM). The ohmic metal was Ti/Al. The specific contact resistance for a n^+-$In_{0.5}Ga_{0.5}As$/graded InGaAs sample grown at 500°C was 2.0×10^{-7} Ω cm^2 (ρ_0 in Table I). The specific contact resistance of n^+-$In_{0.5}Ga_{0.5}As$/graded InGaAs structures changed little between 500 and 530°C, but the surface roughness nearly doubled. Surface roughness did not degrade the ohmic contact characteristics, because it was still smaller than the total InGaAs thickness (100 nm), even for the sample grown at 530°C. The carrier concentration difference was about 15% and hardly affected either.

Table I Dependence of specific contact resistance on growth conditions.

Tg (°C)	y	$d_1=d_2$ (nm)	Roughness (nm)	ρ_c/ρ_0
530	0.50	50	89.2	1.1
500	0.50	50	47.5	1.0
500	0.75	50	139.6	1.3
470	0.75	50	36.8	0.32

Such was not the case for the n^+-$In_{0.75}Ga_{0.25}As$/graded InGaAs structures. The specific contact resistance increased almost 4 times when the growth temperature increased from 470 to 500°C. The surface roughness of the 500°C grown sample was 139.6 nm, exceeding the total InGaAs thickness, and some areas of the InGaAs were too thin to obtain good ohmic contacts. Thus, although the carrier concentration of the 470°C grown sample was lower, a smaller specific contact resistance was obtained. From these results, it is speculated that ohmic characteristics are hardly affected below a critical value of surface roughness.

5. HEMTs Fabrication and Evaluation

We fabricated and evaluated InGaP/$In_{0.2}Ga_{0.8}As$ HEMTs with nonalloyed ohmic contacts (Fig. 7 and Table II). We did not fabricate the same structure with alloyed ohmic contacts, but compared the results obtained from similar structures (Takikawa et al. 1991), our results are as good as those for conventional alloyed ohmic contacts.

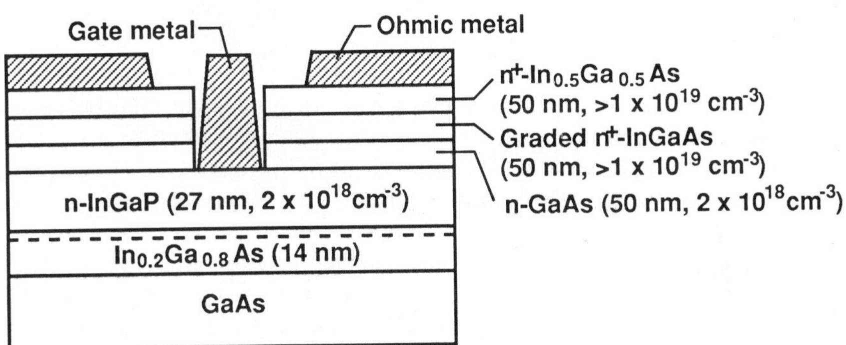

Fig. 7 InGaP/$In_{0.2}Ga_{0.8}As$ HEMTs with nonalloyed ohmic contacts.

Table II Characteristics of InGaP/$In_{0.2}Ga_{0.8}As$ HEMTs with nonalloyed ohmic contacts.

Gate length (μm)	V_{th} (V)	K-value (mS/V mm)	g_m (mS/mm)
0.5	0.039	473.3	352.3
1.0	0.023	383.1	316.5

6. Summary

Heavily Si-doped InGaAs layers for HEMTs nonalloyed ohmic contacts were grown by MOVPE. Using AFM, we studied the dependence of surface roughness on growth temperature and InGaAs thickness. It was found that lower growth temperature and thicker graded InGaAs result in less surface roughness. The specific contact resistance changed with surface roughness. In fabricating and evaluating InGaP/$In_{0.2}Ga_{0.8}As$ HEMTs with nonalloyed ohmic contacts, we found that the characteristics of HEMTs with nonalloyed ohmic contacts are as good as those of conventionally fabricated HEMTs. This means that practical InGaAs nonalloyed ohmic contacts layers for HEMTs can be grown by MOVPE.

7. Acknowledgment

We thank S. Kuroda and T. Ohori for their advice, and N. Kutsuzawa and A. Sasano for their technical support.

8. References

Kikkawa T, Tanaka H, and Komeno J, 1991, J. Cryst. Growth, **107** 370.

Kuroda S, Harada N, Katakami T, Mimura T, and Abe M, 1989, IEEE Electron Devices, **ED-36** 2196.

Kuroda S, Suehiro H, Miyata T, Asai S, Hanyu I, Shima M, Hara N, and Takikawa M, 1992, IEDM Tech. Dig., 323.

Ochimizu H, Kikkawa T, Kuroda S, Kasai K, and Komeno J, 1993, *Extended Abstracts of the Conf. Solid State Devices and Materials, Chiba, Japan* (Business Center for Academic Societies Japan, Tokyo, 1993).

Shiina K, Tanaka H, Ohori T, Kasai K, and Komeno J, 1993, Inst. Phys. Conf. Ser., **129** 61 (*19th Int. Symp. GaAs and Related Compounds, Karuizawa Japan 1992*).

Takikawa M, Ohori T, Takechi M, Suzuki M, and Komeno J, 1991, J. Cryst. Growth, **107** 942.

Woodall J M, Freeouf J L, Pettit G D, Jackson T, and Kirchner P, 1981 J. Vac. Sci. Technol., **19** 626.

Evaluations of V_{th} uniformities and f_T for HEMT/Si fabricated using GaAs/AlGaAs selective dry etching

Takashi Aigo, Mitsuhiko Goto, Aiji Jono, Akiyoshi Tachikawa and Akihiro Moritani

Electronics research laboratories, Nippon Steel Corp., 1618 Ida, Nakahara-ku, Kawasaki 211, Japan.

Abstract: We report on a study of dc characteristics, Vth uniformities and microwave performances of HEMT/Si fabricated using GaAs/AlGaAs selective dry etching. A maximum transconductance of 220 mS/mm is obtained for HEMT/Si with a gate length of 0.8 μm. The standard deviations of Vth for depletion and enhancement mode HEMT/Si with a gate length of 1.2 μm are 36.0 mV and 30.8 mV, respectively. Microwave measurement shows f T of 23 GHz, which demonstrates the highest value for HEMT/Si with a gate length of 0.8 μm. From these results, HEMT/Si is found to be promissing in high speed and large scale GaAs IC applications.

1. Introduction

HEMT/Si technology is very suitable for GaAs IC applications, because GaAs/Si itself is an epitaxially grown material and available for HEMT multi-layer structures, utilizing high thermal conductivity, large diameter and mechanical hardness of Si substrate. Many authors have successfully fabricated HEMT's or MESFET's using GaAs/Si and reported comparable results to GaAs substrates, not only for descrete devices (Fischer et al. 1986 , Aksun et al. 1986, Wang et al. 1989, Charasse et al. 1989), but also for IC applications (Nonaka et al. 1984, Shichijo et al. 1987, Ren et al. 1989).

Applying GaAs/Si to HEMT IC's, it is important to evaluate the influences of high-density dislocations existing in the GaAs/Si and conductive Si substrate. The dislocations can affect the performances of HEMT's and degrade the uniformity of their characteristics in the wafer. Moreover, the conductive Si substrate or doped layer with outdiffused Si along the dislocations at the Si-GaAs interface may act as undesirable factors at high frequency. However, these evaluations have not been carried out so clearly in spite of the demand for the practical use of GaAs/Si.

In this paper, we report on a study of dc characteristics and microwave performances of MOCVD grown HEMT/Si, fabricated using GaAs/AlGaAs selective dry etching, and demonstrate that the standard deviations of threshold voltage (σVth) are 36.0 mV and 30.8 mV for the depletion and enhancement mode HEMT/Si and that the f T over 20 GHz is obtained for HEMT/Si with a gate length of 0.8 μm. From the simulation of equivalent circuit parameters for HEMT/Si, the influences of the doped layer with Si or the conductive Si substrate are also discussed, including the proposal of the suitable equivalent circuit model for HEMT/Si.

2. Epitaxial growth and device fabrication

The epitaxial wafers in this study were grown by MOCVD using standard 3 inch p-type Si (100) substrates misoriented by 3° toward <110>. The growth sequences were conventional steps for GaAs/Si, including the substrate cleaning above 800°C and the two step growth (Akiyama et al. 1986). TMG (trimethylgallium), TMA (trimethylaluminum) and AsH_3 (arsine) were used as source materials with a carrier gas of H_2. Si_2H_6 (disilane) was used for the n-type doping. Thermal cycle annealing was not carried out and EPD's over 10^7 cm^{-2} were estimated with molten KOH etching and X-TEM (cross-sectional transmission electron microscopy).

HEMT structures used in this study are illustrated in Fig.1, which consist of 1.5-μm-thick GaAs buffer layer, 1.5-μm-thick AlGaAs buffer layer, 0.05-μm-thick GaAs channel layer, 0.05-μm-thick doped AlGaAs layer and 0.1-μm-thick doped GaAs cap layer. The carrier concentration of the doped AlGaAs was varied from 7.0 to 12.0 x 10^{17} cm^{-3} in order to find the doping condition for the enhancement mode HEMT/Si and to optimize the doping condition for the depletion mode HEMT/Si, with these layer structures. The fabrication processes of the HEMT's were carried out by mesa chemical etching, AuGe ohmic contacts metallization, a reactive ion etching of the GaAs cap layer using CCl_2F_2/He and an Al gate metallization. The selective ratio of the GaAs/AlGaAs etching by the present system was over 200, and ~1.2 of n-value and ~0.8 V of barrier height for an Al Schottky contact after the etching were obtained, showing little damage of the etching. It should be noted that no special treatment was required to fabricate these HEMT/Si.

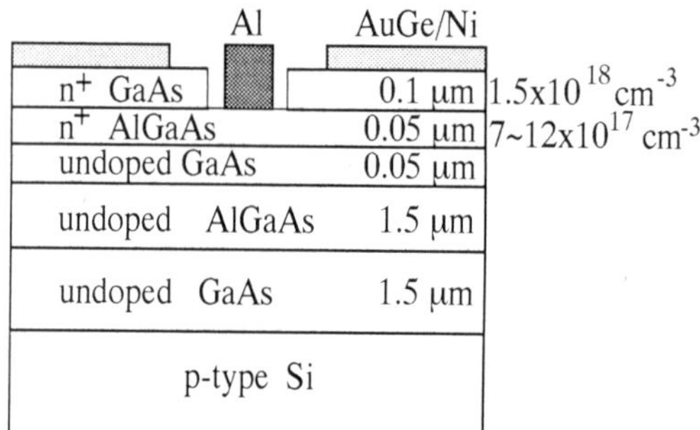

Fig.1 Schematic cross-section of GaAs/AlGaAs HEMT on Si.

3. DC characteristics and Vth uniformities of HEMT/Si

The typical current-voltage characteristics of a depletion mode HEMT/Si are shown in Fig.2 for a gate length of 0.8 μm and width of 20 μm, showing good pinch-off and current

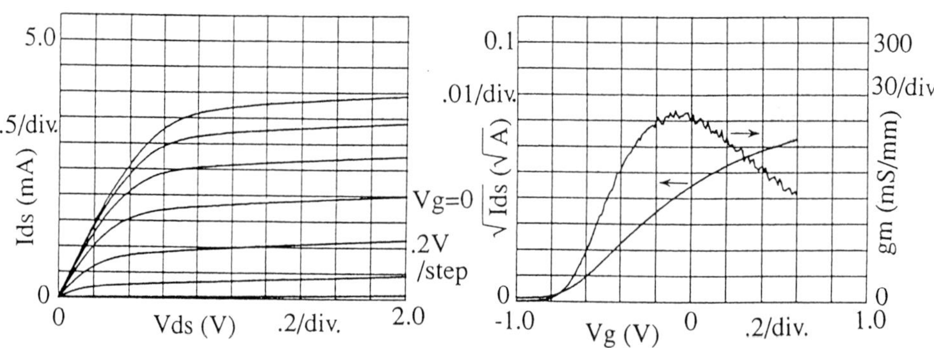

Fig.2 Current-voltage characteristics for a depletion mode HEMT/Si.

Fig.3 Drain current and transconductance vs. gate voltage at a drain bias of 2V for a HEMT/Si.

saturation. The transconductance and drain current as a function of gate voltage are indicated in Fig.3, which shows the maximum transconductance of 220 mS/mm. The comparison of these results and the buffer layer leakage current evaluation using isolated two ohmic contacts to GaAs substrate indicates that the high-density dislocations and the conductive Si substrate for HEMT/Si do not affect the dc characteristics of the devices.

The uniformities of Vth are shown in Fig.4(a) for the depletion and 4(b) for the enhancement mode HEMT/Si with a gate length of 1.2 μm and width of 20 μm, where the standard deviations of the threshold voltage of 36.0 mV and 30.8 mV were obtained respectively. As for the result of the depletion mode, the Vth value of -2.41 V is somewhat deep to evaluate the application of HEMT/Si for IC's. Recently, we have had another result of the Vth uniformity for the depletion mode HEMT/Si, showing the σVth of 27.4 mV with the Vth of -0.094 V. In these results, sudden and random changes of the Vth in the wafer which are possibly attributed to the high-density dislocations were not observed. However, the pattern of the Vth distribution corresponding to the distribution of the carrier concentration or layer thickness for the wafer was observed, suggesting that the uniformties of the Vth for HEMT/Si are not so much affected by the dislocations as the uniformities of the carrier concentration and layer thickness brought by the present MOCVD system.

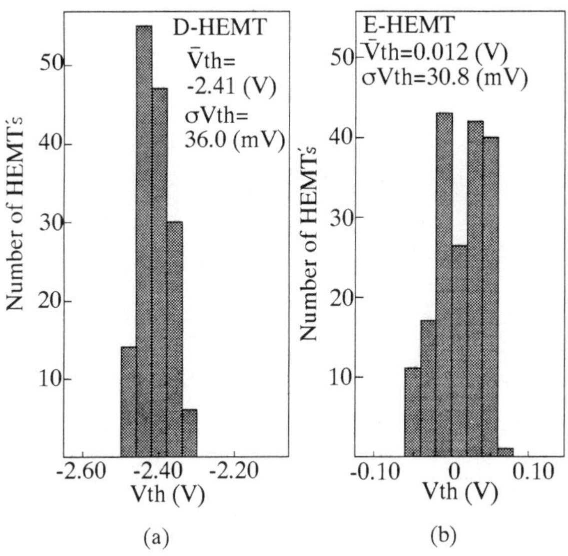

Fig.4 Uniformities of the threshold voltage for the depletion (a) and enhancement mode (b) HEMT/Si.

4. Microwave performances and equivalent circuit parameters

Measurement of microwave performances was carried out with an automatic network analyzer up to 20 GHz using Cascade Microtech wafer probes. Short-circuit current gain, $|h_{21}|$, and unilateral maximum gain, $G_{u(max)}$, are indicated in Fig.5 as a function of frequency for a depletion mode HEMT/Si with a gate length of 0.8 μm. In this measurement, the Vds was 2 V and the Ids was nearly equal to Idss. As can be seen in Fig.5, although the $|h_{21}|$ curve deviates from the typical -6dB/oct. line, a cut-off frequency of 23 GHz is defined as the $|h_{21}| = 0$ dB from the extraporation of the data, which demonstrates the highest value for HEMT/Si with this gate length. However, the maximum frequency of oscillation, f_{max}, is 15 GHz. The reason of the deviation for the $|h_{21}|$ curve and low f_{max} will be discussed later.

Measured S-parameters for HEMT/Si with the f_T of 23GHz are indicated in Fig.6(a) and those for HEMT/GaAs, which was previously fabricated but did not have the identical conditions for the carrier concentration and thickness of the doped AlGaAs layer to the present HEMT/Si, are indicated in Fig.6(b) for comparison. Although the behavior of S 11 for HEMT/Si somewhat deviates from the constant resistance circle for HEMT/GaAs, the deviation is not so large. Moreover, the difference for the magnitude and the phase of S 12 between HEMT/Si and HEMT/GaAs is also observed. However, the difference of the behavior for the S 12 is probably attributed to the point that the HEMT with each substrate did not have the identical conditions for the doped AlGaAs layer structures, because the behavior is mainly determined by the gate-source capacitance, Cgs, and the gate-drain capacitance, Cgd, and they are strongly related to the carrier concentration and thickness of the doped AlGaAs layer under the gate.

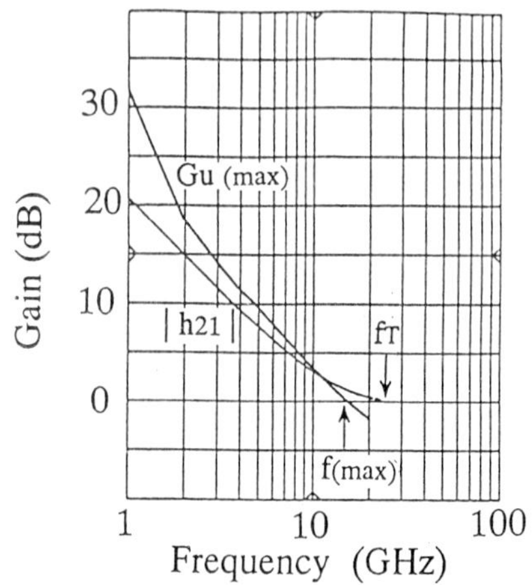

Fig.5 Short-circuit current gain and unilateral maximum gain vs. frequency for a depletion mode HEMT/Si.

On the other hand, the phase shift of S 22 for HEMT/Si is very large, indicating large drain-source capacitance. As the total epitaxial layer thickness for HEMT/Si and HEMT/GaAs is almost the same, the difference of the drain-source capacitance suggests the influences of the thin doped layer with outdiffused Si at the Si-GaAs interface and the conductive Si substrate as mentioned before.

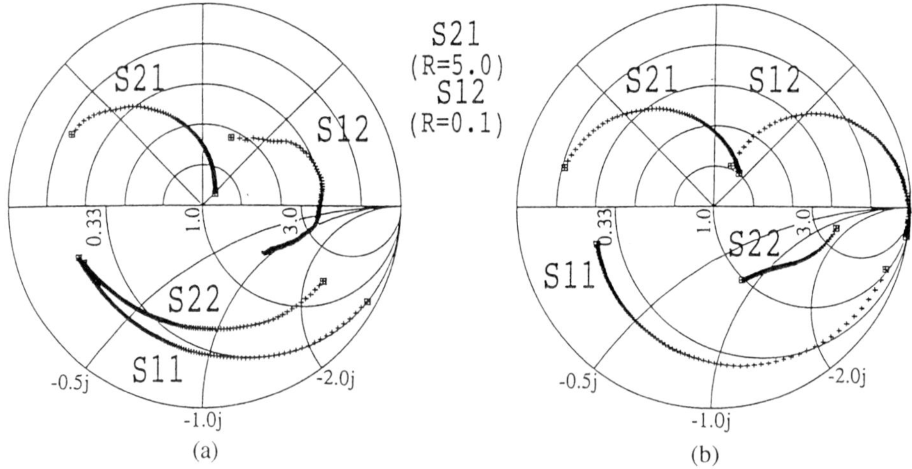

Fig.6 Measured S-parameters from 1 to 20 GHz for a HEMT/Si (a) and a HEMT/GaAs (b).

To verify the influence of the drain-source capacitance observed from the behavior of S-parameters, the small-signal equivalent circuit simulation was carried out. In this simulation, the new parameters, Csub+Rsub, which are parallel to the Cds in the conventional HEMT equivalent circuit model, were introduced as indicated in Fig.7. The Csub means the depletion layer in the GaAs buffer layer formed by the thin doped layer and the Rsub means the conductive path of the Si substrate or the Si-GaAs interface. The extracted parameters are

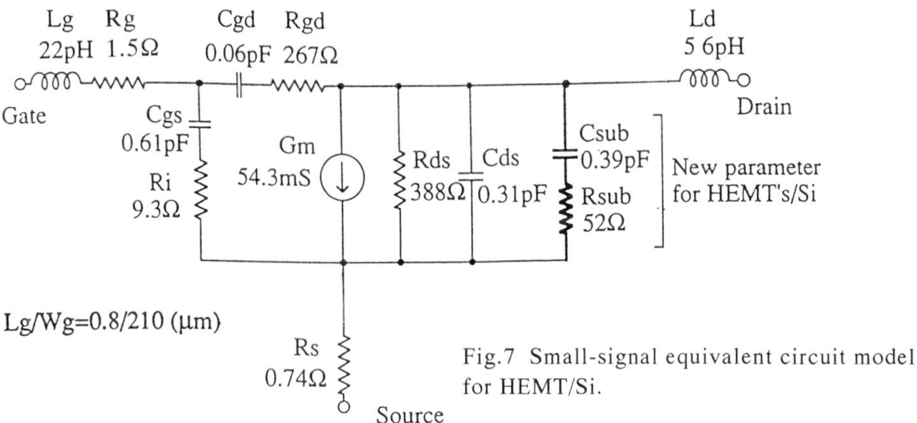

Fig.7 Small-signal equivalent circuit model for HEMT/Si.

also shown in Fig.7 and the simulation error using this new model was about 5% as compared with the measured data, while in the case simulated with the conventional HEMT equivalent circuit model, the error was around 20% as can be seen in Fig.8. This result indicates that this new equivalent circuit model reflects HEMT/Si characteristics well and therefore, the device suffers from the influence of the inherent capacitance and resistance, Csub+Rsub.

Comparing the extracted parameters for HEMT/Si to HEMT/GaAs, the observed capacitance between the source and drain is about a few times larger for HEMT/Si. The capacitance, making matching network with the drain inductance, can affect the output current of HEMT/Si. Therefore, the $|h_{21}|$ curve deviated from the -6dB/oct. may be observed. Moreover, the f_{max} can be also affected by the larger capacitance for HEMT/Si, which causes the low output impedance and would bring smaller f_{max} than the f_T. Further investigation will be needed to make clear the influence of the inherent parameters to f_T and f_{max}.

On the other hand, the Rds, which means the reciprocal of the intrinsic drain conductance, is about two times larger for HEMT/Si, suggesting that the inherent parameters do not affect

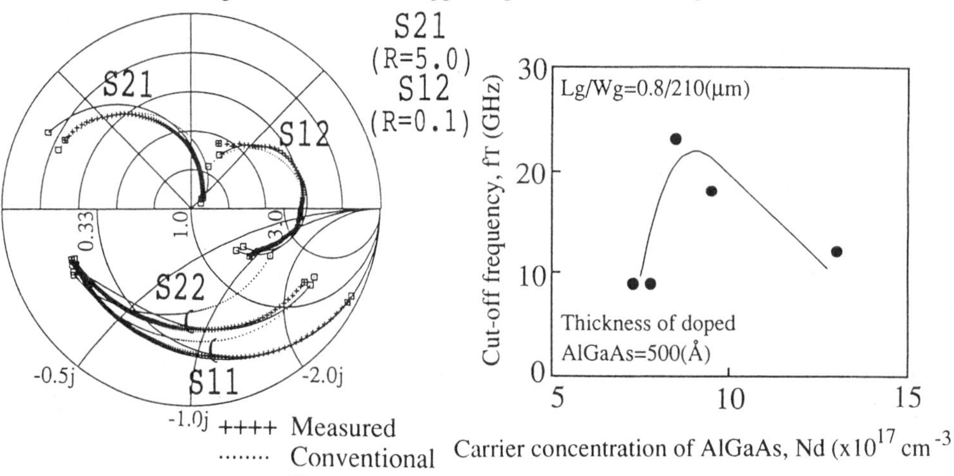

Fig.8 Comparison of simulated S-parameters using the new equivalent circuit model and the conventional model for a HEMT/Si.

Fig.9 Relationship between the carrier concentration of the doped AlGaAs layer and the f_T value for HEMT/Si.

the intrinsic elements of HEMT/Si. In Fig.9, the relationship between the carrier concentration of the doped AlGaAs layer and the f_T for HEMT/Si is shown, which indicates the strong dependence of the f_T on the carrier concentration. This result leads that even in HEMT/Si, the f_T is mainly determined by the intrinsic parameters related to the carrier concentration of the doped AlGaAs layer, namely gm, Cgs and Cgd. The simulation result using the new equivalent circuit model also reveals that the f_T itself is hardly affected by the change of the inherent parameters, although the deviation for the $|h_{21}|$ curve are still observed. Therefore, it is concluded that the high f_T for HEMT/Si obtained here is brought by the optimized condition for the product of the carrier concentration and thickness of the doped AlGaAs layer.

5. Summary

We have reported on a study of dc characteristics, Vth uniformities and microwave performances of HEMT/Si fabricated using GaAs/AlGaAs selective dry etching. The standard deviations of the Vth were 36.0 mV and 30.8 mV for the depletion and enhancement mode HEMT/Si. The influences of the high-density dislocations existing in the GaAs/Si were not observed in the uniformities of the Vth as well as in the dc characteristics. From microwave measurement, we have demonstrated f_T of 23 GHz for the depletion mode HEMT/Si with a gate length of 0.8 μm, which is brought by the optimization for the product of the carrier concentration and thickness in the HEMT layer structures. The new equivalent circuit model including the inherent parameters caused by the thin doped layer with Si at the Si-GaAs interface and the conductive Si substrate has been proposed and simulated HEMT/Si characteristics well, which enables to compare the simulated results for the Si substrate to GaAs substrate and points out the minor dependence of the inherent parameters to the f_T. From these results, HEMT/Si is found to be promissing in high speed and large scale GaAs IC applications.

6. References

Akiyama M., Kawarada Y., Ueda T., Nishi S. and Kaminishi K.
1986, J. Cryst. Growth, **77** 490

Aksun M. I., Morkoç H., Lester L. F., Duh K. H. G., Smith P. M., Chao P. C., Longerbone M. and Erickson L.P.
1986, Appl. Phys. Lett., **49** 1654

Charasse M. N., Bartenlian B., Gerard B., Hirtz J. P., Laviron M., de Parscau A. M., Derevonko M. and Delagebeaudeuf D.
1989, Jpn. J. Appl. Phys., **28** L1896

Fischer R. J., Kopp W. F., Gedymin J. S. and Morkoç H.
1986, IEEE Trans. Electron Devices, **33** 1407

Nonaka T., Akiyama M., Kawarada Y. and Kaminishi K.
1984, Jpn. J. Appl. Phys., **23** L919

Ren F., Chand N., Chen Y. K., Pearton S., Tennant D. M. and Resnic D. J.
1989, IEEE Trans. Electron Device Lett., **10** 559

Shichijo H., Lee J. W., McLevige W. V. and Taddiken A. H.
1987, IEEE Trans. Electron Device Lett., **8** 121

Wang G. W., Ito C., Feng M., Kaliski R., Mclntyre D., Lau C. and Eu V. K.
1989, Appl. Phys. Lett., **55** 1552

Paper presented at the Int. Symp. GaAs and Related Compounds, Freiburg, 1993

An investigation into GaAs-on-InP MESFETs

J.M. Dumas [+], A. Clei [o], P. Audren [✦], M.P. Favennec [✦], R. Azoulay [o], C. Vuchener [+], J. Paugam [+], S. Biblement [o] and C. Joly [o]

[+] FRANCE TELECOM/CNET, BP 40, 22300 LANNION, France
[o] FRANCE TELECOM/CNET, 196 avenue Henri Ravera, 92220 BAGNEUX, France
[✦] Institut Universitaire de Technologie, BP 150, 22300 LANNION, France

ABSTRACT : A study of the electrical performances and degradation mechanisms of GaAs on-InP MESFETs has been carried out. It is evidenced that these devices suffer from (i) low -trap-related parasitic effects and (ii) conventional GaAs MESFET degradations.

1. INTRODUCTION

The GaAs MESFET on InP substrate is an attractive approach for integration with InP-based long wavelength photonic devices as demonstrated by O'Sullivan et al (1991). The GaAs MESFET is a well established technology capable of high levels of integration, but the lattice mismatch between GaAs and InP can introduce anomalous effects in the performances of devices and reliability problems can be expected. After a short description of the device fabrication steps and performances, experimental results on trap-related parasitic effects and degradation mechanisms are reported in this contribution.

2. DEVICE FABRICATION AND PERFORMANCES

A generic technological process has been developped and previously described by Azoulay et al (1991). The GaAs layers are all grown on semi-insulating InP substrate using an MOCVD technique. Once the 100 Å-thick nucleation layer is achieved ; the GaAs buffer, donor and contact layers are grown at a substrate temperature of $\simeq$ 680°C. Then devices are fabricated with the following processes : isolation by boron implantation, AuGeNi ohmic contacts with rapid thermal annealing (RTA), a recessed Ti (500 Å) Pd (800 Å) Au (3000 Å) gate, Ti (500Å) Au (3000 Å) overlayers. A PECVD Si_3N_4 layer ensures the surface passivation.

Because of the 4 %-lattice mismatch between InP and GaAs, high dislocation densities ($10^{10} \sim 10^{11}$ cm^{+2}) are present at the heterointerface. Nevertheless, by optimizing the growth process, this dislocation density at the surface of a 3 µm-tick GaAs buffer layer is reduced to the $10^7 cm^{+2}$ range. The structure is illustrated in Figure 1.

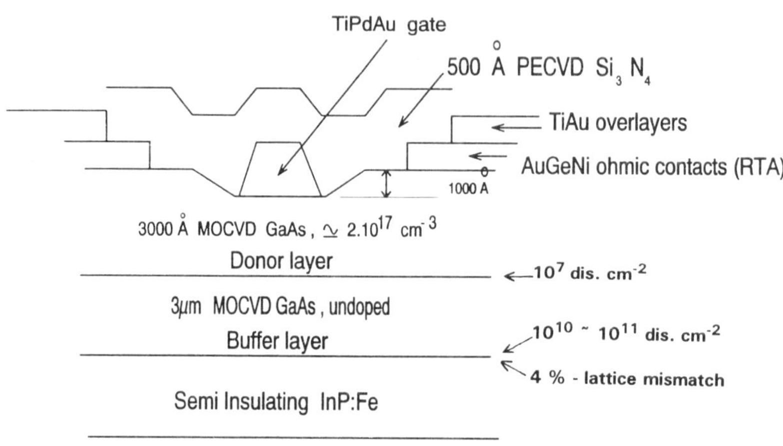

Fig. 1. Cross section of studied devices.

After scribing, devices are soldered with metallic preforms in hermetically sealed microwave packages for electrical characterizations and reliability testing. Typical output characteristics are shown in Figure 2. Extrinsic transconductances, gm's, of 140-150 and 110-115 mS/mm W_G are respectively measured at $V_{GS} = 0$ and $I_{DS} = \frac{I_{DSS}}{2}$; and $V_{DS} = 3$ V. The operating gate current I_{GS} ranges from $\simeq$ 400 to $\simeq$ 500 nA as indicated in Figure 3. This is about one order of magnitude too high for optical receiver application, however the studied devices have been designed for laser modulation.

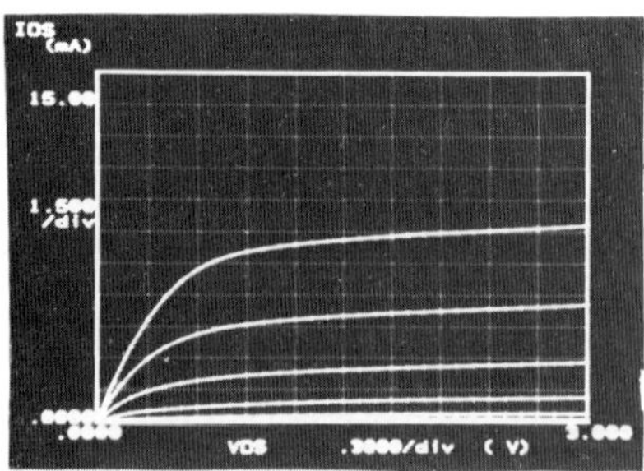

Fig. 2. I_{DS} (V_{DS}, V_{GS}) output characteristics. V_{GS} step = -0.2 V.

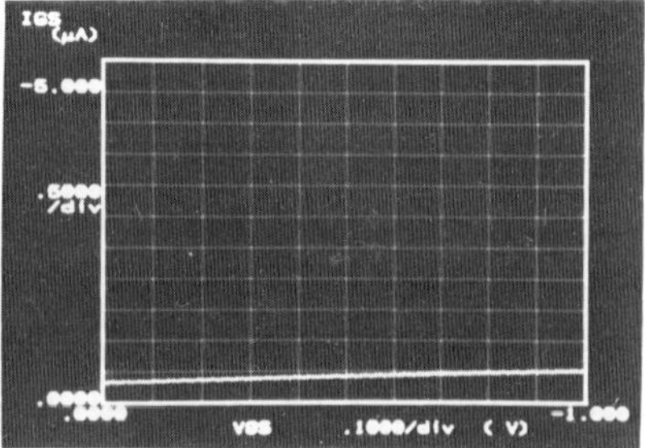

Fig.3. Operating gate current I_{GS} (VGS) at $V_{DS} = 3$ V.

Together with the dislocations, a residual coplanar strain due to the difference in the thermal expansion coefficients between GaAs and InP will remain in the GaAs layers. As a consequence, anomalous effects in the performances of integrated circuits and reliability problems can be expected.

3. TRAP RELATED DRAIN CURRENT TRANSIENTS

Among the anomalies, trap-related drain current transients can have a strong impact on the laser driver operation, depending on mark ratio effects (the mark ratio is the number of "1" in a given sequence of bits). As an example, crossover points of the eye patterns can fluctuate and shifts in logic levels are observed (Dumas et al 1993).

Transients have been studied under gate and drain switching conditions by means of isothermal relaxation experiment. Detailled descriptions of this technique have been previously published by Mottet et al (1991) and Audren et al (1993).

Under gate switching conditions, the device is biased in the saturation region, at $V_{DS} = 2.6$ V ; and the gate voltage switched from $V_{GS} = 0$ ($I_{DS} = I_{DSS}$) to a near pinch-off voltage, V_{GS} off, corresponding to I_{DS}off = 3 % of I_{DSS}. (cf Figure 4). The near pinch-off drain current transient is governed by an electron emission process from a level located in the reverse bias gate space charge region, i.e. in the GaAs donor layer under the gate. The related activation energy and capture cross-section are respectively Ena = 0.19 eV and σna = 1.7×10^{-21} cm^{-2}. However, the total trap concentration is low, $N_T \simeq 10^{-3} \times N_D$ (N_D is the doping concentration of the donor layer) ; and the related current transient amplitude, $\delta\, I_{DS}$off, is not significant.

Under drain switching conditions, the device is biased at $V_{GS} = 0$ and the drain voltage switched in the saturation region from $V_{DS} = 3$V to 1 V (cf Figure 5). The current transient is

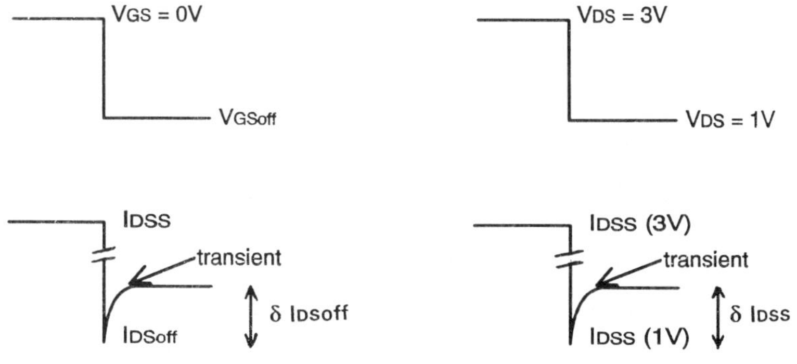

Fig.4. Gate switching conditions at $V_{DS} = 2.6$ V.

Fig.5. Drain switching conditions at $V_{GS} = 0$.

governed by free carrier emission and capture processes from deep levels located in the donor layer / buffer layer interface space charge regions. Two electron emission (Ena, σna) and a "hole-like" (Epa, σpa) processes are measured, as illustrated in Figure 6. Similar signatures obtained by means of deep level transient spectroscopy (DLTS) have been recently reported by Ben Hamida et al (1993) on n-GaAs grown by MOCVD on semi-insulating InP. It was concluded that the results compared to those obtained on homoepitaxial GaAs reference layers grown by MOCVD. One can also point-out that (i) trap concentrations are low, N_T's range from 10^{-3} to $10^{-2} \times N_D$; and (ii) related current transient amplitudes are negligeable : $\delta I_{DSS} < 10^{-2} I_{DSS}$ (1V).

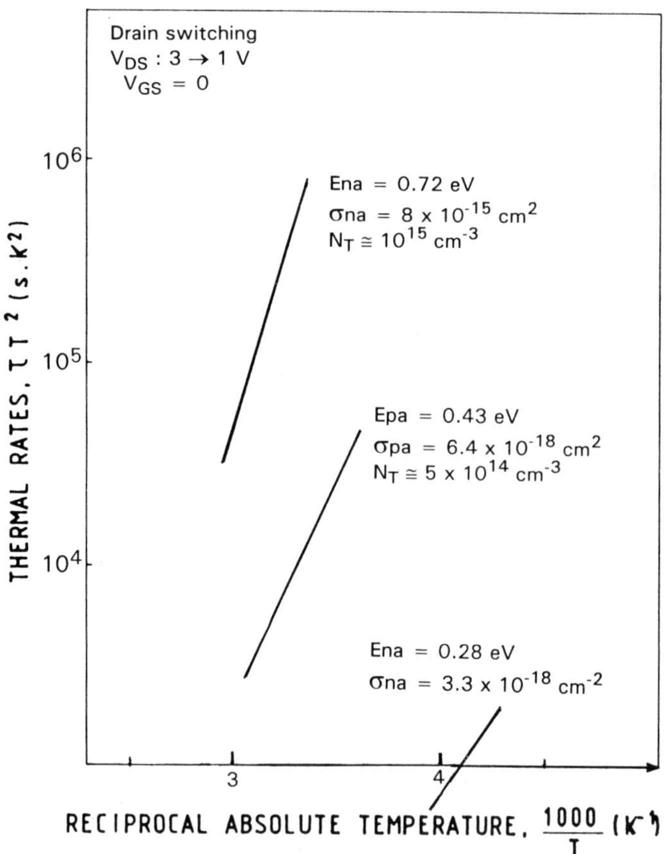

Fig.6. Emission and capture rates versus reciprocal temperature under drain switching conditions.

From the above mentioned results, it clearly appears that trap-related drain current transients will not penalize the device operation.

4. RELIABILITY INVESTIGATION

A set of 20 devices has been carried out into biased aging tests. The structure is the same as illustrated in Figure 1 but a 0.7 µm L_G x 50 µm W_G gate is used. They have been biased at V_{DS} = 2V and $I_{DS} \simeq I_{DSS}/2$ with channel temperatures of $\simeq$ 155 and $\simeq$ 205°C. During aging, a gradual degradation of these components if observed. It consists of (i) a decrease in :
the total drain current, I_{DSS}, as illustrated in Figure 7,
the pinch off voltage, Vp,
the open channel transconductance, gm,
and, (ii) an increase in the open channel resistance, R_{DS} on.

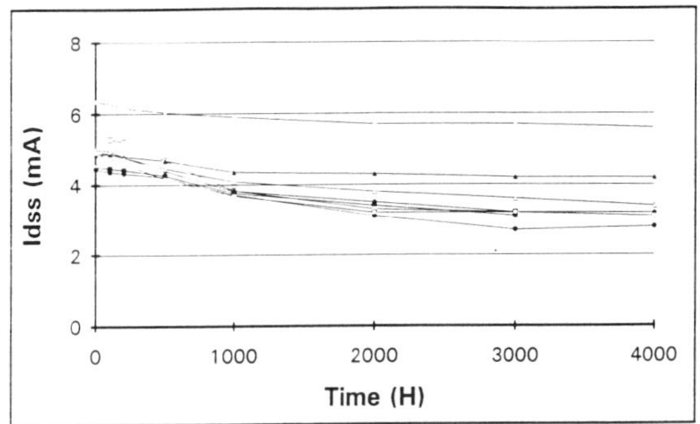

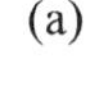

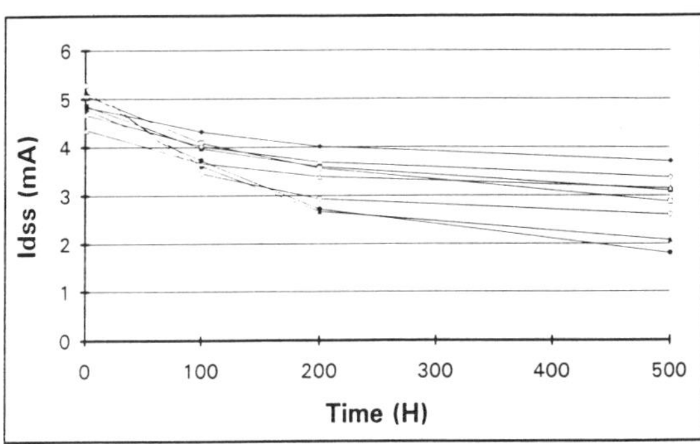

(b)

Fig.7. Total drain current (I_{DSS}) drifts measured at 155°C (a) and 205°C (b). The degradation is of gradual-type and thermally activated.

A comparison between 155 and 205°C aging tests indicates that the drifts are mainly thermally activated. On the other hand, Schottky gate parameters remained practically unchanged but a decrease in the gate current measured at the bias point, I_{Gpf}, has been observed. Regarding the above mentioned parameter drifts, several mechanisms can be involved :

1 - an ohmic contact resistance increase

2 - a reduction of active channel thickness,

3 - a decrease in net donor concentration due to compensation effects.

Ohmic contacts have been analysed using scanning electron microscopy (SEM) and secondary ion mass spectrometry (SIMS). No clear evidence of interdiffusion of the contact with GaAs and/or TiAu overlayers was observed by SIMS. However, a careful SEM investigation together with metallization selective etching evidenced changes in the contact surface morphology with localized collapses (holes) in the Ti layer : a classical balling effect developped during aging modified the size, distribution and ratio of "good" (Ni_2GeAs) and "poor" (Au ?) contact areas, then increasing the contact resistance. High temperature storage tests (Ta = 250°C) later on performed on TLM test vehicules confirmed the ohmic contact degradation.

But, decreases in gm, Vp and $I_G$pf indicate that mechanisms 2 and 3 have also to be taken into account. Trap characterizations have been performed, as described in paragraph 3, on control samples and failed devices. The total trap concentration remained $N_T \leqslant 10^{-2} \times N_D$, both in the GaAs donor layer and at the GaAs donor layer/GaAs buffer layer interface vicinity. This confirms (i) low trap concentrations as previously found and (ii) no significant trap-related compensation mechanism developping during aging. As a consequence, a reduction of the active channel thickness due to the well known "sinking" of the Au-based gate metallization can be postulated (Zanoni et al 1990).

5. CONCLUSION

This investigation into GaAs-on-InP MESFETs evidenced low trap concentrations into the GaAs donor and buffer layers. As a consequence the trap-related drain current transients penalizing the device operation are negligeable. The study also highlighted GaAs MESFET conventional reliability problems such as ohmic contact resistance increase and gate sinking prior to failure mechanisms induced by the lattice mismatch between InP and GaAs. Process improvements have been undertaken.

REFERENCES

Audren P. 1993 J. Phys. III 3 185

Azoulay R. 1991 J. Crystal Growth 107 926

Ben Hamida A. 1993 Proc. Indium Phosphide and Related Materials 150

Dumas J.M. 1993 FRANCE TELECOM/CNET Internal report

Mottet S. 1991 Inst. Phys. Conf. Ser. 120 3 155

O'Sullivan P.J. 1991 J. Elec. Mat. 20 12 1029

Zanoni E. 1990 Quality and Reliability Eng. Int. 6 29.

LT-GaAs MISFET structure for power application

K.-M. Lipka, B. Splingart, U. Erben, E. Kohn

University of Ulm, Abt. Elektronische Bauelemente und Schaltungen,
Oberer Eselsberg, D-89069 Ulm, Germany,
Tel.: +49 731 502-6151, FAX: +49 731 502-6155

ABSTRACT: LT-GaAs MISFETs have been realized indicating a record 2.1W/mm RF power handling capability. To understand the properties of such LT-GaAs power MISFET structures, the MIS-system containing a LT-GaAs insulator and an AlAs interfacial diffusion barrier to the channel has been analysed. A noticeable parallel conductance was found in the insulator leading to a high gate to drain breakdown voltage, however also to a g_m-dispersion in the MHz regime. An electronic equivalent circuit for use in the FET model has been established. At the insulator-semiconductor interface, a low interface potential in the range of 0.3eV below the conduction band is seen, indicating uncommon interface and passivation properties.

1. Introduction

Low Temperature-GaAs layers, grown at extremly low temperatures containing a large amount of excess As, have been successfully used as gate insulator or surface passivation in GaAs FET structures. A detailed overview of the materials properties and applications is given by Smith (1992) and Mishra (1993). In FET structures both high gate-drain and high drain-source breakdown voltages have been shown by Yin et al.(1990) leading to a record power capability of 1.57W/mm for GaAs based FETs demonstrated by Chen et al.(1991). Our recent results, Lipka et al. (1993b), shown in figure 1 indicate 2.1W/mm RF power are feasible.

Using a LT-GaAs insulating layer in GaAs MISFETs an interfacial diffusion barrier has to be introduced to avoid As outdiffusion from the LT-GaAs insulator into the channel. An AlAs diffusion barrier was developed by Yin (1992) and is part of the insulating layer system. Nevertheless, many details concerning the MIS structure, composed of Ti/Pt/Au, LT-GaAs-insulator, AlAs-barrier and the n-doped channel, are still controversal. Therefore important device related parameters have to be investigated in order to obtain information on the MIS-system and its influence on the mode of operation in FET-structures. This study focuses on the electrical properties of the insulator and the characteristics of the insulator-semiconductor interface.

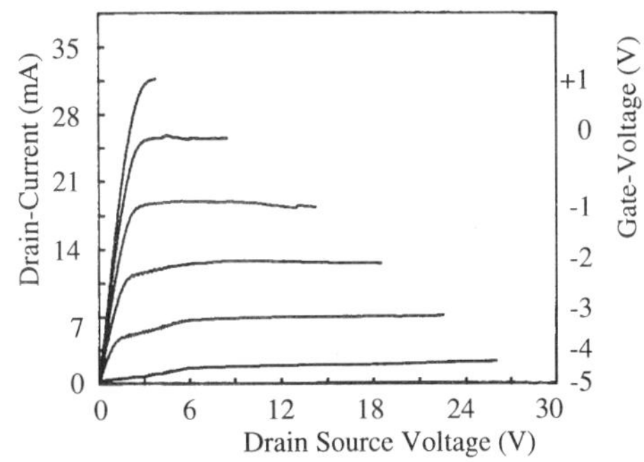

Figure 1: Output characteristics of LT-GaAs power MISFET $w=50\mu m$, $l_g=1\mu m$

MISFETs are of special interest for power applications due to suppressed gate leakage current and therefore high gate breakdown voltage independent of the channel doping. GaAs MISFETs can only operate in the depletion mode due to the insulator interface potential pinning. Therefore, typical problems of these devices are a low transconductance and instabilities due to charging of the interface states, which are insulated from the gate electrode by the dielectric insulator. In the case of the LT-GaAs insulator system a noticeable leakage current is observed like in the case of the Deep Depletion MISFET of Dortu and Kohn (1984). This controlled leakage is also thought to be the key reason for the large gate-drain breakdown voltage shown in figure 2.

2. Device Structure and Fabrication

The investigated MISFET structures contain a silicon doped GaAs-channel and the insulator, consisting of a 20nm AlAs diffusion barrier and a 80nm LT-GaAs top layer. While the sheet carrier concentration in the channel was kept constant at $2.5x10^{12}cm^{-2}$ and $5x10^{12}cm^{-2}$, the channel layer thickness was varied between 30nm and 500nm resulting in $5x10^{16}cm^{-3} < N_D < 1.6x10^{18}cm^{-3}$. The thickness of the insulator was always kept constant.
All structures had been grown in a Riber 32P MBE system on (100) semi-insulating LEC GaAs substrates. The 600nm GaAs-buffer layer and channel layers have been grown under standard MBE conditions. After growing the AlAs layer at 680°C the growth was interrupted and the substrat cooled down to 200°C the growth temperature of the LT-GaAs, determined by thermocouple reading.
Devices have been insulated by wet chemical mesa etching. Drain and source ohmic contacts were made by evaporating and alloying Au/Ge/Ni after removing the LT-GaAs and AlAs layers by selective wet chemical etching and lift-off. Finally Ti/Pt/Au Schottky gate contacts have been deposited by e-beam evaporation. Some structures have been recessed by wet etching in order to obtain a thinner insulator. The gate width of the FET was 50µm and the length between 0.6 and 1µm.

3. Results

3.1 I-V-Characteristics

The two terminal breakdown voltage of the gate-drain MIS-diode, defined by $U_{dg}(I_g= 1mA/mm)$ was found to be between 42V and 51.5V for floating source contacts. These results agree with values reported by Chen et al.(1992) showing a high resistive surface layer between gate and drain smoothing out the gate-drain field and promoting the high breakdown voltage. Due to this conductance a noticeable parallel conductivity to the MIS-capacitance is observed at forward bias gate across the thin insulator. Therefore LT-GaAs MISFETs are expected to act like Deep Depletion MISFETs, but free of interfacial instabilities due to the potential stabilizing conduction path. The metal-LT-GaAs built-in potential determined from the I-V-plot was V_{bi}=0.7V which is in agreement with that of the normal metal-GaAs junction.

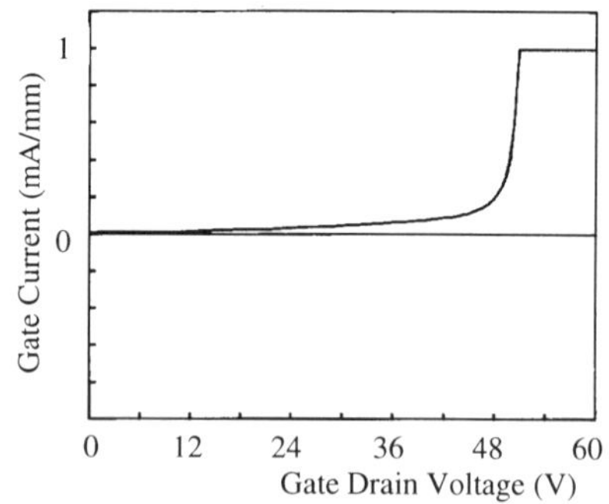

Figure 2: Gate-Drain-breakdown characteristics of MISFET structure

3.2 C-V-Measurements

The MIS capacitance has been measured with a FAT FET structure in a dc-bias range between -5V and +5V at 1MHz with an ac-amplitude of 100mV. The operation mode of the capacitance bridge (HP 4192A) was choosen to be parallel for reverse biased devices and series in case of forward bias.

Lipka et al. (1993c) developed an equivalent circuit shown in figure 4, for the MIS-system by frequency dependent impedance measurements. This circuit can also be described by an effective frequency dependent gate capacitor, leading to a g_m dispersion. At 1MHz the capacitance of the MIS-structure consisting of the bias independent constant capacitance of the insulating layer and the bias dependent space charge layer (SCL) of the channel can be calculated by equation (1). Therefore the measured effective capacitance is compared in figure 3 to the calculated capacitance with different insulator thicknesses d_i. Three cases are of special interest.

(1) no insulator, d_i=0nm

(2) only the AlAs barrier acts as a capacitance, resulting in d_i=20nm

(3) the full insulator acts as a capacitance, resulting in d_i=100nm

In the calculation the dielectric constant of the As-rich insulator was approximated by the value of GaAs.

It is obvious that for the measurement frequency of 1MHz the capacitance of the insulator is determined by the capacitance of the 20nm AlAs diffusion barrier layer whereas the LT-GaAs acts as series resistance. This indicates a larger resistance of the AlAs layer compared to the resistance of the LT-GaAs layer resulting in a reduced low frequency equivalent circuit without LT-GaAs capacitor. Therefore the maximum effective capacitance at forward bias is given by equation (2).

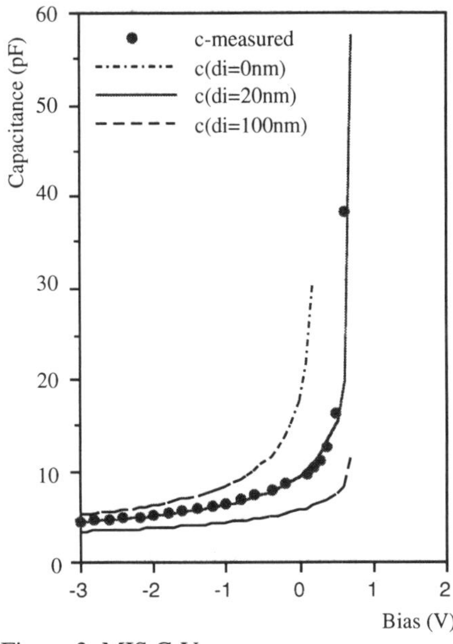

Figure 3: MIS C-V-curve

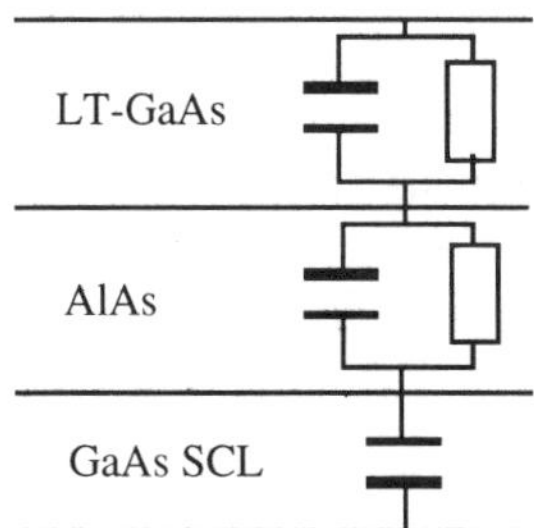

Figure 4: RF-MIS equivalent circuit with channel space charge layer

$$C_{MIS} = \varepsilon \frac{A}{d_i + \sqrt{\frac{2\varepsilon}{q}(V - V_{bi})\frac{1}{N_D}}} \qquad (1)$$

$$C_{eff_{max}} = \frac{1 + \omega^2 C_{AlAs}^2 R_{AlAs}^2}{\omega^2 C_{AlAs} R_{AlAs}^2} \qquad (2)$$

3.3 Transconductance (g_m) of LT-GaAs MISFETs

The transconductance as well as the pinch-off voltage are affected by the parallel resistance of the insulating layer (together with the leakage through the channel-SCL). The dc-pinch-off voltages were compared with the three cases of interest according to section 3.2 confirming the low frequency MIS model, as shown in figure 7 for N_s=2.5x10^{12}cm^{-2}. Corresponding results have been obtained for N_s=5x10^{12}cm^{-2}.

If the g_m for low frequencies is indeed controled by the thickness of the AlAs diffusion barrier layer and the channel-SCL, this parameter should not be affected by the thickness of the LT-GaAs layer. This hypothesis has been confirmed by investigating recessed FET structures with an effective thickness of the LT-GaAs layer between 20nm and 80nm shown in figure 5. In the case of 100nm insulator thickness (including 20nm diffusion barrier) and N_D=1x10^{17}cm^{-3} the low frequency g_m=100mS/mm was reduced to 60mS/mm for frequencies above 45MHz as expected. To limit this dispersion a thin LT-GaAs layer is necessary including a thin effective diffusion barrier. Indeed, a significant increase of the transconductance was recently observed for a diffusion barrier thickness reduced to 5 nm as shown in figure 6 (details about this modified diffusion barrier will be reported elsewhere).

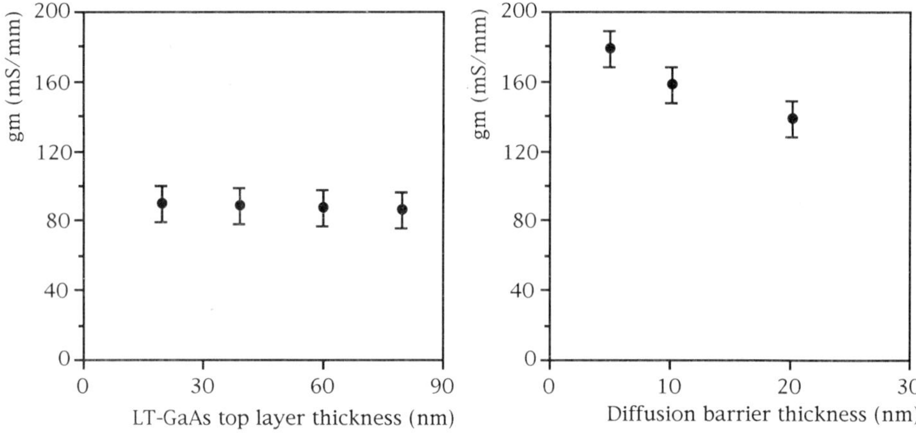

Figure 5: Measured transconductance for varied LT-GaAs top layer thickness N_D=1x10^{17}cm^{-3}

Figure 6: Measured transconductance for varied diffusion barrier thickness N_D=8x10^{17}cm^{-3}

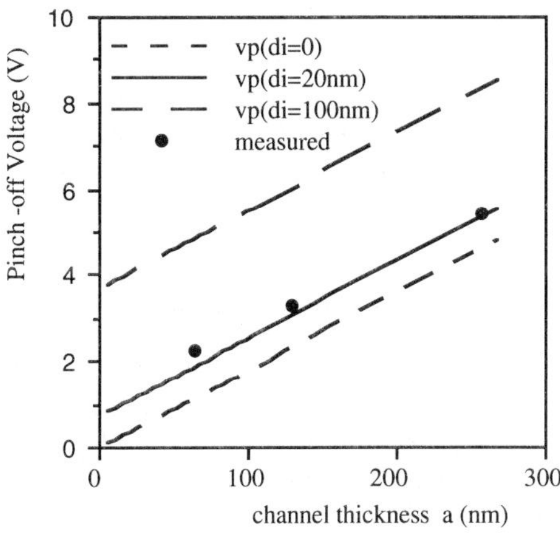

Figure 7: MISFET pinch-off voltage
N_S=2.5x10^{12}cm^{-2}

Saturation Current (mA)
ϕint=0
ϕint=0.3eV
ϕint=0.6eV
measured
channel thickness a (nm)

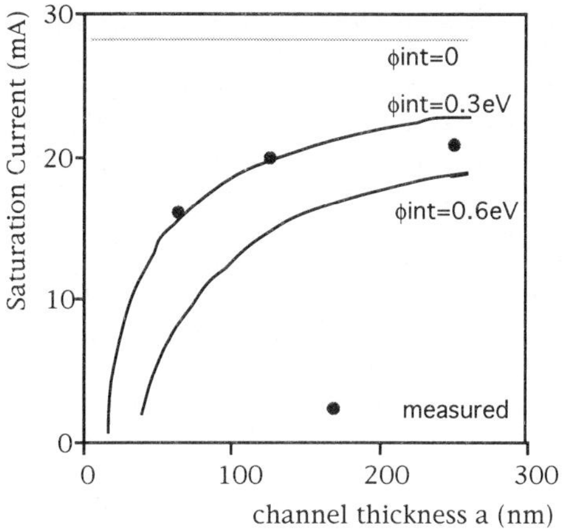

Figure 8: MISFET saturation current
N_S=2.5x10^{12}cm^{-2}

$$V_p = \frac{q}{\varepsilon} N_s \left(\frac{a}{2} + d_i \right) \tag{3}$$

$$I_{DSS} = qN_s V_{sat} w \left(1 - \sqrt{\frac{2\varepsilon}{q} V_{bi} \frac{1}{N_s a}} \right) \tag{4}$$

$$V_{bi} = \phi_{int} - \left[\frac{E_g}{2} - kT \ln \left(\frac{N_s}{n_i a} \right) \right] \tag{5}$$

3.4 Characteristics of the insulator-semiconductor interface

The interface between the LT-GaAs/AlAs insulator and the GaAs channel has been characterised by comparing ungated MISFET structures to ungated MESFET structures. Measuring the channel saturation current in theses structures of identical geometrie and doping concentration, a larger maximum current was observed for the MISFET. This has been studied in the following way:

Two identical FET channels have been grown by MBE but only one contains the LT-GaAs/AlAs insulating layer. After depositing the drain/source ohmic contacts, a clearly higher saturation current was measured for the LT-GaAs/AlAs containing MISFET structure.

This finding has been verified by measuring the saturation current of the ungated MISFET after removing the insulating top layer by selective etching down to the AlAs diffusion barrier layer. Again a lower saturation current was observed after etching.

This has been already qualitatively observed by Yin et al. (1990) and indicates a lower potential of the insulator-semiconductor interface compared to the free surface potential of n-GaAs. Assuming a constant saturation velocity, the active channel thickness can be calculated by the ratio of the saturation currents. Comparing this thickness to the metalurgical thickness, an interfacial potential of ϕ_{int}=0.3eV is extracted. Since the saturation current is already reduced after removing the LT-GaAs layer with the AlAs layer remaining on the channel, it is obvious that a potential at the AlAs/LT-GaAs interface reaches through the diffusion barrier, controling the depletion region in the channel. Therefore, the potential of the insulator-semiconductor interface is due to the fixed Fermi-level in the LT-GaAs layer, as already indicated by Lipka et al. (1993a).

If this picture is correct, a decrease of the saturation current with decreasing channel thickness at constant N_s is expected corresponding to equation (4) and (5). This is actually observed and is shown in figure 8.

4. Discussion and Conclusions

LT-GaAs MISFETs are attractive for power applications, because of their high demonstrated RF-output power of 1.5W/mm and an even higher IV-power product of 2W/mm. High DC-breakdown voltages can be obtained and combined with high output currents. It seems therefore, that with these structures the common GaAs-MESFET breakdown limitation imposed by the channel sheet charge can be overcome (see E. Kohn 1993). However, the MIS-gate configuration introduces a frequency dispersion, which in turn will also translate into a time and temperature instability. To fully explore and utilize the potential of these structures, these instabilities have to be minimized by tailoring the structure. First experiments in this direction are encouraging. A high DC transconductance can be obtained by using a thin diffusion barrier, which is of course a technological challange. First structures with a 5nm modified diffusion barrier have been processed resulting in a maximum low frequency g_m of 180mS/mm. The frequency dispersion is reduced by using a thinner LT-GaAs insulating layer than the 80 nm used in this study. The interface potential between the LT-GaAs MIS system and the channel is approx. 0.3eV, which is clearly lower then the GaAs surface potential of approx. 0.6V. Due to this small interface potential the GaAs channel can be nearly totally opened. Devices with $N_D=8x10^{17}cm^{-3}$ and a=62.5nm have displayed a maximum current densisty of 680mA/mm and the maximum drain current of 28V (see fig.1). This represents a record I-V power product of 16.8W/mm. In class A operation this will translate to $P_{RFmax}=I_{max}(V_{max}-V_{sat})/8$=2.1W/mm. However, the frequency dispersion in the MIS-system means that a larger RF-drive voltage is needed than indicated by the quasi static output characteristics.

Acknowledgement

The authors would like to thank Y. Druelle and G. Salmer from IEMN Lille (France) for many helpful discussions and making their MBE equipment available for the growth of most of the samples. This work was supported by the CEC ESPRIT project 6849 "TAMPFETS".

Literature

Chen C-L, Mahoney L J, Manfra M J, Smith F W, Temme D H and Calawa A R 1992 Electron Device Letters, vol 13, No 6, 335

Chen C-L, Smith F W, Clifton B J, Mahoney L J, Manfra M J, Calawa A R 1991 Electron Device Letters, vol 12, No 6, 306

Dortu J M, Kohn E 1984 Inst. Phys. Conf. Ser. No 74: Chapter 7, 563

Kohn E 1993, 17th WOCSDICE 93, Parma, Italy, Abstracts 128

Lipka K M, Splingart B, Zhang X, Poese M. Panzlaff K, Kohn E 1993a Proc. E-MRS Spring Meeting, Strasbourg, France, paper B-10

Lipka K, Splingart B, Kohn E 1993b Electronics Letters, vol 29, No13, 1170

Lipka K, Splingart B, Erben U, Kohn E 1993c IEEE/Cornell Conference on Advanced Concepts in High Speed Semiconductor Devices, paper 3.2

Mishra U K 1993 Proc. E-MRS Spring Meeting Strasbourg, paper B-15

Smith F W 1992 Mat.Res. Soc. Symp. Proc. Vol 241, 3

Yin L-W, Hwang Y, Lee J H, Kolbas R M, Trew R J, Mishra U K 1990 Electron Device Letters, vol 11, No 12, 561

Yin L-W 1992 University of California, Santa Barbara, ECE Technical Report # 92-23

Measurement and simulation of *p*-buffer MESFETs in impact ionization regime

A Neviani, R Chieu, C Tedesco, E Zanoni

Dipartimento di Elettronica e Informatica, Università di Padova, Via Gradenigo 6A, I-35131 Padova, Italy

W Patrick

Institut für Feldtheorie und Höchstfrequenztechnik, ETH Zürich, Gloriastrasse 35, CH-8092 Zürich, Switzerland

ABSTRACT: We report experimental data on the increase in the output conductance in p-buffer depletion-mode MESFET's showing that this effect occurs at the onset of impact ionization in the channel of the device. This effect was already observed in enhancement-mode MESFET's, and was attributed to a Parasitic Bipolar Effect (PBE) triggered by the injection of impact-ionization-generated holes in the p-buffer. We perform numerical simulations demonstrating that this interpretation is correct, and that the PBE can occur in more general conditions than indicated in previous studies.

1. INTRODUCTION

Hot electron and short channel effects are detrimental consequences of the reduction of the device dimensions in the submicrometric range. The design of optimized power microwave MESFET's requires a detailed evaluation of these effects, in order to improve the output power and obtain a reliable operation of the device. Among hot electron effects, impact ionization plays a major role in limiting the device performances. Avalanche breakdown (often accompanied by burn-out) is only the most dramatic effect of impact ionization: increase in noise power, kinks in the I-V characteristics, increase in the output conductance and increase in negative gate current are other undesired effects which must be controlled in order to obtain a reliable operation of the device.

In this work, we report electrical measurements and numerical simulations performed on a prototype p-buffer depletion-mode MESFET biased at high drain voltages, and investigate the effects of impact ionization on the output characteristics of the device. In section 2 we report the electrical characteristics of the device; the results of the numerical simulations are shown in section 3 and discussed in section 4. Conclusions follow in section 5.

2. SAMPLES AND ELECTRICAL MEASUREMENTS

In this work, we used p-buffer GaAs MESFET's whose structure is reported in Fig. 1. Three side-gate ohmic contacts, like the one reported in the left side of Fig. 1, were introduced to monitor the current due to the possible injection of carriers in the p-buffer. The p-buffer is sufficiently thin to be completely depleted in thermal equilibrium conditions. The drain current I_d (solid lines) versus the drain-to-source voltage V_{ds}, measured at dif-

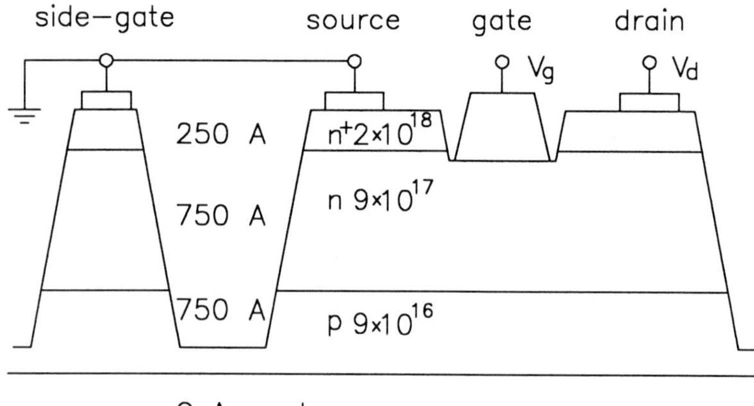

Figure 1: Cross section of the device used in our work. The p-buffer is grown on an unintentionally doped ($\leq 10^{15}$ cm^{-3}) GaAs layer. The gate length is 0.5 μm and the gate-to-source spacing is 1 μm.

ferent gate-to-source voltages V_{gs}, together with the side-gate current I_{sg} (dash-dotted lines) are reported in Fig. 2. The gate current I_g has a behavior similar to that shown by I_{sg}, so it is not reported in the figure. I_{sg} and I_g show a non-monotonic dependence on V_{gs}, which is more evident in Fig. 3, which reports I_g vs V_{gs} for four different V_{ds} values.

When V_{ds} overcomes 4 V, Fig. 2, the drain characteristics show an abrupt increase in the output conductance, which is less evident at V_{gs}=–0.5 V, and disappears at V_{gs}=0 V. This effect was already observed in similar structures (Van Zeghbroeck et al 1987, Fujishiro et al 1989) and was attributed to a mechanism activated by the injection of impact-ionization-generated holes in the p-buffer. Actually, I_{sg} and I_g measured in our devices show a nearly exponential dependence on V_{ds}, as reported in Fig. 2. This behavior is the same as that reported for the gate current in conventional GaAs MESFET's by Hui et al (1990) and by Canali et al (1991), which was demonstrated to come from the collection of impact-ionization-generated holes at the gate contact. This conclusion was supported by numerical simulations (Canali et al 1993).

An explanation of the mechanism responsible for the increase in the output conductance in terms of a Parasitic Bipolar Effect (PBE) was proposed by Van Zeghbroeck et al (1987). They suggested that, due to the onset of impact-ionization generation, part of the generated holes are injected into the p-buffer, raising the electric potential therein, and causing the nearly exponential increase of the side-gate current, see Fig. 2. The increase in the potential forward biases the p-buffer/n-channel junction under the source contact. Consequently, electrons are injected from the source into the p-buffer, where they travel until they reach the high-field region under the drain end of the gate. Here, since the p-buffer/n-channel junction is reverse biased, they are re-injected into the channel, where, due to the high electric field, they contribute to create more electron-hole pairs by impact ionization. This leads to a positive feedback action, which accounts for the increase in the output conductance.

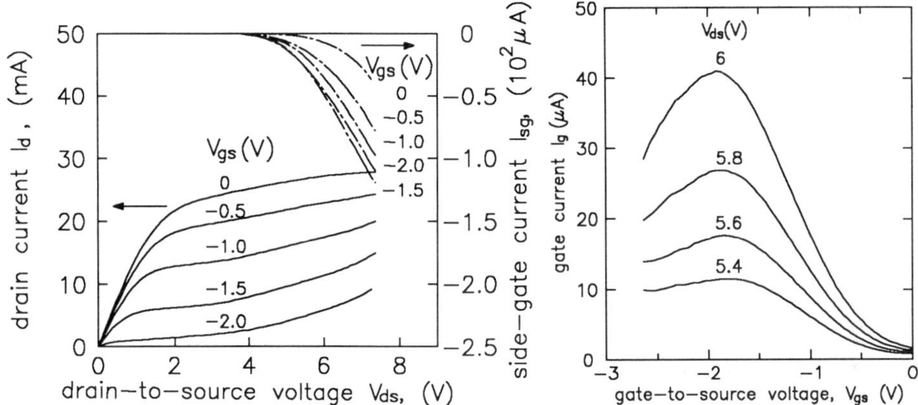

Figure 2: I_d (solid line) and I_{sg} (dash-dotted line) vs V_{ds} measured at different fixed V_{gs}. The side-gate contact is grounded.

Figure 3: Gate current I_g versus gate-to-source voltage V_{gs} at four different V_{ds}.

A similar increase in the output conductance has already been observed in SOI MOSFET's (Kato et al 1985) and in HeteroJunction FET's (HJFET) (Kunihiro et al 1993). Numerical simulations have demonstrated that this effect was originated by a mechanism analogous to the PBE. Nevertheless, in these works, the presence of a hole-confining layer (SiO_2 substrate in SOI MOSFET's, and wide band-gap buffer in HJFET) was thought to be necessary for the occurrence of the effect. Consequently, these results cannot be directly extended to MESFET's. Moreover, Kunihiro et al (1993) suggested that a p-type electrode in the substrate could be useful to suppress the PBE by draining the injected holes. Wada et al (1988) simulated planar n-GaAs MESFET's on n^- substrate, without observing any increase in the output conductance until the breakdown voltage is reached. Nevertheless, they explain breakdown in terms of conductivity modulation in the n^- substrate induced by the injection of holes generated under the drain contact, but their results cannot be applied to recessed geometries, where the generation at the drain is suppressed.

In the next two sections we report the results of numerical simulations demonstrating that the increase in output conductance in our devices is due to a PBE, as suggested by Van Zeghbroeck et al (1987). We also explain the gate bias dependence of this effect. Finally, we analyze the effect of a p-type contact in the substrate.

3. NUMERICAL SIMULATIONS

The numerical simulations of electrically characterized MESFET's were performed using the program HFIELDS (Baccarani et al 1985) implementing a two dimensional drift-diffusion model within a finite element discretization scheme. A mobility model suitable for GaAs based devices (Laux et al 1981) has been included in the program. Thermal generation-recombination is accounted for through the standard Shockley-Read-Hall (SRH) model, while generation by impact ionization is included in a self-consistent manner following the work of Laux et al (1985). Surface states are also taken into account through the introduction of a density $N_t = 4 \times 10^{12}$ cm^{-2} of acceptor levels, modeled according to the SRH theory, with energy 0.7 eV below the conduction band. The ionization rates for electrons α_n and holes α_p are calculated according to the Chynoweth's law:

$$\alpha_{n(p)} = \alpha_\infty \exp\left[-\left(\frac{B}{F_{n(p)}}\right)^m\right]$$

where the values of the parameters $\alpha_\infty = 4 \times 10^6$ cm^{-1}, $B = 2.3 \times 10^6$ V/cm, m=1 for electrons and $\alpha_\infty = 3 \times 10^5$ cm^{-1}, $B = 6 \times 10^5$ V/cm, m=2 for holes are taken from Canali et al (1991) and Sze (1981). The argument $F_{n(p)}$ is defined as $\max\left(0, \mathbf{E} \cdot \mathbf{J}_{n(p)} / \mid \mathbf{J}_{n(p)} \mid\right)$, where $\mathbf{E}$ is the electric field and $\mathbf{J}_{n(p)}$ is the electron (hole) current density.

The drain current I_d and the side-gate current I_{sg} obtained by simulation, as a function of V_{ds} for three different V_{gs}, are shown in Fig. 4. The I_d curves are characterized by an abrupt increase in the output conductance as V_{ds} is raised above approximately 4.5 V. The increase is marked even at V_{gs}=0 V, in contrast to the experimental data. The dashed line is the drain current as obtained by simulation, with the addition of a substrate ohmic contact kept grounded. The side-gate current I_{sg} shows a nearly exponential dependence on V_{ds}, and a non-monotonic dependence on V_{gs}, similar to that observed experimentally.

4. DISCUSSION

The results of the numerical simulations demonstrate that the increase in the output conductance is due to a Parasitic Bipolar Effect. Fig. 5 reports the electric potential and the hole concentration profile under the source contact at different drain bias conditions, with V_{gs}=0 V, as obtained by simulation. At equilibrium (V_{ds}=0 V), the hole concentration is negligible in the depleted p-buffer and $5 \cdot 10^{14}$ cm^{-3} in the unintentionally doped GaAs substrate, while the potential barrier between the n-channel and the substrate is more than 0.6 V. As V_{ds} is moved from 0 to 13 V, the hole concentration increases up to more than 10^{17} cm^{-3} in both the p-buffer and the substrate. Experimentally, a nearly exponential increase of the negative side-gate current is detected, see Fig. 2. At the same time, the potential barrier is reduced to less than 0.2 V at V_{ds}=6 V, and less than 50 mV at V_{ds}=13 V. As a consequence, an increasing number of electrons are injected from the source into the p-buffer, and return into the n-channel under the drain edge of the gate, where the p-buffer/n-channel is reverse biased. This PBE leads to the increase in the output conductance, see Figs. 2 and 4, as discussed in the introduction.

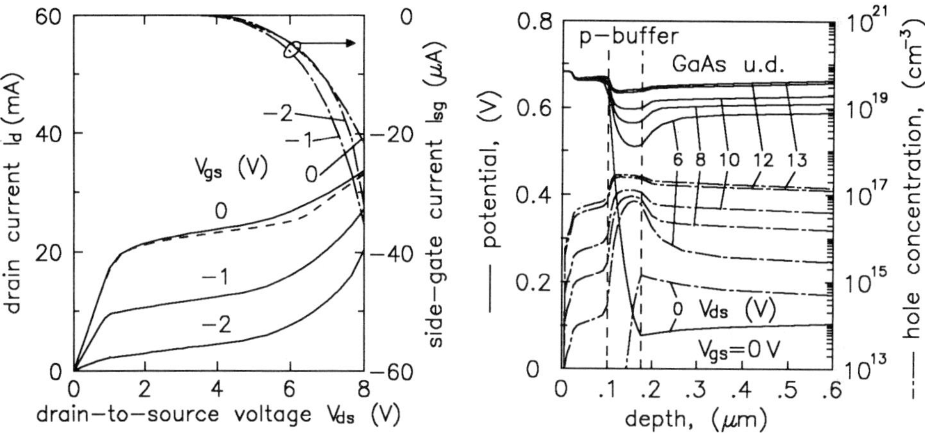

Figure 4: Simulated I_d and I_{sg} vs V_{ds} at V_{gs}=0, −1, −2 V (solid lines), together with I_d obtained adding a substrate contact (dashed line).

Figure 5: Potential (solid lines) and hole concentration (dash-dotted lines) profiles under the source contact for different values of V_{ds} at V_{gs}=0 V. The position of the p-buffer is delimited by dashed lines.

The fact that the increase in the output conductance is not evident in the experimental data taken at V_{gs}=0,−0.5 V is due to (i) the self-heating of the device in open channel conditions; (ii) the gate bias dependence of the impact-ionization generation rate, which, according to Canali et al (1993), is proportional to the gate current I_g. This means that, when V_{gs} approaches 0 V, the generation rate is low, see Fig. 3, while it peaks around V_{gs}=−1.5 V. Consequently, the PBE responsible for the increase in the output conductance is stronger when V_{gs} is around −1.5 V than in open channel conditions.

The addition of an ohmic contact in the substrate in order to collect the injected holes and thus to prevent the occurring of the PBE does not work, as shown in Fig. 4 (dashed line). In fact, even if initially the holes are collected by the substrate contact, they cause a voltage drop along their path. This voltage drop forward biases the p-buffer/n-channel junction under the source. The increase in the output conductance appears at a slightly higher V_{ds} with respect to the previous case.

Despite the positive feedback on the impact-ionization generation due to the PBE, our devices are not subject to avalanche breakdown, but they degradate irreversibly due to the exponential increase of the gate current with V_{ds}. In fact, the positive feedback on I_d induced by the PBE is limited at high V_{ds} by the onset of the high injection regime (the potential barrier in Fig. 5 approaches zero when V_{ds} overcomes 12 V) which reduces the current gain of the parasitic bipolar transistor.

5. CONCLUSIONS

We report experimental data on the increase of the output conductance in p-buffer MESFET's showing, as already reported, that this effect occurs at the onset of impact ionization in the channel of the device. Our numerical simulations demonstrate that the mechanism at the origin of the output conductance increase is the parasitic bipolar effect suggested by Van Zeghbroeck (1987), and that it can occur in more general conditions than indicated in previous studies (Kato 1985, Kunihiro 1993). The exponential increase of the gate current and the onset of the high injection regime for the parasitic bipolar action, which reduces the avalanche generation, cause the device to degradate irreversibly due to the excess gate current flow rather than due to avalanche breakdown.

6. REFERENCES

Baccarani G, Guerrieri R, Ciampolini P and Rudan M 1985 Proc. NASECODE IV Conf., ed Miller J H (Dublin: Boole), pp 3-12

Canali C, Paccagnella A, Zanoni E, Lanzieri C and Cetronio A, 1991 IEEE Elec. Dev. Lett. 12 80

Canali C, Neviani A, Tedesco C, Zanoni E, Lanzieri C and Cetronio A, 1993 IEEE Trans. Elec. Dev. 40 498

Fujishiro H I, Inokuchi K, Nishi S, Sano Y, 1989 Jap. J. Appl. Phys. 28 1734

Hui K, Hu C, George P and Ko P K, 1990 Elec. Dev. Lett. 11 113

Kato K, Wada T and Taniguchi K, 1985 IEEE Trans. Elec. Dev. 40 493

Kunihiro K, Yano H, Goto N and Ohno Y 1993 IEEE Trans. Elec. Dev. 40 493

Laux S E and Lomax R J, 1981 Solid-State Elec. 24 485

Laux S E and Grossman B M, 1985 IEEE Trans. CAD 4 520

Sze S M 1981 Physics of Semiconductor Devices (New York: J Wiley & sons)

Van Zeghbroeck B J, Patrick W, Meier H and Vettiger P, 1987 IEEE Elec. Dev. Lett. 8 188

Wada Y and Tomizawa M, 1988 IEEE Trans. Elec. Dev. 35 1765

About the use of HEMT in front end electronics for radiation detection

G. Bertuccio, G. De Geronimo, E. Gatti, A. Longoni
Dipartimento di Elettronica e informazione, Politecnico di Milano, P.zza Leonardo da Vinci 32, 20133 Milano, Italy.
*J. Ludwig, K. Runge, M. Webel and S. Lauxtermann**
Fakultät für Physik der Universität Freiburg, D-79104 Freiburg, Germany
* Fraunhofer-Institut für Angewandte Festkorperphysik, Tullastrasse 72, D-79108 Freiburg, Germany.

ABSTRACT: The factors affecting the resolution of a detection system for ionising radiation are shortly reviewed, in order to verify the possibility of using heterostructure field effect transistors in these applications.

1. INTRODUCTION.

A radiation detection system is composed of a detector followed by an electronic stage of signal amplification and by a filtering stage for noise reduction. The considered detection system provides an output signal whose amplitude is proportional to the charge delivered by the ionising particle in the detector. In the present work we will consider the system resolution in the measurement of the ionisation charge and its dependence on the characteristics of the amplification stage. This dependence is dominated, in a well designed amplification stage, by the performances of its input transistors.
We will in particular consider the use of heterostructure field effect transistors: High Electron Mobility Transistors (HEMTs). The extremely high bandwidth of these devices determines very low values of their input series white noise. Their input parallel noise can be non negligible, due to the relatively high values of the gate current. The high values of the low frequency noise (e.g. 1/f noise) is generally the true limiting factor to the use of these devices in high resolution applications.
A comparison with the resolution achievable with more conventional input devices will demonstrate the potential of HEMTs in applications where high detection rate is required (such as in the high energy physics experiments with the new accelerators).

2. THE DETECTION SYSTEM.

A schematic diagram of a detection system is shown in Fig.1. The detector is represented by its capacitance C_d. In C_d we also comprises the stray capacitances of the connections between detector and preamplifier and the feedback capacitance of the preamplifier. The current generator $Q\delta(t)$ in parallel to the detector provides the charge Q instantaneously delivered by the ionising particle. The electronics connected to the detector is represented by the input transistor of the preamplifier (with its input capacitance C_{gs}) followed by a 'black box' stage which process the drain current of the input transistor. The step waveform drawn near the drain of the transistor represents the drain current response to a $Q\delta(t)$ current delivered by the detector. The waveform w(t) in the 'black box' represents its response to a current $\delta(t)$. The response of the whole system to a $Q\delta(t)$ current delivered by the detector is a triangular waveform. The results we obtain with this model do not change substantially if the shape of the output pulse is different (e.g. Gaussian output pulse).

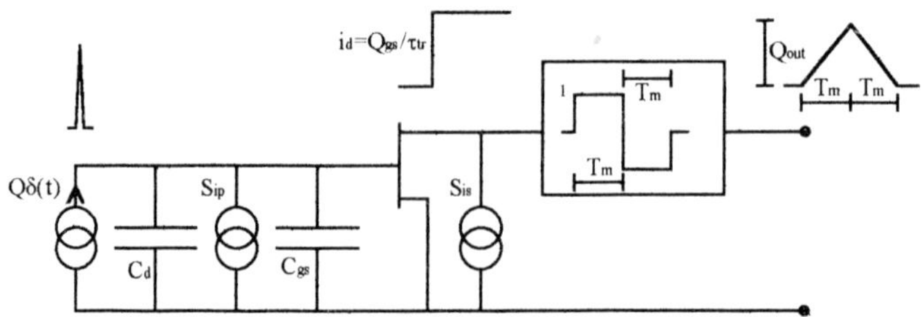

Fig.1 Schematic diagram of a detection system, with relevant signal waveforms.

The current noise generator S_{is} represents the thermal noise of the channel of the FET (or the collector shot noise of a BJT). Its white power spectrum (bilateral) is given by

$$S_{is} = \frac{di_s^2}{df} = \alpha\, 2kTg_m = \alpha\, 2kTC_{gs}\omega_t$$

where $g_m = C_{gs}\omega_t$ is the transconductance of the transistor and ω_t is the cut-off frequency (rad/sec). The coefficient α, of the order of the unity, depends on the transistor type.
The current noise generator S_{ip} has a white spectrum given by

$$S_{ip} = q(I_{leack\ det} + I_{leack\ tr} + I_{th\ res}) = qI_L$$

where $qI_{leack\ det}$ and $qI_{leack\ tr}$ are respectively the shot noise of the detector and transistor leakage current.

$qI_{th\ res}$ is the equivalent shot noise spectrum of the resistors connected to the input

$$S_{i\,th} = \frac{2kT}{R_f} = q\frac{2V_{th}}{R_f} = qI_{th\ res}$$

Non white, low frequency noise components due to the transistor (e.g. 1/f noise) are represented by a current noise generator $S_{i\omega}$ in parallel with S_{is} (the symbol of this generator has been omitted in the Fig.1). The approximating bilateral power spectrum is

$$S_{i\omega} = \frac{c_\omega}{|\omega|} = \alpha 2kTg_m \frac{\omega_1}{|\omega|}$$

where ω_1 is the frequency at which the 1/f noise equals the white thermal noise of the FET channel (or the white collector shot noise in BJT).

We neglect in this work other noise sources which can affect the resolution of the detection system, for instance the dielectric noise. Moreover we suppose, for simplicity, that all the considered noise sources are mutually uncorrelated.

3. The Resolution of a Detection System.

The resolution of the system, in the measurement of the charge delivered by the particle, can be characterised by giving the signal to noise ratio (S/N) at the output of the system or, more used, by giving the equivalent noise charge (ENC) referred to the input. The ENC is the charge, delivered by the detector, which makes the S/N at the output equal to the unity. The lower ENC, the higher is the system resolution. As we suppose that the three considered noise sources are mutually uncorrelated, the squared value of the ENC can be expressed as the sum of three independent contributions [1,2]

$$ENC^2 = ENC^2_{series} + ENC^2_{parall} + ENC^2_{s\omega}$$

$$ENC^2 = C_d\left(\sqrt{\frac{C_{gs}}{C_d}} + \sqrt{\frac{C_d}{C_{gs}}}\right)^2 \frac{\alpha 4kT}{\omega_t T_m} + \frac{2}{3} q I_L T_m + \frac{8}{\pi}\ln(2) C_d\left(\sqrt{\frac{C_{gs}}{C_d}} + \sqrt{\frac{C_d}{C_{gs}}}\right)^2 \alpha k T \frac{\omega_1}{\omega_t^2}$$

where ENC^2_{series} is the contribution of the thermal noise of the channel, ENC^2_{parall} is that of the shot noise at the input and $ENC^2_{s\omega}$ that of the 1/f noise of the transistor.The qualitative diagram of the Fig.2 shows the total ENC^2 and the contributions of the three independent sources, plotted as a function of the shaping time T_m (defined in Fig.1 as the half width of the system output pulse).

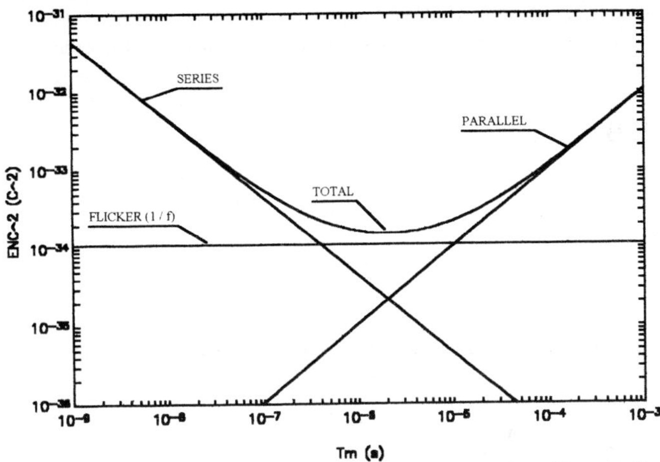

Fig.2 Equivalent Noise Charge (squared) versus filter shaping time. The contributions of the three independent noise sources are indicated.

It is possible to individuate an optimum shaping time T_{mopt} for a minimum ENC^2. The optimum shaping time T_{mopt} is the one for which the white series and parallel contributions are equal.

$$T_{m\,opt} = \sqrt{6}\left[\frac{\alpha kT}{q} C_d\left(\sqrt{\frac{C_{gs}}{C_d}} + \sqrt{\frac{C_d}{C_{gs}}}\right)^2 \frac{1}{I_L \omega_t}\right]^{\frac{1}{2}}$$

The ENC which is obtained for T_m=T_{mopt} represents the best resolution that can be obtained with the considered detection system. The value of ENC^2_{opt} is

$$ENC^2_{opt} = 4\sqrt{\frac{2}{3}}\left[\alpha kTqC_d\left(\sqrt{\frac{C_{gs}}{C_d}} + \sqrt{\frac{C_d}{C_{gs}}}\right)^2 \frac{I_L}{\omega_t}\right]^{\frac{1}{2}} + \frac{8\ln(2)}{\pi} kTC_d\left(\sqrt{\frac{C_{gs}}{C_d}} + \sqrt{\frac{C_d}{C_{gs}}}\right)^2 \frac{\omega_1}{\omega_t}$$

4. Detector and Transistor Parameters Affecting Resolution.

Both T_{mopt} and ENC^2_{opt} decrease by reducing the detector (plus parasitic and feedback) capacitance C_d. We will consider here low capacitance semiconductor vertex detectors, like microstrip and pixel detectors and semiconductor drift chambers. Their output capacitance range from the pF down to several tens of fF. The capacitive matching between detector and input transistor ($C_{gs} = C_d$) minimise both the optimum shaping time and the corresponding ENC.

Both T_{mopt} and ENC^2_{opt} are then reduced by increasing the cut-off frequency ω_t of the transistor. The series noise contribution to ENC^2 is in fact reversely proportional to ω_t. From this point of view the heterostructure FET transistor are the best candidates. They are in fact the highest bandwidth transistor readily available today. While the cut-off frequency of typical Silicon JFETs used in preamplifiers for radiation detection is of the order of 300 MHz, cut-off frequencies of the order of 50÷100 GHz are available with HEMTs and MESFETs. The cut-off frequency of the silicon BJTs reaches also several tenths of GHz, but they suffers for the noise contribution added by the base spreading resistance and by their relatively high base current.

The ENC^2_{opt} term degrades at increasing of the total leakage current I_L. The contribution of the shot noise of the transistor is significant only if its gate leakage current is comparable or greater than the detector leakage current plus the equivalent noise current of the resistor. A leakage current at room temperature of the order of 100 pA or lower per microstrip or per pixel is nowadays typical for silicon detectors. At least one order of magnitude higher current is reported for GaAs diodes. Typical 500 MΩ feedback resistors are equivalent to leakage currents of 100 pA. Therefore, while the leakage current of a typical JFET (of the order of 1 pA or lower) is completely negligible, the gate current of MESFET and HEMT, which ranges around the nA, gives with silicon detector a non negligible noise contribution. The BJT leakage current reaches the μA.

The ENC^2_{opt} is worsened by the presence of the low frequency noise of the transistor. The amplitude of the 1/f noise spectrum has been described in the present work as a function of the ratio ω_l/ω_t. For JFETs typically used in charge preamplifiers the corner frequency $f_l=\omega_l/2\pi$ ranges around a few kHz, therefore ω_l/ω_t is of the order of 10^{-5}. The corner frequency f_l of MESFET and HEMT can be as high as a few hundred of MHz, therefore ω_l/ω_t is of the order of 10^{-2}. This noise contribution is the dominant one with these devices.

5. Noise Measurements.

The HEMTs are usually adopted in high frequency circuits for telecommunications. For these applications the low frequency noise is generally of negligible interest. Therefore this aspect of the device characterisation is generally poor. Nevertheless it is worth remembering that phase noise in high frequency oscillators is affected by the low frequency noise of the devices used.
We present here the performances of the pseudomorphic HEMT ATF35076 from Avantek. Gate length is 0.25 μm, while the gate width is 200 μm. The chosen biasing condition were V_{gs}=0V and V_d=1V. In these conditions the device transconductance reaches a value of 90 mS. In order to precisely evaluate the cut-off frequency, defined as the ratio g_m/C_{gs}, we have estimated the value of C_{gs} on the basis of the reported values of the gate capacitance for these kind of devices, which is of the order of 1.2 pF/mm. We obtain a f_t of the order of 60 GHz. The drain current, in the chosen biasing conditions, is 30 mA, while the gate current is about 1 nA. A lower drain voltage (acceptable down to 0.5 V) could reduce the

gate current to slightly less than 1 nA. A reverse biasing of the gate, which could reduce the drain current and therefore the power dissipation, unfortunately increases rapidly the gate current.
In the chosen biasing conditions the low frequency output resistance R_O is of the order of 100 Ω. The maximum voltage gain G_V of a single stage amplifier, which is given by the product of the transconductance with the output impedance, is of the order of 10. In the frequency range of interest (approximately from zero to 500 MHz) we have verified that this approximation of G_V is reasonable.
A low value of the gain G_V of the input transistor in a multistage preamplifier (e.g. input cascode configuration) is a drawback. In fact in this case the noise contribution of the second stage could result not negligible. For comparison, the G_V with silicon JFETs can reach vales of several tenths.
The low frequency noise of the ATF35076 has been measured in the range from 10 Hz to 100 MHz.
The transistor was biased, in a suitable test fixture in order to avoid high frequency oscillations, through suitably filtered power supplies. The gate was a.c. short circuited. A 600 Ω load resistor was inserted at the drain. The output voltage noise at the drain was further amplified and its spectrum was measured by means of the HP 4195A analyser. The equivalent series noise spectrum at the input of the transistor, at 20 °C room temperature is shown in Figure 3. It has been obtained by dividing the measured output voltage noise spectrum by the squared modulus of the measured voltage gain from the gate to the output. The noise of the measurement system was subtracted. In the same Figure the ideal slope of the 1/f noise is plotted. Two regions of the spectrum with nearly ideal 1/f slope, separated by a region where the slope is lower, are apparent. This behaviour can be explained in terms of superposition of Lorentzian noise due to trapping centres characterised by different time constants [3]. The corner frequency f_l is determined as intercept of the 1/f interpolating curve with the ideal white noise (assuming $\alpha = 0.7$). If the shaping time T_m considered is shorter than 1μs (this is the case of interest), the 1/f interpolating curve to be considered is that related to the spectrum region at frequencies above the shown knee. The corner frequency results to be about 850 MHz.

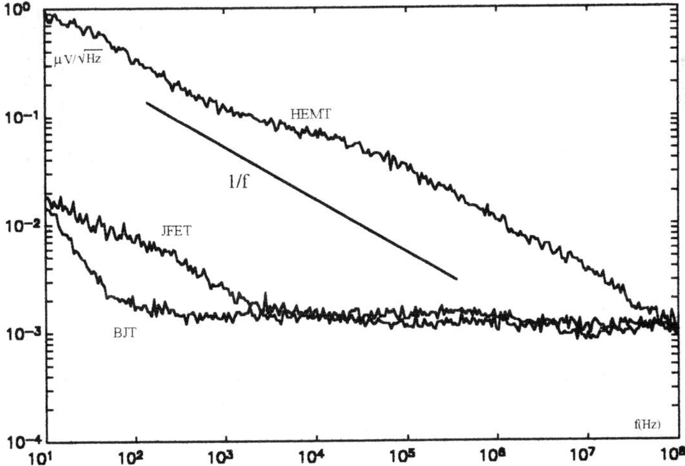

Fig.3 Comparison of the low frequency noise spectra of different transistors (quoted in the text).

For comparison the low frequency spectra of the JFET NJ26, the BJT BFY90 have been measured and shown in the same Figure 3, together with the spectrum of the ATF35076.
For this comparative measurement each transistor was biased in its more reasonable working point. The instrument noise was verified to be negligible in the considered frequency range. On the basis of these noise measurements and of the measurements of the other described relevant transistor parameters

(which we here omit to present) the ENC obtainable with the considered transistors has been evaluated. The results are shown in the Fig.4, for C_d=1pF. We observe the clear advantage in using HEMTs for shaping times shorter than several tens of nanoseconds.

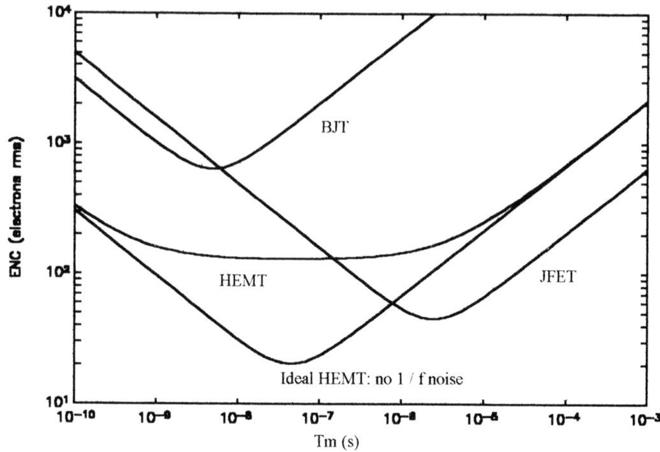

Fig.4 ENC (electrons r.m.s.) obtainable with the transistors considered in Fig.4.

5. Conclusion.

We have verified that heterostructure field effect transistors could be usefully employed in preamplifiers for radiation detectors in applications in which high detection rate is required. In fact for shaping times shorter than several tens of nanoseconds the resolution which can be obtained with HEMTs is higher than that obtainable with conventional silicon JFETs. These values of shaping times are today required for applications with the new high luminosity accelerators. The limiting factors with HEMT is their huge low frequency noise. Efforts in understanding and reducing this kind of noise are therefore welcomed.

References

1 - V.Radeka, *Ann. Rev. Nucl. Part. Sci.*, 38 (1988) 217

2 - E. Gatti, P.F.Manfredi, M.Sampietro and V.Speziali, *Nucl. Instr. & Methods,* NIM A297 (1990) 467-478

3 - A. Van Der Ziel, *Proc. of the IEEE* Vol 76, N.3 (1988), 233-258

Ensemble Monte Carlo simulation of 2D electron transport in a strongly-degenerate pulse-doped GaAs MESFET

Y Yamada and T Tomita

Department of Electrical Engineering and Computer Science, Kumamoto University, Kurokami 2-39-1, Kumamoto City 860, JAPAN

ABSTRACT: The properties of the quantum well structure and the electron transport of a strongly-degenerate 2D electron gas in the pulse-doped GaAs MESFET with a thin doped layer of 100*A* are theoretically studied by a self-consistent calculation of the Poisson's and Schrodinger's equations and by an ensemble Monte Carlo simulation using the exact 2D scattering rates. The theoretical results fairly agree with the experimental ones.

1.INTRODUCTION

Recently a pulse-doped GaAs MESFET(PD-MESFET) with a channel width of about 100*A* has received considerable interests because of its simpler structure than a high electron mobility transistor(HEMT), its fairly good performances and easy fabrication(Nakajima *et.al.*1990a). In contrast to HEMT, the PD-MESFET is based on a homostructure and the channel includes many dopants. So a high electron mobility cannot be expected. However this 2DEG channel can be highly doped to get the high performances owing to the use of a OMVPE process. The maximum electron density and sheet density reported till now exceed $4\times10^{18}cm^{-3}$ and $5\times10^{12}cm^{-2}$, respectively, which are fairly large compared with those in a conventional HEMT(Nakajima *et.al.* 1990b-c). So the electrons are strongly degenerate.

In the present work the quantum well structure of the 2DEG channel and the electron transport are theoretically studied by a self-consistent calculation of the Poisson's and Schrodinger's equations and by an ensemble Monte Carlo simulation(EMC) with the exact 2D scattering rates using a recent techniques for the degenerate system(Goodnick and Lugli 1988).

2.DEVICE STRUCTURE OF THE GaAs PD-MESFET

The device structure of the PD-MESFET used in the present work is shown in Fig. 1(Nakajima *et.al.* 1990a-c). The undoped p-GaAs buffer layer(10000*A*, N_A=$2\times10^{15}cm^{-3}$), the pulse-doped layer(100*A*) and the undoped *n*-GaAs cap layer(300*A*, N_D=$10^{12}cm^{-3}$) are successively grown by OMVPE on a semi-insulating GaAs substrate. The maximum electron density in the doped layer exceeds $4\times10^{18}cm^{-3}$. In addition, the annealing and lithography processes for device fabrication considerably decrease the sheet electron density of the pulse-doped layer by a factor of about two depending on the various process conditions. Thus the sheet density is very sensitive to the processes. So it is important to know how the quantum well structure depends on the activation efficiency of the dopants.

As details of the doping profile have not been published, the uniform and Gaussian profiles are assumed in the present work. The parameters included in the profiles are adjusted to give the sheet density evaluated by the Schubnikov-de Haas measurement.

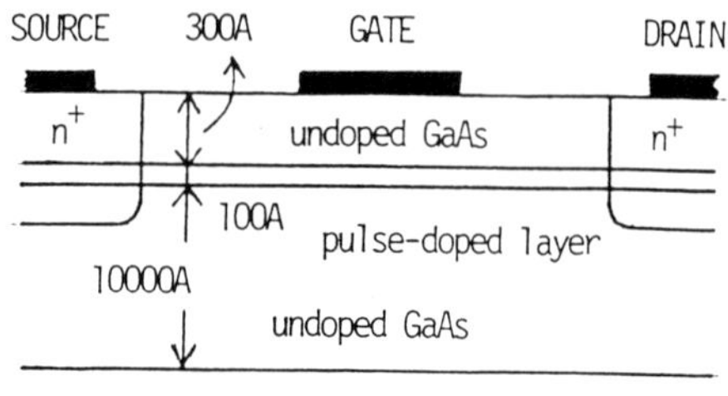

Fig. 1 Schematic drawing of the pulse-doped GaAs MESFET used in the present work.

3. QUANTUM WELL STRUCTURE OF THE 2DEG CHANNEL

The quantum well structure of the pulse-doped channel just grown by OMVPE before the annealing were calculated by a self-consistent solution of the Poisson's and Schrodinger's equations, without considering an impurity band. The effective potential acting on an electron consists of the electrostatic potential and the local exchange correlation potential. It is known from the Schubunikov-de Haas measurement that the sheet electron density up to the third subband is $5.1\times10^{12}cm^{-2}$ at $4.3K$. The parameters included in the doping profile $N_D(x)$ were determined to produce this sheet electron density. The shape of $N_D(x)$ of the uniform doping profile is illustrated in Fig. 2. Its peak density is $5.3\times10^{18}cm^{-3}$. The peak density and standard deviation of the Gaussian doping profile are $11.3\times10^{18}cm^{-3}$ and $1.9nm$, respectively. Table 1 shows the energy separations between the subband energies(E_m) and the Fermi level(E_F) and the subband populations(n_m) at 4.3 and $300K$. The values experimentally obtained by Nakajima *et.al.* (1990a) are also shown. Both at 4.3 and $300K$ at least the three subbands lie under the Fermi level. The fourth subband is very close to the Fermi level. Thus the electrons are strongly degenerate. The present peak density of the uniform profile is larger than the value of $4\times10^{18}cm^{-3}$ presented by Nakajima *et. al*(1990a-c). We confirmed that the present theory using $4\times10^{18}cm^{-3}$ could not reproduce the subband energies and populations experimentally obtained by them. The peak density of the Gaussian profile is unreasonably large considering the OMVPE process(1990a-c). Agreement between the calculated and experimental results is better for the present uniform profile than for the Gaussian one.

	4.3 K			300 K
	Exp.	Uniform	Gaussian	Uniform
n_1	2.9	2.85	3.08	2.61
n_2	1.6	1.48	1.37	1.27
n_3	0.6	0.59	0.55	0.55
n_4		0.07	0.00	0.30
$(\times10^{12}cm^{-2})$				
E_F-E_1	103.0	101.9	110.0	92.5
E_F-E_2	58.5	53.0	49.0	40.4
E_F-E_3	20.7	21.2	19.5	3.29
E_F-E_4		2.4	−2.9	−17.6
(meV)				

Table 1 Comparison between the theoretical and experimental results on the subband population(n_m) and the subband energy(E_m) for the lowest four subbands, where E_F is the Fermi level.

Figure 2 shows the position probability density of finding an electron in the potential well for the lowest three subbands at $4.3K$ for the uniform doping profile. Let us define an average background dopant density, $<N_D>$, given by

$$<N_D> = \sum_m \gamma_m N_{Dm}, \quad (1a)$$

$$N_{Dm} = \int_0^\infty N_D(x)\,|F_m(x)|^2 dx, \quad (1b)$$

where $N_D(x)$ is the doping profile, $F_m(x)$ is the wave function of the mth subband and γ_m is its population ratio. $<N_D>$'s are 3.2×10^{18} and $4.1\times10^{18} cm^{-3}$ for the uniform and Gaussian doping profiles at 300*K*, respectively. Thus it is expected that the impurity scattering rates are larger for the Gaussian doping profile than for the uniform one. N_{Dm} is the average background dopant density for an electron in the mth subband. N_{Dm}'s are 4.75×10^{18}, 2.74×10^{18}, 9.59×10^{17}, 6.54×10^{17}, 3.17×10^{17}, 3.92×10^{17}, $4.38\times10^{17} cm^{-3}$ for the lowest seven subbands at 300*K*, respectively. This indicates that up to the fifth subband the impurity scattering rate monotonically decreases with an increase of the subband index.

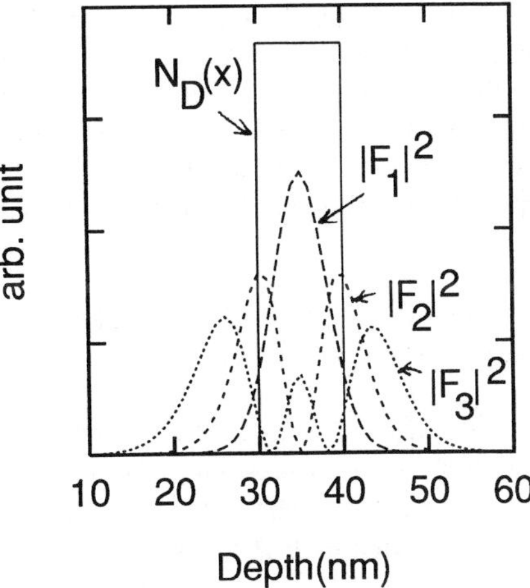

Fig. 2 The assumed uniform doping profile and the position probability density ($|F_m(x)|^2$) (m=1,2,3) of finding an electron in the potential well.

Figure 3 shows the total electron density profile at 300*K*. The solid line shows the experimental results obtained by the $C-V$ measurement(Nakajima *et. al.*1990c) and the others are the theoretical results. The result using the Gaussian profile looks close to the experimental one, while the result using the uniform profile produces the less peak electron density than the experimental one. However it cannot be considered that the Gaussian profile is more reasonable than the uniform one, considering the good results on the subband energies and populations and the unreasonably large peak density of the Gaussian profile. To achieve a better agreement between the theory and the experiment, more detailed analyses of the doping profile and the $C-V$ measurement may be needed.

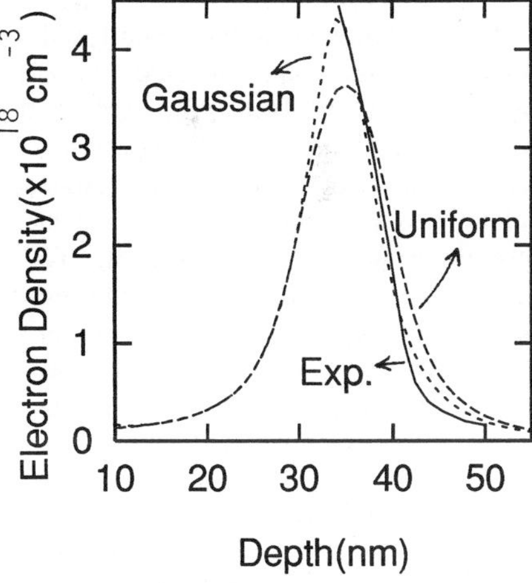

Fig. 3 Electron density profiles at 300*K*.

The dependences of the subband structures on the activation efficiency of the dopants at 300K are shown in Fig. 4. In this calculation $N_D(x)$ is simply multiplied by the activation efficiency. The subband energies and populations monotonically decrease with the activation efficiency. At 50% of the activation efficiency only the lowest subband is degenerate.

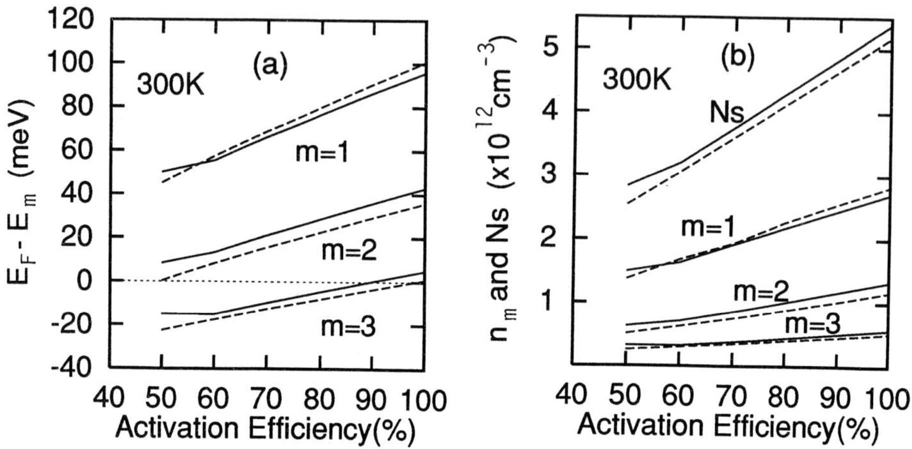

Fig. 4 Dependences of (a) the subband energies(E_F-E_m) measured from the Fermi level, and (b) the subband populations(n_m) and the total sheet density(N_s) upon the activation efficiency at 300*K* for m=1, 2, and 3. The solid and broken lines are obtained for the uniform and Gaussian doping profiles, respectively.

4.ELECTRON TRANSPORT

As the electron density is very high, degenerate effect was taken into account using a rejection technique(Goodnick and Lugli 1988). In addition to the acoustic phonon, polar optical phonon, and ionized impurity scatterings(Yokoyama and Hess 1986), plasmon(Artaki and Hess 1988) and electron-electron scattering(Goodnick and Lugli 1988) were also considered. All the scattering rates and the scattering angle probabilities were numerically calculated using the exact 2D wave functions. To calculate the ionized impurity scattering the Poisson's equation for the change in potential energy of the 2DEG was solved, which included the effects of the screening on the impurity charge by the free carriers with the screening constant S_m. S_m is expressed by n_m, E_F, E_m, and so on. The detailed description is given by Yokoyama and Hess(1986). Figure 5 shows the summations of the intra- and inter-subband scattering rates of the ionized impurity scattering for the lowest four subbands. The scattering rates are a little bit larger for the Gaussian doping profile than for the uniform one. The rate almost decreases with an increase of the subband index. These can be easily understood by the values of $<N_D>$ and

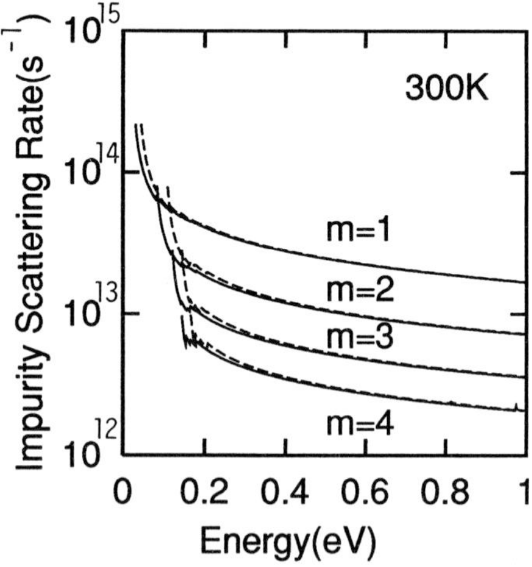

Fig. 5 Energy dependences of the ionized impurity scattering rates for each subband. The solid and broken lines are obtained for the uniform and Gaussian doping profiles, respectively.

N_{Dm} calculated in the last section. The ripples appeared in the lines show the contributions from the inter-subband scatterings. Thus the inter-subband scatterings are negligibly small. Although both the inter- and intra-subband electron-electron scatterings were taken into account, the inter-subband plasmon scatterings were neglected because the intra-subband plasmon scatterings themselves are small compared with the electron-electron scatterings, the ionized impurity scatterings and the polar-optical scatterings. The lowest seven subbands were considered in the ensemble Monte Carlo simulation.

A method using a numerical table of the electron distribution function in **k**-space, $f_m(\mathbf{k})$, for each subband to take the degeneracy effect into EMC may be an exact one in principle. When the ensemble Monte Carlo scheme is applied to the device simulation, however, the method requires much amount of computer resources. So we are particularly interested in an application of a "displaced" Fermi-Dirac distribution function to $f_m(\mathbf{k})$, which was originally proposed for a bulk electron system(Yamada 1991) and didn't need much amount computer resources.

Let us define the following "displaced" Fermi-Dirac distribution functions for each subband.

$$f_m(\mathbf{k}) = \frac{1}{1+\exp\dfrac{\varepsilon(\mathbf{k}-\mathbf{k}_m)-\varepsilon_{fm}}{k_B T_{em}}}, \qquad (2)$$

Here ε_{fm}, T_{em} and $\mathbf{k}_m$ are the effective Fermi level, the effective electron temperature and the average electron wave vector for the mth subband. These parameters can be given by using the following expressions(Yamada 1991).

$$\mathbf{k}_m \approx \frac{1}{N_m}\sum_{j=1}^{N_m}\mathbf{k}_j, \ (3a) \qquad n_m = \int_{E_m}^{\infty} g_m(\varepsilon)f_m(\varepsilon)d\varepsilon, \ (3b) \qquad \varepsilon_m = \frac{1}{n_m}\int_{E_m}^{\infty} \varepsilon g_m(\varepsilon)f_m(\varepsilon)d\varepsilon. \ (3c)$$

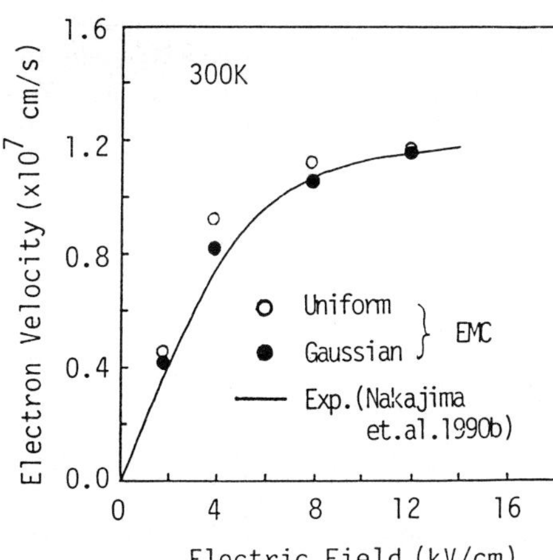

Fig. 6 Velocity-field characteristics. The solid line and circles show the experimental and theoretical results, respectively.

subband or valley	mobility	population ratio
	$cm^2/Vsec$	
1	1804 (1815)	0.469 (0.507)
2	2425 (2359)	0.221 (0.246)
3	3093 (2968)	0.108 (0.107)
4	3159 (2978)	0.070 (0.056)
L		0.021 (0.044)

Table 2 Mobilities of the subbands for the uniform profile evaluated under $2kV/cm$ at $300K$. The values in the parentheses are for the Gaussian profile.

These are obtained every step of the simulation. The ε_m and $f_m(\varepsilon)$ are given by

$$\varepsilon_m \approx \frac{1}{N_m}\sum_{j=1}^{N_m}\varepsilon_j, \qquad (3d) \qquad f_m(\varepsilon) = \frac{1}{1+\exp\frac{\varepsilon-\varepsilon_{fm}}{k_B T_{em}}}, \qquad (3e)$$

where $\mathbf{k}_j$ and ε_j are the electron wave vector and energy of the jth particle, respectively. ε_m and n_m are the average energy and the population of the mth subband. N_m is the number of particles. g_m and f_m are the density of state and the effective Fermi-Dirac distribution function, respectively. For simplicity, the kinetic energy is neglected because it is very small. T_{em} and ε_{fm} are updated every step during the simulation.

Figure 6 shows the velocity-field characteristics at 300*K*. The circles are obtained by the present EMC, while the solid line shows the experimental result(Nakajima *et.al.* 1990b). For a high field larger than $10kV/cm$, agreement between the theoretical and experimental results is good. For a low field the theoretical results are a little bit larger than the experimental one both for the doping profiles. To achieve a better agreement at the low field it may need to take a disorder scattering into the ensemble Monte Carlo simulation. Thus more studies are needed. Table 2 shows the low-field mobilities for electrons in the lowest four subbands and the population ratios evaluated under a uniform electric field at 300*K*. The mobility monotonically increases with the subband index up to the fourth subband. This can be easily understood from the impurity scattering rates shown in Fig. 5.

We are also interested in the temperature dependence of the low field mobility, which will be reported by us elsewhere.

5. CONCLUSIONS

The quantum well structure of the PD-MESFET has been theoretically studied assuming the uniform and Gaussian doping profiles for the pulse-doped layer. Good agreement between the experimental and theoretical results on the subband energies and populations has been achieved for the uniform doping profile. The dependences of the quantum well structure on the activation efficiency have quantitatively studied, which may be useful for the device technology. The transport properties have been evaluated using EMC. The calculated velocity-field characteristics at the high field greater than $10kV/cm$ have well agreed with the experimental ones, while the theoretical results of the low-field mobility are a little bit larger than the experimental one.

REFERENCES

Artaki M and Hess K 1988-II Phys. Rev. **B37** 6 pp2933-45

Goodnick S M and Lugli P 1988-I Phys. Rev. **B37** 5 pp2578-88

Nakajima S, Kuwata N, Nishiyama N, Shiga N, and Hayashi H 1990a Appl. Phys. Lett. **57** 13 pp1316-7

Nakajima S, Otobe K, Kuwata N, Shiga N, Matsuzaki K, and Hayashi H 1990b Proc. of the GaAs IC symposium pp35-40

Nakajima S, Otobe K, Kuwata N, Shiga N, Matsuzaki K, and Hayashi H 1990c IEEE Trans. MTT-s digest pp1081-1084

Yamada Y 1991 Electronics Lett. **27** 8 pp679-80

Yokoyama K and Hess K 1986 Phys. Rev. **B33** 8 pp5595-5606

Inst. Phys. Conf. Ser. No 136: Chapter 2
Paper presented at the Int. Symp. GaAs and Related Compounds, Freiburg, 1993

An analytical model for frequency-dependent drain conductance in HJFET's

Kazuaki Kunihiro, Hitoshi Yano, Hiroshi Nishizawa and Yasuo Ohno
Microelectronics Research Laboratories, NEC Corporation,
34 Miyukigaoka, Tsukuba, Ibaraki 305, Japan

ABSTRACT: An analytical model is developed for frequency dispersion of drain conductance (G_{DS}) in HJFET's. The model, which is based on a backgate-charge-control model and SRH statistics, clarifies the relation between the frequency-dependent G_{DS} characteristics and deep-level parameters. The measured frequency dependence of G_{DS} in an AlGaAs/InGaAs HJFET on MBE-grown layers shows good agreement with the present model. The deep-level properties, which cause the G_{DS} frequency dispersion, are identified by using the model.

1. INTRODUCTION

The frequency dispersion of drain conductance (G_{DS}) is often observed in GaAs MESFET's (Camacho *et al.* 1985) and HJFET's. This has posed serious problems for their analog and digital circuit applications. The phenomenon, which typically occurs in the low-frequency range below 1MHz, has been attributed to the slow response of deep-level traps in the semi-insulating substrate or at the channel/substrate interface. In order to achieve reliable GaAs device and IC performance, it is indispensable to understand the mechanism of G_{DS} frequency dispersion in relation to deep-level characteristics. Numerical device simulation is a practical approach (Li *et al.* 1991), but analytical modeling allows for a more direct understanding of the phenomenon.

In order to describe the frequency dependent G_{DS}, one proposed FET model includes a region that functions like a backgate terminal (Scheinberg *et al.* 1988). This pseudo-backgate terminal modulates channel current through backgate transconductance. Although this model is electrically reasonable, the circuit parameters are not explicitly related to the trap parameters and the device structures. M. Lee *et al.* (1990) developed an FET model for circuit simulation that includes trap parameters, but a nonphysical RC time constant was used to express the slow response of deep-level traps. This is insufficient for extracting deep-level properties from the measured G_{DS} frequency dependence. P. C. Canfield *et al.* (1990a) obtained the frequency dependence of G_{DS} by the Laplace transformation of the exponential response for electron emission. Although their model described the frequency dependence, the amplitude of G_{DS} variation was a parameter. Therefore, the model gave no information as to how to suppress dispersion.

In this paper, we develop an analytical model for frequency dependent G_{DS} based on SRH (Shockley-Read-Hall) statistics and a backgate-charge-control model. In the model, the frequency dependent G_{DS} characteristics are related to the deep-level parameters and the device structures. A comparison with the measured data is made and the trap energy level, which causes G_{DS} dispersion in an AlGaAs/InGaAs pseudomorphic HJFET, is estimated by using the model.

2. ANALYTICAL MODEL FOR G_{DS} FREQUENCY DISPERSION

In considering the band diagram and charge distribution of an ideal HJFET (Fig. 1), we used an undoped Liquid Encapsulated Czochralski (LEC) substrate. The deep-donor level (EL2) compensates for the residual shallow acceptors in the bulk, so the substrate is electrically neutral and exhibits semi-insulating (SI) characteristics. However, almost all the EL2 are neutralized near the epitaxial layer/substrate interface, so a net negative charge appears at the interface. In the backgate model, the edge of the interface charge region is supposed to act as a backgate terminal (Fig. 2), which modulates the channel current with a time constant determined by the capture and emission processes of deep-level traps.

The capture and emission processes are controlled by the rate equation for SRH statistics given by,

$$\frac{df_T}{dt} = nC_n(1 - f_T) - e_n f_T, \qquad (1)$$

where f_T is the electron occupancy ratio for the deep level, C_n is the electron capture coefficient, e_n is the electron emission rate, and n is the free electron concentration. The free electron concentration at the pseudo-backgate, under a small ac voltage application between the source and drain, is expressed as,

$$n = n_0 + \delta n = n_0(1 + \varepsilon e^{j\omega t}), \qquad (2)$$

where ω is an angular frequency of the signal, n_0 is the electron concentration at a steady state, and ε is a small constant. Assuming the Maxwell-Boltzmann distribution for electrons, δn is related to the backgate potential variation $\delta\psi_B$ as follows:

$$\delta n = \frac{e}{k_B T} n_0 \delta\psi_B, \qquad (3)$$

where k_B is the Boltzmann constant and T is temperature. Substituting eq. (2) into eq. (1), we obtain

$$\delta f_T = \frac{f_{T0}(1 - f_{T0})}{1 + j\frac{\omega}{\omega_0}} \frac{\delta n}{n_0}. \qquad (4)$$

Here, ω_0 is the deep-level characteristic angular frequency given by

$$\omega_0 = e_n + n_0 C_n, \qquad (5)$$

and f_{T0} is the electron occupancy ratio for a steady state obtained from $\frac{df_T}{dt} = 0$, as follows:

$$f_{T0} = \frac{n_0 C_n}{e_n + n_0 C_n}. \qquad (6)$$

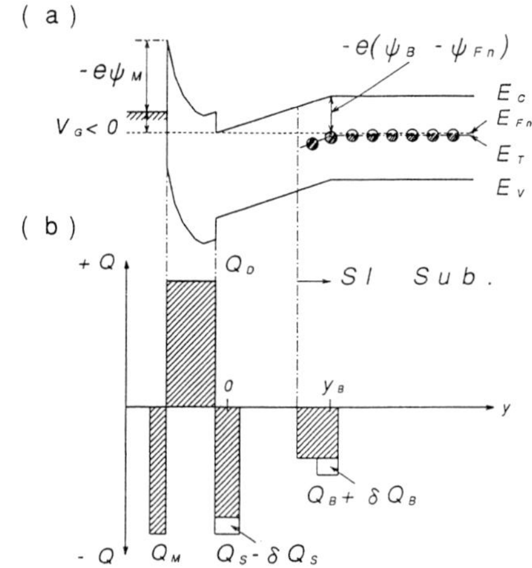

Fig. 1 (a) A band diagram and (b) charge distribution of an ideal HJFET. ψ_B is the backgate potential and y_B is its distance from the channel layer.

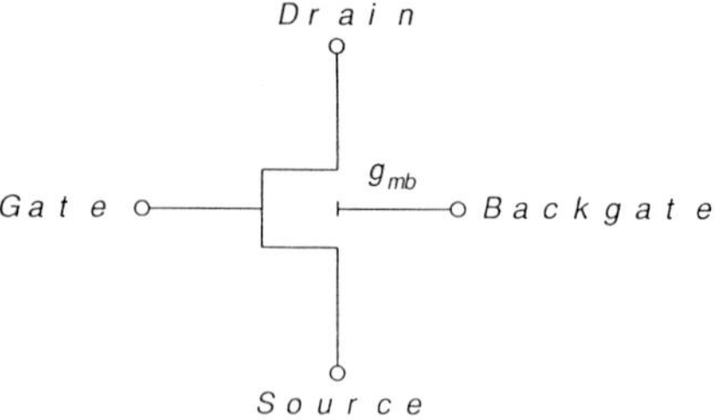

Fig. 2 An FET model with a pseudo-backgate terminal. A channel current is modulated through the backgate transconductance g_{mb}.

The interface charge variation δQ_B is connected with δf_T,

$$\delta Q_B = -eN_{DD}\delta f_T, \tag{7}$$

where N_{DD} is the sheet density of the deep-donor level at the backgate region.

The backgate potential variation has two components: one is two-dimensional modulation by drain voltage variation δV_{DS}, and the other is electrostatic potential variation by the interface charge. Therefore, it is expressed as

$$\delta\psi_B = \alpha\delta V_D + \frac{y_B}{\epsilon_s}\delta Q_B, \tag{8}$$

where α is a device-structure dependent parameter connecting $\delta\psi_B$ with δV_{DS}, y_B is the distance between the channel layer and the backgate and ϵ_s is the permittivity of GaAs.

The channel charge variation δQ_S is related to the backgate potential variation as

$$\delta Q_S = -\frac{\epsilon_s}{y_B}\delta\psi_B. \tag{9}$$

Therefore, the drain current variation δi_D is given by

$$\delta i_D = -v_s W_g \delta Q_S = g_{mb}\delta\psi_B, \tag{10}$$

where v_s is the electron velocity in the channel, W_g is the gate width and g_{mb} is the backgate transconductance defined as

$$g_{mb} = \frac{\epsilon_s v_s W_g}{y_B}. \tag{11}$$

The following discussion considers the drain-current-saturation region where v_s is constant.

From eqs. (3), (4), (7), (8), and (10), admittance between the source and drain Y_{DS} is given by

$$Y_{DS} = \frac{\delta i_D}{\delta V_D} = \alpha g_{mb}\frac{1 + j\frac{\omega}{\omega_0}}{1 + K + j\frac{\omega}{\omega_0}} = G_{DS}(\omega) + jB_{DS}(\omega), \tag{12}$$

where K is defined as

$$K = \frac{e^2 N_{DD} f_{T0}(1 - f_{T0}) y_B}{\epsilon_s k_B T}. \tag{13}$$

In eq. (12), conductance G_{DS} is expressed as

$$G_{DS}(\omega) = \alpha g_{mb}\left\{1 - \frac{K(1+K)\omega_0^2}{(1+K)^2\omega_0^2 + \omega^2}\right\}, \tag{14}$$

and susceptance B_{DS} is expressed as

$$B_{DS}(\omega) = \frac{\alpha g_{mb} K \omega_0 \omega}{(1+K)^2\omega_0^2 + \omega^2}. \tag{15}$$

Using eq. (14), the amplitude of G_{DS} variation from dc ($\omega \to 0$) to high frequency ($\omega \to \infty$) is given by

$$\Delta G_{DS} = G_{DS}(h.f.) - G_{DS}(dc) = \frac{\alpha g_{mb} K}{1+K}. \quad (16)$$

From eqs. (13) and (16), it is predicted that a lower trap density (N_{DD}), a longer distance between the channel layer and the pseudo-backgate (y_B), and a completely occupied or unoccupied deep-level ($f_{T0} = 0, 1$) will suppress the G_{DS} frequency dispersion.

Another method for suppressing the dispersion is to let α in eq. (16) equal zero. In fact, it has been shown that a GaAs MESFET with a fixed potential p-well beneath the channel layer, which corresponds to the case of $\alpha = 0$ in eq. (16), exhibited no G_{DS} frequency dispersion (Canfield *et al.* 1990b).

3. MEASUREMENT

The model was compared with the measured data. The device is a 0.3 μm × 100 μm gate AlGaAs/InGaAs pseudomorphic HJFET fabricated by MBE on an undoped LEC GaAs substrate. The MBE layers consist of a 4500 Å undoped GaAs buffer layer, a 150 Å undoped $In_{0.15}Ga_{0.85}As$ channel layer, a 650 Å Si-doped $Al_{0.2}Ga_{0.8}As$ layer (Si : $2 \times 10^{18} cm^{-3}$) and a 800 Å Si-doped GaAs cap layer for ohmic contact (Si : 3×10^{18} cm^{-3}) as shown in Table I.

Table I Structure of AlGaAs/InGaAs pseudomorphic HJFET epitaxial layers grown by MBE.

n^+-GaAs		3×10^{18} cm^3	800 Å
n^+-$Al_{0.2}Ga_{0.8}As$		2×10^{18} cm^3	650 Å
i-$In_{0.15}Ga_{0.85}As$		-	150 Å
i-GaAs		-	4500 Å
S.I.-GaAs	Sub.	Undoped	LEC

The measurement configuration is shown in Fig. 3. The conductance between the source and drain was measured with an impedance analyzer in the frequency range of 1Hz - 1MHz. The amplitude of the drain-modulation ac signal δV_{DS} combined with the dc voltage is typically 70mV (RMS). The capacitor parallel to the dc power supply blocks the gate voltage V_{GS} variation from the drain ac signal.

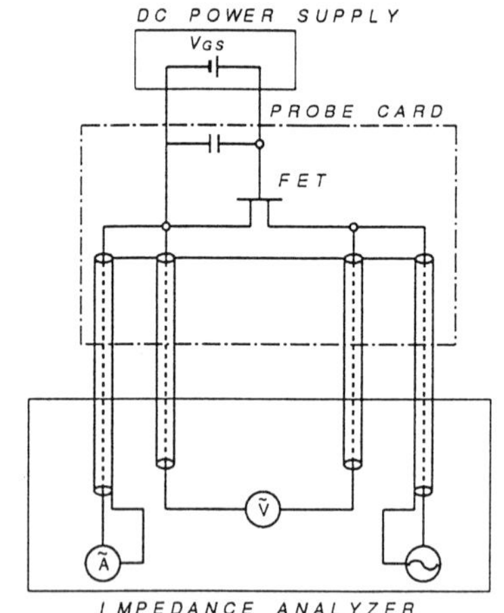

Fig. 3 Measurement configuration used to measure source-drain conductance.

A comparison between the typical measured data and the fitted data is shown in Fig. 4. From eqs. (14) and (15), the analytical expressions should be as follows:

$$G_{DS}(\omega) = a_0 - \sum_i \frac{a_{1i} a_{2i}}{a_{2i}^2 + \omega^2} \quad (17)$$

$$B_{DS}(\omega) = \sum_i \frac{a_{1i} \omega}{a_{2i}^2 + \omega^2}. \quad (18)$$

The subscript i indicates the number of the trap when more than a single trap contributes to G_{DS} frequency dispersion. Here, a_0, a_{1i} and a_{2i} are parameters connected with the device and

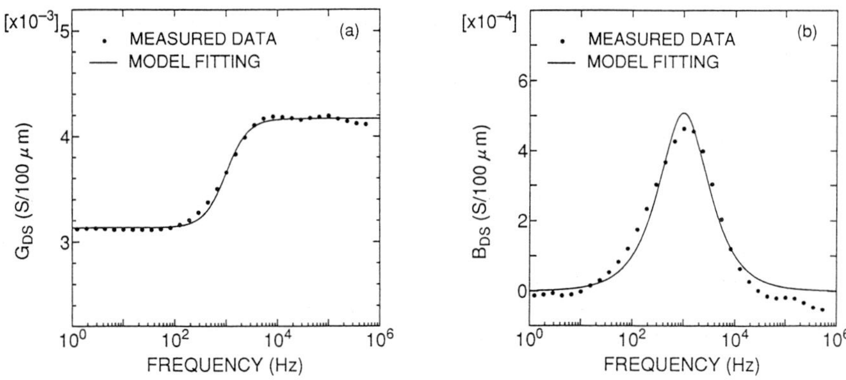

Fig. 4 Comparison of (a) drain conductance G_{DS} and (b) susceptance B_{DS} between the measured data (closed circle) and the results fitted to the present model (solid line). The parameters used are $a_0 = 4.2 \times 10^{-3}[S], a_{11} = 1.0[Ssec^{-1}]$, and $a_{21} = 9.9 \times 10^2[sec^{-1}]$. V_{DS}=2V, V_{GS}=-0.8V.

trap parameters as follows:

$$a_0 = g_{d0} + \sum_i \alpha_i g_{mbi} \quad [S] \tag{19}$$

$$a_{1i} = \alpha_i g_{mbi} K_i \omega_{0i} \quad [Ssec^{-1}] \tag{20}$$

$$a_{2i} = (1 + K_i)\omega_{0i} \quad [sec^{-1}], \tag{21}$$

where g_{d0} indicates a dc drain conductance, which may arise from short channel effects.

The measured frequency dependence of G_{DS} and B_{DS} shows a good agreement with the fitted result by assuming a single trap ($i = 1$) in eqs. (17) and (18). In order to obtain the value of $\alpha_1 g_{mb1}$ by using the relation in eq. (19), we need another technique for distinguishing g_{d0} from $\alpha_1 g_{mb1}$. Instead of using the fitted result of a_0, $\alpha_1 g_{mb1}$ was estimated as 1.2×10^{-3} S from eq. (11) by assuming that $y_B = 0.5$ μm, $v_s = 10^7$ cm/sec and $\alpha_1 = 0.5$. Using this $\alpha_1 g_{mb1}$ value and the relations in eqs. (20) and (21), $K = 7.2$ and $\omega_0 = 1.2 \times 10^2$ sec^{-1} are obtained. From eq. (13), the trap sheet density is evaluated as 1×10^{11} cm^{-2} when $f_{T0} = 0.5$.

Temperature dependence of the G_{DS} dispersion was measured from room temperature to 120°C. The shift of the G_{DS} changing point was observed at higher temperatures as shown in Fig. 5. From eq. (15), the frequency f_0, where susceptance B_{DS} is maximum, is given by $\frac{(1+K)\omega_0}{2\pi}$. Though f_0 is determined not only by emission but also by capture processes, f_0 exhibits almost the same temperature dependence as the emission rate, if the quasi-Fermi level for electrons remains near the deep-level throughout the measured temperature range. Therefore, the activation energy can be determined from the Arrhenius plot for the characteristic frequency. The activation energy is found to be 0.34 eV (Fig. 6), which is different from that for the EL2 level ($E_T = 0.82$ eV) dominant in the undoped LEC substrate. The possible origin of the level is oxygen at the epitaxial layer/substrate interface. This is because a substantial amount of oxygen was detected at the MBE interface by SIMS analysis and oxygen is formed in the deep-level with energy of about 0.4 eV (Arikan *et al.* 1980).

Information about the trap properties is essential for investigating the cause of dispersion and suppressing it. This model can satisfactorily extract the trap parameters, such as energy level and concentration, from the measured data.

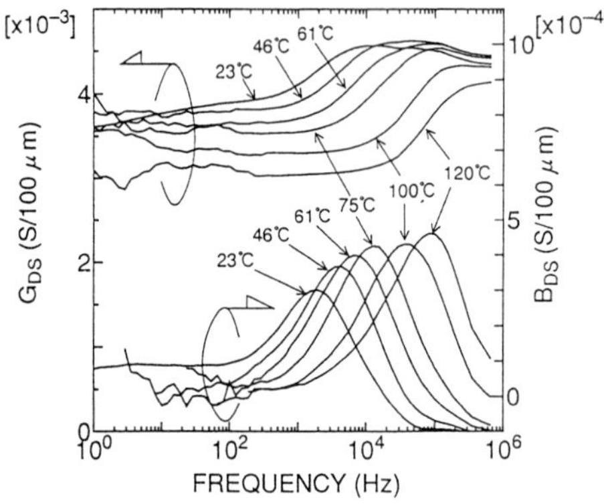

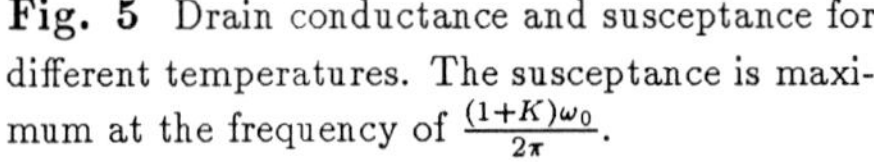

Fig. 5 Drain conductance and susceptance for different temperatures. The susceptance is maximum at the frequency of $\frac{(1+K)\omega_0}{2\pi}$.

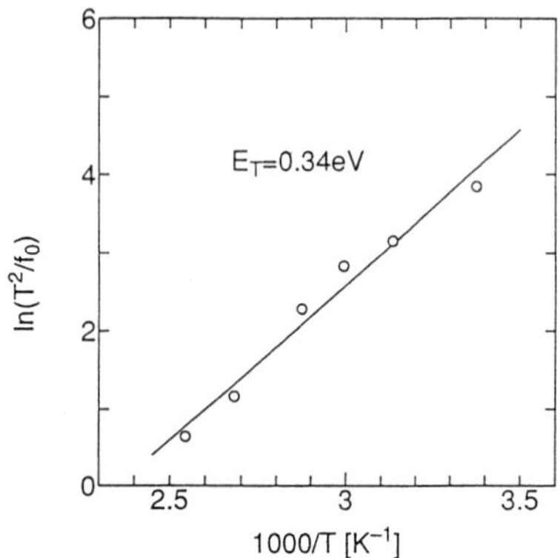

Fig. 6 Arrhenius plot for the characteristic frequency. The activation energy is determined to be 0.34 eV.

4. SUMMARY

An analytical model for the frequency dependent drain conductance in HJFET's was developed. In this model, the frequency-dependent G_{DS} characteristics are related to the deep-level parameters and the device structures. Therefore, the trap properties which cause the frequency dispersion can be extracted by using this model. In fact, it showed good agreement with the measured frequency dependence of G_{DS} in the AlGaAs/InGaAs pseudomorphic HJFET. By fitting the data measured at different temperatures to the model, the deep-level properties were estimated: $N_{DD} \simeq 1 \times 10^{11}$ cm^{-2} and $E_T \simeq 0.34$ eV. The frequency dependent G_{DS} is undesirable for the precise design and the reliable performance of GaAs digital IC's and MMIC's since those applications require uniform performances for wide frequency ranges. This model is useful for designing optimum device structures free from the influence of trap-induced G_{DS} dispersion.

ACKNOWLEDGEMENTS

The authors would like to thank Dr. K. Uetake for providing the samples, and Dr. N. Goto for his valuable discussion. The authors also thank Dr. T. Nozaki for his continuous support and encouragement.

REFERENCES

Arikan M C, Hatch C B and Ridley B K 1980 J. Phys. C: Solid St. Phys., **13** 635
Camacho-Penalosa C and Aitchison C S 1985 Electron. Lett. **21** 528
Canfield P C, Lam S C F and Allstot D J 1990a IEEE J. Solid-State Circuits **25** 299
Canfield P C and Allstot D J 1990b IEEE J. Solid-State Circuits **25** 1544
Lee M and Forbes L 1990 IEEE Trans. Electron Devices **37** 2148
Li Q and Dutton R W 1991 IEEE Trans. Electron Devices **38** 1285
Scheinberg N, Bayruns R and Goyal R 1988 IEEE J. Solid-State Circuits **23** 605

Paper presented at the Int. Symp. GaAs and Related Compounds, Freiburg, 1993

The selectively grown GaAs permeable junction base transistor with a homoepitaxial gate

J. Gräber, G. Mörsch, H. Hardtdegen, M. Hollfelder, M. Pabst, H. Lüth
Institut für Schicht- und Ionentechnik (ISI)
Forschungszentrum Jülich GmbH, Postfach 1913, D-52425 Jülich, FRG

ABSTRACT: Design and processing of the **P**ermeable **J**unction **B**ase **T**ransistor (PJBT) - a novel structure of GaAs-PBT- are presented. Transconductances for this "vertical FET" of 240 mS/mm are achieved both for channels with the usual finger structure and cylindrical channels. Strong dependencies of device characteristics on contact configuration and power densities can be attributed to the velocity overshoot of the electrons. Cut-off frequencies $f_T = 6$ GHz and $f_{max} = 1.5$ GHz have been achieved for 0.6 μm wide channels prepared by optical lithography. By further reduction of channel size and parasitic resistances and capacities a significant improvement of the high frequency performance of this device is to be expected since the transconductance is higher compared to different PJBTs.

1. Introduction

The PJBT consists of vertical conducting channels that penetrate well defined openings of a highly p-doped GaAs grating. This grating serves as gate to control the vertical current of this unipolar device (Fig. 1). The gate and channel length can be defined in the sub-μm range by layer thickness during epitaxial growth. Due to these short distances and the expected velocity overshoot very high frequency operation can be expected.

The PJBT is a new device concept that is based on a modified PBT structure as introduced by Bozler et. al (1979). In case of the PBT a tungsten (W) grating is embedded in GaAs by epitaxial overgrowth. Numerical calculations predict $f_T \approx 200$ GHz and even higher f_{max} for lateral grating periodicities of 320 nm. Experimentally, however, only $f_T \approx 40$ GHz has been achieved by Hollis et al (1987) because effective surface cleaning is difficult due to the different materials and possible underetching. Additionally thermal stress and impurities (Fe, Cr, Cu, Au, Te etc.) from the tungsten diffusing and generating deep traps in the channel during overgrowth reduce the carrier mobility. Thus, technological problems of the hetero-system GaAs/W lead to limitations in device performance.

These problems can be avoided in the PJBT where the homoepitaxial gate (p^{++}-GaAs) and the channel form a pn-junction. Since the diffusion of the p-type dopant carbon is very low and no metallic impurities are present precise doping combined with high mobility can be obtained inside the channel.

The gate of the PJBT can be doped up to $p^{++} \geq 10^{20}$ cm^{-3} using carbon as dopant in MOMBE. Although the specific resistivity (≈ 700 μΩcm) is still higher than for tungsten the RC time constant can be decreased by shorter gate fingers and by imbedding the p^{++}-gate into intentionally undoped layers. Since these layers are totally depleted the extension of the space charge region is enlarged and, hence, the capacity is decreased leading to higher f_T.

Applying selective epitaxy a planar device and a precise control of the channel length can be obtained. Furthermore no etch stop is required for contacting the gate because in the masked areas no deposition takes place and, hence, the depth of the gate remains unaltered. As PJBTs with either n or p-type channels can be built, selective overgrowth enables the sequential growth of n- and p- type PJBTs on one chip. Integration with other devices is also possible.

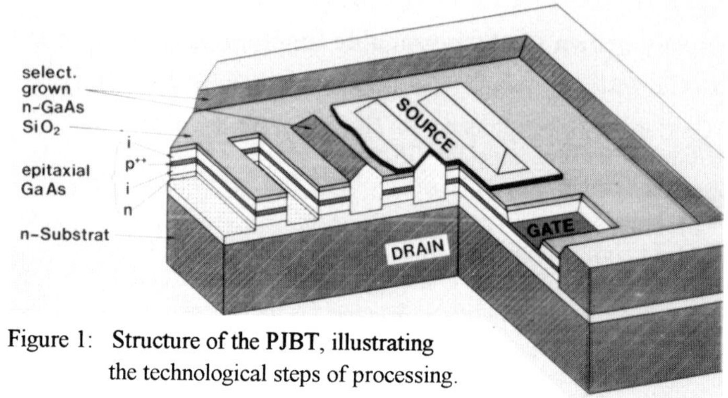

Figure 1: Structure of the PJBT, illustrating the technological steps of processing.

2. Technology of the PJBT

The sequence of processing and the final geometry of the realised structures that are not designed for high frequency operation but for simple processing is illustrated in Fig 1. The single steps of preparation are as follows:

A n^+n i $p^{++}i$ doped sequence of GaAs ($n^+ = 2 \cdot 10^{19}$ cm^{-3}, 2 µm; $n = 3 \cdot 10^{17}$ cm^{-3}, i : $p \leq 10^{15}$ cm^{-3}, $p^{++} = 2 \cdot 10^{20}$ cm^{-3}, 200 nm each) was grown on a semi-insulating GaAs (100) wafer using MOMBE. After deposition of 90 nm SiO_2 the gate structure with openings down to 0.4 µm was transferred into the SiO_2 using optical lithography, reactive ion etching (RIE) with CHF_3 and an O_2-plasma to remove the resist. Subsequently the SiO_2 structure served as a mask for etching pits down into the n-GaAs with CH_4/H_2-RIE followed by a second O_2-plasma to remove the polymer. The wafer was slightly etched in $H_2SO_4:H_2O_2:H_2O$ before the pits were selectively filled ($n \leq 8 \cdot 10^{16}$ cm^{-3}) in MOVPE. For source and drain ohmic contacts Ni/Au/Ge//Ti/Au : 5nm/10nm/20nm//20nm/100nm were evaporated and annealed at 380°C. After opening the gate was contacted by Ti/Au. For high frequency measurements a 1.5 µm thick polyimide layer was structured and coplanar stripes were evaporated for direct probing.

The SiO_2 exhibits a few technological advantages. The pattern generation is improved because the selectivity of the CHF_3-RIE process is high and hence a smooth resist profile can be converted to a sharp profile in the SiO_2. This is a good mask for CH_4/H_2-RIE in contrast to the resist, that is degrading during the etch process at a power ~2 mW/cm^2 and a bias ~ -600V. With the SiO_2 mask we achieved smooth surfaces and vertical walls. Suitable overgrowth conditions were found for selective growth (Fig. 2) by Mörsch et al (1993).

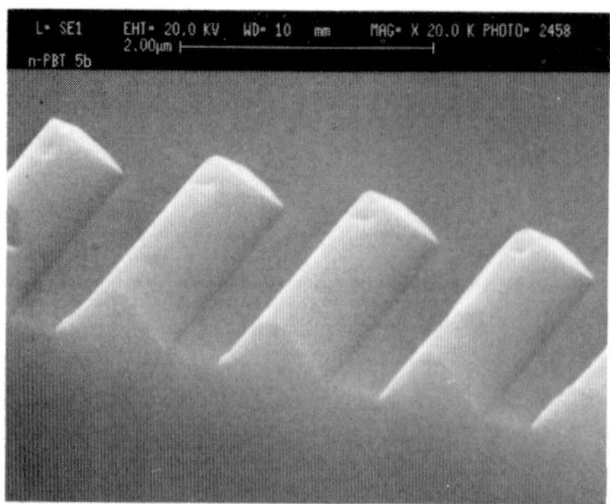

Figure 2: Selectively grown finger structure (left) and cylindrical channels (right)

3. DC-Characteristics

The output characteristic and transconductance measurements have been performed with an automated Semiconductor Parameter Analyser (Hewlett Packard 4145B) by placing probe tips onto the metallization. As the actual doping N_D can not be measured in the channels it has been determined in unstructured areas by CV-profiling.

For high cut-off frequencies it is essential to have a good transconductance that has its maximum when the gate is forward biased. In this case the drain current I_D may be described by the drift saturation model:

$$I_D = q \cdot N_D \cdot v_{sat}(L, U_{DS}, T) \cdot (b_k - 2w) \cdot Z, \qquad \text{(Eq. 1)}$$

where v_{sat} is an effective saturation velocity that depends on an effective gate length L, the drain to source voltage U_{DS} and temperature T. The cross-section of the channel is given by the total gate width Z, i.e. the accumulated width of all gate fingers times b_k-2w(U_{GS}) i.e. the width of the channel b_k minus twice the depletion width w that can be controlled via the gate voltage U_{GS} .

For PJBTs with different doping N_D the drain current has been measured. The change in I_D can be correlated to the change in N_D and w(N_D) only (see Eq. 1). However, this is not the case when the channel width b_k is changed. All PJBTs with equal N_D have been processed under identical conditions.

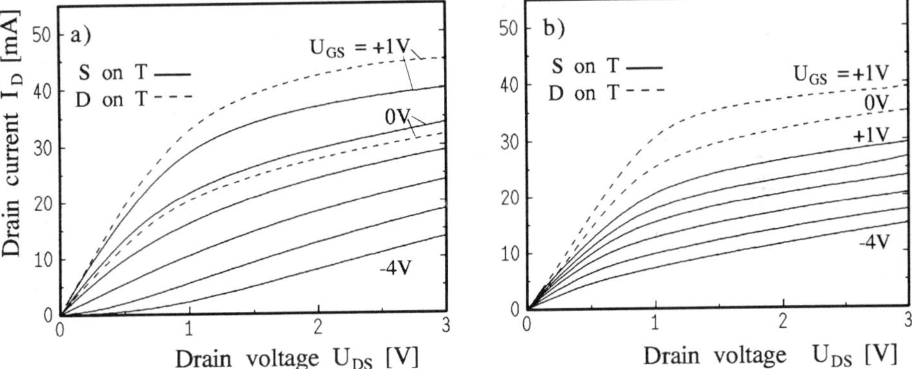

Figure 3: Output characteristics for PJBTs. The channel doping is $N_D = 8 \cdot 10^{16} cm^{-3}$
a) $b_k = 0.6$ µm $Z = 0.168$ mm b) $b_k = 1.0$ µm $Z = 0.096$ mm

Comparing I_D for different PJBTs where N_D and the product $Z \cdot b_k$ are kept constant, experimentally a higher I_D results for smaller channels (see Fig. 3 full lines). According to Eq. 1, however, higher I_D should be expected for wider channels because the term (b_k-2w)·Z is bigger for wider channels. This contradiction can be explained by different effective drift velocities v_{sat} that depend on L. Due to the facets building up during epitaxy (see Fig. 2) the distance between the top contact (source) and gate strongly depends on the channel width. This is important, because at high electric fields ($E \geq 5$ kV/cm) only for short distances (≤ 0.5µm) electrons can move faster due to the velocity overshoot that is discussed later.

When the contact configuration is changed so that the drain is at the top contact (D on T) the distance from source (substrate) to gate does no longer depend on b_k . In this case I_D is increased for D on T and the relative increase is higher for wider channels (see Fig. 3 dotted lines) in accordance to the bigger difference in L.

For a similar device with $Z = 0.240$ mm which is ≈ 1.5 times higher than in Fig. 3 a) the external transconductance per unit length g'_m is plotted in Figure 4 for both contact configurations. Maximum transconductances of g_m=80 mS/mm and 160 mS/mm are achieved.

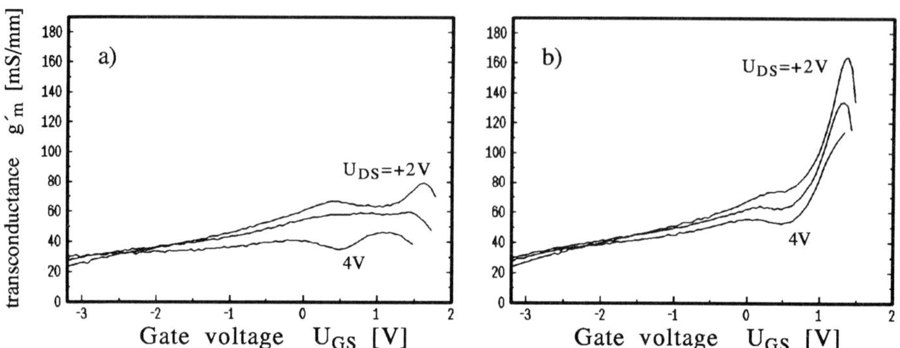

Figure 4: Transconductance per unit length of one PJBT: N_D=$8 \cdot 10^{16}$cm^{-3}, $Z = 0.240$mm, $b_k = 0.6\mu$m
a) S on T; $R_S = 3.5\ \Omega$ b) D on T; $R_S = 1.3\ \Omega$

Due to a voltage drop along the contact resistances R_S and R_D the measurable external transconductance g_m is reduced by R_S and R_D and by the parallel conductance of the channel G_{DS} :

$$\left.\frac{dI_{DS}}{dU_{GS}}\right|_{U_{DS}} = g_m = \frac{1}{\frac{1}{g_{mi}} + R_S + \frac{G_{DS}}{g_{mi}} \cdot (R_S + R_D)} \qquad \text{(Eq. 2)}$$

Thus g_m is limited by the intrinsic transconductance g_{mi} and by contact resistances i.e. especially by R_S as G_{DS}/g_{mi} is small. Since for S on T R_S is three times higher than for D on T a higher transconductance results in the latter case. As R_S, R_D ,G_{DS} and g_m can be measured the intrinsic transconductance that depends on channel properties only can be evaluated. Surprisingly significantly distinct values are found for S on T: $g_{mi} \approx 130$ mS/mm and D on T: $g_{mi} \approx 240$ mS/mm. In the drift saturation model g_{mi} can be derived from Eq. 1 :

$$g_{mi} = \frac{\partial I_{DS}}{\partial U_{GS}} = q \cdot N_D \cdot Z \cdot \left\{ \frac{\partial v_{sat}}{\partial L} \cdot \frac{\partial L}{\partial U_{GS}} \cdot (b_k - 2w) - 2 \cdot v_{sat} \cdot \frac{\partial w}{\partial U_{GS}} \right\} \qquad \text{(Eq. 3)}$$

A higher transconductance for D on T may be explained by higher v_{sat} caused by the velocity overshoot (Hess 1988). For varying L that depends on U_{GS} v_{sat} may increase further. For D on T configuration L is shorter and hence v_{sat} can be higher if L is short enough. Thus the difference in g_{mi} may be caused by the velocity overshoot.

In order to clarify whether the velocity overshoot occurs in the PJBT Monte Carlo Simulations have been performed and the electron drift velocity averaged over the channel cross-section has been calculated. For ($N_D = 1 \cdot 10^{16}$ cm^{-3}) v(x) and the geometry of the channel are plotted in Fig. 5. Under the gate a velocity overshoot with $v \sim 4.5 \cdot 10^7$ cm/s at $\overline{E}$=15kV / cm is reached, that is twice as high as the peak velocity for stationary transport in GaAs. Behind the position $x \sim 0.3\mu$m v(x) is reduced due to scattering. For S on T configuration, where the source to gate distance is increased, the electrons can pass the gate at lower velocities only. Therefore the effective drift velocity is reduced and I_D and g_m are decreased.

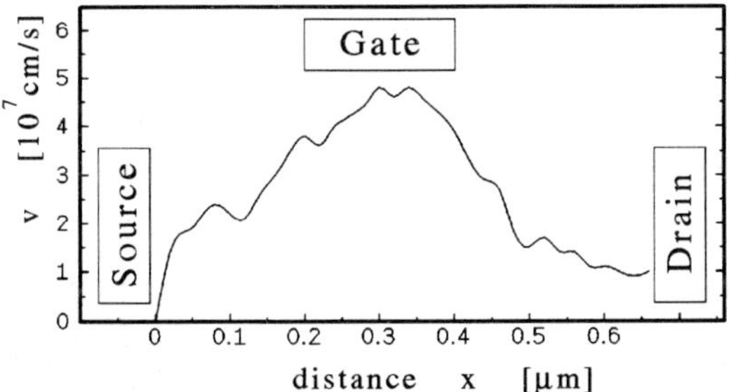

Figure 5: Velocity overshoot calculated with Monte Carlo Simulation for a PJBT. v(x) is averaged over a cross-section at distance x. b_k= 0.6 μm, $N_D = 1 \cdot 10^{16}$ cm^{-3}, U_{DS} = 1 V, U_{GS} = 0 V

The velocity overshoot arises from the smaller effective mass of the electrons and their higher mobility in the Γ-valley compared to the L-valley. At the source most electrons are located in the Γ-valley. But along their way they gain enough energy so that scattering into the "slower" L-valley is possible and the main drift velocity is reduced. However for short distances this process rarely takes place and therefore higher velocities can be obtained than for stationary transport. For increasing temperature the fraction of "slow" electrons is increaseed and additionally phonon scattering is enhenced so that v_{sat} is reduced at higher temperatures.

In some devices with a high power density a negative differential resistance (NDR) could be recorded. For a device (b_k = 0.4 μm, Z = 20·12 μm) a drain current I_D = 63 mA at U_{DS} = 2.5 V decreases to I_D = 54 mA at U_{DS} = 6.0 V . By estimating the thermal resistance an temperature increase of $\Delta T \approx 35$ K is estimated that is confirmed by a thermally induced increase of the gate current (~ exp(U/kT)).While the drain current decreases by 14% only the reduction of the transconductance is bigger than a factor of three.

Vice versa higher transconductances can be obtained for devices with lower power density. When the width of the gate finger is doubled (b_F = 1.2 μm, b_k = 0.6 μm). g'_m is improved to 220 mS/mm (see Fig. 6a, 6c) compared to 160 mS/mm at single gate finger width (Fig. 4). For comparison: Hollis had g'_m = 200 mS/mm and reached f_T = 42 GHz and f_{max}= 197 GHz.

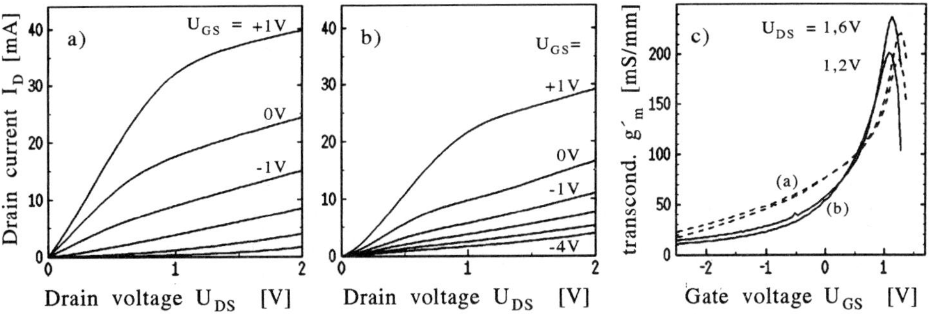

Figure 6: Output characteristics for PJBTs ; $N_D = 8 \cdot 10^{16}$ cm^{-3}; D on T
a) finger structure; b_k =0.156mm, b) cylindrical channels; b_k =0.125mm c) norm. transcond.

Even better, g'_m can be obtained for cylindrical channels of diameter 1 μm where the gate width is defined as Z:= π · R (Fig. 6b, 6c). The cylindrical structures are wider than the finger structures are. Therefore lithography is easier but they cannot be pinched off any more. For both devices the dip in transconductance that occurred at $U_{DS} \approx 1$ V in Fig. 4 is vanished as the power density is reduced. Now for positive gate voltages the transconductance rises continuously as the increase by the velocity overshoot is higher than the negative feedback of the thermal effect.

4 High Frequency Performance

The cut-off frequencies f_T and f_{max} have been determined by S-parameter measurements for devices with varying gate width and gate metallization (grey areas in Fig. 7). The opening of the polyimide i.e. the contact area to the coplanar stripe line is indicated by black rectangulars.

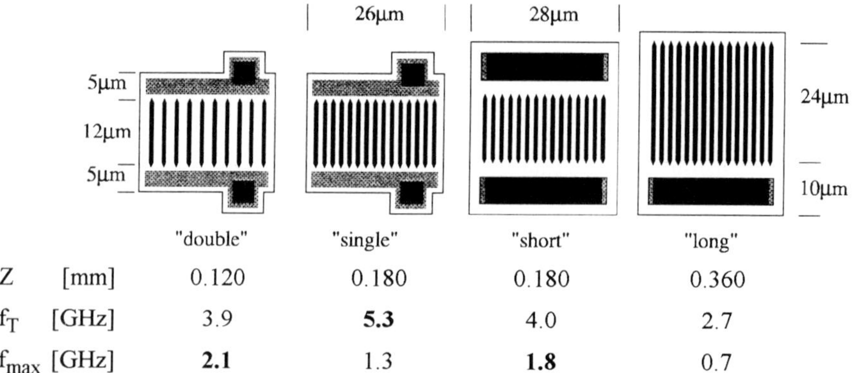

	"double"	"single"	"short"	"long"
Z [mm]	0.120	0.180	0.180	0.360
f_T [GHz]	3.9	**5.3**	4.0	2.7
f_{max} [GHz]	**2.1**	1.3	**1.8**	0.7

Figure 7: Layout and cut-off frequencies of some PJBTs: $b_k = 0.8$ μm; $N_D = 8 \cdot 10^{16}$ cm^{-3}

The highest f_T is obtained for the structure "single" as the ratio of gate width (i.e. high g_m) to parasitic capacities is best besides in "long". But there the thermal effect reduces g_m and thus f_T. The situation is different for f_{max} that also depends on the gate resistance R_G . This consists of resistance of the parallel gate fingers and the contact. The latter is higher for the structure "single" than for "short" and, hence, in this case f_{max} is lower although f_T is higher. When the finger width is doubled the resistances of the finger and even of the contact, because its ratio to gate width has to be taken into account, are reduced leading to $f_{max} = 2.1$ GHz.

For a non optimized structure with 0.6 μm wide channels $f_T = 6$ GHz and $f_{max} = 1.5$ GHz have been achieved. Due to initial contact problems these measurements have been made in the less effective contact configuration S on T.

5 Summary

Due to the homoepitaxial gate structure PJBTs exhibit high transconductances that can be attributed to the velocity overshoot. Although the contact resistances, that limit g_m and lead to additionally heating, and the thermal conductivity are not optimized yet $g_m = 220$ mS/mm have been obtained for both structures defined by optical lithography. Shorter and wider gate fingers will reduce the gate resistance and instantaneously the thermal resistance while the gate capacity does not increase too much because of the two intrinsic layers ($ip^{++}i$). By further reduction of parasitic resistances and capacities higher cut-off frequencies can be expected especially for Drain on Top.

We thank H.P Bochem for taking the SEM micrographs and. A. Fox for performing the high frequency measurements.

Bozler 1979; Bozler C.O., Alley G.D.,et al, IEEE Int. Electr. Dev. Meet. Tech. Dig., (1979), 384-387
Hess 1988; Hess K. and Iafrate G.J., Proceedings of the IEEE **76**(5), (1988), 519-532
Hollis 1987; Hollis M.A., Nichols K.B., Murphy R.A. et Bozler C.O.,
SPIE Advanced Processing of Semiconductor Devices **797**, (1987), 335-347
Mörsch 1993; Mörsch G., Gräber J., Kamp M., Hollfelder M., Lüth H.
contribution to ICCBE-4 (1993) to be published in Journal of Crystal Growth

InP based InGaAs–JFET with δ-doped channel

G. G. Mekonnen, W. Passenberg, C. Schramm, D. Trommer, and G. Unterbörsch

Heinrich-Hertz-Institut für Nachrichtentechnik Berlin GmbH, Einsteinufer 37, D-10587 Berlin, Germany

ABSTRACT: An InP based junction field-effect transistor (JFET) employing a double δ-doped InGaAs channel was designed and fabricated. This new combination of δ-doping and InGaAs material properties promises various advantages over δ-doped GaAs-FETs. Electrical measurements on 1 µm x 50 µm JFETs exhibited a high transconductance of 520 mS/mm being constant over a wide range of gate voltages. A transit frequency and a maximum oscillation frequency of 28 and 39 GHz, respectively, have been obtained.

1. INTRODUCTION

Field-effect transistors (FET) with a δ-doped channel have already been presented in the GaAs/GaAlAs material system (Jeong et al. 1992). In particular, the high transconductance and high frequency performance make these devices very attractive for future applications in monolithic microwave integrated circuits (MMIC) and optoelectronic integrated circuits (OEIC). A transit frequency f_T and a maximum oscillation frequency f_{max} of 7 and 15 GHz, respectively, have been reported. The idea of the δ-doping is to concentrate the carrier transport to an atomic plane similar to a quantum-well structure (Schubert et al. 1986), which enables a very high transconductance in conjunction with a high breakdown voltage. In addition, confining the carriers nearby the gate related junction increases the aspect ratio gate length to channel depth allowing the realisation of short gate length FETs. These facts make the δ-doped FETs competitive to HEMTs (high electron mobility transistors). In contrast to the established δ-doped devices on GaAs substrate, we use InGaAs lattice matched to InP. Due to the higher electron mobility and drift velocity, even further improved device performance is expected. Moreover, these transistors could be monolithically integrated in optoelectronic devices designed for optical communication systems in the long-wavelength range of 1.3/1.55 µm.

2. DEVICE STRUCTURE AND FABRICATION

The cross section of the fabricated JFETs is shown in Figure 1. The ternary layer stack was grown lattice matched on semi-insulating (s.i.) InP substrate by standard MBE technique at a growth temperature of $T_G = 500°$ C. The background impurity concentration of the non-intentionally doped InGaAs was $5{\cdot}10^{15}$ cm^{-3}. The δ-doping is accomplished by depositing Silicon (Si) onto the surface, while the InGaAs growth is shortly interrupted. The Si concentration is adjusted by the Si cell temperature and the deposition time. To reduce the ionized impurity scattering of a single-δ-doped channel we implemented a double-δ configuration

(Passenberg et al. 1991). Thereby, the channel embedded within InGaAs buffer layers consists of two Si planes spaced by a thin undoped InGaAs layer. Both planes are located in the side-walls of the resulting quantum-well. The sheet carrier concentration was $5{\cdot}10^{12}$ cm^{-2}. The optimum gap between the two Si planes was found to be 15 nm for our devices. In this case, the measured sheet resistance $R_\square$ amounted to 160 Ω/□. Reducing the spacing s leads to an increase of $R_\square$ up to 190 Ω/□ in the case of s = 0 (single δ-doped channel), while for s > 30 nm the channel is split in two potential wells.

The gate of the JFET is built by a two-step profile of the Beryllium (Be) doping with a concentration of $5{\cdot}10^{18}$ cm^{-3} (T_G = 400° C) at the pn-junction and $5{\cdot}10^{19}$ cm^{-3} at the gate surface (T_G = 350° C). This yields a low gate contact resistivity and prevents an outdiffusion of the Be into the channel (Trommer et al. 1990). A self-aligned process and standard photolithography were used to form the T-gate mesa structure. Thereby, the gate metallisation Ti/Pt/Au (20 nm/50 nm/200 nm) served as mask for a recess etching in the n-regions (sulphuric acid based solution) and the following metallisation for the source and drain contact. Both ohmic contacts consisting of Ni/AuGe/Ni/Au (5 nm/50 nm/20 nm/50 nm) were annealed at 380° C for 10 sec. The devices were electrically isolated by an additional wet chemical etching process (phosporic acid based solution) building simultaneously an air-bridge from the gate to the corresponding contact pad. The resulting gate length and gate width of the devices were 1 and 50 μm, respectively.

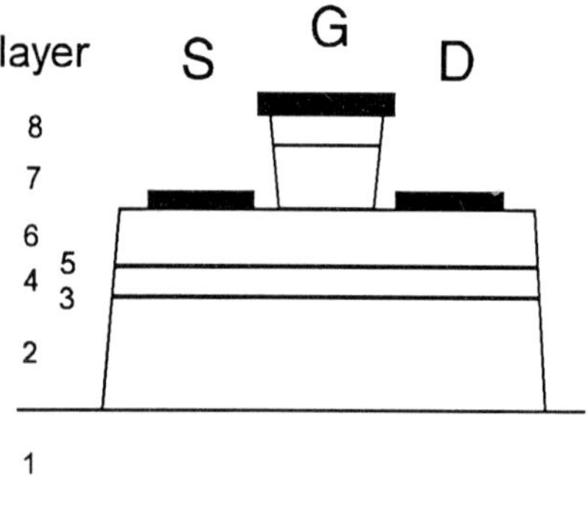

layer	material	doping	thickness
8	InGaAs	p: $5{\cdot}10^{19}cm^{-3}$	80 nm
7	InGaAs	p: $5{\cdot}10^{18}cm^{-3}$	220 nm
6	InGaAs	undoped	26.5 nm
5	Si	n_s: $2.5{\cdot}10^{12}cm^{-2}$	δ
4	InGaAs	undoped	15 nm
3	Si	n_s: $2.5{\cdot}10^{12}cm^{-2}$	δ
2	InGaAs	undoped	200 nm
1	InP substr.	s.i.	

Fig. 1. Schematic cross sectional view of the double-δ-doped InGaAs-JFET with the respective layer parameters given in the table

3. RESULTS AND DISCUSSION

To verify the expected performance of the fabricated devices as well as the possibility of their application in microwave circuits we characterized their DC and AC behaviour.

DC device measurements have been performed on-wafer. Figure 2 shows a typical output characteristics of the double-δ-doped channel JFET. Typical data of these first, not optimized devices were a pinch-off voltage of approximately 1.5 V and a very high tranconductance of around 520 mS/mm. As indicated in Figure 3 this value remains constant over a range of 1.3 V, which is a direct consequence of the δ-doping concept. This is due to the fact that the carrier population in the channel decreases nearly linear with increasing gate voltage, in contrast to homogeneously doped FETs with a square root dependence on V_{GS}. The measured output conductance g_d in the saturation region has a relatively high value of 20 mS ($V_{GS} = 0$ V, $V_{DS} = 1.5$ V) based on a parallel resistance below the channel as indicated in Figure 2. This additional current path was caused by interface state carriers at the substrate surface. The broad plateau of transconductance mentioned above and the high open loop gain g_m/g_d of about 25 open a wide flexibility to design integrated circuits, especially with respect to the choice of the bias point. However, we observed a relatively high gate leakage current of these devices. Excess currents over the thin barrier layer (layer no. 6 of Figure 1) are likely responsible for this finding. It is expected that this effect can be reduced by a more favour design.

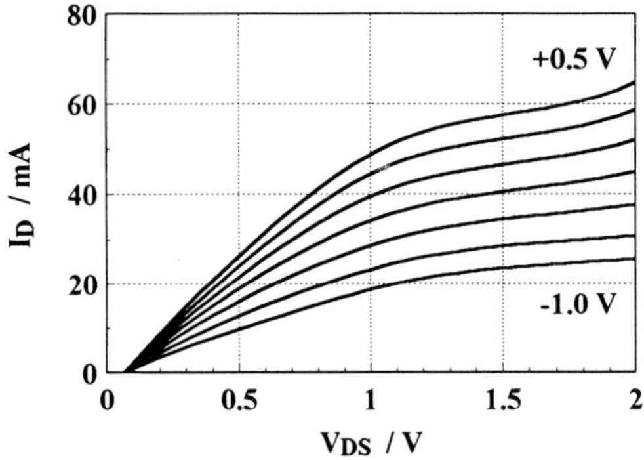

Fig. 2. Output characteristics of a double-δ-doped InGaAs-JFET with a gate length and gate width of 1 and 50 µm, respectively ($V_{GS} = +0.5 \ldots -1$ V, step -0.25 V)

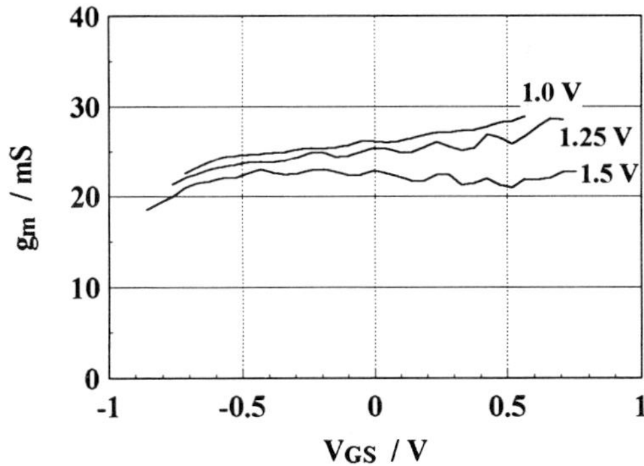

Fig. 3. Extrinsic transconductance versus gate voltage at different drain voltages

On-wafer microwave measurements using 50 Ω probe heads and a network analyzer have been carried out in the frequency range of 50 MHz to 20 GHz. Best performance was obtained under the bias condition of $V_{DS} = 2$ V and $V_{GS} = 0$ V. The resulting current gain as well as the power gain are shown in Figure 4. The extrapolated transit frequency and the maximum oscillation frequency were 28 and 39 GHz, respectively. These are the highest values of δ-doped FETs ever reported, to the best of our knowledge, and they exceed the values observed on corresponding GaAs based devices by more than a factor of 2. Both, f_T and f_{max}, are defined by the intrinsic elements. Due to the used air bridge technique and the achieved low contact resistivity parasitic elements are negligible. Therefore, an increase of the cut-off frequencies is expected by a reduction of the gate length. The measured S-parameter data were utilized to obtain the equivalent circuit elements of the FET. The results of the simulation yield also the existence of a parasitic resistance (110 Ω) parallel to the channel in agreement with the DC data. This reduced the power gain in the low frequency range, as seen in Figure 4.

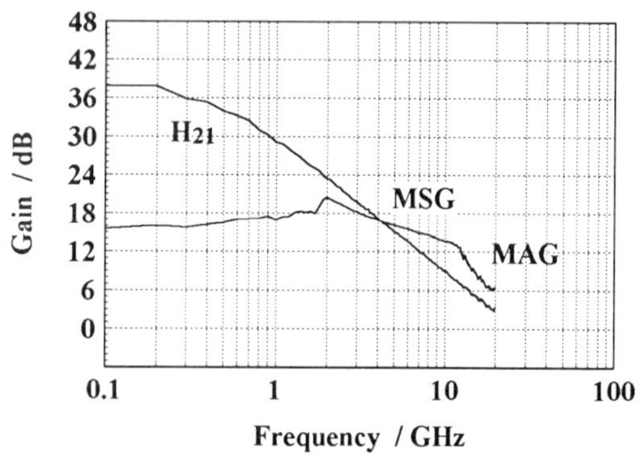

Fig. 4. Current gain H_{21}, maximum stable gain MSG and maximum available gain MAG versus frequency (V_{DS} = 2 V, V_{GS} = 0 V, I_D = 53 mA)

4. CONCLUSIONS

We have demonstrated the realisation of a double δ-doped InGaAs-JFET based on InP with a gate length of 1 μm. A high transconductance of 520 mS/mm over a range of 1.3 V gate voltage and an excellent frequency response with f_T and f_{max} of 28 and 39 GHz, respectively, were obtained. The use of InGaAs lattice matched to InP offers a better performance compared to GaAs based devices. These preliminary results make the device very attractive for high speed applications in MMICs and OEICs.

5. ACKNOWLEDGEMENTS

The authors would like to thank D. Breuer for DC measurements and H.-G. Bach for helpful discussions. The work was funded by the Bundesministerium für Forschung und Technologie, Germany, under contract no. BS 204/6.

6. REFERENCES

Jeong D.-H., Jang K.-S., Lee J.-S., Jeong Y.-H., and Kim B. 1992 *IEEE Electr. Dev. Lett.* **13**, pp 270-2

Passenberg W., Bach H.-G., and Böttcher J. 1991 *J. Electron. Mat.* **20**, pp 989-91

Schubert E. F., Fischer A., and Ploog K. 1986 *IEEE Trans. Electr. Dev.* **ED-33**, pp 625-32

Trommer D., Umbach A., Passenberg W., Mekonnen G., and Unterbörsch G. 1990 *Electron. Lett.* **26**, pp 734-5

Paper presented at the Int. Symp. GaAs and Related Compounds, Freiburg, 1993

Low noise and high gain InAlAs/InGaAs heterojunction FETs with high indium composition channels

K.Onda, A.Fujihara, H.Miyamoto, T.Nakayama, E.Mizuki, N.Samoto and M.Kuzuhara

Kansai Electronics Research Laboratory, NEC Corporation
2-9-1 Seiran Otsu Shiga, Japan 520

ABSTRACT: This paper describes super low noise and high gain performance of a 0.15μm T-shaped gate InAlAs/InGaAs heterojunction FET (HJFET) with a high indium composition channel especially suited for microwave and millimeter wave application with a low drain bias voltage (<1V). The Hall mobility for 70% indium channel was measured to be 14,200cm^2/V·s at 300K. A maximum transconductance at 0.75V drain bias was measured to be 780mS/mm for a 70% indium channel device. The high indium device has shown a peak current gain cut-off frequency (Ft) of 185GHz and 0.39dB noise figure with 16.5dB associated gain at 12GHz. These excellent device performances under low drain bias operation are due to the large conduction band discontinuity and superior electron transport properties of the high indium composition channel.

1.INTRODUCTION

Millimeter wave bands have recently attracted considerable attention for personal communication network, the wireless LAN, mobile communication and so on. In these applications, the development of compact wireless terminal is one of the key issues, and the demand for highly efficient millimeter wave devices has recently increased.

InAlAs/InGaAs HEMTs have demonstrated the excellent device performances at millimeter wave frequencies (Mishra U K et al 1988, Duh K H G et al 1991, Streit D C et al 1991). Several works using strained InGaAs channel (Kuo J M et al 1986, Hong W P et al 1988, Ng G I et al 1988, Akazaki T et al 1992, Yang P et al 1992, Nguyen L D et al 1992) have shown their improved device potential for millimeter wave application, compared to lattice matched HEMTs. However, only a limited number of reports has exhibited noise performance for strained channel InP-based HEMTs (Tan K L et al 1991).

This paper describes excellent device performance of InAlAs/InGaAs heterojunction FETs (HJFETs) with high indium composition channel including noise characteristics under low voltage operation.

2.DEVICE STRUCTURE

The MBE epi-structures were grown on (100) semi-insulating Fe-doped InP substrates. The epi-structure has an InAlAs buffer layer, a 30nm-thick lattice matched $In_{0.53}Ga_{0.47}As$ sub-channel and a 15nm-thick strained $In_{0.7}Ga_{0.3}As$ channel. The growth temperature for the $In_{0.7}Ga_{0.3}As$ layer is chosen at 500°C, which is the same for the $In_{0.53}Ga_{0.47}As$ layer. Si planar doping of $5\times10^{12}cm^{-2}$ is used between a 3nm-thick InAlAs spacer layer and a 20nm-thick undoped InAlAs Schottky layer. The cap layer is a highly-doped lattice matched InGaAs. For comparison, a fully lattice matched device structure with identical Schottky, doping and spacer layers and identical total channel thickness (45nm) was grown. Room temperature Hall measurement data of the high indium channel structure with a 6nm-thick spacer layer exhibit a higher electron mobility of $14{,}200cm^2/V\cdot s$ and larger sheet carrier density of $2.1\times10^{12}cm^{-2}$ as compared to $12{,}200cm^2/V\cdot s$ and $2.0\times10^{12}cm^{-2}$ for the lattice matched $In_{0.53}Ga_{0.47}As$ channel.

Device fabrication process is based on a standard recessed-gate FET technology, which is briefly described as follows: the MBE wafer is isolated by mesa etching using phosphoric acid based etchant. An alloyed AuGe/Ni/Au metal system is used for source and drain ohmic contacts. Typical contact resistances measured are 0.05Ω·mm. After the channel recess step, gates were defined by electron beam lithography and lift-off techniques (Samoto N 1990). Ti/Al was used as gate metallization and Ti/Pt/Au for the contact pads. The fabricated device has a $0.15\times100\mu m^2$ T-shaped gate with two fingers, whose gate resistance is about 1Ω.

3.RESULTS AND DISCUSSION

The typical I-V characteristics of the fabricated devices with 53% and 70% indium channels are respectively shown in Fig.1(a)(b). Both the lattice matched device and the high indium channel device show good pinch-off characteristics. No kink effects are observed for the lattice matched device, but the 70% indium device exhibits an increase in the output conductance (gd) at a drain bias of approximately 0.8V. The increase in gm and gd observed for the high indium device at a drain bias of more than 0.8V is believed to be due to impact ionization in the channel. At a drain voltage of 0.75V, the higher indium device shows a larger transconductance (gm) of 780mS/mm as compared to that for the lattice matched device (gm=570mS/mm). Since the measured source resistance is 0.40-0.45Ω·mm for both devices, we explain the observed gm difference by a difference in electron transport property in the channel layers.

S-parameter measurements were made between 0.06 and 40GHz using a Cascade Microtech microwave probe station and a Wiltron 360 network analyzer. Using measured S-parameters, current gains were calculated versus frequency. Bias dependences of the current gain cut-off frequency (Ft) for the high indium channel device and the lattice matched device were estimated (in Fig.2). The Ft of the high indium channel device shows a higher value

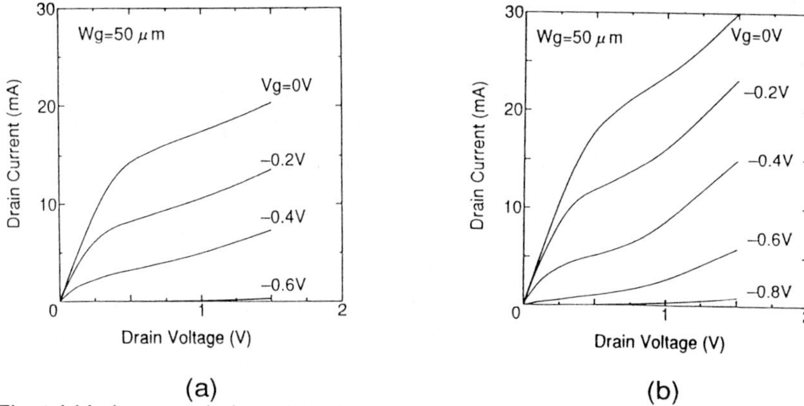

Fig.1 I-V characteristics of (a)the lattice matched channel HJFET and (b)the 70% indium channel HJFET.

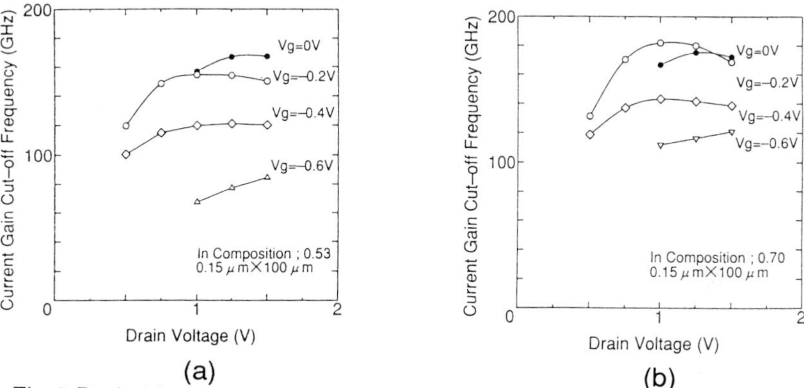

Fig.2 Drain bias voltage dependence of current gain cut-off frequencies for (a)the lattice matched channel HJFET and (b)the 70% indium channel HJFET.

over the whole gate bias range shown in Fig.2 than that of the lattice matched device. A maximum Ft of 185GHz at Vd=1.0V was obtained with the high indium channel device, besides Ft of more than 170GHz were measured at Vd=0.75-1.25V range. On the other hand, the peak of Ft for the lattice matched device was 165GHz at Vd=1.25V. The drain current at Vg=-0.2V of the high indium device is nearly equal to the current at Vg=0 of the lattice matched device. In that gate bias range the high indium device has about 10-15% larger Ft than the lattice matched device.

The noise performances of the fabricated devices with a high indium channel and with a lattice matched channel were measured using on-wafer automatic tuner system over the 2 to 18GHz frequency range. Measured minimum noise figures and associated gains are represented in Fig.3. At 12GHz, the high indium composition channel device shows a noise performance of 0.39dB noise figure with 16.5dB associated gain at a low drain bias voltage of 0.75V. On the other hand, the lattice matched device gives the noise performance of 0.45dB noise figure with 15.5dB associated gain at the same drain bias. Drain current (Id) dependence of noise characteristics for the high indium device is represented in Fig.4, demonstrating a very flat dependence over a wide Id range.

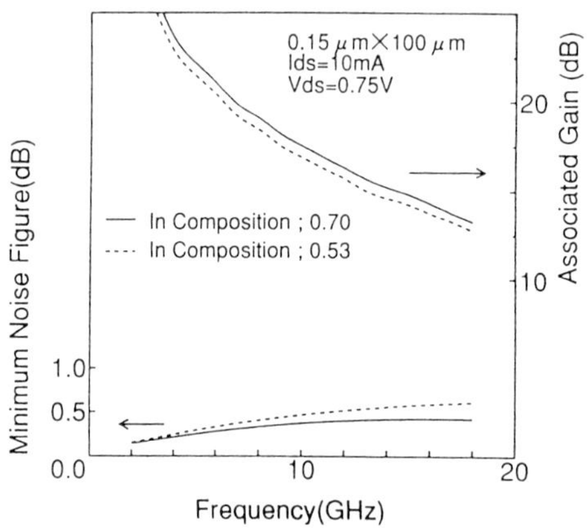

Fig.3 Frequency dependence of minimum noise figure and associated gain for the lattice matched channel HJFET and the 70% indium channel HJFET

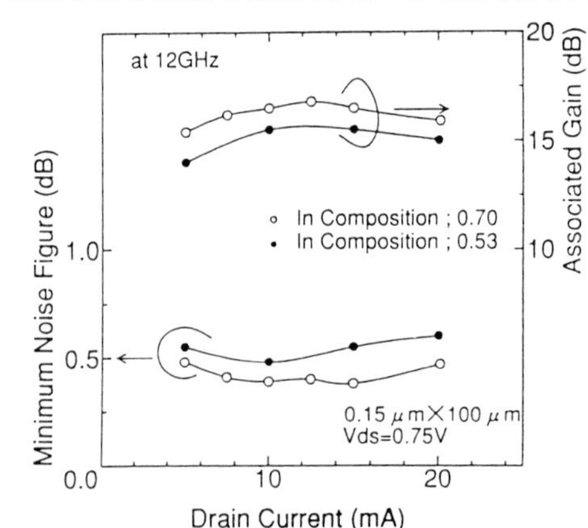

Fig.4 Drain current dependence of minimum noise figure and associated gain for the lattice matched channel HJFET and the 70% indium channel HJFET.

The conduction band discontinuity (ΔEc) between the channel layer and the donor layer (spacer layer) of the fabricated devices were reported to be 0.56eV for the 70% indium device and 0.50eV for the 53% indium device (People R et al 1983). That larger ΔEc of the high indium device contributes to larger sheet carrier density, resulting in larger drain current in the I-V characteristics (Fig.1). Superior electron drift velocity in the high indium channel is contributive to high Ft value for the high indium device compared to that for the lattice matched device. The Ft for the high indium channel device has a peak at lower drain bias than the lattice matched device because of enhanced mobility in the high indium channel. The superior electron transport also contributes to the excellent noise performance of the high indium device.

By using the model by Hughes (Hughes B 1992), the effective noise temperature (Td) and the maximum oscillation frequency (Fmax) were calculated from the noise data to be 625K and 185GHz, respectively. The latter is in reasonable agreement with the value estimated from equivalent circuit parameters fitted to 0.06-40GHz S-parameters. These results show a great potential of the high indium channel device for low noise application both in the microwave bands and in the millimeter wave bands.

4.CONCLUSION

This paper has reported that the high indium composition channel InAlAs/InGaAs HJFETs on InP substrates are superior to lattice matched devices at low drain bias less than in microwave and millimeter wave bands. At a drain bias less than 1.0V, the high indium (70%) channel device with a 0.15μm T-shaped gate has shown a larger transconductance (780mS/mm at Vd=0.75V), a higher cut-off frequency (185GHz at Vd=1.0V) and a superior noise performance (noise figure = 0.39dB, associated gain = 16.5dB at 12GHz). The improved device performance can be explained by larger conduction band discontinuity and larger electron drift velocity compared to the lattice matched channel devices.

ACKNOWLEDGMENTS

The authors would like to thank Dr.H.Abe for his encouragement throughout this work.

REFERENCES

Akazaki T, Arai K, Enoki T and Ishii Y 1992 IEEE Electron Device Lett. **13** 325

Duh K H G, Chao P C, Liu S M J, Ho P, Kao M Y and Ballingal J M 1991 IEEE Microwave and Guided Wave Lett. **1** 114

Hong W P, Ng G I, Bhattacharya P K, Pavlidis D, Willing S and Das B 1988 J.Appl.Phys. **64** 1945

Hughes B 1992 IEEE Trans. Microwave Theory and Tech. **40** 1821

Kuo J M, Lalevic B and Chang T Y 1986 IEDM Technical Digest 460

Ng G I, Weiss M, Pavlidis D, Tutt M, Bhattacharya P K and Chen C Y 1988 Proceedings of Int.Symp.GaAs and Related Compounds 465

Nguyen L D, Brown A S, Thompson M A, and Jelloian L M 1992 IEEE Trans.Electron Devices **39** 2207

People R, Wecht K W, Alavi K,and Cho A Y 1983 Appl.Phys.Lett. **43** 118

Samoto N, Makino Y, Onda K, Mizuki E, and Itoh T 1990 J.Vac.Sci.Techol. **B8** 1335

Streit D C et al 1991 Electronics Lett. **27** 1149

Tan K L, Streit D C, Chow P D, Dia R M, Han A C, Liu P H, Garske D and Lai R 1991 IEDM Technical Digest 239

Yang D, Chen Y C, Brock T and Bhattacharya P K 1992 IEEE Electron Device Lett. **13** 350

Inst. Phys. Conf. Ser. No 136: Chapter 3
Paper presented at the Int. Symp. GaAs and Related Compounds, Freiburg, 1993

High performance GaInP/GaAs hole barrier bipolar transistors (HBBTs)

H. Leier, U. Schaper*, K.H. Bachem**

Daimler Benz AG, Research Center Ulm; Wilhelm Runge Str. 11, 89081 Ulm Germany
* Siemens Reseach Laboratories, Otto-Hahn-Ring 6, 81739 München, Germany
** Fraunhofer Institut für Angewandte Festkörperforschung, Tulla Str. 72, 79108 Freiburg

Abstract: The HBBT may be understood as an advanced variant of the well-known heterojunction bipolar transistor. A thin GaInP film (10-20 nm) embedded between emitter and base layer acts as an efficient hole barrier. Combined with several inherent advantages of GaInP the HBBT leads to an attractive alternative to the conventional AlGaAs/GaAs HBT. We report on HBBT devices with different layer sequences and geometries. HBBT devices exhibit almost constant current gain over a wide range of collector current. Maximum oscillation frequencies above 100 GHz are achieved. Applications in the microwave frequency range will be discussed. GaInP/GaAs HBBTs are well described by a T-like small signal equivalent circuit including bias and geometry scaling. A direct parameter extraction method and first load-pull simulations are reported.

1. INTRODUCTION

The GaInP/GaAs Heterojunction Bipolar Transistor (HBT) has attached considerable interest in the last few years. Impressive dc results with near ideal I-V characteristics have been demonstrated (Liu et al 1992). Current gain cut-off frequencies of 95GHz (Leier et al 1993), power gain cut-off frequencies of 90GHz (Zwicknagl et al 1992) for nonselfaligned devices, 100GHz (Delage et al 1993) and slighly above 100GHz (Liu et al 1993 and Leier et al 1993) for selfaligned devices have been obtained recently.

The considerable effort on GaInP/GaAs HBTs stems from different advantages of the GaInP/GaAs material system compared to the conventional AlGaAs/GaAs heterostructure. From the technological viewpoint the main advantage of the GaInP/GaAs HBT is its easy producibility because selective wet and dry etchants are available in both directions. From the physical viewpoint the band alignment in a npn GaInP/GaAs HBT is favoured compared to an AlGaAs/GaAs HBT due to the higher ratio of $\Delta Ev/\Delta Ec$ for GaInP/GaAs. In addition we have a couple of other benefits from the GaInP compound itself. The GaInP material has a lower reactivity to oxygen, lower concentration of deep levels and a slower surface recombination velocity compared to AlGaAs. Moreover the pn-junction - heterojunction match in a npn HBT is more stable because the p-dopant carbon doesn't form acceptor states in GaInP.

In this paper we report on growth, performance and modelling of npn GaInP/GaAs HBTs which consists almost entirely of GaAs. Only a 10-20nm thick GaInP hole blocking layer is incorporated between emitter and base to improve the emitter injection efficiency. This structure is called Hole Barrier Bipolar Transistor (HBBT) throughout the paper. The introduction of this thin GaInP layer instead of a thick GaInP emitter results in uncritical lattice matching and more reliable processing.

2. MATERIAL GROWTH

The layer structures were grown by low pressure MOCVD on semiinsulating, (100)-oriented, 50mm substrates in a non load locked AIXRON machine. The substrate is etched prior to loading in an ammonia-peroxide solution and does not rotate during deposition. The oxide on the surface of the substrate is removed by annealing at 700°C for 5 minutes in hydrogen atmosphere (8l/min, 100mbar) under arsine overpressure (100ml/min). The different GaAs layers and the GaInP barrier layer have been grown from trimethylgallium, trimethylindium, arsine and phosphine. Silane has been used as dopant source for all n-type layers. All individual layers, except the carbon doped base layer were grown at a deposition temperature of 580°C, a total reactor pressure of 100mbar and 8l/min total flow rate.

The base layer is grown from trimethylgallium and trimethylarsine under slighly different growth conditions (540°C, 150mbar and 2.5l/min). Arsine flow is turned off during the deposition of the base layer. The lower deposition temperature is required for obtaining high hole concentrations (appr. $6x10^{19}cm^{-3}$) in the base layer (Neumann et al 1990). The considerable higher thermal stability of trimethylarsine compared with arsine necessitates higher total reactor pressure and reduced total reactor flow rate for formation of growth sustaining trimethylarsine fragments in the gas phase above the hot susceptor. The proper choice of these parameters is decisive for achieving sufficiently homogenious growth rates and carbon doping over the full diameter of the substrate. The growth behaviour of carbon doped GaAs from trimethylgallium and trimethylarsine is widely controlled by the kinetic of gas phase reactions and therefore strongly influenced by the hydrodynamic of the reactor being used. Different from the wellknown characteristics of gas phase diffusion or surface kinetic controlled growth, the deposition of GaAs from trimethylgallium and trimethylarsine shows some distinctly different features. The growth rate increases with increasing total reactor pressure and decreases with increasing total reactor flow rate. The growth rate is low in upstream and high in downstream position of the susceptor. In a more detailed publication

(Bachem et. al. 1993) we will show that a kinetic parameter (P^2 x L/Q) combining total reactor pressure P, a characteristic length L and the total reactor flow rate Q is the ruling quantity controlling growth rate and carbon doping. The individual layer of the epitaxial layer structure have been characterized by Hall measurements, X-ray diffraction and secondary ion mass spectroscopy.

Fig.1 shows the depth profile of a typical HBBT layer structure. After growth of a 580nm thick GaAs subcollector (n=$5x10^{18}cm^{-3}$) a 10-20nm thin undoped $Ga_{0.5}In_{0.5}P$ etch stop is introduced. This allow an easy access to the subcollector layer during processing. The thickness and doping levels of the next layers from bottom to top are as follows: 250-1000nm (n=$3x10^{16}cm^{-3}$) GaAs collector layer, 60-90nm (p=$6.5x10^{19}cm^{-3}$) GaAs base layer, 10-20 nm (undoped) $Ga_{0.5}In_{0.5}P$ hole barrier, 115nm (n=7 x 10^{17} cm^{-3}) GaAs emitter layer and 115nm (n=$7x10^{18}cm^{-3}$) GaAs cap layer.

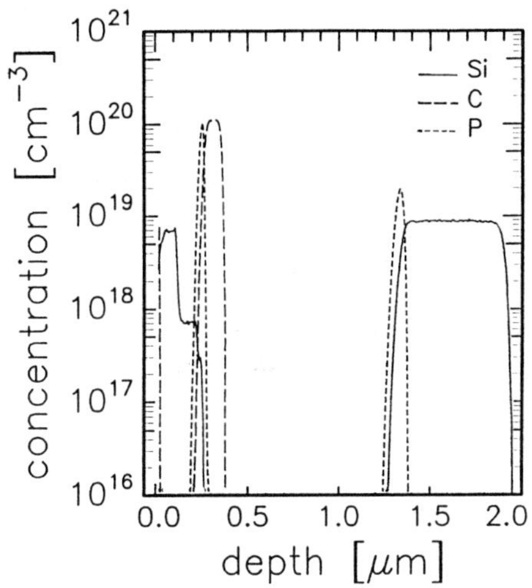

Fig.1: SIMS profile of a HBBT structure.

3. DEVICE FABRICATION AND PERFORMANCE

Selfaligned and nonselfaligned HBBT devices with varying emitter sizes have been fabricated using a triple mesa approach. Selective wet etching techniques and standard optical lithography are applied for device fabrication. GeNiAu and TiPtAu are the n- and p-type contact metallizations, respectively. Device isolation is achieved by a multi-energy proton implantation.

In Fig.2 the dependence of small signal current gain on collector current density for devices with different emitter area/ periphery ratios is shown. Independent of device size and vertical layer structure

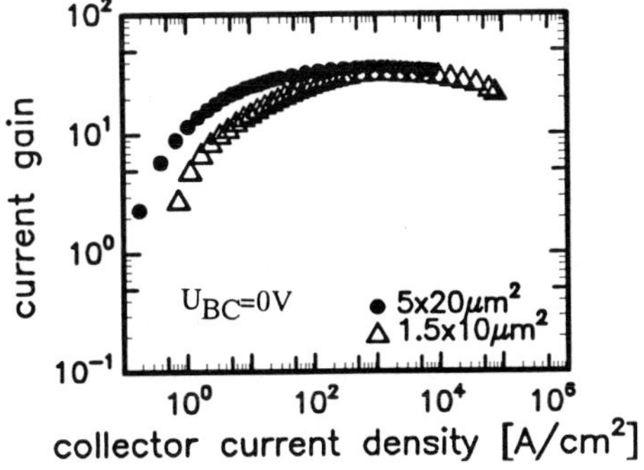

Fig.2: Current gain versus coll. current density for different HBBTs.

we observe a flat decrease of current gain with decreasing collector current density indicating minimum recombination in the emitter-base space charge region.

Despite the high base doping levels and the associated low base sheet resistances of 150 -250Ω (d_b=60-90nm) current gains above 20 are realized. These relatively high current gains are not deteriorated in HBBT devices with GaInP hole barriers as thin as 10nm indicating that even such thin layers effectively block the hole injection.

The high frequency performance and potential of GaInP/GaAs HBBTs is demonstrated in Fig.3a-c. Power gain cut-off frequencies between 100GHz and 110GHz are obtained for two emitter finger devices fabricated from different HBBT structures. As expected the highest current gain cut-off frequency (f_T=80GHz) is obtained with a thin base structure with small base transit time (Fig.3c). The independence of f_{max} on the vertical layer structure can be qualitatively understood if we consider the extrapolation $f_{max} = [f_T/(8\pi R_b C_{BC})]^{1/2}$. A decrease of the base width increases not only f_T but also the base resistance resulting in almost constant f_{max}.

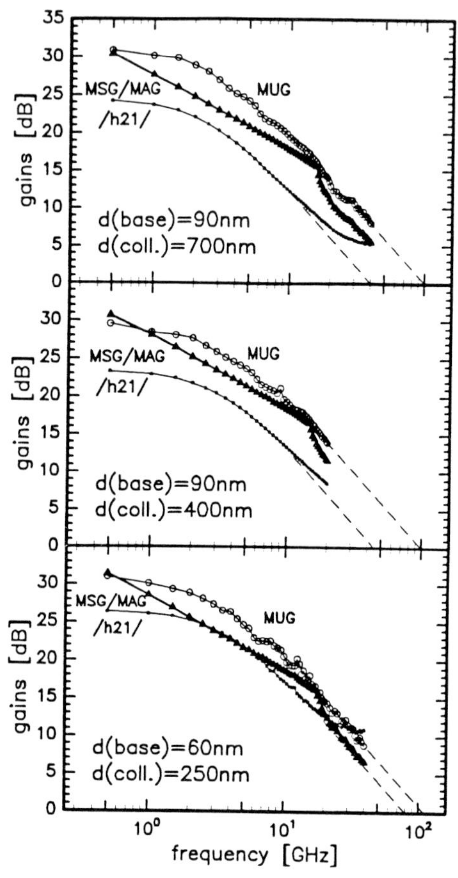

Fig. 3a-c: Current and power gains for HBBTs (2 x 1.5μm x 10μm, Vce=2-2.5V, Ic=20mA).

We are currently engaged to insert the GaInP/GaAs HBBT in our MMIC line for mm-wave oscillators taking advantage of the expected reduced 1/f noise of HBTs in general. Indeed, the phase noise characteristics of already realized oscillators with dielectric resonator at 21GHz are very encouraging (Güttich et al 1993). The fabricated oscillators exhibit an output power of +5dBm with a conversion efficiency of 13%. The phase noise has been measured to be -102.5dBc/GHz at 100kHz and -81.6dBc/Hz at 10kHz off carrier, respectively.

The GaInP/GaAs HBBT is also attractive with respect to power applications. Multifinger HBBT devices in common emitter configuration with four cells each containing 2 fingers with an emitter area of 1.5μm x 15μm per finger were tested under pulsed power conditions in class AB mode (f=10GHz). The pulse width and period are 400ns and 200us, respectively. An output power of 1W and 33% power added efficiency with an associated gain of 4.2dB are achieved at a relatively moderate bias voltage of U_{CE}=7V. This corresponds to power densities of 8.5W/mm or 5.5mW/μm^2. At a somewhat lower power level (0.7W) a power added efficiency of 37% with an associated gain of 6dB is obtained at the same bias conditions. The collector currents varied between 100mA and 270mA during this analysis.

4. MODELLING

A large signal model has to be used for monolithic microwave integrated circuit (MMIC) applications in order to describe the HBBT properties over the operating bias range from DC to 26 GHz. The HBBT is characterized at many bias settings by S-parameter measurements over the frequency range of interest. All these measurement data can be reduced at any bias to a set of 15 frequency-independent parameters using a small-signal equivalent circuit. The modelling results presented here refer to single emitter HBBTs fabricated with a nonselfaligned planarized mesa process by Siemens (Zwicknagl et al 1992).

4.1 Small signal equivalent circuit model

The small signal equivalent circuit with T-like topology is shown in Fig. 4a. The elements of this circuit can be related directly to the physical properties of the device. This topology has been used successfully by several authors (Ramachandran et al 1990, Rodwell et al 1990, Kobayashi et al 1990). The two-port behaviour of this circuit can be described either by impedance ($\underline{\underline{Z}}$) or admittance ($\underline{\underline{Y}}$) matrices which can be converted to the S-parameter matrix. The equivalent circuit is built up of 3 parts like shells. The inner part denoted by $\underline{\underline{Z}}_i$ (dotted box in Fig.4a) represents the intrinsic transistor. The emitter impedance Z_E and collector impedance Z_Q represent the pn-junctions. The current source is described by the DC-currentgain α_o and a roll-off frequency f_α. The internal series resistances of the base and collector layers are R_B and R_C. The extrinsic base collector capacitance C_F is parallel to theinternal transistor. Both build up the second shell of the equivalent circuit which is described by $\underline{\underline{Y}}_k$ (dashed box in Fig. 4a). The equivalent circuit outer shell ($\underline{\underline{Z}}_l$) is built up by adding the external base (Z_1) and collector (Z_2) impedances.

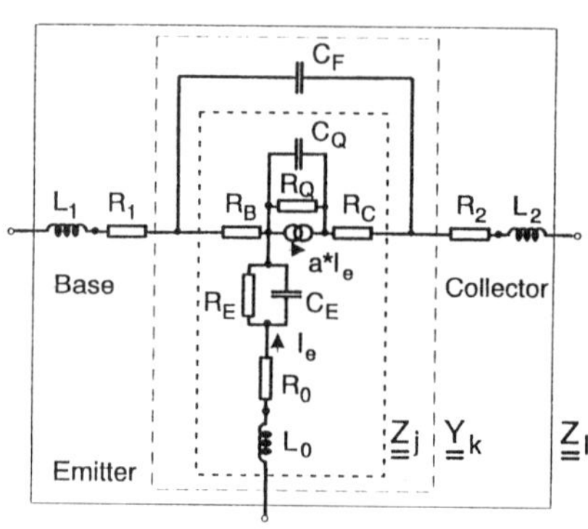

Fig.4a: Small signal equivalent circuit with T-like topology.

$$\underline{\underline{Z}}_i = \underline{\underline{F}}\,(Z_E, Z_Q, \alpha) \qquad Z_E = f(R_0, L_0, R_E, C_E), \quad Z_Q = f(R_Q, C_Q)$$

$$\underline{\underline{Z}}_j = \underline{\underline{Z}}_i + \begin{pmatrix} Z_B & 0 \\ 0 & Z_C \end{pmatrix} \qquad \alpha = f(\alpha_0, f_\alpha)$$

$$Z_B = R_B, \quad Z_C = R_C$$

$$\underline{\underline{Z}}_j \rightarrow \underline{\underline{Y}}_j$$

$$\underline{\underline{Y}}_k = \underline{\underline{Y}}_j + \frac{1}{Z_F}\begin{pmatrix} 1 & -1 \\ -1 & 1 \end{pmatrix} \qquad Z_F = f(C_F)$$

$$\underline{\underline{Y}}_k \rightarrow \underline{\underline{Z}}_k$$

$$Z_1 = f(R_1, L_1)$$

$$\underline{\underline{Z}}_l = \underline{\underline{Z}}_k + \begin{pmatrix} Z_1 & 0 \\ 0 & Z_2 \end{pmatrix} \qquad Z_2 = f(R_2, L_2)$$

Fig.4b: Compositon of the matrices $\underline{\underline{Z}}_j$, $\underline{\underline{Y}}_k$ and $\underline{\underline{Z}}_l$ describing the shells of the equivalent circuit.

Matrices and functions describing the equivalent circuit are shown in Fig.4b.

4.2 Analytical parameter extraction

A method for direct extraction of all equivalent circuit parameters from measured S-parameter data is presented. This method is based on the shell topology of the equivalent circuit. No data fitting is necessary.
The bipolar transistor is measured in the active and cutoff mode. This corresponds to the cold-FET modelling used for the parameter extraction of field-effect-transistors (Berroth et al 1990). An open structure is measured for de-embedding of the transistor S-parameters from the parasitic environment.

The equivalent circuit parameter are calculated directly from the two-port descriptions. The elements of the outer shell have to be determined by an iteration loop due to the coupling of input and output of the transistor by its extrinsic part. Within 2 to 4 iteration steps the frequency independent mean values of all parameters are achieved.

The frequency independent mean values of the extracted model parameters have to describe the measured S-parameters of the transistor. A comparison between measured and modelled S-parameters shows a very good agreement for all 4 complex S-parameters with an error of only 4 %. The highest deviation occurs at the high frequency end where the simple circuit we have used has to be modified.

Applying this equivalent circuit to HBBTs of various emitter length we have demonstrated (Schaper et al 1993) that geometry and bias scaling are well described within this model.

4.4 Large Signal Model

The nonlinear Gummel-Poon model for npn transistors embedded in a linear network accounting for parasitic effects is taken as a basis for the large signal model. The MESFET-based simulator HARPE (OSA 1992) is used for the parameter extraction by optimization.

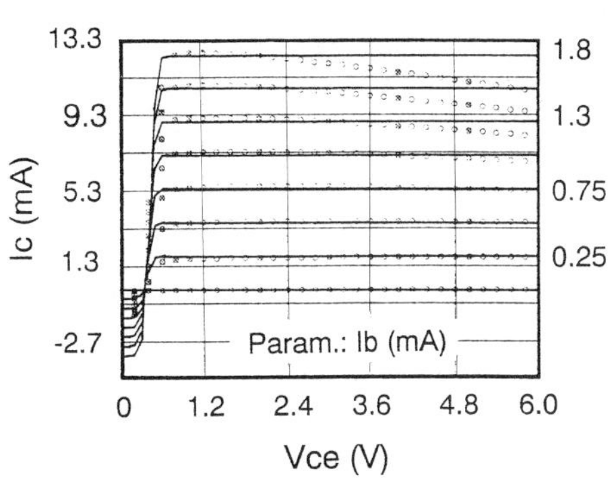

Fig.5: Simulated (-) and measured (o) DC-characteristic of an one-emitter-finger HBBT. ($2.5x21\mu m^2$)

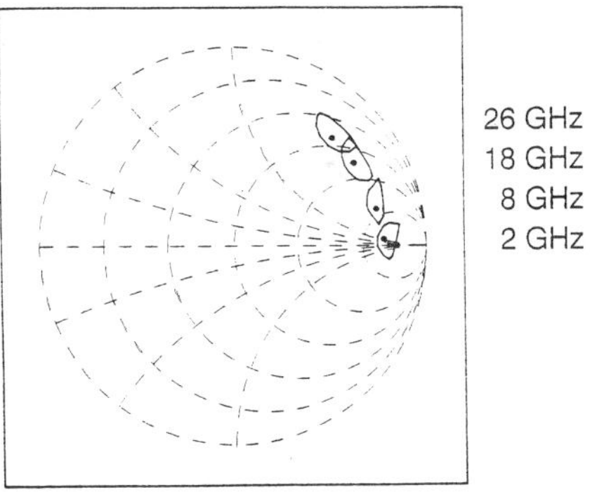

Fig.6: Smith diagram with optimized load impedance (•) for maximum output power at V_{ce}=3V, I_b=0.9mA. The -0.5dB contours are given for each optimum.

With 6 model parameters the DC-characteristics give a good agreement with the measured data (Fig.5) and the measured S-parameters are described very well by this model in the active range $1.5 \text{ V} \leq V_{ce} \leq 3.0 \text{ V}$, $0.3 \text{ mA} \leq I_b \leq 1.5 \text{ mA}$. In this first approach the selfheating effect of the HBBT demonstrated by the negative differential resistance in the DC-output characteristics has been omitted. The aim of this first large signal model of an one-emitter-finger HBBT was to investigate the load impedance for maximum output power (Fig.6). The frequency dependence of the optimal load impedance is explained by a simple circuit with a resistor R and a capacitance C in parallel.

The parameter extraction methods have to be applied and tested now with power transistors consisting of multi-emitter-finger structures. These power transistors are designed to achieve

5W output power for X-band applications. For these devices the selfheating effect have to be included in the model to account for a correct bias description at higher power levels.

5. CONCLUSION

GaInP/GaAs HBBTs with varying lateral dimensions incorporating different vertical layer structures have been analysed. The devices show always reasonable high current gains with small dependence on collector current. Excellent microwave performance with $f_{max} \geq$ 100GHz is demonstrated for HBBTs with different base and collector thicknesses. Examples are given demonstrating the usefulness of GaInP/GaAs HBBTs for oscillator and power applications. A large signal model for single emitter HBBTs is used to simulate load-pull contours. The model can be extended to multi-emitter HBBTs including selfheating effects.

ACKNOWLEDGEMENT

We would like to thank M.Allenson from DRA for expert power analysis. The continuous support of this work by P.Narozny and H.Dämbkes is gratefully acknowledged. The work was partly supported by the ESPRIT Project AIMS and by the German Ministry of Research and Technology.

REFERENCES

Bachem K H, Lauterbach T H (to be published)
Berroth M and Bosch R 1990 IEEE Trans. Microwave Theory Tech. **38**, p.891
Cho H and Burk D 1991 IEEE Trans. Electron Devices **38**, p.1371
Delage S, Forte-poisson M.A.di, Pons D, 1993, Indium Phosphide and Related Materials, Conference Proceedings, p.561
Güttich U, Leier H, Marten A, Riepe K, Pletschen A, Bachem K H (to be published)
Kobayashi K W, Umemoto D K, Esfandiari R, Oki A K, Pawlowicz L M, Hafizi M E,Tran L, Camou J B, Stolt K S, Streit D C and Kim M E 1990 IEEE MTT-S Digest, p.19
Leier H, Marten A, Bachem K H, Pletschen W, Tasker P, 1993, Electron. Lett. **29**, p.868
Liu W and Fan S K 1992 IEEE Electron Device Lett. **13**, p.510
Liu W, Fan S K, Henderson T, Davito D, 1993, IEEE Electron Device Lett., **14**, p.176
Neumann G, Lauterbach T H, Maier M, Bachem K H, Inst. Phys. Conf. Ser. No 112, Chapter 3
Optimization Systems Associates Inc.(OSA) 1992 HARPE User's Manual Version 1.6
Ramachandran R, Nijjar M, Podell A, Stoneham E, Mitchell S, Wang N L, Ho W J, Chang M F, Sullivan G J, Higgins J A and Asbeck P M 1990 GaAs IC Symposium Digest, p 357
Rodwell M, Jensen J F, Stanchina W E, Metzger R A, Rensch D B, Pierce M W, Kargodorian T V and Allen Y K 1990 IEEE Bipolar Circuits and Technology Meeting, p 252
Rodwell M, Jensen J F, Stanchina W E, Metzger R A, Rensch D B, Pierce M W, Kargodorian T V and Allen Y K 1991 IEEE Solid State Circuits **26**, p 1378
Schaper U, Bachem K H, Kärner M and Zwicknagl P 1993 IEEE Trans. Electron Dev. **40**, p.222
Sheng N H, Ho W J, Wang N L, Pierson R L, Asbeck P M and Edwards W L 1991 IEEE MTT-S 1, p.208
Zwicknagl P, Schaper U, Schleicher L, Siweris H, Bachem K H, Lauterbach T, and Pletschen W, 1992, Electron. Lett. **28**, p.327

Paper presented at the Int. Symp. GaAs and Related Compounds, Freiburg, 1993

ALE/MOCVD grown InP/InGaAs HBTs with a highly-Be-doped base layer and suppressed diffusion

H. Shigematsu, H. Yamada, Y. Matsumiya, Y. Sakuma, H. Ohnishi, O. Ueda, T. Fujii, K. Nakajima, and N. Yokoyama

FUJITSU LABORATORIES LTD.
10-1 Morinosato-Wakamiya, Atsugi 243-01, Japan

Abstract: We fabricated InP/InGaAs heterojunction bipolar transistors (HBTs) with a highly doped base region (7×10^{19} cm^{-3}) grown by ALE and other regions grown by MOCVD. We obtained a dc current gain of 30 with a 5x20-μm emitter. An abrupt doping SIMS profile, a low turn-on voltage, and a base sheet resistance of 600 Ω/square indicate that the base dopant of Be didn't diffuse during the growth. Combination ALE of MOCVD is a very promising growth technique for high performance HBTs.

I. Introduction

InGaAs-based heterojunction bipolar transistors (HBTs) lattice-matched to InP have a high-speed performance and low power consumption superior to that of GaAs HBTs because of their high electron mobility and low turn-on voltage. InGaAs-based HBTs are promising for ultrahigh-speed integrated circuits in fiber optic communication systems and high efficiency-microwave devices. Small-scale integrated circuits using InGaAs-based HBTs and high-performance InP/InGaAs HBTs have been reported (Chen *et al.* 1989a, Mishra *et al.* 1989b, Nottenburg *et al.* 1990, Banu *et al.* 1991). A highly doped base layer is known to be vital for high-performance HBTs but, when done using either conventional MBE or MOCVD, could cause the p-type dopant to diffuse from the base into the emitter. To avoid the diffusion, low-temperature growth techniques, such as gas-source MBE (GSMBE) (Hamm *et al.* 1989), have been applied to HBTs and a hole concentration of 1×10^{20} cm^{-3} has been obtained in InGaAs. However, there were few reports on such high base doping using MOCVD, superior to MBE for mass production.

Atomic layer epitaxy (ALE) is promising for high base doping (Bhat *et al.* 1989), because high-quality layers of III-V compounds can be grown even at low temperatures (340°C) due to two-dimensional growth. However, ALE has a low growth rate for its layer by layer growth process. It causes growth time increasing, resulting in a p-type dopant diffusion. We must, therefore, decrease growth time as well as growth temperature to solve this problem. We optimized the growth conditions by combining low-temperature MOCVD with ALE to prevent p-type dopants from diffusing. We also studied the characteristics of InP/InGaAs HBTs grown by this advanced MOCVD system.

II. Growth and Evaluation

We grew an InGaAs layer by ALE using hydrogen as a carrier gas with trimethylindium (TMIn) and triethylgallium (TEGa) mixtures and arsine (AsH_3) as In, Ga, and As sources. They were sent cyclically, separated by 0.5-sec H_2 purges (Fig. 1). In ALE, doping efficiency ordinarily depends on the dopant gas supply sequence. We used four different diethylberyllium (DEBe) doping sequences: (a) doping with AsH_3, (b) doping after AsH_3 supply, (c) doping with TMIn + TEGa, and (d) doping after TMIn + TEGa supply. The carrier concentration for Be-doped InGaAs layers depends on the DEBe supply timing. We obtained a high doping concentration except (c), because Be easily entering sites of V group atoms. Furthermore, we found (a)-type sequence is the best way on surface morphology (Ohtsuka *et al.* 1991).

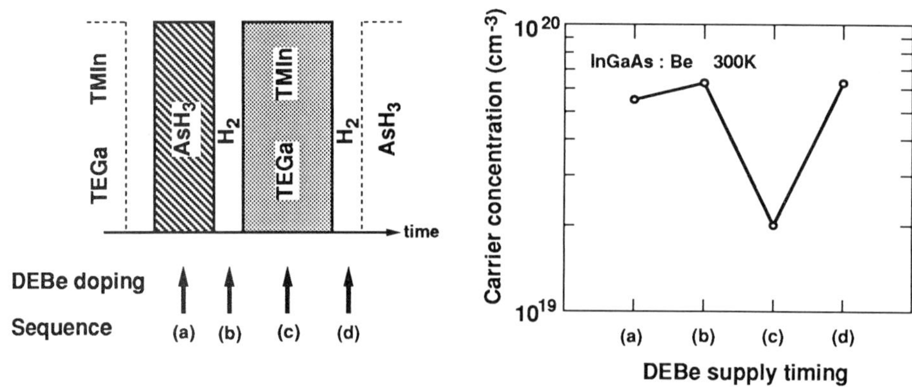

Fig. 1 Gas pulse sequence for p-InGaAs growth

Fig. 2 Dependence of doping concentration on DEBe supply timing

Growth for HBTs' epitaxial structures was done using TMIn, TEGa, PH_3, and AsH_3 as In, Ga, P, and As sources. DEBe and H_2Se were used as p- and n-type dopants. InP/InGaAs HBT device layers were grown lattice-matched to (100) Fe-doped semi-insulating InP substrates by ALE and MOCVD. The device structure utilized a 350-nm-thick n^+-InGaAs subcollector layer doped at 5×10^{18} cm^{-3}, a 300-nm-thick n-InGaAs collector region doped at 1×10^{16} cm^{-3} grown by MOCVD at 520°C. Continuously, we grew a 30-nm-thick p-InGaAs base layer doped at 7×10^{19} cm^{-3} grown by ALE at 340°C. On the base layer, we grew a 20-nm-thick InGaAs undoped spacer layer and a 75-nm-thick InP emitter layer doped at 5×10^{17} cm^{-3}, and two highly-doped ($N_d > 1\times10^{19}$ cm^{-3}) layers of InP and InGaAs to reduce the emitter contact resistance (Fig. 3) under three different growth conditions (Table 1): (sample 1) all regions were grown by MOCVD at 450°C, (sample 2) a spacer layer was grown by ALE at 340°C and other regions were grown by MOCVD at 450°C, and (sample 3) a part of the emitter and a spacer layer were grown by ALE at 340°C and other regions were grown by MOCVD at 450°C.

Layer	Material	Doping(cm^{-3})	Thickness (nm)
Cap	InGaAs	n$^+$: 5E19	50
Emitter	InP	n$^+$: 1E19 n : 5E17	25 50
Spacer	InGaAs	undoped	20
Base	InGaAs	p$^+$: 7E19	30
Collector	InGaAs	n : 1E16	300
Sub-Collector	InGaAs	n$^+$: 5E18	350

Fig. 3 Epitaxial structure

Layer	Sample 1	Sample 2	Sample 3
Cap	MOCVD	MOCVD	MOCVD
Emitter	MOCVD	MOCVD	MOCVD / ALE
Spacer	MOCVD	ALE	ALE
Base	ALE Be doped	ALE Be doped	ALE Be doped
Growth time (min.) (beyond a base layer)	18	40	53

Table 1 Growth conditions

We then measured the SIMS profile of each sample. The SIMS profile of sample 1 shows a highly doped base layer with a peak dopant concentration of $7x10^{19}$ cm^{-3} (Fig. 4 a). The profile also shows that Be has not diffused from the base. In sample 2, however, Be has diffused into the undoped spacer layer, decreasing the base dopant concentration to $5x10^{19}$ cm^{-3} (Fig. 4 b). In sample 3, Be has diffused through the spacer layer into the emitter (Fig. 4 c). The peak concentration in sample 3 is noticeably lower than that of sample 1. These profiles indicate that the base dopant of Be diffuses even at a low growth-temperature of 340°C. This is probably due to the increased growth time, since the growth time of sample 3 is about three times as long as that of sample 1.

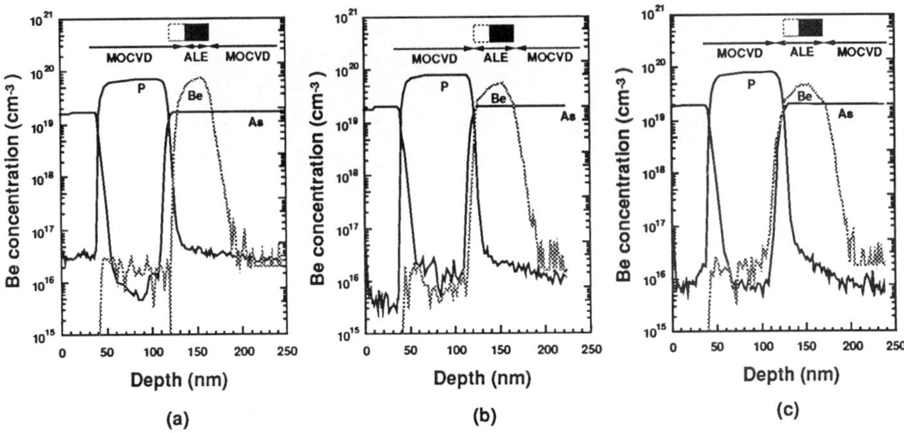

Fig. 4 SIMS profiles of InP/InGaAs HBTs: a) with the spacer layer grown by MOCVD, b) with the spacer layer grown by ALE, and c) with a part of the emitter and the spacer layer grown by ALE

To suppress Be diffusion, we must, therefore, decrease growth time as well as growth temperature. Low temperature-MOCVD is more suitable for growing spacer layers than ALE because of its higher growth rate.

III. Fabrication results and Discussion

We fabricated HBTs using selective wet etching ($HCl:H_3PO_4$, $H_3PO_4:H_2O_2:H_2O$) for the emitter, base, and collector mesas. All the ohmic contacts were formed by evaporating Ti/Pt/Au, followed by lift-off. The nominal emitter areas ranged from 3x5 μm to 250x250 μm. We deposited SiON as the passivation films using P-CVD at 300 °C.

We measured the common-emitter I-V characteristics of an InP/InGaAs HBT grown by ALE/MOCVD with a base layer doped at $7x10^{19}$ cm^{-3} (sample 1) and a 5x20-μm emitter (Fig. 5). We obtained a dc current gain of 30. The output-conductance increased dramatically with an emitter-collector voltage (V_{CE}) because the use of a narrow band-gap collector like InGaAs causes an avalanche collector current at high values of V_{CE}. We are, to our knowledge, the first to show the performance of InP/InGaAs HBTs grown by ALE.

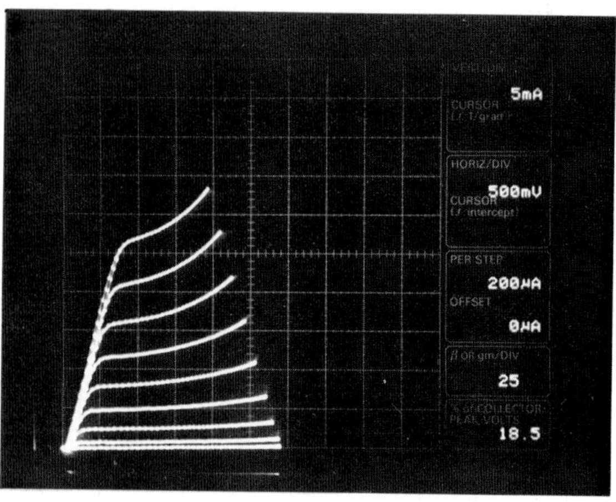

Fig. 5 Common emitter I-V characteristics of a HBT with a base layer doped at $7x10^{19}$ cm^{-3}

We measured the dependence of the small-signal current gain (h_{fe}) on the collector current density in sample 1 at V_{CE} = 1.0 V (Fig. 6). The small-signal current gain is low for small collector current densities but it increases with collector current density, reaching 35 at a density of $5x10^4$ A/cm^2. The ideality factor of the base current calculated from this slope is 1.7. This is probably due to existence of a 20-nm-thick undoped spacer layer at the InP/InGaAs heterointerface, which produces a conduction band potential notch at the heterointerface. And the generation-recombination current related to carrier accumulation in the potential notch at the heterointerface is enhanced and decreases the current gain at low collector current densities. Optimizing of the spacer layer thickness grown by MOCVD will reduce the recombination current at a low emitter bias.

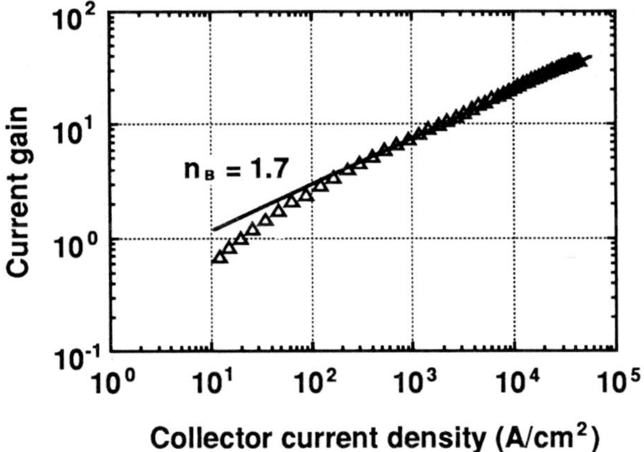

Fig. 6 Dependence of h_{fe} on the collector current density in sample 1 at $V_{CE} = 1.0$V

We drew a Gummel plot of the collector current at $V_{CE} = 1.0$ V. The deviation observed at collector current density over 1×10^3 A/cm^2 is attributed to the voltage drop by the emitter and base resistances. The measured ideality factor for the collector current was 1.2. The deviation from the ideal n = 1.0 for the collector current is probably due to tunneling of electrons at the emitter-base junction. The turn-on voltage at a collector current density of 1×10^2 A/cm^2 is 0.56 V. This value is sufficiently low and indicates that Be has not diffused into the emitter. The existence of a spacer layer also decreases the turn-on voltage. Inserting a spacer layer decreases the barrier height of the spike and the emitter injection current flows at a low emitter bias.

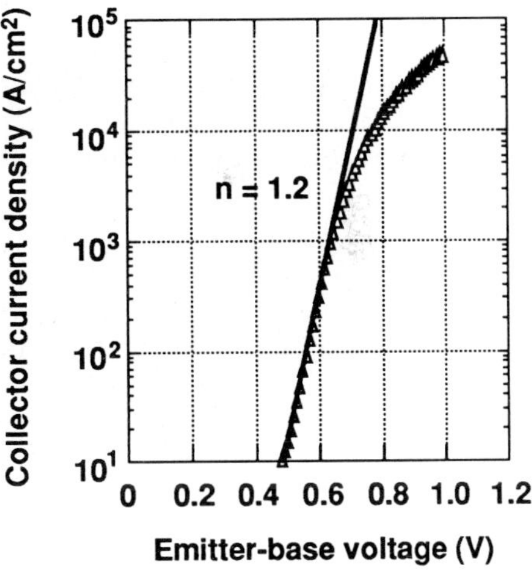

Fig. 7 Gummel plot of sample 1 at $V_{CE} = 1.0$ V

We measured the base sheet resistance across a 2-inch wafer (Fig. 8). We obtained a uniform sheet resistance of 600 Ω/square and contact resistivity of 1.5×10^{-6} Ω/cm^2. The variation of the base sheet resistance is ±3%. These values are also reasonable at this doping level. These results show the excellent epitaxial quality and good control of p-type dopants and layer-thickness achieved across a 2-inch wafer using this growth system.

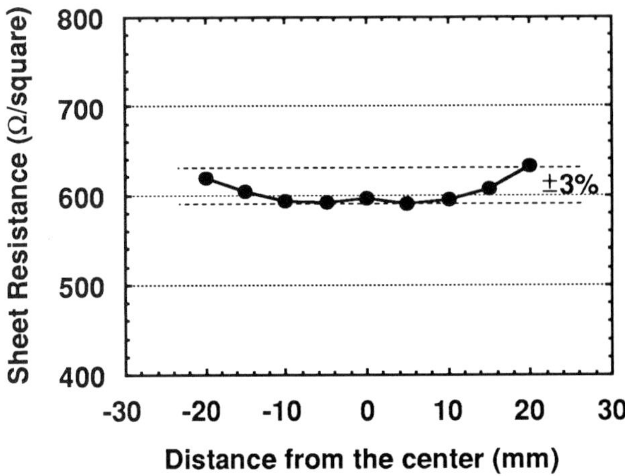

Fig. 8 Base sheet resistance across a 2-inch wafer

IV. Conclusion

We fabricated InP/InGaAs HBTs grown by ALE/MOCVD. Combining conventional MOCVD with ALE gives a highly doped base layer and suppresses Be diffusion. We obtained a dc current gain of 30 at a base dopant concentration of 7×10^{19} cm^{-3} for a HBT with a 5x20-μm emitter. We also obtained a uniform sheet resistance of 600 Ω/square across a 2-inch wafer. These results show ALE/MOCVD is a promising growth-technique for HBTs.

V. Acknowledgments

We thank Messrs T. Iwai, K. Ishii, Y. Yamaguchi, and N. Ohtsuka for their helpful comments and Drs. O. Ohtsuki, and H. Ishikawa for their encouragement during this work.

VI. References

Chen Y K, Nottenburg R N, Panish M B, Hamm R A and Humphrey D A 1989a IEEE Electron. Device Lett. **10** 267

Mishra U K, Jensen J F, Rensch D B, Brown A S, Stanchina W E, Trew R J, Pierce M W and Kargodorian T V 1989b IEEE Electron. Device Lett. **10** 467

Nottenburg R N, Banu M, Jalali B, Humphrey D A, Montgomery R K, Hamm R A and Panish M B 1990 Electron. Lett. **26** (24) 2016

Banu M, Jalai B, Nottenburg R N, Humphrey D A, Montgomery R K, Humm R A and Panish M B 1991 Electron. Lett. **27** (24) 278

Hamm R A, Panish M B, Nottenburg R N, Chen Y K and Humphrey D A 1989 Appl. Phys. Lett. **54** 2586

Bhat R, Hayes J R, Colas E and Esagui R 1988 IEEE Electron. Device Lett. **9** 442

Ohtsuka N, Kodama K, Ozeki M and Sakuma Y 1991 J. Crystal Growth **115** 460

Paper presented at the Int. Symp. GaAs and Related Compounds, Freiburg, 1993

GaAlAs/GaInP/GaAs passivated heterojunction bipolar transistors for high bit rate optical communications

C.Dubon-Chevallier, P.Launay, P.Desrousseaux, J.L.Benchimol, F.Alexandre, J.Dangla, V. Fournier

FRANCE TELECOM, Centre National d'Etudes des Télécommunications, PAB, Laboratoire de Bagneux,
196 avenue Henri Ravera, BP 107, 92225 Bagneux cedex (France)

ABSTRACT: We present a new Passivated Heterojunction Bipolar Transistor structure which yields high performance devices with a very simple processing technology. A thin n-type GaInP layer is inserted between the base layer and the GaAlAs emitter layer and allows, with the same processing step, the passivation of the base layer and the selective etching of the emitter layers. The new structure also gives a low and reproducible p-type ohmic contact resistivity. High bit rate laser driver circuits have been processed using this new structure and are operating at 7 Gbit/s.

1- INTRODUCTION

GaAs/GaAlAs Heterojunction Bipolar Transistors (HBTs) are very attractive devices for a large range of applications from ultra high speed logic to microwave power devices. However, this device requires a complicated fabrication technology and presents several limitations such as the current gain size effect [1], where the current gain decreases with the size of the device due to surface recombinations. Several solutions have been proposed to prevent this problem. Sulphur passivation has been proposed using Na_2S [2] or NH_4S [3] solutions, this process is rather complicated since the devices have to be encapsulated in order to stabilize the passivation process [4]. Another solution is to let a thin GaAlAs layer, part of the GaAlAs emitter, to act as a guard ring [5]. This solution is difficult to use in the HBT fabrication technology, because there is then no selective etch, and also because it is necessary to etch this GaAlAs layer below the p-type ohmic contact in order to obtain the required contact resistivity.

$Ga_{0.51}In_{0.49}P$ alloy lattice matched to GaAs has been presented as an alternative to GaAlAs [6, 7, 8]. Indeed, the GaAs/GaInP heterostructure exhibits a better distribution of the bandgap discontinuities, with a smaller conduction band discontinuity and a larger valence band discontinuity. This structure is then very well adapted to HBT devices, and allows very high DC current gain with abrupt emitter-base interface. Moreover, it has been shown that GaInP is not readily p-type doped with C, and will not suffer from any potential C diffusion during the growth process or during the fabrication technology [9]. The GaInP alloy presents other interesting features such as a high selectivity of chemical etching towards GaAs or GaAlAs.

In this paper, we present a new HBT structure, which takes benefit of the GaInP alloy, and allows us to obtain, with a very simple fabrication technology, the emitter layers selective etch, the passivation of the base surface and a very low p-type ohmic contact resistivity [10,11].

2- PASSIVATED HBT TECHNOLOGY

A schematic cross section of the new HBT structure is presented in Table 1. Its principle is to use 2 layers of different high band gap semiconductor materials to form the emitter: a thin $Ga_{0.51}In_{0.49}P$ layer (E1 30 nm) is introduced between the C-doped GaAs base layer and the $Ga_{0.7}Al_{0.3}As$ emitter layer (E2). The thickness of the GaAlAs layer is typical for HBT structure (150 nm). The GaInP/GaAs and GaAlAs/GaInP heterointerfaces are of the abrupt type.

E"	GaInAs	n^+	$1\ 10^{19}$	50 nm
E'	GaAs	n^+	$4\ 10^{18}$	150 nm
E2	GaAlAs	n	$2\ 10^{17}$	150 nm
E1	GaInP	n	$2\ 10^{17}$	30 nm
B	GaAs	p^+	$3\ 10^{19}$	90 nm
C	GaAs	n	$2\ 10^{16}$	450 nm
C'	GaAs	n^+	$4\ 10^{18}$	500 nm

Table 1: Passivated HBT structure

The fabrication technology to process this structure is a double mesa technology (figure 1). A first mesa is achieved to etch the GaInAs, GaAs and GaAlAs emitter layers. The etch is completed using a chemical selective etch to stop on the GaInP layer. There is then a second mesa to contact the subcollector layer. A multiple implantation process is achieved to insulate the devices and define the active regions. Then the n-type ohmic contacts are deposited, W for the emitter and AuGeNi/Ag/Au for the collector. The AuMn p-type ohmic contact is deposited, without etching the GaInP layer, and then annealed.

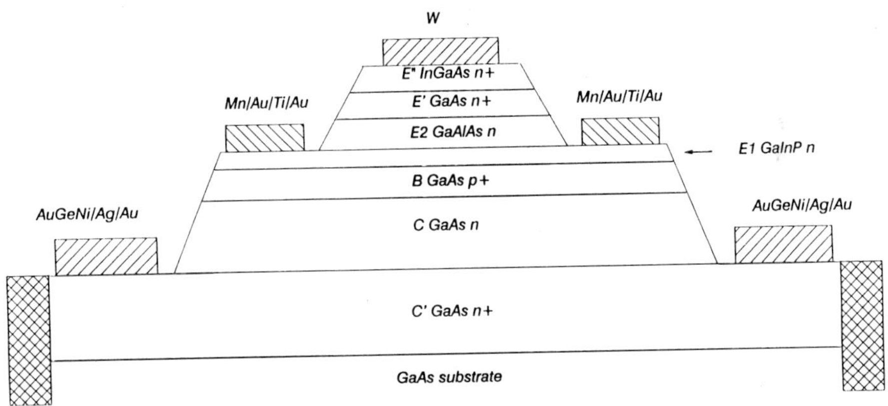

Figure 1: PHBT schematic cross section

Beside the advantages of GaInP/GaAs heterostructure, this structure exhibits other optimised characteristics:

(1) The selective etch of the emitter is easily achieved, since the etch stops on the GaInP layer.
(2) The GaAs base layer is covered by a thin n-type GaInP layer, which acts as a passivation layer [4]. The use of the p-type AuMn ohmic contact [12] permits to keep the GaInP layer even below the contact, and to obtain a very low contact resistivity.
(3) The exposed layer along the mesa is n-type GaAlAs, which acts also as a passivation layer.
(4) The thickness of the GaAlAs layer can be optimized separately, it is then possible to keep convenient emitter thickness.

3- EXPERIMENTAL RESULTS

The PHBT structures were grown by CBE. The CBE growth process and the characterization of the different layers have been presented elsewhere [6,13]. The specific contact resistivity of the AuMn p-type ohmic contact deposited on the thin GaInP layer has been measured using the TLM method. It was found to be as low as 5 10^{-7} $\Omega.cm^2$. Good uniformity and reproducibility of this result has been achieved. HBT structures have been processed using the technological process described before. Small dimension devices, as well as large dimension devices, were fabricated. A very high value of 200, associated with a base sheet resistance of 460 Ω/□, was measured for the DC current gain on large dimension devices, at a collector current density of 110 A/cm^2. This value is similar to that obtained for a conventional HBT structure with a 0.3 μm thick GaInP emitter [6].

Cutoff frequency (f_T) and maximum oscillation frequency (f_{max}) of 21 and 33 GHz, respectively, were measured on 3 x 8 μm^2 non self-aligned HBT devices. These devices, processed with the very simple non self-aligned double mesa technology previously described, were used to design and process circuits for high bit rate optical communication.

4- LASER DRIVER CIRCUIT

The laser driver is a crucial electronic component for lightwave communication systems. Its function is to modulate the laser light intensity given an electrical input stimulus. Its design, though simple in concept, is challenging, because of stringent specifications such as large output current at high bit rate.

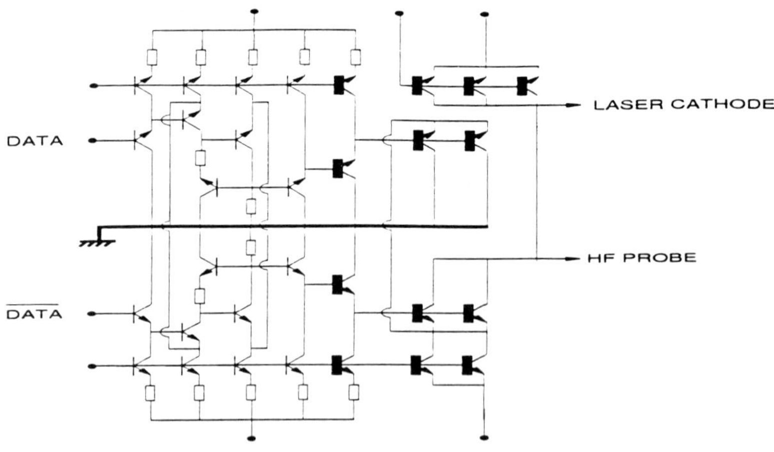

Figure 2: laser driver block diagram

The block diagram of the modulator circuit is shown in figure 2. It consists of a transimpedance amplifier and an output buffer to provide the modulation current. The circuit was processed using the simple non self aligned double mesa technology previously described (figure 3). The laser driver was tested up to 7 Gbit/s, the output eye diagram (25 mA into 50 Ω) is shown in figure 4. The laser driver occupies an area of 1.3 x 1.3 mm^2, requires 6 V supply voltage and dissipates 0.5 W.

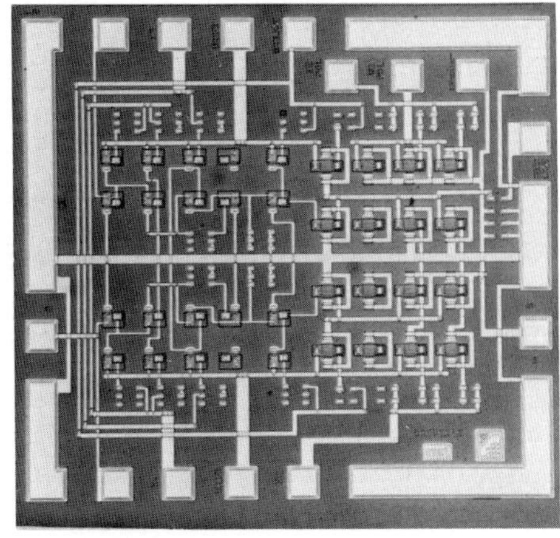

Figure 3: laser driver micrograph

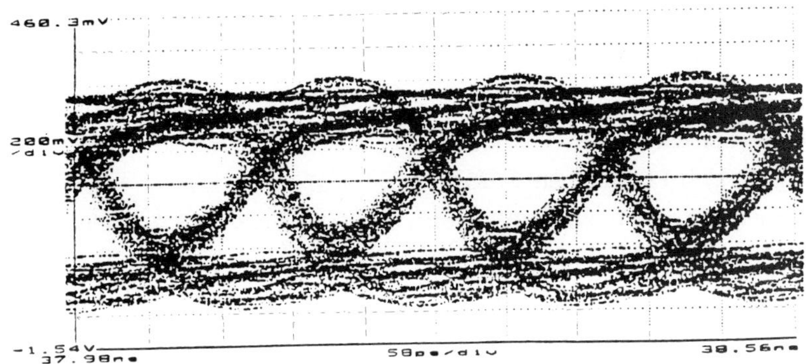

Figure 4: 7 Gbit/s output eye diagram

5- CONCLUSION

We have presented and demonstrated a new Passivated HBT structure which yields high performance devices with a very simple processing technology. This new structure allows to obtain, with the same processing step, the passivation of the base layer and the selective etching of the emitter layers. The new structure also takes advantage of the GaInP/GaAs heterojunction and permits to obtain a low and reproducible p-type ohmic contact resistivity. Laser driver circuits operating at 7 Gbit/s and providing 25 mA modulation current have been obtained using a very simple non-self aligned double mesa technology.

ACKNOWLEDGEMENTS

The authors wish to thank A.M.Duchenois, C.Besombes, L.Bricard, D.Arquey, F. Héliot, J.P. Chandouineau for device processing, J.P.Médus, E. Wawrzynkowski, M.Laporte, L.NGuyen for device characterization, M. Bon for fruitful discussions.

REFERENCES

[1] Nakajima O, Nagata K, Ito H, Ishibashi T and Sugeta T 1985 Jpn.J.Appl.Phys. 24 L596

[2] Nottenburg R, Sandorff C, Humphrey D, Hollenbeck T and Bhat R 1988 Appl.Phys.Lett. 52 218.

[3] Shikata S, Okada H and Hayashi H 1991 Inst.Phys.Ser. 112 251.

[4] Sik H, Amarger V, Riet M, Bourguiga R and Dubon-Chevallier C 1993 Proc 20th GaAs and related compounds symposium.

[5] Lin H and Lee S 1985 Appl.Phys.Lett. 47 839.

[6] Alexandre F, Benchimol JL, Dangla J, Dubon-Chevallier C and Amarger V 1990 Electron.lett. 26 1753.

[7] Zwicknagl P, Schaper U, Scleicher L, Siweris H, Bachem K, Lauterbach T and Pletschen W 1992 Electron.lett. 28 327.

[8] Delage S, 1991 Electron.Lett. 27 253

[9] Abernathy CR, Ren F, Wisk PW, Pearton SJ and Esagui R 1992 Appl.Phys.Lett. 61 1092.

[10] CNET patent.

[11] Dubon-Chevallier C, Alexandre F, Benchimol JL, Dangla J, Amarger V, Héliot F and Bourguiga R 1992 Electron. Lett. 28 2308.

[12] Dubon-Chevallier C, Gauneau M, Bresse JF, Izrael A and Ankri D 1986 J.Appl.Phys. 59 3783.

[13] Benchimol JL, Alexandre F, Dubon-Chevallier C, Héliot F, Bourguiga R, Dangla J and Sermage B 1992 Electron.lett. 28 1344.

Inst. Phys. Conf. Ser. No 136: Chapter 3
Paper presented at the Int. Symp. GaAs and Related Compounds, Freiburg, 1993

Thermal effects and instabilities in AlGaAs/GaAs heterojunction bipolar transistors

Martin Kärner, Helmut Tews, Peter Zwicknagl and Dieter Seitzer*
Siemens Research Laboratories, Otto-Hahn-Ring 6, D-81730 Munich 83, Germany
*Fraunhofer Gesellschaft FhG-IIS, Wetterkreuz 13, D-91058 Erlangen, Germany

ABSTRACT: The interaction between thermal and electrical properties of AlGaAs/GaAs heterojunction bipolar transistors is investigated. Thermal resistance variation has significant influence on the self-heating effect. Multifinger devices exhibit hot spot formation and movement. These effects are attributed to positive electrothermal feedback by the negative temperature coefficient of the base-emitter voltage. A shift of the common-emitter breakdown voltage is observed. This effect occurs due to intrinsic temperature rise by self-heating.

1. INTRODUCTION

The self-heating effect in AlGaAs/GaAs heterojunction bipolar transistors is a major limitation of performance and reliability. Especially power applications are concerned because high current densities and operating voltages are required in order to achieve high output power (Bahl 1988) and high power added efficiency (Sweet 1991).

The heat that is dissipated by transistor action must be removed to maintain low intrinsic temperatures. Lateral heat transfer through the mesa walls however is very small due to the high thermal resistivity of the passivation layer, e.g. silicon nitride (Hirai 1978). Convection and radiation at the wafer surface are negligible. So the substrate is the governing thermal resistance that can be varied by thickness and composition. Local wafer thinning and gold heat sinks were performed by Taylor (1991) to improve the thermal conductivity under the transistors. In the case of compositional variations the GaAs substrate is replaced by Si or InP.

In this paper the interaction between electrical and thermal properties is investigated. In the first part the high sensitivity of AlGaAs/GaAs HBTs against small thermal resistance increase is demonstrated. In the second part three nonidealities in the output characteristics of a large multifinger device are presented and explained.

2. LAYER SPECIFICATIONS AND PROCESS

The HBTs were fabricated with a mesa process in emitter-up configuration. The MOCVD layers were grown on semi-insulating GaAs substrates. Device planarization was done with silicon nitride. The target values for layer thickness and doping are 500 nm and $5\text{x}10^{18}$ cm^{-3} for the subcollector, 500 nm and $2\text{x}10^{16}$ cm^{-3} for the collector, 100 nm and $1\text{x}10^{19}$ cm^{-3} for the base, 150 nm and $3\text{x}10^{17}$ cm^{-3} for the emitter and 200 nm and $5\text{x}10^{18}$ cm^{-3} for the emitter cap. N- and p-type contacts are formed with NiGeAu and CrAu, respectively.

3. SELF-HEATING EFFECT AND THERMAL RESISTANCE VARIATION

The impact of electrical and thermal properties on the self-heating effect and the high sensitivity of AlGaAs/GaAs HBTs against thermal resistance changes shall be demonstrated. For this purpose the HBT layer sequence is deposited on an undoped AlGaAs buffer layer.

The AlGaAs buffer layer has an Al mole fraction of 0.3. This composition exhibits fourfold thermal resistivity compared to GaAs (Afromowitz 1973). After processing the HBT is located on an area with high thermal resistance as shown in Figure 1. This artificial increase of the overall thermal resistance can be varied with the buffer layer thickness. Wafers with AlGaAs buffer layer thickness from 0nm to 1000nm and identical HBT layer sequence were grown.

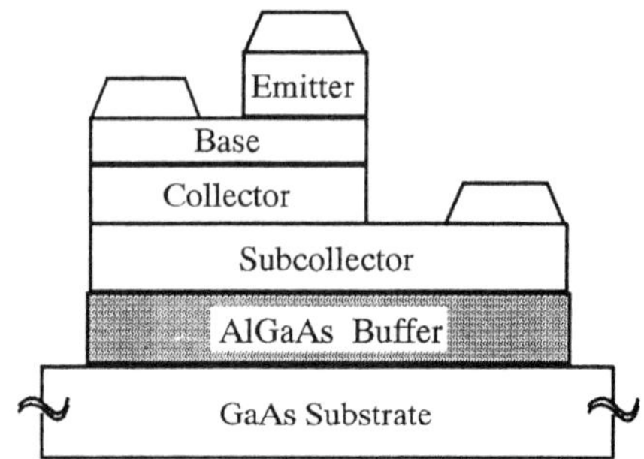

Fig.1 Cross-sectional view of a HBT with incorporated AlGaAs buffer layer

Transistors with similar current gain, offset voltage and on-resistance were selected for comparison. This is necessary as the degree of current gain degradation by self-heating depends on both the electrical and thermal properties of the transistor.

By means of this comparative study the influence of thermal resistivity can be clearly demonstrated. The common emitter output characteristics of HBTs with no buffer layer (solid line) and 1000nm buffer layer (dotted lines) are shown in Figure 2. The current gain degradation is more pronounced for the device with 1000nm AlGaAs buffer. We observed that all devices exhibited significantly stronger current gain degradation with increasing buffer layer thickness. Comparison of the buffer layer thickness (0.2-1μm) to the substrate thickness (500μm) demonstrates the strong dependence of the electrical performance on the thermal properties of the layer sequence.

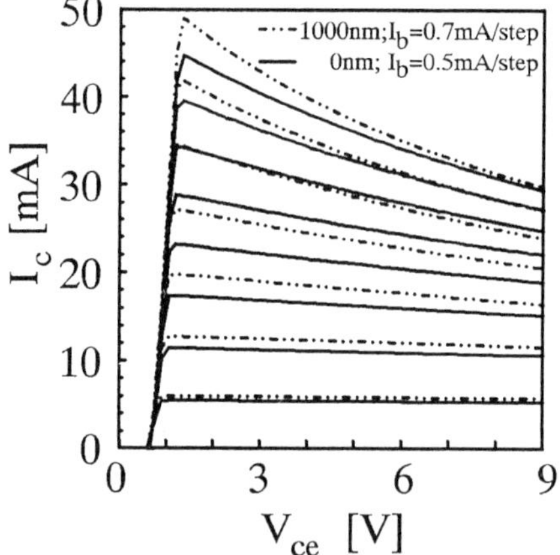

Fig.2 Common emitter output characteristics of HBTs with 0nm and 1000nm buffer

By varying the injection conditions the nature of the current gain degradation can be further discussed. In Figure 3 the typical dependence of the collector current on the dissipated power $P_{diss} = V_{be}I_b + V_{ce}I_c$ is shown. The base current is stepped by 0.7mA. The collector current decreases linearily with increasing dissipated power for constant base current. This suggests a thermal reason for the current gain decrease as the intrinsic temperature is expected to increase proportionally to the dissipated power.

The slope of the collector current however depends strongly on the base current. E.g. no current gain degradation occurs for very low base current ($I_b = 0.7mA$) whereas the slope becomes more pronounced under high injection conditions. Thus the injection conditions play a key role for the self-heating related current gain decrease.

Consequently the self-heating effect must be considered as an electrothermal interaction. Its sensitivity increases drastically with the base current. For very high base currents the slope will approach infinity. The dissipated power will be constant for any collector current.

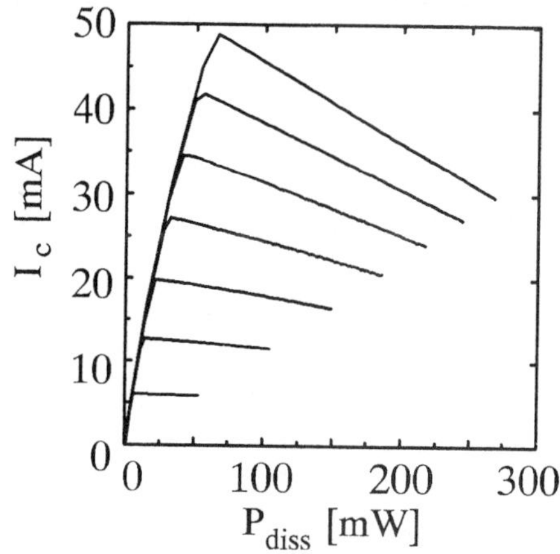

Fig.3 Collector current vs. dissipated power

4. HOT SPOT FORMATION IN LARGE MULTIEMITTERFINGER HBTs

Single emitter HBTs provide collector current densities of about $10^5 A/cm^2$ for the given layer specifications. Thus the current handling capability of a multifinger device with 64 emitter fingers of $2*20\mu m^2$ is theoretically in the order of several amperes. The output characteristics for such a large device however exhibit nonideal effects even at low current levels as shown in Figure 4.

Three nonidealities are observed:

1): current gain degradation for $V_{ce} > V_{ce1} = 6V$,

2): breakdown voltage shift (intersections of curves $I_c(I_b = 1mA)$ and $I_c(I_b = 2mA)$)

3): a steplike current drop in the curve $I_c(I_b = 2mA)$ at $V_{ce3} = 17V$.

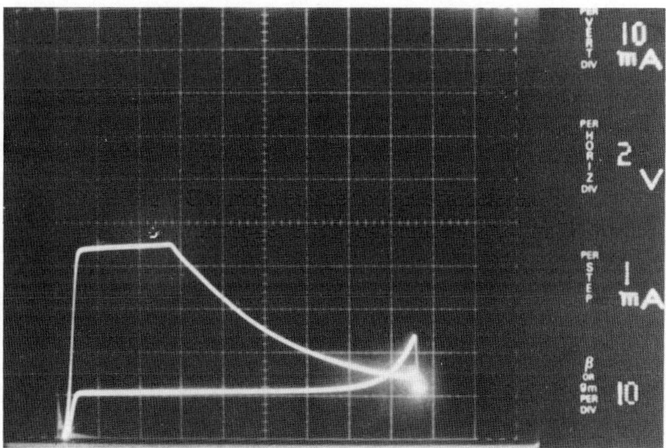

Fig.4 Common emitter ouput characteristics of a multifinger HBT (64 emitters $2*20\mu m^2$)

The collector current remains constant for collector-emitter voltages up to $V_{ce1}=6V$ and degrades strongly for higher V_{ce}. It turns out that the degradation exactly follows a constant dissipated power law as illustrated in Figure 5. This experiment is a key to the explanation of current gain degradation which have to be distinguished from self-heating effects in single emitter HBTs described in the previous chapter. The current gain degradation at constant power is attributed to inhomogeneous current distribution within the multifinger transistor. The so-called hot spot formation occurs due to the negative temperature coefficient of the base-emitter voltage (Adlerstein 1991).

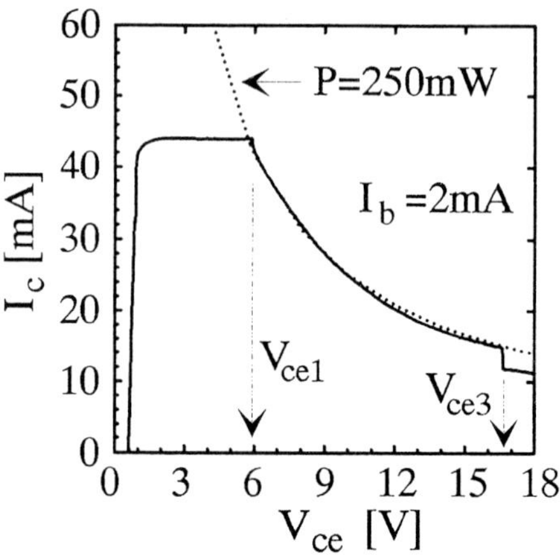

Fig.5 Common emitter ouput characteristics of a multifinger HBT (64 emitters $2*20\mu m^2$)

The decrease of the base-emitter voltage will cause base current concentration in the hottest emitter fingers once a group of emitter fingers is heated up more than others. The hot areas operate at low efficiency as both DC- and high-frequency performance degrade with increasing temperature. On the other hand the cold emitters of the multifinger transistor remain in the cutoff mode due to their higher base-emitter voltage.

In order to confirm this model, direct surface temperature measurement using liquid crystal method has been applied to GaAs microwave structures for the first time. It turns out that generation and disappearance of so-called hot spots (dark zone in Figure 6) can be detected with sufficient resolution to monitor operation conditions and site of appearance simultaneously. The hot spot contour localizes the clearpoint temperature of the liquid crystal (60°C). The liquid-crystal measurement confirms the sudden formation of a hot spot at $V_{ce}=V_{ce1}$. A small region of the multifinger transistor is active and handles the total collector current under these circumstances. Hot spot size and location are constant for increasing collector-emitter voltage.

The intrinsic temperature increase by self-heating causes a current gain decrease in AlGaAs/GaAs HBTs. In the hot spot mode however both base and collector current flow are controlled only by a few emitters within the hot spot area. These emitters instantaneously operate under high injection conditions in the moment of hot spot formation as they use the total base current. This is comparable to infinite base current in Figure 3, where the collector current will follow a curve of constant power. Thus they are heated up to high temperatures that are far beyond the scope of liquid-crystal measurements. Thermal simulations predict an average emitter temperature of about 250°C assuming one active emitter. The current flow is probably controlled by the intrinsic temperature affecting the carrier mobility. Increasing V_{ce} causes power and temperature rise that in turn decreases the carrier mobility. The collector current decreases following $I_c=P_{diss}/V_{ce}$ as the dissipated power and the peak temperature (constant hot spot size) are constant. The hot spot is operating in a power saturated mode. This mechanism prevents the transistor from self destruction by the second or thermal breakdown which is well known for silicon devices. GaAs devices however can operate at high temperatures due to the higher bandgap and the related lower intrinsic carrier density.

Large transistor layouts aid hot spot formation. Firstly, the center areas are heated more than the outer areas due to thermal interaction. Secondly, thermal and electrical properties of the individual emitter fingers may vary due to slightly changing layer properties over the transistor length. As a consequence some emitter fingers may be more sensitive against self-heating than others. In this case the most sensitive emitters, i.e. the emitters with the highest temperature coefficient of the base-emitter voltage will initialize the hot spot formation.

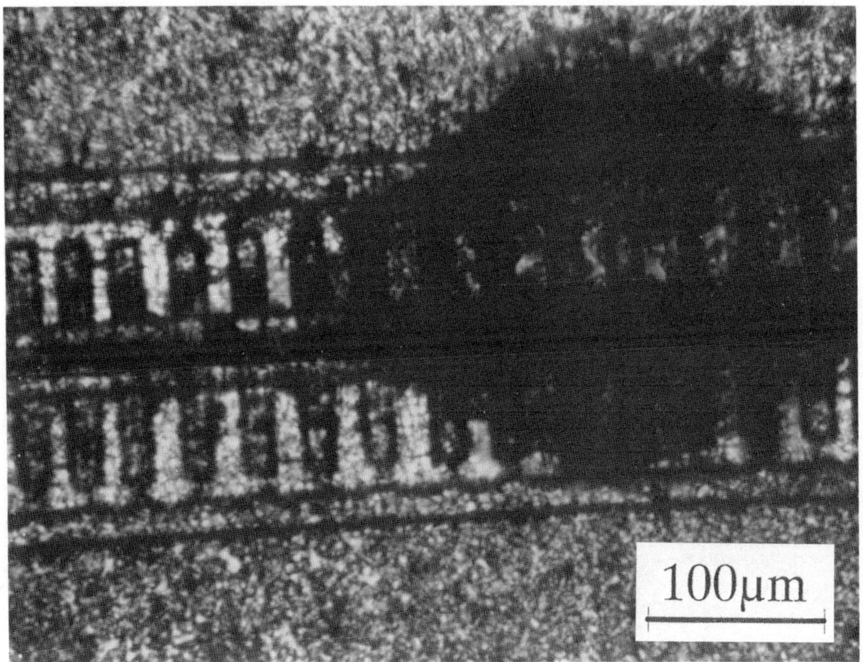

Fig.6 Liquid crystal measurement showing the hot spot in a multifinger HBT (V_{ce}=16V, I_c=2mA)

The observed shift of the breakdown voltage is most probably caused by carrier mobility decrease due to the higher temperature in the hot spot. Breakdown as observed for $I_c(I_b=1mA)$ at $V_{ce}=16V$ is caused by avalanche processes in the collector area. For $I_c(I_b=1mA)$ neither current gain degradation due to self-heating nor hot spot formation occur. All emitters of the multifinger transistor operate near room temperature as has been proven by liquid crystal thermography. For $I_c(I_b=2mA)$, however, hot spot formation takes place. The current flows in a strongly heated area of the transistor. In this operating mode no avalanche breakdown occurs up to $V_{ce}=18V$ and the curve crosses the curve $I_c(I_b=1mA)$. The electron mean free path between two scattering events is reduced at elevated temperatures. Consequently the electrons gain the energy for impact ionization at higher applied electric fields. Temperature dependent breakdown measurements performed by Malik (1992) confirm this finding for AlGaAs/GaAs HBTs.

The steplike current drop at V_{ce3} is closely related to the hot spot formation. The sudden drop of I_c at V_{ce3} = 17V is accompanied by simultaneous formation of another hot spot at a different location while the first hot spot disappears. This I_c-step and the hot spot jumping event are reproducible and only dependent on the bias condition. The switching between two stable operating conditions is a thermal instability described by Shaw (1992). The active area changes depending on a critical parameter, e.g. the electric field or the temperature at a certain location. Local change of such a critical parameter may be caused by any inhomogeneity and can not be measured with standard methods .

These three anomalities in large multiemitterfinger HBTs limit the useful device size without additional thermal stabilization. Emitter ballasting resistors however should be considered cautiously as unavoidable gain decrease is not necessarily traded against thermal stability. The hot spot formation problem is fundamental and arises from electrothermal interaction. For this reason homogeneous epitaxial layers and compact layout using thermal design tools are basic requirements for thermal stability in AlGaAs/GaAs power HBTs.

5. SUMMARY

The interaction between the thermal and electrical properties of AlGaAs/GaAs heterojunction bipolar transistors is experimentally investigated. Thermal resistance variations by the incorporation of thin AlGaAs layers yield strong dependence of the self-heating characteristics on the layer thickness. Large multifinger devices exhibit thermal instability that limits the active transistor area due to hot spot formation. The dissipated power is saturated in the hot spot mode. At high collector-emitter voltages switching of the hot spot areas as well as a shift of the breakdown voltage is observed.

REFERENCES

Adlerstein M G and Zaitlin M P 1991 IEEE Trans.El.Dev. 38 1553

Afromowitz M A 1973 J.Appl.Phys. 44 1292

Bahl I and Bhartia P 1988 Microwave Solid State Circuit Design (New York: Wiley) p.518

Hirai T , Hayashi S and Niihara K 1978 Ceramic Bulletin 57 1126

Malik R.J , Chand N , Nagle J , Ryan R W, Alavi K and Cho A Y 1992 IEEE El.Dev. Letters 13 557

Shaw M P , Mitin V V , Schöll E and Grubin H L 1992 The Physics of Instabilities in Solid State Electron Devices (New York: Plenum Press) pp.404-414

Sweet A 1990 MIC and MMIC Amplifier and Oscillator Circuit Design (Boston:Artech House) p.21

Taylor G C , Bechtle D W , Jozwiak P C , Liu S G and Camisa R L 1991 SPIE Vol.1475 103

Design considerations for the surface passivation layer structure of AlGaAs/GaAs heterojunction bipolar transistors

Hiroshi Ito, Takumi Nittono, and Koichi Nagata
NTT LSI Laboratories, 3-1, Morinosato Wakamiya, Atsugi-shi, Kanagawa 243-01, Japan

Abstract: Surface passivation layer structure using depleted AlGaAs on the extrinsic base of AlGaAs/GaAs HBTs is investigated based on calculations of conduction band edge profiles and experimental evaluations of recombination currents. A fully graded emitter combined with uniform base hardly produces any effective barriers to minority electrons in the extrinsic base when the Fermi level is strongly pinned at surface states near mid-gap. Thus, it is essential to use an abrupt emitter to effectively minimize the surface recombination which is confirmed in a series of fabricated HBTs.

1. Introduction

AlGaAs/GaAs HBTs are one of the most promising electron devices for high speed applications, and operation speeds in the 20 Gb/s range have already been reported (Matsuoka et al. 1991). In order to realize the full potential of the device, the scale down of the device dimensions is required to improve the device speed and suppress the power consumptions. A problem associated with the scale down of HBTs is the current gain reduction with shrinking device area due to the high surface recombination velocity in the extrinsic base region (emitter-size effect) (Nakajima et al. 1985a).

To overcome this, a surface passivation layer on the extrinsic base region (Fig. 1) was proposed (Lin et al. 1985) and its effectiveness for suppressing the surface recombination has been demonstrated (Lin et al. 1985, Lee et al. 1989, Malik et al. 1989, Hayama et al. 1990). This was accomplished by leaving a thin emitter layer which acts as a barrier to the minority carriers in the base. However, the band edge profile of this barrier depends on various parameters. For example, the function of the barrier can be insufficient when the emitter layer is fully compositionally graded because electrostatic potential across the pn junction may compensate the grading potential.

This paper describes design considerations for the surface passivation layer structure based on a band edge profile calculation and measurement of HBT characteristics. HBTs with different emitter/base (E/B) structures were fabricated and evaluated to reveal the influences of the emitter structure on the recombination current characteristics.

2. Band Edge Profile Calculation

First, the conduction band edge profiles in the extrinsic base region of Npn AlGaAs/GaAs HBTs were numerically investigated using the simplified method proposed by Cheung et al. (1975). The energy difference between the surface pinning position of the Fermi level and the conduction band edge for $Al_xGa_{1-x}As$ layer was assumed to be the same as the experimentally obtained Schottky barrier height (Best 1979). Base doping was fixed at 3×10^{19} /cm^3. We used a parabolic compositional grading in the emitter layer at the E/B junction for a thickness of 300 Å.

Figure 2 shows the band edge profiles for the fully graded (x(Al) = 0.3 $\rightarrow$ 0) E/B junctions with different emitter space charge concentrations. Here, surface passivation layer thickness was fixed to be 300 Å. It is clear that there is no barrier to the minority electrons in the base layer at E/B interface even in the case of a very low emitter space charge

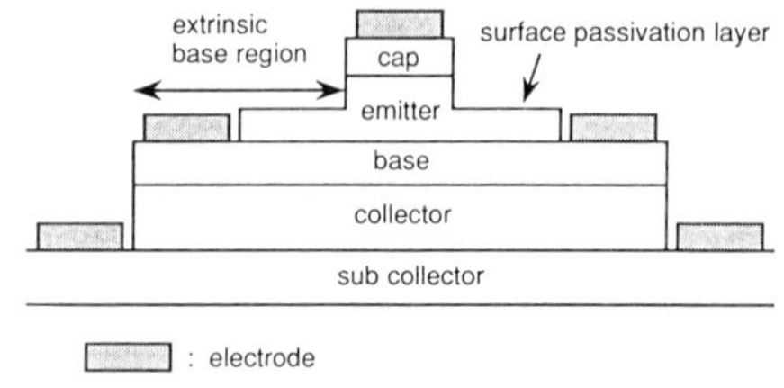

Fig. 1. A cross-sectional view of an HBT with the surface passivation layer on the extrinsic base region.

concentration of 10^{16} /cm^3. The band edge profile can also change with the passivation layer thickness. Figure 3 shows the profiles for the fully graded (x(Al) = 0.3 → 0) E/B junctions with different surface passivation layer thicknesses, where emitter space charge concentration was fixed at 10^{17} /cm^3. Again, there is no effective barrier to the minority electrons in the base layer at the E/B interface even in the case of a very thick passivation layer of 500 Å. Thickening this surface passivation layer can result in decreased surface recombination probability because the thicker layer can, for instance, decrease thermally distributed hole concentration at the extrinsic base surface and a decrease of the tunnelling probability of electrons trapped at surface states into neutral base region. However, these results indicate that surface passivation layer formed using a fully graded emitter layer is not effective for suppressing surface recombination.

Figure 4 shows band edge profiles for three different E/B structures: a fully graded E/B junction ($\Delta x = 0$), a partly graded emitter with an abrupt junction ($\Delta x = 0.1$) and a fully abrupt E/B junction ($\Delta x = 0.3$). Since the base doping level is high, even a small Al compositional discontinuity of 0.1 produces an effective barrier to minority carriers at the E/B interface. Figure 5 shows the band profiles for the partly graded (x(Al) = 0.3 → 0.1) abrupt ($\Delta x = 0.1$) E/B junctions with different space charge concentrations. Here, surface passivation layer thickness was 300 Å. A prominent barrier to minority carriers is formed at E/B hetero-interface regardless of the emitter space charge concentrations. This situation does not change with different surface passivation layer thicknesses. A slight increase of surface recombination current can be expected with increasing emitter space charge concentration and/or decreasing passivation layer thickness. This is because the tunneling probability will increase due to decrease in barrier height and barrier thickness.

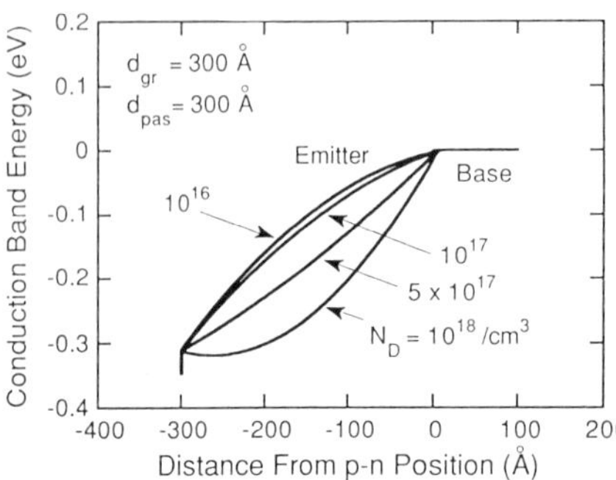

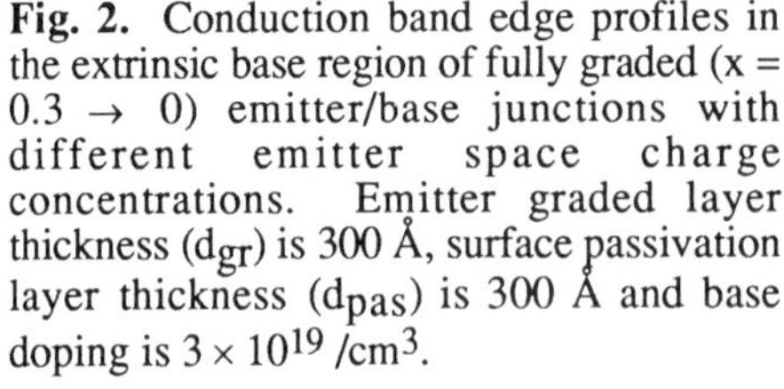
Fig. 2. Conduction band edge profiles in the extrinsic base region of fully graded (x = 0.3 → 0) emitter/base junctions with different emitter space charge concentrations. Emitter graded layer thickness (d_{gr}) is 300 Å, surface passivation layer thickness (d_{pas}) is 300 Å and base doping is 3×10^{19} /cm^3.

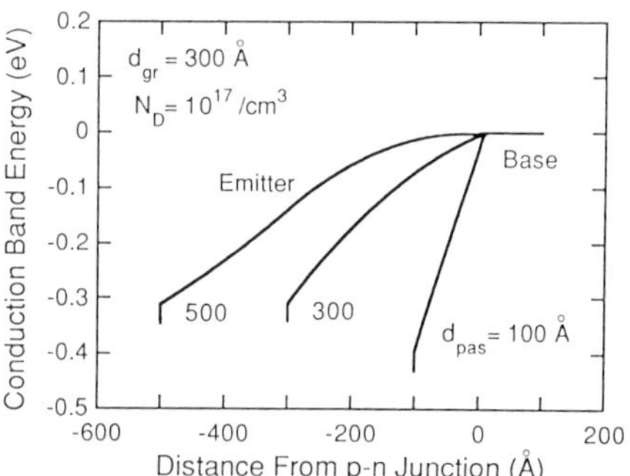

Fig. 3. Conduction band edge profiles in the extrinsic base region of fully graded (x = 0.3 → 0) emitter/base junctions with different surface passivation layer thicknesses. Emitter graded layer thickness is 300 Å, emitter space charge concentration (N_D) is 1×10^{17} /cm^3 and base doping is 3×10^{19} /cm^3.

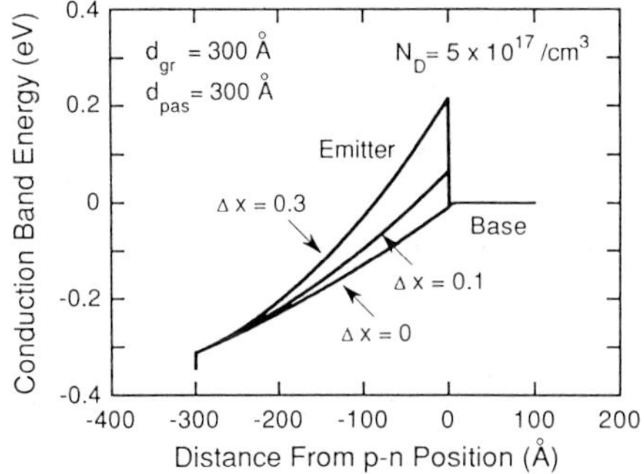

Fig. 4. Conduction band edge profiles in the extrinsic base region of fully graded ($x = 0.3 \rightarrow 0$, $\Delta x = 0$), partly graded with abrupt ($x = 0.3 \rightarrow 0.1$, $\Delta x = 0.1$) and fully abrupt ($\Delta x = 0.3$) emitter/base junctions. Emitter graded layer thickness is 300 Å, surface passivation layer thickness is 300 Å, emitter space charge concentration is 5×10^{17} /cm^3 and base doping is 3×10^{19} /cm^3.

Conduction Band Energy (eV)
0.2
0.1
0
-0.1
-0.2
-0.3
-0.4
d_{gr} = 300 Å
d_{pas} = 300 Å
Emitter
Base
10^{16}
10^{17}
5 x 10^{17}
N_D = 10^{18} /cm^3
-400 -300 -200 -100 0 100 200
Distance From p-n Position (Å)

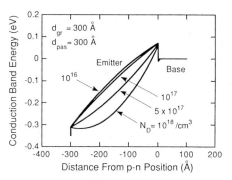

Fig. 5. Conduction band edge profiles in the extrinsic base region of partly graded with abrupt ($x = 0.3 \rightarrow 0.1$, $\Delta x = 0.1$) emitter/base junctions with different emitter space charge concentrations. Emitter graded layer thickness is 300 Å, surface passivation layer thickness is 300 Å and base doping is 3×10^{19} /cm^3.

3. Experimental Results and Discussion

The epitaxial layer structure parameters of HBTs grown by MOCVD on (100) oriented semi-insulating GaAs substrates are shown in Table I. Four different HBTs with different E/B structures were grown. Type I has a fully graded emitter, type II has a partly graded emitter with an abrupt junction, type III has a fully abrupt E/B junction and type IV has a fully graded emitter with a graded base structure. The base layer was doped with C to 3 ×

Table I. Epitaxial layer structure parameters of fabricated HBTs.

Layer	Thickness (Å)	Doping (cm^{-3})	Al Composition
n^+GaAs	2000	5×10^{18}	
N AlGaAs	300	2×10^{18}	0 - 0.3
N AlGaAs	200	2×10^{18}	0.3
N AlGaAs	300	5×10^{17}	0.3
N AlGaAs	300	5×10^{17}	x(emitter)
P^+(Al)GaAs	600	3×10^{19}	x(base)
i GaAs	3500	undoped	
n^+GaAs	5000	5×10^{18}	

	Type I	Type II	Type III	Type IV
x(emitter):	0.3-0	0.3-0.1	0.3	0.3-0.1
x(base):	0	0	0	0.1-0

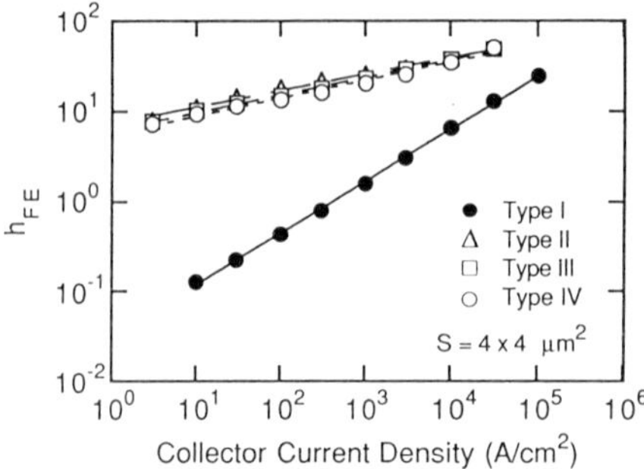

Fig. 6. Dependence of current gain on collector current density for HBTs with different emitter/base structures. The emitter area is $4 \times 4\ \mu m^2$.

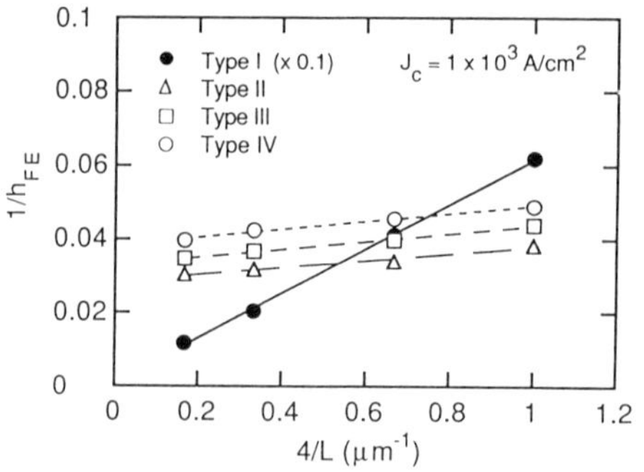

Fig. 7. Dependence of inverse current gain on 4/L at a collector current density of $1 \times 10^3\ A/cm^2$ for HBTs with different emitter/base structures.

10^{19} /cm^3, and the emitter layer to 5×10^{17} /cm^3 with Si. Square shaped mesa-type devices were fabricated by wet chemical etching and lift-off processes. We evaluated recombination current per unit length at the emitter mesa periphery. The thickness and width of the surface passivation layer for each device were 300 Å and 2 μm, respectively.

Figure 6 shows the dependences of current gains on collector current density for the fabricated HBTs with an emitter area of $4 \times 4\ \mu m^2$. The current gain curve only for the type I device has a strong dependence on collector current density. The ideality factors deduced from these curves are about 2.4 for the type I device and 1.2 for the other types.

The inverse current gain (Nagata et al. 1992) of a square shaped emitter mesa transistor can be expressed as

$$\frac{1}{h_{FE}} = \frac{1}{h_{FEi}} + \frac{I_L}{J_c}\left(\frac{4L}{S}\right) = \frac{1}{h_{FEi}} + \frac{I_L}{J_c}\left(\frac{4}{L}\right) ,$$

where h_{FE} is measured current gain, h_{FEi} is the current gain of an infinite size device (intrinsic current gain), I_L is the recombination current per unit length at the emitter mesa periphery , J_c is collector current density, L is the width of the square emitter mesa and S ($= L^2$) is the emitter mesa area. If we plot the inverse current gain of different size devices against 4/L at the same collector current density we can obtain the value of I_L from the gradient of the curve.

Figure 7 shows the dependences of the inverse current gain at a collector current density of $1 \times 10^3\ A/cm^2$ on the value of 4/L. Here, the data for the type I device are multiplied by a factor of 0.1 for ease of comparison. Although the intrinsic current gains are almost the same for all devices, only the inverse current gain of the type I device has a strong dependence

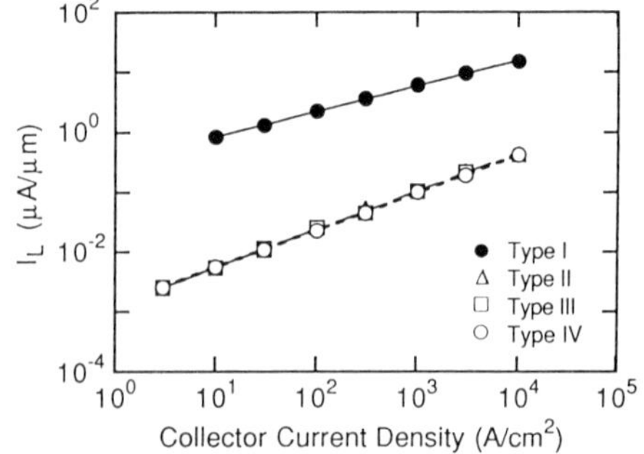

Fig. 8. Dependence of I_L on collector current density for HBTs with different emitter/base structures.

on 4/L, which clearly contrasts to the weak dependence of type II, III, and IV devices. This is due to a very large recombination current at the emitter mesa periphery in the type I device, i. e., a strong emitter-size effect. Figure 8 shows I_L values against the collector current density deduced from series of plots like the one shown in Fig. 7. It is clear that I_L for type II, III, and IV devices is nearly two orders of magnitude smaller than that for the type I device.

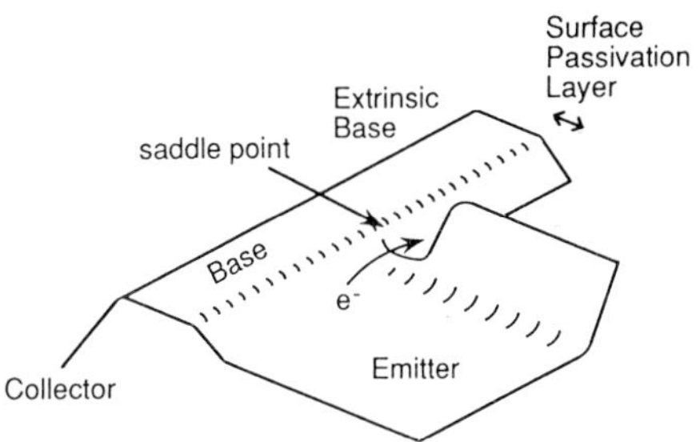

Fig. 9. Schematic drawing of the two dimensional band edge profile of an HBT at around the intersection of the base/emitter junction surfaces with the extrinsic base surface passivation layer. The low barrier height region (the saddle point which allows the electron flow from emitter to the surface depletion region) becomes wider when the surface passivation layer becomes thicker.

When the extrinsic base is covered with AlGaAs, there are two possible paths for the extrinsic base surface recombination. One is the lateral diffusion of minority carriers through the neutral base to the extrinsic base surface and the other is the carrier injection directly into the surface depletion layer through "the potential saddle point" (Fig. 9) which can exist at the intersection of the emitter and the base surfaces (Tiwari and Frank, 1989). The band edge profiles for the graded emitter (type I) and graded base (type IV) devices are nearly the same except for the base built-in field in the type IV device. (In our experiments, the emitter layer for both devices was fully graded.) In the graded base structures, minority carriers injected from the emitter to the base are swept out from the extrinsic base surface towards the collector layer due to the base built-in field (Nakajima et al. 1985b). Therefore, the reduction of I_L in the type IV device compared with that of the type I device is due to the suppression of the lateral diffusion of carriers to the extrinsic base surface. This, at the same time, implies that a surface passivation layer using the fully graded emitter layer is insufficient for preventing surface recombination.

The small I_L values for type II and III devices, which are comparable with that for the type IV device, imply that the potential barrier at the E/B interface is effectively preventing surface recombination in the extrinsic base. The I_L value for the type III device is almost the same as that of the type II device. This means that the Al compositional difference at an E/B junction of 0.1 is high enough to make an effective barrier to electrons in the base.

The ideality factors for I_L deduced form the curves in Fig. 8 are 2.3 for the type I device and range from 1.5-1.6 for the other devices. Because the surface recombination of injected minority electrons in the base layer is dominant in the I_L for the type I device, an ideality factor close to 2 is regarded as the characteristic value for surface recombination on the extrinsic base surface. This is consistent with the reported result that the ideality factor for the surface recombination is 2 when the surface recombination velocity is relatively small (Henry et al. 1978). On the contrary, the intermediate ideality factors of 1.5-1.6 for the others imply that the recombination mechanism dominating the I_L in these devices is different from that of the type I device. Although I_L for the type II - IV devices includes contributions of both recombinations in the neutral region and on the surface of the extrinsic base layer, the ideality factor for the former must be equal to that of emitter current, which is usually close to unity, as shown in Fig. 5. Thus, the intermediate ideality factors for type II - IV devices is attributed to the recombination current through the saddle point because the barrier height of the saddle point for electrons in the emitter layer is smaller than that of the intrinsic p-n junction resulting in a larger ideality factor than unity for the recombination current.

The low barrier height region becomes wider when the surface passivation layer becomes thicker (Fig. 9). This makes the recombination current through the saddle point larger. Therefore, even with nearly ideal surface passivation layers on the extrinsic base layer, a small amount of extrinsic recombination current at the emitter mesa periphery can not be eliminated in addition to the recombination current in the extrinsic neutral base region. The potential profile at around the saddle point may vary with the emitter and base dopings. Consequently, it is important to optimize the compositional profiles and doping profiles in addition to device geometries such as surface passivation layer width and thickness for realizing HBTs with a sufficiently small emitter-size effect.

4. Conclusion

We have investigated the surface passivation layer structure on the extrinsic base layer of AlGaAs/GaAs HBTs based on a conduction band edge profile calculation and an evaluation of device characteristics. Numerical analysis showed the effectiveness of employing an abrupt E/B junction for preventing the surface recombinations of electrons in the base layer. Four types of HBTs with different E/B structures were fabricated and recombination current per unit length at the emitter mesa periphery was compared. It was confirmed that the surface passivation layer with the abrupt E/B structure effectively prevents the surface recombination of injected minority carriers in the extrinsic base region even with an Al compositional difference at an emitter/base interface of 0.1. This is in contrast with the result that a surface passivation layer with a fully compositionally graded emitter structure had an insufficient effect. We have revealed contributions of two different recombination mechanisms at the emitter mesa periphery: surface recombination of laterally diffused electrons through the neutral base and surface recombination of electrons injected through the saddle point.

Acknowledgements

The authors would like to thank T. Koyama and A. Masamoto for assistances in the experiment. They are also grateful to T. Ishibashi, O. Nakajima, T. Makimura, M. Tomizawa, N. Watanabe and K. Hirata for valuable discussions.

References

Best J S 1979 Appl. Phys. Lett. **34** 522
Cheung D T, Chiang S Y and Pearson G L 1975 Solid-State Electron. **18** 263
Hayama N and Honjo K 1990 IEEE Electron Dev. Lett. **11** 388
Henry C H, Logan R A and Merritt F R 1978 J. Appl. Phys. **49** 3530
Lee W-S, Ueda D, Ma T, Pao Y-C and Harris Jr. J S 1989 IEEE Electron Dev. Lett. **10**, 200
Lin H-H and Lee S-C 1985 Appl. Phys. Lett. **47** 839
Malik R J, Lunardi L M, Ryan R W, Shunk S C and Feuer M D 1989 Electron. Lett. **25** 1175
Matsuoka Y, Yamahata S, Ichino H, Sano E and Ishibashi T 1991 Proc. IEDM 797
Nagata K, Nakajima O, Nittono T, Yamauchi Y and Ishibashi T 1992 IEEE Trans. Electron Devices **39** 1786
Nakajima O, Nagata K, Ito H, Ishibashi T and Sugeta T 1985a Jpn. J. Appl. Phys. **24** L596
Nakajima O, Nagata K, Ito H, Ishibashi T and Sugeta T 1985b Jpn. J. Appl. Phys. **24** 1368
Tiwari S and Frank D J 1989 IEEE Trans. Electron Devices **36** 2105

Paper presented at the Int. Symp. GaAs and Related Compounds, Freiburg, 1993

RF-characterization of AlGaAs/GaAs HBT down to 20K

D Peters, W Brockerhoff, R Reuter, H Meschede, A Wiersch, B Becker, W Daumann, U Seiler,[1] E Koenig,[2] FJ Tegude

Duisburg University, SFB 254, Solid-State Electronics Department, Kommandantenstr. 60, 47057 Duisburg, Germany

[1] Daimler Benz Research Center, Ulm, Germany

[2] TH Darmstadt, Department for High Frequency Technology

ABSTRACT: AlGaAs/GaAs Heterostructure Bipolar Transistors have been investigated in the frequency range from 45 MHz up to 50 GHz and in the temperature range from 300 K down to 20 K using an on wafer-measurement setup. A reliable extended equivalent circuit has been developed for the modelling of the s-parameters. Furthermore, the HBT were extensively investigated for the first time in terms of the equivalent circuit elements as functions of temperature and bias conditions. Additional measurements of the ideality factor were carried out to confirm the obtained results.

1. INTRODUCTION

Heterostructure Bipolar Transistors (HBTs) have recently demonstrated significantly improved RF performance by use of self-aligned technology (Asbeck, 1989). For the investigation of the high frequency performance and for further optimization a reliable extended equivalent circuit and a method for appropriate parameter extraction is necessary. Especially for the improvement of the device performance by reduction of parasitics the investigation and interpretation of the temperature dependent HBT parameters measured down to cryogenic temperatures is very effective.

2. HBT STRUCTURE

Self aligned N-p-n AlGaAs/GaAs with a Al mole fraction of about 30 % and an abrupt emitter-base junction were fabricated. Table 1 depicts the epitaxial layers of the HBT with the highly doped base and a spacer between base and emitter to prevent the out-diffusion of beryllium into the emitter, which strongly deteriorates the HBT performance.

Figure 1 schematically shows the triple-mesa-structure of the investigated HBT and the corresponding small-signal equivalent circuit. The active region underneath the emitter contact is represented by the equivalent circuit of two p-n junctions, one forward and one reverse biased, with the dynamic resistances (R_{je}, R_{jc}) and the capacitances C_{je} and C_{jc}

including the depletion and diffusion parts. The base-spreading resistance R_b connects the intrinsic and the extrinsic base. The extrinsic base-collector junction is represented by the feedback capacitance C_{fb}. The active components are described by the current source with the current gain α and the transit time τ. The remaining resistances and inductances account for the parasitics of the HBT.

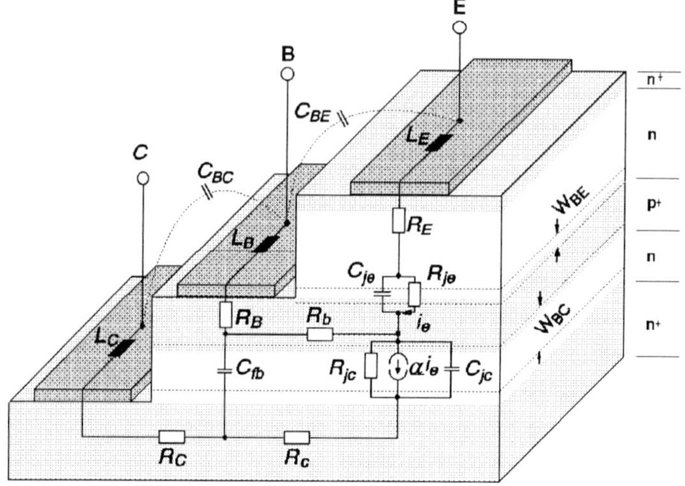

Fig. 1: Triple-mesa-structure of the HBT and the corresponding small-signal equivalent circuit elements.

Layer	Doping (cm^{-3})	Mole Fraction	Thickness
GaAs emitter cap	n = 7.3×10^{18}	0	100 nm
$Al_x Ga_{1-x} As$ grading	N = 2.0×10^{18}	x = 0.0 - 0.3	30 nm
$Al_x Ga_{1-x} As$ emitter	N = 2.0×10^{17}	x = 0.3	100 nm
GaAs spacer	undoped	0	10 nm
GaAs base	p = 8.4×10^{19}	0	90 nm
GaAs collector	n = 2.0×10^{16}	0	1500 nm
GaAs subcollector	n = 5.0×10^{18}	0	1000 nm
Substrate	s.i.		

Tab. 1: Sequence of the epitaxial layers for the investigated HBT. The emitter area is 1×20 μm with 1 emitter finger.

3. EXPERIMENTAL RESULTS

Using an on-wafer measurement setup (Meschede, 1992) scattering parameters were measured in the frequency range from 45 MHz up to 50 GHz and in the extrinsic temperature range from 300 K down to 20 K though the junction temperature of the HBT is not exactly known. Figure 2 demonstrates the measured s-parameters at 300K, 200 K and 20 K, respectively. An increase of the forward transmission S_{21} from 300K down to 200K and a strong reduction to 20K can be found. Furthermore, at low temperatures the phase behaviour of S_{21} and S_{12} changes significantly. However, the observed degradation of the HBT below 60 K demonstrates an increase of the input resistance, correlated to the reduced real part of S_{11}.

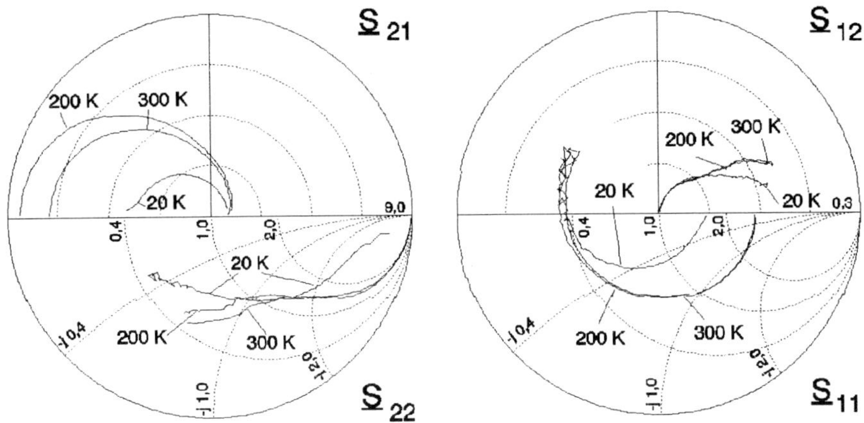

Fig. 2: Measured s-parameters of the AlGaAs/GaAs HBT from 45 MHz up to 50 GHz at 20 K, 200 K and 300 K, respectively (V_{CE} = 3.0 V, I_B = 700 μA).

The increase of the forward transmission S_{21} (Figure 2), correlated to an increase of the current gain β, is in agreement with theoretical estimations. A maximum current gain β can be achieved for HBTs according to the approximation introduced by Kroemer (1982)

$$\beta_{max} = \frac{n_e \, v_{nb}}{p_b \, v_{pe}} \, e^{(\Delta E_v / kT)} \tag{1}$$

where n_e and p_b are the doping levels of the emitter and the base, respectively, and v_{nb} and v_{pe} the mean velocity of the corresponding minority carriers at the end of the two regions. The increase of S_{21} with decreasing temperature can be explained by the enhanced exponential term of equation 1, which gives the factor of suppression of hole injection into the emitter influenced by the value of the valence band offset ΔE_v and the applied temperature.

Figure 3 demonstrates the temperature and bias dependence of the cut-off frequency f_T, which can be directly calculated from the measured s-parameters. The increase of f_T with decreasing temperature is directly correlated to the increase of the forward transmission S_{21} and to the enhanced current gain β (eq. 1). The maximum of the cut-off frequency can be found in the temperature range from 200 K down to 120 K and depends on the base current. The reduction of f_T below these temperatures can not be described by use of the theoretical expression. This deterioration can be explained by additional effects such as space-charge and surface recombination.

For the investigation of the unexpected behaviour at low temperatures the determination of the small-signal equivalent circuit is necessary. A numerical optimization method, based on statistical evolution (Schwefel, 1986), has therefore been developed. The model-generated s-parameters are fitted to the measured s-parameters via the equivalent circuit in order to get physical and device relevant elements.

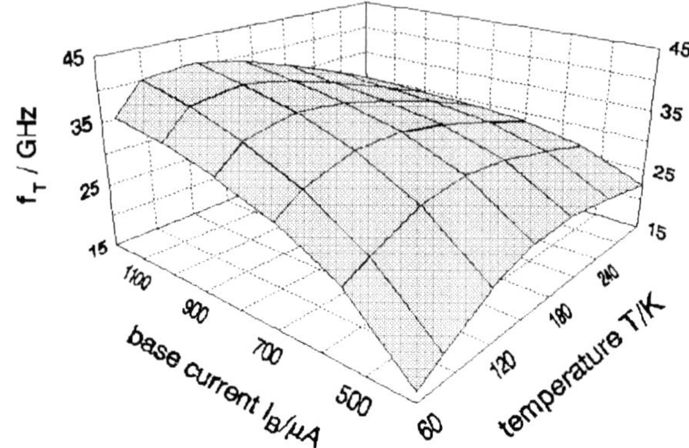

Fig. 3: Cut-off frequency f_T as a function of temperature and base current at V_{CE} =3V.

Figure 4 shows the full equivalent circuit for the HBT at room temperature and active bias conditions.

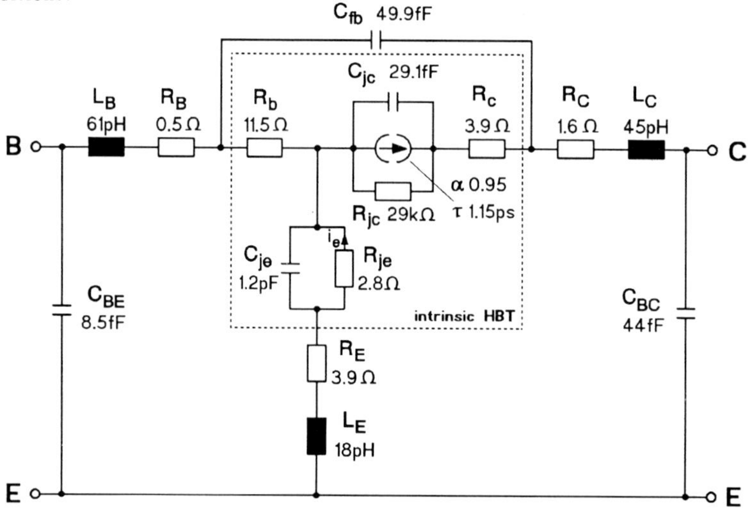

Fig. 4: Full small-signal equivalent circuit (V_{CE} = 3 V, I_B = 700 μA and T = 300 K).

4. DISCUSSION and VERIFICATION

The dynamic emitter resistance R_{je} and the emitter-base capacitance C_{je}, which represent the emitter-base junction, dominate the device performance. Therefore, it is important to investigate the temperature dependence of these two elements of the small-signal equivalent circuit. Figure 5a demonstrates the dynamic resistance R_{je} as a function of temperature and base current. The decrease of R_{je} at constant temperature with increasing base current corresponds to the exponential dependence of the input characteristic $I_B = f(V_{BE})$. However, the dynamic resistance R_{je} is independent of temperature except of very low temperatures and base currents.

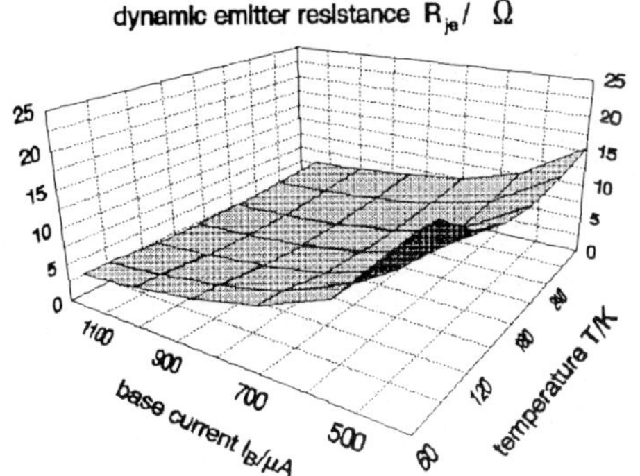

Fig. 5a: Dynamic emitter resistance R_{je} as a function of base current and temperature at V_{CE} = 3V.

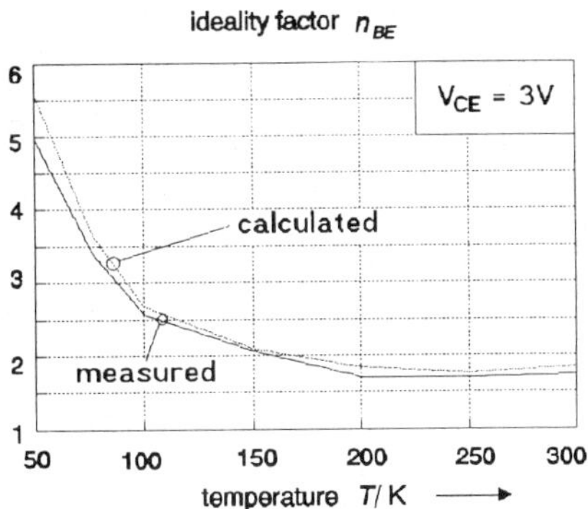

Fig. 5b: Measured and calculated ideality factor n_{BE} of the emitter-base junction as a function of temperature at V_{CE} = 3V and I_B = 700 µA.

For confirmation of this unexpected behaviour the ideality factor n_{BE}, shown in Figure 5b, was both derived from additional dc measurements and extracted from the RF measurements by use of the well-known equation for the dynamic resistance R_{je}, suggested by Sze (1985), with consideration of the temperature dependence of the emitter current I_E :

$$R_{je} = \frac{n_{BE} \cdot k \cdot T}{q \cdot I_E(T)} \tag{2}$$

The comparison of the measured and calculated ideality factor, shows the validity of the results (Figure 5b). Sze (1981) pointed out that an ideality factor close to 1 indicates a diffusion-dominated current transport and for a value close to 2 a pure space-charge recombination current can be assumed. In this case a decrease of the ideality factor from 1.7 at room temperature down to 1.5 at 200 K can be found. The slight reduction of the ideality factor can be explained by the decreasing hole current into the emitter due to the very highly doped base (Chand, 1984). Therefore, the reduction of the space-charge recombination current down to 200 K results in an increase of the current gain β according to equation 1.

On the other hand with decreasing temperature below 200K the recombination currents in the space-charge region and in the bulk material, which increases exponentially with temperature, dominate the charge carrier transport. This leads to the observed increase of the ideality factor (Figure 5b). Moreover, the decrease of the temperature leads to a reduction of the carrier transport across the abrupt N-p junction and tunneling through the spike of the heterojunction dominates. Additionally the extension of the emitter-base depletion region has to be considered caused by the freeze-out effect in the emitter(Chand,1984).

Due to this reduction of the carrier injection from the emitter into the base the extrinsic base surface recombination must be taken into account (Liu, 1992).

Figure 6 demonstrates the bias and temperature dependence of the emitter-base capacitance C_{je}. The enhancement of C_{je} with increasing base current I_B and emitter-base voltage V_{BE}, respectively, corresponds to the reduction of the emitter-base space charge region W_{BE} (Figure 1). Furthermore, at low base currents C_{je} is independent of temperature and at higher base currents the temperature behaviour of C_{je} is similar to the temperature dependence of cut-off frequency f_T (Figure 3) and is in agreement with the temperature behaviour of the forward transmission S_{21} (Figure 2).

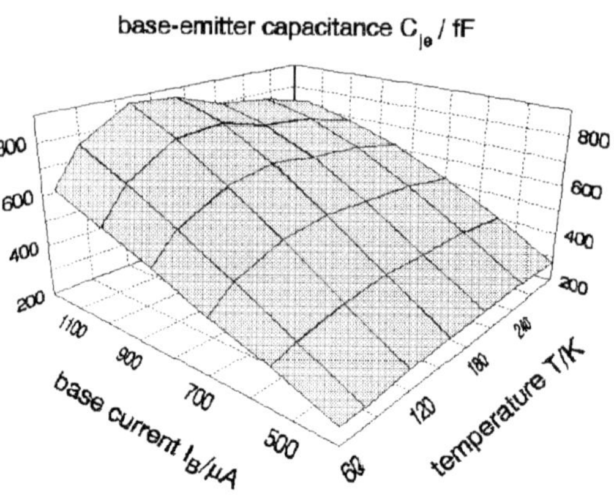

Fig. 6: Base-emitter capacitance C_{je} as a function of temperature and base current (V_{CE} =3V).

5. SUMMARY

The rf performance of AlGaAs/GaAs N-p-n HBTs was investigated in terms of the equivalent circuit elements as functions of temperature and bias conditions. An improvement of the device performance, especially of the current gain β and the cut-off frequency, down to 200K was found. Below this temperature down to cyrogenic temperatures a deteroration of the device performance was observed due to an increase of the recombination effects and the reduction of thermionic carrier transport. Additional dc measurements were carried out to analyse the ideality factor as a function of temperature. The comparison of both the measured and calculated ideality factor have shown the validity of the model.

6. REFERENCES

Asbeck P M et al 1989 IEEE Transactions on Electron Devices ED-36 pp.2032-2042

Chand N, Fischer R, Henderson T, Klem J, Kopp W, Morkoc H 1984 Applied Physics Letters Vol-45 pp. 1086-1088

Fricke K, Hartnagel H L, Leer W Y, Würfl J 1992 IEEE Transactions on Electron Devices ED-39 pp. 1977-1981

Kroemer H 1982 Proceedings of the IEEE Vol. 70 pp.13-25

Liu W 1992 IEEE Transactions on Electron Devices ED-39 pp. 2726-2731

Meschede H et al 1992 IEEE Transactions on Microwave Theory and Techniques Vol. 40 No. 12 pp. 2325-2331

Schwefel H P 1986 Numerische Optimierung von Computermodellen (Basel and Stuttgart: Birkhäuser)

Sze S M 1985 Semiconductor Devices, Physics and Technology (New York: John Wiley & Sons)

Paper presented at the Int. Symp. GaAs and Related Compounds, Freiburg, 1993

Empirical analysis of emitter ballasting resistance effects on stability in power heterojunction bipolar transistors

U. Seiler, E. Koenig*, U. Salz, and P. Narozny

DAIMLERBENZ Research Center, Ulm, Germany
*Department for High Frequency Technology, TH Darmstadt, Germany

ABSTRACT: An empirical DC and RF analysis of the effect of varying the ballasting resistance values for multi-finger Heterojunction Bipolar Transistors (HBTs) has been carried out. A severe disadvantage of HBTs is the positive feedback between the base-emitter junction current and the junction temperature. Non-uniform current distribution among the emitters of a multi-finger power HBT leads to localized overheating and the drastic reduction of the DC gain embodied in the collapse of the collector current. Ballasting resistances can be integrated monolithically in the device to current-couple the emitters and enhance device stability. The phenomenon of the current collapse is discussed, and a practical method for determining the optimal values for the ballasting resistance is presented.

1. INTRODUCTION

Heterojunction Bipolar Transistors (HBTs) have great potential for microwave RF power applications. The benefits of growth-determined critical dimensions, the relatively large minimum feature size (~2 μm) and likely reduced die size give HBTs an advantage over Field Effect Transistors (FETs). The high breakdown voltages, power-added efficiencies and power densities attainable with HBTs make them attractive high-power candidates. The increased power densities result primarily from the high current densities which can be realized implementing an HBT structure. Unfortunately, operating at higher current densities also directly increases the thermal burden on the device. As a bipolar device, the HBT has the enormous drawback of positive feedback between the base-emitter p-n heterojunction current and temperature. This shortcoming was recognized early in bipolar applications [Scarlett 1963]. Without proper thermal considerations in the design, localized hot-spots can develop which are detrimental to device behavior and can escalate to device failure.

In order to sustain large voltages and currents, power HBTs are primarily large-area devices. The area is commonly divided into many thinner emitter fingers to accommodate microwave performance. If the fingers do not heat up uniformly under bias, there is a disparity among the base-emitter junctions of the power device. This results in a non-uniform distribution of current, which due to the positive feedback, effects a further divergence of the finger temperatures. The thermal distribution in Fig. 1 is an example of how the temperature can vary in adjacent fingers of a power HBT.

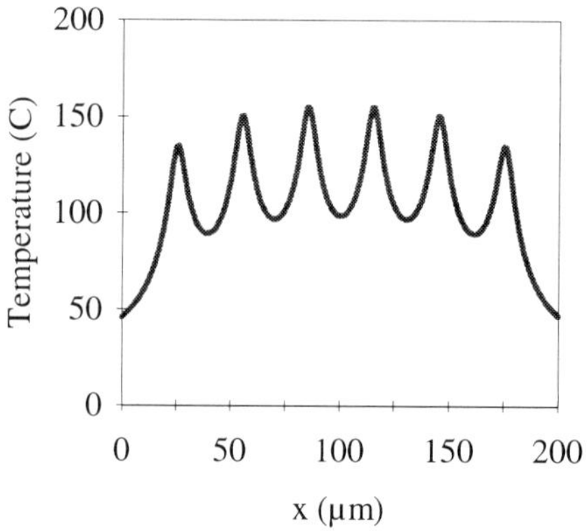

Fig. 1. Calculated thermal distribution in six adjacent emitter fingers shows non-uniform temperature distribution

A thoughtfully considered thermal approach must be implemented for a multi-finger design in order to ensure reliable operation. The inclusion of a stabilizing ballasting resistor in series with each emitter finger facilitates current-coupling and more uniform current distribution [Bergmann 1966]. For high power density Si devices, the effect of varying ballasting resistance values on transistor stability has been investigated theoretically and experimentally [Arnold 1974]. A theoretical investigation of varying resistance's influence on HBT thermal stability exists as well [Liu 1993]. Experimental verification of the stabilizing effect of monolithically integrated ballasting in HBTs has been published [Gao 1989, 1991]. To date, however, no quantitative experimental account of the influence of ballasting resistor value on thermal stability has appeared.

2. EXPERIMENTAL RESULTS

HBTs were designed and fabricated for high power density applications. The material structure used for fabrication is given in Table I. Multi-emitter devices were produced with

Material	Doping (cm^{-3})	Thickness (Å)
InGaAs	n^+ 2x10^{19}	400
graded InGaAs	n^+ 1x10^{19}	400
GaAs	n^+ 5x10^{18}	400
graded AlGaAs	n^+ 1x10^{18}	300
$Al_{0.3}Ga_{0.7}As$	n 2x10^{17}	1000
GaAs	p+ 5x10^{19}	1000
GaAs	n 2x10^{16}	10000
GaAs	n^+ 5x10^{18}	10000
GaAs	S.I.	

Table I. MOCVD grown HBT structure used to fabricate power devices. The n-dopant is Si, the p-dopant is C.

a variety of emitter areas. The layout includes power transistors which can be fabricated with a monolithically integrated metal resistance R_{bal} for ballasting purposes in series with the emitter. A refractory metal can be used for the construction of the ballasting resistors in order to increase the reliability of the devices. A typical power transistor with twelve ballasted emitter fingers is shown in Fig. 2. The device is made up of two parallel rows of six emitter fingers. Each finger has an active emitter area of 60 μm^2. The identical design but without the ballast resistors is also included for control purposes in the implemented mask set. The measured current-voltage (I-V) characteristics from six fingers of an unballasted and a ballasted transistor are shown in Figs. 3a and 3b, respectively. Above a certain magnitude of collector current (I_C), both devices display the negative differential

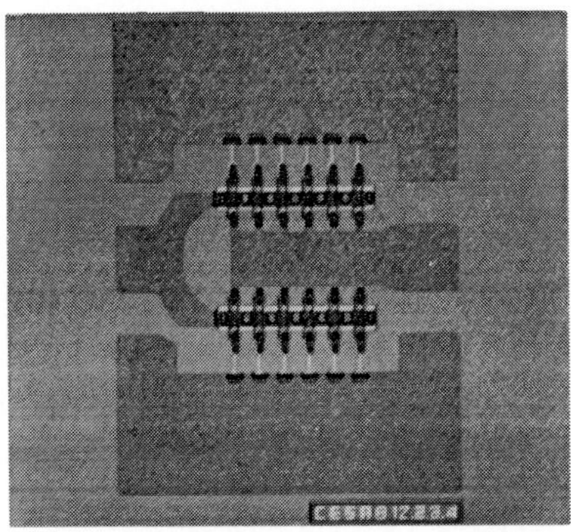

Fig. 2. A twelve-finger power HBT with emitter ballasting resistors.

resistance (NDR) which has been determined to be an inherent trait of HBTs [Chand 1984, Gao 1992]. This current-limiting mechanism stems from the increased injection of holes from the base into the emitter with increasing temperature. The rise in hole injection results

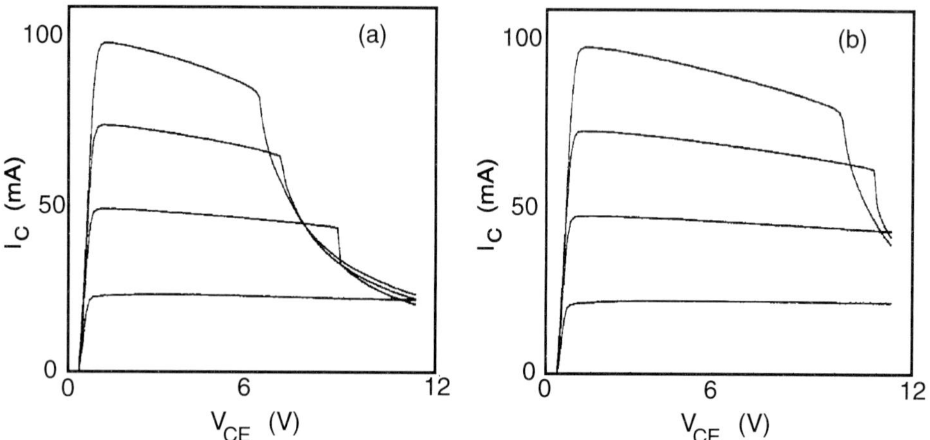

Fig. 3 (a) The I-V characteristic for six unballasted emitter fingers of a power HBT and, (b) the I-V characteristic for six identical fingers but with 2 Ω ballast resistance.

from temperature-induced base band-gap shrinkage and from the emitter-base valence band offset reduction.

A dramatic feature of the I-V characteristics of Fig. 3(a) and 3(b) is the sudden hyperbolic collapse of I_C. This phenomenon is not predicted by any of the known previous work on power HBTs. It is a direct result of the HBTs' negative temperature coefficient and the non-uniform temperature distribution among parallel emitters. The heat resulting from power dissipation in the device is distributed unevenly across the surface, as seen in Fig. 1. This temperature difference among the emitters causes the currents through the fingers to be different. Because of positive current-temperature feedback, the fingers which heat up first continue to draw more current and heat up even more. The hyperbolic collapse results when the power dissipated in the hot emitters reaches a physical limit. This mechanism will be discussed in detail elsewhere [Seiler 1993].

The consequence of the emitter ballasting resistance is evident from the comparison of Figs. 3(a) and 3(b). In the unballasted case, for example, the IC collapse occurs in the top curve at approximately 5.4 V, whereas in the ballasted case the device appears stable until 8.4 V. The enhanced stability is achieved through the current-coupling effect of the ballasting resistance, 2 Ω in this case. The power level at which devices become thermally unstable shifts to higher VCE levels with ballasting. The higher emitter resistance is not introduced without penalty, however. The higher series resistance due to ballasting increases the knee voltage

Vknee of the I-V characteristic. Fig. 4 shows the effect of varying Rbal on IC stability and Vknee for the simplified case of two emitter fingers in parallel and base current IB held constant. The smallest value of Rbal (1 Ω) leads to the lowest Vknee, but at higher VCE values (i.e., >6V), IC shows pronounced thermal influence and drops markedly. The 1 Ω resistance apparently does not sufficiently couple the two fingers under these conditions. The curve from the fingers ballasted with 5 Ω has a slightly higher Vknee than the 1 Ω curve, the thermal degradation though, is significantly reduced at higher VCE. Obviously, the 5 Ω value improves the current distribution among the fingers. An increase of R_{bal} to 15 Ω does not show any improvement over 5 Ω at higher V_{CE}, but V_{knee} is shifted to a higher value. From these measurements, it is evident that an R_{bal} of 5 Ω furnishes an optimal balance between low series resistance and sufficient current-coupling.

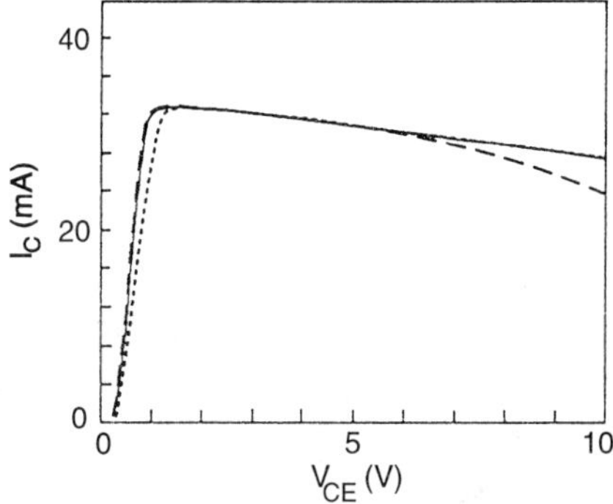

Fig. 4. I-V curves for two emitter fingers with R_{bal}= 1 Ω (---), 5 Ω (—), and 15 Ω (···). I_B= 1mA in all three cases.

The small signal RF performance of a multi-finger HBT is also affected by R_{bal}. The thermal instability of the device brought on by inhomogenous current distribution also causes the fingers to operate under non-uniform bias conditions. The dominance of a few hot emitters, compounded by the bias disparity, brings about severe degradation under RF operation. Ballasting extends the stability of the HBT and thus improves performance. Fig. 5 shows the small-signal RF results for the ballasted twelve-finger HBT such as the one in Fig. 2. The measurements were made at V_{CE} = 3.5 V and I_C = 330 mA with Rbal = 3Ω. For an identical but unballasted device, the conditions at the same bias point were so unstable that the cutoff frequencies were an order of magnitude worse. Too large a R_{bal} was seen to have a negative effect on small-signal performance. The RF gain appeared to diminish with increasing emitter series resistance.

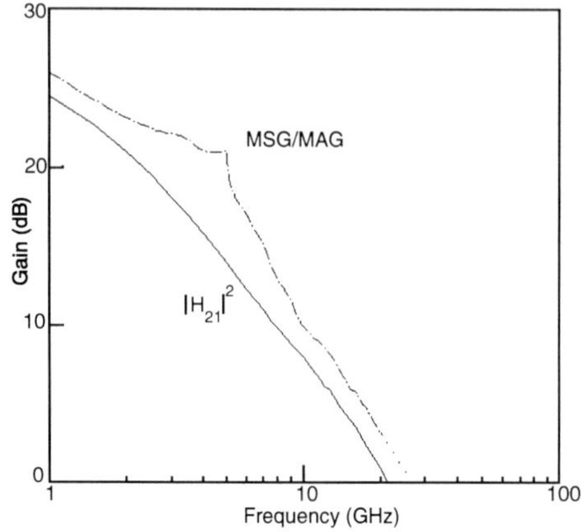

Fig. 5. RF results for a ballasted twelve-finger HBT.

3. CONCLUSION

Non-uniform current distribution in HBTs leads to an extreme collapse in I_C due to the positive current-temperature feedback and negative temperature coefficient of current gain. This collapse results from certain fingers carrying a greatly disproportionate amount of current and becoming excessively hot. Ballast resistors serve to couple the individual emitters of a multi-finger HBT. The optimal ballasting accomplishes uniform current distribution without adversely affecting device performance.

REFERENCES

Scarlett R M, Shockley W and Haitz R H 1963 Physics of Failure in Electronics (Baltimore, Spartan) pp 194-203
Bergmann F and Gerstner D 1966 IEEE Trans. Electron Dev. 13 630
Arnold R P and Zoruglu D S 1974 IEEE Trans. Electron Dev. 21 385
Liu W and Bayraktaroglu B 1993 Solid-State Electron. 36 125
Gao G-B, Wang M-Z, Gui X and Morkoç H 1989 IEEE Trans. Electron Dev. 36 854
Gao G-B, Ünlü M S, Morkoç H, and Blackburn D L 1991 IEEE Trans. Electron Dev. 38 185
Chand N, Fischer R, Henderson T, Klem J, Kopp W and Morkoç H 1984 Appl. Phys. Lett. 45 1086
Gao G-B, Fan Z F and Morkoç H 1992 Appl. Phys. Lett. 61 198
Seiler U, Koenig E, Narozny P and Dämbkes H 1993 BCTM

Paper presented at the Int. Symp. GaAs and Related Compounds, Freiburg, 1993

Infrared devices based on III–V antimonide quantum wells

H. Xie, Y. Zhang, N. Baruch, and W. I. Wang

Department of Electrical Engineering, Columbia University, New York, NY 10027

ABSTRACT: In this paper we present our recent results on normal incidence infrared devices based on GaSb quantum wells (QWs). We have demonstrated a type II p-doped InAs/GaSb QW structure which showed enhanced intervalence subband transitions due to the coupling of the InAs conduction band to the GaSb valence band. The interconduction subband absorption and its growth orientation dependence in GaSb/$Ga_{1-x}Al_xSb$ L-valley QWs are discussed. A new modulation mechanism using electric-field induced direct-indirect transitions in GaSb/$Ga_{1-x}Al_xSb$ QWs is also presented.

1. INTRODUCTION

Intersubband absorption in QWs (West and Eglash 1985) has been attracting a lot of interest recently for applications in various infrared (IR) sensing and switching devices, such as photodetectors and modulators. An advantage of using QW structures for these applications is that the absorption wavelength can be tuned over a wide range by changing the well widths and barrier heights. So far, however, most of the studies in this area have been limited in the n-type direct-gap GaAs/$Ga_{1-x}Al_xAs$ system (Levine *et al* 1987) in which the selection rules forbid interconduction subband absorption for normally incident light, and optical gratings (usually metallic) or waveguides have to be used to couple the radiation into the structure.

For intersubband absorption to be intrinsically allowed at normal incidence, the system must have finite coupling between the electrical field of the normal incidence light (perpendicular to the QW growth direction) and the electron motion during intersubband transitions (parallel to the growth direction). This can be realized in either p-type or indirect-gap n-type QWs. One promising materials system exhibiting such properties is the GaSb-based QWs. In the following sections the intersubband absorption properties in three different GaSb-based QWs will be briefly reviewed.

2. TYPE II P-DOPED InAs/GaSb QWS

In principle, normal incidence intervalence subband absorption is allowed in conventional p-type QWs (Levine *et al* 1991; Katz *et al* 1992a; Xie *et al* 1991a and 1992; Chen *et al* 1992). Unlike s-like conduction-band Bloch states for electrons, the Bloch states for holes are linear combinations of p-like valence-band Bloch states, which can provide nonzero coupling to normally incident radiation. The heavy- and light-hole mixing due to the QW potential can further promote absorption. Due to the inverse relationship between the effective mass and the absorption coefficient, p-type QWs with small heavy-hole effective masses are preferred. Among the III-V semiconductors, GaSb has the smallest heavy-hole effective mass, which leads to large absorption.

We have recently demonstrated that further improvements of the intervalence subband absorption can be achieved in type II p-doped InAs/GaSb QWs (Katz *et al* 1993). In the InAs/GaSb system, the GaSb valence-band edge lies 150 meV higher than the InAs conduction-band edge (Chang *et al* 1981). Due to this unusual band lineup, a strong coupling exists between the InAs conduction and GaSb valence bands (Luo *et al* 1990; Yu *et al* 1991; Chen *et al* 1992). As a result the valence-band structures can be significantly modified and therefore the optical matrix elements for heavy-hole to light-hole intervalence subband transitions can be tremendously enhanced.

Figure 1 shows the intervalence subband absorption spectrum for InAs/GaSb QWs measured at 77K for normal incidence IR radiation. The sample, grown on InAs (100) substrate by molecular beam epitaxy (MBE), has the InAs "barrier" width of 80 Å and the GaSb "well" width of 30 Å, doped p-type with Be to a concentration of 2×10^{18} cm^{-3}. A Fourier-transform infrared (FTIR) spectrometer was used to obtain the

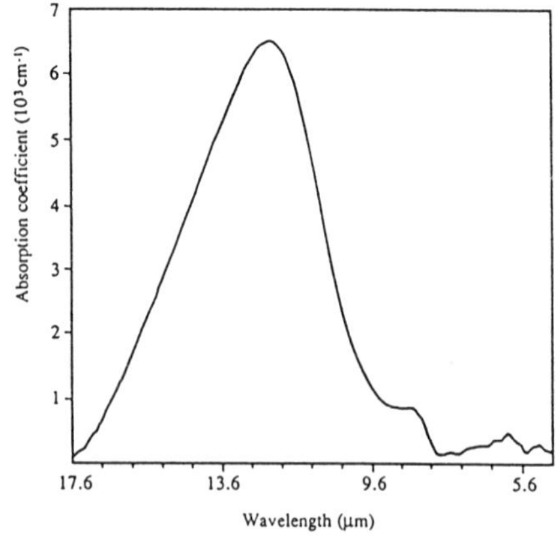

Fig. 1. Intervalence subband absorption spectrum of the InAs/GaSb QWs measured at 77 K for normal incidence IR radiation.

absorption spectrum. Our experimental results show that the absorption peak occurs at 12.3 μm with peak absorption coefficient up to 6500 cm^{-1}. As expected, the observed absorption in the InAs/GaSb system is considerably larger than that in conventional p-type structures (Levine *et al* 1991). This demonstrates the fact that the strong coupling between the InAs conduction band and GaSb valence band plays an important role in enhancing the intervalence subband absorption.

3. GaSb/$Ga_{1-x}Al_xSb$ L-VALLEY QWS

It has been recognized that n-type indirect-gap semiconductor QWs having tilted conduction valleys with respect to the QW growth direction can intrinsically absorb normally incident light (Yang *et al* 1989; Brown and Eglash 1990; Xie *et al* 1991b; Brown *et al* 1992; Katz *et al* 1992b). In an indirect-gap material electrons occupy conduction-band valleys with ellipsoidal constant-energy surfaces (ellipsoids). When the growth direction is misaligned with respect to the principal axes of an ellipsoid, the effective-mass anisotropy of electrons in that ellipsoidal valley can provide nonzero coupling to the normally incident light. Among the III-V indirect-gap heterostructures, the GaSb/$Ga_{1-x}Al_xSb$ L-valley system has the conduction-band offsets large enough to achieve substantial IR absorption. For an L-valley material there are four separate ellipsoids consisting of eight half-ellipsoids located along the eight equivalent <111> axes at the L points of the Brillouin zone. Hence, interconduction subband absorption at normal incidence in L-valley QWs is appreciable as long as the structure is not grown on a (111) substrate.

Although GaSb is a direct-gap material, its L-valley minimum is located only 63 meV above the Γ-minimum (Alibert *et al* 1983). Due to this small energy separation and the much smaller effective mass of the Γ valley than that of the L valleys, the first Γ state in a QW can be easily pushed above the first L state by the quantum-confinement effect. As a result the QW ground state becomes the L state for sufficiently narrow QWs. For the GaSb/$Ga_{0.6}Al_{0.4}Sb_{0.9}As_{0.1}$ QWs designed in our experiments, this Γ-L crossover occurs for a GaSb well thickness of about 50 Å.

As a guide we have evaluated intersubband absorption coefficients and their growth orientation dependence for GaSb/$Ga_{0.6}Al_{0.4}Sb$ QWs with five different growth directions (Zhang *et al* 1993). In Figure 2, we present the intersubband transition wavelength as a function of the well width. For a given well width the transition wavelength decreases in QWs in the following order: [211], [311], [110], [511], and [100]. This is due to the fact that the quantization effective mass, which determines the OW subband energies, becomes

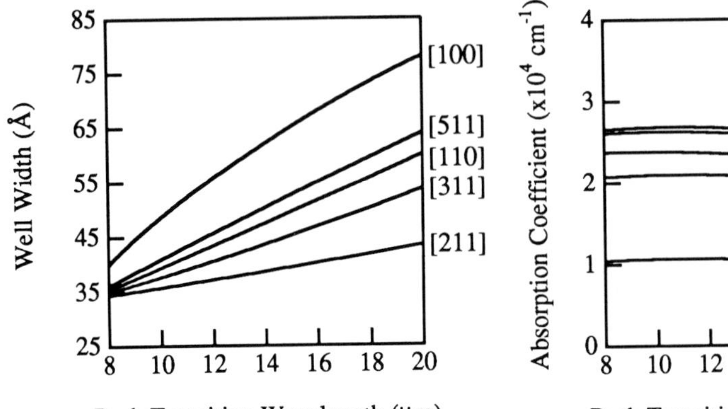

Fig. 2. The peak transition wavelength as a function of the well width.

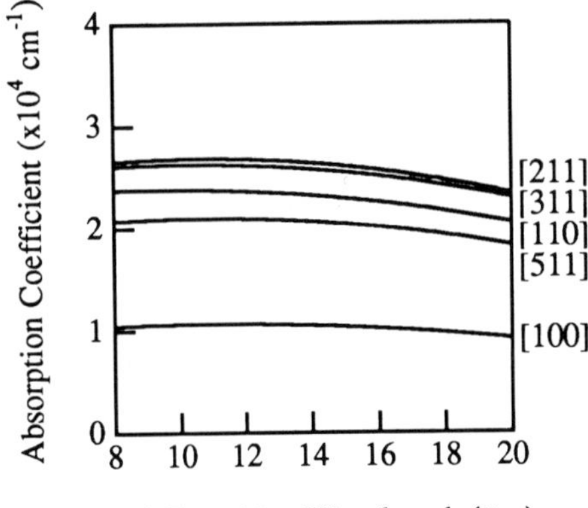

Fig. 3. The peak absorption coefficient at normal incidence as a function of the peak transition wavelength.

smaller for QWs with growth orientations in the above-mentioned order. Figure 3 shows the peak absorption coefficient for normal incidence light as a function of the transition wavelength. As can be seen, absorption coefficients greater than 10^4 cm^{-1} can be easily achieved for normal incidence radiation at the IR wavelength range of 8-20 μm.

Our samples were grown on undoped (100) and (311) GaSb substrates by the MBE. In the QW region, the quaternary $Ga_{0.6}Al_{0.4}Sb_{0.9}As_{0.1}$ was chosen for the barrier with a width of 400Å, the GaSb well width was 40Å and was doped n-type with Te to a concentration of $3x10^{18}$ cm^{-3}. We performed IR normal incidence absorption measurements using FTIR at 68K and room temperature, and observed that the (311) sample exhibited a stronger absorption than the (100) sample. The spectrum shown in Figure 4 for $GaSb/Ga_{0.6}Al_{0.4}Sb_{0.9}As_{0.1}$ (311) QWs at 68K has a peak at 7.8 μm with an absorption coefficient of 9100 cm^{-1}. This is the strongest absorption ever reported in this wavelength range. The spectra of the (100) sample showed a similar

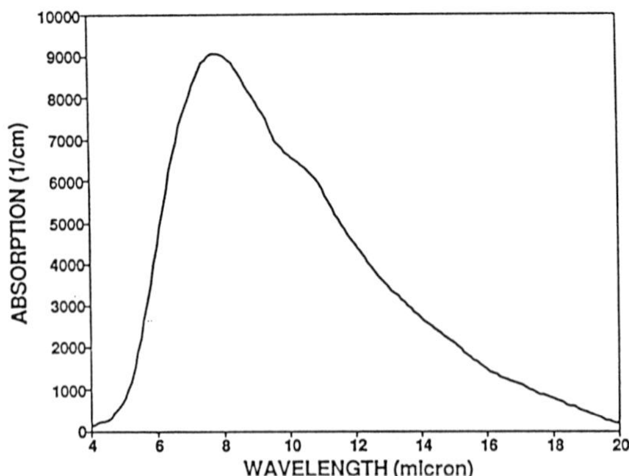

Fig. 4. Normal incidence IR absorption spectrum measured at 68K for (311) $GaSb/Ga_{0.6}Al_{0.4}Sb_{0.9}As_{0.1}$ QWs.

peak at 7.4 μm with an absorption coefficient of 8100 cm^{-1}. At room temperature, we observed a similar spectrum, however, the strength of the peak decreases and the width increases, as expected.Our results show that the experimentally observed spectra are in good agreement with our theoretical calculations. We have also grown photodetector structures and performed responsivity measurements. Good photodetector response covering a wide range of 5-14 μm, with a peak at 8.3 μm and a photoresponsivity of 310 mA/W, was found (Zhang *et al* 1993). Recently a peak absorption coefficient as strong as 8500 cm^{-1} for GaSb/AlSb (100) QWs has been reported by Samoska *et al* (1993).

4. GaSb QW MODULATOR STRUCTURES

Since the initial demonstration of the modulation of optical absorption in QWs using quantum-confined Stark effect (QCSE) in an applied electric field (Miller *et al* 1984), this mechanism has attracted a lot of attention in the area of infrared applications, primarily because there do not exist any efficient IR modulators. The QCSE is applied to IR modulators mainly through the use of intersubband transitions (Harwit and Harris 1987; Pan *et al* 1990; Karunasiri *et al* 1990). In order to have practical applications, the devices must be capable of both achieving a large maximum to minimum absorption ratio under bias and absorbing normally incident light. These can be realized in our proposed modulator which uses the direct-indirect transitions induced by an applied electric field in a $GaSb/Ga_{1-x}Al_xSb$ QW structure (Xie *et al* 1993).

Our device is based on the principles that the quantum-confined Stark shift is proportional to the effective mass (Vina *et al* 1987), and that the inter-conduction subband absorption at

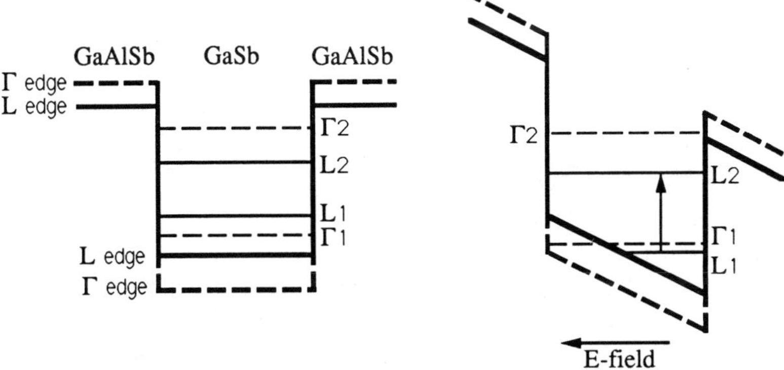

Fig. 5. Schematic band diagram with and without an applied electric field. *Γn* and *Ln* are the *n*th subbands associated with Γ and L states, respectively. The arrow indicates the allowed normal incidence transition.

normal incidence is forbidden in direct-gap QWs but allowed in indirect-gap QWs. Since the effective mass of the L valleys is larger than that of the Γ valley, under an electric field, the L states with heavier masses will exhibit larger Stark shifts than those of the Γ states with lighter masses (the Stark effect lowers the L states further with respect to the Γ states). As the applied electric field is increased, the direct-indirect transition can occur, and thus the first L state becomes the ground state. Our structure is designed such that the ground state is associated with the Γ state at zero voltage and becomes L state under bias. Consequently, the device will switch from being transparent to normal incidence light to strongly absorbing it, as shown in Figure 5. Thus the device acts like a voltage tunable filter, i.e., the modulation of an incident beam.

In Figure 6, the variation of the first energy levels at the center of the QW Brillouin zone for both Γ and L states is shown as a function of the electric field applied perpendicular to a $GaSb/Ga_{0.5}Al_{0.5}Sb$ QW with a well width of 85 Å. The energy zero at zero bias is set at the position of the bulk GaSb conduction-band minimum. It is found that the direct-indirect crossover occurs at a bias voltage of ~200 kV/cm. Near this point, the vast majority of electrons already occupy the first L subband (due to the much larger effective-mass density of states for L-valley electrons) and therefore normally incident light can be absorbed. As a result, the corresponding ratio of the maximum (up to 10^4 cm^{-1} as discussed in Section III) to minimum (almost zero) absorption is the largest ever reported.

Figure 7 exhibits Stark shifts of the L-valley intersubband transition energy as a function of the applied electric filed for the same structure discussed in the Figure 6. As expected, the

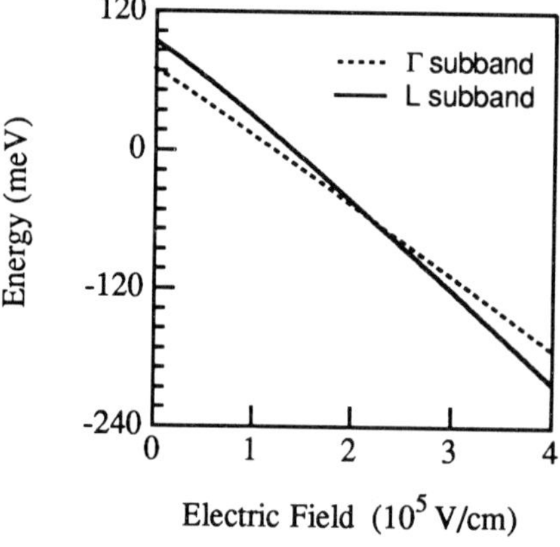

Fig. 6. The first Γ and L levels at the zone center as a function of the applied electric field.

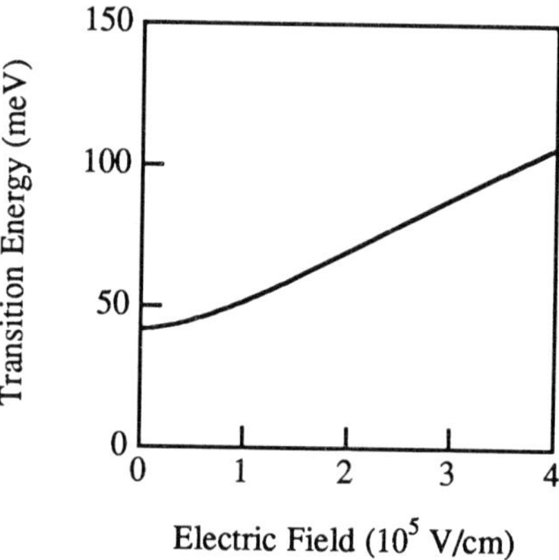

Fig. 7. The intersubband transition energy as a function of the applied electric field.

energy separation between the ground and first excited states increases (blue shift) with the applied bias voltage. For electric fields in the range of 200~300 kV/cm, a significant tunability of about 20 meV for the transition energy (corresponding to the transition wavelength from 18 μm to 14 μm) have been found from our calculation.

It is worth noting that the direct-indirect transition induced by the quantum confinement using a narrow GaSb QW has been reported previously (Wang *et al* 1983, Griffiths *et al* 1983; Voisin *et al* 1985). While the phenomenon is interesting in its own right, the QW thickness is fixed by the crystal growth and can not be tuned after the sample is grown. The mechanism reported here allows for the tuning of the direct-indirect transitions by an applied electric field, and is expected to have a profound impact on many device applications.

5. SUMMARY

In summary, we have presented three types of normal incidence (wide field-of-view) IR detector and modulator structures based on GaSb QWs. The physics principles presented here is quite general and can be applied to a wide range of materials systems such as Si/Ge.

6. ACKNOWLEDGMENTS

This work was supported by the US Office of Naval Research and Air Force Office of Scientific Research.

REFERENCES:

Alibert C, Joullie A, Joullie A M and Ance C 1983 Phys. Rev. B **27** 4946
Brown E R and Eglash S J 1990 Phys. Rev. B **41** 7559
Brown E R, Eglash S J and McIntosh K A 1992 Phys. Rev. B **46** 7244
Chang L L, Sai-Halasz G A, Esaki L and Aggarwal R L 1981 J. Vac. Sci. Techn. **19** 589
Chen H H, Houng M P, Wang Y H and Chang Y C 1992 Appl. Phys. Lett. **61** 509
Griffiths G, Mohammed K, Subbana S, Kroemer H and Merz J 1983 Appl. Phys. Lett. **43** 1059
Harwit A, Harris J S, Jr.1987 Appl. Phys. Lett. **50** 685
Karunasiri R P, Mii Y J and Wang K L 1990 IEEE Electron Dev. Lett. **11** 227
Katz J, Zhang Y and Wang W I 1992a Electronics Letters **28** 932
Katz J, Zhang Y and Wang W I 1992b Appl. Phys. Lett. **61** 1697
Katz J, Zhang Y and Wang W I 1993 Appl. Phys. Lett. **62** 609
Levine B F, Choi K K, Bethea C G, Walker J and Malik R J 1987 Appl. Phys. Lett. **50** 1092

Levine B F, Gunapala S D, Kuo J M, Pei S S and S. Hui 1991 Appl. Phys. Lett. **59** 1864
Luo L F, Beresford R, Longenbach K F and Wang W I 1990 J. Appl. Phys. **68** 2854
Miller D A B, Chemla D S, Damen T C, Gossard A C, Wiegmann W, Wood T H and Burrus C A 1984 Phys. Rev. Lett. **53** 2173
Pan J L, West L C, Walker S J, Malik R J and Walker J F 1990 Appl. Phys. Lett. **57** 366
Samoska L A, Brar B and Kroemer H 1993 Appl. Phys. Lett. **62** 2539
Vina L, Collins R T, Mendez E E and Wang W I 1987 Phys. Rev. Lett. **58** 832
Voisin P, Delalande C, Bastard G, Voos M, Chang L L, Segmuller A, Chang C A and Esaki L 1985 Superlattices and Microstructures Vol **1** 155
Wang W I, Mendez E E, Chang C A, Chang L L and Esaki L 1983 IEEE Trans. on Electron Dev. ED-**30** 1577
West L C and Eglash S J 1985 Appl. Phys. Lett. **46** 1156
Xie H, Katz J and Wang W I 1991a Appl. Phys. Lett. **59** 3601
Xie H, Piao J, Katz J and Wang W I 1991b J. Appl. Phys. **70** 3152
Xie H, Katz J, Wang W I and Chang Y C 1992 J. Appl. Phys. **71** 2844
Xie H and Wang W I 1993 Appl. Phys. Lett. **63**
Yang C L, Pan D S and Somoano R 1989 J. Appl. Phys. **65** 3253
Yu E T, Collins D A, Ting D Z T, Chow D H and McGill T C 1991 Appl. Phys. Lett. **57** 2675
Zhang Y, Baruch N and Wang W I 1993 Appl. Phys. Lett. **63**

Paper presented at the Int. Symp. GaAs and Related Compounds, Freiburg, 1993

A tunneling injection quantum well laser: prospects for a 'cold' device with a large modulation bandwidth

H.C. Sun, L. Davis, Y. Lam, S. Sethi, J. Singh, P.K. Bhattacharya
Solid State Electronics Laboratory,
Department of Electrical Engineering and Computer Science,
The University of Michigan, Ann Arbor, Michigan 48109-2122, USA

ABSTRACT: We have theoretically and experimentally investigated carrier relaxation times in SCH laser structures and have observed the 3-D to 2-D carrier relaxation "bottleneck". We propose and demonstrate a new quantum well laser, in which electrons are directly injected into the lasing quantum well by resonant tunneling, to bypass the long carrier thermalization times in standard MQW lasers. The principle of operation promises a "cold" laser at high injection levels and therefore Auger recombination and chirp are expected to be suppressed. In addition, tunneling of carriers into the active well will ensure very large modulation bandwidths.

I. INTRODUCTION

In recent years, there has been great interest in high speed semiconductor lasers for computer interface links and optical communication systems. Large modulation bandwidths, $\sim$ 80-90 GHz, are predicted for GaAs-based, strained separate confinement heterostructure (SCH) quantum well lasers. However, this can only be achieved at very high photon densities, at which point gain compression, device heating and facet degradation limit the device performance. The gain compression effects are primarily due to a large hot carrier population in conventionally injected lasers. This hot carrier population, which is also responsible for frequency response damping, is largely determined by carrier transport and relaxation mechanisms (Nagarajan et al 1992, Tessler et al 1992, Rideout et al 1991). We have experimentally and theoretically investigated the carrier relaxation times in GaAs/$In_xGa_{1-x}As$ SCH lasers, and have found these times to be long enough to cause a significant degradation in the modulation performance. However, if the carriers were directly injected into the lasing subband by resonant tunneling, then, even under hard driving conditions, there is no significant density of hot carriers. We have demonstrated such a tunneling injection laser (TIL) with strained $In_{0.10}Ga_{0.90}As$ and $In_{0.20}Ga_{0.80}As$ active regions. The principle of operation promises a "cold" laser at high injection levels and therefore Auger recombination and chirp are expected to be suppressed. In addition, tunneling of carriers into the active well will ensure very large modulation bandwidths.

II. CARRIER RELAXATION TIMES: THEORY AND EXPERIMENT

In semiconductor lasers, the drive for low threshold current has made the use of narrow quantum well lasers very attractive. At the same time, since photons have to be strongly confined in the active region, large bandgap cladding layers have to be incorporated. As a result electrons and holes have to be injected into the quantum wells from very high energies and they have to shed $\sim$300–500 meV of energy before lasing through the quantum well states. The carrier relaxation processes to the quantum well, including the carrier diffusion, capture into the quantum well and relaxation in the well, have been studied both theoretically and experimentally.

A Monte Carlo technique is used to study the carrier relaxation in separate confinement heterostructures (SCH) lasers. Optical phonon scattering, carrier-carrier scattering and acoustic phonon scattering have been included for both the 3-dimensional and 2-dimensional systems. Carrier density effects, very important at the high bias levels used in lasers, are included through the Pauli exclusion principle. Spontaneous recombination is included, and is calculated using a 4×4 $\mathbf{k}\cdot\mathbf{p}$ Hamiltonian which incorporates strain through the use of the deformation potential theory (Loehr and Singh 1991).

A series of SCH samples having different well widths (30, 40, 50 and 100 Å) were grown by MBE on semi-insulating GaAs substrates. These structures are schematically shown in Fig. 1, and are the same structures that were theoretically simulated. Time resolved photoluminescence(TRPL) was performed to observe the carrier capture time by the quantum well; the rise time of the luminescence from the quantum well was obtained from a streak camera. The experimental setup is described in detail elsewhere (Davis et al 1993, Norris et al 1989). The output of a mode-locked dye laser, producing a tunable lasing output from 800-870 nm with < 10 ps pulses, is focused on the sample, which is mounted in a cryostat and maintained at a temperature of 200K; the luminescence is collected and directed into a 0.5m Jarrell-Ash spectrometer. The dispersed spectrum is measured with a Hamamatsu streak camera. The temporal resolution of the setup is 40 ps, which is deconvolved from the experimental results.

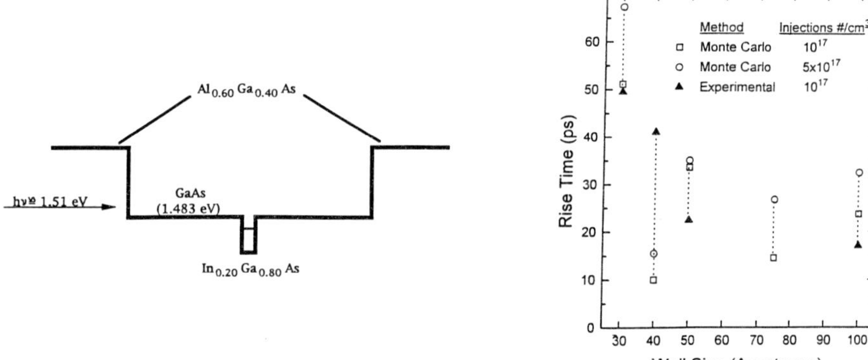

Figure 1 Schematic of the SCH structure analyzed by Monte Carlo and TRPL.

Figure 2 Comparison of experimental and calculated rise time.

The 10-90% rise time derived from both the Monte Carlo calculations and the experiments are compared in Fig. 2. From this figure, we can see reasonable agreement between the experimental and the calculated rise time, in terms of the magnitude of the rise time, as well as in the general trend—decreasing rise time with well size increases. Furthermore, an oscillatory behavior is seen superimposed on the decreasing calculated rise time, which is similar to the theoretical results by Brum and Bastard(1986). These measurements were performed on *unbiased* structures; however, theoretical simulations show that the capture time will decrease to only 12 ps for a 50 Å quantum well under forward injection conditions. If the electron-hole recombination time by stimulated emission for a laser at high injection approaches these relaxation times, the carrier distribution in the quantum well can be "hot" and is not described by a quasi Fermi distribution. This creates serious limitations for the laser performance by introducing gain-compression enhanced Auger rates and other hot-carrier related effects. The problem is schematically described in Fig. 3. The carrier 'bottleneck' caused by relaxation time effects into the quantum well is an intrinsic speed limitation on the present design of semiconductor lasers.

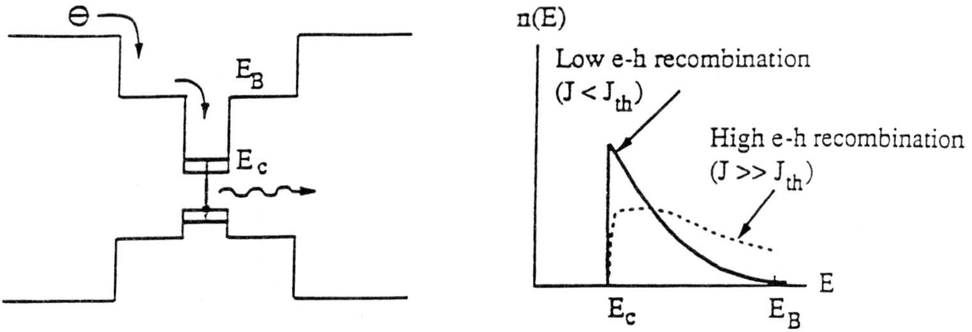

Figure 3 Energy band diagram and carrier distribution in a conventional quantum well laser.

III. SOLUTION: THE TUNNELING INJECTION LASER

If electrons can be injected into the active quantum well of a semiconductor laser at a well-defined energy by tunneling, tremendous advantages are expected at high e-h recombination rates. At low e-h recombination rates, the electron distribution by tunneling will be described by a quasi-Fermi distribution. However, at the high e-h recombination rates, with recombination times approaching a few picoseconds, the carrier distribution will be narrowed, i.e. "cold," as shown in Fig. 4. This will improve the gain at a selected energy and dramatically lower the Auger recombination rates. The important question is whether electrons can be effectively injected into the active lasing region by tunneling. We demonstrate such an injection mechanism in a GaAs-based laser for the first time. We have focused on electron injection since hole thermalization times are expected to be much faster than electron thermalization times. Also, with reference to the structure shown in Fig. 4, the holes injected from the opposite contact remain confined in the lasing well by the large tunneling barriers.

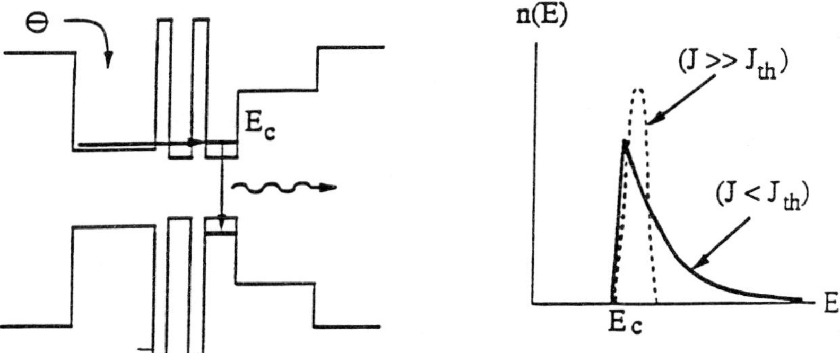

Figure 4 Energy band diagram and carrier distribution in the tunneling injection laser.

The laser structures were grown with pseudomorphic 80Å $In_{0.10}Ga_{0.90}As$ and $In_{0.20}Ga_{0.80}As$ active wells by molecular beam epitaxy; the layer structure is shown in Fig. 5. Broad area and single mode ridge lasers were made by standard photolithography and wet and dry etching. The light-current characteristics of a single-mode laser with cavity length $l = 400$ μm are shown in Fig. 6. From the light-current characteristics the differential quantum efficient η_d is estimated. A plot of $1/\eta_d$ versus cavity length l yields an internal quantum efficiency $\eta_i = 0.56$ and a guide loss of $\gamma = 20$ cm^{-1}. The latter parameter is rather high. The variation of current density with inverse cavity length yields a linear plot, as expected, and typical data are shown in Fig. 7. The transparency current density J_o is obtained from the intercept of this plot on the J-axis to be 80 A/cm^2. The differential gain, $\partial g/\partial n$, of the TIL is estimated to be 6.8×10^{-16} cm^2 from the slope of resonance frequency versus square root of output power.

0.1 μm	CONTACT GaAs	$p^+(5\times10^{18}$ cm$^{-3})$
1.0 μm	OUTER CLAD $Al_{0.60}Ga_{0.40}As$	p $(5\times10^{17}$ cm$^{-3})$
1000 Å	INNER CLAD $Al_{0.30}Ga_{0.70}As$	
80 Å	ACTIVE WELL $In_{0.20}Ga_{0.80}As$	i
20 Å	Resonant Tunneling Barrier AlAs	i
40 Å	Tunneling Well $In_{0.20}Ga_{0.80}As$	i
20 Å	Tunneling Barrier AlAs	i
1000 Å	INNER CLAD GaAs	i
1.0 μm	OUTER CLAD $Al_{0.60}Ga_{0.40}As$	n $(5\times10^{17}$ cm$^{-3})$
0.7 μm	CONTACT GaAs	$n^+(5\times10^{18}$ cm$^{-3})$
	S.I. and n^+ GaAs (100) Substrates	

Figure 5 Layer structure for the tunneling injection laser.

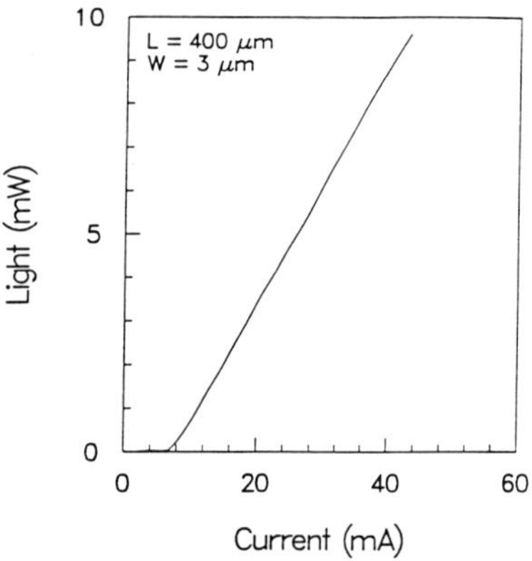

Figure 6 A typical L-I characteristic for a single mode, ridge waveguide laser.

We have also performed low temperature light-current and spectral measurements. We found the threshold current reduced with decreasing temperature and lasing action indeed happened in the 80Å $In_{0.2}Ga_{0.8}As$ active well even at temperatures less than 100K, as shown in Fig. 8. This experimentally proves that thermionic emission in the conduction band is negligible and tunneling is the only possible mechanism for carrier injection.

A broad temperature-controlled carrier distribution in conventional quantum well lasers creates a tremendous waste of injected charge at high injection levels, as reflected by the temperature dependence of the threshold current. This dependence should be minimized in this device. Characteristic temperature T_o measurements of conventional QW lasers and the TIL have been conducted. Higher T_o value has been observed from the TIL comparing with those from conventional QW lasers, as shown in Fig. 9. This higher characteristic temperature of the TIL indicates the reduced temperature dependence and the narrower energy distribution of this device.

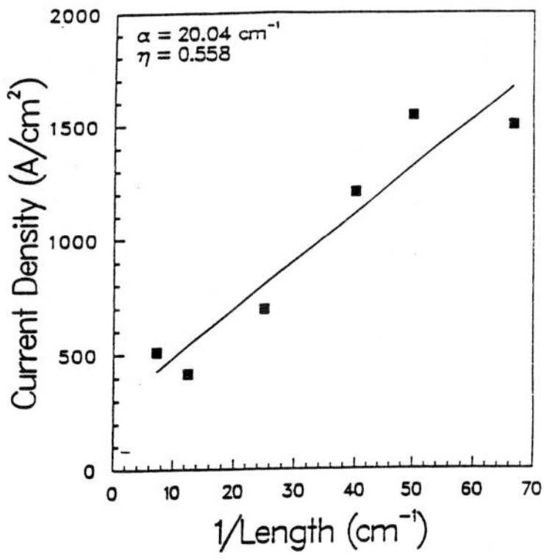

Figure 7 Variation of the current density with inverse cavity length.

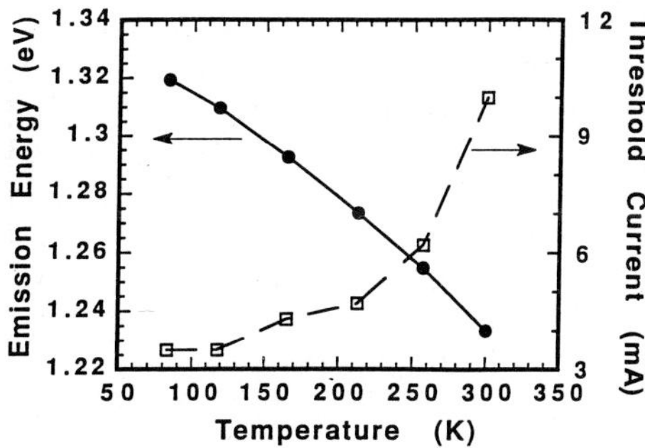

Figure 8 A plot of the temperature dependence of the peak emission energy (at $I = 10I_{th}$) and the threshold current.

The stimulated emission spectrum of the tunneling laser at $I \sim 10I_{th}$ is shown in Fig. 10. Note that it is fairly broad. We believe this is due to the differences in subband energies of the tunneling and lasing InGaAs wells. Obviously, some structure optimization is necessary. The output spectral energy confirms that lasing is out of the 80Å $In_{0.2}Ga_{0.8}As$ well even at low temperature and therefore the expected tunneling is operative. The changing bias under modulation will change the ratio of resonant to sequential tunneling, but will not terminate lasing, as the carriers quickly find the proper energy to tunnel. Carrier injection by tunneling will allow for a smaller modulation signal, as an optimized structure will have a strongly peaked gain curve. Measurement of the small signal bandwidth of the laser shows a characteristic peaking at a specific bias current level, indicating a resonance in the tunneling process.

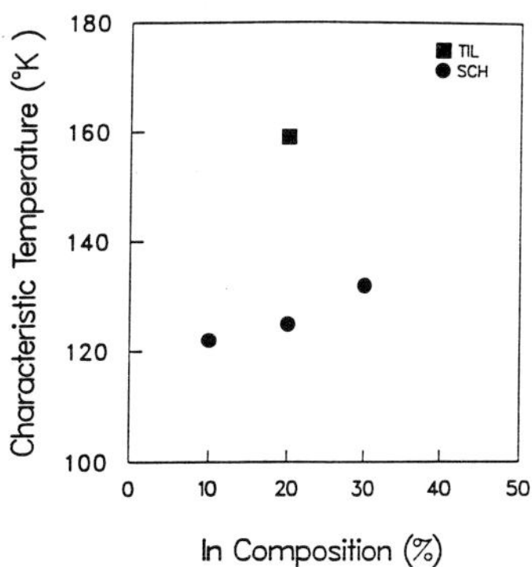

Figure 9 A plot of T_o versus indium composition, showing the superior performance of the tunneling injection laser.

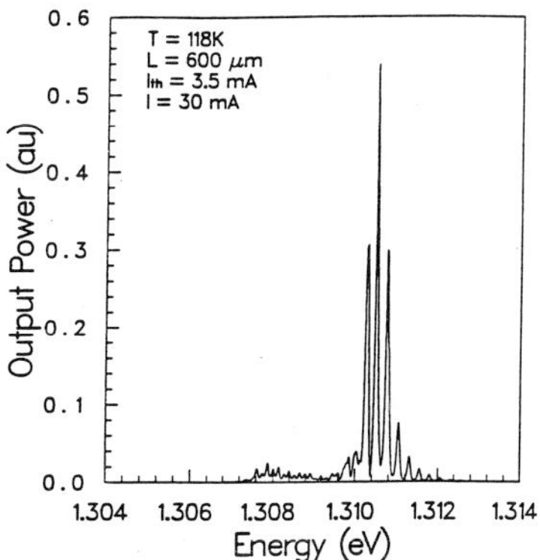

Figure 10 The spectral output of the tunneling injection laser at $I \sim 10I_{th}$.

We have presented preliminary results of a new quantum well laser. Theoretical and experimental work are still in progress to optimize the structure and to study its potential. Some expected characteristics and advantages may be mentioned here. Because the operation of the device does not depend on a broad energy distribution of carriers, the laser is expected to be very mode selective. As stated earlier, Auger recombination, which is a more serious limitation in long wavelength InP-based quantum well devices, and chirp will be suppressed. More importantly, tunneling will ensure an efficient transport of carriers into the lasing subband, bypassing the 3D to 2D capture times, which can be rather long for narrow (< 60Å) wells. By bypassing the relaxation time bottleneck, we hope to achieve small-signal modulation bandwidths larger than the presently achieved values of 30 GHz (GaAs-based laser) (Weisser et al 1992) and 25 GHz (InP-based laser) (Morton et al 1992). Experiments are in progress to optimize the structure.

ACKNOWLEDGEMENTS

The authors acknowledge the help provided by H. Yoon and W. L. Chen. The work is being supported by the Office of Naval Research under Grant N00014-90-J-1831 and the Army Research Office under Grant DAAH04-93-G-0034.

REFERENCES

Brum J A, Bastard G 1986 Phys. Rev. B **33** 1420.

Davis L, Lam Y, Chen Y C, Singh J, Bhattacharya P K, to be published.

Loehr J P, Singh J 1991 J. Quantum Electron. **27**, 708.

Morton P A, Logan R A, Tanbun-Ek T, Sciortino Jr. P F, Sergent A M, Montgomery R K, Lee B T 1992 Electron. Lett. **28** 2156.

Nagarajan R, Ishikawa M, Fukushima T, Geels R S, Bowers J E 1992 J. Quantum Electron. **28** 1990.

Norris T B, Song X J, Schaff W J, Eastman L F, Wicks G, Mourou G A 1989 Appl. Phys. Lett. **54**, 60.

Rideout W, Sharfin W, Koteles E, Vassell M, Elman B 1991 Photonics Tech. Lett. **3** 784.

Tessler N, Nagar R, Eisenstein G 1992 J. Quantum Electron. **28** 1992.

Weisser S, Ralston J D, Larkins E C, Esquivias I, Tasker P J, Fleissner J, Rosenzweig J 1992 Electron. Lett. **28** 2141.

Resonant interband and intraband tunneling in InAs/AlSb/GaSb double barrier diodes

J.L. Huber and M.A. Reed

Department of Electrical Engineering, Yale University
New Haven, Connecticut 06520

G. Kramer and M. Adams

Motorola Inc., Phoenix Corporate Research Laboratories
Tempe, Arizona 85284

C.J.L. Fernando and W.R. Frensley

University of Texas at Dallas
Richardson, Texas 75083

Abstract

We have realized heteroepitaxial tunneling structures in which both interband and intraband tunneling occur, dependent on injection energy. The structures consist of a single InAs well with GaSb barriers that serve as quantum wells for interband tunneling and barriers for intraband tunneling. At low biases, interband tunneling occurs through a coupled double well structure in the GaSb valence band. At higher biases, intraband tunneling occurs through the InAs quantum well.

1. INTRODUCTION

Resonant tunneling diodes (RTDs) usually consist of a double barrier, single quantum well semiconductor heterostructure where the center region has quasi-bound quantum well states accessible to electron (or hole) transport (Chang 1974). Resonant interband tunneling (RIT) devices differ from conventional RTDs in that the confined states, accessible to electron (or hole) transport lie in the valence (or conduction) band rather than the conduction band (Luo 1989, Söderström 1989). Until now, most studies of InAs/AlSb/GaSb structures have investigated tunneling through a single valence band state. We have fabricated a series of structures in which a coupled double well exists in the valence band, and a single well exists in the conduction band. Tunneling through both of these regions was observed. These types of structures may have applications in three terminal operation of interband tunneling structures because the InAs well confined state is separated from the interband states both spatially and in energy. Therefore, it will be possible to vary the potential of the interband tunneling structure by adding charge to the well state.

2. DEVICE STRUCTURE

The structures were grown in a Varian (Intevac) Gen II solid-source MBE system. The growth rates of GaSb, AlSb, and InAs were 1.0, 1.0, and 0.8 μm/h, respectively. Before growth the wafers were heated to 610 °C for 15 minutes to desorb the oxide. The first epitaxial layer, 0.2 μm of GaAs, was deposited at T_{sub} = 580 °C to smooth the wafer surface. A 0.5 μm superlattice buffer region, consisting of a GaSb/AlSb superlattice followed by 0.4 μm of AlGaSb, was then grown at T_{sub} = 530 °C. The T_{sub} was then lowered to 500 °C for the growth of the active region of the structure, which consists of 1μm n+ InAs ($2x10^{18}$ cm^{-3}), 50 nm of lightly doped InAs ($2x10^{16}$ cm^{-3}), the double barrier layer sequence, 50 nm of lightly doped InAs ($2x10^{16}$ cm^{-3}) and then a cap layer of 0.25 μm of n^+ InAs ($2x10^{18}$ cm^{-3}). Growth conditions were such that the InAs/GaSb interfaces were InSb like.

Mesa devices were fabricated using standard lithographic and wet etch techniques with non-alloyed Ni/Ge/Au contacts. The surface of the sample was covered with an insulating layer of silicon nitride; via holes in the nitride were then formed to allow contact between the device contacts and bonding pads.

A series of four structures was investigated. The baseline structure consisted of two 9.5 nm GaSb barriers and a 5.5 nm InAs quantum well (all nominally undoped). A thin AlSb barrier was added, at various positions, to the remaining three structures. Self-consistent band diagrams, not including quantum effects, for the structures are shown in Figure 1a, 2a, 3a, 4a. In structure #2, the AlSb barrier is 1.5 nm thick and each InAs well layer is 3.5 nm thick, while in structure #3 and #4 the AlSb barrier is 2.5 nm thick. A 10 nm undoped InAs spacer layer was placed between both n+ contacts and the double barrier structure for all devices.

At low biases, because of the broken bandgap at the InAs/GaSb interface, the structures are RITs with tunneling occurring through a coupled double well structure in the GaSb valence band. The GaSb layer thickness was chosen such that only one light hole quantum well state exists in the GaSb valence band above the InAs conduction band edge (Yang 1992). The InAs well thickness was chosen such that the lowest energy quantum well state would be above the GaSb valence band edge. Thus, at higher biases, the structures are conventional RTDs, with tunneling occurring through a single InAs quantum well state.

The flatband transmission coefficient for each structure was calculated using a two–band model. The transmission coefficient for structure #1 is shown in Figure 1c. The coefficient was similar for each structure with a broad, double peak at low energies corresponding to transmission through the coupled GaSb valence band wells, and a more narrow single peak (except for structure #2, which had a double peak) corresponding to transmission through the InAs well conduction band state.

3. RESULTS AND DISCUSSION

3.1 Baseline structure

The I-V/G-V curve for structure #1, measured at 77 K, of a 32 μm square device is shown in Figure 1b. The I-V is symmetrical about zero bias, therefore only one bias direction is shown. The NDR region at 0.62 V corresponds to conventional intraband tunneling through the confined state in the InAs well (this resonance occurs at a slightly larger than expected bias, due to parasitic contact resistance). The NDR region near 0.09 V corresponds to interband tunneling through the confined state created in the valence band of the near GaSb barrier. In addition to these NDR regions, a small feature appears in the G-V curve near 0.05 V, corresponding to tunneling into the confined state created in the valence band of the far GaSb barrier.

3.2 AlSb barrier in the center of the InAs well

Structure #2 is similar to the baseline structure (#1), with the addition of a 1.5 nm thick AlSb barrier in the center of the InAs well, creating, in effect, a double well (the total InAs well thickness, including the AlSb barrier, is 0.5 nm larger than that of structure #1). Because the InAs well is actually a double well, two intraband resonances are expected.

The I-V/G-V curve for structure #2, measured at 4 K, of a 16 μm square device, is shown in Figure 2b; again, only one bias direction is shown.

The intraband resonance from the near InAs well is visible at 0.58 V. A weaker feature, assigned to tunneling into the far InAs well, appears at 0.46 V. A strong NDR appears at 0.17 V, consistent with interband tunneling into the near GaSb valence band well. A number of smaller unidentified features in the G-V curve are evident: at 0.26 V, and below 0.17 V. One of these features (occurring below V = 0.17 V) is consistent with interband tunneling into the far GaSb well. The others do not correlate with any intraband resonances predicted by the two–band model.

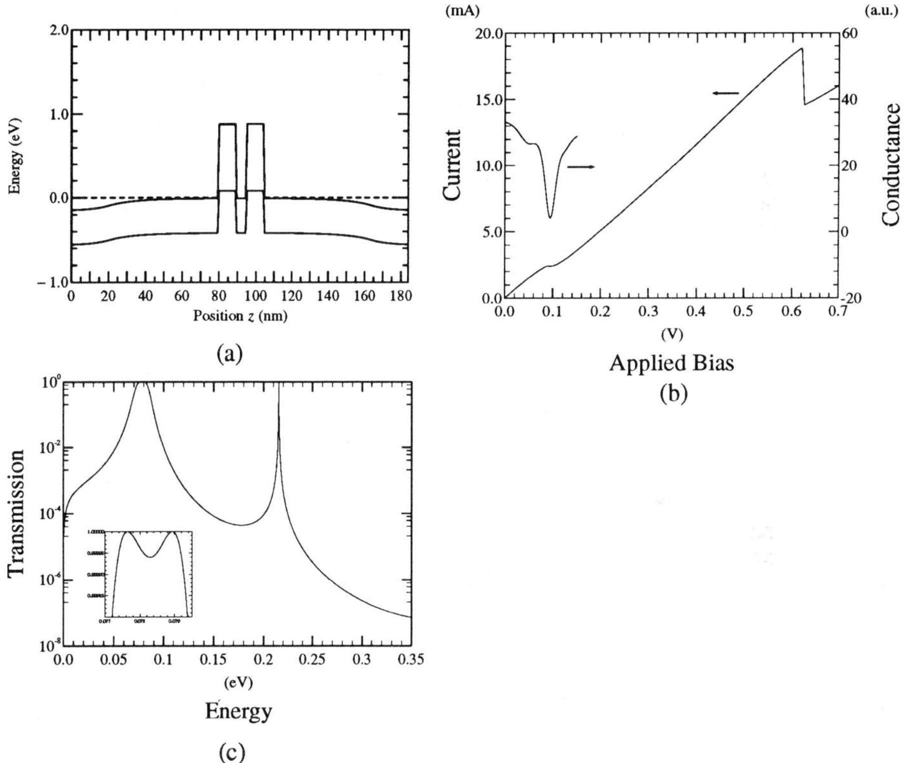

Figure 1. (a) Self-consistent band diagram for structure #1. (b) I-V/G-V for structure #1, T = 77 K. The G-V curve is not shown above V = 0.15 V. (c) Flatband transmission coefficient for structure #1, referenced to the bottom of the InAs conduction band, calculated using a two–band model. The inset is an expanded view of the double peak located at 80 meV.

3.3 AlSb barrier outside the GaSb barrier

Structure #3 is similar to the baseline structure except for the addition of a 2.5 nm AlSb barrier on the outside of one of the GaSb barriers. The I-V/G-V curve from structure #3, measured at 4 K, of a 32 μm square device, is shown in Figure 3b. Negative bias voltages correspond to the case where electrons initially tunnel through the AlSb barrier.

The intraband resonances for the structure appear at -0.90 V and 0.68 V (there are also small features, at -0.96 V and 0.89 V slightly above each of the intraband resonances). The interband resonances into the near GaSb valence band well appear at -0.07 V and 0.18 V. Additional "extra" features occur below the large interband tunneling resonance, similar to those in structure #2. For the negative bias interband resonance, the peak-to-valley ratio is higher and the peak current smaller than for the forward bias resonance. This is expected from, and consistent with, previous observations for other InAs/AlSb/GaSb tunneling structures (Chen 1990) since electrons initially tunnel through the AlSb barrier, which serves as a real tunnel barrier, reducing inelastically scattered valley current. The strengths of the "extra" features are also smaller than for the forward bias direction.

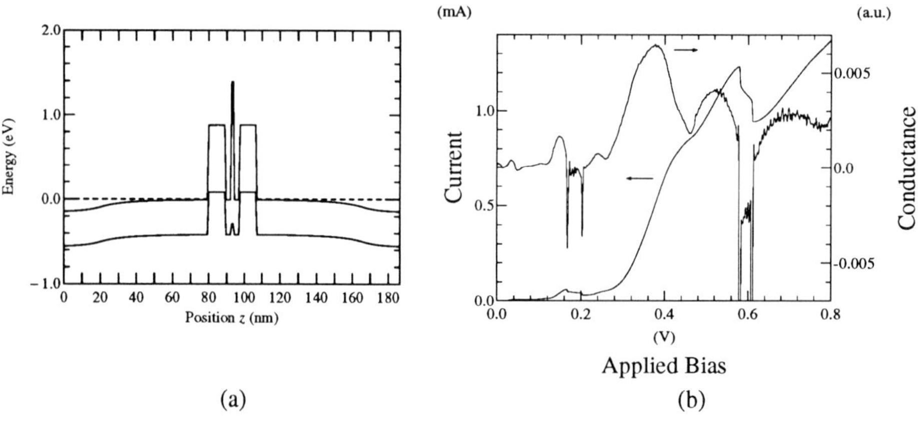

Figure 2. (a) Self-consistent band diagram for structure #2. (b) I-V/G-V for structure #2, T = 4 K.

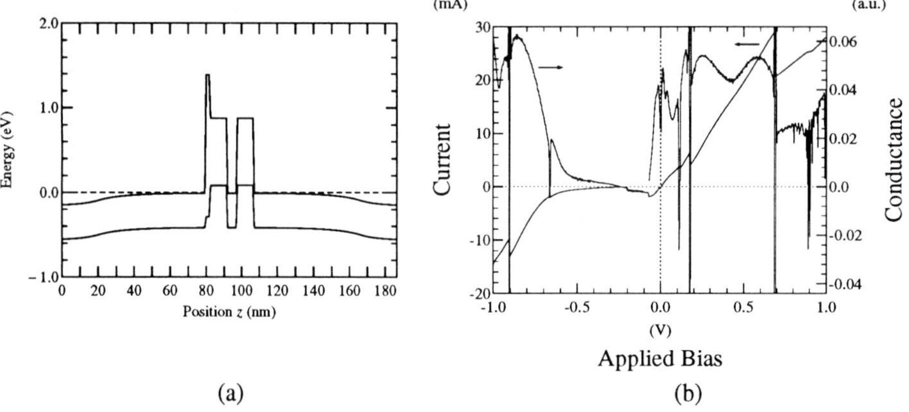

Figure 3. (a) Self-consistent band diagram for structure #3. (b) I-V/G-V for structure #3, T = 4 K. The G-V curve between -0.07 and -0.20 V is not plotted for clarity.

The position and peak current of the intraband resonances are consistent with the asymmetric barriers created by the addition of the AlSb layer. The peak voltage is increased and the peak current reduced for the bias direction where electrons initially tunnel into the thicker barrier (29 mA at 0.68 V, 13 mA at -0.90 V).

3.4 AlSb barrier inside the GaSb barrier

Structure #4 is similar to the baseline structure except for the addition of a 2.5 nm AlSb barrier on the inside of one of the GaSb barriers. The I-V/G-V curve from structure #4, measured at 4 K, of a 32 μm square device is shown in Figure 4b. Positive applied biases correspond to electrons initially encountering the GaSb barrier associated with the AlSb barrier.

In the positive bias direction, the intraband resonance occurs at 0.70 V. Weaker structures occur at 0.21 V, 0.49 V, and 0.96 V, and very weak structures occur at 0.17 V, 0.28 V, and 0.59 V. In the negative bias direction, strong NDR regions occur at -0.04 V and -0.18 V. These two features are qualitatively similar, each with a small peak at a slightly higher bias (-0.13 V and -0.23 V). Another small feature of similar strength is located at -0.31 V. A strong NDR region occurs at -0.36 V, also with a weaker feature at the slightly larger bias (-0.47 V). A number of features appear at reverse biases greater than -0.50 V.

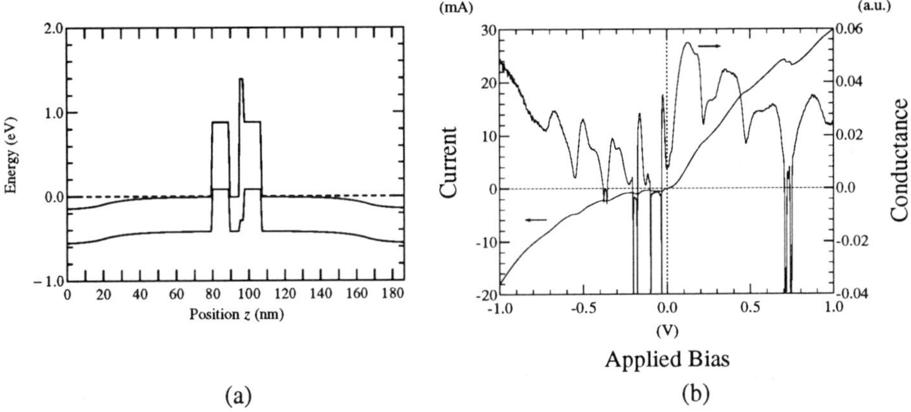

Figure 4. (a) Self-consistent band diagram for structure #4. (b) I-V/G-V for structure #4, T = 4 K.

To assist in the labeling of the various NDR regions, we have used the band profiles at various biases generated by a self-consistent Poisson solver. The energy levels of the InAs quantum well states were then calculated using a single band model. This method is precise for modeling the intraband resonant state energies because the InAs quantum well states lie above the window between the InAs conduction band and the GaSb valence band; little interaction between the various conduction and valence bands is expected. For zero bias, the calculated energy for the lowest InAs well confined state matches reasonably well with the energy of the transmission peak calculated from the two–band model, allowing for fluctuations of one monolayer for each layer in the structure. The interband resonant state energies calculated with a single band model are not precise because of the coupling between the electron-like states in the InAs contacts and the hole-like states in GaSb valence band wells. For these structures, however, the single band model predicts that only one confined state in each GaSb valence band lies above the InAs conduction band edge, and at energies similar to those predicted by the two–band model, and therefore can be used to approximate the resonant voltage.

This method predicts a total of three resonant states for each structure: two interband and one intraband. The two interband resonances result from the localizing of the valence band states as a bias is applied across the structure. For each structure, tunneling through the far well (relative to the emitter) occurs at a lower bias than tunneling through the near well, while intraband tunneling occurs at a much higher bias. For structure #4, it is difficult to predict even the intraband resonance in the negative bias direction because the calculated intraband resonance is not well separated from other measured structure. In addition, there are four relatively strong resonances occurring, where only three were expected.

Clearly, there are more features in the interband tunneling region than are predicted by the simple two–band model. At k = 0, selection rules only allow coupling between the conduction band and the light hole band, thus the justification for using only a two–band model to account for the interband coupling. Away from k = 0, coupling to the heavy hole band is allowed (Ting 1991); thus the additional features may be due to interband tunneling through heavy hole states.

Variable temperature measurements between 1.4 K and 30 K show that the lowest bias conductance peak is formed as a result of current suppression at very low biases, caused by momentum conservation, as has been previously observed (Mendez 1992). The other "extra" features, however, show little temperature dependence over the same temperature range, implying they result from band-to-band coupling rather than phonon-assisted scattering.

4. SUMMARY

We have realized InAs/GaSb/AlSb heteroepitaxial tunneling structures in which both interband and intraband tunneling occur, dependent on injection energy. The structures, which may be adaptable for three terminal operation, consist of a single InAs well with GaSb barriers which serve as quantum wells for interband tunneling and barriers for intraband tunneling. The addition of a single, thin AlSb barrier variably throughout the double barrier structure can greatly affect the nature of the features in the I-V/G-V characteristics.

The number of observed interband features is greater than can be accounted for by simple conduction band—light hole band tunneling. The extra features exhibit little temperature dependence (except the feature at zero bias) and therefore are likely to be a result of band-to-band coupling rather than phonon-assisted scattering into heavy hole states.

We would like thank Saied Tehrani and Herb Goronkin for helpful discussions, and the NNF at Cornell University for assistance with mask fabrication. This work was partially supported by a State of Connecticut Goodyear grant, NSF, and was partially performed under the management of FED (the R&D Association for Future Electron Devices) as a part of the R&D of Basic Technology for Future Industries supported by NEDO (New Energy and Industrial Technology Development Organization, Japan).

5. REFERENCES

Chang L L, Esaki L and Tsu R 1974 *Appl. Phys. Lett.* **24** 593
Chen J F, Yang L, Wu M C Chu S N G and Cho A Y 1990 *J. Appl. Phys.* **68** 3451
Luo L F, Beresford R and Wang W I 1989 *Appl. Phys. Lett.* **55** 2023
Mendez E.E., Nocera J., Wang W.I. 1992 *Phys. Rev. B* **45** 3910
Söderström J R, Chow D H and McGill T C *Appl. Phys. Lett.* **55** 1094
Ting D Z Y, Yu E T and McGill T C 1991 *Appl. Phys. Lett.* **58** 292
Yang R Q and Xu J M 1992 *Phys. Rev. B* **46** 6969

Paper presented at the Int. Symp. GaAs and Related Compounds, Freiburg, 1993

AlAs hole barriers in InAs/GaSb/AlSb interband tunnel diodes

S. Tehrani, J. Shen, H. Goronkin, G. Kramer, M. Hoogstra, and T. X. Zhu

Motorola Inc, Phoenix Corporate Research Laboratories, Tempe, AZ 85284, USA

ABSTRACT: The peak-to-valley current (P/V) ratio in InAs/AlSb/GaSb/AlSb/InAs resonant interband tunnel diodes was increased by inserting monolayers of AlAs adjacent to the AlSb barriers. The highest P/V ratio without the AlAs barriers was 15 while the highest P/V ratio with the AlAs barriers was 30 at room temperature. The AlAs layer acts as a barrier to the parasitic conduction of the holes that are confined in the GaSb quantum well.

1. INTRODUCTION

Resonant tunnel diodes must have sufficient peak current to drive subsequent devices and they must also have low valley current to avoid unnecessary power dissipation in order to be incorporated into transistors for future ULSI. The off-state current must be comparable to MOSFETs used in a DRAMs. For example, in present DRAMs the allowable leakage current density is approaching 2 pA/μm^2 while in existing resonant tunnel diodes, the valley current density is more than 1 nA/μm^2 regardless of the materials used.

There are several possible contributions to valley current. In a GaAs/AlAs resonant tunneling diode, Collins[1] et al observed inelastic excitation of phonons in AlAs barriers which appears as peaks in the conductance derivative at a potential equal to the optical phonon energy. The excess current in AlAs barriers was modeled by Turley and Teitsworth[2] who found that the electron wave function can couple to interface phonons across the barrier and result in a current at voltages beyond the resonance voltage. Other transport mechanisms such as defect[3] and alloy scattering as well as Γ-X band transport can also produce excess current. In this paper we focus on the leakage current arising from unconfined holes in a double barrier resonant interband tunnel diode (RITD). A peak-to-valley current (P/V) ratio as high as 20 at 300K was previously reported[4] in a InAs/AlSb/GaSb/AlSb/InAs RITD. In these type-II structures, resonant tunneling occurs when the occupied conduction band of InAs aligns with the unoccupied valence band of GaSb with electrons in InAs coherently or sequentially tunneling through the GaSb valence band states. When the conduction band in InAs aligns to the bandgap of GaSb, the transmission coefficient becomes minimum resulting in the valley current (Fig. 1) .

There are several ways that parasitic holes can be generated and transported in the double barrier RITD: (1) thermal excitation across the InAs contact layer bandgap, (2) thermal emission from the GaSb quantum well over the AlSb barrier, (3) thermally assisted tunneling through the barrier. As the higher order eigenstates become occupied, holes can tunnel directly through the barrier into the valance band of InAs. In all of these cases, the holes can be collected at the cathode without encountering any potential barriers. In order to reduce the probability of such hole conduction, we have increased the barrier height for holes by replacing a portion of the AlSb barrier with two monolayers of AlAs.

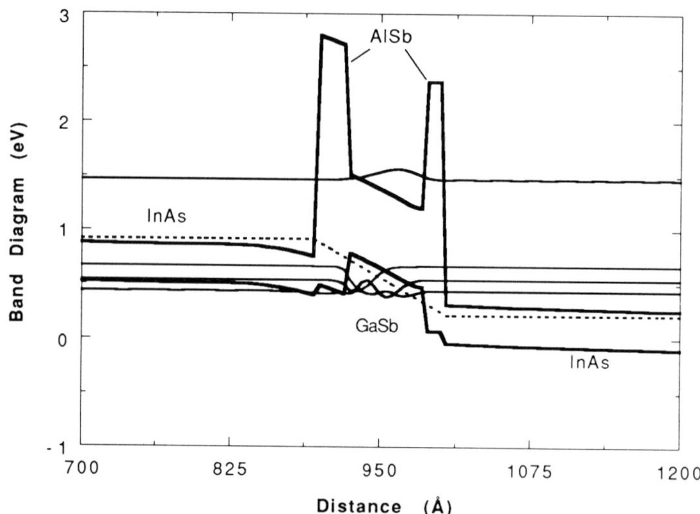

Fig. 1. Calculated band structure for double barrier interband tunneling diode. Also shown are the first three quantized hole states and the first quantized state for electrons in the GaSb quantum well. The quasi-Fermi level (dashed line) is assumed to drop linearly across the double barrier region.

2. RESONANT TUNNELING STRUCTURE

The RITD structures were grown in a Varian (Intevac) Gen II solid-source MBE system. Conventional Sb and As furnaces were utilized to produce tetrameric group-V beams. The substrate temperature (T_{sub}) was measured using an optical pyrometer. Unless otherwise noted, the layers were not intentionally doped. In our MBE system, as-grown layers of InAs, GaSb and AlSb are n-type, p-type and highly resistive, respectively. For the InAs layers, Si was used as the n-type dopant.

The substrates were used as received from the vender. Before growth the wafers were heated to 610 °C for 15 minutes to desorb the oxide. Then, 0.2 μm of GaAs, was deposited at T_{sub} = 580 °C to smooth out the crystal surface. Next, a 0.5 μm superlattice buffer region was grown (T_{sub} at 530 °C) to accommodate the large lattice mismatch (≈ 7%) between the semi-insulating GaAs substrate and the active region of the structure. The buffer region consists of a GaSb/AlSb superlattice followed by 0.4 μm of AlGaSb. T_{sub} was reduced to ≈ 500 °C for the growth of the remaining layers. These consisted of 1 μm of heavily doped InAs (n = 2 x $10^{18}/cm^3$), 50 nm of lightly doped InAs (n = 2 x $10^{16}/cm^3$), the double barrier layer sequence (10 nm of InAs, 1.5 nm of AlSb, 6.5 nm of GaSb, 2.5 nm of AlSb, and 10 nm of InAs), 50 nm of lightly doped InAs (n = 2 x $10^{16}/cm^3$), and finally a cap layer of 0.25 μm of heavily doped InAs (n = 2 x $10^{18}/cm^3$).

The growth rates of GaSb, AlSb, and InAs were 1.0, 1.0 and 0.8 μm/h, respectively. These were determined from reflection high-energy electron diffraction (RHEED) intensity oscillation measurements. The GaSb and AlSb layers were grown using V/III flux ratios that resulted in an Sb-stabilized surface, as evidenced by a (1 x 3) RHEED reconstruction pattern. The InAs layers were grown using a minimal As_4 flux since this has been reported to be the optimal growth condition.[5] Furthermore, the switching technique as reported by Tuttle[6] et al was used for the formation of an "InSb-like" interface at each of the two AlSb/InAs heterointerfaces to improve the material quality.

The structure with the AlAs-inserted barriers is the same as above except the AlSb layer thicknesses were reduced appropriately to accommodate 2 monolayers (ML) of AlAs while maintaining the original number of barrier monolayers. This portion of the structure from bottom to top was 2ML of AlAs, 3 ML of AlSb, 21 ML of GaSb, 2 ML of AlAs, 4 ML of AlSb and 2 ML of AlAs. The transition from AlAs to AlSb was monitored by RHEED. The usual AlSb pattern is perturbed by the addition of 2 ML of lattice mismatched AlAs but returns to normal in a few seconds.

The RITDs were fabricated by first defining and evaporating the top contact layer. The top metal contact layer was used as an etch mask. The top and bottom InAs layers were contacted with Ni/Ge/Au. The diodes were passivated with Si_3N_4 and overlay metal was used to make connection between the top of the diodes and large bond pads.

3. RESULTS AND DISCUSSIONS

Current density-voltage (J-V) characteristics of the two RITD structures are shown in Fig. 2. The forward direction is defined as a positive bias applied to the top contact while the bottom contact is grounded. The RITD with AlAs monolayers had a P/V of 30 at 300K and 85 at 77K when reverse biased (top contact negative) and 5.5 at 300K and 32 at 77K for forward bias. The sample without AlAs barriers had a P/V ratio of 10 at 300K and 20 at 77 k when reverse biased and 14 at 300K and 29 at 77K when forward biased. The P/V ratio (30 at room temperature) achieved in the structures with AlAs monolayers are the highest reported in the Sb-based RITDs.

Two distinctive regions were observed in the temperature dependent valley current as shown in figure 4. At high temperatures the current increases exponentially with 1/T indicating a thermionic emission process. At lower temperatures the current shows no temperature

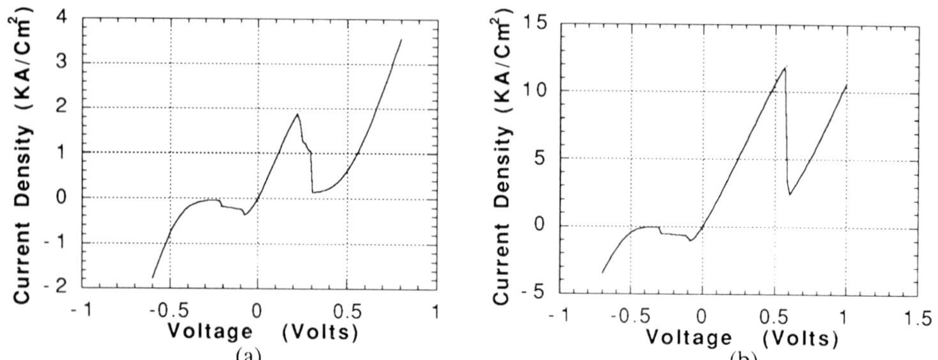

Fig. 2. Room temperature current density-voltage characteristics of RITDs, (a) without AlAs monolayers and (b) with AlAs monolayers.

Fig. 3 shows the temperature dependence of peak and valley current densities as well as the P/V ratios. In general, the peak current increases and the valley current decreases when temperature decreases.

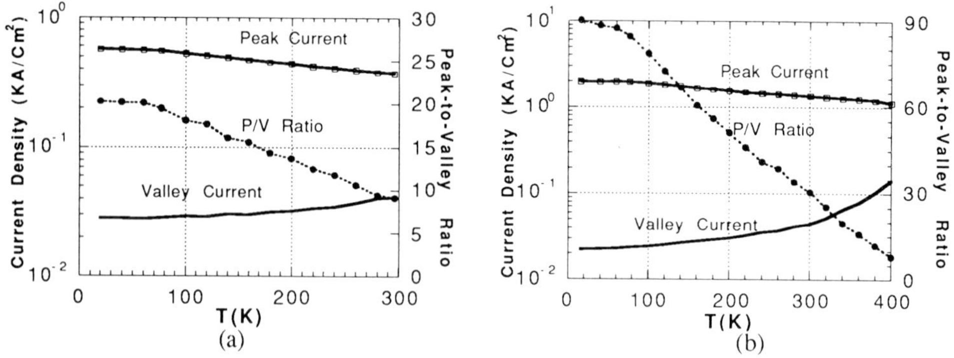

Fig. 3. Temperature dependent measurement of the peak and valley current and P/V ratio, (a) without AlAs monolayers and (b) with AlAs monolayers.

dependence and may be associated with the tunneling process. This can be explained as follows: Figures 1 and 5 show self-consistent Poisson-Schrodinger simulations of the band structures when the diodes are biased at 0.7 volts. The quasi-Fermi potential is assumed to drop across the double barrier tunneling region in these calculations. The figures show that the first three quantum well eigenstates align with the InAs conduction band. In this condition, electrons that tunnel from the GaSb well leave holes behind. These holes can be thermionically excited into the InAs valence band at high temperatures. Inserting the AlAs layers to the barrier increases the energy that is required for holes to escape (Fig. 5a) and this increase in the barrier height reduces the hole current and improve P/V ratio.

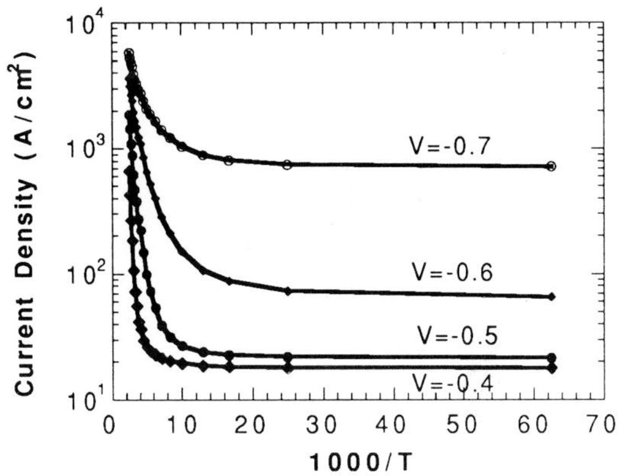

Fig. 4. Reverse-biased temperature dependence of the valley current in RITDs with AlAs barriers.

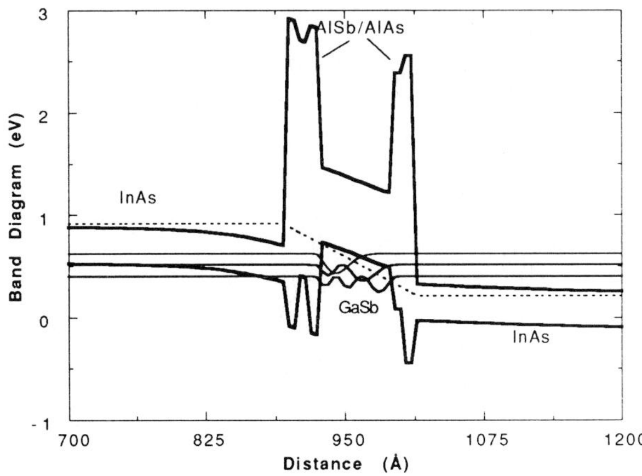

Fig. 5a Reverse biased band structure for the double barrier AlAs inserted interband tunneling diode.

When the diode is biased in the transmissive (peak) region, little voltage is dropped across the double barrier region. The diode resistance is mainly in the contact and the access regions. On the other hand, when the diode is biased in the resistive (valley) region, the primary barrier to carrier transport is the double barrier region. The peak current and the peak voltage in the forward direction (see Fig 2) is higher than in the reverse direction. In figs. 5a and 5b, for 0.7 volt reverse and forward bias respectively, the asymmetry in the thickness of the two barriers alters the potential distributions in the two bias conditions. In the reverse bias case, most of the potential drop is across the double barrier region. In the case of forward bias (Fig. 5b), less potential drop is across the double barrier region and the rest is across the InAs contact layers. Therefore, in order to achieve equal quantum well band bending, more voltage must be applied in the forward direction and this produces a higher peak current.

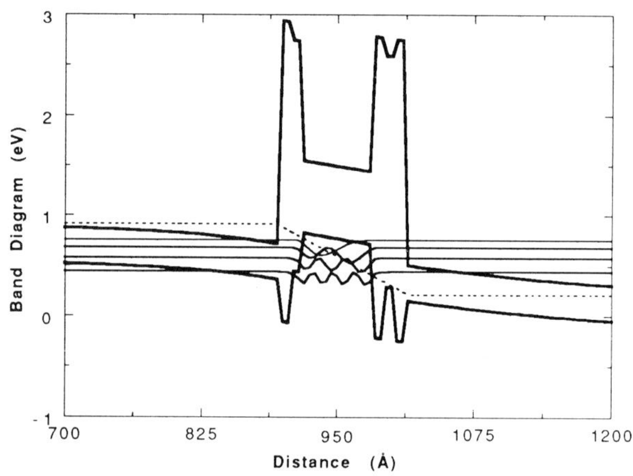

Fig. 5b Forward biased band structure for the double barrier AlAs inserted interband tunneling diode.

ACKNOWLEDGMENTS

This work was partially performed under the management of FED (the R&D Association for Future Electron Devices) as a part of the R&D of Basic Technology for Future Industries supported by NEDO (New Energy and Industrial Technology Development Organization, Japan). We also would like to thank Dr. R. Tsui and M. Adam for useful discussions and assistance in growing the material.

REFERENCES

1. R. T. Collins, J. Lambe, T. C. McGill, R. D. Burnham, Appl. Phys. Lett. **44** (5), 532, 1984.
2. P.J. Turley, S. W. Teitsworth, Phys. Rev. B, **44** (15) 8181, 1991.
3. E. Wolak, K.L. Lear, P.M. Pitner, E.S. Hellman, B.G. Park, T. Weil, J.S. Harris, Jr., D. Thomas, Appl. Phys. Lett. **53** (3), 201, 1988.
4. J. R. Soderstrom, D. H. Chow, and T. C. McGill, Appl. Phys. Lett. **55** (11), 1094, 1989.
5. S. M. Newstead, R. A. A. Kubiak and E. H. C. Parker, J. Crystal Growth **81**, 49 (1987).
6. G. Tuttle, H. Kroemer and J. H. English, J. Appl. Phys. **67**, 3032, (1990).

Inst. Phys. Conf. Ser. No 136: Chapter 4
Paper presented at the Int. Symp. GaAs and Related Compounds, Freiburg, 1993

Room temperature InGaAs coupled-quantum-well base transistor with a graded emitter

Steffen Koch[a], Takao Waho, Takashi Kobayashi, and Takashi Mizutani

NTT LSI Laboratories
3-1 Morinosato Wakamiya, Atsugi, Kanagawa, 243-01 Japan

Abstract: We describe a resonant tunneling transistor with a coupled-quantum-well base operating at room temperature. The molecular beam epitaxy grown InGaAs material system is used. Three thin planar-doped quantum wells separated by strained AlAs barriers constitute the transistor base. A parabolically graded InGaAlAs emitter layer suppresses hole injection into the emitter and enhances the current gain. The DC characteristics show a clear saturation of the collector current with increasing base current, which is evidence for resonant tunneling. Cutoff frequencies are determined employing S-parameter measurements, and the resonant tunneling transit time is estimated.

1. INTRODUCTION

The quantum mechanical phenomenon of resonant tunneling through quantum well structures offers the possibility for functional operation of suitably designed transistors. Such devices have, for example, been realized as bipolar transistors using III-V semiconductor heterostructures with narrow quantum wells in a wide base (the resonant tunneling bipolar transistor (RTBT), Capasso et al. 1986) or with a single quantum well which constitutes the entire base (the bipolar quantum resonant tunneling transistor (BiQuaRTT), Reed et al. 1989). Unipolar devices were studied as well, in particular the resonant tunneling hot electron transistor (RHET, Yokoyama et al 1985). A new approach to the bipolar type was introduced recently (Waho et al 1991) using a *coupled* quantum well (CQW) as the base, where the GaAs material system and AlAs tunneling barriers were employed. In particular, this device showed superior performance in comparison to a reference *single* quantum well transistor. The potential of tunneling transistors for high-speed operation - due to a resonant tunneling time of the order of picoseconds or even less - could also be demonstrated, e.g. using the CQW structure (Waho and Mizutani 1993). However, high-frequency operation was only possible at T=77K in their device.

In the present work, we have realized a resonant tunneling transistor with a CQW base which operates at room temperature. We use the InGaAs material system lattice-matched to InP because it has a large potential for tunneling devices (Inata et al 1987) when $In_xAl_{1-x}As$ barriers are used. This is particularly due to the smaller effective electron mass in InGaAs (leading to higher currents and a larger energy separation of quantum well states) and due to the larger energy difference between the Γ valley in the InGaAs well and the X valley in the $In_xAl_{1-x}As$ barriers (causing smaller valley currents). Here we compare with the GaAs/AlAs system. We employ strained AlAs barriers and a parabolically graded InGaAlAs emitter layer. This device shows current amplification, high-frequency operation and evidence for resonant tunneling at room temperature.

2. EXPERIMENTAL

The transistors discussed in the present work are prepared from a wafer grown by molecular beam epitaxy (MBE). An energy band diagram of the structure is schematically shown in Fig. 1. After Si-doped InGaAs subcollector ($n_{Si} = 3 \cdot 10^{18} cm^{-3}$) and collector ($n_{Si} = 5 \cdot 10^{16} cm^{-3}$) layers the CQW base is grown. It consists of 3 InGaAs quantum wells of 7nm width and 4 strained AlAs barriers. The inner and outer barriers have a width of 3 and 5 monolayers, respectively. The base doping was realized by inserting Be sheet-doping layers into each of the wells, with a sheet concentration of $1 \cdot 10^{13}$ cm^{-2} each. This technique facilitates the confinement of Be in the base. The base is followed by a parabolically graded quaternary InGaAlAs emitter. The Al concentration in the $In_{0.53}(Ga_{1-x}Al_x)_{0.47}As$ emitter follows the law $x(z)=0.10 + 0.65 \cdot (z/35nm)^{1/2}$, where z is the position coordinate in the growth direction, starting with z=0 at the emitter-base interface. We note that this is a parabola as seen from the top of the structure. The graded layer thickness of 35nm is adjusted to the emitter Si-doping concentration of $5 \cdot 10^{17}$ cm^{-3}. A small abrupt part of the emitter has been introduced for facilitating the growth (x=0.10 at the emitter-base interface). Starting from this concentration, the Al flux is increased parabolically, and in the same manner the Ga flux is reduced. Finally, the emitter is linearly graded back to InGaAs to facilitate contact formation, and the doping concentration is increased to $2 \cdot 10^{19}$ cm^{-3}.

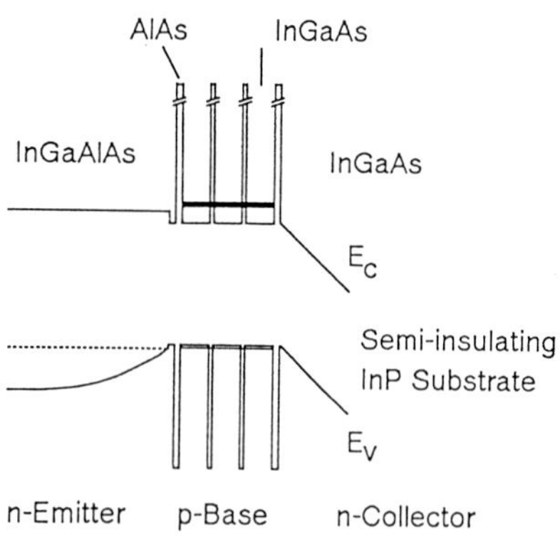

Fig.1: Schematic band diagram of the transistor structure with a parabolically graded quaternary emitter. Dashed line: emitter with constant composition. The lowest electron and hole bands in the CQW are also shown.

For comparison, a wafer with a constant composition $In_{0.53}(Ga_{0.85}Al_{0.15})_{0.47}As$ emitter was also grown. The crystal quality of the base layer was checked employing transmission electron microscopy (TEM). A TEM micrograph is given in Fig. 2, showing a clear CQW structure without any misfit dislocations even though the lattice constants of the wells and the bariers differ by 3.5 percent.

Fig.2: TEM micrograph of the InGaAs/AlAs coupled-quantum-well base, showing good quality strained-layer AlAs tunneling barriers without any misfit dislocations.

Transistor structures were fabricated from these wafers using optical lithography, wet etching and metal evaporation techniques. Typical emitter dimensions are $1x10\mu m^2$, with a spacing of only $0.2\mu m$ between the emitter mesa and the base contact.

3. DC CHARACTERISTICS

DC common-emitter output characteristics are given in Fig. 3. We show the collector current I_C as a function of collector-emitter voltage V_{CE}. The maximum current density j_C of $1 \cdot 10^5$ A/cm^2 is one order of magnitude higher than in the GaAs CQW device (Waho and Mizutani 1993) and leads to improved high-frequency performance. The higher j_C is due to the smaller effective electron mass. In Fig. 3 the differential current gain β at low current is 5; at larger transistor size (with reduced emitter size effect) and at higher measurement resolution (smaller base current steps) the maximum room temperature current gain is β=15. The collector current saturates with increasing base current. This saturation, clearly different from the behaviour in heterobipolar transistors (HBT), indicates the expected quenching of the resonant tunneling current with increasing device bias. The transconductance $g_m=\partial I_C/\partial V_{BE}$ is shown in Fig. 4 . We observe a constant value of 13.5mS at base-emitter voltages V_{BE} larger than 1.1V. The arrow indicates the expected position of the maximum of g_m, taking base resistance into account. This behaviour also clearly differs from the HBT result (which is basically exponential as a function of base-emitter voltage) and thus provides further evidence for resonant tunneling.

The use of a graded emitter is very important for the present device to operate even at room temperature. We note first that it is essential to block the hole current injected into the emitter to reduce the base current. However, it is not possible to use a constant-composition InGaAlAs emitter with a large Al concentration. If the conduction band offset is too large, the lowest resonant states will not contribute to any current because their energy level is below the conduction band edge in the emitter. Thus the device would have to be more highly biased to use higher-order resonant states, with a substantially increased base current. Thus, a graded emitter has to be used, which fulfills the necessity of a small enough conduction band offset ΔE_c at the base-emitter interface and nevertheless provides the advantage of good hole-current rejection. This is experimentally verified by comparing with the reference constant-composition emitter device (with low enough ΔE_c). The graded emitter device shows a current gain about one order of magnitude higher than the constant-composition emitter device. This shows the effectivity of the parabolic grading for suppressing hole injection into the emitter.

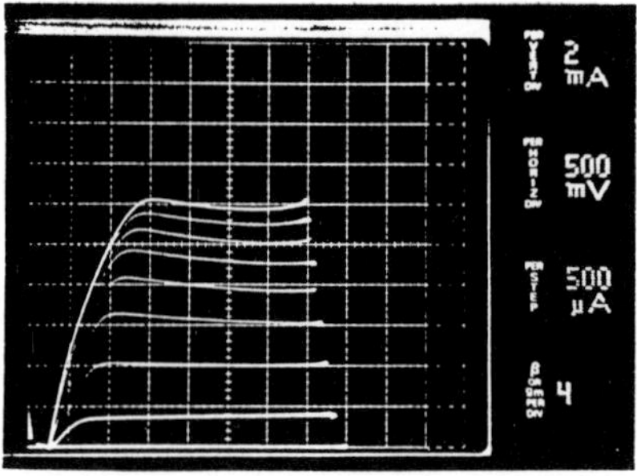

Fig.3: Common-emitter output characteristics for base currents from 0 to 4mA at T = 300 K. Emitter size: $1.2x10\mu m^2$.

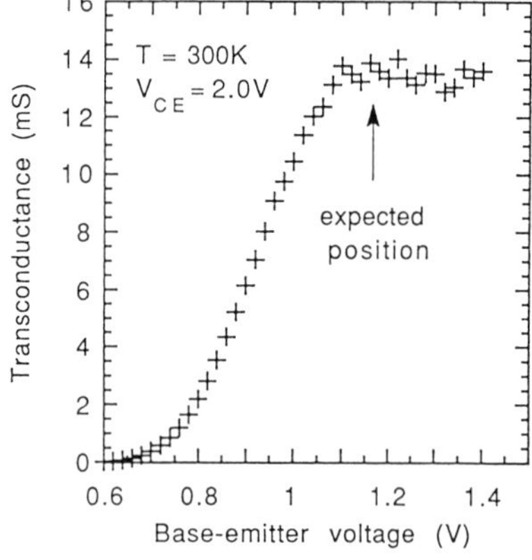

Fig.4: Transconductance as a function of base-emitter voltage. Emitter size: $1.0x10\mu m^2$.

The reason why the resonant tunneling phenomena are not as pronounced as in the similar GaAs device is the larger voltage drop along the base-emitter interface, which is due to both the relatively high base current and the base resistance of about 300Ω. The latter is caused by the thinness of the base for a given doping concentration. We note that - in agreement with this picture - at lower temperature (80K) with reduced base current the transconductance in fact shows a maximum and minimum as a function of V_{BE}. In order to observe an actually negative g_m, this voltage drop must be further reduced. A possibility to obtain this goal may be the use of a larger doping concentration leading to lower base resistance.

4. HIGH-SPEED PERFORMANCE

The high-frequency properties of the device were determined by measuring S-parameters. The results are given in Fig. 5, which shows the current dependence of the unity current-gain cutoff frequency f_T. A maximum value of f_T=19.4 GHz is found at a current density of $2.2 \cdot 10^4$ A/cm^2. At higher currents f_T decreases, which corresponds to the decrease of the differential current gain as shown in Fig. 3. We note that the unilateral gain cutoff frequency was determined as f_{max} =11.5 GHz.

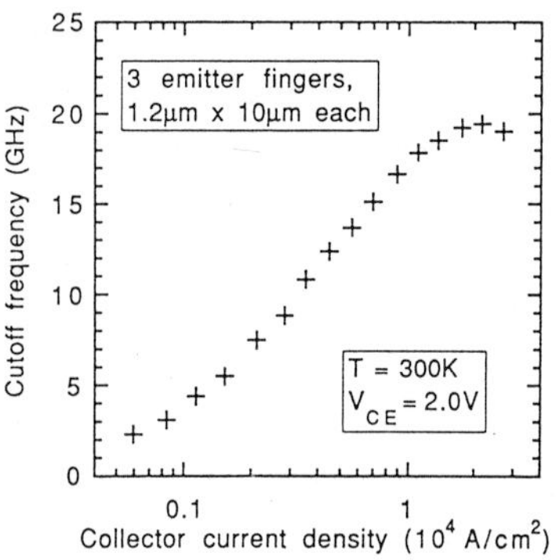

Fig.5: Collector current dependence of cutoff frequency f_T.

The current dependence of the cutoff frequency f_T can be analysed by a procedure originally introduced for HBTs (Yamauchi and Ishibashi 1986). This analysis is based on a decomposition of the total emitter-collector transit time $\tau_{EC}=(2\pi f_T)^{-1}$ into the sum $\tau_E+\tau_B+\tau_C+\tau_{CC}$, where τ_E is the emitter charging time, τ_B and τ_C the base and collector transit time, and τ_{CC} the collector charging time. In Fig. 6 the data of Fig. 5 are re-plotted as the τ_{EC} dependence of the inverse collector current density j_C. Using this representation and extrapolating to infinite collector current (or $1/j_C \rightarrow 0$), we can first eliminate the emitter charging time (which is proportional to $1/j_C$) and find τ_E=2.5±0.3ps at the minimum τ_{EC}=8.2ps. The rest of 5.7±0.3ps is analysed as follows. After measuring collector and emitter resistances and the base-collector capacitance, we obtain τ_{CC}=1.3±0.2ps. An estimate for τ_C yields τ_C=0.6±0.3ps, so that we finally find a base transit time (or resonant tunneling time) of τ_B= 3.8±0.9ps. This can be compared with a numerical determination of the phase time (cf. Waho and Mizutani 1993). The result for the three lowest resonant states (which form a group of levels which are not resolved at room temperature) are 5.1ps, 1.8ps and 2.4ps, respectively. These results are in good agreement with the experimental finding. For further reducing the tunneling time, thinner outer barriers will be useful.

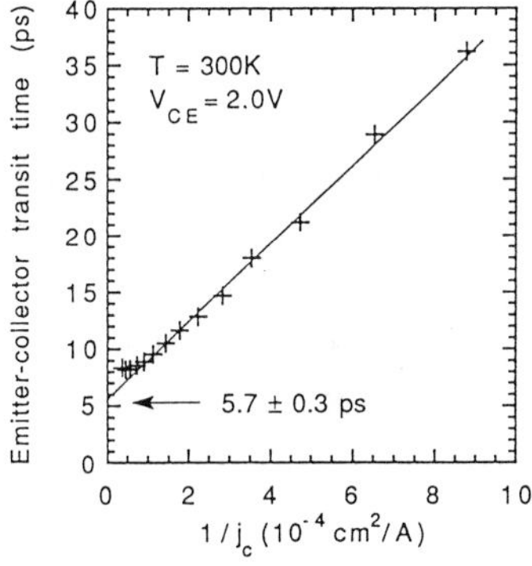

Fig.6: Emitter-collector transit time τ_{EC} as a function of inverse collector current density. The extrapolation to infinite collector current yields $\tau_{CC}+\tau_C+\tau_B$=5.7±0.3 ps.

5. SUMMARY

In conclusion, we report the realization of a coupled-quantum-well base transistor on InGaAs basis using AlAs strained barriers and a parabolically graded emitter. This emitter structure suppresses the hole injection into the emitter and thus allows current amplification at high current density levels even at room temperature. Evidence for resonant tunneling is found as a saturation under increasing device bias of both the collector current and the transconductance. High-frequency operation of the device at room temperature shows current gain cutoff frequencies up to 19 GHz. The resonant tunneling time is estimated as 3.8±0.9ps.

ACKNOWLEDGEMENTS

We would like to thank T. Ishibashi and F. Gueissaz for valuable discussions, T. Ishikawa for wafer growth, H. Takaoka for TEM analysis and K. Hirata for continuous encouragement.

REFERENCES

(a) present address: Max-Planck-Institut für Festkörperforschung, Heisenbergstr. 1, 70569 Stuttgart, Germany.

Capasso F, Sen S, Gossard A C, Hutchinson A L and English J H 1986 IEEE **EDL-7** 573

Inata T, Muto S, Nakata Y, Sasa S, Fujii T and Hiyamizu S 1987 Jpn. J. Appl. Phys. **26** L1332

Reed M A, Frensley W R, Matyi R J, Randall J N and Seabaugh A C 1989 Appl. Phys. Lett. **54** 1034

Waho T, Maezawa K and Mizutani T 1991 Jpn. J. Appl. Phys. **30** L2018

Waho T and Mizutani T 1993 Jpn. J. Appl. Phys. **32** L386

Yamauchi Y and Ishibashi T 1986 IEEE Electron Device Lett. **EDL-7** 655

Yokoyama N, Imamura K, Muto S, Hiyamizu S and Nishi H 1985 Jpn. J. Appl. Phys. **24** L853

Paper presented at the Int. Symp. GaAs and Related Compounds, Freiburg, 1993

Analysis of integrated resonant tunneling devices for millimeter-wave detector applications

B. Landgraf and H. Brugger

Daimler-Benz AG, Forschungszentrum Ulm, D-89081 Ulm, Germany

A. Trasser and H. Schumacher

Universität Ulm, D-89081 Ulm, Germany

ABSTRACT: The millimeter-wave detector performance of biased GaAs-based double barrier resonant tunneling (DBRT) diodes and Schottky-diodes is analysed by a non-linear time domain simulation. The devices are described by their large signal equivalent circuits and a functional approximation of the I/V-curve. In contrast to the Schottky-diode the DBRT-diode has to be stabilized to avoid self-oscillation. If biased at the peak of the I/V-curve, a 300 K DBRT-diode current sensitivity of β_i = 58 µA/µW for an absorbed power level of P_{abs} = -20 dBm (10 µW) at 94 GHz is expected. This value is a factor of three higher than the calculated results of a high performance Schottky-diode simulated at the same absorbed power level. The DBRT-diode detector current depends on the square root of the absorbed power. This leads to Schottky-diode superiority at high power levels. The Schottky-diode´s sensitivity surpasses that of the DBRT-diode for power levels in excess of -16 dBm.

1. INTRODUCTION

Non-linear solid-state two-terminal devices are widely used in communications systems and radar applications, either as mixer or detector diodes. For the detection of millimeter (mm) wave radiation Schottky-diodes are commonly used with a pronounced non-linearity in their current/voltage (I/V) curve and a low parasitic capacitance shunt (Wang et al. 1993). The detector is basically acting as a power measuring device with a DC current or voltage output.

Double-barrier resonant tunneling (DBRT) has attracted considerable attention because it is one of the few solid-state transport phenomena that can provide a fast negative differential resistance (NDR) at room temperature (300 K). Diode oscillators with frequencies above 700 GHz have been realized (Brown et al. 1991), which are among the highest oscillation frequencies reported to date for electronic devices. Furthermore, the pronounced non-linear I/V-characteristic and the NDR-behavior make a DBRT-diode very attracting for detector and mixer applications in the mm-wave frequency range.

Gering et al. (1988) already demonstrated the detector operation of an AlGaAs/GaAs DBRT-diode at 77 K which acts as a full-wave rectifier when biased at the peak current of the I/V-curve. Mehdi et al. (1991) reported about InAlAs/InGaAs/InP DBRT detector diodes with measured sensitivities of 1750 mV/mW in the Ka-Band at 300 K.

We report about a non-linear time domain simulation of the mm-wave detector performance of GaAs-based DBRT-diodes. The results are compared with the calculated detector characteristics of high-performance low-capacitance GaAs Schottky-diodes. The devices are described by their large signal equivalent circuits. Stable operation of the DBRT-diode is achieved through a passive network. A calculated current sensitivity of 58 μA/μW for an absorbed power level of 10 μW at 94 GHz is obtained for a DBRT-diode detector which is significantly higher than the calculated value for the Schottky-diode.

2. BASIC PRINCIPLE OF MILLIMETER-WAVE DETECTOR SIMULATIONS

The mixed-mode simulator ELDO was used for the calculations. It supports a SPICE-like input syntax and has been developed by the French Center of Telecommunication Research (CNET) and by ANACAD GmbH (Ulm, Germany). We used the time domain simulation procedure followed by a Fourier series decomposition to calculate the large signal behavior at an incident mm-wave frequency of 94 GHz. From the spectral information the diode absorbed power is calculated.

Fig. 1 shows the equivalent circuit of a commonly used diode detector. The mm-wave source is represented by a RF generator (94 GHz) with characteristic impedance (Z_G). The DC-bias is supplied via a voltage source (V_b), load resistance (R_L) and low-pass filter network. The detector diode itself is represented by a nonlinear equivalent circuit, which will be discussed in section 3. Separation between the high frequency part and the bias supply was achieved by a second-order low-pass filter, which was treated as concentrated elements (inductances and capacitors) in the simulation. In practise the

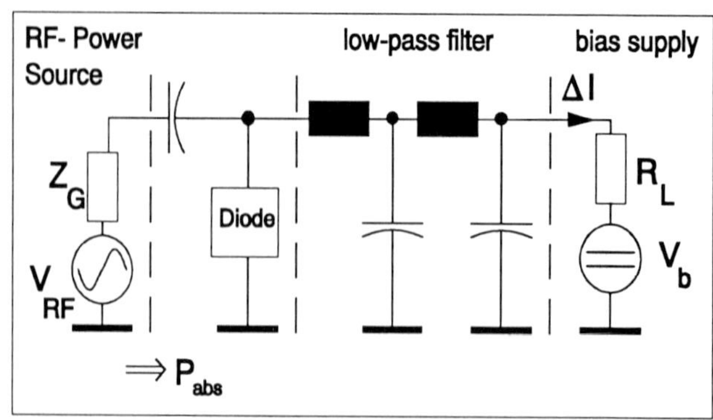

Fig. 1. Equivalent circuit of an integrated mm-wave detector commonly used for Schottky-diodes.

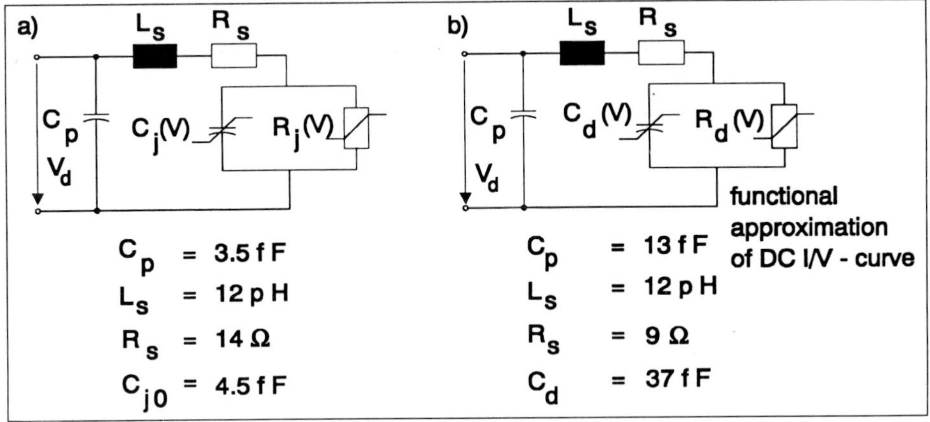

Fig. 2. Large signal equivalent circuits of Schottky- (a) and DBRT (b) mm-wave detector diode.

low-pass filter would be realized by distributed transmission-line structures. The mm-wave signal causes a rectified current component (ΔI) giving rise to a voltage drop across the external load resistance R_L.

For mm-wave detector diodes one figure of merit is the current sensitivity, which is defined as: $\beta_i = \Delta I/P_{abs}$, where ΔI is the induced current and P_{abs} the absorbed high frequency power. The detector dynamic range is limited by the compression point, which is the 1 dB deviation of the functional relationship of the $\log(\Delta I)/\log(P_{abs})$-curve from the linear behavior. An additional important figure of merit for broadband applications is the video resistance which is defined as the small signal resistance on the low-frequency port of the detector.

We use a transient time domain simulation to calculate the high frequency induced rectified current (ΔI). The high frequency power (P_{abs}) absorbed by the device was calculated via the voltage and current values at 94 GHz derived by an FFT. The V_d and I_d values for both devices are taken from the stationary part of our time domain simulation.

3. DIODES LARGE SIGNAL EQUIVALENT CIRCUITS

The diodes used in our simulation are described by their large signal equivalent circuits which are schematically shown in Fig. 2. The data for the DBRT-detector are based on a 15 μm^2 pseudomorphic InGaAs/AlAs/GaAs diodes realized by Lippens et al. (1992) and Brugger et al. (1991). Numerically, a polynomial expression was fitted to the experimental DC I/V-curve.

The negative resistance region (NDR) was assumed to be linear with a resistance value of r_d = - 37.5 Ω. This value was obtained from the corresponding current swing in the NDR-

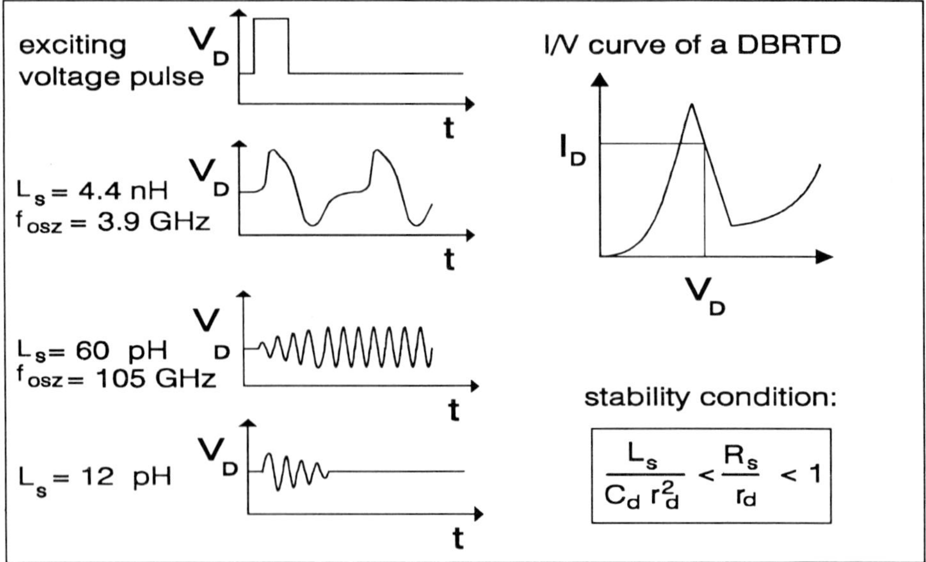

Fig. 3. Intrinsic stability condition required for stable DBRT-diode operation.

region (ΔI = 4 mA) and the difference between peak and valley voltage (ΔU = 0.15 V). The influence of the series inductance (L_S) is discussed separately in section 4. In a first-order approximation, the DBRT-diode capacitance was assumed to be voltage-independent.

For the nonlinear equivalent circuit parameters of the GaAs Schottky-diode, experimental data from Dieudonné et al. (1992) were used for the simulation. The Schottky-contact area of the epitaxial and low-capacitance device is only 1.5 μm^2. The cutoff frequency is about 2 THz.

4. REQUIRED STABILITY CONDITIONS FOR DBRT-DIODES

Due to the negative differential resistance DBRT-diodes tend to oscillate. Kidner et al. (1990) have shown that an intrinsic stability is only obtained if the following equation is fullfilled:

$$\frac{L_s}{C_d r_d^2} < \frac{R_s}{r_d} < 1 \tag{1}$$

Especially the series inductance L_S has to be kept as small as possible to avoid any oscillations. The calculated influence of L_S on the oscillation behavior of the DBRT-diode is shown in Fig. 3 for three different values. The DBRT-diode was excited by a defined voltage pulse of 10 mV, when biased in the NDR region. Stability is only achieved for inductance values $L_S(max) \leq 12$ pH. Therefore L_S = 12 pH is used in our simulations.

To achieve circuit stability the load impedance Z_{load} has to fullfill the following condition:

$$\mathrm{Re}\{Z_{load}\} < |R_S + r_d| \quad (2)$$

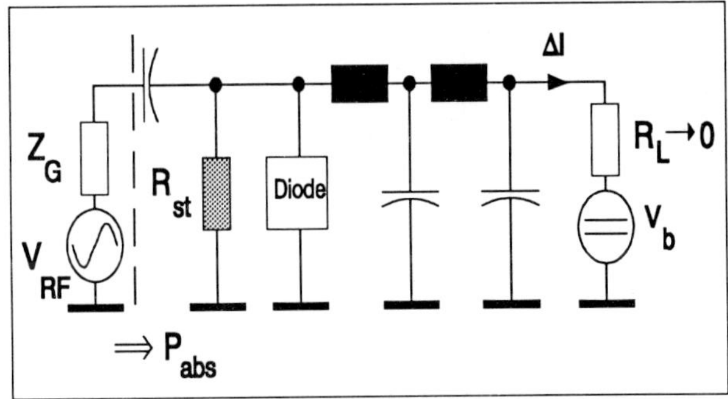

Fig. 4. Equivalent circuit of the integrated mm-wave DBRT-diode detector used in the simulation.

This is realized by a shunt resistor R_{st} in parallel with the DBRT-diode as shown in Fig. 4. $R_{st} = 25\ \Omega < |r_d + R_s|$ was chosen in the following, which is close to the stability limit. This keeps the detector performance as high as possible. Additionally the load resistance was choosen $R_L = 0$, i.e. the rectified short circuit current sensitivity is calculated.

5. DETECTOR SENSITIVITIES

Fig. 5 shows the bias dependent short circuit current sensitivity for both detector devices. The results are obtained for a constant absorbed power of $P_{abs} = 10\ \mu W$ at a frequency of 94 GHz. If biased in the peak current a maximum sensitivity of 58 μA/μW is calculated for the DBRT detector. The Schottky-diode detector shows a maximum sensitivity of 18 μA/μW which is a factor of three lower in comparison with the DBRT-diode.

In Fig. 6 the transfer characteristic is shown for both detector devices. A significant advantage in sensitivity for the DBRT-diode is expected for lower power levels (< -16 dBm). However for P_{abs} > -16 dBm the Schottky-diode appears to be more sensitive. The 1 dB compression point for the Schottky-diode is at -12 dBm. As the DBRT-diode transfer characteristic does not follow a square-law behavior, we do not define any compression point. The DBRT-diode might be advantageous for applications with a wide range of input power. The shunt resistor R_{st} lowers the video resistance for the DBRT-diode circuit to 23.3 Ω, which is useful for broad-band video applications. The corresponding video resistance of the Schottky-diode is 2.8 kΩ.

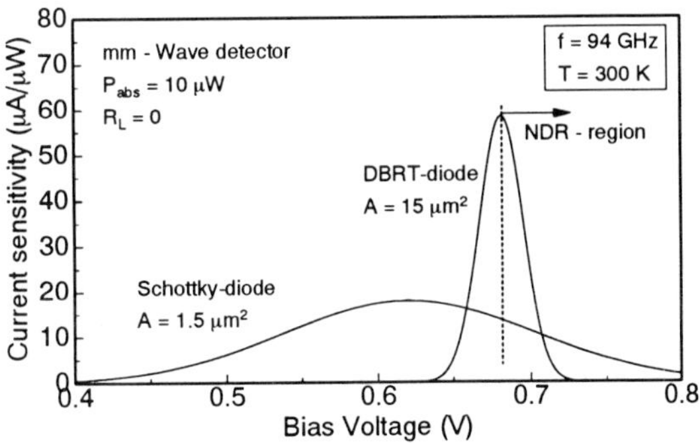

Fig. 5. Detector bias dependence of current sensitivity.

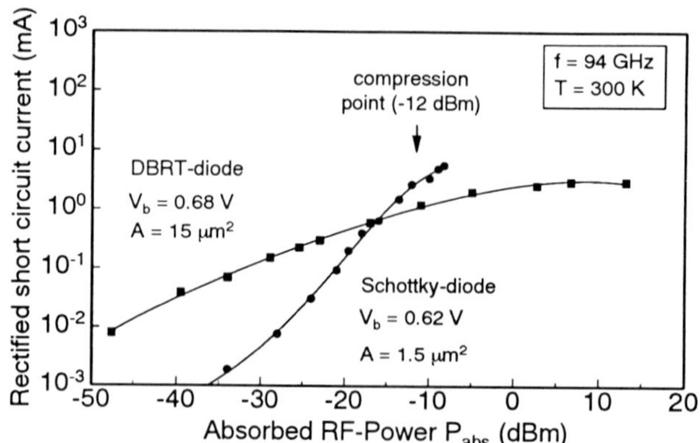

Fig. 6. Transfer characteristica of mm-wave DBRT-diode and Schottky-diode detectors.

6. CONLUSION

A simulation of the mm-wave detector performance of stabilized DBRT-diodes were performed and compared with results from high performance Schottky-diodes. A significantly higher calculated current sensitivity of DBRT-detectors at lower input power levels (< -16 dBm) was obtained for a frequency of 94 GHz. This is very attractive for future mm-wave sensor systems.

ACKNOWLEGDEMENT

This work was partly supported by the Bundesminister für Forschung und Technologie under contract number 01 BM 113/1.

REFERENCES

Brown E R, Söderström J R, Parker C D, Mahoney L J, Molvar K M and McGill T C 1991 Appl. Phys. Lett. 58 2291

Brugger H, Meiners U, Woelk C, Deufel R, Schroth J, Foerster A and Lüth H 1991 Proc. 13th IEEE/Cornell Conf. on Advanced Concepts in High Speed Semicond. Dev. and Circuits Ithaca (U.S.A.) 39

Dieudonné J M, Adelseck B, Schmegner K E, Rittmeyer R and Colquhoun A 1992 IEEE MTT 40 1466

Gering J M, Rudnick T J and Coleman P D 1988 IEEE MTT 36 1145

Kidner C, Mehdi I, East J R and Haddad G I 1990 IEEE MTT 38 864

Lippens D, Nagle J, Grimbert B, Sadaune V, Lheurette E, Vinter B, Tilmant P and Francois M 1992 Microelectronic Engineering Elsevier 19 879

Mehdi I, East J R and G I Haddad 1991 IEEE MTT 39 1876

Wang H, Lam W, Ton T N, Lo D C W, Tan K L, Dow G S, Allen B and Berenz J 1993 IEEE MTT-S Digest 365

Inst. Phys. Conf. Ser. No 136: Chapter 4
Paper presented at the Int. Symp. GaAs and Related Compounds, Freiburg, 1993

Supply and escape mechanisms in $In_{0.1}Ga_{0.9}As$/GaAs/AlAs resonant tunneling heterostructures in a triple well configuration

O. Vanbésien, L. Burgnies, V. Sadaune and D. Lippens
Institut d'Electronique et de Microélectronique du Nord U.M.R C.N.R.S 9929
Département Hyperfréquences et Semiconducteurs
Université des Sciences et Technologies de Lille
59655 VILLENEUVE D'ASCQ Cedex - FRANCE

J. Nagle and B. Vinter
Laboratoire Central de Recherches - Thomson C.S.F
Domaine de Corbeville
91404 ORSAY Cedex - FRANCE

ABSTRACT : Pseudomorphic $In_{0.1}Ga_{0.9}As$/GaAs/AlAs resonant tunneling diodes in a triple well configuration have been fabricated and tested. The devices exhibit good I-V characteristics due to the strong enhancement of the two dimensional character of the carrier injection. The conduction characteristics are analysed in terms of supply and escape mechanisms by means of a self-consistent Schrödinger-Poisson procedure for the size-quantized states whereas extended states are treated in the Thomas-Fermi approximation. Escape times and sheet carrier densities are carefully evaluated for various bias conditions leading to tunneling current characteristics in good agreement with experiment.

1. INTRODUCTION

Resonant Tunneling systems have received a widespread interest for the past few years from the viewpoint of both potential applications and fundamental physics (Chang et al, 1991). The case of the Double Barrier Heterostructure (DBH) is the most elementary example where resonant tunneling occurs. Concerning the theoretical aspects, one of the challenging topic is related to the self-consistent determination of the charge density throughout the structure. Ideally, for such open quantum systems, one should perform a complete quantum kinetic calculation (Frensley 1990, Buot et al 1991) since quantum size effects appear not only within the double barrier region but also in the access regions where strong interference phenomena can be evidenced. Frensley (1991) was thus one of the first to show, in terms of Wigner distribution function, that a complete treatment was particularly difficult to handle and often leads to unphysical band bending profiles. The main difficulty of the pure quantum approach comes from the fact that it is essential to include the inelastic processes in the analysis in order to determine a reasonable self-consistent potential. However, this difficulty can be overcome by means of the Thomas-Fermi (TF) screening model. In addition to its simplicity the main advantage of the TF model is to guarantee charge neutrality at the boundaries and hence to obtain physically acceptable variations of the potential profile. Nevertheless for the carrier concentration, which is assumed to be a local function of the difference between the Fermi level and the conduction band, the TF approximation leads to unphysical jumps which appear as artifacts of the simulation. One elegant way to overcome this problem is to implement the procedure initially proposed by Fiig et al (1991) who assume that the local carrier density can be treated as the superposition of a three dimensional (3D) contribution due to extended states and a two dimensional (2D) one describing the quasi-bound states. In this paper, we applied this procedure to a resonant tunneling structure in a triple well configuration we have also fabricated and tested. Such a structure enforces the 2D character of the injection and thus appears particularly appropriate to test the validity of the theoretical treatment. The device fabrication is described briefly in Section 2 whereas the theoretical aspects are presented in Section 3. In a last Section, we compare theory and experiment on the basis of conduction characteristics.

2 DEVICE FABRICATION AND CURRENT-VOLTAGE CHARACTERISTICS

Recent advances in strained layer epitaxy make it possible to introduce local perturbations in the potential profile of heterostructures aimed at improving both static and dynamic properties. Figure 1 depicts such a structure in a triple well configuration. The basic idea is here to enhance the 2D character of the carrier injection by means of a prewell grown prior to the first barrier. The mid-well controls the escape process whereas the postwell is grown to preserve the symmetry of the structure and hence of the conduction characteristics. In practice, the structure was grown by Molecular Beam Epitaxy starting from a GaAs [100] semi-insulating substrate. The ultra-thin barriers of AlAs with thicknesses of 1.7 nm permit us to achieve high current densities in view of high frequency applications. The prewell and postwell are 5 nm thick whereas the midwell is 4 nm thick. We have intentionally limited the Indium content of InGaAs layers to 0.1 in order to avoid misfit dislocations. For each transition a 0.5 nm thick GaAs spacer was systematically placed to recover good surface states. The triple well-double barrier zone is then sandwiched between two 500 nm thick n^+ GaAs layers acting as electron reservoirs followed by 10 nm thick cladding layers doped 10^{17} cm^{-3}. The device fabrication followed the general procedure of the microwave compatible technology we developed recently and details can be found elsewhere (Lheurette et al, 1992). Measurements were performed at room temperature on a 14 μm^2 diode. A typical I-V characteristic obtained at 300 K is reported Figure 2. State of the art performances were thus obtained for the material system used here (Brugger et al 1991, Riechert et al 1990) with peak current densities as high as 50 $kA.cm^{-2}$ along with a peak to valley current ratio of 7:1. Peak voltage is measured around 0.5 to 0.6 V.

GaAs	$3x10^{18}$ cm^{-3}	5000 Å
GaAs	10^{17} cm^{-3}	100 Å
GaAs	Undoped (UD)	50 Å
$In_{0.1}Ga_{0.9}As$	UD	50 Å
GaAs	UD	5 Å
AlAs	UD	17 Å
GaAs	UD	5 Å
$In_{0.1}Ga_{0.9}As$	UD	40 Å

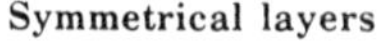

Figure 1: Growth sequence of the pseudomorphic heterostructure in a triple well configuration

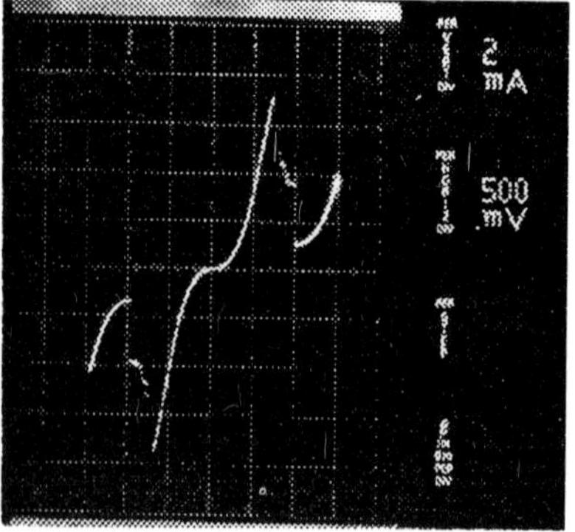

Figure 2: Typical I-V characteristic measured at room temperature

3 THEORETICAL ANALYSIS

The main problem that faces us for the present structure is to describe the way the GaInAs wells are filled up as a consequence of the charge transfer from the adjacent layers. First of all, it is worth mentioning that the spacer layers placed between the highly doped regions and the quantum zone are only 10 nm thick. This means that we have to take into account two interdependent mechanisms i.e. (i) the diffusion due to the doping gradient in the access regions (ii) the charge transfer at the GaAs/InGaAs interfaces as it is observed in a modulation doped heterostructure. The first effect leads to large electrostatic potential barriers which are formed in the emitter and collector regions whereas the second gives strong band bending effects resulting from the charge dipole at the heterointerfaces.

First of all, we suppose that the electron reservoirs are in equilibrium. This zero current approximation permits us to define a constant Fermi level E_F in the access regions. Then we assume that the electron density in the wells are the sum of 2D and 3D contributions respectively with the frontier between these states of different dimensionality defined by the maximum of the conduction band edge ($E_C(z_0)=E_{max}$). The charge density along the growth direction (z) is thus given by :

$$n(z) = N_{c3D}F_{1/2}\left(\frac{E_F - E_C(z)}{kT}\right) + \sum_i \frac{m^*}{\pi\hbar^2} kT \mathrm{Log}\left(1+\exp\left(\frac{E_F - E_i}{kT}\right)\right)\left|\Psi_i(z)\right|^2 \qquad (1)$$

where N_{c3D} is the 3D density of states. $F_{1/2}$ is the 1/2 order Fermi-Dirac integral. $E_c(z)$ represents the conduction band along the growth axis until z_0 and taken equal to E_{max} from z_0 to the DBH. In the second term, $\Psi_i(z)$ is the wave function calculated in the wells for the eigenstates E_i. The other symbols have their usual meaning. To calculate $\Psi(z)$ we used the method proposed by Lassnig et al (1987) based on the determination of the local density of states in a size-quantized region. The advantage of this approach is not only to derive the eigenstates but also to have an estimate of the lifetime of electrons which are temporarily trapped in the well and have the possibility to tunnel through the barrier into the continuum. In the original paper, the method was established under too restrictive assumptions to be applied to the structure under investigation, i.e. a 2D electron gas bounded at one side by an infinite potential and flat band conditions. We have thus extended the formalism with a possible penetration of the wave function on the left hand side and a piecewise potential so that the method can be applied to any arbitrary potential. The numerical procedure is as follows: as initial guess, we started from a potential profile calculated using a self-consistent Poisson solver in the TF approximation throughout the structure. In order to characterise the size-quantized states and hence a more accurate carrier density n(z). The Schrödinger equation is solved by the method outlined above. We assumed evanescent waves (energy $E < E_{max}$) in the diffusion potential barrier and plane waves for the continuum states far from the DBH. At each mesh point of the discretized structure the wave function is calculated and this routine is repeated for each energy value. The next stage is to reconstruct the carrier density using (1) and to use this value of n(z) to update the potential taking the quantum effects into account. The treatment is repeated until convergence is obtained on the potential (absolute error at each mesh point between two trials lower than 10^{-6}).

The last point concerns the current density determination. Basically, we expect two contributions : the first from the extended states resulting from the broadening of the supply function at finite temperature. The second comes from the notch states. For conventional structures this was clearly put in evidence with observation of anomalies in the I-V curves (Mounaix et al, 1992). The conduction current part corresponding to the extended states (J_{3D}) can be evaluated assuming a 3D supply function in the highly doped layer weighted by the transmissivity of the structure taken as a whole and by integrating over energy for $E > E_{max}$. In contrast, for the evaluation of the current part due to the notch states we assumed that the sheet carrier density n_s in the

InGaAs prewell, escaping through the double barrier, gives rise to a current density $J_{2D}(V)=q.n_s(V)/\tau_e(V)$. The escape time τ_e in this expression can be easily deduced from the local density of states D(E) calculated for energies lower than $E_{max.}$ As shown later, D(E) exhibits a sharp peak for each quantum level and τ_e is given by $\tau_e=h/\Gamma$ where Γ is the full width at half maximum of the resonant peak. By this means, we avoid treating in detail the relaxation mechanisms from the extended states to the notch states and the direct supply of the InGaAs well by tunneling through the diffusion barrier.

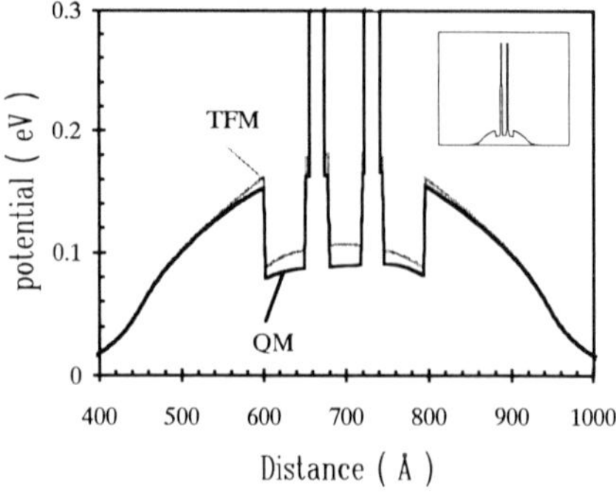

Figure 3: Band bending calculations at equilibrium with a Thomas-Fermi model (TFM) and a with a quantum model (QM) at 300 K

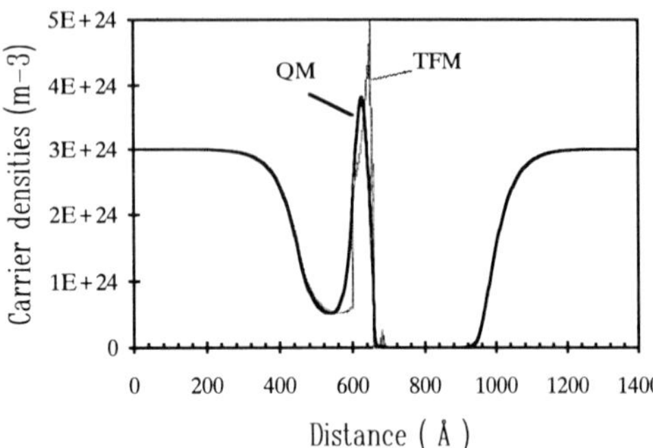

Figure 4: Carrier densities at 300 K under bias with a Thomas-Fermi model (TFM) and with a quantum model (QM)

4 RESULTS AND DISCUSSION

Figure 3 shows a comparison between the potential profile calculated at equilibrium using the TF model (TFM) and the quantum model (QM). The general trends are similar with an overestimation, however, of band bending effects in the TFM. The differences between the two approaches are particularly apparent on the carrier densities reported Figure 4 , under bias in that case. As mentioned in the introduction, unphysical jumps are obtained at each heterointerface using TFM. As expected, under bias the charge is pushed close to the first barrier with a charge density largely overestimated. At the opposite, the QM predicts a continuous variation of n(z) shaped by the presence probability in the prewell. From n(z) behaviour but also quantitatively, it is found here that the 2D filling of the prewell is dominant compared to the extended states. Therefore, the prewell plays entirely its role with a significant improvement of the 2D injection compared to a conventional structure. The variations of n_s versus voltage using the quantum model are given Figure 5. As a general rule, n_s increases monotonously with bias and reaches values rapidly close to 10^{12} cm^{-2}. Let us note that n_s value as a quantity integrated over space turns out to be less sensitive to the model employed. This will not be the case for the current-voltage characteristics as shown later.

As mentioned above, the escape time is deduced from the local density of states D(E). With respect to the situation of an escape process through a single barrier, the fact of introducing a double barrier which acts as an energy filter leads to a strong modulation of τ_e versus bias. This is illustrated Figure 6 showing the variations of D(E) integrated over the prewell region for two bias conditions. At V=0.5 V, two peaks in D(E) separated by an energy offset of around 10 meV are apparent characterised by a large broadening in energy. This situation is very close to the resonance condition which, in a first approximation, should correspond to the anticrossing of the prewell and the midwell levels. Γ takes its maximum value and a high current state is established. At higher bias, the peak in D(E) revealing the quasi-bound state of the prewell becomes very sharp due to the drastic decrease of the DBH transmissivity. From the broadening of D(E) calculated for each bias, we deduced the evolution of $\tau_e(V)$ given in Figure 5 along with $n_s(V)$ previously discussed. A dynamic of two orders of magnitude is obtained for in-and-out resonance conditions with a subpicosecond minimum value for τ_e.

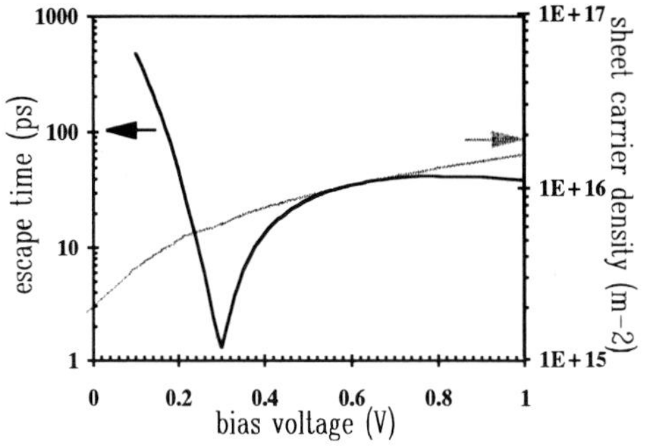

Figure 5: Sheet density of carriers trapped in the prewell and escape time deduced from the local density of states as a function of bias voltage at room temperature.

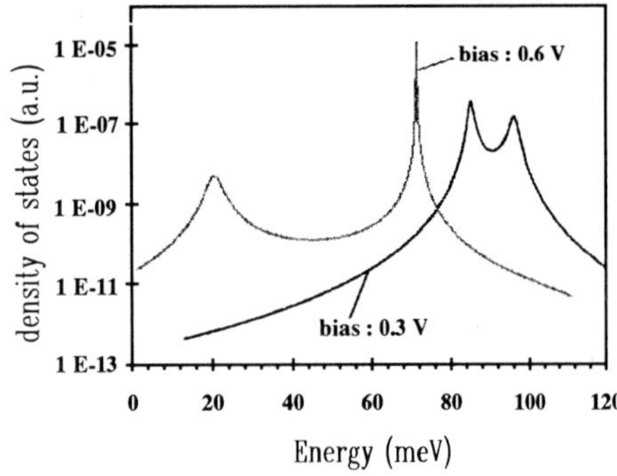

Figure 6: Local density of states integrated over the prewell for two bias voltages

Turning now to the current voltage characteristics, we show in Figure 7 the data we calculated using the two methods. A peak current density of 75 kA.cm^{-2} is obtained by means of the QM comparable to the experimental value Jp=50 kA.cm^{-2} whereas the TFM greatly overestimated Jp with a value of 160 kA.cm^{-2}. One can note that the valley current is underestimated by both methods which assume a coherent transport. In fact, out of resonance conduction is dominated by scattering mechanisms such as phonons or interface roughness (Chevoir et al 1993, Gueret et al 1989) not taken into account in the present models. Both approaches predict comparable peak voltage close to 0.3 V. In fact, the fit of the threshold voltage can be greatly improved by introducing the voltage drop across the access series resistance. Taking a realistic value in the range of 2-4 10^{-6} Ω.cm^2 and a current density of 50 kA.cm^{-2} leads to a shift towards higher voltages of 150 meV for the device under consideration.

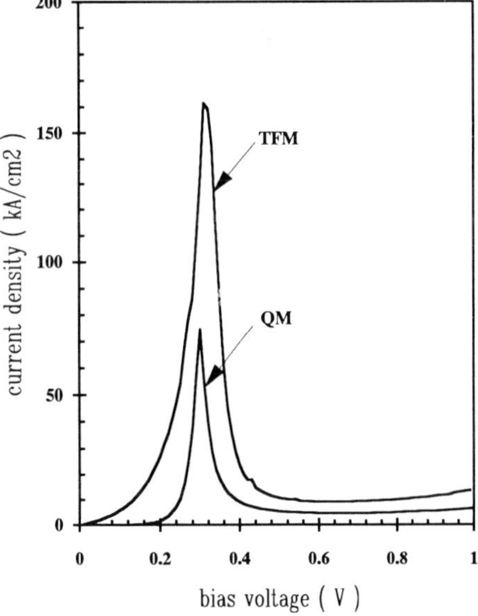

Figure 7: Calculated I-V characteristics at 300 K with a Thomas-Fermi model (TFM) and with a quantum model (QM)

5 CONCLUSION

In summary, the supply and escape mechanisms have been investigated in a double barrier resonant tunneling heterostructure in a triple well configuration. Such a structure exhibits strong band bending effects difficult to handle theoretically in the sense that particles can be exchanged between the wells and their environment. The problem has been solved using a superposition of discrete and continuum states which permits us to calculate the carrier density, the lifetime of electrons in the wells and hence the tunneling current. Comparison between theory and experiment have been carried out on the basis of a pseudomorphic $In_{0.1}Ga_{0.9}As$/AlAs/GaAs structure we have fabricated and tested. The devices exhibit excellent conduction properties which validate the exactness of design rules established by numerical simulation

6 ACKNOWLEDGEMENTS

We would like to thank B. Grimbert, E. Lheurette and P. Mounaix for device processing. The Ministère de la Recherche et de la Technologie (MRT) is acknowledged for financial support.

7 REFERENCES

Brugger H., Meiners U., Wölk C., Deufel R., Schroth J., Förster A. and Lüth H. 1991, Proceedings of the 13th Conference on Advanced Concepts in High Speed Semiconductor Devices and Circuits.

Buot F.A. and Jensen K.L. 1990, Phys. Rev. B 42 9429

Chang L.L., Mendez E.E. and Tejedor C. 1991, NATO ASI Series B277 "Resonant Tunneling in Semiconductors: Physics and Applications", Plenum press

Chevoir F. and Vinter B. 1993, Phys. Rev. B 47 7260

Fiig T. and Jauho A.P. 1991, Appl. Phys. Lett. 59 2245

Frensley W.R. 1989, Solid State Electronics 32 1235

Frensley W.R. 1990, Rev. of Modern Physics 62 745

Gueret P., Rossel C., Schlup W. and Meier H.P. 1989, J. Appl. Phys. 66 4312

Lassnig R. and Boxleitner W. 1987, Solid State Communications 64 979

Lheurette E., Grimbert B., François M., Tilmant P., Lippens D., Nagle J. and Vinter B. 1993, Elec. Lett 28 937

Mounaix P., Vanbésien O. and Lippens D. 1990, Appl Phys. Lett 57 1517

Riechert H., Bernklau D., Reithmaier J.-P. and Schnell R.D. 1990, Elec. Lett 26 341

Investigation of quantum states in V-shaped GaAs quantum wires

R.Rinaldi[a], R.Cingolani[a),b)], F.Rossi[c], L.Rota[d],M.Ferrara[a], P.Lugli[e], E.Molinari[f], U.Marti[g], D.Martin[g], F. Morier-Genoud[g], F.K.Reinhart[g].

[a] Unità GNEQP-Dipartimento di Fisica , Università di Bari, 70100 Bari (Italy)
[b] Dipartimento Scienza dei Materiali, Università di Lecce,73100 Lecce (Italy)
[c] Philipps-Universität Marburg, Fachbereich Physik und Zentrum für Materialwissenchaften, D-35032 Marburg, (Germany)
[d] Department of Physics, Clarendon Laboratory, University of Oxford, Oxford, UK
[e] Dipartimento Ingegneria Elettronica, Università di Roma "Tor Vergata", 00133 Roma (Italy)
[f] Dipartimento di Fisica, Università di Modena, 41100 Modena (Italy)
[g] Ecole Polytechnique Federale de Lausanne, Dep. de Physique, CH-1015 Lausanne

ABSTRACT: The quantized states of GaAs V-shaped quantum wires have been investigated by means of photoluminescence and photoreflectance. The experimental results are succesfully compared with theoretical calculations of eigenstates based on a realistic model for the two-dimensional confining potential.

1.INTRODUCTION

Recently, V-shaped quantum wires fabricated by direct growth of quantum wells on non-planar patterned substrates have attracted much attention (Kapon E. 1989). Unlike the case of rectangular quantum wires, the V-shaped quantum wires can be fabricated with small dimensions ($\leq 20nm$) and exhibit distinct 1D transitions also under low excitation intensity (cw luminescence). In this work we have studied by photoreflectance and photoluminescence the one-dimensional electronic states of V-shaped GaAs/AlAs quantum wires. The observed transitions are compared with the electron and hole confinement energies obtained by solving the full two-dimensional Schödinger equation describing the wires. The spatial localization of the states involved in the transitions is then discussed on the basis of the calculated wavefunctions.

2.RESULTS and DISCUSSION

The investigated GaAs/AlAs quantum wires were fabricated by MBE growth on non-planar GaAs etched substrates (Marti 1991). Two kinds of samples were studied: samples with groove period in the substrate longer than the GaAs carrier diffusion length (pitch 420nm), and samples with groove period of the order of the carrier diffusion length (pitch 250nm). The bending and tapering of the quantum well deposited on the grooves provides an additional lateral confinement with a parabolic-like potential shape. A schematic cross section of the structure and the typical emission spectra of the short pitch sample under ns-pulsed high excitation intensity are reported in

Fig.1. The lines indicate the heterostructure regions from which the emission bands originate. The broad band around 790nm exhibits a fine structure and a fast band filling with respect to the other lines. This band can be attributed to recombination processes between quasi one-dimensional (1D) conduction and valence subbands ($\Delta n = 0$) according to the one-dimensional quantum level calculation reported below. The lines around 750nm and 720nm are due to the flat quantum well, grown as reference outside the patterned region, and to the lateral quantum well on the sidewalls of the V-groove (Kapon K. 1992). The strong emission line centered at 670nm originates from the $(GaAs)_8(AlAs)_4$ superlattices grown as barrier.

To have a clearer evidence of 1D excitonic states, temperature dependent cw-photoluminescence (PL) and photoreflectance measurements have been performed. In Fig.2 we report the temperature dependent PL spectra of the V-shaped quantum wires with the short pitch and in Fig.3 the photoreflectance spectrum of sample with larger pitch. The temperature dependent spectra exhibit the thermal filling of higher energy 1D states up to n=4 at a temperature of 190K. The observed efficient thermal filling of higher energy excitonic states is a consequence of the typical shape of the joint density of states in monodimensional heterostructures (Asada 1986). Samples with the pitch larger than the carrier diffusion length exhibit weak luminescence from the wires, because carriers are not efficiently trapped in the wire regions and recombine in the flat quantum well regions connecting adjacent wires (Christen 1992). Therefore, in these samples photoreflectance measurements were performed to investigate 1D exciton states. In the photoreflectance spectrum the three closely spaced resonances in the energy range between 780nm and 795nm can be attributed to 1D excitonic states associated with the first three subbands, while the resonances around 818nm and 760 nm are due to 1s exciton state of the bulk GaAs substrate and to the flat quantum well connecting the wires, respectively (Cingolani 1993).

To compare the experimentally observed transitions with the theoretically evaluated ones, the quantum wire eigenstates were initially evaluated with a simple approximation. Due to the large extension of the lateral potential with respect to the quantum well thickness, we can separate the confining potential into the two directions. The well widths of the tapered quantum well at several distances (x) from the vertex were measured from TEM pictures. Then, using the envelope function approximation the corresponding confinement energies for electrons and holes were calculated as a function of the distance. The resulting lateral potential was fitted by $U(x) = 1/(cosh^2(Wx))$, where x is the lateral distance and W is a constant (fitting parameter) that gives an estimate of the potential width (Kapon E. 1989). The resulting potential is included in the Schrödinger equation to get the eigenstates (electrons) :

$$E_{e,n} = -\frac{\hbar^2}{32m_e^* W^2}\left[-(1+2n)+\sqrt{1+\frac{32m_e^2 W^2 \Delta E}{\hbar^2}}\right]^2, n = 0,1,2,3... \quad (1)$$

where m_e^* electron effective mass and ΔE is the height of the parabolic-like $U(x)$ confining potential. The energy values of the first five quantized states obtained by eq.(1) are reported in Table 1 (left). As expected, they scale roughly with n. We

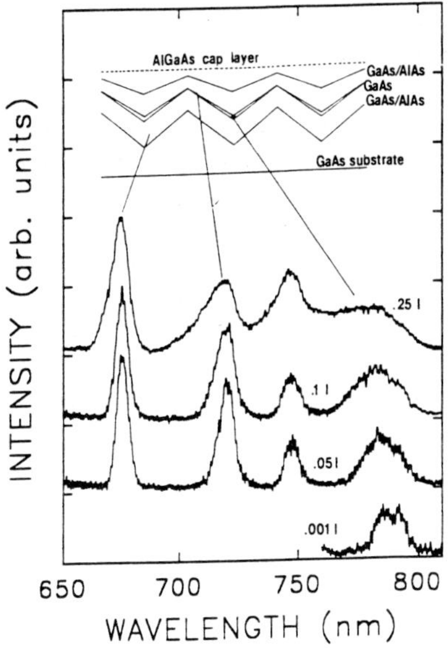

Fig. 1. High excitation intensity PL spectra of the small pitch sample. $I = 50kWcm^{-2}$

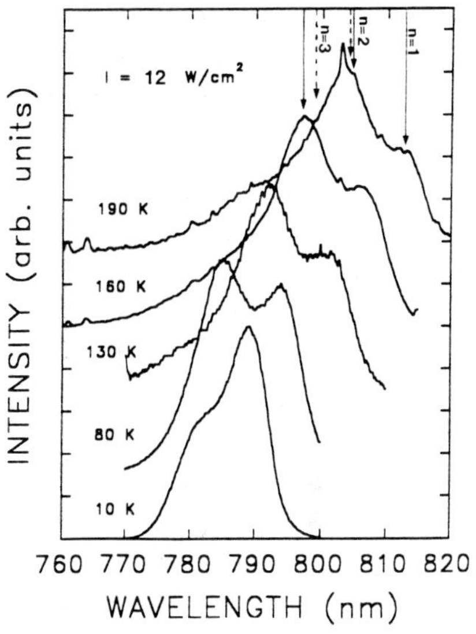

Fig. 2. Temperature dependence of photoluminescence in closely spaced V-shaped quantum wires ($I_{exc} = 12Wcm^{-2}$). In each spectrum the n=0 theoretical transition is fitted to the low energy peak.

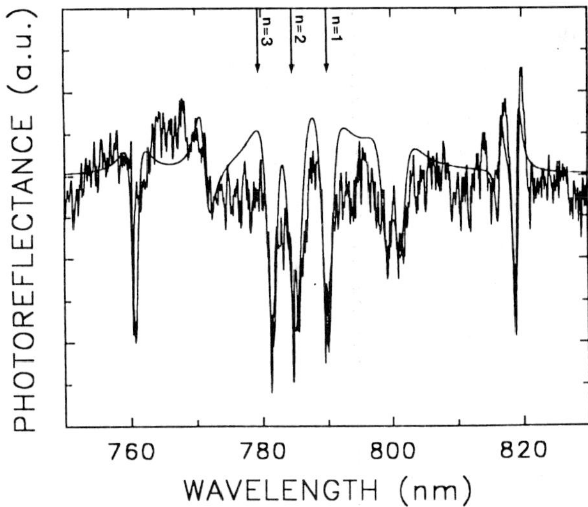

Fig. 3. Photoreflectance of the large pitch sample at 10 K. The continuous line superimposed to the spectrum is a fit performed taking the third derivative Lorentzian model for excitonic resonances.

expect that this approximation is quite good for the first energy levels while it will become worse for the high energy levels, due the penetration of the electronic wave function in the confining potential barriers.

A more precise calculation of the energy levels and of the associated eigenfunctions comes from the solution of a realistic potential in the two-dimensional Schrödinger equation

$$\left(-\frac{\hbar^2\nabla^2}{2m} + V(x,y)\right)\psi(x,y) = \mathcal{E}\psi(x,y)\ . \tag{2}$$

By considering the following set of two-dimensional planewaves over a rectangular domain Ω:

$$\phi(\mathbf{k}_x, \mathbf{k}_y; x, y) = \frac{1}{\sqrt{\Omega}} e^{i(\mathbf{k}_x x + \mathbf{k}_y y)}\ , \tag{3}$$

the Schrödinger equation (2) can be rewritten as

$$\left(\frac{\hbar^2(\mathbf{k}_x^2 + \mathbf{k}_y^2)}{2m} + V(\mathbf{k}_x, \mathbf{k}_y; \mathbf{k}'_x, \mathbf{k}'_y)\right) c(\mathbf{k}'_x, \mathbf{k}'_y) = \mathcal{E}(\mathbf{k}_x, \mathbf{k}_y)\ , \tag{4}$$

where the coefficients $c(\mathbf{k}'_x, \mathbf{k}'_y)$ are the Fourier components of the total wavefunction $\psi(x,y)$ and $V(\mathbf{k}_x, \mathbf{k}_y; \mathbf{k}'_x, \mathbf{k}'_y)$ are the matrix elements of the two-dimensional potential profile in the plane-wave basis given in Eq. (3). By means of a standard numerical procedure, we derive the eigenvalues $\mathcal{E}$ corresponding to the energy levels and the Fourier coefficients $c(\mathbf{k}'_x, \mathbf{k}'_y)$ of the corresponding eigenfunction. Once such coefficients are known, we obtain for each energy level $\mathcal{E}$ the corresponding eigenfunction according to

$$\psi(x,y) = \sum_{\mathbf{k}'_x, \mathbf{k}'_y} c(\mathbf{k}'_x, \mathbf{k}'_y)\phi(x,y). \tag{5}$$

Numerical results obtained for the case of a V-like potential profile are reported in Table 1 (right). Fig. 4 shows a plot of the carrier density corresponding to the first four electronic levels. The vertex of the V is in located at $x = 0$ and $y = 0$. As expected, the ground state exhibits a single maximum in the center of the V-like region, while the excited states, extended over the wings of the V-groove, exhibit an increasing number of maxima as the order of the level increases. Looking at both the electronic wave functions and the energy levels we can understand that the V-grooved structure presents a double behaviour. At low energy it has the typical behaviour of a quantum wire, with well separated 1D energy levels and localised wave functions. At higher energy the spreading of the electronic wave function over the V-groove gives rise to a quasi two-dimensional behaviour. The energy levels computed from the full solution of the two-dimensional Schrödinger equation become closer and closer and the convolution of their 1D density of states almost corresponds to a two-dimensional density of states.

In Figs.2 and 3 arrows corresponding to the first three quantized 1D-level transition energies obtained from eqs.(1) (solid arrows) and (4) (dashed arrows) have been superimposed to the spectra. Neither exciton binding energy nor the thermal shrinkage of the gap have been included in the calculations; therefore, we have shifted the n=0 arrows to coincide with the lowest energy peak in T=190K PL and with the

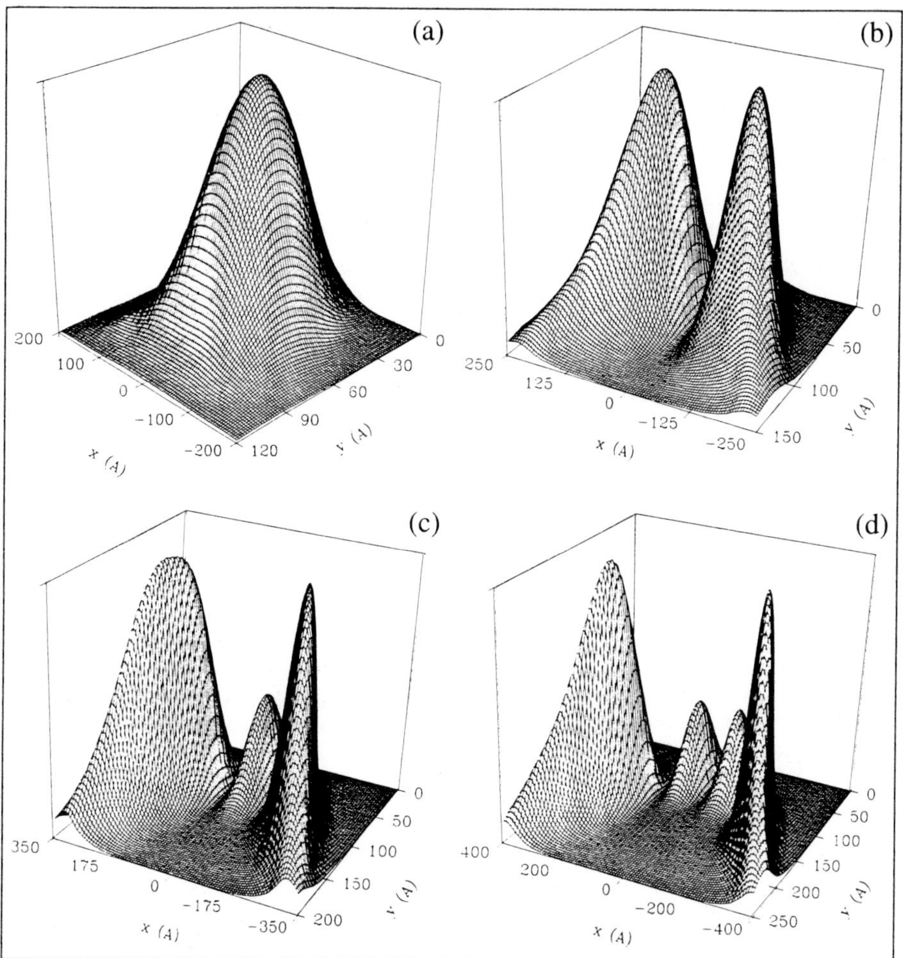

Fig. 4. Three-dimensional plot of the 1D electron density corresponding to the first four energy levels.

Table 1: calculated energy levels of the V-grooved wire

n	energy levels (meV)	
0	1565.5	1574.2
1	1582.0	1592.6
2	1597.0	1603.4
3	1611.3	1610.0
4	1624.6	1617.0
	approx	exact

791nm resonance in PR and scaled the higher energy levels according to the values reported in Table 1. The agreement between theory and experiment is quite good. As shown in Fig. 2 the first two quantized states evaluated by both theoretical models fit well the corresponding PL spectra. For the higher energy levels more refined studies are required.

In conclusion, we have performed a study of one-dimensional exciton states in V-shaped GaAs quantum wires by means of optical methods. The experimental observations are in agreement with the theoretical predictions and show a well defined one-dimensional behaviour of the V-groove in correspondence of the first energy levels

This work was partially supported by EEC commission under the Esprit Basic Science Project NANOPT. The calculations were supported by CNR under grant 92.01598.PF69

REFERENCES

Asada M., Miyamoto Y., and Suematsu Y., 1986, IEEE J. Quantum Electron. **QE22**, 1915.

Christen J., Kapon E., Grudmann M., Hwang D.M., Joschko M., and Bimberg D., 1992, Phys. Stat. Sol. **173**, 307.

Cingolani R., EPS Conference, Regensburg 1993, to appear in Physica Scripta.

Kapon E., Hwang D.M., Bhat R., 1989, Phys. Rev. Lett. **63**, 430.

Kapon K., Kash K., Clausen E.M., Hwang D.M., Colas E., 1992, Appl. Phys. Lett. **60**, 477.

Marti U., Proctor M., Monnard R., Martin D., Morier-Gemoud F., Reinhart F.K., Widmer R., Lehmann H., 1991, APC91, American Vacuum Society Conference Proceedings, **227**, 80.

Possible application to semiconductor devices of one dimensional electron gas (1DEG) systems by periodic bending of *n*-AlGaAs/u-GaAs heterointerfaces

Toshiyuki Usagawa, Akemi Sawada, and Ken'ichi Tominaga
Central Research Laboratory, Hitachi Ltd., Kokubunji, Tokyo 185 Japan

The periodic bending of modulation-doped n-AlGaAs/u-GaAs heterostructure with the bending angle $\theta=\pi/2$ leads to multiple densely packed 1DEG quantum wires with about 2.5 times larger electron concentration compared with the flat n-AlGaAs/ u-GaAs heterostructure. The maximum bending period to keep one-dimensionality is estimated as about 850 Å. The natural extension to the double heterostructures or superlattices will give us new free hand to design devie structures such as to confine electrons one-dimensionally or to separate electrons and holes spatially in the planar structure.

1. INTRODUCTION

Low dimensional electronic materials such as quantum wires(Petroff 1982) or quantum boxes have received growing attention due to expectation to open a new frontier of semiconductor physics and semiconductor devices. Some of the confining nature of electron motions due to quantum effects have already been observed by photoluminescence (PL) measurement(Tsukamoto 1992) or low temperature transport measurement(van Wees 1988) of 1DEG systems.

One of the vehicles to promote such low dimensional material technologies may be quantum wires / quantum boxes semiconductor lasers(Arakawa 1982). For the idealized structures of quantum wires / quantum boxes in Figure 1, detailed theoretical analysis(Arakawa 1982,1984) has been done for the evaluation of laser performances. From the material technology view points, these studies give us serious questions how we realize such artificial superlattices. On the other hand, we do not still have a clear view over application to electron devices by using such low dimensional structures.

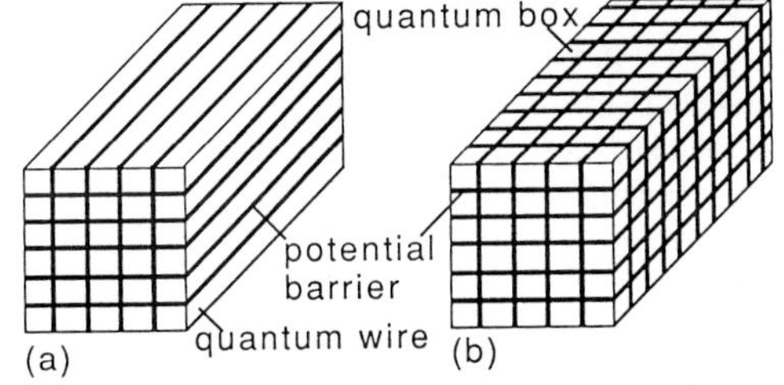

Figure 1 Idealized quantum wires(a) and quantum boxes(b). (Arakawa1982).

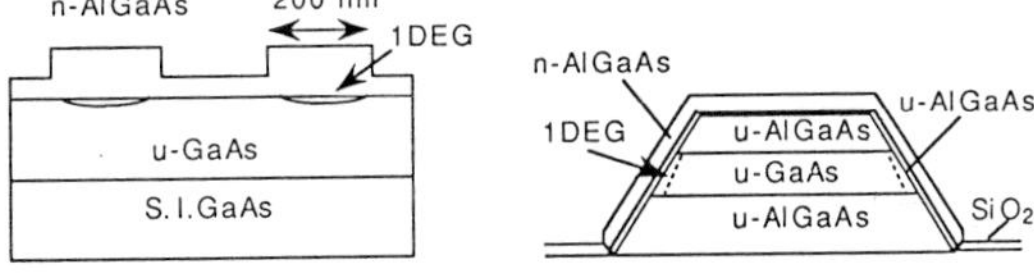

Figure 2 Two typical examples of 1DEG systems. Fine patterning on HEMT structure(a) and MOCVD regrowth on the patterned substrate (Fukui 1989)(b)

In this paper we propose novel 2 or 3

dimensional superstructures based on the previously studied periodic bended structure of modulation doped n-AlGaAs/undoped GaAs heterointerfaces(Sawada 1991,1992a,1992b), which is a natural extension of the modulation doped heterostructures and/or superlattices to 2 or 3 dimensional ways. After reviewing the main characteristics of the periodic bended structure of modulation doped n-AlGaAs/undoped GaAs heterointerfaces, we discuss the new possibitities of device applications.

2. PERIODIC BENDING OF MODULATION DOPED n-AlGaAs/u-GaAs HETEROINTERFACES

Schematic cross sectional view of the two typical 1DEG structures are shown in Figure 2, where one is realized by the fine patterning on the HEMT structures. The other is realized by selective regrowth technique of MOCVD combined with advanced lithography(Fukui 1989).

Common disadvantages of these structures from a device-application viewpoint are

(1) low electron concentration due to the peripheral effects of confining potential which can not exceed the electron concentrations of HEMT structures, and

(2) large dead space area, which prevents the realization of high density of quantum wires.

In order to overcome the above disadvantages, we have introduced periodic bended structure of modulation doped n-AlGaAs/undoped GaAs heterointerfaces. We summarize here the main characteristics and the intuitive undestanding of the 1DEG systems. The schematic cross section of the structure is shown in Figure 3. There, $\lambda \sin(\theta/2)$ is the period of the bending interface and θ is the bending angle.

2.1 Basic Ideas

An intuitive understanding of why much electrons accumulate around the convex corner, point A, of the undoped GaAs in Figure 3 is given below. As the area of electron supplying layer, n-AlGaAs, within the circle around the point A is $[2\pi/\theta - 1]$ times larger than that of electron accumulation layer, u-GaAs, much electrons expect to accumulate at the convex corner. On the other hand, less electrons accumulate around the concave corner, points B and C, of u-GaAs because the area of n-AlGaAs within the circle around the point B is $[2\pi/\theta - 1]$ times less than that of electron accumulation layer, u-GaAs. Simple estimation of the electron concentration at the convex corner is $[2\pi/\theta - 1]^2$ times larger than that of the concave corner, points B and C.

depletion region
n-AlzGa1-zAs
θ
A
B
C
x
y
λsin(θ / 2)
u-GaAs

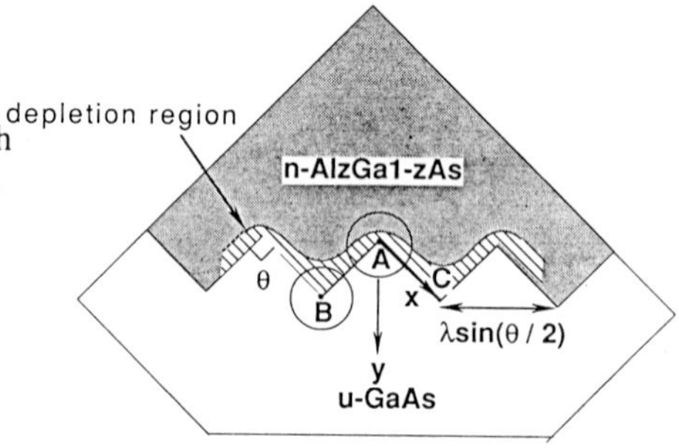

Figure 3 Schematic diagram of the periodic bending structure of n-AlGaAs/u-GaAs heterointerfaces . θ; bending angle, $\lambda \sin(\theta/2)$;bending period

If the bending period is very large, the electron distribution around the mid area between the points A and B will not be influenced by the geometrical effects of bending so that the mid area will be regarded as conventional 2DEG regions.What happens by shrinking the period until this parasitic 2DEG regions disappear? We will then get the multiple 1DEG systems. The threshold period, λs , to emerge such 1DEG systems is an extremely important parameter from the material design point of view.

If λs is estimated to be 100 Å, it meams that it would not be realizable in near future.

2.2 Summary of the 1DEG Characteristics

For the material parameters, we choose the standard values of the conventional two dimensional electron gas (2DEG) structure where the doping concent-ration ND of n-AlzGa1-zAs is 1×10^{18} $/cm^3$, z=0.3, the undoped GaAs is p-type of $1.0 \times 10^{14}/cm^3$. The bending angle θ is fixed as $\pi/2$ for simplicity of numerical calculations. The electro-static potential $\Psi(x,y)$ can be obtained solving the Posson equation based on the classical approach where the electron density n(x,y) is proportional to $\exp(q\psi/\kappa T)$, and the structual potential difference due to the band discontiuity Δ at the heterointerfaceis added as a static potential. The calculation-technique is successfully applied to the device analysis of HEMT structures (Mizuta 1989).

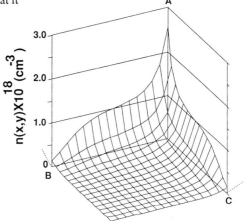

Figure 4 Electron distribution n(x,y) at the convex corner of GaAs (**λ=1200Å**).

The calculated electron density n(x,y) around the convex corner, point A, of the undoped GaAs is shown in Figure 4 for λ = 1200 Å . The maximum electron concentration n(0,0) is 2.5 times larger than that of the conventional flat 2DEG structure, $1 \times 10^{18}/cm^3$. The minimum electron concentration at the concave corner, points B and C, of the undoped GaAs is about a tenth of n(0,0) of the maximum concentration. We find a very close agreement with the approximately expected ratio of $[2\pi/\theta-1]^2 = 9$ given above.

In order to estimate the maximum bending period to keep one-dimensionality without parasitic 2DEG region, we introduce the local sheet density nlocal(x) and the average local sheet density nav as follows,

$$nlocal(x) = \int n(x,y)\,dy,$$

$$nav = \int nlocal(x)\,dx/(\lambda/\sqrt{2}).$$

Among the various calculated results of nlocal(x)versus λ , ND, and z, a typical example of nlocal(x)/n2D for λ= 1000, 1200,2000, and 4000 Å is shown in Figure 5, where n2D is the sheet density of electrons at the flat n-AlGaAs/ u-GaAs heterointerface with the same material parameters. It should be noticed that the nlocal(x)/n2D has a sharp peak around ±20 nm region irrespective of the period $\lambda \sin(\theta/2)$, ND, and the

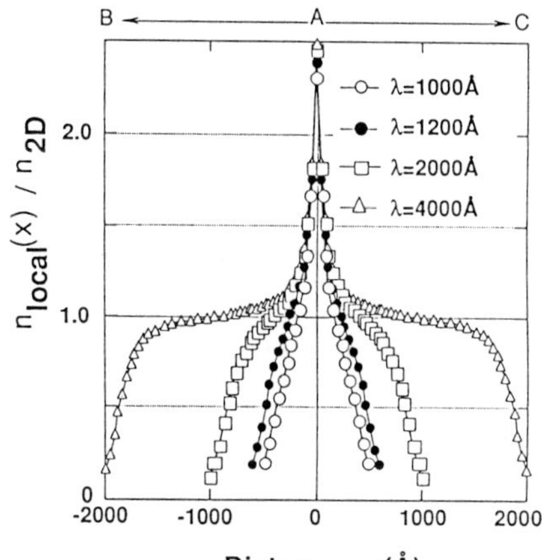

Figure 5 Normalized local sheet density of electrons along the bended interface for various periods; n_{2D} is sheet density for flat n-AlGaAs/u-GaAs structure.

Al mole fraction z. The behavior of nlocal(x) near x=0 is similar for all values of λ and nlocal(0)/n2D ≒2.5.

As λ decreases, the region where nlocal(x)=n2D becomes small, that is, the region of the parasitic 2DEG behavior disappears when λ = 1000 and 1200 Å, and the one-dimensionality is enhanced. From the application view point, the larger sheet density is suitable as active layers for any electron devices and also optical devices so that the maximum period to keep one-dimensionality without parasitic 2DEG region is estimated as about 850 Å (λ = 1200 Å).

2.3 Comparison with conventional 2DEG and/or 1DEG structures

The λ dependence of nav/n2D in Figure 6 shows that nav decreases as λ decreases. For the larger λ , nav/n2D approachs gradually to $\sqrt{2}$, which is expected from the geometrical factor corresponding to $\theta = \pi/2$. The sheet electron density, nav, is about 20 % larger than the flat n-AlGaAs/ u-GaAs 2DEG systems for the case of bending angle $\Theta=\pi/2$ and λ =1200 Å . If the bending angle θ is shrinked to $7\pi/18(70^\circ)$ corresponding to realistic (111) oriented structure on (100) substrate, much more electrons are expected to accumulate according to the intuitive undestanding of the accumulation mechanism given in Section 2.1.

By varying the mesa etching thickness of the n-AlGaAs of the HEMT structure in Figure 2(a) under the constant surface pinning level, the sheet electron density nav is also caluculated in Figure 6. It shows that the sheet electron density is extremely rduced by decreasing the wire width(Tominaga 1993). The periodic bending 1DEG structure has about ten times larger sheet density than that of the conventional 1DEG quantum wires.

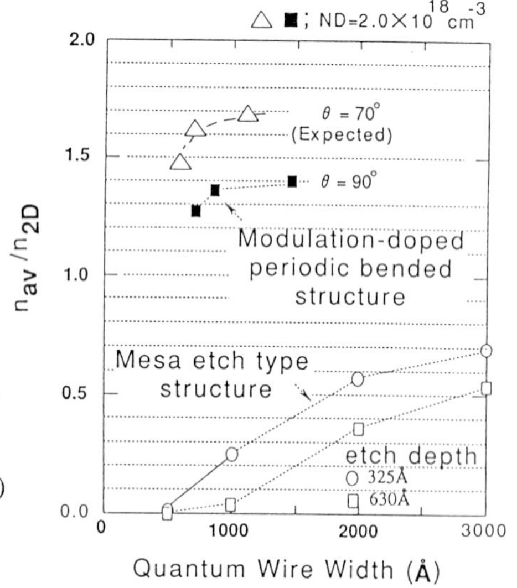

Figure 6 Normalized sheet density of 1DEG for modulation-doped periodic bended structure and mesa etch type structure. n_{2D} is sheet density for flat n-AlGaAs/u-GaAs structure.

3. POSSIBLE APPLICATION TO SEMICONDUCTOR DEVICES

One of the foreseeable device applications is to use as the active layer of FETs schematically represented by Figure 7. This structure will resolve the two main drawbacks of the conventional 1DEG structure mentioned in Section 2. One of the features of the FET-structure is that the electrons are confined in their own 1DEG channel during the device operation and the horizontal difusion of electrons is expected to be extremely small. The conventional FETs have no constraint on the horizontal motion of electrons under the gate electrode. For example, in the case of GaAs/AlGaAs HEMT with the gate length Lg = 0.5 μ m, the diffusion length L =$\sqrt{D\tau}$ is estimated to be 0.3 μ m(for τ = 4 psec, μ =8000 cm /Vs) at room temperature.

It means that the electrons at the source side edge of the gate spreads out averagely 0.3 μ m due to their Brownian motions when they reach the drain side edge. How the constraint of electron motions affects the device performance should be clarified quantitatively.

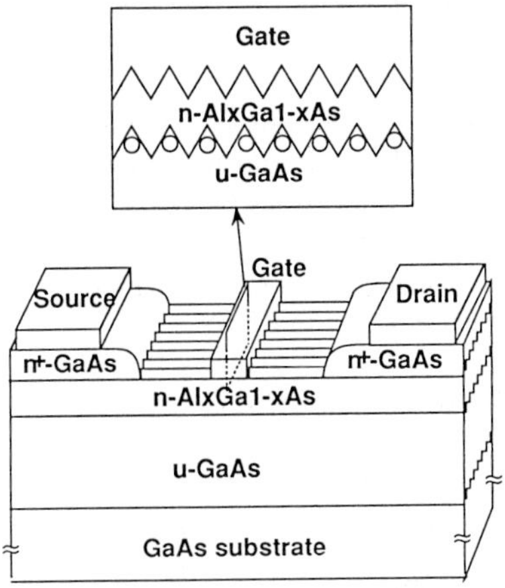

Figure 7 Schematic view of proposed 1DEG-FETs. O ; 1DEG. Electrons are confined in each 1DEG line.

From the material viewpoint, it is easy to extend the original structure to double hetero-structures or superlattices. The band diagrams for the double heterostructure to insert undoped AlzGa1-zAs layer under the u-GaAs layer of Figure 3, are shown in Figure 8, where the horizontal line represents the y direction in Figure 3 through points A and B for λ =1200 Å, respective ly. It shows that the electrons accumulate at the convex corner of u-GaAs/n-AlGaAs. On the other hand, the most stable point of holes is the convex corner of the other u-GaAs/u-AlGaAs interface. It means that electrons and holes are spatially separated (Figure 9(a)) even if electron and hole pairs are generated in the u-GaAs layer. In other words, in the case of photodetectors with FET like structure of Figure 8, spatially separated electrons and holes move antidirectly with each other in the different convex corners. During the cource of motions, electrons and holes rarely meet with each other. It means for this structure to have a new way to enlarge the recombination life time of electrons and holes in the planar structure.The spatial separation within the FET gate-plane makes independent 1DEG and 1DHG(one dimentional hole gas) lines and will gives us high performance of photode tectors. It should be noticed that the spatial separation of electrons and holes in planar structure is a different concept compared with the vertical separation within the multilayers such as nipi superlattices(Dohler 1986), where electrons and holes move freely in each plne of the carrier accumulation layers.

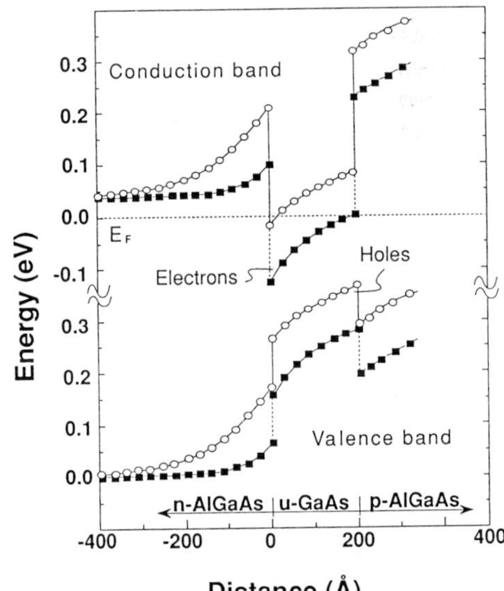

Figure 8 Energy band diagrams for the corrugated double heterostructure through points A(■) and B(○) for λ=1200**Å** in Figure 3. The distance is measured from each hetero-interface. The Fermi level Ef is shown by the dotted line.

In the case of superlattices, multi-layers of 1DEG or 1DHG systems (Figure 9(b)) will be applicable to the active layers to the modulation-doped MQW lasers (Uomi 1990). The confining nature of the accumulated electrons is expected to play an important role in theperformance-improvement of modulation-doped MQW semiconductor lasers. In this way, these novel structures will open a new field of optical devices such as quantum wire lasers or electron devices.

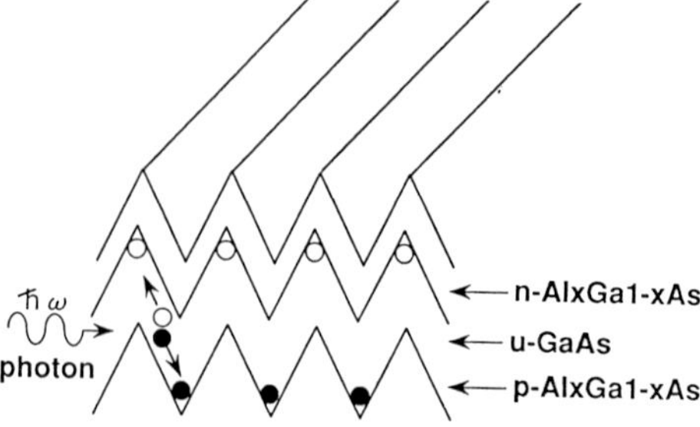

Figure 9(a) Spatial Separation of Electrons and Holes by Modulation-doped Double Corrugate Heterojunction Structure

4. CONCLUSIONS

Periodic bended structures of modulation doped n-AlGaAs/u-GaAs heterostructures give us attractive properties of 1DEG systems such as high electron density, and high packing density of 1DEG wires. It will also give us new ways to confine electrons one-dimensionally or to separate spatially electrons and holes in planar structure. Recent challenge and progress (Tsukamoto 1993) of fine epitaxy technologies combined with advanced lithography will provide the artificial materials technology. We also expect that man-made 2 or 3-dimensional superlattices open the new possibility of the various device applications.

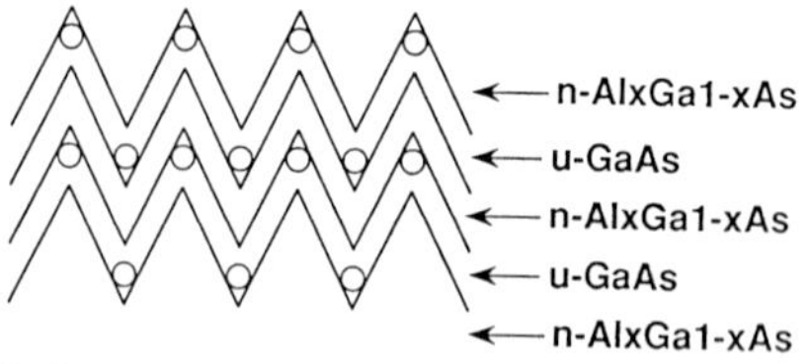

Figure 9(b) Modulation-doped Corrugate MQW Structure

References

Arakawa Y and Sakaki H 1982, Appl.Phys.Lett. 40 939.
Arakawa Y, Vahara K, and Yariv A 1984, Appl.Phys.Lett. 45 950.
Dohler GH 1986, IEEE QE-22 1682.
Fukui T and Ando S 1989, Elect. Lett. 25 410.
Mizuta H, Yamaguchi K, Yamane M, Tanoue T, and Takahashi S 1989, IEEE ED36 2307.
Petroff PM, Gossard AC, Logan RA, and Wiegmann W 1982, Appl.Phys.Lett. 41 635.
Sawada A, Usagawa T, Ho S, and Yamaguchi K 1991, Int. Conf. on Solid State Devices and Materials, Yokohama, p723.
Sawada A, Usagawa T, Ho S, and Yamaguchi K 1992a, Appl.Phys.Lett. 60 1492.
Sawada A, Usagawa T 1992b, Int. Conf. on Solid State Devices and Materials, Tsukuba, p756.
Tominaga K and Usagawa T 1993, private communications.
Tsukamoto S, Nagamune Y, Nishioka M, Arakawa T, Kono T, and.Arakawa Y1992, Int.Symp.GaAs and Related Compounds, Karuizawa, pp929-930.
Tsukamoto S, Nagamune Y, Nishioka M, and Arakawa Y 1993, Appl. Phys. Lett. 61 49.
Uomi K, Mishima T, and Chinone N 1990, Jpn.J.Appl.Phys. 29 88.
Van Wees BJ, Van Houten H, Beenaker CWJ, Williamson JG, Kouwenhove LP, Vander Marel D, and Foxson CT 1988, Phys.Rev.Lett. 60 848.

Inst. Phys. Conf. Ser. No 136: Chapter 4
Paper presented at the Int. Symp. GaAs and Related Compounds, Freiburg, 1993

Control of electron capture in AlGaAs/GaAs quantum wells with tunnel barriers at heterointerfaces

A.Fujiwara, S.Fukatsu, and Y.Shiraki
Research Center for Advanced Science and Technology(RCAST), The University of Tokyo, 4-6-1 Komaba, Meguro-ku, Tokyo 153, Japan

R.Ito
Dept. of Applied Physics, The University of Tokyo, 7-3-1 Hongo, Bunkyo-ku, Tokyo 113, Japan

ABSTRACT: Enhancement of electron capture efficiency is clearly observed in AlGaAs/GaAs quantum wells(QWs) with AlAs tunnel barriers at heterointerfaces. The enhancement occurs when the well width is such that the incoming electron in the conduction-band edge of the AlGaAs barrier resonates with the virtual bound states in QWs. The insertion of only one monolayer(ML) AlAs layer is shown to drastically enhance the resonant effect and the resonant capture is confirmed to be well described as a quantum mechanical interference effect within the framework of the effective mass approximation.

1.INTRODUCTION

Carrier capture in QWs has been intensively investigated since the initial stages of the study of QW and semiconductor heterostructures(Shichijo et al 1976, Tang et al 1982). Some theoretical approaches predicted the enhancement of the carrier capture efficiency through quantum mechanical resonances(Kozyrev and Shik 1985, Brum and Bastard 1986). However, no experimental proof of such resonance had been reported, in spite of several studies of time-resolved photoluminescence (PL) (Deveaud et al 1986, Oberli et al 1989) and PL excitation spectroscopy (Polland et al 1988, Ogasawara et al 1990), until we observed the resonance in the QWs with tunnel barriers inserted at QW interfaces(Fujiwara et al 1992). Resonant capture is a quantum mechanical interference effect which can be understood by regarding a QW as a Fabry-Perot resonator with two interface mirrors for the incident electron wave. The insertion of AlAs tunnel barriers, therefore, corresponds to the "high-reflectance coating" of the interface mirror, leading to the enhanced resonant effect and enabling one to observe the resonant electron capture into QWs. Recently, a time-resolved PL study on this type of structure(Morris et al 1993) and other studies by subpicosecond time-resolved PL (Blom et al 1993, Barros et al 1993) have been reported.

In this paper, we demonstrate the excellent controllability of resonant electron capture via the width of AlAs tunnel barriers and show that the resonant capture is well described within the framework of the effective mass approximation.

2. EXPERIMENTAL PROCEDURE

$Al_xGa_{1-x}As/AlAs/GaAs$ QW structures were grown by the molecular beam epitaxy(MBE) on Semi-insulating GaAs(100) substrates. The sample was rotated during the growth of $Al_xGa_{1-x}As$ and AlAs layers in order to gain homogeneity while we intentionally stopped the rotation during the growth of GaAs well layer in order to obtain a graded well width(L_Z) across the substrate due to the inhomogeneity of Ga beam flux. This allowed us to systematically examine the L_Z dependence of the electron capture by the PL mapping technique. The composition ratio x of the AlGaAs barrier was chosen to be 0.25 and 0.30. AlAs tunnel layers with various thickness(L_{tb}=1ML, 10Å, 20Å) were inserted at both top and bottom QW interfaces.

We evaluated the carrier capture into QWs by means of PL measurements using a cw Ar^+ laser at temperatures 26-200K. The capture efficiency was estimated from the ratio of the PL intensity of QW(I_w) and the AlGaAs barrier(I_b). PL mapping across the sample made us know how the capture efficiency depends on L_Z.

3. EXPERIMENTAL RESULTS

Figure 1 shows the L_Z dependence of the capture efficiency in $Al_xGa_{1-x}As$ (x=0.25) / AlAs / GaAs QWs with various L_{tb}'s. It is seen that the capture efficiency exhibits a resonance at a certain L_Z. This resonance is ascribed to the energy matching of the n=2 electronic state in QW and the conduction-band edge of the barrier. We also found that the resonant well width(L_{res}) increases with increasing L_{tb}. It is remarkable that even the 1ML AlAs tunnel barrier drastically enhances the resonant effect compared to the simple QW structure with no tunnel barriers. That is, the quantum mechanical reflectance of barrier electrons at QW interfaces is significantly increased by the 1ML AlAs tunnel barriers.

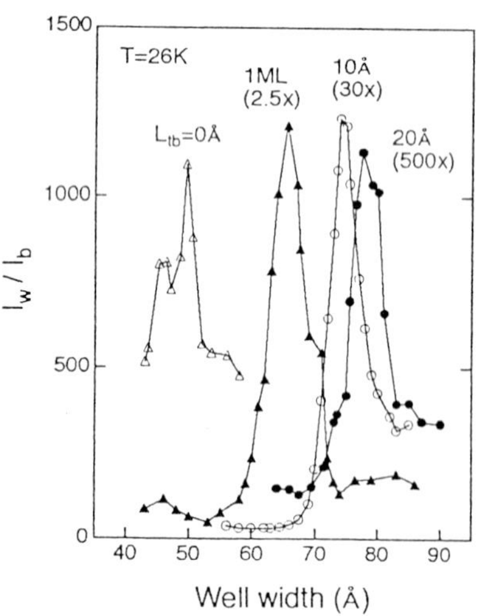

Fig. 1. Well-width dependence of the electron capture efficiency in the QWs with AlAs tunnel barriers of various thickness.

To explain this result, we carried out the calculation of electronic states with the conventional envelope function approach based on the effective mass approximation. If most of the incoming electrons from the $Al_xGa_{1-x}As$ barrier are completely relaxed down to the barrier band-edge and their kinetic energy is negligible, the resonant condition is given as

$$k_w L_{res} + \theta = n\pi \qquad (n=0,1,2\cdots) \tag{1}$$

where k_w represents the wave number of well electrons with the energy of $Al_xGa_{1-x}As$ barrier band-edge, and θ is the phase shift of the well electron with k_w when it is reflected back at one "interface mirror". It should be noted that eq.(1) is quite similar to the resonance condition in the optical Fabry-Perot resonator. θ is expressed as

$$\theta = -2 \tan^{-1}\{\frac{-ik_{tb}m_w}{k_w m_{tb}} \tanh(-ik_{tb}L_{tb})\} \qquad (2)$$

where k_{tb} is the wave number of the electron with k_w in the tunnel barriers(k_{tb} is, therefore, imaginary) and $m_w(m_{tb})$ is the effective mass of the electron in the well (tunnel barrier). Calculated L_{res} as a function of L_{tb} for two different barrier(x=0.25,0.30) is shown in Fig.2 along with experimental results. Nonparabolicity of the effective mass in GaAs well layer was taken into consideration since resonant conditions are strongly affected by the wave number k_w of electrons in QW with high energies equal to conduction-band offset between GaAs well and $Al_xGa_{1-x}As$ barrier. The calculation agrees fairly well with the experimental results, indicating that the effective mass approximation is applicable even in the case of 1ML AlAs layer.

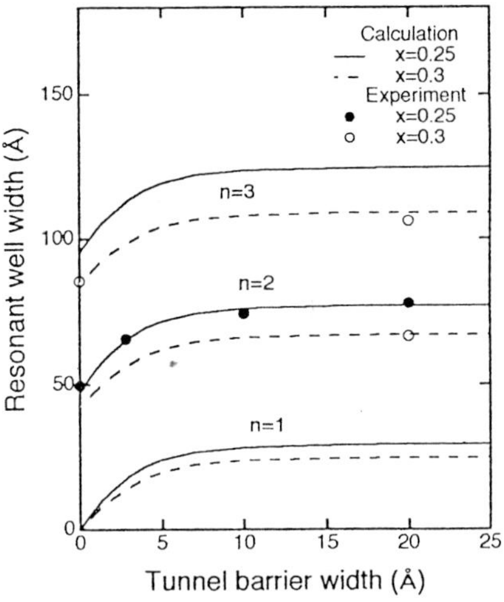

Fig. 2. Resonant well width as a function of AlAs tunnel barrier widths.

For the simple QW(L_{tb}=0), the phase shift θ of AlGaAs/GaAs "interface mirror" is zero as seen in eq.(2) and L_{res} is determined only by k_w. Insertion of tunnel barriers produces the phase shift θ depending on L_{tb} and consequently L_{res} is increased. In this case, both $Al_xGa_{1-x}As$ barrier and AlAs tunnel barrier make up the mirror of the QW resonator. For the larger L_{tb}, L_{res} shows a saturation behavior since the phase shift θ is dictated only by the GaAs/AlAs interface and $Al_xGa_{1-x}As$ barrier plays no large part of the mirror. It is also seen that L_{res} is smaller for the higher $Al_xGa_{1-x}As$ barrier. This is easily understood considering that the energy matching of the QW states and the $Al_xGa_{1-x}As$ barrier band-edge occurs at smaller well widths for higher barriers. These results clearly indicate that the resonant capture in QWs with tunnel barriers is well described by the effective mass approximation and is very much analogous to the optical Fabry-Perot resonator. We further investigated the temperature dependence of the resonant capture. Temperature variation of the resonant curves for samples with L_{tb}=1ML and L_{tb}=20Å is shown in Fig.3. For L_{tb}=1ML the resonant feature is almost smoothed out above 50K, while the resonant behaviour is clearly observed up to 200K for L_{tb}=20Å. The degradation of the resonance with high temperatures can be ascribed to loss of the coherency of the incident electron due to electron-phonon scattering. Therefore, the result may indicate that electrons in virtual bound states in the QW with thinner tunnel barriers are more likely to be scattered by phonons. It is also seen in Fig.3 that the L_{res} decreases with increasing temperature and deviates from the calculated L_{res}.

This is because the kinetic energy of the electron in the $Al_xGa_{1-x}As$ barriers is not negligible at high temperatures and the energy of the virtual bound states becomes larger in order to give rise to resonant capture of energetic electrons in the barriers.

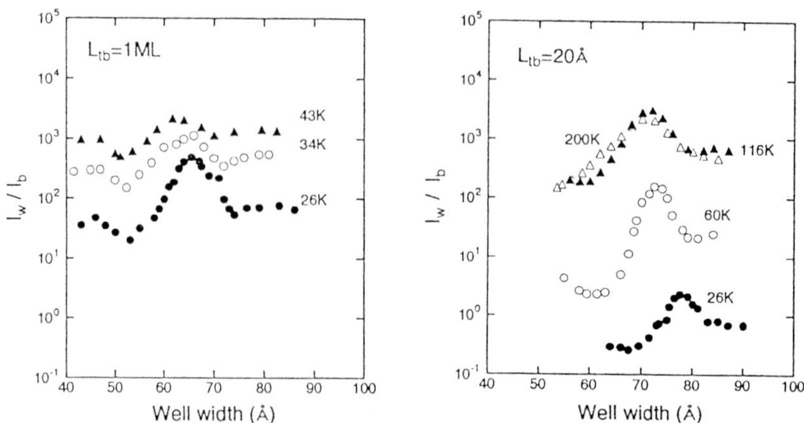

Fig. 3. Temperature dependence of the resonant electron capture in the QWs with tunnel barriers.

4.CONCLUSION

We have investigated the resonant electron capture in $Al_xGa_{1-x}As$ quantum wells with Al tunnel barriers at the heterointerfaces. It was clarified that the resonant electron capture can be controlled by changing the width of tunnel barriers. The role of tunnel barriers was well described by the effective mass approximation and explained by the analogy to the high-reflectance coating of the mirrors in the optical Fairy-Perot resonator.

ACKNOWLEDGMENT

We are grateful to K.Murkai for fruitful discussion and S.Ohtake for the technical support.

REFERENCES

Barros M R X, Becker P C, Morris D, Deveaud B, Regreny A, and Beisser F 1993 Phys. Rev. B 47 10951
Blom P W, Smit C, Haverkort J E M, and Wolter J H 1993 Phys. Rev. B 47 2072
Brum J A and Bastard G 1986 Phys. Rev. B 33 1420
Deveaud B, Shah J, Damen T C, and Tsang W T 1986 Appl. Phys. Lett. 52 1886
Fujiwara F, Fukatsu S, Shiraki Y, and Ito R 1992 Surf. Sci. 263 642
Kozyrev S V and Shik Y Ya 1985 Sov. Phys. Semicond. 19 1024
Oberli D Y, Shah J, Jewell J L, Damen T C, and Chand N 1989 Appl. Phys. Lett. 54 1028
Ogasawara N, Fujiwara A, Ohgushi N, Fukatsu S, Shiraki Y, Katayama Y, and Ito R 1990 Phys. Rev. B 42 9562
Polland H J, Leo K, Rother K, Ploog K, Feldman J, Peter G, Göbel E O, Fujiwara K, Nakayama T, and Ohta Y 1988 Phys. Rev. B 38 7635
Shichijo H, Kolbas R M, Holonyak N,Jr., Dupuis R D, and Dapkus P D 1978 Solid State Comm.27 1029
Tang J Y, Hess K, Holonyak N,Jr., Coleman J J, and Dapkus P D 1982 J. Appl. Phys. 53 6043

Perspective of UV/blue light emitting devices based on column-III nitrides

I.Akasaki and H.Amano

Department of Electrical and Electronic Engineering, Meijo University,
1–501 Shiogamaguchi, Tempaku–ku, Nagoya 468 Japan

ABSTRACT: We have developed p–n GaN homojunction and AlGaN/GaN doubleheterojunction (DH) UV/blue LEDs. Newly developed p–n junction LED showed power efficiency more than 1% at room temperature(RT). By using AlGaN/GaN DH, surface and edge stimulated emissions with low threshold power at RT by optical pumping have been achieved.

1.INTRODUCTION

All the column–III nitrides except boron nitride(BN), that is wurtzite polytypes of indium nitride(InN), gallium nitride(GaN), aluminum nitride(AlN) and their alloys $(Al_xGa_{1-x})_{1-y}In_yN$ $(x\geq0, y\geq0)$ have the direct transition type band structure with the bandgap energy ranging from 1.9eV to 6.2eV. Although they are promising as materials for fabrication of electronic devices such as MISFET(Fujieda et al. 1988) and MESFET(Khan et al. 1993), we believe the most important applications are optical devices, especially short–wavelength light emitters such as light emitting diode(LED) and laser diode(LD) in the blue to ultraviolet (UV) regions. These short wavelength light emitters enable us to realize full–color LED, all solid state flat panel full–color displays, three–dimensional color television system and high–performance optoelectronic systems such as high–density optical storage systems and high–speed full–color printing systems. Realization of such a short–wavelength light emitters is also required for the applications to small medical equipments, biological sciences and physics.

To achieve these desires, it is indispensable 1)to grow high quality column–III nitride semiconductor thin films and 2)to control their conductivity.

In contrast with other III–V compounds such as GaAs and InP, however, it had been fairly difficult to grow high quality epitaxial films with a flat surface free from cracks, because of the large lattice mismatch and the large difference in thermal expansion coefficient between the epitaxial film and the sapphire substrate.

We (Amano et al. 1986, Akasaki et al. 1989) succeeded in overcoming these problems and growing high quality GaN and AlGaN films with a specular surface free from cracks by the prior deposition of a thin AlN buffer layer in MOVPE growth. The electrical and optical properties as well as the crystalline quality can be remarkably improved at the same time. By using such a high quality GaN film, the UV stimulated emission at room temperature(RT) by optical pumping was achieved for the first time(Amano et al. 1990a). Practical bright mis–type blue–LED's have been developed using the same GaN film(Koide et al. 1991). The brightness is about 100mcd typically, and 200mcd maximum at a forward current of 10mA, which is the highest in the blue region to date and comparable with those of commercially available GaAsP red LEDs and GaP green LEDs.

Silicon was found to act as a donor impurity in GaN as well as in AlGaN(Amano et al. 1990b, Koide et al. 1991, Murakami et al. 1991, Nakamura et al. 1992), and the free electron concentration has been controlled from undoped level of less than $10^{15}cm^{-3}$ up to near 10^{19} cm^{-3}.

It has been well known that undoped column−III nitrides show n−type conduction, and p−type films had never been realized. We succeeded, for the first time, in producing p−type GaN in 1989(Amano et al.) and p−type AlGaN in 1991(Akasaki et al. 1991b) by low energy electron beam irradiation (LEEBI) treatment of Mg−doped films. On the basis of these results, we developed the first p−n junction GaN UV/blue LED.

In this paper, (1) the conductivity control of GaN and AlGaN for both n−type and p−type, (2) the performance of the UV/blue LED, and (3) the characteristics of RT stimulated emission by optical pumping from the AlGaN/GaN double heterostructure(DH) are described.

2. MOVPE GROWTH OF GaN AND AlGaN AND THEIR PROPERTIES

A horizontal type MOVPE reactor operated at atmospheric pressure was used for the growth of the GaN and AlGaN films. Trimethylgallium (TMGa), trimethylaluminum (TMAl) and ammonia (NH_3) were used as source gases and hydrogen(H_2) as a carrier gas. Polished sapphire (0001) crystals were used as the substrate. In our process mentioned above, a thin AlN layer about 50nm thick was predeposited at lower temperatures than the growth temperature by feeding TMAl and NH_3 diluted with H_2. Then the temperature was raised to 1050°C, and a single crystalline films of several microns thick was grown.

The surface morphology of GaN films can be remarkably improved by the preceding deposition of the AlN as a buffer layer. GaN films with optically flat surface free from cracks can be grown on the sapphire substrate covered with the AlN buffer layer. On the contrary, island growth occurred in the growth of GaN on the bare sapphire substrate surface.

X−ray rocking curve(XRC) measurements also revealed that GaN grown using the AlN buffer layer has high quality. The full width at half maximum of the GaN film grown with the AlN buffer layer is about 110 arcsec, which is the narrowest up to date in this material.

It was shown that the photoluminescence (PL) properties of the GaN films can be remarkably improved by using the AlN buffer layer. In the PL spectrum at 4.2K of GaN grown with the buffer layer, the free exciton line and the neutral−donor−bound exciton line(I_2−line) clearly appear, while emission bands in the long wavelength region, which may be due to deep−level defects, are scarcely observed. On the other hand, emission bands at long wavelengths dominated in the spectrum of the GaN films grown directly on the sapphire substrate. This indicates that the generation of deep level defects can be suppressed by using the AlN buffer layer.

GaN films grown by the above−mentioned process have electron concentration at RT of about $10^{15}cm^{-3}$, which is four orders of magnitude lower than that of directly grown film. This indicates that our undoped films are very pure. Slightly Si−doped GaN films grown using the AlN buffer layer have n−type conductivity with the electron mobility much higher than that of conventional ones.

Cross sectional transmission electron microscopy (TEM) showed that defects such as dislocations have been markedly decreased by using the AlN buffer layer.

All these results (surface morphology, XRC, PL, electrical properties and TEM) clearly show that by the preceding deposition of the AlN buffer layer, the electrical and optical properties as well as the crystalline quality of GaN film can be remarkably improved.

The AlN buffer layer is effective not only for the growth of GaN but also for the growth of $Al_xGa_{1-x}N$ with a mole fraction x less than 0.4.

We have also succeeded in observing the first RT UV stimulated emission from a GaN film grown using the AlN buffer layer, which was cleaved in a 2.15 mm stripe and excited with a 337.1nm line of a pulsed nitrogen laser. The peak photon energy of the stimulated emission was found to be 3.32eV.

3. CONDUCTIVITY CONTROL OF N–TYPE GaN AND AlGaN

We found that silane is a suitable source gas for Si–doping. The electron concentrations and resistivities of GaN and AlGaN can be easily controlled by changing the silane flow rate from the undoped level of less than $10^{15}cm^{-3}$ up to near $10^{19}cm^{-3}$. Murakami et al.(1991) showed that the intensity of cathodoluminescence(CL) of the near band–edge emission increases with the increase of doping level of Si in the GaN and AlGaN films.

4. REALIZATION OF P–TYPE GaN AND AlGaN

In 1987, we carried out the doping of magnesium(Mg) during the growth of GaN and AlGaN film by supplying biscyclopentadienylmagnesium(Cp_2Mg) as a Mg source gas. Compared with Zn, the vapor pressure of Mg is rather low, and/or the sticking coefficient of Mg at GaN surface is rather high(Amano et al. 1990). Therefore, the Mg concentration in GaN changed linearly with the supply flow rate of Cp_2Mg. This relationship was almost independent of the substrate temperature. Thus we can easily obtain the desired Mg concentration and its profile in GaN by controlling the supply flow rate of Cp_2Mg.

Generally speaking, it is difficult to determine the type of conductivity of as–grown Mg–doped GaN and AlGaN, because the resistivity is too high. We found for the first time, that the Mg–doped GaN and AlGaN tend to show distinct p–type conduction with low resistivity by the LEEBI treatment.

An ohmic contact to the LEEBI treated Mg–doped GaN and AlGaN layer was achieved by depositing Au. Hole concentration at RT up to about $1.4x10^{17}$ cm^{-3} can be achieved. Therefore it can be said that Mg behaves as an acceptor impurity in GaN and AlGaN.

In the PL spectrum at 4.2 K of the Mg–doped GaN with Mg concentration less than $2x10^{19}cm^{-3}$, D–A pair emission and its LO–phonon replica can be clearly observed. On the contrary, in the spectrum of undoped GaN, D–A pair emission did not appear, and the free–exciton line and I_2–line appeared. Therefore, the origin of the D–A pair emission is thought to be residual donor and doped Mg acceptors. This also shows that Mg acts as acceptor impurity in GaN.

From the temperature dependence of the intensity of D–A pair emission, Akasaki et al.(1991a) estimated the activation energy of the Mg acceptor to be about 155~165meV, which is somewhat shallower than that of Zn (210meV).

In the spectrum of Mg–doped GaN with Mg concentration higher than $6x10^{19}cm^{-3}$, a strong blue emission appears and its intensity is high even at RT. Therefore, it should be emphasized that Mg forms blue luminescence centers as well as acts as acceptor in GaN.

By the LEEBI treatment, the intensity of the blue emission is remarkably enhanced, while keeping the shape of the spectrum. This enhancement of blue emission intensity suggests the increase of Mg–related blue luminescence centers, which may be due to the reactivation of Mg atoms by the LEEBI treatment.

5. CHARACTERISTICS OF P–N HOMOJUNCTION UV/BLUE LED

A typical DC–EL spectrum at RT observed from the newly developed diode shows the broad blue emission peaking at 423 nm, which is due to the Mg–associated transition in the p–type GaN layer. A strong and sharp UV emission peaking at 372 nm is also observed, which is thought to originate from band–to–band transition in the n–type GaN layer. An output power at RT more than 1.5mW at a forward current of 30mA and bias voltage of 5.0V have been achieved as shown in Fig.1. The power efficiency is about 1%, which is the highest efficiency ever reported in the LED mode operation of UV/blue LEDs.

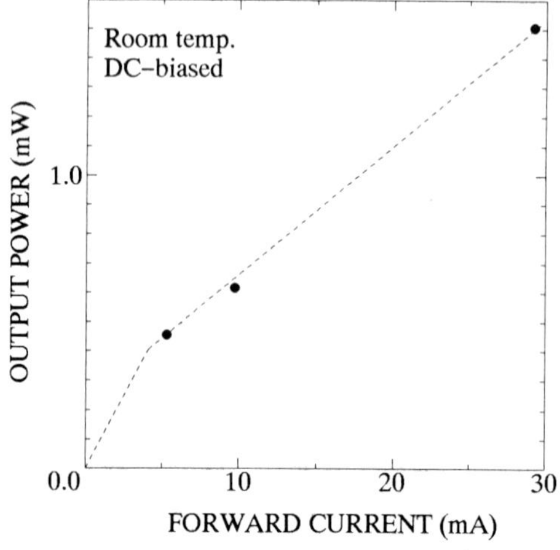

Fig.1 Output power of GaN p–n homojunction LED as a function of forward current taken at RT under DC–biased condition.

6.DEVELOPMENT AND CHARACTERISTICS OF UV/BLUE EMITTING DEVICES WITH AlGaN/GaN DOUBLE HETEROSTRUCTURE

The AlGaN/GaN multi–heterostructure with good crystalline quality, showing quantum size effect, were grown by modulating flow rates of TMAl and TMGa during growth(Itoh et al.1991).

DH was grown on the GaN layer 2μm thick in the same way. Thicknesses of the $Al_{0.1}Ga_{0.9}N$ cladding layer and the GaN active layer were 0.4μm and 0.2μm, respectively. The GaN active layer was slightly Si–doped with a electron concentration of $1x10^{17}cm^{-3}$ at RT. The optical pumping measurements were performed on the DH at RT. The measurement configurations for surface emission mode (Fig.2(a)) and edge emission mode (Fig.2(b)) are schematically depicted. The former is basically the same as that reported by M.A.Khan et al.(1991). A pulsed nitrogen laser with the emission line of 337.1nm, a pulse length of 8nsec with the repetition of 10Hz was used as excitation source. Incident angle of the excitation laser light was about 45° to the sample surface in the case of surface emission mode, and about 90° in the case of edge emission mode. The laser beam was focused on the sample surface yielding a maximum power density of about $0.2MW/cm^2$, which was attenuated by the neutral density filters. It should be noted that the absorption edge of the $Al_{0.1}Ga_{0.9}N$ cladding layer is about 332nm, which means that the cladding layer acts as a window of the 337.1nm excitation.

It was found(Amano et al., 1993) that the refractive index of GaN is larger than that of $Al_{0.1}Ga_{0.9}N$ at around the bandgap energy of GaN by about 0.19, which is promising for the optical confinement of the UV light in the GaN active layer.

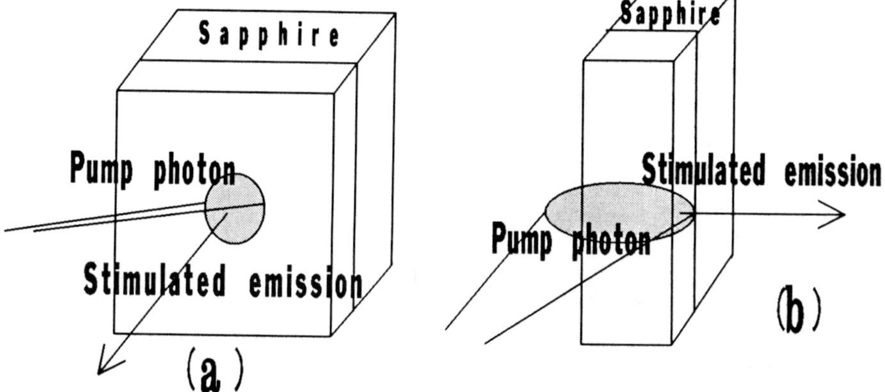

Fig.2 The measurement configurations for surface emission mode (Fig.2(a)) and edge emission mode (Fig.2(b)).

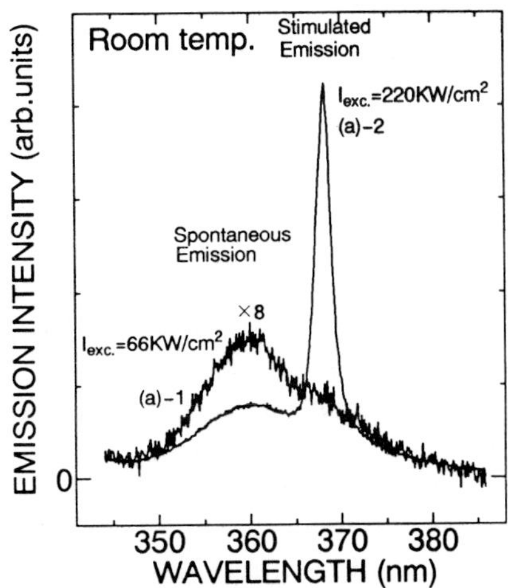

Fig.3(a) Surface emission mode spectra of DH excited below (Fig.3(a)–1) and above (Fig.3(a)–2) threshold input power density.

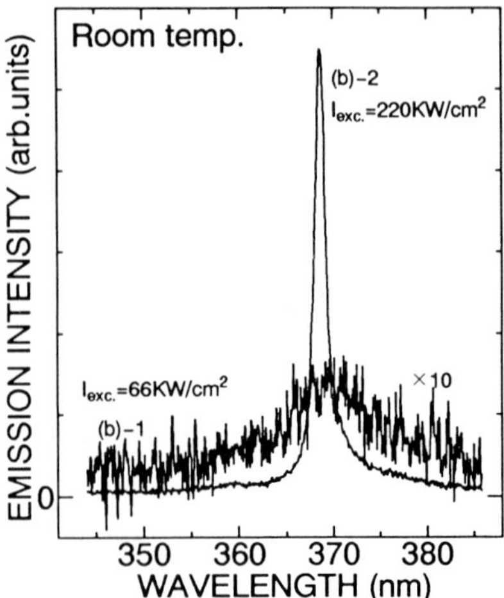

Fig.3(b) Edge emission mode spectra of DH excited below (Fig.3(b)–1) and above (Fig.3(b)–2) threshold input power density.

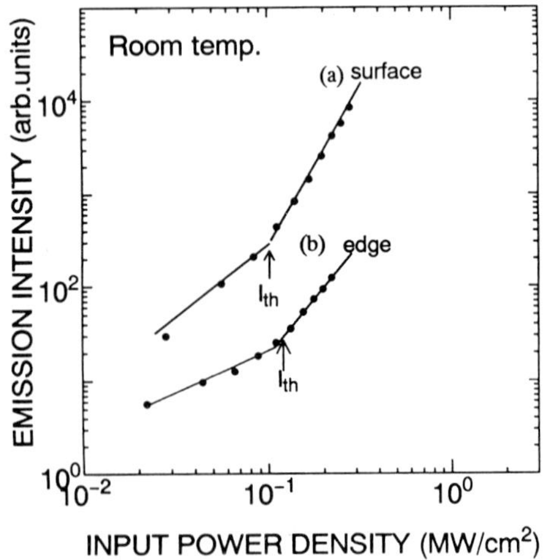

Fig.4 (a)Relationship between surface emission intensity and input power density of DH. (b)Relationship between edge emission intensity and input power density of DH.

Figures 3 and 4 show the surface emission spectra and edge stimulated emission spectra from the DH below (Fig.3(a)–1,Fig.3(b)–1) and above (Fig.3(a)–2,Fig.3(b)–2) a threshold input power density, and the surface emission intensity(Fig.4(a)) and edge emission intensity(Fig.4(b)) as a function of input power density, respectively. Below threshold, only spontaneous emission with a peak wavelength of about 0.36μm was observed. Above threshold, stimulated emission with a peak wavelength of 368.2nm(Fig.3(a)–2) and 368.6nm(Fig.3(b)–2) can be clearly observed along with the weak spontaneous emission. The spectral narrowing and the super–linear dependence of the input–output power above threshold show the onset of stimulated emission. From Fig.4, the threshold powers for stimulated emission in both configurations are found to be around 0.1MW/cm^2, which is about on twentieth(Khan et al. 1991) and one sixth(Amano et al. 1990) those of bulk GaN.

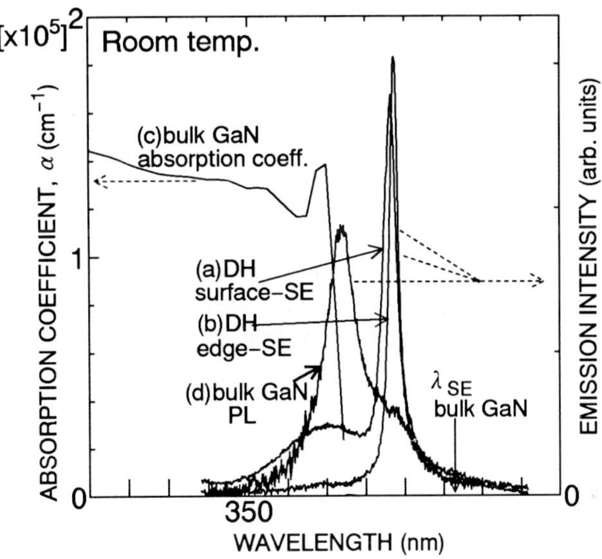

Fig.5 Surface stimulated emission spectrum (Fig.5(a)) and edge stimulated emission (Fig.5(b)) from DH. Absorption spectrum (Fig.5(c)) and weakly excited PL spectrum (Fig.5(d)) of undoped GaN are also shown for comparison. Arrow indicates emission wavelength of stimulated emission from bulk GaN.

Figure 5 shows the spectra of the surface stimulated emission(Fig.5(a)) and edge stimulated emission(Fig.5(b)) from DH. The absorption coefficient(Fig.5(c)) and the weakly excited PL spectrum from bulk GaN 1μm thick(Fig.5(d)) are also shown for comparison. The surface stimulated emission and edge stimulated emission peak wavelengths from the bulk GaN are also indicated by the arrows. It should be noted that the exciton absorption peak can be clearly seen in the absorption spectrum of bulk GaN. Wavelengths of surface and edge stimulated emission from DH are little bit shorter than that of stimulated emission from bulk GaN. This blue shift may be caused by the compressive strain of the GaN active layer, which may be due to the heterostructure. In both the DH case and bulk case, stimulated emission occurs below the absorption edge of the bulk GaN. Further discussion should be necessary to clarify the mechanism of this behavior.

The P-$Al_{0.1}Ga_{0.9}N$/GaN/N-$Al_{0.1}Ga_{0.9}N$ DH diode showed good I-V characteristics, and the similar EL spectrum to that of p-n homojunction LED.

8. SUMMARY

By MOVPE using the AlN buffer layer, the crystalline quality as well as the electrical and optical properties of GaN and AlGaN films can be remarkably improved. Conductivity control for both n-type GaN and AlGaN films has been achieved. GaN film having distinct p-type conduction have been realized for the first time by Mg doping followed by the LEEBI treatment. High performance p-n junction type UV/blue light emitting devices have been achieved. DH is found to be effective for the optical confinement of the UV light. AlGaN/GaN multi-layered structures showed quantum size effects. Further development of the technique for the fabrication of heterostructure together with the understanding of the intrinsic nature of these nitrides will lead to the realization of much higher-performance short wavelength light emitters.

ACKNOWLEDGEMENTS

This work was partly supported by the Grant-in-Aid from the Ministry of Education, Science and Culture of Japan for Scientific Research on Priority Areas "Crystal Growth Mechanism in Atomic Scale" and "Shigaku-Josei". This work was also partly supported by the "The Mitsubishi Foundation". The authors are indebted to staffs of the Toyoda Gosei Co., Ltd. for their help throughout these experiments.

REFERENCES

Akasaki I., Amano H., Koide Y., Hiramatsu K. and Sawaki N. 1989 J.Cryst.Growth **98** 209.

Akasaki I., Amano H., Kito M. and Hiramatsu K. 1991a J.Lumin. **48&49** 666.

Akasaki I. and Amano H. 1991b Proc. Mat.Res.Soc. Symp. **242** 383.

Amano H.,Sawaki N., Akasaki I. and Toyoda Y. Appl.Phys.Lett. 1986 **48** 353.

Amano H.,Kito M., Hiramatsu K. and Akasaki I. 1989 Jpn.J.Appl.Phys. **28** L2112.

Amano H.,Kitoh M.,Hiramatsu K. and Akasaki I. 1990 J.Electrochem.Soc. **137** 1639.

Amano H., Asahi T. and Akasaki I. Jpn.J.Appl.Phys. 1990a **29** L205.

Amano H. and Akasaki I. 1990b Extended abstract of MRS'90 fall meeting **EA-21** 165.

Amano H., Watanabe N.,Koide N. and Akasaki I. Jpn.J.Appl.Phys. 1993 (to be published).

Itoh K., Kawamoto T., Amano H., Hiramatsu K. and Akasaki I. 1991 Jpn.J.Appl.Phys. **30** 1924.

Fujieda S., Mizuta M. and Matsumoto Y. 1988 Jpn.J.Appl.Phys. **27** L296.

Khan M.A., Olson D.T., Van Hove J.H. and Kuznia J.N. 1991 Appl.Phys.Lett. **58** 1515.

Khan M.A., Kuznia J.N., Bhattarai A.R. and Olson D.T. 1993 Appl.Phys.Lett. **62** 1786.

Koide N., Kato H., Sassa M., Yamasaki .S., Manabe K., Amano H., Hiramatsu K. and Akasaki I. 1991 J.Cryst.Growth **115** 639.

Murakami H., Asahi T., Amano H., Hiramatsu K., Sawaki N. and Akasaki I. 1991 J.Cryst.Growth **115** 648.

Nakamura S.,Mukai T. and Senoh M. 1992 Jpn.J.Appl.Phys. **31** 2883.

Paper presented at the Int. Symp. GaAs and Related Compounds, Freiburg, 1993

Recent progress on wavelength tunable laser diodes

Markus-Christian Amann

Siemens AG, Corporate Research and Development, Munich, Germany

ABSTRACT: Single mode laser diodes with an electronically tunable wavelength are among the key components for advanced applications in optical communications, measurement and sensing. The significant progress achieved recently with InGaAsP/InP devices in the 1.55 μm wavelength range is reviewed with particular reference to tuning range, continuous tunability, spectral linewidth and tuning speed. Thereby the different device structures and technological approaches are discussed and compared. Recent concepts and first experimental results on ultra-wide (>50 nm) tunable laser diodes are finally presented and an outlook is given on future developments.

1 INTRODUCTION

Wavelength tunable single mode laser diodes are indispensable key components of advanced photonics applications. Particularly broadband multichannel optical communications, optical switching networks, wavelength dependent measurements and several sensing techniques depend on the availability of laser diodes with an electronically tunable wavelength. With regard to the transmission properties of optical fibers, emission wavelengths around 1.55 μm are usually required in optical communications. Therefore the major development has been performed to date with the InGaAsP/InP material system. Since the performance of broadband communication systems and switching networks is ultimately limited by the wavelength coverage and the number of addressable wavelength channels, the tuning range and spectral resolution play a major role in the laser development.

2 TUNABLE DFB AND DBR LASERS

The longitudinal sections of state-of-the-art tunable laser diode structures [1, 2] are displayed schematically in Fig. 1. The technologically most simple approach is a distributed feedback (DFB) laser, which is the most mature single-mode laser structure, the top contact of which is longitudinally separated into three individually biased sections (Fig. 1a). Wavelength tuning may be performed continuously by a careful mutual adjustment of the laser currents. In this way the carrier density in the active region becomes longitudinally redistributed, which together with thermal heating changes the effective refractive index and the resonance wavelength of the laser cavity. Owing to the random mirror phases or inevitable waveguide perturbations the

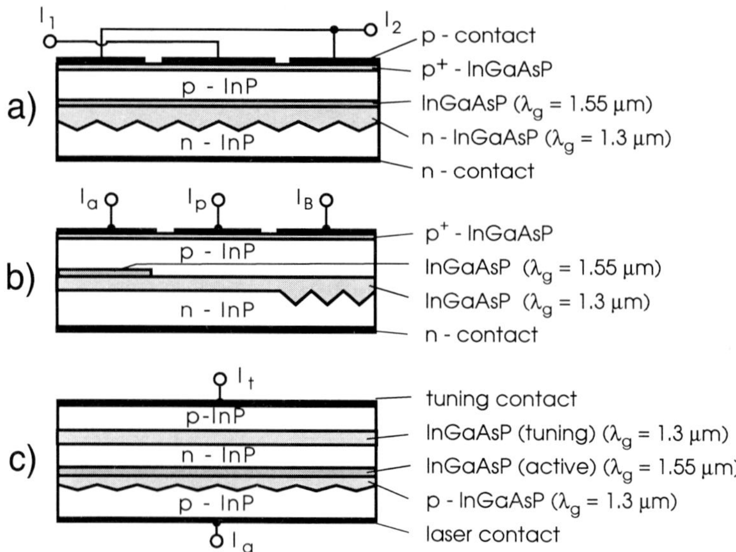

Figure 1: Schematic longitudinal view of tunable three-section DFB laser (a), three-section DBR laser (b) and tunable twin-guide (TTG) DFB laser (c).

longitudinal field and carrier density distributions are different for each DFB laser (even for devices from the same wafer), so that the tuning behaviour, particularly the I_1/I_2-ratio for continuous tuning, differs from device to device. Accordingly a large amount of measurement is required to select and characterize the suited lasers.

An essential disadvantage in the practical application of tunable DFB lasers is that the controls of output power and wavelength are not separated but are both affected similarily by all control currents. With properly selected devices and by exploiting also the wavelength change by thermal heating, tuning ranges up to 6 nm with spectral linewidth below 2 MHz have been reported [3]. Quite recently a spectral linewidth less than 100 kHz has been obtained over a tuning range of 1.3 nm with a corrugation-pitch-modulated multi quantum well (MQW) DFB laser [4].

The three-section distributed Bragg reflector (3S DBR) laser as shown in Fig. 1b provides a more convenient handling since an effective separation between the power control and tuning functions is achieved: Current I_a mainly determines the power while both currents I_p and I_B essentially control the emission wavelength. Changing exclusively I_B or I_p yields a discontinuous tuning by mode jumping. By a proper mutual adjustment of these two currents, however, a continuous wavelength tuning can be obtained. This handling improvement is obtained by separating the laser active region from the passive wavelength selective phase shift and Bragg grating region. This furthermore allows the extension of the tuning range by a strong heating of the Bragg section while keeping the temperature in the gain section constant to maintain the laser action. In this way large tuning ranges up to 22 nm have recently been realized in the quasi-continuous (i. e. stepwise continuous) tuning mode[5], while the maximum continuous tuning range is around 4.4 nm [1].

The purely thermal wavelength tuning via resistive heating of the DBR section yields discontinuous tuning greater than 10 nm with linewidth below 7 MHz [6].

The carrier injection into the passive Bragg and phase control region during tuning introduces shot noise fluctuations that are not damped by the light-carrier interaction as in the tunable DFB laser [7, 8, 9]. The resulting linewidth broadening is typically of the order of several MHz [10], which leads to larger linewidth than with the tunable DFB lasers and limits the applicability of the 3S DBR lasers in linewidth sensitive systems.

Using a four-step MOVPE process and a semi-insulating InP:Fe current-blocking structure an AM modulation bandwidth of 9 GHz and a quasi-continuous tuning range of 9.1 nm were demonstrated with recent 3S DBR lasers [11]. The switching time for the transient between two successive modes can be as small as 500 ps [12].

The most convenient handling with a principally continuous tuning behaviour is achieved with the tunable twin-guide (TTG) laser as shown schematically in Fig. 1c [13, 14, 15]. This laser can be considered as a single mode DFB laser whose effective refractive index is tuned homogeneously along the laser axis by means of current I_t. This is done by carrier injection into the passive tuning region which is collocated below or above the active region. Also in this laser structure the carriers in the tuning region undergo no lifetime shortening by stimulated emission, so that a marked linewidth broadening inevitably occurs in the tuning mode due to the shot-noise [7, 16].

Well designed TTG lasers yield continuous tuning ranges up to more than 7 nm with spectral linewidth below about 30 MHz. For ridge-waveguide TTG lasers 4 MHz linewidth with tuning ranges up to 2 nm have been obtained [14].

Exploiting the Quantum Confined Stark Effect (QCSE) for tuning of the TTG laser [17, 18] no carriers are injected into the tuning region. Correspondingly no shot noise broadening occurs, so that the total spectral linewidth can be kept small. In addition the FM modulation bandwidth can be increased since the carrier lifetime limitation in the tuning region is dropped. On the other hand, however, due to the small optical confinement in the quantum wells, the tuning range is essentially smaller. Improved spectral properties and high-speed FM modulation might also be achieved in future devices using the electron-transfer within an MQW-type tuning region. With the so-called barrier reservoir and quantum-well electron-transfer structure (BRAQWETS) voltage controlled refractive index changes up to 0.02 [19] have been demonstrated and (parasitic free) switching times well below 100 ps have theoretically been predicted [20].

QCSE TTG lasers optimized for broadband FM modulation yield a flat FM response up to more than 2 GHz [17] with a spectral linewidth below 4 MHz, while the tuning range is only of the order of several Å. Using a separate confinement heterostructure quantum well (SCH QW) structure in the tuning region that localizes the holes within the wells while the electrons are distributed over the entire tuning region [21] or applying a MQW twin-active-guide [22] might enable a further simultaneous improvement of optical power and spectral linewidth of the TTG laser.

Employing current induced heating of the top reflector, a tuning range of 2 nm ($\approx$ 600 GHz) at 980 nm has recently been shown for a three-terminal strained layer InGaAs/AlGaAs vertical cavity laser diode with a tuning current as small as 0.7 mA [23].

Owing to the limited refractive index changes, the maximum electronic tuning range (excluding thermal heating) of conventional DFB and DBR type tunable lasers is restricted to values less than about 15 nm at 1.5 μm wavelength [1].

3 WIDELY TUNABLE LASER DIODES

For larger tuning ranges, therefore, completely different laser structures have to be considered. Using the interferometric effect between the two arms of an (asymmetric) Y-coupled integrated laser diode, an extended discontinuous wavelength coverage can be achieved [24, 25]. The principal device structure is displayed in Fig. 2 together with an illustration of its operation. The Y-laser usually consists of an all active waveguide structure with a typical length around 1 mm. By the separation of the top p-contact into 3 to 4 sections, the two interferometer arms A and B can be biased independently. At each bias condition lasing occurs at a wavelength where the two differently spaced comb mode spectra of the two coupled laser cavities exhibit simultaneously an axial mode.

For illustration Figs. 2b and 2c show the comb mode spectra corresponding to arm B and A, respectively, at a certain bias I_a and I_b. Here lasing occurs at λ_1. Increasing I_a slightly shifts the comb mode spectrum of arm A towards shorter wavelengths by $\Delta\lambda_a$ (Fig. 2d). Lasing occurs now at wavelength λ_2, which is shifted by a much larger amount $\Delta\lambda$ towards longer wavelengths with respect to λ_1. As can be seen, the vernier effect of the two differently spaced comb mode spectra yields a significant magnification of the laser wavelength shift as compared with the induced shift of the comb mode spectrum as well the ability to cause either a blue or a red shift of the laser wavelength.

Experimentally, up to 51 nm discontinuous tuning was reported for InGaAsP/InP Y-lasers at 1.55 μm wavelength [25, 26]. Using this laser as a tunable wavelength converter, a conversion of 2.5 Gb/s data streams was demonstrated [26].

Applying a DBR laser structure with two different Bragg reflectors at the rear and front end, that exhibit different comb reflection spectra (similar to the comb mode spectra in Fig. 2b-c), also yields an enhanced tuning range by the vernier effect. The corresponding super-structure grating (SSG) DBR [27, 28] and sampled grating (SG) DBR [29, 30] tunable lasers are described in Fig. 3. The absolute wavelengths of the reflection peaks of each of the two Bragg reflectors R_a and R_b can be controlled by currents I_a and I_b, respectively. In close analogy with the Y-laser, lasing always occurs at a wavelength where the two reflection spectra each exhibit a reflection peak. Consequently, small changes in any of the reflection spectra lead to large changes of the laser wavelength.

The comb reflection spectra are realized by Bragg gratings with side modes in the spatial frequency domain. This can either be achieved by a spatial amplitude modulation of the grating (c. f. Fig. 3b), yielding the SG DBR lasers, or by a spatial frequency modulation (c. f. Fig. 3c), which corresponds to the SSG DBR lasers. The maximum wavelength tuning achieved with the SG DBR laser so far is 57 nm [30], while record tuning ranges of 83 nm and 101 nm

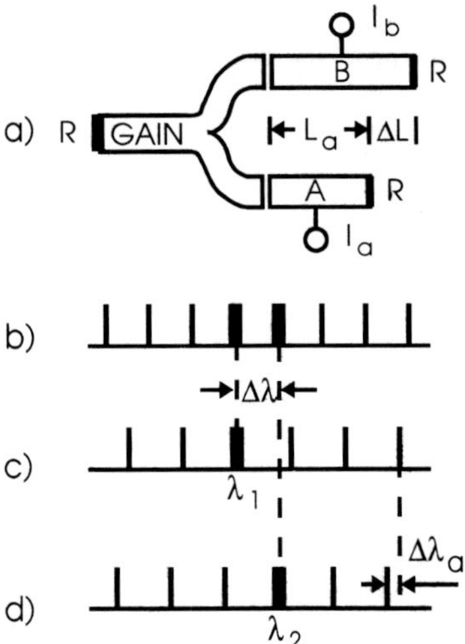

Figure 2: Schematic top view of tunable Y-laser (a), comb mode spectrum of interferometer arm B (b) and arm A at two different currents I_a (c and d).

for single- and multimode operation, respectively, were recently presented for SSG DBR lasers [27]. However, the number of accessible wavelengths within these wavelengths ranges is rather small for both structures ranging below about 12.

A third principal approach for a wide tuning is the application of a codirectionally coupled two-mode twin-waveguide laser structure. The two basic structures presented so far are shown in Fig. 4. Compared with the contradirectionally coupled DFB and DBR lasers the distributed forward coupled (DFC) laser (Fig. 4a) [31] equals the longitudinally quasi-homogeneous DFB laser, while the vertical-coupler filter (VCF) type laser (Fig. 4b) [32, 33, 34] resembles more the DBR laser.

Both devices rely on the interference effect of two codirectionally coupled modes. In the DFC laser periodic absorbers (period $\approx$ 15 μm) are placed within the laser cavity. Only exactly at the phase matching wavelength the superposition of the two waveguide modes yields negligible absorption losses and a low threshold current, since in this case the combined fields almost vanish at the periodic absorbers. In the VCF type lasers, lasing can only occur at a wavelength where the phase matching condition between the two codirectionally propagating waveguide modes applies in the VCF region, since only under this condition a closed feedback loop exists for the light path.

Wavelength tuning is induced in these lasers by shifting the phase matching wavelength. This is done simply with only one control current I_t, by which the refractive index in WG #2 and, as a consequence, the effective refractive index difference between the two waveguide modes can be changed. Owing to the extremely strong dependence of the phase match wavelength on this refractive index difference, a large tuning effect is achieved and the tuning range might ultimately be limited only by the spectral width of the active region gain. With the first DFC lasers tuning ranges around 16 nm were reported with about 9 accessible wavelengths. The more mature VCF type lasers showed tuning ranges between 30 and 57 nm [33, 34] with up to

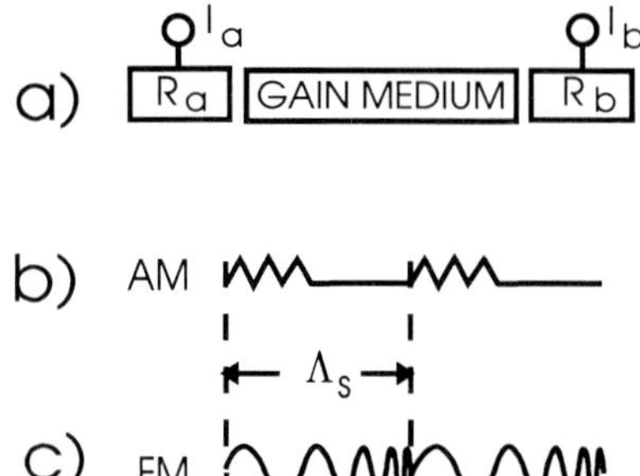

Figure 3: Schematic of sampled grating (SG) and super-structure grating (SSG) tunable lasers (a), with spatially AM modulated (SG) and spatially FM modulated (SSG) multi-wavelength reflectors R_a and R_b. Λ_s is different for R_a and R_b, yielding different wavelength spacing of the comb reflection spectra.

about 15 accessible wavelengths.

It should be stressed, that the underlying operation principles of these widely tunable laser diodes exploit the slight differences of modenumbers or reflection maxima; therefore even rather small variations of these parameters, occuring during the device fabrication, yield large relative changes of the device characteristics, particularly of the laser wavelength. The successful development of these novel components and a reasonable fabrication yield therefore basically require a highly precise and homogeneous fabrication technique; this essentially concerns the crystal growth, for which e. g. a layer thickness control on the 0.01 μm scale is demanded.

A common deficiency of all these approaches for an extended tuning range is the lack of a continuous tunability hindering the access to any wavelength within the tuning range. This still limits the practical applicability of these lasers, particularly as the number of longitudinal modes covered by the wavelength tuning typically is of the order 100. The future development is therefore challenged by improving the wavelength access, e. g. with laser structures exhibiting an enhanced wavelength selectivity or even by the practical realisation of quasi-continuously tunable devices.

4 CONCLUSION

The state-of-the-art of electronically tunable laser diodes has been reviewed, revealing that for tuning ranges below about 10 nm various types of high-performance (single-mode, continuous tuning, linewidth, speed, handling) devices already exist, based on the technologically well developed DFB and DBR laser structures. As with other optoelectronic components, the exploitation of strained-layer quantum well structures has essentially improved the device parameters. The novel device concepts presented so far for wide tuning ranges have been presented together with their relevant characteristics.

Acknowledgement: The author gratefully acknowledges the fruitful discussions with T. Wolf, B. Borchert and S. Illek.

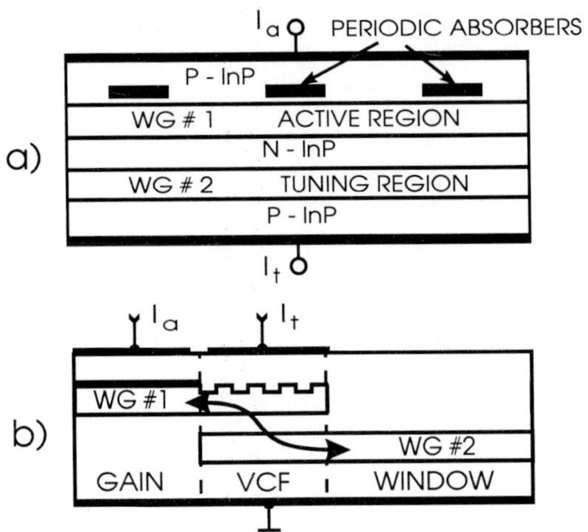

Figure 4: Schematic longitudinal view of tunable distributed forward coupled (DFC) laser (a) and vertical coupler filter (VCF) based lasers. Tuning is performed by only one control current (I_t).

References

[1] Y. Kotaki and H. Ishikawa *IEE Proc. Pt. J.*, vol. 138, pp. 171–177, 1991.

[2] T. L. Koch and U. Koren *J. Lightwave Technol.*, vol. LT-8, pp. 274–293, 1990.

[3] P. I. Kuindersma, W. Scheepers, J. M. H. Cnoops, P. J. A. Thijs, G. L. A. v. d. Hofstad, T. v. Dongen, and J. J. M. Binsma in *Conference Digest of 12th IEEE Semiconductor Laser Conference (Davos, Switzerland)*, pp. 248–249, 1990.

[4] M. Okai and T. Tsuchiya *Electron. Lett.*, vol. 29, pp. 349–351, 1993.

[5] M. Oeberg, S. Nilsson, T. Klinga, and P. Ojala *IEEE Photon. Technol. Lett.*, vol. PTL-3, pp. 299–301, 1991.

[6] S. L. Woodward, U. Koren, B. I. Miller, M. G. Young, M. A. Newkirk, and C. A. Burrus *IEEE Photon. Technol. Lett.*, vol. 4, pp. 1330–1332, 1992.

[7] M. Amann and R. Schimpe *Electron. Lett.*, vol. 26, p. 279, 1990.

[8] B. Tromborg, H. Olesen, and X. Pan *IEEE J. Quantum Electron.*, vol. 27, pp. 178–192, 1991.

[9] M. F. dos Santos Ferreira, J. R. F. da Rocha, and J. de Lemos Pinto *IEEE J. Quantum Electron.*, vol. 28, pp. 833–840, 1992.

[10] Y. Kotaki and H. Ishikawa *IEEE J. Quantum Electron.*, vol. QE-25, pp. 1340–1345, 1989.

[11] B. Stoltz, M. Dasler, and O. Sahlen *Electron. Lett.*, vol. 29, pp. 700–702, 1993.

[12] F. Delorme, G. Gambini, M. Puleo, and S. Slempkes *Electron. Lett.*, vol. 29, pp. 41–43, 1993.

[13] S. Illek, W. Thulke, C. Schanen, H. Lang, and M. C. Amann *Electron. Lett.*, vol. 26, pp. 46–47, 1990.

[14] T. Wolf, H. Westermeier, and M. Amann *Europ. Trans. Telecommun. and Research. Technol.*, vol. 3, pp. 517–522, 1992.

[15] T. Wolf, S. Illek, J. Rieger, B. Borchert, and W. Thulke *IEEE Photon. Technol. Lett.*, vol. PTL-5, pp. 273–275, 1993.

[16] M. Hamada, E. Yamamoto, K. Suda, S. Nogiwa, and T. Oki *Jpn. J. Appl. Phys. (Letters)*, vol. 31, pp. L 1552–L 1555, 1991.

[17] T. Wolf, K. Drögemüller, B. Borchert, H. Westermeier, E. Veuhoff, and H. Baumeister *Appl. Phys. Lett.*, vol. 60, pp. 2472–2474, 1992.

[18] E. Yamamoto, M. Hamada, K. Suda, S. Nogiwa, and T. Oki *Appl. Phys. Lett.*, vol. 59, pp. 2721–2723, 1991.

[19] M. Wegener, T. Y. Chang, I. Bar-Joseph, J. M. Kuo, and D. S. Chemla *Appl. Phys. Lett.*, vol. 55, pp. 583–585, 1989.

[20] J. Wang, J. P. Leburton, and J. L. Educato *J. Appl. Phys.*, vol. 73, pp. 4669–4679, 1993.

[21] Y. Sakata, M. Yamaguchi, S. Takano, J. Shim, T. Sasaki, M. Kitamura, and I. Mito in *Proceedings Optical Fiber Conference (San Jose, USA)*, pp. 9–10, 1993.

[22] E. Yamamoto, K. Suda, M. Hamada, S. Nogiwa, and T. Oki *Jpn. J. Appl. Phys. (Letters)*, vol. 30, pp. L 1884–L 1886, 1991.

[23] T. Wipiejewski, K. Panzlaff, E. Zeeb, and K. J. Ebeling in *Proceedings Optical Fiber Conference (San Jose, USA)*, pp. 122–123, 1993.

[24] M. Schilling, H. Schweitzer, K. Dütting, W. Idler, E. Kühn, A. Nowitzki, and K. Wünstel *Electron. Lett.*, vol. 26, pp. 243–244, 1990.

[25] M. Kuznetsow, P. Verlangieri, A. G. Dentai, C. H. Joyner, and C. A. Burrus *IEEE Photon. Technol. Lett.*, vol. 4, pp. 1093–1095, 1992.

[26] M. Schilling, W. Idler, D. Baums, K. Dütting, G. Laube, K. Wünstel, and O. Hildebrand in *Conference Digest of 13th IEEE Semiconductor Laser Conference (Takamatsu, Japan)*, pp. 272–273, 1992.

[27] Y. Tohmori, Y. Yoshikuni, T. Tamamura, M. Yamamoto, Y. Kondo, and H. Ishii *Electron. Lett.*, vol. 29, pp. 352–354, 1993.

[28] Y. Yoshikuni, Y. Tohmori, T. Tamamura, H. Ishii, Y. Kondo, M. Yamamoto, and F. Kano in *Proceedings Optical Fiber Conference (San Jose, USA)*, pp. 8–9, 1993.

[29] V. Jayaraman, D. A. Cohen, and L. A. Coldren *Appl. Phys. Lett.*, vol. 60, pp. 2321–2323, 1992.

[30] Y. Jayaraman, A. Mathur, L. A. Coldren, and P. D. Dapkus *IEEE Photon. Technol. Lett.*, vol. PTL-5, pp. 489–491, 1993.

[31] M. Amann, B. Borchert, S. Illek, and T. Wolf *Electron. Lett.*, vol. 29, pp. 793–794, 1993.

[32] R. C. Alferness, U. Koren, L. L. Buhl, B. I. Miller, M. G. Young, T. L. Koch, G. Raybon, and C. A. Burrus in *Proceedings Optical Fiber Conference (San Jose, USA)*, pp. 321–324, 1992.

[33] I. Kim, R. C. Alferness, L. L. Buhl, U. Koren, B. I. Miller, M. A. Newkirk, M. G. Young, T. L. Koch, G. Raybon, and C. A. Burrus *Electron. Lett.*, vol. 29, pp. 664–666, 1993.

[34] S. Illek, W. Thulke, and M. Amann *Electron. Lett.*, vol. 27, pp. 2207–2208, 1991.

DC and high-frequency properties of $In_{0.35}GA_{0.65}As/GaAs$ strained-layer MQW laser diodes with *p*-doping

I. Esquivias[*], S. Weisser, A. Schönfelder[**], J. D. Ralston, P. J. Tasker, E. C. Larkins, J. Fleissner, W. Benz, and J. Rosenzweig

Fraunhofer-Institut für Angewandte Festkörperphysik, Tullastrasse 72, D-79108 Freiburg, Germany
[*] *Present address: Dept. Tecnología Electrónica, Univ. Politécnica, Madrid, Spain.*
[**] *Institut für Hochfrequenztechnik und Quantenelektronik, Univ. Karlsruhe, Kaiserstrasse 12, D-76128 Karlsruhe, Germany.*

ABSTRACT: An experimental and modelling investigation concerning the effects of p-doping on the DC and high-frequency properties of MBE grown $In_{0.35}Ga_{0.65}As/GaAs$ strained-layer MQW lasers is presented. P-doping produces a substantial decrease in the non-linear gain coefficient, which can be attributed to a doping-induced decrease in the intraband relaxation time, and a small increase in the differential gain. The factors limiting the maximum measured modulation bandwidth of the p-doped devices (30 GHz) are analyzed and discussed.

1. INTRODUCTION

The modulation dynamics of a single-mode semiconductor laser at microwave frequencies can be well-described with two parameters, the resonance frequency, f_r, and the damping rate γ, which are given by (Su et al 1992):

$$f_r = \frac{1}{2\pi}\sqrt{v_g^2\, g\, \Gamma \frac{\partial g}{\partial n} S} \tag{1}$$

$$\gamma = \frac{1}{\tau_c} + K f_r^2 \; ; \qquad K = \frac{(2\pi)^2}{v_g}\left(\frac{1}{\alpha_i + \alpha_m} + \frac{\epsilon}{\partial g/\partial n}\right) \tag{2}$$

where ν_g denotes the group velocity, Γ the optical confinement factor, g the material gain, $\partial g/\partial n$ the differential gain, S the photon density in the resonator, τ_c the effective carrier lifetime, K the damping factor, α_i the internal losses, α_m the mirror losses and ϵ the nonlinear gain coefficient. The 3 dB modulation bandwidth, f_{3dB}, is proportional to f_r at low power levels ($f_{3dB} \approx 1.55 f_r$) and reaches a maximum value $f_{3dB}^{max} \approx 8.9/K$ (Olshansky et al 1987). The above equations indicate that the main material parameters contributing to f_{3dB} are $\partial g/\partial n$, and ϵ, which should be maximized and minimized, respectively, in order to obtain high modulation bandwidths. In practical devices, the maximum achievable f_{3dB} is determined by other limiting effects, such as ohmic heating, catastrophic failure at high power levels, carrier transport/capture in QW's (Nagarajan et al 1992), multimode behaviour, and electrical parasitics.

P-doping has been shown to enhance the high frequency performance of semiconductor lasers, but the mechanisms leading to this enhancement are still the subject of debate: Uomi (1990) theoretically predicted and experimentally demonstrated (Uomi et al 1990) an increase in f_r due to an enhanced $\partial g/\partial n$ in GaAs/AlGaAs multiple quantum well (MQW) lasers; Su et al (1992) and Lealman et al (1992) reported a doping-induced increase in $\partial g/\partial n$, not counteracted by an increase in ϵ, in InGaAsP bulk lasers and InGaAs/InGaAsP MQW lasers, respectively; Aoki et al (1990) observed a simultaneous increase in $\partial g/\partial n$ and decrease in ϵ in InGaAs/InGaAsP MQW lasers; we have found that p-doping produces only a slight enhancement in $\partial g/\partial n$, but a drastic reduction in ϵ in strained InGaAs/GaAs MQW lasers (Ralston et al 1993a). Differences between material systems, as well as between device structures and doping levels, probably account for these contradictory results.

We have recently demonstrated the first semiconductor laser to achieve 30 GHz direct modulation bandwidth, by means of the addition of p-doping to the active region in InGaAs/GaAs strained-layer MQW devices (Weisser et al 1992, Ralston et al 1993a). In the present work we present more detailed characterization and modelling results concerning the effects of p-doping on the DC and RF properties of such devices.

2. EXPERIMENTAL TECHNIQUES

The design and MBE growth of vertically-compact ungraded SCDH InGaAs laser structures has been presented in more detail elsewhere (Ralston et al 1993a and 1993b). The following is a brief description of the layer structure: an initial 200 nm undoped GaAs buffer layer; 1 μm n^+ GaAs contact layer; 100 nm n-type region graded to $Al_{0.8}Ga_{0.2}As$; 800 nm n-$Al_{0.8}Ga_{0.2}As$ lower cladding region; 48 nm GaAs lower core region; four 5.7 nm $In_{0.35}Ga_{0.65}As$ QWs separated by 20 nm GaAs barriers; 48 nm GaAs upper core region; 800 nm p-doped $Al_{0.8}Ga_{0.2}As$ upper cladding region; 100 nm p-type layer graded to GaAs; 200 nm p^{++} GaAs top contact layer. Structures with five different doping configurations were grown: undoped QW's (sample A); 4.5 nm, 5×10^{18} cm^{-3} Be-doped regions centred in the QW's (sample B), or placed in the GaAs barriers above each of the QW's, separated by a 3 nm undoped GaAs spacer (sample C), 4.5 nm, 2×10^{19} cm^{-3} Be-doped regions centred in the QW's (sample D), or placed above each of the QW's, separated by a 3 nm undoped GaAs spacer (sample E).

Lasers were fabricated with 3-40 μm wide mesas and a coplanar contact geometry suitable for on-wafer RF probing (Offsey et al 1990, Ralston et al 1993b), cleaved to lengths of 200-900 μm, and indium soldered onto copper heat-sinks. The light-output versus current characteristics (P-I) were measured under CW bias and under pulsed conditions using a calibrated large-area photodiode. The devices were characterized in terms of their emission spectra and far-field patterns. Small signal direct modulation response and relative intensity noise (RIN) measurements were carried out using CW bias currents. The high-frequency experimental set-up has been described in detail elsewhere (Ralston et al 1993a). Device testing was performed at a heat-sink temperature of 25°C.

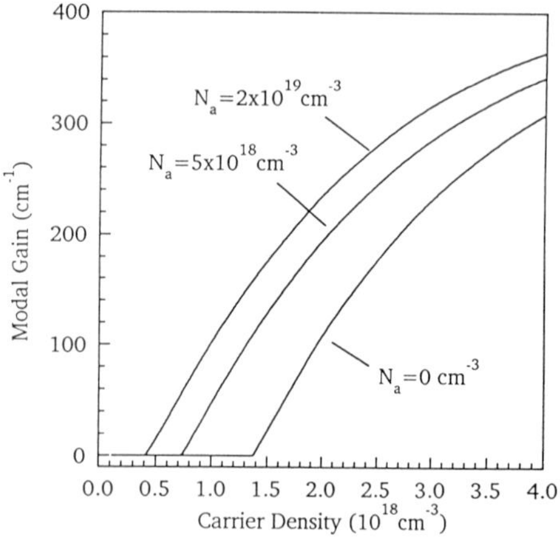

Fig 1 Calculated modal gain as a function of the carrier density.

3. LASER GAIN MODEL

In order to support the laser development, we have simulated the QW gain using a model in which the conduction subbands are approximated as parabolic, whereas nonparabolicity is analytically included in the valence subbands (Ridley 1990). We have considered a carrier-density-dependent bandgap shrinkage

(Kleinman and Miller 1985) and carrier-density-dependent intraband relaxation time τ_{in} (Blood et al 1988), with $\tau_{in,0}$, the intraband relaxation time at a carrier density of 10^{18} cm^{-3}, set to 0.2 ps. A more detailed description of this gain model will be published elsewhere.

Fig. 1 compares the calculated modal gain, $g_{mod} = \Gamma g$, for our laser structure as a function of the carrier density for undoped and p-doped QWs. The calculations predict a monotonic decrease of the transparency carrier density, n_{trans}, with increasing doping level, and an approximately linear dependence of g_{mod} on the carrier density (constant $\partial g/\partial n$) up to $g_{mod} \approx 150$ cm^{-1}. Such a linear behaviour implies that linear gain saturation should not be observed, even in lasers with uncoated facets and relatively short cavity lengths, L (down to L $\approx$ 50 - 200 μm, depending on the optical losses). This fact is a consequence of the high Γ arising from our SCH laser design with $Al_{0.8}Ga_{0.2}As$ cladding layers and MQW active region (Ralston et al 1991). The calculations also predict only a slight enhancement of $\partial g/\partial n$ with doping in the active region, contrary to the large (up to a factor of 4) increase theoretically calculated by Uomi (1990a) in modulation-doped GaAs/AlGaAs MQW lasers. This difference can be explained by the different shape of the valence sub-bands between strained and unstrained materials and by the more realistic inclusion of partial k-selection with a carrier-density-dependent intraband relaxation time in our model.

4. RESULTS AND DISCUSSION

The key DC and RF characterization results for the five structures are summarized in Table I. The first feature to be pointed out is that no significant differences were found between devices with the same doping concentration, but different nominal location. Secondary Ion Mass Spectroscopy (SIMS) profiles of the as-grown epilayer structures reveal that redistribution of the Be doping spikes occurs during the MBE growth (Ralston et al 1993c). This effect leads to very similar doping profiles, and lasing performance, in samples B and C, and in samples D and E.

Sample	A	B	C	D	E
Doping, cm^{-3} (location)	Undoped	$5x10^{18}$ (QWs)	$5x10^{18}$ (barriers)	$2x10^{19}$ (QWs)	$2x10^{19}$ (barriers)
I_{th}^*, best, mA	17	14	13.7	14.2	16.5
λ_{th}, nm	1082	1086	1089	1093	1095
η_i, %	67	57	66	62	63
α_i, cm^{-1}	26	22.5	32	52.5	53
J_{trans}, A/cm^2	360	180	140	290	285
$\partial g/\partial n^*$, cm^2	$2.2x10^{-15}$	$2.2x10^{-15}$	$2.0x10^{-15}$	$2.5x10^{-15}$	$2.5x10^{-15}$
K^*, ns	0.26	0.23	0.23	0.15	0.14
ϵ^*, cm^{-3}	8.6×10^{-17}	7.1×10^{-17}	6.4×10^{-17}	5.0×10^{-17}	4.6×10^{-17}
f_{3dB}^*, best, GHz	21	21	24	30	30

Table I Summary of DC and high frequency properties (* 3×200 μm^2 devices)

Fig. 2 shows the threshold current, I_{th}, for 200 μm long lasers of samples A (undoped), C (5×10^{18} cm^{-3} Be), and E (2×10^{19} cm^{-3} Be), as a function of the mesa width, w. I_{th} shows a linear dependence on w, with a non-zero intercept, I_L, on the vertical axis. A pioneering work on mesa structured lasers (Tsukada et al 1973) attributed a similar dependence to optical loss due to scattering at the mesa sidewalls. We have carried out measurements of the current-voltage characteristics at low current levels and observed the same dependence of the laser current at a given voltage on w as that of I_{th} on w. This result indicates that the offset along the vertical axis I_L is an electrical effect, caused by the high recombination rate at the junction perimeter. The slope of Fig. 1 thus represents the product $J_{th} \times L$, where J_{th} denotes the broad-area threshold current density and L the resonator length. J_{th} initially decreases at a doping level of 5×10^{18} cm^{-3}, but then increases again as the doping level is increased to 2×10^{19} cm^{-3}.

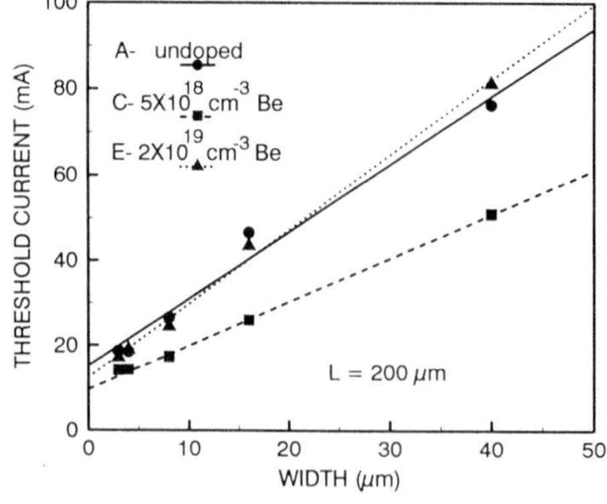

Fig 2 Threshold current vs mesa width for 200 μm long lasers. The experimental points are the average values over 5 devices.

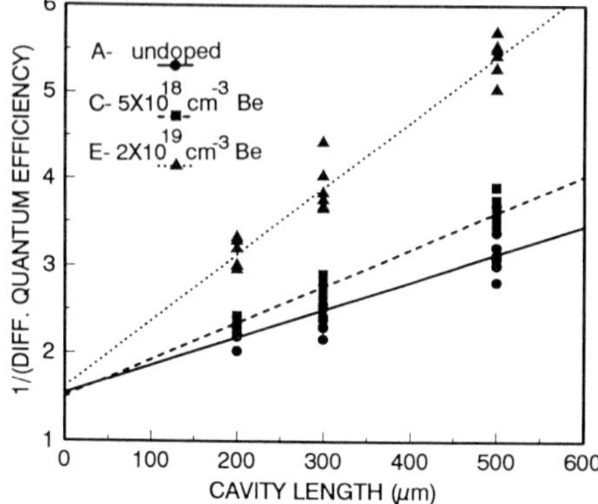

Fig 3 Inverse differential external quantum efficiency vs laser length.

A red shift was observed in the lasing wavelength with increasing doping level (see Table I), similar to that found by Uomi et al (1990) in p-doped GaAs/AlGaAs MQW lasers; this shift can be attributed to the shifting of the QW transition energy due to charge-induced band-bending, and/or bandgap shrinkage. The far-field patterns of the 3×200 μm^2 lasers show a single lobe, with an angular width parallel to the junction of apprioximately 16 °, up to 50-60 mA, indicating a single lateral mode. Higher order lateral modes were observed at higher current levels.

The internal quantum efficiency, η_i, and α_i were estimated from plots of the inverse differential external quantum efficiency as a function of the cavity length (Fig. 3). Within experimental error arising from scatter in the data for different devices, η_i is not strongly affected by the doping level, but α_i increases substantially in the 2×10^{19} cm^{-3} doped devices (see Table I), probably due to free-hole optical absorption.

Fig. 4 shows the modal gain, g_{mod} (calculated from the total losses $g_{mod} = \alpha_i + \alpha_m$), as a function of the experimental J_{th}. The dependence is approximately linear, not showing the onset of linear gain saturation, which confirms the predictions of our theoretical model. The transparency current density, J_{trans}, was estimated by extrapolating to $g_{mod} = 0$ (Table I). As was observed for J_{th}, J_{trans} is first reduced in the 5×10^{18} cm^{-3} doped lasers, but increases again in the 2×10^{19} cm^{-3} doped devices. A similar dependence of J_{trans} on doping level was theoretically predicted by Uomi (1990) for GaAs/AlGaAs MQW lasers, as a consequence of the combined effects of decreases in both n_{trans} and the carrier lifetime at increasing doping concentration. Preliminary measurements of the modulation response at $I < I_{th}$ confirm a decrease in the differential carrier lifetime of the doped devices.

Using parasitic-free RIN measurements, f_r and γ were extracted for 3×200 μm^2 devices as a function of the drive current. Equations (1) and (2) were used to determine $\partial g/\partial n$, the K-factor, and ϵ for the five laser structures (see Table I). The p-doping produces little effect on $\partial g/\partial n$, confirming our calculations, but reduces drastically the K-factor, substantially increasing the maximum intrinsic modulation bandwidth. Fig.

5 shows the non-linear component of the K-factor, K_{NL} (the term including ϵ in equation 2), and ϵ as a function of the doping concentration. The corresponding values, as reported by Su et al (1992) and Lealman et al (1992) for InGaAsP bulk lasers and InGaAs/InGaAsP MQW lasers, respectively, are also included for comparison in Fig. 5. K_{NL} is reduced with doping in all three material systems, but in our case the reduction arises from a drastic decrease in ϵ (almost a factor of two at the higher doping levels), and not from an enhanced $\partial g/\partial n$. The substantially smaller ϵ values for the InP-based devices correlate with the smaller values of $\partial g/\partial n$ (Su et al 1992). According to spectral hole burning theory (Agrawal 1990), ϵ is proportional to the intraband relaxation time τ_{in}. As suggested by Aoki et al (1990), we believe that the observed reduction in ϵ is due to the carrier-density-induced decrease of τ_{in} that we have considered in our gain model. If the dependence of τ_{in} on the carrier density proposed by Blood et al (1988) is valid for doped lasers, the decrease in τ_{in} would only be significant at dopant concentrations much higher than 10^{18} cm^{-3}. Further theoretical and experimental investigations are necessary to clarify this subject.

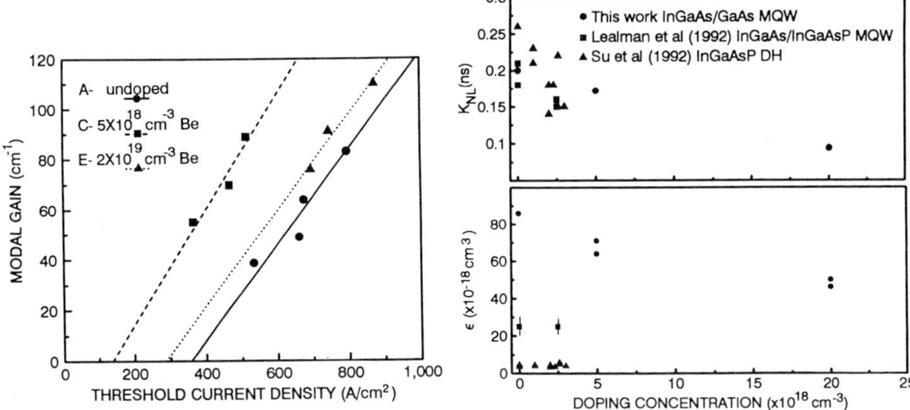

Fig 4 Modal gain vs experimental threshold current density.

Fig 5 Non-linear component of the K factor, K_{NL}, and non-linear gain coefficient ϵ vs p-doping concentration.

Small signal modulation response measurements of the 3×200 μm^2 devices yield f_{3dB} values up to 21 GHz at 60 mA (undoped), 24 GHz at 65 mA (5×10^{18} cm^{-3}), and 30 GHz at 114 mA (2×10^{19} cm^{-3}). Fig. 6 shows the measured f_{3dB} as a function of the square root of the photon density for samples A, C and E. f_{3dB} demonstrates the predicted linear dependence on $S^{1/2}$ up to $S \approx 10^{15}$ cm^{-3}, and saturates at higher photon densities. The higher slope observed in the doped lasers is due to the enhanced $\partial g/\partial n$ and to the higher modal gain at threshold (equation 1). The lines in Fig. 6 have been calculated by fitting the experimental f_{3dB} to equation (1) at low S (considering f_{3dB} proportional to f_r) and by including the effect of damping on f_{3dB} through the value of K determined from the RIN measurements. As it can be observed from the fitted curves, the effect of damping on f_{3dB} is noticeable in the undoped lasers at the measurement frequencies, but should not be significant in the doped devices. We attribute the experimentally observed saturation of f_{3dB} to the combination of two effects: i) the appearance of higher order lateral modes in the strongly index-guided mesa structures, and ii) the existence of a bias dependent cut-off frequency in the modulation response, probably caused by the carrier capture/emission processes into the QWs (Weisser et al 1993).

5. SUMMARY

We have investigated the effect of p-doping on the DC and high-frequency properties of MBE grown $In_{0.35}Ga_{0.65}As/GaAs$ MQW lasers. J_{th} and J_{trans} initially decrease at a doping level of 5×10^{18} cm^{-3}, but then increase again as the doping level is increased to 2×10^{19} cm^{-3}. P-doping produces a red shift in λ_{th}, a substantial increase in α_i, but no important changes in η_i. Both experimental and theoretical results indicate a small increase in $\partial g/\partial n$. The experimentally observed decrease in ϵ is attributed to a doping-induced decrease in τ_{in}. The maximum measured f_{3dB} increases with the doping level, reaching 30 GHz in the most heavily doped devices.

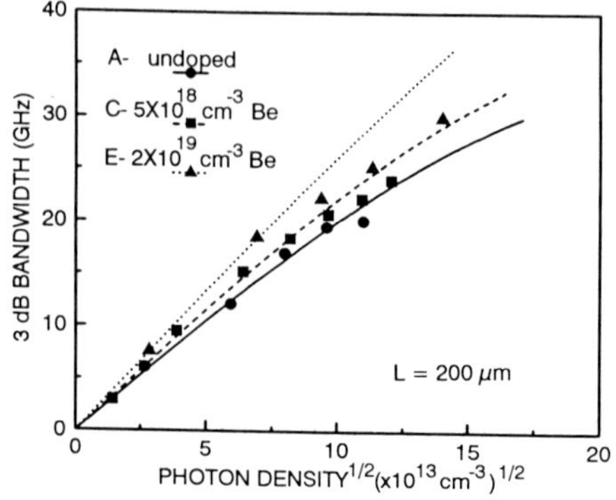

Fig 6 Experimental (symbols) and fitted (lines) 3 dB modulation bandwidth vs square root of the photon density in the laser cavity.

ACKNOWLEDGEMENTS

The authors wish to thank C. Hoffmann, W. Jantz, B. Matthes, and K. Räuber for their assistance with sample preparation and characterization. We also wish to thank G. Grau and H. Rupprecht for their continuing support and encouragement. This work was supported by the Bundesministerium für Forschung and Technologie.

REFERENCES

Agrawal G P, 1990 Apl. Phys. Lett. 57 1.

Aoki M, Uomi K, Tsuchiya T, Suzuki M, and Chinone N, 1990 Electron. Lett. 26 1841.

Blood P, Colak S, and Kucharska A I, 1988 IEEE J. Quantum Electron, QE-24 1539.

Kleinman D A and Miller R C, 1985 Phys. Rev. B 32 2266.

Lealman I F, Cooper D M, Perrin S D, and Harlow M J, 1992 Electron Lett. 28 1032.

Nagarajan R, Fukushima T, Ishikawa M, Bowers J E, Geels R S, and Coldren L A, 1992 IEEE Photonics Technol. Lett. 4 121.

Offsey S D, Schaff W J, Tasker P J, and Eastman L F, 1990 IEEE Photon. Technol. Lett. 2 9.

Olshansky R, Hill P, Lanzisera V, and Powazinik W, 1987 IEEE J. Quantum Electron, QE-23 1410.

Ralston J D, Gallagher D F G, Tasker P J, Zappe H P, Esquivias I, and Fleissner J, 1991 Electron. Lett. 27 1720.

Ralston J D, Weisser S, Esquivias I, Larkins E C, Rosenzweig J, Tasker P J, and Fleissner J, 1993a IEEE J. Quantum Electron, QE-29 1648.

Ralston J D, Larkins EC, Rothemund W, Esquivias I, Weisser S, Rosenzweig J, and Fleissner J, 1993b J. Cryst. Growth 127 19.

Ralston J D, Weisser S, Esquivias I, Schönfelder A, Larkins E C, Rosenzweig J, Tasker P J, Maier M, and Fleissner J, 1993c to be published in Mat. Sci. Engineering B.

Ridley B K, 1990 J. Appl. Phys. 68 4667.

Su C B, Lange C H, Kim C B, Lauer R B, Rideout W C, and LaCourse J S, 1992 IEEE J. Quantum Electron, QE-28 118.

Tsukada T, Ito R, Nakashita H, and Nakada O, 1973 IEEE J. Quantum Electron, QE-9, 356.

Uomi K, 1990 Jpn. J. Appl. Phys. 29 81.

Uomi K, Mishima T, and Chinone N, 1990 Jpn. J. Appl. Phys. 29 88.

Weisser S, Ralston J D, Larkins E C, Esquivias I, Tasker P J, Fleissner J, and Rosenzweig J, 1992 Electron. Lett. 28 2141.

Weisser S, Tasker P, Ralston J R, and Rosenzweig J, 1993 to be published in Proc. IEDM.

High-frequency modulation of a QW diode laser by dual modal gain and pumping current control

Vera B. Gorfinkel and Günter Kompa
Fachgebiet Hochfrequenztechnik, Universtät Kassel, Wilhelmshöher Allee 73, D-34121 Kassel, Germany
Tel.: (0561)8046329, Fax:(0561)8046529

S.A. Gurevich, G.E. Shtengel and I.E. Chebunina
A.F. Ioffe Physico-Technical Institute
Politechnicheskaya 26, 194021 St.-Petersburg, Russia
Tel.: (812)2479391, Fax: (812)2471017

ABSTRACT: A novel dual modulation technique is proposed and realized for a new four terminal QW diode laser structure. The simultaneous output modulation by modal gain and pumping current density **(G&J)** control was investigated experimentally.
Applying a step-like electric signal with a 20 ps rise-time to the side contacts resulted in the laser switching-off time of the same value.
The dynamic laser parameters were extracted from the relaxation oscillation measurements, and simulation of the laser response to the dual **G&J** modulation was carried out.
A 3dB bandwidth as broad as 60 GHz has been attained at moderate laser output powers. In millimeterwave region the laser output response decays as $\mathbf{1/\omega}$.

1. INTRODUCTION

Previously 'V. Gorfinkel and S. Luryi (1993), we considered theoretically a new method of dual modulation of semiconductor lasers. The key idea is to control the laser with an additional high-frequency signal, varied simultaneously with the pumping current $\boldsymbol{I}$. The additional signal can be any one of the physical parameters influencing the optical wave in the laser cavity, such as the gain $\mathbf{g}$, the confinement factor $\mathbf{\Gamma}$, photon lifetime τ_{ph}, etc. Although controlling such parameters may not be as technologically straightforward and natural as modulating the pumping current, technical feasibility of several such schemes is not in doubt, since most of the required elements have been demonstrated in a different context. (Dual laser modulation with a parameter **X** varied together with pumping current $\boldsymbol{I}$ will be referred to as a ***I&X*** dual modulation scheme.) In DFB lasers for coherent optical communications 'M.-C. Amann et al. (1989), it is possible to vary the optical path in the cavity simultaneously with the pumping current, thus implementing the (**I&λ**) dual scheme. Feasibility of the ($\boldsymbol{I\&\tau_{ph}}$) scheme follows from the recently demonstrated electro-optic control of DBR mirror reflectivity in surface-emitting microcavity lasers 'O. Blum et al.(1991). It was shown that for (**I&g**) scheme and ($\boldsymbol{I\&\tau_{ph}}$) scheme 'E. Avrutin et al.(to be published), the dual modulation allows suppressing the relaxation oscillations for an arbitrary shape of the pumping current signal ***I(t)***. Because of that, the rate of information coding can be enhanced to about 80 Gbit/sec. Moreover, dual modulation allows to maintain a *linear* relationship between *I(t)* and the output optical power ***P(t)***, to enhance the modulation frequency, and to achieve pure AM or pure FM modulation regimes of the laser output radiation.

In this paper we shall report on the first experimental realization of a single QW laser structure providing the output modulation by dual modal gain and pumping current density control. This laser is shown to be promising for extremely broad-band operation with the bandwidth exceeding 60 GHz.

2. STEADY-STATE LASER CHARACTERISTICS

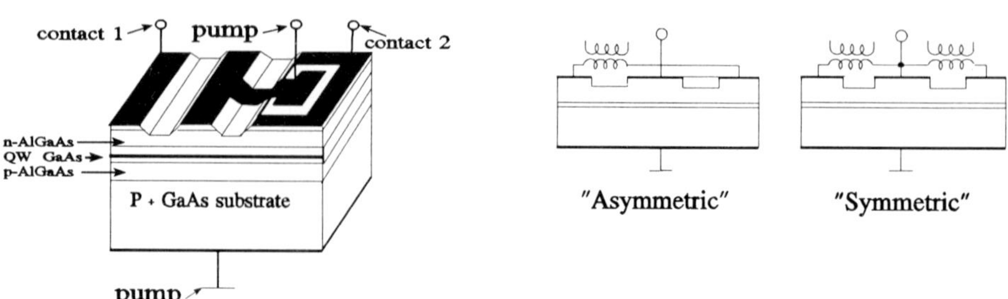

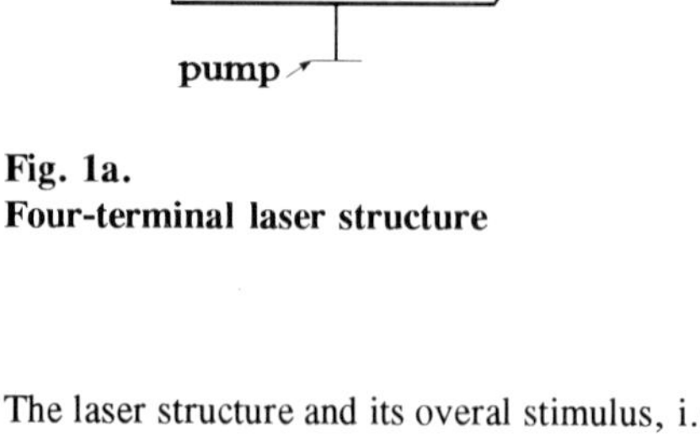

Fig. 1a.
Four-terminal laser structure

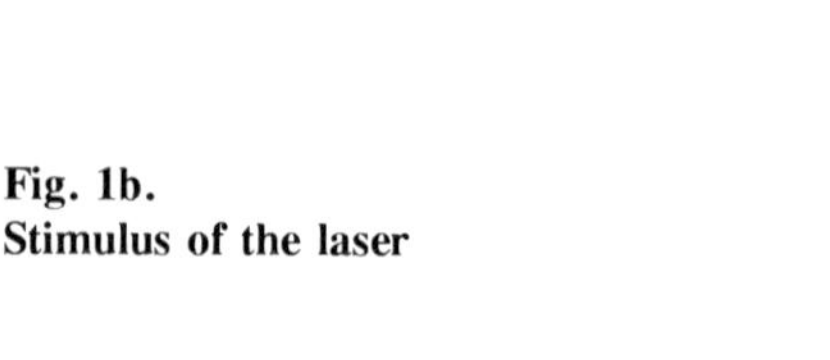

Fig. 1b.
Stimulus of the laser

The laser structure and its overal stimulus, i.e. means for pumping and lateral potential control are shown schematically in Figure. 1. The ridge-guide laser was fabricated on AlGaAs/ GaAs SQW (L_z=100 Å) MBE grown wafer. In the laser fabrication two parallel grooves, 5μm in width, were dry-etched to define a 6 μm ridge-guide. After etching and dielectric layer deposition, the top of the ridge as well as the side areas and the substrate surface were metallized to form a four-terminal device. While the central electrode was used for laser pumping, the side contacts were employed for the modulation. The potentials applied to the side contacts provided additional forward or reverse bias of the areas outside the ridge. To prevent the device from heating damage, all following measurements were at first carried out with 100 ns pumping current pulses at 10 kHz pulse repetition rate. The laser steady-state **P-I** characteristics are shown in Figure 2 for both types of the laser connections ("*asymmetric*" and "*symmetric*"). Varying the side potentials one can influence significantly the laser threshold current. In the "*symmetric*" connection scheme threshold current can be even decreased (see Figure 2, right-hand side picture).

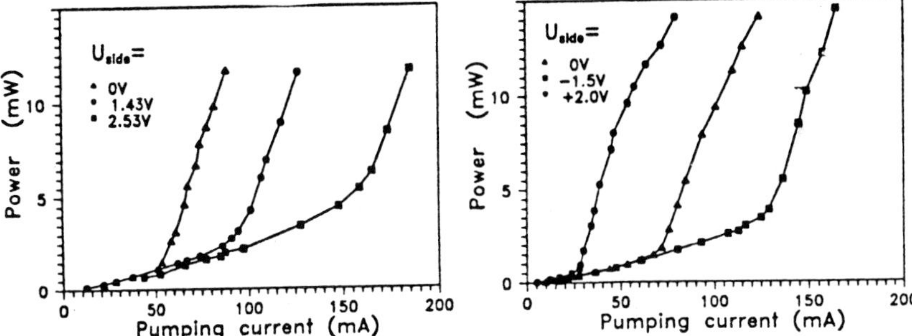

Fig. 2.
Steady-state P-I characteristics of the laser

Physical mechanism of the threshold changes due to the side potential varying consists in a lateral carrier redistribution in the laser active layer. It is demonstrated in a particularly transparent form by the lateral near-field radiation patterns obtained at different potentials at the side contacts with dc pumping below the laser threshold (see Figure 3).

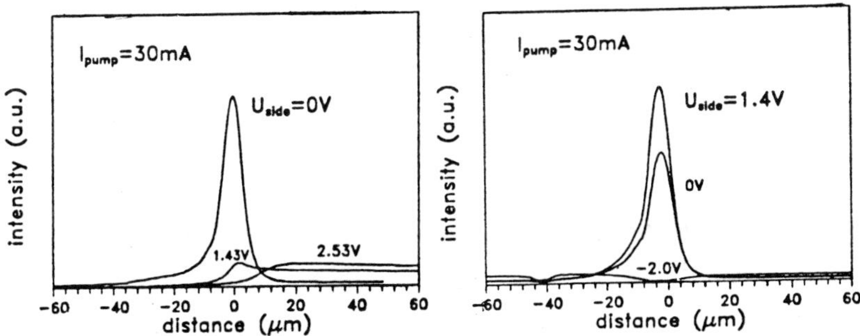

Fig. 3.
Near-field patterns at different side potentials

In the case of the "*symmetric*" connection applying positive potentials to the side contacts decreases the emitting area width and, therefore, reduces the laser threshold current. Thus, varying the side potentials one can control both pumping current density **J(x)** and the optical gain profile **g(x)**. Note that the lateral optical mode profile S(x) determined only by the ridge-guide geometry is fixed. Thus,varying the side potential we change, in fact, the modal gain **G:**

$$G=\left[\int_{-\infty}^{\infty} g(x)\ S(x)\ dx\right] \times \left[\int_{-\infty}^{\infty} S(x)\ dx\right]^{-1}$$

3. DYNAMIC LASER CHARACTERISTICS

3.1 Extraction of the laser intrinsic dynamic parameters

Figure 4 shows the intrinsic laser dynamic parameters at pure pumping current modulation extracted from the laser response to a step-like current switch-on pulse of 120 ps rise-time.

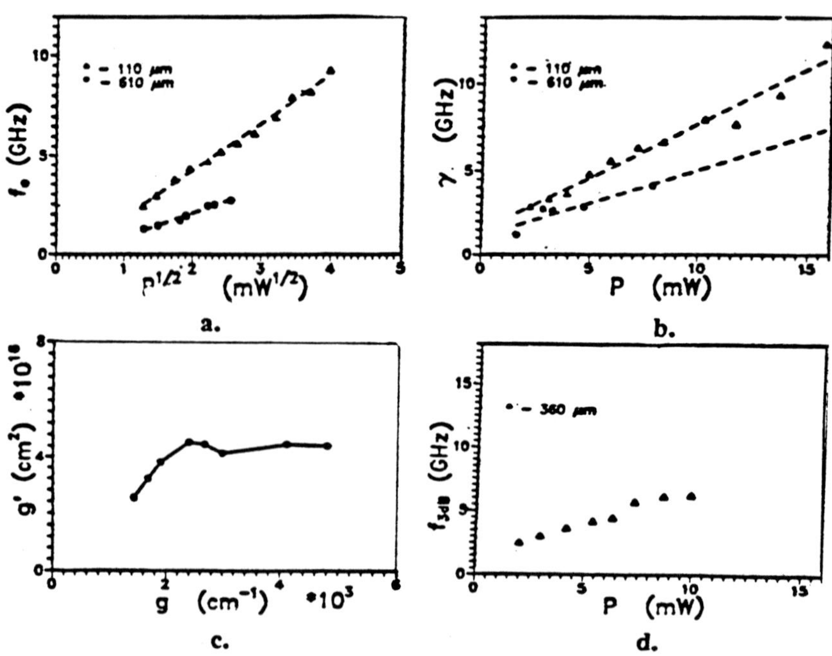

Fig. 4.
Intrinsic laser dynamic parameters at pure pumping current modulation
(resonant frequency (a), damping factor (b), differential gain (c), 3-dB frequency (d))

It is evident that at moderate pumping currents (cw power lower than 10 mW) the relaxation oscillation frequencies do not exceed 5-6 GHz for all cavity lengths being 110 μm up to 610 μm. The 3dB bandwidth for the laser of typical 360 μm cavity length is limited to approximately 5 GHz.

3.2 Laser response to the side potential signal

The dynamic behaviour of the laser subject to the side voltage was studied by applying a step-like electric signal (20 ps rise-time) to the side contacts. The observed switching-off time of the laser output was practically of the same value (**≈20 ps**) (see streak-camera trace, slope A in Figure 5).

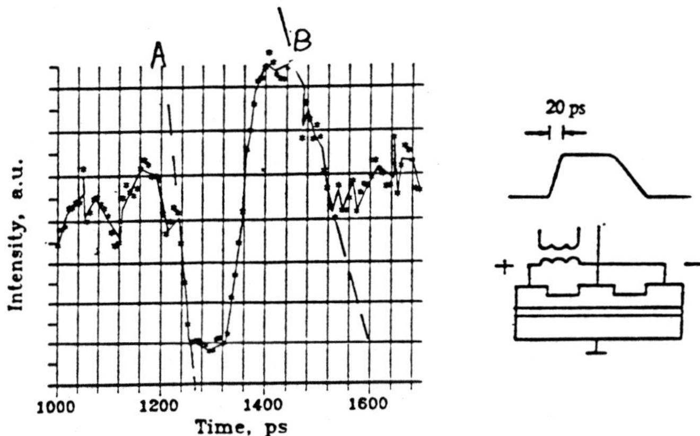

Fig. 5.
Streak-camera trace of the laser response to 20 ps step-like side voltage
(slope A- the laser switching-off subject to the side voltage, slope B- the first relaxation oscillation maximum)

It should be emphasized, that being subjct to the direct modulation (with comparable output power and without any side electrical stimulus) this laser has a relaxation oscillation frequency of only 5 GHz and switching-off time of about 100 ps (see Figure 5, slope B).

3.3 Simulated bandwidth limitation for dual **G&J** modulation

To evaluate principle limitations of the proposed modulation technique we calculated numerically the laser response to the dual **G&J** variation using the dynamic parameters extracted from the relaxation oscillation measurements (see Figure 4). The 3dB bandwidth as broad as 60 GHz has been found for moderate values of the laser output power (Fig.6).

In the millimeterwave range the 3 dB bandwidth is governed by the **1/ω**-like decay of the laser response in contrast to well known **1/ω²**-dependence for direct modulation.

The obtained results demonstrate the considerable advantage of the modal gain modulation with respect to the direct modulation.

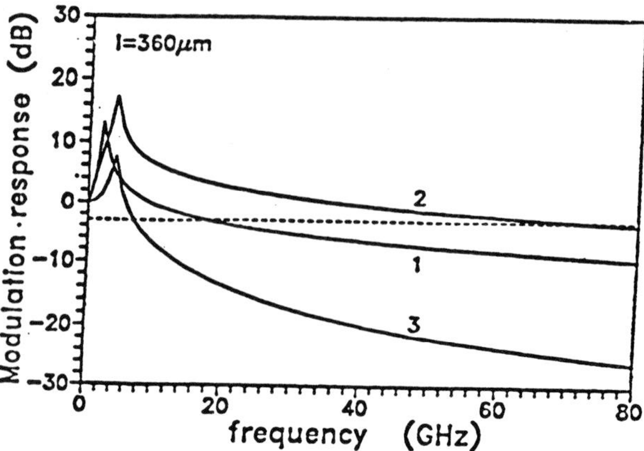

Fig. 6.
Calculated small-signal modulation responses
(modal gain modulation: Po=2 mW (1), 6 mW (2); pure direct modulation: Po=6 mW (3))

References

M.-C. Amann, S. Illek, C. Schanen, W. Thulke, (1989), IEEE Photonics Techn. Lett., 1, pp. 253-254

E.A. Avrutin, V.B. Gorfinkel, S. Luryi, and K.A. Shore, (to be published)

O.Blum, J.E. Zucker, T.H. Chiu, M. Divino, K.L. Jones, S.N.G. Chu, and T.K. Gustafson, (1991), Appl. Phys. Lett., 59, pp 2971-2973

V.B. Gorfinkel, S. Luryi, (1993), *Appl. Phys. Lett.*, ***62***, pp. 2923- 2925

Paper presented at the Int. Symp. GaAs and Related Compounds, Freiburg, 1993

Characteristics of submilliamp tunable three-terminal vertical-cavity laser diodes

T. Wipiejewski[1], K. Panzlaff[1], E. Zeeb[1], B. Weigl[1], H. Leier[2], and K.J. Ebeling[1]

University of Ulm, Department of Optoelectronics, 89069 Ulm, Germany[1]
Daimler Benz Research Center, 89013 Ulm, Germany[2]

ABSTRACT: We have fabricated wavelength tunable vertical-cavity laser diodes in 2D arrays by molecular beam epitaxy, proton implantation, and wet chemical etching. Devices of 8 μm active diameter exhibit threshold currents of 650 μA for cw and 600 μA for pulsed operation. Output power is up to 170 μW cw and 400 μW pulsed. The emission wavelength is about 960 nm. Three terminals for each element supply two separate currents for independent control of output power and emission wavelength. A record wide continuous wavelength tuning range of 5.6 nm is demonstrated with just 1 mA tuning currents.

1. INTRODUCTION

Arrays of vertical-cavity lasers are attractive light sources for multi-channel short distance optical fiber networks due to their potential low fabrication and packaging costs. The very low threshold currents as obtained by Geels et al. (1990) favor simple driving circuits and low power dissipation. The most promising advantage of vertical-cavity laser diodes is the high coupling efficiency of over 90% into single-mode optical fibers with large lateral alignment tolerances as demonstrated by Wipiejewski et al. (1993b) making massive parallel optical interconnects in 2D systems feasible. We have fabricated vertical-cavity laser diode 2D arrays with independent wavelength tuning capability of each element exhibiting threshold currents as low as 650 μA (Wipiejewski et al. 1992). Record wide 5.6 nm continuous wavelength tuning corresponding to 1800 GHz bandwidth is demonstrated by applying separately controllable tuning currents of less than 1 mA to single elements. This figure is higher than previously reported continuous wavelenth tuning ranges in vertical-cavity lasers by Chang-Hasnain et al. (1991) and Yokouchi et al. (1992). Independent control of output power and emission wavelength is provided by two separate currents supplied by three terminals of each element. The wide tuning range makes the devices versatile for 2D parallel wavelength division multiplexing systems and might be also used for wavelength stabilization or equalization within an array.

2. DEVICE FABRICATION

The Nomarski interference micrograph of a 4° angle lapped wafer in Fig. 1 shows the epitaxial layers of the vertical-cavity laser grown by molecular beam epitaxy on n-GaAs substrate. The Bragg reflectors consist of 67 nm GaAs, 5 nm AlGaAs, 80 nm AlAs stacks. The top reflector of 20 periods is p-doped (Be, $p \approx 5 \cdot 10^{18}\text{cm}^{-3}$) and the bottom reflector of 24.5 periods is n-doped (Si, $n \approx 5 \cdot 10^{18}\text{cm}^{-3}$). The one-sided $Al_{0.4}Ga_{0.6}As$ intermediate layers reduce the voltage drop at the Bragg reflector heterojunctions at forward bias by a factor of about three to four compared to abrupt GaAs-AlAs heterojunctions with the same doping levels. A 45 nm GaAs layer is added to the topmost GaAs Bragg reflector layer providing phase matching to the TiAu metal top reflector. The one wavelength thick central region contains three 8 nm thick strained $In_{0.2}Ga_{0.8}As$ quantum wells embedded in 10 nm GaAs barriers and surrounded by 40 nm GaAs and 75 nm $Al_{0.4}Ga_{0.6}As$ layers for efficient carrier confinement. A slight p-type modulation doping with $p \approx 1 \cdot 10^{17}\text{cm}^{-3}$ is applied to the GaAs barriers except for 3-4 monolayers adjacent to the undoped active quantum wells. The p-type doping should increase the normally slow hole diffusion by the electrical field of the ionized acceptor atoms if lateral inhomogeneous carrier distributions due to spatial hole burning effects are present. Therefore differential quantum efficiency and high frequency characteristics should be improved (Ralston et al. 1992).

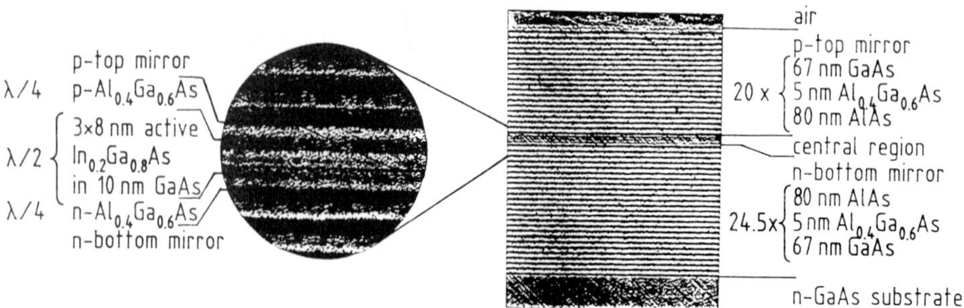

Fig. 1: Normasky micrograph of a 4° angle lapped wafer showing the epitaxial layers of a vertical-cavity laser.

The spacing between individual elements of the 2D laser array is about 60 μm. Fig. 2 illustrates the double-mesa structure of the three-terminal vertical-cavity laser diode. Lateral device formation starts with proton implantation with ion energies of 300, 220, 150, 90, and 40 keV at doses of $1 \cdot 10^{15}$, $8 \cdot 10^{14}$, $6 \cdot 10^{14}$, $5 \cdot 10^{14}$, and $4 \cdot 10^{14}$ cm^{-2}. TiAu contacts are formed over non-implanted areas serving as mask for top reflector mesa etching with 20 NH_4OH (25%) : 2 H_2O_2 (30%) : 100 H_2O. This solution etches AlAs and GaAs layers with comparable etch rates resulting in relatively low lateral underetching. To suppress the underetching effect even more the entire mesa etching is divided into three steps where 5, 6, and 6 periods of the top reflector are removed, respectively. According to the technique described by Wipiejewski et al. (1993a) etching is monitored in-situ providing precise control of etch depth. Resist is spun on the sample between successive etching steps and the TiAu contacts serve as individual self-aligned photolitographic masks in the subsequent flood exposure. The resist protects mesa sidewalls from etching. Mesa etching is interrupted in the AlAs layer of the seventeenth Bragg reflector period which is then selectively removed by 2% HF. This step equalizes latterally inhomogeneous etch depths. A TiAu p-contact is angle evaporated with rotating substrate providing a self-aligned sufficient overlap of metal and unimplanted semiconductor. The fully processed top Bragg reflector mesa can be seen in the scanning electron micrograph of Fig. 3. To define individual laser diodes broad area mesas of about 50 μm diameter are formed by wet chemical etching the TiAu and the semiconductor layers down to the bottom reflector where broad area AuGeNi n-contacts are deposited. A quarter-wavelength thick SiO_x layer is evaporated onto the substrate backside to suppress parasitic reflections.

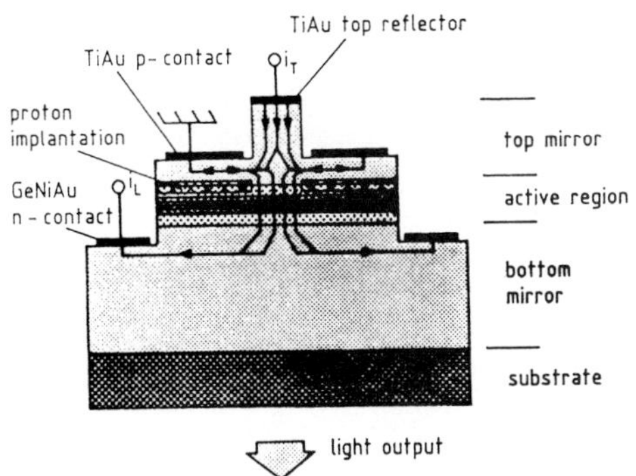

Fig. 2: Schematic of the wavelength tunable vertical-cavity laser diode with laser current i_L and tuning current i_T.

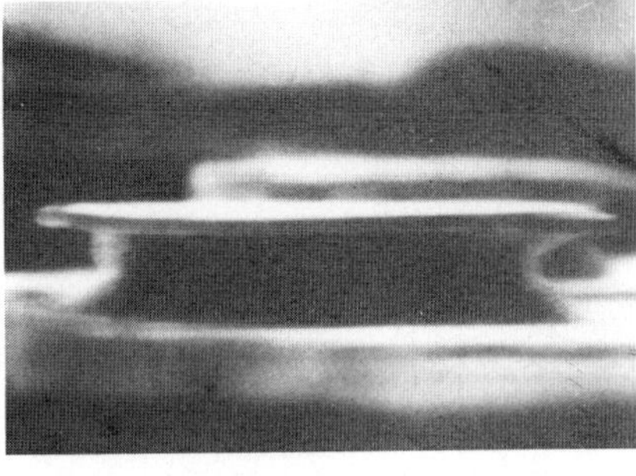

Fig. 3: Scanning electron micrograph of the fully processed top Bragg reflector mesa.

Laser current flow i_L is laterally inwards from the p-contact in the first three periods of the top Bragg reflector and vertical through the non-implanted conductive channel into the active layer and laterally outwards in the n-bottom reflector to the broad area n-contact. The n-contact is shared by all lasing elements of the 2D array. As indicated in Fig. 2 tuning current i_T is supplied by the metal top reflector as third terminal and flows across the heterojunctions of the top reflector mesa to the common p-contact. Laser currents i_L and tuning currents i_T are controlled independently of each other and also separately for each single element in the 2D laser array providing individual control of output power and emission wavelength for all channels operating in parallel.

3. OUTPUT CHARACTERISTICS

Fig. 4 shows the light output versus current characteristics of the 8 μm active diameter device under cw and pulsed (50 ns pulse width, 200 kHz repetition rate) operation for vanishing tuning current. The cw threshold current is as low as 650 μA which is to our knowledge a record value for vertical-cavity lasers. The threshold current density is about 1.3 kA/cm^2. Output power saturates at about 170 μW for a non-heat sinked device. For pulsed excitation the threshold current is about 600 μA and the output power is up to 0.4 mW. The differential quantum efficiency is 5% and 6% for cw and pulsed operation, respectively. The results for pulsed excitation indicate that higher cw output power levels are feasible by applying proper heat sinking to the device. Lasing occurs in the center of the Bragg reflector stop band at about 960 nm wavelength. Emission is single longitudinal mode and stable in the fundamental transversel mode under cw and pulsed operation for all driving currents. Laser current induced heating causes a red-shift of emission at a rate of $\Delta\lambda/\Delta i_L = 0.7$ nm/mA. The bias voltage at threshold of just less than 5 V is relatively large due to non-optimized lateral series resistances and the voltage drop at the heterointerfaces. The latter can be reduced by optimized Bragg reflector fabrication as demonstrated by Chalmers et al. (1993). Despite the non-optimized conditions the applied electrical power at threshold is as low as 3.2 mW.

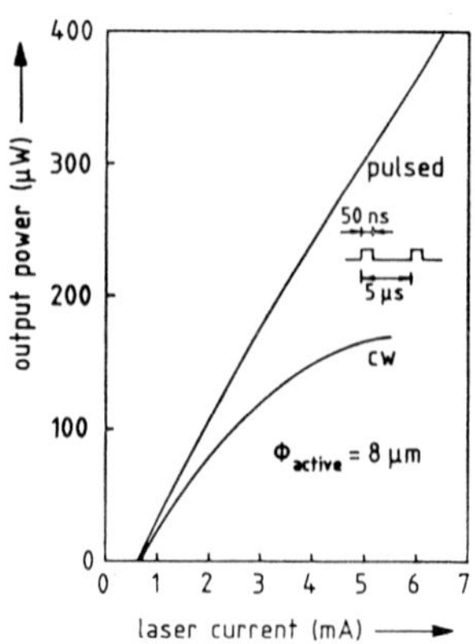

Fig. 4: Cw and pulsed light output characteristics of 8 μm active diameter laser diode for vanishing tuning current.

4. WAVELENGTH TUNING

Tuning current flow in the small size top Bragg reflector mesa causes extra local heating concentrated in the small volume optical laser cavity. Thus emission wavelength is continuously red shifted due to the thermally induced increase of refractive index. Fig. 5 shows the emission wavelengths as functions of tuning current i_T for various fixed laser currents i_L. A record continuous wavelength tuning range of 5.6 nm is achieved with a tuning current of $i_T = 1$ mA and a fixed laser current of $i_L = 1.4$ mA. The corresponding total frequency tuning range is 1800 GHz allowing wavelength division multiplexing of 180 independent channels with 10 GHz bandwidth. The normalized wavelength shift per tuning current is $\Delta\lambda/\Delta i_T = 6.0$ nm/mA indicating efficient heating of the small thermal load. A temperature rise of $\Delta T = 80$ K can be derived for maximum tuning as the emission wavelength shift with temperature is $\partial\lambda/\partial T = 0.7$ Å/K measured at room temperature. No temporal degradation of device performance due to the high operating temperature has been observed in the experiments.

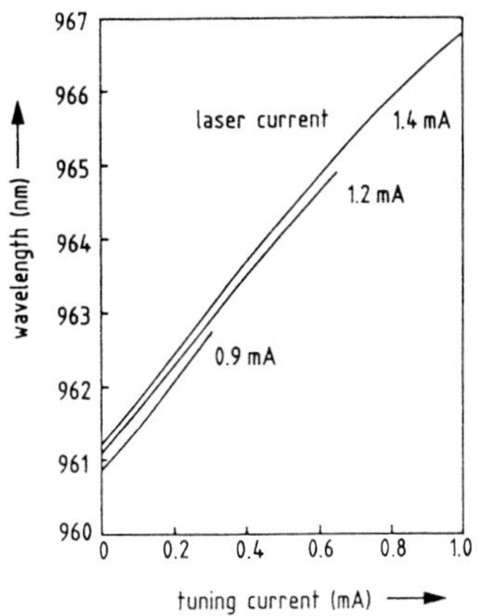

Fig. 5: Emission wavelengths as functions of tuning current for various fixed laser currents.

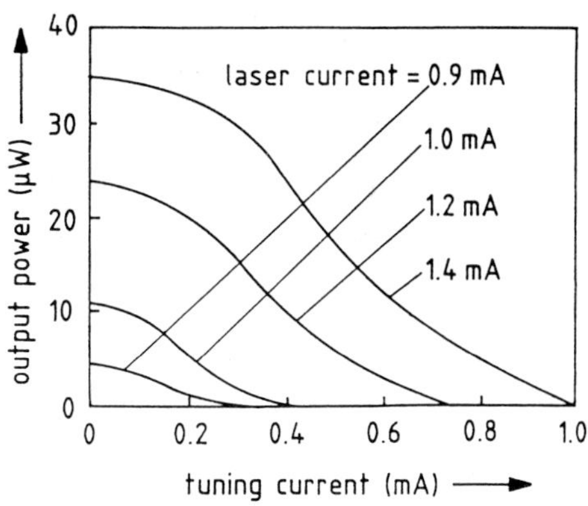

Fig. 6: Cw output power levels of various laser currents with growing tuning current.

With growing tuning current output power may remain constant, increase, or decrease depending on the location of the lasing mode relative to the spectral gain maximum. In Fig. 6 the output power levels decrease with increasing tuning current for fixed laser currents. For many applications this behavior is desirable since the output power decrease can be fully compensated by increasing the laser current which additionally enhances the tuning effect.

In Fig. 7. tuning current pulses of 1 μs duration and various amplitudes are applied to the laser operating at a constant laser current of 1.9 mA. The emission wavelength starts to increase with no observable delay and approaches equilibrium after a few μs. After tuning current turn-off the emission wavelength exponentially decreases to the value for vanishing tuning current. From rise and fall times a $1/e$-time constant of less than 1 μs for the thermally induced tuning is inferred. The time constants for emission red shift due to laser and tuning current heating are approximately equal allowing emission wavelength stabilization fairly independent of the output power level.

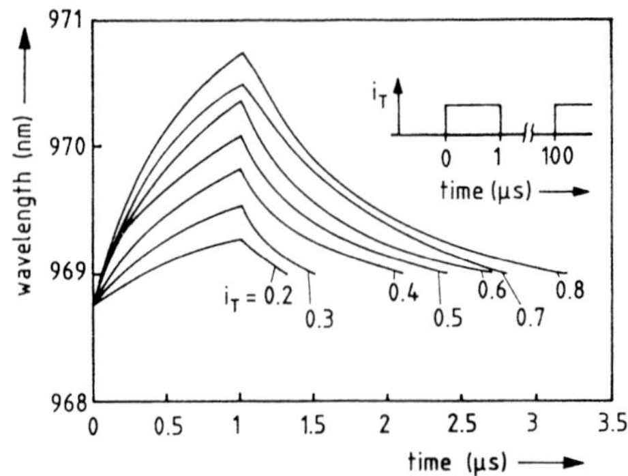

Fig. 7: Tuning dynamics for 1 μs tuning current pulses i_T of various amplitudes.

5. CONCLUSION

We have realized three-terminal vertical cavity laser diodes with record low threshold currents of 650 μA cw and 600 μA pulsed. Maximum output power is 0.4 mW for pulsed excitation and 170 μW for cw operation indicating that proper heat sinking will rise cw output power levels. A record 5.6 nm continuous wavelength tuning range is achieved with tuning currents just below 1 mA. Independent control of output power and emission wavelength is provided for each individual lasing element in 2D arrays. The tuning capability makes the devices very versatile for 2D laser arrays, wavelength division multiplexing systems, fiber networks, or optical interconnects.

ACKNOWLEDGEMENT

Support from the BMFT and DFG is gratefully acknowledged.

REFERENCES

Chalmers S A, Lear K L, Killeen K P, 1993 Appl. Phys. Lett. 62 pp. 1585-1587

Chang-Hasnain C J, Harbison J P, Zah C E, Florez L T, Andreadakis N C, 1991 Electron. Lett. 27 pp. 1002-1003

Geels R S, Coldren L A, 1990 Appl. Phys. Lett. 57 pp. 1605-1607

Ralston J D, Weisser S, Larkins E C, Esquivias I, Tasker P J, Fleissner J, Rosenzweig J, 1992 Proc. 13th IEEE Int. Semicond. Laser Conf., Takamatsu, Japan, PD-5

Wipiejewski T, Panzlaff K, Zeeb E, Ebeling K J, 1992 Proc. 18th Euro. Conf. on Opt. Commun., Berlin, Germany, pp. 903-906

Wipiejewski T, Ebeling K J, 1993a J. Electrochem. Soc. 140 pp. 2028-2033

Wipiejewski T, Panzlaff K, Zeeb E, Weigl B, Ebeling K J, 1993b Proc. 19th Euro. Conf. on Opt. Commun., Montreaux, Switzerland

Yokouchi N, Miyamoto T, Uchida T, Inaba Y, Koyama F, Iga K, 1992 IEEE Phot. Tech. Lett. 4 pp. 701-703

Paper presented at the Int. Symp. GaAs and Related Compounds, Freiburg, 1993

Emission characteristics of proton-implanted vertical cavity laser diodes

B. Möller[1], E. Zeeb[1], R. Michalzik[1], T. Hackbarth[2], H. Leier[2], K.J. Ebeling[1]

University of Ulm, Dept. of Optoelectronics, 89069 Ulm, Germany[1]

Daimler Benz Research Center, 89013 Ulm, Germany[2]

ABSTRACT: We have fabricated and tested planar, proton-implanted InGaAs/GaAs quantum well vertical cavity laser diodes. Minimum threshold currents of 1.4 mA for 10 μm active diameter devices and maximum cw output powers of 20 mW for solder bonded broad area lasers are measured. The modal linewidths linearly increase with inverse modal power at characteristic slopes of about 6 MHz·mW. The residual linewidths due to mainly mode competition noise are 90 MHz for the fundamental and 180 MHz for the first order mode, respectively. The near field patterns and wavelengths of the modes are in good agreement with theoretical calculations.

1. INTRODUCTION

Vertical cavity surface emitting laser diodes (VCSELs) are very attractive for various device applications in future optoelectronic systems. Single longitudinal mode oscillation, alignment tolerant and efficient coupling into single mode fiber (Wipiejewski 1993) and the possibility to produce high bit sequences (Hasnain 1991, Möller 1993) make the devices ideally suited for fiber networks and optical interconnect systems for microelectronic circuits. Two-dimensional arrays are easy to realize (Vakhshoori 1993) and support parallel optoelectronic signal processing. VCSELs with extremely low threshold currents have already been reported (Geels 1990, Wipiejewski 1992) but the maximum output power was limited to several hundred μW.

To enhance the maximum optical output power and the temperature operating range of the devices (Young 1993) an appropriate design of the VCSEL structure and an effective heat sinking of the laser are imperative. For active diameters larger than about 10 μm proton-implanted VCSELs tend to oscillate in several higher transverse modes at driving currents well above threshold.

In this paper we present a simple planar laser structure providing minimum threshold currents of 1.4 mA without heat sinking and maximum cw output powers of 20 mW for broad area devices solder bonded on laser submounts. In addition we perform a detailed experimental and theoretical analysis of emission linewidths and mode patterns for fundamental and higher order transverse mode oscillation.

2. LASER STRUCTURE

The structure under investigation, shown in Fig. 1(a), was grown by molecular beam epitaxy on n-GaAs substrate. The active region contains three 8 nm thick $In_{0.17}Ga_{0.83}As$ quantum wells embedded in 10 nm thick GaAs barriers and surrounded by $Al_{0.3}Ga_{0.7}As$ cladding layers for efficient carrier confinement. The central region is one wavelength thick. The top Bragg reflector is p-doped and consists of 18 quarter wavelength AlAs/GaAs pairs. The bottom one is n-doped and has 22.5 pairs.

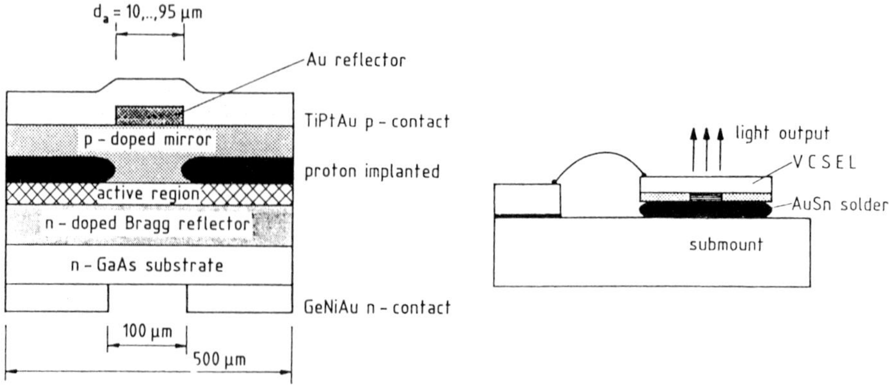

Fig. 1: Structure of VCSEL (a) and VCSEL mount (b).

Both reflectors contain single-sided two 5 nm thick $Al_xGa_{1-x}As$ intermediate layers with Al-contents of x=0.6 and x=0.3, respectively, and are modulation doped with a maximum doping concentration of $9 \cdot 10^{18}$ cm^{-3} to reduce the electrical series resistance at forward bias (Tai 1990, Sugimoto 1992). Lateral current confinement is achieved by proton implantation at varying energies between 40 and 300 keV with a total dose of 10^{15} cm^{-2}. For efficient heat sinking diced broad area VCSELs are solder bonded upside down on molybdenum tungsten laser submounts, as schematically illustrated in Fig. 1(b).

3. OUTPUT CHARACTERISTICS

In Fig. 2 the cw light output characteristics for lasers of different active diameters without heat sinking are shown. Emission wavelength is around 983 nm.

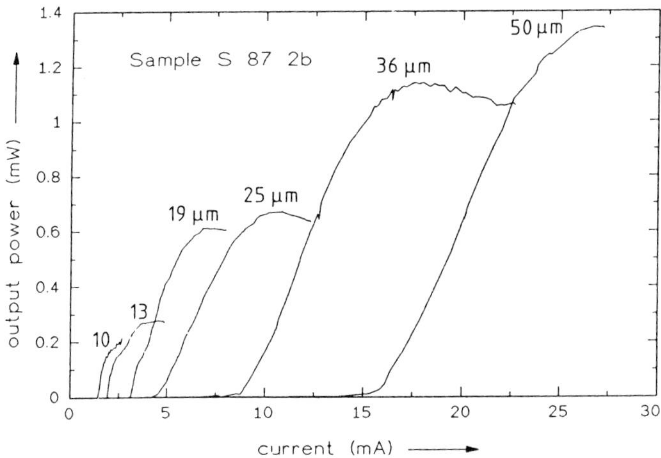

Fig. 2: CW light output characteristics without heat sinking.

Minimum threshold currents of 1.4 mA and a single-sided differential quantum efficiency of 43 % for 10 μm devices are observed. The maximum cw output power of nearly 1.4 mW for 50 μm devices is restricted by a strong heating of the structure mainly due to the still high resistance of the p-doped Bragg reflector. For heat sinked devices the optical power is drastically enhanced as indicated in Fig. 3(a) where we compare the output characteristics of an unbonded and bonded 75 μm device. The maximum power considerably increases from 0.6 mW to 17.4 mW. The observed increase in threshold current for the soldered device is attributed to a deterioration of the upper reflector during soldering. Maximum cw output powers of 18.4 mW at room temperature (20° C) were obtained for 95 μm broad area VCSELs as depicted in Fig. 3(b). An additional slight cooling of the heat sink down to 5° C rises the achievable output power beyond 20 mW.

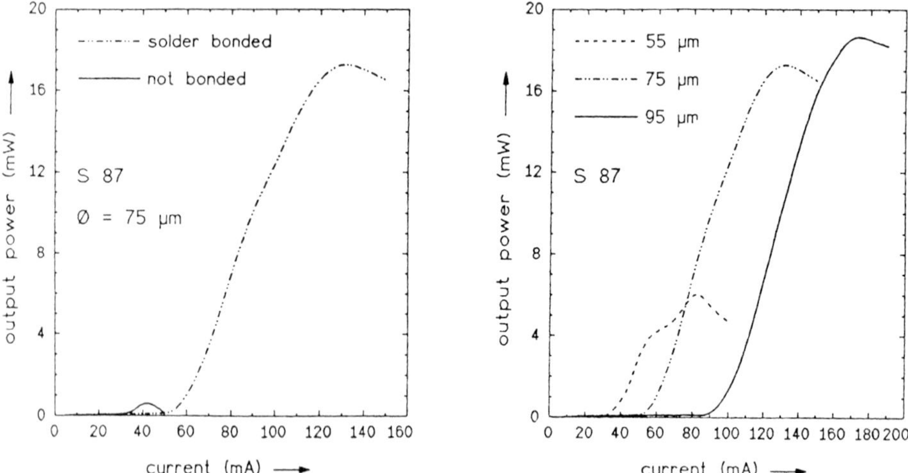

Fig. 3: Comparison of VCSEL output power with and without heat sinking (a). CW output characteristics of heat sinked VCSELs (b).

4. MODE PATTERNS AND LINEWIDTHS

The spectra and corresponding near field patterns for a 25 μm active diameter laser and three different driving currents above threshold are shown in Fig. 4. With increasing current there is a nearly complete transfer of the LP_{01} fundamental mode power into the LP^*_{11} mode. At even higher currents further transverse modes begin to oscillate.

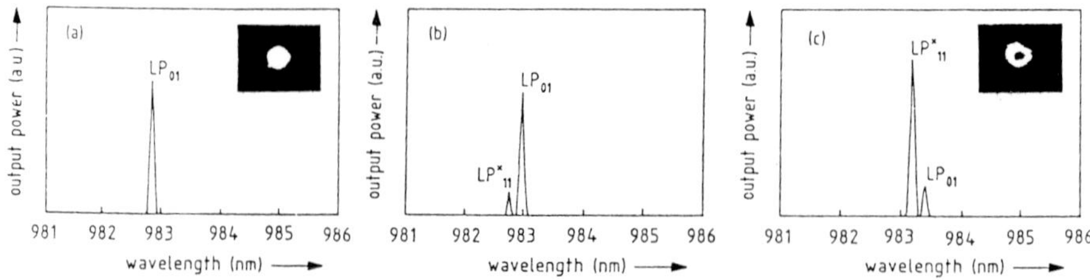

Fig. 4: Spectra and corresponding near field patterns for LP_{01}- and LP^*_{11}-modes at laser currents of 5.2 mA (a), 5.7 mA (b), and 6.3 mA (c) in a 25 μm device.

Fig. 5(a) illustrates the transverse mode spacings relative to the fundamental mode. Theoretical data, obtained with a scalar radial transfer matrix method for an equivalent rectangular gain profile of 14 μm diameter, are in good agreement with the experimentally determined values. The corresponding calculated radial LP_{lp} intensity profiles, normalized to constant optical power, are depicted in Fig. 5(b).

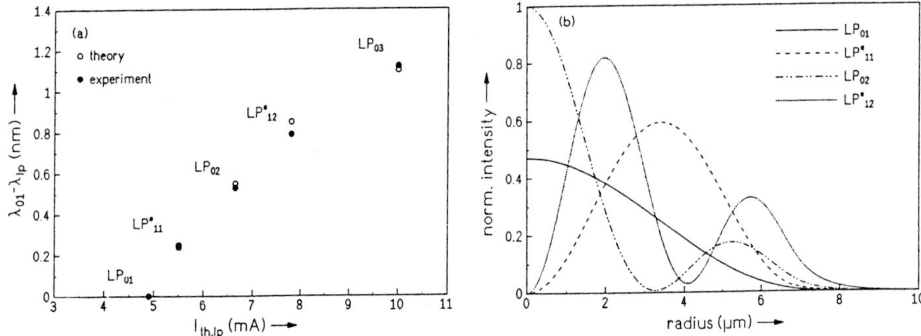

Fig. 5: Threshold currents and emission wavelengths for a 25 μm diameter laser (a) and calculated radial intensity profiles (b).

Measurements of the two lowest order modes emission linewidths in Fig. 6 show a linear increase with inverse modal output power, as theoretically expected. The characteristic slope of about 6 MHz·mW is comparable to DFB laser diodes. In the single mode emission regime a minimum linewidth of less than 200 MHz is achieved for 60 μW output power. However, when the LP^*_{11} mode begins to oscillate, a broadening occurs which is related to the power competition effect illustrated in Fig. 4.

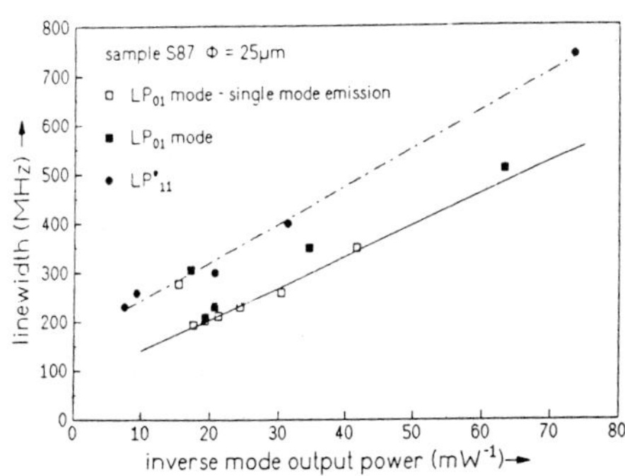

Fig. 6: Linewidths of LP_{01}- and LP^*_{11}-modes for a 25 μm diameter VCSEL.

5. CONCLUSION

We have fabricated planar proton-implanted vertical cavity surface emitting laser diodes with varying active diameters. Threshold currents as low as 1.4 mA and single-sided differential quantum efficiencies of 43 % were measured without heat sinking. Maximum output power exceeding 20 mW was obtained by applying proper heat sinking to the devices. In addition we clarified the mode structure and linewidth behavior of VCSELs. The linewidth shows linear increase with inverse modal output power at a characteristic slope of 6 MHz · mW. The residual linewidths of the fundamental mode and the first order mode are 90 MHz and 180 MHz, respectively. Experimental results are in good agreement with theoretical calculations.

ACKNOWLEDGEMENTS

Financial support from the Bundesministerium für Forschung und Technologie and the Deutsche Forschungsgemeinschaft is gratefully acknowledged.

REFERENCES:

Geels R S and Coldren L A 1990 Appl. Phys. Lett. **57** pp. 1605-1607.

Hasnain G, Tai K, Dutta N K, Wang Y H, Wynn J D, Weit B E, Cho, A Y 1991 Electron. Lett. **27** pp. 915 - 916.

Möller B, Fiedler U, Wipiejewski T, Zeeb E, Panzlaff K and Ebeling K J 1993 ECOC Montreux in press.

Sugimoto M, Kosaka H, Kurihara K, Ogura I, Tai T and Kasahara K 1992 Electron. Lett. **28** pp. 385-387.

Tai T, Yang L, Wang Y H, Wynn J D and Cho A Y 1990 Appl. Phys. Lett. **56** pp. 2496-2498.

Vakhshoori D, Wynn J D, Zydzik G J and Leibenguth R E 1993 Appl. Phys. Lett. **62** pp. 1718-1720.

Wipiejewski T, Panzlaff K, Zeeb E and Ebeling K J 1992 ECOC Berlin pp.903-906.

Wipiejewski T, Panzlaff K, Zeeb E, Weigl B and Ebeling K J 1993 ECOC Montreux in press.

Young D B, Scott J W, Peters F H, Thibeault B J, Corzine S W, Peters M G, Lee S-L and Coldren L A 1993 IEEE Phot. Technol. Lett. **5** pp. 129 - 132.

Inst. Phys. Conf. Ser. No 136: Chapter 5
Paper presented at the Int. Symp. GaAs and Related Compounds, Freiburg, 1993

Novel punch-through heterojunction phototransistors for lightwave communications

Y. Wang, E. S. Yang, and W. I. Wang

Department of Electrical Engineering, Columbia University, New York, NY 10027

Abstract: In this paper, we propose and demonstrate a novel punch-through heterojunction phototransistor (HPT). The base of the transistor is lightly doped and completely depleted under the operating condition. The collector bias current can be applied without the base terminal. The transistors exhibited optical conversion gain as high as 1240 at an incident optical power as low as 0.5 μW. The transient measurements showed that the transistor has a higher response speed than that of conventional two or three terminal HPTs. The results of our simulation show that the punch-through HPTs have a much lower noise characteristics than conventional HPTs. The principle reported here can be applied to low noise HPTs made from other heterostructure material systems, such as AlGaSb/GaSb and InP/InGaAs, for long wavelength optical communications.

Introduction

Photodetectors are very important components in light wave communication systems. In order to achieve satisfactory overall system performance, several specific properties are required for semiconductor photodetectors used in these systems, such as high responsivity at the operating wavelength of the system, minimum introduction of noise by the detection process, and easy to be integrated with amplifier and other functional circuits. In addition to p-i-n photodiodes and avalanche photodiodes, heterojunction phototransistors (HPTs) are also capable of satisfying many of the detector requirements of light wave communication systems. HPTs can provide large optical gain through the transistor action without excess noise, and hence HPTs may have better responsivity and detectivity than both p-i-n diodes and avalanche photodiodes. However, most of the HPTs studied to date were two terminal devices each with a floating base. Due to the low quiescent bias current provided by the incident optical power, the transistor will have a longer charging time for the emitter junction capacitance and a lower f_T. In turn this lower f_T results in inferior signal-to-noise ratio in high speed applications. In order to improve the performance of HPTs, the base terminal was provided and enhanced performance was demonstrated for the first time by Fritzsche et al. [Fritzsche et al, 1981], who reported the operation at 200 MHz with an input optical power of 15 μW. Very recently, Chandrasekher et al reported the thorough study of a three terminal HPT[chandrasekher et al, 1991].and the implementation of a three terminal HPT in an all-biporlar photoreceiver and demonstrated the successful operation at 2 GHz [chandrasekher et al, 1992].

Although there has been some success in the application of electrical bias to improve the gain and the response speed of the HPT, this method is only useful if the thermal noise of the load is a significant fraction of the total noise current [Hata et al,1977]. However, this is not always true. In these cases, the amplified noise associated with the base bias current will contribute a significant noise source to the total noise of the HPT. When the base bias current is much larger than the photogenerated current, the signal to noise ratio will be seriously degraded due to the increased shot noise.

Here we propose and demonstrate a novel HPT based on junction barrier lowering induced through the static induction effect. This transistor can be biased at a proper collector current level without base bias current so that the applified shot noise associated with the base bias current is eliminated. The most important feature of the punch-through HPTs is that the p-type base layer is lightly doped. Unlike the conventional HPTs with a relatively high base doping level, under the normal operating conditions the lightly doped base of PTHPTs is completely depleted. After the transistor reaches punch-through, the barrier height of the base-emitter junction is lowered with increasing V_{CE} through the static induction effect, and collector current increases exponentially with the increase in V_{CE}. Thus under the dark condition, the quiescent bias current I_C is solely controlled by V_{CE}. When the HPT is illuminated, electron-hole pairs are generated in the depleted base and the base-collector junction. The photogenerated electrons contribute a photocurrent (I_{ph}) component to the collector current. In addition, the photogenerated holes are swept to the base-emitter junction. This increases the forward bias of the base-emitter junction, which causes a large electron injection from emitter to collector. The injected electrons are swept to the collector immediately after they pass through the barrier of the base-emitter junction. This novel structure will result in high optical conversion gain and high current cutoff frequency. In the case of weak light detection, punch-through HPTs have larger bandwidths than conventional two terminal HPTs due to the smaller emitter charging time. In comparison with conventional three terminal HPTs, punch-through HPTs have better noise characteristics because the collector quiescent bias currents of the transistors can be provided without the base terminal.

Another advantage of using punch-through HPTs is that its layer structure is compatible with high speed HEMTs. By inserting an undoped spacer between the lightly doped base and the heavily doped emitter, HEMTs and phototransistors can be fabricated on the same wafer grown by molecular beam epitaxy (MBE). This makes it possible to fabricate monolithic integrated high speed photoreceivers for light wave communications.

Results and discussion

For the conventional HPTs, the power signal-to-noise ratio, assuming an input signal of $P_o(1+m\cos\omega t)$ and unity modulation depth, can be expressed as [Milano et al, 1982]

$$\frac{S}{N}=\frac{(1/2)(q\eta P_o/h\nu)^2(\beta+1)^2}{2qB\left(I_{ph}+I_B+I_d\right)\left(h_{FE}+1\right)\left[1+2\beta(\beta+1)/\left(h_{FE}+1\right)\right]+4kTB/R_{eq}}$$

where h is the quantum efficiency, B is bandwidth, I_{ph} is the photogenerated current, I_B is the base bias current, I_d is the dark current, b is the frequency dependent current gain, h_{FE} is the DC current gain, and R_{eq} is the equivalent circuit resistance. For punch-through HPTs, if we neglect the correlation between the noise components associated with the collector quiescent bias current and the amplified photo-generated current, the power signal-to-noise ratio of the punch-through HPT can be written as:

$$\frac{S}{N}=\frac{(1/2)(q\eta P_o/h\nu)^2(\beta+1)^2}{2qB\left(I_{ph}+I_d\right)\left(h_{FE}+1\right)\left[1+2\beta(\beta+1)/\left(h_{FE}+1\right)\right]+2qBI_{cq}+4kTB/R_{eq}}$$

Fig.1 shows the simulated results of signal-to-noise-ratio of both the conventional three terminal HPTs and punch-through HPTs. It can be clearly seen that the punch-through HPTs have much better low noise performance than the conventional HPTs. In the simulation, we made two assumptions for simplicity. First, we assumed that both conventional and punch-through HPTs have the same current gain and same current cutoff frequency f_T. Second, we dropped the term related to the partition noise. In conventional HPTs, the partition noise will be a significant noise component when the transistors work at high frequencies. However, the punch-through HPTs should not suffer from the partition noise as seriously as the conventional HPTs do, due to the fact that there is no neutral base in punch-through HPTs. If we further consider the facts that punch-through HPTs have lower partition noise, higher current gain, and higher current cutoff frequency, the punch-through HPTs will have even higher signal-to-noise ratios than that shown in Fig. 1.

The epitaxial layers of the devices were grown by MBE on n^+ substrates. The collector region consists of a 500 nm n-type GaAs layer doped to 1 x 10^{16} cm^{-3} grown on top of a 600 nm 3 x 10^{18} cm^{-3} n-type buffer layer. After growth of the collector layers, a 500 nm 3 x 10^{16} cm^{-3} Be doped p-type base layer was grown. Finally, a 200 nm 2 x 10^{18} cm^{-3} n-type $Al_{0.3}Ga_{0.7}As$ emitter layer and a 100 nm 5 x 10^{18} cm^{-3} n-type GaAs cap layer were grown. The devices were fabricated by the conventional wet etching method. The area of the devices is 20 x 40 μm^2.

The transistor were probed on wafer for electrical and optical measurements. The punch-through voltages of the transistors were about 2.5 V. After the HPT reached punch-through, collector currents indeed increased exponentially with the voltage applied between the emitter and collector as expected. The optical gain was measured with a Ti:sapphire laser. The wavelength of the laser varies from 790 nm to 850 nm. It can be seen from Fig. 2 that, with a bias voltage of 3 V and quiescent bias current of 1 mA, the optical conversion gain of a punch-through HPT was as high as 1240 at an incident optical power of less than 1µW. The high optical gain at this low incident optical power and low collector bias current is a significant improvement over the conventional HPTs. The transient response of a non-optimally matched punch-through HPT was also measured. The duration of the light pulses produced

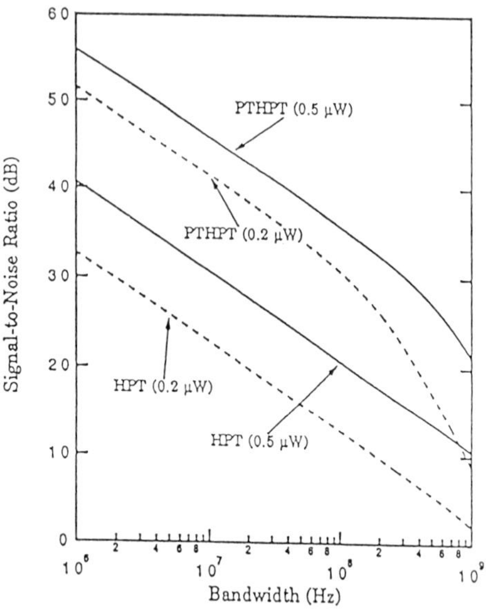

Fig. 1 Simulated signal-to-noise ratios of the punch-through and the conventional HPTs. In the simulation, h_{FE}=500 and f_β=100 MHz were used.

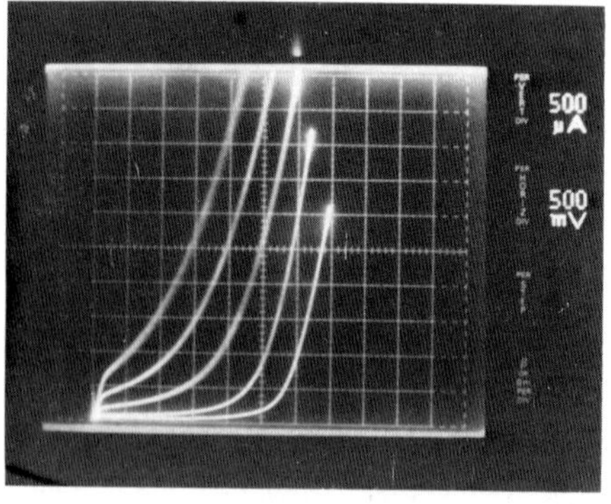

Fig. 2 The photoresponse of a punch-through HPT. the corresponding incident optical powers are 0 µW, 2 µW, 4 µW, 6 µW, and 8 µW.

by the Ti:sapphire mode-locked laser was less than 200 fs and the repetition-rate was 100 MHz. The device, biased at 3 mA, exhibited response pulses with the full width at half maximum (FWHM) of approximately 3.5 ns when the incident power was 5 μW. The estimated 3 dB bandwidth is beyond 100 MHz. This is a significant improvement over the conventional two or three terminal HPTs with similar dimensions, which is typically lower than 50 MHz. The gain-bandwidth product for our devices is beyond 50 GHz.

Conclusion

In conclusion, we have proposed and demonstrated a novel structure heterojunction phototransistor. The simulation has shown that the punch-through HPTs have much better low noise performance than the conventional HPTs. The transistors exhibit optical conversion gain as high as 1240 at an incident optical power lower than 1 μW. The 3-dB bandwidth of the transistors is estimated to be beyond 100 MHz, corresponding to a gain-bandwidth product of more than 50 GHz. In comparison with both the conventional two and three terminal HPTs, punch-through HPTs can offer higher gain, lower noise, and higher speed of response. These advantages make the punch-through HPT an attractive candidate for lightwave communications and weak light detections. The principle reported here is quite general and can be applied to HPTs made from other material systems, such as AlGaSb/GaSb and InP/InGaAs, for long wavelength optical communications.

Acknowledgments

This work was supported by the National Science Foundation and NCIPT/DARPA. We would like to thank Professor M. C. Teich for helpful discussions.

References

Chandraseckha, S. Hoppe, M. K., Dentai, A. G., Joyner, C. H., and Qua, G. J., Demonstration of enhanced performance of an InP/InGaAs heterojunction phototransistor with a base terminal, *IEEE electron Device Lett.*, vol. EDL-12, 550, 1991.

Chandraseckhar, S. et al., *Dig. Device Research Conference*, Boston, MA., 1992.

Fritzsche, D., Kuphal, and Aulbach, Fast response InP/InGaAsP heterojunction phototransistors, *Electron Lett.*, vol. 17, 178, 1981.

Hata, S. Kajiyama, K., and Miyushima, Y., Performance of p-i-n photodiodes campared with APD in the longer-wavelength region 1 to 2 μm, *Electron Lett.*, vol. 13, 668, 1977.

Milano, R. A., Dapkus, P. D., and Stillman, G. E., An Analysis of the performance of heterojunction phototransistors for fiber optical communications, *IEEE Trans. Electron Devices*, vol. ED-29, 266, 1982.

Inst. Phys. Conf. Ser. No 136: Chapter 5
Paper presented at the Int. Symp. GaAs and Related Compounds, Freiburg, 1993

Photovoltaic quantum well intersubband infrared detectors by internal electric fields

H Schneider, S Ehret, E C Larkins, J D Ralston, K Schwarz and P Koidl

Fraunhofer-Institut für Angewandte Festkörperphysik, Tullastrasse 72, D-79108 Freiburg, Federal Republic of Germany

ABSTRACT: We report on quantum well (QW) intersubband infrared detectors which demonstrate a resistive photovoltaic effect, i. e., a *steady state* photocurrent without external bias voltage. Detection relies on a transport mechanism in which the motion of the photoexcited carriers is rectified by built-in electric fields in the barrier layers. This mechanism is demonstrated using asymmetrically doped double-barrier quantum well detectors designed for the 3-5 μm-regime. We also propose a novel detector structure for photovoltaic QW intersubband detection in the 8 - 12 μm-regime.

1. INTRODUCTION

Infrared photodetection in the $3-5\mu$m and $8-12\mu$m atmospheric windows has become a promising application of GaAs-quantum well (QW) structures (Anderson and Lundquist 1992, Kozlowski et al 1991, Levine et al 1991, Martinet et al 1992, Park et al 1992). Photoconductive QW intersubband detectors have been implemented successfully in focal-plane array cameras (Levine et al 1991, Kozlowski et al 1991). Further optimization of these detector structures is still an important issue, with respect to both higher operating temperatures and to application-dependent system requirements, such as lateral homogeneity, detection bandwidth, dynamic range, and noise. In particular, high dynamic range and low noise are expected from photovoltaic QW detector structures. Device concepts for such photovoltaic devices have been reported by Goossen (1988) and by Kastalsky (1988).

In a number of previous publications (Schneider et al 1991a, Schneider et al 1991b, Ralston et al 1992), we have investigated double barrier quantum well (DBQW) structures, i. e., intersubband QW detectors where thin tunnel barriers (AlAs) are incorporated between the wells (GaAs) and barriers (AlGaAs) of a multiple QW, giving rise to a conduction band edge profile similar to those shown in the inset of Fig. 1. The original aim of these structures was to achieve intersubband detection in the $3-5\mu$m regime (Schneider et al 1991a). In a more general context, however, the spatial modulation of the barrier profile provides an additional degree of freedom that can be used in order to separate and to modify the capture and emission processes of the photoexcited carriers.

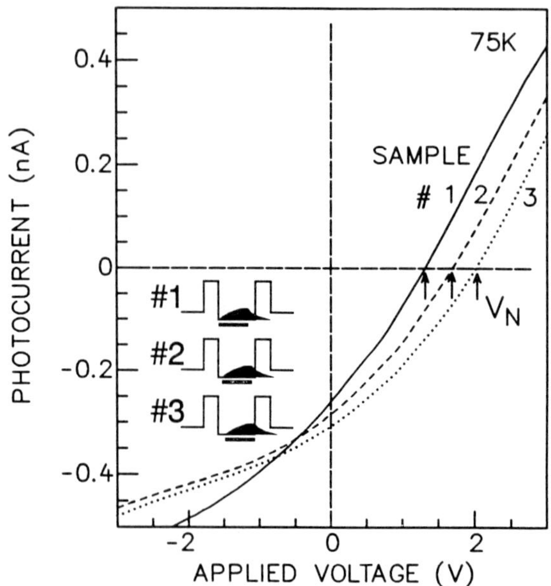

Figure 1: Photocurrent versus applied voltage of the DBQW detectors. Negative photocurrent corresponds to electron transport towards the substrate side. Inset: Spatial dependence of the conduction band-edge. The shaded and black areas represent the intended position of the Si doping spike and a realistic dopant distribution, respectively. Growth direction is from left to right.

In the present work, we focus on an intersubband detection mechanism giving rise to a resistive photovoltaic effect, i. e., a *steady state* photocurrent without external bias voltage. This photovoltaic behavior is achieved by built-in electric fields in the barrier layers, giving rise to a rectification of the motion of the photoexcited carriers.

This transport mechanism is demonstrated experimentally in section 2. In section 3, we discuss an optimized QW detection concept for photovoltaic detection in the $8 - 12\mu m$ regime.

2. PHOTOVOLTAIC 3-5μm DBQW INTERSUBBAND DETECTORS

In order to investigate the influence of internal electric fields in the barrier layers, we realized a series of DBQW intersubband detectors where the location of the doped region relative to the QW layers has been varied systematically, in orde to vary the space charge field across the barrier layers. The samples are 50 period DBQW structures with 5 nm wide Si-doped GaAs QWs, sandwiched between 2 nm thick AlAs tunnel barriers, and further separated by 25 nm $Al_{0.3}Ga_{0.7}As$ layers.

For the three samples under consideration, the intended positions of the nominally 4 nm wide Si

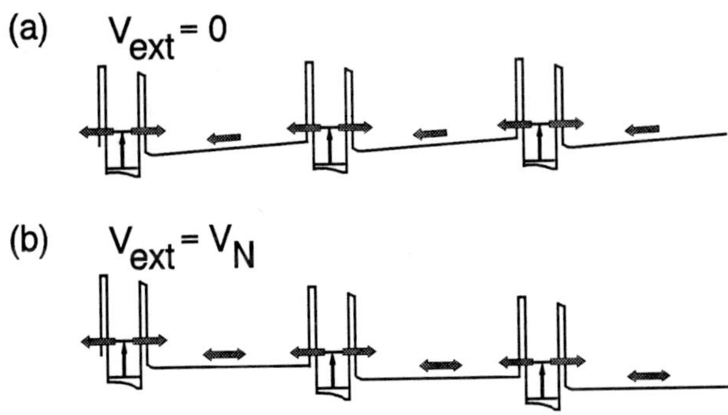

Figure 2: Potential distribution of the conduction band-edge of an asymmetrically doped DBQW structure (a) at vanishing external electric voltage and (b) at an external voltage corresponding to flatband condition in the barrier region. The arrows indicate the processes relevant for detector operation.

doping spikes (dopant concentration of $2 \cdot 10^{18}$ cm^{-3}) in the 5 nm wide GaAs QWs is indicated by the horizontal bars in the inset of Fig. 1. Si-doped GaAs contact layers ($1 \cdot 10^{18}$ cm^{-3}) with thicknesses of 0.5 μm and 1.0 μm were included above and below the active region, respectively. The structures were grown at 580 °C by molecular-beam epitaxy on (100)-oriented semi-insulating GaAs substrates and processed into mesa detectors of 0.04 mm^2 area with ring-shaped ohmic contacts. The photocurrent was measured under broadband illumination using a glowbar and lock-in amplification.

The voltage dependence of the spectrally integrated photocurrent of the three samples is shown in Fig. 1. All three structures show a steady-state photocurrent at 0 V bias, with responsivities of 2.9 ± 0.4 mA/W, and a photovoltage under open-circuit conditions. The polarity of this photovoltage corresponds to a preferential motion of the photoexcited electrons towards the substrate. The photovoltaic behavior is characterized phenomenologically by the compensating voltage V_N, corresponding to the external bias voltage where the photocurrent changes its sign. We can see from Fig. 1 that V_N increases systematically when the doping spike is displaced along the growth direction.

In order to explain the observed photovoltaic behavior, we have plotted in Figs. 2(a) and (b) the schematic potential distribution in an asymmetrically doped multiple DBQW where the dopant atoms are displaced to the right with respect to the QW center. For vanishing external voltage (Fig. 2(a)), the space-charge caused by the partial depletion of the right-hand part of the dopant distribution induce electric fields of opposite signs across the QW layers and across the barrier regions.

The asymmetric potential distribution in such a structure suggests a photoconduction mechanism which is indicated schematically by the arrows in Fig. 2(a). Electrons which are optically

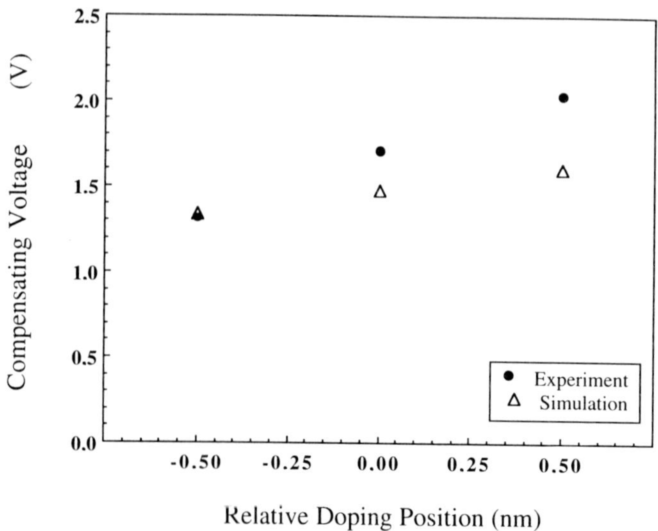

Figure 3: Compensating voltage V_N, as obtained from the experimental data of Fig. 1 (•) and from the theoretical simulation of Si segregation (Δ), plotted as a function of the shift in the intended doping position away from the center of the QW.

excited from the lower to the upper subband of a QW (vertical arrows) have a certain probability (determined by the thickness of the tunnel barriers) to be emitted from the well before intersubband scattering back to the ground state occurs. This photoemission process occurs towards both sides of the QW layer (unless the barriers are made inequivalent). Nevertheless, a net electron current towards the left-hand side is expected, since the asymmetric potential of the AlGaAs barrier layers gives rise to a back-relaxation towards the QW located at the left-hand side of the AlGaAs barrier.

In Fig. 2(b), we have sketched the situation in which the transport asymmetry of the photoexcited electrons is compensated by an external voltage V_{ext}, i. e., $V_{ext} = V_N$. This situation occurs when the built-in field across the $Al_{0.3}Ga_{0.7}As$ layers is approximately compensated by the external field, giving rise to symmetric transport properties in the barrier regions.

The measured compensating voltage V_N versus the relative dopant position is shown in Fig. 3. A simple analysis of the space charge fields, using the one-dimensional Poisson equation (Schneider 1993), predicts an increase of V_N by 0.55 V for the present sample parameters, which compares well with the experimental difference of 0.7 V between sample 1 and sample 3. The absolute value of V_N, however, depends critically on migration of the dopant atoms due to segregation during epitaxial growth. Fig. 3 also shows the results of a recent numerical solution (Larkins et al 1993) of a rate equation model describing the Si dopant incorporation kinetics in the GaAs/AlGaAs system at the experimental growth temperature, without any additional fit parameters. The theoretical results reproduce the experimental values reasonably well. The simulation clearly shows that the photovoltaic transport asymmetry arises predominantly due

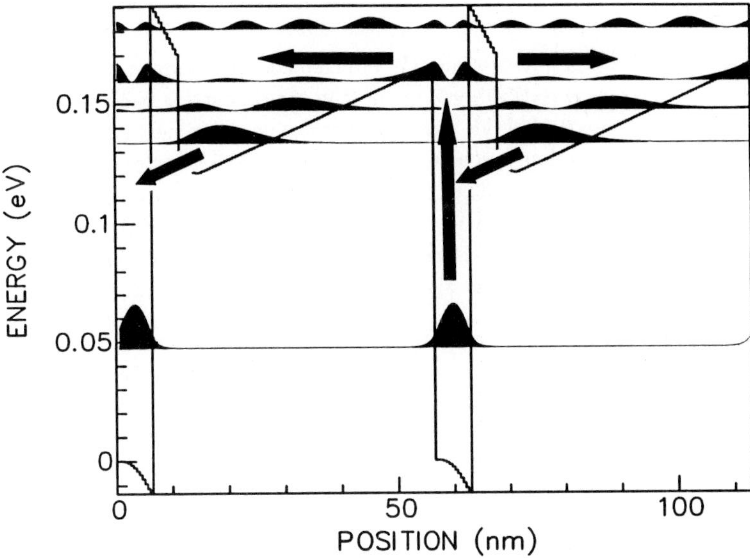

Figure 4: Conduction band edge profile, subband energies, and probability distributions of a novel photovoltaic 8-12 μm QW intersubband detector structure. The proposed device consists of 6.5 nm GaAs, 5.0 nm $Al_{0.25}Ga_{0.75}As$, and 45.0 nm $Al_{0.20}Ga_{0.8}As$, modulation-doped at about 10 nm with respect to the QW center with a sheet carrier density of $4.0 \cdot 10^{11}$ cm^{-2}. Device operation is indicated by arrows.

to dopant redistribution during epitaxial growth. Additional unintentional asymmetries, arising from inequivalent interface roughnesses, compositional asymmetries, or inequivalent tunnel barrier thicknesses, may also contribute to the photovoltaic behavior.

3. PHOTOVOLTAIC DETECTION AT 8-12 μm

The present approach to photovoltaic intersubband detection gives a significant decrease of the noise current, but at the cost of a reduced responsivity. Theoretical considerations indicate that the resulting detectivity of optimized photovoltaic devices should be similar to those of high-quality photoconductive QW intersubband detectors. Therefore, our approach is better suited for infrared detection at 8-12 μm rather than in the 3-5 μm regime, since preamplification is not critical in the long-wavelength infrared range due to the large photon fluxes. In fact, the reduced levels of both the signal and noise currents improve the saturation behavior and the dynamic range of such detectors.

Fig. 5 shows the result of a self-consistent quantum mechanical calculation of the potential profile and subband structure of such an 8-12 μm device. Here, intersubband absorption predominantly occurs between the first and the fourth subband, whereas the second and third subbands are mainly localized in the barrier region. The arrows indicate the expected transport mechanism which should still be analogous to the DBQW case of Fig. 2.

4. CONCLUSIONS

We have investigated a class of photovoltaic QW intersubband detectors in which the motion of the photoexcited electrons is rectified by a gradient in the conduction band potential of the barrier layers. We have demonstrated this transport mechanism experimentally using the space charge fields of asymmetrically doped double-barrier quantum well detector structures operating in the 3-5 μm-regime. The observed photovoltaic behavior is quantitatively explained by the expected space charge fields in the barrier layers. In addition, our experiments and numerical simulations show that these space charge fields depend critically not only on intentional asymetries of the dopant distribution, but also on dopant segregation during epitaxial growth.

Optimized photovoltaic QW intersubband detectors are expected to have smaller noise currents, smaller responsivities, and similar detectivities as compared to photoconductive devices. In particular, the present transport mechanism should give rise to infrared detectors with an extremely high dynamic range, particularly suited for the large photon fluxes occurring in the long-wavelength infrared. We have proposed and simulated such a device which is optimized for the 8-12 μm regime.

ACKNOWLEDGEMENTS: The authors are grateful to H. Rupprecht for his encouragement and continuous support of this work. We would also like to thank J. Fleissner, C. Hoffmann, M. Hoffmann, and K. Räuber for diode processing, and to H. Biebl and B. Dischler for absorption measurements. This work was supported by the BMVg under contract number T/R440/M0149/H1322.

5. REFERENCES

Anderson J Y and Lundqvist L 1992, Intersubband Transitions in Quantum Wells, ed. E Rosencher, B Vinter and B Levine (Plenum London), p. 1

Goossen K W, Lyon S A and Alavi K 1988, Appl. Phys. Lett. **52**, 1701

Kastalsky A, Duffield T, Allen S J and Harbison J 1988, Appl. Phys. Lett. **52**, 1320

Kozlowski L J, Williams G M, Sullivan G J, Craig W F, Anderson R J, Chen J, Cheung D T, Tennant W E and DeWames R E 1991, IEEE Trans. Electron Devices **38**, 1124

Larkins E C, Schneider H, Ehret S, Fleissner J, Dischler B, Koidl P and Ralston J D 1993, submitted for publication

Levine B F, Bethea C G, Glogovsky K G, Stayt J W and Leibenguth R E 1991, Semicond. Sci. Technol. **6**, C114

Martinet E, Luc F, Rosencher E, Bois Ph and Delaitre S 1992, Appl. Phys. Lett. **60**, 895

Park J S, Karunasiri R P G and Wang K L 1992, Appl. Phys. Lett. **60**, 103

Ralston J D, Schneider H, Gallagher D F G, Kheng K, Fuchs F, Bittner P, Dischler B and Koidl P 1992, J. Vac. Sci. Technol. **B 10**, 998

Schneider H, Fuchs F, Dischler B, Ralston J D and Koidl P 1991a, Appl. Phys. Lett. **58**, 2234

Schneider H, Koidl P, Fuchs F, Dischler B, Schwarz K and Ralston J D 1991b, Semicond. Sci. Technol. **6**, C120

Schneider H, Larkins, E C, Ralston J D, Schwarz K, Fuchs F and Koidl P 1993, Appl. Phys. Lett., August 9, to appear

Paper presented at the Int. Symp. GaAs and Related Compounds, Freiburg, 1993

Ultrafast response of metal–semiconductor–metal photodetectors on InGaAs/GaAs-on-GaAs superlattices for 1·3–1·55 mm applications

J. Hugi, C. Dupuy, M. de Fays, R. Sachot and M. Ilegems

Institut de Micro- et Optoélectronique, Ecole Polytechnique Fédérale de Lausanne, CH-1015 Lausanne, Switzerland

ABSTRACT: Lifetime limited ultrafast response (FWHM = 3.3 ps) of Metal-Semiconductor-Metal photodetectors on InGaAs/GaAs-on-GaAs superlattices has been measured by electro-optic sampling. These detectors operate in the 1.3 μm - 1.55 μm wavelength region and show a bias dependent responsivity.

1. INTRODUCTION

The metal-semiconductor-metal photodetector (MSMPD) is a promising device for integration in optoelectronic receivers, due to its ease of fabrication, low capacitance for relatively large area and compatibility with field effect transistors (Sugeta 1980, Ito 1984). For high speed/long distance fiber communication systems, which operate in the low loss and reduced dispersion 1.3 - 1.55 μm wavelength region, InGaAs/InP MSMPDs are normally used (Schumacher 1988). An alternative to the InGaAs/InP material system is to grow the absorbing InGaAs layer lattice mismatched on GaAs substrates to take advantage of the more mature processing techniques of GaAs based devices (Rogers 1988).

Of particular interest is the bandwidth of the device. For the use of the MSMPD as high speed photodetector a major problem is the slow moving holes, producing a long lasting response tail. The two approaches to increase the bandwidth of a MSMPD are to shorten the distance between metallic fingers or to reduce the lifetime of the carriers. Recently, GaAs based MSMPDs with 25 nm finger spacing and width were fabricated (Chou 1992a), and an MSMPD on bulk GaAs with 100 nm finger spacing showed a RC constant limited impulse response with FWHM of 1.5 ps (Chou 1992b). The fastest high-sensitivity MSMPDs were realized on low-temperature grown GaAs (LT-GaAs) where the response is RC constant or carrier lifetime limited and impulse responses with FWHM of 1.2 ps (Chen 1991) and 0.87 ps (Chou 1992b) were measured. Here, we present time resolved measurements of

MSMPDs on an InGaAs/GaAs superlattice grown on GaAs for application in the 1.3 μm - 1.55 μm wavelength region (Zirngibl 1989, 1991). These measurements show lifetime limited ultrafast impulse response of the photodetectors with a FWHM of 3.3 ps.

2. EXPERIMENTAL TECHNIQUES

The active layer of the photodetector consists of a 60-period InGaAs/GaAs superlattice (SL) on a 0.5 μm thick GaAs buffer grown on a semi-insulating GaAs substrate. The superlattice has a total thickness of 0.98 μm, with individual InGaAs layer thicknesses of 8.3 nm and GaAs layer thicknesses of 8.1 nm. Growth rates were 1.5 μm/h for InGaAs and 0.73 μm/h for GaAs. From the growth rates the InAs content in the InGaAs layers is calculated to be 52% in first order approximation (Zirngibl 1991). The active layer is capped by 80 nm GaAs for Schottky-barrier height enhancement. The growth was done by MBE at 450° C and the whole structure is undoped. The metal fingers (Al (50 nm), Pt (20 nm), Au (60 nm)) of the MSMPD were formed by a conventional lift-off process. The active area of the detector is 20 μm x 20 μm and the finger widths and spacings are 1/1, 2/2, 3/3 and 4/4 μm. For the high speed measurements the MSMPDs were integrated into 4 mm long coplanar striplines with linewidths of 100 μm and spacings of 20 μm ($Z_0 = 60\ \Omega$). For comparison MSMPDs were fabricated on epitaxial GaAs layers of 2 μm thickness grown by MBE at 450° C and at 600° C.

The MSMPDs were characterized by external electro-optic sampling (Valdmanis 1987). A 100 fs mode-locked Ti:sapphire laser with 76 MHz repetition rate was used as light source and a $LiTaO_3$ probe tip was placed about 300 μm from the MSMPDs for the sampling. The detectors are biased by a 50 Ω microwave probe head posed on the coplanar striplines, to avoid reflections and make on-wafer-measurement possible. For the spectral response we used a halogen white light source, a monochromator and a lock-in amplifier at 664 Hz.

3. SIMULATION

The transit-time limited current response of a GaAs MSMPD is examined by a two-dimensional simulation based on carrier drift in the electric field. The electrons and holes are represented by 20000 "superparticles" each, where one superparticle stands for a certain number of carriers depending on the energy of the incoming light pulse (Moglestue 1991). The carriers are assumed to move along the electric field lines according to their field dependent steady-state drift velocities. The velocity-field relations for electrons and holes are taken from literature (Shur 1987). The electric field is recalculated after each time step (200 fs) solving Poisson's equation iteratively for a rectangular mesh. The initial vertical electron-hole distribution is calculated with an optical absorption coefficient $\alpha = 1\ \mu m^{-1}$ for 850 nm wavelength. The external current response to Dirac like light pulses is determined by Ramo's theorem (Ramo 1939) and taking a simple RC equivalent circuit for the MSMPD (R = Z_0, C calculated (Soole 1990)). Carrier diffusion and recombination are neglected.

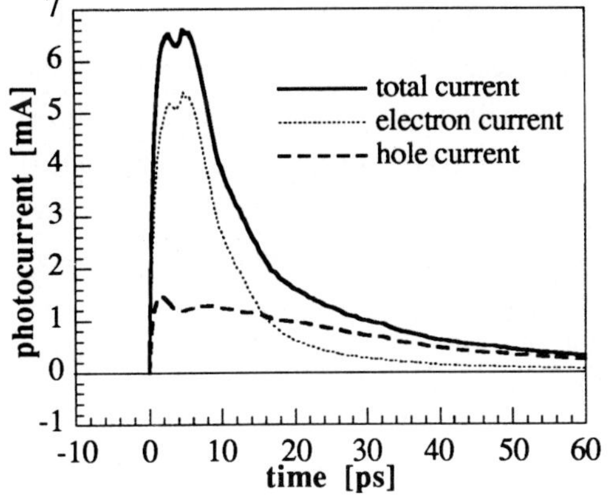

Fig. 1 Simulation of the external current response of a bulk GaAs MSMPD with 1 μm finger width and spacing at 5 V bias. The detector area is 20 μm x 20 μm. The light pulse has 0.8 pJ energy and 850 nm wavelength.
The total external current and the electron and hole current components are shown. Note the difference between the response of the fast moving electrons and the slow holes.

4. EXPERIMENTAL RESULTS

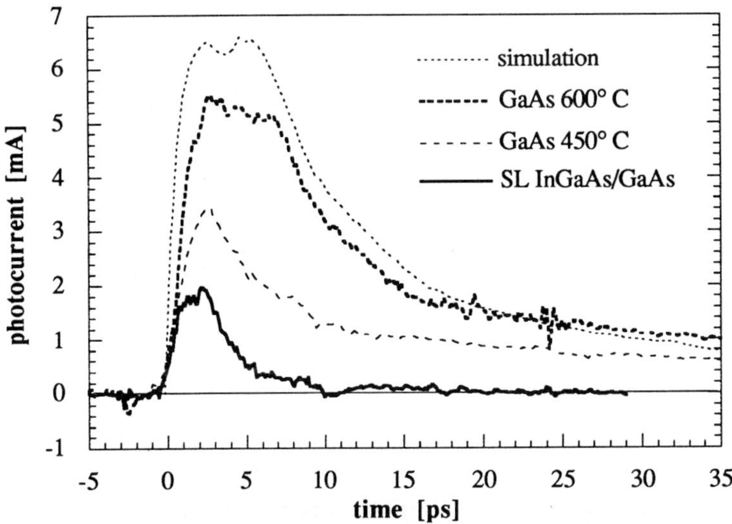

Fig. 2 Temporal response of MSMPDs with 1 μm finger width and spacing.
SL detector: 7 V bias, 2.7 pJ/pulse
GaAs detectors: 5 V bias, 2.7 pJ/pulse
Simulation: 5 V bias, 0.8 pJ/pulse

Figure 2 shows the impulse response of a SL detector compared to detectors on epitaxial GaAs. The SL detector was measured at 895 nm, the GaAs detectors at 850 nm wavelength. The photocurrent was determined by dividing the measured response voltage by the calculated line impedance Z_o. The response of the MSMPD on GaAs grown at 600° C exhibits a FWHM of 10 ps and a long tail from the slow moving holes. The good qualitative agreement with simulation shows that this detector has a transit time limited response. The large difference for the pulse energy between simulation and measurement can be explained partly by the fact that only a fraction of the photons is striking the detector. For the SL-MSMPD the FWHM is 3.3 ps and the tail is significantly reduced. In Figure 3 we see the very similar responses of two SL-MSMPDs, one with 1 μm, the other with 4 μm finger width and spacing, clearly indicating lifetime limited response. The peak signal amplitude is 120 mV for the 1/1 μm and 90 mV for the 4/4 μm detector. We attribute the short carrier lifetime in the order of 3 ps in the SL structure to the low growth temperature and the large lattice-mismatch of the InGaAs layers, producing a high density of deep traps. From the response of the MSMPD on GaAs grown at 450° C in Figure 2 we see that the low growth temperature is partly responsible for the short lifetime of the electrons but does not strongly affect the long response tail.

Figure 4 compares the discrete Fourier transform of the experimental temporal response of the SL-MSMPD with 1 μm finger width and spacing with that of the calculated response for a bulk GaAs MSMPD shown in Figure 1. The 3 dB bandwidth is 48 GHz for the SL-detector and 7 GHz for the transit time limited GaAs detector.

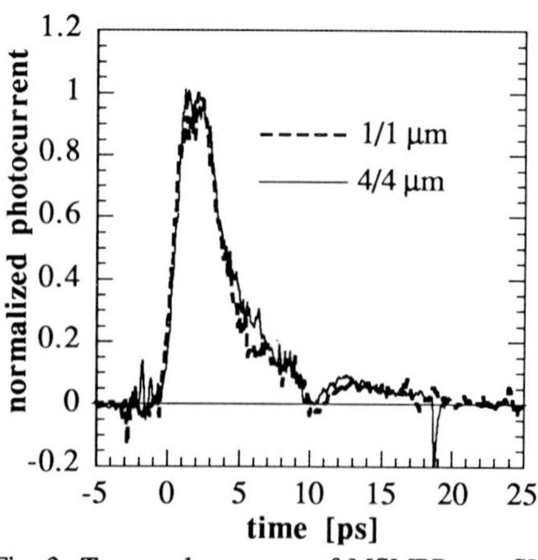

Fig. 3 Temporal response of MSMPDs on SL InGaAs/GaAs with 1 μm and with 4 μm finger width and spacing. 7 V bias, 2.7 pJ/pulse for the 1/1 μm detector, 12 V bias and 3.1 pJ/pulse for the 4/4 μm detector.

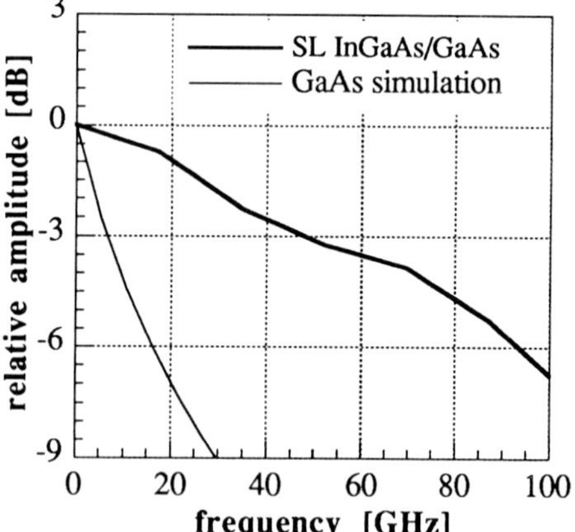

Fig. 4 Frequency response derived from the Fourier transform of the temporal signal of the 1/1 μm SL-detector shown in Figure 3 and of the calculated signal of a transit time limited GaAs detector shown in Figure 1.

The responsivity of a SL-MSMPD in function of the bias is shown in Figure 5. The responsivity is determined by dividing the average photocurrent in the biasing circuit by the average power of the laser beam. The reflective losses from finger shadowing (50 %) and at the GaAs surface (50 % for 50° angle of incidence, TE polarization) amount to 75 %. A strong dependence of the responsivity with bias is also an observed feature of GaAs MSMPDs although the exact reason for this is unknown at the moment. While the maximal amplitude of the time resolved current response increases almost linearly, the responsivity grows exponentially with bias. The FWHM increases from 3.1 ps at 3 V to 4.5 ps at 10 V and to 6.3 ps at 14 V. So at higher bias the increase in responsivity comes essentially from the slow tail which becomes clearly visible above 10 V. This tail decays exponentially with a time constant of about 10 ps. From the bias dependence, we believe charge accumulation within traps near the contacts to be responsible for this tail, but also hopping conductivity has been given as possible explanation (Klingenstein 1992).

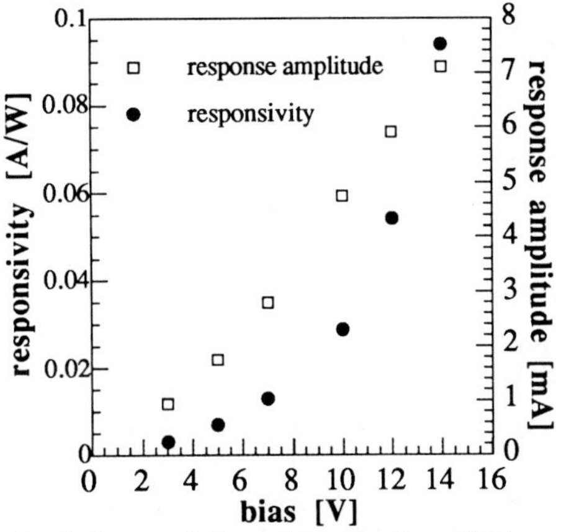

Fig. 5 Responsivity and maximal amplitude of the current response of a 1/1 μm SL-MSMPD in function of bias for 6.1 pJ pulses at 895 nm wavelength.

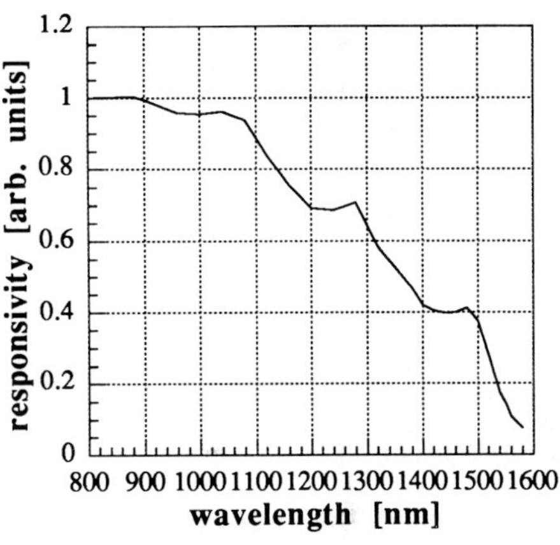

Fig. 6 Spectral response of a 2/2 μm SL-MSMPD at 5 V bias.

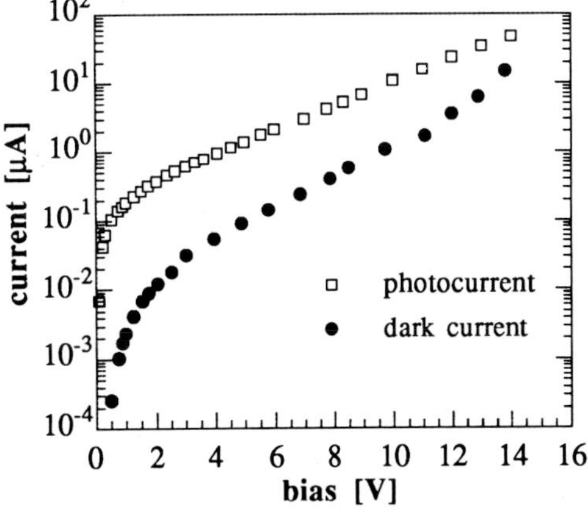

Fig. 7 Dark current and photocurrent for a SL-MSMPD with 1 μm finger width and spacing and 20 μm x 20 μm area. The photocurrent is corrected for dark current. The light pulses have an energy of 2.7 pJ which gives an average power of 200 μW. The wavelength is 895 nm.

From low frequency measurements we see that the responsivity decreases usually at 1300 nm to 60 % and at 1550 nm to 15 % of the value at 895 nm wavelength, as shown in Figure 6.

CONCLUSION

In conclusion, we present ultrafast, lifetime limited response of MSMPDs on InGaAs/GaAs-on-GaAs superlattices with a FWHM down to 3 ps and a very small long lasting tail up to moderate bias. The ultrafast response, the compatibility with GaAs devices and the sensitivity up to 1.55 μm wavelength make this detector a good candidate for optical communication systems.

ACKNOWLEDGMENT

This work was supported by the Swiss Post, Telephone and Telegraph (PTT) research office.

REFERENCES

Chen Y, Williamson S and Brock T, Smith F W and Calawa A R, 1991, Appl. Phys. Lett. 59 (16), pp 1984-1986
Chou S Y, Liu Y and Fischer P B, 1992a, Appl.Phys. Lett. 61 (4), pp 477-479
Chou S Y, Liu Y, and Khalil W, Hsiang T Y and Alexandrou S, 1992b, Appl. Phys. Lett. 61 (7), pp 819-821
Ito M, Wada O, Nakai K and Sakurai T,1984, IEEE Electron Device Lett., EDL-5 (12), pp 531-532
Klingenstein M, Kuhl J, Nötzel R and Ploog K, Rosenzweig J, Moglestue C, Hülsmann A, Schneider J and Köhler K, 1992, Appl. Phys. Lett. 60 (5), pp 627-629
Moglestue C, Rosenzweig J, Kuhl J, Klingenstein M, Lambsdorff M, Axmann A, Schneider Jo and Hülsmann A, 1990, J. Appl. Phys. 70 (4), pp 2435-2448
Ramo S, 1939, Proceedings of the I.R.E., 27, pp 584-585
Rogers D L, Woodall J M, Pettit G D and McInturff D, 1988, IEEE Electron. Device Lett. vol. 9 (10), pp 515-517
Schumacher H, Leblanc H P, Soole J and Bhat R, 1988, IEEE Electron. Device Lett. vol. 9 (11), pp 607-609
Shur M, 1987, GaAs devices and circuits, Plenum Press, New York 1987
Soole J B D, Schumacher H, 1990, IEEE Trans. Electron Devices 37 (11), pp 2285-91
Sugeta T, Urisu T, Sakata S and Mizushima Y, 1980, Japan. J. Appl. Phys., vol. 19 suppl. 19-1, pp 459-464
Valdmanis J A,1987, Electron. Lett. vol. 23 (24), pp. 1308-1310
Zirngibl M, Bischoff J C, Theron D and Ilegems M, 1989, IEEE Electron Device Lett. vol. 10 (7), pp. 336-338
Zirngibl M and Ilegems M, 1991, J. Appl. Phys. 69 (12), pp 8392-8398

Paper presented at the Int. Symp. GaAs and Related Compounds, Freiburg, 1993

Heavily-doped *p*-type GaAs/AlGaAs superlattices for infrared photodetectors

B.W. Kim and A. Majerfeld

Department of Electrical and Computer Engineering. CB425
University of Colorado, Boulder, CO 80309, USA.

ABSTRACT: A theoretical analysis of heavily doped p-type GaAs/AlGaAs superlattices under normal incident light which includes many-body effects is presented. It is shown that many-body effects cause significant changes to the optical absorption coefficient and that the doping level and doping configuration have an important effect on the absorption properties of these superlattices. Peak absorption coefficients for normal light incidence of 3000-5000 cm^{-1} (single polarization) at photon wavelengths of 8 - 9 μm are predicted for p-type superlattices with well doping of 1×10^{19} - 2×10^{19} cm^{-3}.

1. INTRODUCTION

During the last few years there has been considerable theoretical (Chang and James 1989) and experimental (Levine et al 1987, Levine et al 1991) interest on long-wavelength intersubband optical transitions in quantum well structures based on various material systems for application to infrared photodetector devices. In previous theoretical treatments of n-type single quantum well structures it was demonstrated that many-body effects (Hartree and exchange-correlation effects) must be taken into account for electron densities exceeding $n\cong10^{11}$ cm^{-2} in order to reach agreement with experimental observations (Bloss 1989). Although much higher doping levels can be obtained in p-type semiconductors than in direct-gap n-type semiconductors, many-body effects have not been taken into account in preceding analyses of the electrical and optical properties of p-type quantum well structures. In this paper we present a theoretical investigation of many-body effects as well as the effects of doping level and doping configuration on the absorption coefficient for intersubband optical absorption (ISOA) transitions in heavily-doped p-type GaAs/AlGaAs superlattices (SL).

The primary motivation for studying p-type GaAs/AlGaAs SLs rests on their desirable properties for infrared photodetector devices. First, p-type SL structures make device operation possible with normal incident light (polarized normal to the growth direction; the z-direction in this paper). Second, the large effective mass of heavy holes allows the use of much higher doping levels and, consequently, the possibility of larger optical absorption coefficients than for n-type SLs, without the concomitant rapid increase of the thermionic and tunneling components of the dark current. Third, GaAs/AlGaAs SLs are at present preferable to other materials, not only because the III-V semiconductor technology is more mature than the II-VI semiconductor technology but, also, because of the potential for their monolithic integration with high-speed GaAs-based electronics.

For a systematic investigation we use 51 periods of 60 Å well - 140 Å barrier p-type GaAs/$Al_xGa_{1-x}As$ SL as the model structure with 130 meV for the well barrier height. The electronic structure consists of four bound subbands (HH1, LH1, HH2, and HH3) and a miniband (LH2), above the barrier, which contains the final states of the optical transitions. The doping densities investigated in this work are in the range 1×10^{17} - 2×10^{19} cm^{-3} for a structure with doped wells (DW) and 1×10^{17} - 5×10^{18} cm^{-3} for a structure with doped barriers (DB). In both cases the doping width is 60 Å around the center of each well or barrier layer. The computations are performed for a temperature of 300 K. We assume that carbon is the acceptor dopant atom, as it provides excellent doping properties (Hanna et al 1991). In our model, the C atoms are assumed to be fully ionized in the wells and to have the bulk ionization energy in the barriers.

2. OPTICAL PROPERTIES

2.1 Momentum matrix element (MME)

In a SL with a periodic effective perturbation $\mathcal{V}_{eff}(z)$ in the z-direction, the one-particle quantum state can be expressed by the in-plane (the plane perpendicular to the z-direction) wavevector $\mathbf{k_t}$ and an index set (n,q) which designates the quantum state associated with the z-direction, where n is the subband (or miniband) index and q is the SL wavevector. A quantum state of the SL can be expressed by a linear combination of the Bloch wave functions, $|\nu,\mathbf{k}\rangle = e^{i\mathbf{k}\cdot\mathbf{r}}\, u_{\nu,\mathbf{k}}(\mathbf{r})$, where ν is band index in the bulk and $k = (\mathbf{k_t},\mathbf{k_z})$:

$$|\mathbf{k}_t,n,q\rangle = \sum_{\nu,k_z} a(\nu,\mathbf{k}_t,n,q;kz)|\nu,\mathbf{k}\rangle \quad . \tag{1}$$

The effective potential $\mathcal{V}_{eff}(z)$ consists of $\mathcal{V}_{eff}(z)=\mathcal{V}_{sq}(z)+\mathcal{V}_{ht}(z)+\mathcal{V}_{xc}(z)$, where $\mathcal{V}_{sq}$, $\mathcal{V}_{ht}$ and $\mathcal{V}_{xc}$ are the square SL, the Hartree and the exchange-correlation potentials, respectively, and the forms of $\mathcal{V}_{ht}$ and $\mathcal{V}_{xc}$ are obtained from the literature (Ando 1977).

In the dipole approximation, which is appropriate in practical SL structures, the MME of ISOA for direct transitions between the states $|\mathbf{k}_t,n,q\rangle$ and $|\mathbf{k}_t,n',q\rangle$, is given by

$$\langle \mathbf{k}_t,n,q|\hat{\varepsilon}\cdot\mathbf{p}|\mathbf{k}_t,n',q\rangle \equiv \hat{\varepsilon}\cdot\mathbf{p}_{nn'}(\mathbf{k}_t,q) \quad , \tag{2}$$

where p is the momentum operator and $\hat{\varepsilon}$ is the polarization unit vector of light. Upon substituting Eq.(1) into Eq.(2), the MMEs $P^x_{nn'}$ and $P^z_{nn'}$ are obtained and shown in Table Ia and Ib, respectively, for the two valence band system (heavy and light hole bands). In these tables, hh1 (hh2) and lh1 (lh2) represent the m_j=+3/2 (-3/2) and m_j=+1/2 (-1/2) states, respectively, with the total angular momentum J=3/2, γ's are the Luttinger parameters (Lawaetz 1971), and the quantities, Q and R, are defined to be

$$Q^{nn'}_{\nu\nu'} = \int dz\, \varphi(\nu,\mathbf{k}_t,n,q;z)^* \varphi(\nu',\mathbf{k}_t,n',q;z)$$

$$R^{nn'}_{\nu\nu'} = \int dz\, \varphi(\nu,\mathbf{k}_t,n,q;z)^* (-i\partial_z) \varphi(\nu',\mathbf{k}_t,n',q;z) \quad , \tag{3}$$

where $\varphi(;z)$ is the periodic part of the SL envelope function (Fourier transform of $a(;k_z)$ in Eq.(1)).

Table Ia. $P^x_{nn'}$ matrix elements (normal incidence)

	hh1	hh2	lh1	lh2
hh1	$\hbar k_x(\gamma_1+\gamma_2)Q^{nn'}_{h1h1}$	0	$-\sqrt{3}\hbar\gamma_3 R^{nn'}_{h1l1}$	$\sqrt{3}\hbar(\gamma_2 k_x - i\gamma_3 k_y)Q^{nn'}_{h1l2}$
hh2	0	$\hbar k_x(\gamma_1+\gamma_2)Q^{nn'}_{h2h2}$	$\sqrt{3}\hbar(\gamma_2 k_x + i\gamma_3 k_y)Q^{nn'}_{h2l1}$	$\sqrt{3}\hbar\gamma_3 R^{nn'}_{h2l2}$
lh1	$-\sqrt{3}\hbar\gamma_3 R^{nn'}_{l1h1}$	$\sqrt{3}\hbar(\gamma_2 k_x - i\gamma_3 k_y)Q^{nn'}_{l1h2}$	$\hbar k_x(\gamma_1-\gamma_2)Q^{nn'}_{l1l1}$	0
lh2	$\sqrt{3}\hbar(\gamma_2 k_x + i\gamma_3 k_y)Q^{nn'}_{l2h1}$	$\sqrt{3}\hbar\gamma_3 R^{nn'}_{l2h2}$	0	$\hbar k_x(\gamma_1-\gamma_2)Q^{nn'}_{l2l2}$

Table Ib. $P^z_{nn'}$ matrix elements (parallel incidence)

	hh1	hh2	lh1	lh2
hh1	$\hbar(\gamma_1-2\gamma_2)R^{nn'}_{h1h1}$	0	$-\sqrt{3}\hbar\gamma_3(k_x+ik_y)Q^{nn'}_{h1l1}$	0
hh2	0	$\hbar(\gamma_1-2\gamma_2)R^{nn'}_{h2h2}$	0	$\sqrt{3}\hbar\gamma_3(k_x-ik_y)Q^{nn'}_{h2l2}$
lh1	$-\sqrt{3}\hbar\gamma_3(k_x-ik_y)Q^{nn'}_{l1h1}$	0	$\hbar(\gamma_1+2\gamma_2)R^{nn'}_{l1l1}$	0
lh2	0	$\sqrt{3}\hbar\gamma_3(k_x+ik_y)Q^{nn'}_{l2h2}$	0	$\hbar(\gamma_1+2\gamma_2)R^{nn'}_{l2l2}$

The results presented in Table I for inter-valence band optical transitions show that for both normal and parallel light incidence there are non-zero MMEs. However, for normal incident light (Table Ia) HH-LH transitions dominate over HH-HH and LH-LH transitions for small $\mathbf{k}_t$ values. Conversely, for parallel incident light (Table Ib), HH-HH and LH-LH transitions dominate over HH-LH transitions.

2.2. Photo-absorption coefficient (PAC)

Within the dipole approximation, the PAC of the intersubband transition from subband i to j can be expressed as

$$\alpha_{ij}(\hbar\omega)=\frac{\pi e^2}{m^2\omega\varepsilon_s c'\Omega}\Sigma_{\mathbf{k}_t,q}\left[f(\mathbf{k}_t,i,q)-f(\mathbf{k}_t,j,q)\right]\left|\hat{\varepsilon}\cdot\mathbf{p}_{ij}(\mathbf{k}_t,q)\right|^2$$

$$\times\frac{\hbar\Gamma_{ij}(\mathbf{k}_t,q)}{\left[\mathcal{E}_j(\mathbf{k}_t,q)-\mathcal{E}_i(\mathbf{k}_t,q)-\hbar\omega\right]^2+\left[\hbar\Gamma_{ij}(\mathbf{k}_t,q)\right]^2}\,, \tag{4}$$

where ε_S, c', Ω and f are, respectively, dielectric constant, speed of light in the material, volume of the solid and the Fermi-Dirac distribution function, and $\Gamma_{ij}(\mathbf{k}_t,q)$ is the linewidth of the transition i to j for the state $(\mathbf{k}_t,q)$, which accounts for lifetime broadening due to scattering and inhomogeneity of well widths. In this paper $\hbar\,\Gamma_{ij}$=6meV is assumed for numerical computations.

The peak PACs for the HH1-LH2 transition under normal incident light ($\hat{\varepsilon}_x$) for various doping densities at 300 K for the DW and DB doping schemes are shown, respectively, in Figs. 1a and 1b. The squares and triangles indicate PACs obtained taking into account many-body effects and neglecting these effects, respectively. As most of the holes occupy states of the HH1 ground subband, only transitions from HH1 to higher subbands need to be considered. These figures are discussed in the following section.

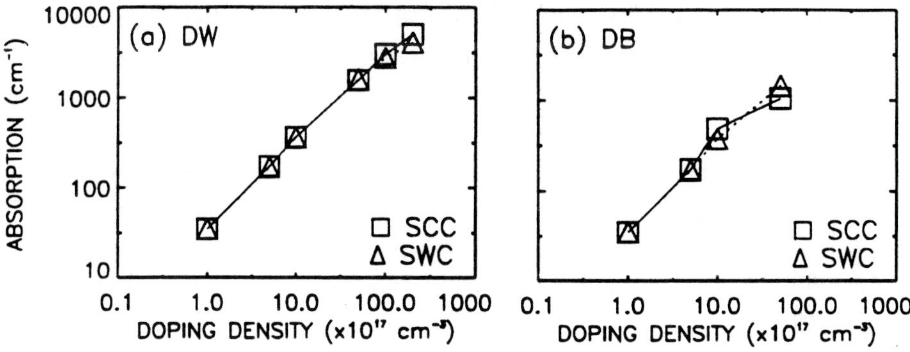

Fig. 1. Doping dependence of the peak absorption coefficients for the HH1-LH2 transition in the DW (a) and DB (b) cases, respectively. Squares show the self-consistent calculation taking into account many-body effects (SCC) and triangles show the results for a square well potential (SWC).

3. DISCUSSION AND CONCLUSIONS

A theoretical investigation of the optical absorption properties of p-type SLs for use in infrared photodetectors was performed. We analyzed the intersubband optical matrix element and the absorption coefficient over a range of doping densities (1×10^{17}-2×10^{19} cm^{-3}), taking into account many-body effects. This work shows that: (1) For the DW structure, the absorption coefficient for normal incident light for HH1-LH2 transitions is enhanced by many-body effects (from α=4000 to α=5000 cm^{-1}) at doping densities around $N_D=2\times10^{19}$ cm^{-3}; (2) The doping density and doping configuration are critical parameters in determining the absorption coefficient, especially for the HH1-LH2 transition. For the DW structure with $N_D=2\times10^{19}$ cm^{-3} the absorption coefficient for this transition, with a photon wavelength of 8.3 μm, reaches $\alpha_{HH1\text{-}LH2}\cong5000$ cm^{-1} (single polarization of light). For the DB structure, even at a moderately high doping density, such as $N_D=5\times10^{18}$ cm^{-3}, the Hartree potential becomes important and produces a large distortion of the original square well-barrier SL structure which, in turn, results in significant changes in the subband energy structure. It becomes increasingly more difficult to pursue numerical computations for the DB case beyond this doping level due to convergence problems related to the Hartree potential.

We demonstrated also that: (1) Many-body effects must be taken into account in order to provide a proper framework for the interpretation of optical and electrical experimental observations and for the design of SL structures; (2) Heavy doping of the wells is required to achieve a high absorption coefficient for normal light incidence.

In conclusion, it should be noted that absorption coefficients of 3,000-5,000 cm^{-1} (single polarization) appear possible for normal incident light by proper structural and doping design of the p-type SL. This absorption value is significantly larger than for n-type SLs and, therefore, may lead to normal incidence photodetectors operating at 77 K.

ACKNOWLEDGMENTS

We thank Z.H. Lu and E. Mao for useful discussions, the Army Research Laboratory, Fort Monmouth, and the Colorado Advanced Technology Institute through the Advanced Materials Institute for partially supporting this work.

REFERENCES

Ando T 1977 Solid State Comm. **21**, 133
Bloss W L 1989 J. Appl. Phys. **66**, 3639
Chang Y C and James R B 1989 Phys. Rev. B **39**, 12672
Hanna M C, Lu Z H, and Majerfeld A 1991 Appl. Phys. Lett. **58**, 164
Lawaetz P 1971 Phys. Rev. B **4**, 3460
Levine B F, Choi K K, Bethea C G, Walker J, and Malik R J 1987 Appl. Phys. Lett. **50**, 273
Levine B F, Gunapala S D, Kuo J M, Pei S S, and Hui S 1991 Appl. Phys. Lett. **59**, 1864

Paper presented at the Int. Symp. GaAs and Related Compounds, Freiburg, 1993

Monolithically integrated transimpedance optical receiver in a planar InGaAs/InP technology

D Römer, Ch Lauterbach, L Hoffmann and G Ebbinghaus

Siemens Research Laboratories, Otto-Hahn-Ring 6, 81739 München , Federal Republic of Germany

ABSTRACT: A monolithically integrated photodiode preamplifier has been fabricated. The integration concept of the OEIC is based on a transimpedance-type preamplifier in a planar InGaAs/InP JFET technology. The new technological approach combines a single step epitaxy on a flat substrate with local doping via ion implantation to keep the surface of the chip planar. The OEIC includes an output buffer stage with 50 Ω impedance matching. A sensitivity of -34 dBm at a bit error rate of 10^{-9} for a data rate of 600 Mbit/s has been derived from signal noise measurements.

1. INTRODUCTION

InP based optoelectronic integrated circuits (OEICs) are promising candidates for future high speed optical communication systems using wavelengths in the near infrared. In recent years, several receiver OEICs in the InGaAs/InP system have been reported. Most of these approaches make use of the integration of pin photodiodes (PDs) for the light detection combined with transimpedance type preamplifiers, based on a junction field-effect transistor (JFET) technology (Lee et al 1990; Newson et al 1991; Uchida et al 1991; Blaser and Melchior 1992; Lauterbach et al 1992). The receiver OEIC reported in this work comprises a planar monolithically integrated pin PD and a transimpedance-type preamplifier in a JFET technology. Multiple epitaxies and/or epitaxy on structured, non planar surfaces have been avoided in order to simplify the technology and enhance the yield and therefore make the OEIC more suitable for mass production. Low contact, source and drain resistances are essential concerning the performance of the OEIC. Therefore a Si contact implantation and an InGaAs p contact layer has been implemented.

2. INTEGRATION CONCEPT

The integration concept is based on a planar InGaAs/InP JFET technology, using a single step metal-organic vapor-phase epitaxy (MOVPE) process on a flat substrate, where the same layer sequence is used for the detector and the JFET. The adaption of these devices is done by local doping via ion implantation. The layer thicknesses and doping concentrations are listed in Table 1. A schematic cross section of a PD, a JFET and a feedback resistor is shown in Figure 1.

Table 1. Epitaxial layer sequence

layer	composition	thickness (μm)	doping concentration (cm^{-3})
substrate	InP:Fe		
buffer	InP:Si	0.2	2×10^{16}
spacer	InP	0.2	undoped
absorption	InGaAs	2.1	undoped
window	InP:Si	0.2	$1\text{-}2\times10^{16}$
contact	InGaAs:Si	0.1	$1\text{-}2\times10^{16}$
cap	InP:Si	0.1	$1\text{-}2\times10^{16}$

On the semi-insulating InP:Fe substrate a layer sequence appropriate for the PD is grown. The doping of the buffer layer is necessary for high speed operation of the device. The doping concentration of the InP buffer has been analyzed by electrochemical capacitance-voltage profiling. With an electron concentration of 2×10^{16} cm^{-3} and a thickness of 0.15 μm the series resistance of the PD is reduced to a value sufficiently low for the operation at the aspired bandwidth. The InGaAs contact layer is implemented for improving the p contacts of the devices due to the small bandgap and compared to InP typically two orders of magnitude higher hole concentration after diffusion. Within the InGaAs absorption layer the channel of the JFET is formed by Si ion implantation, at a dose of 3×10^{12} cm^{-2} and an energy of 480 keV, leading to a maximum electron concentration of 8×10^{16} cm^{-3}. A buried p layer, formed by Be implantation (3×10^{12} cm^{-2}, 800 keV) beneath the channel layer of the JFET, confines the drain current to the channel and suppresses the influence of the InP buffer layer. A second local Si ion implantation beneath the n contacts is used for increasing the n doping to 1×10^{18} cm^{-3}, thus reducing the contact and series resistances of the devices (Römer et al 1992). All implantations were simultaneously annealed at 700°C for 30 s in a PH_3/H_2 atmosphere. The p region of the PD and the level shifting diodes (LSDs) and the gates of the JFETs are formed by Zn diffusion from a spin-on film source at 550°C for 45 s, using SiN_x as a diffusion mask and passivation for the pn junction. The dopant distributions of the JFET structure have been characterized by secondary-ion mass spectrometry (SIMS). The results are depicted in Figure 2. With appropriate parameters for the Zn diffu-

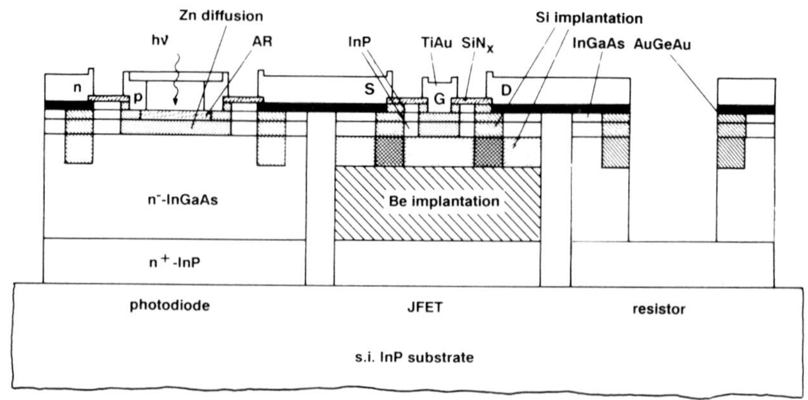

Figure 1 Schematic cross section of the layer structure of the photodiode preamplifier OEIC.

sion, the diffusion front stops beneath the InP/InGaAs heterojunction with a sharp decrease of the atomic Zn concentration from 7×10^{18} to less than 1×10^{16} cm^{-3}. A SIMS profile of the implanted channel and the buried p layer are indicated as well. The maximum electron concentration is **7.5×10^{16}cm^{-3}**. The InP cap layer is removed within the regions of the contacts for exposing the InGaAs contact layer. Later this contact layer is removed within the window of the ring-shaped p contact of the PD to avoid absorption within this layer in case of top illumination. SiN_x is used as an antireflexion coating directly on the InP heterojunction layer. After the deposition of AuGe/Au for the n and Ti/Au for the p contact metallizations, the devices are isolated by mesa etching down to the semi-insulating substrate. The contact resistances of the n and p contacts, determined from transmission line structures, are as low as 0.04 and 0.15 Ωmm, respectively. A feedback resistor is defined by the doped InP buffer layer. Therefore the layers above the resistor are removed.

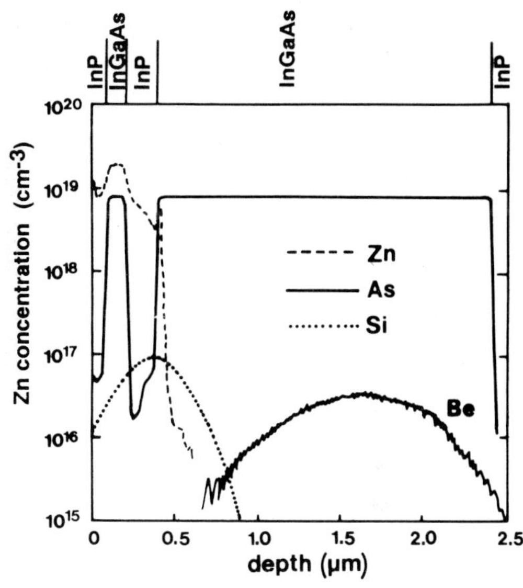

Figure 2. SIMS of the atomic concentration of the diffused Zn profile (the non-calibtated As-signal to distinguish the different epitaxy layers). Atomic concentration of the channel and buried p layer are indicated.

3. CIRCUIT DESIGN

The design of the transimpedance-type photodiode preamplifier, based on the described technology, has been evaluated using the circuit simulator PSPICE and is shown in Figure 3. The detected signal from the PD is amplified at the input stage. A cascode input gain stage reduces the input capacitance resulting from the JFET B1 due to the Miller effect and therefore improves the bandwidth and particularly reduces the circuit noise (Smith and Personick 1982). A source follower with a level shifting stage and a 5.9 kΩ resistor is used to provide negative feedback. In order to enhance the bandwidth of the receiver without deterioration of the transimpedance, 50 Ω impedance matching at the output of the receiver is realized with an output buffer stage, that is connected to the feedback circuit. Biasing of the circuit is achieved with current limiters and LSDs. The receiver comprises 1 PD, 7 JFETs, 7 LSDs and 1 resistor together with 3 power supply nodes. Two power supplies are necessary for biasing the circuit and one for biasing the second JFET at the input stage for minimizing the input capacitance and adjusting the bandwidth. The PD is optimized concerning capacitance, speed and geometry for top illumination by fiber and has a diameter of 30 μm for the active region. The JFETs at the input stage have a gate length of 1.6 μm and a gate width of 100 μm. A micrograph of the chip is shown in Figure 4. The chip size is 1×1 mm^2.

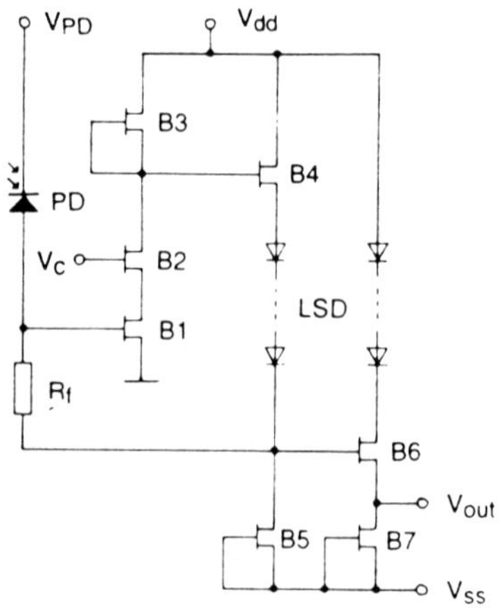

Figure 3. Circuit diagram of the transimpedance-type PD preamplifier with output buffer stage.

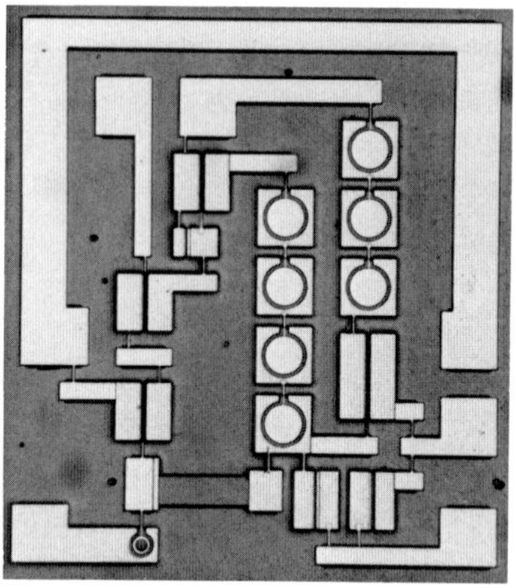

Figure 4. Micrograph of the photodiode-preamplifier.

4. RESULTS

The current voltage characteristics of a forward biased LSD is shown in Figure 5. The series resistance of the diode with a diameter of 66 μm for $V \geq 0.7$ V is as low as 6 Ω. The steep increase of the

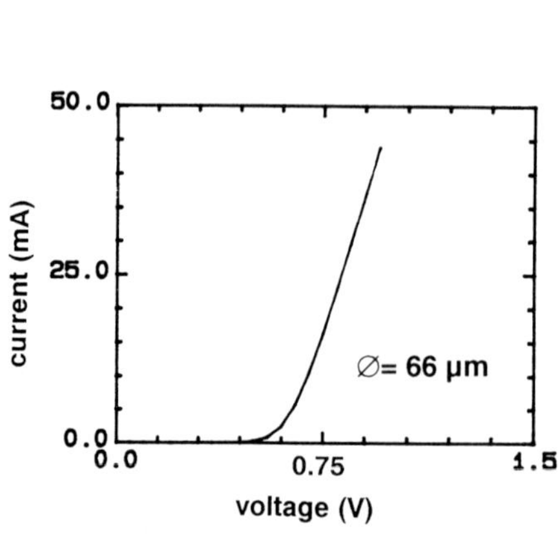

Figure 5 Current voltage characteristics of a forward biased LSD.

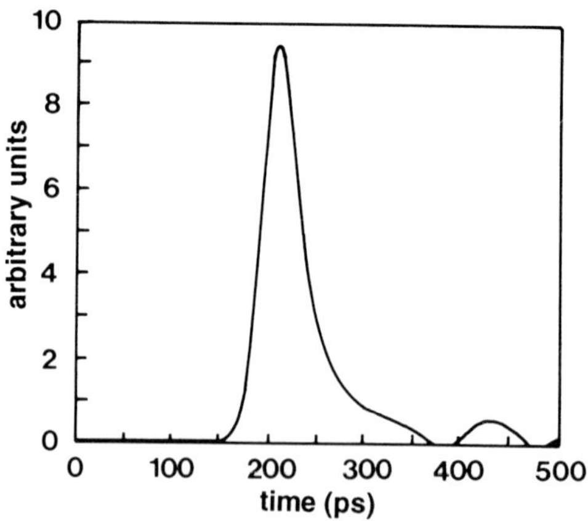

Figure 6. Transient response of a PD illuminated with a short light pulse at 1.3 μm.

diode current leads to a well defined voltage drop across the LSDs in the circuit with an accuracy of better than 0.1 V. From the transient response of the PD, illuminated with light pulses at a wavelength of 1.3 μm, a 3 dB bandwidth of 4.5 GHz is calculated (Figure 6). The quantum efficiency of the PD is 85 %, the leakage current at a reverse bias of 5 V is less than 1 nA. At a drain source voltage of 1.5 V and a gate source voltage of -1 V, the gate leakage of the JFET with a gate length of 100 μm is approximately 40 nA. As a result the shot-noise of the PD and the JFETs does not contribute to the noise at the output of the preamplifier. The output characteristics of a JFET with InGaAs contact layer is shown in Figure 7. Pinch off is achieved at 1 V, the threshold voltage is 2.0 V The maximum transconductance is 150 mS/mm, the output conductance is 10 mS/mm. The current gain cut-off frequency of the JFET, determined from S-parameter measurements, is 10 GHz.

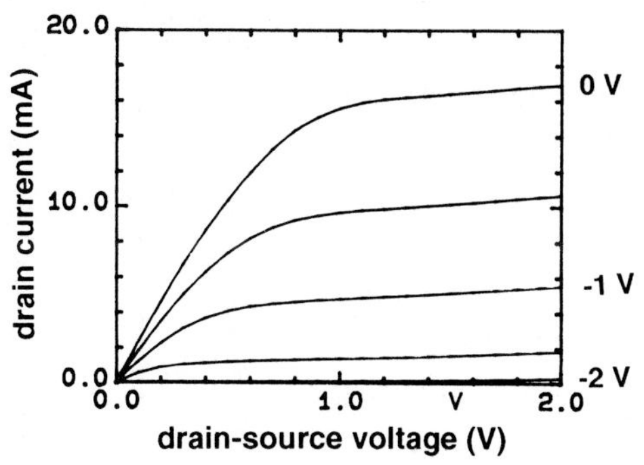

Figure 7. Output characteristics of a JFET with $1.6 \times 100\ \mu m^2$ gate

A measure for a good quality of the ternary absorption layer and the PD design is the correlation of the capacitance with the reverse voltage of the diode (Figure 8). The capacitance at -5 V is less than 0.1 pF. Practically at voltages below -2 V no significant further reduction in the pn junction capacitance can be seen, indicating that the space charge region of the PD reaches through the whole absorption layer at this voltage. The gate source capacitance of a JFET with a gate width of 100 μm is shown as well. At the operation point the capacitance is 0.16 pF.

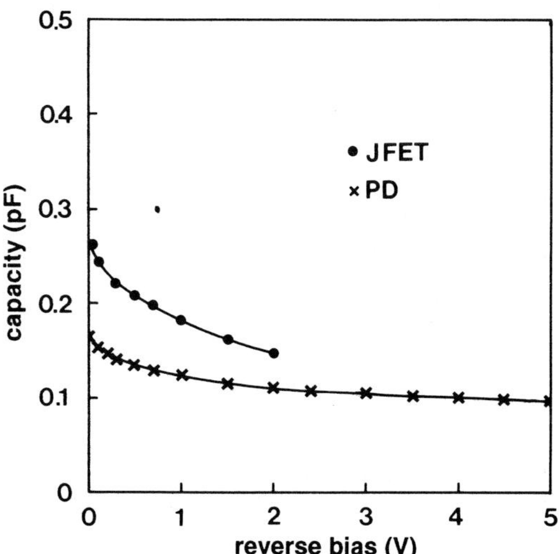

Figure 8. Capacitance of the reverse biased pn junction of the PD and the gate source diode of the JFET

The transimpedance Z_t and the equivalent noise current density deduced from signal and noise measurements of the receiver are shown in Figure 8. The transimpedance of the circuit is 2.1 kΩ, the averaged input noise current density is below $3\ pA/\sqrt{Hz}$. The bandwidth of the receiver is 320 MHz, enabling an operation at a bitrate of 600 Mbit/s. From these data, a sensitivity of -34 dBm of the receiver at a bit error rate of 10^{-9} has been calculated.

CONCLUSION

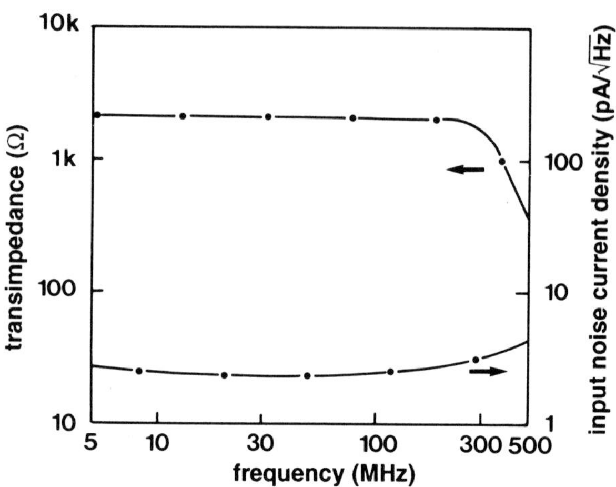

Figure 9. Transimpedance and equivalent noise current density of the OEIC

A photodiode-preamplifier OEIC has been fabricated in a planar InGaAs/InP JFET technology. A single step epitaxy on a flat substrate is combined with local doping via ion implantation in order to keep the surface of the chip planar.
The transimpedance-type photodiode preamplifier array with cascode input gain stage has a bandwidth of 320 MHz, a transimpedance of 2.1 kΩ and an equivalent input noise current density below 3 pA/$\sqrt{Hz}$. The calculated sensitivity at a data rate of 600 Mbit/s is -34 dBm.

ACKNOWLEDGEMENT

We would like to thank J.W. Walter for implantation and annealing, J. Müller for EC-V profiling, Th. Hillmer for SIMS characterization and M. Plihal for discussions. This work has been partly supported by the Deutsche Bundespost Telekom and partly by the RACE project R2070 MUNDI.

REFERENCES

Blaser M and Melchior H 1992 IEEE Photon. Lett. **4**, 1244

Lauterbach Ch, Albrecht H, Römer D, Walter J W, Müller J and Ebbinghaus G 1992 Proc. 4th InP and Rel. Mat. Conf. Newport, Rhode Island, USA, **74**

Lee W S , Spear D A H, Agnew M J, Dawe P J G and Bland S W 1990 Electron. Lett. **26**, 377

Newson D J, Mansfield C, Birdsall P, Quayle J A and MacBean M D A 1991 ECOC, 489

Römer D, Bauer J G, Lauterbach Ch, Müller J and Walter J W 1992 J. Appl. Phys. **72**, 4998

Smith R G and Personick S D 1982 Semiconductor Devices for Optical Communications ed H Kressel (New York: Springer) 89

Uchida N, Akahori Y, Ikeda M, Kohzen A, Yoshida J, Kokubun T and Suto K 1991 IEEE Photon. Technol. Lett **3**, 540

Paper presented at the Int. Symp. GaAs and Related Compounds, Freiburg, 1993

Nonlinear optical absorption due to spatial band bending: a new possibility for the realisation of an optical modulator

C. Väterlein, G. Fuchs, A. Hangleiter, V. Härle, and F. Scholz

4. Physikalisches Institut, Universität Stuttgart,
Pfaffenwaldring 57, D-70550 Stuttgart, Germany

ABSTRACT: We report on nonlinear optical absorption in InGaAs/InGaAsP MQW structures based on charge carrier induced band bending. The basic principle of the nonlinearity is the charge separation after optical pumping due to different depths of the potential wells in conduction and valence band. The resulting band bending leads to a strong change in optical absorption which we observe experimentally. We reach up to 20 % relative change of absorption making the new nonlinearity potentially useful for optical switching devices.

1. INTRODUCTION

Nonlinear optical effects have been a subject of great interest in the last years as a means in the development of optical switching and modulation devices (Peyghambariam 1985). Field screening due to optical excitation (Kan 1989, Ando 1989) as well as state filling (Iannelli 1989, Noda 1990) have been used to modulate the absorption of the device at a certain wavelength. In this paper we present a new effect of nonlinear optical absorption which can be used for the development of an optical modulator.

2. DYNAMIC SPATIAL BAND BENDING

First we will describe the idea on which the new optical nonlinearity relies. In quantum well structures with different effective depth of the potential wells and different effective masses in conduction and valence band, the probability of being outside the wells in thermal equilibrium is different for electrons and holes. In the case of the InGaAs/InGaAsP material system the wells in the valence band are deeper than those in the conduction band. Therefore we expect to find more electrons outside the wells than holes. If we compare electrons with heavy holes which play the most important role at low carrier densities the difference in distribution of the carriers is increased through the difference in the effective masses. The different distribution of electrons and holes over wells and barriers corresponds to a different spatial distribution and therefore to charge separation, which leads to an additional electrostatic potential in the structure even at low carrier densities. For the evaluation of these carrier induced changes of the spatial band structure Fuchs et al. have developed a self consistent calculation method. It has been thoroughly described in (Fuchs 1993) so that in this paper we give only a brief overview. The calculation is based on the alternating solution of Schrödinger and Poisson equation. The initial spatial potential (known from the design data) is used to solve the Schrödinger equation. With the knowledge of the energy states and wavefunctions of electrons and holes, the electrochemical potentials and therefore the carrier distribution for a given number of carriers per unit area can be evaluated. After the solution of the Poisson equation the additional potential due to the inserted carriers is known and can be added to the initial potential. With this new potential the whole process is repeated until the change of potential between two steps is sufficiently small.

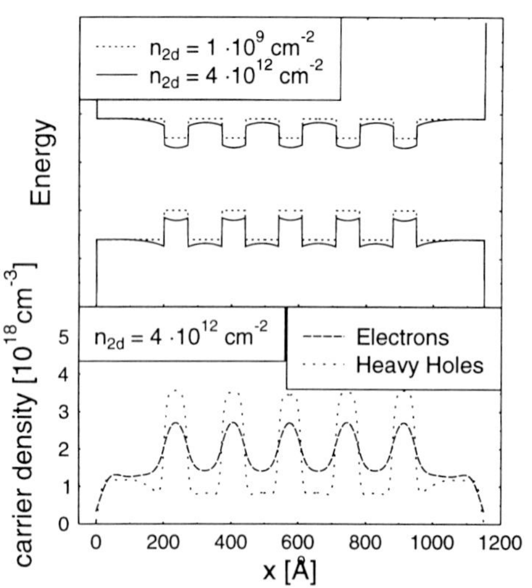

Figure 1: Spatial band diagram of sample 2 with (solid line) and without (dashed line) excited carriers. Shown is the band bending due to carrier separation which leads to the localisation of the n=2 electron subband.

Fig. 1 shows the spatial band diagram of a experimentally investigated sample with (solid line) and without (dashed line) a carrier density of $4 \cdot 10^{12} cm^{-2}$. The important effect happens in the second electronic subband. In the sample we are discussing the second electronic subband is just delocalised in the case of the unbent bandstructure. Due to the band bending however it gets localised so that the overlap between the n=2 electron wavefunction and the corresponding heavy hole wavefunction which is very small in the unbent case rises drastically. Since the optical matrix element for a direct band-to-band transition $M_{i,j}$ is proportional to the normalized overlap integral $F_{i,j} = \leq \varphi_i \mid \varphi_j \geq$ we should be able to observe the band bending through a change in absorption at the corresponding energy. In the next sections we will demonstrate that this indeed is possible and discuss the possibilities of using this effect for optical switching.

3. EXPERIMENT AND SAMPLES

We have investigated two types of samples. The first group are InGaAs/InGaAsP Separate Confinement Multi Quantum Well structures grown by MOVPE. The five InGaAs quantum wells have a width of 7 nm and a gap energy of 0.75 eV at 300 K. The InGaAsP barriers (confinement layers) have a width of 10 nm (200 nm), the gap energy has been varied in different samples between 0.85 eV and 0.95 eV (see table 1).

The second group are tensile strained InGaAs/InP MQW structures with different frac-

Table 1:

Sample	E_g InGaAs	E_g InGaAsP
1	0.75 eV	0.87 eV
2	0.75	0.93
3	0.75	0.97

Table 2:

Sample	x_{Ga}
4	0.47
5	0.7

tions of Ga (see table 2). The experiments with the quarternary samples have been performed at 300 K, those with the strained samples at 2 K.

The experimental setup consists of a halogen lamp as a white light source for the detection of the transmission and a chopped Nd:YAG laser operated at an energy of 1.16 eV which excites carriers into the barriers of the structures. The change of transmission due to the change of carrier density is then detected directly with an lock-in amplifier.

4. InGaAs/InGaAsP STRUCTURES: EXPERIMENTAL RESULTS

Fig. 2 shows a series of differential transmission spectra of sample 2 with different excitation powers. In the region of energy where we expect to find the transition between the excitonic states of the first electron and the first heavy and light hole subbands we find a positive change of transmission. This shows that the oscillator strength of both excitons is reduced through the increase in carrier density. This effect has first been described and explained by Chemla et al. (1985) as a result of carrier induced screening of the Coulomb interaction and phase space filling.

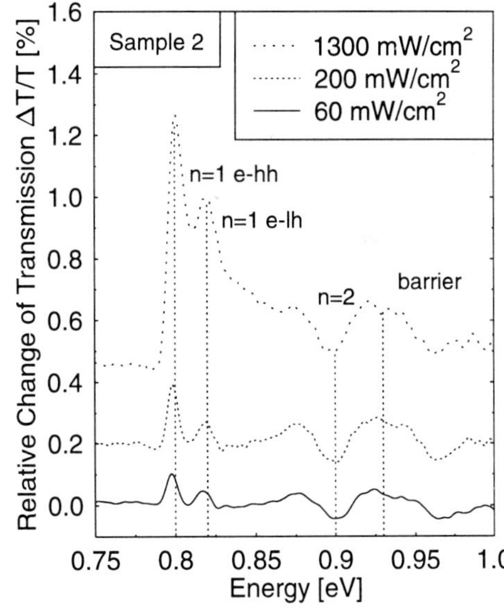

Figure 2: Differential transmission spectra of sample 2 with different excitation intensities. Positive change of transmission at the n=1 subband transition, negative change of transmission at the n=2 subband transition.

Now we will discuss the negative change of transmission at 0.9 eV. This energy lies well below under the gap energy of the barriers which is 0.93 eV so that the change in transmission has to be connected with the n=2 subband transition.

The possibility of localising the n=2 electron subband should depend critically on the effective depth of the potential well in the conduction band. Fig. 3a) shows a comparison of different samples with a different conduction band offset. It can be seen that for a conduction band offset higher than 80 meV it is not possible to achieve an increase of absorption at the n=2 subband transition. In Fig. 3b) the corresponding overlap integrals in dependence of the carrier density are shown. Here it becomes clear why sample 3 shows no negative signal in the differential transmission spectrum – the overlap integral does not change up to the carrier density reached in the experiment.

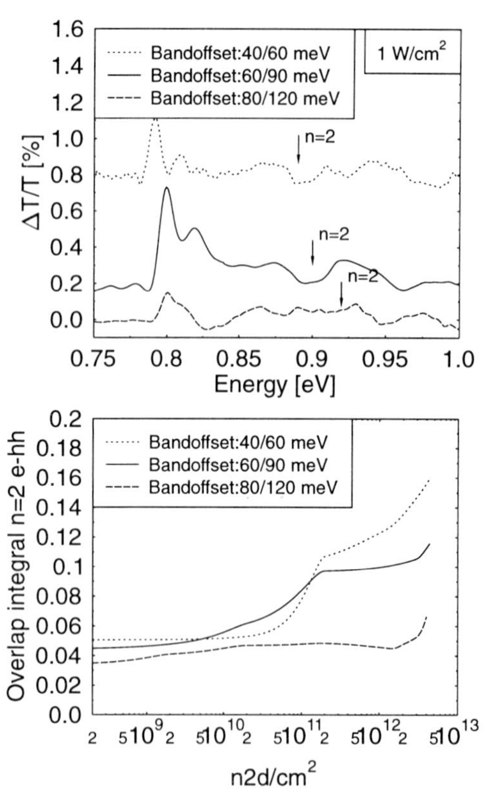

Figure 3: Comparison of three samples with different band discontinuities. a) change of transmission b) evaluated overlap integral. A good agreement between experiment and calculation can be seen.

The relative change of transmission which can be achieved with these samples is rather small. We could improve the situation by using a structure where we can perform the band bending on a n=1 subband. We therefore need a sample with a strongly delocalised first electron subband and a localised first heavy hole subband. This condition is fulfilled by tensile strained InGaAs/InP MQW structures. In the next section our experimental results on these samples are described.

5. TENSILE STRAINED InGaAs/InP STRUCTURES: EXPERIMENTAL RESULTS

Tensile strain in the InGaAs/InP material system leads to a change in the distribution of the band discontinuity. The higher the fraction of Ga is, the shallower the potential wells in the conduction band get. At a Ga fraction of 0.8 we may even get a type II heterostructure (Gershoni 1988).

The change in the distribution of the band discontinuity with the strain can be seen experimentally through the decrease in the overlap of the electron and hole wavefunctions which leads to a decrease in absorption. Fig. 4 shows the absorption spectra of an unstrained (solid line) and a strained (dashed line) sample. With a Ga fraction of 0.7 the absorption edge has vanished completely, we therefore in this case expect to have a nearly flat conduction band.

If we now raise the carrier density in the sample through optical pumping, we get the same effect as described in section 4: due to the different distribution of the carriers in conduction and valence band we get an electrostatic potential which leads to band bending. Our experimental results are shown in Fig. 5. We get the expected result: with the increase in carrier density we get an increase in absorption at the band edge and even see the rise of an excitonic absorption. With the investigated sample we achieved a relative change of absorption of 20 % at a carrier density of $10^{12} cm^{-2}$.

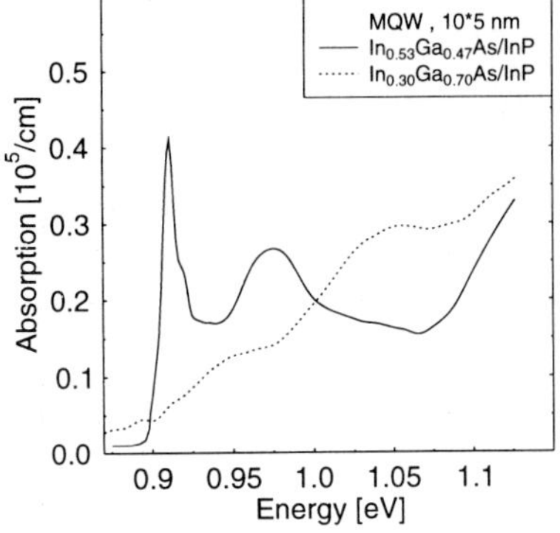

Figure 4: Absorption spectra of sample 4 (solid line) and sample 5 (dashed line).

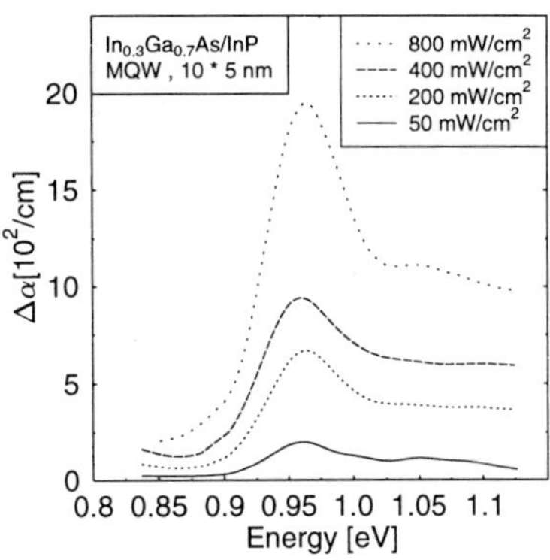

Figure 5: Change of absorption of sample 5 at different carrier densities.

6. DEVICE ASPECTS

For the application of the nonlinearity in an optical device two aspects are important: the total amount of transmission change at low carrier densities and the modulation frequencies which can be reached with the device. We regard the nonlinear absorption due to band bending at the n=1 resp. n=2 subband transition with respect to these two aspects. At the n=2 subband transition the relative change in transmission we can achieve is rather small. On the other hand we do not have any spatially indirect transition due to a large separation of charge carriers in this case. Therefore the modulation frequencies which can be achieved are only limited by the recombination times of the order of 1 ns or below.

The transmission change at the n=1 subband transition reaches much larger values but we expect the frequency limit to be considerably lower because of the spatially indirect transition in this case.

7. CONCLUSION

We have shown a new effect of nonlinear absorption which is based on the change of the overlap integral of the n=2 or n=1 electron - heavy hole subband transition in SCMQW structures. The basic mechanism of this effect is the bending of the spatial band diagram due to carrier separation which leads to the localisation of the formerly unlocalised electron subband. The resulting change in transmission and in the refractive index can be used for optical switching devices which are expected to work with low switching power and at high frequencies.

REFERENCES

Ando H, Iwamura H, Oohashi H, and Kanbe H 1989 IEEE J. Quantum Electronics QE-25 10 p2135.

Chemla D S and Miller A B 1985 J. Opt. Soc. Am.B 2 7 p1155.

Fuchs G, Hörer J, Hangleiter A, and Rudra A 1993 To be published in Phys. Rev. B.

Gershoni D, Temkin H, Vandenberg J M, Chu S N G, Hamm R A, and Panish M B 1988 Phys. Rev. Lett 60 5 p448.

Iannelli J M, Masejian J, Hancock B R, Andersson P O, and Grunthaner F J 1989 Appl. Phys. Lett 54 5 p301.

Kan Y, Obata K, Yamanishi M, Funahashi Y, Sakata Y, Yamaoka Y, and Suemune I 1989 Jap. J. Appl. Phys. 28 9 p1585.

Noda S, Uemura T, Yamashita T, Sasaki A 1990 J. Appl. Phys. 68 12 p6529.

Peyghambariam N and Gibbs H M 1985 J. Opt. Soc. Am. B 2 p1215.

Inst. Phys. Conf. Ser. No 136: Chapter 5
Paper presented at the Int. Symp. GaAs and Related Compounds, Freiburg, 1993

Subnanosecond high-power performance of a bistable optically controlled GaAs switch

David C. Stoudt, Ralf P. Brinkmann[a], and Randy A. Roush

Naval Surface Warfare Center, Dahlgren Division, Pulsed Power Systems and Technology Group (Code B20), Dahlgren, Virginia 22448-5000 USA; [a]Siemens A.G., Munich, Germany

ABSTRACT: Recent subnanosecond results of the Bistable Optically controlled Semiconductor Switch (BOSS) are presented. The processes of persistent photoconductivity followed by photo-quenching have been demonstrated in copper-compensated, silicon-doped, semi-insulating (GaAs:Si:Cu). These processes allow a switch to be developed that can be closed by the application of one laser pulse ($\lambda = 1.06$ μm) and opened by the application of a second laser pulse with a wavelength equal to twice that of the first laser. This report discusses the effects of 1-MeV neutron irradiation on the BOSS material for the purpose of recombination center generation.

1. INTRODUCTION

Photoconductive switches made from semi-insulating (SI) GaAs were proposed in the late 1970's for use as both closing and opening high-power switches (Mourou and Knox 1979). Closing was achieved by exciting electrons from the valence band into the conduction band using a laser with a photon energy greater than that of the bandgap. An alternative method to direct excitation across the bandgap was proposed by Schoenbach, et al. (1988). This concept, which is called the Bistable (or bulk) Optically controlled Semiconductor Switch (BOSS), relies on persistent photoconductivity followed by photo-quenching to provide both switch closing and opening, respectively. Persistent photoconductivity results from the excitation of electrons from the deep copper centers found in copper-compensated, silicon-doped, semi-insulating GaAs (GaAs:Si:Cu). The small cross-section for electron capture back into the Cu centers allows long conduction times (tens of microseconds) after the first laser pulse is terminated. Photo-quenching is accomplished by the application of a second laser pulse of longer wavelength which elevates electrons from the valence band back into the copper levels. This laser pulse floods the valence band with free holes which rapidly recombine with free electrons to quench the photoconductivity over a time scale given by the electron-hole lifetime of the material. These processes allow a switch to be developed which can be closed by the application of one laser pulse ($\lambda \approx 1$ μm) and opened by the application of a second laser pulse with a wavelength about twice that of the turn-on laser.

Recent experimental results have shown that the current through a BOSS switch could not be fully quenched by the application of a 140-ps (FWHM) 2.13-μm laser pulse (Mazzola et al. 1992). A preliminary examination of the semiconductor rate equations indicated that the primary cause for incomplete photo-quenching was that the concentration of the recombination centers was too low. As stated above, the opening transient is the result of a two-step process. The second step is controlled by the electron-hole recombination lifetime which is dominated by the concentration of mid-gap recombination centers (RC) in the bulk material. A defect is considered a RC when the cross-sections for electron and hole capture are approximately equal. Usually, these centers are found near the middle of the bandgap. In this report we examine the effects of varying the RC concentration, by fast-neutron irradiation, on the opening transient of the BOSS switch.

The effects of neutron irradiation on semiconducting materials have been studied for many years (Vavilov 1963), (Borghi et al. 1970). A more recent effort has concentrated on the reduction of the minority-carrier lifetime in GaAs through fast-neutron irradiation (Wang et al. 1989). This work directed us towards the investigation of neutron damage for the purpose of RC enhancement in a BOSS device. The time integral of the neutron flux, called the *fluence*, has the units of neutrons/cm^2. The neutron sources that were used in this work are Sandia National Laboratory's (SNL) SPR-III reactor operating in the pulsed mode and the Annular Core Research Reactor (ACRR). The energy spectrum for these sources is peaked at 1 MeV.

2. SAMPLE PREPARATION

Low resistivity, silicon-doped (n-type) GaAs can be made semi-insulating by the introduction of copper acceptor levels through a thermal-diffusion process (Roush et al. 1993). The sample dimensions were 12 mm by 10 mm by 0.44 mm thick. After processing the material, the samples were irradiated with fast neutrons to increase the RC concentration. Thus far two sets of BOSS devices have been irradiated at SNL. The first set was irradiated in the SPR-III reactor operating in the pulsed mode prior to the fabrication of the electrical contacts. To achieve higher neutron fluences, the second set of devices were irradiated in the ACRR. To avoid excessive heating, the second set of devices were irradiated in three separate exposures. The sample temperature was held below 100°C during both sample runs. The second group of samples were irradiated in the ACRR after the contacts were fabricated. The p^+-i-n^+ devices were manufactured by depositing ohmic contacts that were 1 cm wide and separated by a 5-mm gap on the same side of the sample. After deposition, the contacts were annealed at 450°C for 5 minutes in N_2 at atmospheric pressure.

In the first run (SPR-III) the effect of neutron irradiation on the BOSS material was characterized by using four neutron fluences: 1.74x10^{15} cm^{-2} for Group I; 7.08x10^{14} cm^{-2} for Group II; 3.43x10^{14} cm^{-2} for Group III; and 1.61x10^{14} cm^{-2} for Group IV. The fluence values are given relative to 1-MeV GaAs-equivalent damage (Griffin et al. 1991). Following the irradiation, the sample's radioactivity was allowed to decay, for approximately one month, until it was below the background level. The dc I-V characteristics of the samples, shown in Fig. 1, indicate an increase in the switch resistance from about 60 kΩ for the non-irradiated devices to about 24 MΩ for the devices irradiated with the highest fluence. Figure 1 shows that neutron irradiation moves the Fermi level towards the middle of the bandgap.

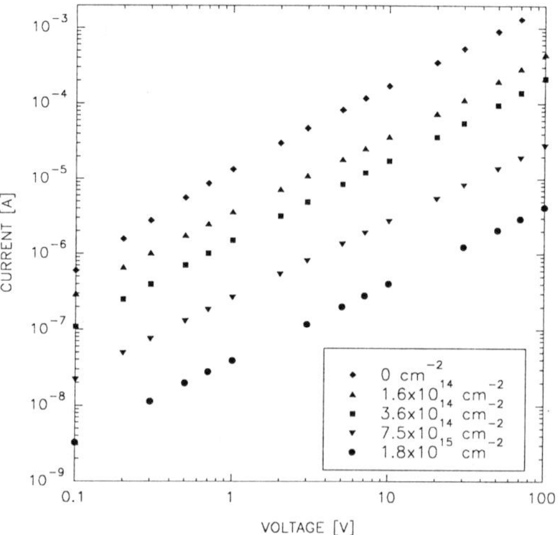

Fig. 1 Effect of neutron irradiation on the dc I-V characteristics of BOSS devices.

The switching results from the SPR-III run indicated that a higher neutron fluence was necessary to achieve the desired opening performance. Therefore, the second sample run was irradiated at the ACRR because it could more readily reach the desired neutron fluence. Two groups of samples were irradiated at the ACRR: 5.3x10^{15} cm^{-2} for Group V, and 1.8x10^{16} cm^{-2} for Group VI. To date only the devices from Group V have been investigated. The dc I-V characteristics of these devices indicated that the switch dark resistance was about 100 MΩ.

3. SWITCHING EXPERIMENTS

The BOSS-switching experiments were conducted with a mode-locked Nd:YAG laser system (1.06 μm), manufactured by Continuum Inc., that was equipped with a optical parametric generator (OPG) that served to double the wavelength (2.13 μm). The laser system produced a Gaussian pulse with a FWHM of about 140 ps. A simple optical delay was then used to adjust the time between switch closure and when the switch was opened. Photoconductivity measurements were performed to evaluate the operation of the neutron irradiated BOSS devices. The BOSS switches were embedded in a 50-Ω transmission line (two-way transit time = 4 ns) that was DC charged. The current through the device was measured by a 50-Ω current-viewing resistor (CVR) placed after the switch in the 50-Ω line. The current waveform was recorded by a Tektronix SCD5000 digitizer with a 4.5 GHz analog/digital bandwidth. The light was delivered to the switch via a 0.5-inch diameter quartz rod that was polished on the effluent end and had a rough-sanded finish on the end that was illuminated by the laser. This served to remove much of the mode structure in the laser pulse; thereby allowing nearly uniform switch illumination. The back face of the switch was illuminated to prevent shadowing of the bulk material located under the contact metalizations.

3.1. SPR-III Devices

The results of the photoconductivity experiments on the devices that were irradiated in the SPR-III are shown in Fig. 2 for three neutron fluences and the non-irradiated material for an applied bias voltage of 50 V. The incident pulse energy for the 1.06-μm laser was 2.1 mJ while the incident pulse energy for the 2.13-μm laser was 5.0 mJ.

As the neutron fluence was increased, the ability of the BOSS device to open was greatly improved. This is direct evidence that neutron irradiation creates recombination centers in GaAs. This also substantiates the earlier findings that indicated the need to increase the recombination center density in order to allow the BOSS switch to open in the subnanosecond regime. For a device in Group I, a curve fit to the switch conductivity during the opening phase, after it was extracted from the load line, indicated a recombination time on the order of 250 ps.

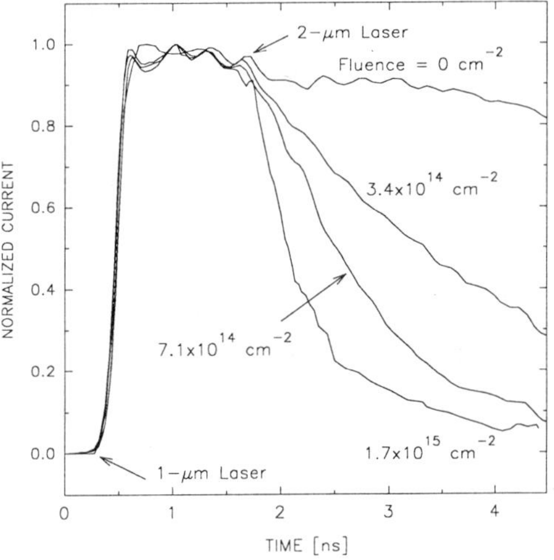

Fig. 2 Effect of neutron irradiation on the opening transient of BOSS devices for several neutron fluences (SPR-III) and non-irradiated material.

3.2. ACRR Devices

As stated above, so far only the devices in Group V have been analyzed. Initial photoconductivity measurements on a device in this group indicated that the RC density was too large because the BOSS device would not remain on after the first laser pulse was terminated. Thermal annealing experiments were then conducted to reduce the RC density by annealing out the damage created by the neutrons (Lang 1977). The results of the photoconductivity/annealing experiments are shown in Fig. 3. The bottom trace in Fig. 3 is the response of the switch after the copper leads were attached with silver epoxy that was annealed at 110°C for 15 minutes. The inability of the switch to remain closed after the 1-μm laser pulse was terminated was an indication that the electrons that were elevated into the conduction band are recombining with holes in the valence band before those holes could be trapped in the Cu_B level.

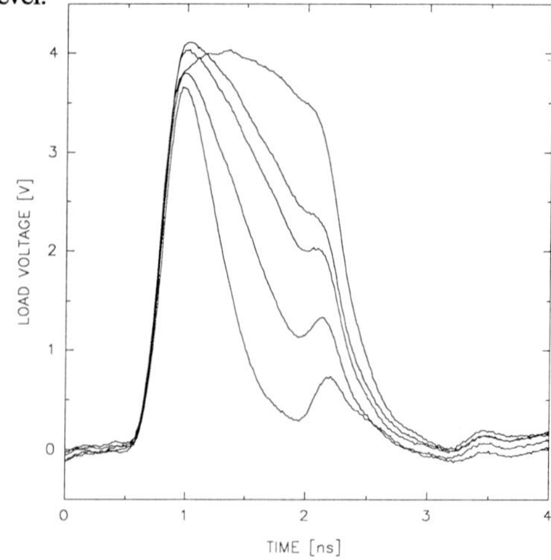

Fig. 3 Voltage across the 50-Ω CVR for an applied bias of 10 V. Effect of 15-minute thermal anneals on the photoconductivity response of BOSS devices irradiated in the ACRR. From the bottom: 110, 207,225,247, and 268°C.

To systematically remove some of the neutron-induced damage, one sample was annealed for 15 minutes at increasing temperatures of 207, 225, 247, and finally 268°C. Both dc I-V and photoconductivity measurements were made between each anneal. The anneals had a negligibly small effect on the dc I-V characteristics. However, the photoconductivity measurements showed a rapid increase in the tail current, or current after the 1-μm laser, with increasing temperature. A curve fit to the switch conductivity, which was extracted from the load line, indicated a recombination time of 180 ps after the 268°C anneal. Initial measurements indicate that the on-state of the switch is in the 10-Ω regime; clearly too high for any practical pulsed-power applications. Efforts are currently being made to reduce the on-state impedance to less than one ohm. Previous nanosecond high-power switching experiments demonstrated an on-state resistance of approximately 3.5 ohms in non-neutron irradiated BOSS devices operating at greater than 5 kV (Stoudt et al. 1993a).

The strong effect that these low-temperature anneals had on the switching performance of BOSS devices indicates that the 450°C contact anneals on the SPR-III samples probably removed some of the neutron damage. Previously reported annealing studies of neutron-irradiated GaAs (Aukerman et al. 1963) show one pronounced annealing step at about 200°C and a second, larger step starting at about 400°C . The effects that we are measuring in the ACRR samples are associated with the first step. However, the contact anneals on the SPR-III samples are in the range of the second annealing step. The actual effect of these steps on the switching mechanism is still under investigation.

It has also been reported that almost all of the neutron-induced defects are annealed out at temperatures between 500°C and 600°C (Coates and Mitchell 1975). This was also seen in our work when neutron-irradiated, n-type GaAs was compensated with copper at a diffusion temperature of 575°C. Although the dark resistance of the sample was in the low megohms, the photoconductivity data indicated that there was no enhancement of the RC density over the non-irradiated material. Therefore, the neutron-induced defects were probably annealed out by the 575°C anneal.

4. DEVICE MODELING

In this section we discuss the semiconductor rate equations which are obtained from the continuity equations for electrons and holes, the rate equations for trapping kinetics, the drift-diffusion current equations, and Possion's equation under the assumptions of spatial uniformity and charge neutrality. These assumptions are valid in the bulk region away from the contacts where the electric field can change rapidly. This is not unreasonable since the samples that were considered in this analysis were fitted with injecting contacts, made from a forward biased p^+-i-n^+ structure, which reduce the fields at the contacts considerably. The rate equations consider only the localized concentration of free and trapped charges, therefore, they are only valid under the conditions of low-carrier injection and moderate electric fields.

The transient development of the electron density in the conduction band (n), the hole density in the valence band (p), and the relative occupation numbers ($0 \le r_i \le 1$) of the various deep levels with a density N_i is given by the following set of rate equations (Stoudt et al. 1993b)

$$\frac{dn}{dt} = \dot{n}_{cv} + \sum_i N_i \dot{t}_{ci} ,$$
$$\frac{dp}{dt} = \dot{n}_{cv} + \sum_i N_i \dot{t}_{vi} , \qquad (1)$$
$$\frac{dr_i}{dt} = \dot{t}_{vi} - \dot{t}_{ci} .$$

The terms on the right hand side denote the various transitions between the deep levels and the conduction and valence bands that are included in the model. Band-to-band transitions ($\dot{n}_{cv}$) consist of direct recombination and thermal emission, and a contribution from the action of two-photon processes induced by the shorter wavelength laser. The terms describing transitions to and from the deep levels contain the effects of stimulated emission due to the external laser irradiation. Also considered are processes of trapping and spontaneous emission.

The rate equation model included two copper levels (Cu_B & Cu_A) which act as deep acceptors, one deep donor level (EL2) which is native to GaAs, and one recombination center (RC). The concentration of the RC level was varied from 4×10^{15} to 1×10^{16} cm^{-3} in the analysis to simulate the effect of increasing the neutron fluence. Because the rate equations are zero dimensional, they do not contain any of the circuit parameters in which the BOSS devices are embedded. To facilitate comparison with the experimental results, the values of the calculated electron and hole concentrations are used to determine the overall switch conductivity (σ) from the following relation

$$\sigma(t) = q\left(n(t)\mu_n + p(t)\mu_p\right) \, [\Omega - cm]^{-1} , \qquad (2)$$

where μ_n (=2900 cm^2/V-s) and μ_p (=400 cm^2/V-s) are the low-field electron and hole mobilities. Once the switch conductivity is determined, we calculate the switch resistance using a length of 0.5 cm and a cross-sectional area of 0.05 cm^2. The switch resistance is then placed into a 100-Ω load line, with an applied voltage of 50 V, to simulate the experiments.

In this section we discuss some of the theoretical results that were obtained with the rate-equation model. Two laser pulses at different wavelengths are used for the optical excitation source. One laser pulse with a photon energy of 1.165 eV (λ=1.064 μm) and a peak photon flux of 1.6×10^{26} cm^{-2}s^{-1} (29.9 MW/cm^2) is used to turn on the switch and a second laser pulse with a photon energy of 0.5825 eV (λ=2.128 μm) and a peak photon flux of 1.6×10^{26} cm^{-2}s^{-1} (14.9 MW/cm^2) is used to turn the switch off. The temporal shape of the laser pulse is assumed to be Gaussian with the standard deviation set equal to 70 ps in the simulation. The results of the simulations are shown in Fig. 4 where the switch current is plotted as a function of time for seven different RC concentrations. Figure 4 clearly shows that by increasing the RC concentration by a factor of 2.5, the ability of the BOSS switch to respond to the 140 ps laser pulse is dramatically improved. The on-state of the switch was not adversely effected until the RC concentration was increased to 1.0×10^{16} cm^{-3} as indicated by a reduction in the on-state current. A higher on-state resistance would be more apparent if the electron and hole mobilities were adjusted with the increase in ionized impurity scattering.

One drawback of an increased RC concentration is that the on-state conductivity will be reduced because electrons in the conduction band will recombine with holes in the valence band before those holes can be trapped in the Cu_B center. This process reduces the number of holes that are trapped in the Cu_B center which, in turn, reduces the available sites to receive electrons from the valence band during the turn-off laser pulse. The benefits of an increased RC concentration are apparent during the simulated turn-off transient of the switch. The primary consequence of these results is that a compromise has to be reached between the benefits of a high RC density for the turn-off transient, and the repercussions of a lower on-state conductivity.

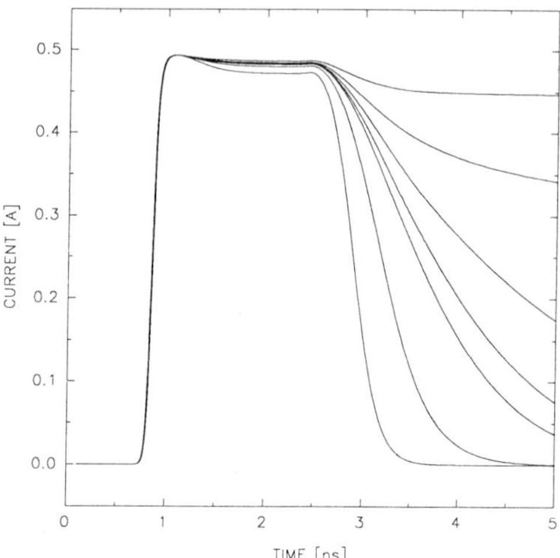

Fig. 4 Simulated current response of a BOSS switch. Recombination center density [x10^{15} cm^{-3}]: (from the top) 4,5,5.5,5.8,6,7,10.

5. CONCLUSION

Experiments were performed to determine the effect of irradiating BOSS material with several different 1-MeV neutron fluences. One effect was that the Fermi level was moved towards the middle of the bandgap, thereby increasing the off-state resistivity of the devices. Simulation studies were performed on the effect of increasing the recombination center density on the BOSS switching cycle. The semiconductor rate equations were solved showing that by increasing the recombination center density, the BOSS switch could be opened in as little as several hundred picoseconds. The theoretical results were in qualitative agreement with the experimental results which showed that by increasing the neutron fluence, the photo-quenching effect could be enhanced. These results support the claim that fast-neutron irradiation creates deep levels in the middle of the bandgap and that these levels function as fast recombination centers. Further work will be necessary to find the optimum set of parameters that result in a BOSS device that can be opened in 100 ps and still have a sufficiently low on-state resistance.

REFERENCES

Aukermann, L.W., Davis, P.W., and Graft, R.D. 1963, J. Appl. Phys., **34**, 3590.

Borghi, L., Stefano, P.De., and Mascheretti, P. 1970, J. Appl. Phys., **41**, 4665.

Coates, R. and Mitchell, E.W. 1975, Adv. Phys., **24**, 593.

Griffin, P.J., Kelly, J.G., Luera, T.F., Barry, A.L., and Lazo, M.S. 1991, IEEE Trans. Nuclear Sci., **38**, 1216.

Lang, D.V. 1977, Inst. Phys. Conf. Ser. No. 31, 70.

Mazzola, M.S., Roush, R.A., Griffiths, S.F., Stoudt, D.C., and Keil, D.H. 1992, Proc. 20th Power Mod. Symp., 266.

Mourou, G. and Knox, W. 1979, Appl. Phys. Lett., **35**, 492.

Roush, R.A., Stoudt, D.C., and Mazzola, M.S. 1993, Appl. Phys. Lett., **62**, 2670.

Schoenbach, K.H., Lakdawala, V.K., Germer, R., and Ko, S.T. 1988, J. Appl. Phys., **63**, 2460.

Stoudt, D.C., Kenney, J.S., Schoenbach, K.H., Roush, R.A., Ludwig, S., and Mazzola, M.S. 1993a, submitted to IEEE Trans. Electron Devices.

Stoudt, D.C., Brinkmann, R.P., Roush, R.A., Mazzola, M.S., Zutavern, F.J., and Loubriel, G.M. 1993b, submitted to IEEE Trans. Electron Devices.

Vavilov, V.S. 1963, Effects of Radiation on Semiconductors (Translation by the Consultants Bureau Enterprises, Inc. New York).

Wang, C.L., Pocha, M.D., and Morse, J.D. 1989, Appl. Phys. Lett., **54**, 1451.

Inst. Phys. Conf. Ser. No 136: Chapter 5
Paper presented at the Int. Symp. GaAs and Related Compounds, Freiburg, 1993

Optoelectronic effects and field distribution in strained (111)B InGaAs/AlGaAs MQW PIN diodes

J.L.Sánchez-Rojas[+], E.Muñoz[+], A.S.Pabla[*], J.P.R.David[*], G.J.Rees[*], J. Woodhead[*] and P.N.Robson[*]

[+]Departamento de Ingeniería Electrónica. E.T.S.I. Telecomunicación. Ciudad Universitaria. 28040 Madrid. Spain.

[*]Department of Electronic and Electrical Engineering, Mappin Street, Sheffield, S13JD. England.

ABSTRACT: A quantitative model describing the influence of the number of quantum wells on the optoelectronic properties and electric field profile, in [111]B-oriented MQW pin diodes, is presented. The implications for the design of optoelectronic and all-optical devices are described.

1. INTRODUCTION

There has been recently a growing interest in the physics and applications of the strong piezoelectric fields within InGaAs layers grown along the [111] orientation (Smith et al 1990). In InGaAs/GaAs single quantum well (SQW) pin structures, photoluminescence (PL) blue shifts due to an external reverse voltage and to carrier screening have been reported (Caridi et al 1991 and Moise et al 1991). In this paper we present a quantitative model describing the field distribution in [111]B-oriented MQW pin diodes, and how the number of quantum wells, nQW, influences these optoelectronic effects.

2. MODEL

We consider a pin diode that contains an InGaAs/GaAs MQW (n quantum wells) in its intrinsic region, V_r being the applied reverse bias. Prior to considering the energy levels, the pin potential profile needs to be determined. For the ideal case (no residual doping), linear potential variations throughout the whole intrinsic region can be assumed

$$V_r + V_{bi} = E_w L_w + E_b L_b \qquad (1)$$

where V_{bi} is the built-in voltage, and L_w (L_b) and E_w (E_b) are the total n-well (barrier) length and field, respectively. We suppose that all the strain is elastically accommodated in the QW's, *inducing an electric field* $E_p = \sqrt{3}\, e_{14}\varepsilon_{ij}/\varepsilon$, with e_{14} being the piezoelectric constant, ε_{ij} the off-diagonal term in the strain tensor, and ε is the dielectric constant of free space times the permittivity of the medium. Then, well and barrier fields differ by just E_p, and the total well field is given by the sum of E_p and E_b. The electric field in the barrier is then

$$E_b = \frac{V_r + V_{bi} - L_w E_p}{L} \quad (2)$$

where $L=L_w+L_b$ is the total length of the intrinsic region.

When the diode is illuminated, the optically generated electrons and holes are stored in the QW's, partially separated by and screening the piezoelectric field. Now the Hartree potential has to be calculated self-consistently and is added to the linear potential defined above. For simplicity, we assume no coupling between QW's, and all of them having the same concentration of electrons and holes (uniform intrinsic region illumination). The free carrier screening produces a potential drop (V_s) across the whole intrinsic region, giving rise to a correction in the barrier electric field, $-V_s/L$, as compared with the empty-well case. This correction has to be introduced in a self-consistent scheme, as V_s changes with each iteration.

In the calculation of electron and hole subbands a self-consistent approach has been developed. For the conduction band, a simple isotropic parabolic band is considered. For the valence band, we use the [111] 4x4 Luttinger-Kohn Hamiltonian, transformed by a unitary operator into a block diagonal form (Ikonic et al 1992):

$$H_p = \begin{bmatrix} X_+ & Y & 0 & 0 \\ Y' & X_- & 0 & 0 \\ 0 & 0 & X_- & Y \\ 0 & 0 & Y' & X_+ \end{bmatrix} + V_h(z) \quad (3)$$

$$X\pm = (\gamma_1 \pm \gamma_3)\, k_t^2 - (\gamma_1 \pm 2\gamma_3)\, k_z^2 \pm \Delta E_{sh}$$

$$Y = (\gamma_2 + 2\gamma_3)/\sqrt{3}\; k_t^2 - 2i(2\gamma_2 + \gamma_3)/\sqrt{3}\; k_t k_z$$

where γ_1, γ_2 and γ_3 are the Luttinger parameters, taken as a linear interpolation between the values for the binary constituents, and $V_h(z)$ is the potential for holes, given below. The shear-strain-induced band splitting ΔE_{sh} is defined in (8), and $k_t^2 = k_x^2 + k_y^2$. The effective mass Hamiltonian for the envelope functions of the electrons in the conduction band is

$$H_n = -\hbar^2\nabla^2/\, 2m_e^* + V_e(z) \quad (4)$$

where m_e^* is the effective electron mass for InGaAs, and $V_e(z)$ is the potential for the electron. The γ parameters, the effective masses and the dielectric constant are assumed to be the same in wells and barriers.

The potentials for electrons and holes are the sum of the offset potential for each band, the external bias, the piezoelectric term (inside the QW's) and the screening potential when optical excitation is considered. We neglect many body effects, and the Hartree potential $\phi(z)$ is the solution of the Poisson equation

$$\nabla^2\Phi = -q/\varepsilon \; (p(z)-n(z)) \tag{5}$$

with the carrier densities for holes and electrons given, respectively, by

$$p(z) = \sum_{\sigma}\sum_{m}\sum_{k_t} |\psi^{\sigma}_{mk_t}(r)|^2 \; [1- f(E_{mk_t} - F_p)] \tag{6}$$

and

$$n(z) = \rho_2 \, k_B \, T \sum_{l} |\Psi_l(z)|^2 \ln\left[1 + \exp\left(\frac{F_n - E_l}{k_B T}\right)\right] \tag{7}$$

where f is the Fermi function, ρ_2 is the 2D density of states for the parabolic conduction band, m and l are the subband index for each band, σ denotes the upper or lower block in H_p, and F_p and F_n are the quasi-Fermi levels for holes and electrons.

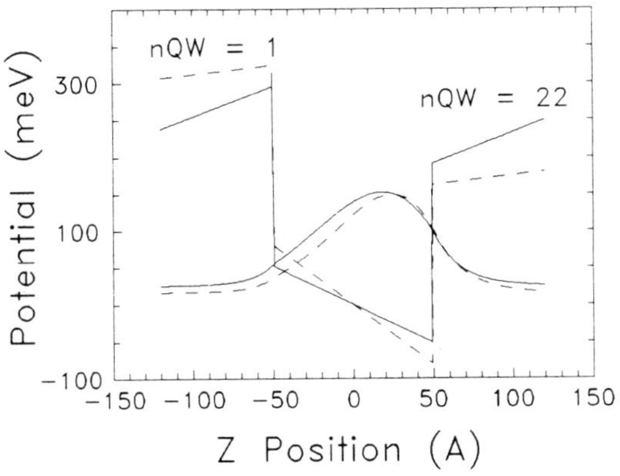

Fig. 1. Conduction band profiles and groundstate wavefunctions.

The self-consistent loop is started with the piecewise-linear potential obtained from (2). Then, E_b is corrected each iteration, as explained above. When convergence is reached, energy levels, wavefunctions, quasi-fermi levels, and charge distributions are obtained. Next, as we are interested in optical transition energies, the effect of biaxial strain on the InGaAs bandgap, E_g, is determined. In a first order approach by considering the shear and hydrostatic contributions,

$$E_g = E_{go}(x) + \frac{1}{2}\Delta E_{sh} - \Delta E_{hy} \tag{8}$$

where

$$\Delta E_{sh} = -\sqrt{3} \;\; \mathbf{d} \; [(C_{11}+2C_{12})/(C_{11}+2C_{12}+4C_{44})]\epsilon$$

$$\Delta E_{hy} = -3\mathbf{a}[4C_{44}/(C_{11}+2C_{12}+4C_{44})]\epsilon$$

ϵ being the lattice mismatch between the epitaxial layer and substrate. The elastic constants, C_{ij}, the deformation potentials, **a** and **d**, are taken as linear interpolation between values for the binary constituents from Landölt-Borstein (1990). The x-dependence of the AlGaAs and of the non-strained InGaAs (E_{go}) band gaps are described as a second order polynomial, with bowing parameters taken from the same reference. Conduction band offset is assumed to be 65% of the bandgap difference.

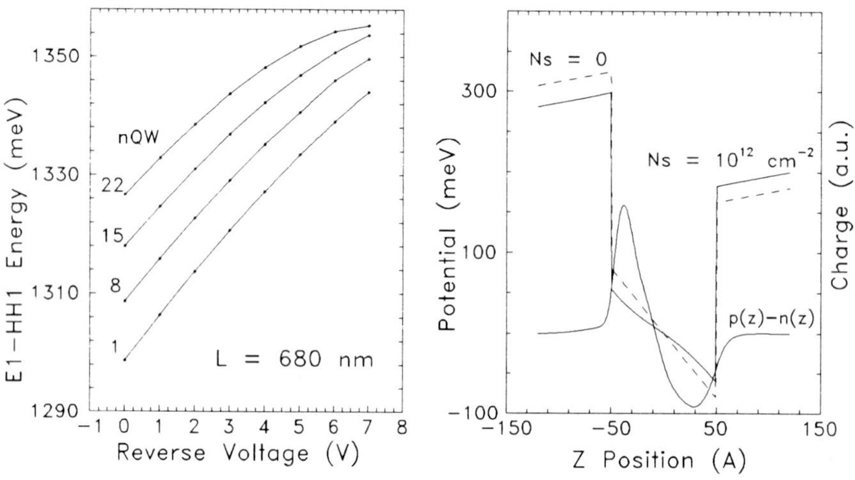

Fig. 2. Transition energies versus reverse bias.

Fig. 3. Conduction band and charge distribution.

Numerical solution of equation (3) is obtained by expanding the wavefunctions in terms of an orthonormal set given by the heavy- and light-hole states in the zone center (Ando 1985). The final differential equations are solved by a finite element technique.

3. RESULTS AND DISCUSSIONS

We have evaluated the band structure changes produced by an external voltage, or by photocarriers (all-optical device), in a pin diode with a MQW in its intrinsic region. This structure plays an important role in designing blue-shifting self-electro-optic effect devices (BS-SEEDs) (Goossen et al 1990). In $In_{.12}GaAs/Al_{.22}GaAs$ structures with one, eight, fifteen, and twenty two QW's (100Å wide), respectively, whith their total intrinsic region length kept constant and equal to 680 nm, have been analyzed. In diodes with one QW (dashed line) and with 22 QW's (solid line), potential profiles, and electron groundstate wavefunctions at the zone center, are shown in figure 1. It is important to note, in these two structures, how nQW influences the electric fields in wells and barriers, giving in turn different energy levels and wavefunctions. Structures with higher nQW should give improved quantum efficiency and absorption, due to the enhanced field in the barriers and to the resulting higher carrier sweep-out rates (tunneling through the triangular barrier) and higher exciton oscillator strength.

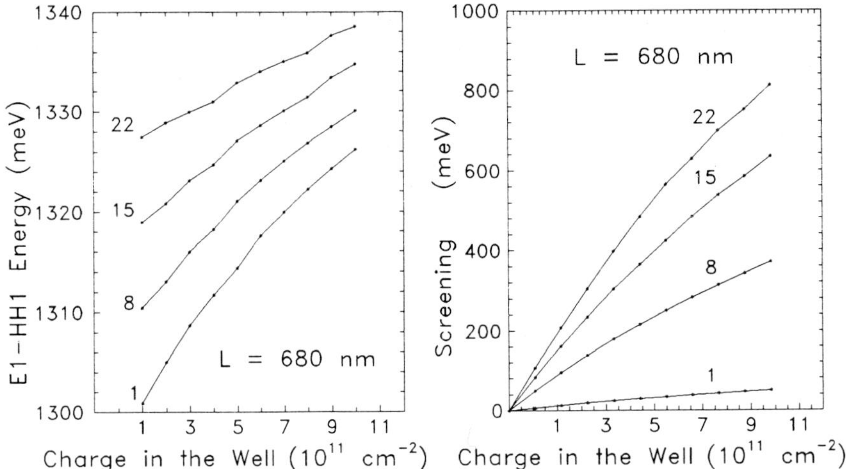

Fig. 4. Transition energies versus carrier density in the QW.

Fig. 5. Screening voltage caused by photocarriers.

For the above four structures, the calculated fundamental optical transitions (first electron level to first heavy hole level), are plotted in figure 2 as a function of the reverse voltage. First, we see that there is nearly 30 meV transition energy difference between the diodes with 22 and with one QW. This is due to the enhanced field in the barriers of the 22 QW sample, i.e. a lower total field in the wells, thus, blue-shifting the transition. Experimental validation of this point has been reported by Pabla et al (1993). Another consequence of this effect is the different flatband voltage in these two samples, leading us to conclude that, for a given intrinsic region thickness, the greater nQW the lower the flat-band voltage required. Therefore, in system applications, a *higher nQW device would reduce its operating voltage*, as the prebias towards flat-band condition has been shown to be an important parameter in BS-SEED design (Goosen et al 1990).

For the structure with one QW, the effects of the free carrier population on the band profiles is shown in figure 3, assuming that $n=p=10^{12}cm^{-2}$ and T=300K. When the light intensity increases, carrier accumulation increases, and screening becomes stronger, producing more and more blue-shifting in the optical transitions. This result is reflected in figure 4. It is important to notice that the sensitivity of the transition energies to the carrier density is higher for the structure with lower nQW. This effect should be even stronger in a real situation, where the carrier generation in each QW is not the same (absorption along the intrinsic region).

These results have practical implications for the operating mode of an all-optical BS-SEED. Higher photon intensities should be used to switch a pin diode with larger nQW, assuming that the rest of conditions are the same.

The total screening voltage, V_s is plotted in figure 5 for the same structures and carrier concentrations as in figure 4. First, we note that V_s can be comparable to V_{bi}, and that the *final correction to the electric field in the barrier is on the order of 10^4 V/cm.* From device design point of view, it is important to point out that V_s is less than linear with nQW: the screening for QW in a structure with 22 QW's is approximately 75% of that in the structure with one QW, indicating again the lower sensitivity of the former, as previously seen in figure 4.

4. CONCLUSIONS

We have shown how the number of QW's influences the electrooptic and the all optical behavior of [111]B-oriented MQW pin diodes. The blue-shifts due to a reverse voltage, and to screening by photocarriers, have been calculated, showing the sensitivity of these effects in terms of the nQW. Structures with higher nQW have a lower flat-band voltage (better for optoelectronic devices), but they have also a smaller sensitivity to optical power (all-optical devices). Band structure calculations have been made selfconsistently, including a [111], 4x4 Hamiltonian for the valence band.

5. ACKNOWLEDGMENTS

We thank Universidad Politécnica de Madrid and CICYT TIC 93-0025 for financial support. Valuable discussions with J.Sánchez-Dehesa are acknowledged. We are also indebted to the Physics Department of the University of Sheffield.

6. REFERENCES

Ando T 1985 J. Phys. Soc. Jpn. **51** pp. 1528-1536

Caridi E A, Chang T Y, Goossen K W and Eastman L F 1990 Appl.Phys.Lett. **56** pp. 659-661

Goossen K W, Caridi E A, Chang T Y, Stark J B, Miller D A B and Morgan R A 1990 Appl. Phys. Lett. **56** pp. 715-717

Ikonic Z, Milanovic V and Tjapkin D 1992 Phys. Rev. B **46** pp. 4285-4288

Landölt-Bornstein 1990 Vol.17 Springer-Verlag Berlin

Moise T S, Guido L J, Barker R C, White J O and Kost A R 1992 Appl. Phys. Lett. **60** pp. 2637-2639

Pabla A S, Sanchez-Rojas J L, Woodhead J, Grey R, David J P R, Rees G J, Hill G, Pate M A, Robson P N, Hogg R A, Fisher T A, Willcox A R K, Whittaker D M, Skolnick M S and Mowbray D J 1993 Appl. Phys. Lett. to be published

Smith D L and Mailhiot C 1990 Rev. Mod. Phys. **62** pp. 173-234

Inst. Phys. Conf. Ser. No 136: Chapter 5
Paper presented at the Int. Symp. GaAs and Related Compounds, Freiburg, 1993

Scaling characteristics of picosecond interdigitated photoconductors

N de B Baynes[1], J Allam[2], J R A Cleaver[1], K Ogawa[2], I Ohbu[3] and T Mishima[3]

[1]Microelectronics Research Centre, Cavendish Laboratory, University of Cambridge, Madingley Road, Cambridge CB3 0HE.
[2]Hitachi Cambridge Laboratory, Hitachi Europe Limited, Cavendish Laboratory, Madingley Road, Cambridge CB3 0HE.
[3]Hitachi Central Research Laboratory, Hitachi Ltd., 1-280, Hitachi-koigakubo, Kokubunjishi, Tokyo 185, Japan.

ABSTRACT: We have measured the bandwidth, responsivity and breakdown voltage of interdigitated photoconductors fabricated on gallium arsenide grown at low temperature, for devices with electrode spacing from 0.5 μm to 8 μm. Contrary to expectation, the pulse height did not increase for decreasing gap width. This is due both to the relation between the initial photocarrier distribution and the field distribution, and to avalanche multiplication effects. Polarisation dependence of the responsivity was also observed.

1. INTRODUCTION

Planar interdigitated photodetectors are promising candidates for opto-electronic integrated circuits due to their high-speed, high-responsivity and simple integration with planar FETs. Metal-semiconductor-metal photo*diodes* with electrode spacing as small as 25 nm have been fabricated using electron beam lithography (Chou et al 1992). Interdigitated photo*conductors* using LT GaAs (grown by molecular beam epitaxy at low temperature and annealed *in situ*) exhibit sub-picosecond response times (Gupta et al 1991), high breakdown fields and high responsivity (Chen et al 1991).

The bandwidth and responsivity of photoconductors depends strongly on the electrode dimensions. For parallel plate electrodes, the responsivity is proportional to the ratio of the recombination time to the transit time and therefore is inversely proportional to the photoconductive gap width (d). For planar interdigitated electrodes the non-uniform field and depth-dependent photogenerated carrier density can be expected to influence the scaling properties. Klingenstein et al (1992) compared planar devices with 0.75 μm and 10 μm gaps at the same nominal field (applied voltage / photoconductive gap width), and found the *same* pulse height. However, Chen et al (1991) report a responsivity of 0.1 $A \cdot W^{-1}$ for photoconductors with 0.2 μm gaps compared with 10^{-3} $A \cdot W^{-1}$ for 20 μm gaps; i.e. responsivity scaling inversely with d. In this paper we study theoretically and experimentally the scaling of the bandwidth and responsivity in interdigitated photoconductors fabricated on LT GaAs and also report two additional effects (avalanche multiplication and polarisation dependence) which influence measurement of the responsivity and response times.

2. CALCULATED SCALING OF PULSE HEIGHT

We have developed a two-dimensional dynamical simulation for the pulse response of interdigitated photoconductors. Photocarriers are generated according to the illumination intensity, wavelength (λ), absorption length ($1/\alpha_0$) and electrode configuration. The field due to the contacts is calculated by conformal mapping (Lim & Moore 1968). The field in one cell of a device with finger width w = d = 1 μm is shown in Figure 1. Also shown is the depth profile of photogenerated carriers. For d < ($1/\alpha_0$) carriers generated away from the surface will experience a significantly reduced field.

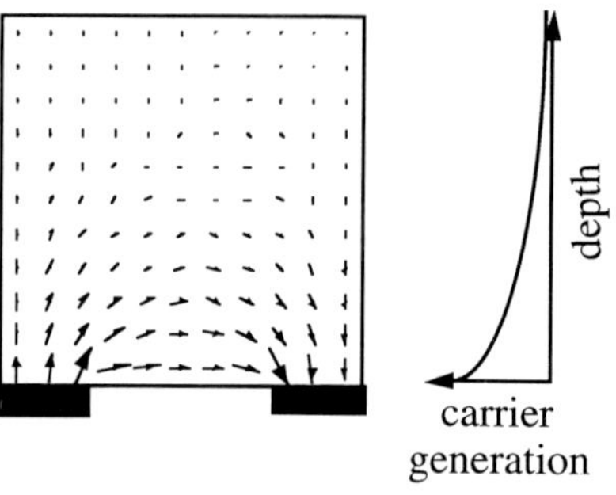

Figure 1. Field distribution in 2 μm x 2 μm unit cell of interdigitated photodetector with 1 μm fingers and gaps. Also shown is the depth profile of the photocarrier generation.

The evolution of the pulse response is studied by modelling the dynamics of individual charge carriers under the applied field, with carrier capture at randomly-distributed traps. The details will be given elsewhere, however the pulse height $I_p(0)$ for delta-function illumination can be calculated without considering the subsequent carrier transport, i.e. before any trapping has occurred. The instantaneous current (including both particle and displacement currents) flowing in the external circuit due to the motion of a carrier of charge q and velocity $\mathbf{v}(\mathbf{r})$ in a field $\mathbf{F}(\mathbf{r})$ is given by Ramo's theorem

$$i = q\mathbf{v}(\mathbf{r}) \cdot \mathbf{F}(\mathbf{r}) / \phi$$

where ϕ is the potential difference between the electrodes (Böttcher et al 1993). Hence in order to calculate the pulse height we calculate the field F_m and number of photogenerated carriers n_m on a (40 x 40) mesh and evaluate

$$I_p(0) = \sum\{n_m F_m v(F_m)/\phi\}.$$

The velocity was assumed to be independent of field (steady-state transport at high field). In the typical operating regime of high fields ($\approx 10^7$ $V \cdot m^{-1}$) and fast capture times (< 1 ps), velocity overshoot effects are important and a more detailed treatment of the transport is required. However the essentially geometrical effect of the planar electrodes and finite absorption length on the scaling of $I_p(0)$ will not depend on the details of the transport.

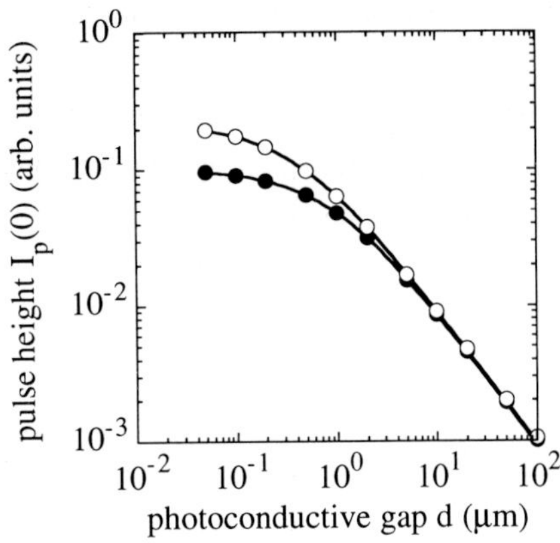

Figure 2. Pulse height calculated for different finger dimensions;
● $\lambda = 0.8\ \mu m$ ($\alpha_0 = 1.8\ \mu m^{-1}$)
○ $\lambda = 0.6\ \mu m$ ($\alpha_0 = 4\ \mu m^{-1}$).

Figure 2 shows the pulse height $I_p(0)$ as a function of photoconductive gap from 0.05 μm to 100 μm, for two different absorption lengths corresponding to $\lambda = 0.8\ \mu m$ and 0.6 μm. The calculation assumes devices with w = d, constant total interdigitated area (A) and constant nominal field $F = \phi/d$. For $d \gg 1\ \mu m$ we find $I_p(0) \propto (1/d)$ whereas for $d \ll 1\ \mu m$ the pulse height approaches a constant. This is a result of the decreasing field for carriers generated away from the semiconductor surface. For $\lambda = 0.6\ \mu m$ and $d = 0.2\ \mu m$ (corresponding to the measurements of Chen et al), the peak height is reduced by a factor of ≈ 3 compared to the value expected for (1/d) scaling.

3. EXPERIMENTAL MEASUREMENT OF BANDWIDTH AND RESPONSIVITY

In order to investigate the scaling experimentally, we have fabricated interdigitated photoconductors on 2 μm thick LT GaAs grown at a nominal temperature of 200 °C and annealed for 10 minutes at 600 °C. The devices were fabricated using electron beam lithography and lift-off of 100 nm thick Ti/Au, with equal finger and gap widths of 0.5 μm to 8 μm and an area of 28 x 32 μm^2. They were contacted by coplanar striplines with 40 μm gaps and 20 μm lines.

The response time was measured by photoconductive sampling using ≈ 60 fs pulses at 800 nm wavelength from a Ti-sapphire laser. Two optical pulses were focussed onto the interdigitated switch, and the time delay between the pulses was varied using a retro-reflector mounted on a fast scanning galvanometer. The photocurrent through the switch was amplified by a current preamp and displayed on a digitising oscilloscope for signal averaging. The photocurrent generated by each of the laser pulses is modified due to the change in the bias caused by current generated by the other laser pulse; hence the current as a function of delay time yields the autocorrelation of the device response.

The full width at half maximum (FWHM) of the autocorrelation of devices with constant area and varying finger widths is shown in figure 3 as a function of the parasitic capacitance calculated by conformal mapping. The capacitance of the 4 μm device, which has only four fingers, is overestimated by the calculation which assumes periodic boundary conditions. A straight line fit (ignoring the data for the 4 μm device) gives an intercept of 2.4 ps, corresponding to an exponential decay time of 1.7 ps in the LT GaAs. The slope is 75 Ω, consistent with a parasitic contribution to the pulse width of ln2•RC, where R = 108 Ω is close to the calculated impedance of the coplanar stripline of 100 Ω.

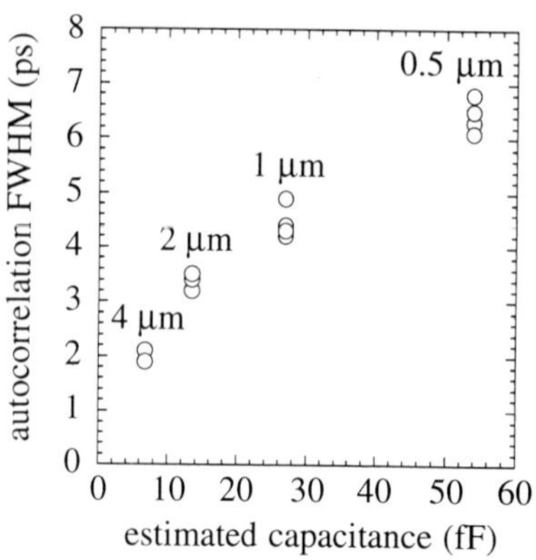

Figure 3. Autocorrelation FWHM for interdigitated photoconductors.

The variation with nominal field of the pulse height and bandwidth (calculated from the numerical Fourier transform of the autocorrelation signal) is shown in Figure 4 for the 1 μm and 2 μm interdigitated structures. The contribution of the parasitic capacitance to the bandwidth and responsivity was removed by calculating the response of the known RC filter to exponential pulses, consistent with Figure 3. The deconvolved data is shown in Figure 5. The 1 μm and 2 μm devices show the same bandwidth at low fields. The pulse height of the 2 μm device is a factor of ≈ 2 *greater* than that of the 1 μm device at the same nominal field, whereas the results of Figure 2 predict a pulse height a factor of ≈ 1.5 *less* for the 2 μm device. The increased pulse height for the 2 μm device, as well as the field dependence of the bandwidth and pulse height, is attributed in the following section to carrier multiplication.

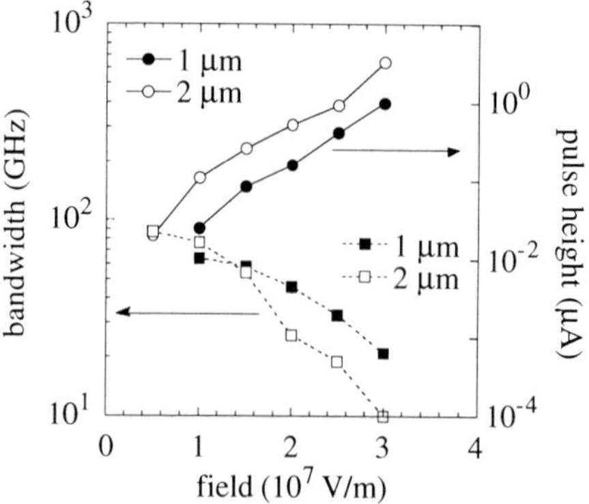

Figure 4. Measured response of detectors with 1 μm and 2 μm fingers and gaps.

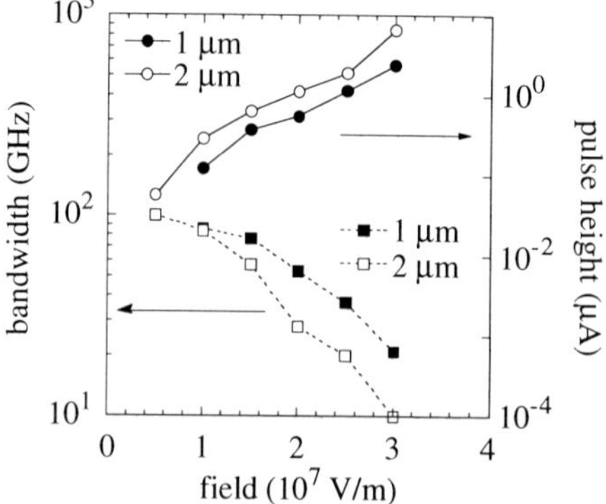

Figure 5. Intrinsic response after deconvolution of the parasitics.

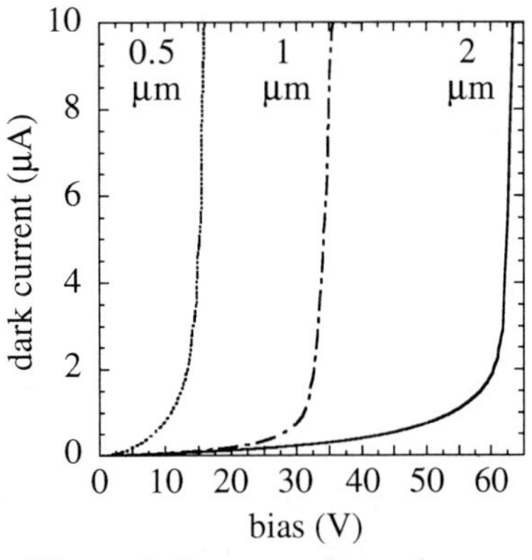

Figure 6. Current-voltage characteristics of 0.5 µm, 1 µm and 2 µm interdigitated devices.

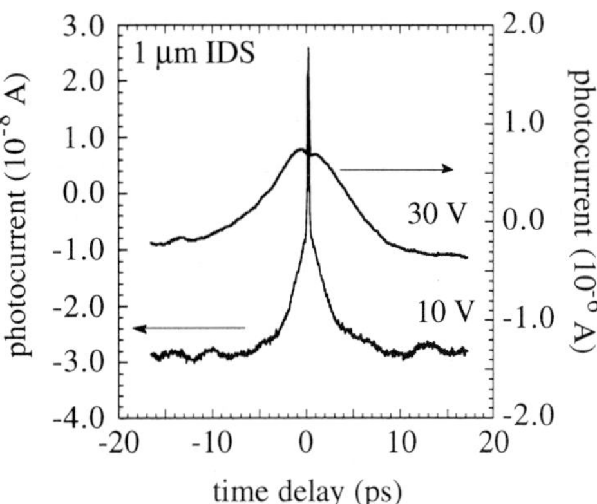

Figure 7. Pulse response of 1 µm interdigitated device at 10 V and 30 V bias.

4. IONISATION AND POLARISATION EFFECTS

Two additional effects were observed in these devices which influence the measurement of the responsivity. All devices fabricated on LT GaAs grown at 200 °C showed a sharp breakdown (Figure 6) at $F \approx 3 \times 10^7$ V•m^{-1}. At fields close to breakdown the pulse response is broadened and the peak response is delayed from zero time. Figure 7 shows the autocorrelation signal; the narrow spike at zero time delay is an artefact due to coherent addition of the laser pulses overlapping in time and position. This behaviour suggests avalanche multiplication due to secondary carriers created by ionisation. Although the breakdown field is consistent with band-to-band impact ionisation in intrinsic GaAs, the exponential increase of the pulse height and decrease in the bandwidth as the field is increased (Figure 5) suggests multiplication of only one carrier type, and impact ionisation or field ionisation of trapped carriers may be important. Such behaviour was not observed in LT GaAs grown at 220 °C (which showed a linear increase in pulse height and constant pulse width up to $\approx 2 \times 10^7$ V•m^{-1}) and 180 °C (which showed photocurrent saturation at $\approx 10^7$ V•m^{-1}).

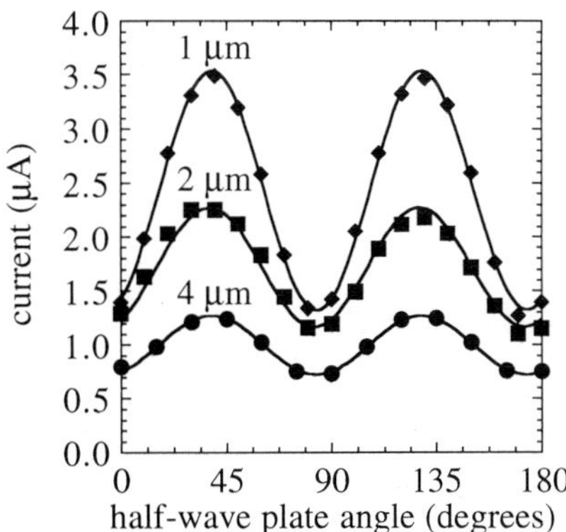

Figure 8. Polarisation dependence of the dc photocurrent for different devices.

We also observed a strong polarisation dependence for the interdigitated structures. Figure 8 shows the dc photocurrent as a function of polarisation angle (varied by rotating a half-wave plate) for devices with d = w = 1, 2 and 4 μm. The modulation depth increases as the finger width approaches the wavelength of the light (0.8 μm). These effects are expected in a wire grid array due to attenuation of the electric field parallel to the wires by induced currents.

5. DISCUSSION AND CONCLUSIONS

Based on a simple model of the initial photocarrier distribution and Ramo's expression for the instantaneous current, we have predicted a failure of the simple scaling rule $I_p(0) \propto (1/d)$ for small devices where d is comparable to the absorption length. This is apparently contrary to the data both of Klingenstein et al (1992) and of Chen et al (1991). We have shown that parasitic pulse broadening and polarisation effects, not discussed by Klingenstein et al, can lead to a reduction in the measured pulse height of small interdigitated photoconductors. There is no advantage to be gained by fabricating devices with d « 1 μm.

The measured pulse broadening due to parasitics was consistent with the estimated capacitance, and could be reduced by a factor of 10 by reducing the area to 10 μm x 10 μm. Even after accounting for the parasitic capacitance we find that the the 2 μm interdigitated structure has a pulse height a factor of 2 greater than the 1 μm device - the opposite behaviour compared to that expected from simple (1/d) scaling. We attribute this to the additional multiplication observed in the wider device at the same nominal field, which also accounts for the reduced bandwidth of the 2 μm device at high fields. Further experimental investigation of scaling of interdigitated structures should be made on material which does not exhibit such multiplication effects.

We observed a strong polarisation dependence when the finger spacing becomes of the same order as the light wavelength. Such behaviour is expected for wire grid polarisers, but somewhat surprisingly we have seen no previous reports of such polarisation dependence. It is evidently important to match the orientation to the polarisation to achieve the maximum responsivity in interdigitated detectors.

6. REFERENCES

Böttcher E H, Hieronymi F, Kuhl D, Dröge E and Bimberg D 1993 Appl Phys Lett 62 p2227
Chen Y, Williamson S, Brock T, Smith F W and Calewa A R 1991 Appl Phys Lett 59 p1984
Chou S Y, Liu Y and Fischer P B 1992 Appl Phy Lett 61 p 477
Gupta S, Frankel M Y, Valdmanis J A, Whittaker J F, Mourou G A, Smith F W and Calewa A R 1991 Appl Phys Lett 59 p3276
Klingenstein M, Kuhl J, Noetzel R, Ploog K, Rozenweig J, Mogelstue C, Huelsmann A, Schneider J and Koehler K 1992 Appl Phys Lett 60 p627
Lim Y C and Moore R A 1968 IEEE ED-15 p173

Inst. Phys. Conf. Ser. No 136: Chapter 5
Paper presented at the Int. Symp. GaAs and Related Compounds, Freiburg, 1993

III–V devices and technology for monolithically integrated optical sensors

Hans P Zappe, Hazel EG Arnot and John E Epler

Paul Scherrer Institute, Badenerstrasse 569, 8048 Zurich, Switzerland

ABSTRACT: Further development of GaAs-based optical circuits will have a significant impact on the field of integrated optical sensors. We present here an integrated interferometer fabricated in GaAs/AlGaAs designed specifically for use as an optical sensor. On a single III-V layer structure, waveguides, lasers, modulators and detectors were fabricated and configured into a Mach-Zehnder interferometer circuit. A modulator and photodetector were monolithically integrated; the laser light source was hybridly coupled. This structure is under further development as an integrated optical medical sensor, whose optics are based solely on III-Vs.

INTRODUCTION

The economies of scale which have so greatly reduced the price while concomitantly increasing the density and functionality of electronic integrated circuits may soon be applicable to integrated optical sensors based on III-V technology. Through complete monolithic integration of a photonic sensor circuit, we may take advantage of the associated increase in robustness, reduction in size and enhancement of processing capabilities. The integration of waveguides with active optical elements such as lasers (Andrew 1992), detectors (Emeir 1992), and modulators (Wickman 1991), has been widely addressed, primarily in the area of telecommunications where speed is of primary interest. However, opportunities for monolithic integration open numerous opportunities for the manufacture of fully self-contained optical sensors, for which simplicity and flexibility often guide a design. It is foreseen that the fabrication of such III-V integrated sensor chips will permit a greater field of application and thus improved marketability of optical sensors.

We present here a GaAs/AlGaAs-based technology for the fabrication of the optoelectronic elements required for an integrated optical sensor circuit, and the use of these as a basis for a Mach-Zehnder interferometer sensor platform. We discuss the technology, individual device elements, and prospects for hybridization and integration.

DEVICE, PROCESS AND CIRCUIT REQUIREMENTS

An optical sensor may measure a change in phase or intensity of an optical mode, measure luminescence intensity or quenching or monitor changes in the coupling efficiency of a grating (Brandenburg 1992). Interferometers are a sensitive means to measure a phase change of an optical signal in response to presence of the analyte. Since interferometers tend to be bulky, complicated, and expensive apparatuses, they lend themselves ideally to, and benefit greatly from, monolithic integration.

A useful configuration is the Mach-Zehnder interferometer shown schematically in Figure 1; it consists of two arms, where the phase of the mode in one arm is shifted due to a change in the refractive index of the material in the sensor pad. Thus the completed sensor consists essentially of two parts, the *III-V-based platform* (which we will discuss here), consisting of waveguides, a laser, detector and modulators, and the environmentally sensitive *sensor pad*, whose optical characteristics change in response to external stimulus. Changes in refractive index on the order to 10^{-5} may be detected with this arrangement. The precise nature of the sensor (whether chemical, biological, or other) depends on the type of film which is used in the sensor pad; a recess is etched into the waveguide and refilled with such a film, which reacts to an analyte (perhaps an ion, an antigen, a gas, etc.) by changing its refractive index.

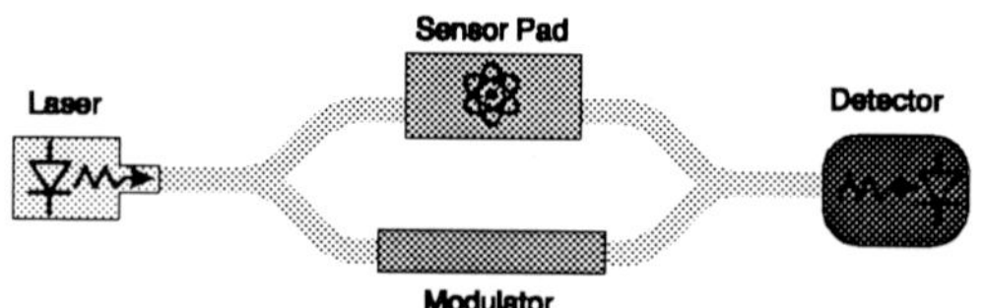

Fig. 1. Schematic of an optical Mach-Zehnder interferometer suitable for integrated optical sensors. All save the sensor pad is fabricated in III-Vs.

The individual devices required for the fabrication of a complete III-V based interferometer platform include low-loss, monomode waveguides, high efficiency phase modulators and waveguide detectors and a suitable light source, preferably a stable, monomode laser with good quantum efficiency. For economy of fabrication and simplicity of integration, we would like to fabricate all of these components in a single waveguide substrate, with no need for evanescent coupling between different waveguide layers. This simplifies growth and layer design, and potentially augments the coupling efficiency between various devices.

Fabrication Technology

The devices were based on dielectric-isolated, dry-etched ridge waveguides formed on a single separate-confinement single quantum well double heterostructure substrate. The layer structure was grown by MOCVD on a Si-doped GaAs substrate and consisted of an 1100 nm n^+ Si-doped $Al_{0.8}Ga_{0.2}As$ lower cladding, a 100 nm not-intentionally-doped $Al_{0.3}Ga_{0.7}As$ waveguide core with a undoped GaAs quantum well, a 980 nm p^+ Mg-doped $Al_{0.8}Ga_{0.2}As$ upper cladding and a 125 nm P^+ Zn-doped GaAs cap layer. Quantum well (QW) thicknesses of 5, 8 and 14 nm were grown on different substrates.

The waveguide ridge was patterned by a standard photoresist subject to a high temperature (130°C for 30 minutes) controlled reflow postbake, which not only improved the etch resistance but also provided a very smooth photoresist edge (Arnot 1992). Ridges were dry etched to a depth of 800 - 900 nm by magnetron reactive ion etching employing 2 sccm of $SiCl_4$ at 0.2 mT driven at 200 W (0.38 W/cm^2); sidewalls were smooth and nearly vertical. The low value of dc bias implied low dry-etch-induced sidewall and substrate damage. The surface was subsequently coated with 200 nm of e-beam evaporated SiO_2 as substrate passivation. After contact opening, the standard Ti/Pt/Au N-metal and Ge/Ni/Au N-metal metallization schemes were employed.

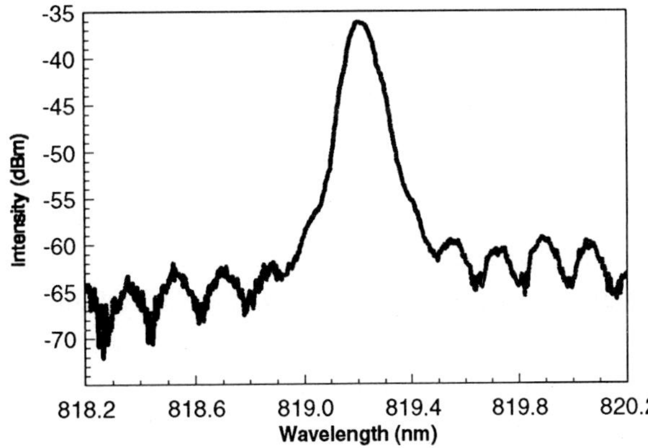

Fig.2. Output spectrum of a 2 μm wide, 500 μm long ridge laser at 30 mA CW drive current.

Device Performance

Since a single substrate was employed for all devices, the lasers, detectors and modulators differed only in metal layout. Electrical and optical characterization was performed by end-fire coupling a temperature-stabilized, optically isolated, polarization-selective external laser source (810 or 830 nm) onto the waveguide, detector, or modulator facets. Waveguide loss was measured by a sweep through the Fabry-Perot cavity resonances stimulated by tuning the external source in wavelength around 830 nm; losses in this material are high (16 dB/cm for 2 μm wide waveguide ridges) due primarily to the high free carrier density.

Laser threshold current densities were below 300 A/cm^2 and 10 mW/facet of single mode CW optical output power was obtained at 30 mA injection current. The monomode spectrum, as shown in Figure 2, had a side-mode suppression ratio of -25 dB at 30 mA CW drive current. Such modal behavior is well suited for the eventual fabrication of integrated interferometer circuits. Output wavelengths of 819, 840 and 850 nm were obtained with the various quantum well thicknesses employed.

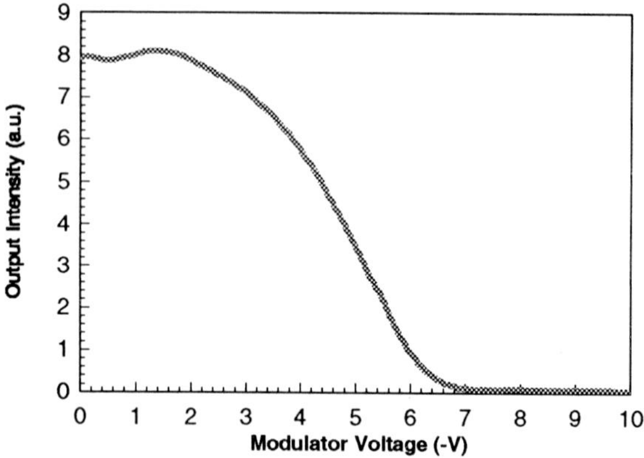

Fig. 3. Intensity modulation characteristic of a waveguide modulator. Laser wavelength was 830 nm, and the modulation depth -22 dB at a modulator bias of -7V.

Waveguide detectors made in the 5 nm QW substrate (absorption edge at 820 nm) were stimulated by an 810 nm external laser and found to have a responsivity of 0.44 A/W implying a quantum efficiency of 67%.

Modulators had the same physical structure, but operated by means of the quantum-confined Stark effect (Miller 1984); the electric field applied normal to the plane of the quantum well causes a red-shift in the absorption edge and will thus drive a nominally transparent device into absorption. Using 830 nm stimulation, the waveguide is thus electrically made absorbing, providing -22 dB of intensity modulation for a bias of -7 V; a typical characteristic is shown in Figure 3. This electrically-induced change in the absorption directly implies a change in the real part of the refractive index, and this will be employed in the interferometer modulator.

INTERFEROMETER PLATFORM, HYBRIDIZATION AND INTEGRATION

The technology used to fabricate these optoelectronic devices has been used to make a Mach-Zehnder interferometer; this includes the phase modulator on one arm of the interferometer and the waveguide detector at the output. A hole is etched into the second interferometer arm, into which waveguiding and the sensor layers will be deposited. Still under development with regard to structure and applications, this arrangement forms the III-V-based platform for many types of sensors.

A light source for the interferometer may be either hybridly attached or monolithically integrated. Cleaved-facet Fabry Perot lasers are easy to make but cannot be monolithically integrated with a photonic circuit of this type. We have circumvented this difficulty by fabricating all the necessary structures at once, cleaving off the lasers, and re-attaching them by a hybridization scheme (Zappe 1993). The laser was bonded to a commercial submount, operated normally, and the emitted light from one facet coupled into the waveguide circuit. This latter component was held by a piezo-driven vacuum pencil, which optimizes and stabilizes the coupling of laser to waveguide by an active feedback network; typical spacings between laser and waveguide facets were 2 to 5 μm. Once optimally coupled, the two components were fixed to a copper submount by UV curable epoxy. Due to epoxy shrinkage, maximal coupling efficiencies of 5% were achievable, resulting in 500 μW of optical power in the waveguide circuit. The development of an integratable (e.g., DBR) laser will remove the need for this somewhat involved step.

Both hybridization and monolithic integration of these components require lasers and detectors to have an emission/absorption wavelength longer than that of the waveguides: the latter are thus transparent. The device substrate employed lends itself to a post-growth shift in the energy gap by means of quantum well intermixing through dielectric cap annealing (Guido 1987). After SiO_2 deposition and a thermal anneal at 960°C, we were able to blue-shift the absorption edge of the waveguides by up to 60 nm, sufficient to render them transparent for the original laser/detector wavelength. After quantum well disordering, the exciton peaks were still clearly seen by photoluminescence and the waveguide losses did not increase with respect to unshifted waveguides. This technique then permits us to define transparent and absorbing regions on a single substrate.

Conclusions and Outlook

Through the development of a GaAs/AlGaAs-based technology for the simultaneous fabrication of low-loss waveguides, monomode lasers, modulators and detectors, we have been able to assemble a GaAs/AlGaAs-based interferometer platform for integrated optical sensors. Application of this interferometer as an actual sensor is a function of the films used in the sensor pad of Figure 1; chemical and biological sensors for medical applications are being developed using this platform, but many other measurands (gasses, chemicals, etc.) may be detected by changing to the appropriate sensing film. The configuration is not intrinsically limited in application and may be applied to many areas of sensor technology.

Acknowledgements

The authors very gratefully acknowledge M Shen, H Brunner, B Graf, HP Schweizer and A Vonlanthen for their valuable contributions to many aspects of this work.

References

Andrew SR, Marsh JH, Holland MC and Kean AH 1992 IEEE Photonics Tech. Lett. **4** 426

Arnot HEG, Zappe HP, Epler JE, Graf B, Widmer R, and Lehmann HW 1993 Electron. Lett. **29** 1131

Brandenburg A, Hinkov V and Konz W 1992 *Sensors* **6**, eds. E. Wagner, R. Dändliker, K. Spenner, (Weinheim: VCH) pp. 399-420

Emeis N, Schier M, Hoffmann L, Heinecke H and Baur B 1992 Electron. Lett. **28** 344

Guido LJ, Holonyak N, Hsieh KC, Kaliski RW, Plano WE, Burnham RD, Thornton RL, Epler JE, and Paoli TL 1987 J. Appl. Phys. **61** 1372

Miller DAB, Chemla DS, Damen TC, Gossard AC, Wiegmann W, Wood TH and Burrus CA 1984 Phys. Rev. Lett. **53** 2173

Wickman RW, Moretti AL, Stair KA and Bird TE 1991 Appl. Phys. Lett. **58** 690

Zappe HP, Shen M, and Arnot HEG 1993 Proc. 6th European Conf. Integrated Optics pp 11.8-11.9

Paper presented at the Int. Symp. GaAs and Related Compounds, Freiburg, 1993

Experience in manufacturing III–V photovoltaic cells

P. A. Iles and F. F. Ho

Applied Solar Energy, City of Industry, California USA

ABSTRACT: In the past ten years, Applied Solar has manufactured and delivered over four million square centimeters of high efficiency GaAs solar cells for use on satellites. GaAs technology was adapted to meet space cell requirements, by continuous operation of high throughput MOCVD reactors, and development of suitable GaAs or Ge substrates. The success with GaAs cells has stimulated development of cells made with other III-V compounds. Several multijunction cells are being developed with projections of near-term production. Possible implications for other III-V devices are discussed.

1.0 INTRODUCTION

Photovoltaic cells made from III-V compounds have attained production status in an important niche market, providing electric power for spacecraft. This paper reviews the properties needed for PV cells operating in space, and outlines the experience gained in adapting III-V technology to manufacture these cells. Although other III-V devices have different requirements, the experience gained in manufacturing solar cells may have some application to these devices.

2.0 REQUIREMENTS FOR PV SPACE CELLS

For effective operation in space, PV cells must have high efficiency (available area limited), and near the Earth, they must also have good tolerance to damage caused by charged particles trapped in the van Allen belts. Individual cells must be fairly large (8 to 60 cm^2) to reduce the number of interconnections, and thinner than 10 mils, to reduce weight. Ohmic contacts must meet stringent conditions. Although not as critical as for terrestrial PV application, cell costs must be as low as possible, a difficult requirement when only a few cells can be formed per slice. To gain user acceptance, cells with validated performance must be producible in quantities sufficient to meet delivery schedules.

Silicon cells have met these requirements for over thirty-five years. However, some III-V materials have the potential for better performance than Si, especially in specific missions. In the past ten years, stimulated by U. S. Air Force MANTECH contracts, and the demand for some specialized applications, GaAs cells have been manufactured on a large scale, and have confirmed their advantages for space use.

3.0 DEVELOPMENT OF III-V SOLAR CELL MANUFACTURING TECHNOLOGY

For space cells, GaAs had shown most-promise for manufacturing, because of its near-optimum bandgap and its maturity of development. Manufacturing methods were required to provide cells with acceptable performance,and the necessary consistency and control using proven processes operated at throughput sufficient to meet delivery schedules. The manufacturing costs should also be as low as possible.

3.1 Technical Goals for GaAs Cell Manufacture

For high efficiency, the GaAs layers must have minimum defects and high minority carrier diffusion lengths. Recombination must be reduced, around the PN junction, in the bulk GaAs and at surfaces. A heteroface structure, using a passivating AlGaAs window was selected. Layer thicknesses and dopant concentrations were optimized for efficiency and radiation resistance by iteratively combining parametric tests with cell modelling.

The processes used to form the Ohmic contacts must meet several requirements. The metallization must have good adhesion, to withstand array bonding conditions, and temperature cycling in operation. Contacts must have low resistance, both at the metal-semiconductor interface and laterally in the grid patterns. The metallization must be fairly thick (4-8μm) to facilitate bonding of interconnects, and to reduce lateral resistance losses. Although very narrow lines are not required, typical line widths are 4-15μm. The contacts are the most important factor determining the operating stability of the solar array. Stable antireflective coatings must be applied, and methods for in-line testing, identification marking and packaging must be developed.

3.2 Cell Structure

Figure 1 shows the heteroface GaAs cell structure. The P/N configuration is shown here; the N/P configuration can also be formed with similar technology.

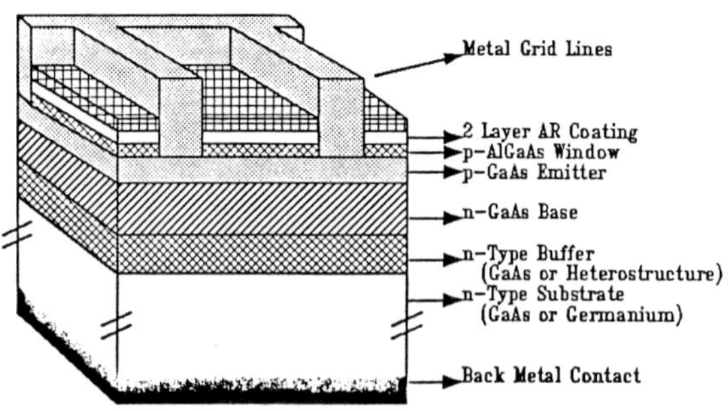

Figure 1. Cross-Section of Heteroface GaAs Cell

3.3 Layer Growth

Metal-Organic-Chemical-Vapor-Deposition (MOCVD) was identified as the growth method with most potential to achieve suitable layer quality and throughput. GaAs substrates were used first, but with a later switch to Ge substrates, the advantages of MOCVD were confirmed. Large capacity MOCVD reactors (processing around 1000 cm^2 of substrates per run) were installed, and it was demonstrated that these large reactors could be operated continuously, to give layers of required quality. Regular maintenance and cleaning procedures were installed, and many reactor modifications were made, including changes of gas flow systems, modified susceptor design, and upgrading of the original computer. The increasingly severe safety and environmental regulations required installation of large scale waste disposal equipment, and a system of detectors and monitors, all under the control of a Safety group.

To adjust to these large barrel type reactors (at least 5 are currently in operation at Applied Solar), gas sources with greater capacity were developed, to reduce frequent replacement while maintaining suitable purity. The quality of these sources has steadily improved, and the costs have decreased.

The MOCVD growth is the main factor in determining layer quality. Many tests were run, to establish the best combination of temperatures, gas flows, layer growth rates, doping concentrations, layer thicknesses, and window layer composition. Although the uniformity requirements for solar cells are not as demanding as for other devices, thickness and doping concentration were controlled to 5-10% for all substrates in the reactor. Several independent assessments by MOCVD experts have shown that the GaAs quality is surprisingly close to the best state-of-the-art. For heteroepitaxial growth of GaAs, even on lattice-matched Ge, considerable modification was required in the growth conditions near the substrate surface. Although an additional photovoltage can be generated at the GaAs/Ge interface, for operation in space, it was preferred to render this interface inactive. Appropriate changes have been made in the growth conditions.

3.4 Substrates

Solar-grade GaAs substrates were developed by both foreign and U. S. suppliers. Ingots were grown in Horizontal-Bridgman furnaces, with dislocation densities less than 5000 cm^{-2}. Square or rectangular shaped boats were used to minimize wastage in forming square or rectangular cells, thereby reducing substrate costs. With experience, and the increased demand for solar cell production, substrate suppliers reduced the cost of solar grade GaAs substrates well below the cost of substrates used for other devices, where substrate cost is less important in overall device costs.
The substrate surface was chemically-mechanically polished to give an epi-ready surface.

While the production effort on GaAs cells was building, an Air Force contract showed that high efficiency GaAs cells could be grown on Ge substrates. The higher mechanical strength of Ge allowed fabrication of GaAs cells which were both larger and thinner. A recent AF MANTECH program demonstrated that GaAs/Ge cells, 36 cm^2 area, and 4 mils thick with efficiencies above 18%, could be produced. The lower cost of Ge substrates further reduced GaAs cell costs.

To ensure good epitaxial growth of GaAs on Ge, a series of tests determined that the Ge slice orientation should be offset several degrees from a $<100>$ plane. The Ge substrate can be thinned (from 8 mils to 4 mils) by etching the substrate after all the front surface processing is completed. Substrate suppliers have developed slicing and polishing methods which can give a substrate 5.5 mils thick at comparable cost to the 8 mil substrates. Substrates 5.5 mils thick can be processed with good yields, and array manufacturers currently prefer 5.5 mil thick substrates as a reasonable balance between array fabrication yields and reduced weight.

3.5 Contacts

Three contact schemes were evaluated, depositing metal on a cap layer of GaAs grown over the window layer (with subsequent removal of the cap layer between gridlines), depositing directly on the window layer, or directly on the GaAs emitter through slots etched in the window layer. Most of the cells manufactured to date have used alloyed Au-based metal stacks evaporated through etched slots in the AlGaAs window, and these contacts have operated successfully. Cells fabricated by direct deposition of refractory metals on the window layer showed minimal degradation after exposure to 500-600°C. Even without the need for high temperature performance, elimination of the Au layers can reduce cell costs. For a limited number of users, cells have been made with front contacts wrapped-around the cell edges, or wrapped-through holes within the cell. The vias required are fairly thick, between 4 and 8 mils.

3.6 Other Processing Steps

A clean processing environment was set-up, including clean rooms for MOCVD substrate preparation, dust-free loading/unloading stations for MOCVD reactors, and dry gas storage cabinets. Standard handling boats were modified for non-round slices, new photolithographic methods were developed, and special fixtures were used to hold the slices during evaporation of metallization and AR coatings. When thin substrates were processed, appropriate modifications were made in handling methods and equipment. Techniques were developed for non-contact marking for identification, for in-line evaluation methods and for cell testing.

4.0 ACHIEVED RESULTS

4.1 Cell Performance and Yields

Median efficiencies around 90% of the best efficiencies reported have been obtained at high yields.

Figure 2 shows I-V data for a 6 cm x 6 cm, 4 mils thick, GaAs/Ge Cell.

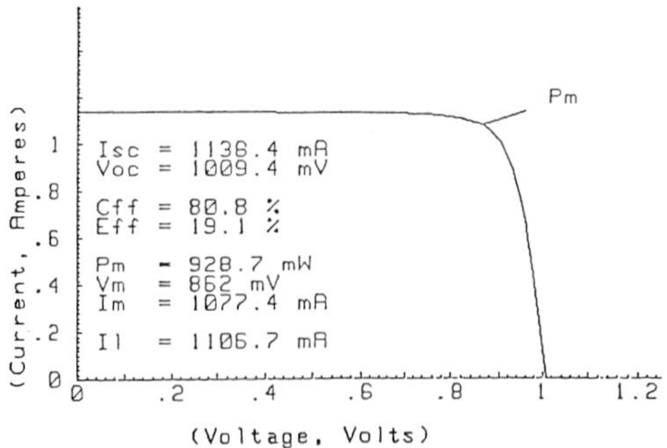

Figure 2. I-V Data for 6 cm x 6 cm, 4 mils thick GaAs/Ge Cell (AMO, 25°C)

Figure 3 shows a histogram plot for part of the weekly output of GaAs/Ge cells. The cells, and arrays made from them have passed space qualification tests specified by users. The temperature coefficients and radiation resistance show good control at state-of-the-art values.

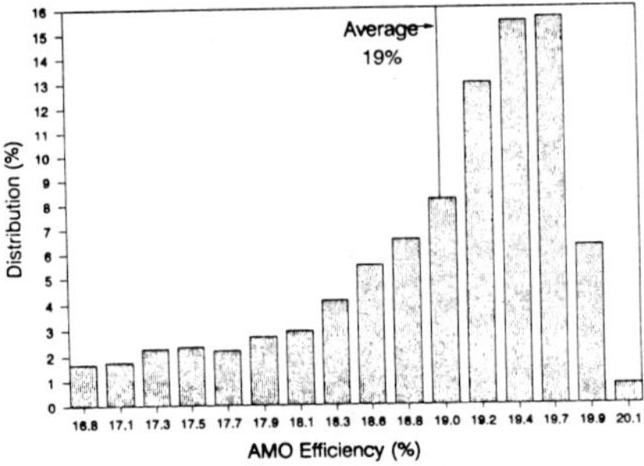

Figure 3. Efficiency Distribution for 1,300 GaAs/Ge Cells, 2 cm x 4 cm.

4.2 Reverse Bias Characteristics

Under reverse bias, simulating the effect of shadowing in a series-string of cells, GaAs/GaAs cells often showed catastrophic failure at isolated defects. GaAs/Ge cells have greater tolerance to reverse bias, and this is attributed to a set of leakage paths originating in antiphase domains formed during growth, and propagating to the GaAs PN junction.

4.3 Stimulating User-Acceptance

When acceptable producibility and performance were demonstrated, only a few users found GaAs cells useful. However, despite the steady decrease in costs, most users could not justify the change from Si to GaAs cells. During the MANTECH programs, analysis of tradeoffs extending to the array level showed that the increased efficiency of GaAs cells could give reduced area and weight, and the life-cycle costs were much lower. Recently, these tradeoff studies have been carried further into the satellite system to include savings in launching fuel costs, and have shown that GaAs/Ge cells can be cost-competitive in more applications. These trade-studies have opened up wider applications of GaAs/Ge cells, on small satellites, on clusters of commercial satellites, or for NASA missions in low Earth orbits.

4.4 Advanced III-V Cells

The success with GaAs production has led to increased activity to produce InP cells which have high radiation resistance, or cascade cells, which combine cells of specific bandgaps to give increased efficiency. The most promising cascade cell structures are monolithic structures, such as the NREL cell shown in Figure 4. Cells using materials lattice-matched to GaAs can be grown on GaAs or Ge substrates. Results to date are that the tunnel diode and top cell can be added to GaAs cells, giving further increase in efficiency (to ~22% AMO average), with further system

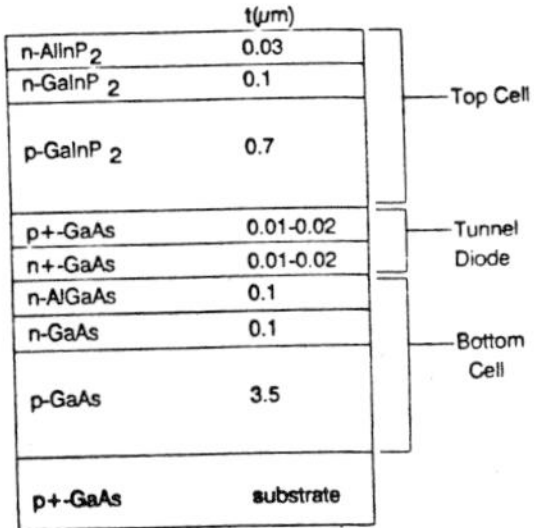

Figure 4. Details of Grown Layers for NREL Cascade Cells. Contacts and Coatings are not shown.

advantages. These extra layers are thin and do not require much additional growth time. Thus the cost increase over GaAs cells should be small. Projections are that production-ready cascade cells will be available by 1995. The success of the heteroepitaxial growth of GaAs/Ge has stimulated tests using Si or other lattice-mismatched substrates. To date, despite extensive annealing schedules or graded lattice layers near the substrate the mechanical and electrical properties are not yet good enough.

4.5 Use of Advanced III-V Technology at Production Levels

(a) Bandgap engineering using ternary or quaternary III-V alloys is being used extensively for solar cells, as well as for thermophotovoltaic converters or laser converters.
(b) Ultrathin GaAs cells have been formed by epitaxial liftoff, from both GaAs and Ge substrates. These cells have high power-to-weight ratios and when a reflecting back contact is applied, the cells can gain from photon recycling.
(c) Advanced surface passivation methods have been applied to production cells.

5.0 SUMMARY

GaAs solar cells have been produced in large quantities with good performance and are finding more applications in a niche market, providing power on spacecraft. The use of Ge substrates has increased cell size, and reduced weight and costs, and systems tradeoffs extended into the spacecraft system show that GaAs cells can be cost competitive. The wider use of GaAs cells is stimulating production efforts on other III-V cells and early production of cascade cells is projected.

Most of the advanced technology reported for GaAs can be applied to production GaAs cells. The experience in solar cell manufacturing may be useful in scaling-up other III-V devices.

Inst. Phys. Conf. Ser. No 136: Chapter 5
Paper presented at the Int. Symp. GaAs and Related Compounds, Freiburg, 1993

Room temperature gallium arsenide radiation detectors

Authors: E.Bauser[1], J.Chen[2], R.Geppert[2], R.Irsigler[2], S.Lauxtermann[2], J.Ludwig[2], M.Kohler[2], M.Rogalla[2], K.Runge[2], F.Schäfer[2], Th.Schmid[2], A.Schöchlin[2], M.Webel[2]

1 MPI FKF Stuttgart
2 Albert-Ludwigs-Universität Freiburg

ABSTRACT: Detectors for registration of ionizing elementary particles and X-ray radiation have been built based on the utilization of semi-insulating GaAs substrates. The commercially available LEC wafers have been processed to feature a Schottky contact at one side and an ohmic at the other. LPE GaAs detectors with 85 μm thick epitaxy layers have been investigated, especially the detection and charge collection efficiency. Noise measurements for various detector-preamplifier configurations have been performed.

1 INTRODUCTION

Semiconductor detectors are widely applied in science and technology. They are not only used in high energy physics, but also for detecting X-ray radiation in medicine. GaAs seems to be a suitable material for these kinds of applications; it has a larger bandgap than Ge, is much faster than CdTe, much more radiation hard than Si and due to its higher Z it has a better detection efficiency for X-ray radiation than Si. The first chapter of this paper deals with characterization of detectors based on LEC and VGF wafers, presented are results of measurements of charge-collection and detection efficiency. In the second chapter we investigate LPE detectors. The last chapter contains noise measurements for various detector- preamplifier configurations.

2 MEASUREMENTS WITH LEC AND VGF GaAs DETECTORS

Semi-insulating wafers with 2 inch diameter and 500 μm thickness, both sides polished have been used. In order to operate the detectors as Schottky diodes, they were successively vacuum deposited and patterned with lift off technique. The front side forms a Schottky contact made of Ti, Pt and Au, the back side features an ohmic contact built up by Ni, Ge, Au, Ni and Au layers. The front side is passivated with silicon nitride to prevent oxidation. The processing of all our detectors has been performed at the IAF.

2.1 Wafer characterization

We have measured the current-voltage-characteristics and α-spectra of 6 Schottky diodes which have been processed on different wafers (table 1).

	Wafer	resistivity 10^7 [ohm*cm]	hall mobility [cm^2/Vs]	carrier concentration 10^7 [cm^{-3}]
1	AXT no.23	>1.0	>5000	<12.5
2	Wacker no.3	4.8	8370	1.5
3	Wacker no.113	3.0	7100	2.9
4	MCP no.4	3.3	7000	2.8
5	MCP no.51	2.9	7150	3.1
6	Wacker no.111	3.0	7100	2.9

Table 1: Material-characteristics of the various wafers

All wafers from Wacker (Wacker Chemitronic) and MCP (MCP Wafer Technology Ltd.) have been fabricated according to the liquid encapsulated Czochralski (LEC) method. The AXT (American XTAL Technology) wafer has been grown using vertical gradient freeze (VGF) method. The current-voltage characteristics have all been obtained at a temperature of 22±1 °C on wafer. The measured diodes are quadratic 4x4 respectively 5x5 mm^2. The current density in reverse direction of the diodes shows two steps, one around −0.2 V to −2 V (see: inlet figure 1), the other starting at about −50 V reverse voltage. The size of the current density in reverse bias direction varies strongly with the origin of the wafers. Diodes from AXT wafer no. 23 have the lowest current density which is 4.1 nA/mm^2 at −300 V. On the other side leakage current densities of the MCP wafers are as high as 30 nA/mm^2.

2.2 Charge collection and detection efficiency

The charge released by each particle is known from the energy required to generate an electron-hole pair, namely 4.2 eV. The charge collection efficiency (CCE) is defined as ratio of the charge, generated by the particle, to the charge detected at the contacts. Alpha-spectra have been recorded using the 8.78 MeV α-line of ^{212}Po for the wafers in tab.1. The CCE of these detectors at a bias voltage between −240 V and −285 V is shown in figure 2. The CCE varies from 18% for the AXT wafer up to 40% on wafer no. 3 of Wacker. These measurements have been performed on wafer.

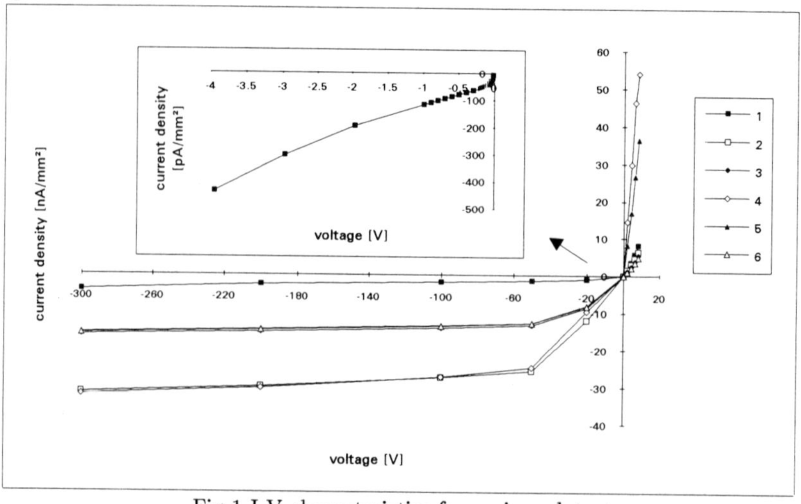

Fig.1 I-V characteristics for various detectors

For a mounted 4x4 mm LEC GaAs detector the CCE and the detection efficiency have been measured as a function of voltage (figures 3, 4) with 70 GeV pions at the SPS accelerator at CERN (Super Proton Synchrotron, European Organization for Nuclear Research) and with 40 MeV electrons at the Freiburg Betatron.

The CCE increases with applied voltage, even at a field of 1V/μm it does not show a plateau. The best value obtained was 45% for electrons for a bias of -600 V.

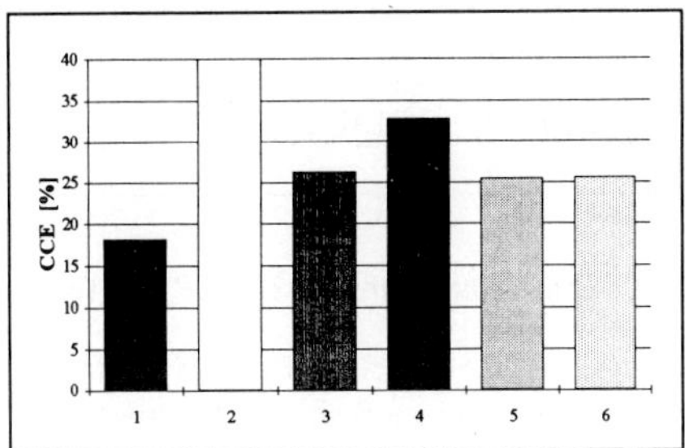

Fig.2 CCE of the detectors

The detection efficiency has been obtained using a conventional scintillator trigger. The two overlapping scintillators have been placed at a small distance behind the detector. The detection efficiency is obtained as the ratio between the number of counts of the GaAs detector to the number of counts in the scintillators. As one can see it shows a constant behaviour for a bias voltage larger than 100 V and reaches a value of 96% for 70 GeV pions and a slightly smaller one, namely 90%, for electrons. The difference stems from setup inaccuracies in the beam.

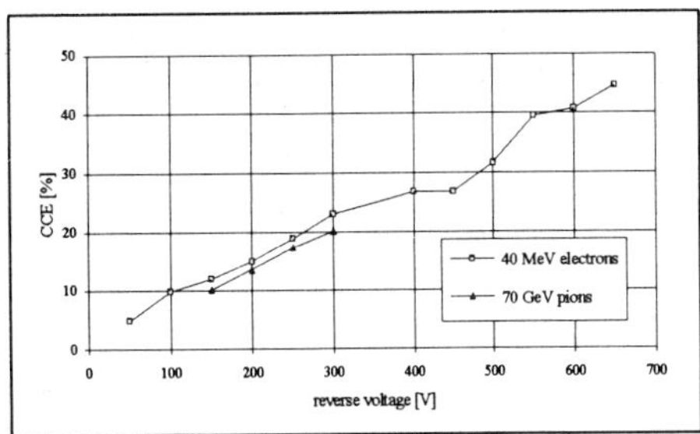

Fig.3 CCE of 70 GeV pions and 40 MeV electrons

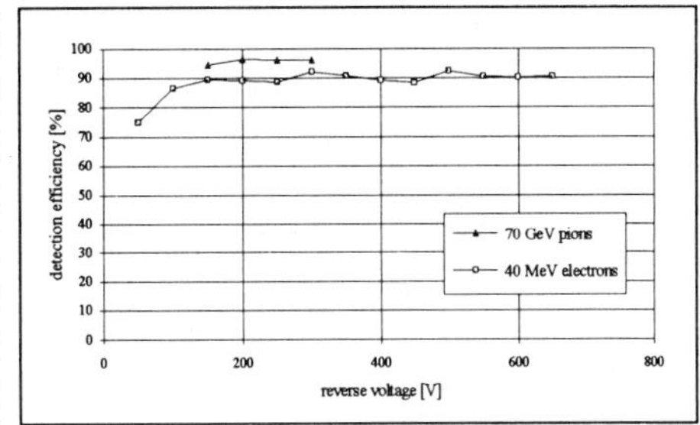

Fig.4 detection efficiency of 70 GeV and 40 MeV electrons

3 MEASUREMENTS WITH LPE GaAs DETECTORS

An epitaxial layer of about 100 μm thickness has been grown with the liquid phase epitaxy technique (LPE) (Bauser 1983). The size of the detectors is about 12 mm^2. The Schottky and ohmic contacts of these detectors are processed with the same technique mentioned for the LEC wafers. A schematic is illustrated in figure 5. The current-voltage characteristic of this detector is shown in figure 6.

	thick-ness [μm]	resis-tivity [Ω cm]	hall-mobility μ [cm^2/Vs]	carrier-concentration η [cm^{-3}]	dopant
substrate	450	4,5*10^7	6790	2,04*10^7	semi-insulating
epitaxial layer	84,5	14,1	7197	6,15*10^{13}	

Table 2: characterization of the LPE detector

The current-voltage characteristic of this detector is shown in figure 7. Spectra of α-particles from ^{241}Am decay have been recorded. Figure 6 shows the spectrum of a measurement in air. The source was placed 5 mm in front of the detector. After crossing the window of the source, 5 mm air and the metallization of the detector, the alpha particles had an energy of 3.7 MeV. The energy resolution is 390 keV FWHM (figure 6). **Even without any bias voltage a CCE of 62% has been obtained.** At –2 V bias the CCE increased to 74%.

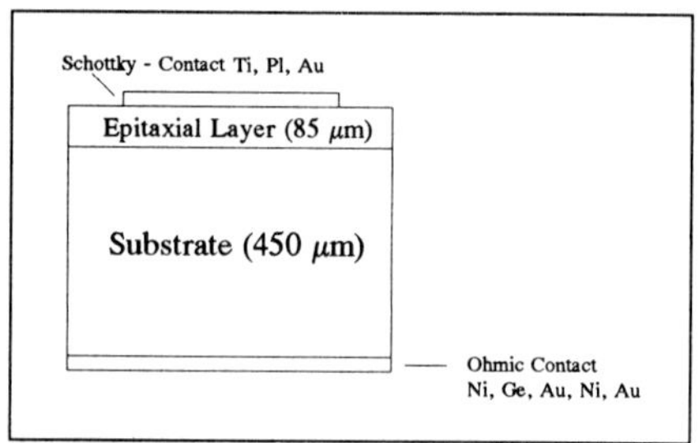

Fig. 5 schematic of the LPE

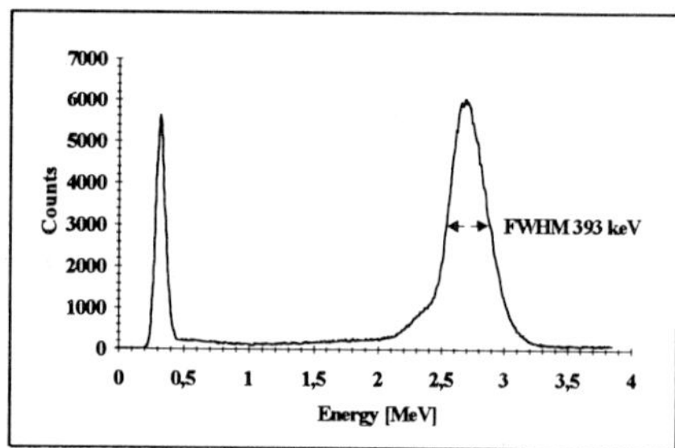

Fig.6 spectrum of 3.7 MeV α particles

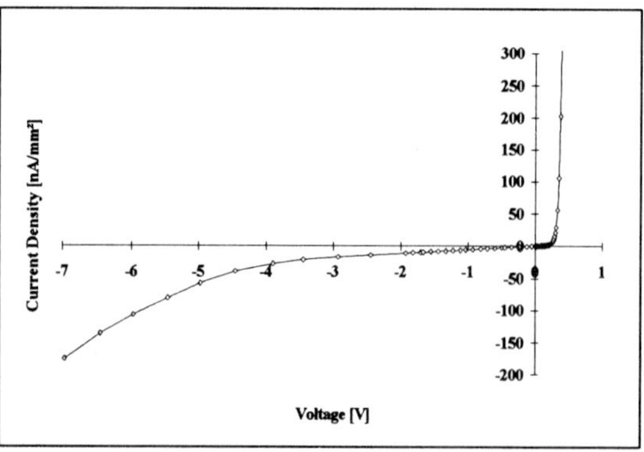

Fig.7 I-V characteristic of the LPE detector

A comparison with LEC GaAs detectors (McGregor 1992, RD8 1992) shows that the CCE of detectors decreases with their thickness (figure 8). This effect is due to the influence of traps.

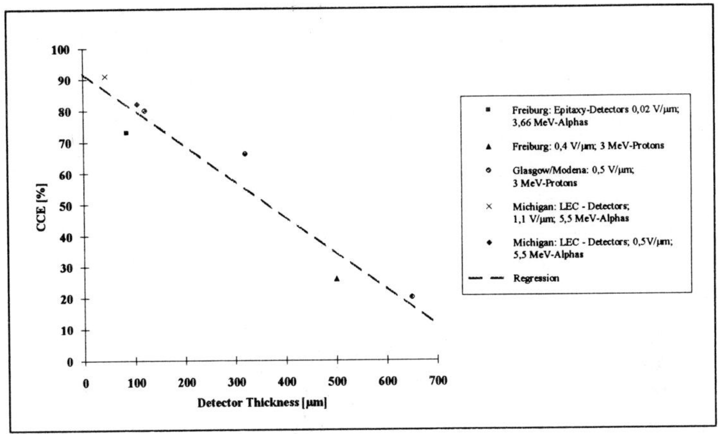

Fig.8 CCE of various detectors

4 NOISE MEASUREMENTS

Figure 9 shows the equivalent noise charge ENC of a 4x4 mm GaAs detector connected to a preamplifier (Ortec 142A) as a function of shaping time. Measurements and calculations for different combinations of detectors and preamplifiers show that with a GaAs setup smaller ENC numbers for shorter shaping times can be obtained. We are therefore developping a special GaAs charge sensitive preamplifier in collaberation with G. Bertucchio, Politecnicio Milano.

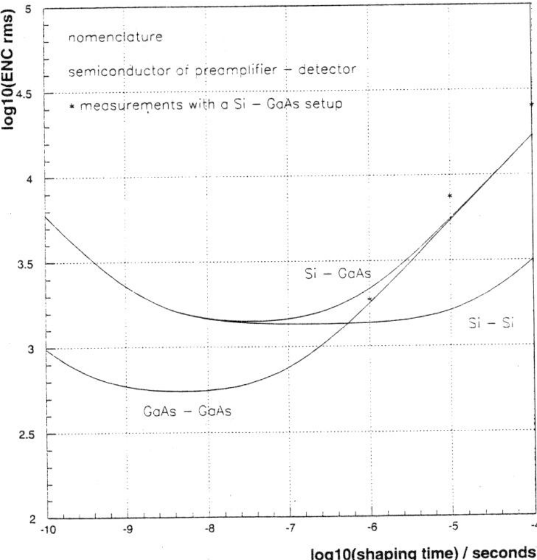

Fig. 9 Noise for different detector-preamplifier configurations

The calculations have been performed with the following formula (Delany 1980). The first part of the formula stands for the serial noise of the FET of the preamplifier, the second for the 1/f series and parallel noise and the third for the parallel noise of the detector.

$$ENC^2 = a_w \frac{1}{T_m} A_1 + 2\pi a_f C_T A_2 + b_w T_m A_3$$

$$b_w = qI_L; a_f = \frac{P_f}{C_G}; a_w = a\frac{2kT_m}{g_m}$$

I_L = *leakage current* P_f = *Power* g_m = *transconductance*
C_G = *gate capacity* C_D = *detector capacity* $C_T = (C_D + C_G)^2$

5 CONCLUSIONS

Detectors for registration of ionizing elementary particles and X-ray radiation have been built. For LPE GaAs detectors no external bias voltage is required to obtain a CCE of 62%. With an applied bias voltage of −2 V the CCE increased to 74%. The current-voltage characteristic for LEC and VGF wafer shows two steps, one arount −0.2 V to −2 V bias, the other starting at about −50 V bias. For LEC GaAs detectors detection efficiencies in excess of 95% have been reached for ionizing particles in the MeV and GeV range. The CCE of the detectors decreases with their thickness.For a GaAs detector with a GaAs preamplifier the lowest ENC is expected at a shaping time of about 10 ns.

6 ACKNOWLEDGEMENTS

We would like to thank Dr. J. Ralston and J. Schneider of the IAF for processing our detectors.

7 REFERENCES

Bauser E 1983 Festkörperprobleme XXII
Delany C 1980 Electronics for the Physicist (New York: Wiley)
McGregor D S et al. 1992 IEEE Trans. Nucl. Sci NS-39(5)
RD8 1992 RD8 Meeting, priv. comm.

Determination of band offsets in GaAsP/GaP strained-layer quantum well structures using photoreflectance and photoluminescence spectroscopy

Y. Hara[†], H. Yaguchi, K. Onabe, Y. Shiraki[*] and R.Ito

Department of Applied Physics, The University of Tokyo,
7-3-1 Hongo, Bunkyo-ku, Tokyo 113 Japan

[*]Research Center for Advanced Science and Technology, The University of Tokyo,
4-6-1 Komaba, Meguro-ku, Tokyo 153 Japan

ABSTRACT: $GaAs_{1-x}P_x$(x=0.81,0.84,0.86) /GaP strained-layer quantum well (SLQW) structures have been studied using photoreflectance(PR) as well as photoluminescence(PL) spectroscopy. The PR measurements have revealed the optical transitions between the quantized subbands in the higher-lying quantum well at the Γ point. From the PR measurements, combined with a square-potential effective-mass calculation, the conduction band offset ratio has been determined to be 0.68±0.1 for x=0.84. This result leads to the type-I quantum well at the X-point conduction band minima. The low-temperature PL spectra are very consistent with the type-I band scheme.

1. INTRODUCTION

The GaAsP strained-layer system is capable of highly flexible band tailoring due to the strain involved (Osbourn *et al.* 1987). In GaAs/GaAsP strained-layer quantum wells (SLQWs) and strained-layer superlattices (SLSs), the type-I band line-ups are realized when either layer is strained (Osbourn 1982, Gourley and Biefeld 1984). The band offsets at the heterointerface and their dependence on the alloy composition or on the strain have been estimated based on photoluminescence and reflectance (Zhang *et al.* 1991) and photoreflectance (Yaguchi *et al.* 1993) measurements. On the other hand, in GaAsP/GaP systems, both type-I and type-II band line-ups are possible depending on the strain configuration (Osbourn *et al.* 1982, Pistol *et al.* 1988). Indirect-gap nature of GaP and GaAsP with P-rich compositions, where the conduction band minima are located at the X points of the Brillouin zone, marks another difference from GaAs/GaAsP systems. Indeed, the crossover behavior from type-I to type-II has been observed in GaAs/GaP SLSs (Recio *et al.* 1990). The band offsets in the GaAsP/GaP system, however, have not been well established to date.

[†]present address: Research and Development Center, Toshiba Corporation,
1 Komukai Toshiba-cho, Saiwai-ku, Kawasaki 210 Japan

In the present study, the band offsets of $GaAs_{1-x}P_x$(x=0.81,0.84,0.86) /GaP SLQWs, in which only GaAsP well layers are strained, have been estimated based on the photoreflectance (PR) measurements combined with a square-potential effective-mass calculation. The PR measurement is extremely useful, as it can reveal the Γ-Γ direct optical transitions between the higher-lying quantized subbands as well as the lowest ones. This allows us a reliable determination of the band offsets by fitting the experimental data with the calculation. Photoluminescence (PL) measurements have also been made to observe the Γ-X indirect transitions. The PL data are compared with the X-point band line-up which is derived from the Γ band offsets.

2. EXPERIMENTAL

The $GaAs_{1-x}P_x$/GaP SLQWs were grown on (100) GaP substrates at 700°C by low-pressure (60 Torr) metalorganic vapor phase epitaxy (MOVPE). Details of the growth conditions have been described elsewhere (Miura *et al.* 1991). The SLQW structure consists of a 5-cycle $GaAs_{1-x}P_x$/GaP multiple quantum well (MQW). The P content (x) was chosen as 0.81, 0.84 and 0.86, corresponding to the biaxial compressive strain of 0.71–0.53%. The thickness of the $GaAs_{1-x}P_x$ well layer (L_z) was 1.6–13nm, and that of the GaP barrier layer was 60–70nm. A 400nm GaP buffer layer was grown before the SLQW structure. The whole structure was determined accurately by double-

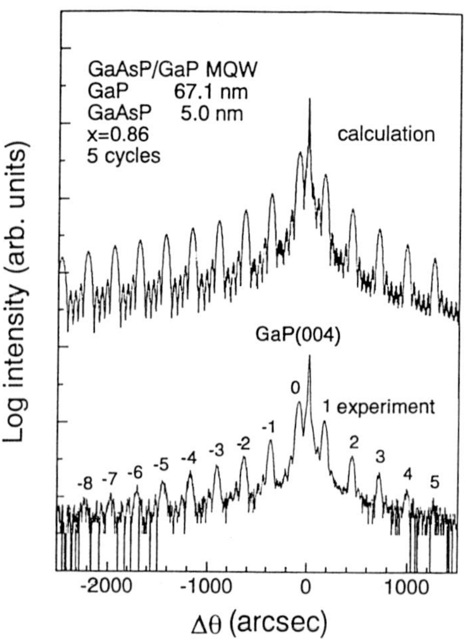

Fig.1. X-ray rocking curve of $GaAs_{1-x}P_x$/GaP SLQW (x=0.86, L_z=5.0nm).

crystal x-ray diffraction combined with a calculation based on the dynamical diffraction theory (Halliwell *et al.* 1987). The x-ray rocking curve was fitted excellently with the calculation as shown in Figure 1.

For the PR measurements, an Ar^+-laser (488 nm) was used as a pump beam (chopped at 210Hz) for surface-field modulation (Yaguchi *et al.* 1993). The Low-temperature PL spectra were obtained using a He-Cd laser (325nm) as the excitation source.

3. RESULTS AND DISCUSSION

Figure 2 shows a typical PR spectrum of the $GaAs_{1-x}P_x$/GaP SLQW with L_z=1.6nm and x=0.81 at 210K. The optical transitions between the n=1 quantized electron and heavy-hole (11H) or light-hole (11L) subbands are clearly observed. The Γ-Γ transitions in the GaP barrier layers, E_0 and $E_0+\Delta_0$, are also observed. The optical transitions between higher-lying subbands (n≥2) were also observed in the SLQWs with a larger L_z, though not shown. The transition energies were derived from the PR spectrum by fitting the experimental data with the third derivative functional form (TDFF) of the dielectric function (Aspnes 1973). Excitonic corrections (Shanabrook *et al.* 1987) were not sensitive in determining the transition energies. In Figure 3, the energy difference between the 11H and the 11L transitons is plotted as a function of the well width (L_z) for the SLQWs with x=0.84. The plots are compared with the

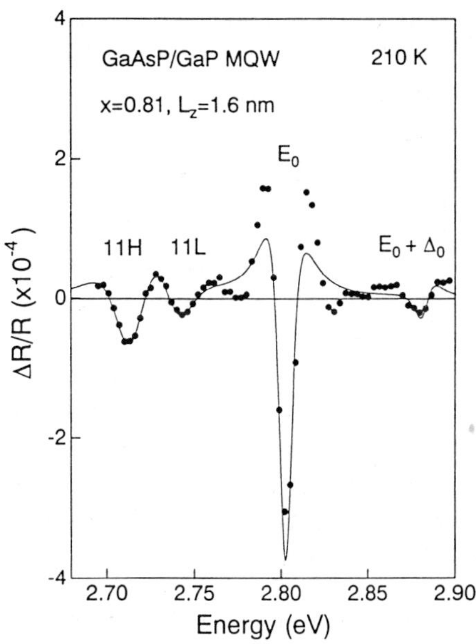

Fig.2. Photoreflectance spectrum of $GaAs_{1-x}P_x$/GaP SLQW (x=0.81, L_z=1.6nm) at 210K. • : Experiment, ——— : TDFF fit.

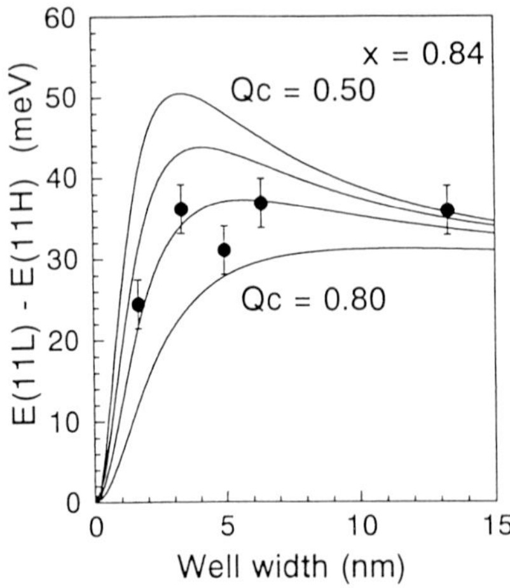

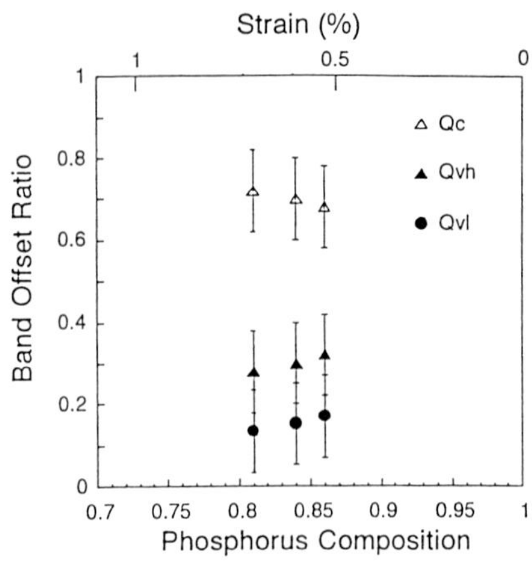

Fig.3. Transition-energy difference between 11H and 11L as a function of the well width (x=0.84). • : Experiment, ——— : calculation for Q_c=0.5, 0.6, 0.7 and 0.8.

Fig.4. Conduction band offset ratio (Q_c), and heavy-hole(Q_{vh}) and light-hole(Q_{vl}) valence band offset ratios at the Γ point for x=0.81–0.86.

square-potential effective-mass calculation where various values of the conduction band offset ratio, $Q_c = \Delta E_c^\Gamma / (\Delta E_c^\Gamma + \Delta E_{vh})$, are assumed. The band offsets, ΔE_c, ΔE_c, ΔE_{vh}, and ΔE_{vl} are defined as shown in Figure 5. The material parameters used in the calculation are found in the literature (Zhang *et al.* 1991). The best fit to the experiment has been obtained with Q_c=0.68±0.1 for x=0.84. This conduction band offset ratio gives the heavy-hole and light-hole valence band offset ratios as Q_{vh}=0.32±0.1 and Q_{vl}=0.16±0.1, respectively, where $Q_{vh} = \Delta E_{vh} / (\Delta E_c^\Gamma + \Delta E_{vh})$ and $Q_{vl} = \Delta E_{vl} / (\Delta E_c^\Gamma + \Delta E_{vh})$.

Similar analyses have been done for x=0.81 and 0.86; the results are collectively shown in Figure 4. It is seen that the band offsets are slightly dependent on the P content or the strain. From these band offsets, using the Γ-X deformation potentials of GaP (Mathieu *et al.* 1979) as an approximation for those of $GaAs_{1-x}P_x$, the band line-ups of the $GaAs_{1-x}P_x$/GaP SLQWs are depicted as shown in Figure 5. It is noted that both the Γ-Γ and the Γ-X gaps are of type-I (*i.e.* both electrons and holes are confined to the $GaAs_{1-x}P_x$ well). The conduction band offset ratio of the X point, $Q_c^X = \Delta E_c^X / (\Delta E_c^X + \Delta E_{vh})$, is calculated to be 0.6 for x=0.84. This value is much larger than Q_c^X=0.37 which was suggested by Pistol *et al.* (1988) for x=0.70.

The Γ-X indirect transitions are observed by the PL measurement. Figure 6 shows the PL spectrum of the $GaAs_{1-x}P_x$/GaP SLQW with x=0.84 and L_z=5.0nm at 18K. No-phonon (NP^X) and LA phonon-assisted (LA^X) emissive transitions are clearly identified. The quantum confinement of electrons and holes is confirmed by the spectral

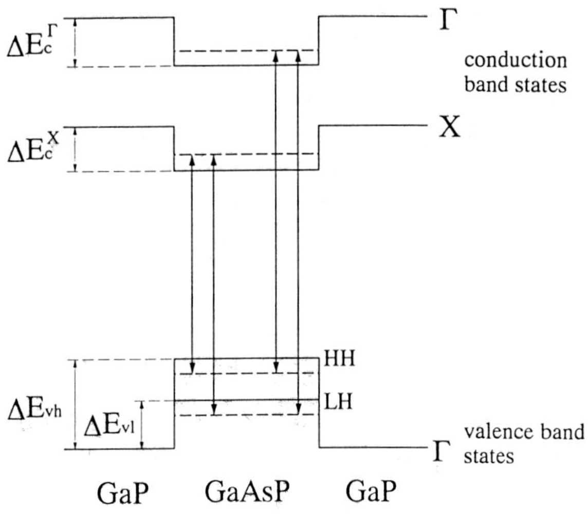

Fig.5. Band line-ups of $GaAs_{1-x}P_x$ (x=0.81–0.86)/GaP SLQWs. Direct (Γ-Γ) and indirect (Γ-X) transitions are shown.

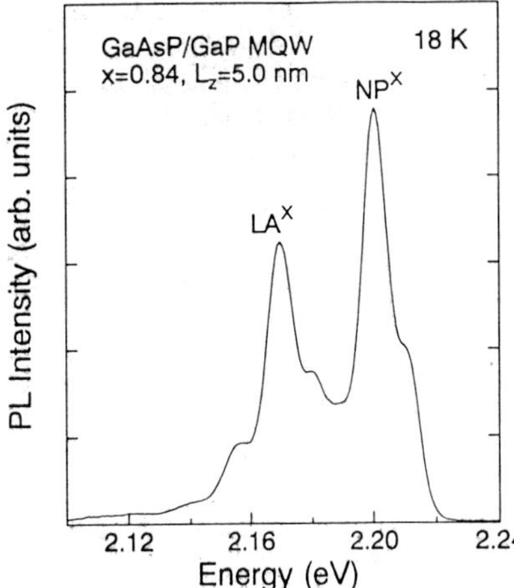

Fig.6. Photoluminescence spectrum of $GaAs_{1-x}P_x$/GaP SLQW (x=0.84, L_z=5.0nm) at 18K. No-phonon (NP^X) and LA phonon-assisted (LA^X) transitions.

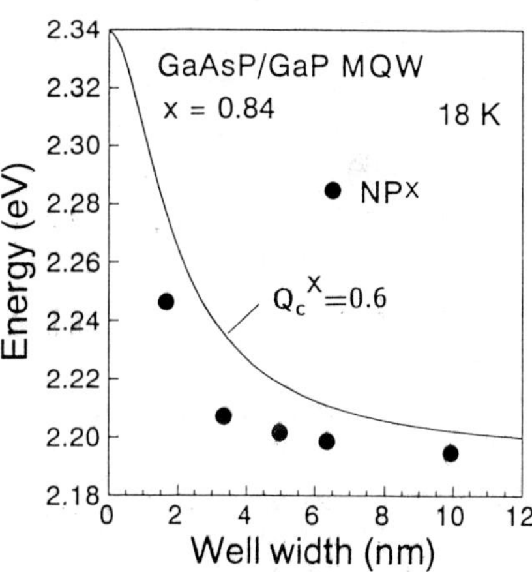

Fig.7. Well width (L_z) dependence of the no-phonon transition (NP^X) energy of $GaAs_{1-x}P_x$/GaP SLQW (x=0.84). • : Experiment, ——— : calculated with the X-point conduction band offset ratio, Q_c^X=0.6.

blue-shift with decreasing the well width as shown in Figure 7. The solid curve is a calculated one with Q_c^X=0.6 at the X point. The exciton binding energy (~20meV for bulk GaP) has not been included in the calculation, so the agreement between the experiment and the calculation is reasonably good. Therefore, it can be concluded that the band line-up of the $GaAs_{1-x}P_x$(x=0.81,0.84,0.86) /GaP SLQWs is of type-I at the X point which forms the fundamental energy gap of the material.

4. CONCLUSION

The conduction band offset ratio Q_c at the Γ point of $GaAs_{1-x}P_x$/GaP SLQWs has been determined to be 0.68±0.1 for x=0.84 from PR measurements combined with the square-potential effective-mass calculation. This leads to the type-I band line-up at the X point whose band offset ratio is 0.6. The low-temperature PL spectra are very consistent with the type-I band scheme.

ACKNOWLEDGEMENTS

The authors would like to thank X. Zhang and S. Miyoshi for their cooperation in MOVPE growth, S. Fukatsu for useful advices, and S. Ohtake for technical support.

REFERENCES

Aspnes D E 1973 Surf. Sci. **37** 418
Gourley P L and Biefeld R M 1984 Appl. Phys. Lett. **45** 749
Halliwell M A, Lyons M H and Hill M J 1987 J. Cryst. Growth **68** 523
Mathieu H, Merle P, Ameziane E L, Archilla B and Camassel J 1979 Phys. Rev. B **19** 2209
Miura Y, Onabe K, Zhang X, Nitta Y, Fukatsu S, Shiraki Y and Ito R 1991 Jpn. J. Appl. Phys. **30** L664
Osbourn G C 1982 J. Appl. Phys. **53** 1586
Osbourn G C, Biefeld R M and Gourley P L 1982 Appl. Phys. Lett. **41** 172
Osbourn G C, Gourley P L, Fritz I J, Biefeld R M, Dawson L R and Zipperian T E 1987 Semiconductors and Semimetals Vol.24 ed Dingle R (Academic) pp 459-503
Pistol M -E, Leys M R and Samuelson L 1988 Phys. Rev. B **37** 4664
Recio M, Armelles G, Meléndez J and Briones F 1990 J. Appl. Phys. **67** 2044
Shanabrook B V, Glembocki O J and Beard W T 1987 Phys. Rev. B **35** 2540
Yaguchi H, Zhang X, Ota K, Nagahara M, Onabe K, Shiraki Y and Ito R 1993 Jpn. J. Appl. Phys. **32** 544
Zhang X, Onabe K, Nitta Y, Zhang B, Fukatsu S, Shiraki Y and Ito R 1991 Jpn. J. Appl. Phys. **30** L1631

Improved structural and transport properties of MBE-grown InAs/AlSb QW's with residual As incorporation eliminated via valved cracker

J. Schmitz, J. Wagner, M. Maier, H. Obloh, P. Hiesinger, P. Koidl, and J.D. Ralston, Fraunhofer-Institut für Angewandte Festkörperphysik, Tullastrasse 72, D-79108 Freiburg, Germany

Abstract: We examine in detail the improvement in the structural and transport properties of MBE-grown InAs/AlSb quantum wells by utilising a valved cracker cell to inhibit unintentional As incorporation in the antimonide layers. Raman scattering and SIMS measurements reveal that the As content of the antimonide layers depends directly on the background As pressure in the growth chamber. Magneto-transport measurements show increases in electron mobilities from 60,000 to 130,000 $cm^2/V{\cdot}s$ due to the elimination of residual arsenic incorporation in such QW's.

1. INTRODUCTION

Antimony-containing III-V compound semiconductor heterostructures have recently become of tremendous technological interest due to their demonstrated application in a wide variety of high-performance electronic and optoelectronic devices, including ultra-high frequency resonant tunneling diodes (Brown 1991), high-transconductance FET's (Werking 1992), high-efficiency mid-infrared photodiodes (Chow 1991), and mid-infrared MQW diode lasers (Choi 1992). The compounds GaSb, AlSb and InAs form a nearly-lattice matched system, with the subsystem InAs/AlSb having the largest conduction bandoffset (1.35 eV) of all fcc-III-V-heterostructures. Combined with the high intrinsic electron mobility of InAs, this system is an interesting alternative to GaAs/AlGaAs heterostructures in high speed applications. However, in terms of materials development and device structures, the antimonides lag behind GaAs. Molecular-beam epitaxy (MBE) is currently the most widely-used growth technique for the above device structures, due to the lack of a stable, high-purity Sb precursor for chemical vapor deposition. Until now, uncontrolled incorporation of arsenic from the MBE background ambient has hindered the reproducible preparation of high-quality antimony-based heterostructures. During conventional molecular beam epitaxy of arsenides, the background pressure (BGP) in the growth chamber varies between the high 10^{-8} torr range and the low 10^{-6} torr range, depending on the As cell temperature, the thickness and location of residual arsenic deposits in the growth chamber, and the pumping system used. Growth of binary antimonides is usually performed with the Sb beam-equivalent pressure (BEP) in the 10^{-7} torr range, so that the As BGP and the Sb BEP are of the same order of magnitude. Unintentional incorporation of As into Sb-containing layers can thus become a severe problem in epitaxial structures containing alternating thin antimonide and arsenide layers, which require that the As evaporation cell remain at its growth temperature during the deposition of the antimony containing layers. In the present study we investigate unintentional arsenic incorporation during the MBE growth of AlSb/InAs/GaSb heterostructures, using both a standard mechanically shuttered As_4 evaporation cell and a valved arsenic cracker. The valved cracker cell was used to accurately control the residual arsenic pressure in the growth chamber. The samples were characterized by Raman spectroscopy, secondary ion mass spectroscopy (SIMS), Hall effect and magneto-transport measurements.

2. MBE GROWTH

All samples were grown on undoped GaAs substrates in a RIBER 2300 system equipped with elemental group III, group V and dopant sources. Arsenic was supplied from both a conventional As_4 sublimator and a valved cracker cell (EPI), the latter used as an As_2 source. Both arsenic cells were equipped with standard mechanical shutters. Reflection high-energy electron diffraction (RHEED) was used both for the calibration of growth rates and for in-situ surface monitoring during growth. Growth rates between 0.8 and 1.0 μm/h were used throughout. To accomodate the misfit between the GaAs substrate and the upper antimonide layers a thick buffer layer was grown, consisting of a 100 nm thick AlSb nucleation layer grown at 570°C, a 1μm thick AlSb layer grown at 530°C and a 10-period smoothing superlattice (25 Å GaSb/25 Å AlSb). On top of this buffer layer an AlSb/InAs/AlSb QW structure, consisting of a 40 nm AlSb bottom barrier, a 15 nm InAs QW and a 10 nm AlSb top barrier, was grown at 500°C. All samples were capped with 10 nm GaSb to prevent oxidation of the AlSb. Four samples (in the following labeled (a) to (d)) with the above structure were grown, with the only intentional difference being the As BGP in the growth chamber during growth. Samples (a) and (d) were grown with a conventional As_4 evaporation cell, producing a high As BGP of 2×10^{-8} torr. Low values of BGP(As) during the growth of samples (b) and (c) were realized by using the valved As cracker with the mechanical shutter closed and the valve opened in different positions during growth of the antimonides. In this way BGP(As_2)=5×10^{-9} torr in sample (b) and BGP(As_2)=5×10^{-10} torr in sample (c) were achieved. Sample (a) and sample (d) were grown under identical conditions; the latter was used for electrical measurements. All samples were grown with "InSb-like" interfaces, leading to superior transport properties in comparison with "AlSb-like" interfaces (Tuttle 1990). The actual formation of an InSb-like interface has been proven by Raman spectroscopy showing an InSb-like interface mode (Sela 1992a, Wagner 1993).

3. RESULTS

Fig. 1 shows SIMS depth profiles (recorded using a Cs^+ sputtering beam) of samples (a), (b) and (c). For clarity only the $CsAl^+$ signal of sample (a) and the $CsAs^+$ signals of all three samples are shown in Fig. 1. The Al profiles in samples (b) and (c) are identical to that of (a). The long and the short plateau in the Al signal correspond to the lower and the upper AlSb barrier of the AlSb/InAs/AlSb structure, respectively, while the peak in the As signal and the dip in the Al signal correspond to the InAs well. The As related levels in the antimonide layers of the three samples decrease with decreasing As BGP (from sample (a) to sample (c)), indicating a reduction in the unintentional incorporation of arsenic into the AlSb barrier layers

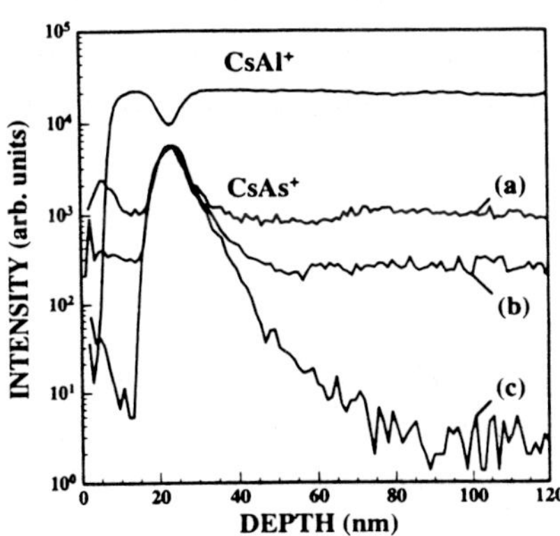

Fig. 1: SIMS depth profiles of three GaSb-capped AlSb/InAs/AlSb QW structures grown with different As background pressures: (a) BGP(As_4)=2x10^{-8} torr, (b) BGP(As_2)=5x10^{-9} torr, (c) BGP(As_2)=5x10^{-10} torr. For clarity only the Al- and the As-related signals are shown.

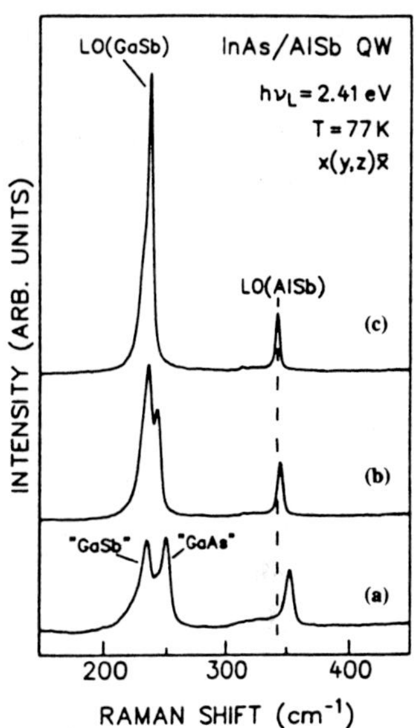

Fig. 2.: Raman spectra of GaSb cap layer and upper AlSb barrier layer of AlSb/InAs/AlSb QWs grown at three different As background pressures: (a) BGP(As_4)=2x10^{-8} torr, (b) BGP(As_2)=5x10^{-9} torr, (c) BGP(As_2)=5x10^{-10} torr. The elimination of residual arsenic incorporation is evident from the disappearance of both the GaAs-like phonon mode ("GaAs") in the GaSb cap layer and the As-induced shift of the AlSb mode.

and the GaSb cap layer. It is evident that this As incorporation scales directly with the As BGP during MBE growth.

The same trend of increasing As incorporation with increasing As BGP was observed in the Raman spectra of the same three samples (Fig. 2). The spectra were excited using ~100 mW from an Ar^+ laser at a wavelength of 514.5 nm (2.41 eV). The absorption depth at this wavelength is estimated to be several tens of nm (Sela 1992a). With increasing As BGP, the AlSb related one-LO phonon peak (from the upper AlSb barrier) is shifted to higher frequencies; at the same time the GaSb related one-LO phonon line (from the GaSb cap layer) splits into a GaSb-like and a GaAs-like mode (Wagner 1993). Such behaviour is characteristic of ternary $GaSb_{1-x}As_x$ and $AlSb_{1-x}As_x$ compounds. $AlSb_{1-x}As_x$ is known to show a single-mode behaviour, with a shift of the one-LO phonon mode to higher frequencies with increasing As content (Sela 1992b), while $GaSb_{1-x}As_x$ shows a mixed mode behaviour (McGlinn 1986). By comparing the measured shift of the AlSb-related phonon mode with the data of Sela et al. (Sela 1992b), As concentrations in the AlSb barriers of 8 and 18 % are estimated for samples (a) and (b). The AlSb LO phonon mode frequency observed in sample (c) corresponds to that of binary AlSb. Comparing the Raman spectra from the GaSb cap layers in samples (a) and (b) with published results for $GaSb_{1-x}As_x$ (McGlinn 1986) indicates As concentrations on the order of 7% for sample (b) and 19% for sample (a). Clearly, unintentional As incorporation of this magnitude must be suppressed in order to achieve reproducible MBE growth of high-quality InAs/AlSb/GaSb heterostructure and QW device layers.

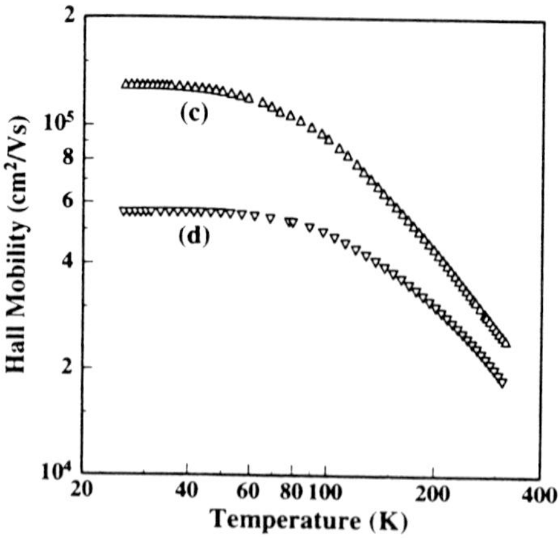

Fig. 3.: Hall mobilities of AlSb/InAs/AlSb QWs with "InSb-like" interfaces. Sample (c) is grown with a low As background pressure using a valved As cracker, sample (d) with a high As background pressure using a conventional As_4 cell.

Despite the fact that the above AlSb/InAs/AlSb QWs are nominally undoped, electron concentrations on the order of 10^{12} cm^{-2} are found in the structures (Tuttle 1990, Nguyen 1992a, Nguyen 1992b, Ideshita 1992). Such large electron concentrations have been proposed to originate from three sources: interface donors associated with the InAs/AlSb interfaces (Tuttle 1990); surface donors (Nguyen 1992a); and deep donors in AlSb bulk material (Nguyen 1992b, Ideshita 1992). In Fig.3 and Fig.4, Hall effect and low-temperature magnetoresistance measurements, respectively, are presented for two samples, grown with the As valved cracker (c) at a low As BGP and with a standard As_4 evaporation cell (d) at a high As BGP. The Hall measurements reveal electron concentrations of $1.4x10^{12}$ cm^{-3} at room temperature for both samples with a very small temperature dependency leading to $1.1x10^{12}$ cm^{-3} (sample c) and $0.9x10^{12}$ cm^{-3} (sample d) at 25 K. Both samples reveal high mobilities, as expected for samples with "InSb-like" interfaces, but the low-temperature and the room-temperature mobilities of the sample grown with a low As BGP are substantially higher ($\mu_{300\,K}$=24,400 $cm^2/V{\cdot}s$, $\mu_{25\,K}$=130,000 $cm^2/V{\cdot}s$) than those of the sample grown at a high As BGP ($\mu_{300\,K}$=20,000 $cm^2/V{\cdot}s$, $\mu_{25\,K}$=56,000 $cm^2/V{\cdot}s$) , although the sample structure and growth parameters are not optimized in terms of electron mobilities. This large increase in the mobility of sample (c) must be attributed to the decrease in the As BGP during growth as compared to sample (d). Similar behaviour is found in the magnetoresistance measurements at 1.6 K (Fig.4). The sample grown using the valved cracker (low As BGP) shows at higher magnetic fields Shubnikov de Haas (SdH) oscillations characteristic for a high-mobility 2D electron gas. A detailed evaluation reveals an electron concentration of N_{2D}= $1x10^{12}$ cm^{-2} and a mobility of μ=114,000 cm^2/Vs. Analysis of the SdH oscillations of the sample grown using the standard As_4 evaporation cell (high As BGP) yield N_{2D}= $9x10^{11}$ cm^{-2} and μ=60,000 cm^2/Vs. Apparent in Fig. 4 is a drift in the baseline

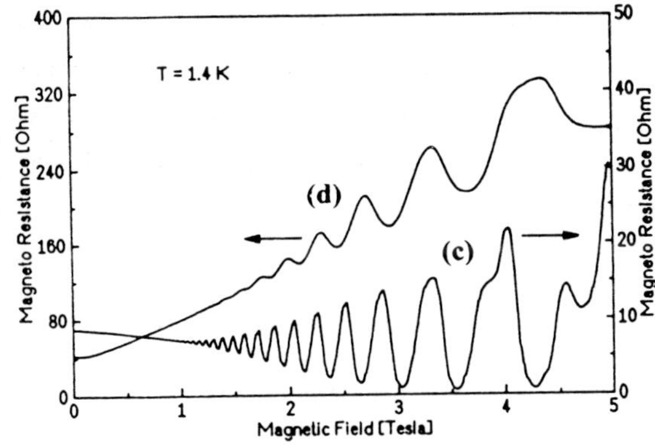

Fig. 4: Magneto-resistance measurements on AlSb/InAs/AlSb QWs. The growth conditions for sample (c) and (d) are the same as in Fig. 3. The drift of the baseline for curve of sample (d) indicates the presence of a 3D conducting channel, in addition to the 2D electron gas in the QW.

of the magnetoresistance of the lower mobility sample, attributed to a 3D channel conducting parallel to the 2D InAs channel. The 3D channel is most likely due to donors in the AlSb barrier (Tuttle 1990, Nguyen 1992a). Since the mobility in the samples is influenced by the As BGP in the MBE machine, it is expected that this donor is related to As incorporated into the AlSb.

CONCLUSIONS

We have demonstrated that unintentional arsenic concentrations as large as 10 - 20 % can be incorporated into AlSb and GaSb films from the background ambient in the MBE growth chamber. Both Raman and SIMS measurements reveal that the As content of the AlSb barriers and the GaSb cap layer is directly dependent on the background As pressure. The As valved cracker facilitates complete suppression of the unintentional As incorporation. Finally, in samples grown with a high As background pressure, magneto-transport measurements reveal the presence of both a 2-D conducting channel in the InAs QW and a 3-D conducting channel in the AlSb barriers. The 3-D channel is most likely due to an As-related deep donor in AlSb. Magneto-transport measurements on samples in which As incorporation has been eliminated with the valved cracker reveal only a 2-D conducting channel, as well as an increase of the low temperature electron mobility from 60,000 to 130,000 $cm^2/V{\cdot}s$.

Acknowledgments

The authors whish to thank G. Bihlmann and T. Fuchs for their assistance with sample preparation and characterization. We also thank H. Rupprecht for his continuing support and encouragement. This work has been financed by the German Bundesministerium für Forschung und Technologie (BMFT) in the framework of the III-V Electronics Research Program.

References

Brown E.R., Söderström J.R., Parker C.D., Mahoney L.J., Molvar K.M. and McGill T.C. 1991 Appl. Phys. Lett. **58** 2291

Choi H.K. and Eglash S.J. 1992 Appl. Phys. Lett. **61** 1154

Chow D.H., Miles R.H., Nieh C.W. and McGill T.C. 1991 J. Cryst. Growth **111** 683

Ideshita S., Furukawa A., Mochizuki Y. and Mizuta M. 1992 Appl. Phys. Lett. **60** 2549

McGlinn T.C., Krabach T.N., Klein M.V., Bajor G., Greene J.E., Kramer B, Barnett S.A., Lastras A. and Gorbatkin S. 1986 Phys. Rev. B **33** 8396

Nguyen C., Brar B., Kroemer H. and English J.H. 1992a Appl. Phys. Lett. **60** 1854

Nguyen C., Brar B., Kroemer H. and English J.H. 1992b J. Vac. Sci. Technol. B **10** 898

Sela I., Bolognesi C.R. and Kroemer H. 1992a Appl. Phys. Lett. **60** 3283

Sela I., Bolognesi C.R. and Kroemer H. 1992b Phys. Rev. B **46** 16142

Tuttle G., Kroemer H. and English J.H. 1990 J. Appl. Phys. **67** 3032

Wagner J., Schmitz J., Maier M., Ralston J.D. and Koidl P. 1993 to appear in Proceedings of MSS-6, Solid State Electron.

Werking J.D., Bolognesi C.R., Chang L.-D, Nguyen Ch., Hu E.L. and Kroemer H. 1992 IEEE Electron. Device Lett. **13** 164

Leakage current mechanisms in strained InGaAs/GaAs MQW structures

J P R David, P Kightley*, Y H Chen, T S Goh, R Grey, G Hill and P N Robson

Department of Electronic and Electrical Engineering, University of Sheffield Mappin Street, Sheffield S1 3JD, U.K.
*Department of Material Science and Engineering, University of Liverpool, PO Box 147, Liverpool L69 3BX, U.K.

ABSTRACT: The reverse leakage current density (J_r) in a range of strained InGaAs/GaAs MQW pin diode structures has been measured. The magnitude of J_r at a particular electric field appears to depend primarily on the factors that determine the average strain of the MQW, thickness of the MQW and the capping layer thickness. Plan view TEM shows that misfit dislocation arrays have formed at both the capping and buffer layer interfaces with the MQW and that the total length of dislocation present correlates with the magnitude of J_r and the strain thickness product of the MQW and capping layer.

1.INTRODUCTION

The use of strain to help improve device performance is now well established by the success of devices such as the pseudomorphic HEMT and the strained quantum well laser. For thin single strained structures such as these, the simple equilibrium energy balance model for dislocation formation of Matthews and Blakeslee (1974) provides a useful guide to the point of the onset of plastic relaxation.

The strain for a bounded pseudomorphic single layer is assumed to be confined to the layer itself. Any capping material which has the same lattice parameter as the substrate material should be free from strain. If the strained layer is thicker than the critical layer thickness, h_c, then dislocation formation occurs at the interface with the substrate and once this has occured, if the capping layer is thick enough, at the top interface also. In strained multi-quantum wells (MQWs) where each strained layer is thinner than h_c and separated by a layer of substrate material "simplistically" no dislocation formation should occur. However in reality dislocations are known to occur at the boundaries of a MQW structure. This suggests that the strain is not confined to the individual strained layers and that the dislocation formation is a characteristic of the overlap of the strain fields for the separate layers comprising the stack. It follows that for MQWs the important features of the stack are, well strain, barrier strain, well thickness (L_W), barrier thickness (L_B) and the number of periods (i.e. the total MQW thickness).

We have characterised a series of InGaAs/GaAs MQW pin diode structures with varying indium composition and MQW dimensions by measuring the reverse leakage current as a function of electric-field. Plan view transmission electron miscroscopy

(TEM) was performed to directly measure the interfacial dislocation content throughout the multilayers and was correlated to the magnitude of the reverse leakage current density.

2.EXPERIMENTAL DETAILS

The samples were grown by MBE using conventional sources on an n^+ (100)GaAs substrate and comprised a 1-2μm n^+GaAs buffer, the undoped InGaAs/GaAs MQW and finally 1-2μm of p^+GaAs capping layer. All except one substrate (layer 353) were mounted indium free. The indium composition of the well was varied from 0.1 to 0.3 with L_W and L_B varying between 65Å and 250Å as detailed in Table I. The number of periods was varied between 30 to 87, giving a total thickness of ~0.6 to 1.2 μm. The average strain of the MQW was determined from the average indium alloy composition of the MQW (David et al 1991). As Table I shows, the MQWs considered in this study varied in average indium fraction from 0.0375 to 0.169, and in total thickness (obtained from capacitance-voltage measurements) from 0.62μm to 1.19μm. The slight discrepancies between the calculated and measured MQW thickness can be attributed to slight variations in the growth rates. Obviously if the MQW behaves as single thick layer with the average indium fraction, it easily exceeds the Matthews and Blakeslee critical layer thickness (1974) and a pronounced misfit dislocation array is expected. Nomarski phase contrast microscopy showed that the cross-hatching that follows dislocation formation was present for all layers and became more pronounced as the average strain and thicknesses of the structure (i.e. numbers of dislocations increased).

TABLE I: Details of the strained MQWs

Layer	indium	L_W(Å)	L_B(Å)	periods	i (μm)	ave. indium	cap (μm)
441	0.10	150	250	30	1.28	0.0375	1
384	0.15	100	200	25	0.75	0.05	2
203	0.30	65	250	32	1.13	0.062	1
353	0.15	100	150	25	0.63	0.06	2
252	0.30	65	50	87	1.03	0.17	1

3. ELECTRICAL CHARACTERISATION

Circular mesa diodes of 100-400μm diameter were fabricated as described previously (David el at 1991) and the electrical characteristics measured using a an HP4140B picoammeter. Capacitance-voltage measurements performed with an HP4275 LCR meter on these devices showed that the background doping of the undoped MQW intrinsic region was low at ~$3\times10^{15}cm^{-3}$ with the devices fully punched through by ~ 1 volt reverse bias. The forward turn on voltage for these layers was typically 1.0 volt and the ideality factor varied from 1.3 - 1.8, with the more highly strained layers having a larger value. The most significant differences were however to be found in the reverse leakage currents which varied by almost four orders of magnitude for a given voltage. Different area devices from each structure gave identical current densities at reverse voltages >3 volts up to the breakdown voltage suggesting that the leakage was due to a bulk mechanism. At very low voltages in the less strained structures there was evidence that the leakage was dominated by edge effects as the current scaled with device radius rather than area.

Since the layers have significantly different total MQW thicknesses, it is essential to compare their current density (J_r) as a function of reverse electric-field (Er) as shown in Fig. 1. With the exception of layer 353, the more highly strained MQWs generally had a higher value of leakage for a given E_r, in agreement with previous observations (David et al 1991). All the structures show an exponential dependence of J_r on E_r, even for relatively low values of E_r. At high values of E_r, a further increase in J_r due to the onset of avalanche multiplication is observed. This behaviour is different to the much flatter J_r vs. E_r characteristics seen in similar lattice matched GaAs/AlGaAs structures.

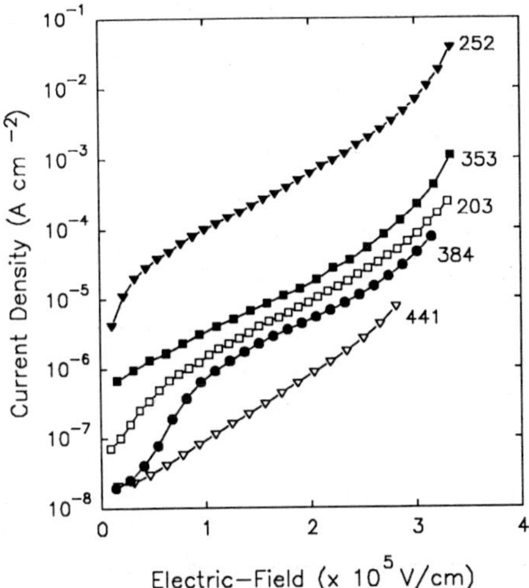

Fig.1 Leakage current density vs. reverse electric-field for the strained MQW pins

Attempts to model the leakage current components using generation-recombination currents, diffusion currents (Forrest 1981) and tunnelling currents (Forrest et al 1980) proved difficult. Tunnelling generally only occurs in narrow band-gap semiconductors and at relatively high values of E_r. In our structures the effective band-gap is still large at ~1.2-1.3eV, however tunnelling appears to be present for E_r as low as 5×10^4 Vcm^{-1}. Good agreement with the experimental could only be obtained when a shunt component term (Forrest et al 1980) and tunnelling via multi-step levels in the bandgap was allowed (Riben and Feucht 1966). Full details of this modelling will be published elsewhere.

4.DISLOCATION CONTENT FROM TEM

The most common form of misfit dislocation formation is that from existing threading dislocations. This is the primary souce of misfit dislocations and it should be noted that the number of layer threading dislocations remains the same, or may even be slightly reduced, over that for the substrate. This mechanism of misfit relief operates adequately provided the rate of relaxation the layer demands as it grows can be satisfied by it. If it cannot be satisfied, then other mechanisms of defect formation (i.e. secondary mechanisms) come into action which usually create a large number of layer threading dislocations.

Provided there is no secondary source of misfit dislocations operating, the relaxation of a thick capped MQW (comprising individual strained layers $<h_c$) is assumed to proceed in the following manner:

i) The MQW behaves as a single thick strained layer and misfit dislocations form at the n^+(GaAs)/MQW interface and not at individual layers within the MQW. Cross section and stereo TEM show this to be the case (Kightley 1991).

ii)The dislocations relax the MQW by an amount *d* which induces a mismatch of *d* into the p^+(GaAs) capping layer that is grown onto the MQW.

iii)Misfit dislocation formation now takes place at the p^+(GaAs)/MQW interface where the final amount of relaxation is approximately in proportion to the relaxation that occurred at the initial n^+(GaAs)/MQW interface and the capping layer thickness.

Secondary nucleation of defects was only seen for layer 252 and so all other layers are assumed to have relaxed in the manner described above. In order to study the microstructure throughout the rather thick MQW structure, portions of the epitaxial layers were selectively etched off so that plan view samples could be used to image dislocation content at different points in the growth. Dislocations were counted and analysed at the MQW/p^+(GaAs)interface (the top dislocation array), n^+(GaAs)/MQW interface (the bottom dislocation array) and for MQW threading dislocations.

Dislocation density measurements for linear arrays of dislocations were initially measured in units of dislocations per micron (measured in an orthogonal <110> direction) and for threading dislocations in units of dislocations cm^{-2}.

For the purposes of this investigation the (a/2)<110> misfit dislocations can be divided into the two simple categories of interfacial 60 degree misfit dislocations and interfacial edge misfit dislocations. As the number of 60 degree dislocations dominate in this comparison, we shall only consider these.

Only layer 252 showed the generation of threading dislocations. These were observed in both MQW and capping layer suggesting that the nucleation and generation occurred during the growth of the MQW. Most of the threading dislocation nucleation appears to be from repetitive nucleation on individual {111} planes. This is shown by both the plan view TEM and Nomarski interference contrast. On the plane itself the dislocation density is massive ($1 \times 10^{10} cm^{-2}$) while away from these planes dislocation density is still very large at approximately 3×10^8 cm^{-2}. It was noted that the threading dislocations formed on the {111} planes that were not exposed as steps by the substrate offcut angle.

5.DISCUSSION

To assess the accuracy of the TEM results we studied the effect of misfit strain present in the layers on the numbers of dislocations present, expecting to find an increase in dislocation numbers for thicker more highly strained MQW's. For both the top and bottom interface arrays the observed dislocation numbers correlated approximately linearly with the products of strain and thickness in the layers above them. The dislocation density in the bottom array was an approximately linear function of the total MQW strain and MQW thickness and the top misfit dislocation array density was an approximate linear function of the capping layer thickness and the mismatch induced in this by the relaxation that had taken place at the bottom interface. This is in good agreement with prediction for the small thickness/strain range studied.

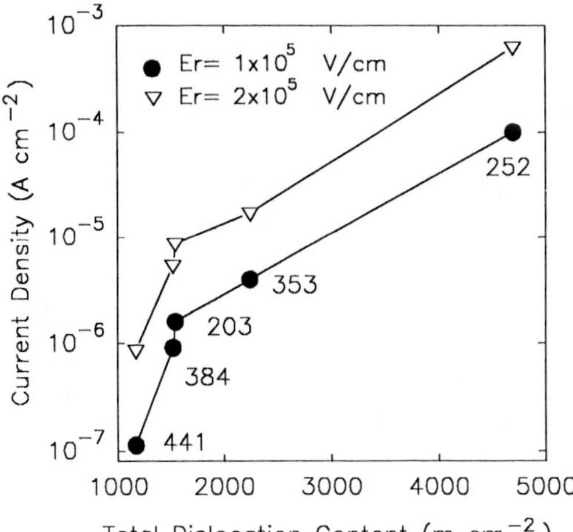

Fig.2 Leakage current density at $E_r = 1 \times 10^5$ and 2×10^5 V/cm vs. the total dislocation content in the strained MQW pin diodes. The line is a guide to the eye.

If a dislocation is regarded as a line of recombination centres (Labusch and Schroter 1980), then the most relevant parameter to compare with the leakage current will be the length of dislocation per unit area of the interface. The number of metres of dislocation cm^{-2} of the interface for the top and bottom interfaces together with the leakage currents at $E_r=1 \times 10^5$ and 2×10^5 V/cm are given in Table II. Both layers 384 and 353 have relatively thin and low average strained MQWs, however the top misfit dislocation content is high because of the thicker capping layer. Layer 353 has a slightly larger than expected J_r possibly due to the indium wetted substrate giving rise to a higher growth temperature. Plotting Jr against the total dislocation length cm^{-2} for the structures examined as shown in Fig.2 demonstrates that there is initially a very rapid increase in J_r from layer 441 to layers 203 and 384. Both layers 203 and 384 have very similar total dislocation length densities and hence have similar values for J_r. Further increases in the average strain and thickness of the structure leads to an increase in the total dislocation length cm^{-2} and J_r but the rate of increase in J_r is not as large.

TABLE II: Details of dislocation content and leakage current densities

Layer	Bottom dislocations (cm^{-2})	Top dislocations (cm^{-2})	Total dislocations (cm^{-2})	J_r (Acm^{-2}) @ 1×10^5V/cm	J_r (Acm^{-2}) @ 2×10^5V/cm
441	839	333	1172	1.13×10^{-7}	8.65×10^{-7}
384	769	756	1525	9.15×10^{-7}	5.45×10^{-6}
203	1034	512	1546	1.6×10^{-6}	8.71×10^{-6}
353	≥1195	1046	2241	4.0×10^{-6}	1.7×10^{-5}
252	2980	1715	4695	1.0×10^{-4}	6.3×10^{-4}

These results show that for strained MQW pins, leakage currents are directly affected by misfit dislocation content. Both the average strain and the thickness of the MQW, and the thickness of the cap determine the magnitude of the misfit dislocation content.

Acknowledgements: This work was partly funded by the Science and Engineering Research Council(U.K.) under grant PB124. We would like to thank D M Carr for help with the material assessment and R.Beanland for the TEM sample preparation.

References:

David J P R,Grey R, Pate M A, Claxton P A and Woodhead J 1991 J.Electron.Mat. 20 295
Forrest S R 1981 IEEE J.Quantum Elect. 17 217
Forrest S R, Kim O K and Smith R G 1983 Solid-State Elect. 26 951
Forrest S R, DiDomenico Jr M, Smith R G and Stocker H J 1980 Appl.Phys.Lett. 36 580
Kightley P 1991 PhD. Thesis, University of Liverpool
Labusch R and Schroter W 1980 Dislocations in solids, ed. Nabarro F R N 5 127
Matthews J W and Blakeslee A E 1974J.Cryst.Growth 27 118
Ribens A R and Feucht D L 1966 Int.J.Elect. 20 583

Inst. Phys. Conf. Ser. No 136: Chapter 6
Paper presented at the Int. Symp. GaAs and Related Compounds, Freiburg, 1993

Conduction-band and valence-band structures in strained $In_{1-x}Ga_xAs$/InP quantum wells on (001) InP substrates

M Sugawara, N Okazaki, T Fujii, and S Yamazaki

Fujitsu Laboratories Ltd., 10-1 Morinosato-Wakamiya, Atsugi 243-01, Japan

ABSTRACT: We calculated conduction-band effective mass of biaxially strained $In_{1-x}Ga_xAs$ and $In_{1-x}Ga_xAs$/InP quantum wells on (001) InP substrates based on $\mathbf{k}\cdot\mathbf{p}$ perturbation approach. By magneto-optical absorption spectra, we determined Luttinger-Kohn parameters for valence bands. Based on the determined band structures, we show that both biaxial compressive and tensile strain can lower the threshold current density of quantum-well lasers by factors 2-3.

1. INTRODUCTION

By introducing homogeneous strain into semiconductor quantum wells, we can design the electronic band structures through the change in the volume and symmetry of the crystal lattice (Pollak 1990, for example). The band engineering produced strained semiconductor quantum-well lasers with much superior performances to lattice-matched lasers (Yablonovitch 1986 and Adams 1986). Note that in-plane band dispersion of strained quantum wells has not been fully established, however. The strain effect on the conduction-band effective mass in quantum wells has received little attention. The valence-band effective-mass parameters (Luttinger-Kohn parameters (Luttinger 1955)), to which the detail of the calculated dispersion is markedly sensitive, have not been well verified, casting doubt on the numerical details of valence-band dispersion.

In this work, we studied the conduction-band and valence-band structures in biaxially strained $In_{1-x}Ga_xAs$/InP quantum wells on (001) InP substrates. We start by discussing band offsets of this quantum well system needed for the dispersion calculation, using our tight-binding model including cation d-orbital bases (Sugawara 1993a). On the basis of the first-order $\mathbf{k}\cdot\mathbf{p}$ perturbation, we calculate conduction-band effective masses both in biaxially strained $In_{1-x}Ga_xAs$ layers and the quantum wells. By magneto-optical absorption spectra, we determined Luttinger-Kohn effective-mass parameters for valence bands as a function of crystal composition. We demonstrate that conduction-band effective mass strongly depends on the well width and strain, and that Luttinger-Kohn parameters are almost half those calculated by Lawaetz (1971). Finally, we discuss the threshold current density of strained quantum-well lasers.

2. BAND OFFSET

Among various theoretical works to predict band-offset values at semiconductor heterojunctions, Harrison's tight-binding model (Harrison 1977) is known to be quite simple and, at the same time, to agree exceptionally well with experiments in III-V semiconductors. In this approach, the valence-band offset between two semiconductors is given as the energy difference between respective valence-band maximum which is simply expressed by cation - anion bonding p-state. The only, but crucial, exception is for systems including aluminum such as $Al_xGa_{1-x}As$/GaAs and $In_{1-x}Al_xAs/In_{1-y}Ga_yAs$ heterojunctions, where the tight-binding model predicts almost vanishing offsets, contrary to experiments. Wei pointed out that this difficulty can be solved by taking into account the repulsion between the bonding p-state and cation d-state (Wei 1988). We improved Harrison's tight-binding model by incorporating cation d-orbitals as basis functions of the tight-binding matrix, and derived a simple equation even applicable to systems including aluminum (Sugawara 1993a). Using the formula, we calculate the band offsets in the present heterojunctions.

Figure 1 (a) shows the calculated band-edge energies at **k** = 0 for free-standing (unstrained) $In_{1-x}Ga_xAs$ as a function of gallium composition, x. As x increases, the conduction-band minimum, $E_{c,o}$, increases monotonously and the valence-band maximum, $E_{v,o}$, decreases. The maximum of the split-off band is also shown by extracting the spin-orbit splitting energy, Δ, from $E_{v,o}$. In Fig. 1 (b), it is assumed that $In_{1-x}Ga_xAs$ layers are grown coherently on (001) InP substrates and under biaxial strain. We added the eigenvalue of orbital-strain Hamiltonian (Pollak 1990) to each band-edge energy. The topmost valence band splits into heavy-hole (HH) and light-hole (LH) states, and the conduction-band edge is less dependent on composition. For both InP and $In_{0.52}Al_{0.48}As$, $E_{c,o}$ and $E_{v,o}$ are shown by dashed horizontal lines at the longitudinal axis. Solid and open circles represent the measured conduction-band edges (Cavicchi 1989 and People 1989), showing a good agreement with the calculation. In $In_{1-x}Ga_xAs$/InP quantum wells, the transition from type-I to type-II potential profile occurs at x = 0.9. $In_{1-x}Ga_xAs/In_{0.52}Al_{0.48}As$ quantum wells are expected to have a type-I potential profile at any composition.

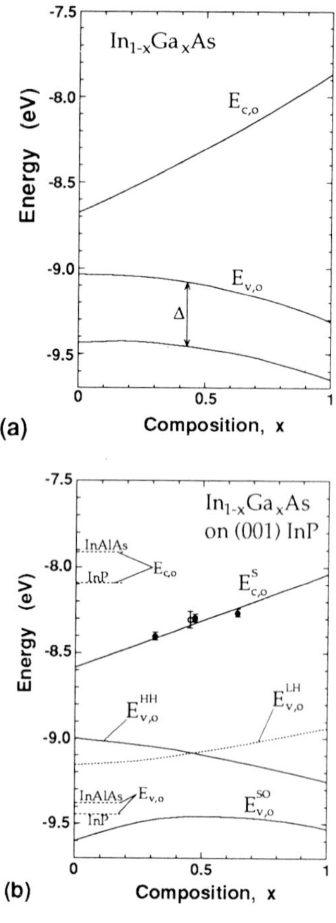

Fig. 1. Band offset.

3. CONDUCTION BAND

Kane's **k•p** perturbation approach revealed that, in bulk semiconductor materials, effective mass and nonparabolic dispersion of the conduction band originate from the first-order interactions between s- and p-state bases (Kane 1957). Bastard (1991) solved the first-order **k•p** matrix in direct-gap semiconductor quantum wells to present the formulation for in-plane conduction-band dispersion. We incorporate the effect of biaxial strain on the **k•p** coupling, using a new set of eight band-edge bases to diagonalize the orbital-strain Hamiltonian. By folding the 8 x 8 first-order **k•p** matrix to 2 x 2 for two s-state bases according to Bastard, we derived a formula for conduction-band dispersion in strained materials (Sugawara 1993c).

We calculated band-edge electron effective mass of $In_{1-x}Ga_xAs$ as a function of the composition, x (Fig. 2(a)). The solid line represents the band-edge mass, m_{Γ_6}, in unstrained $In_{1-x}Ga_xAs$ given by the linear interpolation between InAs and GaAs. The dashed line is for the in-plane mass, $m_e^{\parallel}$, and the dotted line is for the mass perpendicular to the layers, $m_e^{\perp}$, in biaxially strained $In_{1-x}Ga_xAs$ on (001) InP substrates. The mass drops drastically under biaxial tensile strain (x>0.467) and increases under biaxial compressive strain (x<0.467), primarily due to the strain-induced change in the energy difference between valence and conduction bands. Note that the mass shows anisotropy; $m_{\Gamma_6} > m_e^{\parallel} > m_e^{\perp}$ under biaxially tensile strain (x>0.468), and $m_{\Gamma_6} < m_e^{\parallel} < m_e^{\perp}$ under biaxially compressive strain (x<0.468). Fig. 2 (b) shows the

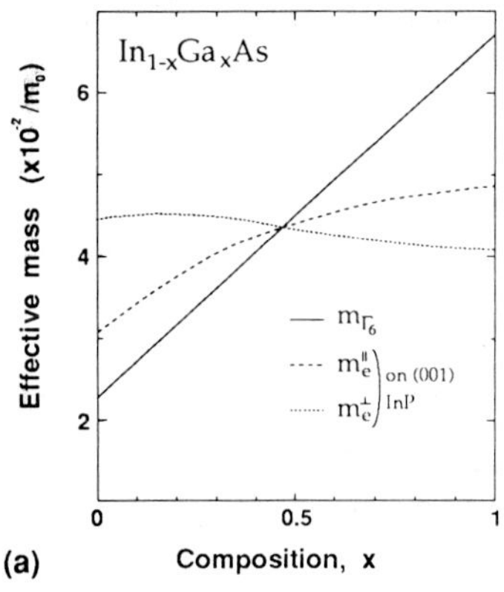

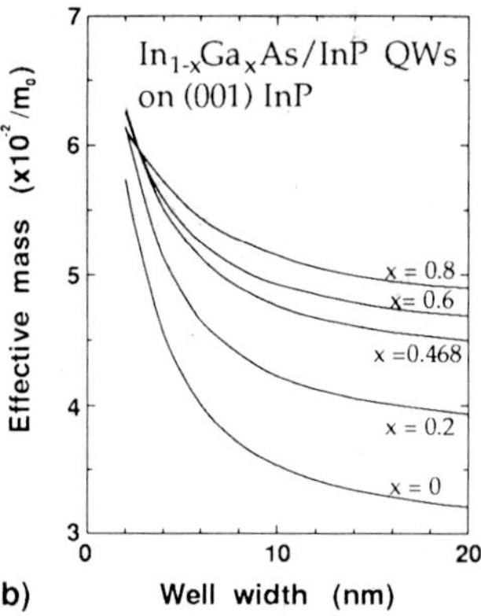

Fig. 2. Band-edge electron effective mass.

in-plane ground-state electron effective mass at the band edge as a function of well width in biaxially strained $In_{1-x}Ga_xAs$/InP quantum wells on (001) InP substrates. As the well width decreases, the mass increases rapidly due to the nonparabolic characteristics of conduction-band dispersion.

4. LUTTINGER-KOHN PARAMETERS FOR VALENCE BANDS

It is well known that the heavy-hole effective mass does not originate from the direct first-order **k•p** coupling, but arises from the indirect second-order **k•p** coupling between p-state bases due to virtual transitions to more remote band edges (Luttinger 1955). The second-order **k•p** matrix with the six p-state band-edge bases is known as the Luttinger-Kohn 6 x 6 matrix. A lot of works have been done to calculate the in-plane valence-band dispersion in quantum wells, especially $GaAs/Al_xGa_{1-x}As$ and $In_{1-x}Ga_xAs$/InP systems (O'Reilly 1989, for example). These studies clarified the significant valence-band nonparabolic characteristics under intersubband mixing and the effect of strain on dispersion. The problem with calculating valence-subband dispersion is, as we pointed out in the introduction, the uncertainty of Luttinger-Kohn parameters. Lawaetz (1971) calculated these parameters for a wide variety of III-V and II-VI semiconductors: γ_1 = 19.67, γ_2= 8.37, γ_3= 9.29, and $\bar{\gamma} = (\gamma_2 + \gamma_3)/2 = 8.83$ in InAs and γ_1 = 7.65, γ_2= 2.41, γ_3= 3.28, and $\bar{\gamma}$ = 2.85 in GaAs ($\bar{\gamma}$ is used when we neglect the anisotropy of valence bands). We next determine a set of Luttinger-Kohn parameters, γ_1 and $\bar{\gamma}$, which properly explain optical absorption spectra of strained quantum wells under a magnetic field.

Under a magnetic field, B, perpendicular to quantum well layers, the exciton resonance energy is written as

$$E_{ex} = E_A + E^z_{c,n} + E^z_{v,n} + E_r + E_{sp}, \tag{1}$$

where E_A is the band gap of well materials, $E^z_{c,n}$ is the quantum confinement energy in the conduction band, $E^z_{v,n}$ is that in valence band, E_{sp} is the spin-splitting term, and E_r is given by the effective mass equation of (Knox 1963)

$$\left[-\frac{\hbar^2}{2\mu}\left(\frac{\partial^2}{\partial r^2} + \frac{1}{r}\frac{\partial}{\partial r}\right) - \frac{e^2}{4\pi\varepsilon\rho} + \frac{e^2B^2}{8\mu}r^2 \right] \Psi_{env} = E_r\Psi_{env} \,. \tag{2}$$

Here, $m = (m^{\parallel}_e{}^{-1} + m^{\parallel}_h{}^{-1})^{-1}$ is the in-plane reduced effective mass, $m^{\parallel}_e$ is the in-plane electron effective mass, $m^{\parallel}_h$ is the in-plane hole effective mass, e is the electron charge, r is the in-plane distance between an electron and a hole, $\rho=[r^2+(z_e-z_h)^2]^{1/2}$, and ε is the static dielectric constant. The third term on the left in Eq. (2), the diamagnetic energy term, forms parabolic in-plane confinement potentials and increases the exciton energy. The envelope wave function is written as

$$\psi_{env} = \frac{1}{\sqrt{D}}\phi_n(\mathbf{r})\varphi_{e,n}(z_e)\varphi_{h,n}(z_h) = \frac{1}{D}\sum_{\mathbf{k}_\parallel} A_n(\mathbf{k}_\parallel)e^{i\mathbf{k}_\parallel\cdot\mathbf{r}}\varphi_{e,n}(z_e)\varphi_{h,n}(z_h) \qquad (3)$$

where $\phi_n(\mathbf{r})$ represents the in-plane relative motion of an electron and a hole, $\varphi_{e,n}(z_e)$ is the electron confined-state wave function, $\varphi_{h,n}(z_h)$ is that of a hole, and $A_n(\mathbf{k}_\parallel)$ is the Fourier coefficient of the relative-motion wave function. We neglected mixing in the confined-state wave functions since excitons are formed primarily from band-edge states. We omitted the terms for the in-plane center-of-mass motion and the angular momentum, taking into account the selection rules of optical transitions under the electric-dipole approximation. Since we found no spin-splitting of the exciton resonances, we neglect E_{sp}. For $\phi_n(\mathbf{r})$, we take the linear combination of the hydrogenic and harmonic-oscillator wave functions with variational parameters (Sugawara 1992).

To include the effect of nonparabolic band dispersion in the calculation, we replace the kinetic energy term in Eq. (2) with the sum in $\mathbf{k}$-space as

$$\left\langle \psi_{env}\right| -\frac{\hbar^2}{2m_i^\parallel}\left(\frac{\partial^2}{\partial r^2} + \frac{1}{r}\frac{\partial}{\partial r}\right)\left|\psi_{env}\right\rangle = \sum_{\mathbf{k}_\parallel} E_{i,n}^\parallel(\mathbf{k}_\parallel)\left| A_n(\mathbf{k}_\parallel)\right|^2 \qquad (i = e,\ h) \qquad (4)$$

where $E^{\parallel}_{e,n}(\mathbf{k}_\parallel)$ is the conduction-band in-plane dispersion calculated by the first-order $\mathbf{k}\cdot\mathbf{p}$ perturbation and $E^{\parallel}_{h,n}(\mathbf{k}_\parallel)$ is that of valence band by the second-order $\mathbf{k}\cdot\mathbf{p}$ matrix including Luttinger-Kohn parameters. By this replacement, we can obtain the reduced effective mass of excitons for a given band dispersion, and calculate E_r as a function of magnetic field using a variational method.

We grew multiple quantum wells consisting of $In_{1-x}Ga_xAs$ well layers and InP barrier layers on (001) InP substrates by metalorganic vapor phase epitaxy. The composition was from x = 0.34 to 0.58, corresponding to in-plane strain of $\varepsilon_\parallel$= -0.88 to 0.78 %, and the well width was from L_W = 6 to 14 nm. The barrier widths were thick enough to neglect the interference of confined-state wave functions between neighboring wells. We measured optical absorption spectra at 2 K. Under biaxial compression (x < 0.468), the splitting between 1e-hh and 1e-lh resonances increased from that of the lattice-matched quantum well as x decreases (Fig. 3). Under biaxial tensile strain (x > 0.468), the positions of 1e-hh and 1e-lh exciton resonances are reversed. We applied the magnetic field of up to 8 tesla in the direction perpendicular to the layers.

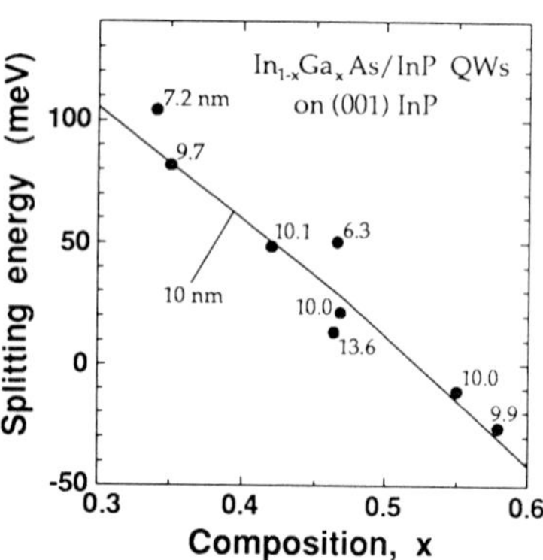

Fig. 3 HH-LH splitting energy.

In magneto-optical absorption spectra of biaxially compressive quantum wells (Fig. 4 (a)), the 1e-hh exciton resonance shows diamagnetic shifts, and its strength remarkably increases. This strength enhancement is caused by the shrinkage of the in-plane relative-motion wave function under parabolic in-plane confinement potential (Sugawara 1992). At the shorter wavelength, a new resonance, which can be attributed to the 1e-hh 2S exciton resonance appears and has larger diamagnetic shifts. In biaxially tensile quantum wells (Fig. 4 (b)), the spectrum of the ground-state 1e-lh resonance hardly changes in either energy or strength. Diamagnetic shifts and increase in strength can be seen in the 1e-hh exciton resonance at shorter wavelengths, and 2S resonance appears. The almost immobile 1e-lh resonance can be attributed to the negative LH effective mass due to the repulsion from lower HH subbands. See our previous work for magneto-

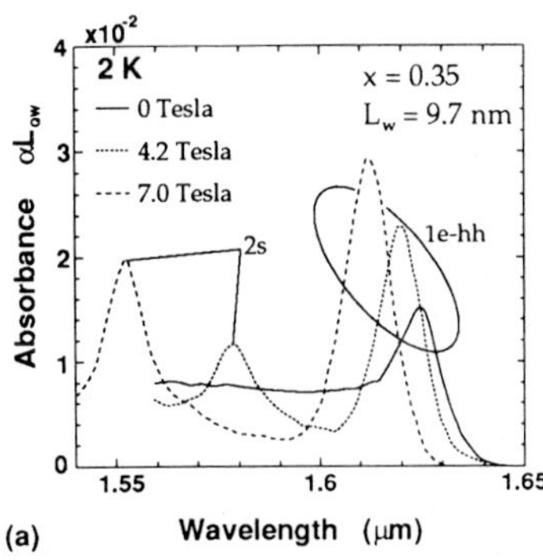

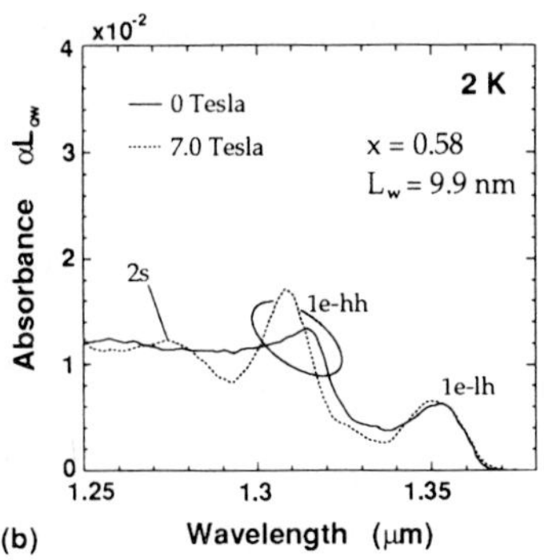

Fig. 4. Magneto-optical absorption spectra under (a) compressive and (b) tensile strain.

optical absorption spectra of the lattice-matched quantum wells (Sugawara 1993b).

In Fig. 5, we plotted the 1e-hh exciton resonance energies for two kinds of quantum wells, including higher-order 2S and 3S resonances, as a function of the magnetic field. The calculation represented by solid lines agree well with measurements. The dashed lines are Landau fans. Note that we must take into account Coulomb interaction, even in higher-order resonances. Detail analyses of 1e-hh and 1e-lh exciton resonance diamagnetic shifts are available (Sugawara 1993c).

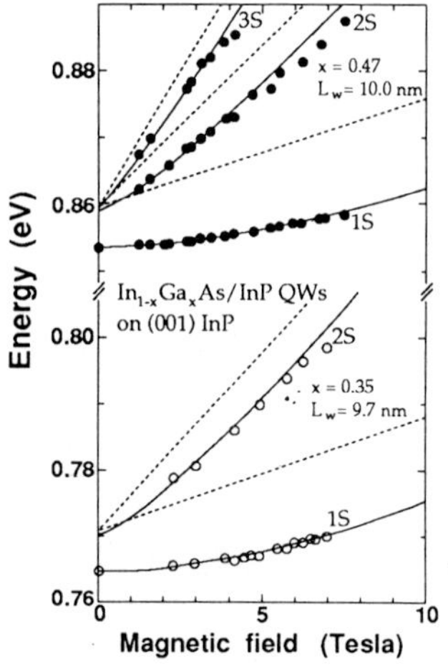

Fig. 5. Diamagnetic shifts.

We plotted Luttinger-Kohn parameters in Fig. 6. Error bars are about ± 1 in γ_1 and ± 0.1 in $\bar{\gamma}$ if we assume $\pm 5\%$ error in the calculated electron effective mass. The solid lines are plotted by multiplying Lawaetz's calculations (linear interpolation between InAs and GaAs) by 0.55 for γ_1 and by 0.5 for $\bar{\gamma}$ (the least square fit); $\gamma_1 = 10.8-6.6x$ and $\bar{\gamma} = 4.4-3.0x$. Note that these values well explain the HH and LH splitting energies (solid line for 10-nm quantum wells in Fig. 3). These values are almost half those calculated by Lawaetz (1971) and significantly modify the calculated in-plane valence-band dispersion through the change in the mass of diagonal terms of Luttinger-Kohn matrix $(m_0/(\gamma_1+\bar{\gamma})$ for HH and $m_0/(\gamma_1-\bar{\gamma})$ for LH) and the change in the degree of intersubband mixing which depends on the splitting energies and the magnitude of nondiagonal terms. To calculate Luttinger-Kohn parameters, we need the interband momentum matrix elements and the energy separations between valence-band edges and remote even-parity band edges. The uncertainties in these quantities, which can hardly be obtained experimentally, may explain the discrepancy between Lawaetz's calculations and our values.

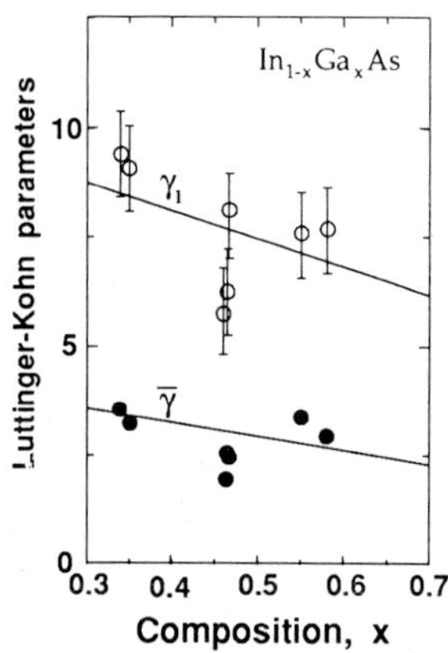

Fig. 6. Luttinger-Kohn parameters.

5. THRESHOLD CURRENT OF STRAINED InGaAs/InGaAsP QUANTUM-WELL LASERS

Based on the determined band structures, we calculated threshold current density in $I_{1-x}Ga_xAs$/InGaAsP separate-heterostructure-confinement quantum-well lasers as a function of composition for 1.5-μm emission (Fig. 7). The right-hand side separated by the longitudinal dotted lines (crossing of HH and LH bands) represent the results for TM-mode electron-LH emission and left hand side for TE-mode electron-HH emission. Solid circles are for the mirror loss of $\xi = 0\ cm^{-1}$ and open circles for $\xi = 20\ cm^{-1}$. Marks on solid lines are for 300 K and those on dashed lines for 360 K. We chose the well width to give 1.5-μm emission and optimized the number of well layers to give minimum threshold. The threshold current density decreases in either side of the crossing composition. We have 1/2 - 1/3 threshold current density in either strain direction, compared with lattice-matched quantum wells. This result agrees well with recent experiments for low-threshold strained quantum well lasers (Thijs 1991 and Yamamoto 1992). A detailed discussion on the mechanism for the threshold current reduction is available (Sugawara 1993d).

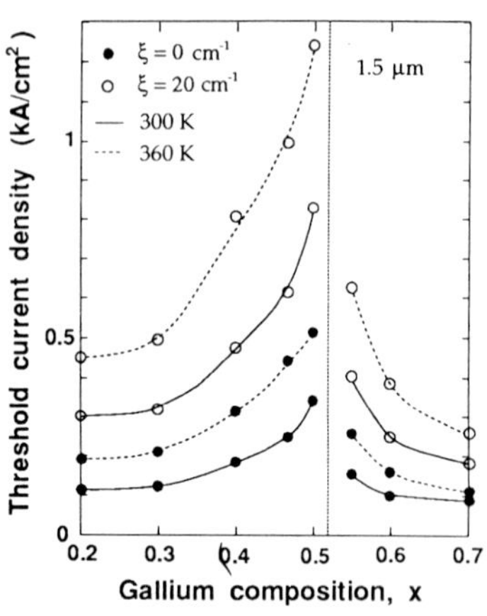

Fig. 7. Threshold current density.

6 CONCLUSION

We clarified conduction-band and valence-band structures in $In_{1-x}Ga_xAs$/InP strained quantum wells on (001) InP substrates using tight-binding model, **k•p** perturbation approach and magneto-optical absorption of exciton resonances. On the basis of the determined band structures, we showed that both biaxial compressive and tensile strain can lower the threshold current density of quantum-well lasers by 1/2-1/3.

REFERENCES

Adams A R 1986, Electron. Lett. 22 249

Bastard G, Brum J A, and Ferreira R 1991 "Electronic States in Semiconductor Heterostructure" Solid State Physics 44 (New York: Academic) pp 229.

Cavicchi R E et al. 1989 Appl. Phys. Lett. 54 739

Harrison W A 1977 J. Vac. Sci. Technol. 14 1016

Kane E O 1957 J. Phys. Chem. Solids 1 249

Knox R S 1963 Theory of Excitons (New York: Academic)

Lawaetz P 1971 Phys. Rev. B 4 3460

Luttinger J M and Kohn W 1955 Phys. Rev. 97 869

O'Reilly E P 1989 Semicond. Sci. Technol. 4, 121

People R et al. 1989 Appl. Phys. Lett. 54, 1457

Pollak F H 1990 Semiconductors and Semimetals Vol. 32 (New York: Academic)Chap.2

Sugawara M 1992 Phys. Rev. B 45 11423

Sugawara M 1933a Phys. Rev. B 47 7588

Sugawara M, Okazaki N, Fujii T and Yamazaki S 1993b Phys.Rev. B (to be published)

Sugawara M, Okazaki N, Fujii T and Yamazaki S 1993c Phys.Rev. B (to be published)

Sugawara M 1993d (submitted to Microwave and Optical Technology Letters)

Thijs P J A, Binsma J M, Tiemeijer L F, and Dongen T V 1991 Technical Digest of ECOC/IOOC (Paris) pp. 31-38

Wei S H and Zunger A 1988 Phys. Rev. B 37 8958

Yablonovitch E and Kane E O 1986 IEEE J. Lightwave Technol. LT-4 504

Yamamoto T, Nobuhara H, Sugawara M, Fujii T and Wakao K 1992 Extended Abstract of the 1992 International Conference on Solid State Devices and Materials, pp. 607

Paper presented at the Int. Symp. GaAs and Related Compounds, Freiburg, 1993

Strain relaxation in $In_{0.2}Ga_{0.8}As$/GaAs MQW structures

G Bender, E C Larkins, H Schneider, J D Ralston and P Koidl.

FhG Institute for Applied Solid State Physics, Tullastr. 72, D-79108 Freiburg, Germany.

ABSTRACT: We study the limits of pseudomorphic strain in MBE grown $In_{0.2}Ga_{0.8}As$/GaAs multiple quantum well structures and the influence of lattice relaxation on the optoelectronic properties of high-speed p-i-n photodetectors with MQWs in the intrinsic region. High-resolution X-ray diffraction yields the degree of lattice relaxation. For the detectors these results are in agreement with photocurrent spectroscopy measurements and subband calculations. The detectors yield a quantum efficiency of unity, in spite of the onset of lattice relaxation.

1. INTRODUCTION

The InGaAs/GaAs material system has gained considerable interest for high-speed optoelectronic device applications since it allows the fabrication of GaAs-compatible devices with operating wavelengths extending beyond 1μm (see also Larkins (1993) and the references therein). The large lattice mismatch between the binary compounds GaAs and InAs is an important feature of this material system, and a potential problem for device applications. Hetero-epitaxial layer systems can accomodate lattice constant variations via elastic strain up to a critical point, above which plastic deformation takes place, and built-in strain is reduced by the formation of dislocations, as reported by Matthews and Blakeslee (1974). Lattice relaxation and the accompanying creation of defects can lead to degraded device performance. In this paper we quantify the degree of lattice relaxation in $In_{0.2}Ga_{0.8}As$/GaAs multiple quantum wells (MQWs) and investigate its influence on the optoelectronic and electronic properties of high-speed p-i-n photodetectors. In spite of the onset of lattice relaxation, we find no degradation of detector sensitivity (i.e. an internal quantum efficiency of unity). This is explained by the fast transport of the photogenerated carriers out of the intrinsic (i) region to the contacts. The transport time of the carriers across the i-region is much smaller than their lifetime due to non-radiative recombination, which can be estimated by time-resolved photoluminescence to be larger than 500ps.

2. STRUCTURAL CHARACTERIZATION

The samples contain $In_{0.2}Ga_{0.8}As/GaAs$ MQW structures grown on semi-insulating GaAs (001) substrates using standard MBE techniques. The samples with 6 and 7 QWs have 5.65nm wells and 20nm barriers. Those with 10, 15 and 20 periods have the MQW structure with 7.6nm wells and 30nm barriers embedded in the i-region of a p-i-n diode structure. For more details about sample structure and processing see Larkins (1993).

High-resolution X-ray diffraction (HRXRD) measurements are performed on a four-circle Eulerian goniometer. A 2-crystal 4-reflection Ge monochromator is used to narrow the primary beam in wavelength and angular divergence. Symmetric 004 reflections and asymmetric 115 reflections (115^{+} at an angle of incidence of 60° and 115^{-} at 30°) are recorded. The simulations are based on a formulation of dynamical theory of X-ray diffraction given by Fewster and Curling (1987). The pseudomorphic deformation of the heterostructure is included according to Seegmüller and Murakami (1988). Lattice relaxation is accounted for by a factor γ in the equations, which link the in-plane components $\epsilon_{\parallel}$ of the strain tensor to the in-plane lattice constants of substrate and film:

$$\varepsilon_{\parallel} = (1 - \gamma) \times (a_{subs} - a_{film}) / a_{subs} \qquad (1)$$

A value of $\gamma=0$ corresponds to pseudomorphic strain accomodation, while $\gamma=1$ indicates complete strain relaxation. The structural parameters are fixed by fitting the angular dependence of the 004 and the 115^{-} reflection. The samples with 6 and 7 QWs are well described in the model of pseudomorphic strain. The samples with 10, 15 and 20 QWs were partially relaxed. The measured and calculated rocking curves near the GaAs 004 reflection of the 10 QWs sample is shown in Figure 1. A measure for the uncertainty of the relaxation factor γ is obtained by also considering the 115^{+} reflection. Here, this uncertainty is the interval in which all three reflections (004, 115^{-} and 115^{+}) are well reproduced. For the partially relaxed samples the values of the relaxation factor γ, and the uncertainties are tabulated in the inset of Figure 1. It is remarkable, that the shape of the MQW reflections near GaAs 004 change with the number of wells as depicted in Figure 2. The sample with 6 QWs shows almost the theoretical line shape. The line form of the sample with 7 QWs broadens for intensities below half maximum of the MQW reflection. The broadening has the character of an exponential decrease, symmetric to the line maximum. The sample with 20 QWs shows this broadening even for intensities up to the line maximum. This symmetric broadening of the theoretical line form with the character of an exponential decrease can be explained with a change of the lattice constant distribution in the MQW structure. In order to quantify the observed broadening, the

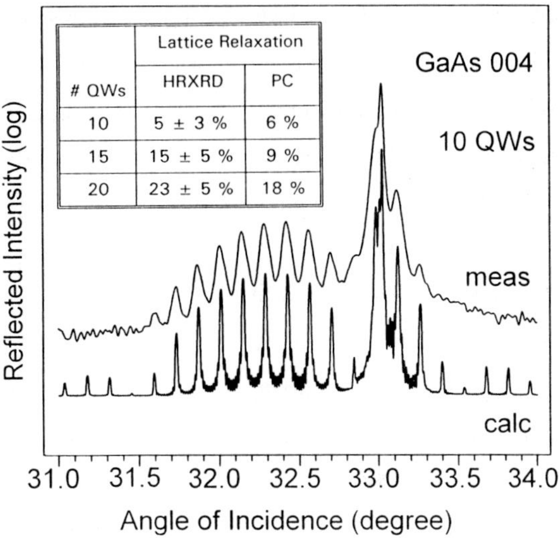

Figure 1: Measured and calculated rocking curve near GaAs 004 for the sample with 10 QWs. The inset tabulates the degree of relaxation defined in equation (1) and obtained from X-ray diffraction and photocurrent data.

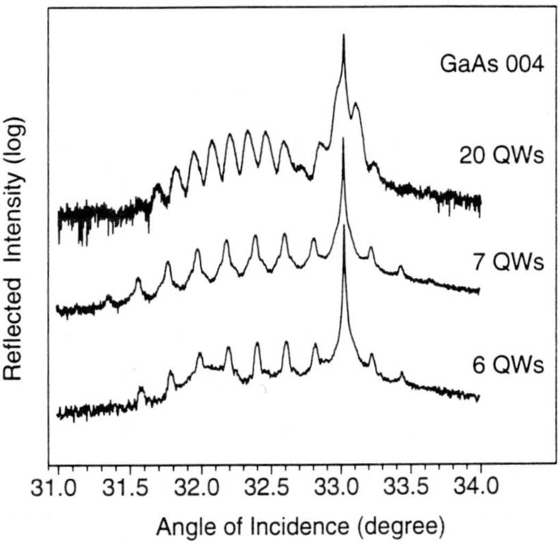

Figure 2: Measured rocking curves near GaAs 004. The MQW features between 31.5° and 32.8° change their line shape and broaden.

difference δFWHM, between measured and calculated FWHM of the MQW reflections near the GaAs 004 reflection are shown in Figure 3. The observed broadening is in qualitative agreement with studies of Kamigaki et al (1986) on strain-relaxed $In_xGa_{1-x}As/GaAs$ single-layer structures with x-values ranging from 5% to 20% at a constant layer thickness of 200nm.

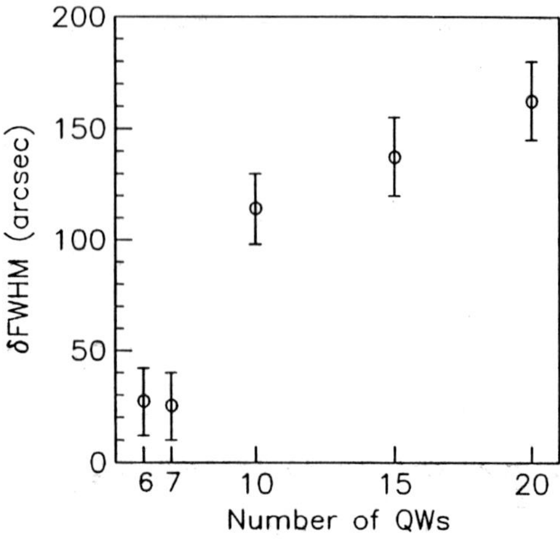

Figure 3: Difference between measured and calculated FWHM of MQW rocking curve features near GaAs 004 vs. the number of QWs.

3. OPTOELECTRONIC PROPERTIES

Photocurrent (PC) spectroscopy measurements are performed at 300K to study the optoelectronic properties of the detector samples (Figure 4). The zero-bias responsivities at the excitonic transition between the first electron and the first heavy hole subband (e1-hh1) scale roughly linearly with the number of wells, as shown in the inset of Figure 4. Together with absorption measurements, the measured responsivities give an internal quantum efficiency of unity for each detector, indicating that all photogenerated carriers reach the contacts. PC spectra with a forward bias of about 0.6V are used to study the subband structure close to flat band voltage. Subband calculations in the envelope function /k•p/ effective mass approximation including strain effects due to tetragonal deformation are carried out using an 8x8 band model described by Cohen and Marques (1990). The excitonic binding energy is set to a constant value of 7meV for all transitions. Lattice relaxation is taken into account using equation (1). Experimentally, the strain sensitive energetic separation between the light hole (lh) and hh excitons is found to decrease with increasing number of QWs (83meV, 77meV and 69meV for 10, 15 and 20 QWs, respectively). The energies of the e1-hh1, e1-lh1 and e2-hh2 transitions are reproduced with relaxation factors γ comparable with those obtained from HRXRD, as tabulated in the inset of Figure 1.

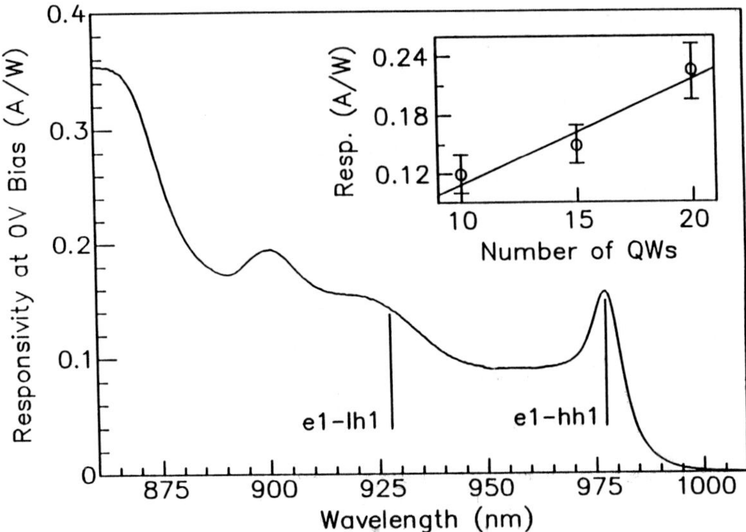

Figure 4: Measured responsivity at 0V applied voltage for the sample with 15 QWs for normal incidence. The inset shows the maximum responsivities at the first heavy hole first electron excitonic resonance versus the number of QWs.

Time-resolved photoluminescence (PL) measurements are performed with a mode-locked Ti:sapphire laser and a spectrometer/streak-camera assembly. The exponential decay constants of the QW luminescence at 300K close to flat band voltage with an excitation wavelength of 837nm are 345ps, 430ps and 570ps for the samples with 10, 15 and 20 QWs, respectively. The decay constants increase with increasing number of wells, which is opposite to the expected influence of the lattice relaxation. The increase of the PL decay constants is therefore attributed to carrier diffusion out of the doped regions and into the intrinsic region. The diffusion current increases as the detectors are biased close to the flat band voltage, and is expected to have decreasing influence for detectors with thicker i-regions. Since the measured decay constants represent lower limits for the lifetimes of the carriers, the lifetime of the photogenerated carriers is larger than 570ps. As published earlier by Schneider et al (1992), for the sample with 15 QWs the vertical transport times are smaller than 20ps for a 2V reverse bias, and on the order of 30ps at zero bias. Thus, carrier transport times across the i-region even without applied voltage are more than an order of magnitude smaller than the carrier lifetime due to non-radiative recombination.

4. CONCLUSION

We have studied the limits of pseudomorphic strain in MBE grown $In_{0.2}Ga_{0.8}As/GaAs$ MQW structures and the influence of lattice relaxation on the optoelectronic properties of high-speed p-i-n photodetectors. The degree of lattice relaxation is quantified using high-resolution X-ray diffraction. An increase in lattice relaxation is observed with increasing - number of wells in the samples with 10 and more QWs. The onset of relaxation leeds to a dramatic change in the line shape and line width of X-ray rocking curves. The degree of relaxation determined from X-ray diffraction is in good agreement with the one determined from photocurrent spectra and subband calculations. Responsivity measurements indicate an internal quantum efficiency of unity, in spite of the onset of lattice relaxation. Time-resolved photoluminescence shows recombination lifetimes of the QW excitons exceeding 500ps. This is at least an order of magnitude larger than the sweep-out time of the photo-generated carriers out of the intrinsic region. This clearly demonstrates, that strain relaxation is not necessarily detrimental for fast photodetectors.

The authors are grateful to acknowledge J. Fleißner and M. Hoffmann for sample processing, M. Pilz and K. Schwarz for technical support, N. Herres for helpful discussions concerning the X-ray measurements, H. Biebl for absorption measurements and H. Rupprecht for his support and encouragement. This work was supported by the BMFT within the framework of the PHOTONIK program.

5. REFERENCES

Cohen A M and Marques G E 1990 Phys. Rev. B41 10608-10621.
Fewster P F and Curling C J 1987 J. Appl. Phys. 62 4154.
Kamigaki K, Sakashita H, Kato H, Nakayama M, Sano N and Terauchi H 1986 Appl. Phys. Lett. 49 1071.
Larkins E C, Bender G, Schneider H, Ralston J D, Wagner J, Rothemund W, Dischler B, Fleissner J and Koidl P 1993 J. Cryst. Growth 127 62-67.
Matthews J W and Blakeslee A E 1974 J. Cryst. Growth 27 118-125.
Schneider H, Larkins E C, Ralston J D, Fleissner J, Bender G and Koidl P 1992 Appl. Phys. Lett. 49 2648.
Seegmüller A and Murakami M 1988 "Analytical techniques for Thin Films" (New York: Academic Press) 27 pp 143-200.

Paper presented at the Int. Symp. GaAs and Related Compounds, Freiburg, 1993

InGaAlAs/InP type II multiple quantum well structures grown by gas source molecular beam epitaxy

Y. Kawamura, H. Kobayashi and H. Iwamura

NTT Opto-electronics Laboratories, Morinosato Wakamiya 3-1,
Atsugi-shi, Kanagawa 243-01, Japan

ABSTRACT: Optical and elctrical properties of InGaAlAs/InP Type II MQW structures grown by gas source molecular beam epitaxy (GS-MBE) are studied. It is found that the transition from the Type I to Type II structures occurs at an Al composition (y) of 0.18, and the properties change drastically between the TypeI and Type II regions. InAlAs/InP Type II MQW diodes are also fabricated, where a clear electro-absorption effect peculiar to Type II structures and an efficient electro-luminescence are observed at room temperature.

I. INTRODUCTION

An $In_{0.52}Ga_{0.48-y}Al_yAs$/InP MQW structure, lattice-matched to InP substrates, has both the Type I structure and Type II structure depending on the Al composition of the InGaAlAs quaternary layers, as schematically shown in Fig. 1. In the limit of y=0, it becomes InGaAs/InP Type I MQWs, which have been studied very extensively, and have been applied to MQW optical devices in the 1.3-1.5 μm wavelength region. On the other hand, in the limit of y=0.48, it becomes InAlAs/InP Type II MQWs. Kroemer and Griffiths (1983) first suggested that the InAlAs/InP hetero-structure has a staggered (Type II) band lineup with an effective band gap of 1.1 eV. In the InAlAs/InP Type II MQW structure, electrons are confined in the InP layers, while holes are confined in the InAlAs layers, and they are spatially separated from each other. Light emission (Aina et al. 1988, Deplon et al. 1992) and the electro-absorption effect (Kobayashi et al. 1993a) were observed for InAlAs/InP

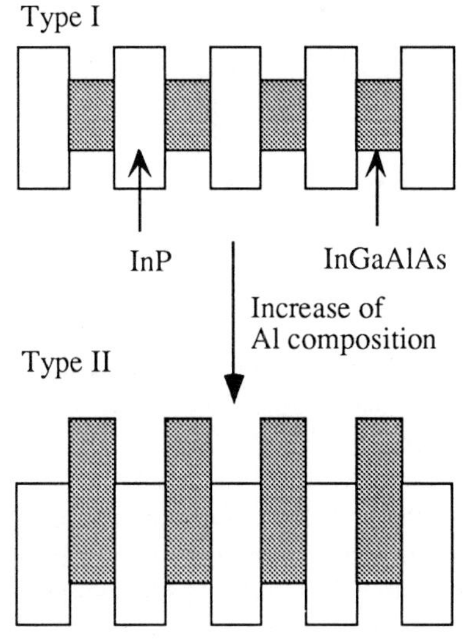

Fig. 1 Change of energy band structure of InGaAlAs/InP MQWs

MQW structures. The resonant tunneling effect was also observed in this material system (Kawamura and Iwamura 1992).

In this paper, Al composition dependence of the optical and elctrical properties of InGaAlAs/ InP Type II MQW structures grown by gas source molecular beam epitaxy (GS-MBE), was studied. It is found that the transition from the Type I to Type II structure occurs at an Al composition of 0.18, and the properties change drastically between the Type I and Type II regions. InAlAs/InP Type II MQW diodes were fabricated, where a clear electro-absorption effect peculiar to Type II structure and an efficient electro-luminescence were observed at room temperature.

II. GS-MBE GROWTH OF InGaAlAs/InP MQW STRUCTURES

InGaAlAs/InP MQW layers (60 periods) were grown on semi-insulating Fe-doped (100) InP substrates by GS-MBE. The growth temperature was 500 °C. The growth rate was 2.6 μm/h for the InGaAlAs layer, and 1.4 μm/h for the InP layer. All samples were not intentionally doped with the residual background electron concentration being $\sim 1\text{x}10^{15}\ \text{cm}^{-3}$. The thick-

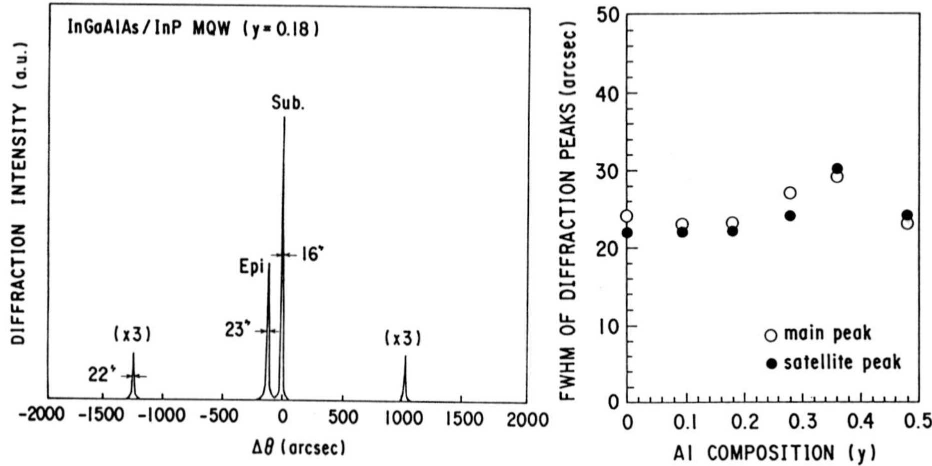

Fig. 2 Double crystal X-ray diffraction pattern of InGaAlAs/InP MQWs (y=0.18)

Fig. 3 Al composition dependence of the FWHM of x-ray diffraction peaks

nesses of the InGaAlAs layer and the InP layer are 85Å and 80Å, respectively. The Al composition (y) varied from y=0.0 ($In_{0.52}Ga_{0.48}As$/InP MQWs) to y=0.48 ($In_{0.52}Al_{0.48}As$/InP MQWs). Figure 2 shows a double crystal X-ray diffraction pattern for the InGaAlAs/InP MQW layer with y = 0.18. The FWHM for the main diffraction peak and first satellite peaks are 23 arcsec and 22 arcsec, respectively. It was confirmed that the FWHM of the diffraction peaks are between 20-30 arcsec for all Al compositon ranges, as shown in Fig. 3. This indicates high uniformity of the MQW layers studied here.

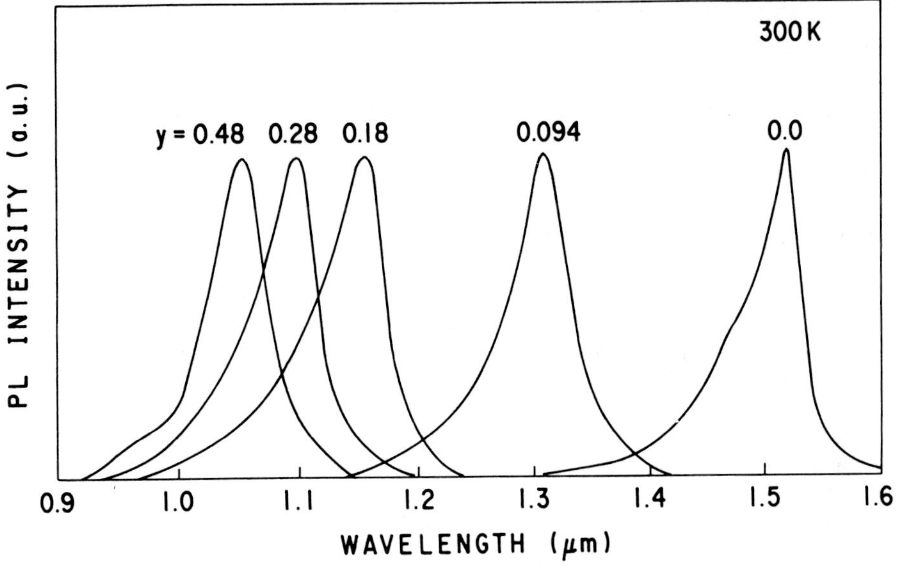

Fig. 4 Photoluminescence spectra for InGaAlAs/InP MQW layers at 300K

III. COMPOSITION DEPENDENCE OF InGaAlAs/InP MQW STRUCTURES

Photoluminescence (PL) measurements were carried out at 300K by using an Ar gas laser (wavelength = 5145Å) as an excitation source. All samples show a clear PL spectrum at room temperature, as shown in Figure 4, where the spectra are normalized at the peak intensity. From the PL peak wavelength, the y dependence of the effective band gap (Eg) at 300K for the InGaAlAs/InP MQW layers is estimated. The result is shown in Fig. 5. It is shown from this figure that dEg/dy changes abruptly at around y=0.18. Below y=0.18, dEg/dy is 1.47 eV, while it decreases to 0.36 eV above y=0.18. The solid line in Fig. 5 is the calculated Eg dependence. In the calculation, the conduction band discontinuity (ΔE_c) of the InAlAs/InP is 0.35 eV (Kawamura and Iwamura 1993), and ΔE_c of the InGaAs/InP is 0.23 eV (Skolnick et al. 1987). It is assumed that ΔE_c of InGaAlAs/InP varies linearly with y. The band gap of InGaAlAs layer is that re-

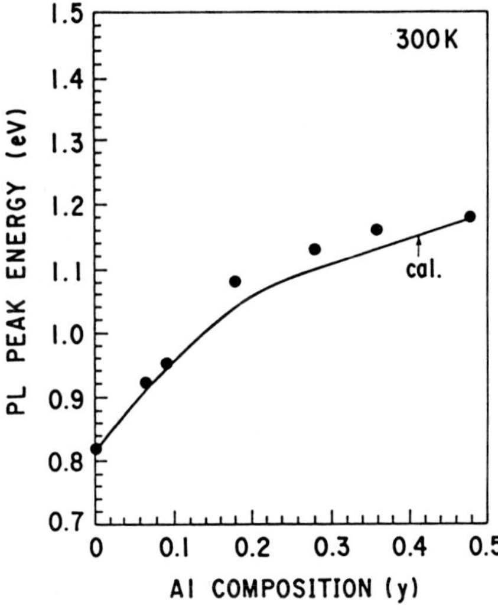

Fig.5 Al composition dependence of the energy gap of InGaAlAs/InP MQW layers at 300K

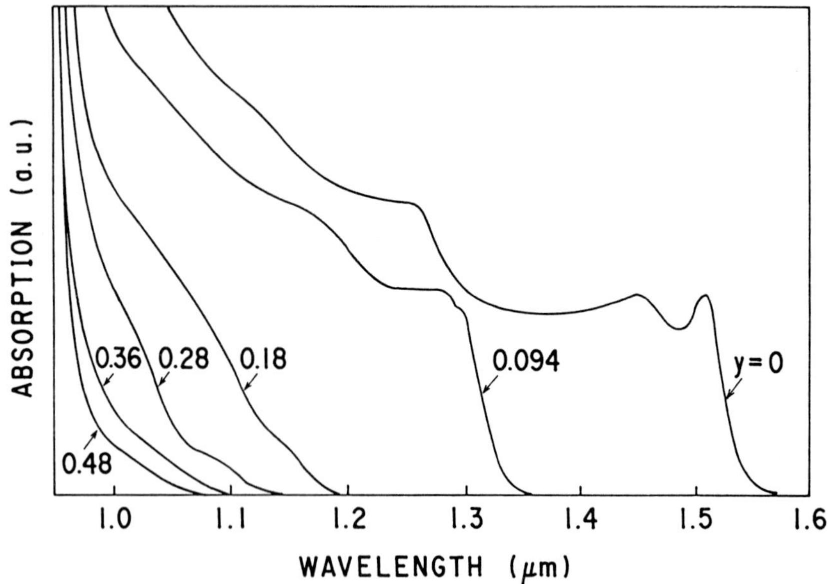

Fig. 6 Absorption spectra of InGaAlAs/InP MQW layers at 300K

ported by Fujii et. al. (1986), and that of InP is 1.35 eV (Turner et al. 1964). The electron effective mass of the InGaAlAs layer is that reported by Olego et al. (1982), and that of InP is 0.08 (Skolnick et al. 1987). The nonparabolicity of the conduction band is taken into account. The calculated value agrees well with the experimental data. This indicates that the change of dEg/dy is due to the transition from the Type I to Type II structure.

To confirm this, the absorption spectra were measured at 300K. Figure 6 shows the absorption spectra with different y at 300K. Clearly, a remarked change is observed at y=0.18. Below y=0.18, a sharp absorption edge with a excitonic resonance peak is observed. On the other hand, the absorption becomes weak above y=0.18. It is known that the absorption coefficient in the Type II MQW structure becomes small due to the decrease in the overlap of the wavefunction between electrons and holes,

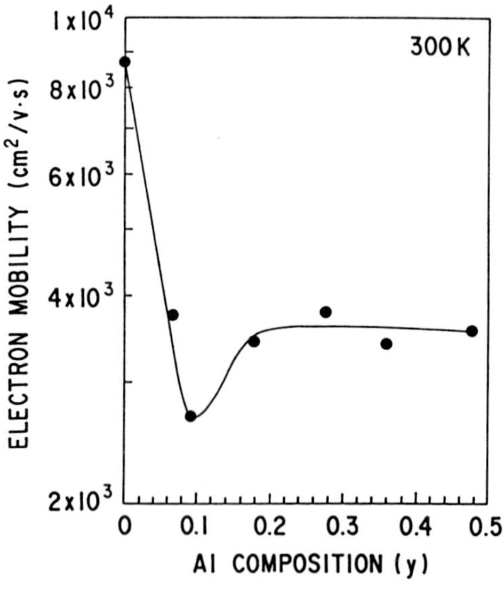

Fig. 7 Al composition dependence of electron mobility at 300K

which are spacially separated. Therefore, the result in Fig. 6 indicates that the MQW has the Type II structure above y=0.18.

The transition between the Type I and Type II structures was also observed in the electrical properties. Figure 7 shows y dependence of the electron mobility (μ) of the InGaAlAs/InP MQW layers. μ was measured by the van der Pauw method. In the region of y<0.18, μ decreases drastically from 8700 to 2700 $cm^2/V{\cdot}s$ with increasing y. However, in the region of y>0.18, μ remains almost constant (3500-3700 $cm^2/v{\cdot}s$). It was confirmed that μ of the InGaAlAs bulk layer is 9700 $cm^2/V{\cdot}s$ at y=0, and decreases rapidly with increasing y, while μ of InP layer is 4500-4700 $cm^2/V{\cdot}s$ at 300K. Taking this into account, the result in Fig. 7 indicates that electrons are confined in the InP layers above y=0.18, while they are confined in the InGaAlAs layers below y=0.18, which is consistent with the results of optical measurements.

IV. PROPERTIES OF InAlAs/InP TYPE II MQW DIODES.

To study the electro-absorption (EA) and electro-luminescence (EL), InAlAs/InP Type II MQW diodes were grown on Sn-doped n-type (100) InP substrates. They are composed of an InAlAs/InP MQW layer (10 or 40 periods) sandwiched between a Si-doped n-type InP cladding layer and a Be-doped p-type InP cladding layer. AuGeNi and AuZnNi were used as n- and p-electrodes, respectively. The MQW diodes were fabricated into a high mesa structure having a 350 μm diameter. A circular electrode with a 200 μm diameter window was evapolated on the upper p-InP cladding layer.

Figure 8 shows the photo-current spectrum under reverse bias for the InAlAs/InP MQW diode with 40 periods at 300K. The InAlAs width is 70Å, and that of InP is 80Å. The input light was introduced through the upper circular electrode. It is clear that the absorption edge shifts to the higher energy side (blue shift), and the excitonic peak structure at the absorption edge is enhanced with the reverse bias. This electric field dependence is opposite to the well-known quantum confined Stark effect (QCSE) observed in the Type I MQW structure, where the absorption edge shows a red shift, and the excitonic peak decreases with the field. The observed electric field dependence is peculiar to the Type II MQW structure (Li and Khurgin 1992). The details of the electro-absorption effect is reported elsewhere (Kobayashi et al. 1993b).

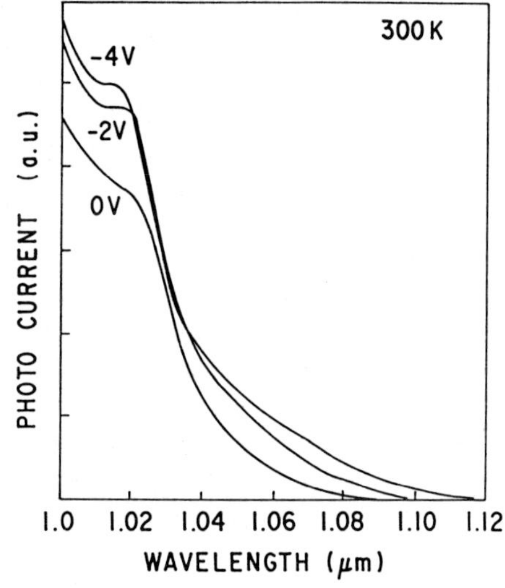

Fig. 8 Photo-current spectrum of the InAlAs/InP MQW diode at 300K under the reverse biased condition.

The EL measurement was also carried out

at 300K. The EL was detected from the top p-side circular window. Figure 9 shows the EL spectrum for the InAlAs/InP Type II MQW diode with 10 periods. The InAlAs layer thickness is 70Å, and the InP thickness is 80Å. The injection current density is 1x10^1 A/cm^2. The emission wavelength is 1.03 μm with a spectral half width of 38 meV. It should be noted that the emission intensity of the InAlAs/InP Type II MQW diodes studied here is comparable to that of the InP homo-junction diodes, which was grown as a reference. This fact suggests that the InAlAs/InP Type II MQW structure can be applied not only for optical modulators, but also for light emitting devices in the 1 μm wavelength region.

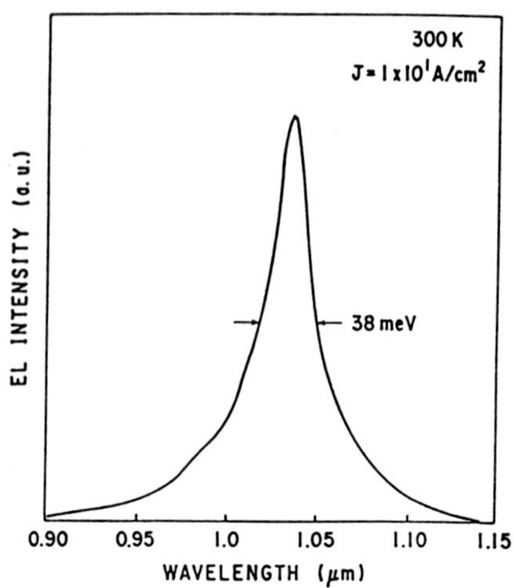

Fig. 9 EL spectrum of the InAlAs/InP MQW diode at 300K under the forward biased condition

V. CONCLUSION

InGaAlAs/InP Type II MQW structures were grown by gas source molecular beam epitaxy (GS-MBE), and Al composition dependence of optical and electrical properties were studied. It was found that the transition from the Type I to Type II structures occured at an Al composition of 0.18, and the properties changed drastically between the TypeI and Type II region. A clear electro-absorption effect characteristic to the Type II structure and an efficient electro-luminescence were observed for the InAlAs/InP Type II MQW diodes at room temperature.

VI. ACKNOWLEDGEMENT

The authors would like to thank Mr. Shin-ichi Iida for processing of the MQW diodes. They are also indebted to Yoshihiro Imamura for his continuous encouragement.

REFERENCES

Aina L, Mattingly M. and Stecker L.; 1988 Appl. Phys. Lett. 53 1620.
Fujii T, Nakata Y., Sugiyama Y., and Hiyamizu S.; 1986 Jpn.J.Apl.Phys. 25 L254.
Kawamura Y. and Iwamura H.; 1992 Jpn.J.Appl.Phys. 31 L1733.
Kobayashi H., Kawamura Y., and Iwamura H.; 1993a Jpn. J. Appl. Phys. 32 548.
Kobayashi H., Kawamura Y., and Iwamura H.; 1993b to be presented at Int. Conf. of Solid State Devices and Materials. Chiba Japan.
Kroemer H. and Griffiths G.; 1983 IEEE Electron Device Lett. EDL-4 20.
Li S. and Khurgin B.; 1992 Appl.Phys.Lett. 60 1969.
Lugagne-Delpon E., Voisin P., Voos M., and André J.P.; 1992 Appl. Phys. Lett. 60 3087.
Olego D., Chang T.Y., Silberg E., Caridi E.A., and Pinkzuk A.; 1982 Appl. Phys. Lett. 41 476.
Skolnick M.S., Taylor L.L, Bass S.J, Pitt A.D., Mowbray D.J., Cullis A.G., and Chew N.G.; 1987 Appl. Phys. Lett. 51 24.
Turner W.J., Reese W.E., and Pettit G.D.; 1964 Phys. Rev. A136 1467.

Paper presented at the Int. Symp. GaAs and Related Compounds, Freiburg, 1993

Determination of minority charge carrier lifetime in non-lattice-matched MOVPE-grown $GaAs_{1-x}P_x/Al_yGa_{1-y}As$ double heterostructures

R.A.J. Thomeer, A. van Geelen, S.M. Olsthoorn, G.J. Bauhuis, M. van Schalkwijk and L.J. Giling

Department of Experimental Solid State Physics, RIM, Faculty of Science, University of Nijmegen, Toernooiveld, 6525 ED Nijmegen, The Netherlands

Abstract

Time-resolved photoluminescence has been used to extract the minority-carrier lifetimes in non-lattice matched MOVPE-grown $GaAs_{1-x}P_x$ ($0.00 \leq x \leq 0.02$) layers. The $Al_{0.2}Ga_{0.8}As$ cladding layers were used for interfacial passivation and carrier confinement.
We report a definite reduction in interface recombination velocity at increasing phosphorus content (S decreasing from 8400 cm/s to 220 cm/s). The bulk recombination was only slightly enhanced.

1 INTRODUCTION

The growth of strained layers (*i.e.* layers with a lattice constant different from that of the substrate) makes it possible to develop III-V based minority-carrier devices with properties not achievable with lattice-matched semiconductor structures. For example, it was shown that the presence of strain in the active region of semiconductor lasers results in a higher performance and reliability as compared to lattice-matched structures (Thijs *et al.* 1991). Furthermore, strain is of great importance in GaAs-based devices grown on economically interesting substrates like germanium and silicon.
The minority-charge carrier lifetime is an important parameter in the determination of the opto-electronic properties of GaAs-based structures. In this paper we have investigated the influence of strain on the carrier lifetimes of $GaAs_{1-x}P_x$ epilayers which were grown by metal-organic vapour phase epitaxy (MOVPE). The phosphorus concentration ranged between x=0.00 and x=0.02. The thicknesses of most active layers were below the critical layer thickness, thereby avoiding the formation of misfit dislocations. Time-resolved photoluminescence (PL) was used to extract the minority-carrier lifetimes, which are a function of bulk lifetime and interface recombination velocity (S). We have used $Al_{0.2}Ga_{0.8}As$ cladding layers for interfacial passivation and confinement of the carriers. The P concentration, and hence the strain, was determined both by X-ray diffraction measurements and by PL spectroscopy. The latter method was also used as a qualitative check of the material.
A marked decrease in S (especially visible at thin layers) has been observed upon enhancing the P concentration in the $Al_{0.2}Ga_{0.8}As/GaAs_{1-x}P_x$ double heterostructures (DHS), whereas the bulk recombination increased only slightly.

2 EXPERIMENTAL DETAILS

The $Al_{0.2}Ga_{0.8}As/GaAs_{1-x}P_x$ DHS were grown on Si-doped (100) 2° off towards (110) GaAs substrates by low-pressure (20 mbar) MOVPE. The source gases were trimethylaluminium (TMA), trimethylgallium (TMG), arsine (AsH_3) and phosphine (PH_3). The V/III ratios and growth rates were 125 and 1.5 μm/h for $Al_{0.2}Ga_{0.8}As$ and 125 and 1.7 μm/h for $GaAs_{1-x}P_x$, respectively. Both the $GaAs_{1-x}P_x$ and $Al_{0.2}Ga_{0.8}As$ layers were grown at 680°C. Sets of samples with varying thicknesses were grown for P concentrations of 0.000, 0.011, 0.015 and 0.019. Both the cladding layers and active layers were nominally undoped and showed n-type conductivity: $N_D = 10^{16}cm^{-3}$ for the former and $N_D = 10^{15}cm^{-3}$ for the latter. A schematical overview of the sample configuration is given in Fig. 1.

BARRIER LAYER	$d = 0.3\mu m$	$Al_{0.2}Ga_{0.8}As$
ACTIVE LAYER	d=variable	$GaAs_{1-x}P_x$
BARRIER LAYER	$d = 0.3\mu m$	$Al_{0.2}Ga_{0.8}As$
BUFFER LAYER	$d = 0.3\mu m$	GaAs
SUBSTRATE		GaAs

Fig. 1: Cross section of the double-hetero structures.

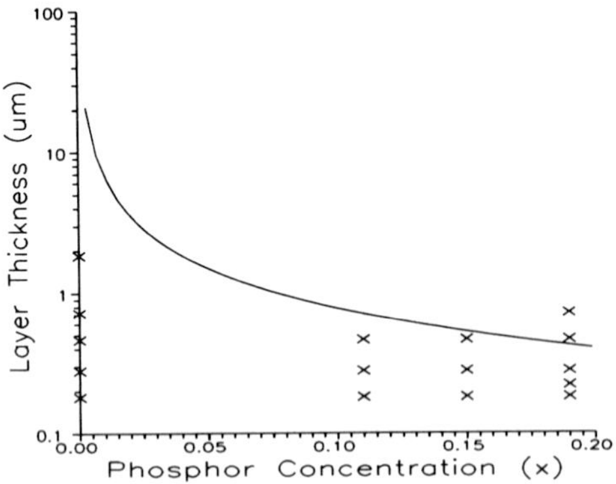

Fig. 2: Plot of critical layer thickness (as determined by van de Leur (1988)) against phosphorus concentration. The crosses indicate the data of our samples.

Most of our samples were thinner than the critical layer thickness (h_{crit}), as can be seen in Fig. 2. Only three of them were thicker than h_{crit} to study the effect of misfit dislocations on the minority-charge carrier lifetime. A phosphorus-spike-doped DHS was grown to investigate if the change in lifetime with respect to P concentration was due to passivation of the interface layer (as is the case at $GaInP_2/GaAs$ layers Olson *et al.* 1989), or to the induced strain at the interfaces: During the switch-over from $Al_{0.2}Ga_{0.8}As$ layer to the active layer (and vice versa) the growth was interrupted and phosphine was let in. In this way the interface was saturated with P, while no strain was introduced in the active layer.

The P concentrations were determined by X-ray diffraction measurements, which give the strain in the active layer, and by PL spectroscopy which measures the effect of both strain and P incorporation. With PL spectroscopy the material quality was checked to control the run-to-run reproducibility of the samples. Time-resolved PL measurements were performed at room-temperature by means of the Time–Correlated Single Photon Counting technique (TCSPC) (O'Connor and Philips 1984). The exciting laser was a dye laser producing 6 ps pulses at variable repetition rate and a wavelength of $\lambda = 600$ nm. Low-level conditions ($\Delta n, \Delta p \simeq 10^{13}cm^{-3}$) were chosen. The activation energies of the interface recombination velocities were studied by varying the temperature (T).

3 RESULTS

In Fig. 3 the PL decay measurements of the DHS with 0.2 μm thick active layers are shown. Because all of these measurements were performed at low level excitation conditions no signs of bimolecular decay are seen. Furthermore only one decay rate is observed, which indicates that no significant majority-charge carrier lifetime is measured (Ahrenkiel 1992). In Fig. 4 two PL measurements of the GaAsP active layer are shown with thicknesses below (Fig. 4a; d=0.25 μm, τ=75.3 ns) and above (Fig. 4b; d=1.0 μm, τ=18 ns) the critical layer thickness. A clear exciton splitting, indicative of good material quality is seen at $E \simeq 1.53$ eV in Fig. 4a. Exciton and acceptor related PL from the GaAs buffer layer are seen around 1.52 and 1.49 eV, respectively. Acceptor related PL from the GaAsP layer appears around 1.505 eV. From the shift of the exciton energies of the strained layer with respect to that of the GaAs buffer the x was determined to be 0.020 and the stress $\epsilon = (a^{\perp}_{GaAsP} - a^{\perp}_{GaAs})/(a^{\perp}_{GaAs}) = -7.2.10^{-4}$. In Fig. 4b no exciton-splitting can be observed. This is caused by a relaxation of the lattice through misfit dislocations in accordance with the low lifetime of this sample. The PL-decay times are shown in Table I. The lifetimes of the samples with $d > h_{crit}$ showed spatial fluctuations. Therefore the underlined decay times are less accurate. Compared with the GaAs/$Al_{0.2}Ga_{0.8}As$ samples of d=1.0 μm and d=3.0 μm the lifetimes of these samples were low. They also exhibited a lower PL-intensity. If we estimate the hole diffusity at D $\simeq$ 10cm^2/s and the minority-charge carrier lifetime at $\tau_{disloc} \simeq 20$ ns, the dislocation density is $N_d \simeq 10^6$cm^{-2} which is a reasonable value for this type of sample (Ahrenkiel *et al.* 1991).

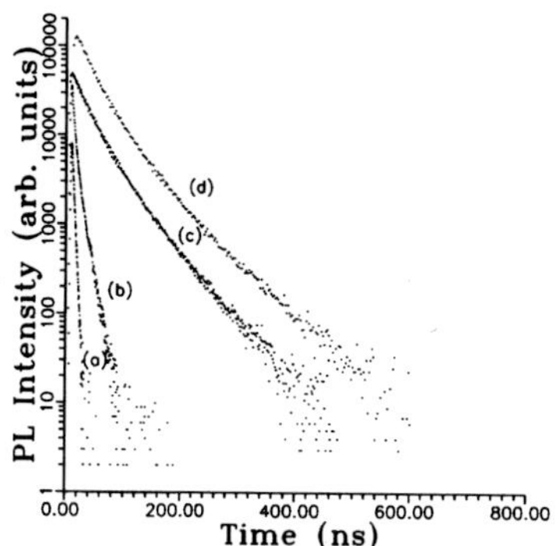

Fig. 3: Time-resolved photoluminescence in four DH-structures with an identical 0.2 μm active layer thickness but different phosphorus concentrations:
(a) x=0.000, τ=3 ns. (b) x=0.011, τ= 11 ns. (c) x= 0.015, τ=55 ns. (d) x=0.019 , τ= 65 ns.

d (μm)	τ (ns)			
3.0	1500			
1.0	1400			<u>19</u>
0.6		279	<u>12</u>	<u>26</u>
0.33	212	158	208	149
0.25				75.3
0.2	3.0	11.0	55.1	65
x	0.000	0.011	0.015	0.019

Table I: Minority charge-carrier lifetimes (τ) of strained $GaAs_{1-x}P_x$ as a function of x and active layers thickness. Data of samples with misfit dislocations are underlined. The accuracy of τ is 5-10%. The values of samples with $d \geq 1.0$ μm at $x = 0.00$ are corrected for photon recycling.

Using the data of Table I, the S were calculated according to:

$$S = \frac{1/d_1 - 1/d_2}{1/S_1 - 1/S_2} \tag{1}$$

The results are shown in Table II. A remarkable decrease in S is found upon enhancing x. This is also seen in Table I for the samples with $d = 0.2$ μm: the lifetime enhances markedly upon increasing x. For larger values of the active layer thickness the lifetime is of the same order, indicating the increasing influence of the bulk-lifetime on the total minority carrier lifetime. An indication of the bulk-GaAs lifetime is obtained via the 1.0 and the 3.0 μm thick DHS's with x=0.000. Their lifetimes of 1.4 μs and 1.5 μs (corrected for photon recycling (Asbeck 1977)) are indicative of excellent material quality, as was also confirmed by the PL spectra. Extrapolation of the lower interface recombination velocities at higher phosphorus concentrations towards higher active layer thicknesses indicates a lower bulk-lifetime at the more strained samples. The phosphorus spike-doped sample showed a lifetime of τ=3 ns, which indicates that the decrease in S is not due to any passivating effect of phosphorus in these samples. Fig. 5 shows the behaviour of decay time as a function of T in a 0.2 μm thick sample (x= 0.019).

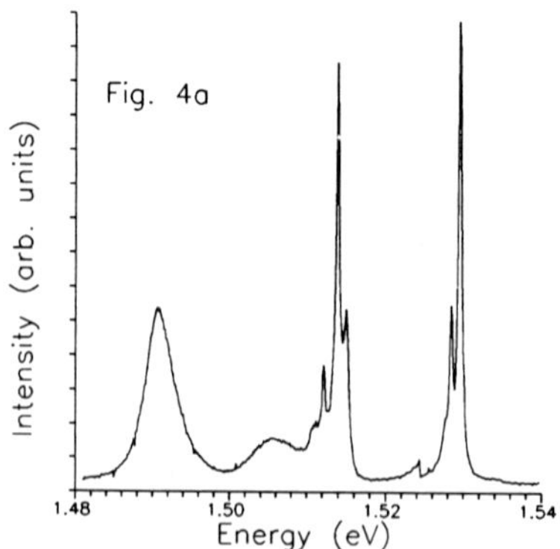

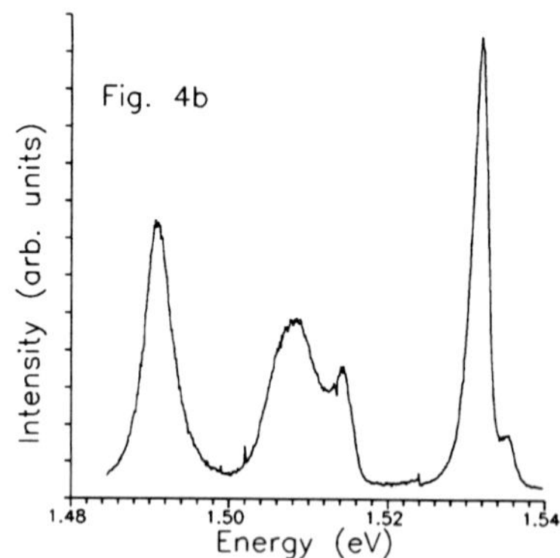

Fig. 4: PL spectra recorded at T= 4.2 K of $GaAs_{1-x}P_x$ layers with thicknesses below (Fig. 4a) and above (Fig. 4b) the critical layer thickness.

x	strain (ϵ)	S	E_{act}
0.000	0.00	8400 cm/s	59±8 meV
0.010	-2.87.10^{-4}	2100 cm/s	131±7 meV
0.015	-4.67.10^{-4}	340 cm/s	227±6 meV
0.019	-7.17.10^{-4}	220 cm/s	286±8 meV

Table II: Interface recombination velocities (S) at different x. The activation energies are the results of the fit according to Eq. 2.

At the interface the charge-carriers predominantly recombine via multiphonon emission (MPE) as described by Henry and Lang (1977) and Lax (1960), who developed a theory of cascade capture of electrons and holes by impurity levels in semiconductors. As the sample temperature is decreased, the recombination rates diminish exponentially with an activation energy $E_{act} = E_c - E_a$, where E_c is the binding energy of the highest excited capture state and E_a the thermal activation energy of the cascade process (Gibb *et al.* 1977 and Ahrenkiel *et al.* 1993):

$$1/\tau = AN_t v_{th} T^{-2} \exp(-E_{act}/kT) \qquad (2)$$

Here A is a constant, N_t the density of impurity states and v_{th} the thermal velocity. Above a certain temperature the radiative recombination dominates. Below this point the temperature dependence of the radiative recombination factor for band-to-band transitions in semiconductors B (Hall 1960), is measured. 't Hooft *et al.* (1985) calculated that $B \propto T^{-n}$ with n= 1.54.

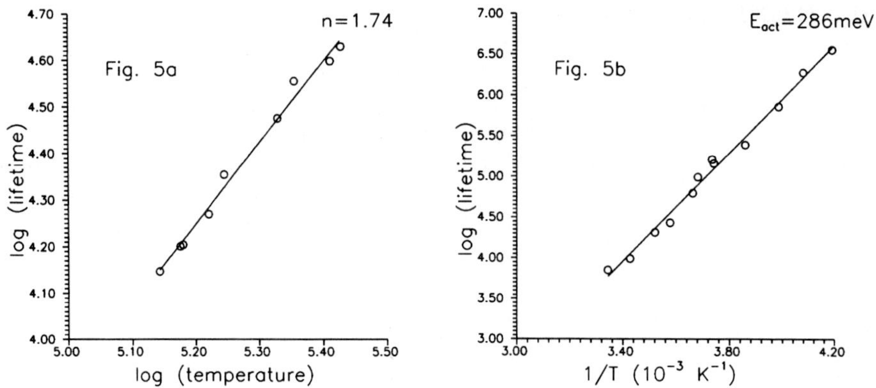

Fig. 5a: Temperature dependence of the radiative recombination.
Fig. 5b: Arrhenius plot of τ for $x = 0.019$; $d = 0.2\,\mu$m showing an activation energy of $E_{act} = 286 \pm 8$ meV.

Freeze out of carriers occurs below $T = 130$ K and band-to-band recombinations are absent: therefore measurements were not performed in this range. From 130 K to ~240 K the lifetime is controlled by radiative recombination as is shown in Fig 5a. The data-points can be reasonably well described by the aforementioned temperature dependence of B with $n = 1.74 \pm 0.07$. The Arrhenius plot of Fig. 5b reveals an activation energy of $E_{act} = 286 \pm 8$ meV. Data of the other $0.2\,\mu$m thick samples showed similar behaviour with different activation energies, which are summarized in Table II. Although the activation energy of the GaAs/AlGaAs interface velocity fits rather poorly with the value of $E_{act} = 27$ meV found by 't Hooft *et al.* (1983) it should be noticed that in their case the aluminium concentration in the barrier layers was different ($x = 0.53$) and their samples were grown under different conditions. If we assume that the density of interface states is the same for all 0.2 μm thick samples, the difference in S is caused by the change in thermal activation energy for the onset of recombination at the interfaces.

4 CONCLUSIONS

We have observed a definite decrease of the interface recombination velocity in strained $GaAs_{1-x}P_x/Al_{0.2}Ga_{0.8}As$ double heterostructures. It is shown that the enhancement of the lifetimes in the thinner DH-structures is not due to the passivating effect of phosphorus at the interfaces. Furthermore, temperature dependent lifetime measurements revealed a change in the activation energy of the strained $GaAs_{1-x}P_x/Al_{0.2}Ga_{0.8}As$ DHS's, responsible for the lowering of S at higher x, *i.e.* if more strain is present.

5 ACKNOWLEDGEMENTS

The authors are grateful to Dr. F.A.J.M. Driessen for critically reading the manuscript. Financial support from the Nederlandse Organisatie voor Energie en Milieu (NOVEM) is gratefully acknowledged.

6 REFERENCES

Ahrenkiel R K, Keyes B M, Dunlavy D J and Kazmierski 1991 Proc. 10th European PV Solar Energy Conf. p 533

Ahrenkiel R K 1992 Solid-State Electronics 35 p 239

Ahrenkiel R K, Zhang J, Keyes B M, Asher S E, Timmons M L 1993 Proc. 23th IEEE PV Specialists Conf. p 15

Asai H and Oe K 1983 J. Appl. Phys. 54 p 2052

Asbeck P 1977 J. Appl. Phys. 48 p 250

Gibb R M, Rees G J, Thomas B W, Wilson B L H, Hamilton B, Wight D R and Mott N F 1977 Phil. Mag. 36 p 1021

Hall R N 1960 Proc. Inst. Elect. Eng. 106B p 983

Henry C H and Lang D V 1977 Phys. Rev. B 15 p 989

't Hooft G W and van Opdorp C 1983 Appl. Phys. Lett. 42 p 813

't Hooft G W, Leys M R and Talen-v.d. Mheen H J 1985 Superl. and Microstr. 1 p 307

Kuo C P, Vong S K, Cohen R M and Stringfellow G B 1985 J. Appl. Phys. 57 p 5428

Lax M 1960 Phys. Rev. 119 p 1502

Leur R H M van de, Schellingerhout A J G , Tuinstra F and Nooij J E 1988 J. Appl. Phys. 64 p 3043

Molenkamp L W and Blik H F J van 't 1988 J. Appl. Phys. 64 p 4253

O'Connor D V and Philips D 1984 Time-Correlated Single Photon Counting (London: Academic Press)

Olson J M, Ahrenkiel R K, Dunlavy D J, Keyes B and Kibbler A E 1989 Appl. Phys. Lett. 55 p 1208

Thijs P J A , Tiemeijer L F, Kuindersma P I, Binsma J J M and Dongen Tvan 1991 IEEE J. Quantum Elect. 27 p 1426

Thermal stability of strained AlGaAs/$In_xGa_{1-x}As$ ($0{\cdot}15 \le x \le 0{\cdot}25$) doped-channel structures

Ming-Ta Yang, Ray-Ming Lin, Yi-Jen Chan, Jia-Lin Shieh and Jen-Inn Chyi
Department of Electrical Engineering
National Central University
Chungli, Taiwan 32054, ROC

Abstract: Thermal stability of strained AlGaAs/$In_xGa_{1-x}As$ ($0.15 \le x \le 0.25$) doped-channel structures was evaluated for different In contents after a high temperature process. Strained channels may be relaxed after heat treatment resulting in carrier decrease, sheet resistance increase, and layers becoming more light sensitive. This degradation is more profound for high In content doped-channel films. Doped-channel FET characteristics confirm this conclusion, and low-fequency noise spectra indicate the existence of traps in the thermally treated devices.

I. Introduction

Strained quantum-well devices, which include a lattice mis-matched active layer, have been extensively applied to electronic and optoelectronic devices. For instance, the excess In content in the GaAs wells can improve carrier transport properties and therefore enhance device speed performance (Rosenberg et. al. 1985, Chan et. al. 1991). The changes of bandgap also help the optimizations in photonic devices.

However, strained layer devices are always limited by the so-called critical thickness. Beyond this thickness, crystal lattices can no longer accommodate stress, and dislocations will form thereafter. This effect will cause a device performance degradation (Pamulapati et. al. 1990, Chan et. al. 1991). In addition to this critical thickness limination, the thermal stability of the strained layer is also another important issue and needs to be investigated. Peercy et. al. (1988), Zipperian et. al. (1988), and Elman et. al. (1990) have demonstrated that this metastable strained layer can be relaxed after high temperature heat treatment, resulting from the generation of dislocations. This thermally unstable strained layer will cause reliability problems which limit the device applications.

In this report, we systematically studied the thermal stability of strained AlGaAs/$In_xGa_{1-x}As$ (x= 0.15, 0.20, 0.25) doped-channel structures, comparing them with the as-grown samples. Doped-channel heterostructure FET's have demonstrated high current handling capability and high breakdown characteristics (Daniel et. al. 1987, Ruden et. al. 1990), which are essential for high power device applications.

II. Device structure and fabrication procedure

These strained AlGaAs/$In_xGa_{1-x}As$ doped-channel structures were grown by molecular beam epitaxy (MBE) with a Riber 32-P system. Fig. 1 shows the device cross-section. The mole fraction of $Al_yGa_{1-y}As$ is y= 0.3. Layers were grown at 620°C, excepting $In_xGa_{1-x}As$ ($0.15 \le x \le 0.25$) channels which were grown at 550°C. In this study we systematically changed the In compositions from 0.15, 0.20 to 0.25, and examined how the strain affects the device performance. For comparison, strained doped InGaAs channels (n= 5E18 cm^{-3}) were all 150 Å thick. The top undoped 200 Å AlGaAs was used to improve Schottky contact, followed by a 200Å n^+-GaAs cap layer.

For these thermal stability studies on strained channels, samples were first subjected to rapid thermal annealing (RTA) and then compared to the as-grown samples. Both types of samples were characterized by Hall effect and photoluminescence (PL) measurements. Device fabrication was realized by conventional optical lithography techniques. The active mesa regions were defined by using $NH_4OH:H_2O_2:H_2O$ chemical etching solutions. Ohmic contacts were carried out by thermal evaporation of the Ge/Ni/Au alloy, followed by a 400°C, 1 min furnace annealing. The separation between source and drain contacts is 3.5 µm. After a gate recess in order to remove the top n^+ GaAs layers, 1 µm-long, 50 µm-wide Al gates were thermally evaporated and defined by the lift-off process. Finally, the devices were completed by the interconnection of the Au metal.

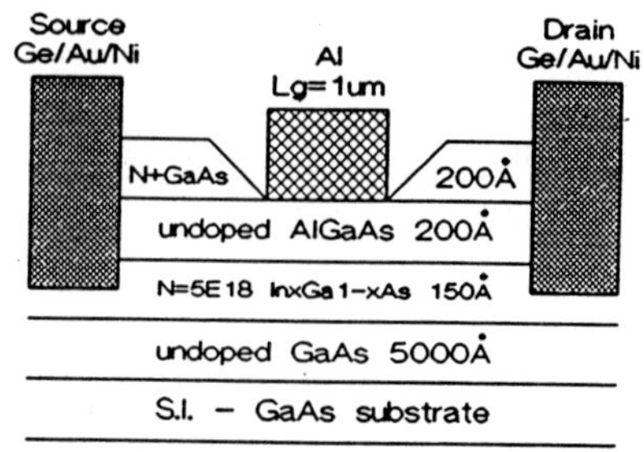

Fig.1 Device cross-section of AlGaAs/$In_xGa_{1-x}As$ ($0.15 \leq x \leq 0.25$) doped-channel FET's.

	$\mu(cm^2/V\text{-}sec)$		$Ns(cm^{-2})$	
	300K	77K	300K	77K
x=0.15	1450	1800	$4.20X10^{12}$	$5.18X10^{12}$
x=0.20	1500	1920	$6.14X10^{12}$	$5.75X10^{12}$
x=0.25	1645	1800	$3.90X10^{12}$	$4.04X10^{12}$

Tab.1 Hall mobility and sheet carrier density for AlGaAs/$In_xGa_{1-x}As$ ($0.15 \leq x \leq 0.25$) doped-channel structures at 300K and 77K respectively.

III. Characterization of thermal effects on strained channels

Tab. 1 summarizes the Hall measurement results for AlGaAs/$In_xGa_{1-x}As$ ($0.15 \leq x \leq 0.25$) doped-channel structures. The mobilities were 1450 cm^2/V-sec for x= 0.15, 1500 cm^2/V-sec for x= 0.20, and 1650 cm^2/V-sec for x= 0.25, corresponding to the sheet charge densities of $4.2x10^{12}$ cm^{-2}, $6.1x10^{12}$ cm^{-2} and $3.9x10^{12}$ cm^{-2}, respectively. Mobilities increased slightly at 77K. The slight decrease of carrier density for the x= 0.25 layer may be associated with the approach to its critical thickness. This assumption is supported by a sensitivity test under illumination conditions for the x= 0.25 sample, and will be discussed later on.

Thermal stress tests were executed by RTA 5 sec annealing, and sheet carrier densities measured by the Hall effect, shown in Fig. 2. 51% of the carriers survived at 850°C for the x= 0.15 layer (58% at 700°C), while this value was only 16% for the x= 0.20 layer (52% at 700°C). As to the x= 0.25 layer, after this thermal stress, carrier densities were not detectable. This suggests that lower In composition strained channels (i.e. x= 0.15, 0.20) are more stable after heat treatment. Highly mis-matched layers are more likely to be thermally relaxed, even the original as-grown layers are stable.

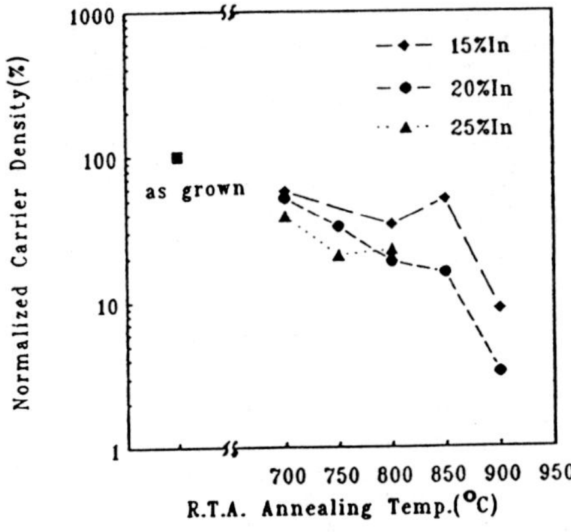

Fig. 2 Normalized sheet carrier density vs RTA temperature for AlGaAs/ $In_xGa_{1-x}As$ ($0.15 \leq x \leq 0.25$) doped-channel structures.

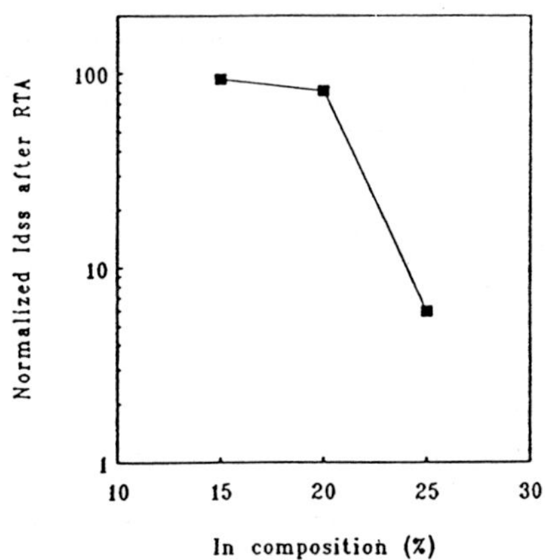

Fig. 3 Normalized Idss after a RTA process at 750°C for AlGaAs/$In_xGa_{1-x}As$ ($0.15 \leq x \leq 0.25$) doped-channel structures.

		x=0	x=0.15		x=0.20		x=0.25	
		as grown	as grown	RTA	as grown	RTA	as grown	RTA
Rs(Ω/□)	light	952	647	1390	869	1046	1189	-
	dark	977	710	1643	1033	1318	1627	-
	light sens.	2.5%	9.7%	18%	19%	26%	37%	-

Tab. 2 Comparisons of sheet resistance vs. different In content channel under illumination conditions at 300K.

Electrical characteristics were also evaluated by the transmission line method (TLM) patterns, in the test cell in our FET mask-set. Fig. 3 presents the results for normalized saturation drain-source current (I_{dss}, measured at V_{ds}= 5V) after RTA 750°C annealing. Only a 6% I_{dss} drop was observed for the x= 0.15 layer and this value increased to 18% for the x= 0.20 and 94% for the x= 0.25. A dramatic I_{dss} drop after heat treatment for x= 0.25 is consistent with the results shown in Fig. 2, which demonstrated an undetectable carrier from Hall measurements.

Sheet resistance (R_s) results evaluated from the TLM patterns indicating the properties of the active layers are listed in Tab. 2. Since we observed previously that thermal relaxation occurred in high In content layers, light sensitivity studies, which are associated with the trapping effects, may provide another tool for the analysis of detail. The as-grown layers were first compared through light sensitivity tests at 300K. The results for lattice-matched layers were also attached. Notice that this sensitivity increased from 2.5% (lattice-matched) to 9.7% (x= 0.15), 19% (x= 0.20), and 37% (x= 0.25) for all as-grown samples. Therefore,

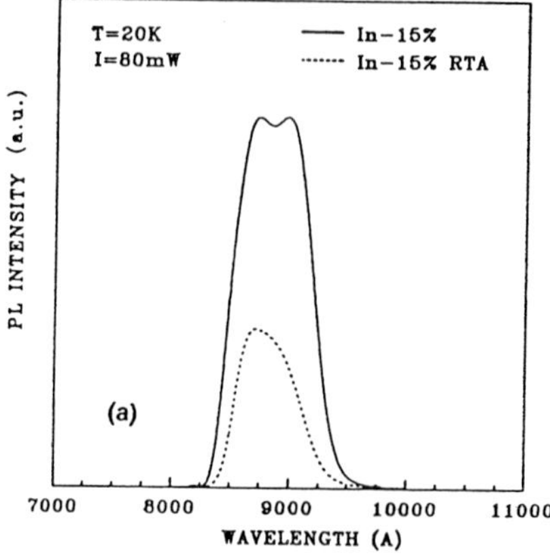

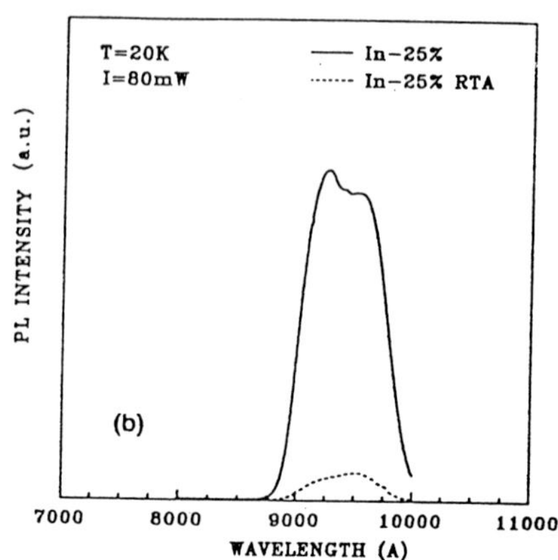

Fig. 4 PL spectra for as-grown and after heat treatment for (a) x= 0.15, (b) x= 0.25 AlGaAs/$In_xGa_{1-x}As$ doped channel structures.

this high sensitivity for the x= 0.25 as-grown layer together with a slight drop in carrier density (Tab. 1) suggests that a 150Å thick channel may be higher than the critical thickness limitation. After RTA 800°C heat treatment, this sensitivity went up to 18% for x= 0.15, and 26% for the x= 0.2, indicating that partial relaxation occurred in both layers. Again, there was no detectable signal for the x= 0.25 after heat treatment.

PL measurements carried out at 20K were compared for as-grown and thermal treatment samples. The spectra are shown in Fig. 4. PL spectra are relatively wider due to the doped-channel properties. Fig. 4(a) illustrates the spectra for x= 0.15. After RTA 800°C annealing, the intensity is lower compared to the as-grown samples. Unfortunately, no significant shift of peak position, reported by Elman et. al. (1990), was found based on this wide spectrum. For the spectra shown in Fig. 4(b), x=0.25 doped-channel structures have a more profound intensity decrease after heat treatment. Based on the previous electrical property analysis, the x= 0.25 layers suffered from thermal relaxation and formed dislocations thereafter. These dislocations would function as non-radiative centers which dramatically reduce the luminescence intensity. In consequence, both electrical and optical investigation indicates that 150Å $In_{0.25}Ga_{0.75}As$ strained channels are thermally unstable, and this instability can be improved by reducing the In content.

IV. Doped-channel FET's characteristics

After fully characterizing the thermal stability of strained films, we started to fabricate the strained doped-channel FET's with a 1.0 μm-long gate. In this section only x= 0.15 layers were chosen. I-V characteristics for as-grown and after RTA 800°C samples subjected to the illumination test are shown in Fig. 5(a) and 5(b), respectively.

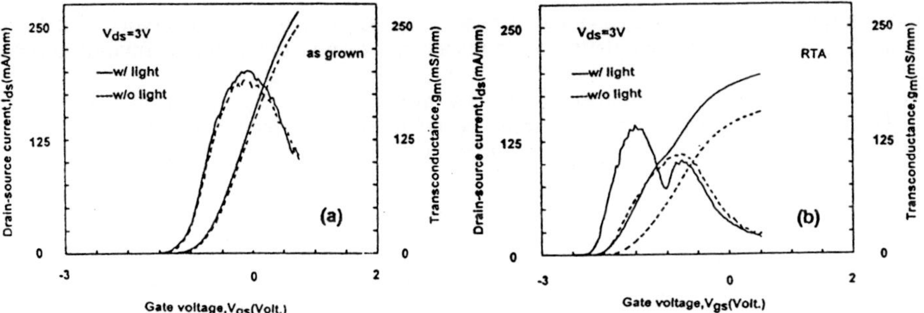

Fig. 5 g_m-I_{ds}-V_{gs} transfer characteristics for (a) as-grown, (b) after heat treatment AlGaAs/ $In_{0.15}Ga_{0.85}As$ doped channel FET's.

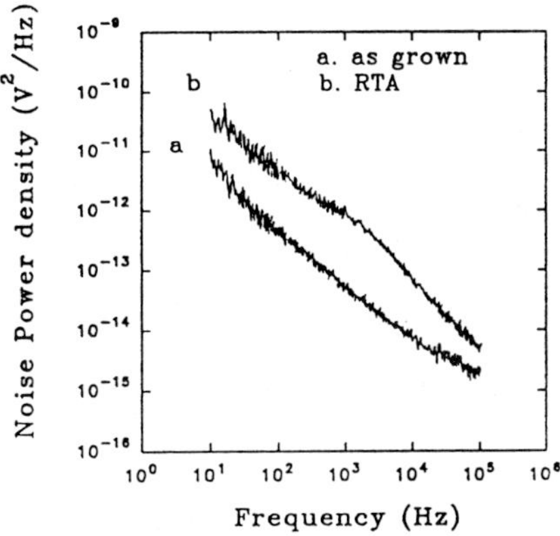

Fig. 6 LF-noise spectra for AlGaAs/ $In_{0.15}Ga_{0.85}As$ doped channel FET's.

The peak transconductance (g_m) was 201 mS/mm under the illumination, and dropped only 5% (191 mS/mm) in the dark at 300K. The associated threshold voltage (V_{th}) shifted only 20 mV. However, for the results shown in Fig. 5(b), not only did the peak g_m drop to 143 mS/mm, but the devices also became more light sensitive. A 24% g_m decrease was found in the dark (109 mS/mm), also causing a V_{th} shift of 360 mV. Device characteristics are responsible for the previous conclusion that strained channels are likely to be relaxed after thermal treatment, causing the generation of dislocations and corresponding degradation of the device.

Low-frequency (LF) noise spectra have demonstrated a very useful technique for device analysis where traps are involved. We compared both as-grown and RTA x= 0.15 doped channel FET's by LF noise measurements. Results are shown in Fig. 6. Both FET's were biased at I_{ds}= 5mA. Pure 1/f noise spectra were found for the as-grown FET's indicating no trapping effect involved in carrier conduction. While for the RTA FET's, the noise floor increases an order of magnitude. Furthermore, Lorentz-shaped spectra were superimposed into 1/f noise. This suggests that generation-recombination (G-R) noise, resulting from the trap center, involves the conduction in RTA doped-channel FET's.

V. Conclusion

In summary, the thermal stability of strained AlGaAs/$In_xGa_{1-x}As$ ($0.15 \le x \le 0.25$)

doped-channel structures was evaluated based on differing In content layers. Through various analysis techniques and device performance, it is demonstrated that thermal relaxation occurrs with a high In content (x= 0.25) resulting in degradation of the performance. However, this thermally unstable effect will also affect, to some degree, samples with even low In contents. Therefore, caution must be taken with strained layer devices, if the fabrication procedures involve high temperature processes.

Acknowledgements:

The authors would like to thank Mr. P.C. Hwang (ITRI, ROC) for low temperature PL measurements, and Mr. C.H. Chen (EE, NCU) for LF noise measurements. This work is supported by the National Science Council, ROC (NSC 82-0404-E-008-103).

REFERENCES:

Chan Y.J., Pavlidis D., IEEE Trans. Electron Devices, 38, p.1999 (1991).
Daniel R.R. et. al., IEDM Tech. Digest, p.921 (1987).
Elman B. et. al., J. Appl. Phys., 68, p.1351 (1990).
Pamulapati J. et. al., J. Appl. Phys., 68, p.347 (1990).
Peercy P.S. et. al., IEEE Electron Device Lett., 9, p.621 (1988).
Rosenberg J.I. et. al., IEEE Electron Device Lett., 6, p.491 (1985).
Ruden P.P. et. al., IEEE Trans. Electron Devices, 37, p.2171 (1990).
Zipperian T.E. et. al., GaAs IC Symposium, 251 (1988).

Paper presented at the Int. Symp. GaAs and Related Compounds, Freiburg, 1993

Optical investigations on strained $Ga_xIn_{1-x}P$ quantum wells

C. Geng, M. Moser, F. Scholz, P. Cygan, P. Michler, A. Hangleiter
4. Phys. Inst., Univers. Stuttgart, Pfaffenwaldring 57, 70550 Stuttgart, Germany

ABSTRACT: Strained GaInP single quantum wells with AlGaInP barriers have been grown by MOVPE and characterized by HRXRD. The influence of strain in these quantum wells has been examined by photoluminescence.
The PL-intensity of single quantum wells at 300K increases for compressive strain. When the critical thickness is reached, the intensity drops drastically. PL-energies at 300K are in good agreement to theory, however at low temperatures we observed a shift below the theoretical values for strained material.
In contrary to unstrained quantum wells, the emission-energy of strained quantum wells shows an anomalous increase of up to 40meV when raising the temperature from 2 to 150K and a shift of 60meV with excitation power. In addition, time-resolved measurements on these samples show a nonexponential decay of emission for temperatures below 150K, indicating various decay mechanisms.

1. INTRODUCTION

GaInP, which is a wide and direct bandgap semiconductor, is steadily gaining interest in optoelectronic applications like visible laser diodes [Itaya 1990, Serreze 1991].
Due to the separation of light- and heavy-hole valence bands and suppression of spontaneous emission, a strong decrease in threshold-current for lasers containing strained quantum wells in the active zone has been theoretically predicted [O'Reilly 1991, Ueno 1993] and experimentally confirmed [Hashimoto 1991, Valster 1992]. This occurrence is very helpful for laser-performance. Nevertheless, many problems with strain induced band structure variations are still unsolved, and further studies are necessary. Besides, the threshold where the built-in strain is reduced by relaxation of the layers, commonly described by the term "critical thickness", is still under discussion [Van der Merve 1988].
Another highly interesting property of GaInP is the spontaneous long-range crystal ordering of the $CuPt_B$-type which can be observed by electron diffraction [Suzuki 1988]. Thermodynamic calculations [Froyen 1991, Osório 1992] show that during epitaxial growth GaInP, the $CuPt_B$-structure is energetically more stable

than any other kind of ordering or the disordered state. According to theory, this structure of alternating monolayers of GaP and InP along $(11\bar{1})$- or $(1\bar{1}1)$-planes leads to a reduction of the band-edge [Kurimoto 1989, Wei 1990]. Depending on epitaxial growth-parameters, the PL-energy has been observed to vary up to 80meV at fixed composition [Kurtz 1990].

This ordering is known to occur also in quantum wells, but the influence of strain on the ordering is not yet investigated. Therefore we have examined strained $Ga_xIn_{1-x}P$ quantum wells in the range of x = 0.28 to 0.66 (+1.76% to -1.06% strain) with different degrees of ordering.

These layers have been grown by low pressure metal organic vapor phase epitaxy (LP-MOVPE). As barrier, $(Al_{0.5}Ga_{0.5})_{0.51}In_{0.49}P$, lattice matched to the GaAs substrate, has been used. The influence of compressive and tensile strain on the optical and structural properties of the material has been examined by photoluminescence (PL).

2. MOVPE GROWTH AND HRXRD

$Ga_xIn_{1-x}P$ strained single quantum wells (SSQWs) and slightly strained (x=0.42 $\hat{=}$ 0.8% strain) MQWs with $(Al_{0.5}Ga_{0.5})_{0.5}In_{0.5}P$ barriers have been grown in a horizontal MOVPE-reactor with a rotating gas foil system (Aixtron) at 100hPa and 750°C. The source materials were trimethylgallium (TMGa), trimethylindium (TMIn), trimethylaluminum (TMAl), phosphine (PH_3) and for the growth of a GaAs-buffer also arsine (AsH_3).

To influence the ordering of the crystal, differently oriented substrates have been used in the same epitaxial run, since increasing misorientation leads to a decrease in crystal ordering. The substrates were misoriented 2° to 6° off the (100)-surface towards the nearest <110>-direction. In bulk material, these growth conditions lead to an ordering-related energy shift of about 25meV for 6° misorientation, which can be considered "quite disordered" and about 55meV for 2° misorientation ("quite ordered").

In order to control composition, thickness, and interface-quality, high resolution x-ray diffraction (HRXRD) has been performed on the MQW-structures. A large number of satellite peaks resulting from the superlattice structure can be found in all samples demonstrating their excellent quality with respect to interface abruptness (fig.1). This has also been observed in the highly misoriented substrates provided that the incidence and exit of the x-ray were exactly symmetric concerning the surface of the sample. By comparing measurements to numerically simulated spectra, the thicknesses (growth rates) and compositions could be evaluated very accurately.

A perfect match could be achieved, indicating excellent homogeneity and good interfacial quality.

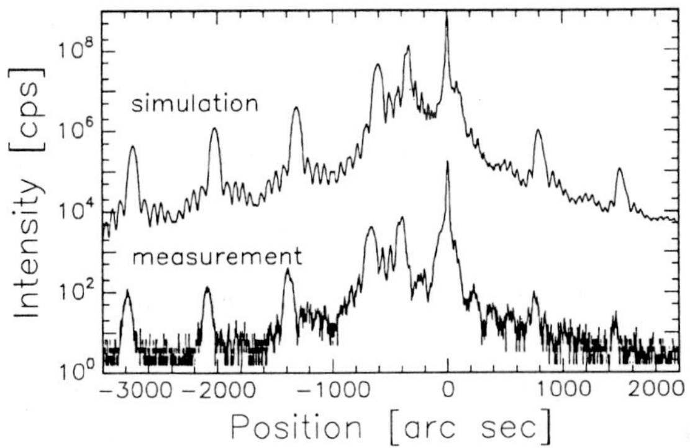

Fig.1: HRXRD measurement and simulation of a MQW-structure of $10x\{Ga_{.42}In_{.58}P/Al_{.25}Ga_{.25}In_{.5}P\}$ *with 4.5nm/22.4nm.*

3. RESULTS AND DISCUSSION

As demonstrated in fig.2, the PL-energy at 300K is in good agreement with the theory of M.P.C.M. Krijn [1991] when the quantization is taken into account.

For a tensile strain of -0.4%, two peaks can be resolved in photoluminescence. This can be explained by the splitting of the light and heavy hole subbands.

Fig.3 shows the PL-intensity of single quantum wells at 300K, it increases strongly for compressive strain. It can be assumed that this is due to the increasing well depth. The assumption is confirmed by the following equation, which describes the thermal emission of the electrons out of the quantum well:

$$I_2/I_1 = \exp\left(-(\Delta E_{C2}-\Delta E_{C1})/kT\right)$$

Where I_i is the PL-intensity and kT the thermal energy at 300K. ΔE_{Ci}, the conduction band offset of GaInP to the AlGaInP barrier has been assumed to be 65% of the band gap difference (Valster 1990).

As the compressive strain exceeds +1.4%, the PL-intensity of 10nm quantum wells drops strongly (fig.3), indicating that the critical thickness is reached. This is in the same order of magnitude as expected from the theory of Matthews and Blackeslee [1974] and in good accordance to the theory of Van der Merwe and Jeser [1988].

The expected energy shift due to strain-release at the critical thickness however, is not observed.

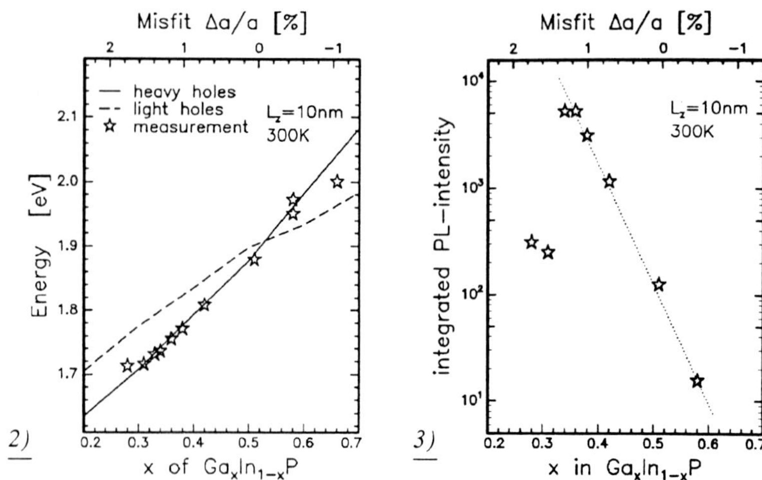

fig.2: PL-peak energy of 10nm strained GaInP quantum wells at room temperature compared to the theory of M.P.C.M. Krijn [1991], where strain and quantization is taken into account.
fig.3: Integrated PL-intensity of 10nm strained GaInP quantum wells at room temperature. The dotted line represents the above mentioned equation, which describes the thermal emission of electrons out of the well.

In contrary to measurements at room-temperature, at low temperature we observed a shift of the PL-peak energies below the expected values from theory, increasing monotonically with strain. A closer look at the temperature-dependence of the PL-peak energy of these samples reveals a very pronounced upshift of up to 40meV with increasing temperature from 2K to 150K. Fig.4b shows this temperature-dependence of the PL-peak energy of a strained quantum well (+1.3%) in comparison to the expected dependence as in the unstrained case.

Also in contrary to the unstrained quantum wells, time-resolved measurements on these samples show a nonexponential decay of emission for temperatures below 150K, this is demonstrated in fig.6.

Furthermore, as showed in fig.5, these low-temperature peaks of highly strained material can be upshifted 62meV by varying the excitation power from $5mW/cm^2$ to $200W/cm^2$.

Such a behavior is well known from doping-superlattices (nipi-structures). In these, the locally modulated band gap, which gives rise to spatially indirect transitions, can be flattened by injection or excitation of excess carriers. This leads to an analogous behavior concerning emission energies and decay times [Döhler 1981].

Similar properties have been found in ordered GaInP bulk material, they could be explained by a staggered band lineup at the boundaries of differently ordered domains [DeLong 1991].

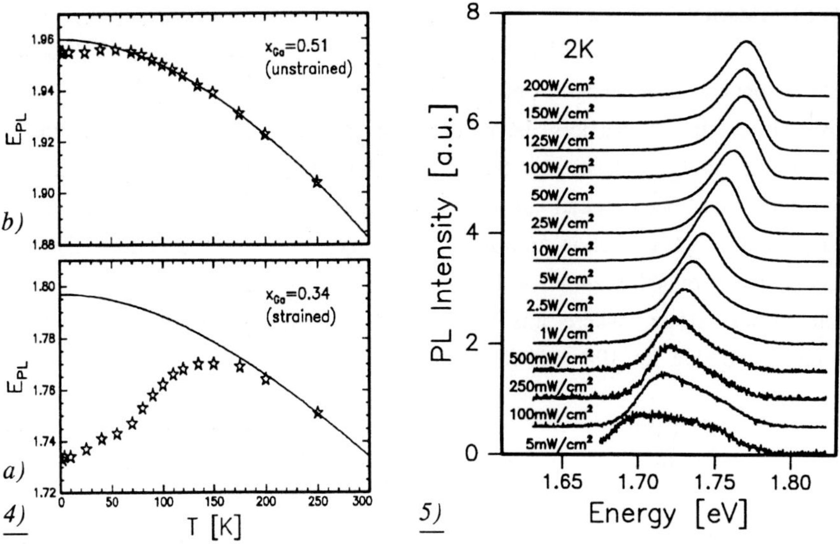

fig.4: PL-peak energy of a) unstrained and b) strained quantum wells depending on temperature. An empirical dependence [Varshini 1967] has been fitted to the unstrained QW and compared to the strained QW.
fig.5: Dependence of PL-peak energy on the excitation power for a QW of +1.3% strain.

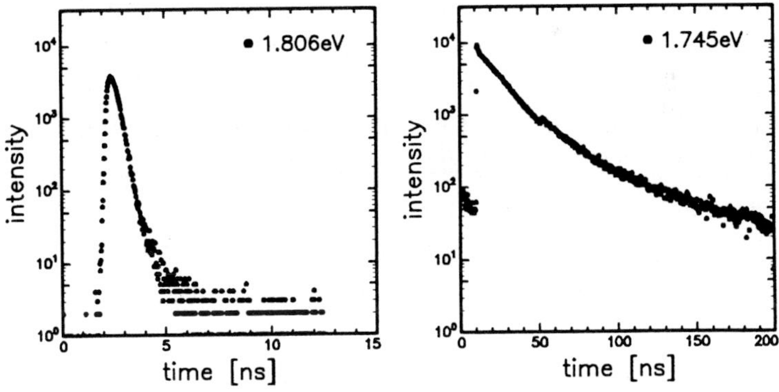

fig.6: Decay of PL-emission for a QW of +1.3% strain at different energies. The temperature is 5K.

These properties have not been found in unstrained quantum wells, which are quite disordered, grown under the above mentioned conditions on 6° misoriented substrates. Although a direct measurement of the degree of ordering in quantum wells was not possible, we deduce from these findings a strong indication, that the formation of ordering is enforced by strain.

4. SUMMARY

The optical attributes of strain in $Ga_xIn_{1-x}P$ quantum wells with $(Al_{0.5}Ga_{0.5})_{0.5}In_{0.5}P$ barriers have been investigated. At room-temperature the PL-energy is in accordance to theory. At low temperature however, a low energy transition has been found in strained quantum wells, which shifts upwards with temperature and excitation power. This is judged as an indication, that ordering is enforced in strained quantum wells.
The critical strain, where misfit-dislocations appear, is found to be +1.4% on the compressive side for 10nm thick layers.

ACKNOWLEDGEMENT

The authors would like to thank in particular E. Kohler and R. Winterhoff for their support in the MOVPE growth and E. Lux for the photoluminescence measurements. We also would like to thank M. Pilkuhn for fruitful discussions. Part of this work has been funded by the EC under contract no. 6134 (HIRED).

REFERENCES

DeLong M C, Ohlsen W D, Viohl I, Taylor P C, Olson J M
1991 Journ. of Appl. Phys. 70 2780
Döhler G H, Künzel H; Olego D, Ploog P, Ruden P, Stolz H J
1981 Phys. Rev. Lett. 47 864
Froyen S, Zunger A 1991 Phys. Rev. Lett. 66 2132
Hashimoto J 1991 Appl. Phys. Lett. 58 879
Itaya K, Hatakoshi G, Watanabe Y, Ishikawa M, Uematsu Y
1990 Electron. Letters 26 214
Krijn M P C M 1991 Sem. Sci. Tech. 6 27
Kurimoto T, Hamada N 1989 Phys. Rev. B 40 3889
Kurtz S R, Olson J M, Kibbler A 1990 Appl. Phys. Lett. 57 1922
Matthews J W, Blakeslee A E 1974 Journ. of Cryst. Growth 27 118
O'Reilly E P, Jones G, Ghiti A, Adams A R Electronics Letters 27 (1991) 1417
Osório R, Bernard J E, Froyen S, Zunger A 1992 Phys. Rev. B 45 11173
Schuster M, Zaus R, Meyer A, Milde A, Cerva H, Sporrer K
1993 Frühjahrstagung der EPS, Regensburg SC16.11
Serreze H B, Harding C M, Waters R G 1991 Electron. Letters 27 2245
Suzuki 1988 Journ. of Cryst. Growth 93
Ueno Y 1993 Appl. Phys. Lett. 62 553
Varshni Y P 1967 Physica 34 149
Liedenbaum C T H F, Valster A, Severens A L G J, 't Hooft G W
1990 Appl. Phys. Lett. 57 2698
Valster A, Van der Poel C J, Finke M N, Boermans M J B
1992 13th IEEE Intern. Semicond. Laser Conf. 152
Van der Merve J H, Jesser W A, 1988 Journal Appl. Phys. 63 1509
Wei S-H, Zunger A 1990 Appl. Phys. Lett. 56 662

Inst. Phys. Conf. Ser. No 136: Chapter 6
Paper presented at the Int. Symp. GaAs and Related Compounds, Freiburg, 1993

Intersubband transition in resonantly coupled asymmetric double quantum well

S.J.Rhee, J.C.Oh, Y.M.Kim, H.S.Ko, W.S.Kim, D.H.Lee, and J.C.Woo,
Department of Physics, Seoul National University, Seoul 151-742, Korea

K.H.Yoo
Department of Physics, Kyung Hee University, Seoul 130-701, Korea

ABSTRACT

The study on the intersubband separation of asymmetric double quantum well (ADQW) is reported. The subband separations in resonantly and non-resonantly coupled ADQWs prepared with AlGaAs-GaAs-InGaAs and AlGaAs-GaAs were measured by photoluminescence excitation spectroscopy. The observed spectra were in good agreement with the subband levels calculated by transfer matrix method. In resonantly coupled ADQW, the interwell transition was significantly enhanced.

1. INTRODUCTION

An asymmetric double quantum well (ADQW), composed of two quantum wells (QWs) of different well-widths, recently attracts the attention due to its optical non-linearity and quantum confined Stark effect. Another reason is because it could be used as infrared (IR) detector. (Berger, 1993) When the intersubband transition is used for IR detection, it provides an advantage of tailoring the bandgap separation precisely tuned to the IR wavelength. The ADQW is an interesting system to study the electron tunneling through the quantum barrier. Electron tunneling for non-resonantly (Sauer, 1988) and resonantly coupled (Gurvitz, 1991) cases, and excitonic tunneling (Clerot, 1990) are reported as coupling mechanisms between two QWs.

In this work, the intersubband transitions of resonantly and non-resonantly coupled ADQWs were studied for two types of samples, one composed of narrow GaAs and wide InGaAs QWs with AlGaAs quantum barrier (QB) and the other composed of two GaAs QWs of different well-width with the same AlGaAs QB. The transitions were observed by photoluminescence excitation spectroscopy (PLE), and the observed spectra were identified by the subband energies obtained from the model calculation. The transfer matrix method using an envelope function approximation was used for the calculation. The low-temperature PLE results show that there is a significant enhancement in the intensity of the interwell transition from the first excited conduction electron (CE) state of the coupled ADQW, which corresponds to the ground state of the narrow QW, to the ground state of heavy hole (HH).

2. EXPERIMENT

Two types of coupled ADQW samples were prepared; one consisted of GaAs and $In_{0.02}Ga_{0.98}As$ QWs with $Al_{0.25}Ga_{0.75}As$ QB, and the other consisted of two GaAs QWs with the same QB. For both types, resonantly and non-resonantly coupled structures were fabricated for comparison. The structures of the samples are summarized in Table 1. All the samples used in this work were grown by molecular beam epitaxy (MBE) using Riber 2300-P system at 620 °C on (100) GaAs substrate. Superlattice buffer layer was introduced to minimize strain. The phase-lock control of reflection high-energy electron diffraction intensity oscillation was adopted in order to obtain the precise QW width, and the growth interruption of 1 minute at each interface were introduced to maximize the flatness of heterointerface in the growth.

In PLE, a tunable Ti-sapphire laser with a birefringent tuning element was used as the excitation source. Typical linewidth of the excitation source was less than 1 Å. The luminescence was filtered by a monochromator to select the light just below the ground state exciton recombination. In some cases, the selection window was varied in order to determine the origin of transition. All PLE spectra were obtained at 20 K with a back-scattering geometry. In the resonant inelastic light scattering

Table 1 : Structures of ADQW samples used in this work. (ML : monolayer)

Sample No.	QW1		QW2		QB		Remarks
	Material	Width	Material	Width	Material	Width	
1	$In_{0.02}Ga_{0.98}As$	62 ML	GaAs	25 ML	$Al_{0.25}Ga_{0.75}As$	12 ML	resonant
2	$In_{0.02}Ga_{0.98}As$	70 ML	GaAs	25 ML	$Al_{0.25}Ga_{0.75}As$	12 ML	non-resonant
3	GaAs	64 ML	GaAs	27 ML	$Al_{0.25}Ga_{0.75}As$	8 ML	resonant
4	GaAs	36 ML	GaAs	18 ML	$Al_{0.25}Ga_{0.75}As$	8 ML	non-resonant

spectroscopy, (Pinczuk, 1981) the same experimental geometry and condition as PLE were used with DCM dye laser, which was tuned at the energy levels higher than the ground subbands separation by the spin-orbit coupling.

3. CALCULATION

In the calculation of electronic energy levels of the coupled ADQW, the transfer matrix method (Ram-Mohan, 1988) using envelope function approximation was adopted. The CE, HH, light hole (LH), and spin-orbit split-off were included in the multi-band Hamiltonian. The input parameters for GaAs and AlGaAs used in the calculation can be found elsewhere, (Lee, 1993) and those for InGaAs in the reference (Madelung, 1991). The strain effect due to lattice mismatch was included for InGaAs.

The calculation shows that E2, H4 and L3 are confined in the narrow GaAs QW, while the others in wide InGaAs QW in Sample 1. In Sample 2, E3, H5 and L3 are confined in GaAs QW. In Samples 3 and 4, E2, H2, L2 and H5 are associated with narrow QW, while the others with wide one. In the above and further on, En, Hn and Ln refer to the sub-levels of CE, HH and LH, respectively. The calculated values are for the electronic states and the exciton binding energy is not included.

4. RESULTS AND ANALYSIS

Typical PLE spectra obtained from Samples 1 and 2 are shown in Fig. 1(a) and (b), and those from Samples 3 and 4 are in Fig. 2(a) and (b), respectively. The peaks observed in PLE are identified by comparing their positions with the separation of electronic subbands.

In Fig. 1(a), the peaks observed at 1,502 and 1,516 meV are identified as E1-H1 and E1-L1 transitions, respectively, which are the ground state transitions localized in QW1, and those at 1,579 and 1,590 meV are as E2-H4 and E2-L3, respectively, which corresponds to, if uncoupled, the ground state transitions in QW2. Between these two pairs of the peaks, the two peaks at 1,546 and 1,550 meV overlapped each other are present. The peak at 1,550 meV is identified as E3-H2 transition, and the only possible justification for the peak at 1,546 meV is the E2-H2 transition.

In Fig. 1(b), the peaks at 1,503 and 1,514 meV are E1-H1 and E1-L1 transitions, and those at 1,582 and 1,598 meV are E3-H5 and E3-L3. These correspond, if not coupled, to the ground state transitions in QW1 and QW2, respectively. The E1-H1 peak is overlapped with another peak at 1,500 meV, which seems to be impurity related. Between these two pairs, a single peak is observed 1,542 meV, which is identified as E2-H2 transition, the excited state transition localized in QW1. Noticed is that the E2-H2 transition, the interwell transition observed in Sample 1, was not observable in this sample. The observed and calculated values are listed in Table 2 for comparison.

Table 2 : Comparison of the calculated and observed separations of the subbands

Sample 1			Sample 2		
Transi.	Cal.	Meas.	Transi.	Cal.	Meas.
E1-H1	1,512.1	1,502	E1-H1	1,519.2	1,503
E1-L1	1,526.8	1,516	E1-L1	1,523.0	1,514
E2-H2	1,556.5	1,546	E2-H2	1,547.2	1,542
E3-H2	1,560.7	1,550	--	--	--
E2-H4	1,581.8	1,579	E3-H5	1,583.4	1,582
E2-L3	1,600.9	1,590	E3-L3	1,599.0	1,598

In Fig. 2(a), the peaks at 1,526 and 1,532 meV are originated from the ground state transitions, E1-H1 and E1-L1, and those near 1,575 and 1,584 meV are E2-H2 and E2-L2. The broad peak around 1,552 meV is nearly the same as the separation beween E2 and H1 levels. In Fig. 2(b), the peaks at 1,545 and 1,554 meV, respectively, are matched with the ground state transitions, E1-H1 and E1-L1, while those at 1,590 and 1,610 meV are E2-H2 and E2-L2, the excited state transitions localized in QW2. A broad background luminescence centered around 1,580 meV with the linewidth of about 40 meV cannot be identified, but this back-

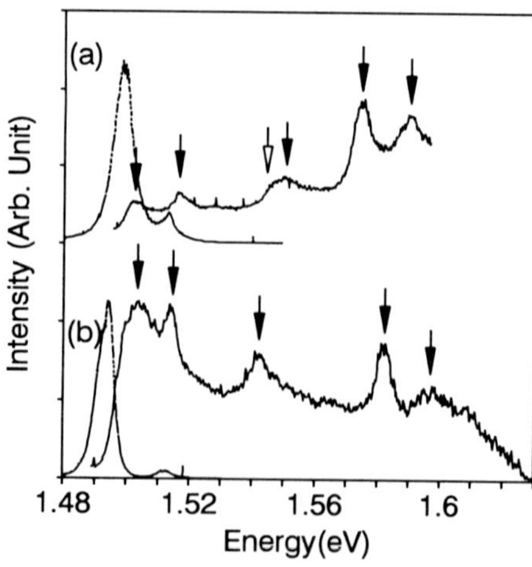

Fig. 1 : PLE (solid) and PL (dotted) of (a) Sample 1 and (b) Sample 2

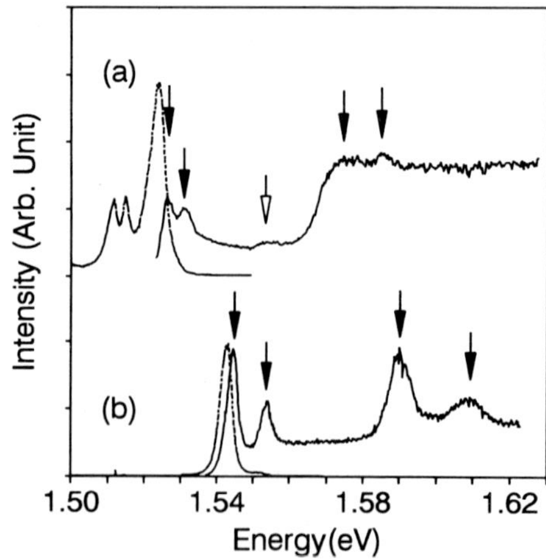

Fig. 2 : PLE (solid) and PL (dotted) of (a) Sample 3 and (b) Sample 4

ground luminescence near E2-H2 and E2-L2 peaks of Sample 3 can be observed in the vicinity of E2-H4 and E2-L3 peaks in Sample 1.

The schematic subband diagrams for Samples 1 to 4 are shown in Fig. 3.

5. DISCUSSION

The transition energies obtained from PLE are smaller than the calculated subband separations by as large as 10 meV. The major source of the discrepancy is caused by the exciton binding energy which is not counted in the calculation. The difference is in comparable magnitude with the exciton binding energy, when estimated from that of narrow GaAs QW. (Kim, 1993) The separations of E1 and E2 of Samples 1 and 4 measured by the resonant inelastic light scattering method are 33 and 47 meV, respectively, which supports the calculated values.

In Samples 1 and 3, the CE ground state of QW2 and the CE first excited state of QW1 are in near resonance. The results confirm that the separation between the first and second excited electron subbands are very small. Noticed is the presence of

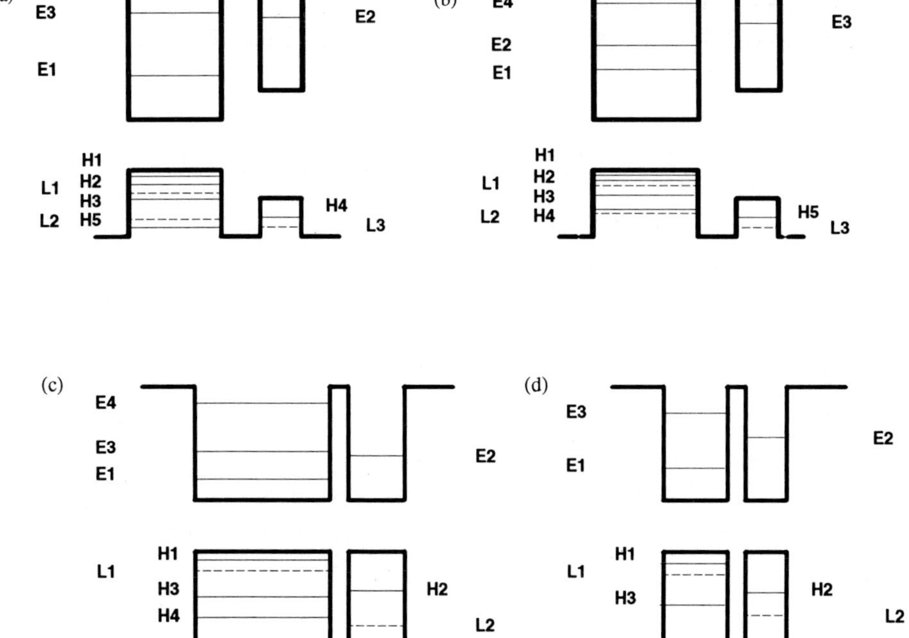

Fig. 3 : Schematic diagrams of the subbands of (a) Sample 1, (b) Sample 2, (c) Sample 3 and (d) Sample 4.

E2-H2 transition in Sample 1 and E2-H1 in Sample 3. These spatially indirect transitions were observed in the resonantly coupled ADQWs, but were not observable in the non-resonantly coupled ones. When there is resonant tunneling between two QWs, the transform from the direct exciton to the cross exciton is possible. (Haacke, 1993) If electron and hole are separated spatially, the eletron hole lifetime is increased, and the photoinduced IR absorption thus is increased. The enhancement of IR absorption was reported for the interwell electron-hole recombination in biased double QW. (Berger, 1993).

More study is needed to learn details of the observed interwell recombination. However, this work may provide a preliminary indication to another possible means of detecting IR.

ACKNOWLEDGEMENT

This work is supported in part by Korea Telecom.

REFERENCES

Berger V, Rosencher E, Vodjdani N, and Costard E, Appl. Phys. Lett., **62**, 378 (1993)

Gurvitz S A, Bar-Joseph I, and Deveaud B, Phys. Rev. **B43**, 14703 (1991)

Haacke S, Pelekanos N T, Mariette H, Zigone M, Heberle A P, and Ruehle W W, Phys. Rev. **B47**, 16643 (1993)

Kim D W, Leem Y A, Yoo S D, Woo D H, Lee D H, and Woo J C, Phys. Rev. **B47**, 2042 (1993)

Lee D H, Kim D W, Leem Y A, Oh J C, Park G H, and Woo J C, J. Appl. Phys. **74**, xxx (1993) (To be published on Sept. 1, 1993)

Madelung O, edited, Semiconductors Group IV Elements and III-V Compounds, Series of Data in Science and Technology (Springer-Verlag, 1991)

Pinczuk A, Shah J, Gossard A C, Wiegmann W, Phys. Rev. Lett. **46**, 1341 (1981)

Ram-Mohan L R, Yoo K H, Aggarwal R L, Phys. Rev., **B38**, 6151 (1988)

Sauer R, Thonke K, and Tsang W T, Phys. Rev. Lett. **61**, 609 (1988)

Paper presented at the Int. Symp. GaAs and Related Compounds, Freiburg, 1993

Modelling α and Δn in strained InGaAs/GaAs quantum wells

A. Simões Baptista, H. Abreu Santos

Electrical and Computer Engineering Department
IST, Technical University of Lisbon, Portugal

ABSTRACT: Simulation results for the dependence of the absorption coefficient α and the nonlinear refractive index Δn on the particle concentration and on the frequency are presented for the $In_xGa_{1-x}As$/GaAs quantum well structure. The effect of strain is also analysed by changing the In content x up to 0.37. This leads to changes in bi-axial compressive strain and further it increases the separation of the light and heavy holes, while reducing the effective mass of the later.
The model includes many-body Coulomb effects and strain induced modifications in electron and hole masses and in the material band-gap. As in unstrained quantum wells, increasing the particle concentration broadens the gain region and increases gain until a saturation behaviour is reached. Increasing the In content and hence the strain is predicted to have a similar effect and deserves experimental verification.

I - INTRODUCTION

With the progress of epitaxial technology (MOCVD, MBE) new possibilities have been opened in the study and development of electronic and optoelectronic devices.

Recently there has been a growing interest in pseudomorphic structures. This results from the technological development of mismatched lattices heteroepitaxy. When pseudomorphic layers are grown on a substrate, if their thickness is less than a critical value, their lattice constant will accommodate to that of the substrate. The ensuing band structure changes offer new ways of improving existing devices and creating new ones. For example, reductions (versus quantum well matched structures) are expected in the threshold current and Auger nonradiative recombination in lasers, accompanied by wider tunability, lower noise factor and higher speed capability, Suemune (1991), Loehr (1991).

An arising interest in ultrafast and coherent phenomena accompanied by applications with high level mobile charge concentrations has led to an ongrowing attention to many body interactions. Namely in lasers, whose gain and refractive index are profoundly affected by the above mentioned effects, Koch (1988), Chow (1990).

In this article the approach developed by Lindberg (1988) and Haug (1989,) for the absorption coefficient α and the nonlinear refractive index variation Δn and the strained layer theory presented by Suemune (1991) are combined into a single model, to be described in Section II.

In Section III results for $In_xGa_{1-x}As/GaAs$ structures are presented and interpreted. Many -body interactions and the influence of the In content on the biaxial compressive strain are analysed.

In Section IV conclusions are stated. Namely, it is predicted that increasing the In content will lead to a broadening of the spectral region with gain, accompanied by gain saturation.

II - MODEL

II.1 - THE INFLUENCE OF STRAIN ON MATERIAL PARAMETERS

Strain modifies the band-structure of a material. So its parameters must be re-calculated as a function of the strain imposed on the lattice. This has been done according to the model of Suemune (1991).

Let us assume an epitaxially grown semiconductor layer that is strained along the QW (quantum well) plane (x-y) by an amount $\varepsilon_{//}$ and uniaxially strained in the growth direction (z) by an amount $\varepsilon_{\perp}$. These quantities are functions of the bulk lattice constants of the substrate α_S and of the layer α_L according to the relations:

$$\varepsilon_{//} = \frac{\alpha_S}{\alpha_L} - 1 \tag{1}$$

$$\varepsilon_{\perp} = -\frac{\varepsilon_{//}}{\sigma} \tag{2}$$

where σ is the Poisson ratio. If the layer is grown on a (001) substrate the strain tensor coefficients are $\varepsilon_{xx}=\varepsilon_{yy}$ and in the growth direction $\varepsilon_{zz} = \varepsilon_{\perp}$. The off-diagonal terms are zero. In this case, assuming an infinite barrier height, the effective valence-band and conduction band masses (in the QW plane) are given by Suemune (1991)

$$\left(\frac{m_v}{m_0}\right)^{-1} = \gamma_1 + \gamma_2 - \frac{3\gamma_3^2}{\gamma_2 + \delta\gamma_a} + 3\left(\gamma_1 - 2\gamma_2 - 4\,\delta\gamma_a\right)\left(\frac{\gamma_3}{\gamma_2 + \delta\gamma_a}\right)^2 \left[\frac{1 + \cos(KL)}{(KL)\times\sin(KL)}\right] \tag{3}$$

$$KL = \pi\left[\left(\gamma_1 - 2\gamma_2 - 4\delta\gamma_a\right)/\left(\gamma_1 + 2\gamma_2\right)\right]^{1/2} \tag{4}$$

$$\delta\gamma_a = \frac{2\,E_U}{\left(\frac{\hbar^2}{2m_0L^2}\right)} \tag{5}$$

$$E_U = -\,b\left(\frac{C_{11} + 2C_{12}}{C_{11}}\right)\varepsilon_{//} \tag{6}$$

$$\frac{m_c}{m_{c0}} = \frac{E_g + E_H - E_U}{E_g} \tag{7}$$

$$E_H = 2a\left(\frac{C_{11} - C_{12}}{C_{11}}\right)\varepsilon_{//} \tag{8}$$

where γ_1, γ_2, γ_3 are the Luttinger mass parameters, b is the shear deformation potential, a is the hydrostatic deformation potential and C_{11} and C_{12} are the stiffness constants. These material parameters, as well as the unstrained conduction band mass m_{c0} and the refractive index n_0 are obtained by linearly interpolating those of the constituent alloys.

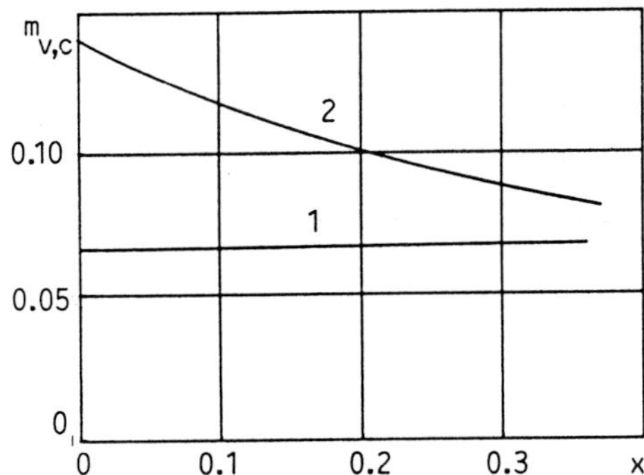

Fig. 1 - Dependence of the effective masses of electrons m_c (curve 1) and holes m_v (curve 2); nomalized to the electron mass at rest m_0, on the In content x.

The effective masses obtained in this way for an $In_xGa_{1-x}As/GaAs$ structure are represented in Fig. 1. m_c is almost constant and m_V decreases with the In concentration.

The band-gap energy is determined, for low concentration of the carriers, from the value without strain taking into account the deformation energies E_U and E_H

$$E_g = E_{g0} + E_U + E_H \tag{9}$$

where E_{g0} is the unstrained band-gap energy. E_{g0} is obtained by quadratically interpolating between the bulk values of the constituent alloys and further corrected for the quantization effect. In these calculations non-parabolicity of the conduction band has been neglected. In two-dimensional structures the degeneracy of heavy and light-hole bands at point Γ is lifted, with changes in their effective masses.

II.2 - ABSORPTION COEFFICIENT α AND NONLINEAR REFRACTIVE INDEX VARIATION Δn

The model for α and Δn includes many body effects. These have three major consequences:

- The reduction of the band-gap with increasing carrier density.
- The screening of electron-hole Coulomb potential.
- The enhancement of optical interband transitions.

The basic theory has been established by Lindberg (1988) and Haug (1989). The nonlinear susceptibility can be described by

$$\chi = \sum_k d_k \chi_k \tag{10}$$

$$\chi = \sum_k d_k \chi_k^0 \left[1 + \frac{\varepsilon_0}{d_k} \sum_{k'} V(k-k') \chi_{k'} \right] \tag{11}$$

$$\chi \cong \sum_k \frac{d_k \chi_k^0}{(1-q_1)} \tag{12}$$

where

$$\chi_k^0 = \frac{d_k (f_{c,k} - f_{v,k})}{\hbar\omega - \varepsilon_{e,k} - \varepsilon_{h,k} + j\hbar\gamma_k} \tag{13}$$

$$q_1 = \frac{\varepsilon_0}{d_k} \sum_{k'} V(k-k') \chi_{k'}^0 \tag{14}$$

and $f_{c,k}$, $f_{v,k}$, are the Fermi-Dirac distributions for electrons in the conduction and valence band, d_k is the dipole transition moment, $\varepsilon_{e,k}$, $\varepsilon_{h,k}$ are the renormalized self-energies of electrons and holes and γ_k describes the relaxation by collisions.

The enhancement of the optical transition by Coulomb effects leads to the denominator $(1-q_1)$ in (12) where $V(k-k')$ is the screened Coulomb potential. The reduction of the band-gap affects the denominator of χ_k^0 given that

$$\varepsilon_{e,k} + \varepsilon_{h,k} = e_{e,k} + e_{h,k} + E_g + \Delta E_g \tag{15}$$

where ΔE_g is due to the exchange and coulomb effects and E_g is the material bandgap for low-concentration of mobile charge carriers. α and Δn are related to the nonlinear susceptibility χ by

$$\alpha \cong \frac{\omega\,\mathrm{Im}(\chi)}{n_0 c} \tag{16}$$

$$\Delta n \cong \frac{\mathrm{Re}(\chi)}{2 n_0} \tag{17}$$

with ω for the light frequency, n_0 for the refractive index and c for the light velocity in vacuum.

III - RESULTS

Curves for the absorption coefficient α and the nonlinear index variation Δn dependence on the normalized energy, fig.2(a), (b), have been obtained according to the model previously described. As expected for the charge carrier densities considered, there is a spectral region with gain. Increasing the In content for fixed carrier densities broadens the spectral gain region and leads to a gain saturation. These effects are similar to those observed when carrier concentrations are increased in a given sample. The consequence is that maximum gain should not increase steadily with the In concentration. In figures 3, 4 the variation of the pseudo-Fermi levels of electrons, μ_e, and holes, μ_h, and of ΔE_g with the carrier densities and the In content are presented.

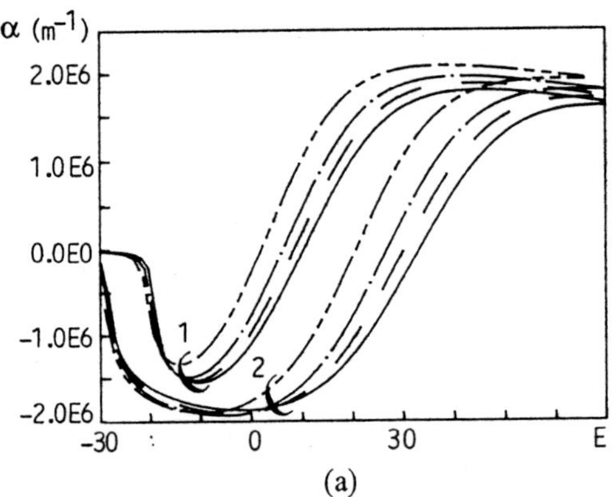

(a)

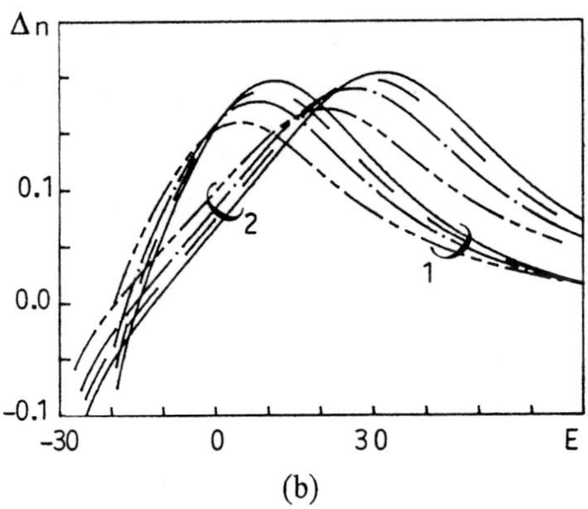

(b)

Fig. 2 - Coefficient of absorption α (a) and nonlinear refractive index variation Δn (b) as a function of normalized energy $E=(\hbar\omega - E_{g0})/E_R$, for two values of carrier density, (1) $N=2\times10^{16}\mathrm{m}^{-2}$, (2) $4\times10^{16}\mathrm{m}^{-2}$. The width of the InGaAs layer is L= 50Å.
—— x=0.37; - - - x=0.28; –·–x=0.18; -··-x=0.
E_R: Rydberg energy of bulk GaAs.

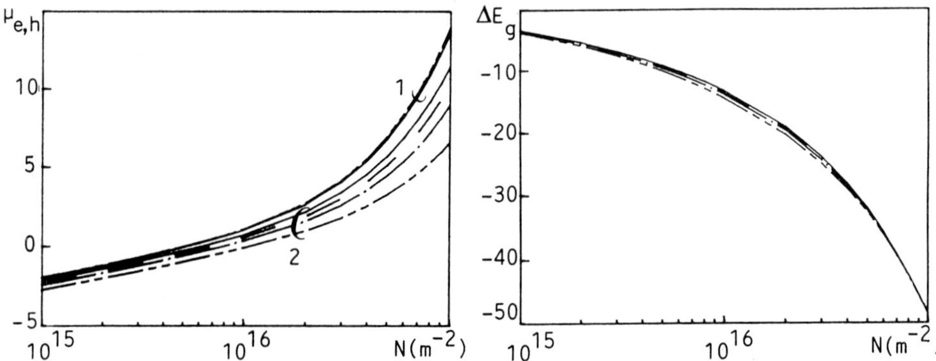

Fig. 3 - Pseudo-Fermi levels of electrons μ_e (1) and holes μ_h (2) normalized to kT, T=300K as a function of the carrier density N. —x=0.37; - - -x=0.28;–·–x=0.18;–··–x=0.

Fig. 4 - Band-gap narrowing ΔE_g normalized to E_R as a function of the carrier density N. — x=0.37;- - - x=0.28;–·–x=0.18;–··–x=0.

Increasing the In content for fixed densities makes the pseudo-Fermi level for holes increase, whereas the pseudo-Fermi level for electrons and ΔE_g both remain practically unaltered.

IV - CONCLUSIONS

Simulation results of the absorption coefficient and the nonlinear refractive index variation of a direct band gap strained semiconductor have been obtained. The model for the two-dimensional electron and hole gas is a generalization of previously published theories Suemune (1991), Haug (1989). It takes into account the coexistence of strain and many body effects. Results are presented for the InGaAs/GaAs quantum well structure with different In contents and hence different strains. It is predicted that increasing the In content results in the broadening of the gain region. This effect becomes more pronounced with increasing particle concentrations. This should be confirmed by experiment.

REFERENCES

Chow W. W., Koch S. W., Murray Sargent III, June 1990, IEEE Journal of Quantum Electronics, Vol. 26, No. 6, pp.1052-1057.

Haug H., Koch S.W., February 1989, Physical Review B, Vol. 39, No. 4, pp. 1887-1898.

Koch S. W., Peyghambarian N., Lindberg M., 1988, J. Phys. C, Solid State Physics, pp. 5229-5249.

Lindberg M., and Koch S.W., 1988, Physical Review B, Vol. 38, No. 5, pp. 3342-3350.

Loehr, P.J. and Singh, J., March 1991, IEEE Journal of Quantum Electronics, Vol. 27, No. 3, pp. 708-716.

Suemune, I., IEEE Journal of Quantum Electronics, May 1991, Vol. 27, No. 5, pp. 1149-1159.

Effect on non-ideal delta doping layers in $Al_{0.3}Ga_{0.7}As/In_{0.3}Ga_{0.7}As$ pseudomorphic heterostructures

S Fernández de Avila[+], JL Sánchez-Rojas[+], P Hiesinger[o], F González-Sanz[+], E Calleja[+], K Köhler[o], W Jantz[o] and E Muñoz[+].

+ Departamento de Ingeniería Electrónica, E.T.S.I.Telecomunicación. Ciudad Universitaria s/n, 28040 Madrid, Spain.
o Fraunhofer-Institut für Angewandte Festkörperphysik Tullastrasse 72, D-79108 Freiburg, Germany.

ABSTRACT: The depth profile of δ–doped layers is investigated by photoluminescence, Hall and DLTS measurements. A GaAs quantum well (QW) embedded into the $Al_{0.3}Ga_{0.7}As$ barrier is δ-doped with Si. The spreading of Si out of the GaAs QW is deduced by comparing the measured charge with self-consistent calculations taking into account variations of the gaussian Si distribution profile and the ionization probability of Si-related DX centers. DLTS measurements confirm the presence of Si atoms in the $Al_{0.3}Ga_{0.7}As$ barrier. It is shown that the influence of charge trapping on the channel conductivity can be neglected for standard MBE growth temperatures.

1. INTRODUCTION

The δ-doping technique can improve the characteristics of selectively doped heterostructures by increasing the two-dimensional electron gas (2DEG) density. δ-doped layers are also used for new devices such as *nipi* superlattices or δ-FETs. The ideal δ-doping, with all dopant atoms concentrated in one monolayer, is difficult to obtain with conventional MBE and MOCVD growth methods. A significant dopant spreading takes place, as shown by SIMS analysis (Clegg J B, 1989).The thickness of the real δ-doping layer embedded in bulk GaAs depends on growth temperature and doping density (Köhler K, 1993).

In this paper, we determine the real spreading of the δ-doping layer and its effects on the 2DEG density in the channel of a series of pseudomorphic heterostructures.

2. EXPERIMENTAL PROCEDURE

The samples were grown in a Varian Gen II MBE system on semi-insulating (100) GaAs substrates at 550°C. The layer sequence consists of 1500Å undoped GaAs buffer, followed by a 35Å GaAs/ 85Å $Al_{0.3}Ga_{0.7}As$ superlattice and a 6000Å undoped GaAs layer. Then the strained $In_{0.3}Ga_{0.7}As$ layer (with thicknesses L_w ranging from 20 to 120Å), a 50Å $Al_{0.3}Ga_{0.7}As$ spacer, a 17Å GaAs QW with a centered nominal $2.45x10^{12}cm^{-2}$ silicon δ-doping layer and a 600Å undoped $Al_{0.3}Ga_{0.7}As$ barrier are grown. Finally, the structure is capped by an undoped 200Å GaAs layer.

PL spectra were taken for all samples from 4K to 300K. Low power excitation (0.5mW) was provided by an Ar^+ laser (5145Å line). Hall measurements were carried out from 20 to 300K in the dark using 6x6mm greek-cross-shaped samples with 0.6mm wide crossbars. Deep Level Transient Spectroscopy (DLTS) was performed on guarded Al Schottky diode structures.

3. SELF-CONSISTENT MODEL

Poisson and Schrödinger equations are solved simultaneously in all the heterostructures taking into account effective mass discontinuity at the interfaces, non-parabolicity effects and the local density approximation to the exchange-correlation interaction. In addition, the DX center behaviour is modeled for δ-doped GaAs/AlGaAs layers. The δ-doping profile is described by a gaussian distribution with standard deviation σ:

$$N_d(z) = \frac{N_\delta}{\sqrt{2\pi\sigma^2}} \exp\left(-\frac{z^2}{2\sigma^2}\right)$$

of a sheet doping density N_δ centered at z=0 (Schubert E F, 1989).

In the GaAs QW full ionization of the dopant (no DX occupancy) is assumed (Ekenberg U, 1983). However, in the present structures one has to consider the possibility that the real δ-layer profile includes Si atoms diffusing out of the GaAs QW into the $Al_{0.3}Ga_{0.7}As$. In this case, the deep levels generated in the $Al_{0.3}Ga_{0.7}As$ barriers have to be taken into account. In our model this is described by an ionization probability depending on the energetical position of the deep level. These DX centers influence the 2DEG density (due to partial donor ionization) as well as the total ionized impurity concentration.

4. PHOTOLUMINESCENCE

The PL spectra show three emission lines.Their energy values at 4K are plotted in Fig. 1 as a function of channel thickness L_w. All spectra show a peak (squares in Fig.1) attributed to the n=1 electron to n=1 heavy hole transition E_{11}. In addition, a higher energy peak (circles in Fig.1) appears in the two samples with thicker channel, which is identified as the transition from n=2 electron to n=1 heavy hole subbands E_{21}. For L_w=120Å the E_{21} peak intensity is higher than that of E_{11} indicating that the Fermi level is above the n=2 electron subband (Colvard C, 1989). However, for L_w=100Å the E_{21} peak intensity is lower but increases with temperature relative to E_{11} peak intensity. Thus the Fermi level is close to the second electron subband, but still below it. This behaviour of the PL spectra with the temperature is shown in figure 2 for the two samples (L_w=100Å and L_w=120Å). The sample with the narrowest channel L_w=20Å shows a low PL intensity as expected from the poor confinement.

Some samples show an additional small PL feature attributed to the Fermi edge singularity (FES) or Mahan exciton (Mahan G D, 1967). This extra structure (triangles in Fig. 1) disappears if either the excitation power or the temperature are slightly increased. Its presence is probably due to hole localization produced by roughness at the $In_{0.3}Ga_{0.7}As/GaAs$ interface. Alloy fluctuations in the channel can be excluded as origin of FES since it is only observed in the samples with narrower channels.

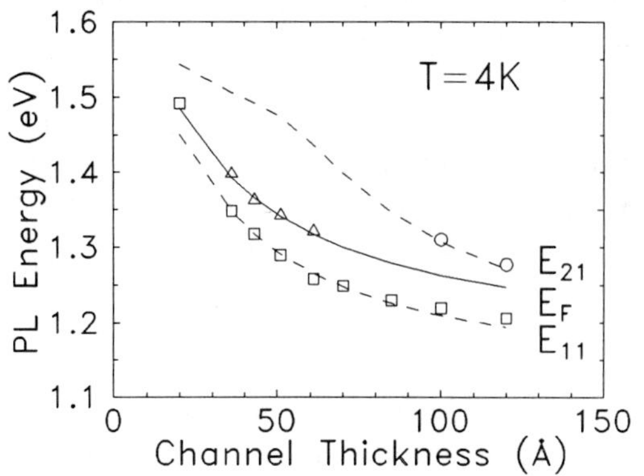

Fig. 1: PL peak positions at 4K (open symbols) as a function of L_w. Squares represent the observed optical transitions E_{11} from n=1 electron to n=1 heavy hole subbands, circles are transitions E_{21} from n=2 electron to n=1 heavy hole subbands and triangles are the observed correlation peaks (FES). The dashed lines are the corresponding theoretical position of the subband transitions and full line is the Fermi level position E_F relative to the n=1 heavy hole subband.

The theoretical fits to the PL energies in Fig. 1 (plotted as lines) are obtained assuming full ionization of the dopant in the whole structure and taking into account the illumination effects on the band-bending (Chaves A S, 1986). An In mole fraction of 0.295 and a δ-doping concentration of $2.5 \times 10^{12} cm^{-2}$ is assumed for all samples. Nominal values of the remaining structure parameters are used. The agreement is excellent for almost all samples. Therefore, the same set of parameters is also used below for the analysis of transport properties. Only for L_w=100Å and L_w=120Å a small discrepancy is observed indicating a lower In mole fraction and a thicker channel. The blue-shift observed in the experimental E_{11} transition for the narrowest sample L_w=20Å could be due to In segregation. For very narrow channels this segregation produces a parabolic profile with higher energy levels as compared to the rectangular profile (Muraki K, 1992).

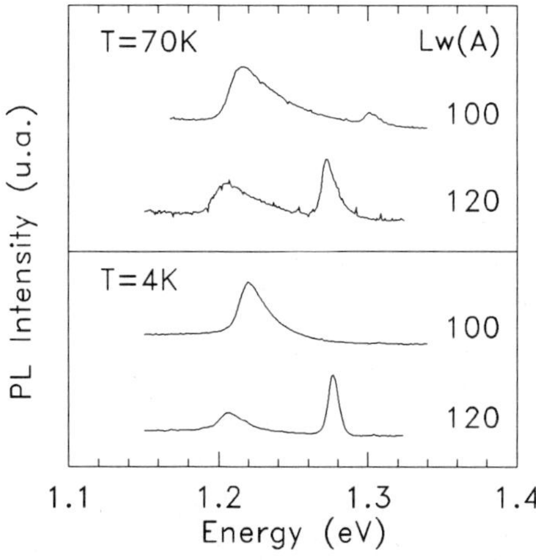

Fig. 2: PL spectra for the two wider channel samples showing the E_{11} and E_{21} peaks. Observe the change in the relative intensity of the two peaks at 4K and 70K for each sample.

5. HALL CARRIER CONCENTRATION

The Hall electron concentration N_H is nearly temperature-independent in all the samples.

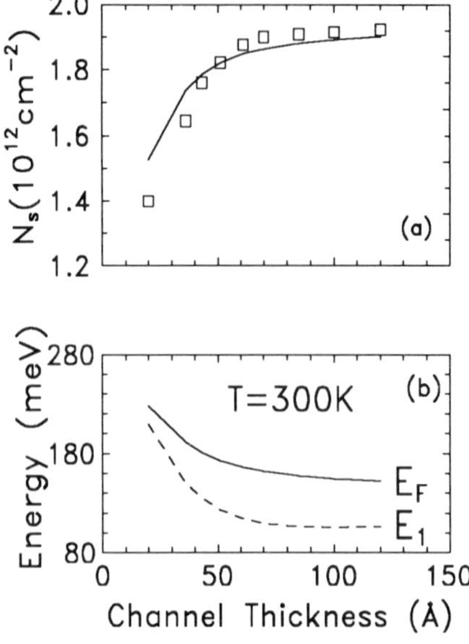

Fig. 3:(a) Hall carrier concentration N_H at room temperature (open squares) as a function of channel thickness L_w. Full line represents the theoretical fitting obtained with σ =15 Å for the gaussian Si distribution. (b) Fermi level E_F and first electron subband E_1 relative to the bottom of the channel conduction band as a function of L_w.

Fig. 4: Theoretical carrier concentration N_S as a function of the standard deviation σ of the Si gaussian distribution with channel thickness L_w as parameter.

The Hall mobility values are ranging from 3600 to 6000 cm^2/Vs at room temperature and from 7020 to 20300 cm^2/Vs at 20K. In Fig. 3a, N_H (open squares) is plotted as a function of L_w. Room temperature values are chosen for comparison with theory in order to minimize metastability effects from DX centers that could affect the measured charge. Because the growth temperature (T_g=550°C) is the same for all samples, the same theoretical Si profile is assumed for the entire series. The width of the gaussian Si distribution is used as a variable parameter to fit N_H. DX centers effects are introduced by assumming a donor level at 115meV below the $Al_{0.3}Ga_{0.7}As$ Γ minimum (Chand N, 1984). The best fit plotted as a solid line in Fig. 3a is obtained with σ=15Å, corresponding to a FWHM of 35Å.

The observed increase in carrier concentration for thicker channels is due to the electron concentration dependence on the difference E_F-E_1 (Fig. 3b), where E_F is the Fermi electron energy and E_1 is the n=1 electron subband energy. For narrow channels (L_w<80Å) the decrease of E_1 (proportional to $1/L_w^3$) with increasing L_w is faster than the decrease of E_F due to band-bending. Thus the total charge increases with the channel thickness. For wider channels (L_w>80Å) the thickness dependence of E_1 and E_F is very small leading to a saturation of the charge as shown in Fig. 3a.

The charge N_s self-consistently calculated as a function of the standard deviation σ of the gaussian Si distribution is shown in Fig. 4 for several samples. For all samples the highest N_s is obtained for a near-ideal δ-distribution (σ=3Å) without any Si atoms spreading out of the GaAs QW. The negative slope of N_s observed for σ<10Å in Fig. 4 is due to the effect of the DX centers. When σ increases some Si atoms penetrate into the $Al_{0.3}Ga_{0.7}As$ barriers and transform into deep traps. The positive slope is produced by the penetration of the Si gaussian distribution into the $In_{0.3}Ga_{0.7}As$ channel where transformation into deep traps does not occur. It should be kept in mind that the beneficial effect of increasing N_s for wider distributions and the deleterious influence of Si atoms in the channel on the electron mobility counteract. Thus the transport properties of these structures cannot be improved by widening the Si distribution.

The simulations also show that for σ values between 10 and 20Å (corresponding to a FWHM between 20 and 50Å approximately) the 2DEG charge in the channel does not change appreciably. Therefore, the model calculations do not allow to determine accurately the real width of the Si distribution but the presence of Si atoms in the $Al_{0.3}Ga_{0.7}As$ barrier is clearly indicated.

6. DEEP LEVEL TRANSIENT SPECTROSCOPY

A further experimental verification of the displacement of some Si atoms out of the GaAs QW is given by DLTS measurements. In this investigation we use the fact that the thermal activation energy of Si-related DX centers in GaAs and $Al_{0.3}Ga_{0.7}As$ differ by about 100 meV.

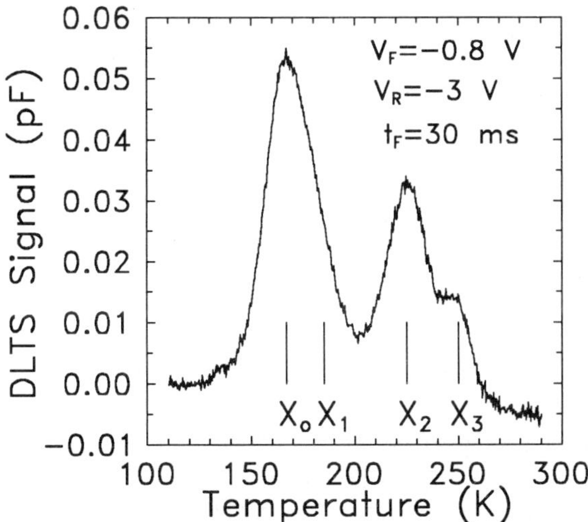

Fig. 5: DLTS spectrum showing the thermal emission peaks of Si-related DX centers.

The occupation of Si-related DX centers in the $Al_{0.3}Ga_{0.7}As$ barrier is rather low because E_F is far below the energy corresponding to the deep donor state, as shown by the comparison of Hall and DLTS measurements. For low reverse voltages only one DLTS peak is observed. It corresponds to the Si-related DX centers present in GaAs and the emission energy obtained, $X_0 = 0.35 \pm 0.02$ eV, agrees with that found in the literature (Calleja E, 1990). A shoulder X_1 is barely detectable at the high temperature side of this peak. To detect clearly the presence of silicon atoms in the $Al_{0.3}Ga_{0.7}As$ barrier, it is necessary to apply an even higher reverse voltage (Fig. 5). Now, besides peak X_0, the peak labeled X_2 with emission energy around 0.47 eV and peak X_3 with 0.46 eV are resolved. Shoulder X_1 and the peaks X_2 and X_3 correspond to Si-related DX centers in $Al_{0.3}Ga_{0.7}As$ with, respectively, one, two and three Al atoms as nearest neighbours (Calleja E, 1990).

The observation of the complete DX configuration corroborates the results about the spreading of the Si δ-layer out of the GaAs QW obtained from the Hall measurements and the self-consistent calculations.

7. CONCLUSIONS

The effects of non-ideal silicon δ-profiles on the charge distribution have been studied in a series of δ-doped strained heterostructures. From the comparison of PL measurements with a self-consistent model the structural and model parameters are confirmed. Hall carrier concentration fitting reveals the spreading of Si atoms out of the GaAs QW and a gaussian distribution with a FWHM of 35 Å is estimated. DLTS measurements give direct proof that Si atoms are incorporated in the barrier layer outside the GaAs QW.

8. ACKNOWLEDGMENTS

This work has been supported by CICYT-Spain, Project TIC-90-0140-C02-01, and by Acciones Integradas Hispano-Alemanas. We thank Dr. M. Maier and Prof. J. Sánchez-Dehesa for helpful discussions.

REFERENCES

Calleja E, Mooney P M, Theis T N and Wright J L 1990 Appl. Phys. Lett. **56** 2102

Chand N, Henderson T, Klem J, Masselink W T, Fischer R, Chang Y C and Morkoç H 1984 Phys. Rev.B **30** 4481

Chaves A S, Penna A F S, Worlock J M, Weimann G and Schlapp W 1986 Surf. Science **140** 618

Clegg J B and Beall R B 1989 Surf. and Interface Analysis **14** 307

Colvard C, Nouri N, Lee H and Ackley D 1989 Phys. Rev B **39** 8033

Ekenberg U and Hess K 1983 Phys. Rev. B **27** 3445

Köhler K, Ganser P and Maier M 1993 J. Cryst. Growth **127** 720

Mahan G D 1967 Phys. Rev. **153** 882

Muraki K, Fukatsu S, Shiraki Y and Ito R 1992 Appl. Phys. Lett. **61** 557

Schubert E F, Tu C W, Kopf R F, Kuo J M and Lunardi L M 1989 Appl. Phys. Lett. **54** 2592

Fabrication of GaAs–AlGaAs nano-heterostructures by through-UHV processing

Y Katayama, T Ishikawa, S Goto, Y Morishita, Y Nomura, M López, N Tanaka and I Matsuyama

Optoelectronics Technology Research Laboratory (OTL), 5-5 Tohkodai, Tsukuba, JAPAN

ABSTRACT: Two kinds of approaches to fabricate GaAs-AlGaAs nano-heterostructure using through-UHV processing are described. One is a method in which quantum-wire structures are formed on the pre-patterned substrates using selective growth on specific crystal orientation of facets. The other is a method which uses a combination of epitaxy and *in situ* pattern formation in an ultra-high vacuum (UHV) multi-chamber system. These two methods are examined, while putting emphasis on the cleanliness of the interfaces obtained and the crystalline defects produced during processing.

1. INTROCUTION

There has been a keen interest in 2- and/or 3-dimensional semiconductor nanostructures motivated by future application to electronic and optoelectronic devices (Sakaki 1980, Arakawa and Yariv 1986, Chang and Esaki 1992, Yokoyama et al. 1992). A variety of methods have been proposed to fabricate such nanostructures. Early attempts along this line include Petroff et al.'s report on the fabrication of GaAs-AlGaAs quantum well-wire structures by a combination of the molecular beam epitaxy (MBE) and photolithography (1982) and Kamon et al.'s trial on the fabrication of the similar structure by selective-area low-pressure organometallic vapor phase epitaxy (LP-OMVPE) using patterned dielectric masks (1986). As a method without the use of specific masks, Petroff et al. proposed a novel idea to fabricate such lateral superlattices using MBE of GaAs-AlAs on the (100) vicinally oriented GaAs substrates (1984). Fukui and Saito (1987) reported the growth of the (AlAs)(GaAs) fractional-layer superlattices on the (100) vicinal surface by OMVPE. The study of electronic and optoelectronic properties and the device application are greatly stimulated by recent works by Kapon et al. (1992) and Tsukamoto et al. (1993).

Summarizing approaches above, we may say that 2- and/or 3-dimensional semiconductor nano-heterostructure are fabricated by combining the technology to form epitaxial layers controlled on the atomic scale and the technology to form superlattice structure with the dimension of nm in the direction parallel to the substrate surface. Since the materials to be dealt with here are compound semiconductors which are very sensitive to the atmosphere and the desired size accuracy in the fabrication is in the order of nm, the chemical change at the surface must be minimized and the cleanliness of the interfaces is essential. Therefore, it will be natural to consider that the entire processes to fabricate semiconductor nano-heterostructure should be carried out in a vacuum chamber without exposing specimen to the air. This is so-called through ultra-high vacuum (UHV) processing or *in situ* processes (Takamori et al. 1985, Akita et al. 1990).

In this context, research on *in situ* patterning using focused-ion beam (FIB) was carried out (Ochiai et al. 1983, Sugimoto et al., 1989, Temkin et al. 1989). However, it turned out that, in the processing using ion beam, ion induced damage extends deeply into the specimen (Sugimoto et al. 1990). So, we have recently developed a lithography process using focused electron-beam which is compatible with an all UHV processing of GaAs-AlGaAs system (Akita et al. 1989). In this lithography which we call *in situ* electron-beam (EB) lithography, an ultra-thin surface oxidized layer of GaAs is used as both the resist film and the etching mask, which can be patterned by EB-induced Cl_2 etching and can be removed by heating. The basic process of in situ EB lithography is the local removal of the surface oxide layer of GaAs by a simultaneous irradiation of electron beam and Cl_2 gas.

In this paper some example of trials conducted at OTL to fabricate semiconductor nano-heterostructures which are compatible with through vacuum processing are described.

2. SELECTIVE GROWTH ON PREPATTERNED SUBSTRATES

We have fabricated quantum-wire structures using two kinds of selective growth on pre-patterned substrates. One is the selective growth of GaAs by metallorganic molecular beam epitaxy (MOMBE) on the facets of specific crystallographic orientation (Nomura et al. 1993) and the other is the use of the difference in the growth rate on the neighboring facets in the molecular beam epitaxy (MBE) of GaAs (López et al. 1993). These methods have the advantage of forming damage- and contamination-free interfaces since these methods do not require post-patterning and/or post removal of the dielectric masks.

2.1 Lateral MOMBE Growth on the Pre-Patterned GaAs (111)B Substrates

This method is based on a finding (Nomura et al. 1991) that the GaAs growth rate on the (111)B surfaces in the MOMBE of GaAs using trimethylgallium (TMGa) and metal arsenic as source materials is extremely low. That is, as shown in Fig.1, GaAs grows only on the sidewalls without any growth on the (111)B surfaces and the arsenic (As_4) pressures higher than 2×10^{-3}Pa, which results in lateral epitaxial growth. Growth of the quantum-wire structure was carried out in an MOMBE system with conventional effusion cells for Ga, Al, As_4 and a gas cell for TMGa. MBE and MOMBE growth were made on the pre-patterned (111)B substrates with mesa-grooves along the $[0\bar{1}1]$ direction with $(1\bar{2}\bar{2})$B sidewalls. The growth sequence was : 1) MBE growth of a 25nm-thick AlAs, 2) lateral MOMBE growth of

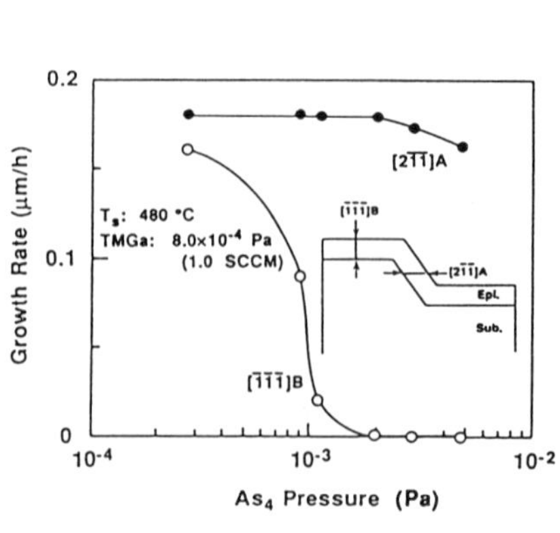

Fig.1 Growth rate of GaAs on (111)B patterned substrate vs. As_4 pressure.

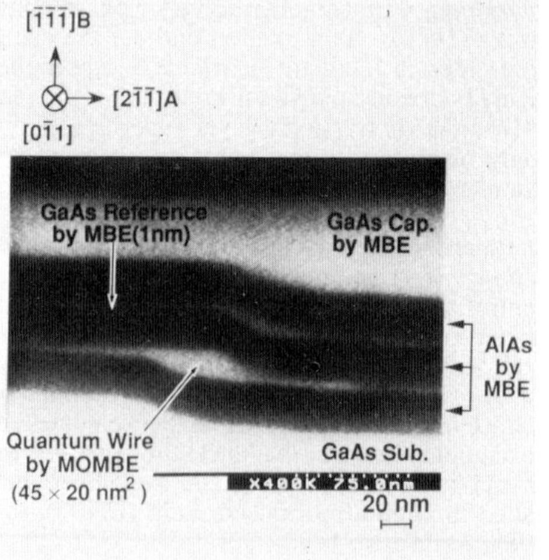

Fig.2 Cross-sectional SEM photograph of grown quantum-wire structure.

GaAs wires, 3) MBE growth of 50nm-thick AlAs overlayer, and 4) MBE growth of 70nm-thick GaAs cap-layer. The growth conditions for lateral MOMBE were, the flow rate of the TMGa was 1.0sccm without any carrier gas, the arsenic (As_4) pressure and the substrate temperature were set at 4.8×10^{-3}Pa and 480°C, respectively. A scanning electron microscope (SEM) photograph of the resulting wire-structure and its schematic illustration are shown in Fig.2.

2.2 MBE growth of GaAs/AlAs Wire-Structures on Mesa Stripes on GaAs (100) Surface

The MBE growth of GaAs-AlGaAs on mesa-stripes oriented along the $[0\bar{1}1]$ and $[0\bar{1}\bar{1}]$ directions of GaAs (100) substrates has been studied widely (for example Yuasa et al. 1987). However, the side walls generated during growth comprised facets with a variety of crystalline orientations and were often rough, which makes the fabrication of quantum-wire structures difficult. López et al. (1993) studied the facet formation during MBE growth of GaAs-AlGaAs on mesa-stripes oriented along [001]-direction and found that the smooth side-walls of (110) and $(1\bar{1}0)$ facets are formed during the growth at appropriate conditions as shown in Fig.3. They found also that the growth rate on the {110} facets is extremely low and that the width of (100) mesa-top facet which is limited by the (110) and $(1\bar{1}0)$ facets is reduced by the MBE growth of GaAs. It is known that, in general, the formation of facets during growth is caused by the difference in the growth rates on the neighboring different crystallographic planes. Facets on which the growth rate is small are formed preferentially. Using these novel characteristics, they have fabricated quantum wire-like structure along the [001] direction as shown in Fig.4. The reduction of the width of the mesa-top (100) facets was made by the GaAs growth using an As_4 pressure of 8×10^{-4}Pa at the growth temperature higher than 600°C. It should be noted that the possibility of quantum-dots fabrication using square-shaped mesa structure surrounded by side walls along [001] and [010] directions. It is also worth while noting that this method can be incorporated as an elemental process in the *in situ* EB lithography described in the following.

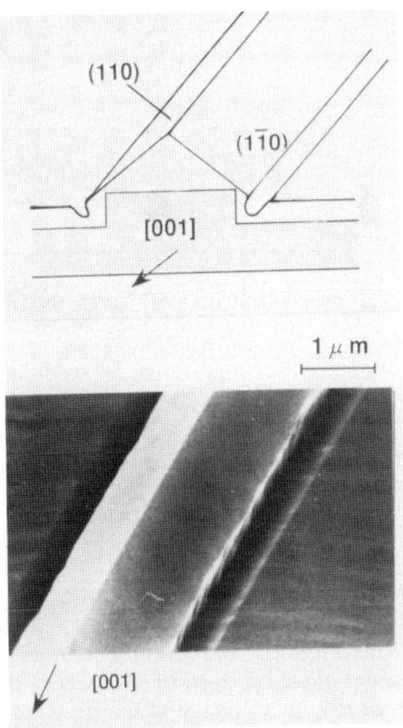

Fig.3 SEM photograph of the surface of GaAs layers grown on [001] mesa stripe and schematic illustration.

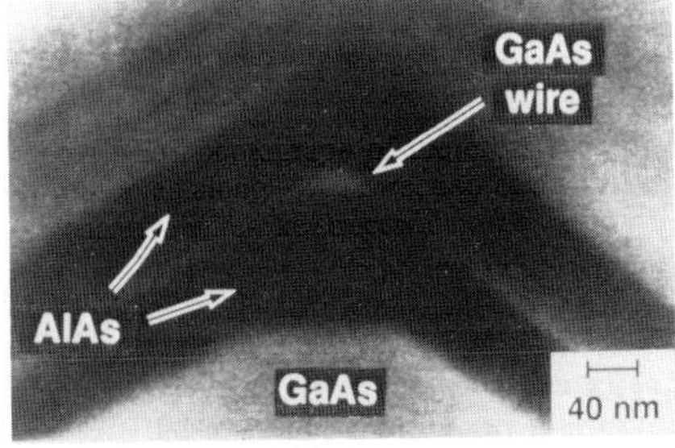

Fig.4 SEM photograph of the cross-section observed along [011] direction of a grown quantum wire-like structure.

3. *IN SITU* ELECTRON-BEAM (EB) LITHOGRAPHY

One of the most important requirements for processing technology in the fabrication of the advanced devices is the fabrication capability of very fine (of the order of nm) and integrated structures which give an additional freedom to design the advanced devices. As briefly stated in the Introduction, neither the conventional lithographic technology which uses organic resist material nor the focused-ion beam (FIB) based through UHV processing do meet the requirements either due to contaminations at the interfaces fabricated or the defects caused by heavy ions. In such a situation, a recently developed *in situ* electron-beam (EB) lithography (Akita et al. 1989) seems to be a hopeful candidate for the fabrication technology of very fine and integrated structures. In the following the outline of this *in situ* EB lithography is described together with the related recent developments.

3.1 *In situ* EB lithography

In *in situ* electron-beam (EB) lithography, an ultra-thin surface oxidized layer of GaAs is used as both the resist film and the etching mask, which can be patterned by EB-induced Cl_2 etching and can be removed by heating. The basic process of in situ EB lithography is the local removal of the surface oxide layer of GaAs by a simultaneous irradiation of electron beam and Cl_2 gas. As illustrated in Fig.5, the procedure of in situ EB lithography comprises the following five steps.1) preparation of a clean GaAs surface by MBE. 2) formation of the thin surface oxide layer as a resist-film by photo-oxidation in a pure oxygen. 3) patterning of the oxide layer by EB-induced Cl_2 etching, and subsequent Cl_2 etching of GaAs as a pattern transfer. 4) removal of the surface layer by heat-treatment under arsenic flux in the MBE chamber. To remove the oxide completely, the irradiation with atomic hydrogen is effective. 5) Overgrowth by MBE.

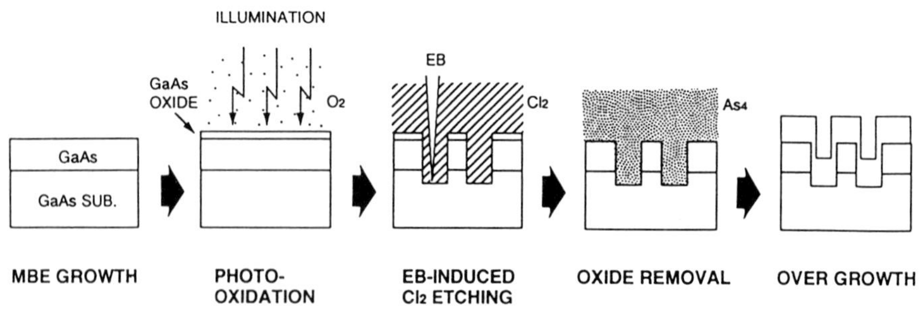

Fig.5 Schematic illustration of *in situ* EB lithography.

All these processes are carried out completely in an ultra-high vacuum chamber system as illustrated in Fig.5. This procedure can be repeated if necessary as described elsewhere (Taneya et al. 1989, 1990, Akita et al.1990). As one important feature in *in situ* EB lithography, the etched depth in the step 3) & 4) is plotted against the total electron dose in Fig.7. A steep rise of etching depth is seen at around the electron dose of $10^{17}cm^{-2}$, under which etching does not proceed effectively. In other words, this dosage is the minium EB-dose to remove the oxide layer completely. This nonlinear etching characteristics is quite favorable for lithography. Furthermore, it has turned out that the electron-beam exposure and the Cl_2 etching can be separated. This means that the electron source can be installed in the

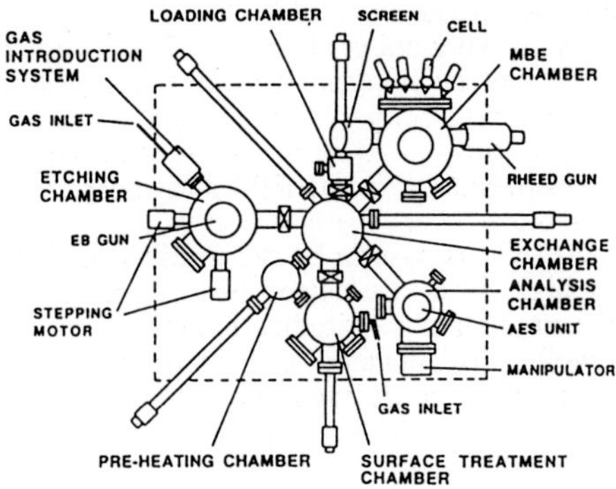

Fig.6 UHV multi-chamber system used for *in situ* EB lithography.

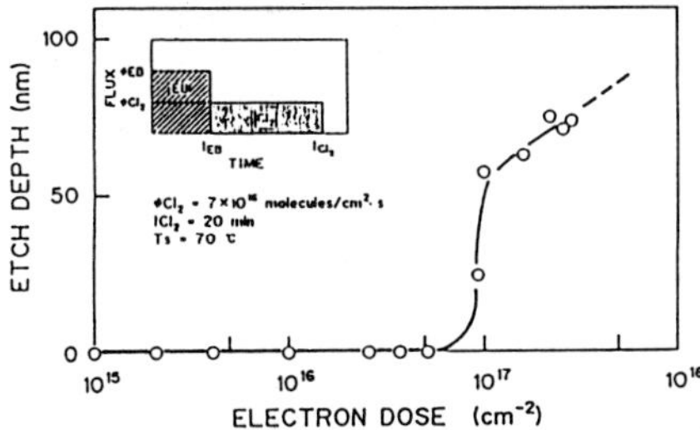

Fig.7 Etched depth vs. total electron dose in the EB-induced etching.

separate chamber from that for Cl_2 etching and will make easier the design of UHV system for *in situ* EB lithography. The details of the five steps and their modifications have already been reported elsewhere (Akita et al. 1989, 1990, Taneya et al. 1989, 1990).

An example of the fine pattern of GaAs produced by this process is shown in Fig.8. The best resolution of patterns obtained is about 50nm, which is considered to be limited by the radius of the electron beam. This indicates that the scattering effect of incident electron is quite small. This seems to be due to the very small thickness of the oxide-layer which acts as both the resist material and etching mask.

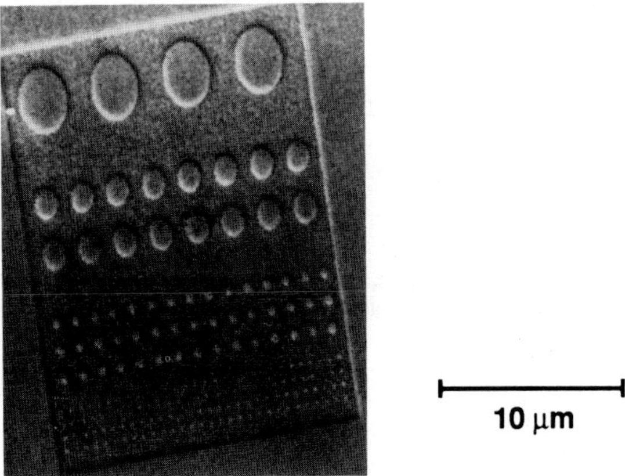

Fig.8 SEM photograph of a fine pattern fabricated by *in situ* EB lithography.

It has also been confirmed that the surface oxide of GaAs can be used as a mask for selective-area growth of GaAs by MOMBE (Hiratani et al. 1990). This means that the selective-area growth of GaAs by MOMBE can be incorporated as a processing unit in *in situ* EB lithography.

3.2 Characterization of Damages Induced in *in situ* EB Lithography

The damage induced in GaAs-AlGaAs system by the irradiation of electron beam must be carefully examined, since a high dose of electron-beam (EB) is used in the *in situ* EB lithography. The effect of EB irradiation was studied by photoluminescence (PL) properties of GaAs-AlGaAs quantum wells (QW) located at the different depth from the surface as shown in Fig.9 (Tanaka et al. 1993). The PL intensity does not change by the irradiation of 10 keV EB at the dose level upto $1{\times}10^{18}$ electrons/cm^2 and the PL intensities from shallower QWs are slightly degraded when the dose is increased to $1{\times}10^{19}$ electrons/cm^2. Even at doses as high as $4.5{\times}10^{19}$electrons/cm^2, the PL intensities from QWs are reduced only by factors of 5 for 2 nm QW and 2 for 5 nm QW. Note that the necessary dose for conventional EB lithography is order of 10^{16}electrons/cm^2 and $1{\times}10^{17}$ - $1{\times}10^{18}$ electrons/cm^2 for *in situ* EB lithography. Thus, the electron-beam induced damage has been shown to be well below the detection limit.

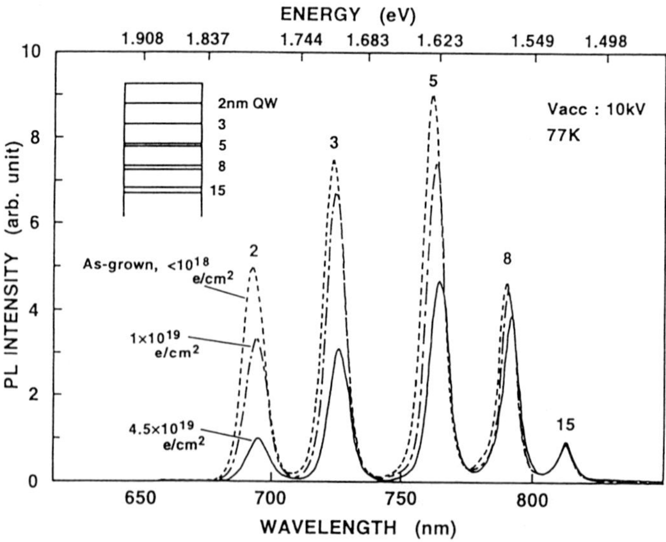

Fig.9 Photoluminescence spectra from quantum wells in the as-grown and the 10keV electron-beam irradiated samples.

Further, the characterization of the buried quantum wells fabricated by *in situ* EB lithography was made by cathode luminescence (CL) and PL topography (Kawanishi et al. 1992) and a substantial intensity of luminescence was recognized from the fine patterns fabricated by *in situ* EB lithography. These results demonstrate that the *in situ* EB lithography which is compatible with through UHV processing is a process technology where the process-induced damage is small.

4. SUMMARY AND FUTURE PROSPECTS

A through-UHV processing in which electron-beam (EB) induced patterning play a central role has been developed as a possible candidate to fabricate semiconductor nano-heterostructures for future electronic and optoelectronic devices. Experimental results show that elecrtron-beam assisted *in situ* processes are the less-damage processes and promising for the future development. However, as the nano-structures desired become smaller and smaller, the interface properties will become more important and more realistic accessments will be necessary. As the requirements for ultra-fine structure become urgent, such other techniques using elecrtron-beam as EB-induced epitaxial growth will become important.

ACKNOWLEDGEMENTS

The authors are sincerely grateful to Dr. Izuo Hayashi for his foreseeing advice and continuous encouragements.

REFERENCES

Akita K, Sugimoto Y, Taneya M, Hiratani Y, Ohki Y, Kawanishi H, and Katayama Y. 1990 SPIE, **1392**, Advanced Techniques for Integrated Circuit Processing, pp.576-587

Akita K, Taneya M, Sugimoto YH, Hidaka H and Katayama Y 1989, J. Vac. Sci. Technol. **B7**, pp.1471-1474

Arakawa Y and Yariv A 1986, IEEE J. Quantum Electron., **QE-22**, pp.1887-99

Chang L L and Esaki L 1992, Physics Today, October, pp.36-43

Fukui T and Saito H 1987, Appl. Phys. Lett. **50**, pp.824-26

Hiratani Y, Ohki Y, Sugiomoto Y, Akita K, Taneya M and Hidaka H 1990, Jpn. J. Appl. Phys. **29**, pp.L1360-L1362

Kamon K, Takagishi S and Mori H 1986, Jpn. J. Appl. Phys. **25**, pp.L10-12

Kapon E, Walther M, Cristen J, Grundmann M, Caneau C, Hwang D M, Colas E, Bhat R, Song G H and Bimberg D 1992, Superlattices and Microstructures **12**, pp.491-499

Kawanishi H, Sugimoto Y, Ishikawa T, Tanaka N, and Hidaka H 1992, MRS Proceedings **267**, pp.147-152

Nomura Y, Morishita Y, Goto S, Katayama Y and Isu T 1991, Jpn. J. Appl. Phys., **30**, pp.3771-73

Nomura Y, Morishita Y, Goto S and Katayama Y 1993, Electronics Letters **29**, pp.163-165

Ochiai Y, Gamo K and Namba S 1983, J. Vac. Sci. Technol. **B1**, p.1047

Petroff P M, Gossard A C, Logan R A and Wiegmann W 1982, Appl. Phys. Lett. **41**, pp.635-8

Petroff P M, Gossard A C, and Wiegmann W 1984, Appl. Phys. Lett. **45**, pp.620-22

Sakaki H 1980, Jpn. J. Appl. Phys. **19** pp.L735-L738

Sugimoto Y, Taneya M, Hidaka H, Akita K, Sawaragi H and Aihara R 1989, Proc SPIE **1098**, pp.52-60

Sugimoto Y, Taneya M, Hidaka H and Akita K 1990, J. Appl. Phys. **68**, pp.2392-2399. López M, Ishikawa T and Nomura Y 1993, Jpn. J. Appl. Phys (in press)

Takamori A, Miyauchi E, Arimoto H, Bamba Y, Morita T, and Hashimoto H. 1985, Jpn. J. Appl. Phys., **24**, 6, pp.L414-L416

Tsukamoto S, Nagamune Y, Nishioka M and Arakawa Y 1993, Appl. Phys. Lett. **62**, pp.49-51

Tanaka N, Kawanishi H and Ishikawa T 1993, Jpn. J. Appl. Phys. **32**, pp.540-543

Taneya M, Sugimoto Y, Hidaka H and Akita K 1989, Jpn. J. Appl. Phys. 28, pp.L515-L517, also 1990, J. Appl. Phys. **67**, pp.4294-4303

Temkin H, Harriott L R, Hamm R A, Weiner J and Panish M B 1989, Appl. Phys. Lett, **54**, pp.1463-1465

Yuasa T, Mannoh M, Yamada T, Naritsuka, Shinozaki K and Ishii M 1987, J. Appl. Phys. **62**, pp.764-770

Yokoyama H, Nishi K, Anan T, Nambu Y, Brorson S D, Ippen E P and Suzuki M 1992, Optical and Quantum Electronics **24**, pp.S245-S272

Inst. Phys. Conf. Ser. No 136: Chapter 7
Paper presented at the Int. Symp. GaAs and Related Compounds, Freiburg, 1993

III–V on dissimilar substrates: epitaxy and alternatives

G. Borghs, J. De Boeck, I. Pollentier*, P. Demeester*, C. Brys*, W. Dobbelaere.
IMEC (Interuniversity Microelectronics Center) Kapeldreef 75, B-3001, Leuven, Belgium
*IMEC -LEA Sint-Pietersnieuwstraat 41, B-9000 Gent, Belgium

Abstract

Heteroepitaxial growth and thin film transfer are two techniques for the integration of III-V technologies on dissimilar substrates. Crystalline quality remains the issue for heteroepitaxial layers of III-V's on Si. Quality improving techniques are discussed. Stress in the thin epitaxial film is another point of concern but recent progress has alleviated this problem to some extend.
Thin film transfer technology is currently tackling more processing and yield related issues and its maturity is making it a competing or complementary alternative to thin-film growth. For some very appealing applications thin-film transfer is the only solution, offering high flexibility and exciting combinations.

1. Introduction

The integration of devices fabricated in different material systems and appropriate technologies on the same substrate offers great advantages in terms of multifunctionality, compactness and speed but also of noise reduction and parasitic impedance control.
Techniques to achieve this can be divided roughly into monolithical and hybrid. Under hybrid techniques we reckon wire bonding, tape automate bonding and "flip-chip". These topics will not be the subject of this paper but will be included in the general comparison between the different approaches. We concentrate here on the monolithical integration obtained by heteroepitaxy and epitaxial lift-off (ELO), the latter also called pseudo-monolithical integration.
In ELO, a homoepitaxially or lattice matched grown III-V film is transferred to its new host substrate after detaching it from its substrate using a highly selective chemical etch on a sacrificial buried layer or by etch-back of the substrate.
The preference for one of the techniques is determined by the application. This paper tries to motivate the choices.

2. Heteroepitaxy

A generic type of heteroepitaxy is the growth of GaAs on Si which illustrates the basic principles of dissimilar growth techniques. In order to obtain the best quality heteroepitaxial GaAs on Si, three major activities must be carefully carried out if possible: Surface preparation and initial layer growth, relaxation of strain due to lattice mismatch and relief of thermal mismatch strain. Although currently the materials problems seem to be much harder to overcome than initially believed, we will see that a combination of innovative ideas in each of these categories can lead to steady progress in materials quality. Moreover, the difficulty to solve GaAs on Si materials problems as reviewed in numerous publications [1,2,3,4,5] has been very instructive for other materials systems.

Cleaning, initial layers

Optimal cleaning of the substrate is of paramount importance. Attempts have been made to keep the thermal budget low during cleaning and outgassing of the wafer to enable growth on already processed Si. Cleaning methods creating a volatile oxide passivation layer, or alternatively a hydrogen passivation after HF dip, made it possible to avoid a high temperature heating step prior to growth. This led however to inferior layer quality due to antiphase domain formation (multivariant film). To reduce this an "even-stepped" surface is necesary which can be obtained by annealing the Si wafer at high temperatures prior to growth. This is opposite to the low thermal budget cleaning scenarios. Moreover, the feasibility of the growth and processing of a 1μm FET ring oscillator GaAs on processed Si while preserving the specifications of the 2μm Si CMOS ring oscillator has been demonstrated [6].

The polar on non-polar growth problem may be resolved by growing on a perfectly ordered Ge (or Si, commercial Si-epitaxial layers on off-cut Si substrates have this even-step structure due to the high growth temperature of the epi-Si [7]) surface consisting entirely of even-atomic surface steps. The closer one can approach this ideal surface, the better the suppression of the multi-variant in the intitial growth stage of GaAs. It was found recently [8] that growth of GaAs on an "even-stepped" Ge surface (realized by epitaxy of Ge on off-oriented Si) always results in multivariant films when the Ge surface is exposed to As_2 prior to GaAs growth. The use of a Ga-prelayer on the freshly grown and cooled Ge surface, however, "freezes" the desirable step configuration and single domain GaAs is realized

On the ideal Si surface the van der Merwe nucleation and layer by layer growth is possible. Much effort has been invested in the initial layer growth and nucleation. Well known is the "2-step" growth technique starting with a thin GaAs layer slowly grown at a low temperature [9]. From epitaxial kinetics it is expected that higher deposition rates favour layer-like growth.[10] It is therefore not surprising that other nucleation mechanisms work equally well like. [11,12] Also in other material combinations like InAs on GaAs [13] low growthrate step is not used.
A special way to improve 2 dimensional growth is obtained by ion beam assisted deposition. Using a 28 eV Ar ion bombardment 3D island growth is suppressed and leads to nearly uniform layer-by-layer growth. [14]

Lattice mismatch, relaxation of strain

The problem of lattice mismatch and threading dislocations is of a general nature, contrary to the cleaning and initial layer growth which asks for a specific solution per substrate. As the growing non lattice matched epilayer reaches a critical thickness, the introduction of misfit dislocatons through strain-induced plastic deformation cannot be avoided. The dislocations created release then part of the unwanted strain in the layer. To avoid dislocations reaching the active layers of the device, mechanisms for elimination of threading segments are of great interest for high mismatch epitaxial growth. Growth conditions which can cause the bending of threading dislocations are important. Through bending one increases the chance meeting of two arms thereby eliminating both or at least one. Moreover the dislocations gliding over a long distance in the horizontal plane cause strain relief and avoid thus creation of new dislocations.

Dislocation nucleation and annihilation for mismatch strain relief can be activated by thermal cycling [15], which gives the dislocations a greater chance to interact. The same is true for high temperature annealing, in situ or ex situ [16]. But the most succesful way to bend dislocations is the inclusion of a step -or continous composition grading of strain during growth. Instead of only an abrupt nucleation of the mismatched layer it is advantage to add after the abrupt nucleation a small deviation in composition thereby varying the strain in the layer.

A typical example of dislocation bending (Fig. 1.) is shown in the cross-section TEM picture of an InAsSb diode grown on GaAs coated Si. [17] The dislocation density is reduced by the gradually strained layer to about 5 10^7 cm^{-2} within 2 μm thickness estimated from etch pit count, plan view TEM and XRD FWHM. This decrease in density can also be obtained by just growing a thicker layer. The evolution of the experimental number of dislocations as a function of distance is shown in Fig 2.a. To appreciate the influence on the performance of an infrared diode, the figure of merit R_0A being the product of the diode resistance at zero bias times the area of the diode is given as a function of distance between the junction and the interface in Fig 2.b..

After 4 μm this value reaches that of a lattice matched sample indicating that an amount of dislocations may be left over without deterioation of the device. Using strained layers, this value can of course be obtained much faster.

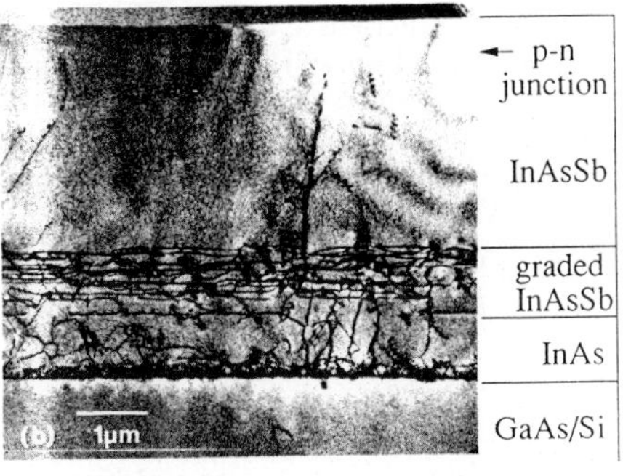

Fig1. TEM Picture of InAsSb diode on GaAs

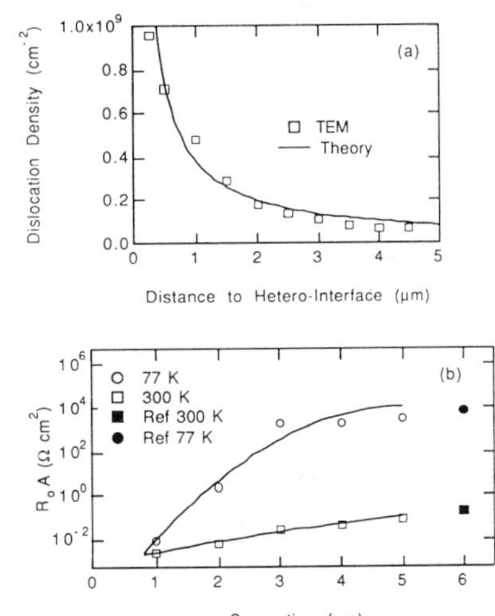

Fig. 2.a. Experimental dislocation density as a function of distance
2.b. R_0A product as function of distance

Experimental data on InGaAs and InAsSb made evident that the dislocation bending using strained layers becomes very complicated for alloy compositions between 30% and 70% . It is believed that local phase transitions produce different lattice defects. Differences in the growth of tensile versus compressive overlayers have also been noticed. When the lattice constant of the overlayer is smaller than the host layer cracks are easily formed [17].

A careful grading in this alloy concentration can nevertheless lead to very good device results as is demonstrated by $In_{0.7}Ga_{0.3}As/InAs_{0.36}P_{0.64}$ near-infrared detectors with low dark current [18].
High mobility HEMTs in $In_xGa_{1-x}As/In_yAl_{1-y}As$ on GaAs with x (and accomodated y values) between 0 and 1 have been reported.[19]. We also mention here the interesting result in the GeSi/Si modulation doped system. Using graded strain by composition variation up to 30% in the Ge_xSi_{1-x} system grown on Si, mobilities at 1.5K of 173 000 $cm^2V^{-1}s^{-1}$ have been obtained in the upper (strained) Si channel [20].

The growth of Ge on Si is also useful for III-V on Si.The issues of lattice mismatch can be overcome by using a fully relaxed, high quality Ge buffer on Si. This buffer can be achieved through epitaxy using compositionally graded GeSi (8) layer, or by a recently reported technique of reducing $Ge_xSi_{1-x}O_2$ in forming gas at 700°C [21]. In both cases good quality Ge can be realized, and the latter technique realizes a stack of GeSi/ GeSiO/Ge/SiO. On the Ge layer GaAs can be grown lattice matched and the problem of heteroepitaxy reduces to the problems associated with polar on non-polar growth and residual thermal strain.

As an alternative for grading , other techniques such as the recently reported atomic hydrogen assisted growth at low temperatures [22] are used. After a 1 μm thick buffer layer grown by a two-step growth technique, the growth front was irradiated with thermally cracked hydrogen, while reducing the growth temperature to 330°C (H_2 backpressure 10^{-6}Torr). The result of the Hydrogen irradiation in combination with the reduced growth temperture is an increased bending of threading dislocations leading to a reduction of the dislocation density to the level of $<10^6$ cm^{-2}. (EPD $3x10^4$ cm^{-2}). A very important factor in this technique is the low GaAs growth temperature leading to lower stress values in the dissimilar material structure.

A second alternative technique that has been recently proposed for possible improvement in heteroepitaxy is based on the preparation of the Si surface in such a way that most of the underlying Si is screened during the growth process and only small GaAs seeding areas have an interface with the Si substrate. The goal of such a process is the reduction of the number of misfit dislocations based on the fact that only the seeding areas can have misfit dislocations. As illustrations for this principle we indicate the studies on conformal growth [23] and GaAs/SiO_2 composite surfaces [24]. In the latter we proposed the use of nanometer-size GaAs seeds surrounded by SiO_2. Single crystal lateral overgrowth is shown to be feasible by MBE, using this technique. A breakthrough in the GaAs-on-Si crystalline quality has been reported using OMCVD regrowth on a sawtooth patterned Si wafer [25]. A close inspection of the interface shows that on the slopes of the grooves a SiO_2 layer are still present and growth is seeded probably through pinholes in the oxide, a situation very comparable to the GaAs/SiO_2 composite surface idea. A thorough understanding (and control) of the defect reduction mechanism is still lacking at present and the reported low defect density seems to disagree with the observation [26] of defect introduction (10^6 cm^{-2}) in the GaAs layer, due to the thermally induced strain.

Thermal mismatch

The difference in thermal expansion coefficient reintroduces stress in the material when cooling down from the growth temperature. There exist many solutions for this problem, the most simple being growth of materials with identical coefficients. But more realistic solutions have been proposed like e.g. low temperature growth, although this not always led to high material quality. Epitaxial lift-off described in a subsequent chapter is another possibility.

A more integrated process to improve materials quality after heteroepitaxial growth uses an AlAs release layer to undercut GaAs mesa structures in HF:H_2O [27,28]. The completely undercut mesas are restrained in their original position by photoresist (or metal) positioners and the residual thermal stress is completely removed by this Mesa Release and Deposition (MRD) technique. Upon removal of the positioners, the III-V mesas settle back to their exact original location on the Si substrate. Successful device processing using MRD has been demonstrated for FET's [29] and strain-free lasers [30]. Regrowth experiments on MRD-regions have successfully shown that strain relief remains after the epitaxy process [27]. An independently developed variant is the undercut GaAs on Si where only part of the GaAs layer is undercut [31]. Using this partly undercut GaAs on Si, LED's have been fabricated [32] and defect reduction after annealing of the these undercut films at 800°C has been reported [33] as well as dark-spot-free AlGaAs grown on undercut GaAs/Si [34].

3. Epitaxial Lift-off and Thin Film Transfer

The selective etching (with respect to GaAs) of a high-quality thin AlAs layer in HF-solutions has triggered the development of the Epitaxial lift-off (ELO) technique [35]. A high-quality epitaxial layer is grown lattice matched with the incorporation of an AlAs release layer. After selective etching of the AlAs layer, the wax-covered epitaxial structure is removed from its original substrate and transferred to its new host substrate. Table I illustrates the evolution of the ELO technique and shows that ELO has clearly demonstrated its power in the fabrication of a wide variety of high quality devices on dissimilar substrates and has now entered the stage of technological development (for a recent review see [36]). Besides the ELO technique, wafer bonding and etch-back and cleavage of lateral epitaxy for transfer (CLEFT) [37] have been developed to realize the thin crystalline film on a dissimilar substrate. In many aspects these related processes have the same characteristics and problems.

In most of the ELO work one relies on the Van der Waals (VdW) interatomic forces [38] of the newly formed interface to assure adhesion of the thin film. Eutectic metal bonding [39] using Au has also been introduced and for solar cells UV-cureable cements are even so suitable. A reliable interlayer is Pd [40]. The GaAs-Pd reaction assures a low-temperature bond with nearly 100% yield and high bond strength.

Either a rectangular area of several cm^2 (for solar cells, e.g.), or an array of several mm wide ribbons (for array-type applications) are transferred to cover the substrate. In this case the processing of the GaAs devices is performed after the transfer and bonding which makes the alignment to the existing circuitry less crucial and well within the achievable accuracy of a few microns. For pre-ELO processed devices, alignment is more critical and other precautions are required such as a moat etch [41] for protection of active layers during undercut. Accuracy of alignment varies with the technique employed [42, 43, 44]: 10μm when using pre-etched surface steps on the Si wafer to locate the ELO film, 5 μm using selective hydrophilisation. Further improvement is expected using more automated ELO-tools and fine tuning of thin film carrier techniques such as polyimide diaphragms. The use of preprocessed devices for ELO is further challenged by the stress that may exist in these fragile structures. Stress is also a more general concern, associated with the existence of bubbles (moisture or particle-related) under the ELO-film and with the step coverage over non-planarities on the fully processed substrate. In addition to immediate device performance degradation, these stresses may also lead to long term reliability problems when heat-cycling occurs during the lifetime of the circuit. Ultra-clean processing with continuous immersion in DI-water, together with adequate surface planarization can lift the hazard of bubble formation. The polyimides used for surface planarisation also improve the adhesion of the transplanted film on the new host substrate.

The deposition of a thin film on a dielectric further enables a higher degree of isolation between the GaAs devices themselves and between the devices and the host substrate, thus reducing side- and backgating effects in amplifier circuits [45]. Also, in the case of the light emitting diodes, the devices can take advantage [46] of the mirror created by the bonding metal layer, using an intermediate dielectric layer to increase their output power by a factor 2 to 3. Recently [47] an array of light emitting diodes with 30% external quantum efficiency has been demonstrated. One of the two key components in the fabrication of LED's beside ELO, is a nanostructuring of the thin film semiconductor surface by "natural lithography" after transplantation on a large dielectric coated gold mirror.

ELO is possible on all kinds of substrates making "exotic" quasi- monolithical integration possible like superconductors and semiconductors on a non-zincblende MgO substrate. The integration of superconducting microwave circuits with III-V semiconductors need an almost monolithic approach to reduce parasitic capacitances. Using ELO the problem of the incompatible processing steps are circumvented and high cut-off frequency HEMTs integrated with ceramic superconductors are fabricated.

As an alternative to pure ELO, where, in principle, the substrate can be reused, a process of a bottom-up wafer adhesion followed by etch-back removal of the substrate has been investigated. Many of the process-related issues discussed above are also relevant to this sacrificial substrate technique. Some remarkable results have been reported using this technique, one example is the film level hybrid integration of AlGaAs lasers with glass waveguides on Si [48].

4. Discussion and Conclusion

Since heteroepitaxy, ELO and hybrid integration are at very different levels of maturity, a comparison of their present merits is not very straightforward. The choice between the different techniques depends on the device, application and the substrate.

One advantage of direct heteroepitaxy is large area uniformity. Thus it serves well for solar cells and as a substrate for other semiconductor devices e.g. $InAs_{0.85}Sb_{0.15}$ infrared photodiodes on GaAs-coated Si [61] and HgCdTe infrared detector arrays that are flip-chip bonded to a Si substrate with read-out circuitry [62]. In the latter case, the thermal match of the detectors and electronics substrates enhances bond reliability needed for the fabrication of large flip-chip arrays.

Direct growth is also unavoidable when no lattice matched substrate exists, as is the case for ternary compositions like high In concentration InGaAs alloys for HEMT applications or infrared detectors. A drawback of heteroepitaxy is that strain and dislocations remain in the active layer being detrimental for some kind of devices. Of all the demonstrated devices, lasers are the most sensitive to structural perfection and residual stress. The 300K CW lifetime of GaAs/AlGaAs QW lasers on sawtooth SiO_2/Si (≈100hr) and on planar Si (≈10hr) [51] is still far below the values (≈10^6 hr) for homoepitaxial GaAs lasers, illustrating the still inferior materials quality. InGaAs/InGaAsP MQW lasers on InP/GaAs/Si seem to suffer less from the defect density and stress, and have exhibited 300K lifetimes of more than 2000 hours [52].

ELO is especially suited for opto-electronic integrated circuits. Also the ability to transplant films to any kind of flat and clean surfaces makes this technique extremely flexible. Very complicated heteroepitaxial growths like ceramic superconductors on Si or III-V can be avoided. The fact that commercially processed chips can be used with ELO is a major advantage and reduces the threshold for its commercial introduction. The flexibility in choice and design of the buffer layer between the III-V and Si is extremely advantageous in the ELO-case for some applications, such as high efficiency LED arrays on Si. These "III-V on Insulator" devices by ELO can outperform their homoepitaxial counterparts.

Typically the OEIC's have a low density of opto-electronic devices and the well established flip-chip technology is presently the primary choice [53,54,55]. Flip-Chip hybrid integration can easily meet today's speed requirements for OEICs and has the advantage of appropriate heat sinking. When dense arrays are envisaged for fast optical interconnects, bond-to-bond capacitance might come into play, and the thin-film approaches, ELO and heteroepitaxy, become more desirable. The boundaries of that cross-over are still to be defined. In addition, crucial alignment between light source and fiber is not straightforward to perform using flip-chip. Furthermore, solder bond reliability is an issue due to power/heat cycles that occur during operation.

It is clear that heteroepitaxy has not (yet) fulfilled its initial promises, leading to increased effort in thin film transfer technology. The ELO-field has progressed rapidly and reaches a high degree of sophistication where the boundaries between monolithic and hybrid integration are vague. While hybrid integration may be the first choice for many applications now, the advantages of a thin film monolithic integration technique will continue to stimulate research efforts.

Acknowledgements

C. Brys acknowledges the support of the Instituut tot Aanmoediging van het Wetenschappelijk onderzoek in Nijverheid en Landbouw.
Support by the ESPRIT BASIC RESEARCH ACTION 6625 is acknowledged

References

[1] R. Fisher, H. Morkoç, D.A. Neumann, H. Zabel, C. Choi, N. Otsuka, M. Longerbone and L.P. Erickson, J. Appl. Phys. 60 (1986) 1640
[2] D. W. Shaw, Mater. Res. Symp. Proc. Vol. 91 (Materials Research, Pittsburgh, PA, 1987) 15
[3] P. Demeester, A. Ackaert, G. Coudenys, I. Moerman, L. Buydens, I. Pollentier and P. Van Daele, Progress in Crystal Growth and Characterization (1991)
[4] M. Van Rossum, J. De Boeck, M. De Potter, G. Borghs in Nato ASI Applied Physics Series:" Novel Si based technologies", ed. R.A. Levy (Kluwer Academic Publishers, 1991) pp. 1 - 24
[5] A. Georgakilas, P. Panayotatos, J. Stoemenos, J.-L. Mourrain and A. Christou, J. Appl. Phys. 71 (1992) 2679
[6] H. Shichijo, R.J. Matyi, A.H. Taddiken, IEEE Electron Device Lett., EDL -9, (1988)444
[7] M. Tachikawa, M. Sugo, Y. Itoh, H. Mori, EMC 1993, Santa Barbara; Paper N4
[8] E. Fitzgerald, J.M. Kuo, Y.H. Xie, P.J. Silverman, EMC 1993, Santa Barbara; Paper N3
[9] M. Akiyama, Y. Kawarada, and K. Kaminishi, Jpn. J. Appl. Phys. 23 (1984) L843
[10] I. Markov, S. Stoyanov, Contemp. Phys., 28(1987)267
[11] See reports of O. Ueda *et. al.* ; R. Venkatasubramanian *et. al.* ; A. Freundlich *et. al.* in Mater. Res. Symp. Proc. Vol. 221, (Materials Research, Pittsburgh, PA, 1991) pp. 263 - 429
[12] R.D. Bringans, D.K. Biegelsen, F.A. Ponce, L.E. Schwartz, J.C. Tramontana, Mater. Res. Symp. Proc. Vol. 198 (Materials Research, Pittsburgh, PA, 1990) 195, and Appl. Phys. Lett. 61 (1992) 195
[13] J.F. Chen, A.Y. Cho, Journ. Electron. Mat., 22 (1993) 259
[14] C-H. Choi, R. Ai, S.A. Barnett, Phys. Rev. Lett. 67 (1991) 2826
[15] M. Yamaguchi, A. Yamamoto, M. Tachikawa, Y. Itoh and M. Sugo, Appl. Phys. Lett. 53 (1989) 24 and M. Yamaguchi, J. Mater. Res. 6 (1991) 382
[16] J. De Boeck, PhD Dissertation, Catholic University of Leuven - IMEC (1991)
[17] W. Dobbelaere, PhD Dissertation, Catholic Universtiy of Leuven-IMEC (1992)
(18) G.H. Olsen, A.M. Joshi, and V.S.Ban, Proc. SPIE,1540 (1991) 596
(19) K. Inoue, J. C. Harmand, and T. Matsuno, J. Cryst. Growth 111 (1991)313
(20) F. Schäffler, D. Többen, H.-J. Herzog, G. Abstreiter, and B. Holländer 7(1992)260
(21) W.S. Liu, M.A. Nicolet, V. Arbet, T. Carns, K.L. Wang,EMC 1993, Santa Barbara; Paper N2
(22) Y. Okada, H. Shimomura, and M. Kawabe, EMC 1993, Santa Barbara; paper N1
[23] D. Pribat, B. Gerard, M. Dupuy, P. Legagneux, Appl. Phys. Lett. 60 (1992) 2144
[24] J. De Boeck, J. Alay, J. Vanhellemont, B. Brijs, W. Vandervorst, G. Borghs, M. Blondeel, C. Vinckier, Mater. Res. Symp. Proc. Vol. 221, (Materials Research Society, Pittsburgh, PA, 1991) 411
[25] K. Ismail, F. Legoues, N.H. Karam, J. Carter, H.I. Smith, Appl. Phys. Lett. 59, 2418 (1991) and N.H. Karam et. at. Mater. Res. Symp. Proc. Vol. 221, (Materials Research Society, Pittsburgh, PA, 1991) 399
[26] M. Tashikawa and H. Mori, Appl. Phys. Lett. 56 (1990) 2225
[27] J. De Boeck, C. Van Hoof, K. Deneffe, G. Borghs , Jpn. J. Appl. Phys. 30 (1991) L423
[28] J. De Boeck , C. Van Hoof, K. Deneffe, R.P. Mertens, G. Borghs, Appl. Phys. Lett. 10 (1991) 1179
[29] J. De Boeck, G. Zou, M. Van Rossum, G. Borghs, Electr. Lett. 27 (1991) 22
[30] G. Burns and C. Fonstad, IEEE Phot. Techn. Lett. 4 (1992) 18
[31] S. Sakai, K. Kawasaki, N. Wada, Jpn. J. Appl. Phys. 29 (1990) L853 and S. Sakai et. al. Jpn. J. Appl. Phys. 29 (1990) 2077
[32] N. Wada, S. Yoshimi, S. Sakai, C.L. Shao, M. Fukui, Jpn. J. Appl. Phys. 31 (1992) L78
[33] S. Sakai, C.L. Shao, N. Wada, T. Yuasa, M. Umeno,App/ Phys. Lett. 60 (1992) 1480

[34] N. Wada, K. Iwabu, S. Sakai, M. Fukui, Appl. Phys. Lett. 60 (1992) 1354
[35] E. Yablonovitch, T. Gmitter, J.P. Harbison, R. Bhat, Appl. Phys. Lett. 51 (1987) 2222
[36] I. Pollentier, P. Demeester, P. Van Daele, Proc. of EFOC/LAN '92 Paris, France (1992), in press.
[37] R.W. McClelland, C.O. Bozler, J.C.C. Fan, Appl. Phys. Lett. 37 (1980) 560
[38] E. Yablonovitch, K. Kash, T.J. Gmitter, L.T. Florez, J.P. Harbison, and E. Colas, Electron. Lett. 25 (1989) 171
[39] R. Venkatasubramanian, M.L. Timmons, T.P. Humphreys, B.M. Keyes, R.K. Ahrenkiel, Appl. Phys. Lett. 60 (1992) 886
[40] E. Yablonovitch, T. Sands, D.M. Hwang, I. Schnitzer, T.J. Gmitter, S.K. Shastry, D.S. Hill, J.C.C. Fan, Appl. Phys. Lett. 59 (1991) 3159
[41] I. Pollentier, L. Buydens, P. Van Daele, and P. Demeester, IEEE Phot.Techn. Lett. 3 (1991) 115
[42] I. Pollentier, P. Demeester, P. Van Daele, unpublished
[43] M. Renaud, I. Pollentier, J.F. Vinchant, P. Demeester, J.A. Cavailles, P. Van Daele, M. Erman, Proc. ECOC/IOOC '91, Paris (1991)
[44] C. Camperi-Ginistet, M. Hargis, N. Jokerst, M. Allen IEEE Phot. Techn. Lett. 3 (1991) 1123
[45] D.Y. Shah PhD Dissertation New Jersey Institute of Technology - Bellcore (1992)
[46] I. Pollentier, A. Ackaert, P. De Dobbelaere, L. Buydens, P. Demeester, P. Van Daele, Proc. SPIE'90 (Optoelectronic Materials and Device Concepts), Aachen, 1990.
[47] I. Schnitzer, E. Yablonovitch, C. Caneau, T.J. Gmitter, and A. Scherer, submitted to Appl. Phys. Lett.
[48] M. Yanagisawa, H. Terui, K. Shuto, T. Miya, M. Kobayashi, IEEE. Phot. Techn. Lett. 4 (1992) 21.
[49] W. Dobbelaere, J. De Boeck, P. Heremans, R. Mertens, G. Borghs, W. Luyten, and J. Van Landuyt, Appl. Phys. Lett. 60 (1992) 3256
[50] R.B. Bailey, L.J. Koslowski, J. Chen, D.Q. Bui, K. Vural, D.D. Edwal, R.V. Gil, A.B. Vanderwyck, E.R. Gertner, M.B. Gubala, IEEE Trans. Electron. Dev, 38 (1991) 1104
[51] D.C. Hall, N. Holonyak, Jr. D.G. Deppe, M.J. Ries, R.J. Matyi, H. Shichijo, J.E. Epler, J. Appl. Phys. 69 (1991) 6844
[52] M. Sugo, H. Mori, Y. Sakai, Y. Itoh, Appl. Phys. Lett. 60 (1992) 472
[53] O. Wada, M. Makiuchi, H. Hamaguchi, T. Kumai, T. Mikawa, IEEE. J. Lightw.Techn.9 (1991) 1200
[54] A. von Lehmen, T.C. Bandwell, R. Cordell, C. Chang-Hasnain, J.W. Mann, J.P. Harbison, L. Florez, Electron. Lett. 27 (1991) 1189
[55] A.J. Mosseley, M.Q. Kearley, R.C. Morris, J. Urquhart, M.J. Goodwin, G. Harris Electron. Lett. 28 (1992) 12

Inst. Phys. Conf. Ser. No 136: Chapter 7
Paper presented at the Int. Symp. GaAs and Related Compounds, Freiburg, 1993

A manufacturable process for HBT circuits

T Lester, R K Surridge, S Eicher, J Hu, G Este, H Nentwich, B MacLaurin, D Kelly, I Jones

Bell Northern Research, P.O.Box 3511, Station C, Ottawa, Ontario, Canada K1Y 4H7
Tel: (613)763 3460, Fax: (613) 763 2404

ABSTRACT: A novel process for the fabrication of high frequency AlGaAs HBT's for future high speed transmission systems has been developed. Particular attention has been paid to circuit manufacturability. All epitaxial material was grown in house by MBE, MOCVD or CBE. Analytical, Monte-Carlo and two dimensional simulation tools were used to optimize the layer structure. Typical devices fabricated on CBE layers gave f_T and f_{max} $\approx$ 80 GHz and β_{max} >100. Details of the process, and examples of modeled and measured device performance will be given.

1. INTRODUCTION

A process for the fabrication of HBT's with cut-off frequencies in excess of 60 GHz has been developed. The primary requirements for the process were reproducibility and reliability. Whilst the main application is for logic circuits with operational speeds in excess of 10 GB/s, analog circuits have also been fabricated with excellent results.

The process is based on the use of a dummy dielectric emitter which is self aligned to the base metal that completely surrounds it. A thick sidewall formed on the dummy emitter provides ample alignment tolerance for the emitter ohmic metal. The base-emitter metal separation is precisely controlled by the magnitude of the lateral etching during the formation of the dummy emitter. To minimize base-collector capacitance, the area beneath the majority of the base metal is rendered isolating by a deep He^+ implant. A simplified diagram of the basic device layout is shown in Figure 1. Circular and rectangular emitters, as well as multiple emitter structures have been fabricated. The minimum allowed emitter size is 1.5x1.5 μm.

An important feature of the process is a selective dry etch to the AlGaAs emitter followed by the deposition of the base metal which is then alloyed through to the base. This facilitates contacting thin graded base layers. The AlGaAs layer between the edge of the alloyed base contact and the edge of the emitter mesa is left in place in order to reduce the surface recombination and increase the current gain of the devices (Malik et al 1989). This method of providing an emitter shelf makes the final device susceptible to high base-emitter leakage if the AlGaAs shelf is not fully depleted. Particular care must be paid to the design of the AlGaAs layer, both in terms of its absolute doping level and thickness in order to avoid excess leakage currents. Alternatively, the shunt current can be decreased by partially etching the shelf, or by a shallow isolation implant.

2. PROCESSING

The process sequence for the fabrication of HBT circuits is outlined in Figure 2. Definition of the device active regions by implant isolation occurs first in the process. A resist mask is used to protect the emitter and collector contact areas from 190keV He^+ ions implanted to a dose of $1.3x10^{15}$ cm^{-2}. Maximum isolation is achieved by annealing at 600°C for 20s. 800 nm of PECVD Si_3N_4, patterned by dry etching in CF_4/O_2 with a photoresist mask, is used to form the dummy emitters (Figure 2a). The photoresist later serves as the lift-off mask for the self aligned p-ohmic metal which completely surrounds the dummy emitter. The Si_3N_4 etch is optically end-pointed and a controlled over-etch makes it possible to achieve reproducible base-emitter separations from 0.2 to 0.5μm. A slightly re-entrant profile for the dummy emitter is obtained by ramping the Si_3N_4 deposition temperature from 240 to 270°C. An $Ar/Cl_2/CF_4$ RIE etch is used to selectively etch off the GaAs emitter cap layer prior to p-ohmic metal deposition (Figure 2b). This etch is anisotropic and the etched hole is almost vertically aligned with the resist edge. Attempts to etch laterally to the foot of the Si_3N_4 in the same process step resulted in uncontrolled undercutting of the dummy emitters and was abandoned in favor of a two step etch process. The second etch is carried out after the p-metal deposition.

In order to obtain a good ohmic contact the p-metal must be alloyed through the AlGaAs to the p^+ base. For reliability reasons it is very important that the metal penetration be self limiting and uniform, with no tendency to spike through to the collector layer. The contact metallization employed in this process consists of Pd-Zn-Pt-Au-Pd and gives good contact resistance (0.2Ω-mm), good morphology, and uniform penetration of the AlGaAs (Figure 3). The penetration depth during alloying can be controlled by the thickness of Pd below the Pt barrier (typically 20 nm). A very thin (2 nm) Zn layer is used in order to limit the extent of the Zn diffusion.

The field Si_3N_4 is removed by masking the emitter regions and dry-etching in CF_4/O_2. After stripping the photoresist, the highly doped GaAs cap layer in the gap between the base contact and the emitter edge, as well as in the field areas, is etched off with a Cl_2/O_2 RIE etch (Figure 2c). This is the second step in the emitter definition and assures that the correct emitter size is achieved. The remaining AlGaAs reduces the surface recombination and, if necessary, may be thinned at this stage by wet chemical etching.

Next a thick SiON sidewall (~1μm) is formed on the dummy emitters. The sidewall protects the gap between emitter and base from the subsequent BCl_3/Cl_2 RIE etch to the sub-collector, and provides an alignment tolerance for the emitter metallization (Figure 2d). The Si_3N_4 dummy emitters are selectively removed by dry-etching in CF_4/O_2. The etch conditions have been optimized to give a Si_3N_4:SiON etch selectivity greater than 5:1. The Ni-Ge-Au emitter and collector contacts are then deposited by e-beam evaporation and bi-layer resist lift-off (Figure 2e).

The back end of the process (Figure 2f) starts with SiON dielectric deposition and via etching followed by the deposition and lift-off of 50Ω NiCr resistors and the first level of Au interconnect. In order to provide high aspect ratio interconnects between first and second level metals, 2.5μm high Au posts are defined by lift-off . The interconnect and post metallizations include a Ti-Pt barrier layer to improve the reliability of the ohmic contacts. A second SiON layer is deposited in order to passivate the NiCr resistors and provide good adhesion for a planarizing layer of benzocyclobutene (BCB), which has been chosen for it's excellent dielectric and planarizing properties. After etching back the BCB using CF_4/Ar the tops of the posts are exposed by dry etching in CF_4/O_2. The interconnect is completed by lifting off a 1.2μm thick layer of Ti-Au. The process is completed by depositing a final passivation layer of PECVD SiON, and exposing the bond pads. A SEM photograph of a cross section of a completed 2x2 μm emitter HBT is shown in Figure 3.

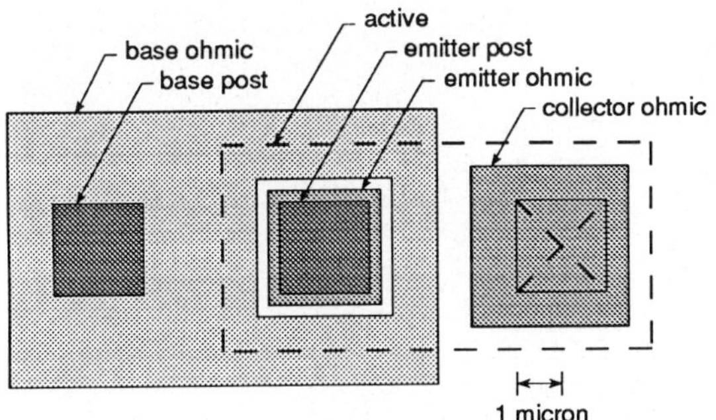

Figure 1. Layout of a 2x2 μm emitter HBT, excluding the interconnect.

Table 1. Standard, fully graded HBT structure for CBE growth.

Layer	% Al	doping (cm^{-3})	thickness (nm)
cap	0	5e18 Si	200
emitter	30-0	5e17 Si	20
emitter	30	5e17 Si	30
emitter	5-30	5e17 Si	15
base	0-5	5e19 C	60
collector	0	5e16 Si	350
sub-coll.	0	4e18 Si	350

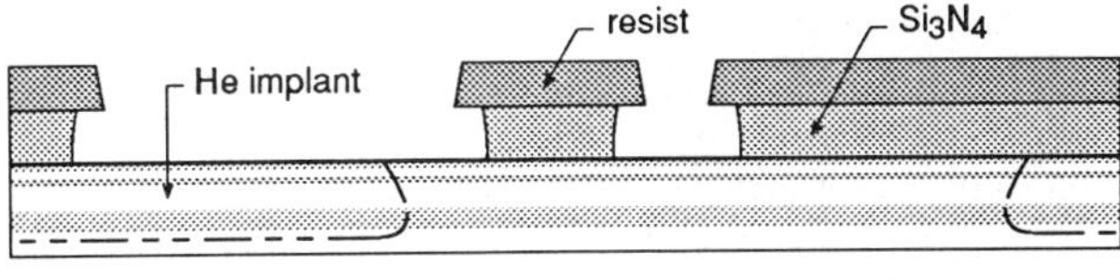

(a) He isolation implant and dummy emitter formation.

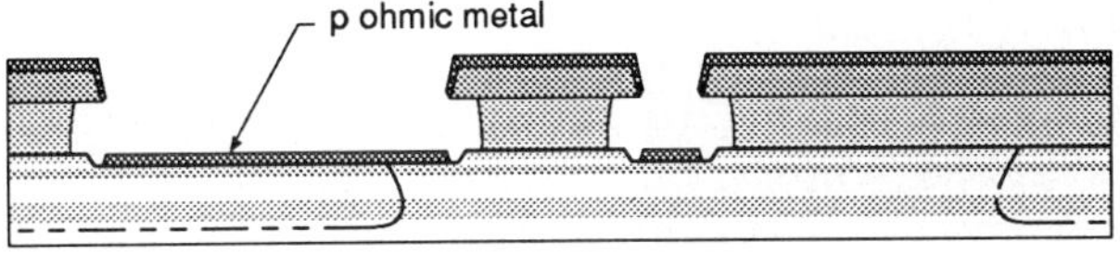

(b) Selective dry etch to AlGaAs emitter and p-ohmic metal deposition.

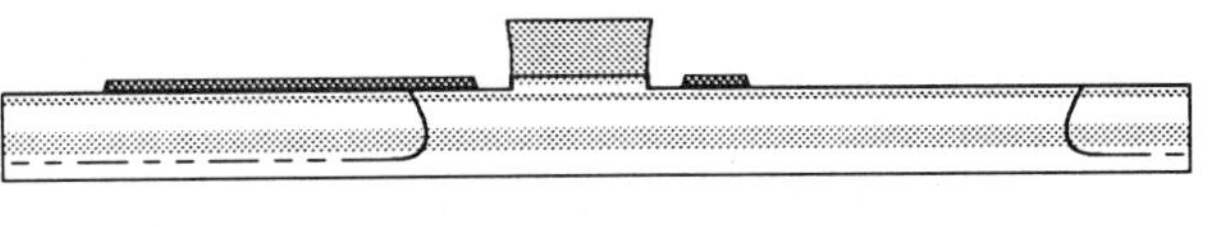

(c) Field Si_3N_4 removal and second etch to AlGaAs emitter.

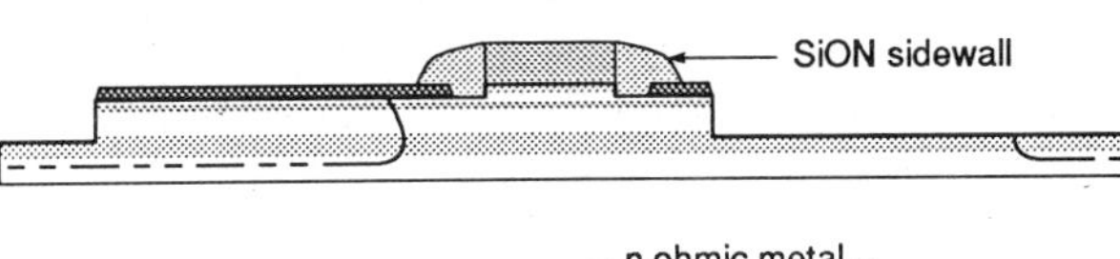

(d) Thick sidewall deposition and RIE etch to sub collector.

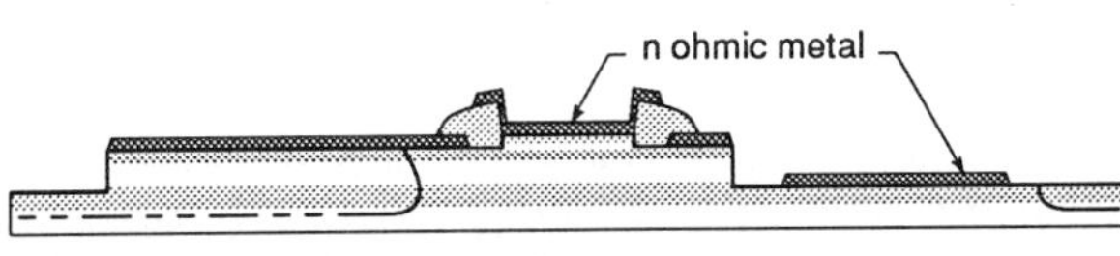

(e) Selective removal of dummy emitter and n-ohmic metal deposition.

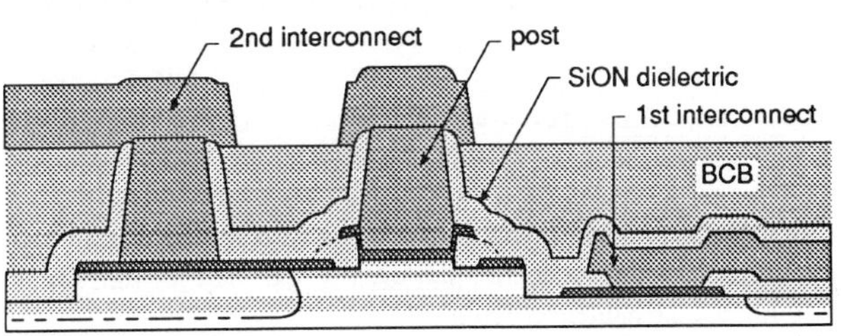

(f) Back end processing:
- SiON deposition and via etch
- NiCr resistor deposition
- ME1 deposition.
- ME1 to ME2 post deposition
- BCB planarization and ME2 deposition

Figure 2. HBT process sequence.

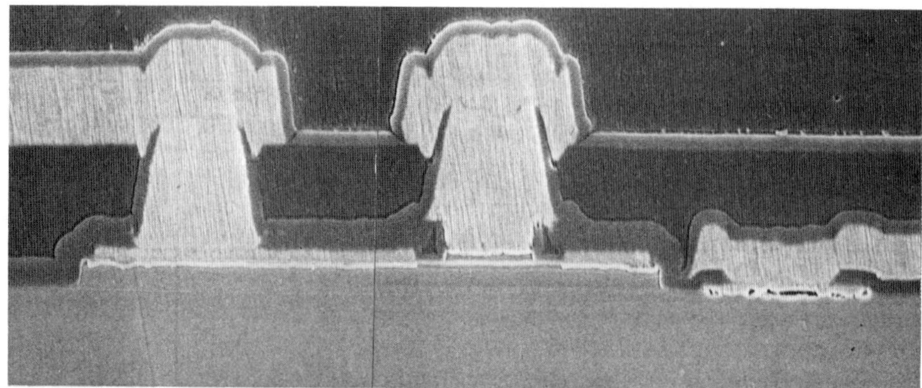

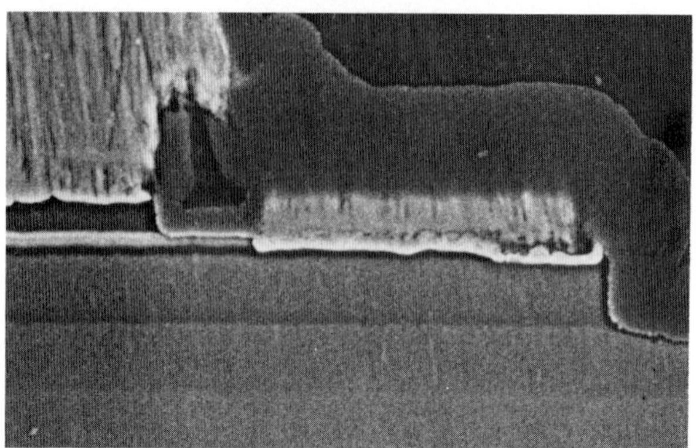

Figure 3. SEM photo of a cross section of a completed 2x2 μm emitter HBT, including a magnified view of the emitter shelf. For identification of individual features refer to Figure 2.

HBT's with cut-off frequencies in excess of 60 GHz have been fabricated from MBE, MOCVD and CBE material. C was the base layer dopant in the CBE and MOCVD material, while Be was used for MBE. The n-type dopant for all three growth methods was Si. Compositional grading of the base and emitter layers, as well as abrupt ungraded structures have been studied. Because of the limitations of space we will restrict our discussion to results obtained with CBE material which generally gave devices with the highest current gain. The standard graded structure grown by CBE (Moore et al 1993) is shown in Table 1.

3. DEVICE SIMULATION

Three separate simulation tools have been used to optimize the HBT structure. A one dimensional analytical program developed at University of British Columbia (Ho and Pulfrey 1989) is used for rapid simulation of DC and RF performance, and has been shown to give good agreement with measured device characteristics. Detailed optimization of the emitter, base and collector layers is accomplished using both a Monte-Carlo program developed at the University of Michigan (Hu et al 1989), and a full two-dimensional simulation program developed at the University of Toronto (Zhang et al 1990). The two-dimensional simulations have been particularly useful for investigating the effects of the emitter shelf parameters on the magnitude of the base-emitter leakage current. Low leakage requires that the AlGaAs shelf region be fully depleted. In 30% AlGaAs the Si donor level is not fully ionized and the atomic concentration is approximately three times the active (Hall) concentration usually specified for the emitter layer. The total dopant concentration determines the depletion depth.

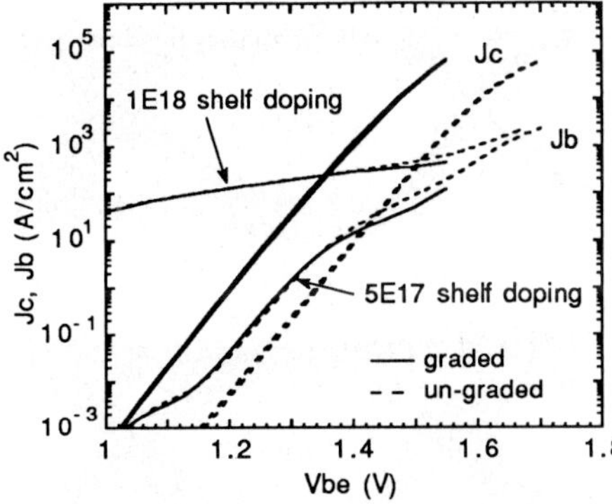

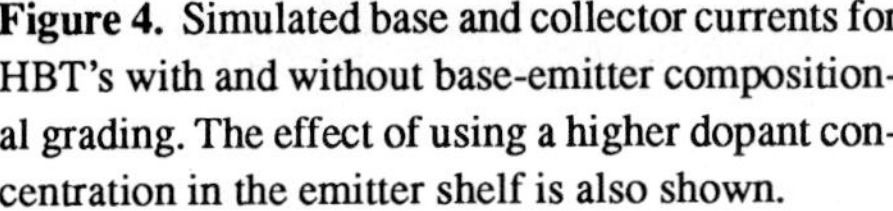

Figure 4. Simulated base and collector currents for HBT's with and without base-emitter compositional grading. The effect of using a higher dopant concentration in the emitter shelf is also shown.

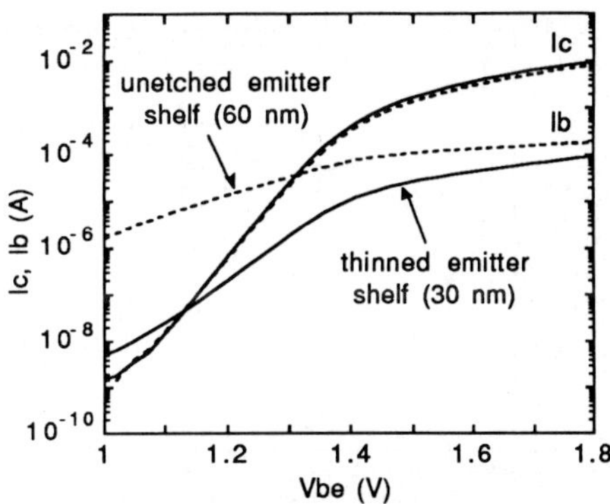

Figure 5. Measured base and collector currents for two 3 μm dia. HBT's, showing the effect of thinning the emitter shelf. The epi-layer structure was similar to that indicated in Table 1.

In Figure 4 the predicted dependence of base current upon V_{be} is compared for shelf doping levels of 5E17 and 1E18 cm^{-3}. In all cases the the doping in the intrinsic emitter region was 5E17, the shelf thickness was 30 nm and the base-emitter spacing was 0.5 μm. The p-ohmic contact was modeled by specifying a p^{++} diffusion through to the base (under the metal) and a region of infinite recombination velocity along the edge of the base ohmic metal. It can be seen that for the higher doped shelf unacceptable leakage currents are predicted. Similarly the predicted leakage current has been shown to depend on shelf thickness as well as base-emitter spacing (Lee et al 1989, Liu and Harris 1992). Also shown in Figure 4 is the dependance of collector current on bias for devices with and without compositional grading at the heterojunction. The higher turn-on voltage for the collector current in the non-graded case results in a lower beta.

4. ELECTRICAL RESULTS

Figure 5 compares the Gummel plots of two 3 μm dia. emitter HBT's from two halves of the same wafer (with a layer structure similar to that in Table 1). One of the devices displays a high base-emitter leakage current, due to conduction through the 60 nm thick emitter shelf, whereas the other exhibits good gain over a large current range. The latter device came from the half of the wafer where the emitter shelf had been thinned to 30 nm by wet chemical etching. The reduction in leakage current is consistent with the results of two-dimensional simulations as discussed above. Low base-emitter shunt currents have also been obtained from material with thinner as grown emitters (30 nm) without any sacrifice of maximum beta. In contrast, we have observed a reduction in beta for devices processed on material without base-emitter compositional grading, due to the increase in collector current turn on voltage.

In Figure 6 we show the effect of device geometry on beta and f_T for HBT's from the half of the wafer with the 30 nm AlGaAs shelf. The results shown were obtained for a variety of square, rectangular and circular emitters with both single and multiple emitters. It can be seen that beta is relatively insensitive to emitter dimensions down to 1.5 μm (highest perimeter/area ratio), showing the effectiveness of the emitter shelf in reducing surface recombination. The maximum beta observed for these devices was 115. Current gains over 200 have been achieved from devices with slightly modified emitters. The dependence of f_T upon emitter size displays the expected decrease for very small emitters due to the increased effect of parasitics, and the expected decrease for larger emitter dimensions. Associated f_{max} values exceeded 60 GHz for devices with emitter dimensions between 2 and 6 μm, with a maximum measured f_{max} of about 100 GHz for devices with 3μm diameter emitters.

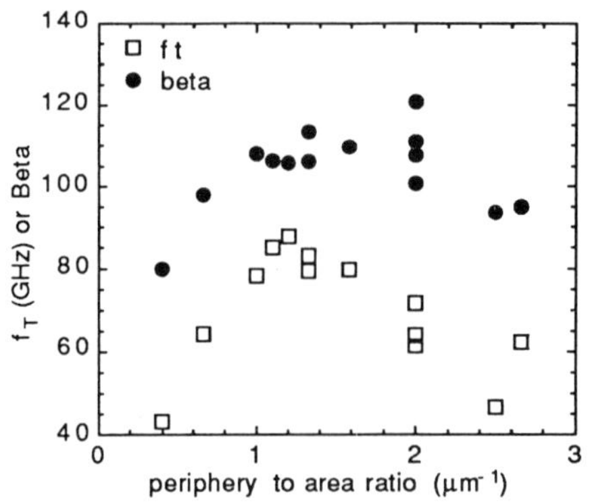

Figure 6. Measured f_T and maximum current gain for single and multiple emitter HBT's.
emitter sizes (µm): 10x10 6x6 4x4 20x2 (2)10x2 3x3 3 dia. 2.5 dia. 2x2 (10)2x2 2 dia. 1.5x1.5

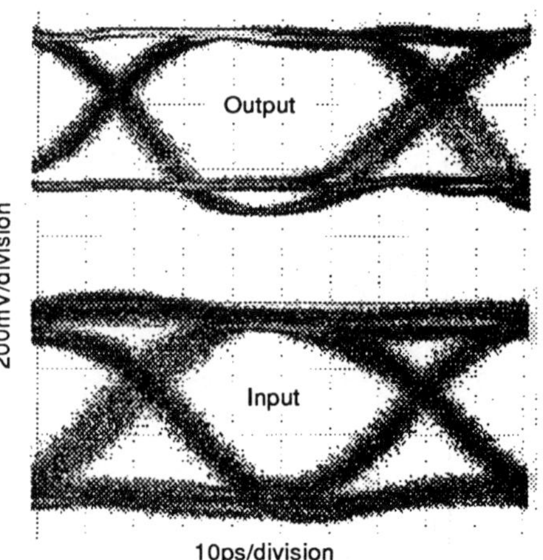

Figure 7. Response of DEMUX/MUX to pseudorandom pattern at 16.4 Gb/s.

5. CIRCUIT PERFORMANCE

A wide range of digital and analog circuits have been designed, built and tested using this technology. A standard pad-out with predefined positions for power, ground and low and high speed signals has allowed high speed testing up to 10 Gb/s using specially designed probe cards incorporating coaxial microwave probes. Among the circuits studied were decision circuits, multiplexers, demultiplexers, frequency dividers, pre-amplifiers and driver amplifiers. All circuits have been shown to perform similar to their design specification. As an example of circuit performance, the eye-diagram shown in Figure 7 was obtained from the output of a demultiplexer-multiplexer pair operating error free at 16 Gb/s. Yields of 83% over 3 inch wafers have been achieved for this circuit. Programmable divide by 4/5 circuits have operated up to 12 Gb/s, while ripple divide by 32 circuits have operated up to 18 GHz.

6. ACKNOWLEDGMENTS

The work described in this paper would not have been possible without the active collaboration of a large number of people, whose assistance we gratefully acknowledge. A. J. SpringThorpe, T. Moore, G. Hillier, M. Majeed and P. Mandeville who grew the epitaxial material; C. Miner, R. Streater, I. Bassignana and D. Macquistan who performed much of the analysis; and J. Sitch, S. Wang, T. Wong, P. Tuok and A. Harrison who performed most of the circuit design and mask layout; as well as our university collaborators J. Xu, Q. M. Zhang, D. Pavlidis and D. Pulfrey.

REFERENCES

Ho S and Pulfrey D 1989 IEEE Trans. El. Dev. 36 2173

Hu J, Tomizawa K and Pavlidis D 1989 IEEE Trans. El. Dev. 36 2138

Lee WS, Ueda D, Ma T, Pao YC and Harris J S 1989 IEEE Electron Device Lett. 10 200

Liu W and Harris J S 1992 Jpn. J. Appl. Phys. 31 2349

Malik R J, Lunardi L M, Ryan R W, Shunk S C and Feuer M D 1989 Electron. Lett. 25 1175

Moore W T, SpringThorpe A J, Lester T P, Eicher S, Surridge R K, Hu J and Miner C J 1993 J. Crystal Growth (to be published)

Zhang QM, Tan G L and Xu J M 1990 User Manual of GPSDA, University of Toronto

Improved *n* and *p* contacts in InP/InGaAs junction field-effect transistors and pin photodiodes for optoelectronic integration

Ch. Lauterbach, D. Römer, L. Hoffmann, J. W. Walter and J. Müller

Siemens Research Laboratories, Otto-Hahn-Ring 6, 81739 München, Federal Republic of Germany

ABSTRACT: Contact resistances of an InP/InGaAs junction field-effect transistor and a pin photodiode have been improved by local n doping via Si ion implantation and implementing an InGaAs p contact layer. AuGe/Au contacts have a contact resistivity R_t of 0.015 Ωmm on $1{*}10^{18}cm^{-3}$ n-doped InP after annealing at 400°C. The Ti/Au metallization shows an ohmic behaviour (R_t=0.09 Ωmm) on $2{*}10^{19}$ cm^{-3} p-doped InGaAs. JFETs with improved contacts show a significant reduction in power dissipation and in the spectral noise density at frequencies below 100 MHz.

1. INTRODUCTION

InP based monolithic optoelectronic integrated circuits (OEICs) offer the potential of small size and high performance components for high-speed and low-cost receivers in the 1300-1600nm spectral range (Uchida et al 1991). To compete with hybrid receivers, the characteristics of the integrated devices have to be optimized. Low-resistance ohmic contacts are essential for avoiding a deterioration of the device parameters. In this work the fabrication of low-ohmic n and p contacts and their influence on the contact resistance of a junction field-effect transistor (JFET) and a pin photodiode (PD) is described. The n contacts of the devices have been improved by increasing the doping concentration of the semiconductor via local Si ion implantation and alloying the n contact material. For the fabrication of ohmic p contacts, an InGaAs contact layer has been implemented.

2. DEVICE STRUCTURE AND FABRICATION

The InP/InGaAs layer sequence is grown lattice-matched in a single step metal-organic vapour-phase epitaxy (MOVPE) on a flat semi-insulating InP:Fe substrate and is designed to meet the demands of the PD. The first epitaxial layer is an 0.3μm thick n^+-doped InP buffer layer to avoid limitation of the transient response of the PD, followed by an 0.3μm thick undoped InP spacer. The 2μm thick undoped InGaAs layer is used as an absorption layer for the PD. This is followed by an 0.2μm InP window layer, an 0.1μm InGaAs contact layer and an 0.1μm InP cap layer for providing a surface passivation. The top-three layers have a moderate n doping of $2{*}10^{16}cm^{-3}$. The JFET is fabricated within this layer structure by local Si ion implantation for the channel doping (Bauer et al 1992) and Be ion implantation for the buried p layer (Lauterbach et al 1990). A schematic cross-section of the JFET and the PD structure is shown in Figure 1.

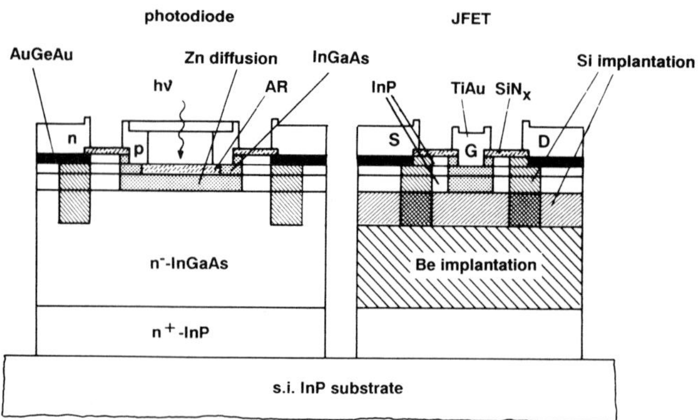

Fig. 1 Schematic cross section of a pin photodiode and an ion-implanted JFET in an InP/InGaAs layer structure with n contact implantation and InGaAs p contact layer

For improving the n contact and series resistance, a four-energy Si implantation (30,70, 150 and 400keV at doses of $2.5*10^{12}$, $6.0*10^{12}$, $1.5*10^{13}$ and $4.4*10^{13}$ cm^{-2}, respectively) was chosen in order to achieve a flat doping profile with an electron concentration of $1*10^{18}cm^{-3}$ within the top-three layers beneath the n contacts. For removing the implantation damage and activating the implanted dopants, the wafer received a thermal capless annealing at 700 ^{o}C for 30 s in a PH_3/H_2 atmosphere. The implantation and annealing conditions were chosen to minimize the degradation of the heterojunction and therefore avoid high leakage currents (Römer et al 1992). The p region of the PD and the gate of the JFET were formed by Zn diffusion, using a SiO_2 spin-on film as a diffusion source and SiN_x as a mask. Two different n contact materials AuGe(12% Ge)/Au (100nm/100nm) and AuGe/NiCr(50%Ni)/Au (100nm/50nm/100nm), deposited by e-gun evaporation and delineated by a lift-off technique, were investigated. Prior to the deposition of the Ti/Au (60nm/1.2µm) metallization for the p contacts, the InP cap layer was selectively removed by HCl to reveal the InGaAs contact layer. Finally the devices were isolated by mesa etching down to the semi-insulating substrate. Simultaneously to the devices, transmission line model (TLM) test structures for the characterization of the contact resistivity (Berger 1972) were fabricated, with a pad size of 50µm*50µm and distances between the pads of 2, 5, 10, 20, 50 and 100µm.

3. DOPING AND CONTACT CHARACTERIZATION

The electron profile of the contact implantation in Figure 2 was measured by electrochemical capacitance-voltage (EC-V) profiling. A constant doping level N_D of $1*10^{18}$ cm^{-3} has been obtained within an overall thickness of 0.7µm, including the top-three layers and the channel layer of the JFET. Using Gaussian distribution to describe the electron profiles, the electrical activation is calculated to 70% in InP and 80% in InGaAs.

The n contacts were characterized directly after the deposition, and after alloying at temperatures in the range from 360°C to 460°C for 10s in a hydrogen atmosphere. The dependence of the contact resistivity R_t on the alloy temperature is shown in Figure 3. In the as-deposited case both contact compositions show a non-ohmic current-voltage characteristic with an R_t of approximately 0.4Ωmm. Ohmic behaviour is achieved at an alloy temperature of approximately 300°C. After alloying at 400°C the contact resistivity of the AuGe/Au contact is 0.015Ωmm, which corresponds to an specific contact resistance of $1*10^{-7}\Omega cm^2$. It decreases to 0.003 Ωmm at an alloy temperature of 460°C. The AuGe/NiCr/Au has a higher contact resistivity with R_t = 0.1Ωmm at 400°C and 0.013Ωmm after alloying at 460°C. It should be noted, that at alloy temperatures of more than 440°C gold spikes were observed within the InP window layer, which may influence the long-term stability of the devices. In contrast to AuGe/Au, the AuGe/NiCr/Au contact has a smoother surface after the heat treatment and does not draw together over oxide layers, which is an important aspect in integrated circuits.

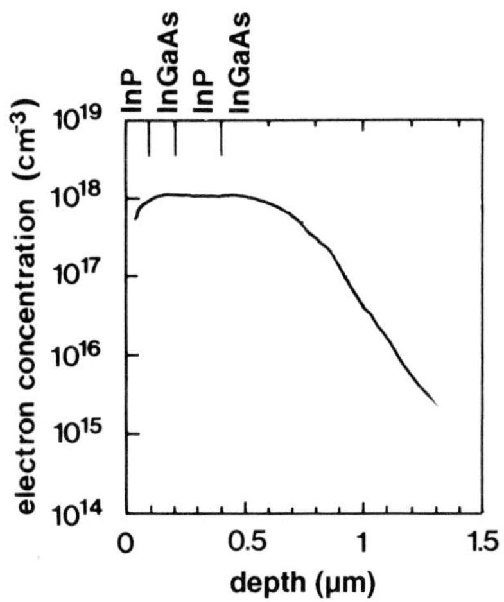

Fig. 2 Electron profile of the Si implanted InP/InGaAs layer sequence after an annealing at 700°C for 30s

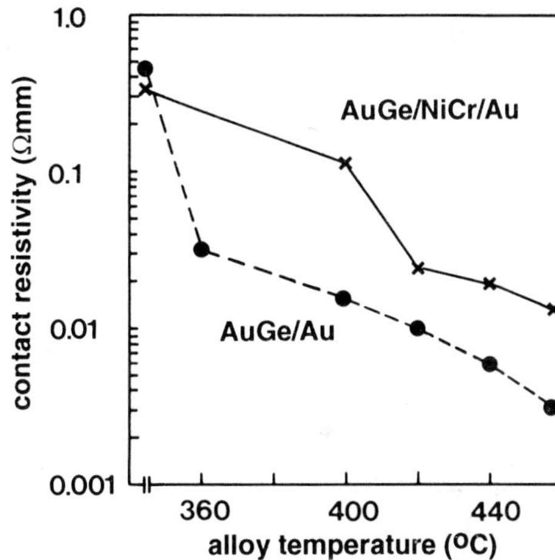

Fig. 3 Contact resistivity of AuGe/Au and AuGe/NiCr/Au contacts on n InP with $N_D=1*10^{18}cm^{-3}$ for different alloying temperatures. The alloying time was 10s.

For the fabrication of an ohmic p contact an InGaAs contact layer has been implemented, which is superior to an InP contact layer due to its lower bandgap, higher segregation coefficient of Zn and its high activation of nearly 100%. For the fabrication of the p region of the PD and the gate of the JFET, the Zn diffusion should stop at the interface between the InP window layer and the InGaAs absorption layer. The suitable Zn diffusion was found to be at a temperature of 550°C for 40s. It was performed on a thermal resistance heated metal stripe under a hydrogen atmosphere.

The Zn distribution within the epitaxial layers was characterized by secondary-ion mass spectrometry (SIMS). The resulting atomic Zn distribution is shown in Figure 4 (solid line). From a Zn concentration of $1*10^{19}cm^{-3}$ at the surface of the InP cap layer, the Zn level drops to $7*10^{18}cm^{-3}$ within the InP and increases to $2*10^{19}cm^{-3}$ in the InGaAs contact layer due to its higher segregation coefficient. Within the InP window layer the Zn concentration decreases to $3*10^{18}cm^{-3}$. The peak in the SIMS profile at a depth of 0.4μm indicates, that the Zn diffusion reaches slightly into the InGaAs absorption layer. The dashed line in Figure 4 shows the according EC-V profile of the hole concentration. Within the InGaAs contact layer a hole concentration of $2*10^{19}cm^{-3}$ is achieved, showing that the Zn is fully activated. In the InP window layer the hole concentration drops to $2*10^{17}cm^{-3}$, which is a typical value for InP diffused under the given conditions.

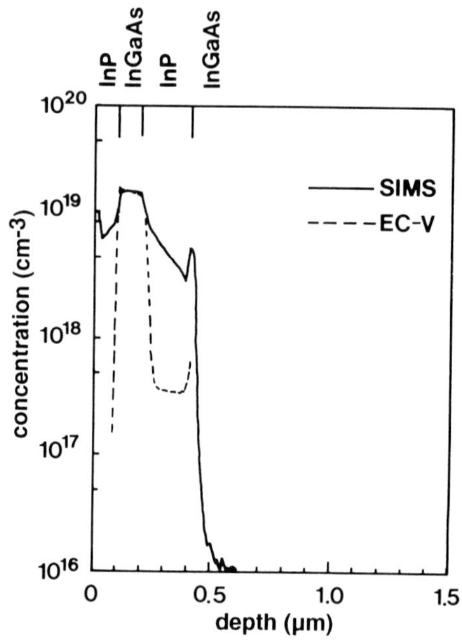

Fig. 4. Comparison of the SIMS depth profile of the Zn distribution and the EC-V profile of the hole concentration. The Zn diffusion was performed at 550°C for 40s.

For the comparison of the quality of the p contact on InP and InGaAs, in a part of the sample the InP cap layer was not removed prior to the Ti/Au deposition assuring, that all changes in device parameters are solely due to the p contact. The Ti/Au metallization on InGaAs shows an ohmic behaviour. As expected due to the low bandgap of InGaAs, the high doping and the usage of an undoped contact material, the dependence of the anneal temperature on the contact resistivity is only weak. It decreases from 0.45Ωmm for the as deposited Ti/Au contact to 0.14Ωmm after annealing at 400°C for 10s. This value corresponds to a specific contact resistance of $6*10^{-7}\Omega cm^2$. The Ti/Au contact on p InP has a non-ohmic characteristic and therefore is not evaluable by TLM measurements. From the forward current of the Schottky contact ($50*50\mu m^2$) a series resistance of approximately 40Ω can be estimated for the as deposited contacts and 15Ω after annealing at 440°C for 10s. In a recent work Malina et al.(1993) have shown by Auger analysis of Ti/Au metallization on InGaAsP, that there is only a limited reaction of the Ti/Au with the semiconductor interface and no detectable indiffusion of the contact material occurs up to an anneal temperature of 400°C, which is an important fact for avoiding degradation in shallow pn junctions. In addition we found, that the adhesion of the Ti/Au contact deteriorates above 400°C.

4. DEVICE CHARACTERIZATION

The impact of contact implantation on the output characteristics of a JFET with a gate length of 1.5 μm and a gate width of 50 μm is shown in Figure 5. With an alloying of the n contacts, the external transconductance increases from 64 to 140mS/mm. In contrast to improving the internal transconductance by e.g. higher channel doping, the contact implantation leads to a higher transconductance without increasing the gate-source capacity, resulting in a higher cut-off frequency and a reduced equivalent input noise current. Furthermore the lower parasitic resistance within the drain-source region enables the operation of the JFET at lower drain-source voltages, leading to a reduction of the power dissipation of the device.

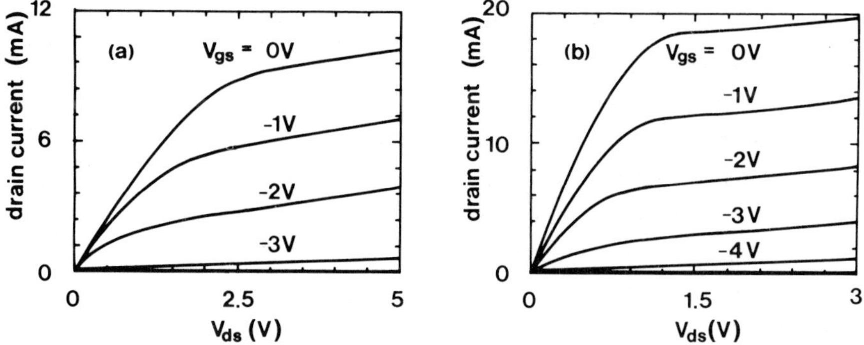

Fig. 5. Output characteristics of InP/InGaAs JFETs: (a) with as grown doping (N_D=2*10^{16}cm^{-3}) and (b) with additional high doping (N_D=1*10^{18}cm^{-3}) by Si ion implantation beneath the n contacts

The threshold voltage of JFETs with an InGaAs p contact is reduced from -3.5 V to -2.2V due to the lower gate contact resistance. The transit frequency increases from 7.5 GHz to 10 GHz. The impact of the p contact on the spectral noise density (non-calibrated) is shown in Figure 6. With an InGaAs contact layer the excess increase of the spectral noise density in the frequency range below 100 MHz is significantly reduced. It is supposed, that the excess noise in case of the InP contact results from the influence of the admittance of the non-ohmic p contact on the intrinsic gate voltage. This p contact can be described as an additional circuit to the pn junction, consisting of a resistance R_c in parallel with a capacity C_c. At low frequencies R_c forms an additional resistance within the gate-circuit of the JFET, leading to a gate noise voltage, that originates from the gate shot noise current.

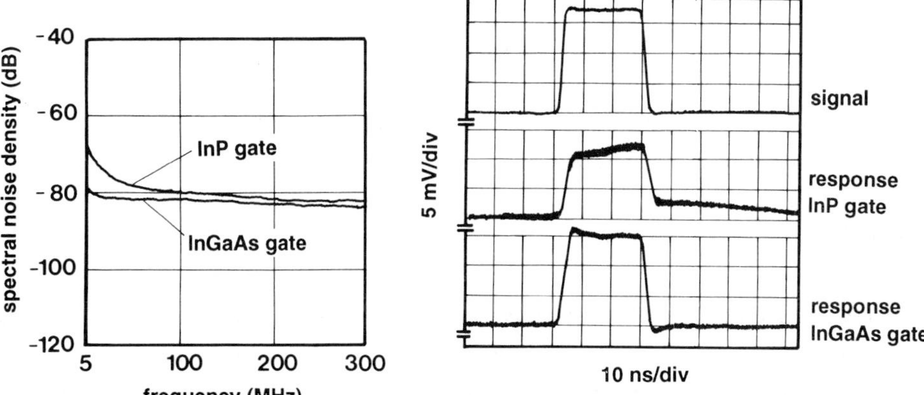

Fig. 6. Comparison of the noise characteristics (non-calibrated)of JFETs with gate contacts on InP and on InGaAs

Fig. 7. Signal and response of JFETs with InP and InGaAs contacts

In Figure 7 the signal and the pulse response of JFETs with p contacts on InP and InGaAs are compared. While the JFET with InP gate contact shows a pulse response with an additional time constant $\tau=R_C*C_C$, the JFET with InGaAs gate contact follows the applied signal with higher amplification and less noise contribution.

The influence of the improved p contact on the PD is shown in Figure 8, where the current-voltage characteristics of forward-biased PDs with InP and InGaAs p contact are compared. The diameter of the p region is 30μm, whereas the contact area of the ring-shaped p contact is only $330\mu m^2$. With InGaAs p contact layer, the series resistance is reduced from approximately 50 Ohm to 9 Ohm. The low-ohmic p contact of the PD offers the possibility to use the same device structure as level shifting diode for adjusting the operation point in integrated circuits.

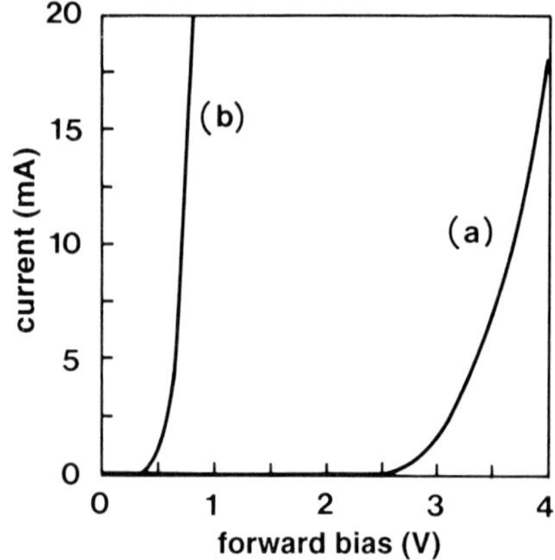

Fig. 8. Comparison of forward-biased PDs with p contacts on (a) InP and (b) InGaAs. The diameter of the p region is 30μm.

CONCLUSION

AuGe/Au and AuGe/NiCr/Au n contacts have been fabricated on $1*10^{18}cm^{-3}$ locally n-doped InP. After alloying at $400^{o}C$ for 10s AuGe/Au and AuGe/NiCr/Au show a contact resistivity of 0.015Ωmm and 0.1Ωmm, respectively. On InGaAs with a p doping of $2*10^{19}cm^{-3}$ Ti/Au has a contact resistivity of 0.14Ωmm. The improved contacts of the JFETs lead to a significant reduction in power dissipation and in the spectral noise density at frequencies below 100 MHz.

ACKNOWLEDGEMENT

We would like to thank G. Ebbinghaus and H. Huber for growing the epitaxial material, Th. Hillmer for the SIMS characterization, as well as J.G. Bauer and M. Plihal for helpful discussions. This work has been partly supported by the Deutsche Bundespost Telekom and by the RACE project R2070 MUNDI.

REFERENCES

Bauer J G, Albrecht H, Hoffmann L, Römer D, Walter J W, 1992, IEEE Photon. Technol. Lett. **4** pp. 253

Berger H H, 1972, Solid State Electronics **12** pp.145

Lauterbach Ch, Römer D, Treichler R, 1990, Appl. Phys. Lett. **57** pp. 481

Malina V, Hájková E, Zenlinka J, 1993, Thin Solid Films **223** pp. 146

Römer D, Bauer J G, Lauterbach Ch, Müller J, Walter J W, 1992, J. Appl. Phys. **72** pp. 4998

Uchida N, Akahori Y, Ikeda M, Kohzen A, Yoshida J, Kokubun T, Suto K, 1991, IEEE Photon. Techn. Lett. **3** pp. 540

Molecular beam epitaxy and technology for the monolithic integration of quantum well lasers and AlGaAs/GaAs/AlGaAs-HEMT electronics

W. Bronner, J. Hornung, K. Köhler, and E. Olander

Fraunhofer-Institut für Angewandte Festkörperphysik, Tullastr.72, 79108 Freiburg, Germany

ABSTRACT: In this presentation the various technology steps for the monolithic integration of GaAs quantum well lasers with Double Pulse Doped AlGaAs/GaAs/AlGaAs Quantum Well E/D HEMT electronics on a single substrate in one process run are described. All layers are grown by molecular beam epitaxy in an Intevac Gen II system. The laser structure, consisting of three 74 Å GaAs quantum wells between two AlGaAs cladding layers, are grown on top of the electronic structure. The laser mesas and contact areas are defined by a combined wet and dry etch process. Apart from the transistor gates which are exposed by electron beam lithography, all photolithography steps are performed using contact printing. To interconnect the devices a two layer metallization is used whereby air-bridges are used to connect the laser mesas to the electronics. Several designs of laser diode drivers and laser diodes were processed in this run. The results showed good performance of the electronic devices, and a high luminescence yield (> 90 %) for the laser diodes measured on-wafer. Some chips were cleaved to obtain laser mirrors. For these chips functional lasers were shown for laser diodes of area 4 x 360 μm^2 with threshold currents in the range of 20 to 30 mA, as well as the operation of laser diode drivers integrated with laser diodes operating at a data rate of 7.4 Gbit/sec.

1. INTRODUCTION

As optical communication systems applications are increasing, there is a growing interest in the monolithic integration of electronic and optical devices due to the demand of high bit rates, high reliability and lower costs.

In the past, components for optical communication systems have been designed and processed in our institute. Examples of these on the transmitter side include a 20 Gbit/sec 2:1 multiplexer [Nowotny et al (1991)], a laser diode driver [Wang et al (1992a)], a 20 Gbit/sec 2:1 multiplexer integrated with a laser driver [Wang et al (1992b)], and a 15 GHz GaAs/AlGaAs multi quantum well laser diode [Ralston et al (1991)]. On the receiver side these are devices such as a monolithically integrated MSM photodiode and optoelectronic receiver with a -3 dB bandwidth of 14 GHz [Hurm et al (1993)] and a 11.6 Gbit/sec 1:4 demultiplexer [Lang et al (1991)]. All electronic devices have been realized using our Double Pulse Doped AlGaAs/GaAs/AlGaAs Quantum Well (DPD-QW) E/D-HEMT process [Köhler et al (1990)].

In this paper we describe the various technology steps to integrate electronic circuits and laser diodes together on one substrate in a single process run. In developing this process it was of great importance that every technological step involved in the process be compatible with every other step. Process steps which are necessary to fabricate the laser diodes must not destroy the electronic structure. On the other hand, it is important that the steps of the HEMT process especially the RIE steps can still be used without damaging the laser diodes.

2. TECHNOLOGY

Molecular beam epitaxy

The AlGaAs/GaAs/AlGaAs hetero structures for the optoelectronic devices were grown by molecular beam epitaxy (MBE) in an Intevac Gen II system on 2-inch wafers. Following a thin undoped buffer layer the HEMT structure is grown at a temperature of 500°C, with the two dimensional electron gas (2DEG) in a 15 nm quantum well (Fig.1). Sufficient electron supply is obtained by growing two Si-δ-doped layers above and below the 2DEG. The vertical structure for the formation of E- and D-FETs follows including two 3 nm AlGaAs etch stop layers. A third 3 nm AlGaAs etch stop layer and a 600 nm GaAs layer are grown above the HEMT structure which serve as the separation of the laser and the HEMT structure and as n-contact for the laser, respectively. The laser structure is then grown at a higher temperature of 700°C. In order to keep the maximum step height on the wafers small a laser structure has been developed which was designed to be vertically compact. It consists of three GaAs/AlGaAs quantum wells, each of thickness 74 Å between two 700 nm thick AlGaAs cladding layers, and ends with a 200 nm thick 8 x 10^{19} cm^{-3} Be-doped GaAs cap layer. This laser structure has been shown to operate at frequencies up to 15 GHz [Ralston et al (1991)].
The growth time for the laser structure is more than two hours. To study the effects of the long time and high temperature growth of the laser structure on the HEMT structure, such structures were grown at 500°C and then the temperature was increased to 700°C for two hours. The electrical properties of devices on structures after this heat treatment compared to those not heat treated were similar. For a detailed description of the laser structure and the HEMT structure see [Ralston et al (1991)] and [Hülsmann et al (1990)] respectively.

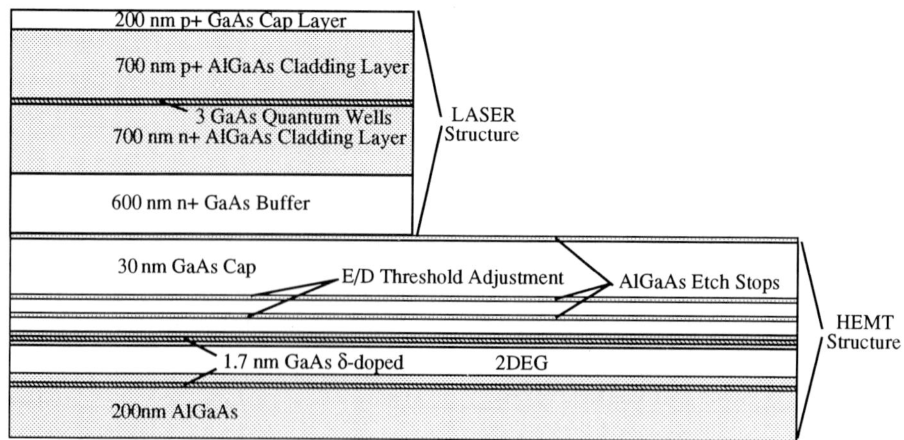

Fig. 1: The complete structure as grown in the MBE system. Note that the scales of the HEMT and the laser structure are not the same

Fabrication of laser mesas

The preparation of the laser diodes was carried out in several subsequent steps (see Fig.2a-c). The first layer of the process is an evaporated Ti/Pt/Au metallization, which serves as the ohmic contact to the p-doped side of the laser diodes. This layer also includes all the necessary alignment marks for the following process steps. Great care has to be taken to align this layer exactly to the wafer flat which corresponds to <110>. This is important to insure that laser mirrors can be obtained by cleaving the wafer perpendicular to the orientation of the mesa structures. After this step the regions for the laser diodes were defined by photolithography. A first wet etching step is done in H_3PO_4:H_2O_2:H_2O at a temperature controlled to 20°C exactly. The etch rate is approximately 300 nm/min with a very good homogeneity across the wafer.

At this etch rate it is possible to interrupt the etching precisely within the n^+-GaAs layer between the laser and the electronic structure. In practice, about 350 - 400 nm of this layer was left as the contact layer for the laser diodes. Then this layer was laterally structured to define the regions which are needed to contact the n-doped side of the laser diodes (Fig.2b). This etch step was performed in a reactive ion etching system (RIE) with CCl_2F_2 as the etching gas at a pressure of 6.0 Pa and a power density of 0.08 W/cm^2, which results in a DC bias of 100 V. This etch step stops at the AlGaAs layer, which separates the laser structure from the electronic structure. The AlGaAs etch stop layer is removed by a short dip in a HCl:H_2O_2:H_2O solution. The samples then have the structure shown in Fig.2c. This is the starting point for the DPD-QW E/D-HEMT electronics process.

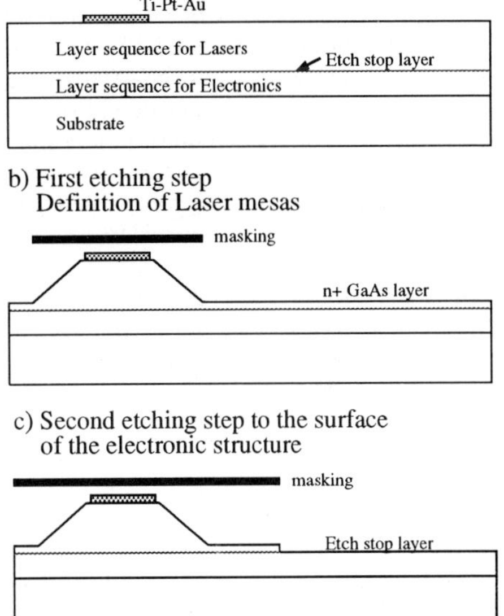

Fig. 2: Etch sequence to define the laser diode mesas and interconnection metallization as described in the text

DPD-QW E/D-HEMT electronics

Our E/D-HEMT process involves three RIE steps, several wet chemical etching steps, and five to seven metal lift-off procedures. The entire process sequence has been described elsewhere [Köhler et al (1990)].

The first layer of the electronic level is the Au-Ge ohmic metal for the source and drain regions of the transistors, which in this case also serves to contact the n-doped side of the laser diodes. This metallization is carried out by optical contact lithography, metal evaporation, and lift-off technique, although the maximum feature step height on the wafers is about 2.2 μm.

The enhancement and depletion transistor gates were defined by electron beam direct write lithography in a PMMA/PMAA two layer resist. This step is followed by two or three RIE steps, depending on the transistor type, and then by the Ti-Pt-Au gate metal evaporation and lift-off sequence.

The electron beam resist has a total thickness of 0.6 μm; for comparision the height of the mesa structures can be up to 2.2 μm. The PMMA resist does not cover the mesas completely due to a planarization effect and, especially at the edge of the mesas, the resist layer thickness is to thin to withstand the etch conditions of the RIE steps, resulting in damage of the mesas. For this reason it was necessary to protect the laser mesas by an additional photo resist layer prior to the etching and metallization sequence. This was done after developing the different electron beam resists for the enhancement and depletion transistors (and in some cases also the MSM photo diodes). We developed a photo resist process during which the small gate structures in the electron beam resist layer are not affected. The protection layer of the laser mesas can be lifted off together with the electron beam resist.

Metallization

The final steps in the process include three metallization layers, namely Ni-Cr for thin film resistors, 200 nm Au for the first interconnection level (first metal), and 3 µm electroplated Au for the second level metal. The two interconnection levels are separated by 200 nm of PECVD-SiON. The second metal deposition can be carried out in airbridge technology. All necessary photolithographic steps are performed by contact printing.
The interconnection between the laser diodes and the electronic level was done with the first metal layer on the n-doped side of the diodes, and by airbridges on the p-doped side. A schematic view of this interconnection is shown in Fig.3.

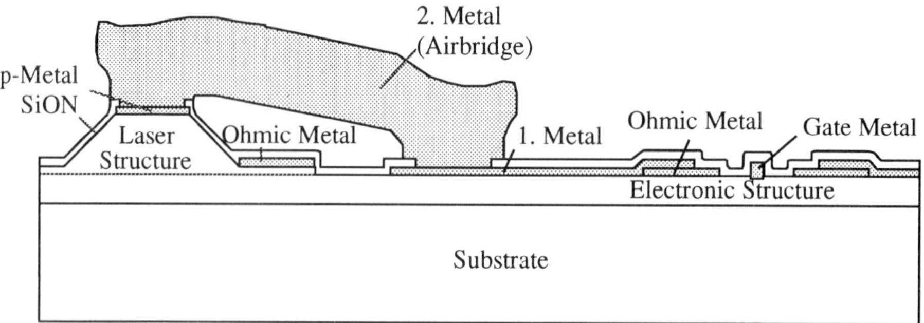

Fig. 3: Schematic cross-section of a laser diode mesa with interconnection metallization. The step height between the laser mesa and the electronic level is about 2.2 µm.

3. RESULTS

Figure 4 shows a photograph of a laser driver circuit with a 4 x 460 µm^2 laser diode mesa. First, the diodes and electronic circuits were characterized by on-wafer measurements. To test the MBE structures, the success of the etch procedures, and the interconnection metallization, the diodes were driven at a constant current of 15 mA while the intensity of the spontanous luminescence was measured. The homogeneineity of the luminescence was very good across the wafer and the yield of the working diodes was better than 90 %. The yield of the laser diode driver circuits was determined only on one wafer and was about 80 %. This slightly lower value was due to some larger etch pits which occured at MBE defects during the mesa etch process.

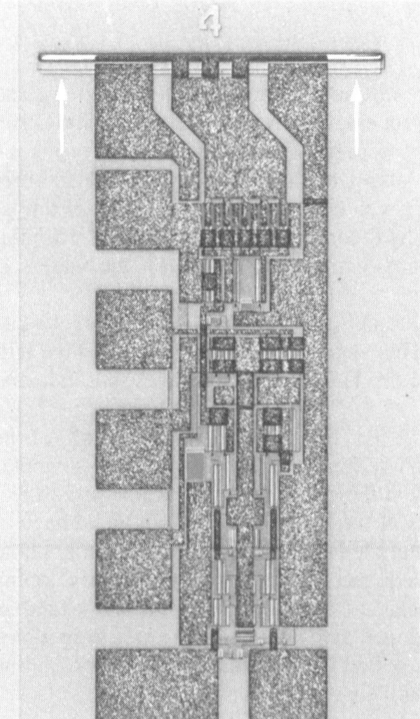

Fig. 4: Photograph of chip with integrated laser diode driver and laser diode mesa. The size of the chip is 0.5 x 1.0 mm^2. The size of the mesa stripe in this case is 4 x 460 µm^2. The arrows mark the regions, where the chip is cleaved to obtain laser mirrors.

Lasers are obtained by cleaving the chips accurately perpendicular to the orientation of the mesa stripes. This was done by dicing the chips from the backside, leaving a remaining thickness of about 80 μm. Then the chips were cleaved and mounted on the edge of a thin sheet metal of copper, which could be fixed in the measurement equipment.
One chip bar with eight isolated laser diodes was characterized. The yield was 100 % in this case and the homogeneity was good. The threshold current was between 23 and 30 mA. A quantum efficiency of 40 % from a single facet was determined.
The integrated structures were characterized by means of an optical transmission system. On the transmitter side this was a pulse generator (Anritsu MP1S012) together with the integrated laser driver and laser diode circuit. A photo diode integrated with a preamplifier and a commercial digital oscilloscope (HP 54120A) was used on the receiver side. Figure 5 shows a pulse diagram at a data rate of 7.4 Gbit/sec. In this diagram the output signal is inverted and the phase is shifted with respect to the input signal.

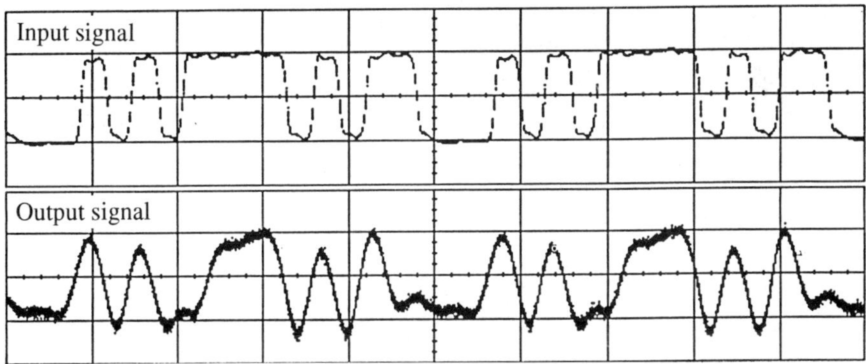

Fig. 5: Pulse diagram of an integrated laser driver and laser diode structure measured with an optical transmission line at a data rate of 7.4 Gbit/s. The output signal is inverted and phase shifted (Timebase: 450 ps/div).

4. CONCLUSION

We have demonstrated a process sequence for the integration of laser diodes and DPD-QW E/D HEMT electronics on a single substrate. The wafers used were grown by MBE with the laser structure on top of the HEMT structure. The laser and electronic regions on the wafer were separated using a combined wet and dry etch process. Our standard HEMT process with some few changes could be subsequently used to further process the chips. Even though the step height on the wafers is as high as 2.2 μm, the process could still be performed by contact photolithography.
The most important result is the fact that neither the extended growth time of the MBE at higher temperature nor the additional etching steps degrade the sensitive HEMT structure.
To further improve the performance of these optoelectronic integrated circuits we plan a few changes for future process runs:

- The laser mirrors were obtained by cleaving the chips perpendicular to the mesa stripes. This is a restriction on the size of the circuits which can be integrated with the laser diodes. To allow further integration (e. g. a multiplexer circuit together with a laser driver and a laser diode) it will be necessary to fabricate the mirrors by an etch process.
- A ridge structure will be etched on the laser diode mesas on the p-contact side to reduce undesirable current flow over the sidewalls.
- The etch stop layer which separates the laser and the HEMT structure is removed using a wet etch dip which causes some larger etch pits on the electronic level. To improve the yield of the electronic devices we will investigate the possibility to remove this layer by a dry etch process.

5. ACKNOWLEDGMENT

This work is a group effort and the authors wish to gratefully thank T. Jakobus and his entire staff of the process technology group. Further we would like to thank Z. G. Wang for designing the laser driver circuits, J. Fleißner, B. Matthes, and J. Rüdiger for cleaving the laser chips, and W. Benz, M. Ludwig, and J. Rosenzweig for measuring the lasers and the laser driver circuits. Special thanks go to H. S. Rupprecht for his directing encouragement. The Bundesministerium für Forschung und Technologie is acknowledged for financial support through the project TK0577.

6. REFERENCES

Hülsmann A, Kaufel G, Köhler K, Raynor B, Schneider Jo, and Jakobus T, Jpn. J. Appl. Phys. 29, 2317 (1990)

Hurm V, Ludwig M, Rosenzweig J, Benz W, Berroth M, Bosch R, Bronner W, Hülsmann A, Köhler K, Raynor B, and Schneider Jo, Electron. Lett. 29, 9 (1993)

Köhler K, Ganser P, Bachem K H, Maier M, Hornung J, and Hülsmann A, Inst. Phys. Conf. Ser. 112, 521 (1990)

Lang M, Nowotny U, and Berroth M, Electron. Lett. 27, 459 (1991)

Nowotny U, Lang M, Berroth M, Hurm V, Hülsmann A, Kaufel G, Köhler K, Raynor B, and Schneider Jo, Microelectr. Engineering 15, 323 (1991)

Ralston R D, Gallagher D F G, Tasker P J, Zappe H P, Esquivias I, and Fleissner J, Electron. Lett. 27, 1720 (1991)

Wang Z G, Berroth M, Nowotny U, Gotzeina W, Hofmann P, Hülsmann A, Kaufel G, Köhler K, Raynor B, and Schneider Jo, Electron. Lett. 28, 222 (1992)

Wang Z G, Nowotny U, Berroth M, Bronner W, Hofmann P, Hülsmann A, Köhler K, Raynor B, Electron. Lett. 28, 1724 (1992)

Paper presented at the Int. Symp. GaAs and Related Compounds, Freiburg, 1993

Fabrication of high speed MMICs and digital ICs using T-Gate technology on pseudomorphic-HEMT structures

A. Hülsmann, W. Bronner, P. Hofmann, K. Köhler, B. Raynor, J. Schneider, J. Braunstein, M. Schlechtweg, P. Tasker, A. Thiede and T. Jakobus

Fraunhofer-Institut für Angewandte Festkörperphysik
Tullastr.72, 79108 Freiburg, Germany

ABSTRACT: A technology for pseudomorphic MODFETs with 160 nm T-gates was developed and combined with a monolithic microwave integrated circuit (MMIC) fabrication process. The realized MODFETs demonstrate transit frequencies of 145 GHz at drain currents of 260 mA and drain bias of 1.2 V. With these MODFETs we were able to fabricate 3-stage 76 GHz amplifiers with a gain of 21 dB, and 5-75 GHz broadband travelling-wave amplifiers with a gain of 9.3 dB and a noise figure less than 4.0 dB in coplanar waveguide (CPW) technology, and dynamic frequency dividers operating from 28 to 51 GHz

1. INTRODUCTION

Monolithic microwave integrated circuits (MMICs) in CPW technology and high speed digital ICs are of growing interest for several commercial applications because of their low cost capabilities compared to hybrid solutions with discrete devices. Pseudomorphic MODFETs on GaAs are attractive devices because of their high electron saturation velocities and large charge transfer from the doped AlGaAs supply layer to the 2DEG in the InGaAs channel.

MMICs require a reliable fabrication process. High yield and good uniformity across the wafer must be achieved for the active and passive devices. The highest uniformity of the threshold voltage can be achieved by a technology which uses a dry etched gate recess in combination with a etch stop layer which is included in the MBE grown pseudomorphic HEMT-structure.

To demonstrate transit frequencies of more than 120 GHz the MODFETs should have gate lengths below 0.2 μm. To reduce the noise figure and input resistance, the gate line resistance should be below 200 Ω/mm which can be achieved by a "T" shaped gate. Therefore a high three-layer-resist structure is convenient which has to be structured by high resolution electron beam direct write lithography.

2. IC FABRICATION

The MMIC fabrication process we have developed starts by growing pseudomorphic (Ψ) HEMT-structures on 2-inch-GaAs wafers by molecular beam exitaxy (MBE). The growth conditions have been published by Schweizer et al. /1/. The layer structure is shown in fig. 1. Under the highly doped GaAs cap layer, the Ψ-HEMT-structure has a thin 3 nm $Al_{0.2}Ga_{0.8}As$ layer which is used as an etch-stop during reactive ion etching (RIE) of the gate recess. This controls the threshold voltage by the exact definition of the gate to channel separation. Before gate deposition the AlGaAs layer is removed by wet chemical etching to obtain a reliable Schottky contact on the 3 nm GaAs layer. A pulse doped 1.7 nm GaAs quantum well between two $Al_{0.2}Ga_{0.8}As$ barrier layers of 5.3 and 5 nm supplies the 2DEG in the 12 nm $In_{0.25}Ga_{0.75}As$ channel. The Ψ-HEMT-structure is grown on a GaAs buffer.

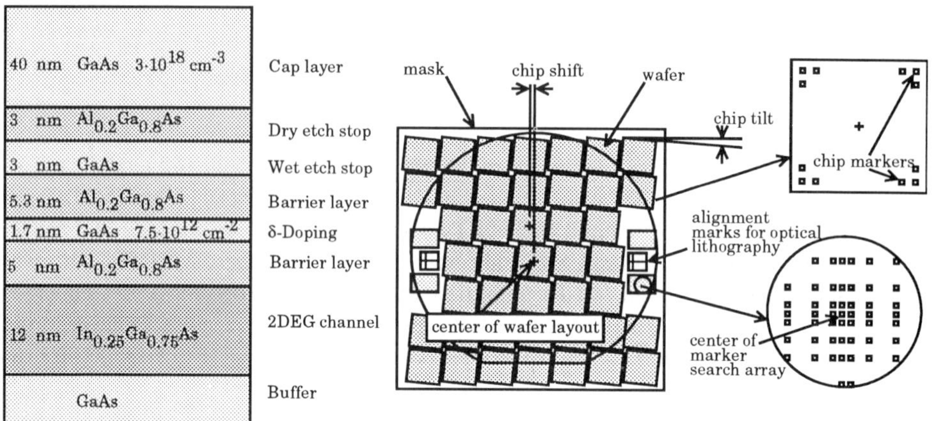

Fig.1 MBE grown Ψ-HEMT structure

Fig. 2 Wafer and chip alignment strategy

After MBE growth the wafer surface is protected against further processing and environment influences by a 20 nm plasma enhanced chemical vapor deposited (PECVD) SiON layer. Ohmic contacts are defined by contact printing using AZ5214 image reversal resist. Before metal deposition the SiON protection layer is removed at the resist openings by RIE in a CF_4 plasma. After evaporating Ni/AuGe/Ni/Au and subsequent lift-off, the contacts are alloyed on a nitrogen purged hotplate for 30 seconds at 390°C. Together with the source and drain contacts, alignment marks comprising 20 µm squares are defined at each corner of the chip to ensure optimal accuracy of the gate positioning for the e-beam direct write.

Device isolation is carried out by oxygen implantation at 80 keV with a dose of 1×10^{12} cm^{-2}. Mesa etched isolation can result in metal interconnection problems of the gate lines at the mesa steps and should be avoided.

After e-beam direct write lithography and subsequent resist development, the gate recess is performed in two steps by a parallel plate plasma reactor operating in RIE-mode. The first etch step removes the 20 nm SiON protection layer and the second step removes the n^+-GaAs cap. For the second etch step Freon 12 is used which forms thin non volatile aluminum fluoride films on AlGaAs surfaces and selectively removes the n^+-GaAs cap on the 3 nm $Al_{0.2}Ga_{0.8}As$ etch-stop layer. Good anisotropy of the gate recess can be achieved at a plasma pressure of 60 μbar and an etch time of one minute. The DC bias of the two parallel plate electrodes in the plasma etching reactor chamber is 55 V to avoid radiation damage of the channel region /2/. The three-layer-resist is based on PMMA and can withstand the RIE step without degradation. Details of the T-gate process are described later. After gate recess etching, Ti/Pt/Au (50/50/350 nm) gate metal is evaporated and subsequently lifted-off in acetone, thus forming the gate electrodes.

Low noise 50 Ω/sq NiCr thin film resistors are fabricated by co-evaporation. Next, a first interconnection layer is formed by evaporating Ti/Au/Ti (20/200/5 nm). Then, a 200 nm SiON layer is deposited by PECVD. This process step encapsulates all active and passive devices and is the dielectric for the MIM capacitors. First and second metal levels are connected by etching via holes into the SiON by RIE in a CF_4 plasma. Finally 3 μm plated gold is grown for the second metal level. Due to the coplanar waveguide MMIC design, backside processing is not required.

3. E-BEAM LITHOGRAPHY

The T-shaped gates are structured by electron beam lithography using a Leica EBPG-5HR operating at 50 keV. We have developed a direct write technology which can expose up to 20 2-inch wafers in one run after a rough manual prealignment. The exposure time is between 0.5 and 3 hours per wafer depending on the exposed area. A special error handling software enables the machine to override problems which may occur during writing, e.g. a defect alignment mark.

An automatic global alignment procedure has been developed to define a coordinate system on the wafer. Fig. 2 shows the principles of the alignment strategy. One of four global alignment marker arrays is used, in which the distance from marker to marker, beginning at the center, increases by 1 μm. By randomly finding three consecutive diagonal markers and measuring the distance to each other, the machine can calculate the exact position of the wafer center. For fine alignment of the gate to source and drain, each chip contains a set of markers in each chip corner. These chip markers are fabricated with the same lithography mask used to define source and drain to achieve maximum alignment accuracy. By finding and measuring the position of these markers the machine can compensate for chip tilt and shift.

The layout detail of a MODFET is shown in the middle of Fig. 3. The first fabrication step is the formation of alloyed ohmic contacts. These contacts are used for source and drain and as a part of the gate pad. The gate layer consists of a 50 nm line with 100 nm patterns on either side which have a distance of 100 nm to the 50 nm line. On the gate pad interconnect area the gate consists of seven parallel 50 nm lines. This gate pattern is converted by the CAT software /3/ to a format which can be read by the EBPG-5HR. This conversion data includes the information of the single line exposure with a dose of 850 μC/cm² (100%) to define the gate length and the side beam exposure with 15% dose to develop only the two top resist layers and thus define the cross-section of the T-gate head. At the lower part of fig. 3 a schematic cross-section of the three-layer-resist after e-beam exposure and development is show. The dose distribution in the three-layer-resist of the described T-gate exposure was simulated and is shown in fig. 4.

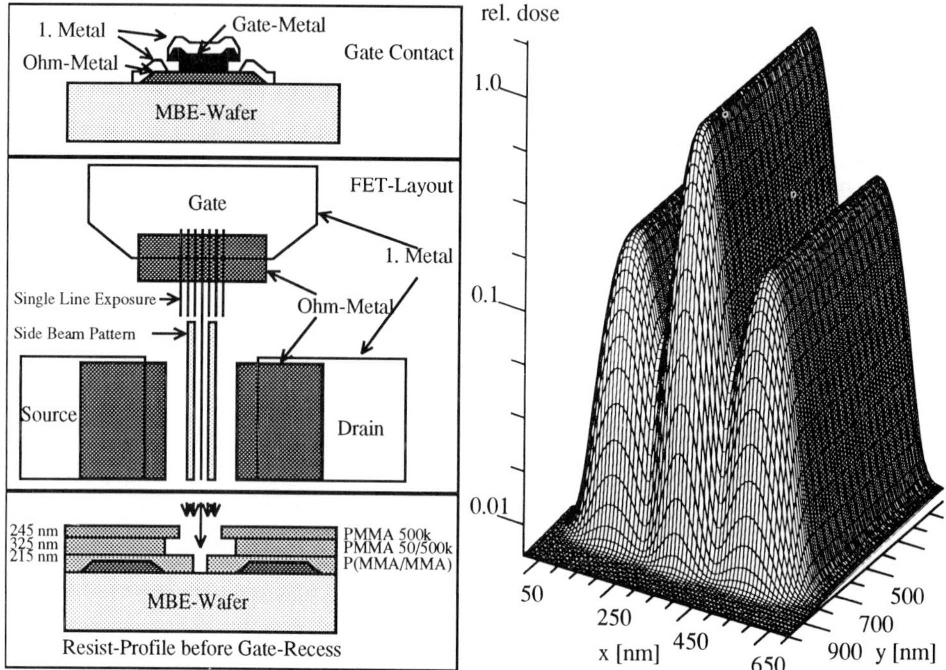

Fig. 3 Schematic of e-beam lithography

Fig. 4 Dose distribution after T-gate exposure

There is always overhanging gate metal even at the gate pad area which can cause bad contacts to the interconnection metal. Therefore it is useful to connect the gate indirectly by an intermediate metal contact which is already formed before the T-gate process. A schematic cross-section of the T-gate contact area is shown at the top of fig. 3.

For a dry etched gate recess a plasma resistant e-beam-resist has to be used to define the gate length. Fig. 5 shows a bargraph which displays the etch rates of several resists in different etch plasmas used for the gate recess. We found that crosslinked P(MMA/MAA) prebaked at 170°C displays the lowest etch rates. Normally P(MMA/MAA) is a high sensitive resist with a low resolution capability when developed in MIBK/IPA. To use P(MMA/MAA) as a high resolution resist with high contrast and low sensitivity, a different developer has to be used. In this case ortho-xylene is a suitable developer. Fig. 6 shows the contrast curves of PMMA 500k, PMMA 50k and the crosslinked P(MMA/MAA). Using the data of the contrast curves and the dose distribution of fig. 4 we simulated the development front at 2, 4, and 6 minutes during xylene development (fig. 7). An electron micrograph shown in fig. 8 confirms the good agreement between simulation and experiment.

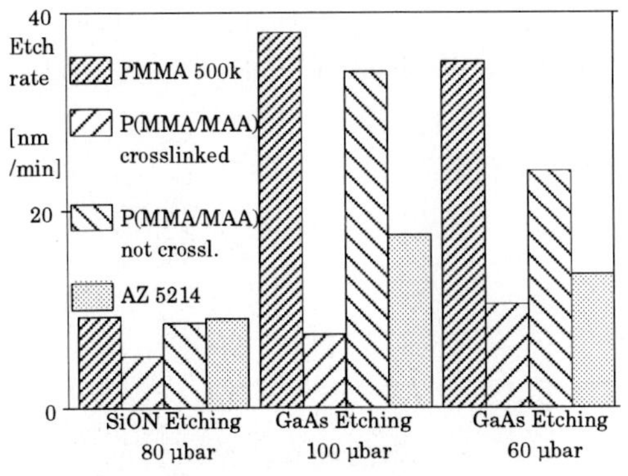

Fig. 5 Etch rates of different resists during gate recess etching

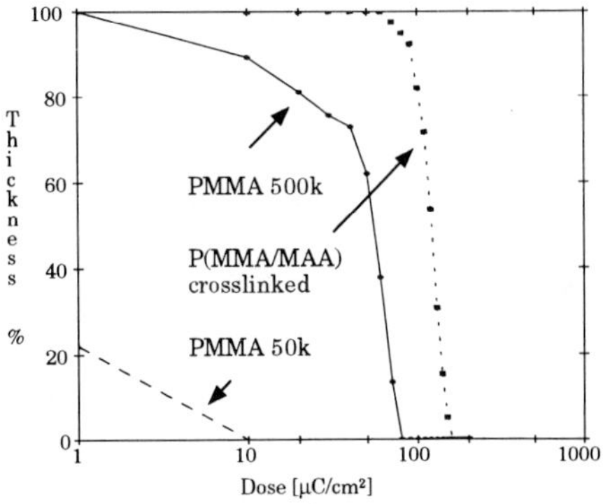

Fig. 6 Contrast curves of the three resists developed in ortho-xylene for 5 min.

4. TECHNOLOGY RESULTS

With the above described technology we were able to fabricate Ψ-MODFETs with 160 nm T-gates. The current gain characteristic is shown in fig. 9. The extrapolated current gain cutoff frequency is 145 GHz. Fig. 10 is a micrograph of a dual gate Ψ-MODFET encapsulated in 200 nm SiON. The sources are connected by airbridge with 3 µm plated gold.

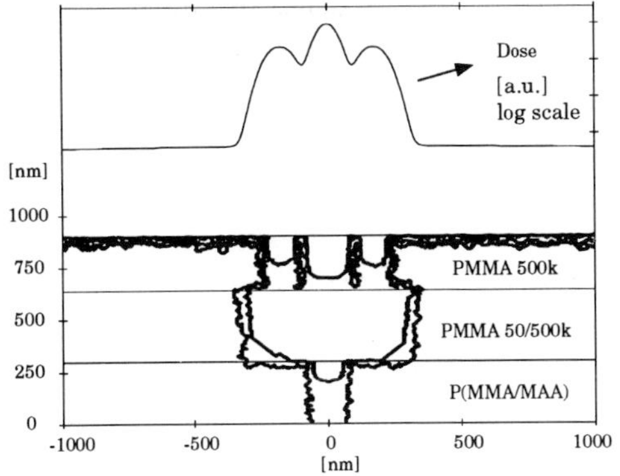

Fig.7 Simulation of a three-layer-resist development

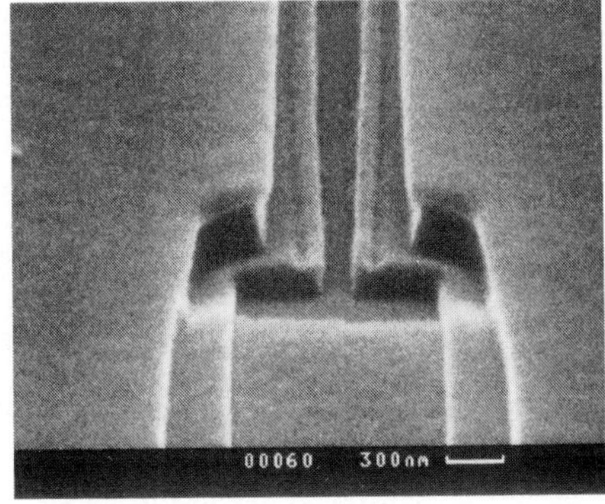

Fig. 8 Micrograph of a three-layer-resist profile

These Ψ-MODFETs were used in several digital and analog circuits. A dynamic frequency divider using NiCr resistor loads operates at a frequency range of 28-51 GHz /4/. A 3 stage 76 GHz amplifier was realized having a gain of 21 dB /5/. Using dual gate FETs in a cascode arrangement we build 5-stage broad band travelling wave amplifiers with a gain of more than 9.3 dB in a range from 5 to 75 GHz. The noise figure is less than 4 dB /6/. As far as we know these circuits demonstrate the highest performance for GaAs chips in their respective field.

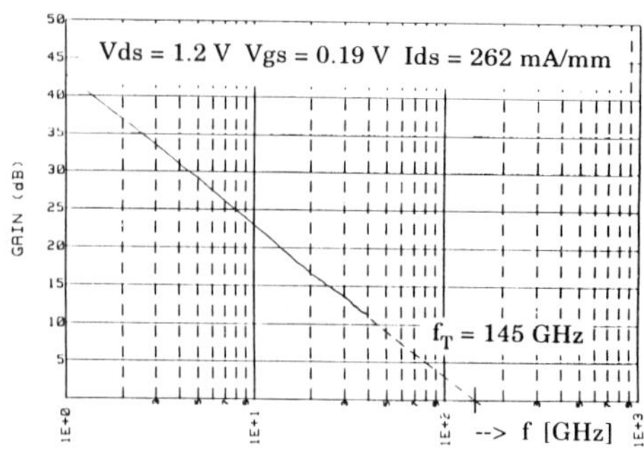

Fig. 9 Current gain versus frequency of a Ψ-MODFET with a 160 nm T-gate

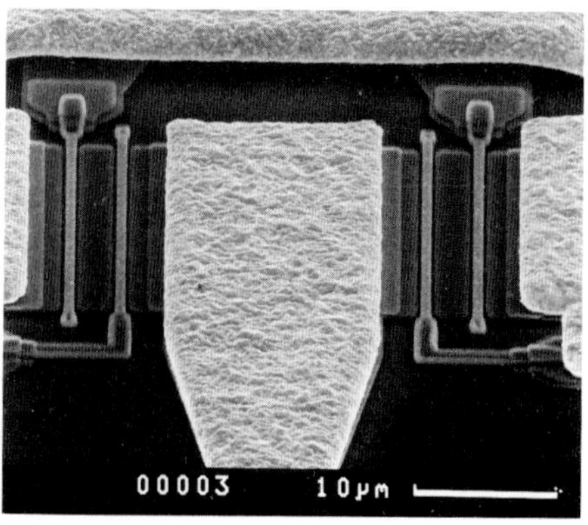

Fig. 10 Micrograph of a dual gate Ψ-MODFET encapsulated in SiON

5. SUMMARY

We have developed a e-beam lithography to fabricate Ψ-MODFETs with 160 nm T-gates. A novel three-layer-resist system was used to withstand degradation during the dry etched gate recess. This resist system requires a development process using xylene. The layout of the FET was improved to ensure failure free connection between gate and pad. The gates were written with a single line exposure and two sidebeams to enlarge the cross-section of the T-gate. A special alignment strategy enables us to expose up to twenty 2-inch GaAs wafers in one batch. The alignment accuracy of the gate to source and drain is less than 100 nm. The cross-section of the T-gate was simulated and is in good agreement with micrographs from the gate area. Realized digital and millimeter wave ICs showed world record performances.

5. ACKNOWLEDGMENT

The authors would like to thank F.Becker, S.Emminger, P.Ganser, K.Glorer, R.Haddad, M.Korobka, M.Krieg, D.Luick, T.Norz, E.Olander, J.Schaub, G.Schilli, and B.Weismann for their help in technology processing, R.Dian for the micrographs, N.Grün, U.Nowotny, J.Rüster and H.Windscheif for the electrical characterization, J.Hornung, M.Sedler and J.Seibel for their help in design, B.Landsberg for his computer support and H.S.Rupprecht for his expert directing and continuous encouragement. This work was supported by the German Federal Ministry of Research and Technology.

/1/ T. Schweizer et al. Applied Physics A 53 1991 pp. 109-113

/2/ D.J. As et al., Journal of Electronics Materials Vol.19 No.7 1990 pp. 747-751

/3/ Transcription Enterprises Ltd. Los Gatos, Ca

/4/ A. Thiede et al., Electronics Letters 13th May 1993 Vol.29 No.10 pp. 933-934

/5/ M.Schlechtweg et al., Electronocs Letters 10th June 1993 Vol.29 No.12 pp. 1119-1120

/6/ J. Braunstein et al., Electronics Letters 13th May 1993 Vol.29 No.10 pp. 851-852

Paper presented at the Int. Symp. GaAs and Related Compounds, Freiburg, 1993

Comparison of two passivation processes for heterojunction bipolar transistors

H.Sik, V.Amarger, M.Riet, R.Bourguiga and C.Dubon-Chevallier

FRANCE TELECOM, Centre National d'Etudes des Télécommunications, Paris B
Laboratoire de Bagneux, 196 avenue Henri Ravera,
BP 107, 92225 Bagneux cedex (FRANCE)
Tel:19 (33 1) 42 31 70 33, Fax:19 (33 1) 47 46 04 17.

Abstract : The current gain dependence on the size of the emitter-base junction for double mesa Heterojunction Bipolar Transistors (HBT) has been investigated, extrinsic base layer being passivated using different passivation techniques. The current gain is improved due to surface recombination reduction by a factor larger than 2 in the extrinsic base region. It has also been found that the base current was dominated by an excess leakage current in the isolation region related to damage created during the Boron-proton implantation. The surface recombination current ideality factor has been found to be 1.17.

I. INTRODUCTION

GaAs Heterojunction Bipolar Transistors (HBTs) have been intensively studied, because of their potential application for ultrahigh speed logic and microwave power devices. However, there are several problems that must be overcome to realize such high speed devices. One of the major obstacles is the DC current gain reduction associated with the scaling down of the transistor size. The excess base leakage current, which contributes to the DC current gain reduction, results from surface recombination at the extrinsic base surface[1]. Since the emitter dimension must be minimized for higher switching speed operation, elimination of this effect is very important. Different approaches have been reported to solve this problem with varying success[2-4].

The purpose of this work is to compare different passivation processes of HBTs to (1)

determine the optimum passivation process, (2) localise the region(s) responsible of the excess leakage current, (3) determine the value of the ideality factor *n*, of the surface recombination current.

II. EXPERIMENTAL PROCESSES

To improve the reliability of the passivation process, we have developed a new structure, the Passivated Heterojunction Bipolar Transistor (PHBT)[5]. The principle of this new structure is to prevent charges from reaching the extrinsic base surface using a thin GaInP layer (high bandgap semiconductor material) between the emitter and the base layers. During all the HBT process, the GaAs base layer is covered by the GaInP which acts as a passivation layer.

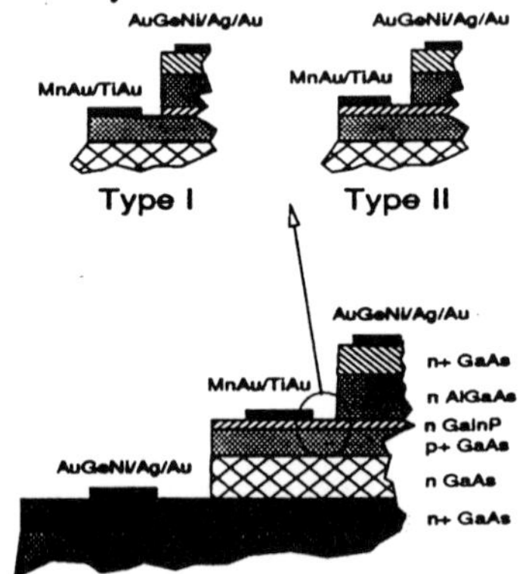

Fig.1. Schematic illustration of the two sample structures.

Table I. Epitaxial layers structure

Layer	Composition	Thickness (nm)	Doping (cm^{-3})
	$Ga_{0.5}In_{0.5}As$	400	Si : 1 10^{19}
Emitter cap	$Ga_{1-y}In_yAs$	400	Si : 1 10^{19}
	GaAs	300	Si : 5 10^{18}
	$Al_{1-y}Ga_yAs$	300	Si : 5 10^{17}
Emitter	$Al_{0.7}Ga_{0.3}As$	1500	Si : 5 10^{17}
	GaInP	300	Si : 2 10^{17}
Base	GaAs	900	C : 3 10^{19}
Collector	GaAs	3500	Si : 2 10^{16}
Subcollector	GaAs	7000	Si : 4 10^{18}

The advantage of this new structure compared to classical fully depleted AlGaAs HBTs, is that the AuMn p-type ohmic contact can be deposited on the GaInP layer, which permits to leave the total GaInP layer on the top of the base layer. Moreover, the GaInP and the GaAs can be selectively etched, leaving the wholeness of the base layer.

On the other hand, a few promising processes involving sulfur treatments have been proposed[6-7] which suppress surface recombination and improve GaAs surface properties. We have investigated the sulfur passivation of HBTs, using an optimised $(NH_4)_2S_x$ solution. In fact, it has been observed that the excess sulfur concentration in $(NH_4)_2S_x$ solution has a strong influence on the treated GaAs surface properties and on the reproducibility of the efficiency of the treatment[8].

The npn HBT device structure was grown by MOCVD, the structure is presented in Table I. In order to compare the different passivation processes, we have processed the same wafer with different processing steps.

The fabrication process is described as follows : First, wet chemical etching is used to expose the base and subcollector layers. Concerning the surface of the extrinsic base, two structures, i.e., with and without the surface passivation layer of 30nm n-type GaInP (type I and type II, respectively), were fabricated on the same wafer as shown in Fig.1. The isolation of the devices is then achieved, using a Boron-proton multi-implantation.

At this step of the process, 'emitter width' and 'emitter length' have been respectively defined by the dimensions of the emitter mesa etch mask and the implantation mask. The two structures were then either S-passivated or not with an $(NH_4)_2S_x$ solution (type Ia, IIa and Ib, IIb, respectively).

Table II

The extrinsic base surface is protected with:	
type Ia	Sulfur
type Ib	nothing
type IIa	GaInP+S
type IIb	GaInP

After that, the S-passivation layer has been protected using an optimised UVCVD silicon nitride overlayer. For ohmic contacts, AuGeNi/Ag/Au and MnAu/TiAu system were used for the n-type and p-type layers, respectively. The emitter mesa and the metallisation were separated by 0.5μm.

In order to investigate the surface recombination effect at the emitter periphery, HBT's with various emitter widths (0.5, 1, 1.5, 2, 3 μm) and emitter lengths (4, 6, 8, 10, 20 μm) are fabricated. Table II summarizes the processes used on the different parts of the wafer.

II. RESULTS AND DISCUSSION

II.1. Passivation process qualification

Considering the fact that the implantation might induce excess leakage current, the base current must be considered as the sum of three components, i.e., base current in the intrinsic base region, which is proportional to the base-emitter junction area, surface recombination current at the emitter mesa edge, which is proportional to the emitter width W_E and excess leakage current or implantation induced recombination current, which is proportional to the emitter length L_E. The relationship between β^{-1}, W_E^{-1} and L_E^{-1} can be approximately expressed as follows:

$$\frac{1}{\beta}=\frac{J_{RV}}{J_C}+\frac{J_{RS}}{J_C}\frac{2}{W_E}+\frac{J_{RI}}{J_C}\frac{2}{L_E} \qquad (E)$$

in accordance with an initial relationship of Hiraoka et al.[9] which did not take into account the implantation effect. J_{RV} is the base current density in the intrinsic base region, J_{RS} is the surface recombination current density in the extrinsic base region, J_{RI} is the recombination current density due to the implantation and J_C is the collector current density.

Fig.2 shows β^{-1} versus L_E^{-1} with various emitter widths for a collector current density J_C of 2.5×10^4 A/cm^2 for structure Ib (i.e. unpassivated structure). It can be observed that the effect of the implantation appears for emitter lengths smaller than 8µm. The excess leakage current density for emitter lengths smaller than 8µm has been estimated to be 38µA/µm which is around 10 times more important than the surface recombination current density J_{RS} for the same structure. When the emitter length becomes larger than 8µm, the DC current gain becomes independent of the emitter length L_E. A similar behaviour is observed for passivated structures (Ia and IIb, i.e. passivated repectively with S and GaInP). Then, if we want to investigate the effect of the surface recombination, it will be necessary to monitor the variation of β^{-1} versus W_E^{-1} for an emitter length larger than 8µm, to eliminate the important effect of the recombination current due to the implantation.

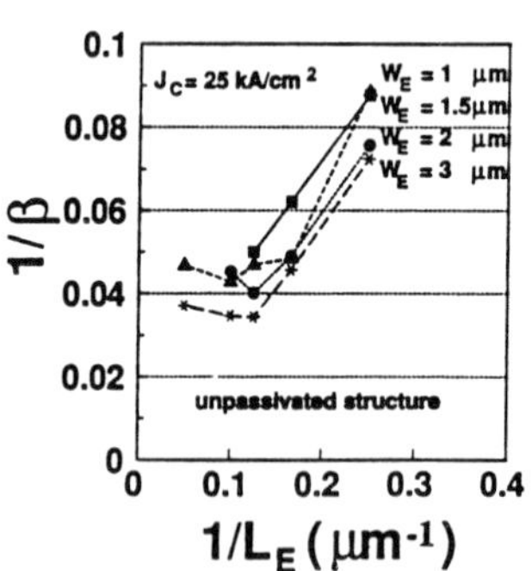

Fig.2 Effect of the B⁺H⁺ implantation on the DC current gain for various emitter lengths.

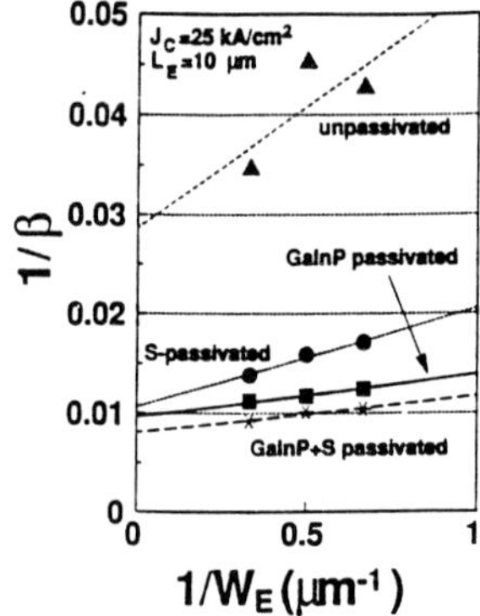

Fig.3 β^{-1} versus W_E^{-1} characteristics for fabricated HBT with an emitter length of 10µm, with different passivation processes.

In this condition, it can be observed in Fig.3 that any passivation of the extrinsic base surface effectively reduces the surface recombination current density J_{RS}. This reduction depends on the type of passivation process. The efficiency of each passivation technique is given by the slope of the different curves in accordance with the relationship (E). Even if the sulfur passivation (Ia S passivated) gives an important reduction of the surface recombination current, (2.4), it is nevertheless less than what observed for IIb (GaInP passivated) which exibits a reduction factor of 6.8. It is also important to observe, that the passivation of the emitter edge on structure IIa, does not have any effect on the surface recombination current J_{RS}. So, the only area concerned by surface recombination is the extrinsic base surface, and more precisely an area defined by the emitter length L_E and a width as small as 0.15μm, close to the emitter edge(10). The fact that the different characteristics do not cross at $W_E^{-1}=0$ indicates that leakage current due to implantation still exists on an area A_I independant of the emitter size. This leakage current depends on the passivation process and the area A_I is believed to be on the emitter edge, the difference between IIa (GaInP+S) and IIb (GaInP) being only related to the emitter edge.

II.2. Recombination surface ideality factor.

The measured Gummel plots of different structures are shown in figure 4. It allows one to observe the change in the base current ideality factor as the contribution of the surface recombination current to the total base current decreases, for the different passivating processes. It is shown that the base current ideality factor for the passivated structure Ia (S) and IIb (GaInP) has a value of *n* around 1.8 whereas it is around 1.5 for unpassivated structure Ib. It has also been observed that *n* increases with W_E for unpassivated structure, whereas it is constant for passivated structures. It indicates that, as the emitter size increases, the contribution of the surface recombination to the total base current decreases for unpassivated devices and that surface recombination current is negligible for passivated structures.

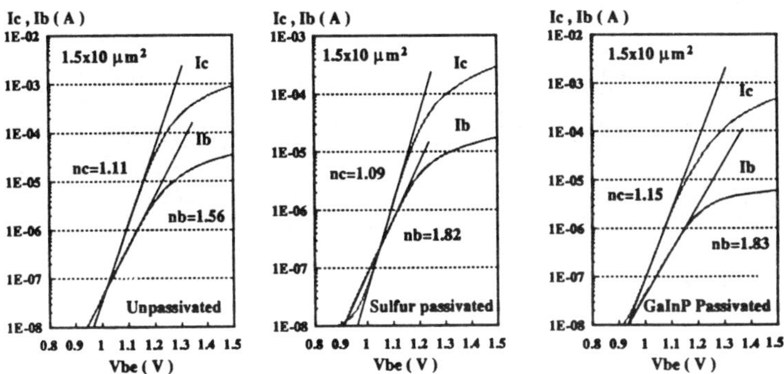

Fig.4 Gummel plots for structures Ib (left), Ia (center) and IIb (right). The emitter area is 1.5x10μm² for all devices.

The difference in base current ΔI_b was calculated in the bias region where the voltage drop due to resistance had no influence on the current-voltage characteristics, for structure IIa (GaInP+S passivated). It is shown that the ideality factor of ΔI_b is 1.17 (very close to 1). ΔI_b

being here essentially the surface current, it can be concluded that the surface recombination current increases exponentially with an ideality factor of 1[4,11]. Figure 5 illustrates ΔI_b as a function of the base-emitter bias V_{be} for $1.5x10\mu m^2$ device.

III. CONCLUSION

In summary, we have experimentally showned that, although the implantation technique is widely used to define the isolation region, alternative techniques , such as mesa etching, will be necessary for devices with an emitter length smaller than 8µm. We have also demonstrated that a drastic reduction of the emitter size effect can be achieved with a passivation of the extrinsic base surface, either with a sulfur passivation or a fully depleted GaInP layer between the emitter and the base layers, the best process being the use of a GaInP layer. It has been shown that the emitter edge surface was not the origin of surface recombination. Finally, it has been found that the surface recombination current has an ideality factor of 1.

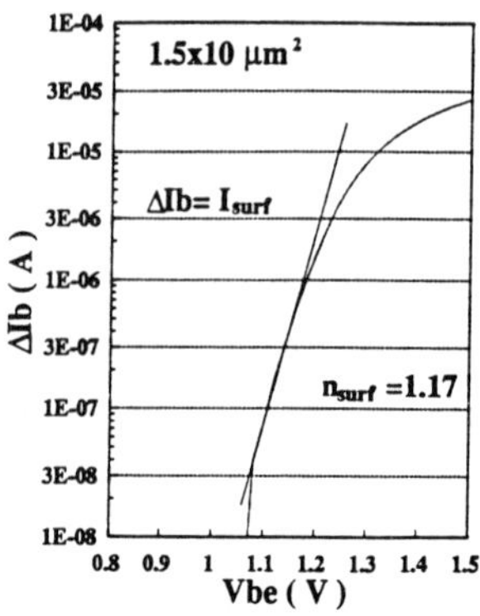

Fig.5 I_{surf} (i.e. difference of the base currents of Ib and IIa) as a function of the base-emitter bias.

REFERENCES

(1) O.Nakajima, K.Nagata, H.Ito, T.Ishibashi and T.Sugeta
Jpn. J. Appl. Phys. **24** (1985) L596.

(2) C.J.Sandroff, R.N.Nottenburg, J.C.Bischoff and R.Bhat
Appl. Phys. Lett. **51**, (33) 1987

(3) S.Noor Mohammad, J.Chen, J.I.Chyi and H.Morkoç
Appl. Phys. Lett. **56**, (10) 1990

(4) K.Mochitzuki, H.Masuda, M.Kawata, K.Mitani and C.Kusano
Jpn. J. Appll. Phys. Vol. 30, n°2B, 1991, L266-268

(5) C.Dubon-Chevallier, F.Alexandre, J.L.Benchimol, J.Dangla, V.Amarger, F.Heliot and R.Bourguiga
Electronic Letters **Vol.28** N° 25 (1992) p2308-2309.

(6) H.Oigawa, J.Fan, Y.Nannichi, K.Ando, K.Saiki and A.Koma
Extended Abstract 20th Conference of Solid State Devices and Material, Tokyo 1988 p263

(7) J.F.Fan, Y.Kurata and Y.Nannichi
Jpn. J. Appl. Phys. **Vol.28**, n°12, 1989, L2255

(8) H.Sik, M.Riet, C.Dubon-Chevallier and B.Sermage
E-MRS Meeting, May 1993 (to be published).

(9) Y.Hiraoka and J.Yoshida
IEEE Trans. Elec. Dev. **ED-34** (1987) 721.

(10)W.Liu and J.S.Harris, Jr.
Jpn. J. Appl. Phys. **Vol.31**, n°12, 1992, pp2349-2351.

(11)W.Liu
Jpn. J. Appl. Phys. **Vol.32**, Part 2 n°5B, 199?, L713-715.

Paper presented at the Int. Symp. GaAs and Related Compounds, Freiburg, 1993

0·2 μm pseudomorphic HEMT technology by conventional optical lithography

C. Lanzieri, M. Peroni, A. Bosacchi *, S. Franchi * and A. Cetronio.

ALENIA un'Azienda Finmeccanica, Direzione Ricerche, 00131 Roma, Italy
*CNR-MASPEC, Via Chiavari 18A, 43100 Parma, Italy

ABSTRACT: A high yield, reproducible 0.2 μm gate length technology utilising conventional optical lithography and highly selective, low damage plasma etching has been developed to produce "T-shaped" Ti/Al gate structures. PM-HEMT devices fabricated with this technique have yielded better than 0.9 dB noise figure and 12 dB associated gain at 12 GHz.

1. INTRODUCTION

To take full advantage of gain and noise performance of pseudomorphic HEMT's, as a result of the superior electron transport properties of this material, i.e. higher electron mobility, sheet carrier density and saturation velocity (Ali and Gupta 1991), extremely short gate length devices with low parasitic resistances and capacitances are required. Said conditions are satisfied by the so-called "T-shaped" gate structures (Chao 1983, Lepore 1988 and Tiberio 1989), fabricated by electron-beam lithography and post metallisation lift-off techniques of multi-layer resists, to yield PM-HEMT's with excellent performance (Lee 1989, Metze 1989 and Plana 1993), but for integrated circuit (IC) applications the latter technology proves to be technically very critical (Enoki 1991). Furthermore the inherent approach of "direct write" on wafer may prove to be too expensive and/or time consuming for medium to large volume IC fabrication.

In this article we will outline a high yield, reproducible 0.2 μm gate length technology which utilises conventional optical lithography and highly selective, low damage plasma etching to produce "T-shaped" Ti/Al gate structures similar to those fabricated by e.b. lithography. In particular we will illustrate that with such a technique the effective gate length of the Ti/AlGaAs Schottky barrier metallisation and thus device input capacitance can be very accurately controlled down to 0.1 μm without affecting the overall gate resistance of the device. The performance of PM-HEMT devices fabricated with this process (i.e. ≈ 0.9 dB noise figure and 12 dB associated gain at 12 GHz) is very interesting for integrated circuit applications considering the low cost, high yield and potentially high volume throughput available with this technique.

2. DEVICE FABRICATION

The epitaxial layers composing the PM-HEMT structure have been grown on 2-inch diameter, semi-insulating LEC GaAs substrates by molecular beam epitaxy. Said structure is composed of a highly silicon doped n+ GaAs cap layer, a step-doped (low-high) AlGaAs donor layer, a thin undoped Al GaAs spacer layer, an undoped In GaAs high mobility channel layer, and an undoped GaAs/Al GaAs superlattice buffer.

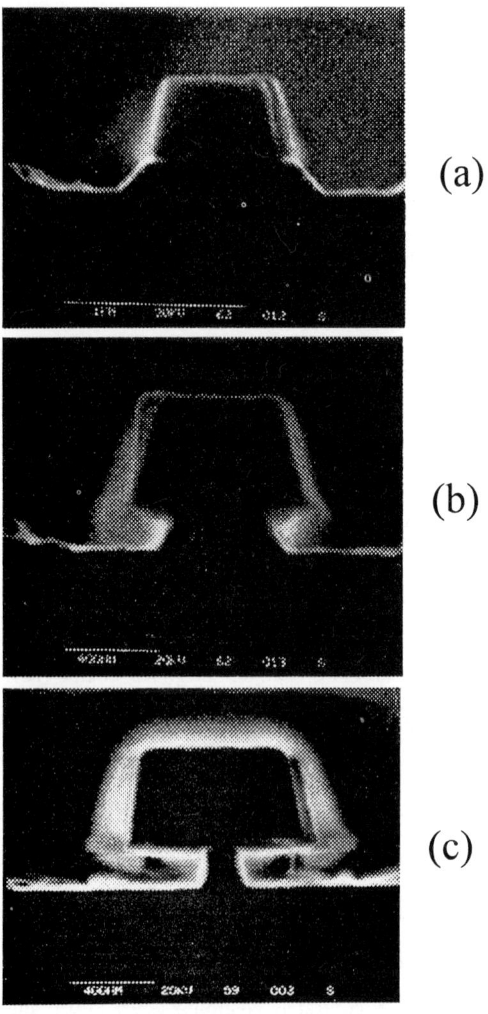

Fig. 1. Scanning electron microscope images of the Ti/Al gate metallisation (a)$L_g \approx 0.8$ µm as-deposited (b) $L_g \approx 0.6$ µm and (c) $L_g \approx 0.2$ µm both after Ti etching.

The fabrication process for the PM-HEMT devices is based on: source-drain ohmic contacts formed by rapid thermal alloying of a Au Ge Ni metallisation scheme, planar isolation by proton implantation, conventional optical lithography for sub-micron gate length, highly selective GaAs/Al GaAs etch solution for channel recessing, Ti/Al gate metallisation, plasma enhanced chemical vapour deposition for device passivation and finally Ti/Pt/Au overlayer metallisation.

After gate metallisation and lift-off, effective device gate length is controlled by means of highly selective, low damage plasma etching of the Ti barrier metallisation. As illustrated in figure 1 with this technique it is possible to reduce the effective gate length from the as-deposited dimension of ≈ 0.8 µm down to less than 0.2 µm without any noticeable deterioration in fabrication yield. The resultant "T-shaped" Ti/Al gate structure, similar to what is obtained by e.b. lithography, is found to have very good on-wafer uniformity and excellent processing reproducibility.

3. DEVICE PERFORMANCE

Typical drain current, transconductance and input capacitance characteristics of 200 μm gate-width PM-HEMT devices are presented in figure 2. To avoid any process variations which may occur

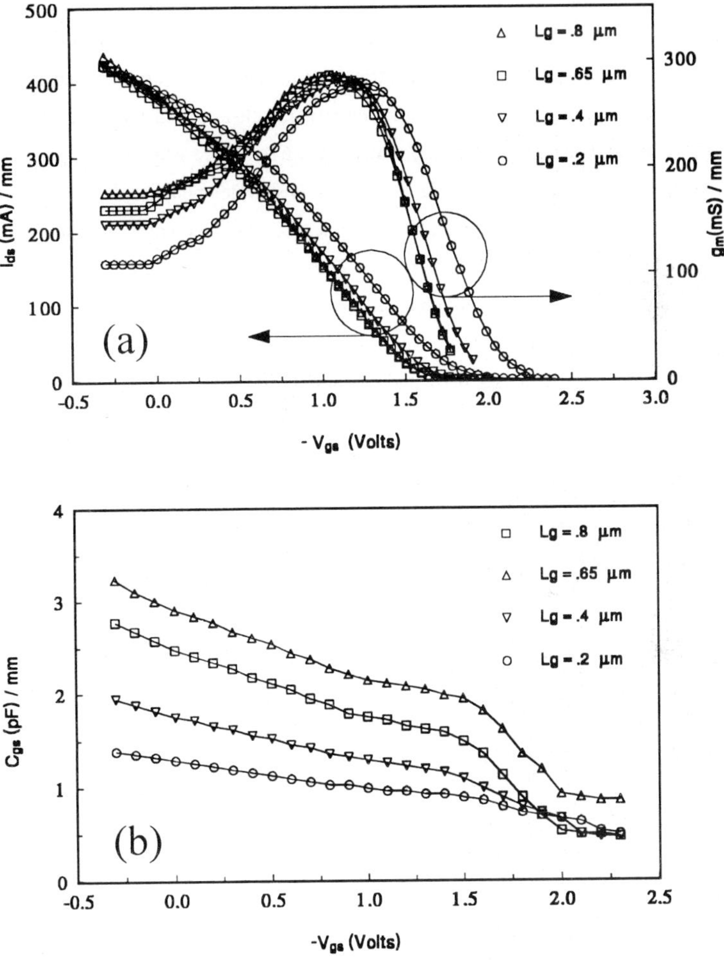

Fig. 2. (a) Drain current, transconductance and (b) input capacitance characteristics of 200 μm gate-width PM-HEMT devices after controlled etching of the Ti barrier metallisation.

during device fabrication the gate structures shown in figure 1 (i.e. as-deposited and under-etched T-gates) where all fabricated on the same wafer by undercutting the same 0.8 μm mask by different amounts. As shown reducing the effective gate length from approximately 0.8 to 0.2 μm (by opportune undercutting) has little effect on device drain current (≈ 350 mA/mm at V_{gs} = 0 V) and maximum transconductance (≈ 280 mS/mm), while the input capacitance is appreciably reduced from approximately 3.0 pF/mm to 1.3 pF/mm at V_{gs} = 0 V without any significant change in gate resistance R_g which is found to increase from approximately 3.4 to 4.1 Ω/mm respectively.

From S-parameter measurements in the frequency range 2 to 20 GHz it is found that all the devices have a maximum intrinsic transconductance of approximately 310 mS/mm at 50% I_{dss} and of approximately 250 mS/mm at ≈ 20% I_{dss} i.e. at minimum noise-figure bias. As shown in figure 3 for the latter bias conditions the key parameters of the intrinsic equivalent circuit are very sensitive to gate length reduction by under-etching of the Ti Schottky barrier metallisation. In particular the intrinsic input resistance R_i is found to increase from approximately 0.6 to 0.9 Ω while the output resistance R_{ds} decreases from approximately 285 to 165 Ω. Said variations, which penalise device performance, can be attributed to the effective ungated recessed channel length which increases from approximately 0.2 to 1.4 μm as the gate length is reduced from 0.8 to 0.2 μm. The negative trend in input and output resistances is compensated by a corresponding reduction in input (C_{gs}) and feedback (C_{gd}) capacitances, both of which tend to improve overall device performance; the former by increasing cut-off frequency and improving the noise-figure and the latter by making the device more stable.

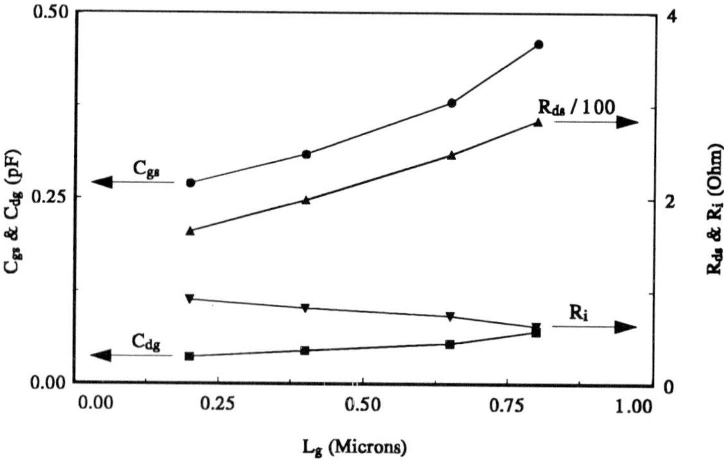

Fig. 3. Key intrinsic equivalent circuit parameters as a function of gate length for minimum noise figure bias conditions (i.e. ≈ 20% I_{dss}).

The overall effect of gate under-etching on device performance is demonstrated by the 12 GHz noise-figure and associated gain characteristics presented in figure 4. As shown the noise performance and associated gain are found to improve continuously from approximately 1.1 to 0.9 dB and 8.5 to 12.0 dB respectively as the gate length is reduced from 0.8 to 0.2 μm. These results confirm that: (a) the influence of input capacitance predominates over the influence of input resistance in determining device noise-figure, as predicted by the Fukui (1979) model, and (b) that the drastic reduction in output resistance impedes the expected high associated gain obtainable with PM-HEMT devices (Ali and Gupta 1991). Nevertheless, by comparing these results with those of commercial HEMT devices fabricated by electron-beam lithography (see figure 4), it is apparent that this approach, which utilises PM-HEMT material and conventional optical lithography for gate definition, can yield results

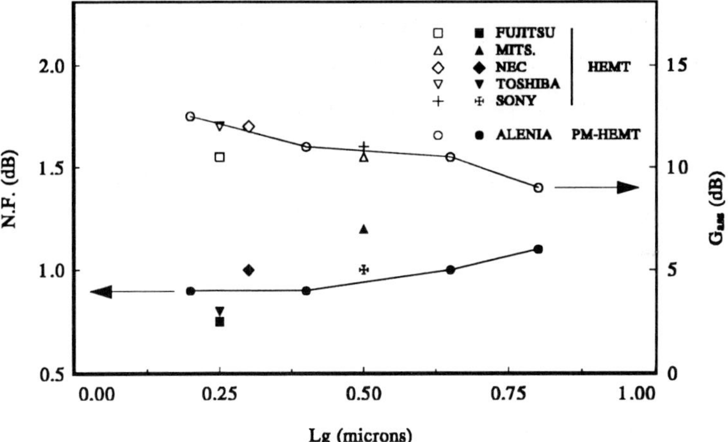

Fig. 4. 12 GHz noise figure and associated gain performance as a function of gate length for commercial HEMT's and PM-HEMT's fabricated with conventional optical lithography.

which are comparable with those obtained with more critical technologies. It is thus potentially very interesting for integrated circuit fabrication where reproducible, high yield and potentially large volume, low cost technologies are a key issue.

4. CONCLUSIONS

A high yield, reproducible 0.2 μm gate length technology has been developed to produce "T-shaped" Ti/Al gate structures, similar to those fabricated by electron-beam lithography by means of conventional optical lithography and a highly selective, low damage plasma etch process. The performance of PM-HEMT devices fabricated with this technique (i.e. ≈ 0.9 dB noise figure and 12.0 dB associated gain at 12 GHz for 200 μm wide devices) are very interesting for integrated circuit applications considering the low cost, high yield and potentially high volume technique used.

5. REFERENCES

Ali F. and Gupta A. 1991 HEMT'S and HBT's: Devices, Fabrication and Circuits (Artech House).

Chao P.C. et al 1983 IEDM Tech. Digest pp. 613.

Fukui H. 1979 IEEE Trans. Microwave Theory Tech. MTT-27 pp. 463

Lee R.E., Beaubien R.S., Norton R.H. and Bacon J.W. 1989 IEEE Trans, Microwave Theory and Tech. 37 (12) pp. 2086.

Lepore A.N. et al 1988 Electronics Letters 24 (6) pp. 364.

Metze G.M. et al 1989 IEEE Electron Devices Letters 10(4) pp. 165.

Plana R. et al 1993 IEEE Trans. Electron Devices 40 (5) pp. 85

Tiberio R.C., Limber J.M., Galvin G.I. and Wolf E.D. 1989 SPIE Electron Beam, X-Ray and Ion Beam Technologies: Submicrometer Lithographies Vol. 1089 pp. 124.

The mechanism for the compositional disordering of InGaAs/InAlAs quantum well structures by silicon ion implantation and annealing

Shin'ichi YAMAMURA, Tadamasa KIMURA, Riichiro SAITO, Shigemi YUGO, Michio MURATA* and Takeshi KAMIYA**
University of Electro-Communications, 1-5-1 Chofugaoka, Chofu-shi, Tokyo 182, JAPAN,
*Sumitomo Electric Industries, Ltd., 1 Taya-cho, Sakae-ku, Yokohama 244, JAPAN,
**University of Tokyo, 7-3-1 Hongo, Bunkyo-ku, Tokyo 113, JAPAN.

ABSTRACT: Interdiffusion at an InGaAs/InAlAs heterointerface due to Si ion implantation was studied by measuring the shift in the photoluminescence peaks from InGaAs quantum wells. The interdiffusion between Ga and Al was found to occur within several seconds and was ascribed to defect-induced interdiffusion. Both direct ion mixing and impurity-induced interdiffusion were found very small. Unimplanted samples showed no interdiffusion except a slight interdiffusion between Ga and In under high-temperature and long-time annealing conditions. Results of B implantation which was found to enhance the interdiffusion of Ga and In were also reported for comparison.

1. INTRODUCTION

An InGaAs/InAlAs quantum well (QW) structure lattice matched to InP is one of the useful systems to construct an opto-electronic integrated circuit (OEIC). InGaAs has a bandgap appropriate for the fiber communication system and also a high electron mobility. The conduction band discontinuity at the InGaAs/InAlAs heterointerface of ~0.5eV is advantageous over the InGaAs/InP heterointerface. The interdiffusion of column III atoms at a heterointerface of quantum wells has been studied for AlGaAs/GaAs (Deppe et al 1988), InP/InGaAs (Julien et al 1991), GaAs/InGaAs (Bradley et al 1993) and InAlAs/InGaAs (Miyazawa et al 1988). Impurity-enhanced interdiffusion of column III atoms has been regarded the major cause and various methods have been reported to incorporate impurities into QWs. Ion implantation into a quantum well structure is one of the popular methods. This method, however, introduces not only impurities into QW but also a lot of defects like vacancies and interstitials.

The interdiffusion at InGaAs/InAlAs is rather complex because InGaAs/InAlAs has three column III atoms; In, Ga and Al. Among them, interdiffusion of In can be neglected if In diffuses uniformly over a QW structure because the composition of In is almost the same between InGaAs and InAlAs lattice matched to InP. Most of the previous papers reported merely on the interdiffusion between Ga and Al (Miyazawa et al 1988, 1989) (Chi et al 1988) (O'Brien et al 1990) (Bryce et al 1991). However, Baird *et al.* (1988a, 1988b) reported on the interdiffusion between Ga and In. They speculated that Ga diffused into InAlAs but Al didn't diffuse from InAlAs and thereafter In diffused into InGaAs. Details of the mechanism are not yet known.

This paper deals with a photoluminescence study on the enhanced interdiffusion at InGaAs/InAlAs interfaces due to Si and B ion implantations and post-implantation annealing. We found the interdiffusion between Ga and Al due to Si ion implantation, and between Ga and In due to B ion implantation. By comparing these results, we discuss possible mechanisms for the interdiffusions.

2. EXPERIMENT

InGaAs/InAlAs QW structures were grown on an Fe-doped InP (100) substrate by metal-organic vapor phase epitaxy. Three non-dope InGaAs wells of 20, 6.4, and 3.5nm widths with 24nm wide non-dope InAlAs barriers were grown successively on a 200nm wide non-dope InAlAs buffer layer and finally capped with a 40nm wide non-dope InP layer. Double energy implantation of Si(80+180keV) or B(30+70keV) was carried out to achieve uniform distributions over QWs. The Si density was varied from 1.8×10^{17} to $3.9 \times 10^{19}cm^{-3}$. The relative depth profile of B was adjusted to become the same as that of Si and the B density was $3.9 \times 10^{18}cm^{-3}$. Post implantation annealing was performed in an infrared lamp furnace. Samples were put into a carbon pill box with an SiO_2 proximity cap, and then annealed at 600~850°C for 15, 180 and 600sec in H_2 flow of 0.5ℓ/min. Interdiffusion was monitored by the energy shift of the photoluminescence (PL) peak from each InGaAs quantum well. PL measurement was performed at 24K using the 514.5nm Ar^+ laser line as an exciter and a liquid-nitrogen-cooled Ge detector.

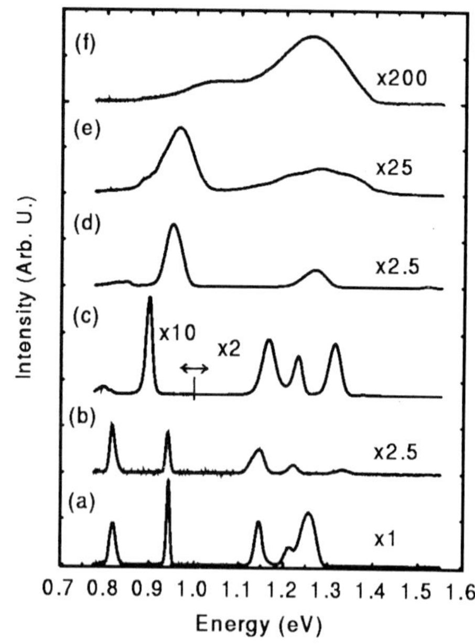

Fig.1 : Typical PL spectra from (a)as-grown, (b)unimplanted, (c)B implanted and (d),(e),(f)Si implanted samples annealed at 700°C for 180sec. The B density is $3.9\times10^{18}cm^{-3}$. The Si densities are (d)$3.9\times10^{17}cm^{-3}$, (e)$3.9\times10^{18}cm^{-3}$ and (f)$1.8\times10^{19}cm^{-3}$.

3. RESULTS

The PL spectrum of an as-grown sample was shown in Fig.1 (a). The PL peaks observed in the as-grown spectrum at 1.145, 0.942 and 0.818eV are from 3.5, 6.4 and 20nm thick wells, respectively. A small peak at 1.213eV which corresponds to the emission from a 2.9nm thick well may come from a well which is by one-monolayer thinner than the 3.5nm thick well. The peak at 1.256eV is speculated as the emission from the InP(cap)/InAlAs(barrier) interface. The PL spectrum of an unimplanted sample which was annealed at 700°C for 180sec was shown in Fig.1 (b). No energy shift of the peaks was observed. However, a slight shift to a lower energy was observed when the sample was annealed at 850°C for 180sec.

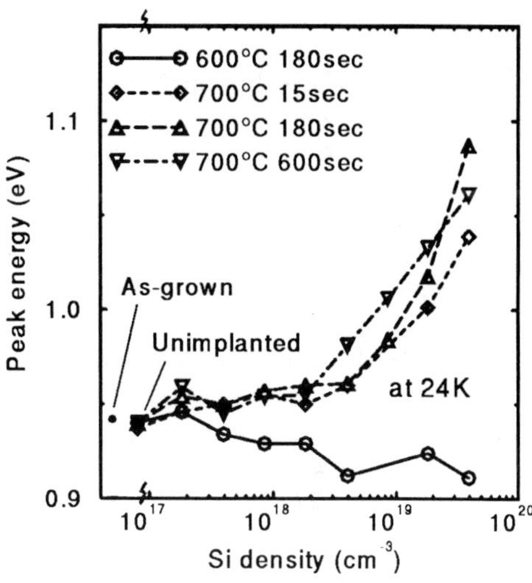

Fig.2 : PL peak energies from the 6.4nm well vs Si densities are shown. Slight red shifts are observed by annealing at 600°C for 180sec.

Typical PL spectra of Si or B implanted samples are also shown in Fig.1. Systematic shifts of the PL peaks to higher energies (blue shift) with increasing Si densities were observed as shown in Fig.1 (d),(e) and (f). On the other hand, shifts to lower energies (red shift) in Fig.1 (c) were found by B implantation. Because unimplanted samples showed no PL peak shift as shown in spectrum (b), the above energy shifts were due to enhanced interdiffusion by Si or B implantations. The blue shift due to Si ion implantation was considered to be caused by interdiffusion between Ga and Al and the red shift due to B implantation by interdiffusion between Ga and In.

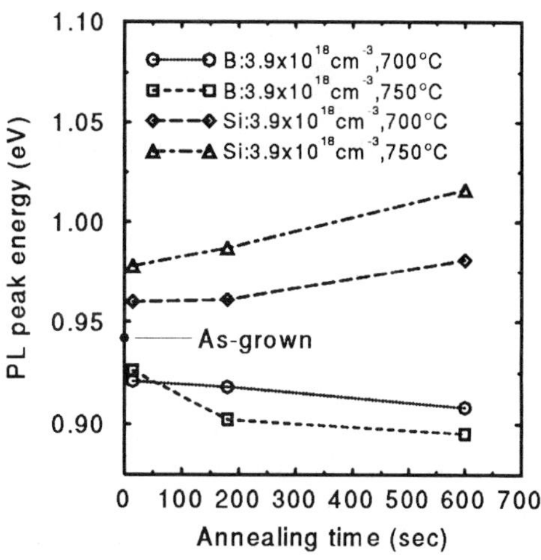

Fig.3 : PL peak energies from the 6.4nm well vs annealing time are shown. The shift finished almost within 15sec either Si or B implantation.

In the following, we analyze mainly the energy shifts from the 6.4nm well, since systematic PL energy shifts in case of Si implantation could be clearly observed in the whole dose range from the 6.4nm well. The PL peak from the 3.5nm well was covered by an implantation induced large broad peak around 1.25eV, and the PL peak from the 20nm well became too weak to detect at high Si densities. The PL peaks were significantly broadened at high Si densities (about 10times larger than the original FWHM at $1.8 \times 10^{19}cm^{-3}$). In case of $3.9 \times 10^{18}cm^{-3}$ B implantation, a red shift was not observed in the emission peak from the 3.5nm well. The PL peaks were not so much broadened (about twice the original value) as those of Si implanted samples. Fig.2 shows the energy shift in the PL peaks from the 6.4nm well as a function of Si density. Significant blue shifts of PL peaks were observed above a critical Si density of $2 \sim 3 \times 10^{18}cm^{-3}$ in samples annealed above 700°C. In contrary, samples annealed under the minimum annealing condition for the PL observation (at 600°C for 180sec), a slight red shift was observed instead of a blue shift. The PL peak energy shift of Si implanted samples as a function of time is shown in Fig.3. The blue shift finished almost within 15sec at above 700°C and then saturated for longer annealing (till 600sec). However, the initial blue shift made a turn towards lower energies after long annealing above 800°C. This indicates that the interdiffusion took place between Ga and In from InGaAs and InAlAs, respectively. In case of B ion implantation, the red shift finished almost within 15~180sec and then saturated for further annealing. The saturated values of both the blue and red shifts mainly depended only on the densities of the impurities and did not almost depend on the annealing temperature, though a slight larger blue shift was observed at higher annealing temperature.

The PL peaks from the 20nm well split into two or three peaks towards higher energies after Si implantation and annealing. It is difficult to consider that these apparent blue shifts were caused by the interdiffusion of Ga and Al, because the shifts were much larger than were expected from our theoretical estimation based on the results for the 6.4nm well. Instead, the split peak values agreed with the possible transition energies between theoretically estimated higher order quantized levels in the conduction bands of the 20nm well. A lot of electrons created by Si impurity are considered to fill the higher quantized energy levels and the transitions related to these levels may appear in the PL like blue shifts. On the other hand, B implantation caused small red shifts in the PL peak from the 20nm well and no peak split was observed. Though B implantation also causes donor type defects (Yamamura et al 1993), its density is not high enough for electrons to fill the higher quantized energy levels.

4. DISCUSSION

Possible mechanisms to cause interdiffusion due to Si ion implantation and annealing are 1) direct mixing during implantation, 2) impurity-enhanced diffusion and 3) defect-induced diffusion. As the most PL energy shift occurred within the minimum annealing time of 15 sec used in this study, we don't know whether the observed PL peak energy shifts are due to direct mixing or not. However, we observed no energy shift in samples annealed under the minimum annealing condition (600°C, 180sec) to measure PL of implanted samples. This result excludes the possibility of direct ion mixing of column III atoms. Impurity-enhanced interdiffusion has been reported by several workers in case of ion implantation of Si (Baird et al 1988a), O (Rao et al 1990), F and B (Bryce et al 1991) into InGaAs/InAlAs MQW. These impurities were considered to make complexes such as Si_{III}-Si_V pairs or O_{As}-V_{III}-O_{As}. Diffusion of these complexes through InGaAs/InAlAs heterointerfaces was followed by interdiffusion between Ga and Al during annealing for above 1 hour or more. However the mechanism described in the previous papers can not explain our results of Si ion implantation because interdiffusion between Ga and Al finished almost only within 15sec of annealing in this study, whereas Si impurity is reported not to diffuse in such a short time (Hailemariam et al 1992). Moreover, the reported critical Si density of $1 \sim 2 \times 10^{19}\text{cm}^{-3}$ is about a decade larger than that obtained in our study. Interdiffusion between Ga and In by B implantation with regard to annealing time showed a similar tendency as that between Ga and Al by Si implantation. The above results suggest defect-induced interdiffusion as the possible mechanism both in Si and B implantations.

Defect-induced interdiffusion was previously reported for GaAs/AlGaAs QW structures by Ga implantation (Vieu et al 1991, 1992). Point defects like vacancies or interstitials can enhance the diffusion of constituents. This kind of enhancement is completed when defects are annealed out by annealing. This defect-enhanced interdiffusion mechanism explains our results. The PL intensity from the QW wells is a good measure of crystal quality and implantation induced defects are almost annealed-out within 15sec, too. The similar behavior between the peak shift and its intensity against annealing supports the defect enhanced interdiffusion mechanism.

Among the three column III atoms in the InGaAs/InAlAs system, Al has the largest binding energy and In has the smallest. We speculate that Al interstitials (I_{Al}) are formed as well as interstitials of other column III atoms and vacancies by Si ion implantation. I_{Al} interstitials may diffuse fastest due to its smallest atomic radius. The diffusion of I_{Al} from InAlAs may be followed by the diffusion of Ga from InGaAs. The diffused I_{Al} interstitials may find column III sites in InGaAs and disappear. The maximum saturated diffusion length evaluated (see appendix) is about 4nm for the Si density of 3.9×10^{19}cm^{-3}. On the contrary, B ion implantation hardly forms I_{Al} interstitials due to its small atomic mass, but forms In interstitials due to its smallest binding energy. Then, Ga atoms of InGaAs move to V_{In} in InAlAs. Unimplanted samples which were annealed at 850°C for 180sec showed the interdiffusion between Ga and In (red shift). A similar phenomenon was reported by Baird *et al.* (1988b) when they annealed non-dope InGaAs/InAlAs heterointerface at 812°C for 10hours. The enhanced diffusion of In from InAlAs into InGaAs due to a chemical potential gradient followed by the diffusion of Ga into InAlAs with no Al diffusion was speculated.

5. CONCLUSION

Enhanced interdiffusion due to Si or B ion implantation into InGaAs/InAlAs interfaces was observed by means of PL measurements from the QWs. Si ion implantation caused blue shifts in the PL peaks, whereas B ion implantation caused red shifts. Both results were explained in terms of defect-induced interdiffusion. In case of Si ion implantation, implantation induced I_{Al} interstitials diffuse fast from InAlAs followed by the diffusion of Ga, leading to the blue shift. The interdiffusion finishes almost within 15sec when these defects are annealed out. In case of B ion implantation, I_{Al} interstitials are not formed and the diffusion of In and Ga takes place, resulting in the red shift.

6. ACKNOWLEDGEMENT

We would like to thank Dr. G. Sasaki and Dr. H. Hayashi of Sumitomo Electric Industries Co. Ltd. and Dr. M. Tsuchiya of Univ. of Tokyo for discussion. We would also like to thank S. Toriihara and K. Suzuki for technical assistant.

APPENDIX

We estimated the atomic diffusion length of the Ga ans Al exchange from the PL peak energy shift. We assumed an error function for the compositional change in Al and Ga. We also assumed that In composition is unchange and Ga and Al similarly interdiffuse each other.

$$x_{\text{Al}} = 0.48 \cdot \left\{ 1 + \frac{1}{2} \cdot \text{erf}\left(\frac{z - d/2}{L}\right) - \frac{1}{2} \cdot \text{erf}\left(\frac{z - d/2}{L}\right) \right\} \tag{1}$$

where d was a well width and z was a direction perpendicular to the well. The center of the well was defined as $z = 0$. The energy gap of the mixed $\text{In}_{0.53}\text{Ga}_{0.42-x}\text{Al}_x\text{As}$ was assumed to change linearly with the Al composition x.

$$Eg(x) = 0.794 + 1.525x \quad (\text{eV}) \quad \text{at 24K.} \tag{2}$$

The discontinuity ratio of the conduction and valence bands was assumed 7:3 (O'Brien et al 1990). The theoretical PL peak energy evaluated as a function of L was calculated by solving one electron state in the potential defined by eqs.(1) and (2) using a method by Cruz Serra *et al.* (1991).

REFERENCES

Deppe D. G. and Holonyak Jr. N. 1988 J. Appl. Phys. 64 R93
Julien F. H., Bradley M. A., Rao E. V. K., Razeghi M. and Goldstein L. 1991 Optical and Quantum Electronics 23 S847
Bradley I. V., Gillin W. P., Homewood K. P. and Webb R. P. 1993 J. Appl. Phys. 73 1686
Miyazawa T., Kawamura Y. and Mikami O. 1988 Jpn. J. Appl. Phys. 27 L1731
Chi J. Y., Koteles E. S. and Holmstrom R. P. 1988 Appl. Phys. Lett. 53 2185
Miyazawa T., Suzuki Y., Kawamura Y., Asai H. and Mikami O. 1989 Jpn. J. Appl. Phys. 28 L730
O'Brien S., Shealy J. R., Chia V. K. F., and Chi J. Y. 1990 J. Appl. Phys. 68 5256
Bryce A. C., Marsh J. H., Gwilliam R. and Glew R. W. 1991 IEE Proc. J 138 87
Baird R. J., Potter T. J., Lai R., Kothiyal G. P. and Bhattacharya P. K. 1988a Appl. Phys. Lett. 53 2302
Baird R. J., Potter T. J., Kothiyal G. P. and Bhattacharya P. K. 1988b Appl. Phys. Lett. 52 2055
Yamamura S., Kimura T., Yugo S., Saito R., Murata M. and Kamiya T. 1993 Proc. of 8th international conference on ion beam modification of materials (to be published)
Rao E. V. K., Ossart P., Thibierge H., Quillec M. and Krauz P. 1990 Appl. Phys. Lett. 57 2190
Hailemariam E., Pearton S. J., Hobson W. S. and Luftman H. S. 1992 J. Appl. Phys. 71 215
Vieu C., Schneider M., Planel R., Launois H., Descouts B. and Gao Y. 1991 J. Appl. Phys. 70 1433
Vieu C., Schneider M., Launois H. and Descouts B. J. Appl. Phys. 71 4833
Cruz Serra A. M. and Abreu Santos H. J. Appl. Phys. 70 2734

Paper presented at the Int. Symp. GaAs and Related Compounds, Freiburg, 1993

Specific role of isoelectronic antimony implants in the disordering of GaAs–AlGaAs multi-quantum well structures

E.V.K. Rao, Ph. Krauz, C. Vieu*, M. Juhel, and H. Thibièrge

France Télécom/CNET-PAB, Laboratoire de Bagneux
196, Avenue Henri Ravera, 92220 - Bagneux, France.

*Laboratoire de Microélectronique et de Microstructures-CNRS
196, Avenue Henri Ravera, 92220 - Bagneux, France.

ABSTRACT: The compositional disorder of GaAs-AlGaAs MQWs promoted by isoelectronic Sb implant damage and furnace anneals is shown to exhibit, in difference with the implants of several other nondopant species, some unique features highly useful for device applications. Each of these features is analysed in comparison with the data on In and Ga implanted MQWs. A satisfactory explanation to this new data on Al/Ga interdiffusion is proposed by considering the group V isoelectronic nature of Sb and the specific character of structural defects generated by its implant in GaAs-AlGaAs MQWs.

1. INTRODUCTION

Since the disordered regions free of charge carriers are highly useful in the development of new opto-electronic and photonic devices, the ion implant-damage induced disorder in GaAs-AlGaAs multi-quantum wells has been extensively investigated in the past (Gavrilovic et al 1985, Hirayama et al 1985, Mei et al 1988, Leier et al 1990). Indeed, while the implant of any nondopant species (lattice constituent or any isoelectronic atom, and either an inert or electrically neutral atom) ensures the absence of free carriers, the use of ion implantation further assures a thorough control on the obtention of spatially localized disordered regions. Of the different nondopants implants investigated in the past in GaAs-AlGaAs MQWs, as will be shown in this paper, the disorder promoted by the (group V) isoelectronic Sb implant damage possesses certain unique features that are highly useful for device applications.

2. EXPERIMENTAL

Nominally undoped GaAs-Al(Ga)As MQWs with different total thicknesses of quantum layers (~0.5 to ~3 μm) grown either by AP-MOCVD or MBE on undoped GaAs buffer over SI GaAs substrates have been investigated in this study. Implants of equal dose (5X10 14 ions.cm-2) and energy (350 keV) of either Sb or In (nearly same mass as Sb) have been performed in these structures at 25 and ~250°C. The implanted and as-grown control samples were furnace annealed at 850°C for periods ranging from 0.5 to 7 h under a continuous gas flow of pure argon with 10% hydrogen. The surface of the samples during heat treatments has been protected by covering face to face with a freshly polished semi-insulating GaAs and lodging the assembly inside a semi-closed graphite boat containing large quantities of pure GaAs powder. The PL spectroscopy and SIMS depth profiling to monitor disorder and cross-sectional TEM measurements to characterize the structural defects have been employed.

3. RESULTS AND DISCUSSION

Several earlier studies on nondopant species implants in GaAs-AlGaAs MQWs have consistently measured a maximum of disorder in as-implanted structures, while subsequent anneals led only to a little or no additional disorder suggesting a saturation tendency of Al/Ga interdiffusion (Hirayama et al 1985, Mei et al 1988, Leier et al 1990). On the other hand, for Sb implants, in addition to the disorder promoted by the collision and relaxation processes during implant as before, we have further measured a significant amount of disorder subsequent to long duration furnace anneals (up to about 7 h) at 850°C. This is illustrated in fig. 1 where we have compared

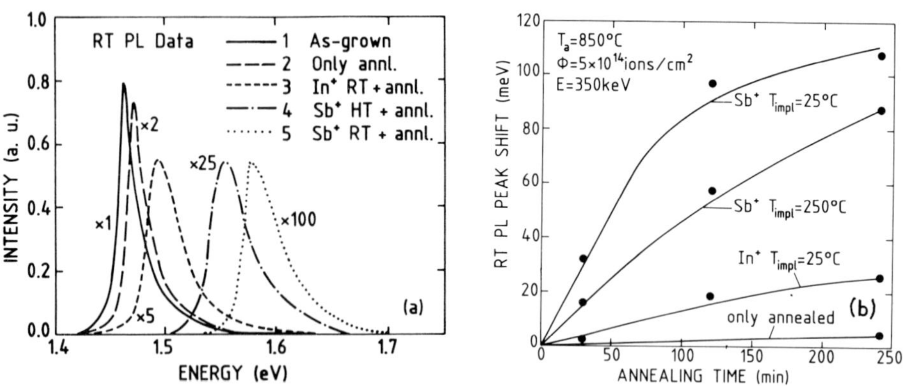

Fig. 1. RTPL spectra (1a) and the corresponding high-energy shift of the PL peaks (1b) with anneal duration at 850°C measured on Sb (at 25 and ~250°C)) and In (at 25°C) implanted MQWs.

the RTPL spectra (fig. 1a) and the high-energy shifts of the corresponding PL peaks (fig. 1b) measured on a given GaAs-AlGaAs MQW subsequent to equal dose implants of Sb (isoelectronic with As) and In (isoelectronic with Ga) followed by long duration furnace anneals at 850°C. Despite a reduced PL quality as expected in implanted and annealed samples, it is clear that the Sb implant damage in contrast to that induced by In implant (of nearly same mass) also at RT, leads to a much severe disorder which exhibits saturation only for prolonged anneal durations (see spectra 3and 5 in fig. 1a). Also, the other noteworthy feature of Sb implant, is that even a hot implant at ~250°C (reduced damage density) causes a significant disorder for prolonged anneal durations as compared to the RT In implant (see spectra 3 and 4 in fig. 1a).

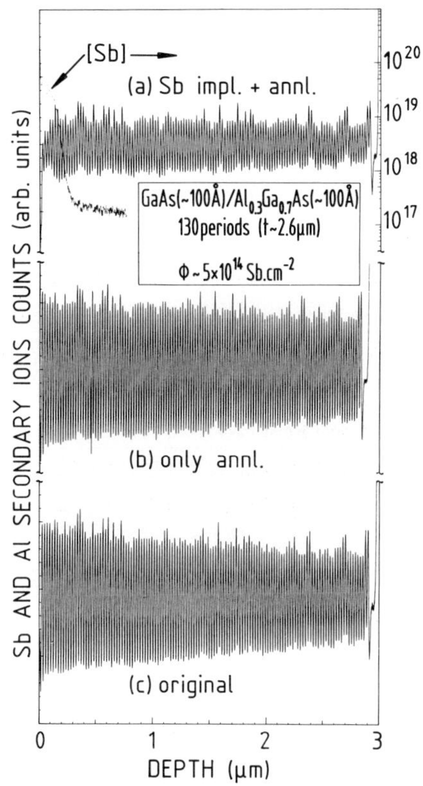

***Fig 2.** SIMS-Al depth profiles recorded on Sb implanted and annealed (2a), only annealed (2b: 850°C-4h) and as-grown (2c) thick MQWs. Also shown in fig. 2a is the Sb distribution which is scarcely affected by anneal.*

The other important feature of the Sb implant is its ability to induce disorder over widths of few microns which property has never been observed before with the implants of other nondopant elements (Rao et al 1993). This is illustrated by the SIMS-Al depth distributions recorded on a ~2.6 μm thick GaAs (~10 nm)-$Al_{0.3}Ga_{0.7}As$ (~10 nm) MQW sample subsequent to a RT Sb implant same as above (corresponding to a R_p and ΔR_p of ~0.09 and ~0.04 μm, respectively) followed by a 4 h anneal at 850°C (fig. 2). Comparing the amplitudes of Al oscillations in as-grown (fig. 2a), only annealed (fig. 2b) and implanted and annealed samples (fig. 2c), it is clear that this shallow Sb implant (~0.17 μm thick when taken as $R_p+2\Delta R_p$) has indeed promoted a *severe and nearly uniform disorder* all over the MQW thickness, ie., ~30 times the depth of the implant projected range (R_p). Furthermore, since the Sb atomic distribution (see fig. 2c) has scarcely shifted from its initial (as-implanted) position, this unusual depth extension of disordered region width must be a consequence of the defects generated by the Sb implant.

For a better knowledge on the properties of implant induced defects, we have undertaken a comparative study of structural defects in Sb and also Ga as-implanted MQWs using TEM measurements. We emphasize here that the Ga implant, much like In and unlike Sb, leads only to a limited disorder which saturates at the early stages of heat treatment (Vieu et al 1991). In addition to the usually observed and often reported (Ralston et al 1986, Arakawa et al 1987) defective zones confined to the implanted regions, we have detected several "spot-like" defects in the depth of the Sb implanted MQW. This can be seen from the micrographs of figures 3a and 3b which respectively represent the X-TEM dark-field images of Sb and Ga implanted MQW samples. These micrographs are deliberately taken underneath the (Sb and Ga) implanted regions to highlight the presence of "spot-like" defects in the depth of the Sb (fig. 3a) but not Ga (fig. 3b) implanted sample. These defects have been identified as precipitates and to a first approximation their size is equal to the well width (~10 nm). During subsequent anneals, a reduction in the size of these defects coincided with the start of quantum well disordering in the depth of the structure. For example, a 30 min anneal at 850°C resulted in a significant reduction of precipitate size while they have completely disappeared for a 4 h anneal also at 850°C (corresponding TEM micrographs are not shown here). This confirms the dissolution of the precipitates for longer anneal durations.

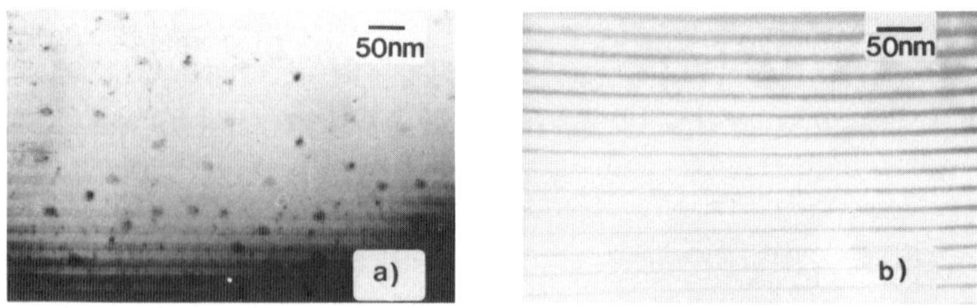

Fig. 3. *Cross-sectional TEM micrographs showing a comparison of defect structure in Sb (3a) and Ga (3b) as-implanted MQWs. Note the presence of precipitates only in the Sb-implanted sample.*

The newly detected precipitates covered nearly ~90% of the Sb implanted MQW thickness (~2.6 μm). Furthermore, as seen from fig. 3a, they are distributed (nearly) uniformly all over the MQW with a majority of them located preferentially in the GaAs well layers while only some are trapped at the hetero-interfaces. Since the Sb atoms are confined to the implanted layer in the near-surface region (from SIMS data), one or all of the lattice constituent atoms (Al, Ga or As) must necessarily participate in the formation of these precipitates. To our knowledge this is the first report of precipitate-like defects in the depth of ion implanted GaAs-AlGaAs MQWs. Incidentally, these defects closely resemble the As precipitates recently detected in GaAs-AlGaAs MQWs grown under special conditions such as, a low growth temperature (~300°C) with excess As, and subsequently annealed at ~700°C for 10s (Mahalingam et al 1992). According to these authors, the excess As present in the form of interstitials (As_i) and anti-site defects (As_{Ga}) in the as-grown material out-diffuses during post-growth anneals to precipitate preferentially in the GaAs well layers. This precipitation behavior of As appears to be much similar to the case Sb implanted MQWs since the precipitates detected here are also preferentially located in the GaAs well layers.

Assuming that the above mentioned As precipitation process is also valid for the formation of defects observed here, we propose to understand the specific properties of Sb implant damage induced disorder in the following framework. Recalling that Sb is a group V isoelectronic element, its substitution on As site (Sb_{As}) during implantation might free As atoms, a fraction of which could get into the interstitial positions (As_i). These inherently mobile interstitial atoms could be further aided by a strain field to migrate deep inside the structure. This strain-field could have different origins : the strain induced by the implant damaged region or the one developed down the Sb concentration gradient (because of size difference between Sb and As atoms) or even the built-in strain prevailing in the as-grown MQW layers. Also, since defects in general are more mobile in AlGaAs (Collins et al 1991) than in GaAs, we expect an accumulation of the As interstitial atoms (As_i) preferentially in the GaAs wells. But, owing to the limited size of wells, the precipitation of As must necessarily liberate some Ga atoms to form As anti-site defects (As_{Ga}). These liberated Ga atoms could be trapped either by the newly formed As-rich precipitates or by the hetero-interfaces. Although speculative at present, this simple model provides a reasonable explanation to the formation of As-rich precipitates all over the thickness (~2.6 μm) of the as-Sb implanted MQW samples. During post-implant annealings, the dissolution of these precipitates might liberate the Ga atoms (bound to the precipitates) and also generate excess Ga vacancies (V_{Ga}) through the annihilation As anti-site defects with the thermally generated As vacancies. These two group III defects in excess would contribute to the enhancement of Al/Ga interdiffusion all over the MQW thickness.

4. SUMMARY AND CONCLUSIONS

In summary, we have shown that the GaAs-AlGaAs MQW disordering achieved with the group V isoelectronic Sb implant damage, in contrast to the implants of several other nondopant species (including group III isoelectronic elements), possesses the following specific features: i). The saturation of Al/Ga interdiffusion only for long duration anneals (typically beyond 4 h at 850°C). ii). Obtention of a long range disorder over distances of few microns in thick MQWs using shallow implants iii). The formation of nearly uniformly distributed precipitates all over the width of the MQW far beyond the implanted region. While the first of these features helps to achieve a high degree of disorder free of charge carriers, the second one permits to extend the widths of disordered regions to few microns depending on the device type. The last of these features permits to predict a preferential direction for defect diffusion, when spatially localized into the depth of MQWs with a minimum lateral spread. Baring these advantages in mind, we have successfully applied this technique for the first time to delineate 2D-array pixels of size down to 2μmx2μm for the fabrication of vertical cavity nonlinear micro-resonators in a ~3 μm thick GaAs-AlGaAs MQW (Sfez et al 1992).

ACKNOWLEDGEMENTS

The authors are grateful to R. Azoulay and F. Alexandre for kindly supplying the different MQW samples investigated in this study. It is also their pleasure to thank B.G Sfez and J.L. Oudar for attempting the first application of this work to fabricate 2D-array vertical cavity optical devices.

REFERENCES

Arakawa Y, Smith J S, Yariv A, Otsuka N, Choi C, Gu B P and Venkatesan T 1987 Appl. Phys. Lett. 50 92

Collins A G, Smith P W, Jacobson D C and Poate J M 1991 J. Appl. Phys. 69 1279

Gavrilovic P, Deppe D G, Meehan K, Holonyak N, Coleman J J and Burnham R D 1985 Appl. Phys. Lett. 47 130

Hirayama Y, Suzuki Y and Okamoto H 1985 Jpn. J. Appl. Phys. 24 1498

Leier H, Forchel A, Hocher G, Hommel J, Bayer S, Rothfritz H, Weimann G and Schlapp W 1990 J. Appl. Phys. 67 1805

Mahalingam K, Otsuka N, Melloch M R and Woodall J M 1992 Appl. Phys. Lett. 60 3253

Mei P, Venkatesan T, Schwarz S A, Stoffel N G, Harbison J P, Hart D L and Florez L A 1988 Appl. Phys. Lett. 52 1487

Ralston J, Wicks G W, Eastman L F, De Cooman B C and Carter C B 1986 J. Appl. Phys. 59 120

Rao E V K, Juhel M, Krauz Ph, Gao Y and Thibierge H 1993 Appl. Phys. Lett. 62 2096

Sfez B G, Rao E V K, Nissim Y I and Oudar J L 1992 Appl. Phys. Lett. 60 607

Vieu C, Schneider M, Descouts B, Gao Y, Planel R and Launois H 1991 J. Appl. Phys. 70 1433

Growth and characterization of huge GaAs crystals

S.Kuma, M.Shibata and T.Inada

Advanced Research Center, Hitachi Cable, Ltd.
5-1-1 Hitaka-cho, Hitachi, Ibaraki 319-14 Japan

Abstract Most of the technology necessary for GaAs device fabrication has now been established: crystal quality, understanding of the activation, processing techniques, etc. The last breakthrough for the widespread use of GaAs devices is the growth of large, long crystals.

Polycrystallization by concentrated dislocations during growth was studied focusing on the curvature of the concave solid-liquid interface, and it was found that the curvature-center must be placed at the outer side of the crystal for single crystal growth. Three-inch GaAs crystal 770 mm long, 4-inch crystal 480 mm long and 6-inch GaAs crystal have been successfully grown.

1. INTRODUCTION

Most of the key technology necessary for GaAs device fabrication has been established and the demand for these devices is spreading. Activation by Si-implantation into SI (semi-insulating) GaAs crystal wafers is the most popular method of fabrication. For uniformity and stability of the activation, the GaAs wafers must be homogeneous, and large, long ingots grown by LEC (liquid encapsulated Czochralski) method are needed for the mass production of devices.
There are several problems in growing these huge GaAs crystals. The most severe one is the occurrence of polycrystallization during growth. Other difficulties are the multiplication of dislocations and the degradation of homogeneity. Theoretical and experimental analysis has now solved these problems, and made possible the stable production of huge GaAs crystals which are also homogeneous.

2. INGOT ANNEALING

2.1 BEHAVIOR OF EXCESS ARSENIC IN GaAs

Semi-insulation and activation are strongly dominated by deep donor EL2 and carbon acceptor. Any deviation of EL2- and C-concentration degrades the homogeneity of the huge GaAs crystals. Impurity C comes from the graphite furnace through the CO formation in the ambient gas. The concentration of CO gas in the chamber is controlled during growth and the C concentration in the crystal becomes homogeneous along its length and across its diameter. The deep donor EL2 forms from excess As in the crystal.

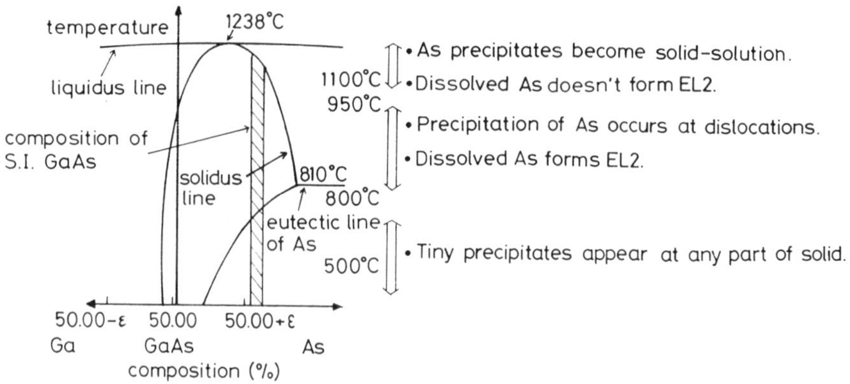

Fig.1 Phase diagram of Ga-As and behavior of excess As in various temperature ranges.

Recent studies (for example, Inada et al. 1989, Otoki et al. 1990) describe the behavior of excess As in GaAs. Figure 1 shows the behavior of excess As at various temperature ranges corresponding to the phase diagram of GaAs. Above 1100 ˚C, the excess As is dissolved in the solid. At 800-950 ˚C, some part of the excess As in the solid-solution precipitates on dislocations. Though this temperature range corresponds to that of solid-solution in the phase diagram, existence of dislocations disturbs the equilibrium state and the precipitation occurs. At this temperature range, part of the excess As in the solid-solution forms EL2. At 500-600 ˚C, far below the temperature corresponding to the solid-solution, As solution in the crystal becomes super saturated, and many tiny precipitates of As appear not only on the dislocations but spontaneously at any part of the crystal.

2.2 ANNEALING PROCESS

Applying the knowledge about excess As described above, a three-step annealing program was developed (Kashiwa et al.1990) and the uniformity of electrical properties in the crystal improved dramatically. The temperature program of the three-step anneal is shown in Fig.2. First, an ingot is annealed at 1100 ˚C or more, and the excess As becomes the solid-solution. Second, the ingot is cooled down rapidly to 500-600 ˚C; rapid cooling creates a large number of nuclei of As precipitates

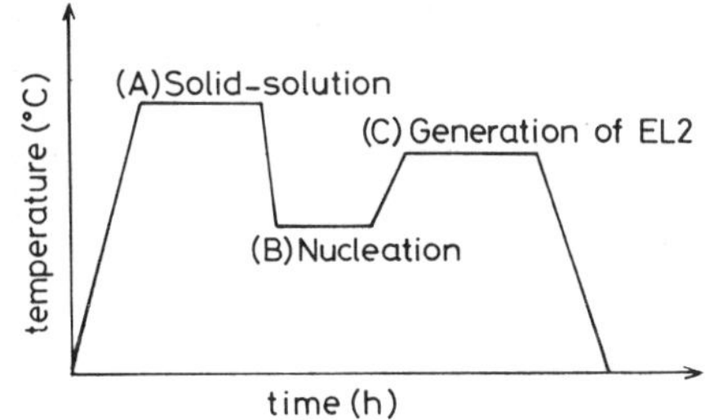

Fig.2 Temperature program of three-step annealing. Solid-solution at high temperature above 1100°C (A), Nucleation at low temperature of 500-600°C (B), Generation of EL2 at 800-1000°C (C).

randomly in the crystal. Finally, the ingot is heated up to 800-950 °C in order to make EL2. Using the three-step anneal method, a distribution of excess As in the crystal is not under the influence of dislocations and becomes uniform.

3. GROWTH OF HUGE SINGLE CRYSTALS

3.1 POLYCRYSTALLIZATION MECHANISM

The difficulty of growing huge single crystals is caused by polycrystallization during growth, and detailed observation of the polycrystallized ingot has shown that the accumulation of grown-in dislocations in the crystal is the cause of this polycrystallization (Shibata et al. 1993). Figure 3 shows KOH etched wafers cut from a polycrystallized ingot; a region of dense dislocations precedes the polycrystalline region. The grown-in dislocations have a tendency to propagate perpendicular to the solid/liquid interface (Scott et al. 1985). When the interface has a concave part towards the melt, grown-in dislocations accumulate and cause polycrystallization. The accumulation of dislocations can reportedly be prevented if the shape of the solid/liquid interface is convex towards the melt. But the interface shape is usually sigmoidal and has a concavity at the periphery. The convex interface is difficult to place during the growth of huge crystal.

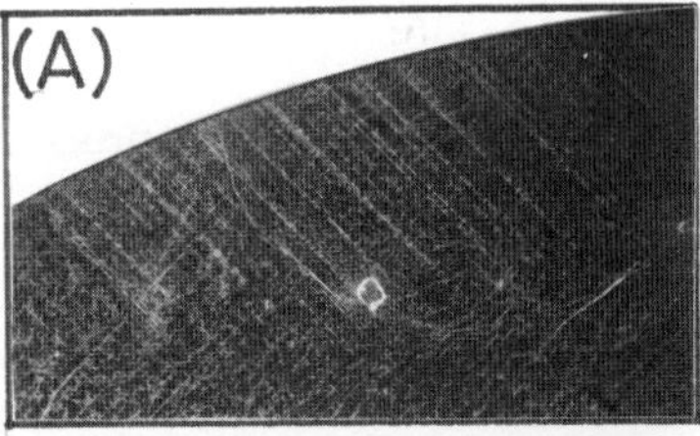

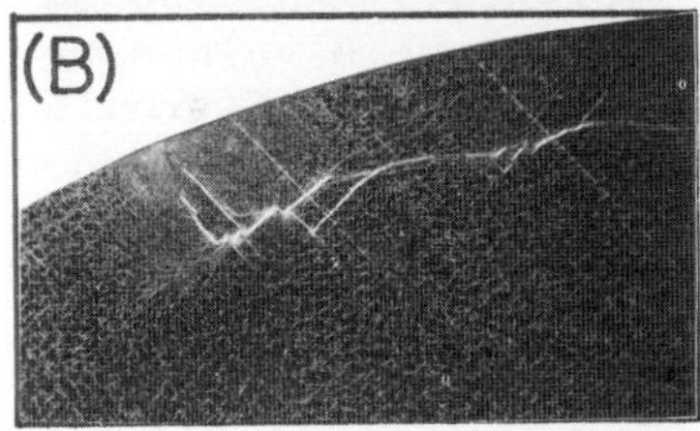

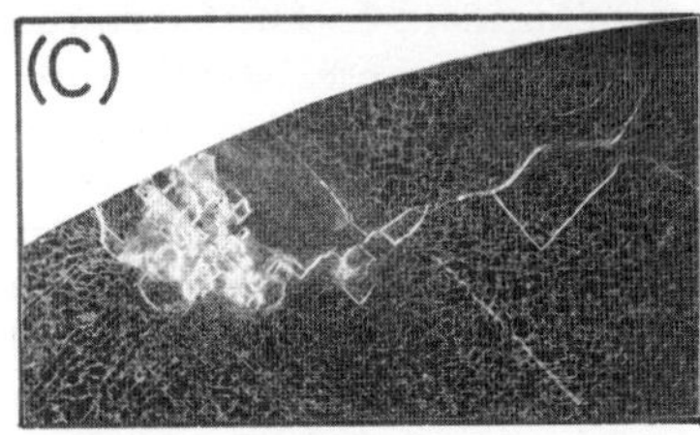

Fig.3 KOH etched (001) wafers cut from a polycrystallized ingot cut 80 mm from the seed end (A), 100 mm (B), 120 mm (C).

3.2 INDEX OF POSSIBILITY OF POLYCRYSTALLIZATION

At the concavity of the interface, the grown-in dislocations would tend to gather aiming at the center of the concave curvature, because they propagate perpendicularly to the interface. If the center of curvature is located within the crystal, dislocations accumulate in the crystal and polycrystallization occurs. When the center is located outside the crystal, the dislocations grow toward the solid/ gas interface of the crystal and disappear. Their disappearance averts polycrystallization. To evaluate the interface shape and quantify the tendency to polycrystallize, index D was proposed (Shibata et al. 1993). D is defined as the distance from the crystal surface to the center of curvature (Fig.4). In investigating index D of single and of polycrystallized 100 mm diameter ingots, the D values were measured from the lines of growth striation. The results are shown in Fig.4. The D value of single crystal was generally very large during growth, while that of polycrystallized crystal was near or below 0.

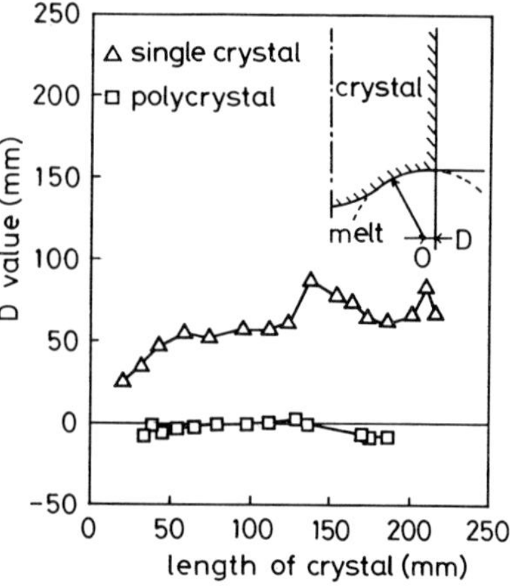

Fig.4 Distance of the center of concavity from crystal periphery (the D-value) as a function of crystal length. The D values were measured for both single and polycrystallized crystals.

3.3 CRYSTAL GROWTH

The shape of the growth interface is formed so that it is nearly the same as the isothermal line in the melt, therefore, this shape can be controlled by changing the temperature profile in the furnace. To set the most suitable temperature profile and to maintain a large D value during growth, the multi-zone heater system was developed which consists of three heaters with the power of each heater controlled independently during growth. Using this system, huge GaAs single crystals were successfully grown. Figure 5 shows the obtained single ingots: a 770 mm long crystal of 75 mm diameter (a), a 480 mm long crystal of 100 mm diameter (b) and a 170 mm long crystal of 150 mm diameter (c). The growth conditions for a 4 inch diameter crystal are shown in Table 1. The maximum 28 kg melt was directly synthesized from Ga and As with 1.6 kg B_2O_3 in a pBN (pyrolytic boron nitride) crucible of 225 mm diameter under 7 MPa of Ar pressure. The ingots were grown in the <001> direction under 2 MPa of Ar pressure.

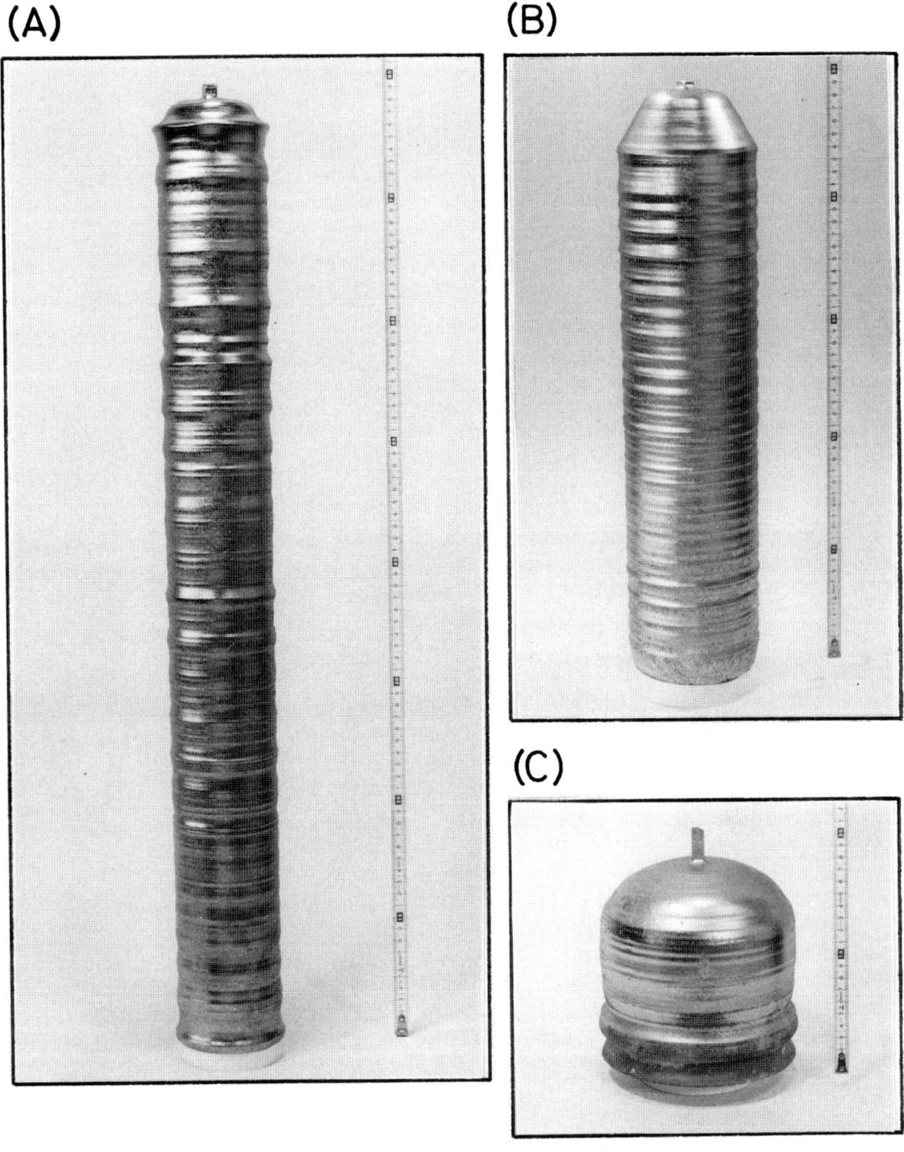

Fig.5 Photographs of 770 mm long, 75 mm diameter GaAs single crystal (A), 480 mm long, 100 mm diameter (B) and 170 mm long, 150 mm diameter (C).
Marker represents 100 mm.

Table 1 Growth conditions for 4 inch crystal

Item	Condition
GaAs weight	28000 g
B_2O_3 weight	1600 g
Crucible	225 mm diameter, pBN
Atmosphere	Ar, 2.0 MPa
Pulling rate	8 mm/h
Seed rotation	5 min^{-1}, clockwise
Crucible rotation	15 min^{-1}, counter-clockwise

4. CHARACTERIZATION

4.1 DISLOCATION DENSITY

There is concern that a long ingot will have more dislocations than a short one, because of the large thermal stress resulting from cooling of the top area of the long ingot. Wafers at the front, middle and tail portions of the 770 mm long ingot of 75 mm diameter were etched by molten KOH and EPD (etch pit density) was measured (Fig.6). The EPD values and profiles are at the same level as those of a conventional ingot, and no increase in EPD was observed. The reason is believed to be the following: the temperature at which the dislocations generate and/or multiply must be above 900 °C (Otoki et al. to be published) and the profile of this temperature range in the multi-heater system does not differ from that in a conventional hot-zone. Thus, dislocation density does not increase even though the top portion of the long ingot is cooled below the short one.

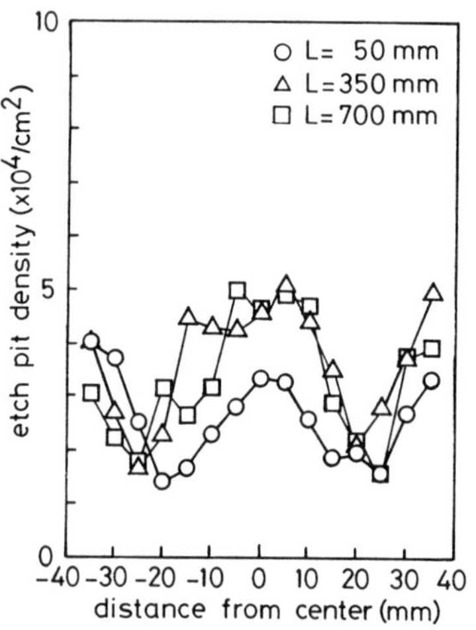

Fig.6 Distribution of EPD along [110] direction in (001) wafers cut from a 75 mm diameter, 770 mm long ingot.

4.2 ELECTRICAL PROPERTIES

After three-step annealing, longitudinal profiles of resistivity in the long and conventional ingots were measured by van der Pauw method at room temperature (RT). The resistivity as a function of crystal length is shown in Fig.7.

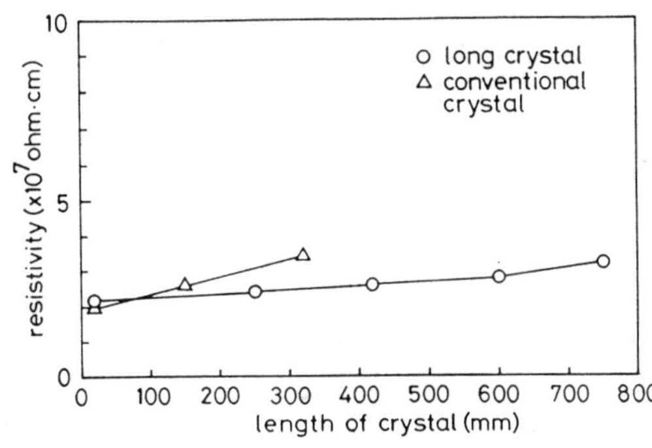

Fig.7 Resistivity of long and conventional ingots as a function of crystal length.

The range of resistivity from the front to the tail of the long ingot is the same as that of the conventional one. The figure suggests that the amounts of EL2 and C concentration were well-controlled in the huge crystal.

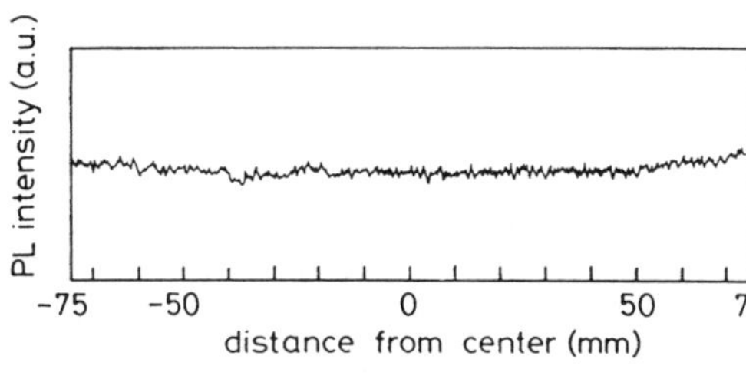

Fig.8 Line scan of PL intensity along [110] direction in a (001) wafer of 150 mm diameter. Measurement pitch is 125 μm across the wafer.

To evaluate the degree of uniformity of a large diameter crystal, the distributions of photoluminescence (PL) intensity of 1.42 eV were measured at RT using a 5145 A Ar laser. Distribution measured in this way is related to that of the threshold voltage (Vth) of FETs fabricated in a wafer (Kuma and Otoki 1987). The sample was a (001) wafer cut from a three-step annealed ingot of 150 mm diameter. The distributions were measured by 0.125 mm pitch along the [110] direction;measurement results are shown in Fig.8. The average standard deviation of PL intensities was less than 7 %. This value is the same level as that of ordinary crystal of 75 mm diameter. Even though the crystal is huge, the desired uniformity can be obtained by the appropriate thermal treatment after growth.

5. SUMMARY

The production technology of huge GaAs single crystals

including annealing has been established theoretically and experimentally. Huge single crystals have been successfully grown by controlling the growth interface shape using the LEC method. The homogeneity of crystal has been improved by controlling the C concentration in the crystal and three-step ingot annealing. The huge crystal has been characterized and found to be satisfactory for fabrication of GaAs ICs. This will allow great advances in GaAs devices.

ACKNOWLEDGEMENTS

The authors wish to thank Mr. T.Suzuki and Mr. M.Wachi for growing the crystals and Mr. Y.Otoki for measuring their properties.

REFERENCE

Inada T. et al., 1989, J. Crystal Growth 96, p.327-332
Kashiwa M. et al., 1990, Hitachi Cable Review No.9, p.55-58
Kuma S. and Otoki Y., 1987 in Defect Recognition and Image Processing in III-V Compounds II, E.R.Weber (Elsevier) p.1
Otoki Y. et al., 1990, J. Crystal Growth 103, p.85-90
Otoki Y. et al., to be published
Scott M.P. et al., 1985, Appl. Phys. Lett. 47 (12), p.1280-1282
Shibata M. et al, 1993, J. Crystal Growth 128, p.439-443

Paper presented at the Int. Symp. GaAs and Related Compounds, Freiburg, 1993

Anomalous increase of residual strains accompanied with slip generation by thermal annealing of LEC-grown GaAs wafers

M. Yamada, T. Shibuya, and M. Fukuzawa

Department of Electronics and Information Science, Kyoto Institute of Technology, Matsugasaki, Sakyoku, Kyoto 606, JAPAN

ABSTRACT: Quantitative characterization of residual strains in commercial 3″-diameter LEC-grown wafers with standard dimensions has been made before and after thermal annealing (20 min at 800°C in AsH_3(2 Torr)+Ar/N_2 atmosphere). It was sometimes observed that slip lines were generated along the ⟨011⟩ in the peripheral region of (100) wafers by the thermal annealing. There was found an anomalous increase of residual strains accompanied with the slip generation. It was discussed on the origins of slip generation and on the anomalous increase of residual strains found here.

1. INTRODUCTION

Semi-insulating GaAs crystals, grown by the liquid-encapsulated Czochralski (LEC) method, are key materials for developing high-speed integrated circuits. During thermal processes such as epitaxial growth and furnace or rapid thermal annealing after ion implantation, unwanted slip lines are sometimes generated in peripheral regions of wafers. Many investigators have known by experience that this slip generation becomes more severe when larger sizes of wafers are used to make devices. Kawase et al (1992) have recently shown that the slip generation during epitaxial growth is less in vapor-pressure-controlled Czochralski (VCZ) wafers than in conventional LEC wafers because the VCZ wafers have a low level of residual strains, compared with the LEC wafers. Although many efforts have been made to suppress the slip generation during thermal processes, the origins of slip generation are not clearly understood at the present stage. However, it may be presumed to be due to residual strains and/or thermal stresses during the thermal processes. Therefore, it is an interesting approach for us to characterize the residual strains quantitatively.

Recently, Yamada et al (1992 and 1993) have developed a high-sensitivity computer-controlled infrared polariscope and quantitatively characterized the residual strains in commercially-supplied GaAs wafers with standard dimensions. To investigate the origins of slip generation, we have quantitatively characterized the residual strains before and after thermal annealing of wafers. As a result of this comparative characterization, we have found that the residual strains are anomalously increased by the slip generation. In this paper, we first explain the quantitative characterization of residual strains and then present the procedure and results of the comparative study on the residual strains before and after thermal annealing. After investigating the origins of slip generation, we discuss on the anomalous increase of residual strains found here.

2. QUANTITATIVE CHARACTERIZATION OF RESIDUAL STRAINS

Since GaAs crystal belongs to the zincblende structure with the crystal symmetry of $\bar{4}3m$, the unstrained crystal is optically isotropic. However, if there are residual strains or external forces are applied, then the crystal exhibits birefringence due to the photoelastic effect. By measuring the strain-induced birefringence and analyzing the photoelastic effect, we may estimate the residual strains or the external forces. Yamada (1985) showed that the residual strain components $|S_{yy} - S_{zz}|$ and $2|S_{yz}|$ can be quantitatively characterized by measuring the principal angle ψ of the birefringence and the phase retardation δ when the infrared probing light with wavelength λ, is incident normal to the (100) GaAs wafer, that is,

$$|S_{yy} - S_{zz}| = k\delta \left| \frac{\cos 2\psi}{p_{11} - p_{12}} \right|, \tag{1}$$

$$2|S_{yz}| = k\delta \left| \frac{\sin 2\psi}{p_{44}} \right|, \tag{2}$$

where p_{ij}'s are the photoelastic constants and $k = (\lambda/\pi d n_0^3)$. Here, n_0 is the refractive index and d is the thickness of wafer. The strain component $|S_{yy} - S_{zz}|$ is the difference of the tensile strains along the crystallographic y and z axes while the strain component $2|S_{yz}|$ is the shear strain between the y and z axes. If we want to separate $|S_{yy} - S_{zz}|$ into S_{yy} and S_{zz} or to characterize the other strain components such as S_{xx}, S_{zx}, and S_{xy}, then we have to direct the probing light along other directions than the [100] direction.

If we use a local cylindrical coordinate system rather than the Cartesian or crystallographic coordinate system, then we may define the following strain component (Yamada 1992 1993):

$$|S_r - S_t| \equiv [(S_{yy} - S_{zz})^2 + (2S_{yz})^2]^{1/2}, \tag{3}$$

where the subscripts of r and t indicate the radial and tangential directions, respectively, in the local cylindrical coordinate system. The strain component $|S_r - S_t|$ describes not only the difference of the tensile strains along the radial and tangential directions in the local cylindrical coordinate system but also the geometrical average of $|S_{yy} - S_{zz}|$ and $2|S_{yz}|$, which may be used as a figure of merit for the total in-plane strains.

In order to measure a small strain-induced birefringence, we used a high-sensitivity computer-controlled infrared polariscope. The light source used was a high-luminosity diode emitting at λ=1.3 μm. The beam diameter and divergence were ~2 mm and ~10 mrad, respectively, at the probing position. The birefringence measurement of a whole wafer was made on a mesh points at intervals of 2.5 mm. In the evaluation of residual strains with eqs.(1)-(3), the refractive index and the photoelastic constants at the probing wavelength were used; that is, n_0=3.40, $|p_{11}-p_{12}|$=0.0463 and $|p_{44}|$=0.0686 (Adachi 1982 1983).

3. EXPERIMENTAL PROCEDURE AND RESULTS

Various LEC-grown GaAs (100) wafers were supplied from several manufacturers. Commercial 3-inch-diameter 600-μm-thick wafers were mainly examined in the present experiment. Their front and rear surfaces were mirror-polished.

Adjacent wafers were selected for the comparative study of residual strains before and after thermal annealing. One of the adjacent wafers was kept as a reference wafer without thermal annealing, which is here denoted by (A). The rest were thermally annealed at the temperature of 800° in the atmosphere of AsH_3(2 Torr)+Ar/N_2 with a furnace. The annealing time was typically 20 min. The wafer-supporting structure was varied to change the temperature distribution of wafer. Some wafers (B) were supported so that the heat conduction between supporting material and wafer may occur in some degree to distort the steady state temperature distribution slightly. The other wafers (C) were supported so that the best homogeneity may be obtained in the steady state temperature distrubution. In the wafers (B), a small number of slip lines were rarely observed along the ⟨011⟩ directions in the peripheral region of wafers. On the other hand, a large number of slip lines were sometimes observed in the wafers (C).

A typical set of two-dimensional distribution maps of $|S_r - S_t|$ measured in the unannealed reference wafer (A) and the annealed wafers (B) and (C) is shown in Fig. 1. The strain distributions are displayed in normarized units using a gray scale. The averaged values and the total variations are also given. The two-dimensional distribution map of the unannealed reference

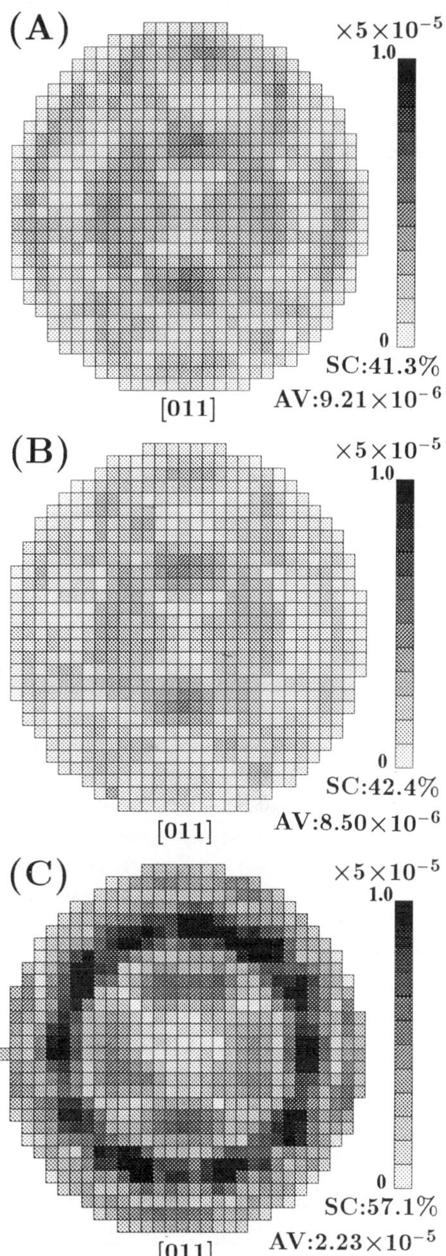

Figure 1. Two-dimensional distribution maps of $|S_r - S_t|$ measured in the unannealed wafer (A) and the annealed wafers (B) and (C) . A large number of slip lines were observed in the annealed wafer (C).

wafer (A) shows a weak fourfold symmetry and its distributions along the ⟨011⟩ directions exhibit a M shape rather than the U shape found frequently in previous works (Yamada 1985 1992 1993). Since the wafers examined here were sliced mainly from ingots which were ingot-annealed after crystal growth, it may be presumed that the residual strains in the peripheral region were relaxed in some degree by the ingot-annealing and hence the distributions along the ⟨011⟩ directions were changed from the U shape to the M shape. The distributions of residual strains vary from ingot to ingot and from wafer to wafer, depending the portion sliced from the ingot.

From the comparison of the annealed wafer (B) with the unannealed reference wafer (A), it is found that their two-dimensional distribution maps of residual strains are almost the same, although their averaged values of residual strains differ slightly. This suggests that the adjacent wafer has almost the same distributions of residual strains.

On the other hand, it is found from the comparison of the annealed wafer (C) with the unannealed reference wafer (A) or with the annealed wafer (B) that extremely large residual strains appear inside of the peripheral region where a large number of slip lines were observed and the averaged value becomes more than 2 times of those in the wafers (A) and (B). Since the distribution and the averaged value of the residual strain in the wafer (B), in which a small number of slip lines were generated by the thermal annealing, are almost the same as those in the unannealed reference wafer (A), the anomalous increase of residual strains observed in the wafer (C) is concluded to be due to the slip generation by the thermal annealing.

Figure 2 shows two-dimensional distribution maps of $|S_{yy}-S_{zz}|$ and $2|S_{yz}|$ measured in the annealed wafers (C), in which a large number of slip lines were observed. It is found that $|S_{yy} - S_{zz}|$ and $2|S_{yz}|$ are distributed in fourfold symmetries with strain maxima distributions rotated 45° against each other. $|S_{yy} - S_{zz}|$ has larger values in the inside of the [010] and [001] peripheral regions while $2|S_{yz}|$ has larger values in the inside of the [011] and [0$\bar{1}$1] peripheral regions. The averaged value of $2|S_{yz}|$ is about two-times larger than that of $|S_{yy} - S_{zz}|$. The reason may be that slip lines were generated mainly in the ⟨011⟩ peripheral regions.

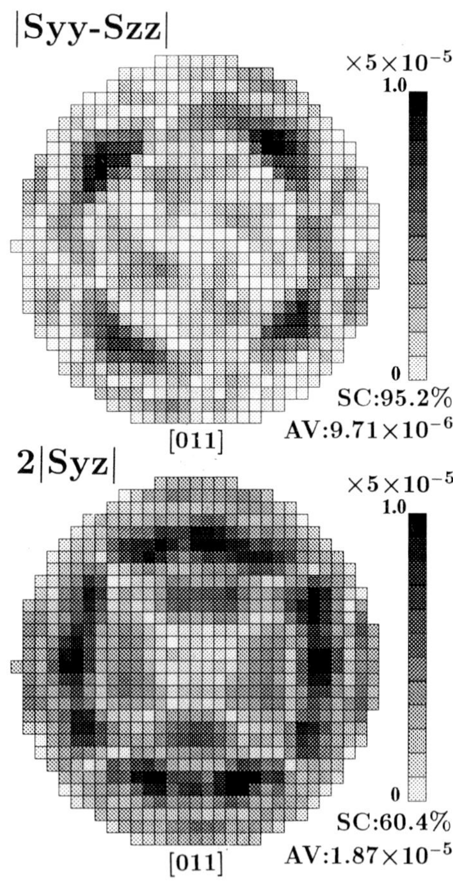

Figure 2. Two-dimensional distribution maps of $|S_{yy} - S_{zz}|$ and $2|S_{yz}|$ measured in the annealed wafers (C), in which a large number of slip lines were observed.

4. DISCUSSIONS

Before discussing the anomalous increase of residual strains accompanied with the slip generation by the thermal annealing, we have investigated the origins of slip generation. The presumable origins searched are as follows:

(a) Stationary temperature gradient due to inhomogeneous heating,
(b) Transient temperature gradient during heating up and down,
(c) Residual strains caused in crystal growth processes,
(d) Deformation layer caused in all wafer processes from slicing to final polishing.

The former two items depend on the furnace structure, the method of wafer supporting, and the rate of heating up and down; that is, how to anneal the wafer. The latter two items are originating in the wafer itself. Although Kawase et al (1992) have recently shown that the slip generation is less in the wafer with a lower level of residual strains, the items (c) and (d) are presumably not the main origin in the present experiment, because the residual strain level in the wafers used are almost the same as stated previously. Therefore, the slip lines may not be generated if there is no gradient in temperature distribution during thermal annealing. The deformations due to the residual strains as well as the deformations induced by the wafer processes may play a trigger and/or an additional rule in the deformation caused by the stationary and/or transient temperature gradient. At the present stage, we have not experimentally investigated further where the temperature gradient comes from.

In most cases, the stationary temperature distribution can easily be made homogeneous. However, even if it is homogeneous, the transient temperature distribution during heating and cooling may not be always homogeneous. Therefore, we have tried to calculate the transient temperature gradient under the condition that a wafer is cooled immediately after it is kept at the homogeneous temperature in an ideal furnace.

Figure 3 shows calculated transient temperature gradients during natural cooling from 800°C. In the calculation, the thermal diffusivity of GaAs used was 0.04 cm^2/s and the heat transfer coefficient the wafer to the surrounding ambient was 0.05 cm^{-1}. The detail of formulation on thermal strains or stresses as well as transient temperature distribution will be reported elsewhere. It is clearly seen that a large temperature gradient is caused in the peripheral region of wafer. It can be, therefore, understood that the peripheral region of wafer is strongly deformed during the ideal natural cooling.

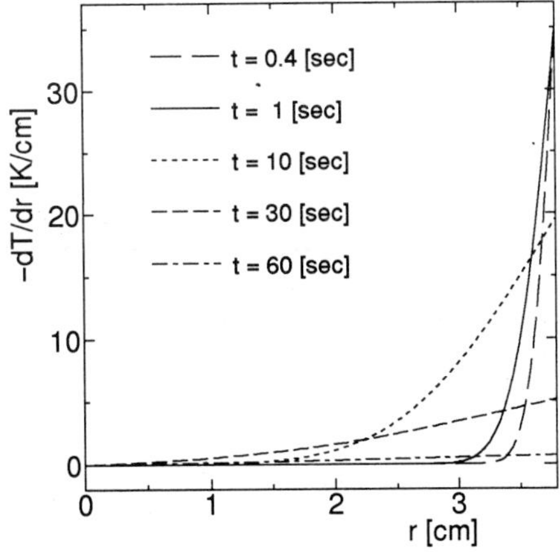

Figure 3. Transient temperature gradient during natural cooling from 800°C.

In the actual thermal annealing, the transient temperature gradient may be produced in the peripheral region of wafer, although it depends on the rate of heating and cooling. Also, the stationary temperature gradient may be caused by the heat flow through the wafer-supporting material

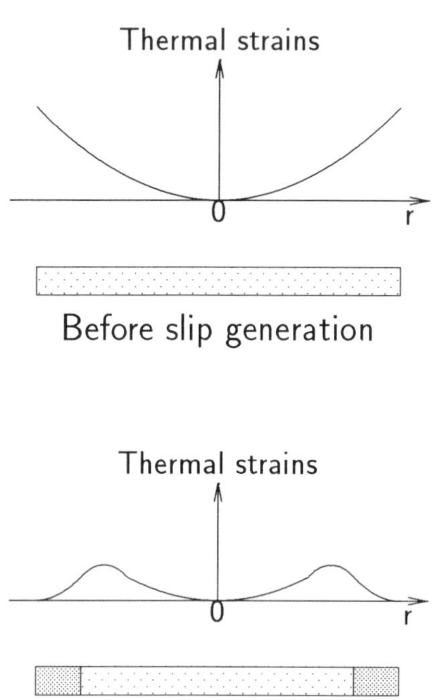

Figure 4. A thermoelastic model for the anomalous increase of residual strains found here.

even if the wafer is homogeneously heated. If there is a temperature gradient during thermal annealing, then it causes thermal strains or stresses. If the thermal strains or stresses do not exceed a yield value generating slip lines, then the residual strain distribution may not be changed after thermal annealing, as observed in the annealed wafer (B). On the other hand, if the thermal strains exceed the yield value in some places, then they are relaxed there by generating slip lines, subsequently they are redistributed to satisfy the new boundary conditions produced by the slip lines, and finally a part of them are frozen into the wafer after thermal annealing. Therefore, the residual strains may be increased after the thermal annealing accompanying with slip line generation, as observed in the annealed wafer (C). This situation is sketched in Fig. 4.

5. CONCLUSION

We have made the quantitative characterization of residual strains before and after thermal annealing of standard 3″-diameter LEC-grown wafers and found an anomalous increase of residual strains accompanied with the slip generation by the thermal annealing. This anomalous increase of residual strains is fairly well explained with a thermoelastic model based on the temperature gradient during thermal annealing.

Adachi S 1982 J. Appl. Phys. 53 5863
Adachi S and Oe K 1983 J. Appl. Phys. 54 6620
Kawase T, Wakamiya T, Fujiwara S, Kimura K, Tatsumi M, Shirakawa T, Tada T and Yamada M 1992 Proc. 7th Conf. Semi-insulating III-V Materials (Ixtapa Mexico) ed Carla J Miner in press
Yamada M 1985 Appl. Phys. Lett. 47 365
Yamada M, Fukuzawa M, Kimura N, Kaminaka K and Yokogawa M 1992 Proc. 7th Conf. Semi-insulating III-V Materials (Ixtapa Mexico) ed Carla J Miner in press
Yamada M 1993 Rev. Sci. Instrum. in press

Paper presented at the Int. Symp. GaAs and Related Compounds, Freiburg, 1993

Direct MBE growth of low-dimensional GaAs/AlGaAs–heterostructures on RIE patterned substrates

M Walther, T Röhr, H Kratzer, G Böhm, W Klein, G Tränkle and G Weimann

Walter Schottky Institut, Technische Universität München, D-85747 Garching

ABSTRACT: Facetted MBE growth on ridges oriented in [001] direction and etched by RIE was used for the direct formation of GaAs/AlGaAs nanostructures. On deep etched ridges (etch depths > 300 nm) facets appear, which are isolated from the surrounding area by distinct trenches. Narrow ridges form surfaces exhibiting only $(01\overline{1})$ facets at the sidewalls and (111)B facets on the top, self-adjusting a triangular surface shape. These facets with high migration lengths and low sticking coefficient for Ga atoms ('non growth surfaces') are remarkably smooth; inaccuracies of the patterning process (roughness of mask) are completely compensated during regrowth. Ga moves to the vertex of the triangular shaped structure formed by (111)B facets, Al incorporates on all facets; hence a thin GaAs layer growing at the vertex is enclosed completely between AlGaAs layers. Quasi one-dimensional quantum wires with high cathodoluminescence yield are obtained.

1. INTRODUCTION

New concepts for devices, realized by epitaxial regrowth on patterned substrates have attracted much attention in the last years (Cho 1991). It has been shown, that this technique is well suited for the growth of buried structures (Lievin et al 1992), the direct growth of lasers (Kapon et al 1990), butt joints and nanostructures (Ando et al 1989). Molecular beam epitaxial (MBE) regrowth on nonplanar substrates has been reported using various patterning techniques (Smith et al 1985, Choquette et al 1992). As shown in a previous paper (Walther et al 1993), MBE regrowth on nonplanar substrates patterned by optical lithography and reactive ion etching (RIE) yields GaAs/AlGaAs epilayers with unchanged high quality. In this paper, we report on the growth of GaAs/AlGaAs heterostructures on small ridges in [011] orientation. Secondary electron microscopy (SEM) investigations and spatially resolved cathodoluminescence (CL) measurements reveal, that the facetted growth on patterned substrates allows the direct growth of quantum wire structures.

2. MBE REGROWTH ON DRY ETCHED SUBSTRATES

To investigate the facetted growth on nonplanar substrates and GaAs/AlGaAs epilayers narrow ridges in various crystallographic directions were formed through a photoresist mask by RIE with $SiCl_4$. After dry etching, the photoresist mask was removed in organic solvents. Prior to loading the wafer into the MBE system they were cleaned in an oxygen plasma and ammonium hydroxide solution (NH_4OH 2 %, H_2O 98 %).

The structures were directly grown on patterned AlGaAs after thermal desorption of thin GaAs passivation layers in the UHV (Tanaka et al 1991). The second epitaxy was performed at a substrate temperature $T_S = 680^oC$ with continuous wafer rotation at 11.7 rpm. A detailed description of the patterning process and the thermal desorption is given in a previous paper (Walther et al 1993).

Facetted growth on ridges and grooves is determined by the migration lengths of the adatoms on different crystallographic planes and by shadowing effects (Meier et al 1989). The shape of the regrown structures depend on the orientation of the ridge, the lateral size, the thickness of the regrown material, the adatom species (Al,Ga) and the growth parameters.

An example for the different growth of GaAs and AlGaAs on a small ridge is shown in Fig. 1. This Figure shows the stain etched cross section of a 0.45 μm high and 0.75 μm wide ridge oriented in [011] direction after regrowth. The overgrown structure consists of several GaAs and $Al_{0.3}Ga_{0.7}As$ layers of nominal thickness of 150 nm each. Brighter areas show GaAs and darker ones indicate $Al_{0.3}Ga_{0.7}As$. It is obvious in Fig. 1, that these deep etched ridges in [011] direction are isolated from the surrounding structure by sharp trenches due to shadowing effects during MBE regrowth.

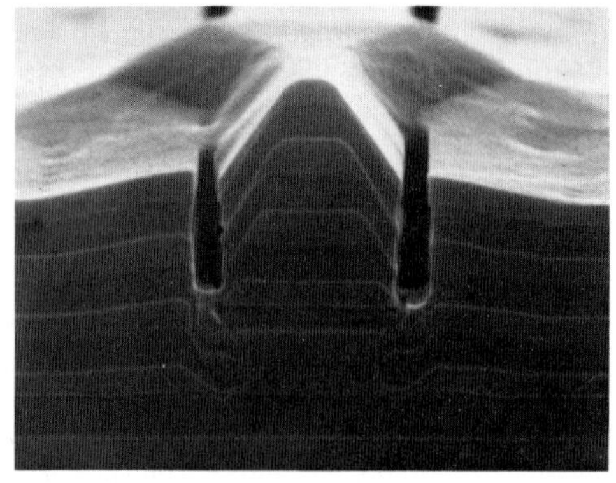

Fig. 1. Facet formation during regrowth for a 0.75 μm wide ridge in [011] direction after the alternating regrowth of 150 nm thick GaAs and $Al_{0.4}Ga_{0.7}As$ layers. Bright areas are showing GaAs, darker ones indicate $Al_{0.3}Ga_{0.7}As$.

(111)B facets on both sides of the ridge are clearly visible as well as the (100) planes in the center of the ridge. The growth rate of GaAs on (111)B planes is drastically reduced due to the enhanced migration of Ga atoms to the (100) facets in the center of the ridge. The migration of Ga atoms towards the (100) planes results in thicker GaAs layers in the central part of the ridge. In contrast, the growth rate of $Al_{0.3}Ga_{0.7}As$ is roughly the same on all crystallographic planes due to the much lower migration length of Al atoms. The shape of the structure remains therefore nearly unchanged during regrowth of AlGaAs. As a result, the GaAs layers grown on the upper part of the ridge are completely embedded in AlGaAs. Decreasing the lateral dimension of the ridge or increasing the thickness of the regrown GaAs layer leads to triangular shaped structures, as the (111)B facets on both sides of the ridge restrict the formation of the central (100) plane (Walther et al 1993).

3. DIRECT MBE GROWTH OF LOW-DIMENSIONAL STRUCTURES

The facetted growth on small ridges and the occurrence of sharp trenches on both sides of the ridges allow the direct growth of isolated nanostructures.

In Fig. 2 the SEM picture of a GaAs structure regrown at $T_S = 680^oC$ on a 0.75 μm wide and 1 μm high ridge in [011] direction is shown. On top of the 1.5 μm thick GaAs layer a 3 nm thick single quantum well (SQW) embedded in $Al_{0.3}Ga_{0.7}As$ barriers was grown. The shape of the overgrown ridge is only determined by (111)B planes on top of the ridge and $(01\bar{1})$ facets at the sides. These surfaces with a high migration length for Ga atoms ('non growth surfaces') are remarkably smooth so that inaccuracies of the patterning process (roughness of mask) are completely compensated during regrowth. The self-adjusted shape of the structure, determined by planes with a high migration length and the formation of sharp trenches at the sidewalls are the key issues for the direct growth of isolated nanostructures. Directly grown quantum wires on these structures are obtained by embedding the GaAs SQW between the AlGaAs layers. Due to the high migration length of Ga on the (111)B surfaces, the growth rate on these planes is drastically reduced as compared to (100) planes. GaAs grows therefore preferentially at the vertex of the structure, resulting in a quasi one-dimensional GaAs quantum wire enclosed in AlGaAs (Roehr et al 1993).

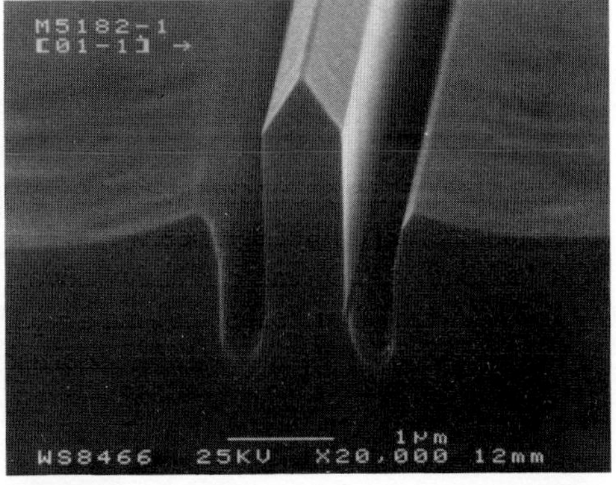

Fig. 2. SEM picture of a 1 μm high and 0.75 mm wide ridge in [011] orientation, overgrown with 1.5 μm GaAs and a SQW ($L_z = 3$ nm) embedded in 30 nm $Al_{0.3}Ga_{0.7}As$ barriers.

The SEM picture in Fig. 2, taken at the center of the 2"-wafer, shows a symmetrical shape of the (111)B facets on top of small ridges. With increasing distance from the center of the wafer, the dimensional asymmetry of the (111)B facets on the left and the right side of the regrown structure increases. The inner (111)B plane is smaller due to changes of the impinging fluxes caused by the geometrical arrangement of the effusion cells. In a Intevac Gen II the flux on a (111) facet varies during sample rotation between 0 % and 115 % of the value for the (100) plane. In addition, the sources of the group III and group V elements are located at different places, leading to a variation of the V/III-ratio on tilted facets. It is obvious, that small variations in the flux distribution are much more pronounced on tilted surfaces, resulting in an increasingly asymmetric shape towards the perimeter of the wafer.

4. CATHODOLUMINESCENCE MEASUREMENTS ON NANOSTRUCTURES

For a detailed investigation of the growth behavior of SQWs on this triangular shaped structures CL measurements have been performed. The spatially resolved CL images and the spectra were taken at T = 85 K from cleaved samples as shown in the SEM picture in Fig. 2.

Fig. 3 shows a spatially resolved panchromatic CL-image of the sample structure shown in Fig. 2. The CL signal of the quantum well on each side of the ridge (etched area) can be seen as well as an emission with comparable intensity from the top of the ridge. The CL-signal of the 1.5 μm thick GaAs layer is not resolved due to the detector characteristics and the exposure conditions of the photography.

The CL spectra of an overgrown SQW is shown in Fig. 4 for an etched area of the sample (left) and for a 1.2 μm wide ridge (right) with a slightly asymmetric shape. Additionally to the emission line at 1.689 eV from the unpatterned part of the sample, two other lines with comparable intensities and linewidths are observed at 1.631 eV and 1.767 eV.

Spatially resolved CL images of each of the three peaks in Fig. 4 are shown in Fig. 5. The emission at 1.689 eV is associated to the surrounding region of the ridge. The blue shifted line at 1.767 eV is

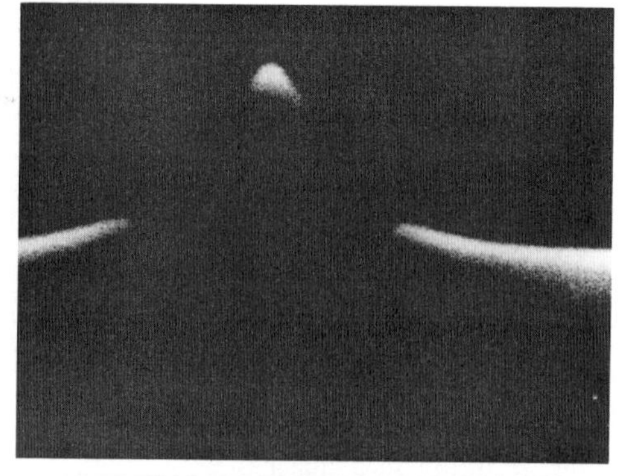

Fig. 3. Panchromatic CL image of the sample structure shown in Fig. 2 at T = 85 K.

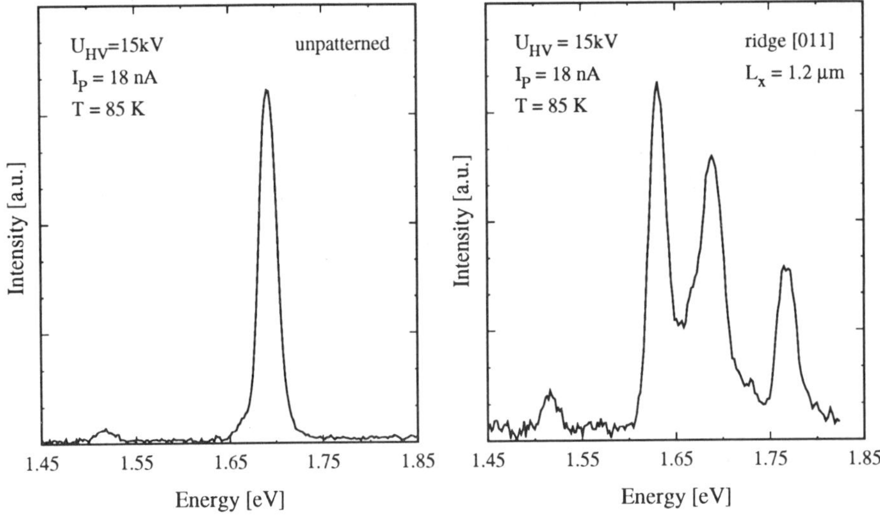

Fig. 4. CL spectra at T = 85 K of the SQW with L_z = 3 nm after regrowth. Left: Unpatterned region. Right: Emission lines for a 1.2 μm wide ridge with a slightly asymmetric shape.

caused by the SQW grown on the (111)B facets at the right side of the ridge. Due to the slight asymmetry of the structure two emission lines from the (111)B planes are observed in the CL image. The red shifted line at 1.631 eV is associated with the emission from the quantum wire, grown at the vertex of the triangular shaped structure. CL-scanning along 350 μm long ridges reveal no changes in the spectral position of these peaks. The CL yield of the quantum wire structure grown on top of the triangular ridge is comparable to the emission of the SQW in unpatterned regions.

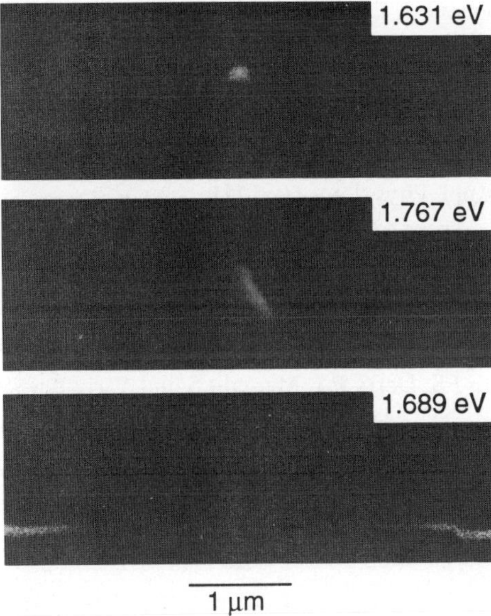

Fig. 5. Spatially resolved CL imaging of the spectrum shown on the right side in Fig. 4. The asymmetric growth of the (111)B results in two emission lines from the (111)B facets.

5. SUMMARY

MBE regrowth of GaAs/AlGaAs heterostructures on dry etched wafers yields epilayers with high morphological and optical quality. Isolated nanostructures with a triangular shape are obtained during regrowth on high and narrow ridges aligned in [011] direction. Facetted growth of SQWs on this triangular shaped structures allows the direct fabrication of quantum wire structures.

6. ACKNOWLEDGEMENT

The financial support by the German Federal Ministry of Research and Technology (BMFT) under contract BM 117/5 is gratefully acknowledged.
We thank A. Hülsmann and B. Raynor (Fraunhofer Institut für Angewandte Festkörperphysik, Freiburg) for the ebeam masks.

7. REFERENCES

Ando S and Fukui T 1989 J. Crystal Growth 98 646

Cho A Y 1991 J. Crystal Growth 111 1

Choquette K D, Hong M, Freund R S, Chu S N G, Mannaerts J P, Wetzel R C, Leibenguth R E 1992 Appl. Phys. Lett. 60 1738

Kapon E, Simhony S, Harbison L T, Florez L T and Worland P 1990 Appl. Phys. Lett. 56 1825

Lievin J L, Le Gouezigou D, Bonnevie D, Gaborit F, Poingt F and Brillouet F 1992 Appl. Phys. Lett. 60, 1211

Meier H P, Van Gieson E, Walter W, Harder C, Krahl M and Bimberg D 1989 Appl. Phys. Lett. 54 433

Roehr T, Walther M, Rochus S, Boehm G, Klein W, Traenkle G and Weimann G 1993 to be published in J. Mater. Sci. Eng. B

Smith J S, Derry P L Margalit S and Yariv A 1985 Appl. Phys. Lett. 47 712

Tanaka H and Mushiage M 1991 J. Crystal Growth 111 1043

Walther M, Roehr T, Boehm G, Traenkle G and Weimann G 1993 J. Crystal Growth 127 1045

Quantitative study of oxygen incorporation on MBE-grown AlAs surfaces during growth interruption and its effect on nonradiative recombination in GaAs/AlAs quantum wells

T. Someya[1], H. Akiyama[1], Y. Kadoya[2], and H. Sakaki[1,2]

1) Institute of Industrial Science, University of Tokyo, Roppongi 7-22-1, Minatoku, Tokyo 106, Japan
2) Quantum Wave Project, ERATO, JRDC, Keyaki House 302, Komaba 4-3-24, Meguroku, Tokyo 153, Japan

We report on our systematic study to quantify the concentration of oxygen accumulated at MBE-grown GaAs-on-AlAs heterointerfaces during the growth interruption from SIMS measurements and to clarify its effects on the photoluminescence (PL) efficiency. When the growth is interrupted at AlAs surfaces for 5, 15, and 30 minutes, the PL efficiency of GaAs(7.4nm)/AlAs quantum wells decreases at 77K to 36%, 7%, and 1% of its full value, while the adsorbed oxygen concentration increases to $2\times10^{10}cm^{-2}$, $5\times10^{10}cm^{-2}$, and $1\times10^{11}cm^{-2}$, respectively. These values provide a new insight on the nonradiative decay of carriers and a significant guideline for advancement of UHV in-situ processing.

1. Introduction

It has been suggested that AlAs (or AlGaAs) surfaces are easily degraded by the adsorption of residual gas such as oxygen- and carbon-containing molecules (Foxon 1985, Achtnich 1987) when the growth is interrupted in molecular beam epitaxy (MBE). For those devices fabricated in ultra-high vacuum (UHV) processing systems, the surface degradation of similar nature is expected. Hence it is important to characterize it quantitatively as well as to minimize it. In this work we evaluated via secondary ion mass spectrometry (SIMS) the concentration of oxygen accumulated at inverted (GaAs-on-AlAs) heterointerfaces and systematically studied its effect on the photoluminescence (PL) efficiency of GaAs/AlAs quantum wells (QWs). It is found that the number N_{ox} of oxygen atoms adsorbed on AlAs surfaces during growth interruption is proportional to the growth interruption time. Then we have shown that the recombination rate $1/\tau$ in QWs increases linearly with N_{ox} incorporated at the inverted interface. This implies that the lifetime measurement is quite effective to evaluate quantitatively the oxygen concentration N_{ox} in QW samples.

2. Oxygen Incorporation on MBE-grown AlAs Surfaces

To quantify the oxygen incorporation process, a sample having four different GaAs-on-AlAs interfaces was grown by MBE (Riber-2300) on a (001) semi-insulating GaAs substrate at the substrate temperature of 580°C. The growth rates of GaAs and AlAs were 0.65μm/hour and 0.22μm/hour, respectively. The sample contains five periods of 1nm thick AlAs layers and 200nm thick GaAs spacer layers alternately. AlAs surfaces were purposely degraded by the adsorption of oxygen during growth interruption. Growth interruption time at inverted interfaces was 4 hours, 1 hour, 30 minutes, 10 minutes, and 5 minutes, respectively. Then we analyzed the concentration of oxygen in this sample by SIMS (Perkin Elmer 6600) using a Cs^+ source operated at a potential of 5kV. To detect oxygen atoms, ^{16}O was monitored. The absolute levels of the oxygen concentration was calibrated from the reference sample prepared with oxygen implantation. Figure 1 shows the SIMS depth profile of oxygen accumulated during the very long growth interruption and indicates that the oxygen concentration increases with the length of the growth interruption. The oxygen signals were reduced to noise level at AlAs surfaces when the growth interruption is 10 and 5 minutes.

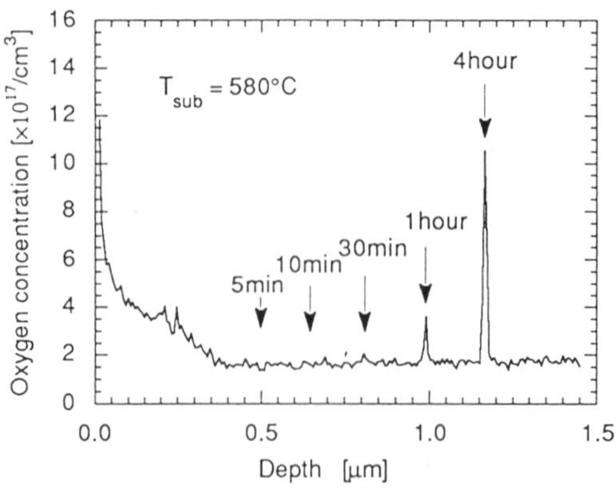

Figure 1 SIMS depth profile of GaAs/AlAs multilayer. AlAs surfaces are degraded by the adsorption of oxygen during growth interruption.

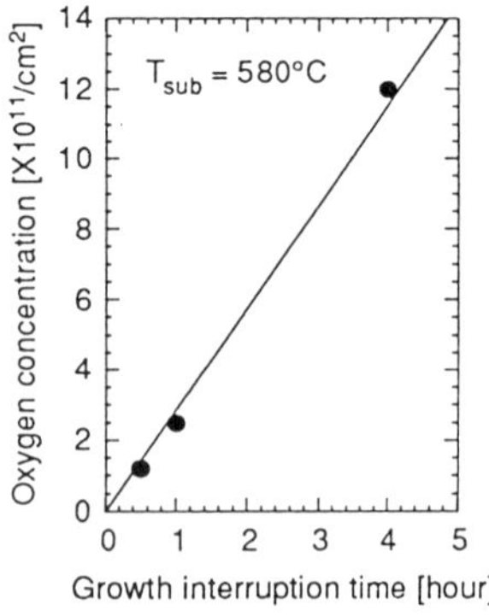

Figure 2 The number of oxygen atoms adsorbed on AlAs surfaces during growth interruption is proportional to the growth interruption time.

As shown in Fig. 2, the number of oxygen atoms adsorbed on AlAs surfaces is proportional to the growth interruption time, suggesting that the desorption rate of oxygen from AlAs surface is very low. Note here that the incorporation of very small oxygen concentration can be easily evaluated by adopting the long interruption time (Someya 1993). Although the oxygen incorporation rate depends on the vacuum quality of MBE systems, it is found to be almost constant (2×10^{11}/cm^{2}hour) in our well-outgassed system throughout the period of our study.

3. Nonradiative Recombination in GaAs/AlAs Quantum Wells

We have grown four different GaAs(7.4nm)/AlAs single quantum well (QW) structures for the PL measurements and a reference inverted interface for SIMS analysis. The growth was interrupted at inverted interfaces of QWs for 30, 15, 5, and 1 minute for samples A, B, C, and D, respectively. The oxygen concentrations N_{ox} at inverted interfaces of these QWs were carefully estimated by performing the SIMS measurement on AlAs surfaces prepared with one hour interruption in the same growth run. From this data, we find that N_{OX} is 1×10^{11}cm^{-2}, 5×10^{10}cm^{-2}, 2×10^{10}cm^{-2}, and 3×10^{9}cm^{-2} for samples A, B, C, and D, respectively.

The PL measurements were done at 77K by using 514.5nm Ar^+ laser light with 2mW focused in the spot size of 200μm. As shown in Fig. 3, the PL efficiency η of these QWs decreased to 36%, 7%, and 1% of its full value when the growth was interrupted for 5, 15, and 30 minutes in our well out-gassed MBE system. The PL efficiency decreases drastically with the increase of the concentration of incorporated oxygen. This means that the surface cleanliness of AlAs is a crucial issue for the device fabrication in UHV processing systems because it takes typically a few ten minutes to transfer or to process samples.

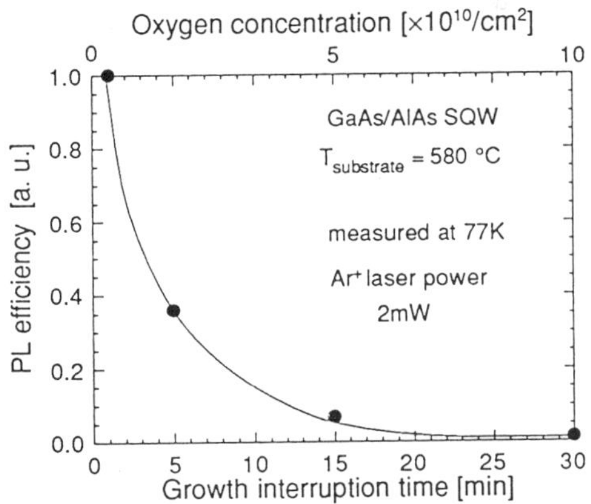

Figure 3 The PL efficiency of QWs decreases drastically with the increase of the oxygen concentration.

The drastic decrease in PL efficiency η due to the oxygen incorporation demonstrates that PL is a sensitive characterization method(Tsang 1980). However 1/η does not increase linearly with the oxygen concentration. In order to provide quantitative information on the nonradiative recombination, we have also measured the time-resolved PL at 77K . The output of a cw mode-locked yttrium-lithium-fluoride (YLF) laser was frequency doubled in a KTP ($KTiOPO_4$) crystals which operated with the average power of 10mW, a repetition rate of 75.4MHz. The luminescence was detected by a cooled microchannel tube, or MCP-PMT (Hamamatsu Photonics R2809-11).

The PL spectrum and the time resolved PL of these samples is shown in Fig. 4 and Fig. 5, respectively. PL intensity decays rapidly when oxygen concentration at inverted interfaces increases. The recombination rate $1/\tau$ increases linearly with the growth interruption time, or the oxygen concentration at the inverted interfaces as shown in Fig. 6. The recombination rate of photoexcited carriers is described as

$$\frac{1}{\tau} = \frac{1}{\tau_{rad}} + \frac{1}{\tau_{nonrad}}$$

where τ_{rad} is the radiative lifetime and τ_{nonrad} is the nonradiative lifetime. Our experimental results show that nonradiative recombination rate $1/\tau_{nonrad}$ is proportional to the oxygen concentration N_{ox}. The time resolved PL measurements are quite effective to characterize N_{ox} in the samples quantitatively.

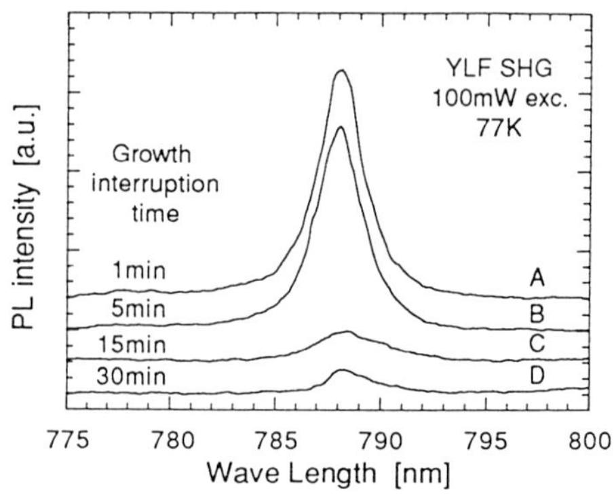

Figure 4 PL spectra of QWs (sample A-D). The growth was interrupted at inverted heterointerfaces of QWs for 30, 15, 5, and one minute.

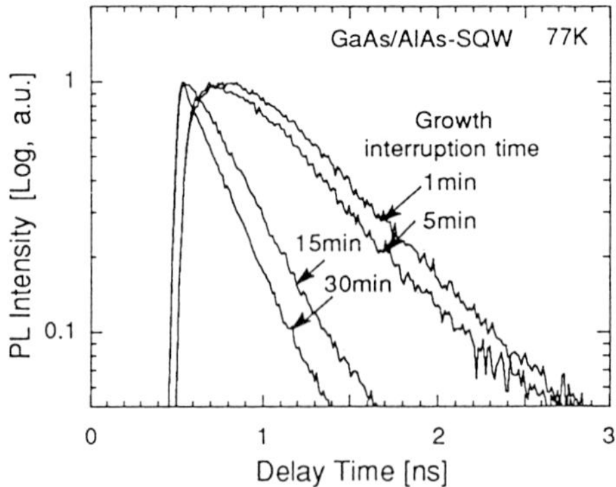

Figure 5 Time resolved PL of GaAs/AlAs QWs. PL intensity decays rapidly when the growth interruption time increases.

In addition, we comment on the relation between the PL efficiency η and the PL decay time τ. The PL efficiency η is described as

$$\eta = \frac{1/\tau_{rad}}{1/\tau} \propto \tau$$

if τ_{rad} is constant (Feldmann 1986). Though η decreases with N_{ox}, η is not proportional to τ as shown in Fig. 7. The inconsistency is likely to be due to the complexity of initial energy relaxation processes of photogenerated carriers. Indeed, effects of these processes are found to be important in separate PL measurements in which the resonant and nonresonant excitation of QWs are compared. We will discuss this subject elsewhere.

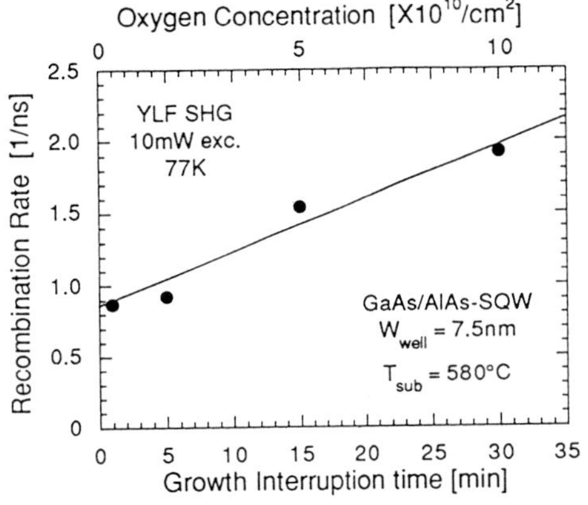

Figure 6
The recombination rate $1/\tau$ increases linearly with the oxygen concentration N_{ox} at the inverted interfaces.

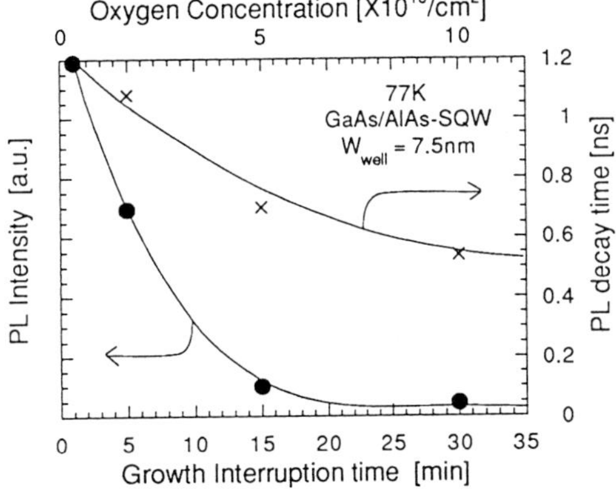

Figure 7
The oxygen concentration dependence of PL efficiency η and PL decay time τ.

4. Conclusions

PL studies show that, when the oxygen concentration N_{ox} at the GaAs-on-AlAs interface increases, the PL efficiency η of QWs decreases drastically and at the same time the recombination rate $1/\tau$ increases linearly with N_{ox}. The time resolved PL measurements are quite effective to characterize N_{ox} in the samples.

Since a strong correlation is found between the oxygen concentration and the PL efficiency η degradation, we conclude that tolerable oxygen concentration at inverted heterointerfaces for photonic-QW devices is $2\times10^{10}cm^{-2}$. This is a typical value adsorbed during growth interruption for 5 minutes in our MBE system. These values serve as the significant guideline for advancement of UHV in-situ processing.

Acknowledgements

We acknowledge T. Matsusue, T. Noda, H. Noguchi, and Y. Ohno, University of Tokyo, for the technical supports of MBE growth and stimulating discussions. We wish to express our sincere thanks to the SIMS staff of MST Foundation for valuable discussion concerning SIMS analysis. This work is mostly supported by a Grant-in-Aid from the Ministry of Education, Science, and Culture, Japan.

References

T. Achtnich, G. Burri, M. A. Py, and M. Ilegems, Appl. Phys. Lett. 50, 1730 (1987).

J. Feldmann, G. Peter, O. Göbel, P, Dawson, K, Moore, C. Foxon, and R. J. Elliot, Phys. Rev. Lett. 59 2337 (1987).

C. T. Foxon, J. B. Clegg, K. Woodbridge, D. Hilton, P. Dawson, and P. Blood, J. Vac. Sci. Technol. B3(2), 703 (1985).

W. T. Tsang, F. K. Reinhart, and J. A. Ditzenberger, Appl. Phys. Lett. 36, 118 (1980).

T. Someya, H. Akiyama, Y. Kadoya, T. Noda, T. Matsusue, H. Noge, and H. Sakaki, Appl. Phys. Lett. 63 (1993)

Paper presented at the Int. Symp. GaAs and Related Compounds, Freiburg, 1993

MBE growth of $In_{0.35}Ga_{0.65}As$/GaAs MWQs for high-speed lasers: relaxation limits and factors influencing dislocation glide

E.C. Larkins, M. Baeumler, J. Wagner, G. Bender, N. Herres, M. Maier, W. Rothemund, J. Fleißner, W. Jantz, J.D. Ralston, G. Flemig* and R. Brenn*

Fraunhofer Institut für Angewandte Festkörperphysik, Tullastraße 72, 79108 Freiburg, Germany
**Fakultät für Physik, Universität Freiburg, Hermann-Herder-Straße 3, 79104 Freiburg, Germany*

ABSTRACT: The structural limitations of $In_{0.35}Ga_{0.65}As$/GaAs MQWs for high-speed laser structures are investigated. The gradual onset of defect formation and strain relaxation are studied in test structures and lasers with i) different numbers of QWs; ii) Be-doped MQWs, and; iii) different substrates. The gradual onset of strain relaxation is observed by PL microscopy. Prior to the onset of strain relaxation, resonant Raman scattering and Rutherford backscattering measurements show increasing defect densities with increasing numbers of QWs. Defect formation and dislocation glide are shown to be strongly influenced by the growth temperature, substrate material and dopant impurities.

1. INTRODUCTION

The optimization of $In_{0.35}Ga_{0.65}As$/GaAs multiple quantum wells (MQWs) for high-speed laser applications requires the optimization of both the molecular beam epitaxial (MBE) growth parameters and the MQW structural parameters. The influence of the MBE growth parameters on the optical and structural properties of $In_yGa_{1-y}As$/GaAs 3 QW structures with y=0.20 and y=0.35 has been reported in a previous publication (Larkins et al 1993). It was observed that lower growth temperatures (450°C) resulted in significantly lower interface roughness than higher growth temperatures (≥480°C). At the same time, it was shown that the lower growth temperatures did not degrade the photoluminescence (PL) efficiency - a fact which can be attributed to the careful preparation of a clean growth environment. Growth stops, either in the GaAs barriers or directly in the $In_yGa_{1-y}As$ QWs, improved the planarity of the heterojunction interfaces, indicating that planar growth is thermodynamically favored for $In_yGa_{1-y}As$/ GaAs MQWs with up to (at least) 3 QWs and y≤0.35.

Optimization of the structural parameters of $In_yGa_{1-y}As$/GaAs MQWs is highly application dependent and is often limited by the formation of lattice defects and relaxation of the pseudomorphic strain (Matthews and Blakeslee 1974). In this work, we are interested in optimizing and determining the ultimate structural limitations of pseudomorphic $In_{0.35}Ga_{0.65}As$/GaAs MQWs for high-speed lasers. For high-speed lasers, it is desirable to maximize both the optical confinement factor (Γ) and the number of QWs (n). To maximize Γ, $Al_{0.8}Ga_{0.2}As$ cladding layers are used to obtain a large refractive index difference between the core and cladding regions. This refractive index difference determines the optimum thickness of the core region. In order to fit as many QWs as possible in the core region and to maximize Γ, the thickness of the GaAs QW barrier layers must be minimized. This paper summarizes the epitaxial structures and materials characterization used to optimize the MQW structure of high-speed, pseudomorphic $In_{0.35}Ga_{0.65}As$/GaAs MQW lasers. Lasers fabricated with the optimized MQW structures developed in this study have already achieved modulation bandwidths of 30GHz.

2. SAMPLES

The test structures and lasers investigated in this study were all grown by MBE using cracked arsenic (beam equivalent pressure=1.5×10^{-5} torr). The MQW test structures were grown on (001) liquid-encapsulated Czochralski (LEC) GaAs substrates (etch pit density (EPD) $>5 \times 10^4$ cm^{-2}) as follows: a buffer consisting of 14 periods of 25nm i-GaAs followed with a 10s growth stops for smoothing; a 160s growth stop to allow for the substrate temperature change; an n-period (3<n<9) MQW structure consisting of 20nm i-GaAs barriers and 5.7nm i-$In_{0.35}Ga_{0.65}As$ QWs, including a 20nm i-GaAs cap layer. The buffer layers were grown at 580°C and the MQW structures were grown at 450°C. The growth temperatures were measured with a radiatively coupled thermocouple, calibrated to the 640°C GaAs congruent sublimation temperature (Larkins et al 1993).

The lasers share a common epitaxial structure, differing only in the details of the MQW structure. The lasers consist of: 1μm n^+ GaAs, [Si]=4x10^{18} cm^{-3}; 0.1μm graded n^+ $Al_xGa_{1-x}As$ ($0 \leq x \leq 0.8$), [Si]=3x10^{18} cm^{-3}; 0.8μm $Al_{0.8}Ga_{0.2}As$, [Si]=3x10^{18} cm^{-3}; the n-period ($3<n<6$) MQW structure; 0.8μm $Al_{0.8}Ga_{0.2}As$, [Be]=1x10^{18} cm^{-3}; 0.1μm graded p^+ $Al_xGa_{1-x}As$ ($0.8 \geq x \geq 0$), [Be]=8x10^{18} cm^{-3}; 0.2μm p^{++} GaAs, [Be]=8x10^{19} cm^{-3}. The MQW structures are composed of 5.7nm $In_{0.35}Ga_{0.65}As$ QWs separated by 20nm GaAs barriers. These MQWs are sandwiched symmetrically in undoped GaAs to obtain an optimized waveguide core thickness of 0.177-0.185μm. To minimize interface roughness and non-radiative defect incorporation (Ralston et al 1993a), the $Al_xGa_{1-x}As$ layers are grown as AlAs/GaAs binary short-period superlattices. The p^{++} GaAs cap layer was grown at 540°C, while the $In_{0.35}Ga_{0.65}As$ QWs, the remaining GaAs layers and the $Al_xGa_{1-x}As$ layers were grown at 450°C, 580°C and 700°C, respectively. Four of the 4 QW lasers were grown with Be doping in the MQW structure (4.5nm wide regions of [Be]=6.3x10^{18} cm^{-3} or [Be]=2.5x10^{19} cm^{-3} either centered in the QWs or placed above the QWs in the GaAs with a 3nm spacer). Additional lasers were grown on (001) vertical Bridgman (VB) GaAs substrates with low etch pit densities ($<10^3$ cm^{-2}).

3. CHARACTERIZATION

All of the above lasers and test structures were characterized with photoluminescence (PL) and photoluminescence microscopy (PLM). The test structures were also characterized with resonant Raman scattering, Rutherford backscattering (RBS) and X-ray diffraction (XRD). High-speed mesa-geometry lasers with coplanar electrodes were fabricated from the laser materials. The dc and ac performances of these lasers were extensively characterized, as described by Weisser et al 1992, Ralston et al 1993b and Esquivias et al 1993. The heavily p-doped lasers achieved -3dB modulation bandwidths of 30GHz for drive current of 114mA and -3dB modulation bandwidths of 20GHz for pump currents as low as 50mA.

The PL spectra were excited at 10K using the 457.9nm Ar+ ion laser line with a power of 1mW. The $In_{0.35}Ga_{0.65}As$ MQW luminescence was dispersed with a 1.0m spectrometer and detected with a cooled Ge detector. The PL spectra from the laser structures with $3<n<6$ QWs are shown in Figure 1. The integrated PL intensities and linewidths are similar to within ±20%. There are no clear trends in either the intensity or the linewidth with respect to the number of QWs. This is particularly interesting, since the PLM images clearly show the onset of lattice relaxation, as discussed below.

The PLM images were obtained at 80K using an improved version of the experimental apparatus described by Wang et al 1993. The spatial resolution of this technique is ~1μm. The luminescence was excited with a tungsten lamp using a high-pass filter with a -3dB cutoff wavelength of 650nm and detected with a Hamamatsu C1000 infrared video camera. The images were formed with the MQW luminescence, using lowpass filters with -3dB cutoff wavelengths of 850nm and 900nm to remove the reflected light and the signal from the GaAs substrate. No significant differences were observed between images obtained with the 850nm and 900nm filters, indicating that the defect features are

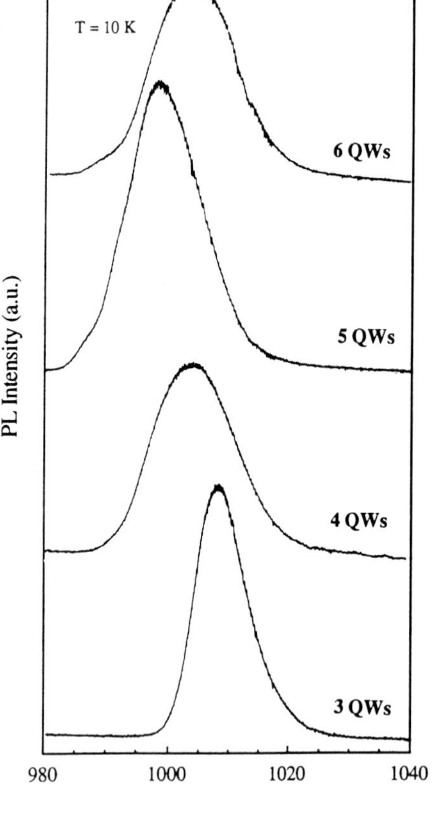

Figure 1: Photoluminescence spectra (10K) from $In_{0.35}Ga_{0.65}As$/GaAs MQW lasers grown on LEC substrates. The undoped MQW regions contain n=3, 4, 5 and 6 QWs.

due to quenching of the MQW luminescence. Figure 2a-d are images from the as-grown, undoped MQW lasers with 3, 4, 5 and 6 QWs, respectively. The images from the 3QW laser structure (Fig. 2a) show no evidence of defects or dislocations. However, the 4 and 5 QW laser structures show the formation of defects which are preferentially oriented along the [100] and [010] directions. In the 4 QW laser structure (Fig. 2b), these defects are 2-40μm long and tend to be isolated from each other. In the 5 QW laser structure (Fig. 2c), the average length of these defects increases, as does the defect density. In addition, there is a growing tendency for these defects to intersect. In the 6 QW laser structure (Fig. 2d), the average length and density of these defects continues to increase and misfit dislocations with a <110>-oriented dislocation line also begin to appear. (Due to the extremely small misfit dislocation densities, we are unable to observe a pronounced asymmetry and do not distinguish between the inequivalent [110]- and [1-10]-oriented misfit dislocation lines. We also do not distinguish between the equivalent [100]- and [010]-oriented defects.)

No defects were observed in the images from the as-grown MQW test structures with less than 6 QWs. The 6≤n≤9 MQW test structures show very small numbers of misfit dislocations with <110> dislocation lines. In the 6≤n≤8 MQW test structures, the dislocation line densities (dislocation density x average length) are very small ($<10^{-4}$ μm^{-1}) and relatively constant, though the dislocation line density increases noticably for the 9 QW sample. Many of these dislocations appeared to have nucleated at oval defects, suggesting that the 6≤n≤8 MQW structures may be metastable. The difference between the laser structures and the MQW test structures probably results from the high-temperature (700°C) growth of the 0.8μm $Al_{0.8}Ga_{0.2}As$ cladding layer on top of the MQW active region of the lasers. This apparent dependence on post-growth annealing supports the supposition that the as-grown MQW structures with 6≤n≤8 QWs are in a metastable strain state. PLM also shows the presence of <110>-oriented misfit dislocation lines extending from Ga-droplets on the surface of one highly strained MQW sample. The presence of <100>-oriented defects extending from an oval defect has also been observed.

The authors are only aware of two observations of such <100>-oriented defects in GaAs-based semiconductor materials. In the first observation, Qin and Roberts 1989 were able to create these defects through

a) 3 QWs

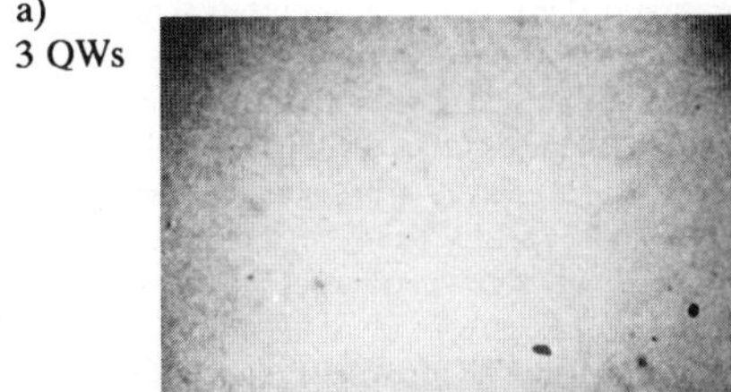

b) 4 QWs

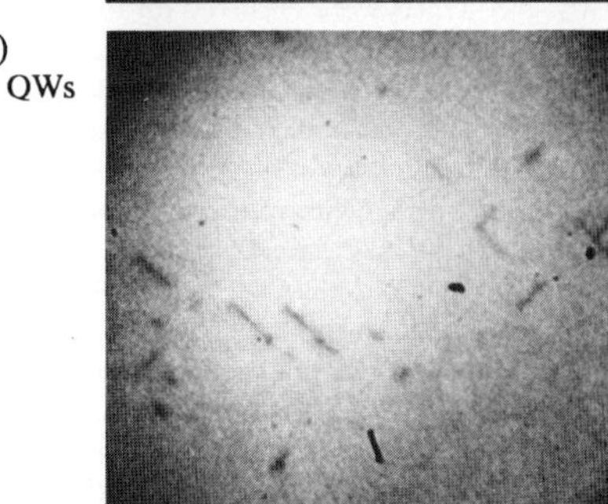

c) 5 QWs

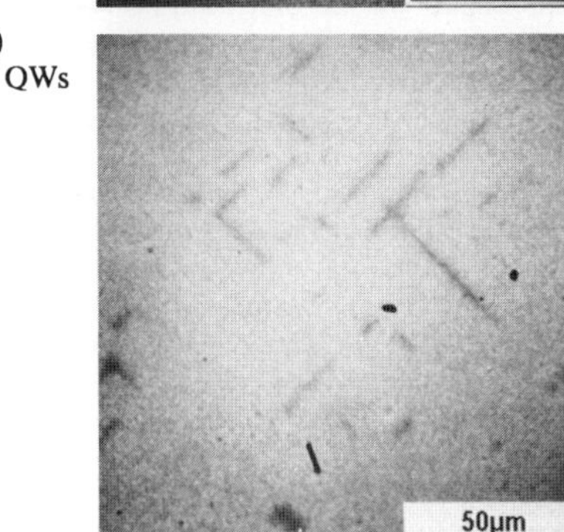

d) 6 QWs

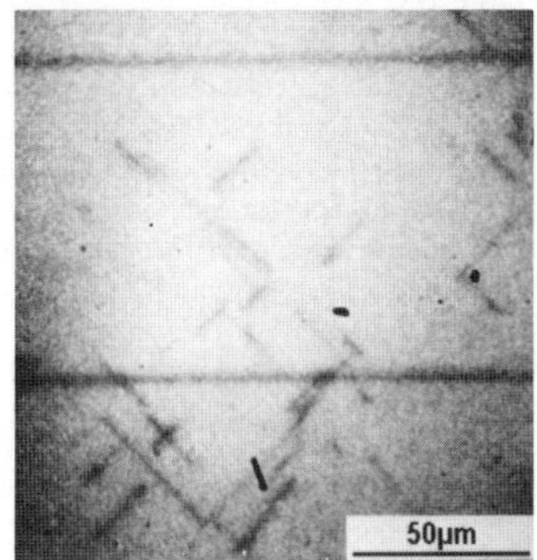

Figure 2: Photoluminescence (PLM) images from high-speed $In_{0.35}Ga_{0.65}As$/GaAs MQW laser structures grown on LEC substrates with a) 3 QWs; b) 4 QWs; c) 5 QWs; and d) 6 QWs. The scale bars are oriented along the [110] direction.

indentation of GaAs. In the other reference, Bonar et al 1992 observed these defects while studying the relaxation of 20nm $In_yGa_{1-y}As/GaAs$ ($0 \leq y \leq 0.5$) QWs with transmission electron microscopy. They suggested that the behavior of these defects was consistent with dislocation glide along the {110} planes inclined to the interface. We find that the density of these <100> defects depends strongly on the substrate material. After comparing many PLM images from identical 5 QW laser structures grown on semi-insulating LEC and VB GaAs substrates, it was estimated that the structure grown on the LEC material had a defect density which was a factor of $\sim 10^6$ larger than that grown on the VB material (Baeumler et al 1993). This difference is particularly significant, since the ratio of the threading dislocation densities of the LEC and VB substrates is only $\sim 10^2$-10^3, suggesting the new possibility that dislocation interaction *may* be important for the nucleation and/or propagation of these <100>-oriented defects.

PLM images from the laser structures with Be-doped MQWs had much lower defect densities, both for the <100>-oriented defects and the <110>-oriented misfit dislocation lines. This defect suppression was more effective for the samples with Be-doped GaAs barrier layers than for those with Be-doped $In_{0.35}Ga_{0.65}As$ QWs. No defects were observed for the laser structure with [Be]=$2x10^{19}$ cm^{-3} doped barriers - even though it was grown on an LEC substrate. These results suggest that the Be pins the movement of the dislocations in the GaAs barrier layers. The defect reduction observed in the lasers with doped $In_{0.35}Ga_{0.65}As$ QWs can be attributed to diffusion/segregation of the Be into the adjacent GaAs layers. This Be redistribution has been corroborated by subsequent SIMS measurements (Ralston et al 1993c). Schweizer et al 1992 have observed a similar (but only partial) dislocation pinning with Si doping in growth studies of pseudomorphic MODFETs. Kirkby 1975 has estimated the elastic pinning energies of impurities on dislocations in GaAs, predicting that In impurities should be even more effective than dopant impurities. Therefore, it is reasonable to expect that impurities in the GaAs will have a greater dislocation pinning effect than doping in the $In_{0.35}Ga_{0.65}As$ wells, where the movement of the dislocations is already pinned by the In. It is also reasonable to expect that impurity-induced hardening of the GaAs barrier layers can be used to increase the critical layer thickness (CLT) at which strain-relaxation begins to occur through the formation of lattice defects and misfit dislocations.

Raman scattering measurements were performed at 77K with an excitation energy of 1.915eV in resonance with the $E_o+\Delta_o$ bandgap of the GaAs barriers. The polarizations of the incident and scattered light were parallel to the same (100) crystallographic direction [x(y,y)x] as described by Wagner et al 1993. In this measurement geometry, the ratio of the one-LO/two-LO phonon scattering intensities is very sensitive to point defects and defect clusters. Figure 3 shows the corresponding spectra from the MQW test structures with n=3, 5, 7 and 9 QWs. For n>4, the intensity of the symmetry forbidden one-LO Raman line increases rapidly with respect to the intensity of the allowed two-LO Raman line. Under resonant excitation conditions, defect-induced one-phonon scattering can be observed, so that this rapid increase in the one-LO/two-LO phonon scattering intensity ratio reflects an increase in the density of defects or defect clusters due to increased stress in the structure. Such defects may be nucleation sites for and/or participate in the propagation of the <100>-oriented defects and misfit dislocations.

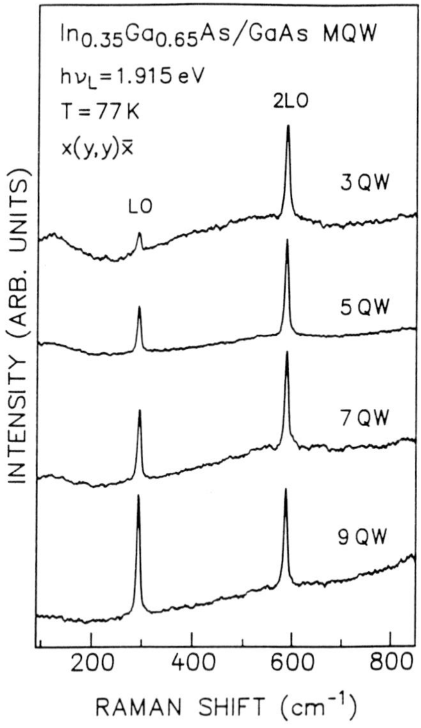

Figure 3: Raman scattering spectra (77K) from $In_{0.35}Ga_{0.65}As/GaAs$ MQW test structures with n=3, 5, 7 and 9 QWs, excited in resonance with the $E_o+\Delta_o$ GaAs bandgap. The increase in the intensity ratio of the one-LO/two-LO Raman lines with number of QWs, reflects an increase in the lattice defect density.

RBS channeling data was obtained with a 10±3nA 2.0MeV $^4He^+$ beam, taking great care to measure and control the beam current. The elimination of the dead time of the analog to digital signal converter was particularly important for obtaining both high sensitivity and good reproducibility. Backscattering yields were obtained from the {011} planar channeling measurements. These channeling yields were normalized to the backscattering yields obtained with a randomized sample orientation. The channeling and random backscattering measurements were performed temporally as close to each other as possible to minimize errors introduced by slow drift in the measurement system. To compare different samples, the random backscattering yields were also normalized to each other. These normalization procedures effectively suppressed errors introduced by drift in the RBS system and by slight differences in the condition of the sample surfaces. Figure 4 shows the normalized backscattering yields obtained from the MQW test structures with 3<n<9 QWs. The normalized backscattering yield clearly increases as the number of QWs increases. This is believed to be caused by an increase in the density of lattice defects. Very little asymmetry was observed between the {011} and the {0-11} planar channeling measurements, suggesting that the defects responsible for the increased backscattering affect both planes equally. The kink angles determined from the RBS measurements following the method of Flagmeyer 1992 are all 0.94°±0.08° and show no clear trend with respect to the number of QWs. In this case, it is clear that the planar channeling measurements are much more sensitive to defects and lattice relaxation than are more traditional measurements based on the determination of the kink angle.

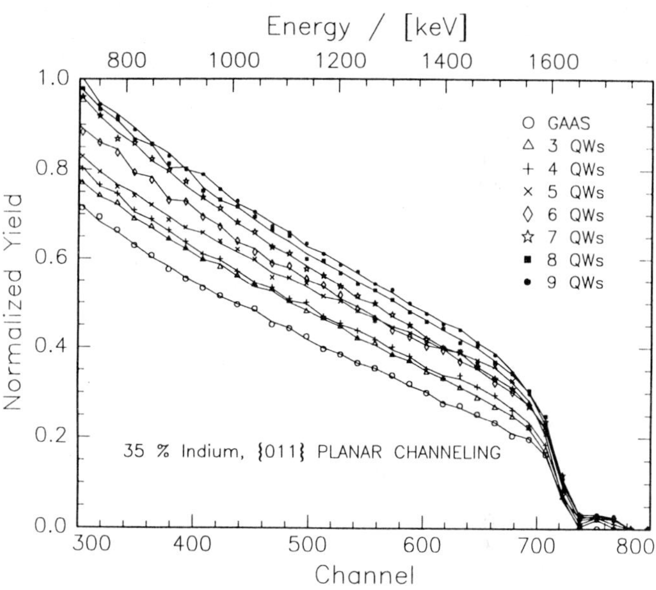

Figure 4: Normalized backscattering yields from {110} planar channeling measurements performed on $In_{0.35}Ga_{0.65}As$/GaAs MQW test structures with 3≤n≤9 QWs, using a 10±3nA, 2.0MeV $^4He^+$ ion beam. The increase in the backscattering yield with increased number of QWs reflects an increase in the lattice defect density.

High-resolution X-ray rocking curves were obtained near the GaAs 004 and 115 reflections. The difference ($\delta_{FWHM}=FWHM_{meas}-FWHM_{cal}$) in the measured and calculated MQW reflection linewidths near the GaAs 004 reflection were typically $+3" < \delta_{FWHM} < +7"$. Similar rocking curve simulations are described by Bender et al 1993. The weak fringes between the MQW reflection maxima were clearly resolved-even for the 8 and 9 QW test structures, which showed evidence of relaxation in the PLM images. The nearly ideal reflection linewidths and the observation of the weak fringes are generally considered an indication of very high quality epitaxial structures. Careful analysis of the X-ray rocking curve data suggests that there is an increase in the tendency of the indium to segregate toward the surface as the number of quantum wells increases.

4. CONCLUDING REMARKS

Numerous techniques have been used to determine the critical layer thickness (CLT) pseudomorphically strained semiconductor structures, including XRD, PL, RBS and PLM. The acccuracy of CLT measurements depends strongly on the sensitivity of the measurement technique employed (Fritz et al 1987 and Gourley et al 1988). The PLM technique appears to be the most suitable technique for $In_yGa_{1-y}As$/

GaAs heterostructures, since PLM can be used to observe the very onset of lattice relaxation at which the dislocation line densities are $<<10^{-5}$ μm^{-1}. Although changes in the PL spectra have been widely used to determine the CLT, we observe no significant change in the PL intensity, linewidth or energy - even for samples which have dislocation line densities $>3x10^{-2}$ μm^{-1}. This is technologically significant, since practical device integration requires dislocation line densities of $<<10^{-5}$ μm^{-1}. PLM images show the onset of strain relaxation in MQW laser structures with n>4 $In_{0.35}Ga_{0.65}As/GaAs$ (5.7nm/20nm) QWs through the formation of <100>-oriented defects. The density of these defects appears to increase strongly with increased annealing temperatures and with increasing substrate dislocation density. RBS planar channeling measurements and resonant Raman scattering measurements show what appears to be an increase in the density of lattice defects - *even before the onset of lattice relaxation is observed by PLM.* X-ray diffraction suggests that increasing misfit strain may increase indium surface segregation. The use of Be-doping in the GaAs barrier layers of the MQW appears to inhibit the formation and/or propagation of both misfit dislocations and <100>-oriented defects. These results suggest that a 4 QW $In_{0.35}Ga_{0.65}As/GaAs$ (5.7nm/20nm) active region is close to the practical limit for high-speed MQW lasers with an In composition of 35%. Small decreases in the barrier thickness and/or the introduction of a fifth QW may be possible using impurity-hardened GaAs QW barriers, lower growth temperatures for the $Al_xGa_{1-x}As$ waveguide cladding layers and low EPD substrates.

Acknowledgements

The authors wish to thank H. Thaden for his assistance with sample preparation. They also wish to thank H. Rupprecht for his continuing support and encouragement. This work was supported by the German Bundesministerium für Forschung und Technologie (BMFT) within the framework of the PHOTONIK program, and by the Freiburger Materialforschungszentrum (FMF).

5. REFERENCES

Baeumler M, Larkins E C, Köhler K, Bernklau D, Riechert H, Ralston J D and Jantz W 1993 to be presented at the Fifth International Conference on Defect Recognition and Image Processing in Semiconductors and Devices (DRIP 5) and to be published in the Inst. of Phys. Conf. Ser.

Bender G, Larkins E C, Schneider H, Ralston J D and Koidl P 1993 to be presented at this conference.

Bonar J M, Hull R, Walker J F and Malik R 1992 Appl. Phys. Lett. 60 1327.

Esquivias I, Weisser S, Schönfelder A, Ralston J D, Tasker P J, Larkins E C, Fleißner J, Benz W and Rosenzweig J, to be presented at this conference.

Flagmeyer R 1992 Nucl. Instr. and Meth. in Phys. Res. B68 190.

Fritz I J, Gourley P L and Dawson L R 1987 Appl. Phys. Lett 51 1004.

Gourley P L, Fritz I J and Dawson L R 1988 Appl. Phys. Lett. 52 377.

Kirkby P A 1975 IEEE J. Quantum Electron. QE-11 562.

Larkins E C, Rothemund W, Maier M, Wang Z M, Ralston J D, Jantz W 1993 J. Crystal Growth 127 541.

Matthews J W and Blakeslee A E 1974 J. Crystal Growth 27 118.

Qin C-D and Roberts S G 1989 Inst. Phys. Conf. Ser. No. 104 321.

Ralston J D, Larkins E C, Rothemund W, Esquivias I, Weisser S, Rosenzweig J and Fleißner J 1993a J. Crystal Growth 127 19.

Ralston J D, Weisser S, Esquivias I, Larkins E C, Rosenzweig J, Tasker P J and Fleißner J 1993b IEEE J. Quantum Electron. June 1993.

Ralston J D, Weisser S, Esquivias I, Schönfelder A, Larkins E C, Rosenzweig J, Tasker P J, Maier M and Fleißner J 1993c to be published in Mat. Sci. and Engr. B.

Schweizer T, Köhler K, Ganser P, Hiesinger P and Rothemund W 1992 Mat. Res. Soc. Symp. Proc. Vol. 263 323.

Wagner J, Larkins E C, Herres N, Ralston J D and Koidl P 1993 submitted for publication.

Wang Z M, Baeumler M, Jantz W, Bachem K H, Larkins E C and Ralston J D 1993 J. Crystal Growth 126 205.

Weisser S, Ralston J D, Larkins E C, Esquivias I, Tasker P J, Fleißner and Rosenzweig J 1992 Electron. Lett. 28 2141.

Paper presented at the Int. Symp. GaAs and Related Compounds, Freiburg, 1993

Selective high-temperature-stable oxygen implantation and MBE-overgrowth technique

H. Muessig, C. Woelk, and H. Brugger

Daimler-Benz AG, Forschungszentrum, Wilhelm-Runge-Straße 11, D-89081 Ulm, Germany

A. Forchel

Technische Physik, Universität Würzburg, Am Hubland, D-97074 Würzburg, Germany

ABSTRACT: A selective isolation implantation technique by use of single charged oxygen ions is applied to convert highly doped GaAs material with doping concentrations in the range between $1x10^{18}cm^{-3}$ and $1x10^{19}cm^{-3}$ into highly resistive regions (>10^9 Ω/sq). The isolation behavior is observed to be thermally stable up to temperatures above 750°C. This allows subsequent layers to be grown by molecular beam epitaxy (MBE) at substrate temperatures of 600°C. After a surface cleaning procedure modulation-doped pseudomorphic GaAs/InGaAs/AlGaAs heterostructures are grown on the implanted wafers. An excellent surface morphology and a strong photoluminescence response is observed on the overgrown samples. A two-dimensional electron gas mobility of 29 500 cm^2/Vs (6 650 cm^2/Vs) and a carrier density of $1.4x10^{12}cm^{-2}$ ($1.9x10^{12}cm^{-2}$) are measured by Hall technique at 80 K (300 K) in the dark.

1. INTRODUCTION

Ion implantation is firmly established in III/V-technology, either to create doped layers or to produce resistive regions for device isolation. In GaAs material a high resistance behavior is achieved by damage-induced compensation of carriers as a consequence of ion bombardment. Most commonly, protons and boron ions are used for implant isolation (Clauwaert et al. 1987). However, a long-term stability of a high resistance behavior is only observed for temperatures well below 600°C (e.g., Pearton 1990). In the last decade III/V molecular beam epitaxy (MBE) has been established as a standard technique for the fabrication of high quality heterostructure device layers, which are normally grown in one epitaxial run. A technology based on a combination of a selective implantation process for lateral patterning and a following large-area MBE-overgrowth is a promising way for a monolithic integration of different device components on one single chip. This allows a minimization of parasitic losses and a reduction of fabrication cost of future millimetre-wave application electronic systems (Brugger et al. 1992).

A high-temperature-stable isolation implantation process by use of oxygen ions implanted into Si-doped and MBE-grown GaAs is reported. Initially highly conducting (< 20 Ω/sq) layers are converted into highly resistive layers with a sheet resistance of $R_{SH} > 10^9$ Ω/sq. The crystal defects responsible for damage-induced isolation anneal out for temperatures

$T_A > 600^oC$. However, a highly resistive and thermally stable isolation behavior is achieved after an anneal process at $T_A > 750^oC$. This allows an overgrowth of subsequent layers by MBE on selectively implanted material at elevated temperatures between 500°C and 700°C without a degradation of the isolation. The overgrown layers show an excellent surface morphology with defect densities comparable with standard epitaxially grown samples. Pseudomorphic (PM) modulation-doped GaAs/InGaAs/AlGaAs heterojunction field-effect transistor (HFET) structures on selectively implanted areas exhibit a strong photoluminescence (PL) response and a two-dimensional electron-gas (2DEG) mobility $\mu_H = 29\,500\ cm^2/Vs$ ($6\,650\ cm^2/Vs$) and carrier density $n_s = 1.4x10^{12}cm^{-2}$ ($1.9x10^{12}cm^{-2}$) at 80 K (300 K).

2. SELECTIVE ISOLATION BY OXYGEN IMPLANTATION

The implant experiments are performed on highly doped GaAs material on semi-insulating (s.i.) (100) oriented GaAs substrates. The epitaxial material is grown by MBE and consists of the following layer sequence: 100 nm undoped GaAs buffer layer, 40 nm undoped short-period AlAs/GaAs superlattice (SL), 100 nm undoped GaAs and a final 0.4 μm thick and Si-doped GaAs layer ($N_D = 1x10^{18}cm^{-3}$ - $1x10^{19}cm^{-3}$). The samples are covered with a 2 μm thick SiO_2 dielectric layer by a plasma-enhanced chemical vapor deposition (PECVD) process at 300°C. For the selective implantation the dielectric layer is patterned by photolithography and openings are fabricated by a reactive ion etching (RIE) $CHCl_3/CF_4$ plasma process. The implantation is carried out in a VARIAN 500 XP implanter at room temperature with the wafers normal axis inclined during ion bombardment by 7° off.

Single charged oxygen ions ($^{16}O^+$) are used as the implant to convert the highly conducting GaAs into highly resistive regions. Energy and dose values are calculated on the basis of a TRIM-91 Monte-Carlo simulation (Ziegler et al. 1985) to match the oxygen profile to the Si-doping distribution. A multi-energy implantation is performed with the following dose values: 250 keV ($3x10^{14}cm^{-2}$), 115 keV ($1.7x10^{14}cm^{-2}$), and 40 keV ($2.4x10^{14}cm^{-2}$). This corresponds to a total dose of $7x10^{14}cm^{-2}$. The total dose is varied keeping the individual energies and dose ratios constant.

After implantation both sides of the wafer are capped with 200 nm thick PECVD-deposited SiO_xN_y followed by a rapid thermal anneal (RTA) for 30 sec. T_A is varied between 500°C and 850°C. For the electrical characterization on-mesa and off-mesa test structures are realized by standard photolithography methods. Ohmic contacts are fabricated by alloyed GeNiAu metallization on n^+-GaAs. Transmission line measurements (TLM) and Hall experiments are performed to investigate the isolation characteristics of the implanted material. A schematic cross-sectional view of the sample geometry is shown in Figure 1.

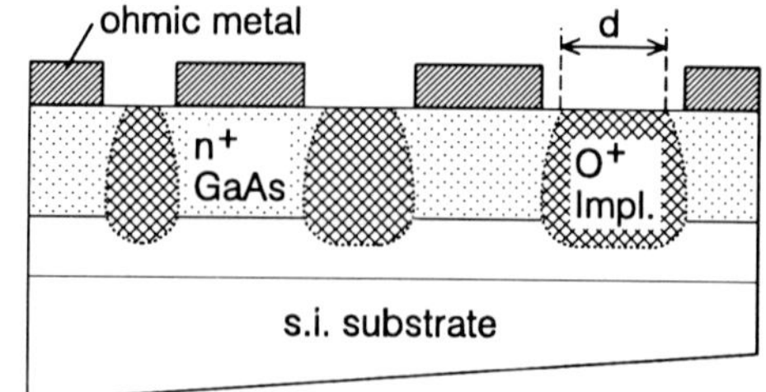

Figure 1. Schematic cross-section of samples used for oxygen implantation investigations.

In Figure 2 a series of room temperature current/voltage (I/V) curves are shown from selectively implanted and annealed samples. The results are obtained from a measurement on a TLM contact configuration with different isolation widths (d). Due to the very high sheet resistance of $R_{SH} > 10^9$ Ω/sq the measured current levels are in the nA-range. The experiments are performed inside a shielded box in the dark. Quantitative values of R_{SH} are deduced from the d - dependent slopes (δV/δI) at low voltages, where the electrical characteristic shows a nearly ohmic behavior. The reliability of the ohmic contacts is proven by separate test structures on the same wafers. Reference measurements on s.i. substrates, which are specified by a specific resistance of > 10^7 Ωcm, yield values of $R_{SH} \approx 10^9$ Ω/sq. Therefore, the oxygen implanted regions in Figure 2, which are annealed at 750°C, exhibit an isolation behavior comparable with presently available GaAs substrates used for millimetre wave applications.

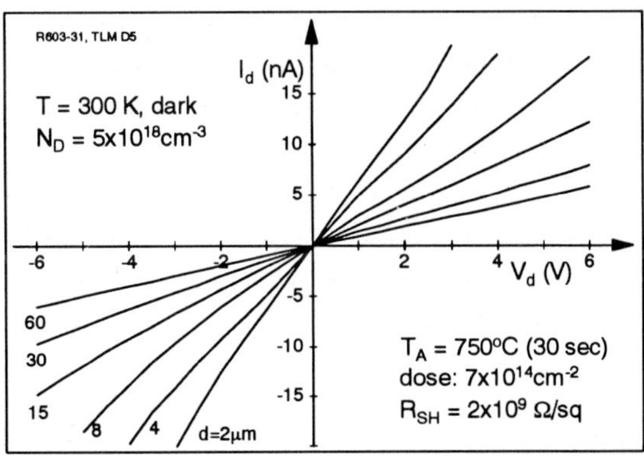

Figure 2. I/V-curves of an implanted and annealed sample measured on TLM test patterns with oxygen isolation widths d. The initial sheet resistance was 17 Ω/sq.

In Figure 3 the oxygen implant isolation characteristic is shown as a function of anneal temperature and for two different total oxygen dose values. In the low temperature region ($T_A < 600$°C) damage-related carrier trapping is the dominating isolation process. For $T_A > 600$°C the crystal defects anneal out and the conductivity is continuously recovered. A similar isolation characteristic is observed for protons and boron ions at lower temperatures (Pearton 1990). For the lower dose curve R_{SH} increases again for $T_A > 670$°C up to a maximum value of $R_{SH} > 10^9$ Ω/sq at 750°C. For even higher annealing temperatures the resistance drops again. A similar behavior is observed on an identical sample with a significantly higher dose value of 2×10^{15}cm^{-2} which is also shown in Figure 3. However, the increase of R_{SH} starts at higher temperatures ($T_A > 750$°C).

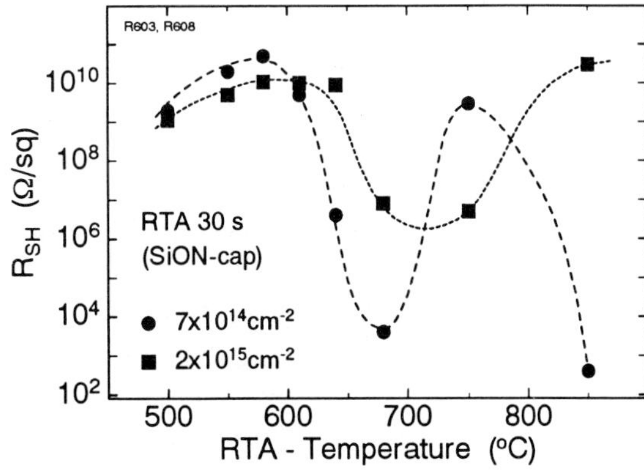

Figure 3. Measured R_{SH} as a function of T_A for samples with $N_D = 5 \times 10^{18}$cm^{-3} and an initial R_{SH} of 17 Ω/sq.

In addition to DC I/V experiments high frequency measurements (0.5 GHz - 26 GHz) are performed by use of coplanar test structures with different geometries on oxygen implanted areas with $R_{SH} = 5x10^8$ Ω/sq. Experiments on n^+-doped regions with low $R_{SH} \leq 10^2$ Ω/sq and on s.i. GaAs substrates without MBE-layers are performed for comparison. From the characteristic impedance and the measured damping factors there is further evidence of a nearly identical isolation characteristic of the oxygen implanted areas and the used s.i. GaAs substrates in accordance with the DC results, but in contrast to results obtained from structures on the n^+-doped regions.

The reported results demonstrate an implant isolation by oxygen ions in highly doped GaAs with $N_D = 5x10^{18}cm^{-3}$ for $T_A \geq 750°C$. A similar behavior is observed on samples with N_D up to $10^{19}cm^{-3}$. The microscopic mechanism responsible for the carrier compensation is not clarified yet. Experiments on the dose dependent isolation characteristics clearly show that a threshold concentration of oxygen ions close to $10^{15}cm^{-2}$ is necessary to achieve high resistivity behavior for $T_A > 700°C$ (Muessig and Brugger 1993). This value is expected to be close to the amorphization limit of the crystal. In comparison a damage related resistance value of $R_{SH} > 10^9$ Ω/sq for $T_A \leq 600°C$ is achievable at significantly lower dose values. Further analytical investigations are under way to clarify the compensation mechanism of oxygen in highly doped n-type GaAs. Favennec (1976) suggested, that oxygen forms a deep double electron trap for T_A up to 800°C. Schnell et al. (1991) argued, that the increase of R_{SH} at higher temperatures is due to the presence of electrically active oxygen. From spectroscopic studies by Alt (1990) there is strong evidence of a double electron trap caused by substitutional oxygen due to a midgap Ga-O-Ga defect complex.

3. MBE OVERGROWTH ON IMPLANTED STRUCTURES

After the selective implantation and RTA process the wafers are regrown by a second MBE run. Prior to the epitaxial growth the wafers are ex-situ wet chemically cleaned by a H_2SO_4 dip followed by a $KOH/H_2O_2/H_2O$ etching procedure. Inside the MBE chamber a small amount of GaAs is thermally desorbed under As_4 pressure. A series of PM HFET test structures is grown with the following layer sequence: undoped 0.1 µm thick GaAs buffer, undoped 12 nm PM $In_{0.2}Ga_{0.8}As$ quantum well , 5 nm $GaAs/Al_{0.25}Ga_{0.75}As$ spacer layer, 40 nm pulse-doped $Al_{0.25}Ga_{0.75}As$ electron supply layer and a final 40 nm thick GaAs cap layer. In Figure 4 the schematic layered structure of a regrown

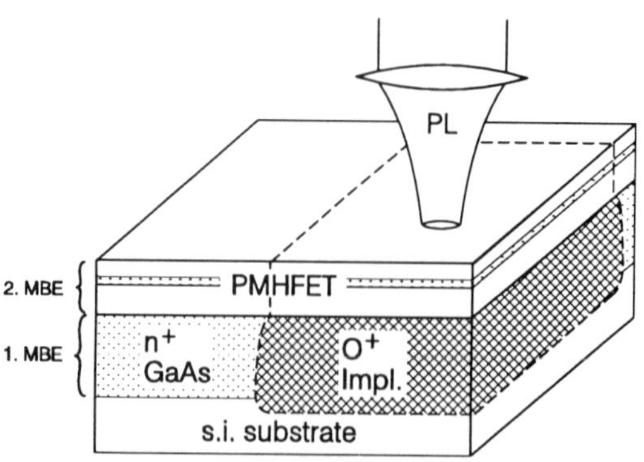

Figure 4. Schematics of the sample layered structure for PL measurements on selectively implanted wafers with an overgrown PM HFET layer sequence.

sample is shown. The material quality of the overgrown layers is investigated by a visual inspection of the wafers surface morphology, micro-PL measurements and Hall characterization of the 2DEG transport properties.

The achieved surface defect density on overgrown samples is comparable with conventionally grown samples on s.i. substrates. The optical material quality is demonstrated by the PL response shown in Figure 5. Strong spectral bands are observed arising from recombinations of the 2DEG carriers from the n=1 (E_{11}) and n=2 (E_{21}) electron subbands, respectively, with photo-excited holes in the n=1 subband. Curves c) and d) are from overgrown PM HFET layers with a 0.1 μm thick GaAs buffer layer located on top of a selectively oxygen implanted region and on an unimplanted n^+-doped area, respectively. For comparison, under the same conditions a PL spectrum is recorded of an identical PM HFET layer sequence grown on a s.i. substrate in one epitaxial run (curve b). The PL intensities of curves (b) and (c) are nearly identical. This demonstrates the high optical quality of the overgrown material. A more intensive PL response is observed on a reference sample on s.i. substrate (curve a), which is mainly attributed to the additional SL and the thicker buffer layer (0.8 μm).

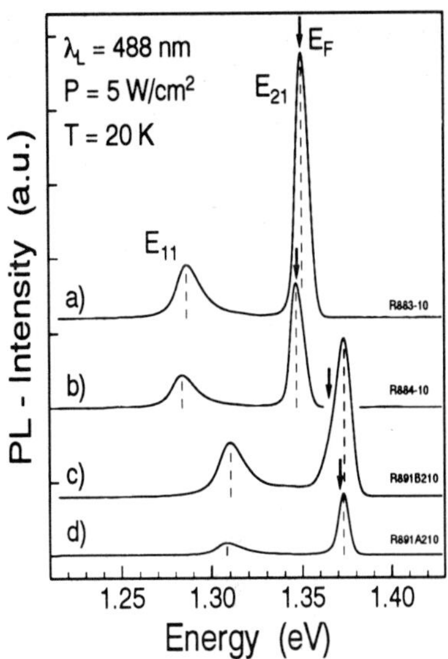

Figure 5. PL spectra of PM HFET layers on
- s.i. GaAs substrates with
  a) SL and 0.8 μm thick GaAs buffer
  b) 0.1 μm thick GaAs buffer
- selectively implanted wafers with a 0.1 μm thick GaAs buffer on
  c) oxygen implanted areas
  d) n^+ doped regions.

From the PL line shape information about the carrier density (n_s) is deduced (Brugger et al. 1991). The arrows indicate the Fermi energy (E_F). From the spectral width ($E_F - E_{11}$) a 2DEG $n_s \approx 1.5 \times 10^{12} cm^{-2}$ is deduced for the overgrown sample on oxygen implanted areas. This value is about 20% lower in comparison with the n_s-values on reference samples (a, b).

Table I. 2DEG transport properties of the PM HFET structures grown on a s.i. GaAs substrate (reference samples) and fabricated on selectively oxygen implanted wafers by a MBE overgrowth technique.

	PM HFET layer on implanted wafers		PM HFET layer on reference substrates	
T (K)	n_s ($10^{12}cm^{-2}$)	μ_H (cm^2/Vs)	n_s ($10^{12}cm^{-2}$)	μ_H (cm^2/Vs)
80	1.4	29 500	1.7	27 900
300	1.9	6 650	2.1	7 000

The electronic properties are investigated by a magnetic-field-dependent Hall measurement which allows a selective determination of the 2DEG carrier density and mobility in the presence of parallel conducting cap-layers (Koser et al. 1993). Especially the low-temperature mobility behavior is expected to be sensitively dependent on the material quality of the overgrown samples. In Table I the measured 2DEG n_s and μ_H-values are listed for PM HFET layers grown on large-area oxygen implanted and annealed structures together with the data obtained from reference samples. The 2DEG transport properties achieved are close to the data obtained from reference material. The results are a promising basis for future work on multi-functional integration of different device layers by the presented technology.

4. CONCLUSION

A selective isolation process with oxygen ions implanted in highly doped n-type MBE-grown GaAs material is presented. For annealing temperatures $T_A > 750^oC$ a high resistive isolation characteristics with $R_{SH} > 10^9$ Ω/sq is observed. This allows MBE-layers to be grown in a second epitaxial run at elevated temperatures without isolation degradation of the underlying regions. The overgrown PM HFET layers on implanted GaAs material exhibit a low-defect-density surface morphology. The achieved 2DEG mobility values of 29 500 cm^2/Vs at 80 K are comparable with results obtained from reference samples.

ACKNOWLEDGEMENT

The authors would like to thank Th. Hackbarth, H. Haspeklo, S. Heuthe, H. Koser, H. Leier and U. Salz for expert technical help. This work was partially supported by the Bundesministerium für Forschung und Technologie (Bonn, Germany) under contract number 01 BM 113/1.

REFERENCES

Alt H Ch, 1990, Phys. Rev. Lett. 65, 3421

Brugger H, Woelk C, Muessig H, 1992, Proc. of Material Research Society MRS Fall Meeting Symposium D ed. by Houghton D C et al. Boston (USA), in press

Brugger H, Muessig H, Woelk C, Kern K, Heitmann D, 1991, Appl. Phys. Lett. 59, 2739

Clauwaert F, Van Daele P, Baets R, Lagasse P, 1987, J. Electroch. Soc., 711

Favennec P N, 1976, J. Appl Phys. 47, 2532

Koser H, Voellinger O, Brugger H, 1993, Proc. of the 20th Int. Conf. on Gallium Arsenide and Related Compounds, Freiburg (Germany), Aug 29 - Sept 02, in press

Muessig H and Brugger H, 1993 (unpublished)

Pearton S J, 1990, Mat. Science Report 4, 313

Schnell R D, Gisdakis S, Alt H Ch, 1991, Appl. Phys. Lett. 59, 668

Ziegler J, Biersack J, Littmark V, 1985, Doping and Range of Solids (New York: Pergamon)

Importance of V/III supply ratio in low temperature epitaxial growth of InAs

T. Hamada*, T. Hariu and S. Ono**

Department of Electronic Engineering,
**Research Institute of Electrical Communication,
Tohoku University, Sendai 980 Japan
*Present address: Oki Electric Industry Co., Hachioji 193, Japan

ABSTRACT: Crucial effects of V/III supply ratio in low-temperature epitaxial growth of InAs on GaAs are described. The incorporation rate of In is less than 1 even at 300 °C and increases with the supply rate of As. Optimum supply ratio shifts to a lower value as the growth temperature is reduced and as the rf power to excite the plasma is increased. It is indicated that the excited atomic As supplied by plasma-cracking of As_4 assists the low-temperature epitaxial growth and the optimization leads to high quality InAs layers on GaAs at low temperatures (μ_n=1.1x10^4 cm^2/Vs for a layer with thickness 2.0 μm grown at 300 °C, μ_n=1.5x10^4 cm^2/Vs for a layer with thickness 2.5 μm at 400 °C).

1. Introduction

In order to expand the flexibility in the choice of III-V semiconductor materials for higher performance of electronic and optoelectronic devices, InAs, InSb and their alloys have been attracting more attention. In our efforts devoted to the low temperature epitaxial growth of these materials in view of the requirement to produce advanced device structures, more pronounced importance of V/III supply ratio has been noticed at lower growth temperature to obtain high quality layers as well as for the control of alloy composition (Hamada T et al. 1992). The purpose of this paper is to describe the low temperature epitaxial growth of InAs on GaAs down to 270 °C by paying particular attention to the crucial effects of V/III supply ratio, with some comparison with InSb.

A particular epitaxial growth method in hydrogen plasma, which we call plasma-assisted epitaxy (PAE) (Hariu T et al. 1990), was employed here to reduce the growth temperature in comparison with the conventional molecular beam epitaxy (MBE). One of the important advantages of PAE among others is the supply of atomic group-V elements in the excited state which are produced through dissociation in discharging plasma after they are evaporated from heated crucibles, as in MBE, in the form of typically tetramers (As_4, Sb_4) (Hariu T et al. 1989). These chemically active species, which are detected by optical emission spectroscopy (OES) from plasma, assist the low temperature epitaxial growth with less and wider V/III supply ratio, compared with the conventional MBE.

2. Experimental

A similar PAE apparatus as described elsewhere (Matsushita K et al. 1984) was used for the growth of InAs films. Elemental metallic In (99.9999 % pure) and As (99.99999 % pure) were evaporated by resistive heating and supplied toward the growing surface through plasma which is excited by rf power at 13.56 MHz through inductive coupling. Hydrogen gas of 99.9999 % purity was used as a discharging gas after further purification through a palladium diffuser. Semi-insulating GaAs (100) substrates were chemically etched in $3H_2SO_4:1H_2O_2:1H_2O$ and immediately introduced into the growth chamber. The growth temperature was varied between 270 °C and 400 °C. The total pressure during the growth was 20 mTorr. The growth rate of InAs ranged from 0.6 to 0.8 μm/h. The electrical property was measured by van der Pauw method at room temperature.

3. Results and discussions

Figure 1 shows the dependence of the growth rate on As/In supply ratio at the growth temperature of 300 °C and 400 °C. Although the growth rate for the growth temperature of 400 °C was almost constant in spite of the increasing As/In supply ratio, the growth rate for the growth temperature of 300 °C was slower than that for 400 °C, and increased with the increased As/In supply ratio. It was then noted that the incorporation rate of In is less than 1 even at 300 °C, as was also observed in MBE at higher temperatures (Ferguson I T et al.

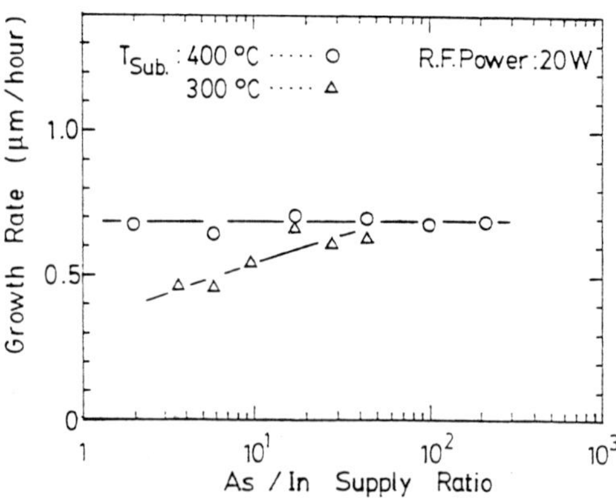

Figure 1 The dependence of the growth rate of InAs on As/In supply ratio at the substrate temperature of 300 °C and 400 °C.

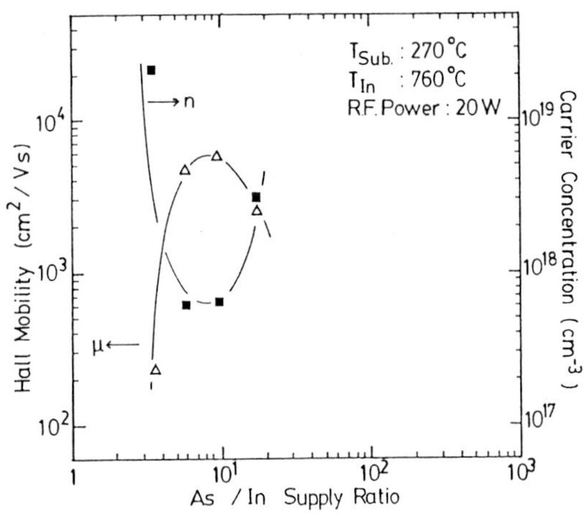

(a) the growth temperature of 270 °C

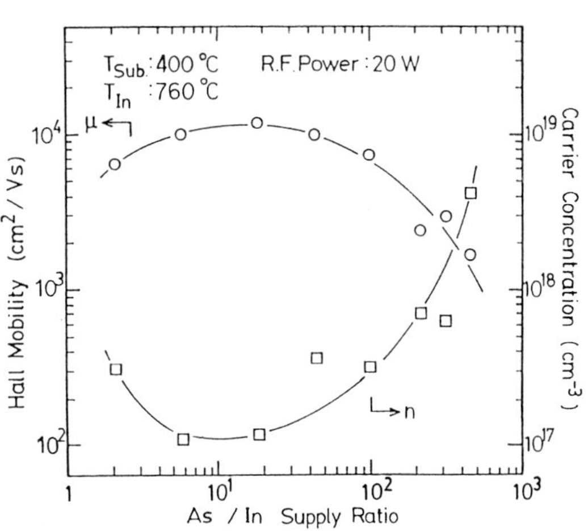

(b) the growth temperature of 400 °C

Figure 2 The effect of As/In supply ratio on the electrical properties of InAs layers grown on GaAs at different temperatures. (The substrate temperature of (a) 270 °C and (b) 400 °C)

1992), and increases with the supply rate of As. This phenomenon is opposite to the case of InSb (Ohshima T et al. 1989).

Figure 2 shows how the electrical properties varied with As/In supply ratio for InAs layers grown at 270 °C (Figure 2(a)) and 400 °C (Figure 2(b)) with 20 W rf power applied to excite the plasma. It is noticed here that the electrical property at 270 °C depends much more critically upon the As/In supply ratio than those grown at 400 °C and also that, as the growth temperature is lowered, the optimum As/In supply ratio shifts to a lower value. The electronic property as well as surface morphology depends more critically on As/In supply ratio and then more precise control of supply ratio is required at lower growth temperature.

Figure 3 shows the dependence of the optimum As/In supply ratio on the growth temperature, where open circles are the optimum As/In supply ratios at which the largest Hall mobility is obtained and solid circles are the As/In supply ratios at which the Hall mobility reduces to one fourth of maximum value. This more critical dependence of electrical property on As/In supply ratio at a lower temperature should come from the decreased reevaporation of As (compared with the slight change of sticking coefficient of In) from the

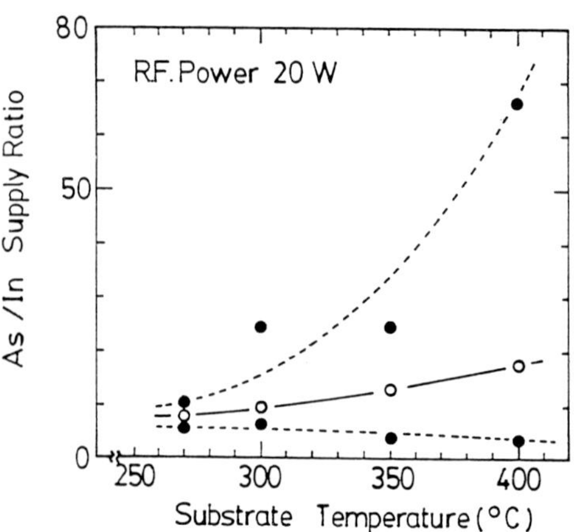

Figure 3 The dependence of the optimum As/In supply ratio on the growth temperature. (Open circles are the optimum As/In supply ratios at which the largest Hall mobility is obtained and solid circles are As/In supply ratios at which the Hall mobility reduces to one forth of maximum value.)

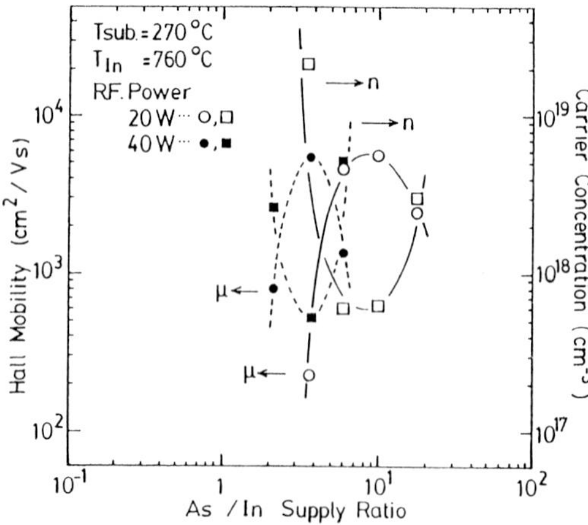

Figure 4 The electrical properties of InAs layers grown at the growth temperature of 270 °C with rf power of 20 W and 40 W as a function of As/In supply ratio.

growing surface. It is here concluded that As/In supply ratio should be more precisely controlled to grow good quality epitaxial layers at a lower growth temperature. Optimum supply ratio shifts to a lower value from 12.0 at 350 °C to 6.8 at 270 °C, compared with 1.8 to 0.72 for InSb (Hamada T et al. 1992), as the growth temperature is reduced.

Figure 4 shows the electrical properties of InAs layers grown at a growth temperature of 270 °C with rf power of 20 W and 40 W as a function of As/In supply ratio. It is noted here that, as rf power is increased, the electrical property depends more critically upon the As/In supply ratio and the optimum ratio shifts to a lower value. This result can be explained by the increased dissociation of As molecules through enhanced plasma-cracking at 40 W compared with 20 W. The same tendency was also confirmed in the case of PAE-InSb growth (Ohshima T et al. 1989).

Figure 5 shows the electron mobility of InAs epitaxial layers on GaAs by PAE in comparison with other methods as a function of a growth temperature, where solid circles are for thickness around 2.0 to 2.5 μm and open circles for around 0.5 to 0.7 μm. This optimization leads to high quality InAs layers on GaAs ($\mu_n=1.1\times10^4$ cm^2/Vs for a layer with thickness 2.0 μm grown at

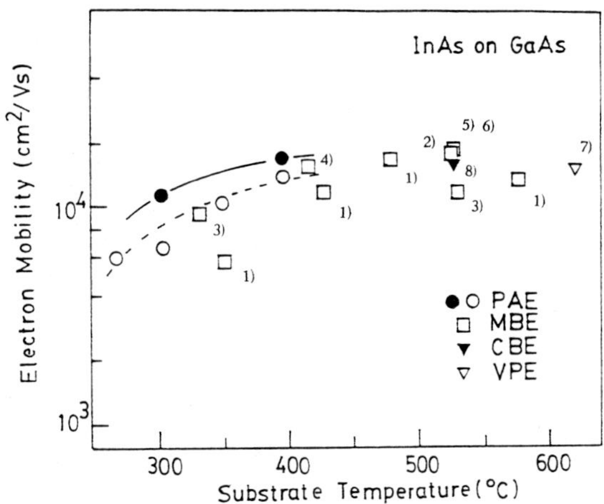

Figure 5 The electron mobility of InAs epitaxial layers on GaAs by PAE in comparison with other methods as a function of a growth temperature. (Solid circles for thickness around 2.0 to 2.5 μm and open circles for around 0.5 to 0.7 μm.)
< 1) Yano M et al. 1977, 2) Kubiak R A A et al. 1984, 3) Meggitt B T et al. 1978, 4) Chiu T H et al. 1990, 5) Kalem S et al. 1988a and 1988b, 6) Kalem S. 1989, 7) Godhino N et al. 1970, 8) Kamp M et al. 1990 >

300 °C, μ_n=1.5x10^4 cm^2/Vs for a layer with thickness 2.5 μm at 400 °C) comparable to those obtained at higher temperatures by other methods.

4. Conclusions

The plasma-assisted epitaxy (PAE) has been successfully applied to low temperature epitaxial growth of InAs on GaAs down to 270 °C by paying particular attention to the crucial effects of V/III supply ratio. The electrical property of the films grown at 270 °C depends much more critically upon the As/In supply ratio than those grown at 400 °C. It is noticed also that, as the growth temperature is lowered and as rf power to excite the plasma is increased, the optimum As/In supply ratio shifts to a lower value. The optimization leads to high quality InAs layers on GaAs at low temperatures (μ_n=1.1x10^4 cm^2/Vs for a layer with thickness 2.0 μm grown at 300 °C, μ_n=1.5x10^4 cm^2/Vs for a layer with thickness 2.5 μm at 400 °C).

Acknowledgement

The authors would like to thank Dr. Y. Sano of Oki Electric Industry Co., Ltd. for partial support of the work and supply of GaAs wafers, and also Murata Foundation for encouraging support of this work.

References

Chiu T H and Ditzenberger J A, 1990 Appl. Phys. Lett. 56, 2219
Ferguson I T, de Oliveria A G and Joyce B A, 1992 J. Cryst. Growth 121, 267.
Godhino N and Brunnschweiler A, 1970 Solid State Electronics 13, 47
Hamada T, Ohshima T, Hariu T and Ono S, 1992 Proc. 19th Int. Symp. GaAs and Related Compounds, Inst. Phys. Conf. Ser. No.129, 163
Hariu T, Fang S F and Ohshima T, 1990 Proc. 17th Int. Symp. GaAs and Related Compounds, Inst. Phys. Conf. Ser. No.112, 137
Kalem S, Chyi J, Litton C W, Morkoc H, Kan S C and Yariv A, 1988a Appl. Phys. Lett. 53, 562
Kalem S, Chyi J I, Morkoc H, Bean R and Zanio K, 1988b Appl. Phys. Lett. 53, 1647
Kalem S, 1989 J. Appl. Phys. 66, 3097
Kamp M, Weyers M, Heineeke H, Lueth H and Balk P, 1990 J. Cryst. Growth 105, 178
Kubiak R A A, Parker E H C and Newstoad S, 1984 Appl. Phys. Lett. A35, 61
Matsushita K, Sato T, Sato Y, Sugiyama Y, Hariu T and Shibata Y, 1984 IEEE Trans. Electron Devices ED-31, 1092.
Meggitt B T, Parker E H C and King R M, 1978 Appl. Phys. Lett. 33, 528
Ohshima T, Yamauchi S and Hariu T, 1989 Proc. 16th Symp. GaAs and Related Compounds, Inst. Phys. Conf. Ser. No.106, 241
Yano M, Nogami M, Matsushita Y and Kimata M, 1977 Jpn. J. Appl. Phys. 16, 2131

Re-evaporation and sub-oxide transport in molecular beam epitaxy

Colin E.C. Wood, Richard A. Wilson, Seyed A. Tabatabaei
Joint Program for Advanced Electronic Materials, Electrical Engineering Department.University of Maryland and Laboratory for Physical Sciences, College Park, MD. 20740

Peter Sheldon
National Renewable Energy Laboratory, Golden, CO. 80401

ABSTRACT

Silicon, germanium and aluminium, can be transported to substrate interfaces as sub-oxides. Impurities can also be re-evaporated from surface close to effusion cells.

INTRODUCTION

Unintentional impurities are the single biggest problem of epitaxial films and interfaces. In this paper we consider two processes that can introduce unwanted concentrations of Si, Ge, Be, Ga, Al, and In.

EXPERIMENTAL

Ozone treated GaAs substrates, were heated to 723K for 1 hour to remove CO_2, H_2O, etc. then to ~973K under a dimeric arsenic flux to sublime native oxides in the growth chamber. Effusion cells were simultaneously 'outgassed'. Elemental concentrations of grown films and interfaces were profiled by secondary ion mass spectrometry, and net ionized impurity concentrations were determined by 'Polaron' electro-chemical profiling.

SUB-OXIDE TRANSPORT

Suboxide transport, was indicated first in parabolic quantum well structures, where sheet electron concentrations (Hall effect) were often anomolously high. $3x10^{13}cm^{-2}$ Si was found in SIMS profiles of these samples (see figure 1). In samples heated for longer in the preparation chamber, silicon signals were reduced. We propose that Si is transported as sub-oxides. We considered first native oxide desorption: Arsenic oxide As_2O_3 in 'ozone oxides' desorbs at ~720K toward the cell shroud, and the gallium oxide desorbs decompositionally above 913K:

$$Ga_2O_3^c + As_2^g \iff Ga_2O^g + As_2O_3^g \qquad \ldots\ldots\ 1$$

where superscripts c and g denote condensed, gaseous states respectively.

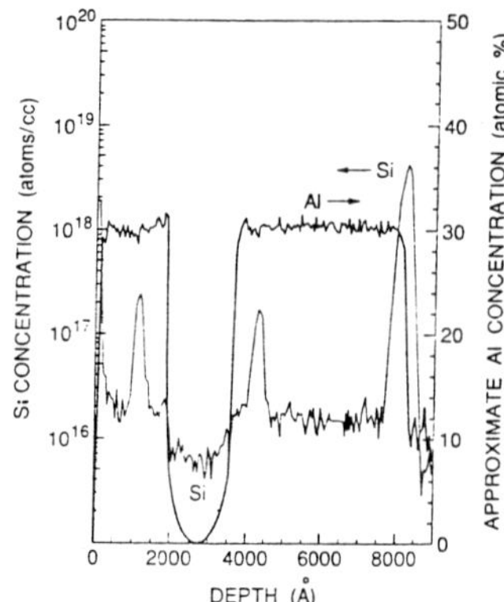

Fig. 1. SIMS depth profiles for a double Si planar doped parabolic quantum well showing ~$1.4\mathrm{x}10^{13}\mathrm{cm}^{-2}$ unintentional silicon at the interface.

Next we consider the various reactions which could produce sub-oxides, for example, with water, carbon monoxide, arsenic oxide and gallium oxide:

$$H_2O^g + Si^c \iff SiO^g + H_2^g \qquad \ldots\ldots\ 2$$

$$CO^g + Si^c \iff SiO^g + C^c \qquad \ldots\ldots\ 3$$

$$1/3As_2O_3^g + Si^c \iff SiO^g + 1/3As_2^g \qquad \ldots\ldots\ 4$$

$$Ga_2O^g + Si^c \iff SiO^g + 2Ga^l \qquad \ldots\ldots\ 5$$

Free energies[1] ΔG_o for reactions 2 through 5 are shown in figure 2.

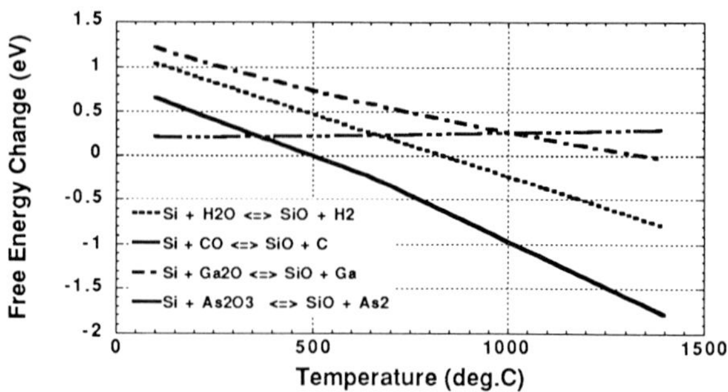

Fig. 2. Free energy change vs. temperature for Si monoxide formation by reaction with gaseous oxygen containing species. Note that carbon monoxide does not readilly oxidize silicon.

It should be noted that arsenic oxide is the most reactive. Suboxides can be reduced by gallium and, or arsenic, according to equations 4 and 5 at the substrate oxide desorption temperature. Making the reasonable assumption that the free energy of surface gallium atoms is similar to that of liquid gallium at 973K, reduction by gallium is more favorable than by arsenic.

Next, Ge and Al impurity spikes were found at interfaces of AlGaAs/Ge/AlGaAs quantum well structures on GaAs substrates The energetics for germanium and aluminum redox reactions analogous to silicon are shown in fig. 3. As GeO is more volatile than SiO, interfacial germanium probably has the same origin as silicon. Similarly reduction and incorporation is favorable below 973K with gallium and dimeric arsenic.

Al_2O can also be formed by interactions with the same species, e.g.

$$1/3As_2O_3^{g} + 2Al^{l} \Leftrightarrow Al_2O^{g} + 1/3As_2^{g} \qquad \text{....... 6}$$

however, reduction is not favorable under MBE conditions

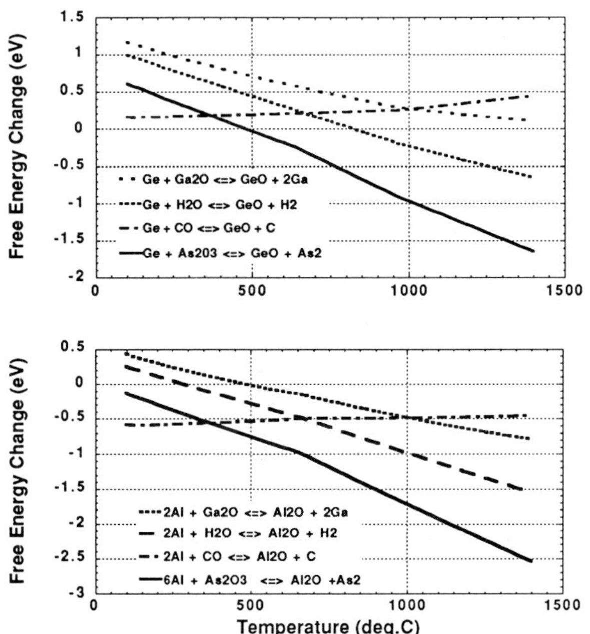

Figure 3. Free energy change for formation of Ge and Al suboxides by reactions analogous to fig. 2.

From the arguments above suboxide transport can be reduced by lowering cell temperatures during the native oxide desorption. However subliming As oxides in the preparation chamber is most effective in reducing sub-oxide transport.

RE-EVAPORATION

To demonstrate re-evaporation by heating from the effusion cell furnaces, a GaAs film was grown with only the gallium and, or arsenic shutters open. Other cells were rapidly heated to outgassing temperatures K_o and cooled to quiescent temperatures as indicated in figure 4a.

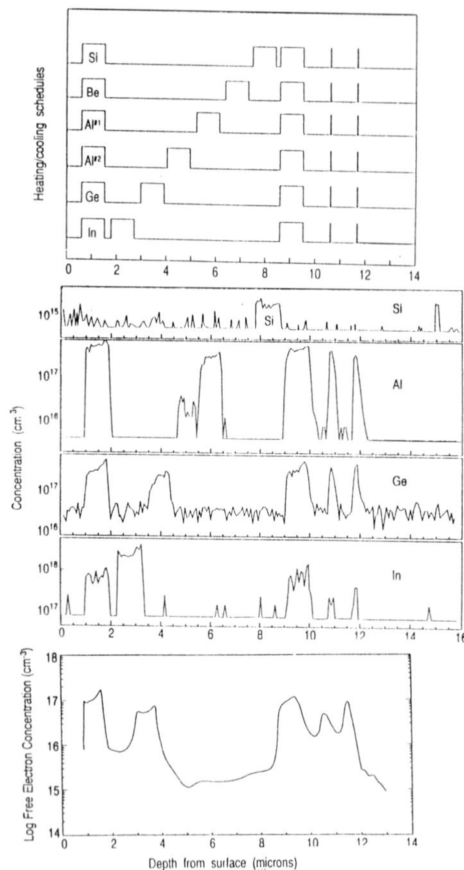

Fig. 4a. Time vs. cell temperatures for a nominally undoped GaAs film. b. corresponding Al, Si, Ge and In depth profiles, and c. ionized impurity profile.

High donor cell temperatures K_{Ge} and K_{Si} correlate with high n_e values of 4b.Moreover high Ge, In, Si and Al cell temperatures coincide with increased SIMS yields of fig. 4c. Temperatures and fluxes for unshuttered J_o and shuttered cells J_S and J_{Po} respectively are listed in table 1. J_o values are from calibrations. J_S and J_{Po} are inferred from n_i, and n_e depth profiles. J_S can be seen to be > 10^{-4} J_o. We explain these observations by re-evaporation from surfaces, see figure 5. Vanes are heated by effusion furnaces, and present direct trajectories toward substrates.

Cell	Temperature (Kelvin)	Jo x10¹² $cm^{-2}s^{-1}$	JS x10⁸ $cm^{-2}s^{-1}$	JPo x10⁸ $cm^{-2}s^{-1}$	JS/Jo x10⁻⁶
Al#1	1473	~450	140	0.42	31
Al#2	1423	~600	5.6	- - -	1
In	1273	~1200	560	2.2	47
Ge	1603	~115	70	42	61
Si	1473	2	0.56	0.84	28
Be	1173	1.7	- - -	---	- - -

Table 1. Flux data for open and shuttered cells at outgassing temperatures

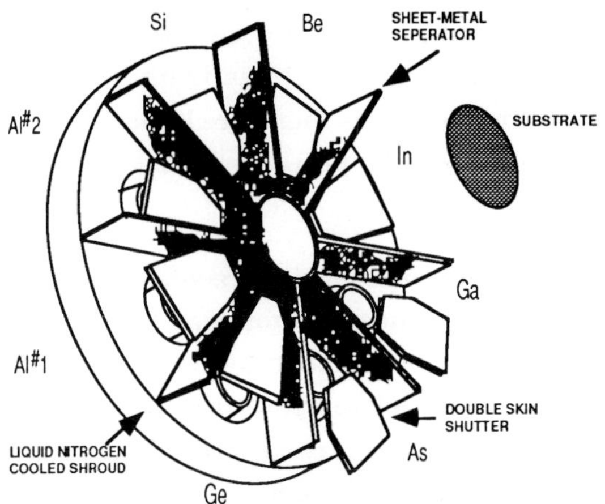

Fig. 5. Effusion cell, shutter and seperator vane arrangement

If re-evaporation from vanes is the source of impurities, then effective vane temperatures K_{eff} can be estimated from J_S, see table 2. K_{eff} should not be greater than the temperature predicted from radiative power balance considerations K_v. That is, at equilibrium, radiative losses must sum to or,

Cell	Kc	Kv	Keff	Kc/Keff
Al#1	1473	1239	1133	1.3
Al#2	1423	1197	978	1.46
In	1273	1070	968	1.32
Ge	1603	1348	1233	1.3
Si	1473	1239	1183	1.25

Table 2. Cell temperatures (K_c), calculated vane temperatures (K_v) and effective re-evaporating surface temperatures (K_{eff}).

be lower than radiation incident from furnaces etc.:

$$2A_v\sigma\varepsilon_v K_v^4 = A_c\sigma\varepsilon_c K_c^4 \qquad \text{...... 7}$$

where σ is Stefan's constant, and K_i absolute temperature. Ignoring differences in emissivity ε and geometric factors, etc. the relationship simplifies to:

$$K_v = K_c/1.19 \qquad \text{...... 8}$$

This estimation makes the reasonable approximation that re-evaporating surface areas A_v are on the order of the radiant surfaces A_c.
In support of this model, K_v values derived from equation 8 are higher than K_{eff} inferred from re-evaporation fluxes. Re-evaporation will also be aggravated by heat from adjacent cells. This effect is seen by higher n_{Ge} when Al#1 cell was hot, and the relatively higher increases in n_{Ge} in the two regions with all cells hot.

Finally, $5x10^{18}cm^{-3}$ re-evaporated aluminum found in a GaAs layer grown with the hot shuttered aluminum cell was shown to degrade characteristic photoluminescence lifetimes from ~270ns to ~95ns. by introducing non-radiative centers. It is therefore important to minimize re-evaporation effects for minority carrier property devices.

CONCLUSIONS

Suboxides are formed by reaction with arsenic oxide, and are reduced at the substrate. This problem is most pronounced for oxides produced by ozone exposure. The effect can be reduced by removing arsenic oxide in preparation chambers, and lower cell temperatures whilst remaining oxides are sublimed in the growth chamber. Al, Ge, Si, In, and Ga, etc. re-evaporation can be minimized by cooling cells when not in use, double plate shutters and increased cooling of separator vanes, etc.

ACKNOWLEDGMENTS

The NREL portion of this work was performed under contract #DE-ACO2-03CH10093 to the US Department of Energy.

REFERENCES

1 Handbook of Chemistry and Physics, CRC Press Inc. Boca Raton, USA. (1986)

Inst. Phys. Conf. Ser. No 136: Chapter 9
Paper presented at the Int. Symp. GaAs and Related Compounds, Freiburg, 1993

Electrical properties of heavily Si-doped GaAs grown on (311)A GaAs surfaces by molecular beam epitaxy

K. Agawa, Y. Hashimoto, K. Hirakawa, and T. Ikoma
Institute of Industrial Science, University of Tokyo
7-22-1 Roppongi, Minato-ku, Tokyo 106, JAPAN

Abstract: We have systematically studied the doping characteristics of Si into GaAs grown on the (311)*A* GaAs substrates by molecular beam epitaxy in order to realize highly conductive *p*-type GaAs layers. The highest hole density obtained for the uniformly doped layers was 4.2×10^{19} cm^{-3}, while for the δ-doped layers the sheet hole density as high as 2.6×10^{13} cm^{-2} was achieved. Furthermore, the growth temperature dependence of Si doping has been investigated. It is found that the conduction-type sharply changes from *p*-type to *n*-type with decreasing growth temperature at a critical temperature of 430 - 480°C.

1. Introduction

Realization of highly conductive and stable *p*-type GaAs layers is technologically quite important, in particular, in forming low-resistance base regions of *npn* heterobipolar transistors (HBTs). Recently, Be (for example, Ilegems 1977) and C (for example, Konagai et al 1989) have been extensively investigated as candidates for acceptors. However, it is pointed out that the high diffusivity of Be may induce instability during the device operation. Also, lattice distortion due to the large difference between C and As atomic radii might deteriorate the crystalline quality when the carbon density is very high. Thus, exploration for new acceptor species which have low diffusion constants and lattice constants similar to that of GaAs is highly desirable.

On the (100) GaAs surfaces, Si is the most widely used *n*-type dopant in molecular beam epitaxy (MBE). However, it has been shown by the earlier work (Wang et al 1985) that Si acts as an acceptor in GaAs grown on (*N*11)*A* GaAs substrates when $N\leq3$, while it behaves as a donor on (*N*11)*A* GaAs for $N\geq5$ and (*N*11)*B* GaAs for all the values of *N*. It has also been reported that at high growth temperatures (≥600°C) Si acts as an acceptor on the (311)*A* orientation, while it becomes a donor at low growth temperatures (≤500°C) (Meier et al 1988, Li et al 1992). Therefore, there is a possibility to control the conduction type of heavily Si-doped GaAs intentionally. The main factors which control the amphoteric nature of Si are the substrate orientation, the substrate temperature, and the V/III ratio. On the (111)*A* and (211)*A* GaAs surfaces, wet chemical etching is not well performed, while for the (311)*A* orientation excellent surface morphology is available even at high doping levels (Takamori et al 1987).

Thus, the (311)*A* GaAs substrate is considered to be the most suitable orientation from the viewpoint of doping control.

In this work, we have systematically studied the electrical properties of heavily Si-doped GaAs layers grown on the (311)*A* GaAs substrates by MBE. It is shown that very high hole density which is comparable to those in Be- or C-doped GaAs can be achieved by Si-doping. Furthermore, we will also show results on the growth temperature (T_S) dependence of the conduction-type of Si-doped GaAs.

2. *p*-type conduction in heavily Si-doped GaAs (uniform doping)

The (311)*A*-oriented semi-insulating GaAs substrates were used in the present experiments. The crystal orientation is accurate within ±0.5°. The substrates were prepared by standard wet chemical etching in $8H_2SO_4$:$1H_2O_2$:$1H_2O$ solution. The (311)*A* GaAs substrates were then mounted on the Mo-blocks with In solder, together with the (100) GaAs substrates for reference.

The uniformly doped samples were prepared by successively growing a 3000Å-thick undoped GaAs buffer layer, a 0.5µm-thick uniformly Si-doped GaAs layer, and a 1000Å-thick undoped GaAs capping layer by MBE. The doping density was controlled by adjusting the Si flux while keeping the growth rates (0.7 µm/h) and the V/III ratio (~5) constant. The growth temperature T_S was monitored by a pyrometer.

First, we studied the electrical properties of uniformly Si-doped GaAs grown on the (311)*A* and conventional (100) GaAs substrates. For this series of the samples, T_S was set to be 600°C. Figure 1 shows the carrier densities in these samples determined by Hall measurements. On the (100) orientation Si-doped layers become *n*-type. Although the electron density initially increases with the doping level, it saturates at around 4×10^{18} cm^{-3} and, then, starts to decrease with increasing dopant density due to the auto-compensation effect.

On the other hand, Si-doped layers on the (311)*A* surfaces exhibit *p*-type conduction. The hole density monotonically increases with increasing doping level and reaches the density as high as 4×10^{19} cm^{-3}. The doping efficiency is about 60% when the

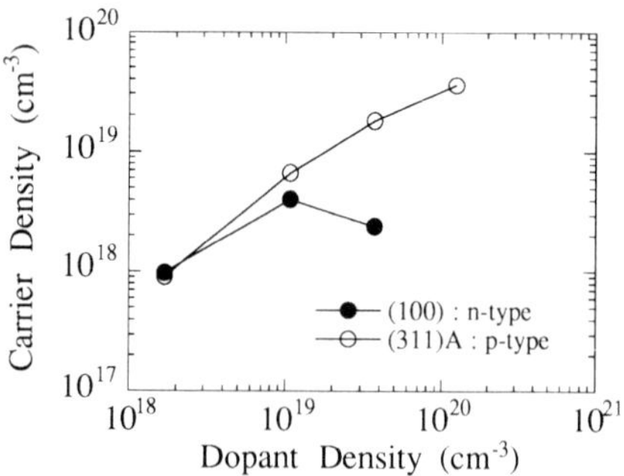

Fig. 1. Carrier density of uniformly doped GaAs layers measured by Hall measurements at room temperature. The doping efficiency of *p*-type layers on (311)*A* GaAs is much higher than *n*-type layers on (100) GaAs at more than 1×10^{19}cm^{-3}.

doping density N_{Si} is up to $1\times10^{19}cm^{-3}$, and it is 30% even when N_{Si} is as high as $1\times10^{20}\ cm^{-3}$.

In Fig. 2, the hole mobilities in these *p*-type layers grown on (311)*A* GaAs substrates are plotted as a function of the hole density, together with the results for Be- and C-doped layers grown on (100) GaAs substrates (Konagai et al 1989, Malik et al 1988). The hole mobilities in Si-doped GaAs are comparable to those in Be- and C-doped layers. This fact suggests the high crystalline quality of heavily Si-doped GaAs layers grown on (311)*A* substrates and their usefulness in realizing highly conductive metallic *p*-type base regions in *npn* HBTs. In Fig. 3, the measured resistivities of Si-doped layers are plotted as functions of temperature. It should be noted that a GaAs layer with resistivity as low as 3.8 mΩcm at 300 K can be obtained by Si-doping. As shown in the figure, *semiconductive p*-type conduction gradually changes to *metallic* conduction at hole density of $\sim4\times10^{18}\ cm^{-3}$.

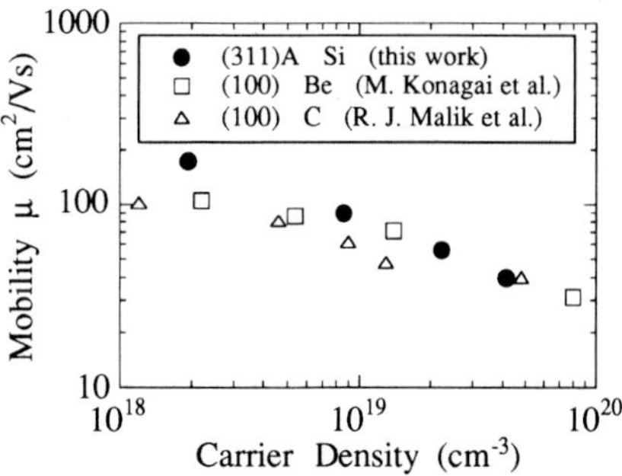

Fig. 2. Room temperature hole mobilities in uniformly doped GaAs layers are plotted as a function of the hole density. Data for Be- and C-doped layers are also shown.

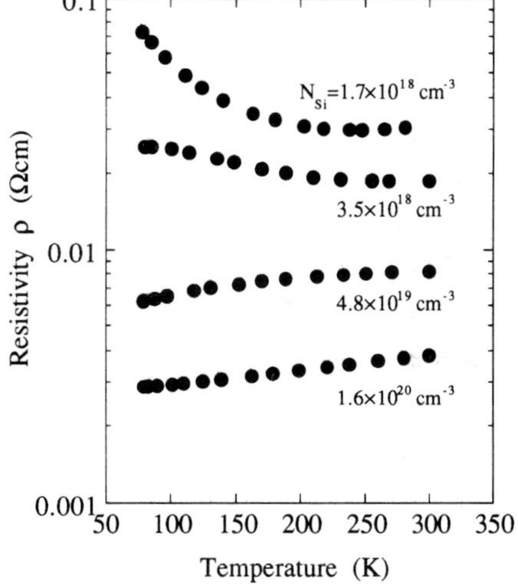

Fig. 3. Temperature dependence of resistivities of uniformly Si-doped *p*-type GaAs layers grown on (311)*A* GaAs.

3. δ-doping on the (311)*A* GaAs surfaces

In the δ-doping case, the uniformly doped GaAs layers described in the previous section were replaced with δ-doped layers. During the δ-doping process, a constant Si flux of $2.6\times10^{11}\ cm^{-2}s^{-1}$ was supplied for various intervals under the constant background As_4 flux of $2\times10^{15}\ cm^{-2}s^{-1}$. The sheet doping density of Si was varied from 0.01 to 0.3 monolayers (MLs), equivalent to $6.3\times10^{12}\ cm^{-3}$ to $1.9\times10^{14}\ cm^{-2}$. T_s was kept at 500°C. The doping profiles were investigated by the secondary ion mass spectrometry (SIMS).

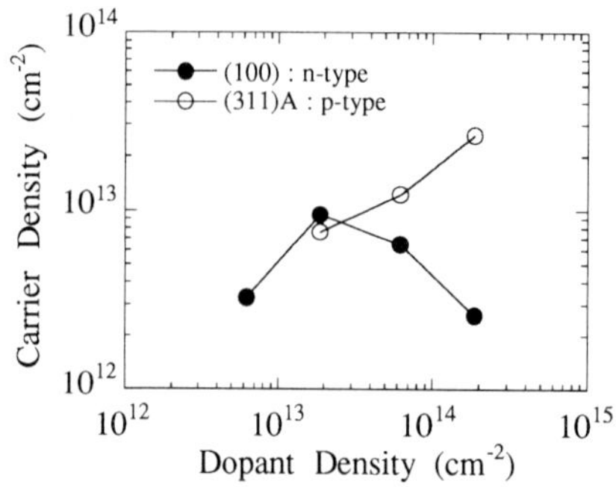

Fig. 4. Carrier densities in δ-doped GaAs layers measured by Hall measurements at room temperature are plotted as functions of sheet doping density.

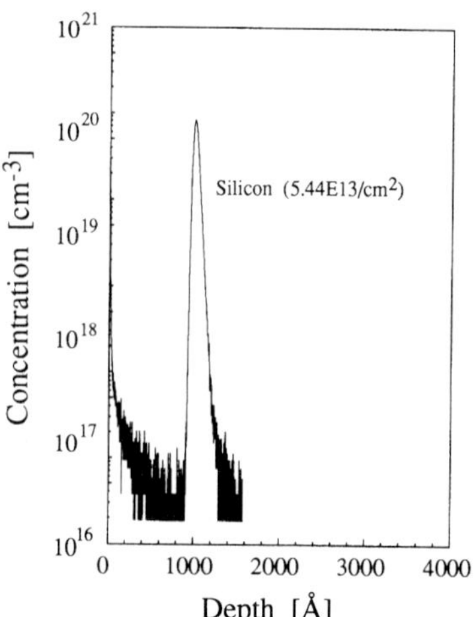

Fig. 5. Si dopant distribution of Si-doped GaAs layers grown on (311)*A* GaAs measured by SIMS measurements.

In Fig. 4, the hole densities in δ-doped samples determined by Hall measurements at room temperature are plotted as functions of the sheet doping density of Si. As in the case for uniform doping, the conduction type is *p*-type for the (311)*A* orientation, while it is *n*-type on the (100) surfaces. For the (100) orientation, the electron density initially increases with increasing doping density and reaches 9×10^{12} cm^{-2} and, then, starts to decrease due to the auto-compensation effect. On the other hand, for the (311)*A* orientation the hole density keeps increasing with doping density and does not show saturation. Note that the highest hole density achieved in the present experiment is 2.6×10^{13} cm^{-2}, which is seven times higher than the highest hole density ever reported for the δ-doped layers with carbon (Malik et al 1988). We think even higher densities could be obtained by optimizing the growth conditions.

Figure 5 shows the Si profile in the δ-doped sample with sheet doping density of 5.4×10^{13} cm^{-2} measured by SIMS. The full width at half maximum (FWHM) of Si peak is as narrow as 60Å even for such a high doping density. This value is comparable to one of the best results on the δ-doped *n*-type GaAs grown at 490°C on the (100) substrates (Harris et al 1989). It is also noted that the Si doping profile is quite symmetric and the segregation of Si atoms is suppressed at T_S = 500°C. These facts confirm that δ-doping profiles which are as sharp as those for the (100) *n*-type samples can be obtained for the (311)*A* *p*-type samples.

(a)

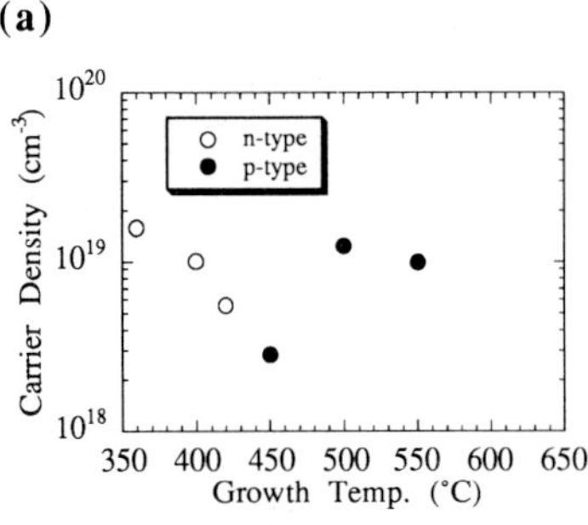

(b)

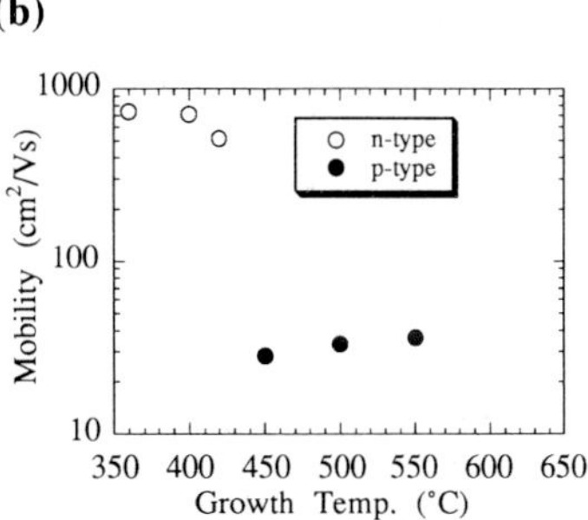

Fig. 6. Growth temperature dependence of uniformly doped layers. Carrier density (**a**) and mobility (**b**) determined by Hall measurements at room temperature are shown.

(a)

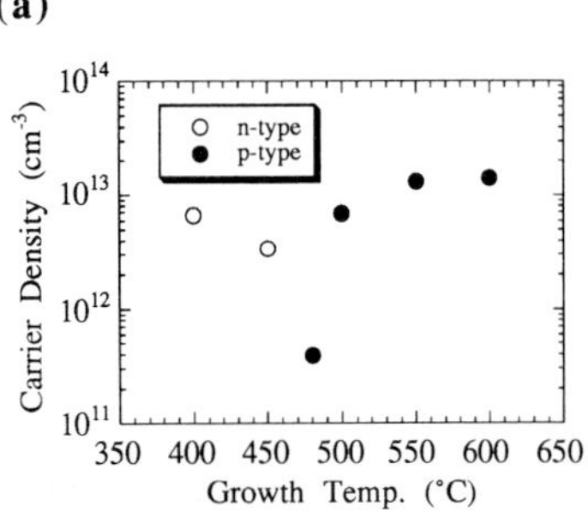

(b)

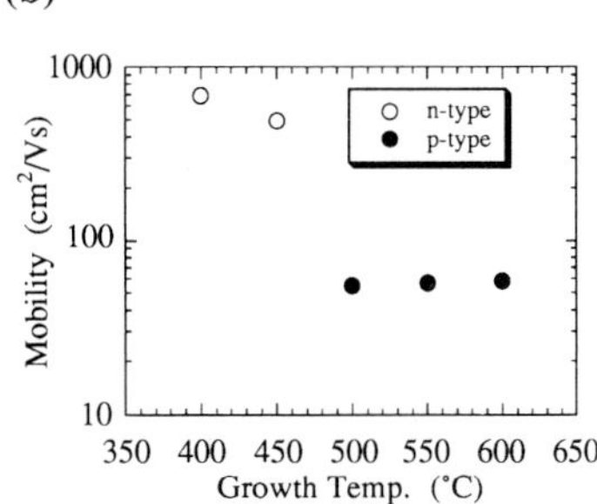

Fig. 7. Growth temperature dependence of δ-doped layers. Carrier density (**a**) and mobility (**b**) determined by Hall measurements at room temperature are shown.

4. Growth temperature dependence of conduction type

Furthermore, we have systematically studied the T_S-dependence of the conduction type in both uniformly doped and δ-doped GaAs layers grown on the (311)*A* oriented wafers. In case of uniform doping, the growth rate of GaAs and the V/III flux ratio was fixed to be 0.7 μm/h and ~5, respectively, during the growth. The Si doping density was set to be 1.6×10^{19} cm^{-3}. For δ-doping case, a constant As_4 flux of 2×10^{15} cm^{-2}s^{-1} was supplied during the δ-doping process. The sheet doping density of Si was fixed at 6.3×10^{13} cm^{-2}.

Figure 6 shows the T_S-dependence of the carrier density and mobility in uniformly doped samples. When T_S is above 500°C, *p*-type conductivity is obtained, and the hole density and mobility are almost independent of T_S. With decreasing T_S below 500°C, however, the hole density sharply drops by one order of magnitude. This drastic reduction of carrier density is attributed to the auto-compensation effect. By further lowering T_S, the conduction type suddenly changes to *n*-type and the electron density recovers up to the value almost identical to that of the hole density for T_S >500°C. The critical temperature T_C for the conduction type change is 430°C and the temperature width of the transition is as narrow as ~ 50°C.

A very similar behavior was observed for δ-doped samples, as shown in Fig. 7, except for the fact that T_c for δ-doping is 480°C, 50°C-higher than that for uniform doping case.

Although the mechanism for the drastic change in the conduction type is not clear at present, it might be related with the temperature dependence of the microstructural morphology of the (311)*A* surfaces, which was recently reported by Notzel et al.(Notzel et al 1992). Further investigation is necessary to clarify the mechanism.

As seen in Fig. 7, the critical temperature T_c for δ-doping is higher than that for uniform doping by ~50°C. This difference is probably due to the difference in the V/III flux ratio, which is 5 in the uniform doping case and infinite in the δ-doping; since under high V/III ratio Si atoms preferentially occupy Ga sites, δ-doped samples tend to become more *n*-type than uniformly doped samples at the same T_S.

5. Summary

In summary, we have systematically studied the doping characteristics of Si into GaAs grown on (311)*A* GaAs substrates. The highest hole density obtained in the present work for the uniformly doped layers was 4.2×10^{19} cm^{-3}, while for the δ-doped layers a sheet hole density as high as 2.6×10^{13} cm^{-2} was achieved. Furthermore, the growth temperature (T_S) dependence of Si doping has been investigated. It is found that the conduction type suddenly changes from *p*-type to *n*-type with decreasing T_S at a critical temperature of 430 - 480°C. The results suggest that by simply controlling the growth temperature *pn* doping can be done with only one dopant element (Si).

Acknowledgments

We thank Prof. T. Okumura and S. -Q. Shao and M. Noguchi for helpful discussions and J. Nakagawa for supplying (311)*A* GaAs substrates. This work is supported by the Grant-in-Aid for Scientific Research from Monbu-sho and by the Industry-University Joint Research Program "Mesoscopic Electronics."

References

Harris J J, Beall R B, Clegg J B, Foxon C T, Battersby S J, Lacklison D E, Duggan G, and Hellon C M 1989 J. Cryst. Growth **95** 257

Ilegems M 1977 J. Appl. Phys. **48** 1278

Konagai M, Yamada T, Akatsuka T, Saito K, Tokumitsu E, and Takahashi K 1989 J. Cryst. Growth **98** 167

Li W Q, Bhattacharya P K, Kwok S H, and Merlin R 1992 J. Appl. Phys. **72** 3129

Malik R J, Nottenberg R N, Schubert E F, Walker J F, and Ryan R W 1988 Appl. Phys. Lett. **53** 2661

Meier H P, Broom R F, Epperlein P W, van Gieson E, Harder Ch, Jackel H, Walter W, and Webb D J 1988 J. Vac. Sci. Technol. **B6** 692

Nagle J, Malik R J, and Gershoni D 1991 J. Cryst. Growth **111** 264

Notzel R, Ledentsov N N, Daweritz L, Hohenstein M, and Ploog K 1992 Phys. Rev. **B46** 3507

Takamori T, Fukunaga T, Kobayashi J, Ishida K, and Nakashima H 1987 J. J. Appl. Phys. **26** 1097

Wang W I, Mendez E E, Kuan T S, and Esaki L 1985 Appl. Phys. Lett. **47** 826

Paper presented at the Int. Symp. GaAs and Related Compounds, Freiburg, 1993

Planar GaInP/GaAs HBT technology achieved by CBE selective collector contact regrowth

D.Zerguine, F.Alexandre, P.Launay, R.Driad, P.Legay, J.L.Benchimol

FRANCE TELECOM, Centre National d'Etudes des Télécommunications, Paris B
Laboratoire de Bagneux, 196, avenue Henri Ravera, BP 107, 92225 Bagneux Cedex (France)

ABSTRACT : A new planar GaInP / GaAs Heterojunction Bipolar Transistor technology has been developed. The process is achieved by a selective collector contact regrowth from the bottom subcollector layer to the top of the device. Both the multilayer HBT structure and the selective collector regrowth are achieved by Chemical Beam Epitaxy (CBE). Thus, emitter, base and collector are contacted on the top surface of the device. A cutoff frequency and a maximum oscillation frequency of 30 GHz and 25 GHz respectively have been obtained on devices with a $2x15\ \mu m^2$ emitter-base junction area.

1. INTRODUCTION

Planar HBT technologies are of great interest in order to increase the reliability of metal connections and the circuits scale integration. The multilayer HBT structure is vertical and leads to different technological possibilities to contact the subcollector buried layer. Among them, the multimesa technology to contact both the base and the subcollector buried layers is the most useful one, thanks to its simplicity. But this technology presents the inconvenient of a non planar topology, with step height higher than 1 µm which becomes critical as the devices sizes are reduced. Such a technology should not allow a large scale integration.

To overcome these difficulties, we have developed a new HBT technology in which the main process step is a selective regrowth of the n^{++}GaAs collector contact layer, from the bottom subcollector layer to the top of the device. The final step height of the HBTs is then reduced to less than 300 nm and the emitter, base and collector are contacted on the top surface of the device. The reliability of the device metal connexions and a wide reduction of the device sizes are enhanced which gives an opportunity to increase the circuit scale integration. Another advantage of this technological process is the ability to grow, on top of the regrown subcollector layer, a low Schottky barrier $n^{++}Ga_{0.35}In_{0.65}As$ contact layer which allows us to use a tungsten contact with a very low associated ohmic contact resistivity ($10^{-6}\ \Omega cm^2$). In our technology, both the emitter and the collector ohmic contacts are made with the same refractory metal.

Several planar HBT technologies have already been developed with different approaches and static characterisations have been reported[1,2,3,4]. In the present publication, we report the first microwave characterisation of a planar GaInP / GaAs HBT with a selective regrown collector contact layer grown by CBE.

2. MATERIAL GROWTH

The initial N-p-n HBT epitaxial structure described on table 1, is grown by CBE on a 2 inch semi-insulating (100) GaAs substrate. The specific GaAlAs / GaInP emitter bilayer structure offers many advantages. In particular, the thin GaInP layer between the GaAlAs emmitter and the GaAs base layers acts as an etch stop layer allowing to contact very easily the base layer through this thin GaInP layer[5].

All the group V and III precursors are gaseous. AsH_3 and PH_3 are used for Arsenic and Phosphorus respectively, Triethylgallium (TEG) and Trimethylindium (TMI) are used as group III precursors. Carbon provided from Trimethylgallium (TMG) and solid silicon are used as p-type and n-type dopant precursors respectively. The CBE selective growth conditions of GaAs and GaInAs in terms of growth temperature, which is the main parameter to control the growth selectivity, V/III ratio and growth rate have already been reported[6]

	n^+ $Ga_{0.35}In_{0.65}As$	Si : 10^{19} cm^{-3}	50 nm
Emitter cap	n^+ $Ga_{1-x}In_xAs$	Si : 10^{19} cm^{-3}	50 nm
	n^+ GaAs	Si : 2.10^{18} cm^{-3}	50 nm
Emitter	n GaAlAs	Si : 4.10^{17} cm^{-3}	170 nm
	n GaInP	Si : 4.10^{17} cm^{-3}	30 nm
Base	p^{++} GaAs	C : 6.10^{19} cm^{-3}	110 nm
Collector	n^- GaAs	Si : 2.10^{16} cm^{-3}	490 nm
Subcollector	n^+ GaAs	Si : $2.5.10^{18}$ cm^{-3}	510 nm
GaAs SI substrate			

Table 1. Initial N-p-n HBT multilayer structure grown by CBE

For a high 2.5 µm/h growth rate, the selective growth of GaAs and GaInAs are achieved respectively at 640°C and 550°C growth temperature. Above these temperatures, neither GaAs nor GaInAs poly-crystaline deposition are observed on the Si_3N_4 film which is used as a mask to ensure the growth selectivity. Prior to the subcollector regrowth, a specific surface preparation is achieved in order to avoid C surface contamination, and to ensure the n^+ type continuity at the regrowth interface. This preparation consists in a wet chemical etch (100 Å) followed by an UV assisted ozone treatment (1 mn) and a silicon predeposition under high precracked arsine flow.

3. PLANAR HBT PROCESS DESCRIPTION

Figure 1 shows the schematic cross-section view of the planar HBT. The first step of the technological process is the patterning of the dielectric mask for the selective subcollector contact regrowth. A thin Si_3N_4 film (1000 Å) is deposited by PECVD and etched by an SF_6 reactive ion etching. Then the vertical etching of the epitaxial layers, from the emitter cap layer down to the subcollector layer, is performed by $SiCl_4$ reactive ion etching.

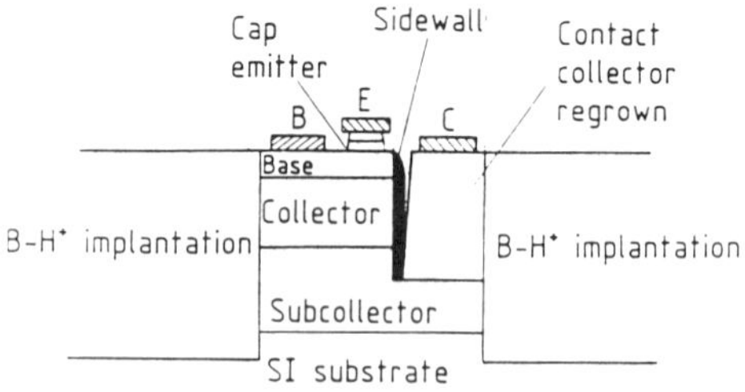

Fig 1. Cross-section of the planar GaInP / GaAs HBT

At last the selective regrowth of a 1 µm thick n^+GaAs layer and a 100 nm thick $n^{++}Ga_{0.35}In_{0.65}As$ layer is achieved on the whole 2" substrate to obtain a planar surface. PECVD Si_3N_4 spacers ensure the lateral isolation between the subcollector regrown layers and the other HBT active layers. The other technological steps proceed as follows. The HBT isolation is achieved by a deep Boron and Proton implantation (2 μm). Then the sputter deposited tungsten emitter ohmic contact is etched by reactive ion etching in a SF_6-O_2 gas mixture.

Then the GaInAs and GaAs emitter cap and GaAlAs emitter layers are etched selectively over the GaInP layer to delineate the emitter mesa. Thereafter, the tungsten collector ohmic contact is deposited by sputtering and etched by reactive ion etching in a SF_6-O_2 gas mixture. The C-doped GaAs base is contacted by a p-type Mn-Au-Ti-Au ohmic contact through the n-GaInP layer (300 Å). Figure 2 shows a scanning electron microscope view of the planar GaInP / GaAs HBT.

Fig 2. SEM view of the planar GaInP / GaAs HBT

4. DEVICE ELECTRICAL CHARACTERISATIONS AND DISCUSSION

Transmission line model measurements (TLM) have been used to determine the N type and P type ohmic contact resistivities which are about 10^{-6} Ω.cm^2. In the case of the subcollector regrowth, a very low additional resistance (1 Ω) to the HBT collector access resistance (10 Ω) has been measured. The base sheet resistance is 140 Ω.
The breakdown voltage of the collector-emitter junction BV_{ceo} is above 15 V and the offset voltage is about 400 mV. Nevertheless, a low current gain (Hfe=3) is obtained, as compared to the current gain of 23 obtained on the same HBT epitaxial multilayer structure processed with a mesa technology without subcollector regrowth. This degradation is not related to carbon diffusion from the base layer to the emitter as indicated by SIMS measurements. On an other hand, the same decrease of the current gain is observed on HBT processed with a multimesa technology, after an annealing (640°C, 60 mn) similar to the thermal treatment during the subcollector regrowth.
This poor value has been correlated to a decrease of the minority carrier lifetime τ_e in the C doped base layer[7,8], after the thermal treatment (640°C / 60 mn).
Electron lifetime τ_e has been measured at 300 K by time-resolved photoluminescence in carbon doped GaAs layers grown by CBE, for three C doping levels: 5 10^{18}, 4 10^{19} and 8 10^{19} cm^{-3}. The results displayed on the figure 3 show that the minority carrier lifetime decreases with the carbon doping level in the as grown layers (curve a), and there is chiefly a drastic decrease of the lifetime when the samples are annealed at 640°C. The degradation ratio $(\tau_{eo} - \tau_e)/\tau_{eo}$, where τ_{eo} is the minority carrier lifetime in the as grown layers, is 95 % for a 8 10^{19} cm^{-3} C doping level while it is bellow 30 % for a 5 10^{18} cm^{-3} doping level. As shown in figure 4 the ratio $(\tau_{eo} - \tau_e)/\tau_{eo}$ increases when the selective regrowth temperature increases.

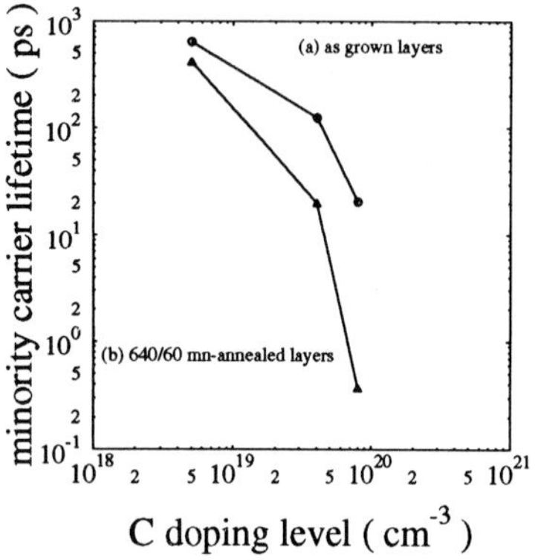

Fig 3. Minority carrier lifetime as a function of the carbon doping level in the GaAs base layers. (a) Before annealing, (b) after annealing (640°C, 60 mn).

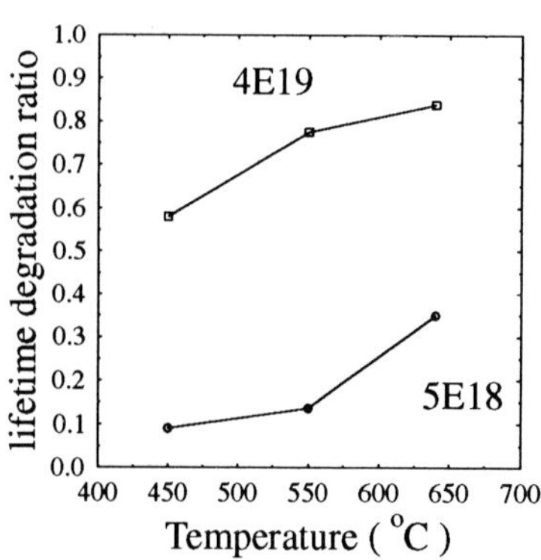

Fig 4. lifetime degradation ratio $(\tau_{eo} - \tau_e)/\tau_{eo}$) as a function of the regrowth temperature.

Microwave measurements have been carried out on a Wiltron 360 network analyser. The S parameters of the HBTs are measured up to 40 GHz. The transistors exhibit a cutoff frequency, f_T, of 30 GHz and a maximum oscillation frequency, f_{max}, above 25 GHz for a current density of 4.5 10^4 A / cm^2 and for the following HBTs bias conditions: V_{EB} = 1.5 V and V_{CE} = 3 V. The microwave current and power gains of the transistors are displayed in figure 5 (a). Figure 5 (b) shows the evolution of the frequencies f_T and f_{max} as a function of the collector current.

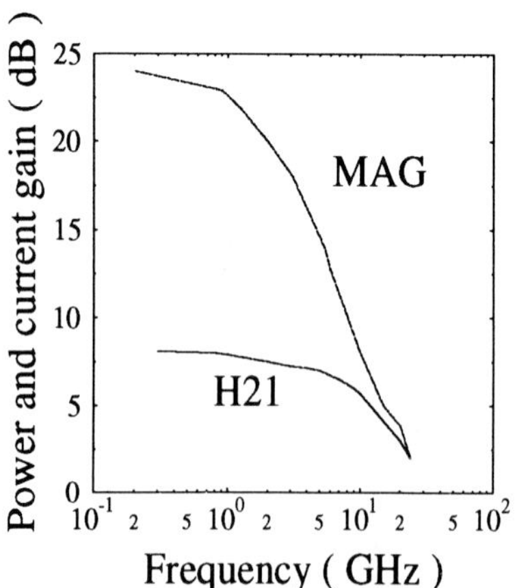

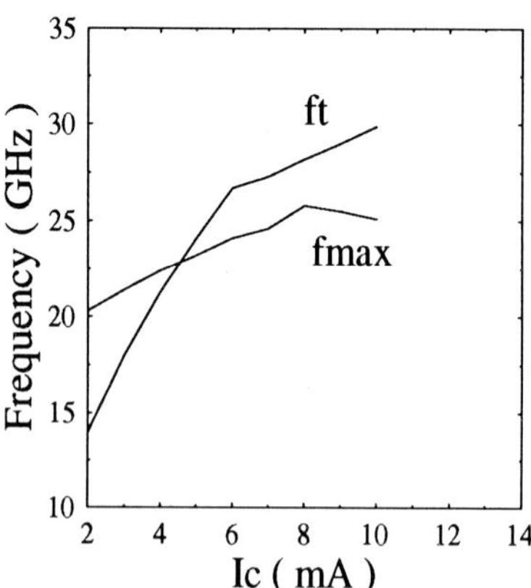

Fig 5. RF characteristics of the GaInP / GaAs HBTs : (a) current and power gain as function of the frequency, (b) f_t and f_{max} frequencies as function of the collector current.

Based on these earlier results and on our discussion about the drastic minority carrier lifetime decrease for very high base doping level, we have processed HBTs with a lower base doping level (4 10^{19} cm^{-3}).

The same technology has been used after an optimization of the regrowth conditions to insure the selectivity at a lower temperature (550°C)[9]. Our first results indicates a current gain of 75 (figure 6) with the same microwave performances. Further improvements are under development to optimize these results for base doping level higher than 4 10^{19} cm^{-3}.

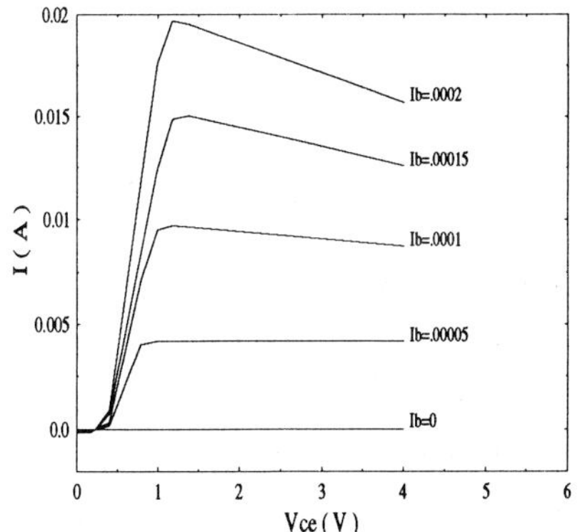

Fig 6: Common emitter DC characteristics for planar HBT with a C base doping level of 4 10^{19} cm^{-3}.

5. CONCLUSION :

A novel planar GaInP / GaAs HBT technology has been presented. The main technological step is a selective CBE subcollector contact regrowth from the bottom subcollector layer to the emitter layer of the HBT, which allows to obtain a planar device. This enhances the reliability of the device metal connections and enables a wide reduction of device size, thus allows an increase of the circuits scale integration. Those planar transistors exhibit f_T and f_{max} values of 30 GHz and 25 GHz, respectively. These results are already comparable to the conventional multimesa technology, with the same initial HBT epitaxial structure grown by CBE. For very high base doping level (> 5 10^{19} cm^{-3}), a drastic decrease of the HBT current gain is observed. This is related to a decrease of the minority carrier lifetime τ_e in the heavily carbon doped base layer after the post growth annealing. For lower base doping level (4 10^{19} cm^{-3} instead of 6 10^{19} cm^{-3}) and optimized selective growth conditions (550 °C instead of 640 °C), current gain as high as 75 has been obtained with the same microwave performances.

ACKNOWLEDGMENT :

The authors are grateful to C.Besombes, A.M.Duchenois, F.Héliot, D.Arquey and L.Bricard for support in technology, to M.Juhel for SIMS analysis, to B.Sermage for minority carrier life-time PL-measurements and to J.P.Médus for DC and RF measurements.

REFERENCES :

1 YANG Y, PLUMTON D L, and WHITE, 1989 Elect. Lett. vol 25 n°4 pp. 282-3

2 MITANI K, MASUDA H, MOCHIZUKI K, and KUSANO C 1992 IEEE Elect. Dev. Lett vol EDL-13 pp. 209-10

3 KRÄUTLE H, 1986 Elect. Lett. vol 22 n°22 pp.1191-3

4 PLUMTON D L and al, 1990 IEEE trans. Elect. Dev. vol 37 n°5 pp.1187-92

5 DUBON-CHEVALIER C, ALEXANDRE F, BENCHIMOL J L, DANGLA J, AMARGER V, HELIOT F and BOURGUIGA R, 1992 Elect. Lett. vol 28 n°25 pp. 2308-9

6 ALEXANDRE F, ZERGUINE D, LAUNAY P, BENCHIMOL J L and ETRILLARD J, 1993 J. Cryst. Growth vol 127 n°1-4 pp.221-5

7 STRAUSS U, HEBERLE A P, ZHOU X Q, RÜHLE W W, LAUTERBACH T, BACHEM K H and HAEGEL N M 1993 Jpn. J. Appl. Phys. vol 32 pp. 495-7

8 HAN W Y and al 1992 Appl. Phys. Lett vol 61 n°1 pp.87-9

9 LEGAY P, ALEXANDFRE F, NUNEZ M, ZERGUINE D, SAPRIEL J to be published in J. Cryst. Growth.

Paper presented at the Int. Symp. GaAs and Related Compounds, Freiburg, 1993

Growth interruption effects on GaAs *p–n* structures grown on GaAs(111)A using only silicon dopant

K. Fujita, M. Inai, T. Yamamoto, T. Takebe and T. Watanabe

ATR Optical and Radio Communications Research Laboratories, 2–2 Hikaridai, Seika–cho, Soraku–gun, Kyoto 619–02, Japan

ABSTRACT: A study has been made on the effects of growth interruption on GaAs p–n structures grown on GaAs(111)A using only Si dopant by MBE. CL spectra and I–V characteristics of the samples confirm the p–n structure. However, the quality of p–n structure is strongly affected by pausing after changing the growth condition; as the pausing time is decreased, the breakdown voltage becomes higher and the peak intensity of a sample's emission spectrum increases drastically. Furthermore, for a pausing time of 1 min, emission in the p–layer appears with increasing forward current; this indicates a decrease in the donor and acceptor levels. As a result, strong emission of 875 nm is observed even at room temperature.

1. INTRODUCTION

Silicon is widely used as an n–type dopant in GaAs growth using molecular beam epitaxy (MBE). It has been reported, however, that in GaAs growth on GaAs(n11)A substrates, Si atoms incorporate as an acceptor when $n \leq 3$ (Ballingall et al. 1982; Wang et al. 1985; Subbanna et al. 1986; Takamori et al. 1987). Furthermore, due to the unique bonding structure on the GaAs(311)A surface, the Si–doping type can be controlled by controlling the MBE growth conditions (Meier et al. 1988). Li et al. (1992) have demonstrated the MBE growth of AlGaAs/GaAs heterojunction bipolar transistors on GaAs(311)A substrates using controlled all–Si doping. We previously reported that the Si–doping type in GaAs growth on GaAs(111)A can be controlled by the growth conditions, i.e., the V/III flux ratio and the off–angle of the substrates (Shigeta et al. 1991; Okano et al. 1991). Furthermore, we demonstrated GaAs p–n junction LEDs as an approach to the production of p–n junctions on GaAs(111)A substrates using only Si dopant (Fujita et al. 1993). However, the device quality of such LEDs is strongly affected by interruption caused by changing the growth condition.

In this paper, we report the effects of growth interruption on p–n structures grown on GaAs(111)A using only silicon dopant by MBE.

2. EXPERIMENTAL

Silicon–doped n–type GaAs(111)A misoriented 5° toward the [100] direction was used as the substrate. Before the substrate was introduced into the MBE chamber, it was degreased and etched with an etchant ($NH_4OH : H_2O_2 : H_2O$ = 2 : 1 : 96) for 2 minutes; the substrate was subjected to thermal etching at a temperature of 700°C and an As pressure of $2x10^{-3}$ Pa. Then, a 1–μm n–type GaAs layer was grown at 540°C and a flux ratio J_{As4}/J_{Ga} of 7. After changing the substrate temperature from 540°C to 620°C in 5 minutes, the temperature was held constant for a time before re–starting growth. This time is referred to as "pausing time" in this paper. Then, a 1–μm p–type GaAs layer was grown at a flux ratio of 2. We used two As cells to change the As pressure abruptly. After growing the n–layer using the two cells, one As cell's shutter was closed when the substrate temperature became 620°C. The pausing time was varied from 1 minute to 20 minutes to investigate the effects of As pressure change on the p–n structures. The flux intensity of Ga was fixed. The growth rate was 1.0 μm/hr for both the n–type and p–type GaAs layers. Only Si was used as the dopant and the Si cell temperature was varied to control the carrier concentration: $1x10^{18}cm^{-3}$ for both the n–type and p–type GaAs layers. After growing the n–type GaAs layer, the p–type GaAs layer was also grown using Be.

The p–n structures were evaluated by cathodoluminescence (CL) and current–voltage (I–V) measurement. The LED chips were fabricated as follows: the ohmic contact (0.4 mm x 0.4 mm) onto the p–layer was made using evaporated Au/Mn alloyed at 430°C for 1 minute. A mesa etch was used for device isolation (1 mm x 1 mm). The chip was mounted onto a Au–evaporated substrate using Indium solder. The electroluminescence (EL) spectra of the LEDs were measured under DC–biased conditions at 14 K and 300 K.

3. RESULTS AND DISCUSSION

Figure 1 shows the cross–sectional CL spectra of a LED grown with a pausing time of 5 minutes. The peak wavelength of the p–layer is 825 nm and that of the n–layer is 810 nm. These peak wavelengths correspond to those of Si–doped GaAs layers whose conductivity type can be confirmed by Hall measurement. The same two peaks were also observed in the CL spectra of samples with a pausing time of 1 minute and 20 minutes. The results indicate that p–n structures can be grown using only Si dopant by controlling the growth condition, independent of the pausing time.

Figure 2 shows the I–V characteristics of the LEDs as a function of pausing time.

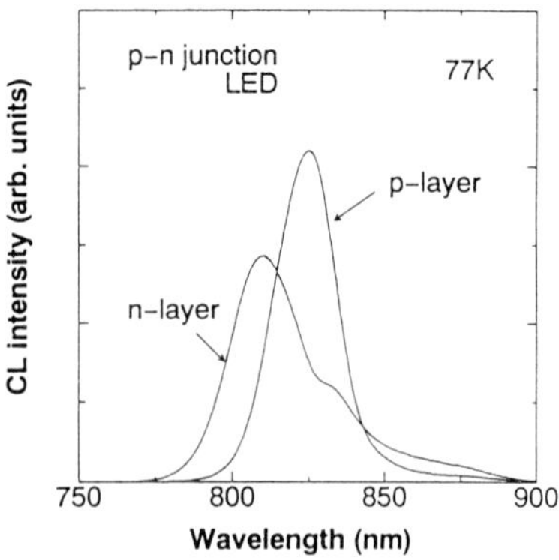

Fig.1. Cross–sectional CL spectra of a p–n junction GaAs LED. Labels "p–layer" and "n–layer" indicate spectra from the p–type layer and n–type layer, respectively.

There is no difference in the line shape in the forward bias region; the ideality factors are approximately 2 for these samples from 0.2 V to 0.6 V. This means that carrier recombination at the junction is dominant in this region (Sze 1981). By contrast, the breakdown voltage is strongly affected by the pausing time; as the time is decreased, the breakdown voltage becomes higher. The difference in breakdown voltage between these samples can be attributed to the different interface structures.

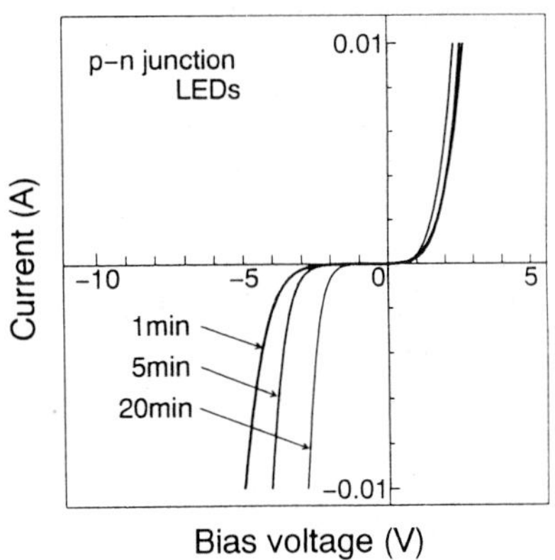

Fig.2. Dependence of I–V characteristics of p–n junction GaAs LEDs on pausing time during growth interruption.

To investigate the interface structure, the EL spectra of the LEDs were observed at a low temperature of 14 K. When the pausing time was 20 minutes, no EL spectra were observed. However, the peak intensity increased as the pausing time was decreased. Figure 3 shows the EL spectra of LEDs grown with a pausing time of (a) 1 minute and (b) 5 minutes. For (b), a weak EL peak is observed at 860 nm. The origin of the peak may be ascribed to the donor–acceptor (DA) pair recombination at the p–n interface, where the electron is injected from the n–layer to the p–layer. As the forward current is increased, the peak intensity increases without a peak shift. At 60 mA, a weak EL peak (885 nm) is observed. This longer–wavelength emission can be ascribed to the DA pair reconstruction by the hole injection from the p–layer to the n–layer, because a CL measurement of the n–GaAs layer shows a deep–level emission (Fig. 1). This peak is also observed in the sample with a pausing time of 1 minute when the forward current is 20 mA. However, the peak intensity increases drastically and shifts to a

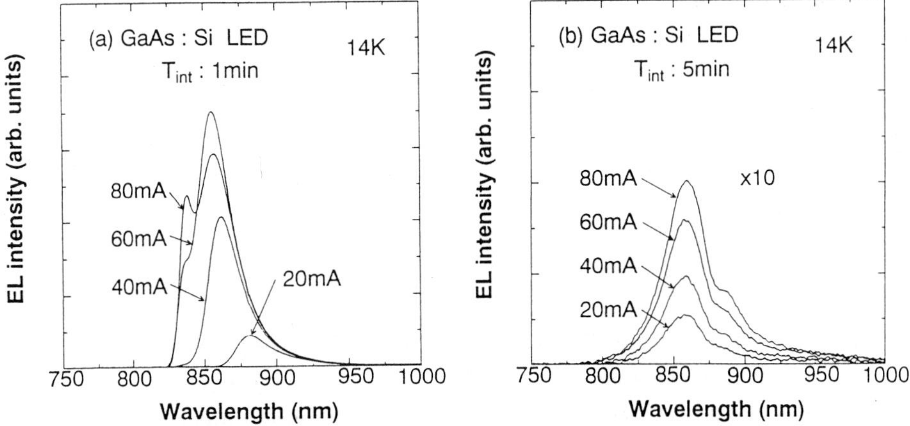

Fig.3. Emission spectra of p–n junction GaAs LEDs grown with a pausing time of (a) 1 min and (b) 5 min.

longer wavelength as the forward current is increased. When the forward current is 80 mA, emission at 830 nm can be observed. This can be ascribed to emission in the p–layer because a CL measurement of the p–layer shows 830 nm emission (Fig. 1). The results suggest a decrease in the donor and acceptor levels which cause emission at around 860 nm. As a result, strong emission is observed even at room temperature as shown in Fig. 4. Figure 4 shows the spectra of the sample grown with a pausing time of 1 minute. The peak wavelength is 875 nm. The origin of the peak can be ascribed to the emission in the p–layer. The peak intensity increases as the forward current is increased without a peak shift.

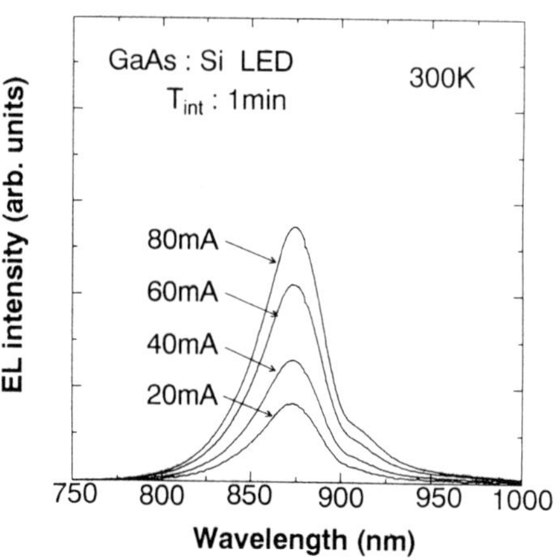

Fig.4. Emission spectra of p–n junction GaAs LEDs grown with a pausing time of 1 min.

To evaluate the quality of p–n junction LEDs grown using only Si dopant, the EL spectra of LEDs grown using Be as the p–type dopant were observed. Figure 5 shows EL spectra of LEDs grown with a pausing time of 1 minute. In the spectra at 300 K, emission in the p–layer (875 nm) can be observed. The peak intensity increases as the forward current is increased without a peak shift. The intensity at 80 mA is equivalent to that of the LED grown using only Si dopant. In the spectra at 14 K, a strong peak of 860 nm is observed at 20 mA. The origin of the peak may be ascribed to the DA pair recombination. As the forward current is increased to 30 mA, emission in the p–layer (830 nm) appears. However, the

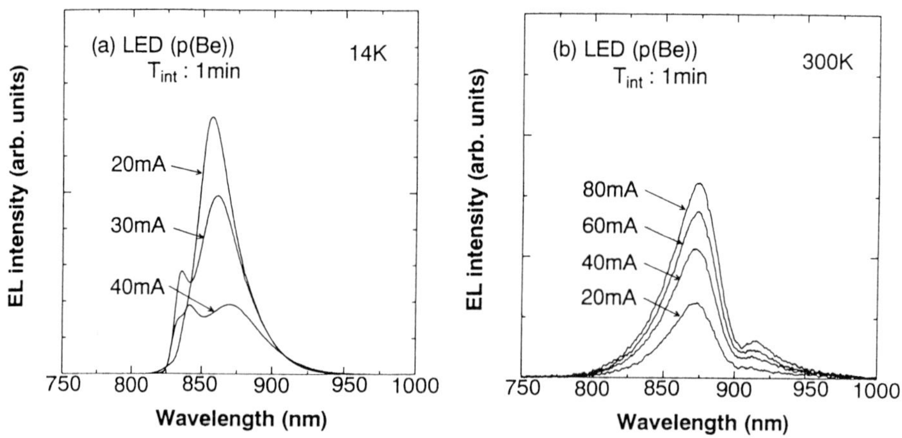

Fig.5. Emission spectra of p–n junction GaAs LEDs grown with a pausing time of 1 min.

peak intensity of the emission is weaker than that in the EL spectrum of the LED grown using only Si dopant. Furthermore, the intensity decreases drastically as the forward current is increased to 40 mA. This results from the characteristics of the p–layer. One possibility is the diffusion of Be due to annealing caused by heat induced by the current injection (Fujita et al. 1993). This indicates that Si is more stable as a p–type dopant on GaAs(111)A as compared to Be. These results show that the MBE growth of GaAs p–n junctions on GaAs(111)A using only Si dopant is effective in improving the quality of devices with a p–n structure such as LEDs.

It is remarkable that the quality of LEDs is more affected by the pausing time than by the interruption caused by changing the substrate temperature. We consider the pausing time dependence of the quality of LEDs to be related to the amount of As atoms supplied on the surface. As the substrate temperature is being changed, an arsenic stabilized surface may be formed because of a high As pressure. However, when the As pressure is decreased to grow the p–layer with a flux ratio of 2, the amount of As atoms supplied to the surface decreases drastically. As a result, a desorption of As atoms from the surface is expected to occur, and thus, As vacancies are formed. Therefore, a long pausing time under low As pressure will produce many donor and acceptor levels due to the As vacancies at the p–n structure interface. This is consistent with the result that no emission spectra were observed for the sample with a long pausing time of 20 minutes.

From these results, we find that a control of the pausing time during growth interruption caused by changing the growth condition, is an important factor in improving the quality of LEDs grown on GaAs(111)A using only silicon dopant.

4. CONCLUSION

The effects of growth interruption on GaAs p–n structures grown on GaAs(111)A using only Si dopant by MBE were studied. CL spectra and I–V characteristics of the samples confirm the p–n structure. However, the quality of p–n structure is strongly affected by pausing after changing the growth condition; as the pausing time is decreased, the breakdown voltage becomes higher and the peak intensity of a sample's emission spectrum increases drastically. Furthermore, for a pausing time of 1 minute, emission in the p–layer appears with increasing forward current. This indicates a decrease in the donor and acceptor levels. As a result, strong emission of 875 nm is observed even at room temperature.

ACKNOWLEDGEMENT

The authors would like to thank Drs. Y. Furuhama and H. Inomata for their encouragement throughout this work.

REFERENCES

Ballingall J.M. and Wood C.E.C. 1982 Appl. Phys. Lett. **41** 947

Fujita K., Shinoda A., Inai M., Yamamoto T., Fujii M., Lovell D., Takebe T. and Kobayashi K. 1993 J. Cryst. Growth **127** 50

Li W. Q. and Bhattacharya P. K. 1992 IEEE Electron Device Lett. **13** 29

Meier H. P., Broom R. F., Epperlein P. W., Gieson E.van, Harder Ch., Jäckel H., Walter W. and Webb D.J. 1988 J. Vac. Sci. Technol. **B6(2)** 692

Okano Y., Shigeta M., Seto H., Katahama H., Nishine S. and Fujimoto I. 1991 Jpn. J. Appl. Phys. **29** L1359

Shigeta M., Okano Y., Seto H., Katahama H., Nishine S. and Kobayashi K. 1991 J. Cryst. Growth **111** 248

Subbanna S., Kroemer H. and Merz J. L. 1986 J. Appl. Phys. **59** 488

Sze S. M. 1981 Physics of Semiconductor Devices (John Wiley & Sons, Inc., New York) 2nd ed., Chap.2, p.81

Takamori T., Fukunaga T., Kobayashi J., Ishida K. and Nakashima H. 1987 Jpn. J. Appl. Phys. **26** 1097

Wang W.I., Mendez E.E., Kuan T.S. and Esaki L. 1985 Appl. Phys. Lett. **47** 826

Paper presented at the Int. Symp. GaAs and Related Compounds, Freiburg, 1993

GaAs growth on Si(111) using a two-chamber MBE system

K. Fujita, A. Shinoda, T. Yamamoto, T. Takebe and T. Watanabe

ATR Optical and Radio Communications Research Laboratories, 2–2 Hikaridai, Seika–cho, Soraku–gun, Kyoto 619–02, Japan

ABSTRACT: The effects of As pre–deposition on the quality of GaAs grown on Si(111) were investigated using a two–chamber MBE system. This system improved the crystalline quality of GaAs, independent of the As pre–deposition temperature. However, the surface morphology became drastically smoother at the temperature of 600°C. SIMS measurement of the sample indicated that the improvement in morphology is caused by a stable As–Si bond which is formed at an initial stage of GaAs growth on Si. This study shows how using a two–chamber MBE system is effective in improving the quality of Si substrates and thus the crystalline quality of GaAs layers.

1. INTRODUCTION

In spite of the large mismatch in lattice constants and thermal expansion, remarkable progress has been made in the epitaxial growth of GaAs on Si substrates (Fang et al. 1990). Almost all studies reported to date have employed Si(100) wafers as substrates because of the well–established Si(100) technology. Nonetheless, GaAs growth on Si(111) substrates has also been investigated intensively because of interest in the growth mechanism of GaAs on a Si surface with a lower atomic density than that of the (100)plane (Bringans et al. 1987; Radhakrishnan et al. 1988; Maehashi et al. 1990; Sobiesierski et al. 1991).

Shigeta et al. (1991) reported that the conduction type of Si–doped GaAs layers can be controlled through the growth conditions in MBE growth on GaAs(111)A substrate. This technique is attractive because if a GaAs(111)A layer can be grown on a Si(111) substrate, a p–n junction can be grown using only Si dopant. Takano et al. (1990) reported that the polarity of GaAs(111) grown on Si(111) substrate depends on the substrate temperature at which the As prelayer is deposited onto Si; GaAs(111)A was obtained with As pre–deposition at 20°C and 350°C, while GaAs(111)B was obtained with As pre–deposition at 580°C and 700°C. This dependence could be useful for controlling the polarity of GaAs(111) on Si(111). However, it is known that the quality of GaAs grown on Si(100) is strongly affected by the contamination of Si during thermal treatment of the Si surface; as a result, pre–deposition at a high temperature degrades the quality of the GaAs layer (Nishimura et al. 1991).

In this paper, we report the effects of As pre-deposition on the quality of GaAs grown on Si(111) (GaAs/Si) using a two-chamber molecular beam epitaxy (MBE) system. This system was designed to avoid contamination of Si during thermal treatment of the Si surface.

2. EXPERIMENTAL

N-type Si(111) misoriented 3° toward the $[1\bar{1}0]$ direction was used as the substrate. Before the substrate was introduced into the pre-heating chamber in the two-chamber MBE system, it was degreased and etched in etchants based on the method developed by Ishizaka et al. (1986); the substrate was subjected to pre-growth annealing at 1000°C for 5 minutes with a pressure of less than $8x10^{-6}$Pa. Then, the substrate was transferred to the growth chamber under a pressure of less than $8x10^{-8}$Pa. This was followed by As pre-deposition at $1x10^{-3}$Pa for 5 minutes. The effects of As pre-deposition were examined by varying the substrate temperature from 250°C to 850°C. The GaAs layer was grown by a two-step growth method (Akiyama et al. 1984); a thin (130 nm) GaAs initial layer was grown at 250°C prior to the growth of a thick (1.4 μm) GaAs top layer at 505°C. The As_4/Ga flux ratio was 2. The As_4 beam flux intensity was fixed at $1x10^{-3}$Pa. The growth rates for the initial layer and the top layer were 0.5 μm/h and 0.7 μm/h, respectively. The Si substrate was also annealed in the growth chamber at 1000°C for 5 minutes.

The surface after pre-growth annealing was investigated with a scanning tunneling microscope (STM) (JSTM-4500VS, JEOL Ltd.) which was combined with the pre-heating chamber. The surface morphology of GaAs layers were observed by Nomarski microscopy. The crystalline quality of a GaAs layer was evaluated by double crystal X-ray diffraction using a CuKα_1 radiation source and a Si(111) first crystal; the full width at half-maximum (FWHM) of (111) reflection from the GaAs layer was measured. The GaAs/Si interface structure was examined by secondary ion mass spectroscopy (SIMS). The SIMS depth profiles were measured using Cs^+ primary ions at normal incident, and an energy of 14.5 keV. The polarity of GaAs on Si(111) was estimated using selective etching with the etchant (HF : H_2O_2 : H_2O = 2 : 32 : 40) developed by Takebe et al. (1993).

3. RESULTS AND DISCUSSION

Figure 1 shows the STM image of a Si(111) surface after pre-growth annealing in the pre-heating chamber. This image clearly shows a 7 x 7 structure. The result indicates that the native oxide on the Si surface was removed. However, it can also be seen that the surface morphology of the GaAs layer was strongly affected by the As pre-

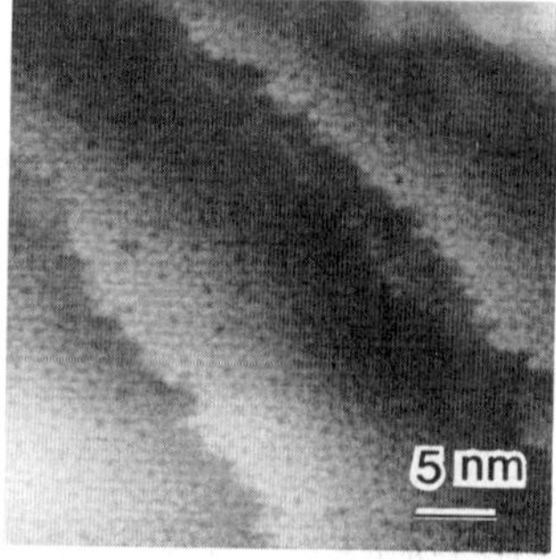

Fig. 1. STM image of Si(111) surface after heating at 1000°C.

deposition temperature in the growth chamber; it became smooth above 600°C. Figure 2 shows the surface morphology of samples subjected to As pre-deposition at (a) 250°C and (b) 750°C; the morphology of sample (b) was improved drastically, compared with the morphology of sample (a). This indicates that As pre-deposition at a high temperature is effective in improving the quality of the GaAs layer grown on Si(111).

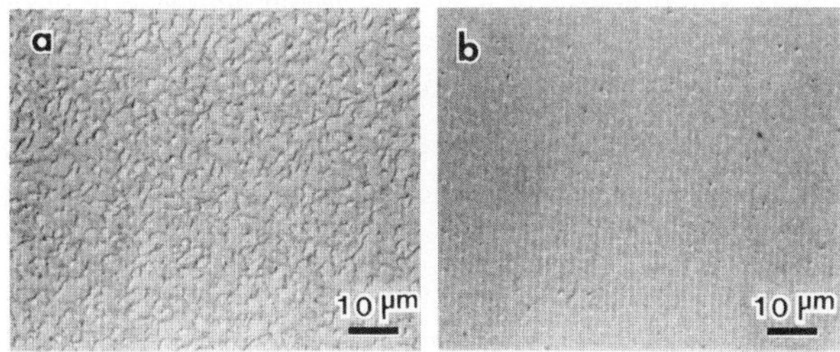

Fig. 2. Surface morphology of GaAs layers grown on Si(111) with As pre-deposition at (a) 250°C and (b) 750°C. The substrate was annealed in the pre-heating chamber.

By contrast, the surface morphology of GaAs layers grown on Si(111), with pre-growth annealing in the growth chamber, were rough, even with As pre-deposition at a high temperature. Figure 3 shows the surface morphology of samples subjected to As pre-deposition at (a) 250°C and (b) 750°C; the morphology is rough for both samples. This roughness may have been caused by contamination of the Si substrate during pre-growth annealing in the growth chamber, because the pressure in the growth chamber was ten times as high as that in the pre-heating chamber for pre-growth annealing. These results indicate that the quality of Si substrates and the subsequent As pre-deposition at a high temperature are important factors for improving the morphology of GaAs grown on Si(111).

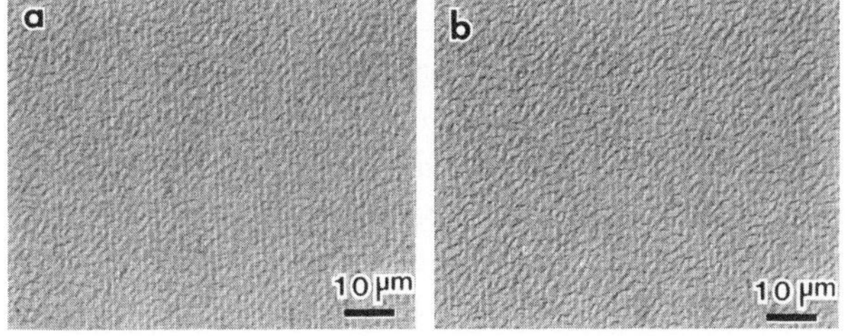

Fig. 3. Surface morphology of GaAs layers grown on Si(111) with As pre-deposition at (a) 250°C and (b) 750°C. The substrate was annealed in the growth chamber.

The crystalline quality of each sample was evaluated by X-ray FWHM. Figure 4 shows the relationship between X-ray FWHM of GaAs/Si and As pre-deposition temperature. The results indicate that the FWHMs for the

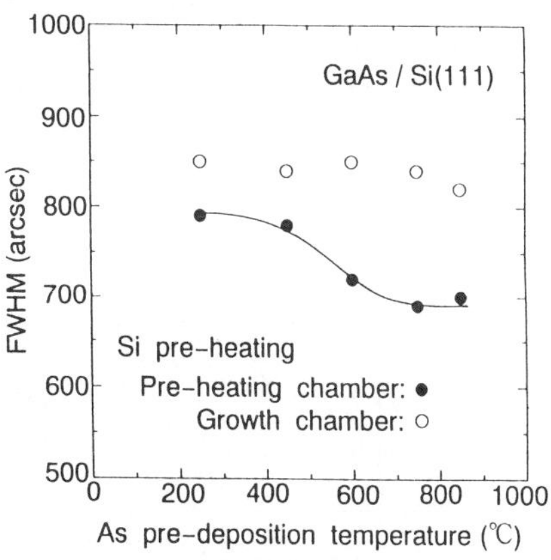

Fig. 4. Relationship between X-ray FWHM of GaAs/Si and As pre-deposition temperature.

samples with pre–growth annealing in the pre–heating chamber become narrow, compared to those for the samples with pre–growth annealing in the growth chamber. Furthermore, the FWHMs become narrow at temperatures above 600°C, corresponding to the smoother surface morphology. These results indicate that the quality of GaAs can be improved by controlling the As pre–deposition temperature in addition to the pre–growth annealing in the pre–heating chamber.

To understand the reason for the quality improvement, the GaAs/Si interface structure was investigated by SIMS. Figure 5 shows SIMS depth profiles for Ga, As and Si at the GaAs/Si interface: (a) and (b) show profiles for As pre–deposition at 250°C and 750°C, respectively. The ion intensity of Si changes more sharply around the interface for sample (b) than that for sample (a). This indicates that the improvement of the quality is caused by a stable As–Si bond formed at an initial stage of GaAs/Si.

It is remarkable that As pre–deposition at a high temperature improves the quality of a GaAs layer grown on Si(111). Stolz et al. (1989) reported that As pre–deposition at a high temperature degrades the quality of GaAs grown on Si(100) by MBE. They explained that the degradation is caused by a stable As–layer on the Si surface formed at a high temperature; the stable layer hinders a rearrangement required for charge neutrality at the polar (GaAs) and nonpolar (Si) interface. We consider the quality dependence on the As pre–deposition temperature to be related to the stable As–layer on the Si(111) substrate. The Si(111) surface has a lower atomic density than that of the (100)plane, namely only one dangling bond. Therefore, a stable As–layer on Si(111) might be effective in incorporating Ga adatoms into the lattice sites, rather than hindering the rearrangement required for charge neutrality. As a result, As pre–deposition at a high temperature improves the quality of GaAs grown subsequently. The result is consistent with the consideration that the ion intensity of Si in the SIMS depth profile of the sample with As pre–deposition at 750°C changed more sharply around the interface than that of the sample with As pre–deposition at 250°C.

The polarity of GaAs on Si(111) was estimated using selective etching characteristics of an etchant (HF : H_2O_2 : H_2O = 2 : 32 : 40); the etching rate was 1.72 μm/min for GaAs(111)A and 4.46 μm/min for GaAs(111)B. The results showed the etching rates to be

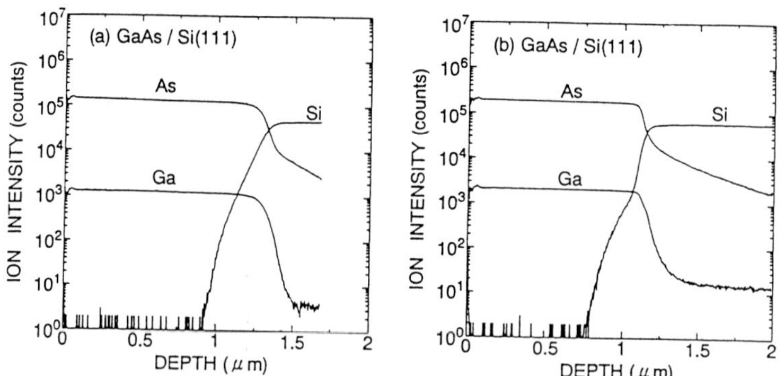

Fig. 5. SIMS depth profiles for Ga, As and Si at the GaAs/Si interface: (a) and (b) show profiles for As pre–deposition at 250°C and 750°C, respectively.

between 4.0 μm/min and 4.5 μm/min, indicating a GaAs(111)B layer. This tells us that the GaAs(111)A layer is not observed even for As pre–deposition at a low temperature. This may be caused by the poor interface structure, as shown in Fig. 5. To obtain the GaAs(111)A layer, the change from As to Ga on the Si surface is required at an initial stage of GaAs/Si. We will investigate the dependence of initial layer on growth condition of GaAs on Si(111) in future work.

These results indicate that using a two–chamber MBE system is effective in improving the quality of Si substrates and thus the quality of GaAs layers. Consequently, the controllability of the polarity of GaAs(111) grown on Si(111) can be improved.

4. CONCLUSION

This paper presents the effects of As pre–deposition on the quality of GaAs grown on Si(111) using a two–chamber MBE system. The crystalline quality of GaAs was improved, independent of the As pre–deposition temperature. However, the surface morphology became drastically smoother at the temperature of 600°C. SIMS measurement of the sample indicated that the improvement in morphology is caused by a stable As–Si bond formed at an initial stage of GaAs growth on Si. A two–chamber MBE system is effective in improving the quality of Si substrates and thus the crystalline quality of GaAs layers.

ACKNOWLEDGEMENT

The authors would like to thank Drs. Y. Furuhama and H. Inomata for their encouragement throughout this work.

REFERENCES

Akiyama M., Kawarada Y. and Kaminishi K. 1984 Jpn. J. Appl. Phys. **23** L843

Bringans R.D., Olmstead M.A., Uhrberg R.I.G. and Bachrach R.Z. 1987 Rev. B, **36** 9569

Fang S.F., Adomi K., Lyer S., Morkoç H., Zabel H., Choi C. and Otsuka N. 1990 J. Appl. Phys. **68** R31

Ishizaka A. and Shiraki Y. 1986 J. Electrochem. Soc. **133** 666

Maehashi K., Hasegawa S., Sato M. and Nakashima H. 1990 Jpn. J. Appl. Phys. **29** L13

Nishimura T, Kadoiwa K., Hayafuji N., Miyashita M., Mitsui K., Kumabe H. and Murotani T. 1991 J. Cryst. Growth **107** 468

Radhakrishnan G., Liu J., Grunthaner F., Katz J., Morkoç H. and Mazur J. 1988 J. Appl. Phys. **64** 1596

Shigeta M., Okano Y., Seto H., Katahama H., Nishine S. and Kobayashi K. 1991 J. Cryst. Growth **111** 248

Sobiesierski Z., Woolf D.A., Westwood D.I. and Williams R.H. 1991 Appl. Phys. Lett. **58** 628

Stolz W., Naganuma M. and Horikoshi Y. 1988 Jpn. J. Appl. Phys. Phys. **27** L283

Takano Y., Kanaya Y., Kawai T., Torihata T., Pak K. and Yonezu H. 1990 Appl. Phys. Lett. **56** 1664

Takebe T., Yamamoto T., Fujii M. and Kobayashi K. 1993 J. Electrochem. Soc. **140** 1169

Quantitative analysis of Be diffusion in δ-doped AlInAs and GaInAs during MBE growth

W Passenberg and P Harde

Heinrich-Hertz-Institut für Nachrichtentechnik Berlin GmbH
Einsteinufer 37, D-10587 Berlin, Germany

ABSTRACT: Diffusion of Be acceptors in MBE grown multiple δ-doped AlInAs and GaInAs layers was investigated. Over the investigated concentration range the diffusion coefficient in AlInAs appears to be higher than in GaInAs. In agreement with theory a quadratic relationship with dopant concentration was verified. The temperature dependence in the investigated range from 400 to 530 °C can be described by an Arrhenius plot with activation energies of 0.6 eV and 1.1 eV for AlInAs and GaInAs, respectively. The diffusion behaviour appears to be unaffected by the V/III beam equivalent pressure ratio.

1. INTRODUCTION

Beryllium (Be) is the most commonly used p-type dopant in molecular beam epitaxy (MBE) of III-V semiconductor compounds. As with any acceptor species, diffusion of this dopant during growth represents a major issue of practical relevance as this undesired effect may severely degrade the performance of devices (cf. Scott et al. 1989; Passenberg et al. 1990). This is particularly true for very high Be-doping concentrations as required for applications such as the doping of base layers in hetero bipolar transistors, ohmic contact layers, or p-surface doping for enhanced Schottky barriers. So far, only relatively few quantitative results on the diffusion behaviour of Be have been reported, mainly concentrating on the most widely investigated Ga(Al)As material system. In this work we present an investigation on the MBE grown ternary alloys AlInAs and GaInAs lattice matched to InP. In particular, the influence of the doping concentration and of the main MBE process parameters, i.e. growth temperature, As pressure, and growth time, was evaluated utilizing multiple δ-doped samples.

In previous related studies a quadratic dependence of the Be diffusion coefficient on doping concentration, as expected from basic diffusion theory, was measured in GaAs and AlGaAs layers subjected to post-growth annealing (Ilegems et al. 1977; Miller et al. 1985) as well as for closed ampoule diffusion in GaAs (Yu et al. 1991). Discordantly, Pao (1986) claimed the diffusion of Be in GaAs to be exponentially dependent on doping concentration. In the case of GaInAs, Scott et al. (1989) have found only very little concentration dependence of the Be diffusion coefficient.

As regards the temperature dependence, acceptor diffusion in III-V semiconductors generally proves to be thermally activated following $D \propto \exp(-E_A/kT)$ (Schubert et al. 1989, Cunningham et al. 1990, D: diffusion coefficient, E_A: activaton energy, k: Boltzmann constant, T: temperature). The relevance of the As partial pressure during growth of

GaAs: Be is described controversially: a weak $p_{As4}^{-1/4}$ dependence (Ilegems et al. 1977) on the one hand and a drastical variation of over more than three orders of magnitude at a threefold p_{As4} increase (Pao et al. 1989) on the other hand.

2. MBE GROWTH AND SIMS MEASUREMENTS

Experimentally, multiple Be δ-doped AlInAs and GaInAs test layers comprising five nominally identical δ-planes with an equal spacing of 200 nm were employed in this study. In contrast with the single δ-doped samples as used previously by Schubert et al. (1989) when investigating Be diffusion in GaAs, multiple δ-doping not only allows to directly assess the time dependence of dopant diffusion induced during growth but also helps improve the accuracy of the analysis as a set of multiple profiles is contained in each sample. The layers were grown lattice-matched onto InP:Fe substrates by elemental-source MBE using a standard V/III beam equivalent pressure ratio of 15 (As_4 vapour pressure = $2 \cdot 10^{-5}$ mbar) and a growth rate of 1 μm/hr. The substrate temperature was calibrated by monitoring the RHEED patterns. Because after-growth annealing yielded divergent results solely diffusion whilst MBE growth was investigated.

The Be concentration profiles were measured by SIMS (Atomika 6500, quadrupole instrument). In GaInAs samples $^{9}Be^{+}$ ions were detected utilizing a 3 keV O^{+}_{2} primary ion beam at an angle of incidence of 45 ° and with a spot diameter smaller than 20 μm (FWHM). With these experimental parameters the detection limit of the Be concentration in GaInAs was estimated to be below 10^{14} cm^{-3}. In the case of AlInAs, the $^{9}Be^{+}$ signal interferes with that of $^{27}Al^{3+}$ ions which necessitates the use of Cs^{+} primary ions (15 keV; 30°) and detection of the $^{84}AsBe^{-}$ cluster, under which conditions its detection limit is worsened by two orders of magnitude and also the depth resolution is somewhat deteriorated.

3. Be δ-DOPING CONCENTRATION

For a given Be flux, i.e. at constant Be cell temperature (850 °C), SIMS measurements confirmed a linear relationship between the integral count rate across each δ-plane and the corresponding Be deposition time, which was varied up to 280 seconds. This behaviour indicates that even at longer periods of δ-doping the quantitative incorporation of the deposited Be into the crystal is not noticeably impaired.

Hall measurements were performed to correlate the sheet δ-doping density with the resultant sheet carrier densities. In AlInAs which appears to be of high resistivity when not intentionally doped, sheet carrier densities were nevertheless not measurable below $1 \cdot 10^{13}$ cm^{-2} per δ-plane. Above this limit, again a linear dependence of the sheet carrier density on deposition time was observed (figure 1), which implies an almost complete electrical activation of the incorporated Be. The doping concentration was virtually independent of the growth temperature over the range from 400 °C to 500 °C exceeding the aforementioned densities, as reported elsewhere (Shih et al. 1992). In contrast with AlInAs, measurements on GaInAs samples gave rather irreproducible results, which may be attributed to electrical compensation. The same has been reported to occur in bulk doped GaInAs layers at low Be-concentration and is believed to be due to the presence of oxygen (Le Corre et al. 1987). Therefore, in GaInAs the Be δ-doping densities could only be indirectly estimated via a calibration of the Hall sheet carrier densities measured in bulk doped layers where the amount of incorporated Be dopants is known.

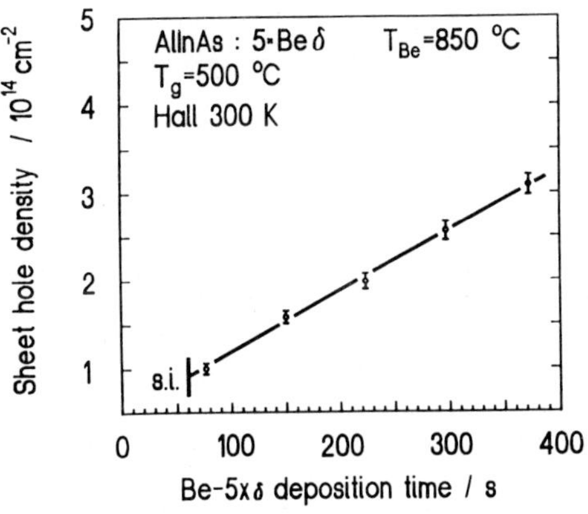

Fig.1: Integral sheet carrier density as obtained from Hall measurements in 5 x Be *δ-doped AlInAs layers as function of the total Be deposition time.*

4. DIFFUSION CHARACTERISTICS

Figure 2 shows a typical Be concentration profile (^{84}AsBe$^-$) as measured by SIMS in a 5 x δ-doped AlInAs layer, grown in this instance at 425 °C. Broadening of the δ-spikes along with hyperbolically decreasing peak concentrations due to diffusion during growth can clearly be noticed. From such SIMS-profiles the diffusion properties were quantitatively evaluated. As the individual δ-planes undergo diffusion for different lengths of deposition time, each sample provides a plot of diffusion length squared vs. time from which a diffusion coefficient can be derived. For the diffusion length the half width at half maximum of the SIMS-profiles was taken. As demonstrated in figure 3 the diffusion behaviour for δ-doping follows the basic $x^2 \propto t$ law in accordance with one-dimensional thermal diffusion in the absence of an additional driving force. In experiments reported elsewhere samples with only single δ-doping were used and diffusion coefficients were determined from a single diffusion profile. A similiar evaluation of our data would yield different results for the diffusion coefficients as could be derived from the dashed lines in figure 3. The intercept of the diffusion length at t = 0 probably reflects the influence of the ion mixing effects during SIMS analysis and a possible slight roughness of the surface.

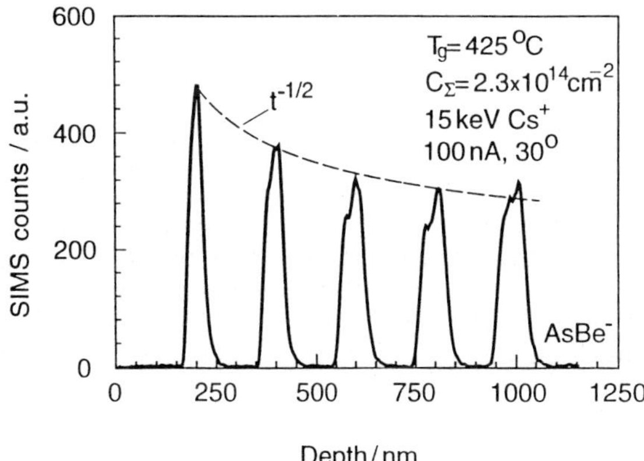

Fig. 2: SIMS concentration profile of Be in a 5 x δ-doped AlInAs layer illustrating the time dependent broadening of the doping spikes and the associated decrease in peak concentration due to dopant diffusion during growth.

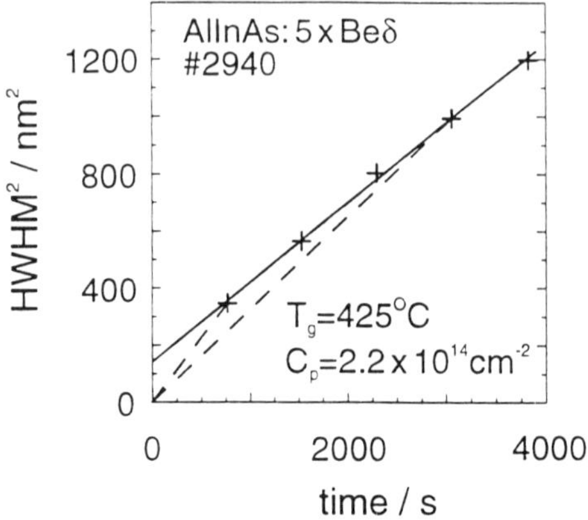

Fig. 3: Diffusion width as characterized by the half-width-at-half-maximum (HWHM) values of the SIMS δ-profiles in quadratic dependence of the diffusion time. The dashed lines illustrate the evaluation method if only single δ-doped layers were used.

To assess the influence of the temperature on the Be diffusion coefficient various AlInAs and GaInAs layers with five nominally identical δ-doping densities were grown in the temperature range between 400 °C and 530 °C. For each temperature the diffusion coefficient was determined as described above. The Arrhenius plot of the values obtained is depicted in figure 4, from which activation energies of 0.6 eV and 1.1 eV are derived for AlInAs and GaInAs, respectively. The latter value is to be compared with $E_A = 0.45$ eV as given by (Scott et al., 1989), a discrepancy which may be attributed to the different experimental conditions and evaluation methods applied.

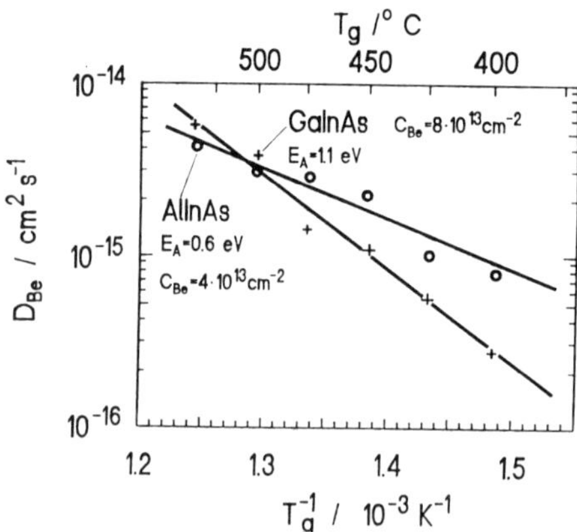

Fig. 4: Arrhenius plot of the Be diffusion coefficient in δ-doped AlInAs and GaInAs.

We also studied the concentration dependence of the Be diffusion coefficient in the two ternary materials. In both cases we found, over a range of one order of magnitude, a quadratic concentration dependence as depicted in figure 5. This behaviour is characteristic of acceptor diffusion in III-V compound semiconductors (cf. Casey et al. 1967) and was also observed by Yu (1991) for Be diffusion in GaAs. In the case of δ-doping where the diffusion coefficient D depends on the square of the doping density C (figure 2) we conclude that the maximum achievable doping concentration is primarily limited by the duration and the temperature of growth. Figure 5 also shows that the Be diffusion is greater in AlInAs than in GaInAs.

To interpret the $D \propto C^2$ behaviour in the case of MBE-growth given here we prefer to follow the model proposed by Frank and Turnbull (1956) rather than the "kick-out" mechanism suggested by Gösele (1981). This is because we assume that the Be dopants are incorporated on lattice sites during MBE growth which is also reflected by the high degree of their electrical activation.

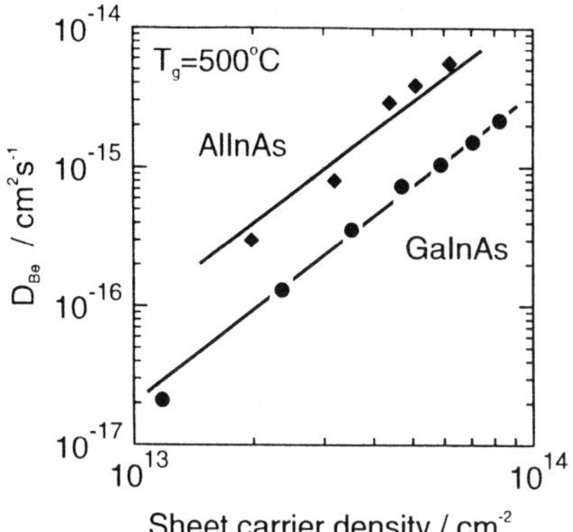

Fig. 5: Quadratic dependence of the Be diffusion coefficient in AlInAs and GaInAs on the δ-doping concentration given in terms of the Hall measured sheet carrier density is shown.

In addition, the influence of the V/III beam equivalent pressure ratio was investigated. We found no measurable change in diffusion behaviour when the As vapour pressure was varied by a factor up to six at otherwise constant MBE growth conditions, in accordance with findings by Ilegems (1977).

5. SUMMARY

The diffusion behaviour of Be in MBE grown AlInAs and GaInAs layers during growth was assessed using multiple δ-doped structures. Evaluation of the samples was mainly done by means of SIMS. Basically, the diffusion properties were found to obey the well-known fundamental diffusion equations. In particular, a quadratic dependence of diffusion length on diffusion time was verified. The Be diffusion coefficient proved to be greater in AlInAs than in GaInAs showing a quadratic dependence on dopant density. An Arrhenius-like temperature dependence with activation energies of 0.6 eV and 1.1 eV for AlInAs and GaInAs, respectively, was found. The As vapour pressure at MBE growth appeared to have no significant effect on the diffusion behaviour.

6. ACKNOWLEDGEMENT

This work was partially funded by the Deutsche Bundespost Telekom under the OEIC project.

7. REFERENCES

Casey H C, Panish M B, Physical Review Vol. 162, 3, (1967) 660

Cunningham J E, Chiu T H, Tell B, and Jan W, J. Vac. Sci. Technol. B 8 (2), (1990) 157

Frank F C, Turnbull D, Physical Review Vol. 104, 3 (1956) 617

Gösele U, Morehead F, J. Appl. Phys. 52 (7) (1981) 4617

Ilegems M, J. Appl. Phys., Vol. 48, No. 3, (1977) 1279

LeCorre A, Caulet J, Gauneau M, Loualiche S, L'Haridon H, Lecrosnier D, Roizes A, David J P, Appl. Phys. Lett. 51 (20), (1987) 1597

Miller J N, Colins D M, Moll N J, Appl. Phys. Lett. 46 (10), (1985) 960

Pao Y C, Hierl T, Cooper T, J. Appl. Phys. 60 (1) (1986) 201

Pao Y C, Franklin J, J. Crystal Growth 95 (1989) 301

Passenberg W, Harde P, Künzel H, Trommer D, 2nd Intern. Conf. on InP and Related Materials, Denver USA, 1990

Schubert E F, Kuo J M, Kopf R F, Luftman H S, Hopkins L C, Sauer N J, J. Appl. Phys. 67 (4) (1990) 1969

Scott E G, Wake D, Spiller G D T, Davies G J, J. Appl. Phys. 66 (11) (1989) 5344

Shih Y C, Block T R, Streetman B G, J. Vac. Sci. Technol. B 10(2) (1992) 863

Yu S, Tan T Y, Gösele U, J. Appl. Phys. 69 (6) (1991)3547

Inst. Phys. Conf. Ser. No 136: Chapter 9
Paper presented at the Int. Symp. GaAs and Related Compounds, Freiburg, 1993

Orientation-dependent growth behavior of GaAs(111)A and (001) patterned substrates in molecular beam epitaxy

T. Takebe, M. Fujii[a], T. Yamamoto[b], K. Fujita, and T. Watanabe

ATR Optical and Radio Communications Research Laboratories
2-2 Hikaridai, Seika-cho, Soraku-gun, Kyoto 619-02, Japan

ABSTRACT: Extra facet generation on ridge-type triangles with (001)-, (110)-, and (201)-related equivalent slopes on GaAs (111)A substrates and stripes running in the $[\bar{1}10]$, [110], and [100] directions on (001) substrates during molecular beam epitaxy of GaAs/AlGaAs multilayers was investigated. Growth of extra (114)A, (110), and $(\bar{1}\bar{1}\bar{1})$B facets was common to the (111)A and (001) patterned substrates. By investigating local variation in layer thickness in the regions adjacent to these facets and extra facets specific to the respective substrates, the orientation-dependent Ga surface diffusion length, λ_{Ga}, was elucidated as $\lambda_{Ga}(001) \approx \lambda_{Ga}(\bar{1}\bar{1}\bar{3})B < \{\lambda_{Ga}(\bar{1}\bar{1}\bar{1})B, \lambda_{Ga}(\bar{3}\bar{3}\bar{1})B, \lambda_{Ga}(013), \lambda_{Ga}(113)A\} < \lambda_{Ga}(159) \approx \lambda_{Ga}(114)A \approx \lambda_{Ga}(111)A \leqq \lambda_{Ga}(110)$.

1. INTRODUCTION

The gallium arsenide (GaAs) (111)A surface has several unique properties. Most notably, the surface has three-fold rotational symmetry and Si doping in molecular beam epitaxial (MBE) growth results in p-type conduction. Taking advantage of these properties, we realized lateral carrier confinement in a triangular (111)A p-type region surrounded by three equivalent (113)A n-type sidewalls by MBE growth of Si-doped GaAs on (111)A substrates patterned with ridge-type triangles (Fujii *et al.* 1992). We also successfully fabricated "lateral p-n subband junctions" that combined vertical quantum confinements with lateral p-n junctions on stripes with the (113)A sidewall and observed intersubband recombination emissions and tunneling currents under foward bias (Yamamoto *et al.* 1993). Control of the extra facet generation on the sidewalls and corners of the triangle during MBE growth was of critical importance for obtaining uniform sidewall layers and firm carrier confinement (Takebe *et al.* 1993a).

In this paper, extra facet generation behavior during MBE growth of GaAs/AlGaAs multilayers on ridge-type triangles on (111)A substrates and stripes on (001) substrates is investigated. Orientation dependence of the growth rate and the Ga surface diffusion length is discussed in the substrate plane - facet - sidewall system for extra facets common to the two patterned substrates and those specific to the respective substrates.

2. EXPERIMENTAL

Ridge-type equilateral triangles with heights of 5 - 7 μm whose three crystallographically equivalent sidewalls are composed of (001)-related, (110)-related, and (201)-related surfaces, briefly designated as "(001) triangle", "(110) triangle", and "(201) triangle", respectively, were formed on (111)A substrates using photolithography and selective etching techniques. Stripes running in the $[\bar{1}10]$, [110], and [100] directions with heights of 5 - 7 μm whose sidewalls are composed of (111)A-related, $(\bar{1}\bar{1}1)$B-related, and (010)-related surfaces, briefly designated as "$[\bar{1}10]$ stripe", "[110] stripe", and "[100] stripe", respectively, were formed on the (001) substrates. These patterns are schematically shown in Figure 1.

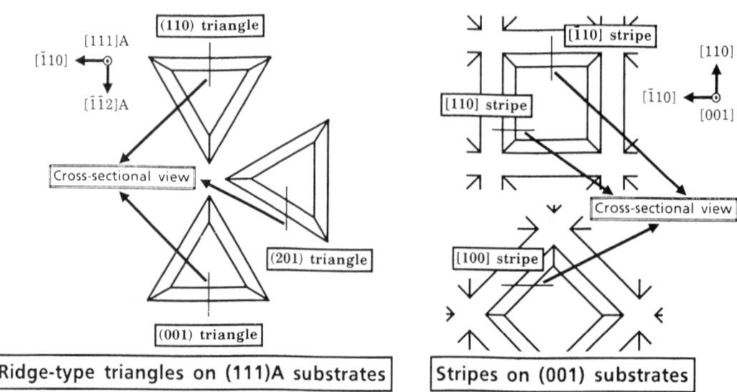

Figure 1 Schematic presentation of patterns on (111)A and (001) substrates.

The intersection angles θ of the sidewalls to the substrate plane were varied over wide ranges using H_2O_2-excess $HF+H_2O_2+H_2O$ mixtures of various compositions (Takebe *et al.* 1993b). These θ ranges include various low- and high-index planes for the side walls.

Undoped 0.2 μm thick GaAs / 0.2 μm thick $Al_{0.3}Ga_{0.7}As$ multilayers were grown simultaneously on the (111)A and (001) patterned substrates at a substrate temperature of 620 °C, a V/III flux ratio of 7.4 for GaAs and 6.2 for AlGaAs, and a substrate rotation speed of 60 rpm. The growth rate was 0.77 μm/h for GaAs and 1.10 μm/h for AlGaAs. The AlGaAs layers were used as "markers" in order to observe how the growth proceeded.

The grown layers were closely observed by scanning electron microscopy (SEM) at an acceleration voltage of 10 kV and a probe current of 1 nA in cross-sections shown in Figure 1. No appreciable variation of growth behavior across each sample was observed.

3. RESULTS AND DISCUSSION

The layers grown simultaneously on the (111)A and (001) patterned substrates exihibited the same growth rate, R, shown in the previous section, on the substrate planes far from the patterns, that is, R[111]A=R[001]. The orientation-dependent growth rate, therefore, has meaning for systems where there exist interactions of adatoms among two or more coexisting different surfaces, especially for the adjacent substrate plane - facet - sidewall system.

From the orientation, intersection angle to the substrate plane, and surface morphology, growth of extra (114)A, (110), and $(\bar{1}\bar{1}\bar{1})$B facets was confirmed as common to the (111)A and (001) patterned substrates. Extra (113)A and (159) facets on the (111)A patterned substrates and extra $(\bar{1}\bar{1}\bar{3})$B and (013) facets on the (001) patterned substrates also were confirmed as main facets. Generation behavior of these facets with respect to θ will be discussed elsewhere.

The relative growth rate, R'[lmn], of an extra (lmn) facet to the growth rate of the layer on the substrate plane was evaluated graphically from the cross-sectional SEM images. Accurate values of R's could be obtained by averaging over several points on a sample. One simple expression for the variation of the effective molecular beam fluxes with θ is the "cosine law" and if no interactions of adatoms between two adjacent surfaces or orientation-dependent incorporation of adatoms are assumed, the relative growth rate, R"[lmn], of the facet with an intersection angle θ_f is expressed as $R''=\cos\theta_f$. A deviation of R'/R" from unity represents a simple measure of the strength of interactions between the adjacent surfaces and the orientation-dependent incorporation of adatoms. Table 1 summarizes the values of R', R", and R'/R" for the observed facets.

In the substrate plane - facet - sidewall system, exponential thickness variations of the layers grown on the substrate plane and/or on the sidewall towards the boundary with the facet reflect the orientation-dependent anisotropy of the Ga surface diffusion length, λ_{Ga}. Since Ga adatoms incident on a substrate where variously oriented surfaces are exposed migrate towards a surface with the minimum λ_{Ga} and are incorporated there, the surface showing an exponential thickness variation of the layers towards the adjacent surfaces has a shorter λ_{Ga} than the adjacent surfaces. By studying this point for the observed systems of extra facets and their adjacent surfaces, orientation dependence of λ_{Ga} was deduced.

Figure 2 shows cross-sectional profiles of the (111)A sidewall of the [$\bar{1}$10] stripe and the (001) sidewall of the (001) triangle. Generation of the (114)A facet is confirmed on both patterns. In (a), an exponential thickness variation is observed for the layers grown on the (001) substrate plane towards the boundary with the (114) facet, while there is no such variation on the (111)A sidewall. The same result has been reported in many papers studying the growth on [$\bar{1}$10] stripes on (001) substrates (for example, Shen *et al.* 1991). In (b), an exponential thickness variation of the layers grown on the (001) sidewall is observed towards the boundary with the (114) facet, while there is no such variation on the (111)A substrate plane.

Table 1 Relative growth rates of the facets.

Facet	Substrate	θ_f	R'[lmn]	R''[lmn]	R'/R''
(114)A	(111)A	33°	0.81	0.84	0.96
	(001)	21°	0.86	0.93	0.92
(110)	(111)A	35°	0.73	0.82	0.89
(011)	(001)	45°	0.47	0.71	0.67
(11$\bar{1}$)B	(111)A	71°	0.37	0.33	1.13
($\bar{1}$11)B	(001)	53°	0.46	0.60	0.77
($\bar{1}\bar{1}$3)A	(111)A	80°	0.33	0.17	1.91
(113)A		30°	1.05	0.86	1.22
(159)	(111)A	34°	0.85	0.83	1.02
			0.69*		0.83*
($\bar{1}$13)B	(001)	24°	0.87	0.91	0.95
(013)	(001)	19°	0.90	0.94	0.96

*In coexistence with the ($\bar{2}$38) facet.

These results consistently imply that some of the Ga atoms incident on the (111)A surface migrated to the (001) surface through the (114)A surface and finally were incorporated in the lattice. The interchange of the roles of substrate plane and sidewall between the (111)A and (001) surfaces, however, might lead to different results since the effective beam fluxes and flux ratio are different between the substrate plane and sidewall, and λ_{Ga} is affected by such a growth ambience. The significance of the present results is that we have clearly confirmed for the first time that the experimental evidence obtained on (001) patterned substrates that the Ga surface diffusion length on the (111)A surface, λ_{Ga}(111)A, is longer than λ_{Ga}(001) essentially originates from the crystalline anisotropy and applies to all (111)A-(001) systems.

(114)A facet

(a) (114)A / θ_f = 21°

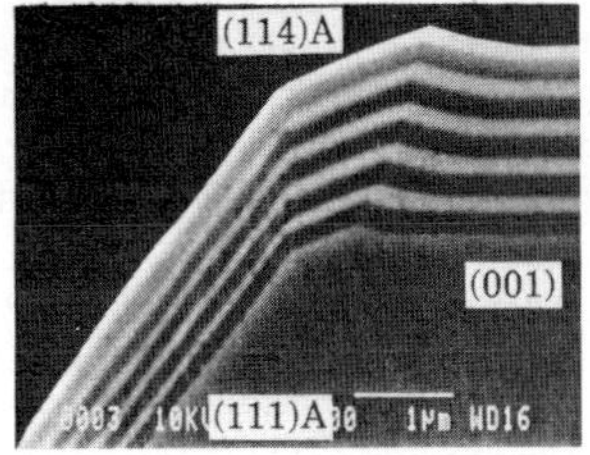

"[$\bar{1}$10] stripe : θ = 54° / (111)A"

(b) (114)A / θ_f = 33°

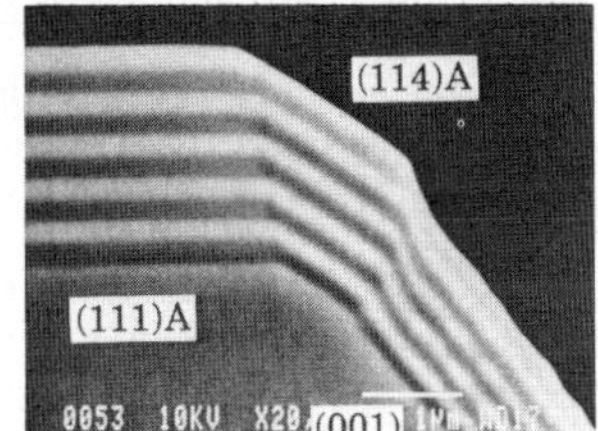

"(001) triangle : θ = 54° / (001)"

Figure 2 The (114)A facet on (a) (001) and (b) (111)A patterned substrates.

Since the (11N)A surface (N=1, 2, 3, - - -) is composed of the (111)A and (001) elements, λ_{Ga}(11N)A will have a value intermediate between λ_{Ga}(111)A and λ_{Ga}(001). The generation of the ($\bar{1}\bar{1}$3)A facet with θ_f=80° adjacent to the (114)A facet on the ($\bar{2}\bar{2}$5)A sidewall of the (001) triangle (Takebe *et al.* 1993a) with R'/R" much larger than unity (Table 1) shows the Ga migration from the (111)A substrate plane to the ($\bar{1}\bar{1}$3)A facet. The generation of the (113)A facet with θ_f=30° adjacent to the (110) facet on the (110) triangles (shown below) with R enhanced even over R[111]A (Table 1) indicates the preferential incorporation into the (113)A facet of Ga adatoms migrating from the (110) facet to the (111)A substrate plane. These results are strong evidence that λ_{Ga}(113)A is shorter than λ_{Ga}(111)A. The R'/R" for the (114)A facet is slightly lower than unity on both (111)A and (001) patterned substrates, which is responsible for the appearance of the facet itself and the exponential thickness variation of the layers on the (001) and (11N)A surfaces adjacent to the facet. This suggests that λ_{Ga}(114)A is comparable to λ_{Ga}(111)A. Thus, the dependence of λ_{Ga} on N is not simple.

Figure 3 shows cross-sectional profiles of the (010) sidewall of the [100] stripe and the (33$\bar{1}$)B sidewall of the (001) triangle. Generation of the (110) facet is confirmed on both patterns. The R'/R" for the (110) facet, as listed in Table 1, is extremely low on both (001) and (111)A patterned substrates, which suggests that a large number of the Ga atoms incident on the (110) facet migrated to the adjacent sidewall and substrate plane and were incorporated there. In (a), an exponential thickness variation of the layers grown on the (001) substrate plane is observed towards the boundary with the (011) facet. This means that λ_{Ga}(110) is longer than λ_{Ga}(001). In (b), an exponential thickness variation of the layers grown on the (33$\bar{1}$)B sidewall is observed towards the boundary with the (110) facet and the (113)A facet is formed on the (111)A substrate plane adjacent to the (110) facet. Since the generation of the (113)A facet instead of an exponential thickness variation is due to preferential incorporation of Ga atoms into the (113)A surface, as discussed earlier, the result leads to a conclusion that λ_{Ga}(110) is longer than λ_{Ga}($\bar{3}\bar{3}\bar{1}$)B (more generally ($\bar{N}\bar{N}\bar{1}$)B composed of the ($\bar{1}\bar{1}$0) and ($\bar{1}\bar{1}\bar{1}$)B elements) and λ_{Ga}(113)A, and is comparable to or longer than λ_{Ga}(111)A.

(110) facet

(a) (011) / θ_f = 45 °

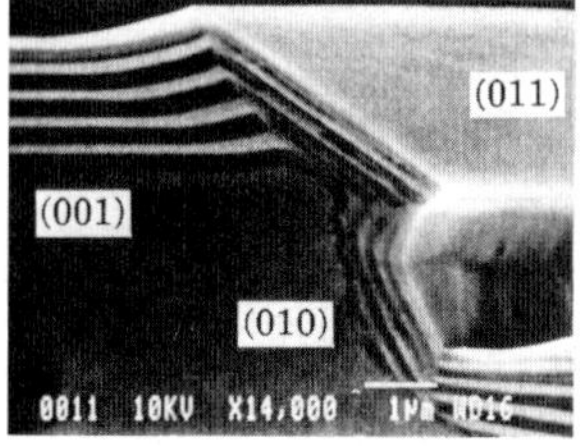

"[100] stripe : θ = 89 ° / (010)"

(b) (110) / θ_f = 35 °

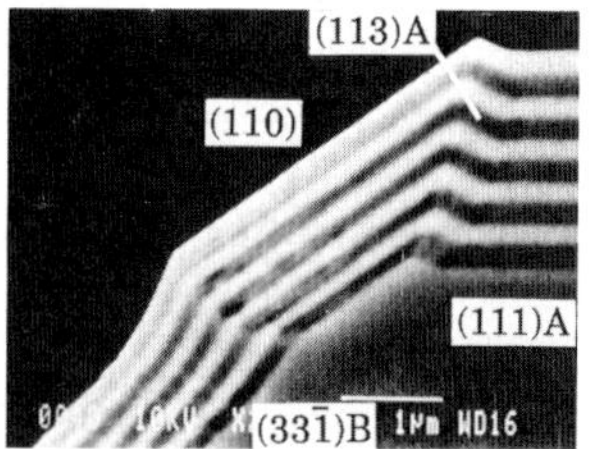

"(110) triangle : θ = 49 ° / (33$\bar{1}$)B"

Figure 3 The (110) facet on (a) (001) and (b) (111)A patterned substrates.

Figure 4 shows cross-sectional profiles of the ($\bar{1}$11)B sidewall with an inverted mesa of the [110] stripe and the (77$\bar{8}$)B sidewall of the (110) triangle. Generation of the ($\bar{1}\bar{1}\bar{1}$)B facet is confirmed on both patterns. In (a), an exponential thickness variation of the layers grown on the (001) substrate plane is observed towards the boundary with the ($\bar{1}$11)B facet. This means that λ_{Ga}($\bar{1}\bar{1}\bar{1}$)B is longer than λ_{Ga}(001). The R'/R" for the ($\bar{1}$11)B facet on the (001) patterned substrate is much lower than unity as listed in Table 1. This strongly suggests a lateral flow to the (001) substrate plane of excess Ga adatoms not incorporated on the initially formed ($\bar{1}$11)B plane. Similar results have been presented in previous reports (for example, Tsang and Cho 1977). The R'/R" for the (11$\bar{1}$)B facet on the (111)A

($\bar{1}\bar{1}\bar{1}$)B facet

(a) ($\bar{1}$11)B / θ_f = 53 °

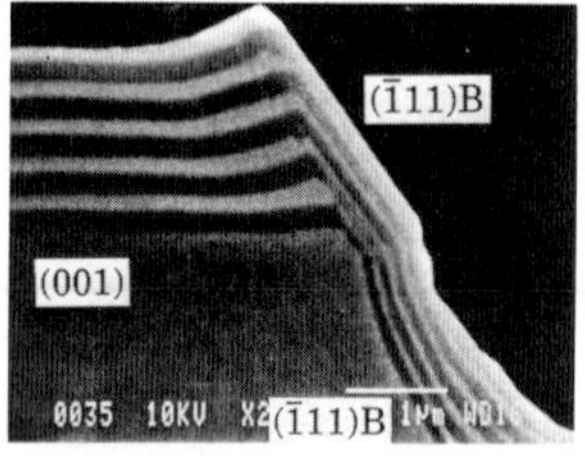

"[110] stripe : θ = 54 ° / ($\bar{1}$11)B"

(b) (11$\bar{1}$)B / θ_f = 71 °

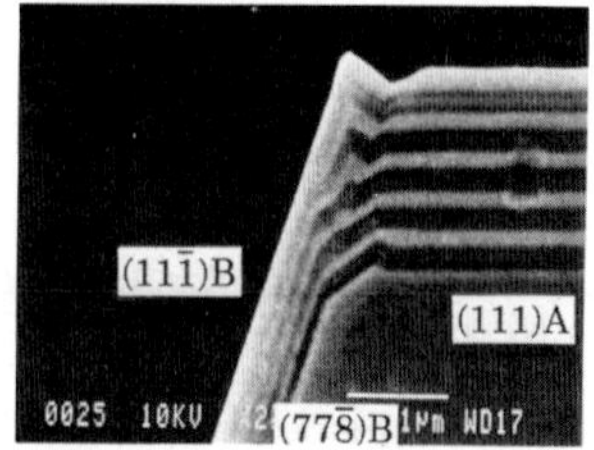

"(110) triangle : θ = 74 ° / (77$\bar{8}$)B"

Figure 4 The ($\bar{1}\bar{1}\bar{1}$)B facet on (a) (001) and (b) (111)A patterned substrates.

patterned substrate is larger than unity as listed in Table 1. Since the growth rate on $(\bar{1}\bar{1}\bar{1})$B substrates is suppressed as the As_4 pressure increases (Nomura *et al.* 1991), this is strong evidence of Ga migration from the (111)A substrate surface to the sidewall. This excess Ga supply from the (111)A substrate surface reduced the effective V/III flux ratio, hence favored the growth of the $(11\bar{1})$B facet. Thus, it can be safely concluded that λ_{Ga}(111)A is longer than $\lambda_{Ga}(\bar{1}\bar{1}\bar{1})$B although no exponential thickness variations on the facet were observed in (b).

Figure 5 shows a cross-sectional profile of the $(\bar{1}12)$B sidewall of the [110] stripe. Generation of the $(\bar{1}\bar{1}\bar{3})$B facet is confirmed. The generation of the $(\bar{1}\bar{1}\bar{3})$B facet has also been reported in the lateral growth of GaAs on $(\bar{1}\bar{1}\bar{1})$B patterned substrates under high As_4 pressures by metalorganic MBE (MOMBE) (Isu *et al.* 1993). The R'/R" for the $(\bar{1}13)$B facet is slightly lower than unity, which is responsible for the appearance of the facet itself. It should be noted that no exponential thickness variations of the layers grown on the (001) substrate plane are observed towards the boundary with the $(\bar{1}13)$B facet in contrast to the $(\bar{1}11)$B facet. This means that $\lambda_{Ga}(\bar{1}\bar{1}\bar{3})$B is comparable to λ_{Ga}(001). Therefore, λ_{Ga} on the $(\bar{1}\bar{1}\bar{N})$B surface composed of the $(\bar{1}\bar{1}\bar{1})$B and $(00\bar{1})$ elements decreases as N increases in accordance with the (11N)A cases.

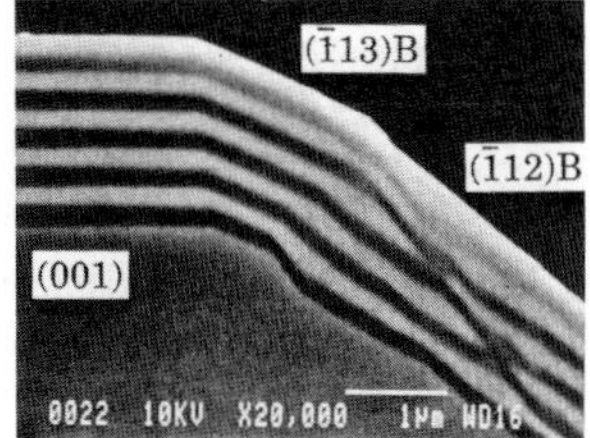

Figure 5 The $(\bar{1}\bar{1}\bar{3})$B facet on (001) patterned substrates.

Figure 6 shows a cross-sectional profile of the (045) sidewall of the [100] stripe. Generation of the (013) facet is confirmed. An exponential thickness variation of the layers grown on the (001) substrate plane is observed towards the boundary with the (013) facet. This means that λ_{Ga}(013) is longer than λ_{Ga}(001). The R'/R" for the (013) facet is slightly lower than unity, which is responsible for the appearance of the facet itself and the exponential thickness variation of the layers on the (001) substrate plane adjacent to the facet.

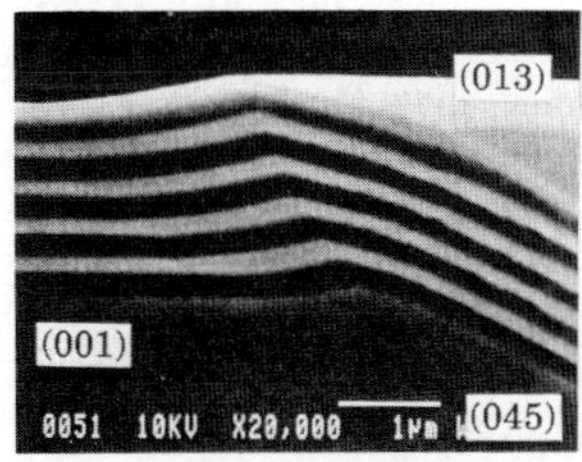

Figure 6 The (013) facet on (001) patterned substrates.

Figure 7 shows a cross-sectional profile of the $(\bar{3}27)$ sidewall of the (201) triangle. Generation of the (159) facet is confirmed. No exponential thickness variations of the layers grown on the (111)A substrate plane are observed towards the boundary with the (159) facet and the R'/R" for the (159) facet is close to unity. Thus, λ_{Ga}(159) is comparable to λ_{Ga}(111)A.

Table 2 summarizes the orientation dependences of λ_{Ga} discussed above. These orientation dependences of λ_{Ga} can be consistently unified to a single inequality as follows.:

$\lambda_{Ga}(001) \approx \lambda_{Ga}(\bar{1}\bar{1}\bar{3})B < \{\lambda_{Ga}(\bar{1}\bar{1}\bar{1})B,\ \lambda_{Ga}(\bar{3}\bar{3}\bar{1})B,$
$\lambda_{Ga}(013),\ \lambda_{Ga}(113)A\} < \lambda_{Ga}(159) \approx \lambda_{Ga}(114)A \approx$
$\lambda_{Ga}(111)A \leqq \lambda_{Ga}(110).$

That is, λ_{Ga} increases in the order of the (001), $(\bar{1}\bar{1}\bar{1})$B-related, (111)A-related, and (110) surfaces. Since the (013) facet is composed of the (011) and (001) elements and the (159) facet is composed of the $(\bar{1}01)$ and (111)A elements, the locations of λ_{Ga}(013) and λ_{Ga}(159) in the above inequality are reasonable.

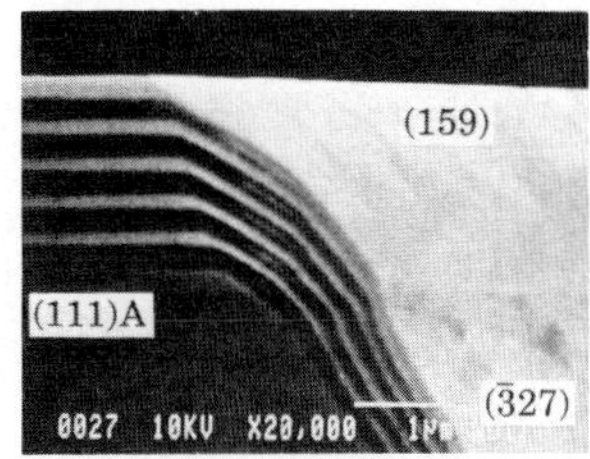

Figure 7 The (159) facet on (111)A patterned substrates.

Table 2 Orientation dependences of the Ga surface diffusion length.

Substrate	Facet	Side wall	Orientation dependence
(111)A	(114)A	(001)	$\lambda_{Ga}(001) < \lambda_{Ga}(114)A \approx \lambda_{Ga}(111)A$
(001)		(111)A	
(111)A	(110)	$(33\bar{1})$B	$\lambda_{Ga}(001) < \lambda_{Ga}(\bar{3}\bar{3}\bar{1})B < \lambda_{Ga}(111)A \leqq \lambda_{Ga}(110)$
(001)	(011)		
(111)A	$(11\bar{1})$B		$\lambda_{Ga}(001) < \lambda_{Ga}(\bar{1}\bar{1}\bar{1})B < \lambda_{Ga}(111)A$
(001)	$(\bar{1}11)$B	Inverted mesa	
(111)A	$(\bar{1}\bar{1}3)$A		$\lambda_{Ga}(113)A < \lambda_{Ga}(111)A$
	(113)A		
(111)A	(159)		$\lambda_{Ga}(111)A \approx \lambda_{Ga}(159)$
(001)	$(\bar{1}13)$B		$\lambda_{Ga}(001) \approx \lambda_{Ga}(\bar{1}\bar{1}\bar{3})B$
(001)	(013)		$\lambda_{Ga}(001) < \lambda_{Ga}(013)$

Considering device application of patterned substrates, such as lateral p-n junctions, the present result that $\lambda_{Ga}(001)$ is the shortest may limit the use of (001) substrates because an exponential variation of the layer thickness always appears on the (001) substrate plane at the boundary with the sidewall and hampers the formation of simple lateral junctions. In contrast, (111)A substrates with a relatively long λ_{Ga} are free from such a limitation, which led to the successful formation of lateral carrier confinement structures (Fujii *et al.* 1992) and lateral p-n subband junctions (Yamamoto *et al.* 1993).

4. SUMMARY AND CONCLUSION

Extra facet generation on ridge-type triangles with (001)-, (110)-, and (201)-related equivalent slopes on GaAs (111)A substrates and stripes running in the $[\bar{1}10]$, [110], and [100] directions on (001) substrates during molecular beam epitaxy of GaAs/AlGaAs multilayers has been studied. By investigating local variation in layer thickness in the regions adjacent to the facets common to the (111)A and (001) patterned substrates and extra facets specific to the respective substrates together with the relative growth rates of the facets, the orientation-dependent Ga surface diffusion length, λ_{Ga}, has been elucidated as $\lambda_{Ga}(001) \approx \lambda_{Ga}(\bar{1}\bar{1}\bar{3})B < \{\lambda_{Ga}(\bar{1}\bar{1}\bar{1})B, \lambda_{Ga}(\bar{3}\bar{3}\bar{1})B, \lambda_{Ga}(013), \lambda_{Ga}(113)A\} < \lambda_{Ga}(159) \approx \lambda_{Ga}(114)A \approx \lambda_{Ga}(111)A \leqq \lambda_{Ga}(110)$.

ACKNOWLEDGMENTS

The authors would like to thank Dr. Y. Furuhama for his encouragement throughout this work.

[a] Present affiliation : Mitsubishi Cable Industries, Ltd.
[b] Present affiliation : Murata Manufacturing Co., Ltd.

REFERENCES

Fujii M, Yamamoto T, Shigeta M, Takebe T, Kobayashi K, Hiyamizu S, and Fujimoto I 1992 Surface Science 267 26.
Isu T, Morishita Y, Goto S, Nomura Y, and Katayama Y 1993 J. Crystal Growth 127 942.
Nomura Y, Morishita Y, Goto S, Katayama Y, and Isu T 1991 Jpn. J. Appl. Phys. 30 3771.
Shen X Q, Tanaka M, Nishinaga T 1991 10th Symp. Rec. Alloy Semicond. Phys. Electron. 65.
Takebe T, Fujii M, Yamamoto T, Fujita K, and Kobayashi K 1993a J. Crystal Growth 127 937.
Takebe T, Yamamoto T, Fujii M, and Kobayashi K 1993b J. Electrochem. Soc. 140 1169.
Tsang W T and Cho A Y 1977 Appl. Phys. Lett. 30 293.
Yamamoto T, Inai M, Takebe T, and Watanabe T 1993 Jpn. J. Appl. Phys. 32 L28.

Paper presented at the Int. Symp. GaAs and Related Compounds, Freiburg, 1993

Differences in the growth mechanism of $In_xGa_{1-x}As$ on GaAs studied by the electrical properties of $Al_{0\cdot3}Ga_{0\cdot7}As/In_xGa_{1-x}As$ heterostructures ($0\cdot2 \leq x \leq 0\cdot4$)

K Köhler, T Schweizer, P Ganser, P Hiesinger and W Rothemund
Fraunhofer-Institut für Angewandte Festkörperphysik, Tullastraße 72, 79108 Freiburg, FRG

ABSTRACT: The growth modes, three dimensional (3d) or two dimensional (2d), of InGaAs layers have been studied by their influences on electrical properties of InGaAs composed heterostructures. We have grown AlGaAs/InGaAs heterostructures by MBE with InAs mole fractions (x) ranging form 0.2 to 0.4. For x<0.25 the layers grow in 2d growth mode. Exceeding the critical layer thickness (CLT) inverted HEMT structures show highly anisotropic electron mobilities with higher mobility in <011> direction. For x>0.25 the layers grow in 3d growth mode which degrades the transport properties of AlGaAs/InGaAs/GaAs HEMT structures stronger than of inverted GaAs/InGaAs/AlGaAs HEMT structures. Higher electron mobility is observed in <01-1> direction.

1. INTRODUCTION

Heterostructures of the system $Al_yGa_{1-y}As/In_xGa_{1-x}As$ grown on GaAs are of great interest for electronic devices. $In_xGa_{1-x}As$ based high electron mobility transistor (HEMT) structures for high speed devices offer better properties than devices using $Al_yGa_{1-y}As$/GaAs HEMT structures. However the lattice mismatch between GaAs substrate and $In_xGa_{1-x}As$ layer can only be accomodated within the layer by elastic strain when the $In_xGa_{1-x}As$ layer thickness is below the critical layer thickness (CLT) (Matthews and Blakeslee 1974). If the layer thickness exceeds the CLT strain between the layers forces the formation of misfit dislocations which degrades the electrical (Fritz et al 1985) and optical (Anderson et al 1987) properties. For the growth of $In_xGa_{1-x}As$ layers with an InAs mole fraction x>0.25 (Price 1988), it was shown using reflection high energy electron diffration (RHEED) (Berger et al 1988, Price 1988) and transmission electron microscopy (TEM) (Yao et al 1991), that a change from a two dimensional (2d) to a tree dimensional (3d) growth mode occurs. We will discuss the influences of different growth modes of $In_xGa_{1-x}As$ on the electrical properties of $Al_{0.3}Ga_{0.7}As/In_xGa_{1-x}As$/GaAs heterostructures.

2. EXPERIMENTAL

The layer sequence of the normal $Al_{0.3}Ga_{0.7}As/In_xGa_{1-x}As$/GaAs HEMT structures starts with a superlattice and a 600 nm undoped GaAs buffer layer. The structure continues with the $In_xGa_{1-x}As$ quantum well (QW) which was varied in thickness and InAs mole fraction.

The following layer sequence for the electron supply consists of a 5 nm wide $Al_{0.3}Ga_{0.7}As$ spacer and a 1.7 nm wide GaAs QW with a δ-doping of $3.5x10^{12}cm^{-2}$ followed by 60 nm $Al_{0.3}Ga_{0.7}As$. The structure is capped by 20 nm GaAs.

The layer sequence of the $GaAs/In_xGa_{1-x}As/Al_{0.3}Ga_{0.7}As$ inverted HEMT (I-HEMT) structures starts with an undoped 300 nm GaAs buffer layer followed by a superlattice and 200 nm $Al_{0.3}Ga_{0.7}As$. The electron supply consists of a 1.7 nm wide GaAs QW which is δ-doped with a Si-concentration of $2.8x10^{12}cm^{-2}$. The spacer is 5 nm thick. On top of the $In_xGa_{1-x}As$ QW which was also varied in thickness and InAs mole fraction, an undoped GaAs layer was grown. The structure is capped by a 30 nm $1.0x10^{18}cm^{-3}$ highly Si doped GaAs layer. This cap layer is designed so as to be just depleted by the surface potential, and to ensure that the conduction band edge at the heterojunction is below the Fermi energy.

The heterostructures were grown by molecular beam epitaxy in an Intevac Gen II on LEC grown 2" GaAs substrates. The growth rate of GaAs was 1.25μm/h. Growth rates, InAs and AlAs mole fractions were controlled by RHEED intensity oscillations. Both HEMT and I-HEMT structures were grown under identical growth conditions at 550°C. Mobility measurements at 300K and 77K in the dark were performed on Hall bar samples with 6 contacts for the orientation dependent Hall mobility measurements. The CL wavelength selective topograms were carried out at 20 K with an electron beam focused to a diameter of about 1 μm.

3. RESULTS AND DISCUSSION

Differences in the growth mechanism of $In_xGa_{1-x}As$ were studied by electrical properties of $Al_{0.3}Ga_{0.7}As/In_xGa_{1-x}As$ heterostructures. For a 2d growth mode we found anisotropical behaviour of $GaAs/In_{0.2}Ga_{0.8}As/Al_{0.3}Ga_{0.7}As$ I-HEMT structures. Electron mobilities of these structures measured in <011> and <01-1> at 300 K are shown in Fig. 1 as a function of the $In_{0.2}Ga_{0.8}As$ QW width. Open circles represent the measurements in <011> direction, and full circles in <01-1> direction. The QW width was varied from 12 nm to 30 nm. Results of the Hall effect measurements on the same samples carried out at 77 K show similar behaviour. Up to an $In_{0.2}Ga_{0.8}As$ QW width of 17.5 nm the electron mobility in both directions is constant and the electron mobility in the <01-1> is slightly higher in comparison to the <011> direction. Increasing the QW width results in a decrease in the electron mobility in both directions. For the <011> direction a slower decrease was observed in comparison to the electron mobility measured in the <01-1> direction. The mobility ratio $\mu_{<011>}/\mu_{<01-1>}$ at 300 K increases from 2.7 to 23 for QW widths of 20 and 30 nm respectively. With decreasing temperature this ratio increases and for the structure with a QW width of 20 nm,

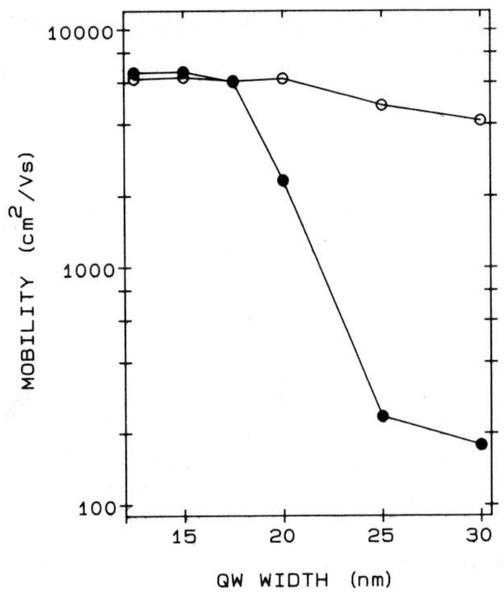

Fig. 1 Electron mobilities of I-HEMT structures measured at 300 K in <011> (open circles) and <01-1> (full circles) direction versus the QW width of $In_{0.2}Ga_{0.8}As$.

the ratio of 2.7 at 300 K is 54 at 77 K. It was not possible to determine the electron mobilities for the I-HEMT structures with QW widths of 25 nm and 30 nm at 77 K, because the resistance in the <01-1> direction is > 10^6 Ω.

To study the temperature behaviour we have measured the resistance of (6x6)mm^2 samples. The resistance in the <011> and <01-1> direction of the I-HEMT structures with a QW width of 20 nm, 25 nm and 30 nm as a function of the inverse temperature is shown in Fig. 2. Open symbols represent the measurements in the <011> and full symbols in the <01-1> direction. Resistance in both directions increases with increasing QW width. However, the increase in the <01-1> direction is clearly higher than in the <011> direction. With decreasing temperature the resistance in the <01-1> direction increases exponentially, as shown for the samples with QW widths of 25 and 30 nm by about 5 orders of magnitude. Below 50K the resistance remains nearly constant. For the I-HEMT structures with QW widths of 25 nm and 30 nm, the resistance ratio $r_{<01-1>}/r_{<011>}$is>10^5, measured at 30K.

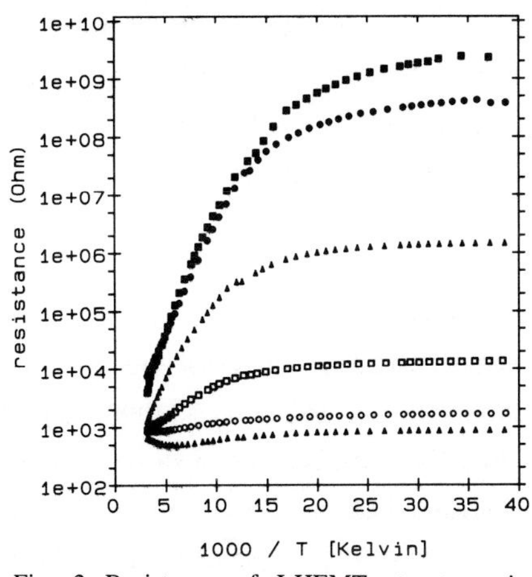

Fig. 2 Resistance of I-HEMT structures in <01-1> (full symbols) and <011> (open symbols) direction versus inverse temperature. The I-HEMT structures with an $In_{0.2}Ga_{0.8}As$ QW width of 30 nm, 25 nm, and 20 nm are represented by squares, circles, and triangles.

The occurence of the anisotropy in the <011> and <01-1> direction can be correlated to the occurrence of misfit dislocations which are asymmetrically oriented. This is shown in Fig. 3 by CL topograms of the investigated I-HEMT structures, with QW widths shown below the CL topograms. The onset of misfit dislocations is observed for the I-HEMT structure with a QW width of 17.5 nm, in which case we attribute the dark straight lines to misfit dislocations, because for CL image measurements, misfit dislocations appear as dark lines due to nonradiative recombination centers (Petroff et al 1980). The images were obtained by wavelength selective measurements (920 nm - 940 nm) to ensure that the CL arises from the $In_{0.2}Ga_{0.8}As$ layer.

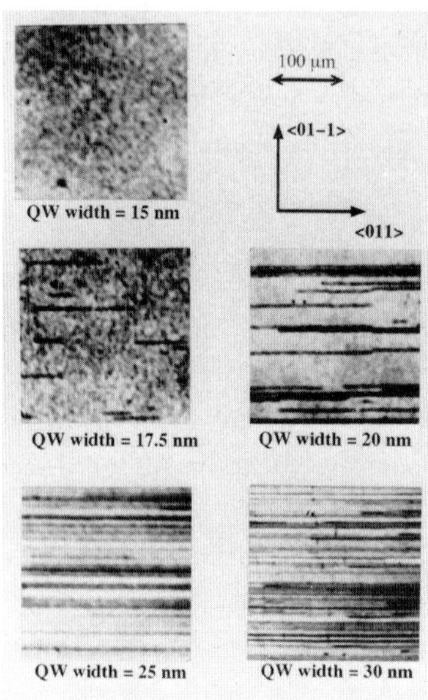

Fig. 3 Wavelength selective CL topograms of I-HEMT structures with $In_{0.2}Ga_{0.8}As$ QWs.

Increasing the QW width results in an increase in the misfit dislocation density, preferentially oriented parallel to the <011> direction. Only a few misfit dislocations perpendicular to the <011> direction could be observed for a QW width of 30 nm. Similar asymmetrical dislocation densities have been observed by Watson et al (1990).

The temperature dependent resistance, shown in Fig. 2 could be explained using a model of Sheng (1980). This model allows to describe the temperature dependent resistance of disorderd materials which are characterized by large conducting regions separated by small insulating barriers. The electrical conduction can be described by a simple tunneling process through a barrier at low temperatures, and as a thermally activated process at higher temperatures. In our case the barriers are formed due to insulating regions around the dislocations (Woodall et al 1983). The misfit dislocations form a line of deep traps which pin the Fermi level, depleting a radial region around them. We used the model of Sheng (1980) to calculate the temperature dependent resistance in the <01-1> which showed good agreement to the experimental results (Hiesinger et al 1992). The equation proposed by Sheng was modified by a factor NxL which is the total number of barriers in the investigated sample. N is the dislocation density and L the lenght of the sample. The dislocation density is determined from CL topograms. In our calculations other parameters were kept constant and only the dislocation density was varied for the different samples.

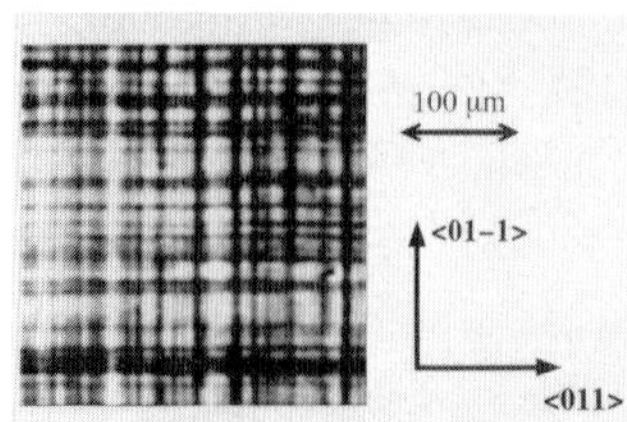

Fig. 4 Wavelength selective CL topogram of I-HEMT structures with $In_{0.2}Ga_{0.8}As$ QW and without Si δ-doping.

A highly asymmetrical dislocation density in the <110> and <-110> direction was also found for the material system GaAsP/GaAs by Fox and Jesser (1990). They correlated the asymmetry to differences of the two types of Ga and As dislocations. Further they showed that the differences are more pronounced if the material is n-type as against p-type. In order to determine if such behaviour is responsible for the asymmetrical dislocation density for the material system $Al_{0.3}Ga_{0.7}As/In_xGa_{1-x}As$, an I-HEMT structure without Si-doping and a QW width of

25 nm was grown. This heterostructure is slightly p-type, due to background doping in the MBE system. The CL topogram of this structure is shown in Fig. 4. As can be seen, the build up of the dislocations parallel to the <01-1> and <011> directions are of similar density.

For $In_xGa_{1-x}As$ based heterostructures where 3d growth occurs, anisotropic electron mobilities were also measured. This anisotropy was studied on HEMT structures, and differs from the anisotropy observed for the above described I-HEMT structures. For the HEMT structures the higher electron mobility is observed in the <01-1> direction.

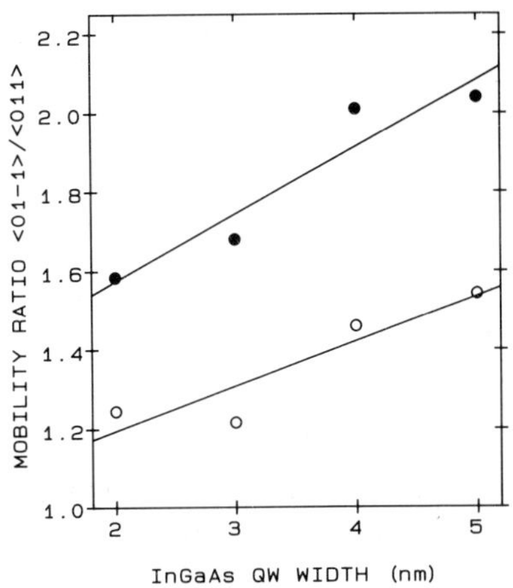

Fig. 5 Mobility ratio $\mu_{<01-1>}/\mu_{<011>}$ of HEMT structures versus $In_{0.38}Ga_{0.62}As$ QW width at 300 K (open circles) and 77 K (full circles).

The effect of variing QW width on the anisotropy in the electron mobility of HEMT structures with a fixed InAs mole fraction is shown in Fig. 5 where the mobility ratio $\mu_{<01-1>}/\mu_{<011>}$ is plotted against the $In_{0.38}Ga_{0.62}As$ QW width. Hall effect measurements carried out at 300 K and 77 K are indicated by open and full circles respectively. As seen from Fig. 5 an increase in the mobility ratio with increasing $In_{0.38}Ga_{0.62}As$ QW width is obtained at both temperatures. The anisotropy is less pronounced compared to the one described above. The anisotropic electron mobility increases above an In mole fraction of 0.3 while the effect is in the range of $\mu_{<01-1>}/\mu_{<011>}=1.1$ at 300K. This anisotropy of the electron mobility can be explained by anisotropic interface scattering of the electrons due to an asymmetric extension of the growth islands. This assumption is confirmed by results of Ohta et al (1989) and Sugaya et al (1989) who found anisotropic surface migration of Ga and Al atoms. A higher migration in the <1-10> direction was measured in comparison to the <110> direction.

Differences in the growth mechanism of $In_xGa_{1-x}As$ layers on the electrical properties of HEMT and I-HEMT structures, which have been already indicated in the discussion of 2d and 3d growth, are demonstrated in Fig. 6. The ratio of the measured mobility at 77 K of HEMT and I-HEMT structures for different InAs mole fractions is plotted versus the $In_xGa_{1-x}As$ QW width. The electron mobilities presented in Fig. 6 are averaged mobilities in the <011> and <01-1> direction. Up to a QW width of 15 nm, the mobility ratio remains constant at 1 for 2d growth (x=0.2), which means that the mobility of both kinds of structures is similar. Increasing the QW width results in a strong decrease of the mobility ratio. This can be correlated to the onset of misfit dislocations which degrade the electronical properties.

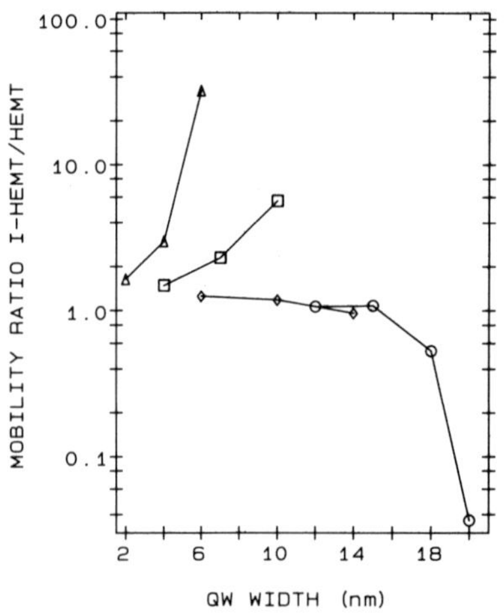

Fig. 6 Mobility ratio I-HEMT/HEMT versus QW width at 77 K with InAs mole fractions of 0.2, 0.3, 0.35, 0.38 indicated by circles, diamonds, squares, and triangles.

The decrease in the electron mobility for I-HEMT structures is more pronounced than for HEMT structures, because the misfit dislocations are at the lower interface where the two dimensional electron gas of the I-HEMT structure is localized. For a 3d growth mode (x=0.38) the electron mobilities of the I-HEMT structures are higher in comparison to the HEMT structure. This behaviour can be explained by 3d-growth, because for a 3d growth mode the upper interface shows interface roughness where the two dimensional electron gas of the HEMT structure is localized.

In conclusion, the influences of the different growth modes and structual properties of $In_xGa_{1-x}As$ layers grown on GaAs were studied by electrical properties of $Al_{0.3}Ga_{0.7}As/In_xGa_{1-x}As$ modulation doped heterostructures. For 2d growth a highly anisotropic electron mobility was measured for inverted heterostructures caused by misfit dislocations. For normal heterostructures with InAs mole fractions where 3d growth occurs, a second anisotropy in the electron mobility was observed, caused by oriented growth.

REFERENCES

Anderson T G, Chen Z G, Kulakovskii V D, Uddin A and Vallin J T 1987 Appl. Phys. Lett. 51 752
Berger P R, Chang K, Bhattacharya P, Singh J and Bajaj K K 1988 Appl. Phys. Lett. 53 684
Fox B A and Jesser W A 1990 J. Appl. Phys. 68 2739
Fritz I J, Picraux S, Dawson L R, Drummond T J, Laidig W D and Anderson N G 1985 Appl. Phys. Lett. 46 967
Hiesinger P, Schweizer T, Köhler K, Ganser P, Rothemund W and Jantz W 1992 J. Appl. Phys. 72 2941
Matthews J W and Blakeslee A E 1974 J. Crystal Growth 27 118
Ohta K, Kojima T and Nakagawa T 1989 J. Crystal Growth 95 71
Petroff P M, Logan R A and Savage A 1980 J. Microscopy 118 255
Price G L 1988 Appl. Phys. Lett. 53 1288
Sheng P 1980 Phys. Rev. B21 2180
Sugaya T, Yokoyama S and Kawabe M 1990 Inst. Phys. Conf. Ser. 106 147
Watson G P, Thompson M O, Ast D G, Fischer-Colbrie A and Miller J 1990 J. Electr. Mater. 19 957
Woodall J M, Pettit G D, Jackson T N, Lanza C, Kavanagh K L and Mayer J W 1983 Phys. Rev. Lett. 51 1783
Yao J Y, Andersson T G and Dunlop G L 1991 J. Appl. Phys. 69 2224

n-type doping of GaAs(111)A with tin using MBE

M.R. Fahy, K. Sato*, C. Roberts and B.A. Joyce.

IRC Semiconductor Materials, The Blackett Laboratory, Imperial College, Prince Consort Rd, London, SW7 2AZ, United Kingdom.

* Permanent Address: Nikko Kyodo Co. Ltd., 10-1, Toranomon 2-chome, Minato-ku Tokyo 105, Japan.

ABSTRACT: We have investigated the tin doping of (111)A oriented GaAs grown using MBE. The surface morphology and the measured carrier concentration were found to be strongly dependent on both growth temperature and As:Ga flux ratio. SIMS showed that the tin distribution within the layers was similar to that for the growth on (001) substrates, but with rather less surface segregation, resulting in a 50% greater concentration in the film. In contrast to silicon doping of GaAs (111)A, under all growth conditions the epi-layers were n-type. At high As_4:Ga flux ratios layers were weakly compensated, though highly compensated at lower As_4:Ga ratios.

1. INTRODUCTION

In recent years there has been considerable interest in the growth of the GaAs on substrates other than (001) (Kadoya et al. 1991, Okano et al 1989). This has developed as attempts have been made to grow quantum wire-like structures on etched and misoriented surfaces (Miller 1985, Notzel et al. 1991). The use of the {111} oriented substrates may also allow fabrication of devices using strong piezoelectrically generated electric fields (Smith et al. 1986). To achieve this requires an ability to grow high quality epitaxial layers, but, the current level of understanding of growth on {111} surfaces still lags far behind that of the (001). Of particular importance is the ability to dope , both n and p type, in a well controlled fashion. Beryllium is a well behaved p-type dopant for both (001) and (111)A growth, but. silicon, which is the most common n-type dopant when growing GaAs on (001), behaves differently during growth on (111)A oriented substrates. The Ga terminated nature of the (111)A surface greatly increases the chances for Si to be incorporated as an acceptor. Thus, under conditions typical for high quality (001) growth (580°C, As_4:Ga flux ratio of 2:1), on (111)A Si is an acceptor (Wang et al. 1985), occupying As sites. The distribution of Si between Ga and As sites depends on growth conditions, but only at low growth temperatures and under high As fluxes does Si behave as a net donor, and even then it is highly compensated. There is, therefore, still a need for a well behaved donor species for use when growing on (111)A substrates. We have examined the use of tin as such a possibility. Tin was the original choice of n-type dopant in growth of GaAs using, MBE, but it has been replaced by Si because of its tendency to surface segregate. However, Sn does have the advantage that it is possible to achieve much higher n-type doping levels than with Si. There is little indication that Sn can act as a substitutional acceptor (Harris et al. 1982), although there is some evidence of a tin-related deep acceptor level in GaAs about 220meV above the valence band edge (Zemon et al. 1986). To our knowledge there has been no report of tin doping of GaAs(111)A using MBE, although Bhat et al. (1991) have used tetraethyltin with organometallic chemical vapour deposition (OMCVD). They report growing compensated n-type material. We have made a preliminary study of the tin doping of (111)A oriented GaAs using MBE by doping with tin over a range of growth temperatures and As_4:Ga flux ratios and analysing the layers using Secondary Ion Mass Spectroscopy, (SIMS), and Hall effect measurements.

2. EXPERIMENTAL

Samples were grown in a conventional solid source VG V80H MBE machine. Semi-insulating (111)A GaAs substrates (Sumitomo Electric Co.) were etched for 10 minutes in 100:1:1 H_20:NH_3OH:H_2O_2, then rinsed in deionised water before being blown dry with N_2. Pieces approximately 1cm^2 were mounted side by side with epi-ready undoped (001) substrates (Showa Denko Co.) onto molybdenum blocks using indium. Ga and As_4 fluxes were calibrated using RHEED oscillations. Temperatures were measured using a pyrometer which was calibrated using the oxide desorption temperature (600°C) and the (2x4) to c(4x4) transition (530°C) on the (001) surface. Due to the strong dependence of tin behaviour on growth conditions, all cells and substrates were given 2-3 hours to reach a stable condition before starting growth.

1μm thick samples were grown at substrate temperatures of 420, 460, 500, 520, 540 and

580°C. At 580°C layers were also grown under three different As_4:Ga flux ratios (2:1, 5:1 and 10:1). In all cases the Ga flux was set at 6.2 x 10^{14}atoms.$cm^{-2}.s^{-1}$ (equivalent to 1ML/s on (100) and 0.88ML/s on (111)A). Observation of the RHEED pattern during cleaning of the etched (111)A surface showed a clear streaky pattern above 450°C but to ensure complete cleaning the substrates were heated to 650°C in an As_4 flux. Under all growth conditions both facetting and some evidence of 3 dimensional growth were observed by RHEED.

Hall measurements were made at room temperature using a Polaron Hall measurement system with a 0.5T magnetic field. The distribution of tin within the grown layers was determined using a Cameca IMS 4f Secondary Ion Mass Spectrometer with a 12.5KeV oxygen beam and concentrations were obtained by using a reference of Sn^+ implanted GaAs.

3 RESULTS AND DISCUSSION

Under all growth conditions layers grown on (001) substrates were highly reflective with a very low density of surface defects. No tin droplets were observed. However, morphology of the layers grown on the (111)A was strongly affected by the growth conditions. With an As_4:Ga flux ratio of 3:1, layers grown at temperatures below 500°C were reflective by eye, but showed a high density of small, non-crystallographic defects on a smooth background when viewed by Normaski interference contrast microscopy. At higher growth temperatures samples appeared matt, but increasing the As_4:Ga flux ratio to 10:1 improved the morphology and the surface again became reflective.

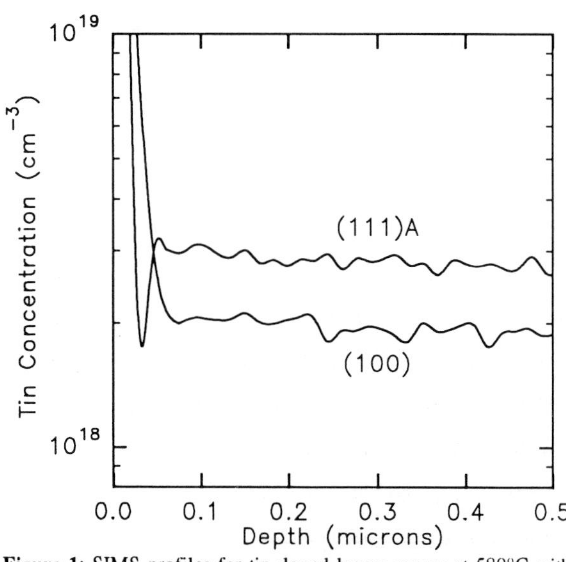

Figure 1: SIMS profiles for tin doped layers grown at 580°C with a As_4:Ga flux ratio of 3:1. The concentration of tin incorporated is higher on the (111)A than the (001)

3.1 Tin distribution

SIMS measurements were performed on layers obtained

under each of the growth conditions. A typical result, for (001) and (111)A layers grown simultaneously at 580°C with an As_4:Ga flux ratio of 3:1, is shown in figure 1. Under all conditions the tin is incorporated uniformly throughout the bulk of the grown layer with approximately 50% more tin being incorporated during growth on (111)A than on (001). It does, however, surface segregate strongly on both orientations, but the extent is lower on the (111)A than the (001). This not only gives a higher incorporation rate, but also a lower surface concentration of Sn on the (111)A, as shown in fig. 1.

3.2 Electrical measurements

The carrier concentrations for layers grown between 420°C and 580°C with an As_4:Ga flux ratio of 3:1 are shown in figure 2. For layers grown on (001) the carrier concentration remains approximately constant with growth temperature, the small dip at 580°C probably being due to surface segregation. The carrier concentration is in very good agreement with the tin concentration measured by SIMS, suggesting full electrical activation of the tin. The tin doped (111)A layers are n-type under all growth conditions, but with varying degrees of compensation. Increasing the As_4:Ga flux ratio dramatically increases the carrier concentration for layers grown on (111)A (figure 3). A similar, but much smaller effect is seen for Si doping, where the layers become more compensated as the high As_4 flux pushes Si back onto Ga sites. The nature of the compensating centre is not clear although simple comparison with Si would suggest it is Sn atoms occupying As sites (Fahy et al. 1993). However, 4.2K photoluminescence measurements show a broad deep level at about 250meV above the valence band edge, similar to that observed by Bamba et al (1983) in compensated (001)

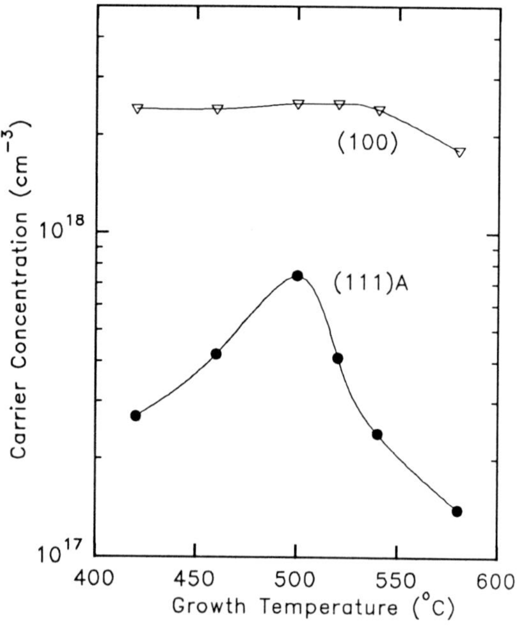

Figure 2: The effect of growth temperature on carrier concentration for tin doped GaAs (001) and (111)A. The As_4:Ga flux ratio is 3:1 and the growth rate is 1μm/hour.

GaAs layers doped during growth with a tin ion beam. They assigned this to a large number of tin related complexes such as (Sn_{Ga}-V_{Ga}) and (Sn_{As}-V_{As}) caused by ion beam damage. Similar studies of melt-grown Sn doped GaAs (Panish, 1973) have also produced highly compensated layers, again assigned to tin related defects. Given the difficulties in growing high quality layers on (111)A and with the added complication of the strain caused by the large atomic size of tin, it is likely that a large number of vacancy related defects would also be introduced during the MBE growth. Further, it seems unreasonable that a substitutional tin acceptor would be as deep as 250meV when compared to carbon at 27meV and silicon at 33meV. The large decrease in compensating centres on increasing the As_4 flux and the lack of a similar effect for Si suggests that the compensation is, at least in part, related to a tin-arsenic vacancy complex.

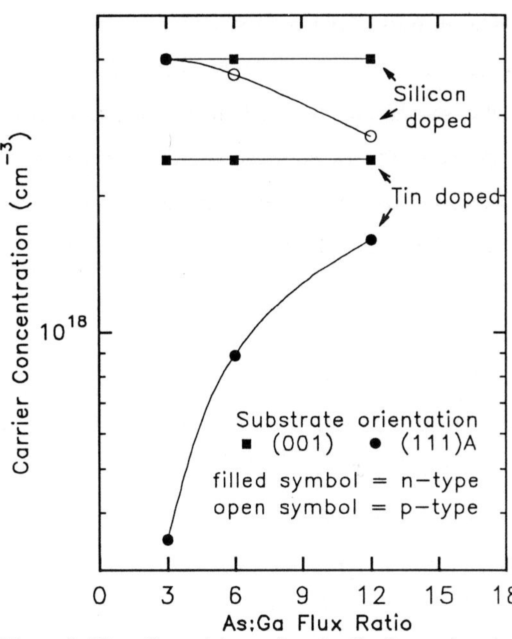

Figure 3: The effect of increasing As_4:Ga flux ratio when doping (001) and (111)A GaAs with Si and Sn. It should be noted that Si dopes (111)A GaAs p-type.

4 CONCLUSION

We have doped GaAs (111)A n-type with tin under growth conditions where Si is a p-type dopant. The carrier concentration is very strongly dependant on the growth conditions, in particular the As_4:Ga flux ratio. Sn also surface segregates, although to a lesser extent than on (001), so sharp doping interfaces would be problematical. Nevertheless, it basically behaves only as a donor and represents the best option for an n-type dopant at this stage. Further work is certainly needed to improve the quality of layers grown on (111)A GaAs, to clarify the nature of the defect centre which compensates the tin so that its effect can be minimised and to devise optimum growth conditions to restrict the surface segregation.

ACKNOWLEDGEMENTS

We should like to thank Paul Mookherjee for photoluminescence measurements. The support of Imperial College and the Research Development Corporation of Japan under the auspices of the "Atomic Arrangements; Design and Control for New Materials" Joint Research Program is gratefully acknowledged.

REFERENCES

R. Bhat, C. Caneau, C. Zah, M. Koza, W. Bonner, D. Hwang, S. Schwarz, S. Menocal and F. Favire, J. Cryst. Growth 107 (1991) 772.
Harris J, Joyce B, Gowers and Neave J 1982 Appl. Phys. A 28 63
Kadoya Y, Sato, A, Kano, H and Sakaki H 1991 J. Cryst. Growth 111 28
Miller D 1985 App. Phys. Let. 47(12) 1309
Notzel R, Daweritz L and Ploog K 1991 J. Cryst. Growth 115 318
Okano Y, Seto H, Katahama H, Nishine S, Fujimoto I and Suzuki T 1989 Jap. J. Appl. Phys. 28,2 L151
Smith D 1986 Solid State Commun. 57 919
Wang W, Mendez E, Kuan T and Esaki L 1985 Phys. Rev. Lett. 47 826
Zemon S, Vassell M, Lambert G and Bartram R 1986 J. Appl. Phys. 60(12) 4253
Bamba Y, Miyauchi E, Kuramoto K, Takamori A and Furuya T 1983 Jap. J. Appl. Phys. 22(6) L331
Fahy M, Neave J, Ashwin M, Murray R, Newman R, Joyce B, Kadoya Y and Sakaki H 1993 J. Cryst. Growth 127 871
Panish M 1973 J. Appl. Phys. 44(6) 2659

Decomposition of AsH_3 and PH_3 in the epitaxial growth of III–V compounds

A. S. Jordan* and A. Robertson**

*AT&T Bell Laboratories, Murray Hill, NJ 07974
**AT&T Bell Laboratories, Engineering Research Center, Princeton, NJ 08540

ABSTRACT: The gas-phase composition for the pyrolysis of AsH_3 and PH_3 in thermodynamic equilibrium has been determined over a wide range of temperatures and pressures (p_{tot}). Although the calculations are applicable to both MOMBE and MOVPE, here the MOMBE aspects are emphasized. The required data for AsH_3 and its subhydrides (AsH and AsH_2) as well as the mixed species AsP, As_2P_2, AsP_3 and As_3P were derived from statistical thermodynamics. A robust free-energy minimization technique with material balance constraints specific to MOMBE cracking or MOVPE provide the product distributions. The major subhydrides in MOMBE cracking are AsH and PH, while AsH_2 and PH_2 are significant in MOVPE. Furthermore, in MOMBE growth the dominant species are As_2 and P_2 but in MOVPE the tetramers As_4 and P_4 become more important. If in the growth of InGaAsP, AsH_3 and PH_3 are combined into a single stream before entering the MOMBE cracker, we show that AsP replaces much of As_2 and a lesser amount of P_2.

1. Introduction

AsH_3 and PH_3 are the most widely used gaseous sources for As and P in the epitaxial growth of GaAs, InP and their various ternary and quaternary alloys by metal-organic molecular beam epitaxy (MOMBE or CBE), hydride-source MBE (HS-MBE) and metal-organic vapor-phase epitaxy (MOVPE). In MOVPE growth the substrate temperature and the system's total pressure influence the hydride pyrolysis if one neglects catalytic factors. On the other hand, in molecular beam technologies, hydride decomposition is independently controlled by the temperature of cracker cells which produce all the volatile species impinging on the substrate surface in the UHV chamber. Consequently, it is likely that the control of stoichiometry, especially of the group V sites in quaternary material, is more readily realized by MBE-related techniques.

The stoichiometry and purity of epitaxial films are in part related to the product distributions of AsH_3 and PH_3 pyrolysis. One can determine the equilibrium species concentrations in hydride decomposition by the methods of chemical thermodynamics. Equilibrium calculations provide an essential baseline even if the kinetics is relatively slow. Moreover, under certain reaction conditions when high temperatures, long residence times or catalysts are employed, kinetic limitations may be unimportant.[1] Clearly, to perform a thermochemical analysis of hydride pyrolysis there is a need for reliable data. Recently, we have evaluated the thermodynamic properties (free-energy function, standard entropy, enthalpy and heat capacity) of AsH_3 and its subhydrides (AsH, AsH_2) as a function of temperature[2] and derived their standard enthalpies of formation[2] together with the phosphorus analogs from bond energy measurements.[1] Subsequently, using a Gibbs energy minimization technique the equilibrium gas-phase concentrations of As, As_2, As_4, AsH,

AsH_2, AsH_3, H and H_2 and the corresponding P-related species were determined over a wide range of temperatures and total pressures.[1,2] While the main thrust of the investigations was MOMBE, the calculations were also applicable to low pressure MOVPE.

In this paper, the main focus is on MOMBE crackers and in applying the mass conservation constraints we take into account the effusive flow through an orifice. This permits the generation of charts for the dimer/tetramer ratio as a function of AsH_3 or PH_3 input flowrate and cracker temperature. Then, we outline the method to estimate the thermodynamic properties of the mixed gaseous species AsP, As_2P_2, AsP_3 and As_3P by means of statistical thermodynamics. These molecules are generated in a single cracker that combines the AsH_3 and PH_3 streams before entering the reaction zone, a configuration used in the MOMBE growth of InGaAsP, InAsP, and GaAsP. The equilibrium product distribution of a single cracker is then contrasted with that of separate AsH_3 and PH_3 crackers. Finally, we discuss the implications of the analysis to MOMBE growth and in comparing the results with experimental findings we confirm that the mixed species are stable and not an artifact of the mass spectrometric measurement.

2. Thermodynamic Methodology

We have employed a free-energy minimization technique to calculate the concentration of species in AsH_3 or PH_3 pyrolysis.[1,3] Accordingly, for each gaseous molecule we determined the Gibbs free-energy, F_i, from the relation

$$F_i = -(FEF_i)T + \Delta H^{\circ}_{fi}\ (298) \qquad (1)$$

where ΔH°_{fi} is the enthalpy of formation and FEF_i is the free-energy function ($FEF_i = -(F^{\circ}_i - H^{\circ}_{298,i})/T$). Then, from F°_i the chemical potential is evaluated in the ideal gas state at T and total pressure, p_{tot}. The chemical system is constrained by mass balance which will be demonstrated for AsH_3 decomposition in a MOMBE cracker. Subsequently, the total free-energy of the system ($F_{tot} = \Sigma n_i\ F^{\circ}_i$ where n_i is the number of moles of species i) is minimized and the constraint is removed by Lagrange multipliers. This procedure leads to two algebraic equations in the multipliers that can be solved by the Newton-Raphson iteration technique.[4]

In formulating the constraint for a cracker with orifice area A we must couple chemical reactions and the Knudsen effusion law.[5] Consequently, for a flowrate of pure AsH_3, f, expressed in molecules/sec, we can write the following mass balance with the effusive flow:

$$f = \frac{p_{tot}A}{(2\pi kT)^{0.5}} \sum_{i=1}^{8} \frac{\alpha_i^{As} x_i}{(M_i)^{0.5}} \quad (2a) \quad \text{and} \quad 3f = \frac{p_{tot}A}{(2\pi kT)^{0.5}} \sum_{i=1}^{8} \frac{\alpha_i^{H} x_i}{(M_i)^{0.5}} \quad (2b)$$

where x_i, M_i, and α_i^{As} (or α_i^{H}) are the mole fraction, molecular weight and the number of As (or H) atoms contained in species i. The species included in Eq. 2a and Eq. 2b, respectively, are As, As_2, As_4, AsH, AsH_2, AsH_3 and AsH, AsH_2, AsH_3, H, H_2.

The MOMBE cracker finds its steady-state pressure obeying Eqs. 2. In contrast, in a MOVPE flow system at a constant T and p_{tot} the input number of moles of AsH_3/H_2 (expressed as As and H) balance a simple linear combination of the moles of products in the output. Consequently, the numerical calculations for a cracker are more complex than for pyrolysis in a MOVPE flow system. The procedure we have adopted to obtain the species concentrations starts with an initial estimate of the atomic H/As ratio and p_{tot} assuming complete dissociation to H_2 and As_2. Second, the free-energy minimization technique is applied to the sum of As atoms and the sum of H atoms in all the molecular species to obtain the equilibrium x_i. The calculation yields an updated p_{tot} in comparison with Eqs. 2. Then, using the H/As ratio as a parameter we perform several iterations to satisfy the constraints in Eqs. 2 at a converging p_{tot}.

Generalization to a single cracker, combining AsH_3 and PH_3 flows at a flowrate of f_{AsH_3} and f_{PH_3}, respectively, is in principle straightforward. In this case, Eqs. 2 are replaced by three equations for the constraints (f_{AsH_3}, f_{PH_3} and 3 (f_{AsH_3} + f_{PH_3})) and the Knudsen effusion terms here include up to 18 species. Besides the molecules relevant in separate crackers (As, As_2, As_4, AsH, AsH_2, AsH_3, H, H_2 and their phosphorus analogs) we also consider the mixed gaseous species AsP, As_2P_2, As_3P and AsP_3. The numerical procedure is more complicated then for AsH_3 or PH_3 pyrolysis alone because the system consists of three different atoms.

There is reliable thermodynamic data in the literature for AsH_3 [2] and PH_3 [1,6] and their subhydrides as well as H and H_2 [6] and the monatomic and polyatomic forms of As[7,8] and P[6]. However, the thermodynamic properties of the mixed species have not yet been investigated. We have estimated the thermodynamic functions of AsP, As_2P_2, As_3P and AsP_3 on the basis of statistical thermodynamics. AsP is a diatomic molecule for which the interatomic distance (1.99Å), spectroscopic vibrational and rotational constants have been tabulated by Huber and Herzberg.[9] By standard methods one can evaluate the translational, vibrational, rotational and anharmonicity contributions to the thermodynamic functions[2,10] (free-energy function, standard entropy, etc.) for AsP(g). The normal modes of vibration for As_2P_2, As_3P and AsP_3 have been determined by Ozin employing laser Raman spectroscopy.[11] The interatomic distances for the four-atomic species were estimated for a tetrahedral geometry on the basis of data for As_4 [8] and P_4 [6]. The enthalpy of formation for AsP at 298K was derived from the bond dissociation energy of the molecule (4.46eV).[12] For the four-atomic species the enthalpy of formation was based on the average pair bond energies in As_4 and P_4.

3. Results and Discussion

In Figures 1 and 2, we present the dimer/tetramer ratio for AsH_3 and PH_3 pyrolysis, respectively, as a function of MOMBE cracker temperature and input flowrate. Note that the cracker reaches a steady-state p_{tot} that is a function of flowrate and T. For example, at 10 sccm of AsH_3 flow with a cracker orifice of $2cm^2$, p_{tot} varies between 1.6×10^{-5} and 2.5×10^{-5} atm in the temperature range 900-1500K. In the same temperature interval in a PH_3 cracker p_{tot} increases from 1.2×10^{-5} to 1.8×10^{-5} atm. Under MOMBE conditions the prominent decomposition products are the dimers P_2 and As_2 and their abundance compared to the tetramers rises with increasing temperature. Furthermore, the dimer-tetramer ratio diminishes at higher hydride flowrates. The dimer-tetramer ratios are in good accord with the experimental data. At a cracker T of 1000°C and a flowrate of 10 sccm we predict ratios of 1300 and 3700 for AsH_3 and PH_3 decomposition, respectively. At the same temperature Huet et al.[13] have determined the product distribution from AsH_3 and PH_3 cracker cells employed in HS-MBE growth experiments by a modulated beam mass spectrometric

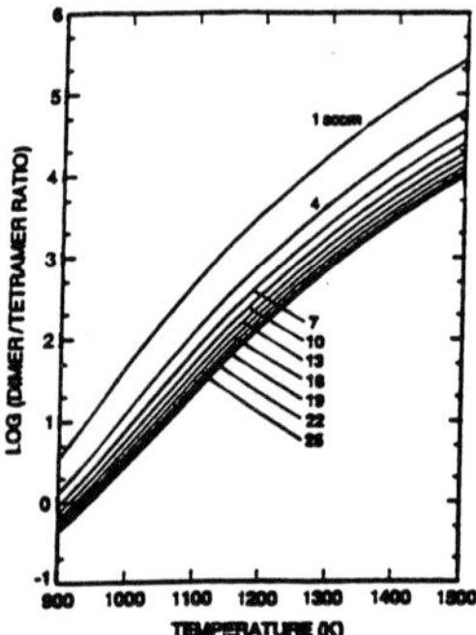

Fig. 1-Dimer/tetramer ratio as a function of temperature and arsine flowrate in a cracker.

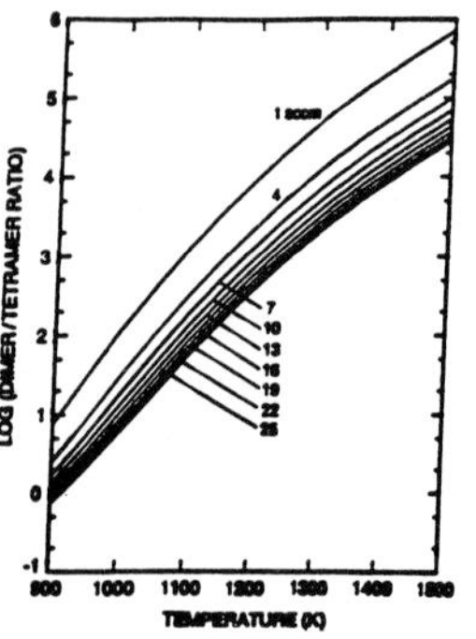

Fig. 2-Dimer/tetramer ratio as a function of temperature and phosphine flowrate in a cracker.

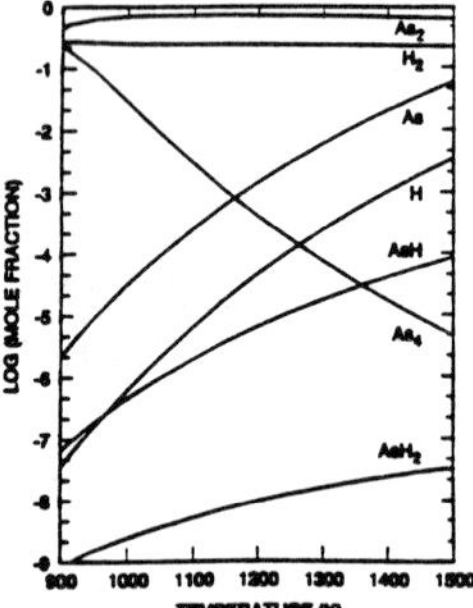

Fig. 3 The equilibrium gas phase composition as a function of temperature for cracking 2 sccm AsH_3.

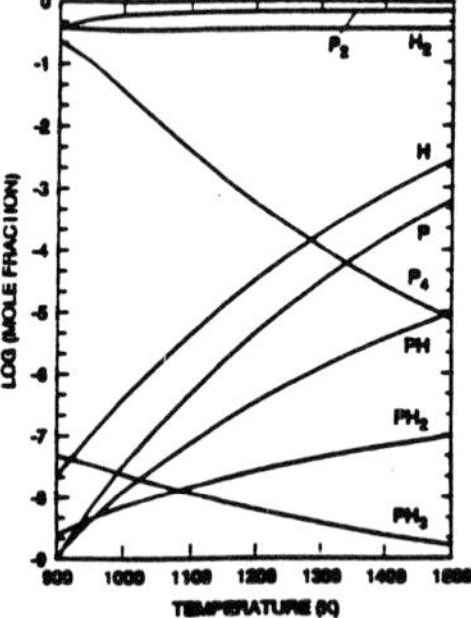

Fig. 4-The equilibrium gas phase composition as a function of temperature for cracking 8 sccm PH_3.

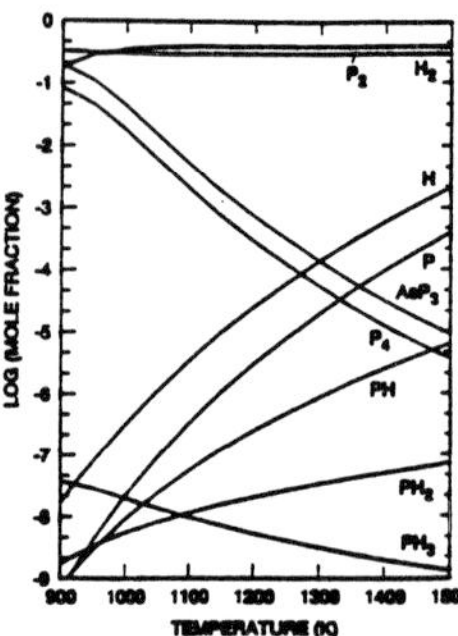

Fig. 5-The equilibrium gas phase composition as a function of temperature for cracking 2 sccm AsH_3 and 8 sccm PH_3 in a single cracker.

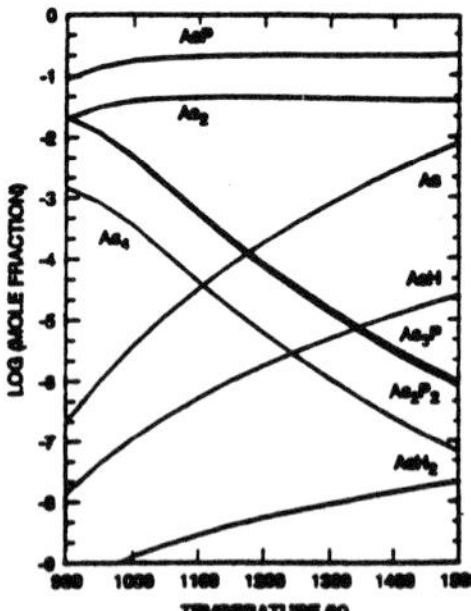

Fig. 6-The equilibrium gas phase composition as a function of temperature for cracking 2 sccm AsH_3 and 8 sccm PH_3 in a single cracker.

technique and reported dimer-tetramer ratios of 1000 and 2000, respectively. Whether the difference between the measurements and predictions is due to the experimental method, kinetic factors or slight error in the thermodynamic data (i.e., ~ 2kcal) is difficult to tell.

The calculations can be readily extended to MOVPE conditions. Here the H/As or H/P input ratios are very large due to the dilution of AsH_3 or PH_3 with the H_2 carrier gas. Moreover, kinetics may play a more significant role at the typical deposition temperature of 900K. In low pressure MOVPE ($p_{tot} \approx 0.1$ atm) the tetramers dominate over the dimers, though their concentration is by no means negligible. At atmospheric pressures tetramers become the major species, while the concentration of dimers is minuscule.[1,2] Recently, the uniformity of group V sites in quaternary (Q) InGaAsP epitaxial layers ($\lambda = 1.35\mu m$) grown by low pressure MOVPE has been successfully interpreted in the framework of this thermodynamic analysis.[14]

In the preparation of Q material by MOMBE or HS-MBE the AsH_3 and PH_3 flows are frequently combined before they reach a single cracker cell.[15] Another alternative is to employ two separate crackers for AsH_3 and PH_3. In Figs. 3 and 4 we show the equilibrium product distribution as a function of cracker temperature for AsH_3 (2 sccm) and PH_3 (8 sccm), respectively. With separate crackers the dominant species reaching the surface of the substrate are As_2 and P_2 at all temperatures. One should also note that there is ample atomic H present in the MBE chamber and that the dominant subhydrides are AsH and PH which may play an important role in reducing carbon contamination.[1] In contrast, under MOVPE growth conditions AsH_2 and PH_2 are the thermodynamically favored subhydrides.[1,2] Indeed, recently the presence of AsH_2 and PH_2 have been identified by Raman spectroscopy[16] and AsH by mass spectroscopy[17] in MOVPE and MOMBE systems, respectively.

In Figs. 5 and 6 the species distribution as a function of temperature are presented when AsH_3 (2 sccm) and PH_3 (8 sccm) streams are combined in a single cracker. The steady-state total pressure in the cracker rises from 1.28×10^{-5} to 1.84×10^{-5} atm in the temperature interval between 900 and 1500K. Here the calculations show that in addition to the normal volatile products of separate AsH_3 and PH_3 crackers, the mixed species AsP, As_2P_2, As_3P and AsP_3 are also generated. The most prominent mixed species is AsP with a concentration intermediate between P_2 and As_2.

Mixed species of As and P in the VPE literature were first reported by Ban.[18] However, one could not be certain that the formation of these molecules in the pyrolysis of AsH_3 and PH_3 mixtures is not an artifact associated with the mass-spectrometric technique. Subsequently, Panish and Hamm, employing a single cracker for AsH_3 and PH_3 pyrolysis in a HS-MBE apparatus, observed the mixed species by modulated beam mass spectrometry.[15] They concluded that the concentration of AsP much exceeds that of As_2P_2, As_3P and AsP_3. The thermodynamic calculations presented in Figs. 5 and 6 confirm that the mixed species indeed exist and they are not fragments formed in the detector of the mass spectrometer. In view of the fact that the sticking coefficient of AsP is not known, it remains to be seen whether the presence of AsP, in addition to As_2 and P_2, compromises the compositional uniformity on group V sites and morphology in the MOMBE and HS-MBE growth of InGaAsP and InAsP. Therefore, we suggest that experiments should be performed using single and separate crackers to evaluate the effect of the mixed species on epitaxial growth.

REFERENCES

[1] A. S. Jordan and A. Robertson, Submitted to J. Vacuum Sci. Technol. B.

[2] A. S. Jordan and A. Robertson, In press, Journal of Materials Science - Materials in Electronics.

[3] W. R. Smith and R. W. Missen, "Chemical Equilibrium Analysis: Theory and Algorithms," John Wiley and Sons, Inc., New York (1982).

[4] G. A. Corn and T. M. Corn, "Mathematical Handbook for Scientists and Engineers," second edition, McGraw-Hill Book Co., New York (1968).

[5] See, for example, S. Dushman, "Scientific Foundations of Vacuum Technique," New York, John Wiley and Sons (1949).

[6] D. R. Stull and H. Prophet, "JANAF Thermochemical Tables," second edition, National Bureau of Standards, Washington, DC (1971).

[7] D. R. Stull and G. C. Sinke, "Thermodynamic Properties of the Elements," Advances in Chemistry Series No. 18, American Chemical Society (1956).

[8] R. J. Capwell, Jr. and G. M. Rosenblatt, Journal of Molecular Spectroscopy, *33*, 525 (1970).

[9] K. P. Huber and G. Herzberg, "Molecular Spectra and Molecular Structure, IV Constants of Diatomic Molecules," Van Nostrand Reinhold Co., New York (1979).

[10] G. N. Lewis and M. Randall, Revised by K. S. Pitzer and L. Brewer, "Thermodynamics," second edition, McGraw-Hill Book Co. Inc., New York (1961).

[11] G. A. Ozin, Journal Chem. Soc. *(A)*, 2307 (1970).

[12] N. Rajamanickam, U. D. Prahllad and B. Narasimhamurthy, Spectroscopy Letters, *15*, 557 (1982).

[13] D. Huet, M. Lambert, D. Bonnerie and D. Dufresne, J. Vacuum Sci. Technol., *B3*, 823 (1985).

[14] J. L. Zilko, P. S. Davisson, L. Luther and K. D. C. Trapp, J. Crystal Growth, *124*, 112 (1992).

[15] M. B. Panish and R. A. Hamm, J. Crystal Growth, *78*, 445 (1986).

[16] P. Abraham, A. Bekkaoui, V. Souliere, J. Bouix and Y. Monteil, J. Crystal Growth, *107*, 26 (1991).

[17] C. R. Abernathy, J. Crystal Growth, *107*, 982 (1991).

[18] V. S. Ban, J. Electrochem. Soc., *118*, 1473 (1971).

Inst. Phys. Conf. Ser. No 136: Chapter 9
Paper presented at the Int. Symp. GaAs and Related Compounds, Freiburg, 1993

Light emission from lateral *p–n* junctions on patterned GaAs(111)A substrates

N. Saito, M. Yamaga, F. Sato, I. Fujimoto, M. Inai*, T. Yamamoto* and T. Watanabe*

NHK Science and Technical Research Laboratories, Setagaya, Tokyo 157, JAPAN
*ATR Optical and Communications Research Laboratories, Seika-cho, Kyoto 619-02, JAPAN

ABSTRACT: Light emission is observed from the lateral p-n junctions on patterned GaAs (111)A substrates. The dependences of output power and emission spectra on Si concentration suggest that the emission characteristics are governed by the hole concentration in p-region. When Si concentration is $1\times10^{19}cm^{-3}$, the light output from the lower junctions is higher than that from the upper ones. This is because the lower junctions have round shape and are formed on (111)A surface, where the hole concentration is higher than on (311)A slope on which the upper ones are formed.

1. INTRODUCTION

Si is an n-type dopant most commonly used in epitaxially grown GaAs on (100) substrate. Recently, however, doping characteristics of Si on non-(100) surfaces has been investigated. The conduction type depends on growth conditions and crystallographic orientation of the substrates (Ballingall and Wood 1982, Wang *et al* 1985). Using this amphoteric nature of Si in molecular beam epitaxial (MBE) GaAs, both n- and p-type regions can be grown simultaneously on a patterned substrate, and lateral p-n junctions are formed at the boundary of the two regions. Fujii *et al* (1992a) have made lateral p-n junctions on patterned GaAs (111)A substrates (Figure 1(a)). Because of three-fold symmetry of the (111) surface, three equivalent junctions encircle the (111) surface. Hence, they can be used in vertical-cavity surface emitting lasers as current blocking layers.

Inai *et al* (1993) have studied the electrical characteristics of lateral p-n junctions grown on (111)A substrates. They have shown that the upper junctions are graded junctions whose width depends on the growth temperature. The structure of the upper junctions is determined by Ga adatom migration from the top (111)A surface to the (311)A slope surface. The junctions are formed not on the ridge between the (111)A surface and the (311)A slope, but on the (311)A slope surface (Figure 1(b)). On the other hand, the structure of the lower junctions has not been fully investigated since they are less abrupt than the upper ones, because of the rounded shape of the lower junctions caused by mesa etching (Fujii *et al* 1992b).

In this paper, we report light emission from the lateral p-n junctions formed on patterned GaAs (111)A substrates in the direction perpendicular to the substrates. We have compared the light output and emission spectra of the lower junctions and those of the upper ones. When Si concentration is $1\times10^{19}cm^{-3}$, the output power from the lower junctions is higher than that from the upper ones. The different dependences of light output and emission spectra on Si concentration are discussed in relation to the difference in the shape of the junctions mentioned above.

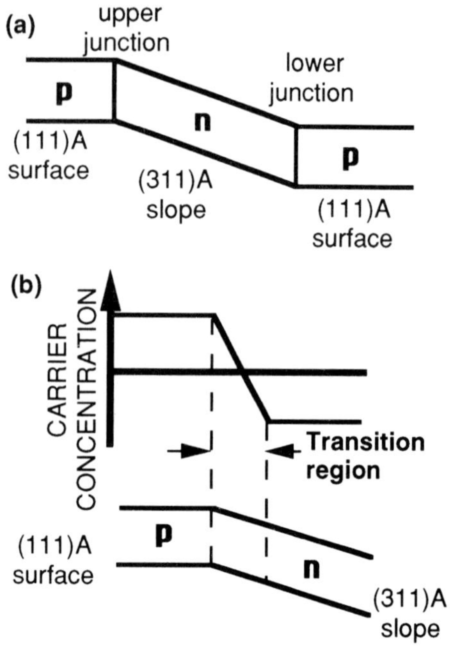

Fig.1.Schematic structure of the lateral p-n junction: (a) simple model and (b) the model proposed by Inai *et al.*

2. EXPERIMENTAL

The preparation of the samples, that is, patterning of substrates and MBE growth of Si doped GaAs layers, followed the procedure employed by Fujii *et al* (1992a). Semi-insulating (111)A on-axis substrates were etched in a H_3PO_4:H_2O_2:H_2O=3:1:50 solution at 40℃ using conventional photolithography technique to obtain a triangular (111)A surface surrounded by three (311)A slopes. After degreasing with organic solvents and final etching in an NH_4OH:H_2O_2:H_2O=2:1:96 solution, the substrates were loaded into the MBE chamber, with unpatterned (111)A and (100) wafers. Thermal cleaning was made at a temperature of 680℃ and an As pressure of 1.5 mPa. Si-doped GaAs layers 1 μ m thick were grown at a substrate temperature of 600℃ and a growth rate of 1.3 μ m/h with an As_4/Ga flux ratio of 2. Silicon concentration was varied through the temperature of Si K-cell.

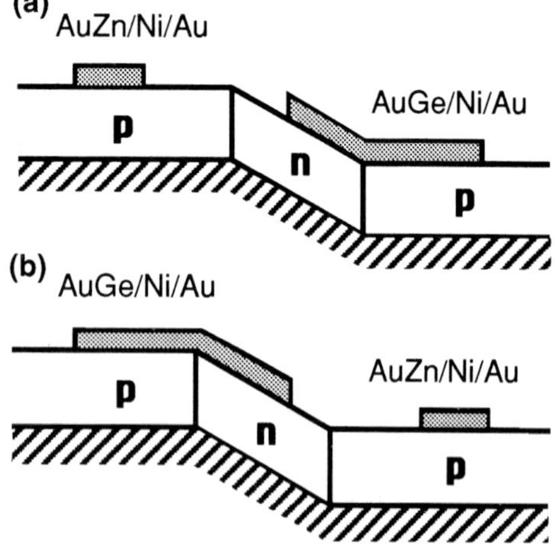

Fig.2 Two types of electrodes to observe light emission from (a) the upper junction and (b) the lower junction.

Two types of electrodes were deposited to measure the power and the spectrum of the light emitted from the upper and the lower junctions (Figure 2). Output power and emission spectra were measured at room temperature with a DC current injected using tungsten probe needle. Cathodoluminescence (CL) spectra were also measured at room temperature to investigate conduction types and carrier concentrations of small regions near p-n junctions. The reference data were taken from capacitance-voltage measurements made on the flat samples.

3. RESULTS AND DISCUSSION

Figure 3(a) shows cross-sectional view of lateral p-n junctions. The (311)A slope gradually changes into (111)A surface with a transition width of 3 μ m at the lower junctions. This is consistent with the results of CL measurements (Figure 3(b)). The CL peak wavelength of the bottom (111)A surface, which is the same as that of flat (111)A (p-type), gradually changes into the n-type (flat (100)) value on the (311)A slope. On the other hand, the upper end of the (311)A slope makes a clear ridge with the top (111)A surface. At the upper junction, the CL peak wavelength changes with a transition width of 1 μm. Since the conduction type and carrier concentration depend on tilt angle of the substrate from (111)A surface at a given MBE growth condition (Okano *et al* 1990), the gradual change in carrier concentration occurs at the lower junctions.

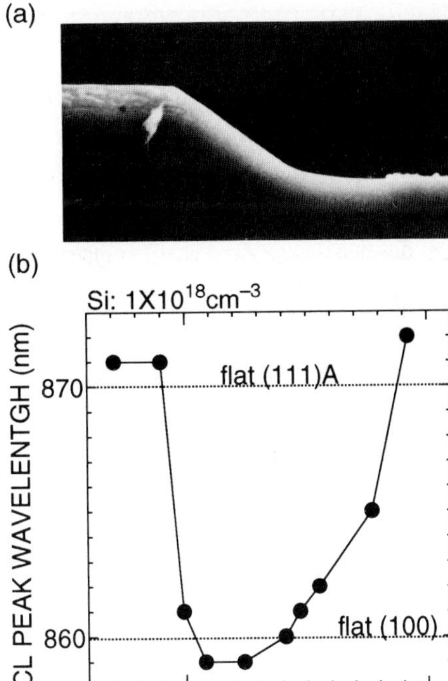

Fig. 3. (a) Cross-sectional view of the lateral junction and (b) Peak wavelength of cathode luminescence spectrum as a function of distance from the upper interface.

Fig. 4. Near field pattern of lateral p-n junctions.

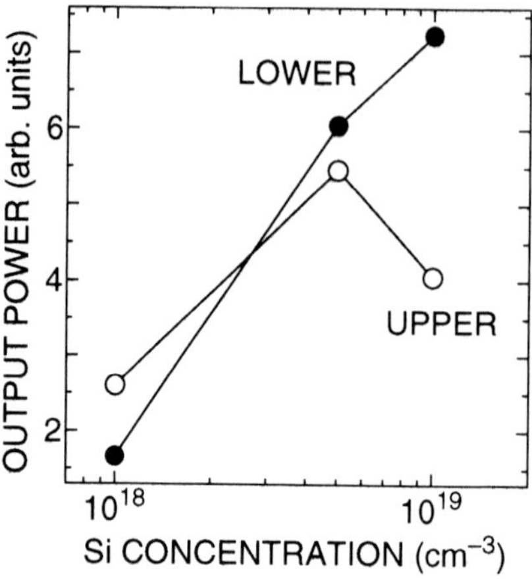

Fig. 5. Si concentration dependence of light output from the lateral p-n junctions at 100mA.

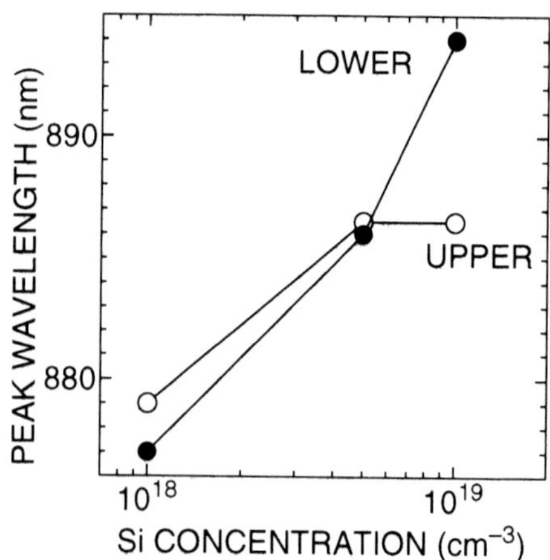

Fig. 6. Si concentration dependence of peak wavelength from the lateral p-n junctions at 50mA.

Light output was observed in the near infrared as shown in Figure 4. Although the left corner is hidden by the probe needle, the three sides emit light equally. Figure 5 shows the light output at 100mA forward current through a junction as functions of Si concentration. Light output from the lower junction increases as Si concentration increases. On the other hand, the light output from the upper junction with a Si concentration of $1\times10^{19}cm^{-3}$ is less than that with a Si concentration of $5\times10^{18}cm^{-3}$.

The peak wavelength of emission spectra shows similar dependence on Si concentration (Figure 6). The peak wavelength from the lower junction increases as Si concentration increases, while that from the upper junction with a Si concentration of $1\times10^{19}cm^{-3}$ is almost the same as that with a Si concentration of $5\times10^{18}cm^{-3}$.

The dependences of the light output and the peak wavelength from the lower junctions on Si concentration agree with those observed in CL spectra of p-type samples (Cusano 1964). That is, as carrier concentration increases, the light output (efficiency) increases and the peak shifts to the longer wavelengths. Hence, the light output and emission spectra are governed by the effective carrier concentration in the p side of the junction. The light output and the peak wavelength from the upper junctions with Si concentration up to $5\times10^{18}cm^{-3}$ show that the effective carrier concentration near the upper junctions is almost the same as that at the lower junctions at least up to this doping level. The decrease in output power from the upper junction at a doping level of $1\times10^{19}cm^{-3}$ is attributed to the decrease in carrier concentration at the upper junction. This is probably due to the difference in doping characteristics of Si acceptor on nearly (111)A surface and (311)A surface, on which the lower and the upper junctions are formed, respectively. In other words, carriers are more compensated at the upper junctions that is formed on (311)A slope than at the lower junctions on nearly (111)A surface.

However, the dependence of the peak wavelength on injected current (Figure 7) showed no evidence of compensation. The peak wavelength from the lower junctions did not depend on current, while that from the upper junctions shifted to longer wavelength. The peak wavelength should shift to high-energy in donor-acceptor (D-A) pair recombination. A slight increase with Si concentration $1\times10^{18}cm^{-3}$ may be due to temperature rise of the junctions caused by the resistance of p- and n- regions. The difference of light emission mechanism from the lateral p-n junctions should be investigated further.

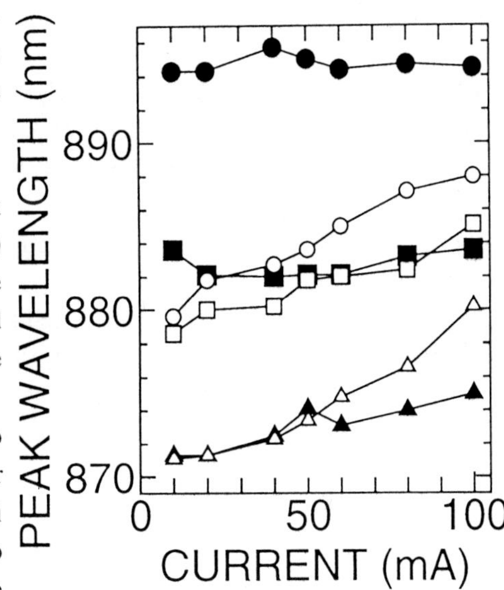

Fig. 7. Dependence of peak wavelength of light emitted from the lateral p-n junctions on injected current. Si concentration is ○ ●:$1\times10^{19}cm^{-3}$, □■:$5\times10^{18}cm^{-3}$, △▲: $1\times10^{18}cm^{-3}$. Solid symbols denote the peak wavelength from the lower junctions and open symbols that from the upper ones.

4. CONCLUSION

We have observed light emission from the lateral p-n junctions formed at the boundary of p- and n-type regions simultaneously grown on patterned GaAs (111)A substrates that have (111)A surface and (311)A slope. The dependences of output power and emission spectra on Si concentration suggest that the emission characteristics is governed by the hole concentration in p-type region. The light output from the lower junctions is higher than that from the upper ones when Si concentration is $1\times10^{19}cm^{-3}$. This is because the lower junctions have round shapes and are formed on (111)A surface, where the hole concentration is higher than on (311)A slope on which the upper ones are formed.

ACKNOWLEDGMENTS

The authors would like to acknowledge K. Kobayashi, T. Tajima and K. Goto for their useful discussions. We also thank N. Hinomoto, A. Kuroe and O. Nanba for their help in sample preparation and experiments.

REFERENCES

Ballingall J M and Wood C E C 1982 Appl. Phys. Lett. **41** 947

Cusano D A 1964 Solid State Commun. **2** 353

Fujii M, Yamamoto T, Shigeta M, Takebe T, Kobayashi K, Hiyamizu S and Fujimoto I 1992a Surf. Sci. **267** 26

Fujii M, Takebe T, Yamamoto T, Inai M and Kobayashi K 1992b Superlattices and Microstructures **12** 167

Inai M, Yamamoto T, Fujii M, Takebe T and Kobayashi K 1993 Jpn. J. Appl. Phys. **32** 523

Okano Y, Shigeta M, Seto H, Katahama H, Nishine S and Fujimoto I 1990 Jpn. J. Appl. Phys. **29** L1357

Wang W I, Mendez E E, Kuan T S and Esaki L 1985 Appl. Phys. Lett. **47** 826

Inst. Phys. Conf. Ser. No 136: Chapter 10
Paper presented at the Int. Symp. GaAs and Related Compounds, Freiburg, 1993

Nitrogen doping in GaP layer grown by OMVPE using TBP

Akihiro WAKAHARA, Kotaro HIRANO, Xue–Lun WANG, and Akio SASAKI
Department of Electrical Engineering, Kyoto University
Kyoto 606–01, Japan

ABSTRACT: Nitrogen–doped GaP layer is grown at 720°C by OMVPE using tertiarybutylphosphine (TBP) and NH_3. The NH_3 partial pressure is varied from $5x10^{-3}$ to $4x10^{-2}$ atm. High quality GaP layers with excellent surface morphology are obtained. Strong luminescence caused by the nitrogen isoelectronic trap is observed in the photoluminescence spectra. The nitrogen concentration calculated from the PL intensity is the order of $10^{18}cm^{-3}$. Large incorporation efficiency is achieved by using TBP as compared with that by PH_3.

1. INTRODUCTION

AlGaP alloy semiconductor has been paid much attention to application in light–emitting devices operating in a visible light region. Although the band gap of AlGaP is large (Eg=2.26–2.45eV), the band structure of AlGaP is indirect. Therefore, the improvement of luminescence capability is required to realize light–emitting devices. Recent theoretical study predicts that an AlP/GaP short period superlattice can be the quasi–direct band structure by zone–folding and band mixing effects (Schuurmans 1987, Kumagai 1988). However, the oscillator strength of AlP/GaP short period superlattices is theoretically expected to be the order of 10^{-4}, which is about 10^{-3} times compared with that of the Γ–Γ transition in GaP (Schuurmans 1987). Even if an ideal AlP/GaP super–lattice can be obtained, further improvement of luminescence capability would require to actualize the light–emitting devices using AlP/GaP superlattice. Isoelectronic centers such as nitrogen also have studied in this alloy system to improve the radiative recombi–nation rate. Although strong luminescence can be achieved with heavy nitrogen doping, the luminescence peak shifts longer wavelength because NN pair recombination centers are made at high nitrogen concentration. In order to obtain pure green light–emitting devices, it is effective to introduce the nitrogen into the AlGaP layers and/or AlP/GaP superlattices.

OMVPE is a powerful technique for the growth of superlattices. The epitaxial growth of P–containing materials has been carried out by using PH_3. Since the decomposition temperature of PH_3 is relatively high, high temperature growth is required to obtain a high quality AlP/GaP superlattices. We have demonstrated high quality $Al_xGa_{1-x}P$ ($0 \leq x \leq 1$) layers and AlP/GaP short period superlattices grown by OMVPE using $(NH_4)_2S_x$–treated substrate and tertiarybutylphosphine (TBP), simultaneously (Wakahara 1992, Wang 1993). However, few studies have made for the nitrogen doping of the GaP OMVPE layer using TBP. In this paper, we study the basic characteristics of the nitrogen doping into the GaP layer grown by OMVPE using TBP.

2. EXPERIMENT

The epitaxial growth was carried out in a conventional atmospheric pressure OMVPE system with RF–heated horizontal reactor. Trimethylgallium (TMGa) was used as the group III source, and TBP and NH_3 were used as the group V sources. Pd–diffused H_2 was used as the carrier gas. Sulfur–doped n^+–GaP(100) wafers were used as substrates in the experiment. After the degreasing process, the substrates were etched in a hot HNO_3–HCl–H_2O solution (1:2:2 at 60°C) for 2 min. Then the substrates were dipped in the $(NH_4)_2S_x$ solution. With the $(NH_4)_2S_x$ treatment, the oxide film can be removed from chemically etched substrates and the surface is protected from further oxidation by the formation of Ga–S bonds (Fan 1988). The detail of the $(NH_4)_2S_x$ treatment was previously described (Wakahara 1992, Wang 1993). The growth conditions are summarized in table 1. The growth temperature was 720°C. The partial pressure of NH_3 was varied from 5×10^{-3} to 4×10^{-2} atm.

The epitaxial layers were characterized by Nomarski–interference microscopy observation and photoluminescence (PL) measurement. In the PL measurement, Ar^+ laser operating in an ultraviolet region was used as the excitation source to suppress the luminescence from the substrate. The samples were cooled by using He gas flow cryostat.

Table 1. Epitaxial growth conditions of nitrogen–doped GaP.

SUBSTRATES	GaP:S (100)
TOTAL GAS FLOW RATE	0.22 mol/min.
TMGa FLOW RATE	10 μmol/min.
TBP FLOW RATE	50~200 μmol/min.
V/III RATIO	5~20
NH_3 FLOW RATE	1~9 mmol/min.
GROWTH TEMPERATURE	720°C

3. RESULTS

Figure 1 shows the Nomarski–interference photomicrographs of the nitrogen–doped GaP epitaxial layer. The surface morphology is specular at the entire growth conditions. In the case of the hydride VPE, it has been reported that the growth rate of nitrogen–doped GaP layer is decreased by introducing NH_3 (Stringfellow 1975). Similar behavior has been reported for the PH_3–used OMVPE (Roehle 1981). However, the growth rate was not affected by the presence of NH_3 in this experiment. Our results are different from that for the OMVPE using PH_3.

Figure 2 shows the low temperature (10K) PL spectra of the nitrogen–doped and undoped GaP layers. For the undoped layer, a very week luminescence which may be assigned as the S–C related donor–acceptor pair (DAP) recombination was observed. On the other hand, some strong sharp lines labeled A and NN_x (x=1,3,..) were observed for the nitrogen doped layer. The A line and NN_x lines correspond to the luminescence from excitons bound at an isolated nitrogen atom and nitrogen pairs of the xth nearest on P lattice site, respectively (Thomas 1966). Luminescence intensity of the NN_1 line was larger than that of the A line, which is expected to be indicative for a high nitrogen concentration. Observed full width at half maximum (FWHM) of the A line is 3.5meV which is almost same as the resolution of the spectrometer (ΔE=4.3meV). These results suggest that the high quality nitrogen–doped GaP layer has been obtained.

Figure 3 shows the nitrogen concentration as a function of NH_3 partial pressure. The nitrogen concentration was calculated by using the intensity ratio of I_A/I_{NN1} (Thierry–Mieg, 1983). As can be seen from Fig.3, the doping concentration of nitrogen atom in the GaP layer is the order of $10^{18}cm^{-3}$. The nitrogen concentration increases with NH_3 partial pressure but tends to saturate. The nitrogen concentration also increases with decreasing V/III ratio. The maximum nitrogen concentration obtained in this experiment

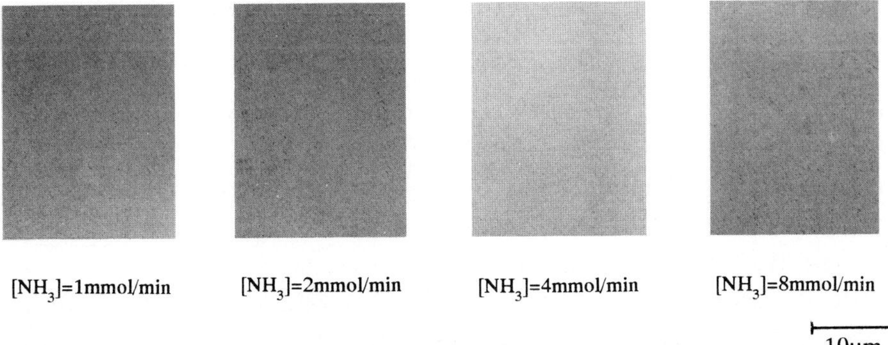

Fig.1. Surface morphology of the nitrogen–doped GaP epitaxial layer. V/III=10.

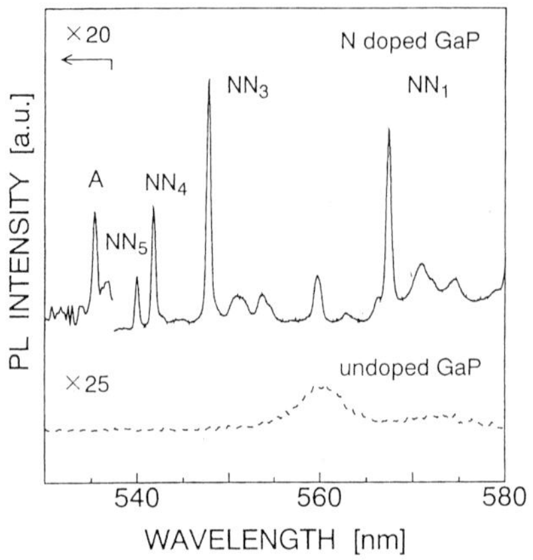

Fig.2. PL spectra of undoped and nitrogen-doped GaP layers. The nitrogen doped layer exhibits much stronger luminescence compared with the undoped layer.

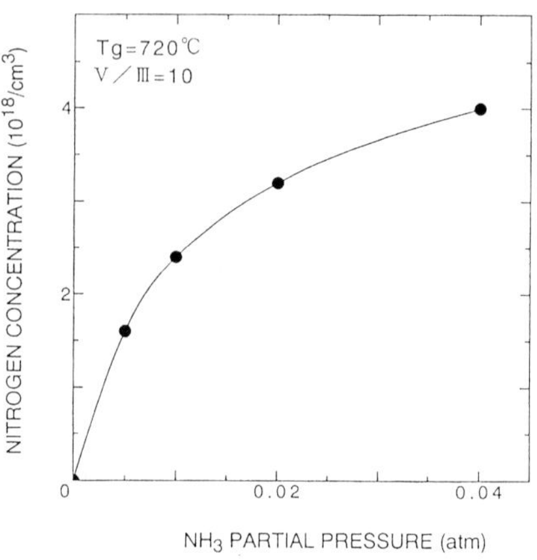

Fig.3. NH_3 partial pressure dependence of nitrogen concentration. V/III ratio was 10.

was $6x10^{18}$ cm^{-3} at V/III ratio of 5. Figure 4 shows the nitrogen distribution coefficient. The nitrogen distribution coefficient was defined as X_N^S/X_{NH3}^g. Here, X_N^S is nitrogen content in GaP layer and X_{NH3}^g is NH_3 composition in vapor phase. The circles indicate the present results. The squares indicate the distribution coefficients for PH_3-used OMVPE which have reported by Rohele and Beneking (Roehle 1981). They grew a nitrogen-doped GaP layer by OMVPE using pre-cracked PH_3 and achieved a high quality GaP layer grown at low V/III ratio. Their growth conditions, especially low V/III ratio, are similar to that used in this experiment. The nitrogen incorporation efficiency increases by decreasing V/III ratio. Similar behavior has also observed for the nitrogen doping of GaP grown by using PH_3 (Roehle 1981). However, the nitrogen incorporation efficiency for TBP OMVPE is about 100 times larger than that for PH_3

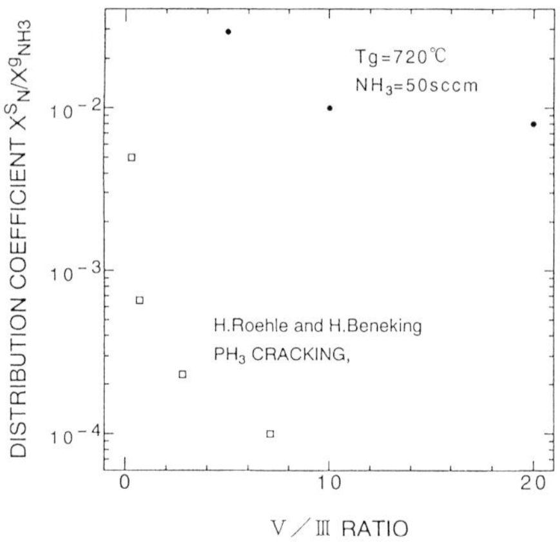

Fig.4. V/III ratio dependence on the nitrogen concentration. The NH_3 flow rate was fixed as 2.2 mmol/min. The nitrogen distribution coefficient was defined as X_N^S/X_{NH3}^g. Here, X_N^S is nitrogen content in GaP layer and X_{NH3}^g is NH_3 composition in vapor phase.

OMVPE.

4. DISCUSSION

As described above, the nitrogen incorporation efficiency of GaP layer becomes larger by using TBP compared with PH_3. This effect may be attributed to that the growth reaction in TBP–used OMVPE is different from that in PH_3–used OMVPE. Since the decomposition temperature of PH_3 in vapor phase is higher than 800°C, the dominant decomposition process of PH_3 around 720°C is catalytic reaction at the substrate surface (Stringfellow 1989). The decomposition process of NH_3 may also be catalytic reaction because NH_3 is thermally stable material. The decreasing of the growth rate observed in PH_3–used VPE and OMVPE is explained with blocking effect by NH_3 molecules adsorbed on growth sites (Stringfellow 1975, Roehle 1981). On the basis of this model, decomposition rate of NH_3 is considered to be lower than that of PH_3 because of low nitrogen incorporation efficiency. That is, lower decomposition rate leads to lower incorporation efficiency. On the other hand, TBP is decomposed in vapor phase about 500°C. By this vapor phase pyrolysis, many radicals such as PH_x(x=1,2) are generated from TBP (Stringfellow 1989). In the OMVPE using TBP, PH_x radicals are considered as the P species of the growth reaction. Generally, radicals indicate a high chemical activity. Therefore, these radicals may assist the catalytic reaction of NH_3 by which reactive nitrogen species are supplied. Once reactive nitrogen species are generated, nitrogen incorporation efficiency could be improved.

4. CONCLUSIONS

We have studied the nitrogen–doping of GaP layer by OMVPE using TBP. The high quality nitrogen–doped GaP epitaxial layer with a specular surface has been obtained. Photoluminescence spectra have indicated the sharp A and NN_x lines related with nitrogen isoelectronic center. The nitrogen concentration calculated from the intensity ratio I_A/I_{NN1} has been the order of $10^{18}cm^{-3}$, and the large incorporation efficiency has been achieved by using TBP as compared with that by PH_3.

ACKNOWLEDGMENT

The authors wish thank Professors Y.Takeda and S.Noda for helpful suggestion. This work was supported in part by the Kurata Research Grant, and it was also supported in part by Scientific Research Grant–in–Aid No.04452175 from the ministry of education, Science and Culture of Japan.

REFERENCES:

Fan, J.F., Oigawa,H., and Nannichi, Y., 1988, Jpn. J. Appl. Phys. **27** L1331.

Kumagai, M., Takagahara, T., and Hanamura, E., 1988, Phys. Rev. **B37** 898.

Roehle, H., and Beneking, H., 1982, Inst. Phys. Conf. Ser. No.**63** 119.

Schuumans, M.F.H., Pompa, H.W.A.M., and Eppenda, R., 1987, J. Luminescence **37** 269.

Stringfellow, G.B., 1989, Organometallic Vapor Phase Epitaxy, Academic Press, Inc.

Stringfellow, G. B., and Weiner, M. E., and Burmeister, R. A., 1975, J. Electron. Mater. **4** 363.

Thierry–Mieg, V., Marbeuf, A., Chevallier, J., Mariette, H., Bugajski, M., and Kazmierski, K., 1983, J. Appl. Phys. **54** 5358.

Thomas, D.G., and Hopfield, 1966, Phys. Rev. **150** 680.

Wakahara, A., Wang, X–L., and Sasaki, A., 1992, J. Cryst. Growth **124** 118.

Wang, X–L., Wakahara, A., and Sasaki, A., 1993, J. Cryst. Growth **129** 289.

Metalorganic vapour phase epitaxy of III/V-semiconductors using alternative metalorganic-group-V-compounds decomposing under *in-situ* formation of group-V-H-functions

G Zimmermann[a], Z Spika[a], W Stolz[a], E O Göbel[a], P Gimmnich[b], J Lorberth[b], A Greiling[c], A Salzmann[c]

Wiss. Zentrum für Materialwissenschaften (WZMW) und Fachbereiche Physik[a] und Chemie[b], Philipps-Universität, D-35032 Marburg, Germany
[c]sgs mochem products GmbH, D-35032 Marburg, Germany

ABSTRACT: The key feature of the novel trialkyl-group-V-precursors is the in-situ formation of group-V-H-functions in the hot temperature zone of the MOVPE reactor by the ß-elimination process, as verified by decomposition studies of diethyltertiarybutyl-arsine (DEtBAs) under MOVPE growth conditions. The results for the optimization of the MOVPE growth of GaAs using several batches of different compounds in combination with standard TMGa as determined by use of the temperature-dependent Hall- and photoluminescence (PL) technique are presented and discussed. Reproducible residual p-type-doping levels of 3-5*10^{15} cm^{-3} with hole mobilities of up to 390 cm^2/Vsec at 300 K are observed, Mg being the dominant acceptor impurity at present as determined by PL spectroscopy. Deep level photoluminescence studies show a reduction of the deep defect EL2 as compared to AsH_3-grown layers, due to the smaller V/III ratio.

1. INTRODUCTION

The high toxicity of AsH_3 and PH_3, standardly used as group-V-precursors in metalorganic vapour phase epitaxy (MOVPE) causes severe safety risks and restrictions in particular for large scale production facilities for III/V-semiconductor based devices. A significant reduction of the toxicity of metalorganic arsenic precursors is observed, if all group-V-H functions are replaced by organic groups (Stringfellow 1990). However, the use of the As-trialkyl sources trimethylarsine (TMAs) and triethylarsine (TEAs) leads to a high carbon incorporation into the grown layers (see for example Kuo (1983), Lum (1988a) and Lum (1988b)) and the use of trisdimethylaminoarsine (TDMAAs) is limited to aluminum-free structures, because an enormous amount of nitrogen incorporation is reported for the growth of (AlGa)As layers by Zimmermann (1993a).
Recently, a novel class of specific liquid trialkyl-group-V-compounds has been proposed by Zimmermann (1992). The key feature of these molecules with a principle molecular structure of $As(R_{rad})_n(R_ß)_{3-n}$ (n=0,1,2; R_{rad}= alkyl group decomposing under radical formation, $R_ß$= alkyl group decomposing under ß-elimination) is the in-situ formation of group-V-H-functions only in the hot temperature zone of the MOVPE reactor. Thus, this new class of precursor materials combines the lowest possible toxicity of any MO-group-V-compound with the capability of forming group-V-hydride functions, needed for high quality MOVPE growth, in-situ in the MOVPE reactor by thermal decomposition under ß-elimination. Therefore, the usage of these compounds will

tremendously reduce the safety risks and related costs for chemical synthesis, transport, handling and storage for the MOVPE process.
A ß-elimination decomposition of the source diethyltertiarybutyl-arsine (DEtBAs) as a model precursor for this class of molecules was reported by Zimmermann (1992) for increased temperatures in a high vacuum ersatz reactor system, by the detection of the cracking products diethyl-arsine (DEAsH) and isobutene (C_4H_8). To clarify, that this mechanism also takes place under normal MOVPE growth conditions, thermal decomposition studies of various As- as well as P-precursors in the gas phase of the MOVPE reactor have been performed. In this paper, we will discuss the decomposition behaviour of the DEtBAs source under normal growth parameters. Furthermore, the influence of isobutene on the growth process of GaAs epilayers has been studied. The electrical and optical properties of GaAs epilayers, grown by using several batches of the compounds DEtBAs and diethylisopropyl-arsine (DEiPrAs), as well as the result of deep level photoluminescence studies on DEtBAs-grown layers are presented and discussed.

2. EXPERIMENTAL

Both, thermal decomposition studies and MOVPE growth experiments have been performed in a commercial horizontal low pressure MOVPE equipment (Aix 200, Aixtron Corp.) at 50 mbar reactor pressure by using palladium-diffused hydrogen as carrier gas. The decomposition of the precursors has been studied in the temperature range of 30 to 650°C by a quadrupole mass spectrometer (QMS) (BALZERS QMG 511, mass range: 4 to 1024 amu) with cross beam ion source, operating at 40 eV electron ionization energy and a faraday cup as detection system. This setup could be evacuated by a diffusion pump system down to a base pressure of $3*10^{-6}$ mbar. The reactor gas was sampled at a height of 1 cm above the susceptor and introduced into the ionization chamber via a pipe of 6 mm diameter, consisting of quartz in the heated region of the MOVPE reactor and electro-polished stainless steel outside the reactor chamber. The pressure in the QMS system could be controlled by a needle valve.
For MOVPE growth experiments the As-precursors DEtBAs, DEiPrAs and DEAsH were synthesized and used in combination with standard TMGa in the temperature range of 600 to 675°C by varying the V/III ratio between 8 and 30. The influence of great amounts of isobutene in the gas phase on the layer quality has been investigated by using a ratio of isobutene to AsH_3 in the range of up to 2:1. The epilayers were grown on semi-insulating (100) 2° off [011] oriented GaAs wafers with a growth rate of 1μm/h. The GaAs substrates, prepared according to a procedure described by Fronius (1985) and Contour (1988), are cleaned prior to growth in an AsH_3 atmosphere at a temperature of 750°C for 5 min, in order to establish comparable substrate surface conditions for the various tested alternative precursor compounds.
The epitaxial layer quality has been characterized electrically by means of the standard temperature-dependent Hall technique, using a correction for the surface and interface depletion of the GaAs samples. Photoluminescence (PL) spectroscopy has been performed, using an Ar ion laser (E = 2.41 eV) as excitation source and a GaAs cathode photomultiplier. The luminescence of deep levels in the layers was detected with a PbS-photodiode.

3. RESULTS AND DISCUSSION

3.1 Decomposition studies

In this section, the thermal decomposition of the As-precursor DEtBAs is presented in detail to verify, that the ß-elimination mechanism is dominant under MOVPE conditions.

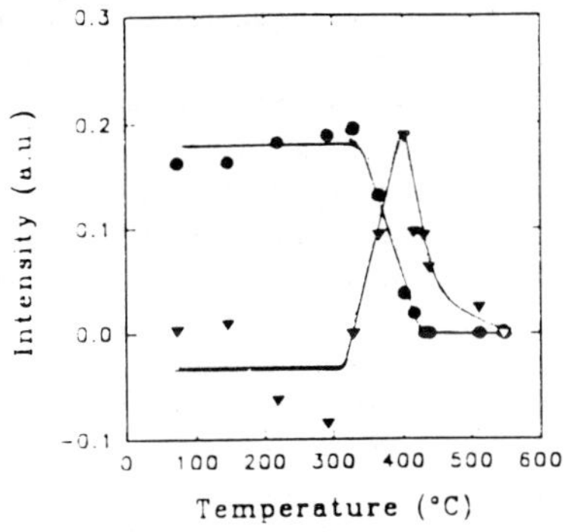

Fig. 1 Mass peak intensity of DEtBAs (•) and the main cracking product DEAsH (▼) versus decomposition temperature.

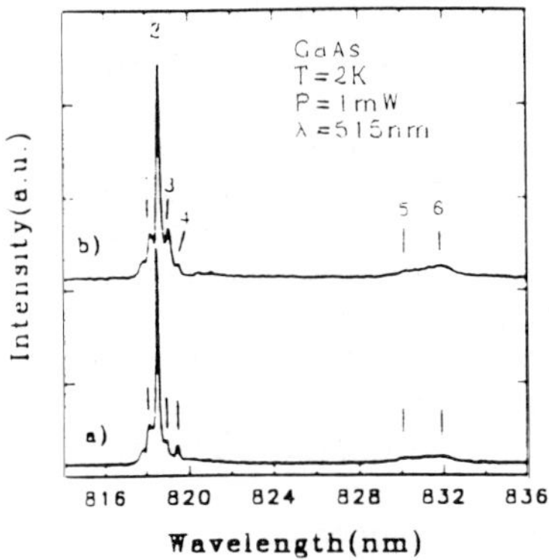

Fig. 2 Photoluminescence spectra at 2K with an excitation power of 1mW for an isobutene/AsH_3-grown (a) and a standardly grown layer (b). The transitions are: 1-(X), 2-(D°X), 3-(D^+X), 4-(A°X), 5-(e,$A°_C$), 6-(D°,$A°_C$)

The DEtBAs molecule mass intensity at m/e = 190 amu and the mass intensity of the dominant cracking product DEAsH (m/e = 134) are depicted in Figure 1 with respect to the temperature. To have a correct interpretation of the mass spectra, the part of the signals has been substracted in Figure 1, that is only due to the formation of cracking products after electron bombardement in the QMS by taking into account the room temperature spectra of the molecule. A significant decrease of the DEtBAs mass intensity and an increase of the DEAsH- and isobutene- (not shown in Figure 1) related signals is clearly observed for temperatures higher than 350°C, as is expected for a ß-elimination decomposition. At temperatures above 400°C, a decrease of the DEAsH-peak indicates a thermal decomposition of this molecule. A radical abstraction of the tertiarybutyl-, as well as of one ethyl group has to be assumed also, because of an increase of the mass intensities for the DEAs• (m/e = 133)- and ETBAs• (m/e = 161)-radical (not shown in Figure 1). However, the ß-elimination process dominates the radical decomposition pathways by a ratio of about 3:1, as can be estimated from the observed mass intensities. A similar decomposition process is also observed for diethyltertiarybutylphosphin (DEtBuP), leading to diethylphosphine (DEPH) and isobutene. Further detailed experiments, performed also with other As- and P-compounds out of the class of ß-eliminating trialkyl-V compounds will be presented elsewhere (Zimmermann 1993b).
These results, that prove the dominant formation of DEAsH and isobutene by thermal ß-elimination decomposition of the DEtBAs molecule at temperatures above 350°C under MOVPE conditions are in good agreement with calculations, reported by Foster (1993), indicating, that for the primary arsine $TBAsH_2$ the ß-elimination process has a lower activation energy than the radical abstraction.

3.2 Influence of isobutene on the epitaxial growth

MOVPE experiments at 650°C have been performed using TMGa, AsH_3 and isobutene (C_4H_8) to examine, whether or not the formation of isobutene influence the crystal growth mechanism. As a result, the isobutene/AsH_3 grown layers exhibit excellent surface morphologies and no change of the growth rate could be observed, as compared to standardly grown layers. In Figure 2 the PL spectrum at 2K of an isobutene/AsH_3-

grown (ratio 2:1) GaAs layer (a) is compared to a spectrum of a standardly grown layer (b). The identification of the observed peaks, as can be obtained from the literature (Hamilton 1990) is as follows: 1) free exciton (X), 2) neutral donor bound exciton (D°X), 3) charged donor bound exciton (D^+X), 4) neutral acceptor bound exciton (A°X), 5) band to acceptor carbon ($e,A°_C$) and 6) donor to acceptor carbon ($D°,A°_C$) transition. Both layers show no significant difference in the PL spectrum. A dominant sharp line at the D°X transition is observed in the near gap region of GaAs. There are only weak intensities for the acceptor-related transitions (A°X), ($e,A°_C$) and ($D°,A°_C$). This indicates, that only small amounts of carbon are incorporated in both layers. The electrical properties of these samples confirm the results, obtained from the PL spectra. Therefore, no influence of isobutene on the epitaxial layer quality can be detected.

3.3 Electrical properties

The background doping level of DEtBAs- and DEiPrAs-grown GaAs layers have been improved by changing the chemical synthesis route of these precursors (Gimmnich 1993) from values of $1*10^{18}$ cm^{-3}, n-type to $3*10^{15}$ cm^{-3}, p-type. In this study, the influence of different As-precursors, synthesized by using similar chemical routes, on the properties of the epilayers has been investigated. The mobility μ (Figure 3a) and the residual carrier concentration p (Figure 3b) at 300 K for GaAs-layers, grown by using DEtBAs(o), DEiPrAs(o) and DEAsH(o) in combination with TMGa are shown as a function of the growth temperature. There are no significant differences in the electrical properties by changing the As-precursor or the substrate temperature. All layers exhibit p-type behaviour with hole mobilities between 350-460 cm^2/Vs and carrier concentrations around $3\text{-}6*10^{15}$ cm^{-3}, indicating a small compensation ratio N_D/N_A. For 77K, the hole mobilities ranged between $\mu=1300\text{-}3800$ cm^2/Vs and the carrier concentration was around $p=1\text{-}3*10^{15}$ cm^{-3}.

3.4 Optical properties

PL studies at 8K in the region between 817 and 836 nm have been carried out to clarify, which kind of impurities are responsible for the detected p-type doping level of the grown layers. The PL intensity of a GaAs layer, grown with DEtBAs and TMGa at 650°C with a V/III ratio of 16 is depicted in Figure 4 on a logarithmic scale versus wavelength for the excitation power of 5, 10 and 20 mW, respectively. The observed transitions are numbered, indicating the transitions (Hamilton 1990) 1) free exciton (X), 2) neutral donor bound exciton (D°X), 3) charged donor bound exciton (D^+X) and 4) neutral acceptor bound exciton (A°X), as well as 5) band to acceptor carbon ($e,A°_C$). Independent of excitation power, the dominating peaks are the acceptor-related transitions 5) and 6). The transition at 831.6 nm (line 6) is assigned to a band to acceptor magnesium recombination ($e,A°_{Mg}$), based on Mg-detection in the DEtBAs source by atom absorption spectroscopy (AAS) and a comparative study of intentionally doped standardly grown GaAs:Mg using bis(methylcyclopentadienyl)-Mg as a doping source. In the excitonic region only the (A°X)-related peak is well resolved. These PL results indicate a dominating acceptor incorporation into the epilayers, in agreement to the Hall transport studies. Carbon is assumed to result from the decomposition of TMGa and magnesium is an impurity of the ß-eliminating compounds, because in the present synthesis route Grignard-Mg-reagencies are used (Gimmnich 1993). We observe similar spectra for DEiPrAs-grown layers. For DEAsH a significantly lower content of Mg, due to the used different synthesis route for DEAsH was detected by PL. Additional purification steps of the ß-eliminating precursors with respect to Mg are underway, to further reduce the impurity level in the epitaxial layers.

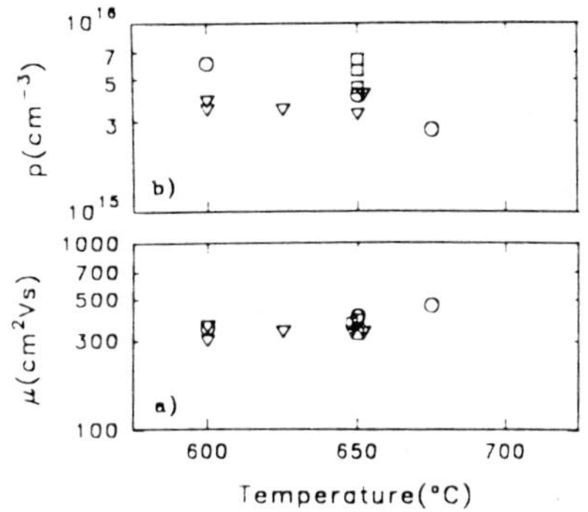

Fig. 3 Mobility μ (a) and net carrier concentration p (b) versus growth temperature at 300 K for GaAs layers, grown by using DEAsH (**o**), DEtBAs (▼) and DEiPrAs (**□**) at different V/III ratios.

Fig. 4 Photoluminescence spectra of a DEtBAs-grown GaAs layer (T=650°C, V/III = 16) at the excitation power of a) 5 mW, b) 10 mW and c) 20 mW.

3.5 Photoluminescence of deep levels in GaAs

PL studies in the wavelength range between 900 and 2400 nm have been performed to investigate the influence of the growth conditions on the formation of deep traps in the epilayers by using the ß-eliminating compounds. The results are compared to those of standardly grown layers with similar background doping levels. The EL2-PL band at around 1900 nm is detected to be the dominant deep defect. No significant differences in the formation of the EL2 defect could be observed for samples, grown at different temperatures. However, defect concentrations are reduced significantly, as compared to standard layers grown by using AsH_3. This behaviour is explained by the smaller V/III ratio (Watanabe 1983), used for the MOVPE growth with the alternative sources. The decrease of the EL2 PL intensity with decreasing V/III ratio is shown in Figure 5 for DEtBAs-grown layers at V/III ratios of 30, 15 and 7.5. The integrated intensity of the EL2-PL band at 1900 nm is depicted versus V/III ratio in Figure 6 for DEtBAs- and AsH_3-grown layers, also proving this clear trend. From deep level transient fourier spectroscopy (DLTFS) studies on these layers, EL2 concentrations around $1*10^{12}$ cm^{-3} for DEtBAs-grown layers are observed (Spika 1993).

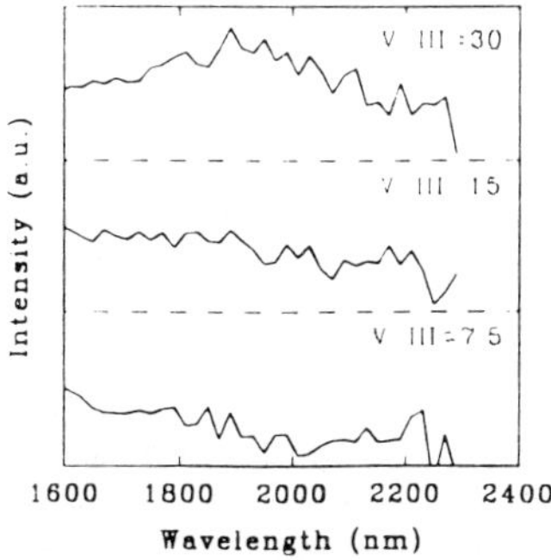

Fig. 5 Deep level photoluminescence spectra of GaAs layers, grown by using DEtBAs and TMGa at a substrate temperature of 650°C and with different V/III ratios of 7.5 up to 30.

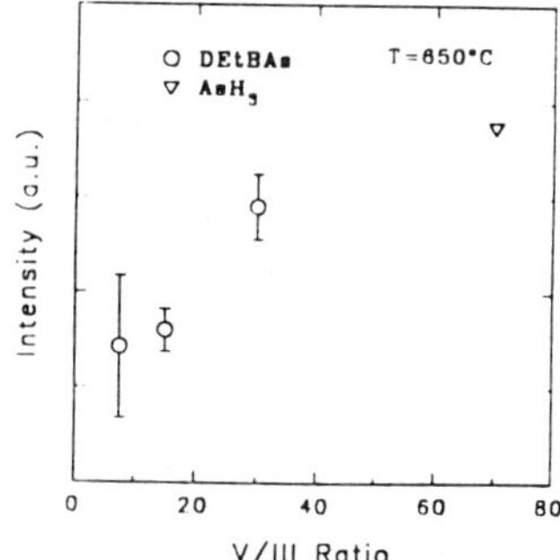

Fig. 6 Integrated intensity of the EL2-related luminescence band at 1900 nm as a function of V/III ratio for layers, grown by DEtBAs and AsH_3, respectively.

4. CONCLUSION

Alternative group V compounds, that can decompose under ß-elimination at increased temperatures have been used for MOVPE growth studies. The thermal decomposition for the compound diethyltertiarybutyl-arsine (DEtBAs) clearly show the dominant ß-elimination decomposition under MOVPE growth conditions. Thus, the cracking product DEAsH is used as actual precursor in the MOVPE. Isobutene, the other cracking product, has no detectable influence on the growth process. The electrical properties of GaAs layers, grown by several batches of different precursors, show reproducible p-type-doping levels around $4*10^{15}$ cm^{-3} with hole mobilities of up to 390 cm^2/Vs at 300K, independent from the growth temperature in the range of 600 to 700°C. The dominant impurities are detected by PL to be Mg, due to the present chemical synthesis route of the As-compounds and C, supposed to result from the TMGa decomposition. Additional purification steps are underway to further reduce the impurity level. A lower concentration of the deep level EL2, compared to standardly grown layers is observed by PL- and DLTFS studies, when using the alternative As-sources, because of the chosen lower V/III ratio.

These results clearly prove the great potential of this novel class of group-V-compounds as substitutes for the highly toxic group-V-hydrides. Already at the present stage of development, the achieved crystalline quality should allow the application of these compounds for the growth of light-emitting diode (LED) device structures.

5. ACKNOWLEDGEMENTS

The authors would like to thank T. Ochs for expert technical support during MOVPE. This work has been sponsored by the "Bundesministerium für Forschung und Technologie (BMFT)" of the Federal Republic of Germany under contract number TK 0456/4.

REFERENCES

Contour J P, Massies J, Fronius H and Ploog K 1988 Japan. J. Appl. Phys. **27** L167.

Foster D F, Glidewell C and Cole-Hamilton D J 1993, unpublished.

Fronius H, Fischer A and Ploog K 1985 Japan. J. Appl. Phys. **25** L137.

Gimmnich P, Greiling A, Lorberth J, Thalmann C, Rademann K, Zimmermann G, Protzmann H, Stolz W and Göbel E O 1993 Mater. Science and Engineering **B17** 21.

Hamilton B in Properties of Gallium Arsenide, 2nd edition, INSPEC, London, New York 1990 246 and references therein.

Kuo C P, Cohen R M and Stringfellow G B 1983 J. Crystal Growth **64** 461.

Lum R M, Klingert J K, Kisker D W, Tennant D M, Morris M D, Malm D L, Kovalchik J and Heimbrook L A 1988a J. Electron. Mater. **17** 101.

Lum R M, Klingert J K and Lamont M G 1988b J. Crystal Growth **89** 137.

Spika Z, Zimmermann G, Stolz W, Göbel E O, Weiss S, unpublished

Stringfellow G B 1990 J. Crystal Growth **105** 260.

Watanabe M O, Tanaka A, Udagawa T, Nakanisi T and Zohta Y 1983 J. J. Appl. Phys. **22** 923

Zimmermann G, Protzmann H, Marschner T, Zsebök O, Stolz W, Göbel E O, Gimmnich P, Lorberth J, Filz T, Kurpas P and Richter W 1993a J. Crystal Growth **129** 37.

Zimmermann G, Protzmann H, Stolz W, Göbel E O, Gimmnich P, Greiling A, Lorberth J, Thalmann C and Rademann K 1992 J. Cryst. Growth **124** 136.

Zimmermann G et al. 1993b, unpublished.

Coherency limits of tetragonal III–V In-containing alloys on GaAs and InP substrates

B.L. Pitts*, M.J. Matragrano†, D.T. Emerson*, B. Sun*, D.G. Ast† and J.R. Shealy*

OMVPE Facility, Cornell University, Ithaca, N.Y. 14853, USA

* School of Electrical Engineering, Cornell University

† Department of Materials Science and Engineering, Cornell University

Abstract. The coherency range for the flow modulation OMVPE growth of GaInP on GaAs was investigated. Excellent surface morphology was realized over a 10% In mole fraction range with films exceeding the Matthews-Blakeslee critical thickness. X-ray diffraction was used to measure the tetragonal distortion of epitaxial films. High quality photoluminescence spectra and low densities of dark line defects associated with misfit dislocations indicate that these tetragonal films are of device quality. Similar behavior of GaInAs films on InP was observed over a 6% In composition range. Tetragonal distortion of GaInAsP/GaInP structures on GaAs was also investigated and compared to GaInP.

I. Introduction

High purity III-V compound semiconductor materials systems have supported many high speed and Optoelectronic device applications. Generally, for GaAs and InP based material systems, strict lattice matching conditions must be enforced in order to eliminate misfit dislocations, which degrade device performance. The epitaxial film's lattice parameter must be carefully controlled to either match that of the substrate, for the case of thick films, or nearly match the substrate lattice parameter, for the case of thin pseudomorphic films. It appears, however, that the pseudomorphic limit can be relaxed for several In-containing compounds using organometallic based epitaxial processes (Ozasa et al. 1990, Lee et al. 1992 and Kamei et al. 1988). In this paper, we report the coherency range of OMVPE grown GaInP and GaInAsP/GaInP structures on GaAs, and GaInAs on InP. For GaInP, we observe excellent surface morphology over a 10% Indium mole fraction range for films greater than the critical thickness (Matthews 1975). Dark line density dependency on lattice mismatch was examined over the entire coherency range. The tetragonal distortion of GaInAsP/GaInP structures and GaInP on GaAs was also investigated. It was observed that GaInAsP structures can accommodate a much wider range of tetragonal distortion than GaInP.

II. Experimental

Epitaxial layers were produced by Flow Modulation Epitaxy (Kobayashi et al. 1989) using a vertical barrel multichamber OMVPE system described elsewhere (Pitts et al. 1993). Substrates were rotated through group III spatially separated zones in a group V background without valve switching. The substrate exposure cycle, based on experimental thickness data, for GaInP growth is shown in Figure 1. The exposure cycles varied from 2 to 5 monolayers/cycle. Film thickness ranged from 0.4 to 1.0 μm. Triethylgallium, trimethylindium (TMI), arsine and phosphine were used as sources and the substrates were (100) GaAs and (100) InP. The growth temperature for all experiments ranged from 550 to 675 °C. The reactor chamber was cleaned by heating the quartz reaction cell to 900 °C (hot wall), while injecting HCl into the chamber. For the growth of GaInAsP on GaAs, a 400 Å GaInP buffer layer was necessary to achieve good surface morphology. An ultrasonic analyzer, placed downstream from the TMI bubbler, was used to monitor and regulate the TMI bulk flow for GaInAsP films. Double crystal X-ray diffraction was used to determine the lattice deformation. Measurements were performed on (004) and {224} planes to calculate the normal and parallel lattice mismatch. Rutherford Backscattering Spectrometry (RBS) was used to determine alloy composition, as well as layer thickness. The mobility and carrier concentration were measured using the standard van der Pauw method. Optical quality was assessed using both low temperature (1 K) and room temperature photoluminescence (PL). Dark line and oval-like defects were observed using a combination of Scanning Electron Microscopy (SEM) and Cathodoluminscence (CL).

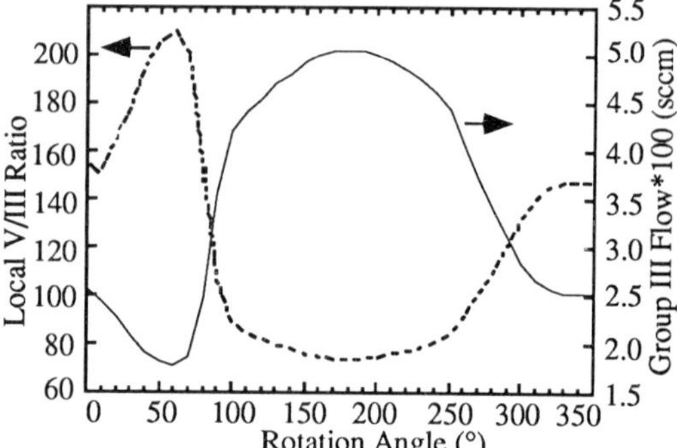

Figure 1. The substrate exposure cycle for GaInP in the multichamber reaction cell. PH_3 is uniformly injected in the cell, while substrates are rotated through a spatially separated group III rich zone.

III. Results and Discussion

The parallel lattice mismatch $\Delta a_{\|}/a_{sub}$ (where $\Delta a_{\|}=a_{\|}-a_{sub}$) dependency on In composition is shown in Figure 2a. The GaInP coherency range extends from 0.40 to 0.52 Indium composition, where lattice mismatch is accommodated by tetragonal distortion. Also shown, the relaxation occurs earlier for compressive stress than for tensile stress. This differs from Ozasa et al. who reported the opposite effect. This discrepancy may be attributed to the difference in material preparation (Ozasa et al. used Chemical Beam Epitaxy). For GaInAs on InP, preliminary data, provided in Figure 2a, shows that the coherency extends over a 6% In composition range. Low temperature PL linewidth as low as 1.7 meV and room temperature mobilities exceeding 12,500 cm^2/V s have been observed on GaInAs/InP structures. A more thorough study of the coherency range of GaInAs on InP system is currently under investigation. Figure 2b shows the dark line densities, measured using CL, over the entire range investigated. The thickness of each sample is given in terms of critical thickness (h_c). As shown in the figure, the dark line defect density increases dramatically when the system is in compression. Low dark line defect densities ($<500\ cm^{-1}$) were realized for In mole fractions ranging from 0.45 to 0.51 and the absence of dark lines were observed over a 3% composition range (see figure). Beyond the coherency range, high density striations along the (011) direction were observed

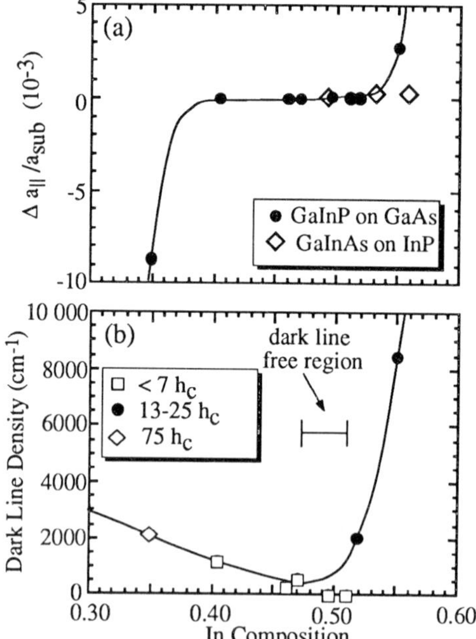

Figure 2. (a) Parallel lattice mismatch dependency on In mole fraction of GaInP on GaAs and GaInAs on InP. **(b)** Dark line defect density (using CL) of GaInP on GaAs versus In mole fraction. The film thickness are indicated in units of critical thickness. Solid curves are drawn empirically to suggest trend in data. The bar is drawn to denote the range of compositions free of dark lines.

on the $Ga_{0.45}In_{0.55}P$ layer and cross hatching was seen on the $Ga_{0.65}In_{0.35}P$ surface. The dark line density on these samples were difficult to calculate due to resolution of the electron microscope. This is evidence that these structures are partially relaxed and misfit can not be accommodated elastically. Both samples showed weak PL and no electrical conduction was observed which can be attributed to deep level traps associated with misfit dislocations (Schaus et al. 1986). The CL images of both samples are presented in Figures 3a and 3b.

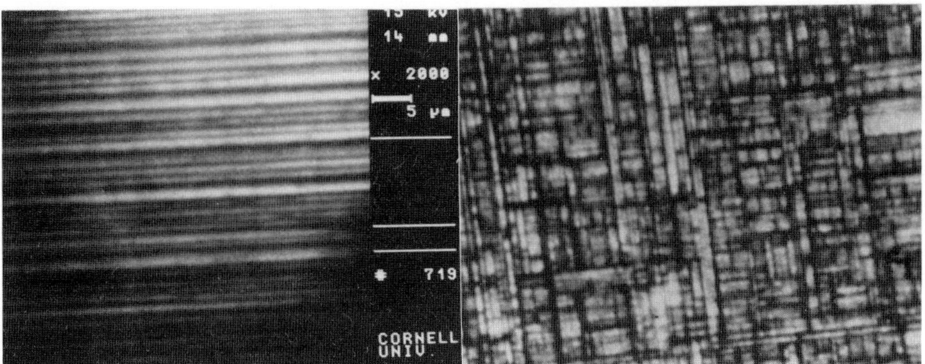

Figure 3. Cathodoluminscence images of **(a)** $Ga_{0.45}In_{0.55}P$ and **(b)** $Ga_{0.65}In_{0.35}P$ film's surface. The magnification is 2000X for each micrograph.

Figure 4a shows a SEM micrograph of a $Ga_{0.55}In_{0.45}P$ surface with oval-like defects present. These defects have been postulated by others as In rich and polycrystalline (Hageman et al. 1990). Oval defects appear after several runs, as a result of deposit build up in the reaction chamber. This effect is believed to be a result of In desorption from the reaction cell wall. The density of these oval-like defects can be as high as a few thousand per cm^2, depending on the condition of the reaction cell. This effect is even more dramatic for GaInAs, requiring that the tube be cleaned before each run. The corresponding CL image (Figure 4b) shows that the only dark lines that are present appears to be those that nucleate from the oval-like defects.

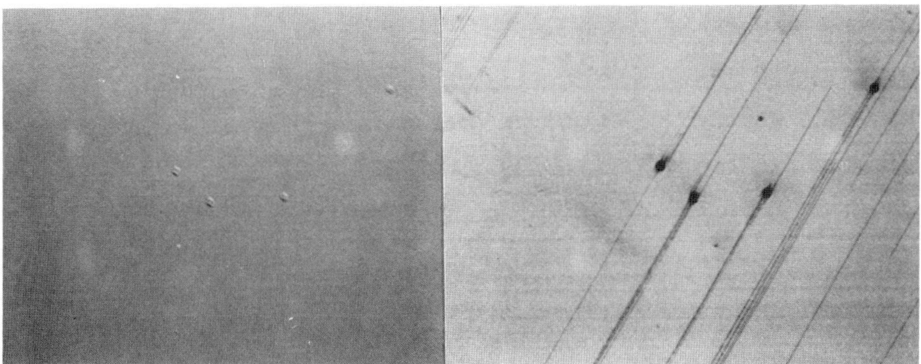

Figure 4. **(a)** SEM image and **(b)** the corresponding room temperature CL image of a coherent $Ga_{0.55}In_{0.45}P$ film's surface illustrating the absence of dark line defects associated with misfit dislocations. The dark lines present originate from oval-like defects (In rich regions at a density of roughly $1000/cm^2$ in this case. The magnification is 200X.

However, after the cell was cleaned (hot wall HCl treatment) no dark lines originating from these defects were observed on samples that were closely lattice matched to GaAs. It is speculated that if these oval defects are eliminated, no dark lines would be present in films of the same thickness and composition as the sample whose image appears in figure 4.

Room temperature PL spectra of three different compositions of GaInP within the coherency range are shown in Figure 5. Transmission electron diffraction revealed the presence of ordering in these films, which had 300 K energy gaps ranging from 1.82 to 1.92 eV. As shown in the figure, although linewidths are approximately the same, the intensity depends on alloy composition. The PL intensity decreases only slightly as the lattice mismatch is increased. This loss in radiative efficiency is most likely due to the increased density of dislocations which generate deep levels. Two dominant electron traps associated with misfit have been reported (Schaus et al. 1986). The PL linewidths are typically 30 meV for ordered alloys (26 meV is the lowest value observed, but on random GaInP) without significant broadening due to misfit as shown, and low temperature (1K) linewidths are as narrow as 10 meV. All conducting samples had carrier concentrations in the low 10^{16} cm^{-3} range and room temperature mobilities ranging from 2500 to 3200 cm^2/V s. Low temperature mobilities (77K) were as high as 14,000 cm^2/V s on undoped GaInP.

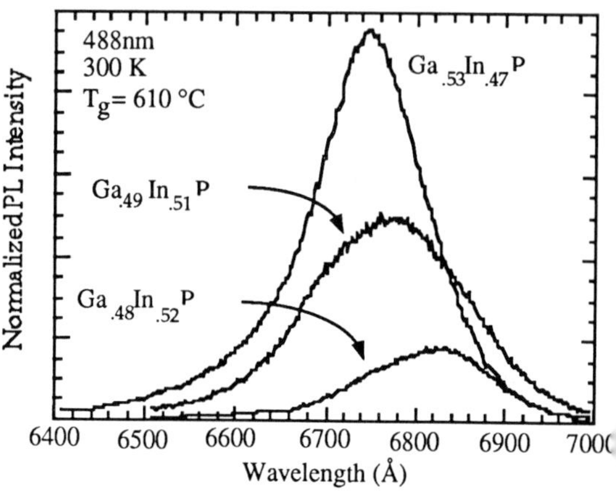

Figure 5. Room temperature PL spectra of ordered GaInP layers within the coherency range. The alloy compositions and experimental conditions are provided in the figure.

Following Lee J et al., the parallel lattice parameters of the of GaInAsP and GaInP are plotted versus their perpendicular lattice parameters in Figure 6. All GaInAsP films had thicknesses greater than 5000 Å. The degree of tetragonal distortion is represented by the difference between the distance of the data point from the $GaAs_{||}$ line and the distance of the data point from the $GaAs_{\perp}$. The GaInAsP samples had 300 K energy gaps ranging from 1.6 to 1.82 eV. We are unable to make a direct comparison to simple GaInP/GaAs structures due to presence of the GaInP buffer under the GaInAsP. The extended range of coherency for GaInAsP is very useful due to the difficulty in controlling its composition. We identify the alloy composition of GaInAsP using a combination of PL and Raman Spectroscopies. The Raman spectrum of GaInAsP is compared to that of GaInP (lattice matched to GaAs) in figure 7. We observe 3 dominant LO mode behavior with an additional feature appearing due to the GaAs sub-lattice. The position of the GaP like vibrational mode is an indicator of the degree of lattice matching, where as, the relative intensity of the GaAs like vibrational mode is a sensitive measure of the As mole fraction. Each of the features in these spectra are identified as LO phonons by their polarization behavior. We attempted to study the coher-

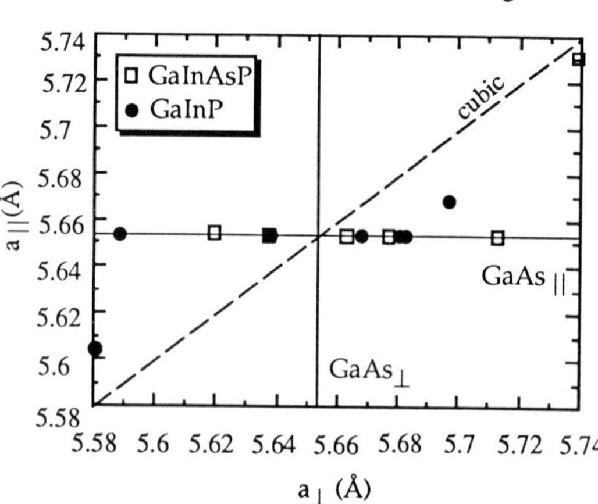

Figure 6. Comparison of normal and parallel lattice parameters of GaInP and GaInAsP/GaInP structures. The dashed line is shown to represent a cubic lattice ($a_{\perp}=a_{||}$).

ency range of GaP/InP (100) superlattices. In addition to the flow modulation scheme previously mentioned, the phosphine was also pulsed into the chamber during the GaP cycle to further enhance the surface mobility (Lee M et al. 1992). However, due to the dramatic differences in the growth temperature requirement for GaP and InP using OMVPE, we were unable to produce structures with high quality surfaces. These samples exhibit room temperature PL characterized by a broad linewidth (175 meV) and relatively weak intensity at 700 nm. The (004) symmetric X-ray diffraction showed the zeroth order diffraction peak at 1000 arc sec from the substrate peak. Based on the X-ray and growth rate data, the period of this superlattice is believed to be $(GaP)_3/(InP)_2$. However, the X-ray intensity was too weak to measure asymmetric reflections, and as a result, we were unable to determine the coherency range of GaP/InP superlattices.

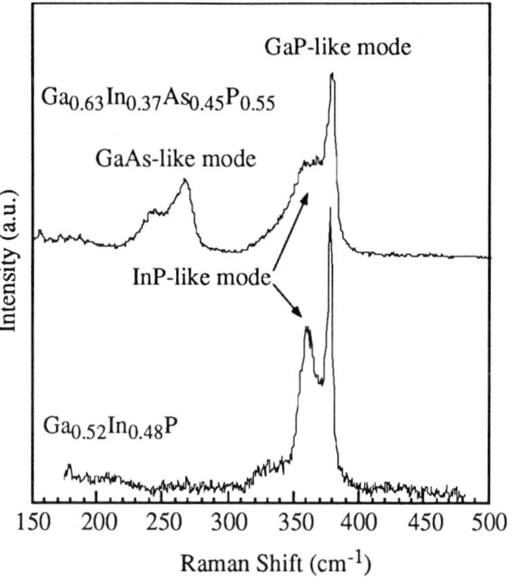

Figure 7. Raman Spectra of $Ga_{0.63}In_{0.37}As_{0.45}P_{0.55}$ and $Ga_{0.52}In_{0.48}P$ on GaAs.

VI. Conclusion

In summary, we have investigated the coherency range of GaInP and GaInAsP on GaAs and GaInAs on InP produced by Flow Modulation Epitaxy. These films were characterized by X-ray diffraction, RBS, CL, PL, and Hall measurements. The results indicate that asymmetric X-Ray diffraction is sensitive to complete lattice relaxation, but not useful in identifying whether a film is of device quality. The CL technique was used in combination with RBS to establish the range of alloy compositions in GaInP where defect free films may be produced. High quality GaInP alloys have been prepared over a 3% Indium mole fraction range which are free from both darkline and oval-like defects. The removal of the oval-like defects require that the OMVPE reaction cell be cleaned between each run. We have demonstrated that in-situ HCl, hot wall cleaning between runs is an effective means to eliminate these defects. Although both ordered and random alloys of GaInP were studied, a systematic trend on the influence of atomic ordering on the lattice matching conditions was not observed. Future studies investigating this and the coherency range of GaInAsP alloys on InP is of primary importance prior to the development of devices based on thick tetragonal epitaxial films. Finally, as a result of these studies, device quality GaInP/GaAs epitaxial structures with greater range of alloy compositions can be exploited for new electronic and Optoelectronic devices.

V. Acknowledgements

The authors wish to thank G.F. Redinbo for the CL measurements, and K.L Whittingham and B.P. Butterfield for technical assistance. This work was supported by the Joint Services Electronics Program under Grant No. F49620-90-C-0039, the BMDO/IST Electronic/Optical Materials Program under contract No. N00014-89-J-1311, and the Advanced Research Projects Agency-Optoelectronics Technology Center under contract No. MDA97290C0058.

VI. References

Hageman P R, van Geelen, Gabielse W., Bauhuis L.J. Giling 1992 J. of Crystal Growth 125 336.
Kamei H, Hashizume K, Murata M, Kuwata N, Ono K, and Yoshida K 1988 J. Crystal Growth 93 329.
Kobayashi N, Makimoto T, Yamauchi Y, and Horikoshi 1989 J. Appl. Phys. 66 640.
Lee J, Mayo W E, and Tsakalakos T 1992 J Electronic Materials 21 867.
Lee M K, Horng R H and Haung L C 1992 J of Crystal Growth 124 358.

Matthews J 1975 J. Vac. Sci. Tech. 12 126.
Pitts B L, Emerson D T, and Shealy J R 1993 Appl. Phys. Lett. 62 1821.
Ozasa K, Yuri M., Tanaka S, and Matsunami H 1990 J. Appl. Phys. 68 107.
Schaus C F, Schaff W J, and Shealy J R 1986 J. of Crystal Growth 77 360.

Suitability of N_2 as carrier in LP-MOVPE of (AlGa)As/GaAs

Hilde Hardtdegen, M. Hollfelder, Chr. Ungermanns, K. Wirtz, R. Carius, D. Guggi and H. Lüth

Institut für Schicht- und Ionentechnik, Forschungszentrum Jülich,
P.O. Box 1913, 52425 Jülich, Germany

ABSTRACT: The influence of the carrier gas N_2 on the optical and electrical properties of (AlGa)As/GaAs in LP-MOVPE was studied with respect to high frequency device application (HBTs and HEMTs). The suitability of the growth process was demonstrated by a HEMT structure exhibiting a 4 K Hall mobility of 249,000 cm^2/Vs for a sheet carrier concentration of $4.9 * 10^{11}$ cm^{-2}. Intentional p-type doping in GaAs with C using TMAs was studied for HBT applications. A carrier concentration of $1.3 * 10^{20}$ cm^{-3} and a mobility of 57 cm^2/Vs was achieved.

1. INTRODUCTION

The metalorganic chemical vapor phase epitaxy (MOVPE) growth process has become well known for its compatibility to industrial application. Nevertheless there are some aspects concerning the safety of the process, which need to be looked after. First of all the toxicity of the hydrides, AsH_3 and PH_3, together with their high vapor pressure presents a problem. Secondly the commonly used carrier gas H_2 forms explosive mixtures with the air. In this study we will devote our attention to this second problem. Our aim is to improve the growth process by using N_2 as an inert carrier gas. We will test the suitability of N_2 - a cheap and safe replacement for H_2 - in LP-MOVPE.

A set of growth parameters was developed using nitrogen as the carrier gas, with which heterostructures for high frequency device application (hetero bipolar transistors (HBTs) and high electron mobility transistors (HEMTs)) can be obtained, to compare its suitability to the carrier hydrogen. The material system under investigation for these studies was $Al_xGa_{1-x}As/GaAs$ - a system that is extremely sensitive to oxygen and moisture and which therefore puts the growth process including the carrier to the test.

2. EXPERIMENTAL

The experiments were carried out in a horizontal low pressure MOVPE reactor (AIXTRON) using TMGa, TMAl and AsH_3. For n-type doping SiH_4 (1% in H_2) was employed, whereas p-type doping was achieved by replacing AsH_3 with TMAs. The ambients used, N_2 and H_2, were purified using a getter column and a Pd-cell, respectively. All epilayers were grown on semi-insulating GaAs substrates oriented 2° off (100) toward <110> and were cleaned in concentrated H_2SO_4 prior to growth.

The growth rates were determined by weighing and surface profiling after selective chemical etching for homoepitaxial GaAs and $Al_xGa_{1-x}As$, respectively. Surface morphology was studied with a Nomarski interference microscope.

Electrical characterization was carried out measuring the Hall effect at 77 K and 300 K. The $Al_xGa_{1-x}As$ layers under investigation were deposited on 100 nm thick AlAs buffer layers. In addition the HEMT-type samples were studied with the temperature dependent Hall effect in the temperature range from 4 to 300 K.

Raman spectroscopy of GaAs was studied at room temperature to evaluate the layer quality. High resolution photoluminescence (PL) measurements were carried out at 2 K for the same purpose and for the determination of the $Al_xGa_{1-x}As$ composition. The Al- fraction of the samples was calculated using the relation for the excitonic transition as proposed by Bosio et al. (1988). In addition the composition was determined by high resolution X-ray diffractometry (HR-XRD). HR-XRD was also employed to determine the carbon concentration in intentionally p-doped layers.

3. OPTIMIZATION OF (AlGa)As/GaAs GROWTH

A set of growth parameters needed to be developed with which high quality GaAs as well as (AlGa)As could be deposited. First GaAs growth was optimized. The results of this optimization have been published by Hardtdegen et al. (1992) in detail elsewhere. Nevertheless a short evaluation shall be given here with respect to the growth of heterostructures.

Although Raman studies indicate that 650 °C is the best deposition temperature, good quality GaAs can also be deposited at 50 °C higher and lower growth temperatures without leaving the diffusion controlled growth regime - the regime in which homogeneous growth can be performed in MOVPE. The V/III ratio is also not a very critical growth parameter. As long as ratios of 50 and above are employed, lowly compensated n-type layers are obtained. The gas velocity was varied from 0.3 to 1.5 m/s, but no significant influence could be seen on the electrical and optical quality of the material. The reactor pressure, however, is the most important growth parameter. As seen in Figure 1 from photoluminescence studies on samples grown at 20 and 100 hPa, the intensity of the excitonic transitions - a measure for the crystalline quality and low impurity concentration of the grown layers - is definitely higher for the lower reactor pressure. We believe that at higher reactor pressures laminar gas flow cannot be established and that turbulences occur due to the higher viscosity and lower thermal conductivity of nitrogen in comparison to hydrogen. A further comparison between the spectra for N_2 and H_2 shows, that the carbon concentration as seen from the ratio of the intensity of the conduction band to acceptor (CA) transition to the donor acceptor transition (DA) is reduced in the case of nitrogen whereas the donor concentration as seen from the intensity of the exciton bound to neutral donors is increased. All in all the electrical data of the layers - as has been previously reported - is comparable.

The growth parameter we will hold on to in the optimization of (AlGa)As growth will therefore be the reactor pressure. 20 hPa reactor pressure seems to be the precondition for excellent crystal growth with nitrogen as the carrier. The growth temperature was varied between 600 and 750 °C - this temperature region certifying diffusion controlled growth. A factor of merit for the optical quality of the (AlGa)As layers grown is on the one hand the intensity ratio of the bound exciton (BE) to donor acceptor (DA) recombination (BE/DA) which indirectly is a measure for the carbon and deep level impurity concentration (the higher this ratio the better) and on the other hand the full width at half maximum of the bound exciton transition, the broadening of which for a given Al-concentration is related to macroscopic inhomogeneities and inferior crystal quality. Photoluminescence experiments done on this sample series which contained an aluminum concentration x = 29 % shows a sharp increase in the BE/DA ratio until it is equal to about 1 when the growth temperature is increased from 600 to 700 °C. At 750 °C this ratio drops off again to about 0.5. The full widths at half maximum of the BE transitions follow the pattern inversely so that they are at their narrowest at 700 °C. This means that the best $Al_{0.29}Ga_{0.71}As$ optical quality in terms of low impurity concentration and superior crystal quality is achieved at 700 °C growth temperature.

Next the influence of V/III ratio in the range from 40 to 160 on optical $Al_{0.29}Ga_{0.71}As$ layer quality was studied. For this purpose photoluminescence spectroscopy was again employed. Here an increase in BE/DA intensity with V/III ratio was observed. Therefore V/III ratio of 160 was used for further investigations.

At last the gas velocity was varied for $Al_{0.24}Ga_{0.76}As$ between 0.4 and 1.9 m/s. Photoluminescence studies reveal narrow FWHMs for the BE transition at 0.9 and 1.4 m/s, whereas the highest BE/DA ratios were observed at 0.4 and 0.9 m/s. Best electrical data, however, was achieved for 0.9 m/s gas velocity: 4,470 and 2,260 cm^2/Vs at 77 and 300 K, respectively, were obtained for a carrier concentration of $1.3 * 10^{16}$ cm^{-3}, making this the optimum gas velocity.

A comparison of photoluminescence spectra for $Al_{0.29}Ga_{0.71}As$ grown with the N_2 process and with our standard growth process (using hydrogen) is presented in Figure 2. Clearly the deep level and carbon impurity concentration is lower for the carrier gas nitrogen!

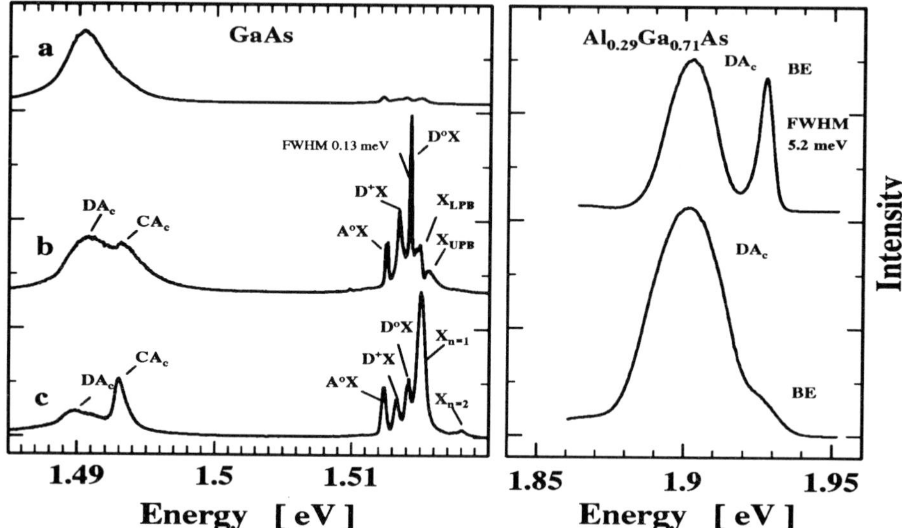

Fig. 1: 2 K photoluminescence spectra for GaAs grown with N_2 at 100 hPa (a) and 20 hPa (b) and with H_2 at 20 hPa (c)

Fig. 2: 2 K photoluminescence spectra for $Al_{0.29}Ga_{0.71}As$ with N_2 (upper curve) and H_2 (lower curve).

Whereas the optical characteristics are unambiguously different for (AlGa)As grown with N_2 and H_2, the electrical data is comparable for both (AlGa)As and GaAs grown under optimized conditions. The data is presented in Table 1. The GaAs samples were highly resistive and needed to be slightly doped with Si. The (AlGaAs) samples are not intentionally doped.

carrier	material	n_{300} [10^{14} cm^{-3}]	μ_{77} [cm^2/Vs]	μ_{300} [cm^2/Vs]
H_2	GaAs	5.2	48,980	7,620
N_2	GaAs	5.3	58,340	7,530
H_2	$Al_{0.24}Ga_{0.76}As$	140	3,830	2,180
N_2	$Al_{0.24}Ga_{0.76}As$	130	4,470	2,260

Table 1: electrical data for bulk (AlGa)As and GaAs material for both N_2 and H_2 carriers showing that the electrical quality is comparable.

Last the uniformity of samples deposited at optimized conditions without substrate rotation was investigated with HR-XRD on AlAs/GaAs superlattices. The thickness uniformity was increased from $\pm$ 7 % to $\pm$ 4 % and the compositional uniformity from $\pm$ 0.8 % to $\pm$ 0.2% for H_2 and N_2, respectively. The increase in uniformity is due to the lower thermal conductivity of N_2 compared to H_2, so that less reactor side wall and predeposition is observed.

4. CARBON DOPING

One major prerequisite for the deposition of semiconductor device structures is to be able to control intentional n- as well as p-type doping levels and profiles. In particular for

(AlGa)As/GaAs HBT and HEMT structures the control of n-type doping in GaAs and (AlGa)As as well as p-type doping in GaAs is necessary. N-type doping usually doesn't present a problem. It was possible to dope GaAs with Si up to 5 * 10^{18} cm^{-3} and $Al_{0.3}Ga_{0.7}As$ up to 3 * 10^{17} cm^{-3} (Hall measurements). P-type doping needed for HBT structures in particular, however, remains the bigger problem and is the focus of our attention in this paper. Carbon, with which high doping levels and sharp doping profiles was obtained for example by Konogai et al. (1989), was chosen as the p-type dopant, the intention being the employment in HBT structures. For the incorporation of carbon, growth was performed with TMAs instead of AsH_3 as the group V source and TMGa as the group III source. What is required for HBT structures with respect to p-type doping? We need to strive for heavy doping in GaAs at a deposition temperature as compatible as possible to (AlGa)As growth with excellent suface morphology. The carrier gases H_2 and N_2 again were compared in their influence on doping behavior.

Following observations hold for both carrier gases: As shown in Figure 3 the Arrhenius-plots of GaAs grown with TMAs and TMGa have the same slopes for both H_2 and N_2. The diffusion controlled growth regime shifts to a 100 K higher temperature in comparison to GaAs grown using AsH_3 as the group V source. Good sample morphology is found for deposition temperatures at and below 600 °C, the morphology being excellent for the N_2 grown samples. The electrically active carbon concentration increases as the growth temperature decreases as seen in Figure 4. In total, however, a lower carrier concentration is observed for H_2 than for N_2 as the carrier. Probably due to lower diffusion constants in nitrogen, the removal of C-containing species is less complete than for hydrogen. At

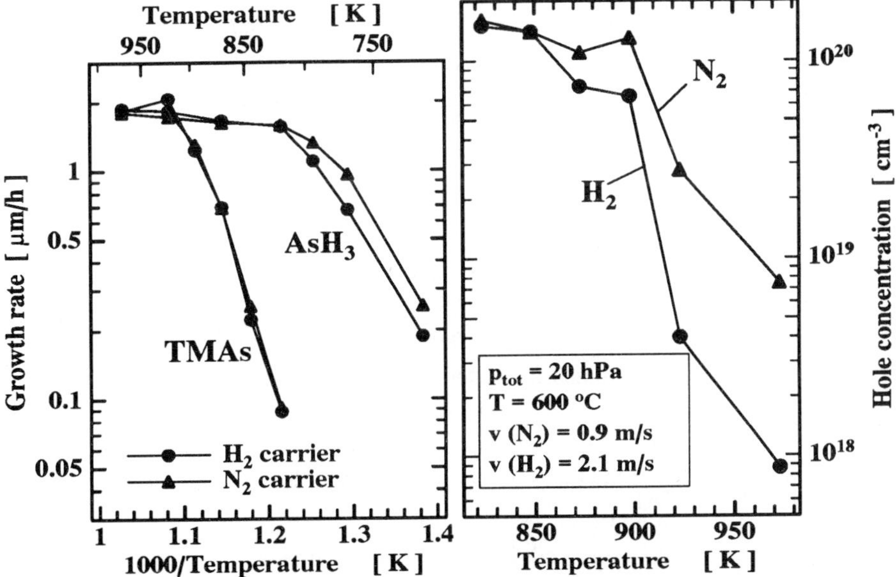

Fig. 3: GaAs growth rate dependence on reciprocal temperature using AsH_3 or TMAs (group V source), and H_2 or N_2 (carrier)

Fig. 4: Growth temperature dependence of hole concentration in GaAs using TMAs (group V source)

600 °C - the highest growth temperature at which morphologically perfect layers are grown for N_2 as the carrier - at last the V/III ratio was varied between 20 and 80. A V/III ratio of 40 is optimal with respect to surface morphology and high carrier concentration without compensation (the carbon concentration measured X-ray equaling the acceptor concentration measured by Hall-effect). At a carrier concentration of $1.3 * 10^{20}$ cm^{-3} a mobility of 57 cm^2/Vs was obtained. Under the same conditions only 70 cm^2/Vs at a carrier concentration of $6.3 * 10^{19}$ cm^{-3} was achieved for H_2 as the carrier.

5. GROWTH OF DEVICE STRUCTURES

Using the optimized set of growth parameters a HEMT structure was deposited to demonstrate the suitability of the N_2 growth process for the fabrication of HEMT structures. Table 2 presents the electrical data for both carriers. It is apparent that the N_2 growth process is as suitable for growth of HEMT type structures as H_2.

carrier	n_4 [$10^{11}cm^{-2}$]	μ_4 [cm^2/Vs]	μ_{300} [cm^2/Vs])
H_2	4.1	240,000	8,150
N_2	4.9	249,000	7,000

Table 2: comparison of electrical data for HEMT-type structures for both carriers

HBT structures were also deposited under the optimized growth conditions described above. Such preliminary results as can be given for HEMT structures are, however, not as easily obtained. Device fabrication is in progress. The results will be published later on.

6. CONCLUSIONS

Growth of the (AlGa)As/GaAs material system was optimized using nitrogen as the carrier. The optical quality of the layers is greatly improved particularly for $Al_xGa_{1-x}As$ whereas the electrical data is comparable to that of layers deposited with our standard growth process. Thickness and compositional uniformity is increased for nitrogen grown lyers. N-type doping was investigated in GaAs as well as in $Al_xGa_{1-x}As$. Heavy carbon doping with a concentration of $1.3 * 10^{20}$ cm^{-3}, which was all electrically active, was achieved in GaAs. All in all it has been shown that the N_2 growth process is very suitable for growth of $Al_xGa_{1-x}As/GaAs$ heterostructures with respect to high frequency device application and not only a safer but also a better alternative to hydrogen.

7. REFERENCES

Bosio C, Staehli J L, Guzzi M, Burri G, and Logan R A 1988, Phys. Rev. B 38, 3263

Hardtdegen H, Hollfelder M, Meyer R, Carius R, Münder H, Frohnhoff S, Szynka D, and Lüth H 1992, J. Cryst. Growth 124, 420

Konagai M, Yamada T, Akatsuka T, Saito K, Tokumitsu E, Takahashi K 1989, J. Cryst. Growth 98, 167

Anomalous photoluminescence behaviour for GaInP/AlGaInP quantum wells grown by MOVPE on misoriented (001) substrates

H. Hotta, A. Gomyo, F. Miyasaka, K. Tada, T. Suzuki and Ke. Kobayashi

Opto-Electronics Res. Labs., NEC Corp., 34, Miyukigaoka, Tsukuba 305, Japan

ABSTRACT: Anomalous photoluminescence (PL) behavior and "step-bunching" as the origin of the anomaly, are reported for GaInP/AlGaInP-quantum wells (QWs) grown by MOVPE on the (001) GaAs substrates misoriented toward [110]. The PL peak energies of these QWs are significantly smaller than the calculated values. This anomalous behavior is attributed to step-bunching observed by transmission electron microscopy. Remarkable step-bunching is observed at substrate-misorientations of 6° and 10°, while it is significantly reduced at 0° and 15.8°.

1. INTRODUCTION

Multi-quantum-well (MQW) structures for AlGaInP visible-light emission laser diodes (LDs) have played an important role in achieving short-wavelength emission and high-power operation. To shorten operating wavelengths, MQW structures have often been grown on (001) GaAs substrates misoriented toward [110], because these substrates tend to suppress "natural superlattice" formations which cause a decrease of the band gap energy of (Al)GaInP-bulk (Gomyo et al. 1988, 1989). Crystal qualities for QW-structures grown on the misoriented substrates have been evaluated by photoluminescence (PL) measurement (Valster et al. 1991, Wang et al. 1992, Yanagisawa et al. 1992, Watanabe et al. 1993). As far as we know, no one has pointed out that an anomalous behavior of PL peak energy appears for GaInP/AlGaInP QWs. With regard to heterointerfaces grown on (001) vicinal substrates, so-called "step-bunching" or "multi-atomic step" formation has been reported for AlGaAs alloy systems (Fukui and Saito 1990, Kasu and Fukui 1992, Kasu and Kobayashi 1993), but there have been no reports of step-bunching for AlGaInP alloy systems and of the relation between step-bunching and PL. This paper reports for the first time anomalous PL behavior, and step-bunching as the origin of the anomaly, for GaInP/AlGaInP-QWs grown on (001) GaAs substrates misoriented toward [110].

2. MOVPE GROWTH FOR GaInP AND AlGaInP ON MISORIENTED SUBSTRATES

Two types of GaInP- and AlGaInP-layer structures are examined. One, as shown in Fig. 1(a), consists of 100 nm-thick GaInP-bulk and GaInP-QWs with well width Lz=6, 3, 1, and 0.5 nm. The other, as shown in Fig. 1(b), consists of the GaInP layers into which 1 nm-thick AlGaInP marker layers are inserted every 20 nm. The composition of the AlGaInP layers is $(Al_{0.6}Ga_{0.4})_{0.5}In_{0.5}P$.

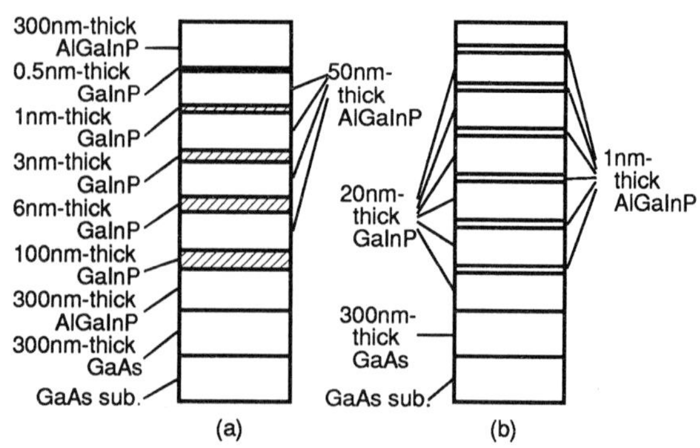

Fig. 1. Cross-sectional structures of (a) QWs and (b) AlGaInP-marker-inserted GaInP layers.

The GaInP- and AlGaInP-layers were grown by metalorganic vapor phase epitaxy (MOVPE). Trimethylaluminum (TMAl), triethylgallium (TEGa), trimethylindium (TMIn), phosphine (PH_3) and arsine (AsH_3) were used as the MOVPE-sources. The carrier gas was hydrogen. The reactor pressure and the substrate temperature were kept constant at 70 Torr and 660°C, respectively. The ratio between source gas flow rates for column V and III atoms (V/III ratio), was 500 for the AlGaInP growth. The growth rate for the GaInP and AlGaInP layers was 1.8 μm/h. The lattice mismatch of AlGaInP to GaAs was controlled within ±0.1%, which was determined by X-ray diffraction measurements. The substrates were semi-insulating (001) GaAs with misorientaions (hereafter called "off-angle") of 0°, 6°, 10° and 15.8° toward [110]. The nominal accuracy of the substrate orientation was ±0.5°.

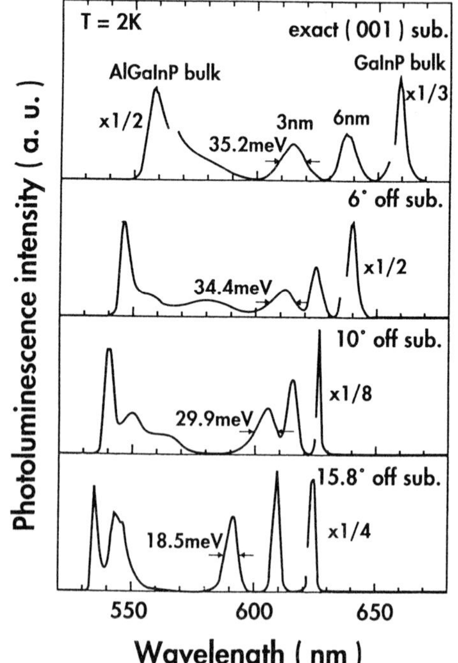

Fig. 2. PL spectra for GaInP-SQWs, 100nm-thick GaInP-bulk, and AlGaInP-barrier (bulk) on the substrate misoriented (off) toward [110]. The thicknesses of QWs were designed as 0.5, 1, 3 and 6 nm.

3. PL AND TEM MEASUREMENTS

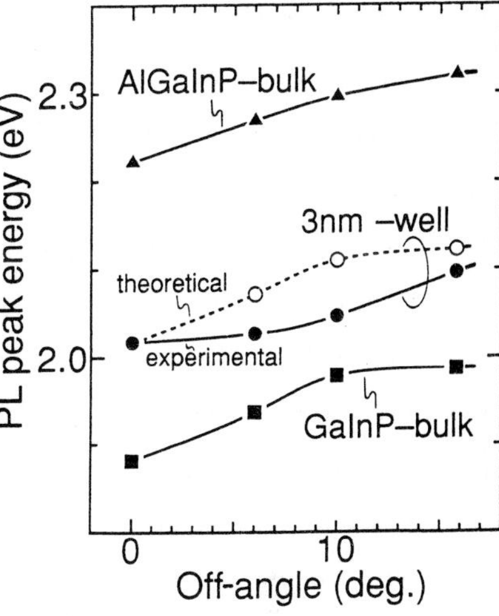

Fig. 3. Misoriented (off) angle dependence of PL peak energies for 3 nm-thick QWs and bulks. Open circles indicate the calculated energies for a 3 nm-thick QW, on the basis of the bulk-energies.

PL spectra were measured at 2 K by using an argon ion laser for excitation. The excitation intensity was approximately 1.3 W/cm^2. Transmission electron microscopy (TEM) observation of the two structures was carried out for the ($\bar{1}$10) cross-section with a 200 kV acceleration voltage. Samples for the TEM observation were prepared by cleaving, mechanical polishing, and argon ion milling.

Figure 2 shows PL spectra for the QW structure. All the PL peak energies for 3 and 6 nm-thick QWs and GaInP and AlGaInP bulks are blue-shifted with the off-angle increase. These blue-shifts were thought to be due to disordering of the natural superlattice. However, Fig. 3 shows that the PL blue-shifts for QWs are significantly smaller than those for the energies calculated for a 3 nm-thick QW on the basis of the bulk-energies, at 6° and 10° off-angles.

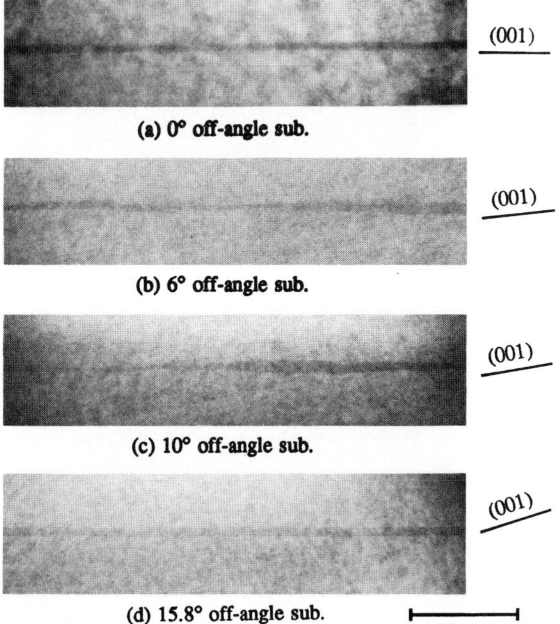

Fig. 4. ($\bar{1}$10) cross-sectional TEM images for QWs with a nominal thickness of 3 nm.

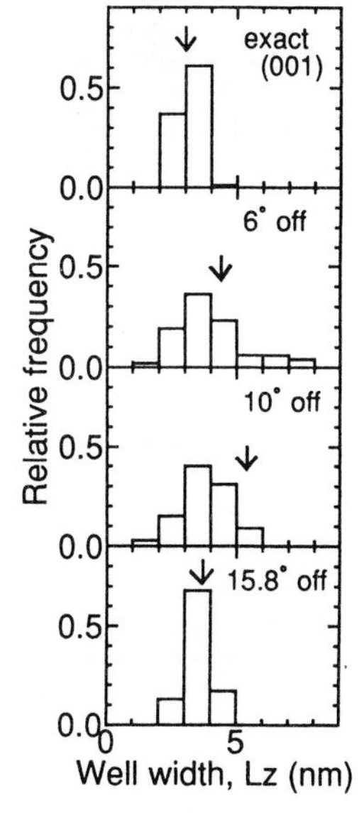

Fig. 5. Actual thickness distributions for 3 nm-thick QWs. Arrows indicate well-widths calculated from PL peak energies.

The TEM images for 3 nm-thick QWs are shown in Fig. 4. Waving in the QW heterointerfaces and variations in QW width appear at 6° and 10° off-angle substrates. But the degrees of waving and variations are significantly reduced at 0° and 15.8° off-angle substrates. The periods of these variations are a few hundreds of nanometers. As shown in Fig. 5, the actual thickness distribution for a 3 nm-thick QW is small at 0° and 15.8° off-angle substrates and large at 6° and 10° off-angle substrates. These QW thicknesses were measured at every 8 nm separated position. The number of measurement points is from 46 to 116 for each off-angle substrate. For each off-angle substrate, each well-width calculated from the PL peak energies is in the range of this distribution. TEM images at the heterointerfaces for the 0° substrate reveal that there are both slightly ascending and descending bunches of steps, while there are only descending bunches of steps toward the substrate misoriented direction for 6°, 10° and 15.8° off-angle substrates. From these, it is concluded that the variations of QW thickness at 6° and 10° off-angle substrates were caused by significant step-bunching. The number of bunched steps, as defined

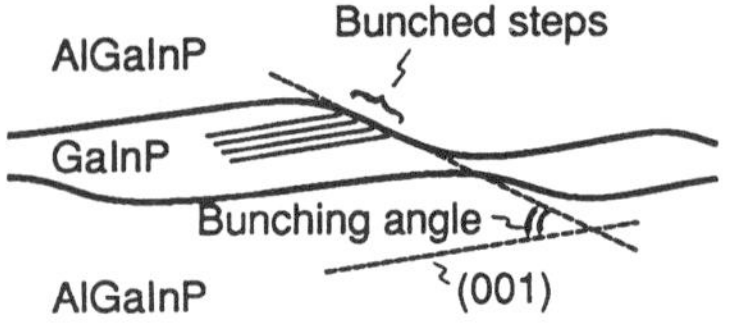

Fig. 6. Defenitions for (i) the number of bunched steps and (ii) the bunching angle of the step-bunched surface with a (001) surface.

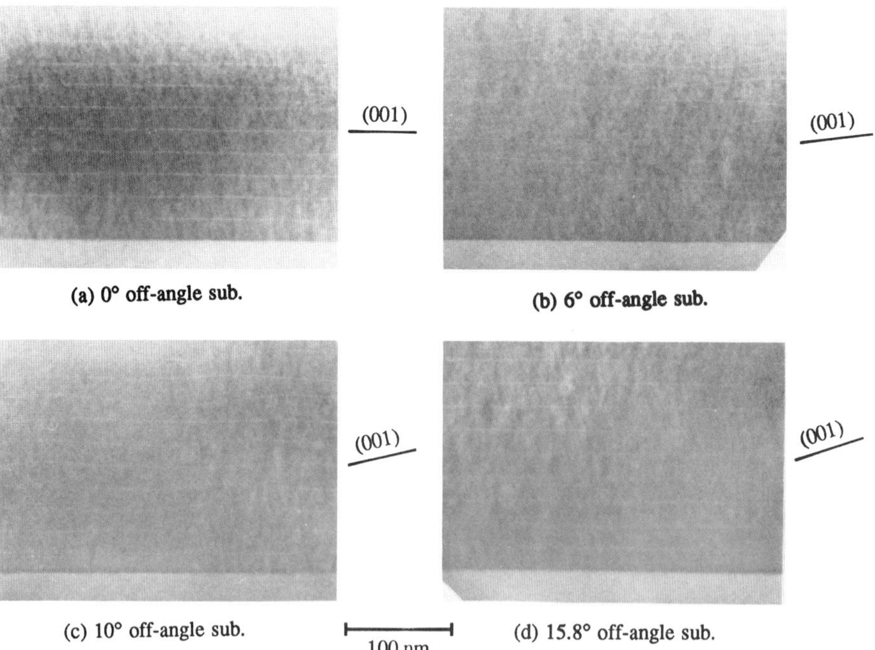

Fig. 7. ($\bar{1}$10) cross-sectional TEM images for the GaInP layers into which 1 nm-thick AlGaInP layers were inserted every 20 nm.

in Fig. 6, was from 15 to 18 monolayers. The maximum bunching angle of the step-bunched surface with the (001) surface lay between 20° and 25° for off-angles of 6°, 10°, 15.8°, although it was 3° for a 0° off-angle.

The dependence of step-bunching on the off-angle is clearly seen in TEM images (Fig. 7) for samples with the structure shown in Fig. 1(b). At 6° and 10° off-angle substrates, significant step-bunching was observed even at the first heterointerface between GaInP and AlGaInP grown on a flat GaAs buffer layer. However, at 0° and 15.8° off-angle substrates, the degree of step-bunching at any of the heterointerfaces was significantly reduced.

4. DISCUSSION

As mentioned in section 3, the anomalous reduction in PL peak energy for QWs grown on the misoriented substrates corresponds to the remarkable variation of the QW thickness which is caused by step-bunching. A QW with a thicker well width has a lower quantum energy level than a QW with a thinner well width. Thus, when the QW thickness variation occurs, the argon ion laser-excited carriers, electrons and holes, move from the thin well-width regions to the thick well-width regions. In turn, the PL from the thick well-width regions dominates the observed PL spectrum. This is believed to be the cause of the anomalous PL peak energy reduction. This interpretation is consistent with the result that every well width calculated from the measured PL peak energies for all off-angles is in the range of the actual distribution of QW thicknesses in Fig. 5. Therefore, the anomalous PL behavior is attributed to the QW thickness variation due to step-bunching. The full width of half maximum (FWHM) of PL spectra for the QW designed with a thickness of 3 nm, as shown in Fig. 2, becomes narrower with the off-angle increase. This dependence does not correspond to that for the QW thickness variation caused by step-bunching. It is thought that the FWHM of PL for a QW is also affected by the degree of ordering of natural superlattice formation.

The cause of the strong dependence of step-bunching on the examined off-angle range can be discussed from two viewpoints. The first is the growth mode. According to TEM observation of the heterointerfaces, both ascending and descending bunched steps are observed for the 0° case, while only descending bunched steps are observed for the 6°, 10° and 15.8° cases. This means that "2-dimensional nucleation mode growth" is occurring at 0° and "step flow mode growth" is dominant at 6°, 10° and 15.8°. The reduction of step-bunching at 0° is attributed to uni-directional step flow mode growth, which is necessary for step-bunching formation, being not established on the exact (001) substrate. When the off-angle is increased (e.g., 6°), the step flow mode growth tends to become dominant, and this promotes step-bunching under the present growth conditions.

The second viewpoint is the existence of the maximum bunching angle, as defined in Fig. 6. As observed in this study, the maximum angles for any off-angle case range from 20° to 25° and most of the bunching angles are less than approximately 20°. This indicates that no step-bunched surface can have an angle as large as that of the (113)A surface (25.2°), which has the minimum terrace width (0.4nm) of (001). Because of this singularity, it is probable that in the (113)A surface, some reconstruction occurs that prevents column III atoms from adsorbing to its steps. This reconstruction possibly limits the step-bunching for the 15.8° case.

5. CONCLUSION

It was observed that step-bunching occurs for AlGaInP material grown by MOVPE on (001) GaAs substrates misoriented toward [110]. The step-bunching caused the anomalous PL peak behavior of a QW. The PL peak energies of QWs for misorientations of 6° and 10° were significantly smaller than the calculated values. This anomalous PL behavior was attributed to the variation of QW thickness caused by the step-bunching. Remarkable step-bunching appeared at 6° and 10°, while it was significantly reduced at 0° and 15.8°. This off-angle dependence was explained by taking into consideration both the growth mode and the existence of the maximum inclination angle of a step-bunched surface from the (001) surface.

Acknowledgements

We would like to thank Kohroh Kobayashi, Ikuo Mito, Kenji Endo, Isao Hino, and Hiroaki Fujii for their encouragement and useful discussions, and Emiko Saito for technical support.

References

Fukui T and Saito H 1990 Jpn. J. Appl. Phys. **29** L483
Gomyo A, Suzuki T, Iijima S, Hotta H, Fujii H, Kawata S, Kobayashi K, Ueno Y and Hino I 1988 Jpn. J. Appl. Phys. **27** L2370
Gomyo A, Kawata S, Suzuki T, Iijima S and Hino I 1989 Jpn. J. Appl. Phys. **28** L1728
Kasu M and Fukui T 1992 Jpn. J. Appl. Phys. **31** L864
Kasu M and Kobayashi N 1993 Appl. Phys. Lett. **62** 1262
Valster A, Liedenbaum C T H F, Finke M N, Severens A L G, Boermans M J B, Vandenhoudt D E W and Bulle-Lieuwma C W T 1991 J. Cryst. Growth **107** 403
Wang T Y, Welch D F, Scifres D R, Treat D W, Bringans R D, Street R A and Anderson G B 1992 Appl. Phys. Lett. **60** 1007
Watanabe M, Rennie J, Okajima M and Hatakoshi G 1993 Electron. Lett. **29** 250
Yanagisawa H, Tanaka T, Minagawa S and Yano S 1992 Extended Abstracts of 11th Record of Alloy Semiconductor Phys. and Electron. Symp. pp 491

Paper presented at the Int. Symp. GaAs and Related Compounds, Freiburg, 1993

MOVPE growth of strained $GaP_{1-x}N_x$ and $GaP_{1-x}N_x$/GaP quantum wells

S. Miyoshi, H. Yaguchi, K. Onabe, Y. Shiraki*, and R. Ito

*Department of Applied Physics, The University of Tokyo,
7-3-1 Hongo, Bunkyo-ku, Tokyo 113, Japan*

**Research Center for Advanced Science and Technology (RCAST),
The University of Tokyo, 4-6-1 Komaba, Meguro-ku, Tokyo 153, Japan*

ABSTRACT: We report the successful growth of $GaP_{1-x}N_x$ ($x < 0.04$) alloys on GaP by metalorganic vapor phase epitaxy (MOVPE). In spite of the large miscibility gap, which is calculated to extend from x = 0.00001 to 0.99999 at 700°C, $GaP_{1-x}N_x$ alloys have been obtained at the growth temperature of 630 ~ 700°C. The peak energy of low-temperature photoluminescence (PL) spectra shows a redshift with increasing *x*. We have made strained $GaP_{1-x}N_x$ (well) / GaP (barrier) multiple quantum wells (MQWs) and observed strong PL emission from these MQWs.

1. INTRODUCTION

The $GaP_{1-x}N_x$ alloy is a candidate for a blue and ultraviolet optical material due to the possibility of its large variation of bandgap, from 2.3 eV (GaP) up to 3.4 eV (GaN) at room temperature. A limiting factor for obtaining this alloy is the difference in the lattice structure between the two end materials, GaP (zincblende) and GaN (wurtzite). The differences in the lattice structure and the lattice constant cause a large miscibility gap (Stringfellow 1972, 1974), which is calculated to extend from $x = 0.00001$ to 0.99999 at 700°C, and has prevented a successful growth of this alloy system. It is true that a large amount of efforts have been dedicated to the study of N isoelectronic trap in N-doped GaP (for instance, Thomas and Hopfield 1966, Faulkner *et al.* 1968, Masselink *et al.* 1983, Zhang *et al.* 1993). However, there have been only two demonstrations of $GaP_{1-x}N_x$ alloy growth up to now, with $x < 0.08$ by MBE (Baillargeon *et al.* 1992) and $x > 0.91$ by CVD (Igarashi 1992). Besides, there has been no report on quantum well structures using this alloy system.

In the present paper, we report the first successful growth of $GaP_{1-x}N_x$ (x < 0.04) alloys on GaP by MOVPE. The alloys with a mirror-like surface were obtained. The N incorporation in the MOVPE growth was examined. The peak energy of the low-temperature PL spectra of the alloy showed a redshift with increasing x. We made $GaP_{1-x}N_x$/GaP MQW structures with the alloy serving as a well for the first time and observed strong PL emission from these MQWs.

2. EXPERIMENTAL

Samples were grown on nominally on-axis (100) GaP substrates with a conventional low-pressure (60 Torr) MOVPE system with H_2 carrier. Trimethylgallium (TMG), PH_3, and dimethylhydrazine (DMHy) were used as the Ga, P, and N sources, respectively. Prior to the growth of $GaP_{1-x}N_x$ alloys or the MQWs, a 0.3 μm-thick GaP buffer layer was grown at 750°C with TMG and PH_3. Then the substrate temperature was lowered to the growth temperature; 600~700°C. In the growth of $GaP_{1-x}N_x$ alloy, TMG, PH_3, and DMHy were supplied simultaneously. The total film thickness was 0.15 ~ 0.8 μm. In the growth of the MQWs (5 periods), we turned the DMHy flow on and off to grow $GaP_{1-x}N_x$ and GaP, respectively. The $GaP_{1-x}N_x$ well width was 10 ~ 110 Å, and the thickness of GaP barrier was 450 Å. TMG and DMHy were kept in an isothermal bath with a constant temperature of -10 and 10°C, respectively, bubbled and carried into the reactor with H_2. The V/III ratio was 20 ~ 200. The growth rate was 2.2 μm / hour at 650°C with the TMG flow rate of 2.0 sccm (= 4.5 μmol / min). Solid composition (x) of the samples was determined from a double crystal x-ray rocking curve of cubic (511) reflection (All our samples had a zincblende structure). We assumed Vegard's law between x and the lattice constant a ; $x = (a - a_{GaP}) / (a_{c\text{-}GaN} - a_{GaP})$, where a_{GaP} (= 5.451 Å) and $a_{c\text{-}GaN}$ (= 4.51 Å) was the lattice constant of GaP (taken from the data of Straumanis *et al.* 1967) and cubic GaN (taken from the data of Miyoshi *et al.* 1992), respectively. We also assumed Vegard's law for the elastic constants. These of GaP and cubic GaN were taken from the work of Madelung (1987) and that of Sherwin and Drummond (1991), respectively. Low-temperature photoluminescence measurements were performed using a He-Cd laser operating at 325 nm to excite the samples.

3. RESULTS AND DISCUSSION

In the MOVPE growth of the $GaP_{1-x}N_x$ alloys, we varied the relative molar flow (x_v) of DMHy in the total group V gas flow; $x_v = f_{DMHy} / (f_{PH_3} + f_{DMHy})$. The x_v versus x relationship at 650°C is shown in Fig. 1. The V/III ratio was around 100 and f_{TMG} = 2.0 sccm. When x_v < 0.63, the alloys (x < 3 %) with a mirror-like surface were obtained. The x-ray diffraction curve has a single peak; the solid composition was uniform. However, when x_v ~ 0.75, the diffraction curve had double peaks; its solid composition was not uniform (4.0 % and 3.0 %). When x_v

was increased to 0.80, the film surface covered with black powder and no x-ray signal was detected. We were not able to determine the nitrogen content of the deposited powder. In order to obtain the alloys with larger x, we tried several growth conditions, changing growth temperature and growth rate, but 3.2 % was the maximum x where we obtained a epitaxial film with uniform composition (which was grown at 630°C and f_{TMG} = 2.0 sccm). It is considered that the strain energy involved in the interface between the alloy and the substrate, which increases with increasing x, prevents the epitaxial growth of the alloy.

The relationship between x and x_V is described in the same way as the case of $GaAs_{1-x}P_x$ (Samuelson *et al.* 1982). We define the distribution coefficient of N as $k_N = x / x_V$, and that of P as $k_P = (1-x) / (1-x_V)$. Then we obtain

$$(1 - x) / x = C_{PN} (1 - x_V) / x_V, \qquad (1)$$

where C_{PN} (= k_P / k_N) is the relative distribution coefficient between P and N. We fitted the data to Eq. (1) taking C_{PN} = 80 (the solid curve in Fig. 1). This means that P is 80 times easier to incorporate than N at 650 °C.

Figure 2 shows the growth temperature dependence of x of the alloy with constant source gas flow rate (f_{TMG} = 2.0 sccm, V/III = 80, and x_V = 0.625). As the temperature was lowered from 700°C down to 630°C, x increased from 0.39 % to 3.2 %. However, when the temperature was as low as 600°C, the grown surface appeared powder-like and no x-ray signal was detected. With

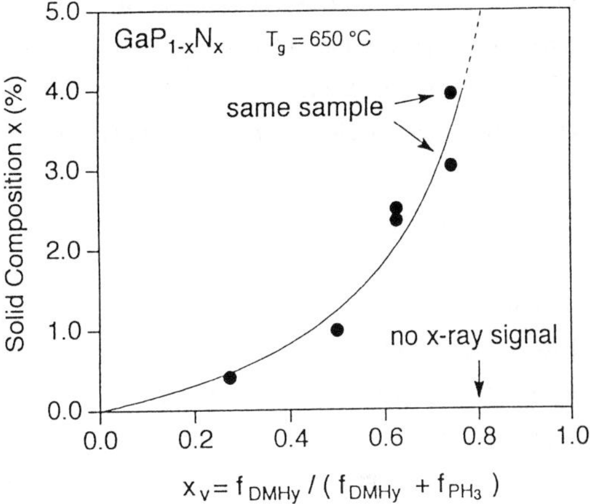

FIG. 1. Solid composition x versus vapor composition x_V for the alloy $GaP_{1-x}N_x$ grown at 650°C. The solid curve is a fitting curve to Eq. (1) taking C_{PN} = 80.

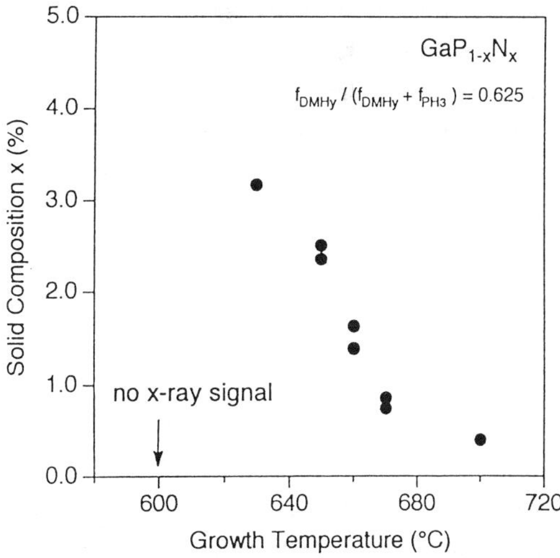

FIG. 2. Solid composition x versus growth temperature for the alloy $GaP_{1-x}N_x$ with constant x_V.

increasing growth temperature, C_{PN} increases from 50 to 420. This decreasing nitrogen incorporation with increasing temperature is considered to be due to increasing PH_3 decomposition and nitrogen desorption from film surface during growth.

The low-temperature PL spectra of the alloys are shown in Fig. 3. The alloys have a strong luminescence intensity (it is apparent when compared to that of GaP buffer layer). As the bandgap of GaN (3.4 eV) is larger than GaP (2.3 eV), the bandgap of $GaP_{1-x}N_x$ is expected to increase with increasing x. However, the luminescence peak shows a redshift with increasing x. This redshift is in accordance with the 77 K PL data measured by Baillargeon *et al.* (1992). Their calculation predicts that the bandgap decreases with increasing x when $x < 50$ % and it increases with larger x. It is a well-known fact that a pair of two N atoms makes a deep radiative center (named NN_i) in GaP even at small nitrogen concentration of ~10^{19} cm^{-3} due to their large electronegativity (Thomas and Hopfield 1966), thus the bandgap reduction is possible when x is small. As shown in Fig. 3, the spectrum for GaP (x = 0) consists of nitrogen trapped A-exciton (2.317 eV) and its phonon replicas. As the nitrogen content increases to 0.13 %, the emission from the nitrogen pairs, *i.e.*, NN_1 (2.18 eV), NN_3 (2.26 eV), and NN_4 (2.29 eV) and their phonon replicas, dominates the spectrum. When x is increased to 0.34 %, the deep NN_1 line and its phonon replicas

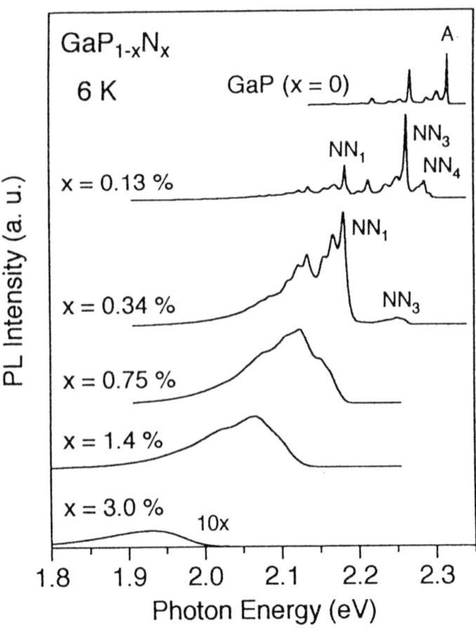

FIG. 3. Low-temperature PL spectra of $GaP_{1-x}N_x$ alloys with different x.

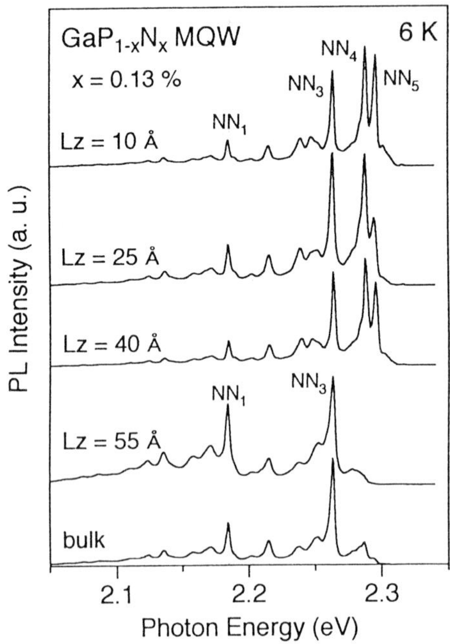

FIG. 4. Low-temperature PL spectra of $GaP_{1-x}N_x$/GaP MQWs (x = 0.13 %).

become dominant, the NN_3 line gets weak, and the NN_i lines with larger i (in other words, higher emission energy) disappear. When x gets larger than 0.34 %, the NN_1 emission and its phonon replicas overlap, giving a broader line shape, and the peak energy shows a redshift with increasing x. We interpret this broad peak (in the spectra for $x \geq$ 0.75 %) as the band edge of the alloy. Thus the bandgap decreases with increasing x when $x \leq 3.0$ %.

Figures 4 ~ 6 show the PL spectra of the $GaP_{1-x}N_x$/GaP MQWs with different N content in $GaP_{1-x}N_x$ well: 0.13 % (Fig. 4), 0.75 % (Fig. 5), and 1.4 % (Fig. 6). In all these spectra, the emission shows a blueshift with decreasing well thickness (Lz). For the samples with $x \geq 0.75$ % (Fig. 5 and Fig. 6), the emission peak of the spectra shows a blueshift with decreasing Lz. However, for the samples with $x = 0.13$ % (Fig. 4), the line shape and the peak energy of the spectra do not change with decreasing Lz. Only the relative intensity of each peak changes. The spectra for $x = 0.13$ % (Fig. 4) consist of NN_i lines (i = 1, 3, 4, 5) and its phonon replicas. In the alloy with x = 0.13 %, it is considered that N atoms only give NN_i lines in GaP host crystal, and that the energy bands characteristic to the alloy are not formed. Thus there is no quantum confinement effect in the MQWs with $x = 0.13$ %, so there is only a small difference in the spectrum with different Lz. In the other two samples ($x = 0.75$ % in Fig. 4 and $x = 1.4$ % in Fig. 5), on the other hand, the energy bands

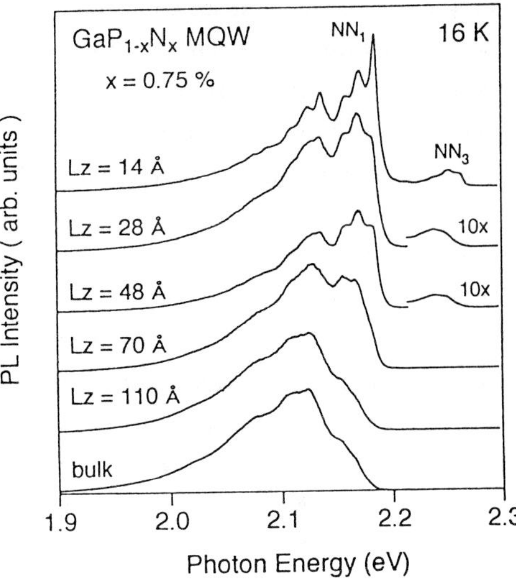

FIG. 5. Low-temperature PL spectra of $GaP_{1-x}N_x$/GaP MQWs (x = 0.75 %).

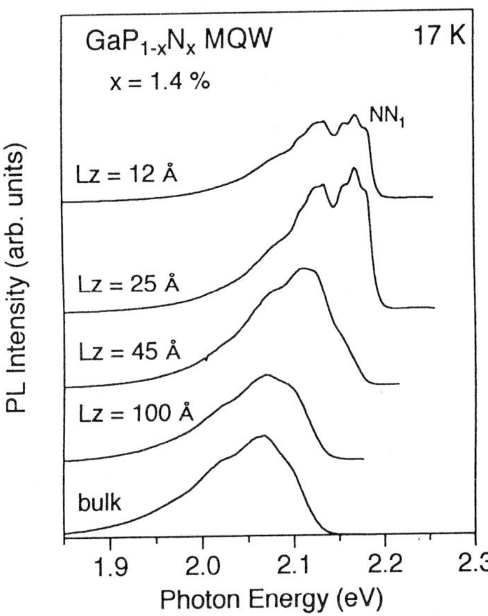

FIG. 6. Low-temperature PL spectra of $GaP_{1-x}N_x$/GaP MQWs (x = 1.4 %).

characteristic to the alloys are formed in the alloy, thus the quantum confinement is efficient, and the spectra move to higher energy considerably with decreasing Lz. In the MQWs with $Lz <$ 30 Å, as shown in Fig. 5 and 6, the wave functions spread into the GaP barrier, so the emission from the NN_1 and/or NN_3 line in the GaP layer gets dominant in the spectra (as shown in Fig. 3, a few nitrogen atoms are incorporated into GaP epitaxial layer and form radiative centers).

4. CONCLUSION

We have grown $GaP_{1-x}N_x$ ($x < 0.04$) alloys on GaP (100) substrates by MOVPE. TMG, PH_3, and DMHy were used as Ga, P, and N sources, respectively. Alloys with a zincblende structure and a mirror-like surface have been obtained. Nitrogen incorporation increases with decreasing growth temperature. The peak energy of low-temperature PL spectra shows a redshift with increasing x. We have made $GaP_{1-x}N_x$/GaP MQW structures with the alloy serving as a well for the first time and observed strong PL emission from these MQWs.

ACKNOWLEDGEMENTS

The authors would like to express their appreciation to M. Nagahara, K. Ota, W. Pan, and X. Ma for assistance in MOVPE growth and to S. Fukatsu for useful discussion, and acknowledge S. Ohtake for his technical support. We greatly appreciate Sumitomo Chemical Co., Ltd. and Kimmon Electric Co., Ltd. for supplying the TMG source and the He-Cd laser, respectively, which were used in this work.

REFERENCES

Baillargeon J N, Cheng K Y, Hofler G E, Pearah P J and Hsieh K C 1992 Appl. Phys. Lett. **60**, 2540
Faulkner R A 1968 Phys. Rev. **175**, 991
Igarashi O 1992 Jpn. J. Appl. Phys. **31**, 3791
Madelung O (editor) 1987 *Intrinsic Properties of Group IV Elements and III-V, II-VI and I-VII Compounds* (Landolt-Börnstein Vol. 22, Springer-Verlag, Berlin)
Masselink W T and Chang Y -C 1983 Phys. Rev. Lett. **51**, 509
Miyoshi S, Onabe K, Ohkouchi N, Yaguchi H, Ito R, Fukatsu S and Shiraki Y 1992 J. Cryst. Growth **124**, 439
Samuelson L, Omling P, Titze H and Grimmeiss H G 1982 J. Physique **43**, C5-323
Sherwin M E and Drummond T J 1991 J. Appl. Phys. **69**, 8423
Straumanis M E, Krumme J -P and Rubenstein M 1967 J. Electrochem. Soc. **114**, 640
Stringfellow G B 1972 J. Electrochem. Soc. **119**, 1780
Stringfellow G B 1974 J. Cryst. Growth **27**, 21
Thomas D G and Hopfield J J 1966 Phys. Rev. **150**, 680
Zhang Y, Ge W, Sturge M D, Zheng J and Wu B 1993 Phys. Rev. **B47**, 6330

Inst. Phys. Conf. Ser. No 136: Chapter 10
Paper presented at the Int. Symp. GaAs and Related Compounds, Freiburg, 1993

GaAs crystallographic selective growth by atomic layer epitaxy and its application to fabrication of quantum wire structures

Hideo Isshiki, Yoshinobu Aoyagi, and Takuo Sugano

Frontier Research Program, The Institute of Physical and Chemical Research(RIKEN), 2–1 Hirosawa, Wako–shi, Saitama, 351–01, JAPAN

Sohachi Iwai, and Takashi Meguro

The Institute of Physical and Chemical Research(RIKEN), 2–1 Hirosawa, Wako–shi, Saitama, 351–01, JAPAN

Abstract

Crystallographic selective growth using atomic layer epitaxy (ALE) has been developed for the fabrication of low–dimensional quantum structures. In addition, ALE growth mode change between isotropic (side wall) and anisotropic (selective) growth, without growth temperature change, was achieved by control of the hydrogen purge time after AsH_3 supply. Due to the self limiting effect and high selectivity of ALE growth, trapezoidal shaped GaAs/$(AlAs)_1(GaAs)_1$ quantum wire structures were successfully realized.

1.Introduction

Selective growth is one of the most attractive techniques for providing direct control of the *lateral* dimensions of semiconductor structures(Fukui 1990). This control is very important in realizing low–dimensional quantum structures, such as quantum wires and quantum dots(Kapon 1989, Tsukumoto 1992) . Recently, we have reported that selective growth between several facets is possible using atomic layer epitaxy(ALE)(Isshiki 1993). It is expected to be very useful for vertical and/or lateral dimensional control in nano fabrication, because single monolayer control for several facets can be achieved(Ide 1988, Usui 1990) and because uniform growth over the selective growth area(Iwai 1988) is possible due to the self limiting effect.

In this paper, we report the crystallographic selective growth using ALE and its application to fabrication of quantum wire structures. In addition, we demonstrate an experiment of ALE growth mode change between side wall and selective growth without growth temperature change.

2.Experiment

A horizontal quartz reactor was used here for ALE growth experiments. GaAs growth was carried out using trimethylgallium (TMG) as the Ga source and AsH_3 as the As source. These were alter–

nately fed into the growth chamber for 1 s and 2 s respectively, with a hydrogen purge time between each gas supply. The flow rates of TMG and AsH_3 were $6x10^{-7}$ mol/s (0.4Pa partial pressure) and $1.8x10^{-5}$ mol/s (12Pa partial pressure) respectively. Hydrogen carrier gas was fed continuously into the chamber with a flow rate of 2500 cc/min, and the total pressure was $1.2x10^3$ Pa.

3.Results and discussion

A. Crystallographic selectivity of GaAs ALE growth

Figure 1 shows the GaAs growth rate on differently oriented GaAs substrates as a function of growth temperature. There the source gases were fed into the growth chamber with a 1 s hydrogen purge time between each gas supply. GaAs ALE growth was achieved on (100) GaAs substrate in the temperature range from 530°C to 570°C, the so-called "ALE window". However ALE growth on the other substrates was not observed. The growth rate on GaAs (111)B substrate increased with increasing growth temperature. On the other hand the growth rate on GaAs(111)A substrate began to increase at about 480°C, decreasing rapidly above 545°C, and finally the GaAs growth rate was nearly zero over 560°C. The growth temperature dependence of growth rate on GaAs (110) substrate is very similar to that of growth rate on GaAs (111)A substrate.

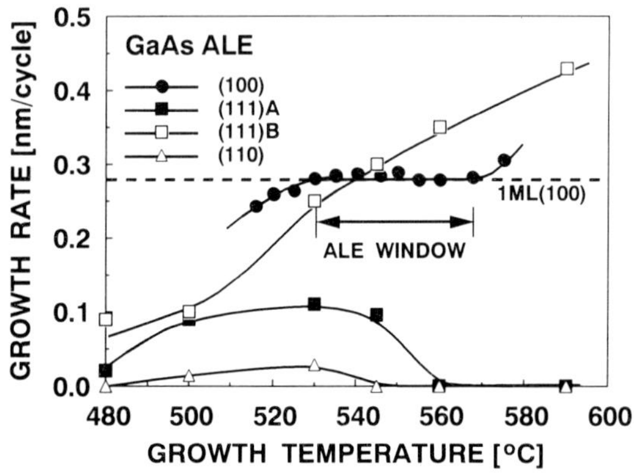

Figure 1 ALE growth rate of GaAs on variously oriented GaAs substrates as a function of growth temperature.

Figure 2 shows the growth rate of GaAs on each substrate as a function of hydrogen purge time after AsH_3 supply at 570°C. There the hydrogen purge time after TMG supply was 1 s. The growth rates on GaAs (100) and (111)B substrates were almost independent of the purge time, but those on GaAs (111)A and (110) substrates decreased *exponentially* with increasing the purge time, with time constants of less than 0.2s.

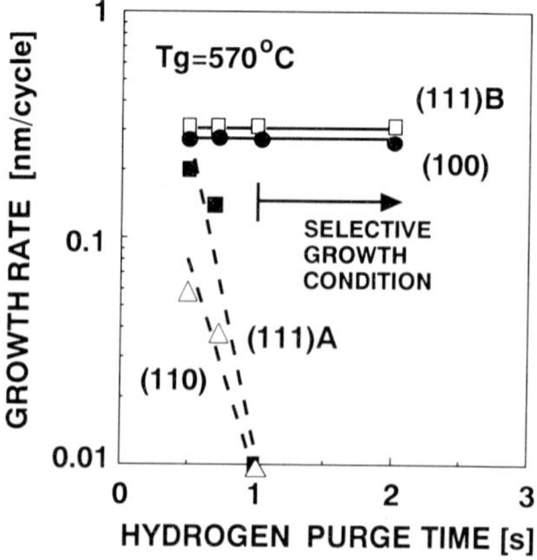

Figure 2 ALE growth rate of GaAs on variously oriented GaAs substrates at 570°C as a function of hydrogen purge time after AsH_3 supply.

For the mechanism of growth selectivity it is speculated that the limitation of growth rates on GaAs (111)A and (110) planes may be caused by limiting effect of the adsorption of a Ga precursor at a As site, which is formed on the surface by As desorption in the hydrogen purge time after AsH_3 supply(Isshiki 1993).

B. Application to fabrication of GaAs quantum wire structures

As shown in Fig.1, GaAs ALE growth with high selectivity between GaAs (100) and (111)A surfaces is possible in the temperature range from 560°C to 570°C. Moreover it is notable that the temperature dependence of the GaAs growth rate shows the possibility of GaAs selective growth between GaAs (100) surface and (110) surface. Also shown in Fig.2, using control of the hydrogen purge time after AsH_3 supply, change of the growth mode between isotropic (side wall) and anisotropic (selective) growth is possible without changing in growth temperature or flow rates of source gases. These features of ALE growth are very useful for fabrication of low–dimensional quantum structures.

The GaAs/$(AlAs)_1(GaAs)_1$ multi quantum well layers were grown on V–grooved GaAs (100)substrate with (111)A sidewalls, and on a GaAs(100) substrate with <011> direction SiO_2 stripe mask to observe crystallographic selective growth by ALE. GaAs/$(AlAs)_1(GaAs)_1$ MQW layers were grown at 570°C, using dimethylalminumhydride as the Al source (Ishizaki 1991). The flow rate was $3x10^{-7}$ mol/s and other growth conditions were the same as for GaAs growth.

Figure 3 shows the scanning electron microscope (SEM) cross–sectional photograph of the GaAs/$(AlAs)_1(GaAs)_1$ quantum wire structures grown on V–grooved substrate. GaAs and $(AlAs)_1(GaAs)_1$ layers consisted of 75 and 200 growth cycles respectively. The source gases were fed into the growth chamber with a 1 s hydrogen purge time between each gas supply. The $(AlAs)_1(GaAs)_1$ layers appear as dark lines by stain etching. As shown in fig.3 GaAs/$(AlAs)_1(GaAs)_1$ quantum wires with 20nm thickness

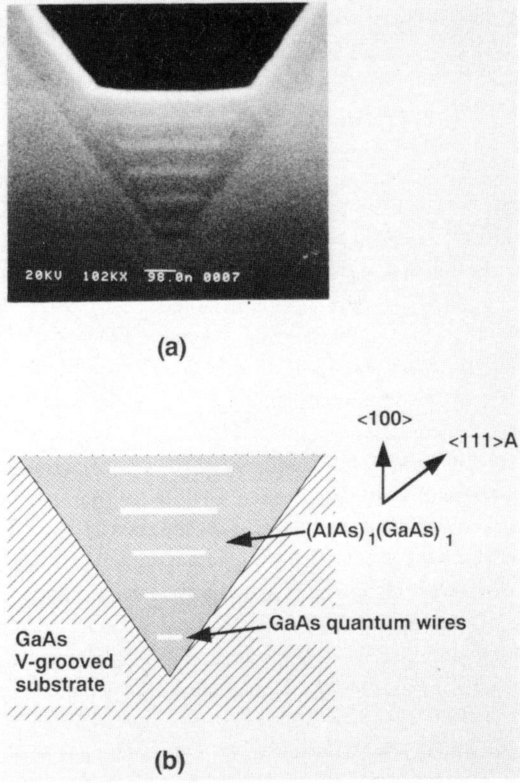

Figure 3 (a) SEM cross–sectional photograph, and (b) schematic structure of GaAs/$(AlAs)_1(GaAs)_1$ quantum wire structures grown on V–grooved substrate.

and 50nm width were obtained. GaAs epitaxial layers are grown in the <100> direction of V-grooves formed on the GaAs substrate, and no GaAs growth in the <111>A direction is observed. This growth selectivity, between GaAs (100) and (111)A surfaces, is consistent with the crystallographic orientation dependence of GaAs growth rate shown in Fig.1. Under this condition, the AlAs ALE growth on (100) GaAs surface has been achieved, however, the high contrast of growth rate between GaAs (100) and (111)A surfaces such as the GaAs ALE growth was not observed. In addition, an excess growth on the boundary region between GaAs (100) and (111)A surface, which is usually observed in conventional MOVPE growth on V-grooved GaAs substrates(Tsukumoto 1992), is not observed so that the epitaxial layer is very uniform. The reason for the absence of excess growth may be that excess source gases supplied by non-reaction source gas migration from the (111)A to the (100) region do not contribute to growth in the <100> direction due to the self limiting effect, which is particular to ALE.

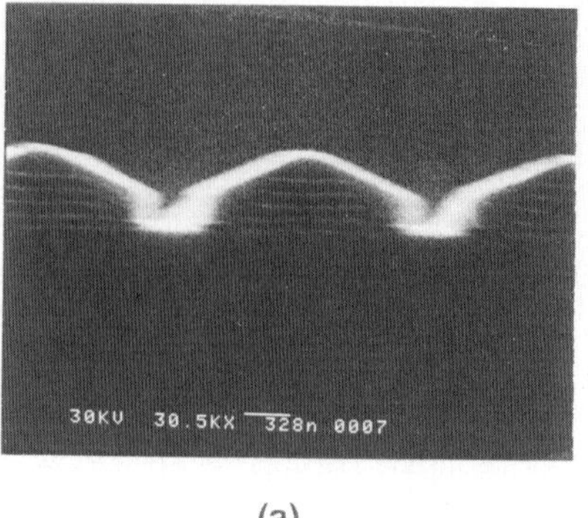

(a)

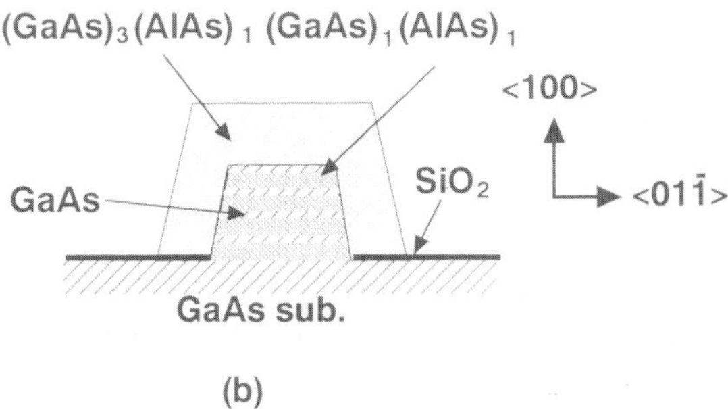

(b)

Figure 4 (a) SEM cross-sectional photograph, and (b) schematic structure of GaAs/$(AlAs)_1(GaAs)_1$ quantum well grown on (100) GaAs substrate with <011> direction SiO_2 mask.

Figure 4 shows the SEM cross-sectional photograph of the GaAs/ $(AlAs)_1(GaAs)_1$ MQW structures grown on <011> direction SiO_2 masked substrate. GaAs and $(AlAs)_1(GaAs)_1$ layers (MQW) consisted of 100 and 150 growth cycles respectively, and $(AlAs)_1(GaAs)_3$ top layer consisted of 300 growth cycles. The hydrogen purge times after AsH_3 supply in the MQW growth and the top layer growth were 1.5s and 0.3s respectively, and both growth temperatures were 570°C. The $(AlAs)_1(GaAs)_1$ layers appear as dark lines. GaAs and AlAs ALE growth is achieved in the <100> direction so that the epitaxial layer is very uniform. However, no growth in the <011> direction (lateral growth) is observed under the MQW growth condition. This growth selectivity, between the GaAs (100) and (110) surfaces, is consistent with the crystallographic orientation dependence for GaAs growth rate, shown in Fig.1. Furthermore, it is notable that no excess edge growth is observed, which is due to the self limiting effect. In contrast, side wall epitaxy is observed under the top layer growth condition, which can be achieved using only control of the hydrogen purge time after AsH_3 supply, without growth temperature or gas flow rate change. It is demonstrated that the fabrication of ideal low dimensional quantum structures, such as rectangular quantum wire arrays and quantum boxes, may be possible by crystallographic selective growth using ALE.

4.Conclusion

Crystallographic selective growth using atomic layer epitaxy (ALE) has been developed for the fabrication of low-dimensional quantum structures. In the temperature dependence of the GaAs growth rate, no GaAs growth on (111)A and (110) planes was observed in the high temperature range under the condition of GaAs ALE growth on GaAs (100) plane. Also it is shown that ALE growth mode change between isotropic and anisotropic growth, without growth temperature change, is possible by the control of hydrogen purge time. Due to the self limiting effect and these special features of ALE growth, trapezoidal shaped GaAs/$(AlAs)_1(GaAs)_1$ quantum wire structures were successfully realized. It is expected that the fabrications of ideal low dimensional quantum structures can be realized by the crystallographic selective growth using ALE.

References

Fukui T, Ando S and Fukai Y K 1990 Appl.Phys.Lett. **57**, 1209

Kapon E, Simhony S, Bhat R and Hwang D M 1989 Appl.Phys.Lett. **55**, 2715

Tsukamoto S, Nagamune Y, Nishioka M and Arakawa Y 1992 J.Appl.Phys. **71**, 533

Isshiki H, Iwai S, Meguro T, Aoyagi Y and Sugano T 1993, accepted for publication in Appl. Phys. Lett.

Ide Y, McDermott B T , Hashemi M, Bedair S M and Goodhue W D 1988 Appl.Phys.Lett. **53**, 2314

Usui A, Sunakawa H, Stutzler FJ and Ishida K 1990 Appl.Phys.Lett. **56**, 289

Iwai S, Meguro T, Doi A, Aoyagi Y and Namba S 1988 Thin Solid Filmes **163**, 405

Ishizaki M, Kano N, Yoshino J and Kukimoto H 1991 Jpn.J.Appl.Phys. **30**, L428

New method for maskless selective growth of InP wires on planar GaAs substrates

J. Ahopelto,[a] *H. Lezec, A. Usui[b] and H. Sakaki[c]

Quantum Wave Project, ERATO, Research Development Corporation of Japan (JRDC), 34 Miyukigaoka, Tsukuba, Ibaraki, 305 Japan.

*Fundamental Research Laboratories, NEC Corp., 34 Miyukigaoka, Tsukuba, Ibaraki, 305 Japan.

ABSTRACT: Maskless growth of InP wires on planar (100) GaAs substrates by hydride vapor phase epitaxy (VPE) is demonstrated. The selectivity of the growth was achieved by using focused ion beam (FIB) to modify locally the substrate surface. The nucleation of deposited InP is enhanced on the exposed areas leading to the selectivity. Continuous 200 μm long wires with sub-micron cross-sectional dimensions were obtained in a single growth process. TEM micrographs show that the number of dislocations in the wires is low, indicating the suitability of the present method for fabrication of nanoscale structures, *e.g.* quantum wires.

1. INTRODUCTION

The fabrication of nanoscale semiconductor structures, *i.e.* quantum boxes and quantum wires, has attracted attention due to the predicted improvement in the device performance. The fabrication of these structures typically consist of complicated lithographical, etching and regrowth steps. The simplification of the fabrication and the reduction of the number of the process steps would be desirable. Besides laser assisted growth [1-3], there are only few reports of maskless selective growth on planar substrates. Selectivity in the growth of diamond was achieved on ion-implanted silicon [4] and nanoscale InSb islands have been grown by droplet-epitaxy [5]. Recently the authors reported the growth of nanoscale InP islands on GaAs [6]. Here we report a new selective growth method with which well-defined wire-structures can be fabricated on planar substrates in one growth step. The method is based on the selectivity in the growth caused by local modification of the substrate surface by bombarding the surface by focused ion beam (FIB) prior to the growth. Using this method, sub-micron InP wires on planar GaAs substrates have been fabricated.

---

[a] Permanent address: VTT Semiconductor Laboratory, Otakaari 7 B, FIN-02150 Espoo, Finland.

[b] Permanent address: Fundamental Research Laboratories, NEC Corp., 34 Miyukigaoka, Tsukuba, Ibaraki, 305 Japan.

[c] Permanent address: Institute of Industrial Science, University of Tokyo, 7-22-1 Roppongi, Minato-ku, Tokyo 106, Japan.

2. EXPERIMENTAL

The fabrication process is shown schematically in Fig. 1. Semi-insulating GaAs (100) 2°-off substrates were bombarded by FIB, Fig. 1(a), followed by deposition of InP using atmospheric hydride vapor phase epitaxy (VPE), Fig. 1(b). The implanted patterns consisted of 200 μm long lines in [011] and [01$\bar{1}$] directions, separated by 0.5, 1.0, 2.0 or 5.0 μm spacings. The patterns were implanted at room temperature using JEOL JBL 150 FIB system with a base pressure of 10^{-7} Torr. Silicon ions with energies of 20 keV, 60 keV and 260 keV were used for implantation. The corresponding beam spot diameters were estimated to be 0.5 μm, 0.2 μm and 0.1 μm, respectively. The ions were extracted from a Au-Si-Be -liquid metal source. All the lines were scanned in one pass, controlling the dose by changing the dwell time on each pixel, spaced by 0.02 μm. Prior to the FIB exposure, the wafers were degreased and etched in H_2SO_4:H_2O_2:H_2O (3:1:1) solution. Just before loading into the FIB apparatus, the wafers were dipped in HCl and rinsed in deionized water. InP was grown from PH_3 and InCl in H_2 carrier. The partial pressure of InCl was 1.4×10^{-3} atm and the V/III ratio 7. The implanted substrates were dipped in HF:ethanol (1:9) solution, rinsed in ethanol and blown dry by N_2 before loading into the growth apparatus. Before the deposition of InP, the samples were heated up to 665 °C under As_4 overpressure and kept at that temperature for about 20 minutes. The deposition temperature for InP was 620 °C or 665 °C and the deposition time was 30 seconds. To explore the effect of the above thermal treatment, an implanted reference sample was fabricated using the same conditions but without InP deposition. Details of the growth apparatus and the InP deposition procedure are given in [6]. The grown structures were characterized by scanning electron microscope (SEM), crosssectional transmission electron microscope (XTEM) and atomic force microscope (AFM).

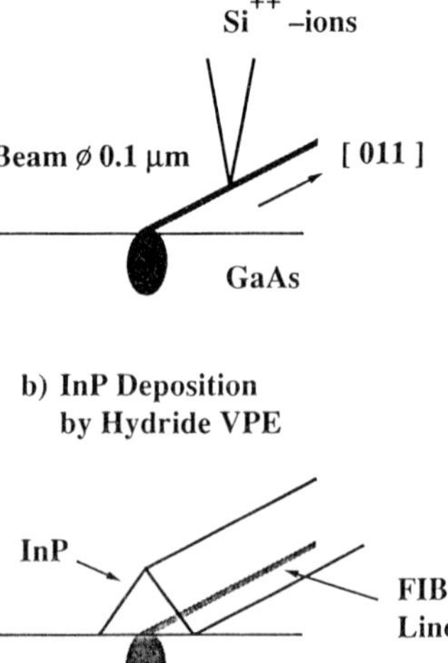

Fig. 1. The selective growth process shown schematically. (a) Lines are first drawn on the substrate by FIB. (b) InP is deposited on the exposed substrate by hydride VPE.

3. RESULTS AND DISCUSSION

Continuous wires were only obtained on lines implanted in [011] direction. In [01$\bar{1}$] direction, rows of detached islands were formed. The different behaviour of the growth in these two directions is demonstrated in Fig. 2.

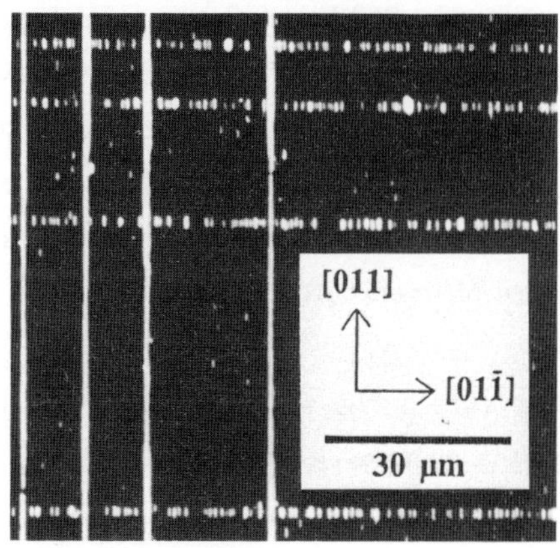

Fig. 2. Topview SEM micrograph showing the difference in growth of InP deposited on lines implanted in [011] and [01$\bar{1}$] directions. Continuous lines were obtained in the [011] direction but in the [01$\bar{1}$] direction rows of detached lines were formed.

The selectivity of the growth showed strong dependence on the implanted dose. Based on SEM micrographs from samples grown on substrates implanted with various doses, a minimum dose required to enhance the nucleation was determined. An example is shown in Fig. 3(a). The wires obtained at the threshold dose were not continuous but formed a row of elongated islands and some extra nucleation occurred between the implanted lines. The threshold line doses for wires grown at 665 °C were 3×10^9-1×10^{10} cm^{-1} being higher at lower energies. The doses correspond to average areal doses of 2×10^{14}-3×10^{14} cm^{-2} at all energies used. When the growth temperature was lowered to 620 °C, the threshold doses increased about tenfold. To obtain continuous wires, a dose of one order of magnitude higher than the threshold was needed. Best wires were

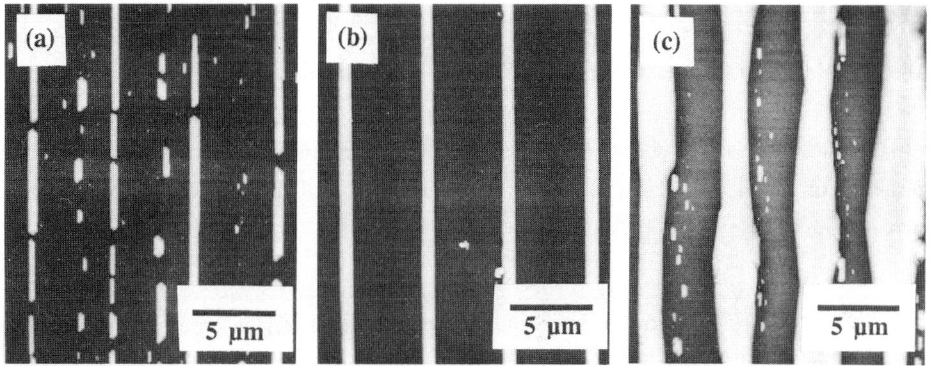

Fig. 3. Topview SEM micrographs showing the InP wire formation dependence on the dose. The lines were implanted using 260 keV Si^{++}-ions at the average areal doses of 3×10^{14} cm^{-2} (a), 3×10^{15} cm^{-2} (b) and 1×10^{16} cm^{-2} (c). The lines were in [011] direction with a 5 µm spacing between the lines. The samples were grown in the same run at 665 °C.

obtained by 260 keV Si^{++}-ions at line dose of 3×10^{10} cm^{-1}, corresponding an average areal dose of 3×10^{15} cm^{-2}, Fig. 3(b). At this dose the spacing of 0.5 μm was too narrow and InP grew also between the implanted lines. At the optimum dose the selectivity was very high for the spacings of 1.0, 2.0 and 5.0 μm, and continuous 200 μm long lines were obtained. Further increase in the dose resulted in rapid increase in the width of the wires due to the effect of beam tails [7], Fig. 3(c).

The wires grown at 665 °C on lines implanted with 260 keV Si^{++}-ions at line dose of 3×10^{10} cm^{-1}were 700 nm wide and 500 nm high. By lowering the growth temperature, smaller wires were obtained. At 620 °C, 230 nm wide and 160 nm high wires were obtained using the same implantation parameters, shown in Fig. 4.

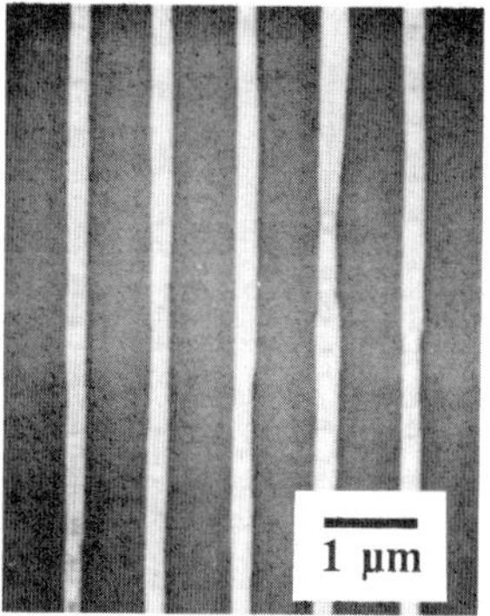

Fig. 4. Topview SEM micrograph of InP wires grown at 620 °C on lines implanted in [011] direction by 260 keV Si^{++}-ions at a dose of 3×10^{15} cm^{-2}.

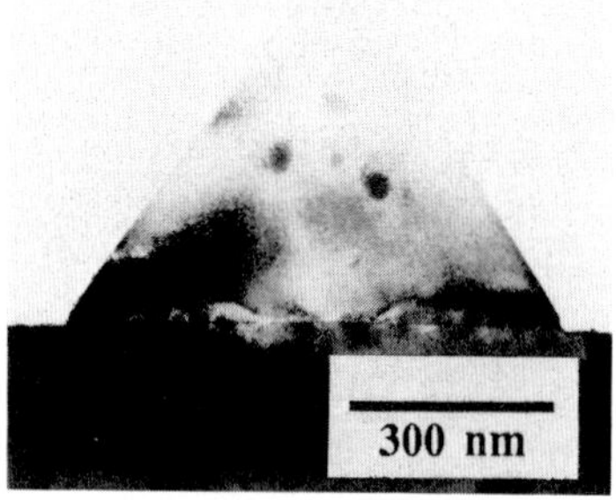

Fig. 5. XTEM micrograph of a wire in [011] direction showing the relatively low defect density in the interface. The implanted dose was 1×10^{15} cm^{-2} and the growth temperature 665 °C.

An XTEM micrograph of a wire grown along the [011] direction is shown in Fig. 5. The line dose was 1×10^{10} cm^{-1} corresponding an average areal dose of 1×10^{15} cm^{-2}. The cross-section is a triangle formed by the substrate surface and (111)B sideplanes. Despite of the large lattice mismatch of 3.8 %, the defect density in the center region of the wire is vanishingly small, increasing to a value of 10^4-10^5 cm^{-1} in the edges of the wire.

The threshold dose found in this work corresponds the value for the onset of rapid damage accumulation in Si^+-implanted GaAs [8]. The damage inside the crystal probably does not affect the nucleation on the surface directly. The damage is associated by introduction of strain in the lattice and this can modify the surface. Besides damage inside the crystal, FIB bombardment effectively sputters material from the substrate surface [9,10]. In Fig. 6.

is shown an AFM image of an FIB implanted square after a thermal treatment similar to one the grown samples underwent. The implanted species was 260 keV Si^{++}-ions at a dose of 1×10^{15} cm^{-1}. Some amount of material is removed from the surface resulting in surface roughness. Introduced steps and kinks on the surface enhance the nucleation probability of the deposited material. In the previous work the density of InP islands on GaAs was found to be a strong function of the surface step density [6], and this suggests that the surface steps play an important role also in the selective growth observed. Although the possible effect of residual strain cannot be excluded, the surface sputtering seems to be responsible for the locally enhanced nucleation of InP. The formed nuclei act as effective sink for the migrating surface species, leading to the selectivity.

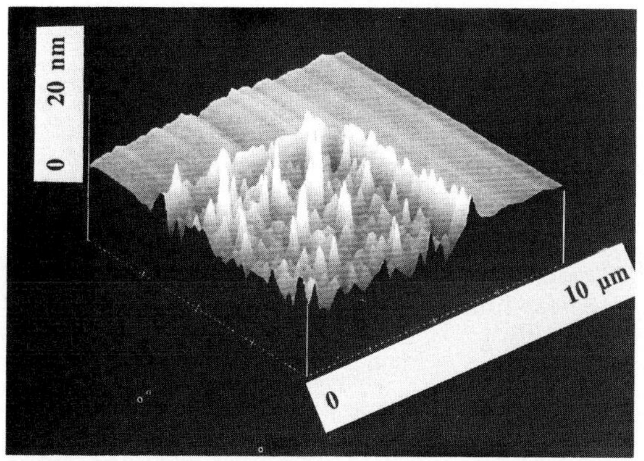

Fig. 6. An AFM image of a square implanted with 260 Si^{++}-ions at a dose of 1×10^{15} cm^{-1} after annealing 20 minutes at 665 °C. Notice the different horizontal and vertical scales.

The different behaviour of InP growth on lines implanted along the [011] and [01$\bar{1}$] directions is due to the slow growth rate of the {111}B planes [11]. In the [011] direction faster growing planes of the nuclei coalesce easily to form a continuous wire but in the [01$\bar{1}$] direction the formation of the slow growing (111)B planes prevents the coalescense and rows of detached islands are formed on the implanted lines. The slow growth rate of the {111}B planes can be used to control the size of the cross-section of wires grown in [011] direction. By decreasing the growth rate by lowering the growth temperature, smaller wires can be obtained. The reason for the increase in the threshold dose induced by lowering the deposition temperature is not clear. It may be due to slightly different temperature cycles during the growth.

The defect density in the wires is relatively low as compared to large area growth of InP on GaAs [12]. This may partly be due to the small growth area [13] and partly due to possible remaining strain in the wires.

4. SUMMARY

In summary, a new method for maskless selective growth of well-defined structures on planar substrates is reported. The selectivity is achieved by modifying locally the substrate surface by FIB bombardment prior to growth. The nucleation probability of the deposited material is enhanced on the bombarded areas, leading to the selectivity of the growth. Using the method, sub-micron InP wires on planar GaAs substrates was fabricated in one growth step by hydride VPE. Due to the simplicity, the present method is promising for the fabrication of various nanoscale structures, *e.g.* quantum wires.

ACKNOWLEDGMENTS

The authors would like to thank Y. Ochiai for the FIB arrangement and A. A. Yamaguchi, K. Nishi, H. Sunakawa and T. Baba for fruitful discussions.

REFERENCES

[1] S. Iwai, T. Meguro, A. Doi, Y. Aoyagi and S. Namba, Thin Solid Films **163**, 405 (1988).
[2] N. H. Karam, H. Liu, and S. M. Bedair, Appl. Phys. Lett. **52**, 1144 (1988).
[3] Q. Chen, J. S. Osinski, and P. D. Dapkus, Appl. Phys. Lett. **57**, 1437 (1990).
[4] S. J. Lin, S. L. Lee, J. Hwang and T. S. Lin, J. Electrochem. Soc. **139**, 3255 (1922).
[5] N. Koguchi, S. Takahashi and T. Chikyow, J. Cryst. Growth **111**, 688 (1991).
[6] J. Ahopelto, A. A. Yamaguchi, K. Nishi, A. Usui and H. Sakaki, Jpn. J. Appl. Phys. **32**, L32 (1993).
[7] T. Bever, G. Jäger-Waldau, M. Eckberg, E. T. Heyen, H. Lage, A. D. Wieck, and K. Ploog, J. Appl. Phys. **72**, 1858 (1992).
[8] T. E. Haynes and O. W. Holland, Appl. Phys. Lett. **58**, 62 (1991).
[9] R. L. Kubena, R. L. Seliger and E. H. Stevens, Thin Solid Films **92**, 165 (1982).
[10] H. Morimoto, Y. Sasaki, Y. Watakabe, and T. Kato, J. Appl. Phys. **57**, 159 (1984).
[11] D. W. Shaw, J. Cryst. Growth **31**, 130 (1975).
[12] M. Tamura, D. Olego, Y. Okuno and T. Kawano, in *Proc. 19th Int. Symp. Gallium Arsenide and Related Compounds*, Karuizawa, 1992, ed. by T. Ikegami, F. Hasegawa and Y. Takeda (IOP Publishing Ltd, Bristol, 1993), p. 151.
[13] E. A. Fitzgerald, G. P. Watson, R. E. Proano, D. G. Ast, P. D. Kirchner, G. D. Pettit, and J. M. Woodall, J. Appl. Phys. **65**, 2220 (1989).

Paper presented at the Int. Symp. GaAs and Related Compounds, Freiburg, 1993

Growth of InAlAs/InGaAs modulation doped structures on low temperature InAlAs buffer layers using trimethylarsenic and arsine by metalorganic chemical vapour deposition

N. Pan, J. Elliott, J. Carter, H. Hendriks, and L. Aucoin
Raytheon Company, Research Division, 131 Spring Street, Lexington, MA 02173

ABSTRACT: Modulation doped InAlAs/InGaAs structures grown on InP substrates incorporating a highly resistive InAlAs buffer layer were consistently demonstrated without a conductive impurity spike at the epitaxial/substrate interface. The highly resistive buffer layer (2 x 10^5 ohm-cm) was obtained at a growth temperature of 475°C using a combination of trimethylarsenic and arsine as the arsenic sources. Excellent surface morphology was obtained for the low temperature (LT) InAlAs buffer layer and the modulation doped structures. Two terminal breakdown voltage measurements performed on the LT InAlAs buffer layer showed a breakdown voltage exceeding 50 V confirming the high resistivity at high electric fields. The presence of a two dimensional electron gas in the modulation doped structure was confirmed by the observation of clear plateaus in the quantum Hall-effect at magnetic fields up to 17 T.

1. INTRODUCTION

Lattice matched InGaAs/InAlAs high electron mobility transistors (HEMT) have demonstrated excellent low noise performance at high frequencies. A noise figure as low as 1.2 dB at 94 GHz has been demonstrated (Duh, et al. 1991). High current gain cutoff frequencies exceeding 200 GHz have been shown (Enoki, et al. 1990, Mishra, et al. 1989, Nguyen, et al. 1992). However, the progress in device performance of InGaAs/InAlAs HEMT grown by the metalorganic chemical vapor deposition technique (MOCVD) has not been as rapid. One of the factors limiting the performance of MOCVD grown HEMT structures has been the quality of the epitaxial/substrate interface. A conductive impurity spike is often observed at the epitaxial/substrate interface. A number of researchers had associated the conductive spike to Si since it was frequently observed at the epitaxial/substrate interface by SIMS measurements (Bass, et al. 1984, Briggs, et al. 1987, Huber, et al. 1984, Ishikawa, et al. 1992, Svensson, et al. 1991). The highest current gain cutoff frequency achieved by a MOCVD grown HEMT was 187 GHz (Adesida, et al. 1993). Fe doping in both the InP and InAlAs buffer layers was used to suppress the conductivity of the epitaxial/substrate interface.

In this work, InGaAs/InAlAs HEMT structures incorporating a highly resistive InAlAs buffer layer were routinely grown without a conductive impurity spike at the epitaxial/substrate interface. The high resistivity could only be realized by growing the InAlAs at a low growth temperature (LT) of 475°C using a combination of trimethylarsenic (TMAs) and arsine as the arsenic sources. The breakdown voltage of the LT InAlAs layer as determined by two terminal measurements exceeded 50 V confirming the high resistivity at high electric fields. The high quality of a HEMT structure grown on top of a LT InAlAs buffer layer was verified by the presence of clear plateaus in the quantum Hall-effect.

2. EXPERIMENTAL RESULTS AND DISCUSSION

The structures were grown in an atmospheric pressure MOCVD reactor. Trimethylindium (TMI), trimethylaluminum (TMA), trimethylarsenic (TMAs), and pure arsine were used as the source materials for the LT InAlAs layers. TMI, TMA, and arsine were used for the growth of conventional InAlAs layers. TMI, trimethylgallium (TMG), and arsine were used to grow InGaAs layers. Tertiarybutylphosphine (TBP) was used as a phosphorous overpressure during the oxide removal step prior to epitaxial growth. The arsine was purified using an In-Ga-Al eutectic for oxygen and water vapor removal. Disilane diluted in high purity hydrogen (50 ppm) was used as the dopant. The TMG, TMAs, TMA and TMI bubbler temperatures were maintained at -10.0°C, 10.0°C, 18.1°C and 19.9°C respectively. Fe doped InP substrates (100) with 2° off orientation towards the (110) direction were used. High magnetic field Hall-effect measurements were performed on a conventional Hall-bar pattern at 4.2 K in a Bitter solenoid with the magnetic field applied perpendicularly to the sample surface. A maximum field of 17 T was used. Ohmic contacts were formed by alloying pure In at 350°C for 2.5 minutes in a N_2/H_2 atmosphere. Electrochemical capacitance voltage (C-V) measurements were performed using a Polaron Profiler. Secondary ion mass spectroscopy (SIMS) was performed using a 14.5 kV Cs primary ion beam. The negative ions were collected from an image area of 30 mm in diameter. The primary current was 0.065 nA resulting in an average sputter rate of 7 Å/s.

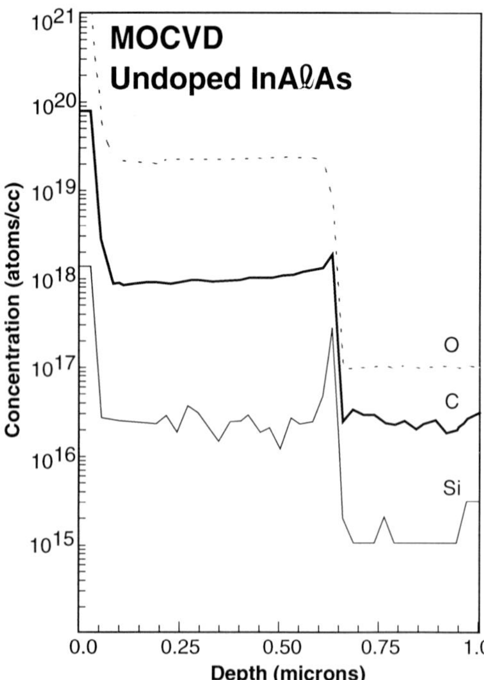

Figure 1. SIMS profile of a conventional undoped InAlAs layer is shown. High levels of C, O, and Si are detected at the epitaxial/ substrate interface. C, O, and Si are present in the layer.

The HEMT structure consisted of a 850 Å LT InAlAs buffer layer, a 750 Å undoped conventional InAlAs buffer layer, a 400 Å undoped InGaAs channel layer, a 30 Å undoped InAlAs spacer layer, a Si delta-doped layer, a 200 Å undoped InAlAs Schottky layer, and a 100 Å undoped InGaAs cap layer. The growth temperature for conventional InAlAs and InGaAs layers was 650 °C and the growth temperature for the LT InAlAs layer was 475°C.

The presence of a conductive impurity spike at the epitaxial/substrate interface has been commonly observed in structures grown on InP substrates as shown by an n-type conductive spike in the C-V profile. Figure 1 shows a typical SIMS profile of a conventional undoped InAlAs layer. The presence of Si, C, and O are clearly revealed at the epitaxial/substrate interface. It was believed that Si rather than C or O is responsible for the n-type impurity spike. Since an n-type impurity spike was evident, it was desired to electrically compensate the spike with a p-type buffer layer or a highly resistive buffer layer. The introduction of p-type or semi-insulating dopants onto the MOCVD system was not attempted due to the uncertain memory effects associated with these dopants.

The growth of p-type InAlAs buffer layers using TMAs as the sole arsenic source was initially investigated in an attempt to achieve high carbon doping. TMAs was chosen as the arsenic source since it had no memory effect and it was previously shown to result in very high carbon incorporation in the growth of GaAs yielding heavily doped p-type GaAs (Kuech, et al. 1988, Neumann, et al. 1990, Shimazu, et al. 1990). However, undoped InAlAs grown using only TMAs as the arsenic source showed n-type (5×10^{15} cm^{-3} - 3×10^{16} cm^{-3}) rather than p-type conductivity with poor surface morphologies in the growth temperature range of 550°C to 650°C. Undoped InAlAs layers grown using only arsine as the arsenic source showed very good surface morphologies with n-type conductivity ($1\text{-}3 \times 10^{16} cm^{-3}$) in the same temperature range. Attempts in growing InAlAs at temperatures below 550°C using either arsine or TMAs resulted in extremely poor surface morphologies due to insufficient arsenic overpressure. Since it was not possible to realize a p-type InAlAs or a highly resistive InAlAs buffer layer using either TMAs or arsine in the growth temperature range of 550°C to 650°C, both of the sources were simultaneously used at lower growth temperatures (< 550°C) in an attempt to improve the cracking efficiency. The combination of TMAs and arsine was previously used to control the impurity concentration in GaAs and InGaAs but this combination had not been investigated for LT growth (Kuech, et al. 1988, Dietze, et al. 1981).

Sample	T^G	TMAs (SCCM)	A_sH_3 (SCCM)	ρ_{300} (Ω-cm)
A	550°C	10	2	1.05×10^{-1}
B	500°C	20	2	3.80×10^{1}
C	475°C	40	3	2.00×10^{5}

Table 1. 300 K resistivities as a function of growth temperature and TMAs and arsine flow are shown.

Table I shows the resistivities of LT InAlAs layers (≈ 1.0 μm) as a function of the growth temperature using a combination of TMAs and arsine. A growth temperature of 475°C was found to yield the highest resistivity (2×10^5 ohm-cm) InAlAs layer. The decrease in the growth temperature was mostly responsible for the high resistivity since changes in the As mole fraction at a particular growth temperature did not change the resistivity of the sample. The resistivity of this layer did not change appreciably (1×10^5 ohm-cm) after a 600°C anneal under an arsine overpressure confirming the stability of this layer. The As mole fraction was increased at lower growth temperatures because of the decreased cracking efficiency of both As sources. The surface morphologies of the InAlAs layers under an optical microscope at low (10 X) and high magnifications (250 X) resembled the InP substrate with the exception of the edges. The fact that excellent surface morphologies could be obtained at a growth temperature as low as 475°C suggested that the cracking efficiency of both arsine and TMAs was significantly enhanced when the sources were used in combination.

The mechanism for enhanced cracking efficiency when both As sources are simultaneously used will require future investigation. The thickness uniformity of this layer across a two inch InP wafer was ±3%. Low temperature photoluminescence measurement performed at 4.2 K revealed no optical emission at the expected InAlAs transition. Double crystal X-ray diffraction of the LT InAlAs layer showed a weak and broad shoulder located at +100 arcseconds from the InP substrate peak confirming that the LT InAlAs layer was lattice matched. Evaporated ohmic contacts with 2.0 μm spacing was formed and the measured breakdown voltages (100 μA) exceeded 50 V confirming the high resistivity under high electric fields (250 kV/cm). It is essential for a buffer layer in a HEMT structure to have a high breakdown voltage in order obtain good pinch-off characteristics. A SIMS profile of the LT InAlAs layer is shown in Figure 2. The concentration levels of C, O, and Si in the layers are not much different in comparison to a conventional InAlAs layer (Figure 1). Impurity spikes located at the epitaxial/substrate interface are not observed in this particular sample. The addition of TMAs did not increase the C concentration in InAlAs in contrast to the increased C incorporation observed in the growth of GaAs (Kuech, et al. 1988). The high resistivity of the LT InAlAs layer is probably due to the presence of deep levels as a result of the lower growth temperature rather than the compensation of shallow impurities.

The effectiveness of the LT InAlAs buffer layer was demonstrated in the growth of InAlAs/InGaAs HEMT structures. The surface morphologies of the HEMT structures grown with the LT InAlAs buffer layer were excellent. Figure 3 shows the SIMS profile of the HEMT structure. The SIMS profile was performed at a lower primary current (100 nA) than the previous ones in order to monitor the Si delta-doped region located near the surface. The presence of Si is again evident at the epitaxial/substrate interface. The Si peak located at about 300 Å below the surface corresponds to the Si delta-doped region. The Al trace indicates the regions of InAlAs and the drop in the Al level corresponds to the undoped InGaAs channel layer. Figure 4 shows the Polaron profile of the same HEMT structure. The peak in the C-V profile corresponds to the Si delta-doped region in close agreement to that measured

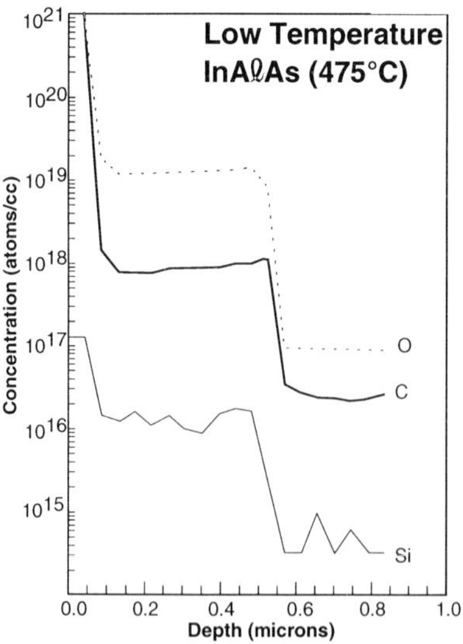

Figure 2. SIMS profile of a LT InAlAs grown using TMAs and arsine is shown. C, O, and Si are detected throughout the epitaxial layer.

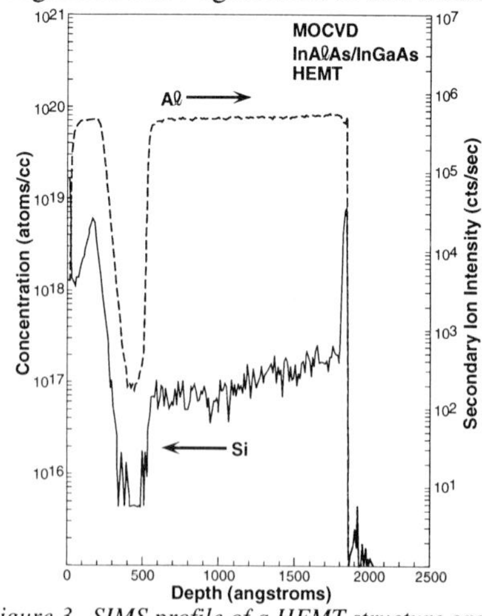

Figure 3. SIMS profile of a HEMT structure grown on top of a LT InAlAs buffer layer is shown. Si is present at the epi/sub interface. The Si peak at about 200 Å below the surface corresponds to the Si delta-doped region. The Al trace corresponds to the InAlAs regions.

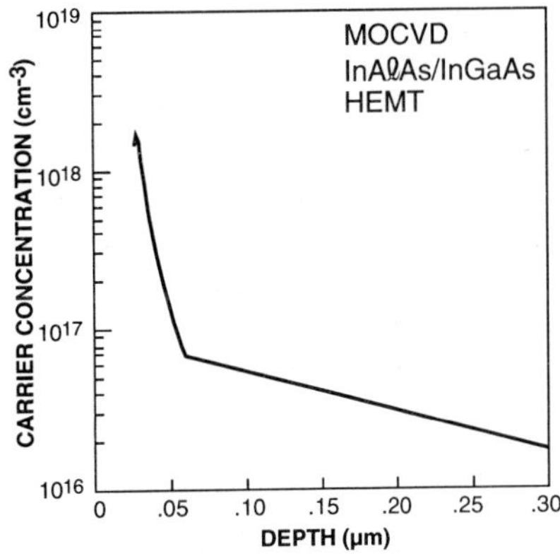

Figure 4. C-V profile of a HEMT structure grown on top of a LT InAlAs is shown. The peak in the profile corresponds to the Si delta-doped region. A conductive impurity spike situated at the epitaxial/substrate interface is clearly absent.

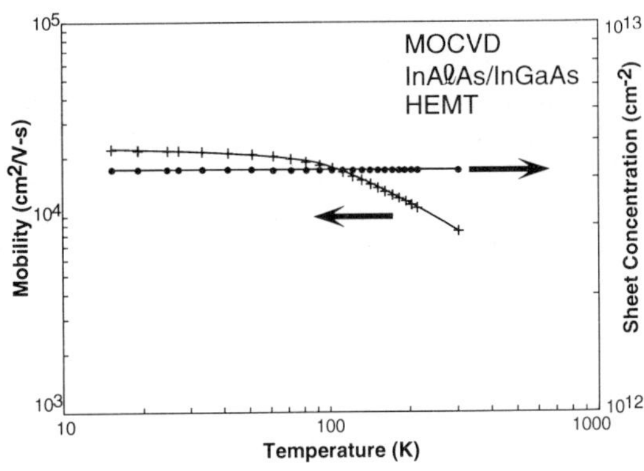

Figure 5. Variable temperature Hall-effect measurements performed on a HEMT structure grown on top of a LT InAlAs buffer layer is shown. The mobility at low temperatures saturates at 22,000 cm^2/V-s. The sheet carrier concentration (4.2 x 10^{12} cm^{-2}) remained constant throughout the entire temperature range.

by SIMS. However, there was no evidence in the C-V profile of an n-type impurity spike at the epitaxial/substrate interface. HEMT structures that were previously grown without the incorporation of a LT InAlAs buffer layer frequently showed a conductive impurity spike at the epitaxial/substrate interface. The presence of deep levels in the LT InAlAs buffer layer is suggested to compensate the conductivity of Si impurities at the epitaxial/substrate interface.

Variable temperature Hall-effect measurements performed on a HEMT structure is shown in Figure 5. The mobilities at 300 K, 77 K, and 10 K are 8,600 cm^2/V-s, 20,000 cm^2/V-s, and 22,000 cm^2/V-s respectively. These measurements are similar to HEMT structures that were grown on conventional undoped InP buffer layers. The mobility as the temperature was lowered down to 10 K showed a plateau indicating a reduction in ionized impurity scattering as would be expected in high quality two dimensional electron gas (2DEG). The sheet density shows no carrier freeze-out and it remains constant throughout the entire temperature range at 4.2 x 10^{12}cm^{-2}. High magnetic field Hall-effect measurements performed on the same sample is shown in Figure 6. Clear plateaus (ρ_{xy}) in the quantum Hall-effect is observed confirming the existence of a 2DEG. The Fast Fourier Power Transform of the magnetoresistance oscillations (ρ_{xx}) revealed the occupation of two subbands with a combined sheet density of 3.6 x 10^{12} cm^{-2} in reasonable agreement to the sheet density obtained from variable temperature Hall-effect measurements. The sheet density value obtained from the magnetoresistance measurements is more accurate since the mobility of the subbands are separated in the determination of the sheet density.

3.0 CONCLUSIONS

The growth of InAlAs using a combination of TMAs and arsine permitted a lower growth temperature resulting in a highly resistive InAlAs buffer layer. The maximum resistivity achieved was 2 x10^5 ohm-cm at a growth temperature of 475°C. The high resistivity of the LT InAlAs buffer layer was maintained at an electric field as high as 250 kV/cm. InAlAs/InGaAs HEMT structures grown with the incorporation of the LT InAlAs buffer layer consis-

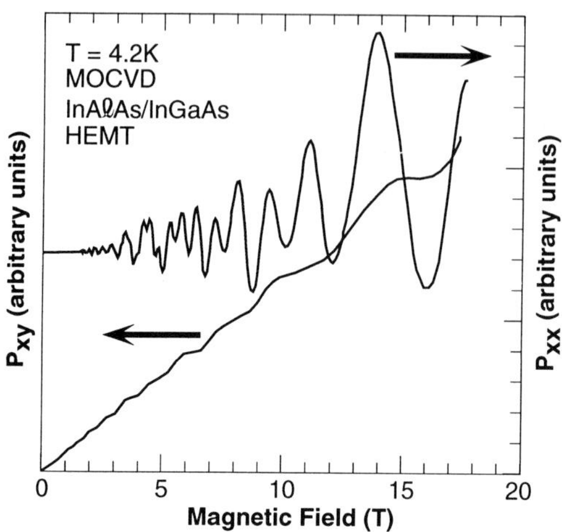

Figure 6. High magnetic field Hall-effect measurements performed on a HEMT structure grown on top of a LT InAlAs buffer layer is shown. The plateaus in the quantum Hall-effect (ρ_{xy}) are very clear. The magnetoresistance oscillations (ρ_{xx}) correspond to occupation of two electron subbands with a combined sheet density of 3.6 x 10^{12} cm^{-2}.

tently showed the absence of conductive impurity spikes at the epitaxial/substrate interface. Excellent transport properties were verified by variable temperature Hall-effect measurements and the presence of clear plateaus in the quantum Hall-effect. The ability of the MOCVD growth technique to deposit layers at a low growth temperature with highly resistive electrical properties by using a combination of group V sources opens up a new growth regime which has not been previously explored.

ACKNOWLEDGEMENTS

The authors wish to thank H. Statz, D. Massé, B. Hoke, and S. Shanfield for their unconditional support of the MOCVD HEMT research effort. Assistance in the preparation of the manuscript from M. Stock is greatly appreciated.

REFERENCES

Adesida I, Nummila K, Tong K, Caneau C, and Bhat R 1993 in InP and Related Materials Conf. Proc. Paris, France (IEEE, Piscataway, NJ, 1993) 405
Bass S J and Young M L 1984 J. Cryst. Growth **68** 311
Briggs A T R, and Butler B R, 1987 J. Cryst. Growth **85** 535
Dietze W T, Ludowise M J, Cooper C B, 1981 Electron. Lett. **17** 698
Duh K H G, Chao P C, Liu S M J, Ho P, Kao M Y, and Ballingal J M 1991 IEEE Microwave and Guided Wave Lett. **1** 104
Enoki T, Arai K, Ishii Y, and Tamamura T 1991 Electron. Lett. **27** 115
Huber A M, Razeghi M, Morillot G 1984 in Proc. Intl. Symp. on GaAs and Related Compounds Inst. Phys. Conf. Ser. **74** 223
Ishikawa H, Miwa S, Maruyama T, and Kamada M, 1992 J. Appl. Phys. **71** 3898
Kuech T K, Tischler M A, Wang P J, Scilla G, Potemski R, Cardone F, 1988 Appl. Phys. Lett. **53** 1317
Mishra U K, Brown A S, Jelloian L M, Thompson M, Nguyen L D, 1989 IEDM Tech.Dig. 101
Neumann G, Lauterbach Th, Maier M, and Bachem K H, 1990 in Proc. Intl. Symp. on GaAs and Related Compounds Inst. Phys. Conf. Ser. **112** 167
Nguyen L D, Brown A S, Thompson M A, Jelloian L M, Larson L E, and Matloubian M 1992 IEEE Electron Dev. Lett. **13** 143
Shimazu M, Kimura H, Kamon K, Shirakawa T, Murai S, and Tada K 1990 in Proc. Intl. Symp. on GaAs and Related Compounds Inst. Phys. Conf. Ser. **112** 173
Svensson S P, Beck W A, Mautel D C, Uppal P N, and Cooke D C, 1991 J. Cryst. Growth **111** 450

Intramolecular and intermolecular alane-adducts for the growth of $Al_xGa_{1-x}As$ by atmospheric and low pressure MOVPE

B.P. Keller, R. Franzheld, G. Franke, V. Gottschalch, R. Schwabe*, S. Keller
Universität Leipzig, Fachbereiche Chemie und Physik *
Linnéstr. 3-5, D-04103 Leipzig, Germany

U. Dümichen
Martin-Luther-Universität Halle / Wittenberg, Fachbereich Chemie
Geusaer Str., D-06217 Merseburg, Germany

ABSTRACT: We present results of the atmospheric and low pressure MOVPE growth of AlAs/GaAs and $Al_xGa_{1-x}As$ heterostructures using the precursors methylalane-trimethylamine ($MAlH_2$-TMN) and dimethylaminopropylalane ($DAAlH_2$). The alanes were used in combination with trimethylgallium (TMGa), triethylgallium (TEGa), and arsine as group-III and Group-V precursors, respectively. The reactivity and the mechanism of the decompositon for the growth of AlAs and $Al_xGa_{1-x}As$ are discussed. Electrical and optical characterisation of the $Al_xGa_{1-x}As$ layers revealed the intramolecular alane as an interesting group-III precursor.

1. INTRODUCTION

The problem of high carbon and oxygen uptake into epitaxial layers of Al-containing III-V semiconductors grown by metalorganic vapor phase epitaxy (MOVPE) has been tried to overcome with increasing effort by the application of hydrogen instead of methyl-substituted Al-metalorganics (Roberts et al. 1990,Hobson et al. 1991, Protzmann et al. 1991, Olsthoorn et al. 1992). Using the intermolecular adduct trimethylamine alane (TMAA) the residual carbon incorporation into AlAs and $Al_xGa_{1-x}As$ (if combined with triethylgallium, TEGa) could be largely reduced (Hobson et al. 1991). However, these alane precursors also show disadvantageous effects under MOVPE-growth conditions like homogeneous prereactions in the gas phase resulting for example in methyl-exchange with trimethylgallium (Grady et al. 1990) and a nonuniform composition of the layers in the case of $Al_xGa_{1-x}As$ epitaxy at higher reactor pressures (Roberts et al. 1990).
We present for the first time, to our knowledge, results of the MOVPE growth of AlAs/GaAs and $Al_xGa_{1-x}As$ heterostructures using the precursors methylalane-trimethylamine ($MAlH_2$-TMN) and

dimethylaminopropylalane ($DAAlH_2$) (fig.1). Our attempt was to modify the molecular structure of the alane in such a way, that a resonably stable molecular configuration is achieved with the Al-atom bonded mainly to hydrogen.

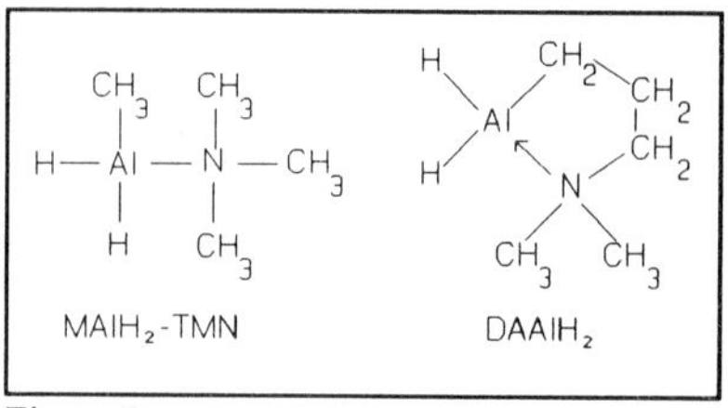

Figure 1.

2. EXPERIMENTAL

The alane compounds were prepared in our laboratory by the reaction of methylaluminiumdichloride-trimethylamine with litiumchloride and lithiumaluminiumchloride with N,N-dimethylallylamine, respectively. The physical properties of the alane adducts are summarized in table 1. The chemical structure and the composition of the adducts were verified by nuclear magnetic resonance spectroscopy (NMR) and x-ray diffraction analysis.

The MOVPE experiments were performed in horizontal reactor systems one operating at normal atmospheric pressure (AP) and the other at reduced pressure (20 mbar, LP). TEGa, TMGa, and arsine (100%) were used as additional group-III and Group-V precursors, respectively.

Table 1

	$DAAlH_2$	$MAlH_2$-TMN
p_{eq} (20°C) / mbar	0.2	2.5
melting point / °C	37	- 35

The total gas hydrogen flow amounted to 6 l/min (AP) and 7 l/min (LP) yielding gas velocities of about 100 and 900 cm/s at growth temperature in the AP and LP reactor tubes, respectively. We investigated heterostructures of AlAs / GaAs and $Al_xGa_{1-x}As$ / GaAs to evaluate the properties of the alanes for the MOVPE grow that temperatures t_{gr} between 450 - 800 °C. The layers were investigated using high resolution X-ray diffraction (HRXRD), low temperature photoluminescence (PL), and Hall-measurements (standard van der Pauw-technique).

3. RESULTS AND DISCUSSION

The normalized growth rate of AlAs (related to the alane input partial pressure, $p°_{III}$) as a function of the growth temperature is shown in figure 2. for AP and LP conditions. Generally, the growth rate using the alane precursors is diffusion controlled even at temperatures as low as 450°C. Upstream of the susceptor we found no deposition on the cold parts of reactor tubes neither at atmospheric nor at reduced pressure. At AP we observed at t_{gr} > 700°C an enhanced deposition on the hot walls of the quartz tube opposite to the susceptor. This deposition is more pronounced

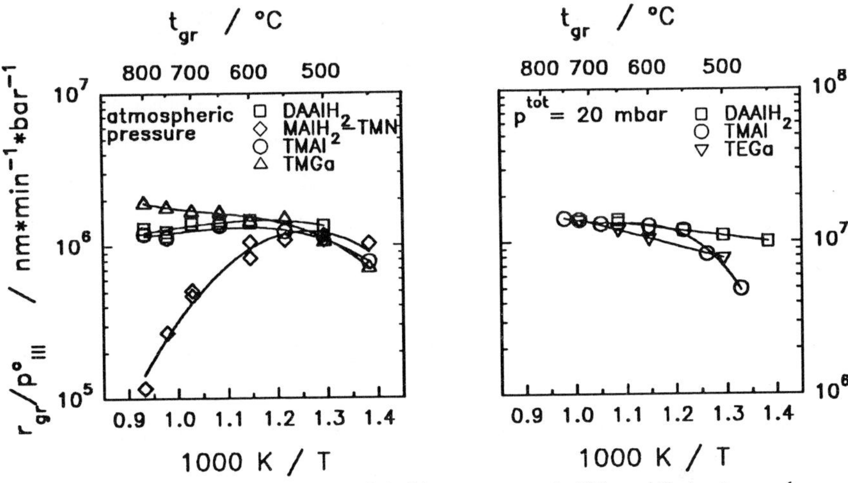

Fig. 2. Normalized growth rate of the binary compounds AlAs and GaAs at normal atmospheric and low pressure using different precursors

in the case of $MAlH_2$-TMN than in the case of $DAAlH_2$ explaining the decrease of the AlAs growth rate using $MAlH_2$-TMN. However, we could not find any of indication of a further increased predeposition in the case of $Al_xGa_{1-x}As$ due to alkyl-exchange with the gallium precursor. Comparing this with the results of Roberts et al. (1990) one can establish a relation of thermal and chemical stability of the alanes as follows: TMAA < $MAlH_2$-TMN < $DAAlH_2$ < TMAl.

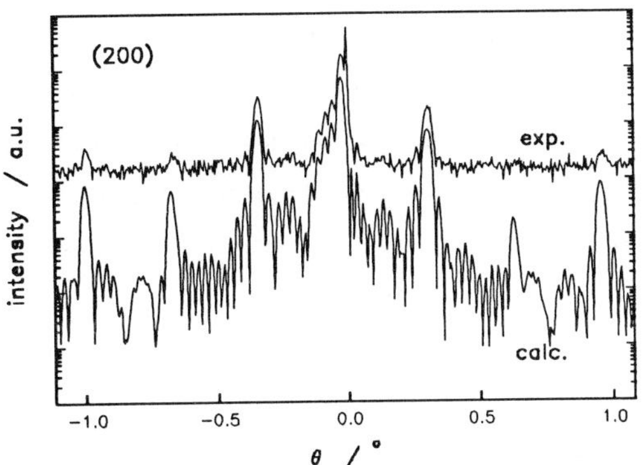

Fig. 3. Experimental and calculated (200) HRXRD-curves of a 10 period AlAs-GaAs superstructure

The experimental and simulated (200) HRXRD rocking curves of 10 period superstructures of (8 nm AlAs - 7 nm GaAs) with 50 nm AlAs and 240 nm GaAs cladding layers grown with $DAAlH_2$ and TEGa are shown in fig. 3. In addition to the main superlattice peak, we observed satellite peaks up to the 3rd order indicating a very good lateral and vertical homogenity of the layer structure. The small full width of half maximum (FWHM) of the satellites and the Pendellösung fringes around the satellite peaks point at sharp AlAs-GaAs heterointerfaces in the dimension of one monolayer. Using the above mentioned layer parameters, which were confirmed by cross edge transmission electron microscopy, the theoretical curve corresponds excellently with the HRXRD experiment.

Next, we investigated the growth of $Al_xGa_{1-x}As$ using the more stable alane $DAAlH_2$. The alloy

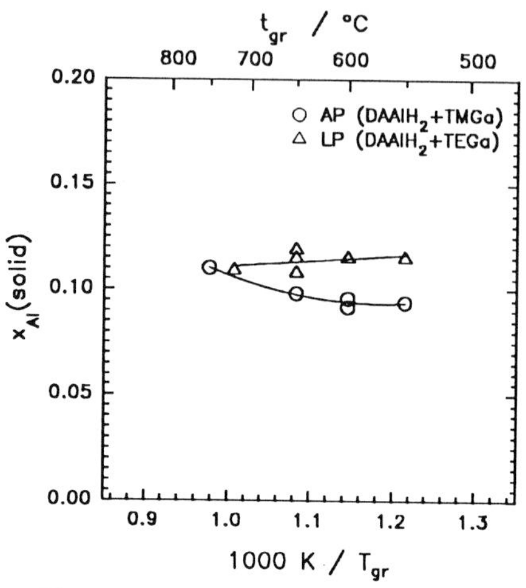

Fig. 4. Composition of $Al_xGa_{1-x}As$ grown at atmospheric and low pressure using $DAAlH_2$ as function of the growth temperature

composition at fixed gas phase mole fractions of 0.10 and 0.12 for AP and LP, respectively, as a function of the growth temperature is shown in fig. 4. Using the alane in combination with TMGa as well as with TEGa $Al_xGa_{1-x}As$ layers can be grown in a wide temperature range. The alloy compositions correspond well with the expected values calculated from the growth rates of the binary compounds AlAs and GaAs. Only at 750°C we found a slight increase of the Al solid mole fraction in the case of atmosheric pressure growth. This could be an indication of an enhanced depletion of gallium species due to prereactions with the alane (alkyl-exchange) as reported also in the case of the TMAA-TMGa and TMAA-TEGa combinations by Roberts et al.(1990), Schneider et al.(1992), and Pitts et al.(1993) respectively. Using the alane $DAAlH_2$ these effects were diminished under low pressure growth conditions (see fig. 4.) due to the reduced collision probability of the gas molecules.

The results of the electrical characterisation of the $Al_xGa_{1-x}As$ layers grown at low pressure are summarized in table 2. All investigated $Al_xGa_{1-x}As$ samples showed n-type conductivity. Obviously, the relatively high net doping level results from the not yet optimized preparation procedure of the alane. Similar results were obtained for atmospheric and low pressure growth. We observed increasing compensation ratios at higher Al solid mole fractions which originate from residual impurities.

Table 2

	x_{Al}(solid)	μ_{300} / cm^2/Vs	μ_{77K} / cm^2/Vs	n_{300K} / cm^{-3}
$DAAlH_2$/TMGa	0.18	1840	2740	$5.3*10^{16}$
$DAAlH_2$/TEGa	0.12	2040	3510	$1.5*10^{16}$

Typical $Al_xGa_{1-x}As$ photoluminescence spectra recorded at 4.2 K are shown in fig. 5. The excitation densities amounted 1 W/cm^2 (spectrum (a)) and 19 W/cm^2 (samples (b+c). The recombination of bound excitons (BE) and donor-acceptor-pair luminescence were observed. Samples grown with the combination $DAAlH_2$/TEGa at reduced pressure exhibited FWHM of 5 meV at $x_{Al}=0.17$. In comparison with previously reported values using TMAA we attribute the somewhat increased FWHM

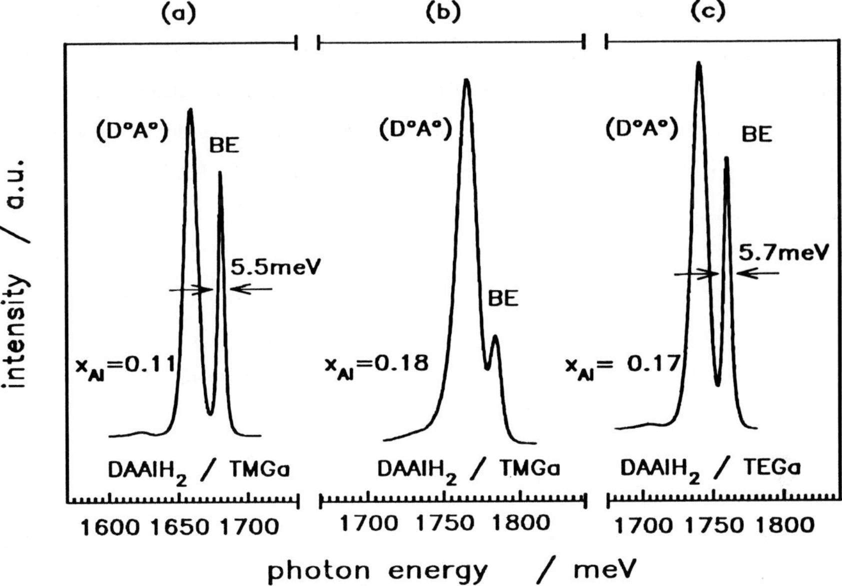

Fig. 5. 4.2 K PL of $Al_xGa_{1-x}As$ grown at (a) atmospheric pressure and (b+c) low pressure using the alane $DAALH_2$ in combination with TMGa and TEGa, respectively

to the residual impurity contamination of our alane.

Investigating samples grown with $DAAlH_2$ at reduced pressure an increased intensity of the impurity pair luminescence was observed in the case of using TMGa instead of TEGa. Additionally, the relative intensity of the bound exciton decreased and the exciton FWHM increased using TMGa. This again points to the affinity of methyl- groups from TMGa to Al as observed also for the TMAA adduct (Hobson et al.1991). However, we could not differentiate exactly between intrinsic carbon or residual impurities due to the not optimized preparation as impurity source. This investigations will be the subject of forthcoming investigations.

4. CONCLUSIONS

Results of the MOVPE growth using the new alane precursors $DAAlH_2$ and $MAlH_2$-TMN were presented. Layers of AlAs and $Al_xGa_{1-x}As$ could be grown in a wide range of growth temperatures at atmospheric and low pressure. The thermal and chemical stability of the alanes as follows the relation TMAA < $MAlH_2$-TMN < $DAAlH_2$ < TMAl. Electrical and optical characterisation of the $Al_xGa_{1-x}As$ layers revealed the intramolecular alane $DAAlH_2$ as a promising group-III precursor for atmospheric and low pressure MOVPE.

ACKNOWLEDGMENTS

The authors wish to thank G. Benndorf, E. Kramer, S. Kriegel, and G. Wagner for the help in the physical characterisation of the AlGaAs. Parts of this work were supported by the Bundesministerium für Forschung und Technologie under contract No. 412-4001-01 BT 106/3 and by the Deutsche Forschungsgemeinschaft.

REFERENCES

Grady A S, Markwell R D, Russell and Jones A C 1990 J.Cryst.Growth **106** 139

Hobson W S, Harris T D, Abernathy C R, Pearton S J 1991 Appl.Phys.Lett. **58** 77

Pitts B L, Emerson D T, Shealy J R 1993 Appl.Phys.Lett. **62** 1821

Protzmann H, Marschner T, Zsebök O, Stolz W, Göbel E O, Dorn R, Lorberth J 1991 J.Cryst.Growth **115** 248

Roberts J S , Button C C, David J P R, Jones A C, Rushworth S A 1990 J.Cryst.Growth **104** 857

Schneider jr. R P, Bryan R P, Jones E D, Biefield R M, Olbright G R 1992 J.Cryst.Growth **123** 487

Growth parameter optimization for multiwafer production of GaAs/$Al_xGa_{1-x}As$ solar cells on 4" substrates

B Marheineke, J Knauf*, D Schmitz*, H Jürgensen*, M Heuken and K Heime

Institut für Halbleitertechnik, RWTH Aachen, Templergraben 55, 52056 Aachen, FRG
*Aixtron Semiconductor Technologies GmbH, Kackertstr. 15-17, 52072 Aachen, FRG

ABSTRACT: The growth conditions in a commercial PLANETARY multiwafer MOVPE reactor were optimized for large scale production of GaAs/$Al_xGa_{1-x}As$ solar cells to reduce the costs of epitaxy. This reactor type shows growth properties similar to common horizontal reactors making transfer of process technology easy. The growth efficiency of the group III precursors is as high as 42.2 %. The variation of GaAs film thickness is less than 1.25 % across 4" wafer area. A GaAs solar cell without anti-reflection coating shows a promising efficiency of 12.3 %.

1. INTRODUCTION

GaAs solar cells have become of major interest for spacecraft power supply. Providing high efficiencies, thermal stability and radiation hardness they are preferred to silicon solar cells. The major drawbacks of GaAs solar cells are weight, fragility and high costs. To overcome the problems of weight and fragility Dieter et. al. (1992) have successfully deposited GaAs/AlGaAs solar cell structures on Si substrates. Accordingly Flores et. al. (1992) have grown GaAs/AlGaAs solar cells with high efficiencies on Ge substrates. A reduction of costs can also be achieved by using highly efficient multiwafer production technology. PM Frijlink (1988) introduced a multiwafer planetary MOVPE reactor capable of up to 5x3" substrates in which about 30% of the available group III material depositis from the gas phase (Frijlink 1991). In this work a reactor of this type has been investigated for its use in the mass production of GaAs based solar cells.

2. EPITAXIAL GROWTH

The structures were grown in a commercial AIXTRON AIX 2400 planetary multiwafer MOVPE reactor described elsewhere (Hergeth 1993). This horizontal reactor with radial geometry has a capacity of up to 5x4" or 15 x(4.5x4.5 cm^2) wafers. TMGa, TMAl and AsH_3 were used as precursors. N-type and p-type doping was obtained by using SiH_4 and DMZn, respectively. The structures were grown on semiinsulating 4" GaAs substrates except for the solar cell structures wich were grown on 4.5x4.5 cm^2 GaAs:Si substrates.

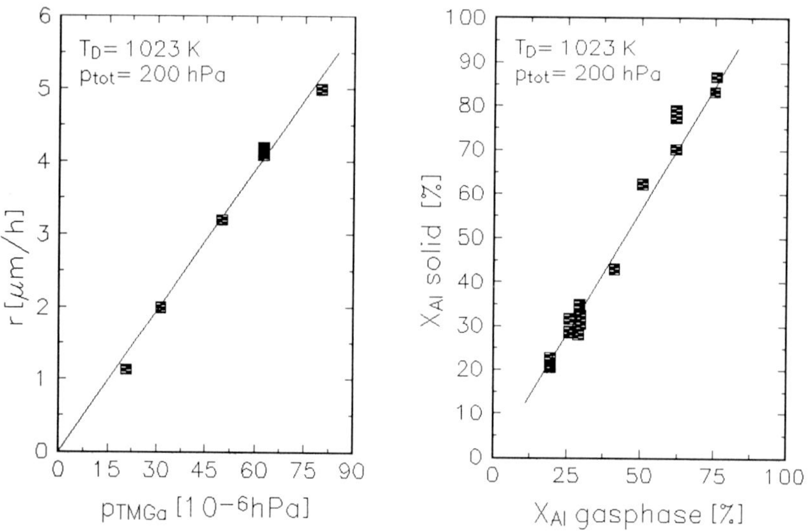

Fig. 1 GaAs growth rate and $Al_xGa_{1-x}As$ compostition.

The layers were deposited at a growth pressure of 200 hPa and a growth temperature of 1023 K. The variation of the TMGa partial pressure from $20x10^{-6}$ hPa to $80x10^{-6}$ hPa resulted in GaAs growth rates of 1 to 5 μm/h showing a linear dependence (figure 1). The variation of the AsH_3 partial pressure did not show any effect on the growth rate. Specular surfaces were obtained using a V-III ratio as low as 25. The composition of $Al_xGa_{1-x}As$ was only dependent on the TMAl content in the gas phase (figure 1) and the growth rate was not affected by the TMAl/TMGa mole fraction ratio. DCXD spectra show the high crystal quality of the deposited $Al_xGa_{1-x}As$. The FWHM of the layer peak is less than 24 arcsec compared to 28 arcsec of the substrate. The variation of the V-III ratio results in p-conducting material for V/III ratios below 22 and n-type conduction was obtained for higher ratios. The optimized V-III ratio of 56 results in nominally undoped GaAs with a background doping level of $n=1.4\times10^{14}$ cm^{-3} and electron mobilities as high as 90,000 cm^2/Vs at T=77 K on 4" wafers. N-type doping results in free electron concentrations in the range from 10^{15} cm^{-3} to 2×10^{18} cm^{-3} being linearly dependent on the SiH_4 partial pressure. Intentio-

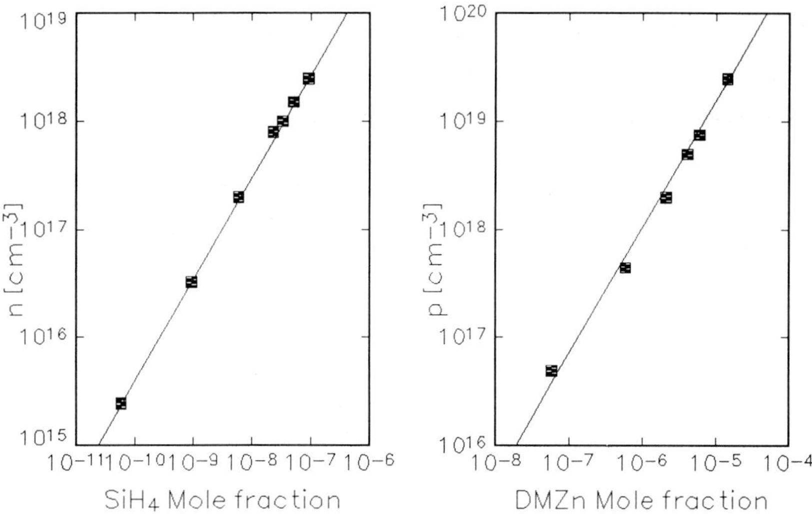

Fig. 2 N-type and p-type doping of GaAs: free carrier concentration vs. dopant mole fraction.

nally n-type doped GaAs structures show electron mobilities which are in very good agreement with theoretically determined data ($\Theta=0.3$, Walukiewicz 1982). For p-type doping the free hole concentration also increases from 3×10^{16} cm^{-3} to 2×10^{19} cm^{-3} linearly with the DMZn partial pressure. Figure 2 shows the free carrier concentrations of n-type and p-type doped GaAs vs. SiH_4 and DMZn mole fraction, respectively. $Al_{0.85}Ga_{0.15}As$ was p-type doped in the range of 10^{16} cm^{-3} to 10^{19} cm^{-3} for solar cell window layers. So far the general properties of GaAs and $Al_xGa_{1-x}As$ growth with the planetary reactor revealed that it is very similar to growth in common horizontal MOVPE reactors (Schmitz 1988). Thus process technology can be easily transferred.

For the mass production of GaAs based solar cells the efficiency of the precursors is of major interest. At an H_2 flow rate of 30 sml/min through the TMGa source kept at $T_s=273K$ and $p_s=500$ hPa the GaAs growth rate was determined to be 5.1 $\mu m/h$. Thus about 42.2% of the available group III material deposits on the five 4" wafers. This efficiency is about 7 times the efficiency of common horizontal MOVPE reactors (Kuech 86). Consequently the planetary reactor is designated for industrial use in large scale production of GaAs based compounds.

Hydrides and metalorganyles were led separately into the reactor. Keeping the total flow constant it is possible to vary the flows through each inlet. Therefore the mixing point of the gases can be varied and the behaviour of gas phase depletion can be changed. It was found that for a certain ratio of the flow through the metalorganic inlet to the flow through the hydride inlet an optimized uniformity of film thickness of ± 1.25 % across 4" wafer area is obtained.

In accordance with theory the optimization of flow conditions allows to adjust a linear depletion profile, thus a linear decrease in growth rate across the 4" wafer diameter is obtained. This was verified by High Resolution X-Ray diffraction (HRXD) of an unrotated 4" AlAs/GaAs Bragg-reflector structure. Figure 3 shows a typical HRXD spectrum of such a Bragg-reflector at a single point. Together with the rotation of the wafer a variation of

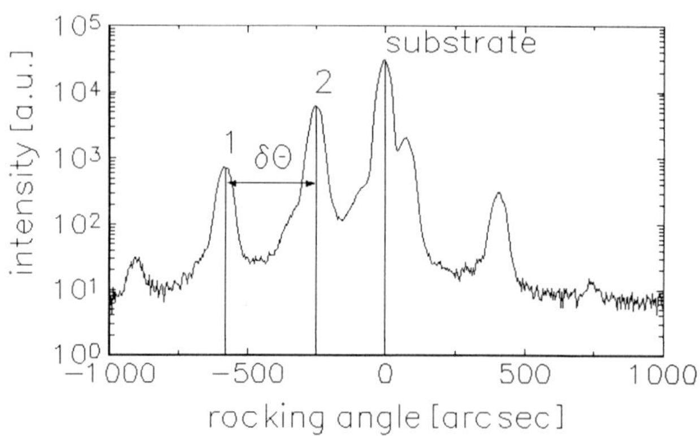

Fig. 3 Typical HRXD spectrum of a GaAs/AlAs Bragg reflector at a single point. The film thickness can be derived from the splitting $\delta\Theta$ of the satellite peaks (1,2).

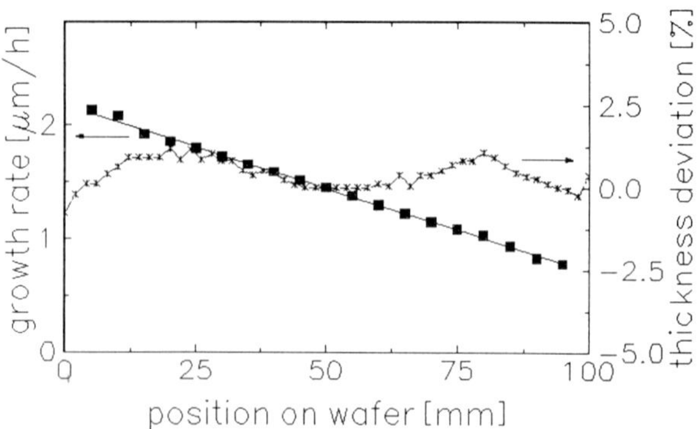

Fig. 4 GaAs growth rate (■) across unrotated and thickness deviation (x) across rotated 4" wafers.

GaAs film thickness of less than $\pm 1.25\%$ across 4" wafer area was achieved. In figure 4 the decrease in growth rate of the unrotated structure and the film thickness variation across a rotated GaAs structure on a 4" wafer are combined. In addition the variation of dopant concentration and $Al_XGa_{1-X}As$ composition are better than $\pm 1.8\%$ and $\pm 1.85\%$ across 4" wafer area respectively.

3. SOLAR CELLS

Based on these results a set of GaAs/$Al_XGa_{1-X}As$ solar cell structures was grown. The structures were grown on 4.5x4.5 cm^2 GaAs:Si substrates with (100) orientation. They mainly consist of a GaAs(n^+) buffer layer doped to a level of $2x10^{18}$ cm^{-3}, a GaAs(n) base with $n=2x10^{17}$ cm^{-3}, a shallow GaAs (p^+) emitter with $p=2x10^{18}$ cm^{-3} and a very thin AlGaAs(p^+) window layer with an Al-content of 85 % and a doping level of $p=2x10^{18}$ cm^{-3}. The structure is presented in figure 5. From the grown structures 2x2 cm^2 pieces were cut and solar cell devices were fabricated. The solar cells were tested under AM0 conditions (135 mW/cm^2). In figure 6 the I-V curve of the best solar cell is shown. The short circuit current density is 22.7 mA/cm^2, the open circuit voltage is 1.005 V, and the fill factor is 73%. The efficiency is 12.3 % without anti-reflection coating (ARC). This is a promising approach for growing high efficiency solar cells. Further optimization of the process especially concerning the p-n junction and the heterointerface will result in higher efficiencies. Since transfer of process technology is easy the substitution of the GaAs substrates by Si or Ge substrates will be possible.

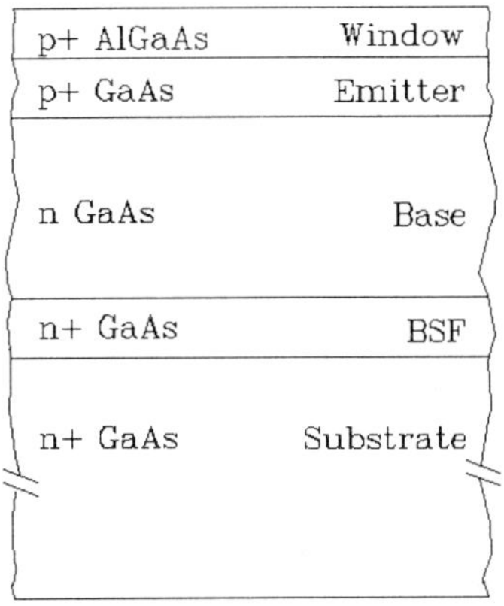

Fig. 5 GaAs/AlGaAs solar cell structure

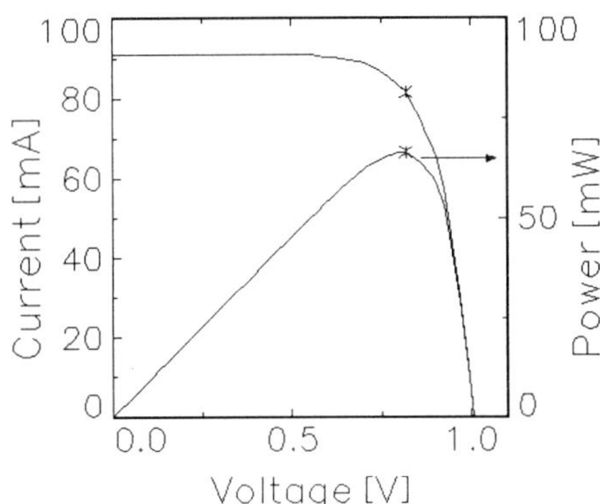

Fig. 6 I-V curve of the GaAs/AlGaAs solar cell without ARC under AM 0 illumination.

4. CONCLUSION

The process technology of the used planetary reactor turned out to be very similar to conventional horizontal MOVPE reactors. Thus transfer of technology is quite easy. The possibility of growing simultaniously on 5x4" or 15x(4.5x4.5)cm^2 substrates and the extremely high efficiency of the precursors drastically reduce the costs of epitaxy. The grown GaAs and $Al_XGa_{1-X}As$ show good electrical properties comparable to material grown in conventional MOVPE reactors. Under optimized flow conditions the variation in film thickness, dopant concentration and $Al_XGa_{1-X}As$ composition could be reduced to less than 1.25%, 1.8% and 1.85% across 4" wafer area, respectively. The presented solar cell structure shows and efficiency of 12.3% under AM0 illumination. This is a promising approach for growing high efficiency GaAs/AlGaAs solar cells.

ACKNOWLEDGEMENT

We would like to thank Mr. P. Wiedemann from SEL Alcatel, Stuttgart for performing the HRXD measurements and Dr. C. Flores from CISE Spa for processing and testing the solar cell devices.

REFERENCES

Dieter RJ, Scholz F, Martin W, Hangleiter A, Dörnen A, Michler P, Kürner W, Lu B, Frese V, Hilgarth J and Rasch K-D 1992 Microelectronic Engineering 18 189

Flores C, Bollani B, Campesato R, Passoni D and Timó GL 1992 Microelectronic Engineering 18 175

Frijlink PM 1988 J Crystal Growth 93 207

Frijlink PM, Nicolas JL and Suchet P 1991 J Crystal Growth 107 166

Hergeth J, Marheineke B, Knauf J, Strauch G, Schmitz D and Jürgensen H 1993 Inst. Phys. Conf. Ser. No. 129 109

Kuech TF, Veuhoff E, Kuan TS, Deline V and Potemski R 1986 J Crystal Growth 77 257

Schmitz D, Strauch D, Knauf J, Jürgensen H, Heyen M and Wolter K 1992 J Crystal Growth 93 312

Walukiewicz W, Lagowski J and Gatos HC 1982 J Appl Phys 53(1) 769

The LP-MOVPE of GaAs/$Al_xGa_{1-x}As$ with DEAlH-NMe_3 as Al source

R.Hövel, E.Steimetz and K.Heime

Institut für Halbleitertechnik, RWTH Aachen, Templergraben 55, D-52056 Aachen, Germany

ABSTRACT: High quality GaAs/$Al_xGa_{1-x}As$ was grown by low pressure MOVPE (20 hPa) with the Al-source DEAlH-NMe_3 (Diethylaluminiumhydride-trimethylamineadduct), TEGa and AsH_3. The growth behaviour was studied with regard to relevant growth parameters such as growth temperature and V/III ratio. The high optical quality of the $Al_xGa_{1-x}As$ is proved by a narrow photoluminescence linewidth at low temperature which is comparable to the best values reported in literature. The background carrier concentration for all grown layers ($0<x<0.6$) was n-type ($n \approx 10^{16}$ cm^{-3}). The high carrier mobility and the high PL-intensity indicate a low oxygen and carbon incorporation.

1. INTRODUCTION

In MOVPE the growth of AlGaAs with the commonly used group III precursors (trimethylaluminium, TMAl; trimethylgallium, TMGa) leads to layers with a high carbon content. Due to this C incorporation layers with an Al content $x>0.3$ are p-type (Kuech et al, 1987). To overcome this problem ethyl based group III precursors were used. The combination of TEAl and TEGa has indeed been shown to result in AlGaAs with no discernible C signal in the low temperature PL spectrum (Kuech et al, 1986). A further problem, associated with the use of highly reactive metalorganics, is the incorporation of oxygen. In an effort to make the precursors stable against reactions with oxygen and moisture saturated Al-alkyles were investigated. The oxygen content could be lowered by two orders of magnitude when using the coordinatively saturated Al-alkyl 1,3dimethyl-aminopropyl-1-ala-cyclohexane and the homologeous Ga precursor (Hövel et al, 1992). However, the low vapour pressure makes these precursors suitable only for a small regime of applications. Hobson et al (1991) used the solid compound trimethylamine alane (TMAAl) with either TEGa or TMGa to grow $Al_xGa_{1-x}As$ by MOVPE with encouraging properties. This alane does not contain Al-C bonds and does not form volatile alkoxides. Like trimethylindium, the TMAAl is a solid and changes in the evaporation rate may occure. In order to overcome the negative aspects of the alternative Al precursors mentioned above we studied the growth of $Al_xGa_{1-x}As$ with the ethyl based group III precursors diethylaluminiumhydride-trimethylamine (DEAlH-NMe_3) and TEGa in combination with AsH_3, since this Al precursor is a liquid with an adequate vapour pressure for MOVPE applications (38 Pa at 20°C). It has successfully been used in combination with TMGa and AsH_3 to grow $Al_xGa_{1-x}As$ by atmospheric pressure MOVPE (Jones et al, 1989). The resulting layers were of good quality, but nevertheless carbon remained the dominant impurity in low Al content AlGaAs layers. This implies that TMGa is the supplier of the carbon and the most promising route to grow high quality AlGaAs is to use TEGa.

2. EXPERIMENTAL PROCEDURE

The experiments were carried out in a horizontal rectangular reactor with a capacity of one two inch wafer. The pressure was 20 hPa and the average gas velocity was 1.2 m/sec. Starting materials were TEGa (Morton International) and the Al-source DEAlH-NMe_3 (EPICHEM). The AsH_3 was dried with a cleaning stage (Waferpure); the H_2 was Pd-diffused. A N_2 purged glove box for loading the samples was used to keep oxygen and water vapour away from the reactor. Substrates were (100) GaAs, misoriented by 2°off towards (110). After cleaning and etching the samples were loaded into the reactor and annealed for 5 min at 1020 K. After a 50 nm undoped GaAs buffer layer the 1-2 μm thick AlGaAs layer was grown and covered with a 10-20 nm undoped GaAs toplayer.

3. RESULTS AND DISCUSSION

In order to study the Al incorporation we carried out growth experiments with different gas phase compositions while the total group III partial pressure ($p_{Al}+p_{Ga}$=0.2 Pa), temperature T=1023 K and V/III ratio (V/III=100) were kept constant. A linear dependence of the Al content in the AlGaAs layer on the Al content in the gas phase was found. This simplifies the adjustment of the solid composition. The Al distribution coefficient as a function of the partial pressures in the gas phase was 1.7 while the growth rate of AlAs is a factor of 1.5 higher than the rate of GaAs. Comparison with a simple mass transport model showed that the results are hardly consistent with the assumption that the diffusing species are TEGa and DEAlH-NMe_3. Two explanations are possible: either fragments of the Al-precursor diffuse through the gas phase or the Ga diffusion species is a TEGa-AsH_3 adduct.

By investigating the temperature dependence of the growth rate and solid composition, parasitic gas phase reactions might be detected and the suitability of precursors and precursor combinations respectively can be checked. In Figure 1 the growth rates of GaAs, AlAs and $Al_xGa_{1-x}As$ for Al contents of approximately x=30 % are plotted. For the binaries and the ternary the growth is diffusion controlled. To obtain an easy process control a diffusion controlled growth is desirable. Then small fluctuations in temperature will have no influence on growth rate and solid composition. To exclude errors when estimating the Al-content of layers with a thickness of about 1 μm with DXRD (C.R. Wie, 1989), the Al-content obtained from room temperature PL is added in figure 1. The solid composition is nearly temperature independent. Such a behaviour points to the absence of undesired gas phase reactions. To exclude the presence of gas phase reactions with AsH_3, the V/III ratio was varied while the group III partial pressure was kept constant. The Al content decreases from x=25% to x=28% if the V/III ratio is varied from 200 to 25. Nevertheless parasitic reactions cannot be excluded because some dark deposits occur at the reactor inlet during growth. To check that the deposit does not lead to a depletion of the gas phase we studied the homogeneity of the Al content on a two inch wafer. In the direction of the gas flow a variation of $\Delta x/x=\pm4\%$ and vertical avariation of $\Delta x/x=\pm0.5\%$ was found. Taking into account that a medium gas velocity of v=1.2 m/sec and non-rotating susceptor was used, this does not point to undesired gas phase depletion due to gas phase reactions.

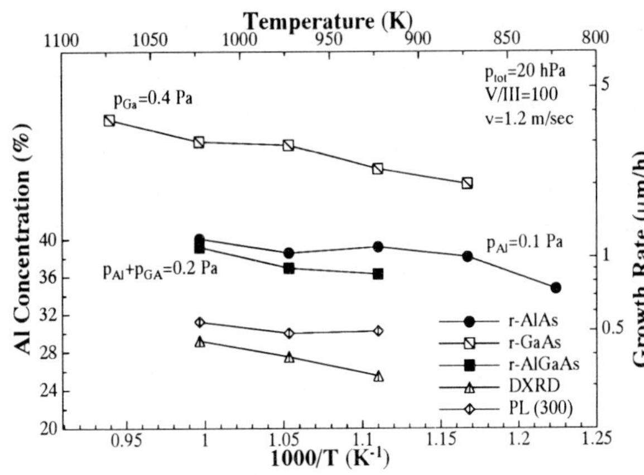

Fig. 1 Temperature dependence of GaAs, AlAs, AlGaAs growth rate and Al-composition.

The electrical properties were estimated with van der Pauw measurements. The residual background carrier concentration for all Al contents (0<x<0.6) and all growth parameters was n-type. In GaAs layers the electrical properties are $n = 3 * 10^{14}$ cm^{-3} with a correspondinng 77 K mobility of 80.000 cm^2/Vs. The background carrier concentration in $Al_xGa_{1-x}As$ is more than one order of magnitude higher. It increases with increasing Al content. After exceeding an Al content of x=30% a small decrease was found. The value of $n=7.9*10^{16}$ cm^{-3} for $Al_xGa_{1-x}As$ with x=30% can be reduced by decreasing temperature

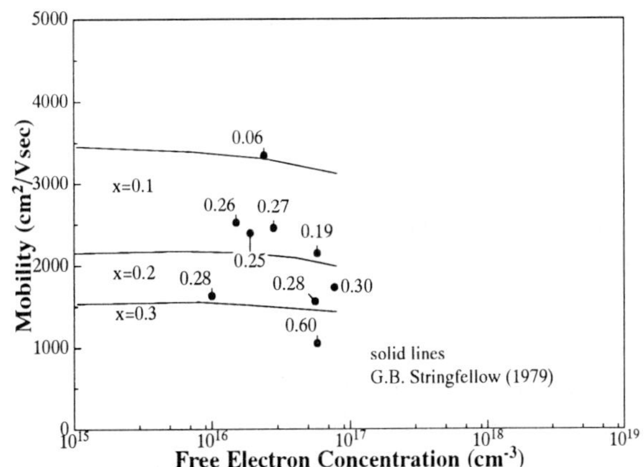

Fig. 2 Room temperature mobility versus free carrier concentration, solid lines theoretical calculations after Stringfellow (1979)

and increasing V/III ratio. Such a behaviour is typical when ethyl based group III precursors are used. In Figure 2 the mobility versus carrier concentration from layers with different Al content and different growth parameters is plotted. Theoretical calculations after G.B. Stringfellow (1979) of the mobility for three Al contents as a function of the free carrier concentration are added as solid lines. The measurements of mobility and carrier concentration were carried out with illumination and in the dark and no difference was found. This indicates a low level of traps. All layers showed mobilities higher than the calculated ones and point to a low compensation level. These good electrical properties prove that only a low concentration of acceptors, probably carbon, is present.

Figure 3 shows the 12 K PL spectrum of an $Al_xGa_{1-x}As$ x=30% sample at an excitation density of 0.1 W/cm^2. The band edge emission is seen on the high energy side of the spectrum, whereas the acceptor related emission is seen on the low energy side. The inset shows the band edge emission with a higher resolution. The FWHM of 5.3 meV is comparable to the best values reported in literature. The intensity of the band edge to the acceptor related peak is a measure for the acceptor incorporation. The band edge intensity is here three times higher than the acceptor related response. Usually the acceptor peak is caused by carbon incorporation. Analysing the impurity related response in a GaAs PL spectrum, the main impurities were the acceptor magnesium and the donor silicon. This implies that not carbon but magnesium is the main acceptor in the $Al_xGa_{1-x}As$ samples, too.

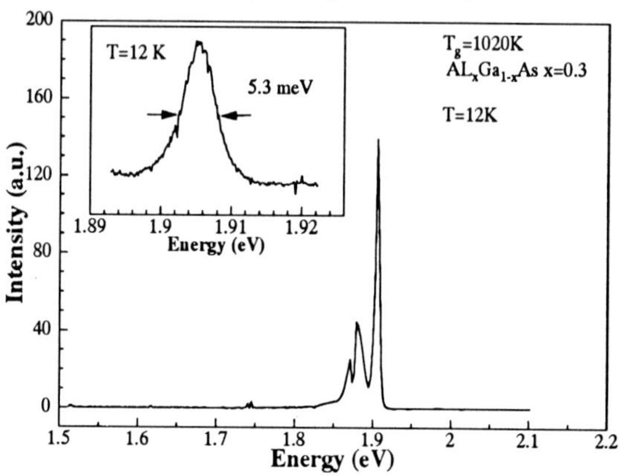

Fig. 3 Low temperature PL spectra (12K) of an $Al_xGa_{1-x}As$ layer with x=0.3 grown at 1023 K.

The high intensity of the band edge transition indicates a low oxygen incorporation. This was found for optically direct and indirect AlGaAs. AlGaAs layers grown at temperatures of 970 K and 1020 K show nearly the same intensity of the band edge transition which point to a low oxygen incorporation. The high quality is also demonstrated by the PL spectrum of a sample with a high Al (x=60%) content (Fig. 4).

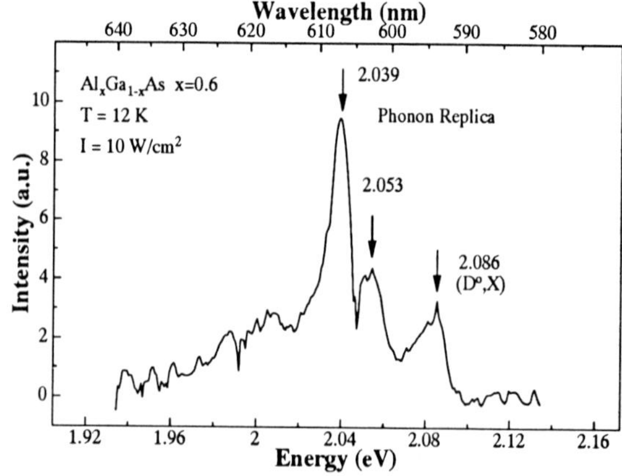

Fig. 4 Photoluminescence spectra of an $Al_xGa_{1-x}As$ sample with x=0.6 (indirect energy gap)

Their shape is typical for good quality indirect gap $Al_xGa_{1-x}As$ (R.Dingle et al, 1979). The peak at the highest energy is attributed to the recombination of bound excitons associated mainly with donors (D°,X). The (D°,X) recombination is a zero-phonon process that becomes allowed in the indirect gap alloy due to alloy scattering. The two other structures, distance ~33 meV and ~47 meV from the (D°,X) transition, are ascribed to phonon replica of the donor bound excitation recombination. The small structure appearing at low energies may be associated with impurities; they are possibly due to donor acceptor pair recombinations and their phonon replica. Calculation of the Al fraction x of the sample after Guzzi et al (1991) leads to an Al content of 0.55±0.04 instead of 0.6±0.01 estimated by DXRD.

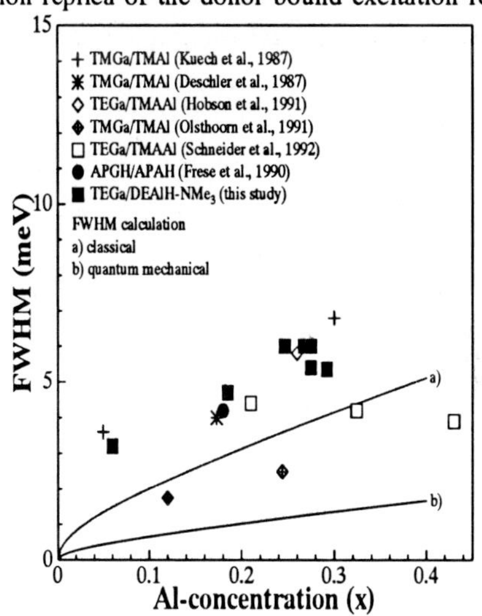

Fig. 5 Comparison of FWHM with values of literature. Solid line theoretical calculations due to alloy broadening a) Schubert et al (1984) b) Singh et al (1986)

The FWHMs of all grown AlGaAs samples are shown in fig. 5. The high optical quality is proven by a FWHM which is comparable to most of the best values reported in literature. All the measurements were carried out with an excitation density of 0.1 W/cm^2. The solid lines represent theoretical calculations of the alloy broadening due to compositional disorder. The dispersion of the values is due to different growth parameters.

4. CONCLUSION

The alternative Al precursor $DEAlH\text{-}NMe_3$ was studied in combination with TEGa and AsH_3. The Al precursor has an adequate vapour pressure for MOVPE applications and is a liquid. Dark deposits at the reactor inlet point to gas phase reactions. However, due to our results these gas phase reactions do not lead to any negative effect of the growth behaviour. Neither the investigation of temperature dependence of the growth rate and Al composition nor the variation of the V/III ratio gives any indication for undesired gas phase reactions. All grown samples with a Al fraction of $0<x<0.6$ showed n-type conductivity and a high mobility. The high optical quality is proved by a high Pl emission intensity and by a narrow FWHM which is comparable to the best values quoted in literature and a high PL emission intensity. The electrical and optical results indicate a low amount of carbon and oxygen in the samples. These results prove $DEAlH\text{-}NME_3$ to be a powerful substitute to TMAl.

5. ACKNOWLEDGEMENTS

The authors would like to thank Professor C. Whitehouse and Dr. T. Martin of the Defence Research Agency, Malvern for their support and fruitful discussions. Parts of this work were financially supported by the EC under the ESPRIT contract MORSE.

6. REFERENCES

M.Deschler, M.Cuppers, A.Brauers, M.Heyen and P.Balk 1987 J. Cryst. Growth **82** 628.

R.Dingle, R.A.Loganand, R.J.Nelson 1979 Solid State Commun. **29** 171.

V.Frese, G.K.Regel, H.Hartdegen, A.Brauers, P.Balk, M.Hostalek, M.Lokai, L.Pohl, A.Miklis and K.Werner 1990 J. Electron. Mater. **19** 305.

M.Guzzi, E.Grilli, S.Oggioni, J.L.Staehli, C.Bossi and L.Pavesi 1992 Phys. Rev. B **45** 10951.

W.S.Hobson, T.D.Harris, C.R.Abernathy and S.J.Pearton 1991 Appl. Phys. Lett **58** 77.

R.Hövel, W.Brysch, N.Neumann, K.Heime and L.Pohl 1992 J. Cryst. Growth **124** 106.

A.C.Jones, S.A.Rushworth, J.S.Roberts, C.C.Button and J.P.R.David 1989 Chemtronics **4** 235.

T.F.Kuech, D.J.Wolford,E.Veuhoff, V.Deline, P.M.Mooney, R.Potemski and J.Bradley 1987 J. Appl. Phys. **62** 632.

S.M.Olsthoorn, F.A.J.M.Driessen, L.J.Gilling, F.M.Frigo and C.J.Smit 1991 Appl. Phys. Lett. **58** 1274.

R.P.Schneider, Jr.,R.P.Bryan, E.D.Jones, R.M. Biefeld and G.R.Olbright 1992 J. Cryst. Growth **123** 487.

E.F.Schubert, E.O.Göbel, Y.Horikoshi, K.Ploog and H.J.Queisser 1984 Phys. Rev. B **30** 813.

J.Singh and K.K.Bajaj 1986 Appl. Phys. Lett **48** 1077.

Inst. Phys. Conf. Ser. No 136: Chapter 10
Paper presented at the Int. Symp. GaAs and Related Compounds, Freiburg, 1993

Strain and strain relaxation in selectively grown GaAs on silicon

G. Frankowsky, A. Hangleiter, K. Zieger, F. Scholz

4. Physikalisches Institut, Universität Stuttgart, Pfaffenwaldring 57,
D-70550 Stuttgart 80, Germany

ABSTRACT: We report on strain and strain relaxation of selectively grown GaAs on Si. Photoluminescence microscopy (PLM) images demonstrate that the formation of cracks in the GaAs layers can be avoided. We show that it is possible to grow patterns up to a size of 1x1 mm^2 without any cracks. The reason for that is a better control of the pattern edges. Cathodoluminescence (CL) spectra show that the patterns are not homogeneously strained. The strain relaxes within 75 μm towards the pattern edges.

1. INTRODUCTION

The heteroepitaxial growth of GaAs on Si substrates has been extensively studied in the recent years. There is a strong interest to combine the high speed and optoelectronic properties of GaAs with the highly developed Si technology. One very interesting application is the development of highly efficient and cheap solar cells. Unfortunately, the thermal and lattice properties of GaAs and Si are quite different. The quality of the layers is limited mainly by the large lattice mismatch (4.1 %) and the big difference in the thermal expansion coefficients. The different lattice constants cause high dislocation densities (10^6 - 10^8 cm^{-2}) (Roedel (1991), Wada (1992)) while the different thermal coefficient leads to the presence of a tensile biaxial strain in the GaAs layer. The stress is responsible for the formation of micro cracks in the epitaxial layers. From investigations on planar grown samples we found all cracks starting at the edge of the Si wafer. We have never observed cracks starting and ending inside the wafer. Ackaert et al. (1989) have shown that cracks can be induced in selectively grown patterns when specially designed masks were used. The mask produces very high strained spots in the layer which induce cracks. The reason for the formation of micro cracks in planar grown samples is the undefined structure of the Si wafer edge. Small inhomogeneities at the wafer edge cause small very highly strained spots in the GaAs layer. The cracks are induced by these spots. We show that selective heteroepitaxy offers a better control over the pattern edges. The highly strained spots can be completely avoided and therefore it is possible to grow large GaAs patterns without any micro cracks.

2. EXPERIMENTAL

The samples were grown on misoriented (6°) Si substrates. In a first step a 100 nm SiO_2 layer is deposited on the substrates. The patterns were defined by conventional photolithography and wet chemical etching. Our standard mask consists of 2x2 mm^2 openings in the SiO_2 layer containing patterns with a size between 20x20 μm^2 and 2x2 mm^2. After the processing of the SiO_2 mask, a 100 nm GaAs nucleation layer is grown. The final layer with a nominal thickness of 7 μm is applied with a thermal cyclic growth (TCG) process (Dieter (1992)).

The presence of strain modifies the band structure of the GaAs layers. According to Van de Walle (1989) the band gap shifts with the strain ϵ. The electron – heavy hole transition shows a linear shift with the strain ϵ. For the electron – light hole transition the relation between the strain ϵ and the energy shift ΔE_{lh} is more complicated. In the case of tensile strain, the transition with the lowest energy is the electron – light hole transition.

We can use the luminescence as a probe for the strain. For a tensile strained layer we expect a shift to lower energies. Assuming that the GaAs layer is lattice matched at growth temperature T_g (Fang (1990)), we can calculate the strain ϵ in the layer as

$$\epsilon = \int_{T_g}^{T_m} \left[\alpha_{Si}(T) - \alpha_{GaAs}(T)\right] dT$$

α_{Si} and α_{GaAs} are the thermal expansion coefficients of Si and GaAs, T_g is the growth temperature and T_m is the temperature where the measurements were performed. For T_m = 7 K and T_g = 970 K we obtain a tensile strain $\epsilon = 2.98 \cdot 10^{-3}$. This leads to a band gap shift of ΔE_g = 18 meV. For the investigation of the micro crack formation we are using spectrally resolved photoluminescence microscopy (PLM). The samples were mounted in a continuous flow cryostat and excited by an Ar^+ ion laser with a typical spot size of 4 - 8 mm. The spectral resolution is realized using a tunable interference filter. The luminescence images of the sample were recorded using a standard CCD-camera. The bandwidth of our interference filter is about $\Delta\lambda \approx 12$ nm, which is sufficient for the detection of cracks in the GaAs layers.

For spatially resolved strain measurements a higher spectral resolution is needed. Therefore we are using the spectrally and spatially resolved cathodoluminescence (CL). The CL offers high spatial (limited by carrier diffusion) and spectral resolution. We performed our measurements at T = 7 K.

3. RESULTS AND DISCUSSION

Fig. 1 shows a PLM image of a planar grown GaAs layer. The interference filter was tuned to a shorter wavelength for this image compared to the luminescence of the layer. The bright vertical lines are indicating micro cracks in the GaAs layer. Due to the relaxation of the strain close to the cracks the luminescence is shifted to higher energy and these regions appear as bright parallel stripes in the image. Typical distances between the cracks were found between 30 μm and 200 μm.

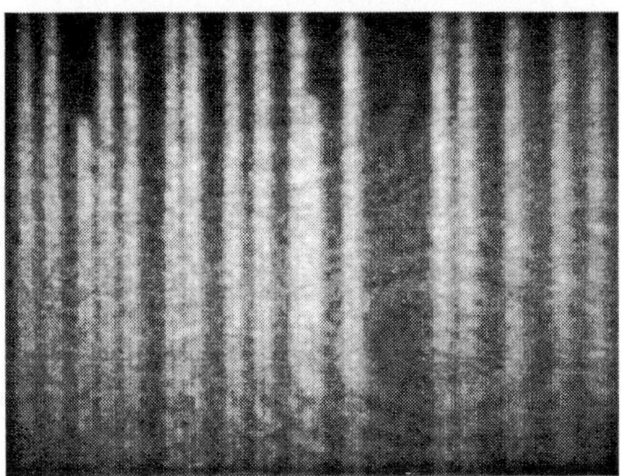

Figure 1: PLM-image of a planar grown GaAs layer taken at a $T_m = 7K$ and $\lambda = 826$ nm. The image width is 900 μm.

A PLM image of four 1x1 mm^2 sized selectively grown GaAs patterns is shown in Fig. 2. No cracks can be found in these patterns. The increased intensity at the pattern edges

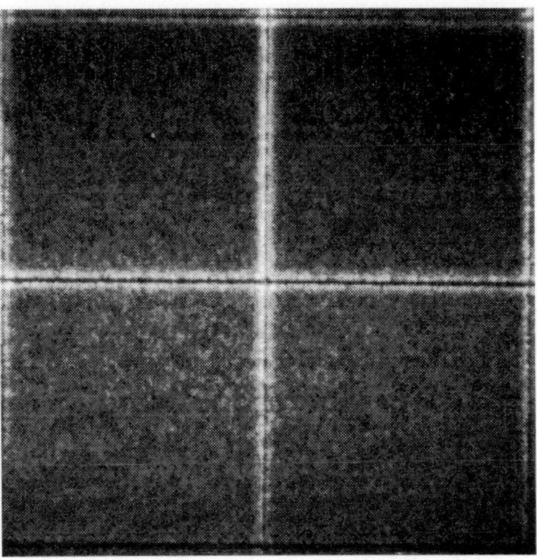

Figure 2: PLM-image of selectively grown GaAs pattern at a $T_m = 7$ K. The pattern size is 1x1 mm^2. The image was recorded at a wavelength of $\lambda = 832$ nm

indicates strain relaxation.

In Fig. 3 a single 200x200 μm^2 sized pattern can be seen. Like in Fig. 2 the intensity is increased at the pattern edges. The intensity fluctuations inside the pattern are caused by the high dislocation density. Several authors have reported strain relaxation of GaAs patterns on Si obtained by post growth patterning (Lee 1988, Linguis 1990, Tsukamoto 1992). They used conventional photoluminescence spectroscopy and found a shift of the band gap up to 12 meV to higher energies for pattern sizes between 4 μm and 100 μm. These measurements are not spatially resolved and reflect only an average strain in the patterns. We have measured cathodoluminescence spectra at T_m =7K with excitation only in the center of the pattern to avoid the influence of the less strained pattern edges. The size of the patterns varies from 20x20 μm^2 up to 2x2 mm^2. The result of these measurements is shown in Fig. 4. The dashed line indicates the band gap we observe for planar grown GaAs layers. For the selective grown GaAs we obtain nearly the same value of the band gap in the center for pattern sizes above 200 μm. Smaller patterns appear to be less strained in the center and we obtain a shift of 17 meV between patterns of 20 μm and 200 μm. The strain increases from $\epsilon = 1.2 \cdot 10^{-3}$ for 20x20 μm^2 sized patterns to $\epsilon = 2.8 \cdot 10^{-3}$ for patterns larger than 200 μm.

The spatial distribution of the strain was investigated using CL linescans. The electron beam is scanned along a line perpendicular to the pattern edges. At each point the electron beam reaches a complete spectrum is recorded. From the evaluation of the CL line scan spectra we obtain the lateral distribution of the strain in the layer. In Fig. 5 the results for four differently sized patterns are plotted. We find a characteristic relaxation of the strain towards the pattern edges within a distance of 75 μm, which is surprisingly large compared to the thickness of the GaAs layer (7 μm). For large pattern we obtain

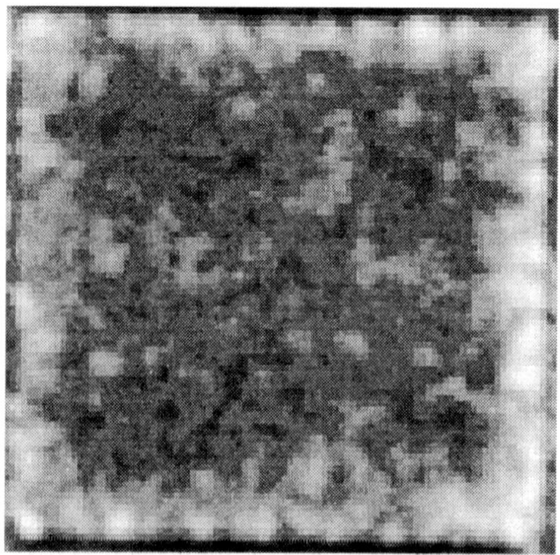

Figure 3: PLM-image of a selectively grown 200x200 μm sized GaAs pattern at a T_m = 7 K and λ = 830 nm

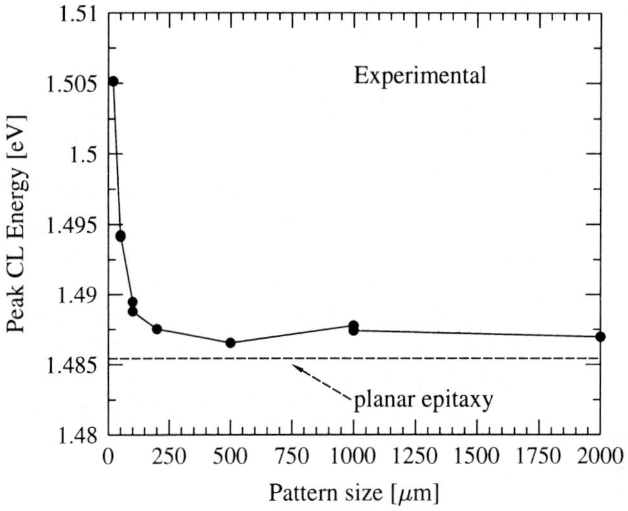

Figure 4: Dependence of the band gap energy in the center of selectively grown GaAs layers on the pattern size at $T_m = 7$ K. The dashed line indicates the band gap for planar grown samples.

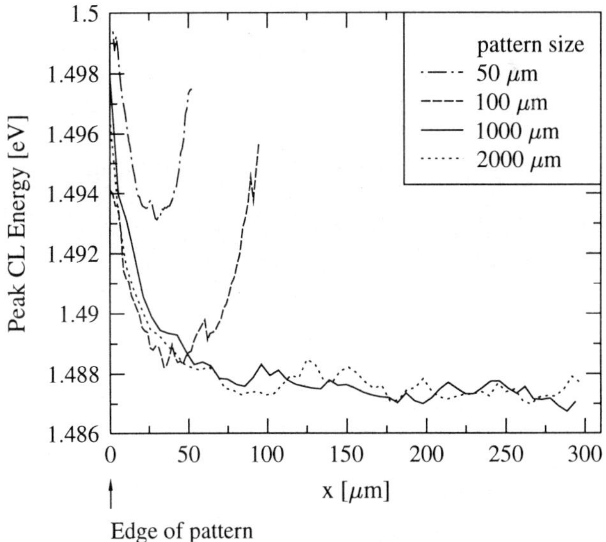

Figure 5: Variation of the band gap energy due to the position within the patterns measured with spatially and spectrally resolved cathodoluminescence line scans at $T_m = 7$ K. The respective minima of the Peak CL Energy for the different pattern sizes correspond to the data points plotted in Fig. 4.

the following behavior: The strain increases from the edge towards the center of the pattern. In a distance of $\approx 75\ \mu m$ from the edge the strain reaches a maximum value of $\epsilon = 2.6 \cdot 10^{-3}$ comparable to planar grown samples and remains constant. At the pattern edges we obtain a strain of $\epsilon = 1.6 \cdot 10^{-3}$. For patterns with a size below 150 μm the behavior is very similar. We obtain the maximum strain in the center of the patterns, but dependent on the pattern size the strain does not reach the value for planar samples. The maximum strain reached in the center of the patterns decreases with the pattern size. For a 50 μm sized pattern we obtain a strain of $\epsilon \approx 2.1 \cdot 10^{-3}$ in the center and $\epsilon \approx 1.6 \cdot 10^{-3}$ at the pattern edges.The patterns are not homogeneously strained.

4. CONCLUSION

With selective epitaxy it is possible to grow GaAs patterns up to a size of at least 1x1 mm^2 without any cracks. The main advantage of selective epitaxy is a better control of the strain at the edges of the GaAs patterns. Highly strained spots responsible for the formation of the micro cracks can be avoided. PLM images exhibit an increased intensity at the pattern edges if the interference filter is tuned to shorter wavelengths, indicating strain relaxation. From spatially resolved CL measurements we obtain a relaxation of the strain within 75 μm from the pattern edges which is nearly 11 times the layer thickness. The distribution of the strain depends on the lateral size of the patterns. Strain comparable to the value observed for planar grown samples is reached in the center of patterns with size above 200 μm.

5. REFERENCES

Ackaert A. et al. 1989 Appl. Phys. Lett. 55 2187
Dieter R. D. et al. 1992 11. E.C. Photovoltaic Solar Energy Conference
Fang S. F. et al. 1990 J. Appl. Phys. 68 (7) R31
Lee H. P. et al. 1988 Appl. Phys. Lett. 53 2394
Linguis A. et al. 1990 Solid State Commun. 76 303
Roedel R. J. et al. 1991 J. Electrochem. Soc. 138 3120
Tsukamoto N. et al. 1992 Appl. Phys. Lett. 61 810
Wada a. et al. 1992 Appl. Phys. Lett. 60 1354
Van de Walle C. G. 1989 Phys. Rev. B 39 1871

Minority carrier lifetime of III–V compound semiconductors

R K Ahrenkiel

National Renewable Energy Laboratory, Golden, CO 80401.

ABSTRACT: The status of minority-carrier lifetime improvement in popular III-V materials will be reviewed. The most important recombination mechanisms in semiconductors including GaAs, $Al_xGa_{1-x}As$, InP, and GaInP will be discussed.

1. INTRODUCTION

Minority-carrier lifetimes are fundamental to the performance of a wide range of optoelectronic devices. With the continuous improvement in epitaxial growth, many record minority-carrier lifetimes have been reported in the last few years. This paper reviews the current status of lifetime improvement for various popular materials.

The lifetime is determined by a combination of both intrinsic and extrinsic or defect related mechanisms. The intrinsic mechanisms include both radiative and Auger recombination. Radiative recombination (τ_R) is the dominant effect in doped GaAs as will be shown. The most common extrinsic mechanism is Shockley-Read-Hall (SRH) recombination at both bulk defects, surfaces and heterointerfaces.

The standard technique for measuring minority-carrier lifetime in III-V compounds is time-resolved photoluminescence (TRPL). This technique has been described extensively in the literature (Ahrenkiel, 1992). For planar geometry, the low-injection PL lifetime related to the other recombination lifetimes through the relationship

$$\frac{1}{\tau_{pl}} = BN + \frac{1}{\tau_{SRH}} + \frac{1}{\tau_A} + \frac{S_1 + S_2}{d} \tag{1}$$

Here B is the recombination constant for a given semiconductor, N is the majority carrier density, and S_1 and S_2 are the surface/interface recombination velocities at the two surfaces. Also τ_A is the Auger lifetime and τ_{SRH} are the bulk SRH lifetimes.

2. MINORITY-CARRIER LIFETIME IN GaAs

Early measurements by Hwang on bulk, n-type GaAs wafers [1971] found room temperature lifetimes in the range of 10 to 20 nanoseconds (ns) at doping levels below about 1 x 10^{17} cm^{-3}. These data are shown in Figure 1. At higher doping levels, the lifetimes decrease approximately as 1/N where N is the donor concentration. A recent wafer measurement on AlGaAs passivated wafers by Ehrhardt and coworkers [1991] is comparable to the earlier measurement of Hwang. Also shown by the solid line is the intrinsic, radiative lifetime τ_R. Here τ_R is calculated assuming that B= 2 x 10^{-10} cm^3/s. At the lower doping levels, the lifetimes are also much smaller than the radiative lifetime and are controlled by deep level defects in the wafers; i.e. by Shockley-Read-Hall recombination.

Nelson and Sobers (1978 a,b) measured the minority-carrier lifetime in epitaxial, p-type AlGaAs/GaAs isotype double heterostructures and found quite different results. Their data (●) are shown in Figure 1 and indicate lifetimes that are comparable to the intrinsic radiative recombination rate (τ_R). One sees that the PL lifetime was larger than the predicted radiative lifetime over a range of doping levels from about 1 x 10^{17} cm^{-3} to 1 x 10^{19} cm^{-3} because of photon recycling. The thickness of the active layer of these devices varied from several microns to about 15 microns.

Recent TRPL data on MOCVD-grown n-type GaAs obtained by Lush and coworkers [1992a] is shown by the (▲) data points. These data were measured on thin (1.0 μm or less) DH structures for which the

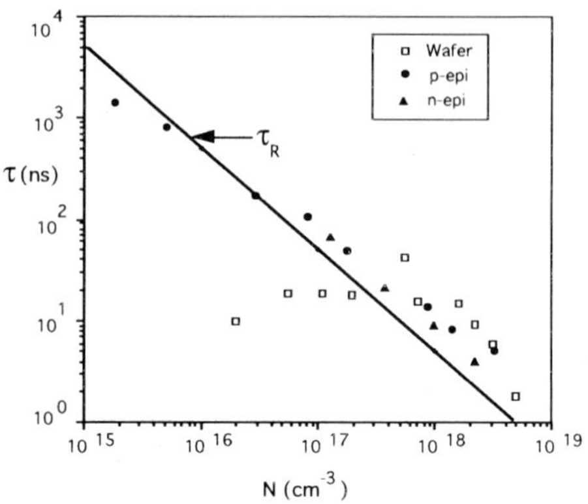

Figure 1: Minority carrier lifetime in GaAs.

photon recycling effects are minimal. In summary, these data show that the GaAs bulk lifetime in epitaxial material, is primarily controlled by the intrinsic, radiative recombination. By contrast, the lifetime of wafer material is controlled by bulk SRH recombination except at very high carrier concentrations.

3. PHOTON RECYCLING EFFECTS MEASURED IN GaAs

Measurements [Ahrenkiel et al 1992] on n-type DH structures were analyzed to find the photon recycling factor. These data were compared with the values of ϕ calculated by Asbeck [1977] for similar DH structures at various doping concentrations. The factor ϕ is about 12 for a 10 μm DH structure when the doping level is below 1 x 10^{18} cm^{-3}. A decrease in ϕ is observed at the higher doping levels. This decrease results from a smaller overlap between the internal emission spectrum with the absorption spectrum.

A remarkable enhancement of the photon recycling effect was obtained [Lush et al 1992b] by substrate removal from selected area. TRPL measurements were made on the sample DH device in etched regions and compared with unetched regions. The PL lifetime for the d = 5.0 μm device (substrate etched) is 1.07 μs and is 28 times the radiative lifetime.

4. GaAs GROWN HETEROEPITAXIALLY ON SILICON

In recent years, there have been many reports of GaAs photovoltaic (PV) devices grown on silicon substrates. The minority-carrier lifetime is greatly degraded by dislocation recombination in the GaAs:Si system. Measurements by Ahrenkiel and coworkers [1988a,] combined with a recombination theory developed by Yamaguchi and coworkers [1986] at Nippon Telephone and Telegraph (NTT) find the lifetime in GaAs as a function of dislocation density. The dislocation contribution to the lifetime is given by:

$$\frac{1}{\tau_d} = \frac{\pi^3 D N_d}{4} \tag{2}$$

Here N_d is the dislocation density in lines/cm^2, D is the minority-carrier diffusivity, and τ_d is the lifetime produced by the dislocations. Recent data show that the lowest dislocation densities for GaAs:Si devices are in the range of 10^7 cm^{-2}. The corresponding maximum minority-carrier lifetimes are in the range of 1 to 2 ns at doping densities of 2 x 10^{17} cm^{-3}. Comparable devices grown on GaAs substrates typically show lifetimes of 100 to 400 ns.

5. RECOMBINATION VELOCITY OF THE $Al_xGa_{1-x}As$/GaAs INTERFACE

5.1 Doped GaAs

The measurement of interface recombination in $Al_xGa_{1-x}As$/GaAs DHs was first shown in the measurements of Nelson [1978] using the TRPL technique. Here S increased with the GaAs doping level and is about 350 cm/s at 3 x 10^{16} cm^{-3} compared to 550 cm/s at 2 x 10^{17} cm^{-3}. Recent studies [Ahrenkiel et al 1990] found that the recombination velocity at the MOCVD n-$Al_{0.30}Ga_{0.80}As$/GaAs interface changed by orders of magnitude, depending on growth temperature. These data show that S drops a factor of 40 for growth temperatures greater than 740°C. The interface recombination velocity varies from a maximum of 20,000 cm/s (700°C growth) to a minimum of about 500 cm/s (775°C growth). In general, the interface recombination velocity of MOCVD-grown $Al_xGa_{1-x}As$/GaAs improves with growth at temperatures greater than 740°C. The smallest measured values of S on doped DHs were obtained by Lush and coworkers [1992b]. These DH devices were grown by MOCVD at 740°C. For an active layer doping of 1.3 x 10^{17} cm^{-3}, they found S less than 11 cm/s using a $Al_{0.30}Ga_{0.70}As$ window layer.

5.2 Undoped GaAs

Undoped GaAs DH structures are useful diagnostic structures for the analysis of SRH recombination, both in the bulk and at interfaces. In these structures, the radiative recombination contribution is often negligible in effect on the total lifetime. The PL lifetime in this limit is approximately:

$$\frac{1}{\tau_{pl}} = \frac{1}{\tau_{SRH}} + \frac{2S}{d} \tag{3}$$

To allow for the variation in thickness as it effects the surface lifetime, one may compute a figure of merit $F = \tau_{PL}/d$ where:

$$F = \frac{\tau_{PL}}{d} = \frac{1}{d/\tau_{SRH} + 2S} \tag{4}$$

For very thin DH structures, the figure of merit is approximately 1/2S.

An alternative window layer is GaInP that is lattice-matched to GaAs. Recent measurements on undoped GaInP/GaAs DH devices have shown record photoluminescence (PL) lifetimes. The best reported undoped device of the series has a lifetime of 14.2 µs with d = 1.0 µm, which is a world record for GaAs [Olson et al 1989].

Figure 2 shows the figure of merit, τ_{PL}/d, versus temperature of a number of undoped GaAs DH structures that have been reported in the literature. Curve A is the data for the 3 µm thick GaInP/GaAs DH that was recently reported (Olson et al, 1989, Ahrenkiel et al, 1990). The PL lifetime of this device varied as $T^{1.59}$

between 77 K to 300 K. The data of curve B is for a 0.12 μm, undoped $Al_{0.3}Ga_{0.7}As/GaAs$ structure and shows thermally actuated temperature dependence (Molenkamp et al, 1988). In this device, the lifetime is dominated by the interface recombination and the authors calculate a room temperature S of 18 cm/s. Curve C is the data of Wolford and coworkers (1991) from an undoped MOCVD-grown $Al_{0.30}Ga_{0.70}As/GaAs$ DH with a thickness of 0.30 μm. The lifetime is weakly dependent on temperature from about 150 K to 300 K, indicating that the dominant recombination mechanism is nonradiative. The temperature dependence of the lifetime is not indicative of dominant radiative recombination in any temperature range. Between 50 K and 100 K, the lifetime is thermally activated with a slope similar to Curve B. This behavior is indicative of dominant interface recombination.

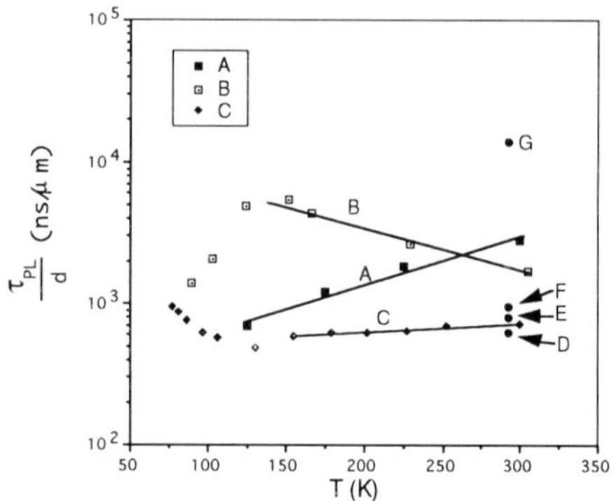

Figure 2: Figure of merit for undoped GaAs DH devices.

Other room temperature measurements referred to in the above text are plotted as single points in Figure 3. The datum of Dawson and Woodbridge (1984) is shown as point D. Other data shown are point E (Hummel et al, 1990), point F ('t Hooft et al, 1985), and point G (Olson et al, 1989).

6. $Al_xGa_{1-x}As$ LIFETIME

The lifetime in $Al_xGa_{1-x}As$ can be measured by the TRPL technique on DH devices with the structure $Al_yGa_{1-y}As/Al_xGa_{1-x}As$ (y > x). A review of lifetime measurements in $Al_xGa_{1-x}As$ can be found in some recent literature [Ahrenkiel 1992b, 1993]. In contrast to GaAs, the lifetime in $Al_xGa_{1-x}As$ has been shown to be SRH limited over a wide range of doping levels. One expects a smaller B-value for increasing aluminum content. Early measurements on $Al_xGa_{1-x}As$ lifetime by van Opdorp and 't Hooft [1981] showed DH structures that were grown both by LPE and by MOCVD. The composition range examined was $0.4 < y < 0.62$ and $0.10 < x < 0.17$. These data indicate that the $Al_xGa_{1-x}As$ lifetime drops sharply with increasing aluminum concentration. Zarem and coworkers [1989] measured the PL lifetime in MBE-grown, undoped $Al_xGa_{1-x}As$ DHs over the range $0 < x < 0.38$. The measured lifetimes were less than instrumental response or about 4 ns for $x < 0.35$ and injection densities of 3 x 10^{18} cm^{-3}. At x = 0.35, the PL lifetime increases to 20 ns, and at x = 0.38, it increases to about 30 ns. The lifetime is attributed to effects related the direct-indirect crossover.

The largest lifetimes in high aluminum (x~0.40) compositions were grown by liquid phase epitaxy (LPE). These were DHs [Ahrenkiel et al, 1988b] with active layers of composition n-Al0.38Ga0.62As and window layers of n-$Al_{0.8}Ga_{0.2}As$. A record low-injection lifetime of 20.2 ns was measured for a 5.0-mm device doped to 4 x10^{15} cm^{-3}. Here SRH recombination is dominant and no dependence on majority-carrier concentration was observed.

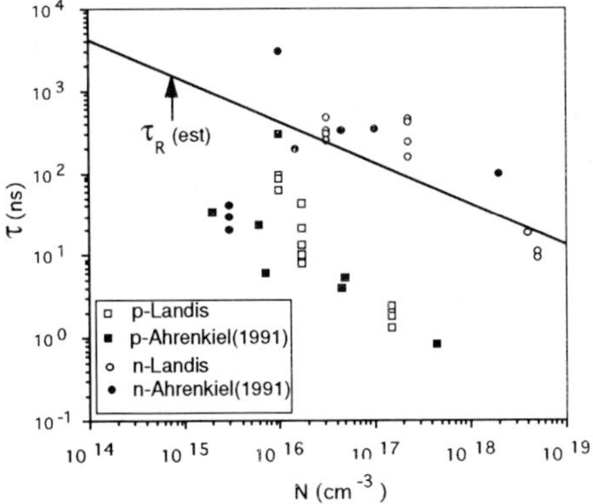

Figure 3. Composite of lifetime measurements in InP

Recent studies of MOCVD $Al_xGa_{1-x}As$, indicate that the dominant SRH defects are of double acceptor, oxygen complexes [Zhang et al 1993]. Because of their large capture cross-sections for holes ($\sigma \sim 10^{-12}$ cm^{-3}), a fairly low concentration ($\sim 10^{14}$ cm^{-3}) effectively "kills" the minority-carrier lifetime in n-type material. These centers are either deactivated or removed by growth above 740°C.

7. GaInP

Several DH diagnostic devices have been fabricated using the ternary AlInP as the confinement layer. The bandgap of these GaInP alloys is about 1.85 eV and the lifetimes have ranged from 40 to 409 ns. These data are markedly larger than $Al_{0.40}Ga_{0.60}As$ films that have a comparabale bandgap.

8. InP LIFETIMES

There is relatively little known about the minority-carrier properties of InP as compared to GaAs. By application of the van Roosbroeck-Shockley relationship to the absorption spectra, a B-coefficient is calculated that is similar to that of GaAs. Data on InP has been marked by inconsistency and a large lifetime discrepancy between n- and p-type material. Also, in contrast to GaAs, bulk crystals have produced better lifetimes than epitaxial films.

Figure 3 is a plot of available data on n- and p-type InP from a variety of sources [Ahrenkiel 1991, Landis et al 1992]. There is a great deal of scatter in the data. The solid line is an estimate of the radiative lifetime assuming that $B = 2 \times 10^{-10}$ cm^3/s; i.e. that B has the same value as for GaAs. Most of the data for p-InP lie at least a factor of 10 below the estimated τ_R. One must suspect that some SRH defects are contaminants in p-InP but there is not definitive evidence for that speculation.

Figure 3 also shows a plot of available data on n-type InP from several different sources. Again the estimated radiative lifetime is represented by the solid line and most of the data lies above the line for n-type material. This observation has been noted in numerous reports; i.e. that n-type InP has much longer lifetimes than p-type InP. These data certainly have important consequences.

9. CONCLUSIONS

The maximization of minority-carrier lifetimes in III-V materials is an important component of the materials research and development. Significant improvements in materials technology have been made in the last half decade of research. The TRPL technique is extremely useful for research as well as quality control.

10. ACKNOWLEDGMENT

The author wishes to thank his colleagues B. M. Keyes and D. H. Levi for their contribution to this work. This work is performed for the US Department of Energy under contract DE-AC02-83CH10093.

11. REFERENCES

Ahrenkiel R K, Al-Jassim M M, Dunlavy D J, Jones K M, Vernon S M, Tobin S P, and Haven V E, 1988a, Appl. Phys. Lett. 53, 222.
Ahrenkiel R K, Dunlavy D J, Loo R Y, and Kamath G S, 1988b, J. Appl. Phys. 63, 5174.
Ahrenkiel R K, Olson J M, Dunlavy D J, Keyes B M, Kibbler A E, 1990, J. Vac. Sci. Technol. A8, 3002.
Ahrenkiel R K, 1991, In "Properties of Indium Phosphide", (INSPEC, IEE, 1991), p. 77.
Ahrenkiel R K, 1992, Solid-St. Electron. 35, 239.
Ahrenkiel R K, 1993, In Properties of Aluminum Gallium Arsenide", EMIS Data review, (INSPEC, IEE, 1993), p. 221.
Ahrenkiel R K, and Dunlavy D J, 1989, J. Vac. Sci. Technol. A7, 822.
Ahrenkiel R K, Dunlavy D J, Keyes B M, Vernon, S M, Tobin S P, and Dixon T M, 1990, 20th IEEE Photovoltaic Specialists Conference, 1990, (IEEE, New York), p. 432.
Ahrenkiel R K, Keyes B M, Lush G B, Melloch M R, and Lundstrom M S, and MacMillan H F, 1992, J. Vac. Sci. Technol. A 10, 990.
Asbeci P, 1977, J. appl. Phys. 48, 820.
Dawson P and Woodbridge K, 1984, Appl. Phys. Lett. 45, 1227.
Ehrhardt A, Wettling W, and Bett A, 1991 Appl. Phys. A53, 123
'tHooft G W, Leys M R, and Thalen-van der Mheen H J, 1985, Superlattices and Microstructures 1, 307.
Hummel S F, Beyler C A, Zou Y, Grodzinski P, and Dapkus P D, 1990, Appl. Phys. Lett. 57, 695.
Hwang C J, 1971 J. Appl. Phys. 42, 4408
Landis G A, Jenkins P, and Weinberg I, 1992, In "Proceedings of Third International Conference on InP and Related Compounds", (IEEE, New York), p. 636.
Lush G B, MacMillan H F, Keyes B M, Levi D J, Melloch M R, Ahrenkiel R L, andLundstrom M S, 1992 J. Appl. Phys. 72, 1436.
Lush G B, Melloch M R, Lundstrom M S, Levi D H, Ahrenkiel R K, and MacMillan H F, 1992, Appl. Phys. Lett. 61, 2441.
Molenkamp L W and van Blik H F, 1988, J. Appl. Phys. 64, 4253.
Nelson, R J and Sobers, R G, 1978a J. Appl. Phys. 49, 6103.
Nelson R J and Sobers R G, 1978b Appl. Phys. Lett. **32**, 761.
Nelson R J, 1978, J. Vac. Sci. Technol. 15, 1475.
Olson J M, Ahrenkiel R K, Dunlavy D J, Keyes B M, and Kibbler A E, 1989, Appl. Phys. Lett. 55, 1208.
van Opdorp C, and 't Hooft G W, 1981, J. Appl. Phys. 52, 3827.
Wolford D J, Gilliland G D, Juech T F, Smith L M, Martinsen J, Bradley J A, Tsang C F, Venkatasubramanian R, Ghandi S K, and Hjalmarson H P, 1991, J. Vac. Sci. Technol. B9, 2369.
Yamaguchi M, Yamamoto A, and Itho Y, 1986, J. Appl. Phys. 59, 1751.
Zarem H A, Lebens J A, Nordstrom, K B, Sercel P C, Sanders S, Eng L E, Yariv A, and Vahala K, 1989, Appl. Phys. Lett. 55, 2622.
Zhang J, Ahrenkiel R K, Keyes B M, Asher S E, and Timmons M L, 1993, Appl. Phys. Lett. (In press).

Inst. Phys. Conf. Ser. No 136: Chapter 11
Paper presented at the Int. Symp. GaAs and Related Compounds, Freiburg, 1993

The ordered $GaInP_2$ alloy: reasons for its 'anomalous' optical properties

F.A.J.M. Driessen, G.J. Bauhuis, S.M. Olsthoorn, and L.J. Giling.

Department of Experimental Solid State Physics, RIM, Faculty of Science, University of Nijmegen, Toernooiveld, 6525 ED Nijmegen, The Netherlands

Abstract

We report properties of ordered $GaInP_2$ alloys as a function of temperature T by a combination of photoreflectance (PR), photoluminescence (PL), PL excitation and electrical measurements. A variety of evidence for localization at confined donors was obtained for $T < 30$ K, such as: a first-derivative PR line shape, hopping conduction, and a consistent description of the T dependence of PL intensity assuming localized states. Along with transport data and properties of the moving emission, it is proposed that the origin of the inverted-S shape of PL energy as a function of T is connected with thermal population of the D^- band.

1 INTRODUCTION

The ternary semiconductor $Ga_xIn_{1-x}P$ forms an attractive alternative to $Al_xGa_{1-x}As$ in optoelectronic devices, such as $Ga_xIn_{1-x}P/GaAs$ tandem junction solar cells and (Al,Ga,In)P laser diodes, because of the very low value of the surface recombination velocity of the $Ga_xIn_{1-x}P/GaAs$ interface, and the absence of carrier-trapping DX centres in $Ga_{0.52}In_{0.48}P$ (hereafter abbreviated as $GaInP_2$), which is the composition at which lattice matching with GaAs is achieved. In spite of its successful application, various fundamental material properties of $GaInP_2$ are not well understood. The reason for this is that optical and transport properties of the equimolar $GaInP_2$ are complicated by the occurrence of regions of spontaneous long-range order of the CuPt type, namely as monolayer $(GaP)_1(InP)_1$ superlattices in the $[\bar{1}\ 1\ 1]$ and $[1\ \bar{1}\ 1]$ directions. As a consequence of this ordering the symmetry of the crystal is lowered in these regions, which leads to a splitting of the valence band and a decrease of the band gap (Mascarenhas and Olson 1990). For an infinitely large, ordered $GaInP_2$ crystal a reduction of 260 meV was calculated (Wei and Zunger 1990) whereas values around 100 meV (*e.g.* DeLong *et al* 1990) and recently even 190 meV (Lee *et al* 1992) have been found experimentally. The degree of ordering -size, shape, homogeneity and density of the ordered domains- depends on growth kinetics, particularly on surface mobility of group III adatoms (Stringfellow and Chen 1991). This surface mobility is in turn determined by growth conditions such as temperature of growth, V/III ratio and substrate orientation. The optical properties of ordered $GaInP_2$ show several anomalies: the so-called inverted-S shaped behaviour of PL energy as a function of T (Kondow *et al* 1989); the strong shift of the dominant PL peak towards higher energy upon increasing the excitation density P; and the long carrier lifetimes, which also depend strongly on P. DeLong *et al* (1990) concluded from the last two properties that spatially indirect recombination took place in the alloy.

In this paper we present data on ordered $GaInP_2$ from photoreflectance (PR), photoluminescence (PL) and PL excitation spectroscopy (PLE). The transport properties of these samples are reported on in a separate contribution (Bauhuis*et al* 1993) We show that optical and transport properties are greatly influenced by localization on relatively deep donor states that are confined in the ordered domains. Anomalous properties of $GaInP_2$, including its inverted S-shaped behaviour, are discussed.

2 EXPERIMENTAL DETAILS

For details on the PR, PL and electrical techniques we refer to Driessen *et al* (1993). A scanning standing-wave dye laser on DCM was used to record the PLE spectra. The experiments were carried out on $GaInP_2$ epilayers grown by metal organic vapour phase epitaxy at a pressure of 20 mbar. The 2.1-μm thick epilayers were grown lattice matched on a (100) 2^0 off towards (110) semi-insulating GaAs substrate at $T = 700^0$ C and V/III ratios of 400 (sample 1) and 870 (sample 2). The epilayers were nominally undoped and showed n-type conductivity. At room T the values for carrier concentration and mobility were 4.2×10^{15} cm^{-3} and 2500 cm^2/Vs for sample 1, and 3.4×10^{15} cm^{-3} and 2250 cm^2/Vs for sample 2. The aforementioned growth conditions are known to produce samples with long-range regions of monolayer superlattice ordering. The existence of the ordered structure in these samples was confirmed by transmission electron microscopy (TEM) which showed extra diffraction spots for the $[\bar{1}\ 1\ 1]$ and $[1\ \bar{1}\ 1]$ directions corresponding to long-range ordering of the CuPt-type. The TEM diffraction patterns also showed diffused streaks in the $\langle 0\ 0\ 1 \rangle$ directions, which provides evidence for the quasi two-dimensionality of the ordered domains: thin in the two ordering directions and extended in the directions perpendicular thereto. Furthermore, qualitative measures of the degree of ordering of $GaInP_2$ epilayers are the values for both the anomalously strong shift of PL energy as a function of P (the so-called "moving emission" (Fouquet *et al* 1990) and for the carrier lifetime, which is extremely long ($\sim$ μs-ms (DeLong *et al* 1991)) compared with random alloys ($\sim$ ns). The moving components of the 4K-PL spectra of samples 1 and 2 increased 4.7 meV and 6.1 meV per order of magnitude increase in P, and their lifetimes were 80 and 110 μs, respectively. These values are, under the given growth conditions, in good agreement with other published data (DeLong *et al* 1990). Almost all measurements we perfomed gave comparable results on both samples; unless specifically mentioned we report the results of sample 1.

3 RESULTS AND DISCUSSION

PR spectra recorded at various T are shown in fig.1; the upper spectrum is the 4.2-K PL spectrum. In those recorded at 240 and 293 K, a third-derivative line shape (TDLS) is observed. For semiconductors with a homogeneous intrinsic field distribution this TDLS occurs for low values of the modulation field (Aspnes and Rowe 1971), such as it is the case under our experimental conditions. In the T interval 40-190 K the spectra show a markedly different PR line shape from those at higher T. This complex line shape corresponds nearly to that observed in the high field limit (Aspnes and Frova 1969); the low energy side of the line shape may be affected by impurities. Because the value of the modulating field was not increased while lowering T, the high-field line shape can only be explained by field inhomogeneities in the sample. These inhomogeneities are presumably caused by confinement of electrons in the ordered domains, which have lower band gaps. A second interesting change in PR line shape is observed in spectra recorded at T between 4.2 K and 22 K: the spectra then show a stepped line shape. The effects of electric field modulation on the PR line shapes of isolated confined systems, such as quantum wells and superlattices, have been calculated by Glembocki and Shanabrook (1989), and by Enderlein, Djiang and Tang (1988, 1992). In all cases a first-derivative line shape (FDLS) was found resulting from the Stark effect of minibands and sublevels. Therefore, our observation of a pronounced FDLS at $T \lesssim 22$ K

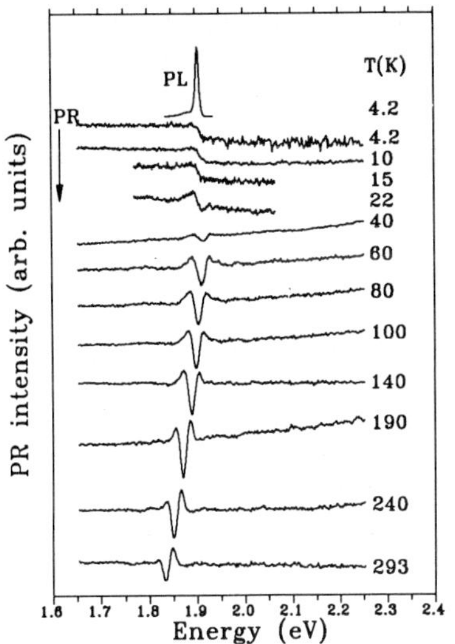

Figure 1: PR spectra as a function of T, and 4.2-K PL spectrum.

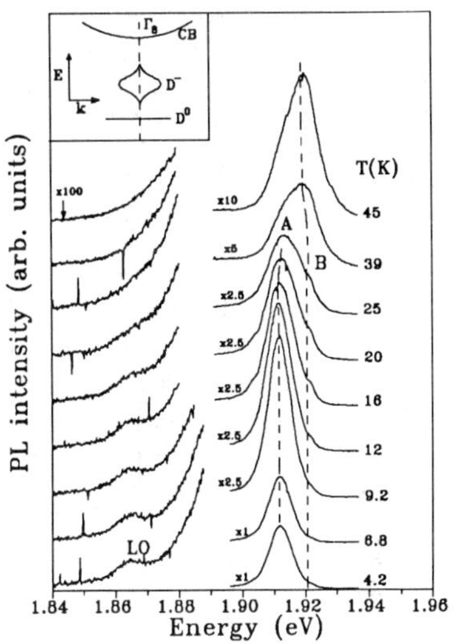

Figure 2: T dependence of the PL spectra from 4.2-45 K. The inset shows the band structure near the conduction-band minimum.

is a first proof that the carriers are strongly localized at low T. Before discussion of the 4.2-K PL spectrum we turn to some of the transport properties of these samples, the details of which we refer to Bauhuis *et al* (1993). Firstly, a donor binding energy of 36 meV was found, which is significantly larger than that of 11.1 meV calculated with effective mass theory. This is caused by compression of the donor wave function in the ordered domains by the disordered barrier material. Secondly, hopping conduction was clearly identified as the dominant conduction mechamism below $T = 30\ K$; consequently, electrons are then bound by donors. Therefore, the FDLS of PR below 20-40 K will be caused by localization at donors in the ordered domains. Thirdly, ϵ_2-conduction was identified as the dominant conduction mechanism in the intermediate T-range of 30-80 K. This ϵ_2-conduction is associated with conduction in an impurity band that is believed to arise from a resonance between negatively charged donors, the D^- ions (Nishimura 1965). Owing to the great spatial extent of these D^- states the interaction between them is strong, which results in a large D^- band width. Upon increasing the donor concentration n_D this width increases and ϵ_2, which is the energy difference between the bottom of the D^- band and the D^0 level, is consequently reduced. The resulting band structure near the Γ_8 point of the conduction band is shown schematically in the inset of fig.2: the broad D^- band is located between donor level and conduction band.

We now return to the low excitation density PL and PR spectra at $T = 4.2$ K in fig.1. The most important feature is that the dominant PL signal is emitted at the same energy as the step in the PR signal. This is a remarkable result because PR more strongly reflects the band structure and PL comes preferentially from impurity states in the band gap. Therefore, the absence of Franz-Keldysh oscillations and the clear first-derivative line shape at the impurity edge shows that the PR line shape is entirely determined by electron localization at these impurities in our samples. On the basis of transport data and the long lifetimes, we attribute the low-T PL to spatially separated recombination between electrons at donors in the ordered regions and photoexcited holes, *i.e.* (D^0_{ord}, h).

Fig.2 shows PL spectra for T below 45 K recorded at $P = 1$ Wcm^{-2}. An LO phonon replica is now observed in the lowest T spectra 47 meV below the dominant PL (labelled 'A'). According to selection rules for scattering of excitons by LO phonons, the first LO replica is forbidden owing to momentum conservation (Permogorov 1982). However, this selection rule is relaxed in cases of strong spatial localization, for which the uncertainty in momentum is comparable with the phonon momentum. Our observation of an LO phonon replica below $T = 25$ K is therefore further evidence for the localization that appears in $GaInP_2$ alloys at low T. At the high-energy side a second shoulder (labelled 'B') is present, which dominates the spectrum above $T > 25$ K. Peak B is not accompanied by a phonon replica, indicating a non-localized nature of that PL.

Table I: Rates of emission shift dE/dP (meV per order of magnitude of P) and exponents x in $I \propto P^x$ for peaks A and B at different T.

	peak A		peak B	
T(K)	dE/dP	x	dE/dP	x
4.2	4.7	0.98	†	†
25	5.2	1.01	0	1.45
40	7.2	1.08	0	1.49
70	†	†	0	1.61

†not observable

Table I summarizes the rate of emission shift dE/dP and the dependence of PL intensity I upon P for peaks A and B at various T. The I of peak 'A' depends virtually linearly on P. This behaviour is in accordance with the expected behaviour of (D^0_{ord},h): I proportional to the hole density [h], which in turn depends linearly on P. The I of peak 'B' depends superlinearly on P. An exponent between 1 and 2 is typical for PL processes in which both electrons and holes are photocreated (*e.g.* bound-excitons, free excitons), and in which parallel radiationless processes participate also. The dE/dP was found to be zero for peak 'B' which shows that electrons and holes are not spatially separated. The rate at which the PL energy of peak 'A' shifts is seen to increase if peak 'B' increases. This can be rationalized by assuming that peak 'B' has a short lifetime, which results in a reduction of the valence band filling. The accompanying increase in strength of the internal depletion fields will then cause a larger dE/dP of peak 'A'. The P dependence of PLE spectra showed that the PLE edge stayed at constant energy so that the P dependence of the Stokes shift (*i.e.* the energy difference between PLE and PL) was identical to that of PL, as can be seen in fig.3 for sample 2.

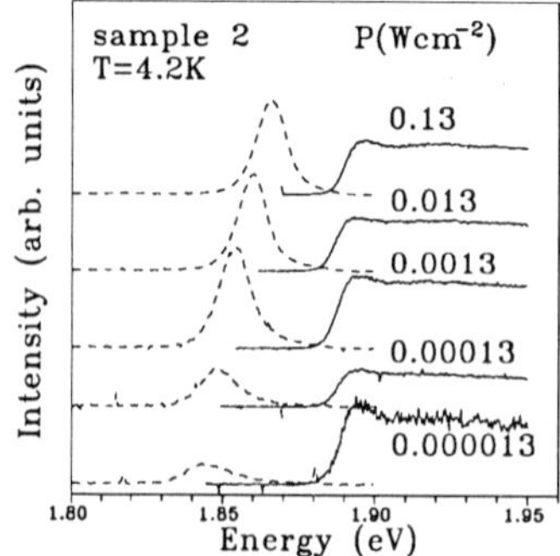

Figure 3: PLE (solid) and PL (dashed) spectra as a function of excitation density recorded at $T = 4.2$ K.

The anomalous behaviour of photon energy as a function of T in ordered $GaInP_2$(Kondow *et al* 1989) is shown in fig.4; where possible, energies of the moving and non-moving components are shown separately. The dashed curve visualizes the energy of the PL maximum: the inverted-S shape. The energies of the band gap E_0 as deduced from PR using Aspnes and Rowe's formalism (1971) are also shown in fig.4 by triangles. The complex line shape between $T = 40$ and 190 K hindered the precise determination of E_0 (Driessen *et al* 1993) and data points are therefore not given here.

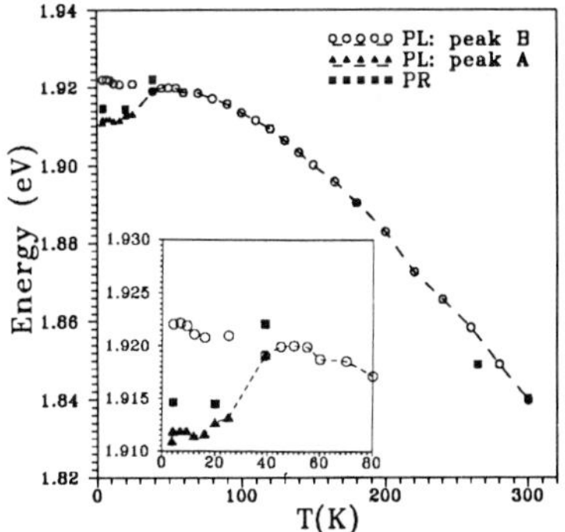

Figure 4: Energy of PL emissions as a function of T. The dashed line (inverted S) shows the maximum of PL intensity. The values of the band gap as determined by PR are shown by triangles.

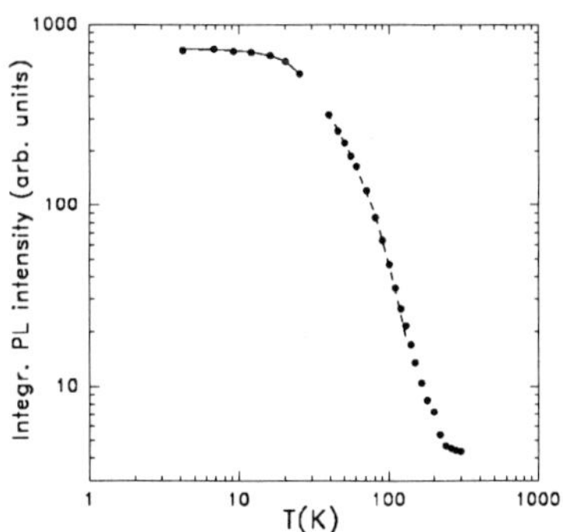

Figure 5: Numerically integrated PL intensity versus T. The solid and broken curves are best fits described in the text.

Because the inverted S-shape was also reported in disordered superlattices (Kasu *et al* 1990) and $Al_xIn_{1-x}As$ (Olsthoorn *et al* 1993), which both showed carrier localization, it appears that the inverted S-shaped behaviour occurs only together with carrier localization. For our sample the anomalous increase of PL energy starts if T exceeds 25 K, which is virtually the same T below which PR spectra showed evidence for strong localization through FDLS. Furthermore, the conduction mechanism changed from ϵ_3 conduction to ϵ_2 conduction at this T. The origin of the inverted S-shaped dependence of photon energy on T may therefore be the following. At low T (D^0_{ord},h) emission occurs (moving peak 'A'). Because the gap between the bottom of the D^- band and the donor states is small, the lowest lying states in the D^- band can be occupied at relatively low T and this wide band will then be filled. The observed increase in PL energy upon increasing T may therefore be due to PL from the D^- band: peak 'B' (note its consistent superlinear relationship between I and P). The additional, usual band gap reduction (Varshni 1967) upon increasing T leads to a decrease in PL energy at higher T.

Yamamoto *et al.* (1990) also reported PL intensity as a function of T for their disordered superlattices. It appeared that the data could be well fitted to a relationship valid for amorphous semiconductors because of the existence of localized states therein (Street *et al* 1974): $I_{PL}(T) = I_0/[1 + A\exp(T/T_0)]$, where I_{PL} is the PL intensity, T_0 is a characteristic T corresponding to the energy depth of localized states, A is the tunneling factor, and I_0 scales the PL intensity at the low-T limit. In fig.5 we show the numerically integrated PL intensity of the $GaInP_2$ PL as a function of T. Our data could be well described by the above equation for $T < 30$ K; the solid curve represents a best fit yielding $T_0 = 7.5$ K, $A = 0.01$ and $I_0 = 753.1$. The fit became less good if data points above $T = 30$ K were added showing this data was not consistent with the equation. However, an Arrhenius equation $I_{PL}(T) = I_0\exp(-T/T_0)$, with $T_0 = 32.0$ K, described these data perfectly, as is shown by the broken curve in fig.4. This result is in view with the onset of PL from non-localized D^- states above $T = 30$ K. Furthermore, a discontinuity is seen around 200 K. Above this T PR data indicated a homogeneous electron distribution. Therefore, the discontinuity in the PL intensity will most likely be caused by the onset of luminescence from the disordered regions above 200 K.

4 CONCLUSIONS

We have performed PR, PL, PLE and electrical measurements as a function of T on long-range ordered $GaInP_2$ obtained by MOVPE. The data provide information on the origin of the anomalous properties of the ordered alloy. Below $T = 200$ K we found that optical and transport properties were strongly influenced by a donor state which, owing to confinement in the ordered domains, had a high binding energy of 36 meV. Evidence for strong localization was obtained for $T < 30$ K: the PR spectra showed a pronounced first-derivative line shape, the conductivity was dominated by hopping conduction, the relationship between PL intensity and T could be well fitted assuming localized states, and an LO phonon replica was observed. Energies of the PL and PR features were found to be identical at these low T. The PLE data showed excitation-density-dependent Stokes shifts; the PLE edge stayed constant. On the basis of transport data the low-T moving PL transition was attributed to spatially separated recombination between a photo-excited hole and an electron that is localized on a donor in an ordered superlattice domain. All properties of both moving and non-moving emissions strongly suggest that the non-moving emission is to be attributed to PL from the D^- band. Consistently, the inverted-S shape of PL energy as a function of T would then be connected with thermal population of the D^- band. In the T interval 30-200 K the PR showed a complex line shape owing to the absence of carriers in the disordered regions. Above $T \approx 200$ K the PR showed third-derivative line shapes, which indicates a uniform electric field distribution. The discontinuity in the PL intensity at 200 K is therefore attributed to the contribution to PL from randomly alloyed regions above this T.

The authors are grateful to P.R. Hageman and A. van Geelen for growing the samples, to A.J. Bons for performing the TEM measurements, and to D.M. Frigo for his critical reading of the manuscript. Financial support from the Stichting voor Fundamenteel Onderzoek der Materie (FOM), and the Nederlandse Organisatie voor Energie en Milieu (NOVEM) is gratefully acknowledged.

Aspnes D E and Frova A 1969 Solid State Commun. 7 155
Aspnes D E and Rowe J E 1971 Phys. Rev. Lett. 27 188
Bauhuis G J, Driessen F A J M, Olsthoorn S M and Giling L J *this conference*
DeLong M C, Taylor P C and Olson J M 1990 Appl. Phys. Lett. 57 620
DeLong M C, Ohlsen W D, Viohl I, Taylor P C and Olson J M 1991 J. Appl. Phys. 70 2780
Driessen F A J M, Bauhuis G J, Olsthoorn S M and Giling L J 1993 Phys Rev. B *accepted*
Enderlein R, Jiang D and Tang Y 1988 Phys. Stat. Sol. B 145 167
Enderlein R 1992 Proc. 20th Int. Conf. on the Physics of Semiconductors ed E M Anastassakis and J D Joannopoulos (Singapore: World Scientific), p1089
Fouquet J E, Robbins V M, Rosner J and Blum O 1990 Appl. Phys. Lett. 57 1566
Glembocki O J and Shanabrook B V 1989 Superlattices and Microstructures 5 603
Kasu M, Yamamoto T, Noda S and Sasaki A 1990 Jap. J. Appl. Phys. 29 828
Kondow M, Minagawa S, Inoue Y, Nishino T and Hamakawa Y 1989 Appl. Phys. Lett. 54 1760
Lee M K, Horng R H and Haung L C 1992 J. Cryst. Growth 124 358
Mascarenhas A and Olson J M 1990 Phys. Rev. B 41 9947
Nishimura H 1965 Phys. Rev. 138 A815
Olsthoorn S M, Driessen F A J M, Eijkelenboom A P A M and Giling L J 1993 J. Appl. Phys. 73 7798
Permogorov S 1982 Excitons ed E I Rashba and M D Sturge (Amsterdam: North-Holland) p 177
Street R A, Searle T M and Augustin I G 1974 Amorphous and Liquid Semiconductors ed J Stuke and W Brenig (London: Taylor and Francis) p 953
Stringfellow G B and Chen G S 1991 J. Vac. Sci. Technol. B 9 2182
Varshni Y P 1967 Physica 34 149
Wei S-H and Zunger A 1990 Appl. Phys. Lett. 56 662
Yamamoto T, Kasu M, Noda S and Sasaki A 1990 J. Appl. Phys. 68 5318

Paper presented at the Int. Symp. GaAs and Related Compounds, Freiburg, 1993

Regeneration of the EL2 defect in hydrogen passivated GaAs

C A B Ball, A B Conibear, and A W R Leitch

Department of Physics, University of Port Elizabeth, P.O. Box 1600, Port Elizabeth 6000, Republic of South Africa

ABSTRACT: The EL2 defect in GaAs is passivated by exposure to a hydrogen plasma forming an electrically inactive EL2-H complex. The electrically active defect is restored to approximately 30 percent of its original concentration by annealing at 384 K for 2 hours in the presence of an electric field. The passivation of the EL2 is therefore concluded to be less stable than previously reported. The thermal dissociation of the EL2-H complex is characterised by two first-order processes, one significantly more rapid than the other.

1. INTRODUCTION

Exposure of GaAs to a hydrogen plasma has been shown to passivate shallow dopants and to render some defects deep in the bandgap electrically inactive (Chevallier 1991). The passivation of deep levels has been believed to be more stable to annealing (Dautremont-Smith 1986) than that of the shallow dopants and has enjoyed relatively little attention. Among the defects passivated by hydrogen is the extensively studied EL2 defect. This deep level is considered to be primarily responsible for the compensation of shallow acceptors in nominally undoped semi-insulating GaAs. The passivation of the EL2 defect by hydrogen, and the thermal stability of the passivation are therefore of considerable interest. Lagowski and coworkers (1982) first reported the passivation of the EL2 in horizontal Bridgman (HB) GaAs by exposure to a hydrogen plasma. They concluded that hydrogen passivation constitutes an effective low-temperature process for controlling the EL2 concentration. Omel'yanovski et. al. (1987) reported approximately 10 percent re-activation of the EL2 defect in liquid encapsulated Czochralski (LEC) GaAs after 30 minutes annealing at 623 K. This treatment entirely re-activated the shallow dopants in the material. Rapid thermal annealing at 830 K for 10 s has been reported by Cho et al (1988) to regenerate the EL2 in passivated HB GaAs. There has been, to our knowledge, no work done in this regard on metal-organic vapour-phase epitaxial (MOVPE) material. The indications have been that the passivation of the EL2 is at least as thermally stable as that of the shallow dopants and that relatively high temperatures are required to effect any reactivation. In this paper we show that the EL2 in MOVPE-grown GaAs is measurably reactivated by annealing at temperatures lower than those required for the reactivation of shallow dopants.

2. EXPERIMENT

The GaAs used in this work was grown by MOVPE on Si doped ($n=1\times10^{18}$ cm^{-3}) substrates. Samples with free carrier densities of $n=1\times10^{15}$ cm^{-3} (nominally undoped) and $n=2\times10^{16}$ cm^{-3} (Si doped) were used. After Ni-AuGe-Ni Ohmic contacts had been formed on the back of the substrates, the samples were exposed to a dc hydrogen plasma in a parallel plate system similar to that described by Stutzmann et al (1990). One of the plates was grounded and the other biased at 600 V. The sample was mounted 10 cm downstream on a holder biased at -250 V. The hydrogen pressure was 0.4 mBar and the power dissipated by the plasma was less than 1.2 W. After passivation for 1 hour at 180 °C, the sample was lightly etched to remove 0.2 μm of the material so as to prevent any surface damage created by the plasma treatment from affecting the electrical measurements. Gold Schottky contacts were evaporated onto the samples. Ohmic and Schottky contacts were also evaporated onto reference samples which had not been exposed to the plasma. Capacitance-voltage (C-V) profiling at 1MHz showed hydrogen penetration to a depth of 2 μm in the case of the 1×10^{15} cm^{-3} material and 0.7 μm for the 2×10^{16} cm^{-3} material.

3. RESULTS

Figure 1 shows Deep Level Transient Spectroscopy (DLTS) spectra of the reference and passivated samples for the Si-doped material. The data was collected at a very low pulsing frequency using a digital system (Ball et al 1991) in order to bring the DLTS peaks to temperatures at which little reactivation of passivated samples occurred. The reverse bias was 2 V and the pulse height 1.8 V for the passivated samples. The reverse bias and pulse height were appropriately adjusted for the reference so as to sample to a corresponding depth in the material. Spectrum (a) is for the reference sample in which the EL2 concentration was 1.6×10^{14} cm^{-3}. It may be seen that the EL2 level is greatly reduced in the spectrum (b) for the passivated sample. The peak heights are in arbitrary units, but those for the passivated samples have been scaled to account for the fact that the free carrier density is reduced. Peak heights shown are thus proportional to trap concentration. Spectrum (c) is for a passivated sample subjected to 2 hours annealing at 384 K under a reverse bias of 3 V. The EL2 is restored to approximately 30 percent of its concentration in the unpassivated material. This is significantly more than previously reported for annealing at higher temperatures (Omel'yanovski et al 1987).

Calculations based on data (Roos et al 1991) for the dissociation of the Si-H complex in the depletion region of a Schottky barrier diode indicate that the reactivation of shallow donors is negligible for the annealing in this study.This was verified by carrying out C-V measurements on passivated samples before and after annealing. No change in the free carrier profiles was detected. It may thus be accepted that any changes in DLTS peak heights are not related to reactivation of shallow dopants.

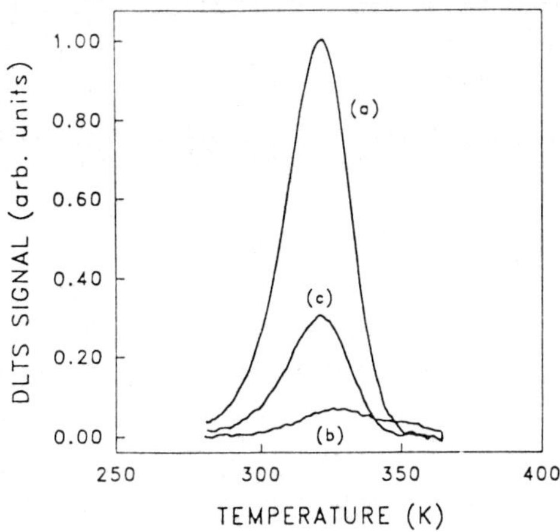

Fig. 1. DLTS spectra for Si-doped GaAs: a)reference, b) as-passivated, c) passivated sample annealed for 2 hours at 384 K.

A nominally undoped sample was passivated and then subjected to annealing under 0.5 V reverse bias at 384 K. The EL2 peak height is plotted as a function of annealing time in Figure 2. The data was collected at the peak temperature of the chosen DLTS rate window and annealing was carried out at the same temperature (Conibear et al 1993). It is apparent that there is a rapid intial reactivation over the first approximately 1.5 hours (5400 s) after which the recovery of the EL2 proceeds more slowly.

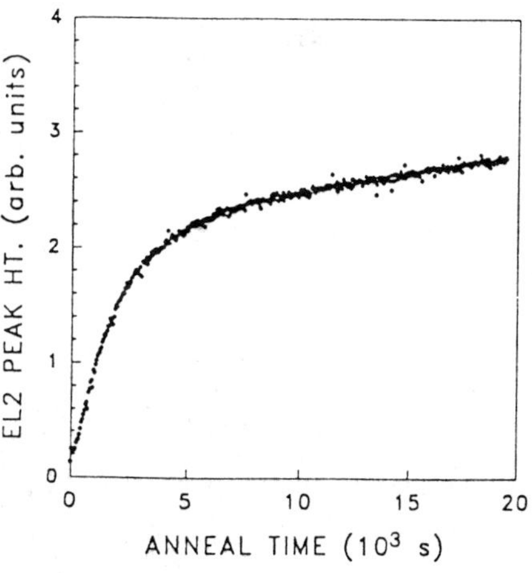

Fig. 2. EL2 peak height in passivated undoped GaAs as a function of annealing time at 384 K.

The concentration of the passivated EL2 or EL2-H complex may be derived from this data since [EL2-H] = $[EL2]_{total}$ - [EL2], where [EL2-H] is the concentration of the EL2-H complex in the passivated sample, [EL2] is the electrically active EL2 concentration in the passivated sample, and the total EL2 concentration, $[EL2]_{total}$ is measured in the reference sample.

A first-order dissociation process for the EL2-H complex is characterised by the expression $N(t)=N_0\exp[-\nu(T)t]$, where N_0 is the concentration of the complex at the commencement of annealing and $\nu(T)$ is the reactivation rate at the specific annealing temperature. A plot of the natural logarithm of the concentration against anneal time is linear for a first-order reactivation. We shall show that the dissociation of the EL2-H complex may be characterised by two first-order processes, one significantly faster than the other. This implies two components to the EL2-H concentration, one which decays rapidly which we shall denote N_1 and a slowly decaying component denoted N_2.

The total time-varying EL2-H concentration may be written as

$$N(t) = N_{10}\exp[-\nu_1(T)t] + N_{20}\exp[-\nu_2(T)t]$$

where N_{10} and N_{20} are the initial concentrations of the two components and ν_1 and ν_2 the respective reaction rates. Figure 3 shows the natural logarithm of the complex concentration plotted against time. The solid line is a fit to data for annealing times greater than 12500 s after which the rapid reaction is completed. The linearity of this portion of the curve indicates that the slow reaction is first-order. With ν_2 and N_{20} determined from this fit, the contribution of the slow decay was subtracted from the experimental data for the shorter annealing times.

The resulting data representing the rapid reactivation process is plotted logarithmically in Figure 4. The linearity again indicates a first-order process. The reaction rates for the fast and slow processes at 384 K were determined to be $\nu_1=(5.3\pm2.1)\times10^{-4}\ s^{-1}$ and $\nu_2=(4.6\pm1.8)\times10^{-6}\ s^{-1}$ respectively. The rather large uncertainties are due to possible under-rating of the EL2 concentrations because of the proximity of the trap and the Fermi level (Ma et al 1988). N_{10} and N_{20} were determined from equation 3. The rapidly reactivating component constituted of the order of thirty percent of the total starting EL2-H concentration in our samples.

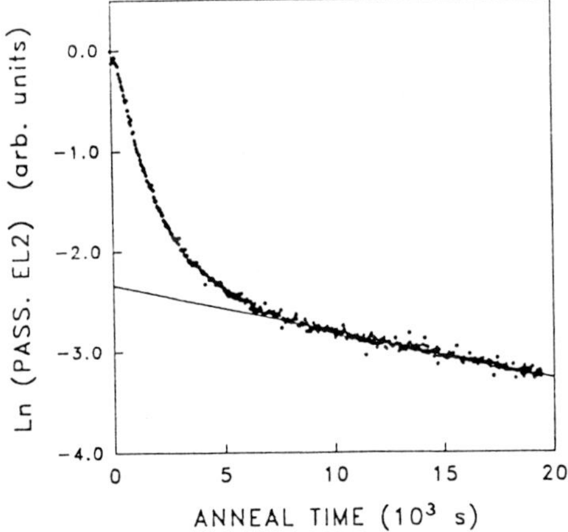

Fig. 3. Natural logarithm of the EL2-H complex concentration as a function of annealing time at 384 K. Sample: passivated undoped GaAs. The fitted line indicates the slow component of the EL2 reactivation.

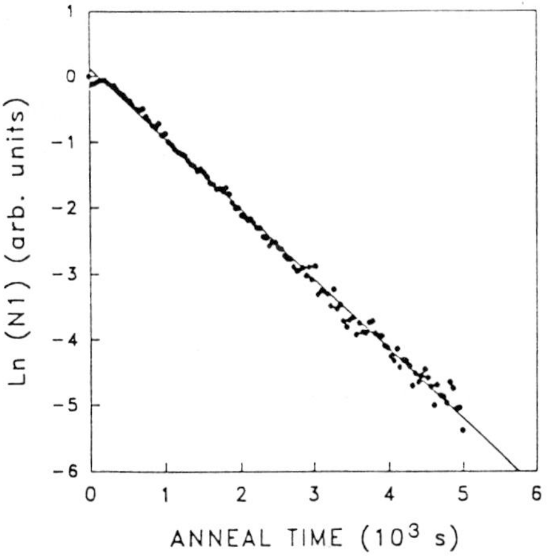

Fig. 4. Natural logarithm of the rapidly decaying EL2-H component N_1 after subtraction of the slow component N_2.

4. DISCUSSION AND CONCLUSIONS

The EL2 defect is believed by some to be a family of related defects (Cho et al 1989, Taniguchi et al 1983), rather than a single level. There are clearly at least two components in our samples which interact differently with hydrogen. The trap energy (0.82 eV) and capture cross section (1×10^{13} cm^2) of the reactivated EL2 were measured by DLTS. These parameters were not significantly different from those measured for the EL2 in the reference samples.

Annealing carried out for reverse biases ranging from 2 V to 8 V showed very similar reactivation rates to those we have reported. Annealing under zero bias however resulted in only a slow reactivation rate of the order of 10^{-7} s^{-1}. It may therefore be tentatively suggested that the rapid reactivation involves thermal emission of an electron from the El2-H complex, followed by dissociation of the complex. In the zero bias case, there are free electrons available for capture, thus precluding dissociation. Under reverse bias the emitted electron is swept away by the electic field in the depletion region. The greater passivation stability reported by other workers may possibly be ascribed to differences in passivation conditions and to the fact that no bias was applied during annealing. The possibility that the relative concentrations of the rapidly and slowly reactivating species could depend on growth conditions should be investigated.

In summary, we have shown that the EL2 defect significantly reactivates by reverse bias annealing at temperatures as low as 384 K. The reactivation may be described by two simultaneous first-order processes which appear to be dependent on the application of an electric field to deplete the region of interest. While early reports indicated that there could be a "temperature window" (Pearton et al 1987) in which reactivation of shallow dopants could be effected, while leaving deep levels passivated, this is evidently not the case for the EL2 defect.

5. ACKNOWLEDGEMENTS

This work was partially funded by the Foundation for Research and Development. We express our appreciation to Mr J.R. Botha who grew the samples used in this study.

6. REFERENCES

Ball C A B and Conibear A B 1991 Rev. Sci. Instrum. 62 2831
Chevallier J, Clerjaud B, and Pajot B 1991 Hydrogen in Semiconductors, Semiconductors and Semimetals 34 ed Pankove J I and Johnson N M (Academic Press) p 447
Cho H Y, Kim E K, Min S K, Kim J B, and Jang J 1988 Appl. Phys. Lett. 53 856
Cho H Y, Kim E K, and Min S K 1989 Phys. Rev. B 39 10376
Conibear A B, Leitch A W R, and Ball C A B 1993 Phys. Rev. B 47 1846
Dautremont-Smith J C, Nabity J C, Swaminathan V, Stavola M, Chevallier J, Tu C W, and Pearton S J 1986 Appl. Phys. Lett. 49 1098
Lagowski J, Kaminska M, Parsey Jr. J M, Gatos H C, and Lichtensteiger M 1982 Appl. Phys. Lett. 41 1078
Ma Q Y, Schmidt M T, Wu X, Evans H L, and Yang E S 1988 J. Appl. Phys. 64 2469
Omel'yanovski E M, Pakhomov A V, Polykov A Y 1987 Sov. Phys. Semicond. 21 514
Pearton S J Sorbett J W, and Shi T S 1987 L. Appl.Phys A 43 153
Roos G, Johnson N M, Herring C, and Harris S. J 1991 Appl. Phys. Lett. 59 461
Stutzmann M and Herrero C P, 1990 20th Int. Conf. on the Physics of Semiconductors, Thessaloniki Greece, ed Anastassakis E M, and Joannopoulos J D (Singapore: World Publishing Co.) p 783
Taniguchi M and Ikoma T 1983 J. Appl. Phys. 6448

Electrical conduction in ordered $GaInP_2$ epilayers

G.J. Bauhuis, F.A.J.M. Driessen, S.M. Olsthoorn, and L.J. Giling

Department of Experimental Solid State Physics, RIM, University of Nijmegen, Toernooiveld, NL 6525 ED Nijmegen, The Netherlands

Abstract

A Hall-Van der Pauw and high-field magnetotransport study is reported on $GaInP_2$ epilayers which contain naturally ordered regions with lower band gap than that of the embedding disordered regions. Below $T = 200$ K three conductivity mechanisms are distinguished in this material: hopping conduction (activation energy ϵ_3) dominates for $T < 30$ K, band conduction (ϵ_1) for $T > 85$ K, and ϵ_2-conduction at intermediate temperatures. The behaviour of the conductivity σ in a transverse magnetic field B is found to be in good agreement with theory. From mobility data above $T = 200$ K a radius for ordered regions of 3.5 nm was calculated.

1 INTRODUCTION

The direct band gap semiconductor $GaInP_2$, which is lattice matched to GaAs, is a promising material for optoelectronic devices such as laser diodes and tandem solar cells (Valster 1991, Roentgen 1991, Ikeda 1986, Olson 1990). An interesting feature of $GaInP_2$ is that it contains regions that exhibit CuPt-type ordering on the group III sublattice, which leads to a decrease in band gap (Wei 1990, DeLong 1990, McDermott 1991). This decrease depends on the degree of ordering, which is in turn determined by growth conditions; theoretically this reduction of E_g can be as high as 260 meV (Wei 1990). Understanding of the electrical properties of this material is of great importance for device purposes. To our knowledge no experimental data are available on the transport properties in ordered $GaInP_2$ at temperatures below 200 K. For higher T it is well known that electrical conduction of III-V ternary and quaternary alloys may be influenced by ordered clusters (Bhattacharya 1985, Lee 1991). By treating spherically ordered domains as clusters, Friedman, Kibbler and Olson (1991) calculated that scattering at clusters in $GaInP_2$ strongly affects the Hall mobility μ_H. Their model predicts that, for samples with large cluster radius r_c , μ_H should even increase with increasing T. However, experimental verification of the effect of cluster scattering on μ_H at high temperatures could not be made from their measurements on ordered $GaInP_2$ samples; they could only conclude that $r_c < 2$ nm. In this paper we present Hall-Van der Pauw and high-field magnetotransport measurements on two ordered $GaInP_2$ layers in the temperature interval from $T = 2.1$ K to 450 K. At T below 200 K we identify the prescence of hopping, D^--band conduction and carrier freeze-out, all donors are confined in the ordered regions. Furthermore, the first experimental proof of cluster scattering in $GaInP_2$ is demonstrated for $T > 200$ K.

2 EXPERIMENTAL DETAILS

The experiments were carried out on two nominally undoped $Ga_xIn_{1-x}P$ epilayers grown by metalorganic vapour phase epitaxy (MOVPE), both showing n-type conductivity. Experimental details of growth parameters and sample characteristics have are given by Driessen (1993, this conference). Rocking curves were measured using high-resolution X-ray diffraction. From these curves gallium contents x of 0.471 and 0.468 were calculated for samples 1 and 2, respectively. Using the equation given by Auvergne (1977) this corresponds to bandgaps of 1.924 and 1.920 eV for the disordered regions. Photoluminescence measurements performed at very low excitation densities showed an E_g for the ordered regions of 1.893 eV, for sample 1, corresponding to a band gap reduction ΔE_g of 31 meV. For sample 2 this ΔE_g amounted 82 meV.

Hall-van der Pauw measurements were performed for T ranging from 20 to 600 K. For T above 450 K we found that intrinsic conduction through the GaAs substrate influenced the measurements. From the T-dependence of the Hall mobility no evidence was found for the presence of a two-dimensional electron-gas at the $GaInP_2$ /GaAs interface, as was reported by Friedman et al. (1991). Measurements of the transverse conductivity in a magnetic field (up to 20 T) were performed on sample 2 in the High-Field Magnet Laboratory, Nijmegen for T's in the range 2.1 - 140 K.

3 RESULTS AND DISCUSSION

3.1 Transport properties at low temperatures

For T below 200 K previously reported optical data (Driessen 1993) showed that carriers are absent in the disordered regions. In fig.1 the inverse T-dependence of the Hall coefficient R_H of each sample is shown. The sharp drop in R_H at $T < 70\ K$ (sample 1) or $T < 85\ K$

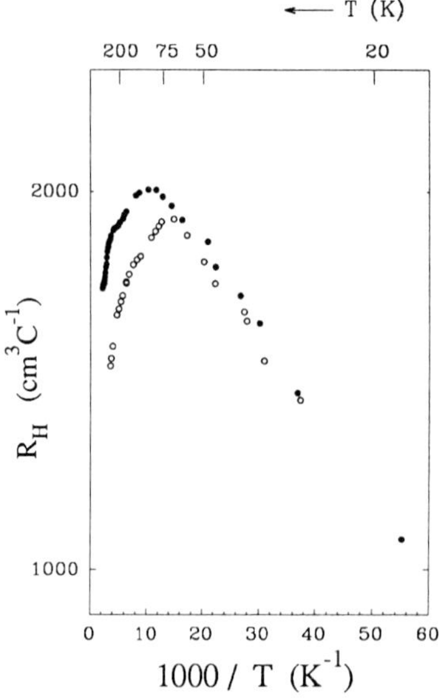

Figure 1: Hall coefficient R_H as a function of temperature for samples 1 (•) and 2 (o).

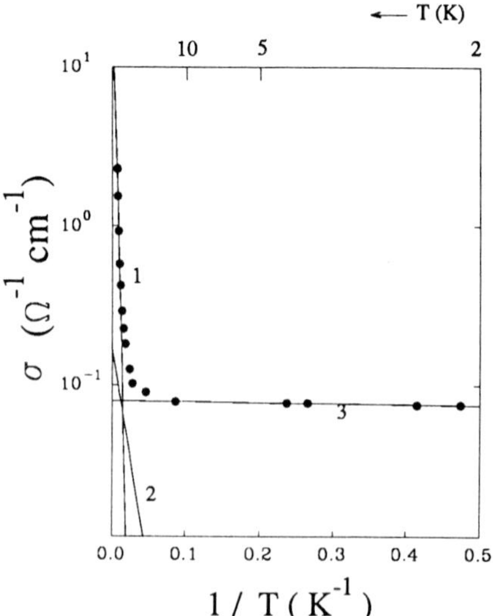

Figure 2: $\sigma(1/T)$ for temperatures below 140 K. The solid lines show the results from the fitting procedure: lines labelled 1, 2 and 3 correspond to the contribution of ϵ_1-, ϵ_2- and ϵ_3-conduction, respectivily.

(sample 2) indicates a change in the dominant conduction mechanism (Shklovskii 1984a). In fig.2 the inverse T- dependence of the conductivity σ is shown for sample 2 in the range corresponding to 2 - 140 K. The conductivity can be expressed as (Shklovskii,1984a)

$$\sigma \;=\; \sigma_1 e^{-\epsilon_1/kT} \;+\; \sigma_2 e^{-\epsilon_2/kT} \;+\; \sigma_3 e^{-\epsilon_3/kT} \tag{1}$$

with activation energies $\epsilon_1 > \epsilon_2 > \epsilon_3$ and conductivity pre-factors $\sigma_1 > \sigma_2 > \sigma_3$.

The three conductivity terms correspond to different conduction mechanisms. The first term corresponds to band conduction and yields the donor binding energy ϵ_1. The third term corresponds to hopping conduction with ϵ_3 being caused by dispersion of the donor states. In the intermediate temperature range, between the band and hopping regimes, ϵ_2-conduction dominates. The ϵ_2-conduction mechanism is connected to a resonance of electrons that are weakly bound to neutral donors, the D^- states (Shklovskii 1984a).

The data of fig.2 are well described by equation (1) with activation energies of: $\epsilon_1 = 36$ meV, $\epsilon_2 = 5.5$ meV and $\epsilon_3 = 0.011$ meV. The dominating mechanism for $T > 85$ K is ϵ_1-conduction, which is caused by electrons excited from donor levels to the conduction band with binding energy ϵ_1. The value of 36 meV found here is much higher than that expected on the basis of effective mass theory of shallow donors in a random alloy (Driessen 1993). We believe this is caused by constriction of the donor wave functions to the ordered domains, in which the conduction-band offset between ordered and disordered regions acts as a potential barrier.

Between $T \approx 30$ and 85 K ϵ_2-conduction is observed, a characteristic of which is the quadratic dependence of activation energy ϵ_2 on magnetic field B (Yamanouchi 1965):

$$\epsilon_2(B) \;=\; \epsilon_2(0) \;+\; \gamma B^2 \tag{2}$$

where γ is a constant. In order to determine whether this relationship is valid for our $GaInP_2$ layers, we performed magnetoconductivity measurements on sample 2 at T's between 2.1 and 140 K. The values of ϵ_2, as determined from the temperature dependence of σ, are shown as a function of B^2 in fig.3. For $B^2 > 30$ T^2 the quadratic dependence of expression (2) is clearly present. A value of $5.4 \cdot 10^{-6}$ eV T^{-2} was found for γ, which is in the same

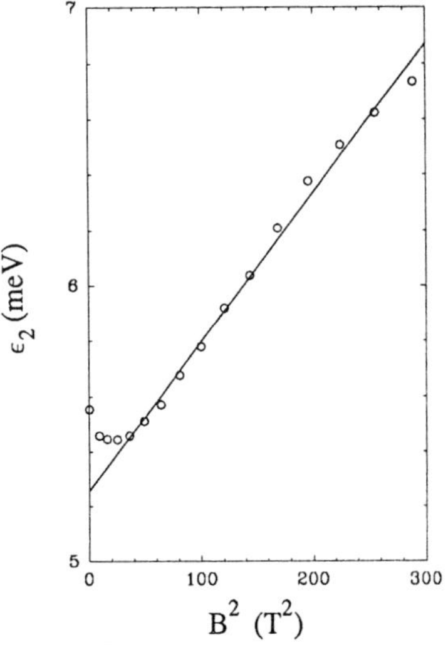

Figure 3: Activation energy ϵ_2 as a function of the quadratic magnetic field B.

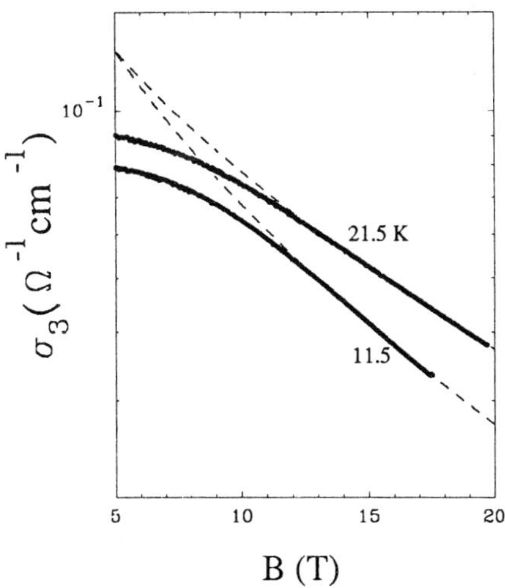

Figure 4: Pre-exponential factor σ_3 as a function of magnetic field B at two different temperatures. Dashed lines represent fits to expression (3).

order of magnitude as the values found for germanium (Sadasiv 1962).

Below $T \approx 30$ K ϵ_3-conduction is the dominating conduction mechanism (fig.2), wherein the electrons are localized on the donor cores and conduction occurs by hopping. Evidence for localization in our samples was already found via optical measurements in this temperature range (Driessen 1993). A characteristic feature of ϵ_3-conduction is a large negative magnetoconductivity that depends exponentially on the magnetic field. This behaviour is due to squeezing of the impurity wavefunctions in directions perpendicular to the field, leading to a sharp decrease in wavefunction overlap for an average pair of neighbouring impurities. Since the wavefunction overlap determines the probability of a 'hop' between two impurities, this squeezing strongly reduces the conductivity. According to Shklovskii and Efros (Shklovskii 1984b), there are two formulas that describe the dependence of σ_3 on the strength of the magnetic field. For strong fields this has the form

$$\sigma_3(B) \propto e^{-const \cdot B^m} \tag{3}$$

where m is predicted to be slightly greater than 0.5 (Shklovskii 1984b). The transition between the strong-field and weak-field case is given by the critical field $B_c = \hbar N_D^{1/3}/(a_B e)$ where a_B is the Bohr radius and N_D the donor concentration. Using $a_B = \hbar/(\sqrt{2m^* E_D})$ with effective electron mass $m^*/m_0 = 0.108$ and $E_D = 36$ meV, a value of $B_c = 3.2$ T is calculated. Fig.4 shows the dependence of σ_3 for two temperatures in the range where ϵ_3-conduction predominates. The σ_3 data at five temperatures were fitted to expression (3) for $B \gg 3.2$ T and gave $m = 0.54 \pm 0.04$, which is in good agreement with theory.

3.2 Transport properties at high temperatures

The effect of ordered clusters on the Hall mobility μ_H of III-V alloys can by accounted for by introducing an additional scattering mechanism. Friedman, Kibbler and Olson (1991) used the Harrison-Hauser/Marsh (HHM) formalism (Marsh 1982, Harrison 1976a 1976b) to analyse the temperature dependence of μ_H in their ordered $GaInP_2$ samples. With this HHM model they calculated the cluster Hall mobility μ_H^{CL} as a function of the cluster radius r_c,

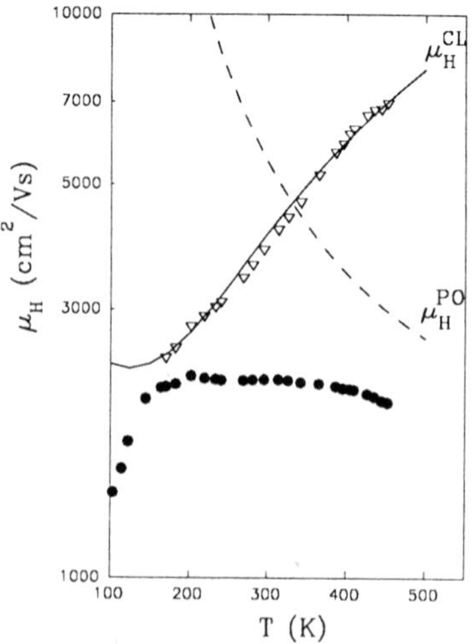

Figure 5: Temperature dependence of the Hall mobility: experimental values of μ_H (•); calculated values for μ_H^{CL} (∇). The solid curve represents the best fit with r_c= 3.5 nm and $V_0^2 f = 3.8 \cdot 10^{-4}$ $(eV)^2$.

the fraction f of the crystal volume occupied by clusters, and the energy difference V_0 resulting from alloy fluctuations. According to this model, $\mu_H^{CL} \propto (r_c^3 V_0^2 f)^{-1} T^{-0.5}$ for $r_c < 1.5$ nm, whereas for larger r_c the μ_H^{CL} increases with increasing temperature. For larger r_c, $V_0^2 f$ and r_c can be determined independently: a larger r_c gives a larger increase in mobility with increasing T, whereas a larger $V_0^2 f$ shifts the $\mu_H(T)$ curve downwards.

Fig.5 shows the Hall mobility of sample 2 above 200 K, where carriers are present in the entire crystal. In this temperature range the most important scattering mechanisms in $GaInP_2$ are polar optical scattering (PO), acoustic deformation potential scattering (AD) and ionized impurity scattering (II) (Friedman 1991, Seeger 1988). Only the Hall mobility (μ_H^{PO}) due to PO scattering is of the same order of magnitude as the measured Hall mobilities. Matthiessen's rule was used to calculate the cluster Hall mobility μ_H^{CL}:

$$\frac{1}{\mu_H^{CL}} = \frac{1}{\mu_H^{meas}} - \frac{1}{\mu_H^{PO}} - \frac{1}{\mu_H^{AD}} - \frac{1}{\mu_H^{II}} \tag{4}$$

The μ_H^{CL} data were fitted by variation of r_c and $V_0^2 f$ using the HHM model. It is seen from fig.5 that μ_H^{CL} increases with increasing T, which is consistent with $r_c > 1.5$ nm. The best fit was obtained for $r_c = 3.5$ nm and $V_0^2 f = 3.8 \cdot 10^{-4}$ (eV)^2, and is shown by the solid line in fig.5. Assuming V_0 to equal ΔE_g, the value for $V_0^2 f$ corresponds to a fraction f of 0.06. Similarly, using the mobility versus temperature curve for sample 1, we found $r_c = 3.5$ nm and $f = 0.31$.

It should be noted that in the HHM model the shape of the ordered domains is assumed to be spherical, whereas in reality the domains are thinner in the ($\bar{1}11$) and ($1\bar{1}1$)-direction than perpendicular thereto. However, although the spherical description is an approximation, we believe that these values give a good indication of the degree of ordering. To illustrate the strong effect of clustering on the Hall mobility, the room temperature Hall mobilities found here are compared with those of samples grown under different conditions in the same MOVPE reactor (Hageman 1992). For the less ordered $GaInP_2$ layers, the room temperature Hall mobilities were in the range 3000-6000 cm^2/Vs. Considerably lower values of 2500 and 2250 cm^2/Vs were found for samples 1 and 2, respectively.

4 CONCLUSIONS

We have performed Hall-Van der Pauw and magnetotransport measurements on ordered $GaInP_2$ epilayers grown by MOVPE. The ordering appears to have a large effect on the electrical properties of the alloy.

Below $T = 200$ K conductivity measurements confirm previous results obtained by optical measurements: electrons are confined in the ordered regions where donor binding energies are relatively high (36 meV). This leads to ϵ_2- and ϵ_3- (hopping) conduction up to temperatures as high as 85 K. The prescence of ϵ_2-conduction is confirmed by the quadratic dependence of ϵ_2 on magnetic field. Where hopping predominates, the magnetic field dependence of the pre-exponential factor σ_3 agrees excellently with the theoretical expression for strong magnetic fields.

For temperatures above 200 K the Hall mobility is reduced as a result of cluster scattering. In our samples this effect could be calculated for the first time. Using the Harrison-Hauser/Marsh model on cluster scattering, cluster radii of 3.5 nm were calculated from the temperature dependence of the Hall mobility. The fractions of the layers occupied by clusters are 0.06 and 0.31 for the two samples investigated.

Acknowledgements

The authors are grateful to P.J. van der Wel for providing the computer program for cluster scattering, to P.R. Hageman and A. van Geelen for growing the samples, to P. van der Linden for assistance in the High-Field Magnet Laboratory and to D.M. Frigo for critical reading of the manuscript. This work was financed by the Nederlandse Organisatie voor Energie en Milieu (NOVEM).

References

Auvergne D., Merle P., Mathieu H. 1977 Solid State Commun. **21** 437
Bhattacharya P.K., Ku J.W. 1985 J. Appl. Phys. **58** 1410
DeLong M.C., Taylor P.C., Olson J.M. 1990 Appl. Phys. Lett. **57** 620
Driessen F.A.J.M., Bauhuis G.J., Olsthoorn S.M., Giling L.J. 1993 *this conference*
Friedman D.J., Kibbler A.E., Olson J.M. 1991 Appl. Phys. Lett. **59** 2998
Hageman P.R., Van Geelen A., Gabriëlse W., Bauhuis G.J., Giling L.J. 1992 J. Cryst. Growth **125** 336
Harrison J.W., Hauser J.R. 1976a Phys. Rev. B **13** 5347
Harrison J.W., Hauser J.R. 1976b J. Appl. Phys. **47** 292
Ikeda M., Nakano K., Mori Y., Kaneko K., Watanabe N. 1986 Appl. Phys. Lett. **48** 89
Lee M.K., Horng R.H., Haung L.C. 1991 Appl. Phys. Lett. **59** 3261
Marsh J.H. 1982 Appl. Phys. Lett. **41** 732
McDermott B.T., El-Masry N.A., Jiang B.L., Hyuga F., Bedair S.M., Duncan W.M. 1991 J. Cryst. Growth **107** 96
Olson J.M., Kurtz S.R., Kibbler A.E., Faine P. 1990 Appl. Phys. Lett. **56** 623
Roentgen P., Heuberger W., Bona G.L., Unger P. 1991 J. Cryst. Growth **107** 724
Sadasiv G. 1962 Phys. Rev. **128** 1131
Seeger K. 1988 *Semiconductor Physics, An Introduction* (Springer, Berlin) chapter 6
Shklovskii B.I., Efros A.L. 1984a *Electronic Properties of Doped Semiconductors* (Springer, Berlin) chapter 4
Shklovskii B.I., Efros A.L. 1984b *Electronic Properties of Doped Semiconductors* (Springer, Berlin) chapter 7
Valster A., Liedenbaum C.T.H.F., Finke M.N., Severens A.L.G., Boermans M.J.B., Vandenhoudt D.E.W., Bulle-Lieuwma C.W.T. 1991 J. Cryst. Growth **107** 403
Wei S.H., Zunger A. 1990 Appl. Phys. Lett. **56** 662
Yamanouchi C. 1965 J. Phys. Soc. Japan **20** 1029

Interface formation and surface Fermi level pinning in GaSb and InSb grown on GaAs by molecular beam epitaxy

J. Wagner, A.-L. Alvarez [a], *J. Schmitz, J.D. Ralston, and P. Koidl*

Fraunhofer-Institut für Angewandte Festkörperphysik, Tullastrasse 72, D-79108 Freiburg, Federal Republic of Germany
[a] *permanent address: ETSI Telecomunicación, Universidad Politécnica, E-28040 Madrid, Spain*

Abstract: We have used resonant Raman scattering by longitudinal optical (LO) phonons to analyze GaSb/GaAs and InSb/GaAs interfaces for GaSb and InSb grown on (100) GaAs by molecular-beam epitaxy. Striking differences with respect to the abruptness of the interface and the minimum layer thickness required to achieve a good crystalline quality were found. Further, the position of the Fermi level at the surface of epitaxial InSb has been studied by electric-field-induced LO phonon scattering.

1. INTRODUCTION

GaSb and InSb based heterostructures are currently of considerable interest for both fundamental studies (Stradling 1991) and device applications (Aardvark et al. 1993). The fabrication of such devices requires epitaxial growth of these III-V materials. For technological reasons heteroepitaxy on GaAs substrates is advantageous (Yano et al. 1989, Parker et al. 1989). In the case of GaSb on GaAs, only very thin GaSb layers, i.e. with thicknesses of a few monolayers (ML), are required to achieve two-dimensional growth in spite of the 7.8 % lattice mismatch (Bourret and Fuoss 1992, Brandt et al. 1993). However, depending on the growth conditions, there is a strong tendency for an exchange between the two different anions leading to the formation of $GaSb_{1-x}As_x$ rather than binary GaSb (Yano et al. 1989, Yano et al. 1991, Brandt et al. 1993). The critical layer thickness for the growth of InSb on GaAs is below two monolayers, which reflects the large lattice mismatch of 14.6 % between these two materials. It has been shown that heteroepitaxial growth of InSb on GaAs starts with the nucleation of three-dimensional islands, and that a complete coverage of the GaAs substrate is only obtained for equivalent layer thicknesses exceeding 100 ML (Zhang et al. 1990).

An important issue, for example in the formation of ohmic contacts, is the position of the Fermi level at the surface. For bulk InSb (110) surfaces exposed to air it has been concluded that the surface Fermi level is pinned close to the valence band edge, causing a p-type surface accumulation layer (Mead and Spitzer 1964, Geurts et al. 1985). For epitaxial InSb layers grown on (100) GaAs, there is a recent report of the formation of an electron accumulation layer at the surface; such behaviour implies a Fermi level pinning at the conduction band edge (Söderström et al. 1992). However, as pointed out by Söderström et al. (1992), the position at which the Fermi level is pinned may depend on the details of layer growth.

2. RESULTS AND DISCUSSION

In the present study we have used resonant Raman scattering by longitudinal optical (LO) phonons to analyze the heteroepitaxial growth of GaSb and InSb on GaAs, including the Fermi level position at the InSb surface. The samples were grown by solid-source molecular beam epitaxy (MBE) on (100) GaAs substrates. After the growth of a GaAs buffer layer the substrate temperature was lowered to 520 °C for the growth of GaSb, and the As_2 molecular beam was interrupted by closing the valve of the valved-cracker As effusion cell. Then the deposition of GaSb was started by opening the shutters for Ga and Sb_2 simultaneously. The As background pressure decreased by one order of magnitude within $\simeq 10$ sec, which corresponds to the growth of less than 10 ML of GaSb. InSb was grown using also an Sb_2 molecular beam, supplied by a non-valved cracker cell. The substrate temperatures were in the range 400-450 °C. Si and Be were used as n- and p-type dopants, respectively. Undoped InSb layers showed residual p-type conductivity at low temperatures, with a hole concentration of $\simeq 10^{16}$ cm^{-3}. The thickness of the layers varied from 10 ML to 1 μm. The equivalent layer thicknesses for both InSb and GaSb were determined by the deposition time, based on growth rate calibrations using reflection high-energy electron diffraction (RHEED) oscillations.

Raman spectra were recorded in backscattering from the (100) growth surface, with the polarization of the incident and scattered light parallel to the same (100) crystallographic direction $[x(y,y)\bar{x}]$. Optical excitation was at 2.015 or 2.18 eV, in resonance with the E_1 band gap of InSb or GaSb, respectively (Dreybrodt et al. 1972, Kauschke et al. 1987). For the present scattering configuration intrinsic two-LO phonon scattering is allowed, whereas one-LO phonon scattering by the deformation potential mechanism is symmetry-forbidden. For resonant excitation, however, intrinsic one-LO phonon scattering via the Fröhlich mechanism, as well as defect-induced and electric-field-induced one-LO phonon scattering, contribute to the Raman spectrum (Menéndez and Cardona 1985, Menéndez et al. 1985).

Fig. 1 shows low-temperature Raman spectra of GaSb grown on GaAs with equivalent layer thicknesses of 10, 40, and 100 ML. For comparison, the spectrum of a 0.5 μm thick heteroepitaxial GaSb layer is also displayed. At a GaSb layer thickness of 10 ML, well defined peaks from scattering by GaSb LO and transverse optical (TO) phonons are

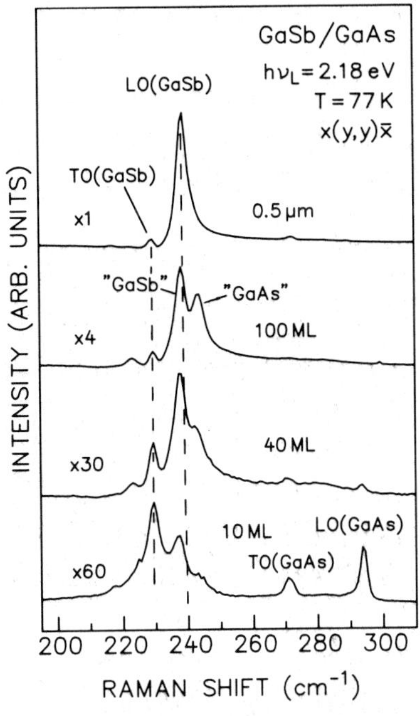

Fig. 1 Low-temperature (77 K) Raman spectra of GaSb grown on (100) GaAs for equivalent GaSb layer thicknesses ranging from 10 monolayers (ML) to 0.5 μm. The spectra were excited at 2.18 eV in resonance with the E_1 band gap of GaSb.

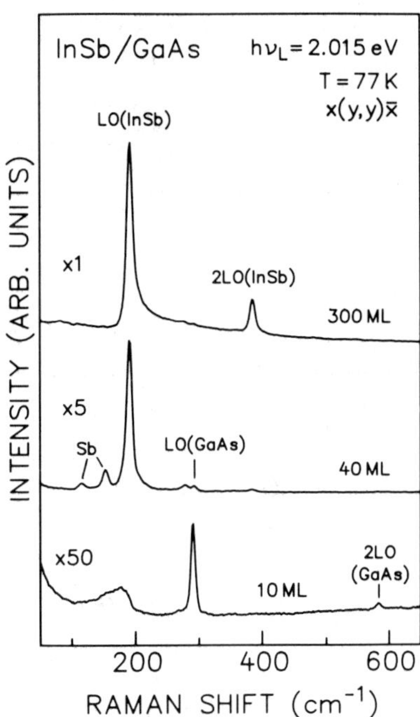

Fig. 2 Low-temperature (77 K) Raman spectra of InSb grown on (100) GaAs for equivalent InSb layer thicknesses ranging from 10 to 300 monolayers (ML). The spectra were excited at 2.015 eV in resonance with the InSb E_1 band gap energy.

already observed, in addition to scattering by LO and TO phonons from the GaAs buffer layer. The GaSb TO phonon peak, which is symmetry-forbidden for backscattering from a perfect (100) surface, is dominant for the 10 ML sample but decreases considerably in relative strength when the layer thickness is increased to 40 and 100 ML. At the high-energy side of the GaSb one-LO phonon peak a shoulder is resolved which develops into a separate peak with increasing layer thickness up to at least 100 ML. The splitting of the GaSb one-LO phonon line into two separate peaks indicates the presence of $GaSb_xAs_{1-x}$ rather than binary GaSb (McGlinn et al. 1986). From the energy separation between the GaSb-like ("GaSb") and the GaAs-like LO phonon ("GaAs") the average As concentration is estimated to about 3 % for the 100 ML thick layer (McGlinn et al. 1986).

The present finding of an exchange between As and Sb for GaSb grown on GaAs with layer thicknesses up to at least 100 ML is qualitatively consistent with previous reports on this material system (Yano et al. 1989, Yano et al. 1991, Brandt et al. 1993). However, the amount of As incorporated in the present GaSb layers, which certainly depends on the actual growth conditions, is much smaller than that of $\geq$70 % reported for GaAs/GaSb/GaAs double-heterostructures (Brandt et al. 1993). In spite of the formation of $GaSb_xAs_{1-x}$ instead of binary GaSb, resonantly enhanced two-LO phonon scattering is observed already for an equivalent layer thickness of 40 ML (not shown in Fig. 1). This obervation indicates a good crystalline quality of the grown material (Wagner 1988, Lusson et al. 1989), despite the considerable lattice mismatch of 7.8 %, even for very thin layers of GaSb grown on GaAs.

Low-temperature Raman spectra of InSb layers grown on GaAs substrates with equivalent layer thicknesses of 10, 40, and 300 ML show that equivalent layer thicknesses of several tens of monolayers are required to produce films which exhibit the one-LO phonon line of crystalline InSb (Fig. 2). In addition, scattering by crystalline Sb is observed, with two lines resolved at 116 and 153 cm^{-1} (Hünermann et al. 1987), for an equivalent layer thickness of 40 ML. For a film thickness of 10 ML, only a broad peak at 150 - 200 cm^{-1} is observed, other than the one-LO and two-LO phonon scattering from the GaAs buffer layer. This broad peak can be assigned to scattering by highly disordered or amorphous InSb, and possibly by amorphous Sb. For a thickness of 300 ML both the InSb one-LO and the two-LO phonon line are well resolved and there is no further change in the Raman spectrum when the InSb layer thickness is increased further.

If we combine the present findings with results reported by Zhang et al. (1990), obtained by transmission electron microscopy and RHEED, the following picture emerges. For equivalent InSb layer thicknesses on the order of 10 ML, small islands about 10 nm in height are present with a surface coverage of about 25 %. These islands give rise to the broad Raman peak observed in the lowest curve of Fig. 2. With increasing equivalent layer thickness the surface coverage and the island size both increase, forming a connected InSb network (Zhang et al. 1990). As a consequence, the phonon coherence length also increases, leading to the appearence of mainly defect-induced one-LO phonon scattering from crystalline InSb. At this stage crystalline Sb is also present. Equivalent layer thicknesses exceeding 100-300 ML are required to achieve a complete coverage with InSb (Zhang et al. 1990) and to observe intrinsic InSb two-LO phonon scattering along with dominantly intrinsic one-LO phonon scattering; both of the latter indicate good crystalline quality in the fully-relaxed InSb film (Wagner 1988, Lusson et al. 1989) despite the large dislocation density arising from the 14.6 % lattice mismatch to the GaAs substrate. In summary, for InSb on GaAs no indication was found for an exchange between the anions, but layer thicknesses exceeding 100 ML are required to achieve two-dimensional growth and a good crystalline perfection. These findings are in sharp contrast to the above results for GaSb on GaAs.

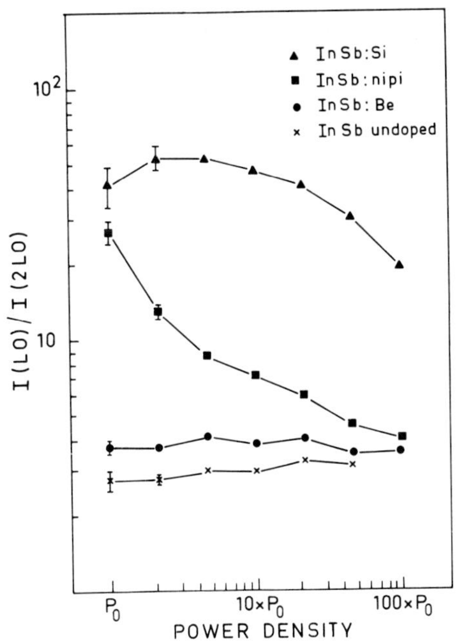

Fig. 3 Normalized LO phonon scattering strength I(LO)/I(2LO) versus optical power density. P_0 corresponds to a power density of $\simeq 10$ W/cm^2. Data are shown for undoped InSb (x), p-type InSb:Be (●) with a low-temperature hole concentration of 9.1×10^{17} cm^{-3}, an InSb nipi doping superlattice (■), and n-type InSb:Si (▲) with a low-temperature electron concentration of 1.6×10^{17} cm^{-3}. Error bars are indicated for the data points recorded at the lowest power densities.

In Fig. 3 the intensity of one-LO phonon scattering, normalized to the strength of the purely intrinsic two-LO phonon signal, is plotted versus the optical power density for a series of $\simeq$ 1 μm thick n- and p-type doped InSb layers. The data were recorded in the $x(y,y)\bar{x}$ scattering configuration. Both the nominally undoped, but residually p-type, and the intentionally p-type doped layers show a comparatively low one-LO phonon intensity which shows very little variation with power density. The homogeneously n-type doped sample exhibits a much larger one-LO phonon intensity at low power densities along with a marked decrease in intensity with increasing excitation power density. An InSb nipi structure was also analysed, consisting of 10 nm thick Si- and Be-doped layers separated by 40 nm wide undoped spacers. Proceeding from the epilayer surface back towards the substrate, the layer sequence starts with a 40 nm thick undoped InSb cap layer followed by the first Be-doped layer. The donor (N_D) and acceptor (N_A) concentrations were in the 10^{18} cm^{-3} range with $N_D > 2N_A$. Similar to the homogeneously n-type doped layer, the residual n-type nipi structure also shows a large relative one-LO phonon intensity at low power densities, but the signal decreases rapidly in intensity for increasing incident power density.

The above findings can be explained by the presence of a surface electric field in the n-type samples, which leads to electric-field-induced one-LO phonon Raman scattering. The intensity of this scattering is proportional to the square of the strength of the surface electric field averaged over the probing depth in the Raman experiment, estimated to be about

14 nm (Wagner et al. 1993). This field is progressively screened by photogenerated carriers as the optical power density is increased. The surface electric field, the presence of which has been confirmed by the observation of interference effects between dipole-allowed and electric-field-induced LO phonon scattering (Wagner et al. 1993), is caused by a pinning of the surface Fermi level at the valence band edge (Geurts et al. 1985).

3. CONCLUSIONS

We have shown how resonant Raman scattering by LO phonons can be used to analyse GaSb/GaAs and the InSb/GaAs heterointerfaces, as well as to study surface Fermi level pinning in epitaxial InSb. Striking differences were found between GaSb/GaAs and InSb/GaAs heterostructures with respect to the abruptness of the interface and the minimum layer thickness necessary to achieve a good crystalline quality.

ACKNOWLEDGEMENTS: We would like to thank G. Bihlmann for technical assistance in the MBE growth and P. Hiesinger for performing the Hall effect measurements. A.-L.A. wishes to thank the Comunidad Autónoma de Madrid for financial support.

REFERENCES

Aardvark A et al. 1993 Semicond. Sci. Technol. 8 S380
Bourret A and Fuoss P H 1992 Appl. Phys. Lett. 61 1034
Brandt O, Tournié E, Tapfer L and Ploog K 1993 J. Cryst. Growth 127 503
Dreybrodt W, Richter W and Cardona M 1972 Solid State Commun 11 1127
Geurts J, Pletschen W and Richter W 1985 Surf. Sci. 152/153 1123
Hünermann M et al. 1987 Surf. Sci. 189/190 322
Kauschke W, Mestres N and Cardona M 1987 Phys. Rev. B 36 7469
Lusson A, Wagner J and Ramsteiner M 1989 Appl. Phys. Lett. 54 1787
McGlinn T C et al. 1986 Phys. Rev. B 33 8396
Mead C A and Spitzer W G 1964 Phys. Rev. 134 A713
Menéndez J and Cardona M 1985 Phys. Rev. B 31 3696
Menéndez J, Vina L, Cardona M and Anastassakis E 1985 Phys. Rev. B 32 3966
Parker S D et al. 1989 Semicond. Sci. Technol. 4 663
Söderström J R et al. 1992 Semicond. Sci. Technol. 7 337
Stradling R A 1991 Semicond. Sci. Technol. 6 C52
Wagner J 1988 Appl. Phys. Lett. 52 1158
Wagner J, Alvarez A-L, Schmitz J, Ralston J D and Koidl P, 1993 Appl. Phys. Lett. 63
Yano M, Ashida M, Kawaguchi A, Iwai Y and Inoue M 1989 J. Vac. Sci. Technol. B7 199
Yano M, Yokose H, Iwai Y and Inoue M 1991 J. Cryst. Growth 111 609
Zhang X et al. 1990 J. Appl. Phys. 67 800

Paper presented at the Int. Symp. GaAs and Related Compounds, Freiburg, 1993

Impact ionization and associated light emission phenomena in GaAs devices: a Monte Carlo study

G. Zandler, A Di Carlo, P. Vogl and P. Lugli*

Physik Department and Walter Schottky Institut, TU München, D-85747 Garching, FRG

ABSTRACT: A theoretical investigation of impact ionization and associated light emission in GaAs MESFETs, GaAs/AlGaAs HEMTs and HBTs is presented. Based on band structure and Monte Carlo calculations, we show that light emission by hot carriers in submicron GaAs devices can be explained by direct radiative interband transitions, both below and above the gap energy. In agreement with experimental findings, a strong correlation is found between the currents associated to minority carriers and light emission.

1. INTRODUCTION

Breakdown phenomena are one of the limiting factors in the performance of microwave devices, especially for power applications. One of the main causes leading to device breakdown is the generation of electron-hole pairs via impact ionization processes by hot carriers in the high field regions of the device. The characterization of such phenomena and the study of its onset through electrical measurements are in general a quite difficult task, both because of the inhomogeneities of the electric field distribution and because of the strong influence of technological parameters. Recently, light emission from devices has been extensively used as a tool for the investigation of near breakdown phenomena (Herzog and Koch, 1988; Zappe and As, 1990; Lanzoni et al., 1991; Zanoni et al., 1992a) and a direct correlation between the onset of avalanche processes (evidenced by the collection of minority carriers at the gate or base terminals, for FET or HBT respectively) and that of light emission has been found (Canali et al., 1993; Zanoni et al., 1993).

Accurate modelling of light emission in semiconductor devices requires a spatially resolved detailed knowledge of the carrier distribution function (d.f.) and of the dynamics of minority carriers. Furthermore, it relies on theoretical models for the emission mechanism. Despite the considerable amount of experimental work available, no consensus exists on the microscopic mechanism responsible for light emission. Electron-hole recombination, interband direct transitions, or impurity-assisted intraband transitions have been invoked to explain the measured spectra. Recently, Bude et al. (1992) suggested that the main contribution to light emission in Si devices could come from vertical transitions between different conduction or valence bands that are populated because of the very high electric fields.

In this paper, we present realistic calculations of the optical response functions and radiative emission rates of III-V compounds in the presence of very hot carriers. The absorption and emission spectra due to direct interband transitions are calculated in the framework of relativistic non-local empirical pseudopotential theory (Chelikowski and Cohen, 1976). In Zandler et al (1993) we have shown that, when carriers have effective temperatures of several thousand degrees, the luminescence is dominated by transitions between conduction bands (cc) and valence bands (vv) . In such condition, electrons occupy not only the band edge states but also the first and second conduction band across a substantial fraction of the Brillouin zone. Similarly, holes extend over the heavy and light hole band up to the Brillouin zone edge. The most significant optical transition occur between the second and first conduction band at the X-point, with additional weaker contributions from higher lying states along the Δ and Σ axis, and from light to heavy hole bands. When one or both type of carriers are near equilibrium, then conduction-to-valence (cv) contribution becomes important.

The hot electron and hole d.f.'s to be used in the calculation of the emission rate are obtained from self consistent ensemble Monte Carlo (MC) simulations for GaAs MESFET's and GaAs/AlGaAs HBT's. It should be stressed, the MC is the only method which, at present, can provide an accurate physical description of hot carrier effects in devices, together with the full details of the carrier distribution function (Jacoboni and Lugli, 1989). The transport model for our simulation is based on a three-valley description of the conduction band and on three valence bands, with non parabolic isotropic dispersions. The carrier interaction with polar optical, acoustic, equivalent and nonequivalent intervalley, intraband and interband phonons, and ionized impurities is considered. For impact ionization, Kane's model is adopted. A list of all relevant parameters and more details on the MC algorithm are presented in Lugli (1992). In the following sections we will briefly discuss the MC results for GaAs MESFETs, HEMTs and HBTs, and present a comparison with available experiments. A more detail discussion will be presented in a forthcoming publication.

2. GaAs MESFETs

We have simulated the ion-implanted n-type GaAs MESFET described and characterized in Canali et al (1993). At high drain voltages, a high field region forms between the gate and the end of the recess. As electrons heat up to average kinetic energies of few tenth's of eV, impact ionization processes occur. This causes minority carriers to be produced which leave the device at the gate contact or move towards the substrate. At a drain voltage of 7 V, the simulation gives a gate current of about 70 μA (at zero gate voltage), in fair agreement with the experimental results. In addition, the collection efficiency (that is, the ratio between the number of holes collected at the gate and the total number of generated holes) is calculated to be about 60 percent. The rate of hole generation and hole collection of course depends on the terminal biases.

Figure 1 compares the electron average kinetic energies for two drain biases of 4 and 7 Volts respectively (at a fixed gate bias of 0 Volts), in the high field region between gate and drain. We can notice that electrons reach higher average energies, and correspondingly a hotter d.f., at the higher voltage. Consequently, the ionization rate is enhanced, producing about three orders of magnitude more holes at 7 V than at 4 V, as can clearly be seen from the comparison of the hole d.f.'s.

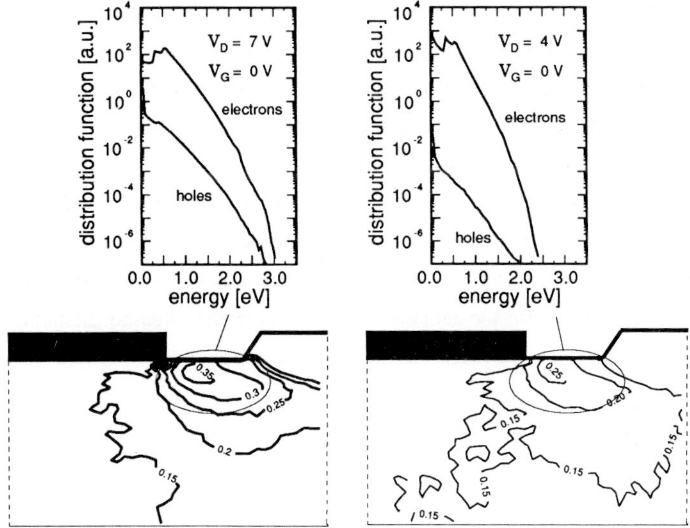

FIG 1. Contour lines of electron kinetic energy (in eV) in the high field region of the simulated MESFET. The top part shows the electron and hole distribution functions.

The shape of the distribution functions differs from the ones depicted in Fig. 1 in other regions of the device. For instance, holes are accelerated by the built-in field of the gate Schottky barrier, thus heating up near the gate electrode, while, on the contrary, they cool down as they penetrate into the substrate. Thus, also the calculated emission spectra are strongly position dependent. Figure 2 presents the calculated total spectra (solid lines) at 7 Volts drain bias when the whole device (left) or only the region between gate and drain (right) are considered. Light reabsorption by the top layer has been taken into account by an energy dependent weighting factor proportional to the

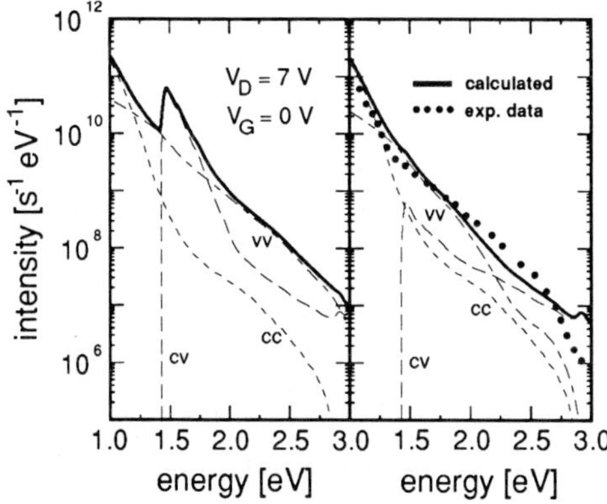

FIG 2. Calculated electroluminescence spectra, indicating the contribution from different transitions when the whole device is considered (left) and the comparison with measurements when the region under the gate is neglected (right)

absorption coefficient determined directly from our pseudopotential calculations. Our analysis shows that at 7 V the *below-gap* emission is controlled by **cc** transitions, while *above gap* **vv** transitions dominate (left figure). The high energy tail above 2 eV originates from very hot holes under the gate corner. The pronounced recombination peak at gap energy is due to the presence of warm electrons and cold holes under the gate. Since the gate electrode is opaque, these two contributions may be screened and should not be found in the measured spectra. In particular, because of the reduction of vv contributions, the high energy part of the spectra is dominated by cv processes. Indeed, good agreement with experimental data (dots) is found by neglecting the region beneath the gate. It should be noticed though that the hole contribution to the spectra is very sensitive to the device geometry and doping profile, as well as to the adopted valence band model. Additional studies are needed to clarify the influence of such parameters on the calculated spectra.

3. GaAs/AlGaAs HEMTs

We have simulated the GaAs/AlGaAs submicron HEMT depicted in Fig. 3a. At a drain bias of 3 V (with 0 Volts on the gate) the strongest longitudinal electric field is found in the AlGaAs layer between the end of the gate and the beginning of the highly doped GaAs cap layer. Somewhat lower field are found in the channel. As a consequence, electrons heat up considerably both in the GaAs channel and in the AlGaAs layer. Because of the smaller band gap, ionizations occur mainly in the GaAs channel as indicated in Fig. 3b. Gate current and electroluminescence spectra for this device are currently being calculated.

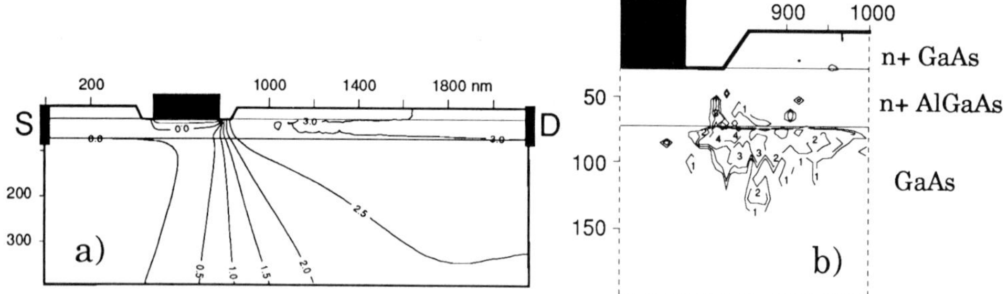

FIG 3. Equipotential lines for the simulated HEMT (a) and contour lines for the common logarithm of the ionization rate, in units of $(\mu m^3\ ps)^{-1}$ (b)

4. GaAs/AlGaAs HBTs

Our theoretical investigation has focused on carrier multiplication phenomena occurring at the base-collector (BC) region. The structure of the studied HBT has been described in Zanoni et al. (1992b). Our MC simulation has shown that electric fields of several hundreds kV/cm exist in the collector, which lead to sizable multiplication effects as soon as V_{CB} exceeds 8 V (Di Carlo and Lugli, 1993). Such multiplication originates from impact ionization processes by hot electrons inside the collector. The generated holes drift towards the base, gaining energy from the junction field. This is clear for instance from the d.f.'s of electrons and holes at the beginning (left panel) and well inside (right panel) of the collector shown in Figure 4. Due to the high electric field the d.f.'s are heavily heated and cannot be fitted by maxwellian distributions. The structures in these distributions are due to the density of state. We found that the holes populate the valence band up to the zone boundary. Around 1.7 eV umklapp processes strongly influence the carrier dynamics in the heavy hole band, resulting in pronounced drop of the d.f. at that energy. While the electron distribution does not change significantly in the collector, holes are strongly accelerated by the junction field and their distribution gets hotter as they approach the base.

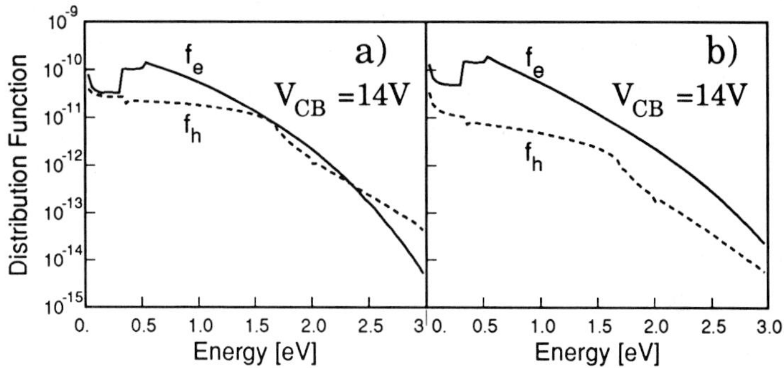

FIG 4. Electron (solid lines) and hole (dashed lines) distribution functions at the beginning (a) and well inside (b) the collector of the HBT.

The calculated spectra are shown in Fig. 5, together with the individual contributions. From the comparison of the left and right panels, we see that reabsorption by the top layers drastically cuts the cc contribution coming from hot electrons deep into the collector. As a result, vv transitions dominate, except for the strong band-edge peak which is due to the recombination of cold electrons and holes in the base.

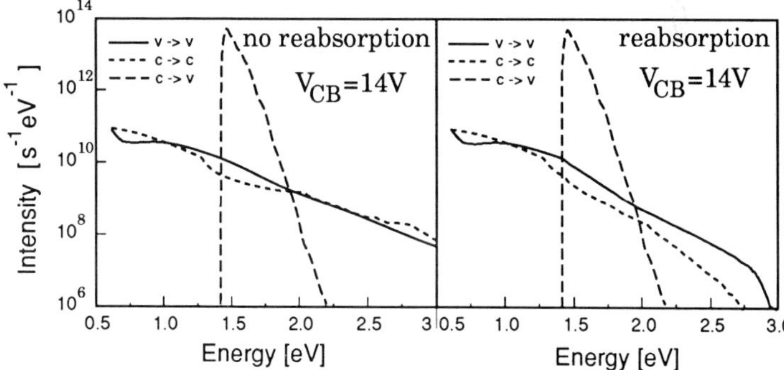

FIG 5. Contribution of the different radiative transitions to the calculated spectra of the HBT, with (right) and without (left) reabsorption by the top layers.

The bias dependence of the spectra is shown in Fig. 6a. As expected, the band-edge contribution does not change, as the emitter current is kept constant. On the contrary, the hot-carrier related luminescence becomes stronger at increasing V_{CB}, an indication of enhanced ionization processes. In fact, Fig. 6b demonstrates that the integrated intensity both above and below gap follows closely the calculated multiplication factor, in perfect agreement with the experiment of Zanoni et al. (1993). The correlation comes from the fact that the multiplication factor and the light emission are directly controlled by the number of impact generated holes and by their d.f.'s, respectively.

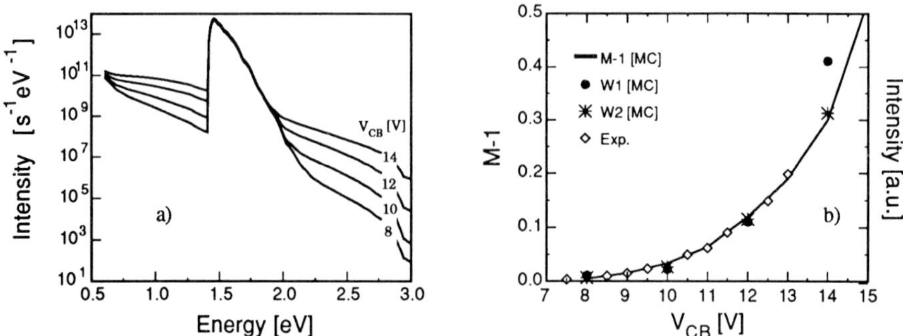

FIG 6. Dependence on base-collector voltage of the calculated spectra (a); correlation of the integrated light intensity in the region 0.8–1.4 eV (W2) and 2–2.7 eV (W1) with the calculated multiplication factor M-1. The experimental results refer to the integrated visible light and to the variation of base current

5. CONCLUSIONS

We have presented a MC study of hot carrier effects in GaAs devices, showing that light emission gives a characteristic signature of each specific device which is strongly correlated with impact ionization processes. The spectral analysis of the electroluminescence can be an excellent tool for the characterization of hot carrier effects in devices provided that sophisticated modeling tools are combined to careful experimental investigations.

This work was partially supported by the DFG and by the Siemens project SFE.

*Permanent Address: Dipartimento di Ingegneria Elettronica, Universita' di Roma "Tor Vergata", 0013 Roma, Italy

6. REFERENCES

Bude J. , Sano N. and Yoshii A. 1992 Phys. Rev. B **45**, p. 5848

Canali C, Neviani A., Tedesco C., Zanoni, E. Cetronio, A. and Lanzieri C. 1993 IEEE Trans. Electr. Dev., **40**, p. 49

Chelikowski J. R. and Cohen M. L. 1976 Phys. Rev. B**14**,p. 556

Di Carlo A. , Lugli P., Pavan P. , Zanoni E., and Malik R. 1992 Microelectr. Eng., **19**, p. 135

Di Carlo A. , Lugli P. 1993 IEEE Electr. Dev. Lett., **14**, p. 103

Herzog M. and Koch F. 1988 Appl. Phys. Lett. **53**, p. 2620

Jacoboni C. , and Lugli P. 1989 *The Monte Carlo Method for Semiconductor Device Simulation*, Springer Verlag, Wien

Lanzoni M. , Sangiorgi E., Fiegna C. , Manfredi M. and Ricco' B. 1991 IEEE El. Dev. Lett. **12**, p. 341

Lugli P. 1992 Microelectr. Eng., **19**, p. 275

Toriumi A. 1989 Solid State Electronics **32**, p. 1519

Zandler G., Lugli P., and Vogl, P., 1993a, in Proc. Int. Conf. Phys. Sem., Eds. P. Jang and H. Zheng, p. 329

Zanoni E. , Tedesco C., Manfredi M., Saraniti M. and Lugli P. 1992a Semicond. Sci. Technol. **7**, p. B543

Zanoni E. , Malik R. , Pavan P., Nagle J., Paccagnella A. and Canali C. 1992b IEEE Electr. Dev. Lett., **13**, p. 253

Zanoni E. , Vendrame L. , Pavan P., Manfredi M. , Bigliardi S. , Malik R. , and Canali C. 1993 Appl. Phys. Lett. **62**, p. 402

Zappe H. P. and As D. J. 1990 Appl. Phys. Lett. **57**, p. 2919

Avalanche breakdown in GaAs/Al_xGa_{1-x}As multilayers and alloys

J P R David[1], J Allam[2,3], J S Roberts[1], R Grey[1], G Rees[1] and P N Robson[1]

[1]Department of Electronic and Electrical Engineering, University of Sheffield, S1 3JD
[2]Department of Physics, University of Surrey, Guildford GU2 5XH
[3]Permanent address: Hitachi Cambridge Lab., Cavendish Lab., Cambridge CB3 0HE.

ABSTRACT: We have made a systematic study of the avalanche breakdown voltage (V_b) in Al_xGa_{1-x}As / GaAs alloys and multilayers. The alloys show linear dependence of V_b on x up to at least x = 60 %. For multilayers with thin layers (≤ 200 Å), V_b is close to that of the "equivalent alloy". For thick layers (≥ 500 Å), the breakdown voltage tends towards the bulk value. The transition from pseudo-alloy to bulk behaviour is interpreted in terms of the momentum and energy relaxation lengths calculated from a Monte Carlo simulation.

1. INTRODUCTION

Avalanche breakdown due to impact ionisation limits the electric field which can be applied across GaAs/Al_xGa_{1-x}As multilayer structures such as modulators and photodiodes. The avalanche breakdown voltage (V_b) is determined by the electron and hole impact ionization coefficients (α and β respectiveiy). The ability to tailor the ionization coefficients for a given electric field would allow control of V_b and the noise performance of avalanche photodiodes.

The dependence of the ionisation rates on the bandstructure is not well understood even for bulk GaAs (Allam et al 1990). Claims have been made for enhancement of ionisation rates in certain multilayer structures (Chin et al 1980) due to the effect of the heterojunction discontinuities on the carrier energy. However, it was recently shown (Czajkowski et al 1990) that this mechanism does not lead to enhancement of α in GaAs/Al_xGa_{1-x}As multilayers due to the effects of intervalley transfer and the small conduction band discontinuities in the X and L valleys. Experimental measurements of GaAs/Al_xGa_{1-x}As multilayers have yielded conflicting results with several groups finding an enhanced α/β ratio compared to bulk GaAs (Capasso et al 1981; Juang et al 1985; Kagawa et al 1989) while others observed no significant change (Susa & Okamoto 1984; Franks 1990; Salokatve et al 1992). However, detailed measurements of the ionisation rates for a wide range of layer dimensions have not been previously reported.

We have investigated the ionization process in a range of bulk $Al_xGa_{1-x}As$ alloys and $Al_xGa_{1-x}As$ /GaAs multilayers by measuring V_b directly. V_b is the parameter of interest for devices limited by avalanche breakdown; it is critically dependent on α and β and can be measured accurately in low dark current diodes without the errors which affect photomultiplication measurements used to determine α and β directly.

2. EXPERIMENTAL MEASUREMENT OF BREAKDOWN VOLTAGES

The breakdown voltage was measured in p-i-n diodes with nominally 1 μm i-regions, with the bulk $Al_xGa_{1-x}As$ layer or $Al_xGa_{1-x}As$/GaAs multilayer in the i-region. The structures were grown by atmospheric pressure MOVPE or by conventional MBE on a slightly misoriented (100) n^+ GaAs substrate. Circular mesa diodes of 100 μm - 400 μm diameter with optical access were fabricated using wet chemical etching.

Photoluminescence and x-ray rocking curve measurements were performed on the structures to determine the alloy composition and layer thicknesses. The i-region doping and thickness were obtained from capacitance-voltage profiling of the fabricated devices. The background doping was typically $< 2 \times 10^{15}$ cm^{-3} and hence the field in the i-region is very uniform. Dark current measurements were carried out using an automated HP4140B picoammeter to determine the leakage currents and V_b. All of the devices studied exhibited low dark current and sharp breakdown. By measuring the leakage current as a function of device area it was shown that edge leakage was low and the breakdown was due to a bulk process. Photomultiplication measurements were also carried out to confirm that avalanche multiplication was responsible for the breakdown observed. The definition of V_b for these studies was the voltage at which the reverse leakage current increased by several orders of magnitude for a small (< 0.2 V) increase in reverse bias. Because of small variations in the

Table I. Details of bulk $Al_xGa_{1-x}As$ diodes.

Layer	x (%)	i (μm)	V_b (V)	V_b(GaAs) (V)	$\Delta V_b/V_b$(GaAs) (%)
RMB305	0	1.02	33.5	33.6	00.3
CB74	15	0.86	32	27.6	15.9
CB80	30	0.90	40	30.2	32.5
RMB365	40	0.75	38.2	26.2	45.8
RMB366	50	0.73	40	25.6	56.0
RMB367	60	0.75	43.8	26.2	67.2

i-region thickness and doping between structures, it is necessary to normalise V_b to that of a bulk GaAs diode of the same thickness and doping. These values V_b(GaAs) were obtained by using published data for ionisation rates in GaAs (Bulman et al 1985) and solving for the breakdown condition (Moll 1964) numerically. For devices with higher doping than 2×10^{15} cm^{-3} the appropriate formula for a "reach-through diode" was used. The normalised breakdown voltage of a multilayer "ML" is given by $\Delta V_b/V_b$(GaAs) where $\Delta V_b = V_b(ML) - V_b(GaAs)$.

Table I shows the characteristics and measured breakdown voltage of the bulk $Al_xGa_{1-x}As$ layers. The normalised breakdown voltage $\Delta V_b/V_b$(GaAs) increased linearly with x up to at least x = 60 %, as shown by the circles in Figure 1. No change in slope due to the transition from direct to indirect bandgap at x ~ 45 % was observed. This is consistent with recent hydrostatic pressure measurements (Allam et al, 1990) which indicated that the ionisation process involves the X valleys even in GaAs. A linear fit to the data (solid line in Figure 1) gives the relationship $\Delta V_b/V_b(GaAs) \approx 1.1$ x.

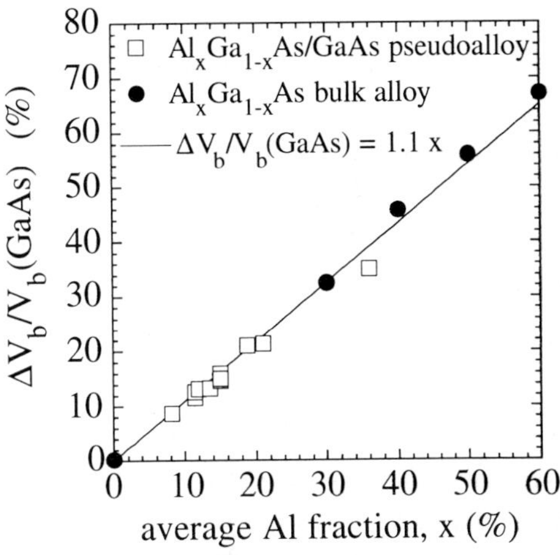

Figure 1. Normalised breakdown voltage of $Al_xGa_{1-x}As$ bulk alloys (circles) and $Al_xGa_{1-x}As/GaAs$ multilayer pseudoalloys (squares) as function of average Al fraction x.

Table II. Details of $Al_xGa_{1-x}As/GaAs$ multilayers with equal well and barriers widths.

Layer	$L_w=L_b$ (Å)	period	i (μm)	V_b (V)	V_b(GaAs) (V)	$\Delta V_b/V_b$(GaAs) (%)
RMB736	5000	1	1.1	39.2	35.7	9.8
CB557	2500	2	0.826	28.8	26.2	9.9
CB144	500	10	1.0	36.2	33.0	9.7
CB130	500	10	1.02	37.0	33.6	10.1
CB77	250	20	0.9	36.5	32.5	12.3
CB556	100	50	0.9	32.4	28.0	15.0
MV251	60	75	0.8	32.0	27.6	15.9
MV256	60	37	0.43	19.5	17.0	14.7
CPM418	30	160	1.04	39.0	34.0	14.7
CB558	15	333	1.05	36.4	31.6	15.0

The breakdown voltage was measured in two different series of $Al_xGa_{1-x}As/GaAs$ multilayers. The first series consisted of $Al_xGa_{1-x}As/GaAs$ multilayers grown with x = 30 % and equal well (L_w) and barrier (L_b) widths in the range 15 Å to 5000 Å. The details of the structures and the measured breakdown voltages are shown in Table II. In all cases the measured V_b was larger than the calculated V_b(GaAs), showing that the presence of the barriers was suppressing the ionization coefficients. Figure 2 shows $\Delta V_b/V_b$(GaAs) against log(L_w). Three different regions are observed. For layer dimensions ≤ 100 Å, V_b is increased by ~ 15%. This is in close agreement with the value of 16 % for the "equivalent alloy" with the average Al content of the multilayer. For thick layers (≥ 500 Å), V_b is increased by ~ 10 %. Intermediate behaviour is observed for structures with 250 Å layers.

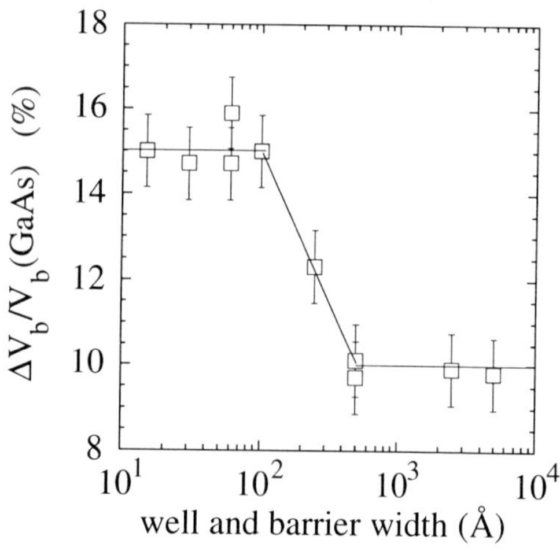

Figure 2. Breakdown voltage in $Al_xGa_{1-x}As/GaAs$ multilayers with equal well and barrier widths. The straight lines are intended as a guide to the eye. The error bars indicate approximate experimental errors.

The pseudo-alloy behaviour of thin multilayers was confirmed by studying a second series of structures with layer thickness ≤ 200 Å. The composition of the barriers was x = 30 % (except in one case with x = 60 %), but the well and barrier widths were varied as shown in Table III

Table III. Details of $Al_xGa_{1-x}As/GaAs$ superlattices.

Layer	x (%)	L_w (Å)	L_b (Å)	period	i (µm)	V_b (V)	V_b (GaAs) (V)	Average Al (%)	ΔV_b / V_b(GaAs) (%)
MV245	30	160	60	45	0.85	31.5	29.0	8.2	8.6
CPM408	30	160	100	38	1.02	37.5	33.6	11.5	11.6
CPM419	30	160	100	38	1.125	41.0	36.4	11.5	11.6
CPM429	30	random	60	14	1.02	38.0	33.6	12.0	13.1
CPM244	30	60	50	100	1.08*	34.0	30.0	13.6	13.3
CPM405	30	60	100	60	0.91	37.0	30.6	18.75	21.0
CB79	30	85	200	35	1.06	42.0	34.6	21.0	21.4
CB114	30	85	200	35	1.35	51.0	42.0	21.0	21.4
CPM427	60	40	60	40	0.39*	21.0	15.8	36.0	35.0

*reach-through structures

to achieve varying "equivalent alloy" composition. In one structure (CPM429) the barrier width was varied randomly with a mean value of 90 Å. In all cases, the measured breakdown voltages of the multilayers (squares in Figure 1) were in good agreement with those of the bulk alloys (circles in Figure 1).

3. MONTE CARLO CALCULATION OF RELAXATION LENGTHS

We can expect a change in the high field transport when the layer widths are greater or less than the characteristic length for relaxation of the carrier. In the spirit of the "lucky drift" concept of high field transport and impact ionisation (Ridley 1983), we consider the energy relaxation length (l_E) and the momentum relaxation length (l_m) where $l_E \gg l_m$ in the drift mode. The quantities l_E, l_m and $l_{ii} = 1/\alpha$ were calculated for electrons in bulk GaAs using a Monte Carlo simulation, similar to that previously described (Czajkowski et al 1990). The material parameters of Shichijo and Hess (1981) were used, with an isotropic threshold energy $E_{th} = 2.0$ eV and a Keldysh "softness" factor of P = 0.1. The momentum relaxation length was assumed to be the same as the mean free path for scattering, since the scattering is dominated by intervalley phonons. The energy relaxation length was calculated from the energy balance equation

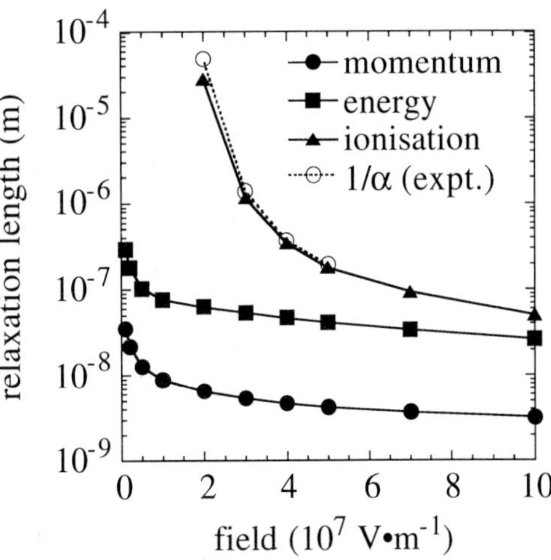

Figure 3. Relaxation lengths for momentum, energy and impact ionisation, calculated by Monte Carlo simulation. The experimental data for α of Bulman et al (1985) is also shown.

$$l_E = \langle E \rangle / eF$$

where $\langle E \rangle$ is the mean electron energy and F is the electric field. The results are shown in Figure 3 together with the experimental data for α of Bulman et al (1985). The calculated ionisation rate was in good agreement with the experiment. l_E and l_m do not vary greatly in the field range 2 - 5 x 10^7 V•m^{-1} and at a field of 3 x 10^7 V•m^{-1} take the values $l_E = 530$ Å and $l_m = 53$ Å.

4. DISCUSSION

For layer widths less than or comparable to the mean free path l_m, a carrier will typically sample both well and barrier layers during a free flight, and we therefore expect some average or pseudo-alloy behaviour consistent with our experimental results for thin multilayers:

$$\alpha(ML) = \alpha(Al_yGa_{1-y}As) \text{ where } y = \{L_b/(L_b+L_w)\}\ x.$$

For layer widths much greater than l_E and l_m, the energy and momentum are randomised within the initial portion of each layer and the ionisation rate is the geoemetric mean:

$$\alpha(ML) = \{L_w/(L_b+L_w)\}\alpha(GaAs) + \{L_b/(L_b+L_w)\}\ \alpha(Al_xGa_{1-x}As)$$

where $\alpha(GaAs)$ and $\alpha(Al_xGa_{1-x}As)$ are the ionisation rates in bulk layers.

For layer widths comparable to l_E enhancement of the ionisation rate is possible for carriers which ionise within a length l_E of the heterojuncion. Although previous work (Czajkowski et al 1990) indicated no enhancement due to the heterojuncion discontinuities, this does not preclude enhancement by increased carrier energies due to reduced phonon scattering or ionisation in the barrier layers. However, photomultiplication measurements showed no enhancement of the *ratio* of electron to hole ionisation rates in these samples (Franks, 1990).

5. ACKNOWLEDGEMENTS

We acknowledge G. Hill and M.A. Pate for device fabrication, I. Czajkowski for contributions to the Monte Carlo programme, and the Science and Engineering Research Council for financial support. One of us (JA) would like to thank the University of Surrey for providing a Visiting Fellowship and Hitachi Cambridge Laboratory for use of computer resources.

6. REFERENCES

Allam J, Adams A R, Pate M A and Roberts J S 1990 *GaAs and Related Compounds 1990* (Inst. Phys. Conf. Ser. No. 112) 375

Bulman G E, Robbins V M and Stillman G E 1985 IEEE Trans. Electron. Dev. 32 2454

Capasso F, Tsang W T, Hutchinson A L and Williams G F 1982 Appl. Phys. Lett. 40 38

Chin R, Holonyak N and Stillman G E Electronics Letters 1980 16 468

Czajkowski I K, Allam J and Adams A R 1990 Electronics Letters 26 1311

Franks R B 1990 Solid State Elect. 33 1235

Juang F-Y, Das U, Nashimoto Y and Bhattacharya P K 1985 Appl. Phys. Lett. 47 972

Kagawa T, Iwamura H and Mikami O 1989 Appl. Phys. Lett. 54 33

Moll J L 1964 Physics of Semiconductor Devices (McGraw-Hill) Chapter 11

Ridley B K 1983 J Phys C: Solid State Physics 16 3373

Salokatve A, Toivonen M and Hovinen M Electronics Letter 1992 28 416

Shichijo H and Hess K 1981 Phys. Rev. B23 4197

Susa N and Okamoto H 1984 Jap. J. Appl. Phys. 23 317

Characterization of dislocation reduction in MBE-grown (Al, Ga)Sb/GaAs by TEM

G. D. Kramer†, M. S. Adam†, R. K .Tsui † and N. D. Theodore*

†Motorola Inc, Phoenix Corporate Research Laboratories, Tempe, AZ 85284
*Motorola Inc, Materials Technology Center, Mesa, AZ 85202

ABSTRACT High dislocation densities result when InAs or (Al,Ga)Sb epitaxial layers are grown on GaAs substrates due to the large lattice mismatch (~8%). Cross-sectional Transmission Electron Microscopy (TEM) has been used to study the distribution of these dislocations. For InAs and AlSb layers the dislocation density decreases quickly for the first 2 μm, then slowly decreases to 3 x $10^8/cm^2$ after 10 μm of epi growth. In a thin-element superlattice (GaSb/AlSb, 10 nm periods) the dislocation densities are comparable to those of the bulk InAs and AlSb layers. When the layer thickness of the same periodic structure is increased to 100 nm the dislocation density is reduced by an order of magnitude to 4 x $10^7/cm^2$ after 10 μm of epi.

1. INTRODUCTION

The (Al,Ga)Sb/InAs/GaAs material system is being explored for applications in high-speed field-effect transistors (Luo *et al* 1989, Werking *et al* 1990) and quantum effect devices because of the favorable material properties. In the case of InAs/AlSb, there is near lattice-match and a large conduction-band discontinuity across the heterojunction. InAs has an electron mobility as high as 30,000 cm^2/V-sec and a saturation velocity of 2 x 10^7cm^2/sec at room temperature due to the small effective mass of electrons and the large intervalley energy difference in the system (Ideshita *et al* 1992). Furthermore, the large conduction-band discontinuity between InAs and (Al,Ga)Sb leads to the strong confinement of electrons in InAs quantum wells sandwiched between (Al,Ga)Sb barrier layers. One difficulty however is the lattice mismatch to available semi-insulating substrates.

High dislocation densities result when InAs or (Al,Ga)Sb epitaxial layers (epi) are grown on GaAs substrates due to the large lattice mismatch (approximately, 6.8% and 8.5%). These defects decrease in density as the distance from the substrate increases. Even though it is typical to grow 2 μm or more of buffer in order to reduce the density of defects, devices fabricated in the InAs/(Al,Ga)Sb/GaAs system have ~$10^9/cm^2$ dislocations at the surface of the wafer according to Soderstrom *et al* (1991). For the fabrication of effective and reliable devices, it is necessary to understand the behavior of extended dislocations and then to minimize and control their formation. As a step in that direction, this study investigates the behavior of dislocations in >10 μm thick layers by the use of TEM. The InAs and AlSb layers were investigated, as well as a thin-element superlattice (1000 periods of 5nm GaSb/5nm AlSb) and a thick-element periodic structure (100nm GaSb/100nm AlSb).

2. EXPERIMENT

The samples studied were grown in a Varian (Intevac) Gen II solid source MBE system. Semi-insulating (100) GaAs substrates were used as received from the vendor. They were first outgassed at 420°C in vacuum for 30 minutes and then heated to 610°C for 15 minutes under an arsenic flux just prior to growth. The substrate temperature was measured using an optical pyrometer calibrated to the congruent sublimation temperature of GaAs. The growth rates of GaSb, AlSb and InAs were 1.0, 1.0 and 0.7μm/h, respectively. GaSb and AlSb layers were grown using V/III flux ratios that resulted in an Sb-stabilized surface, as evidenced by a (1 x 3) RHEED reconstruction pattern. The InAs layers were grown using the minimum As_4 flux that still results in an As-stabilized RHEED pattern. None of the layers were intentionally doped.

The first epi layer is 200nm of GaAs grown at 580°C to smooth the surface of the wafer. Next, 50nm of AlSb is grown to nucleate the mismatched material. During initial growth of the AlSb layer, the temperature is reduced from 580°C to 530°C and the RHEED pattern goes from the (2 x 4) streaks characteristic of GaAs to a spotty pattern indicative of three-dimensional (3-D) island growth. The spots begin to elongate and within 30 monolayers the (1 x 3) pattern of AlSb begins to appear. The RHEED pattern continues to improve and after ~50nm the RHEED streaks are smooth. This portion of the growth is common to all of the structures.

All of the samples were a minimum of 10μm thick. The AlSb sample is 15.6μm of AlSb followed by a 5nm GaSb cap to inhibit oxidation of the AlSb. The InAs sample is 10μm thick and was grown at 500°C. The SL structure is 5nm of GaSb/5nm of AlSb repeated 1000 times. The periodic structure is repeated layers of 100nm GaSb/100nm AlSb.

A modification of the usual TEM sample preparation technique was necessary because of the propensity of AlSb to oxidize. By preparing the sample using liquid nitrogen cooled ion-milling and then immediately storing the sample in vacuum, oxidation was sufficiently suppressed. The TEM imaging and analysis was then performed using a JEOL 200CX operating at 200keV.

3. RESULTS AND DISCUSSION

All of these heterojunction structures show dislocations in the TEM micrographs. The micrograph of the AlSb epitaxial layer grown on GaAs is shown in Figure 1. The lattice

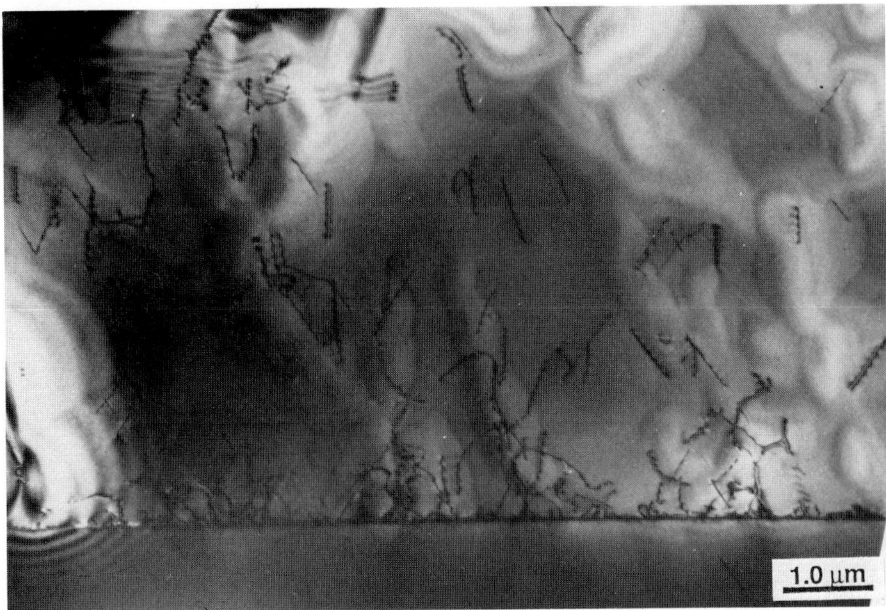

Figure 1. The TEM micrograph of AlSb epi grown on GaAs substrate shows threading dislocations that decrease in density with distance from the substrate.

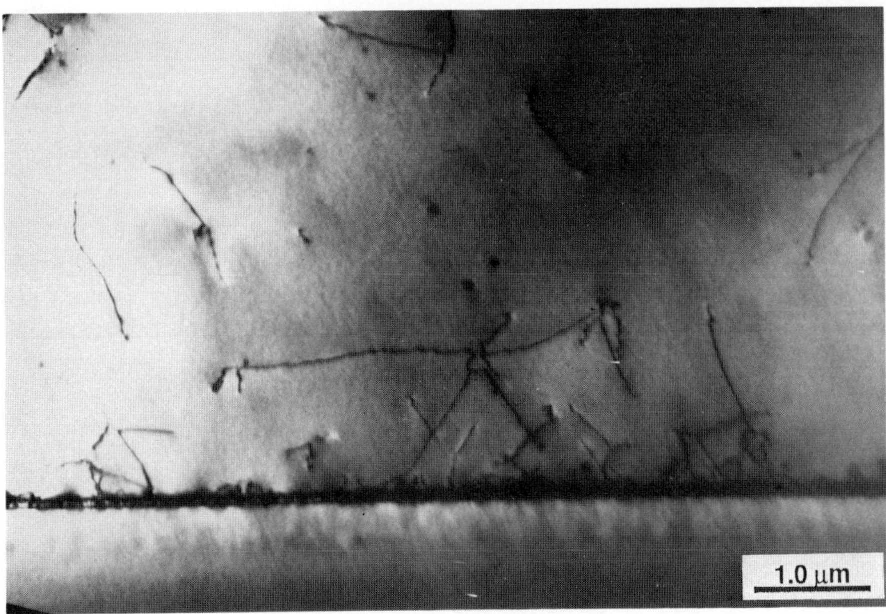

Figure 2. This TEM micrograph shows InAs grown epitaxially on GaAs. A comparison with the AlSb layer suggests that the dislocation density is more quickly reduced in InAs compared to AlSb.

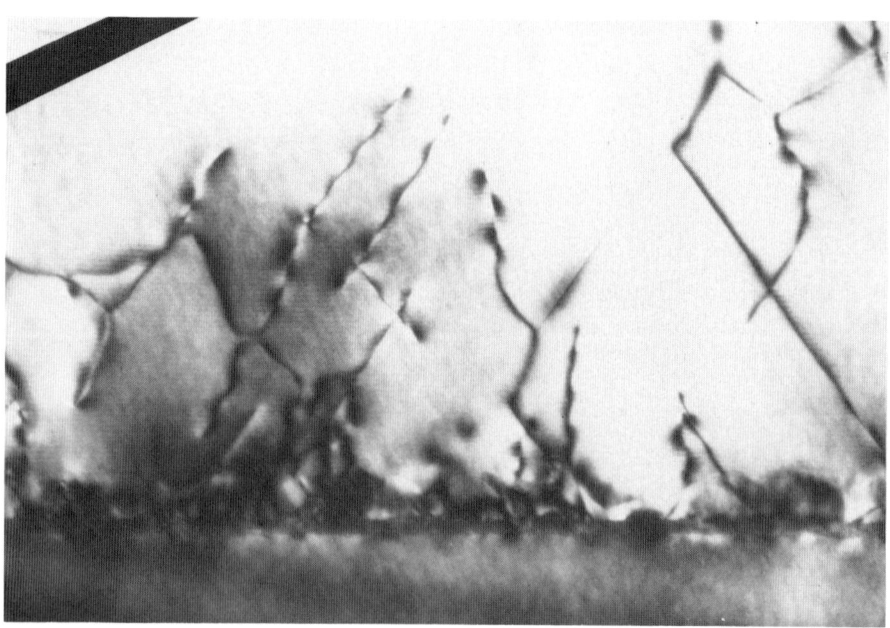

Figure 3. This TEM micrograph shows that the thin-element superlattice composed of 5nm AlSb/5nm GaSb layers has no apparent effect on dislocation reduction.

Figure 4. In this TEM micrograph the thicker periodic structure with 100nm AlSb/100nm GaSb layers has bent many dislocations parallel to the epitaxial layers, which results in a reduction of dislocation density.

mismatch is 8.5%. This large mismatch and the 3-D nucleation result in high dislocation densities. The threading dislocation density at the start of epi growth is ~10^{10}/cm^2, this falls 2 orders of magnitude to ~2 x 10^8/cm^2 after 15.6 μm. The defects drop more rapidly in the first 2 μm and then nearly level out, suggesting there is a lower limit of ~10^8/cm^2 that is reached when we try to lower threading dislocations densities only by growing thicker layers. The defect density drops because threading dislocations with opposite Burgers vectors are attracted and then annihilate one another. The more rapid decrease in density at the beginning is at least in part explained by the higher density of dislocations resulting in an increased probability of annihilation.

The TEM micrograph of the InAs epi layer grown on GaAs (6.8% mismatch) is shown in Figure 2. This 10.5μm layer has a dislocation density of ~10^{11}/cm^2 at the epi/substrate interface that falls to 3 x 10^8/cm^2 at the surface, this is very similar to the AlSb example. The defect density decreases more quickly at the start of growth for InAs compared to AlSb. This could be due to a higher mobility of dislocations in InAs.

The 5nm AlSb/5nm GaSb superlattice TEM micrograph is shown in Figure 3. The dislocation density is high (~10^{11}/cm^2) at the start of growth and falls to ~4 x 10^8/cm^2 at the surface. Although this structure is effective in smoothing the growth surface as indicated by the RHEED pattern, there is no evidence that this superlattice is at all useful in causing dislocations to bend toward the growth plane and interact with other dislocations. The dislocation density and distribution are very similar to the AlSb and InAs example. A summary of the defect distributions for all of the samples is presented in Figure 5.

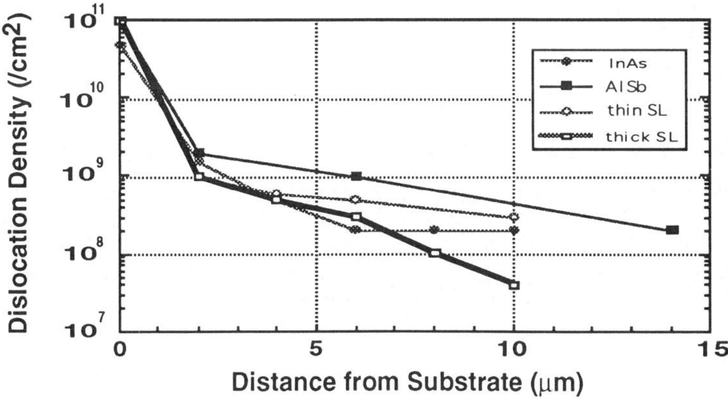

Figure 5. This graph shows the reduction of dislocations with distance from the substrate. The drop in dislocation density is higher nearest the substrate. This can be explained by the the reduced probability of dislocations with opposite Burgers vectors annihilating each other as the density is reduced.

The TEM micrograph of the periodic 100nm AlSb/100nm GaSb structure is shown in Figure 4. The defect density is $\sim 10^{11}/cm^2$ at the beginning of growth and after 12μm of epi falls to $\sim 5 \times 10^7/cm^2$ at the surface. The critical thickness for GaSb/AlSb (~0.6% mismatch) is ~40nm. In this case the epi layer thickness has exceeded the critical thickness of GaSb/AlSb. This results in relaxed layers with high strain at the interfaces. It can be seen from the micrograph that the structure is bending many of the defects parallel to the growth plane and this improves the probability of annihilation. This results in the order-of-magnitude decrease in threading dislocations at the surface. It can also be seen that not all of the defects interact with the heterointerface. Those whose Burgers vectors are ~60° to 90° from the interface have a lower driving force for interaction with the heterointerface.

4. SUMMARY

The formation and behavior of dislocations in (Ga,Al)Sb/GaAs heterostructures was investigated using cross-sectional TEM. Four structures (>10μm thick) were studied: AlSb, InAs, 5nm GaSb/5nm AlSb superlattice and a periodic structure with 100nm GaSb/100nm AlSb layers. Dislocations were present in all of these heterosturctures that decrease in the epitaxial growth direction. In the superlattice the dislocation densities are very comparable to those of the bulk InAs and AlSb layers, *i.e.*, $\sim 10^{10}/cm^2$ at the start of growth decreasing to $\sim 10^8/cm^2$ in the top layer. When the layer thickness of the periodic structure is increased to 200nm (100nm GaSb/100nm AlSb) the dislocation density is reduced by an order of magnitude, to $4 \times 10^7/cm^2$ after 10μm of epi growth. The probability that dislocations will be annihilated is increased when the higher strain in thicker layers is sufficient to cause threading dislocations to bend over to form misfit dislocations at the superlattice interfaces.

5. ACKNOWLEDGMENTS

This work was partially performed under the management of FED (the R&D Association for Future Electron Devices) as a part of the R&D of Basic Technology for Future Industries supported by NEDO (New Energy and Industrial Technology Development Organization, Japan).

REFERENCES

Ideshita S, Furukawa A, Mochizuki Y, Mizuta M, Appl. Phys. Lett. **60,** 2549 (1992).
Luo L F, Beresford R, Wang W I, Munekata H, Appl. Phys. Lett. **55**, 789 (1989).
Soderstrom J R, Brown E R, Parker C D, Mahoney L J, Yao J Y, Andersson T G, McGill T C, Appl. Phys. Lett. **58,** 275 (1991).
Werking J, Tuttle G, Nguyen C, Hu E L, Kroemer H, Appl. Phys, Lett. **57**, 905 (1990).

Inst. Phys. Conf. Ser. No 136: Chapter 11
Paper presented at the Int. Symp. GaAs and Related Compounds, Freiburg, 1993

Shallow and deep levels in GaAs grown by atomic layer MBE

A Bosacchi, E Gombia, M Madella,* R Mosca and S Franchi
CNR - MASPEC Institute, Via Chiavari, 18a, I-43100, Italy

ABSTRACT: We report on a study of the incorporation of shallow and deep levels in Si doped (n ~ 10^{16} cm^{-3}) GaAs grown by Atomic Layer MBE at temperatures between 370 °C and 530 °C. The free carrier concentration depends on whether Si is supplied during: a) both the Ga and As subcycles, b) the As subcycles, or c) the Ga subcycles. As for deep levels, the occurrence of M1, M3 and M4 levels and their concentrations allow us to relate the low-temperature grown ALMBE-GaAs to MBE-GaAs prepared at conventional temperatures.

1. INTRODUCTION

Atomic Layer Molecular Beam Epitaxy (ALMBE (Briones and Ruiz 1991), also termed Migration Enhanced Epitaxy, MEE (Horikoshi et al 1988)) is a variant of Molecular Beam Epitaxy (MBE) where the group-III and the group-V beams are supplied to the growing surface in alternate cation and anion subcycles, instead of simultaneously as in MBE. If the number of cations supplied during each ALMBE cycle equals the concentration of surface sites, the growth rate is a monolayer per cycle. During cation subcycles, the composition of the surface becomes less rich in As; this results in an enhanced cation migration rate (compared to the case of MBE), which, in turn, makes the two-dimensional atomic layer-by-atomic layer growth more likely. These features are prerequisites for the growth of interfaces smooth on an atomic scale, and, then, of epitaxial structures where the physical properties are determined by the carriers confined close to heterointerfaces. The interesting consequence of the enhancement of cation surface migration is that the growth temperature can be significantly reduced (Horikoshi et al 1988), with the consequent reduction of thermally activated phenomena, such as diffusion and/or surface segregation of constituent atoms or dopants. Moreover, ALMBE is attracting increasing attention also for: i) the preparation of mismatched structures (Gerard et al

---

* present address: CSELT, Via G Reiss Romoli 274, I-10148 Torino, Italy

1991), where growth tends to take place according to 3D mechanisms, unless a 2D mode is forced by ALMBE, and ii) the growth of continuously and abruptly composition-graded structures, with an unprecedented flexibility (Madella et al 1993).

Most of the studies concerning ALMBE materials and structures have dealt so far with their optical properties, while the electrical ones are by far less investigated. p-type and n^+-type doping has been reported in ALMBE GaAs by Tadayon et al (1989) and by Tadayon et al (1990) and Ramsteiner et al (1991), respectively. The present work reports on a study on the incorporation of shallow and, for the first time, of deep electronic levels in $n \sim 10^{16}$ cm^{-3} ALMBE GaAs doped with Si, under different conditions. Our results on shallow and deep levels allow us to relate GaAs grown by ALMBE at low temperatures to GaAs prepared by MBE at conventional temperatures instead of to material deposited by low temperature MBE.

2. EXPERIMENTAL PROCEDURES

The ALMBE growth has been carried out in a Varian Gen II Modular system, with shutters controlled by especially developed hardware and software, which give a resolution of the molecular-beam supply times of <10 ms. The substrates were non-In-bonded and radiatively heated. The growth temperatures were in the range 370 - 530 °C. Most of the samples were grown at a temperature of 450 °C, which is in the middle of the optimum range for the preparation of AlGaAs/GaAs quantum well structures, as deduced by the intensities and the linewidths of the photoluminescence recombinations (Madella et al 1993). The substrate temperatures T_s quoted in this paper were measured by an optical pyrometer for $T_s \geq 450$ °C and by a thermocouple (TC) not in direct contact with the substrate for $T_s < 450$ °C. The TC readings were corrected by the difference (50-80 °C) between the TC and the optical pyrometer values measured at $T_s > 450$ °C. The Ga fluxes have been chosen by calibrating the GaAs MBE growth rates at about 0.9 μm/h by RHEED oscillations. The As_4 beam equivalent pressure was ~8×10^{-6} Torr; the As supply times were selected according to the time evolution of the RHEED intensity signal. In different experiments, the As supply times τ_{As} were chosen in the range 0.5-1.5 s, while the supply time of Ga was 1.1 s. The shortest τ_{As} was sufficient to give a (2x4) As-stabilized surface reconstruction at the end of the As subcycle. A monolayer of GaAs is deposited in each ALMBE cycle.

The samples consist of: i) n^+ (100)-GaAs substrates, ii) GaAs:Si ($n = 1 \times 10^{17}$ cm^{-3}) buffer layers grown by MBE at 600 °C and iii) the Si doped ALMBE layers grown under the different conditions. The doping levels were chosen so as to allow the measurements of the net donor concentration profiles over ~ 2 μm ranges with reverse biases of up to 14 V. Unless stated otherwise, the Si/Ga ratio was kept constant during the growth of all the

ALMBE layers.

C-V measurements were carried out at 300 K by a HP 4192A impedance analyzer on Al Schottky diodes fabricated by photolithography. DLTS measurements were performed by using a high sensitivity lock-in type spectrometer, with pulse durations of 550 μs and rate windows of 55 s^{-1}. The deep level concentrations were evaluated so as to take into account the so-called λ effect.

3. SHALLOW LEVELS

The carrier concentration profiles of Si doped GaAs grown by ALMBE at 450 °C are shown in Figure 1. The curves (a), (b) and (c) refer to samples where Si was supplied during: a) both the As and Ga subcycles, b) the As subcycles, and c) the Ga subcycles, respectively. In order to rule out any possible artifact due to fluctuations of growth conditions in different runs, multi-layer structures were grown, where layers of the three types (a)-(c) (with thicknesses of ~0.5, ~1.0 and ~1.0 μm, respectively) were stacked in sequence; the carrier concentration profile of one of these samples is shown by the curve (d) in Figure 1. The net donor concentration (N_d-N_a) profiles measured in multi-layer structures are definitely consistent with the results measured in the individual layers (a)-(c).

Figure 1 shows that the net donor concentration (N_d-N_a) is systematically larger in the case of curve (b) than in the case of curve (c), while the number of Si atoms supplied to the surface is the same. Since the Si sticking coefficient is likely independent of the way (a)-(c) used to supply Si to the surface, the different net donor concentration of the layers (b) and (c) can be related to the electrical compensation of Si shallow donors; the compensation may be due to the amphoteric nature of Si in GaAs, which may have different probability of being incorporated as an acceptor, depending on whether Si is supplied during the As- or Ga- subcycles. This explanation is confirmed by the observation that the same ratio between (N_d-N_a) measured in type (b) and (c) layers has been found also in samples doped with a Si flux 10 times higher than that of the samples shown in Figure 1. It is worth mentioning, however, that for

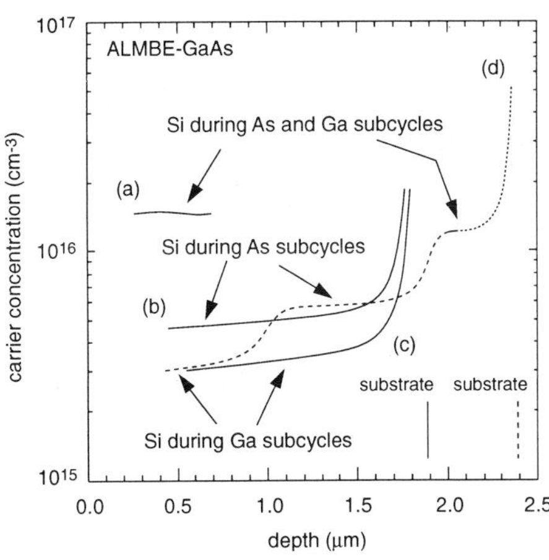

Fig. 1. Carrier concentration profiles of Si doped GaAs grown by ALMBE at 450 °C. The curve (d) refers to a sample where type-(a), -(b) and -(c) layers were stacked in sequence. The dotted part of curve (d) was obtained after etching 2.1 μm of material. The growths were carried out with $\tau_{As} = \tau_{Ga} = 1.1$ s

doping levels in the low 10^{18} cm^{-3} range, Tayadon et al (1990) have measured a higher carrier concentration when Si is supplied during the Ga subcycle, rather than during the As one; this finding, obtained in samples grown at 580 °C, is opposite to our well reproducible results. It should be noted that the ALMBE growth at relatively high temperatures results in AlGaAs/GaAs quantum well structures with degraded properties (Foxon et al 1990), unless the growth is carried out by depositing fractions of monolayer in each ALMBE cycle (Madella et al 1993).

4. DEEP LEVELS

Figure 2 shows the DLTS spectra of Si doped GaAs (with a net carrier concentration in the range shown in Figure 1) grown by ALMBE at 450 °C. The curves refer to samples where Si is supplied to the surface during: a) both the As and Ga subcycles, b) the As subcycles, and c) the Ga subcycles. The low temperature side of the DLTS spectra show three main peaks having the same features of the M1, M3 and M4 levels observed in MBE GaAs (Lang et al 1984); the peak at about 330 K, that we detect in all of the ALMBE samples, corresponds to a level (termed here as M(330)) with an activation energy of 0.70 eV and a capture cross section of 8.1×10^{-14} cm^2, that can be hardly ascribed either to M6 or M7 (Lang et al 1984). We note that the concentrations of the M1, M3 and M4 levels are higher in samples grown with the Si shutter open during both the As and Ga subcycles ($7\text{-}13 \times 10^{12}$ cm^{-3}) than in samples of type-(b) and (c) ($2\text{-}4 \times 10^{12}$ cm^{-3}).

In order to study the influence of the growth temperature on the deep level density, ALMBE GaAs have been grown at 370, 450 and 530 °C respectively. The results on the concentration of deep levels are summarized in Figure 3, for samples with Si supplied during the As subcycle. Within the explored temperature range, the concentration of M1, M3 and M4 levels weakly depends on the substrate temperature T_s. This is in contrast with the case of MBE GaAs, where the trap density decreases by about two orders of magnitude for growth temperatures increasing from 520 to 630 °C (Blood and Harris 1984).

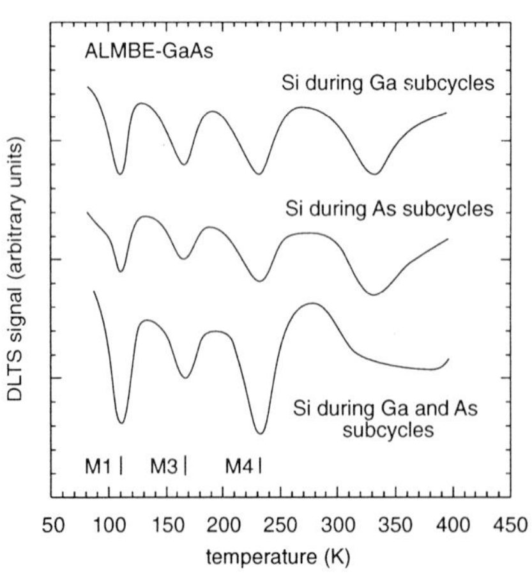

Fig. 2. DLTS spectra of GaAs:Si grown by ALMBE at 450 °C with different Si-supply schemes. The carrier concentrations are the same as those of samples shown in Figure 1.

It is worth noting that the deep level densities in ALMBE GaAs grown at $370 < T_s < 530$ °C

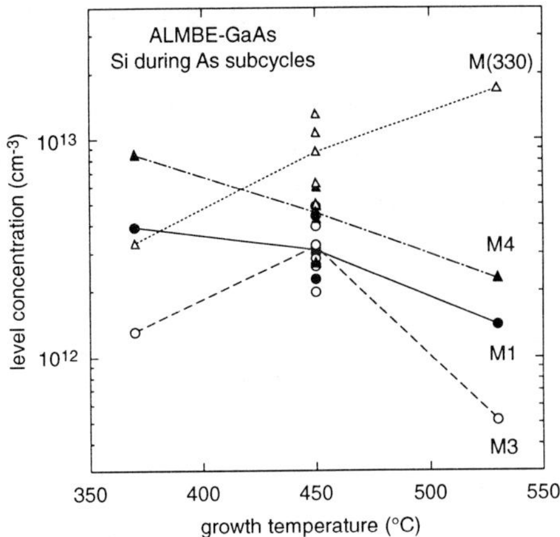

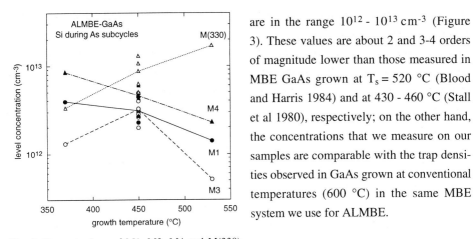

Fig. 3. Concentrations of M1, M3, M4 and M(330) deep levels as functions of substrate temperatures of Si:GaAs samples grown by ALMBE. Si was supplied during the 1.1 s long As subcycles. The net carrier concentrations are the same as that of curve (b) in Figure 1. The lines are guides for the eye.

are in the range 10^{12} - 10^{13} cm^{-3} (Figure 3). These values are about 2 and 3-4 orders of magnitude lower than those measured in MBE GaAs grown at T_s = 520 °C (Blood and Harris 1984) and at 430 - 460 °C (Stall et al 1980), respectively; on the other hand, the concentrations that we measure on our samples are comparable with the trap densities observed in GaAs grown at conventional temperatures (600 °C) in the same MBE system we use for ALMBE.

We have also studied the effect of the length of the As supply times on the deep level density. Figure 4 shows the DLTS spectra of three ALMBE GaAs samples grown with As supply times τ_{As} of 0.5, 1.1 and 1.5 s respectively, while τ_{Ga} was 1.1 s. Since levels such as M0 or M5 do not occur reproducibly, we focus our attention on the M1, M3 and M4 levels; their concentrations slightly increase when the As supply time increases. It should be noted that in MBE GaAs the M1 and M4 concentrations decrease as As/Ga ratios increase, while the M3 density has the opposite behaviour (Blood and Harris 1984). As for the M(330) trap, our results do not allow us to identify definite trends in the τ_{As} dependence of the level concentration.

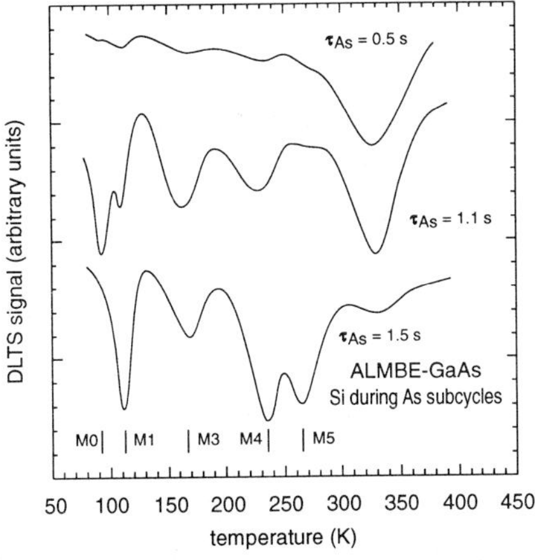

Fig. 4. DLTS spectra of GaAs:Si grown by ALMBE at 450 °C, using different the As supply times. Si was supplied to the surface during the As subcycles. The marks show the positions of the M0, M1, M3, M4 and M5 deep levels. The concentrations of the M1, M3 and M4 levels in samples grown with As supply times of 0.5 s, 1.1 s and 1.5 s are in the ranges 1 - 2 x 10^{12} cm^{-3}, 3 - 5 x 10^{12} cm^{-3}, and 6 - 8 x 10^{12} cm^{-3}, respectively. The carrier concentrations are the same as that of curve (b) in Figure 1.

5. CONCLUSIONS

We have studied the incorporation of shallow and, for the first time, of deep levels in GaAs grown by ALMBE. We conclude that the net donor concentration of Si doped GaAs, and, then, the compensation ratio of shallow Si donors depend on the Si-supply scheme which is used, i.e. on whether Si is supplied during: a) both the Ga and As subcycles, b) the As subcycles, or c) the Ga subcycles. Our DLTS data show that GaAs grown by ALMBE at 370 - 530 °C is characterized by the same M1, M3 and M4 deep levels, which are typical of GaAs prepared by MBE at 520 - 650 °C; apart from a few differences between the dependence of the deep level concentrations on growth conditions in the material deposited by MBE and ALMBE, we note that the trap concentrations of ALMBE GaAs is comparable to that of MBE material grown at conventional temperatures (~600 °C), and it is lower by orders of magnitude than that of MBE GaAs grown at temperatures close to those used in the present work.

ACKNOWLEDGEMENTS

The technical assistance of P Allegri, V Avanzini and A Motta is acknowledged. The work is partially supported by the CNR PF MSTA.

REFERENCES

Blood P and Harris JJ 1984 J. Appl. Phys. 56 993

Briones F and Ruiz A 1991 J. Crystal Growth 111 194

Foxon CT, Hilton D, Dawson P, Moore KJ, Fewster P, Andrew NL and Olson JW 1990 Semicond. Sci. Technol. 5 721

Gerard JM, Marzin JY and Jusserand B 1991 J. Crystal Growth 111 205

Horikoshi Y, Kawashima M, and Yamaguchi H 1988 Japan J. Appl. Phys. 27 169

Lang DV, Cho AY, Gossard AC, Ilegems M, and Wiegmann W 1984 J. Appl. Phys. 47 2558

Madella M, Bosacchi A, Franchi S, Allegri P and Avanzini V 1993 J. Crystal Growth 127 270

Ramsteiner M, Wagner J, Silveira JP and Briones F 1991 GaAs and Related Compounds 1990 ed KE Singer Inst. of Phys. Conf. Series 112 (Bristol: Institute of Physics) pp 85-9

Stall RA, Wood CEC, Kirchner PD and Eastman LF 1980 Electron. Lett. 16 171

Tadayon B, Tadayon S, Schaff WJ, Spencer MG, Harris GL, Tasker PJ, Wood CEC and Eastman LF 1989 Appl. Phys. Lett. 55 59

Tadayon B, Tadayon S, Spencer MG, Harris GL, Griffin J and Eastman LF 1990 J. Appl. Phys. 67 589

Influence of annealing on electron lifetimes in transistor base-layers on GaAs:C

U. Strauss, A. P. Heberle, W. W. Rühle, H. Tews[1)], T. Lauterbach[2)] and K. H. Bachem[2)]

Max-Planck-Institut für Festkörperforschung, Heisenbergstr.1, D-70506 Stuttgart, FRG

[1)] Siemens AG, Otto-Hahn-Ring 6, D-81730 München, FRG

[2)] Fraunhofer-Institut für Angewandte Festkörperphysik, Tullastr. 72, D-79108 Freiburg, FRG

ABSTRACT: The minority carrier lifetimes in heavily p-doped GaAs epitaxial layers for the bases of heterobipolar transistors are investigated by time-resolved photoluminescence. The samples are grown with carbon dopings from $4x10^{18}$ to $1.5x10^{20}$ cm^{-3}. A post-growth annealing process at 600 °C for 10 min guarantees activation of carbon as an acceptor to more than 90 %. The electron lifetimes of annealed samples are comparable with the lifetimes of as-grown samples with same free hole densities. The lifetimes are limited by Auger recombination at densities higher than $5x10^{19}$ cm^{-3}.

1. INTRODUCTION

Carbon is an important doping material for the base of fast GaAs heterobipolar transistors because of its small diffusivity. Very high hole concentrations of more than 10^{20} cm^{-3} can be achieved (Pena-Sierra 1992, Wagner 1992, and Aitchison 1990). The activation of carbon as an acceptor depends on the growth conditions. A post-growth annealing process of samples with low activation at carbon densities up to 10^{20} cm^{-3} increases the electrical activation to more than 90 % (Watanabe 1992 and Han 1992).

The electron lifetime in the p-base is an important parameter for device performance and modelling. We investigate the influence of post-growth annealing on the electron lifetimes. We determine these lifetimes by measuring the luminescence decay after excitation with ultrashort laser pulses. We further discuss microscopic models for the non-active carbon.

2. EXPERIMENTAL

Our samples are grown by metal-organic vapor deposition (MOCVD) at temperatures between 560 and 600 °C. Epitaxial layers of GaAs:C with thicknesses of 150 nm or 700 to 1100 nm are deposited on GaAs substrates. The structures with 150 nm thin layers contain additionally diffusion barriers of $Al_{0.6}Ga_{0.4}As$ towards the substrate. The carbon concentrations p_C are measured by secondary ion mass spectrometry (SIMS). We achieve densities from $p_C = 4x10^{18}$ to $1.5x10^{20}$ cm^{-3}. The free hole concentrations p_0 are determined without electrical contacts by resonant plasma reflection, and a calibration is made by Hall measurements on similar samples. The activation of carbon ranges between 50 and 100 %. The higher values are obtained for lower p_C. The samples are annealed at 600 °C for 10 min in arsenic atmosphere. After annealing, the activation of carbon is between 90 and 100% .

The samples are optically excited at 600 nm (2.07 eV). We use a dye laser, which is synchronously pumped by a mode-locked Nd:YAG laser with glass-fiber grating pulse compressor and subsequent frequency doubler. Pulsewidth is as short as 300 fs. The excitation densities are at least two orders of magnitude lower than the doping concentrations. The time-resolved luminescence is detected by photoluminescence up-conversion with a time-resolution of 300 fs. This high time-resolution permits precise measurements of the luminescence decay times, and the temporal resolution is three orders of magnitude better than in experiments with single photon counting technique (de Lyon 1992).

3. RESULTS AND DISCUSSION

The recombination of nonequilibrium carriers in p_0-doped GaAs occurs via trapping of electrons, via emission of radiation, or via valence-band Auger effect. These three processes are characterized by the recombination coefficients A, B, and C_p, respectively. The recombination kinetics of the electrons is described by

$$\frac{dn}{dt} = -An - Bp_0 n - C_p p_0^2 n, \qquad (1)$$

if the electron density is small compared with p_0. The luminescence intensity I is proportional to the minority carrier density, and Eq. (1) yields

$$I \propto \exp(-t/\tau). \qquad (2)$$

The luminescence decay time is

$$\tau = (A + Bp_0 + C_p p_0^2)^{-1}, \qquad (3)$$

and is equal to the minority carrier lifetime. A typical exponential luminescence decay is shown in Fig. 1. We investigate lifetimes of as-grown and annealed GaAs:C layers as a function of the majority carrier densities at 300 and 50 K. All data are compiled in Fig. 2.

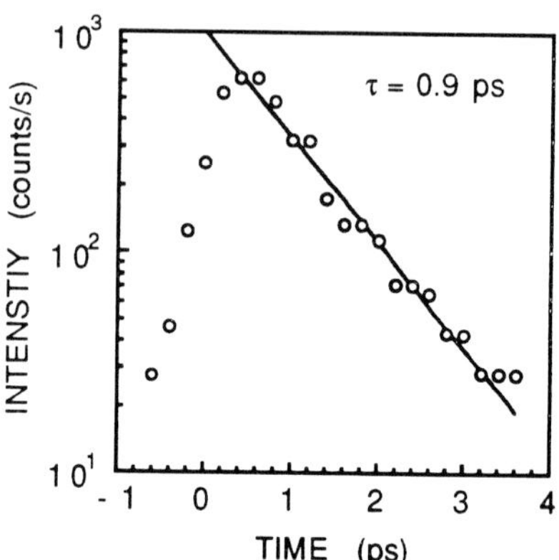

Fig. 1: Luminescence decay of annealed GaAs:C with p_0=1.1x10^{20} cm^{-3} at 50 K.

The lifetimes at room temperature (Fig. 2a) are independent of p_0 at carrier densities smaller than 1x10^{19} cm^{-3}. In this doping regime, the lifetime is limited by diffusion of the electrons to the GaAs:C-GaAs interface and by trapping at defects. For p_0 > 5x10^{19} cm^{-3}, the lifetime varies like $\tau \propto p_0^{-2}$, which is the characteristic dependence for Auger recombination. Radiative recombination is not dominant in our samples at room temperature, since no dependence $\tau \propto p_0^{-1}$ is observed. Therefore we simply assume a value of 1.5x10^{-10} cm^{3}s^{-1} for B. This value was calculated by Stern (1976) and experimentally verified (Casey 1976 and Strauss 1993a). The overall dependence of the lifetime is well described with A = 3x10^{10} s^{-1}, B = 1.5x10^{-10} cm^{3}s^{-1}, C_p = (2±1)x 10^{-29} cm^{6}s^{-1}.

Figure 2b depicts the results obtained at 50 K. Trapping together with diffusion as well as Auger recombination are obviously nearly the same as at 300 K, since lifetimes at low and high p_0 depend only weakly on temperature. The weakly temperature-dependent Auger recombination is typical for an impurity or phonon assisted process.(Haug 1988 and Bardyszewski 1985) However, the lifetimes at medium concentrations are shortened at 50 K. Obviously the influence of radiative recombination becomes visible. The overall variation of lifetime at 50 K is well described taking A = 3x10^{10} s^{-1}, B = 2.2x10^{-9} cm^{3}s^{-1} and C_p = (2±1)x10^{-29} cm^{6}s^{-1}. The values for A and C_p are similar to those obtained at 300 K. The value for B is higher at 50 K, the increase being consistent with calculations by Stern (1976).

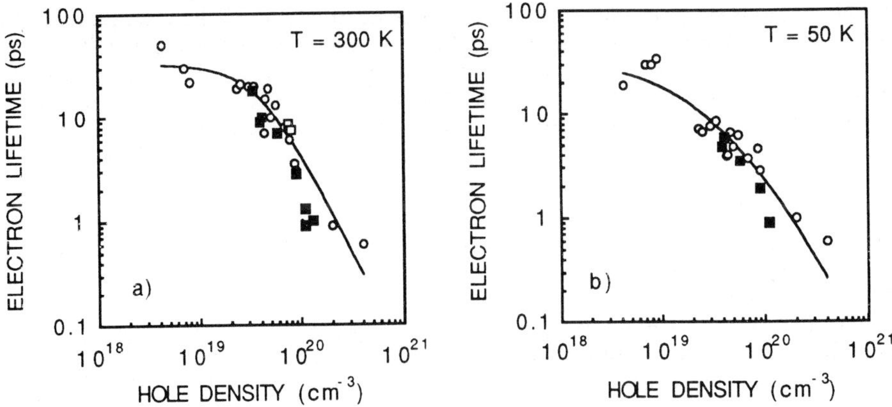

Fig. 2: Minority carrier lifetimes of GaAs:C at 300 and 50 K. Open symbols: samples as grown, closed symbols: samples annealed at 600 °C, squares: this work, circles: earlier data (Strauss 1993b), solid line: fit to as-grown samples a) with $A = 3x10^{10}\ s^{-1}$, $B = 1.5x10^{-10} cm^3s^{-1}$ *and* $C_p = 2.2x10^{-29}\ cm^6s^{-1}$, *b) with* $A = 3x10^{10}\ s^{-1}$, $B = 2.2x10^{-9} cm^3s^{-1}$ *and* $C_p = 1.8\ x\ 10^{-29}\ cm^6s^{-1}$.

We now compare as-grown samples with annealed samples. The post-growth annealing increases the free hole densities without changing the carbon concentrations. Activations of carbon between 90 and 100% are achieved in our samples with carbon concentrations up to 10^{20} cm^{-3}. Fig. 2 shows, that the minority carrier lifetimes are related to the free hole concentrations rather than to the carbon concentrations, since we observe similar lifetimes at similar p_0 with and without annealing. The lifetimes of annealed samples close to $p_0=10^{20}$ cm^{-3}, where no as-grown samples are available, seem to be slightly shorter than described with the coefficients A, B and C_p above. However, this deviation is still within the error limits given for these values.

The recombination at high hole densities seems to be faster than measured in intrinsic GaAs at similar photoexcited hole densities (Strauss 1993a). Therefore an about 2.5 times larger Auger coefficient is obtained in the doped samples. This increase is attributed to impurity assisted Auger effect (Strauss 1993b). However, the lifetimes of annealed samples are not longer than those of as-grown samples with same p_0. This result shows, that non-active carbon, present in non-annealed samples only, does not shorten the lifetime.

The origin of the reduced activation of carbon is not yet completely understood. Three models are discussed: self-compensation of carbon by substituing Ga as well as As (de Lyon 1990 and Giannini 1992), passivation of carbon by hydrogen (Watanabe 1992, Wagner 1992 and Kozuch 1993) and interstitial carbon (Höfler 1992 and Giannini 1993). Determination of the lattice constant by x-ray diffraction gives useful information to choose between these models, since carbon doping leads to contraction of the lattice. X-ray measurements of de Lyon et al. (1990) and Giannini et al. (1992) are interpreted assuming, that C is exclusively on As and Ga sites.

We measure the lattice contraction by x-ray diffraction before and after annealing for a layer with $p_c = 1.1x10^{20}$ cm^{-3}, thus two layers with same p_c but with different p_0 can be compared. We observe an increase of p_0 from $6.9x10^{19}$ to $1.1x10^{20}$ cm^{-3}. The x-ray peak-position of the doped layer shifts from 0.70 to 0.85 mrad relative to the peak of the GaAs substrate. The shift shows an increase in lattice contraction from 0.108 to 0.131 %. The larger contraction in the layer with larger p_0 but same p_c cannot be explained by assuming that the reduced doping efficiency is only due to donor-like C atoms on Ga sites; since C on Ga as well as on As sites leads to nearly the same contraction. We conclude that interstitial carbon and/or hydrogen passivation of substitutional carbon additionally reduce doping efficiency. Interstitial C leads to less contraction than substitutional C (Höfler 1992), and hydrogen passivation could possibly increase the size of the substitutional C.

4. CONCLUSIONS

Reduced activation of carbon as an acceptor in layers of GaAs:C is due to carbon on interstitial sites or hydrogen passivation. Additional self-compensation might be possible. Post-growth annealing guarantees an acceptor activation as high as 90% for carbon doping up to 10^{20} cm^{-3}. Therefore, no further increase of the free hole density p_0 is expected in these layers during operation in devices. The minority carrier lifetimes of annealed and as-grown layers are comparable for similar p_0.

ACKNOWLEDGMENTS

We are grateful to K. Eberl for helpful discussion and experimental support. We wish to thank A. Breitschwerdt, H. Klann, K. Rother and S. Tippmann for expert technical assistance and E. Tournié for critical reading of the manuscript. This work is supported by the Bundesminister für Forschung und Technologie.

REFERENCES

Aitchison B J , Haegel N M, Abernathy C R and Pearton S J 1990 Appl. Phys. Lett. 56 1154
Bardyszewski W and Yelnik D 1985 J. Appl. Phys. 57 4820
Casey H C and Stern F 1976 J. Appl. Phys. 47 631
Giannini C, Fischer A, Lange C, Ploog K and Tapfer L 1992 Appl. Phys. Lett. 61 183
Giannini C, Brandt O, Fischer A, Ploog K and Tapfer L 1993 J. Crystal Growth 127 724
Han W Y and Lu Y 1992 Appl. Phys. Lett. 61 87
Haug A 1988 J. Phys. Chem. Solids 49 599
Höfler G E and Hsieh K C 1992 Appl. Phys. Lett. 61 1992 327
Kozuch D M, Stavola M, Pearton S J, Abernathy C R, and Hobson W S (1993) J. Appl. Phys. 73 3716
de Lyon T J, Woodall J M, Goorski M S and Kirchner P D 1990 Appl. Phys. Lett. 56 1040
de Lyon T J, Woodall J M, Kash J A, McInturff D T, Bates R J S, Kirchner P D and Cardone F 1992 J. Vac. Sci. Technol. B 10 846
Pena-Sierra R, Escobosa A and Sanchez-R V M 1992 Appl. Phys. Lett. 62 2359
Stern F 1976 J. Appl. Phys. 47 5382
Strauss U, Rühle W W and Köhler K 1993a Appl. Phys. Lett. 62 55
Strauss U, Heberle A P, Zhou X Q, Rühle W W, Lauterbach T, Bachem K. H and Haegel N. M. 1993b Jpn. J. Appl. Phys. 32 495
Wagner J, Maier M, Lauterbach T, Bachem K H, Fischer A, Ploog K, Mörsch G and Kamp M 1992 Phys. Rev. B 45 9120
Watanabe K and Yamazaki H 1992 Appl. Phys. Lett. 60 847

Si-doping characteristics and deep levels in MBE-AlInAs layers

H. Hoenow*, H.-G. Bach, H. Künzel, and C. Schramm

* Humboldt-Universität zu Berlin, Inst. für Werkstoffe und Verfahrenstechnik
Invalidenstr. 110, D-10099 Berlin, Germany
Heinrich-Hertz-Institut für Nachrichtentechnik Berlin GmbH
Einsteinufer 37, D-10587 Berlin, Germany

ABSTRACT: AlInAs layers lattice matched to InP are widely used in opto-electronic and microwave devices. To achieve n-type material, silicon is the most important donor. We report on the Si-doping characteristics in conjunction with deep electron trap properties of AlInAs grown by MBE at different temperatures and doping levels. Temperature resolved Hall, C-V and DLTS measurements were carried out for characterisation. A clear reduction of the free carrier concentration was found for decreased growth temperatures despite of constant Si-doping. To explain the doping characteristics, we discuss the donor ionisation, the role of electron traps, and an additional compensation mechanism.

1. INTRODUCTION

AlInAs layers grown on InP-based materials either lattice matched or strained are widely used for both optical and microwave applications. Low-temperature MBE (Molecular Beam Epitaxy) growth with high crystalline quality and regrowth capability is very interesting for opto-electronic integrated devices. Under these conditions AlInAs growth yields high-resistivity layers and low leakage currents, which are attractive for new device concepts.
Recent investigations by Künzel et al. (1991) revealed that AlInAs grown in the low-temperature range of $T_g = 200...400°$ C exhibits high-resistivity behaviour even with a silicon (Si)-doping up to 10^{17} cm^{-3}. To clarify the apparent reduction of free carrier concentration we investigated the Si-doping characteristics and deep electron trap properties of AlInAs layers grown by MBE at different temperatures and doping levels. Thus, our investigation is mainly directed to lower doped AlInAs layers, aimed for Schottky barrier improvement in MSM (Metal Semiconductor Metal) photodiodes and AlInAs/GaInAs MeSFETs (Metal Semiconductor Field-Effect Transistors) as well as for device isolation.

2. SAMPLE FABRICATION

The MBE growth was performed in an ISA/Riber 32 P system. The AlInAs layers were grown simultaneously on untreated epi-ready (100)-oriented semi-insulating InP:Fe substrates aimed for Hall samples and on n^+-InP:S substrates for capacitance-voltage (C-V) and deep-level transient spectroscopy (DLTS) analyses. The growth rate was about 1 μm/h and the V/III BEP (Beam Equivalent Pressure) ratio was kept at 10. We investigated the Si-doping behaviour in the range of $5*10^{16}$ to $4*10^{17}$ cm^{-3}, and we have grown samples in the growth temperature range from 350 to 550° C. For Hall measurements we processed rectangular samples with alloyed In-ball contacts. The C-V and DLTS measurements were performed on Schottky diodes achieved by a Ti/Pt/Au e-beam metal evaporation in conjunction with a standard lift-off procedure. Previously, a slight etching of the semiconductor surface in a phosphoric acid based solution was applied to obtain reproducibly low-leakage current devices.

3. MEASUREMENTS

From **room temperature Hall measurements** the reduction of the free carrier concentration for lower growth temperatures is obvious, in agreement with data of Higuchi et al. (1991). For comparison, we measured the chemical Si-concentration incorporated into these layers by SIMS (Secondary Ion Mass Spectroscopy) and found a nearly T_g-independent concentration. As shown in Fig. 1, samples with a Si-doping level of $1.2*10^{17}$ cm^{-3} exhibit a room temperature free carrier concentration down to a value of $5*10^{16}$ cm^{-3} for T_g = 350° C grown layer.

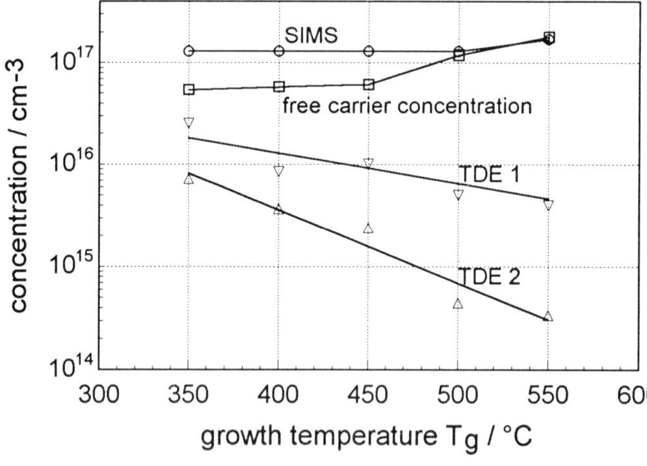

Fig. 1. Si-doping level (measured by SIMS), room temperature free carrier concentration (measured by Hall), and the trap concentration of the two main deep levels TDE 1 and TDE 2 (measured by DLTS) in AlInAs layers versus MBE growth temperature

In order to get more detailed knowledge about the Si-doping behaviour (e.g. dopant ionisation) we performed **temperature dependent Hall measurements** in the range of 77 to 350 K on samples with different doping concentrations. For a growth temperature of 350° C the results are given by the full lines in Fig. 2. As seen, all layers exhibit only a slight temperature dependence of the free carrier concentration. This fact clearly proves the shallow donor character of Si as dopant in AlInAs, also in low-T_g grown layers, and an approximately 100% ionisation of the respective donor atoms at all temperatures investigated. However, a distinct difference between the chemical Si and measured carrier concentration is noted. This difference vanishes for samples grown at $T_g \geq 500°$ C as shown by Hoenow et al. (1992).

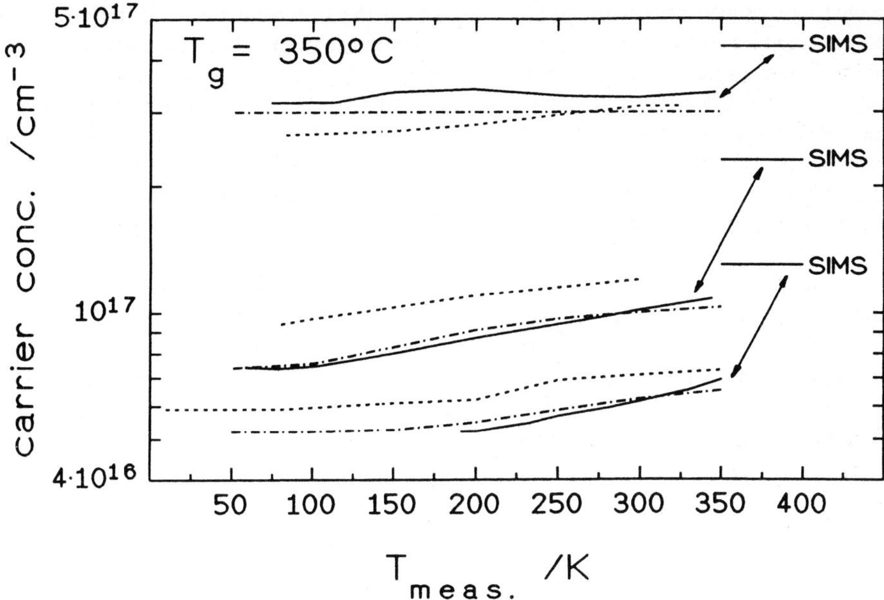

Fig. 2. Free carrier concentration in AlInAs:Si ($T_g = 350°$ C) obtained from temperature resolved Hall measurements (full lines) compared with C-V derived effective dopant concentrations (dashed lines) and Hall-simulating data (dashed-dotted lines) as a function of the measurement temperature $T_{meas.}$ for three different doping levels verified by SIMS

To investigate the low temperature net effective dopant concentration (N_D-N_A), we have made **temperature resolved C-V measurements**. Fig. 3 shows a typical set of C-V curves for temperatures from 10 to 325 K. The samples were grown at 350° C and the Si-concentrations measured by SIMS were 1.3 and 4.3*10^{17} cm^{-3}. No carrier freeze-out was detected by

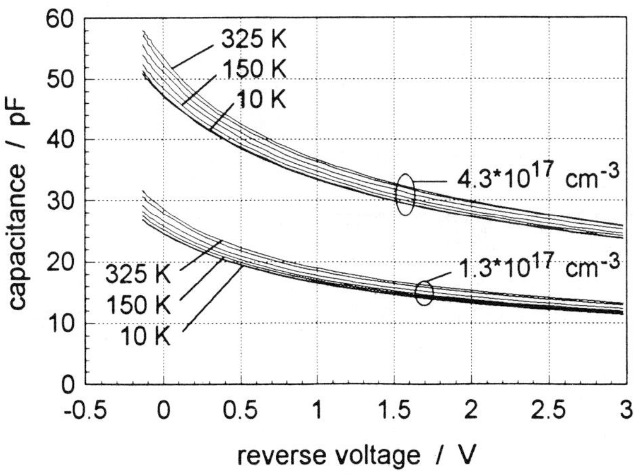

Fig. 3. C-V curves of samples grown at 350° C with different Si-doping concentrations; curves were measured at 10, 80, 150, 200, 250, 300, and 325 K

the related capacitance behaviour down to 10 K. On the one hand the C-V derived total ionised net dopant concentration is slightly larger than the corresponding free carrier concentration for the definitely non-degenerate samples (see Fig. 2), but on the other hand an amount of 6...10*10^{16} cm^{-3} is still missing compared to the SIMS detected chemical Si-concentrations in all samples (T_g=350° C).

In order to resolve this apparent discrepancy we analysed the properties of deep levels in the AlInAs layers by **DLTS measurements**. The existance of two dominant electron traps in AlInAs, called TDE 1 and TDE 2, were already reported by Hoenow et al. (1992). The trap parameters vary with growth temperature. Especially, the concentration of the levels increases with decreasing growth temperature (Fig. 1). Compared to layers grown at 500 °C, in low-T_g (350° C) grown samples the trap concentration is higher by one order of magnitude (space-charge edge correction and peak-broadening was considered). However, the trap concentration (acceptor type assumed) in all layers is still too low to explain quantitatively the full discrepancy between measured free carrier and Si-doping concentration.

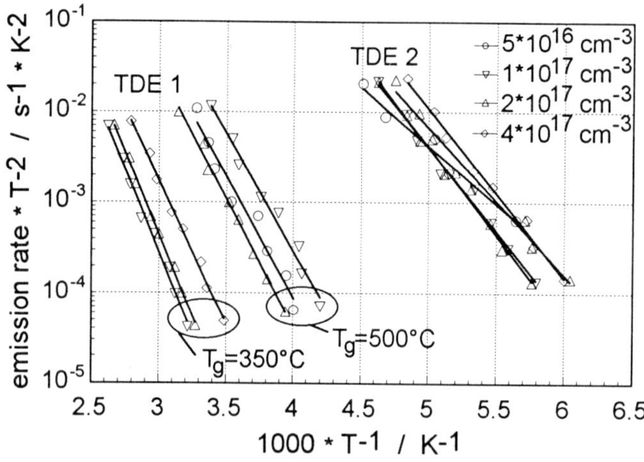

Fig. 4. DLTS Arrhenius plot for AlInAs:Si samples grown at different growth temperatures comprising different Si-doping concentrations

The Arrhenius plot shows a clear dependence of the emission behaviour of TDE 1 on the growth temperature, but no significant influence of the TDE 2 properties. These results are in agreement with data from Higuchi et al. (1991), which described a peak temperature variation with growth temperature for a trap level, which is comparable to TDE 1.
The variation of the Si-doping concentration leads only to a small variation in activation energy and capture cross section of both trap levels TDE 1 and TDE 2. The trap concentration increases slightly with increasing Si-doping concentration.

4. DISCUSSION

As described above, especially in low-temperature (T_g=350° C) grown MBE AlInAs a distinct reduction of the free carrier concentration compared to the SIMS-measured Si incorporated concentration is observed. In AlInAs the measured carrier concentration should equal the chemical (Si) donor concentration, if Si exhibits a shallow donor energy level. This was verified e.g. by Passenberg et al. (1993), which show a well behaved Arrhenius activation of electron concentration versus Si cell temperature up to a carrier concentration exceeding the effective density of states by more than one order of magnitude. In our opinion, the total carrier deficiency observed in lower doped AlInAs is attributed to two different mechanisms.

For addressing the first order effect, Hoenow et al. (1992) have shown that the amount of carrier reduction depends on the decay of the growth temperature (referenced to T_g=500° C). Thus, it is proposed that the observed carrier deficit is related mainly to the compensating effect of an assumed deep acceptor, which reduces the electron concentration for a constant amount, independently from the measurement temperature. The energetic depth of the acceptor is assumed in the lower half of the AlInAs bandgap, because its charge state does not change with temperature in the range investigated. The relation to the DLTS detected level TDE 1 can actually not be assured due to a lack of absolute trap concentration by a reasonable amount. The deep acceptor concentration depends on the MBE growth temperature. This effect may be related to bulk AlInAs lattice imperfections.

Two additional experimentally observed findings of second order will be explained in the following way. This work established that the DLTS-measured trap concentration (acceptor-type of majority carrier trap assumed) increases slightly with increasing Si-concentration at fixed growth temperatures. Furthermore, the free carrier concentration increases moderately with the measurement temperature, see Fig. 2. The latter finding will not be related to any varying partial donor ionisation, because the Hall-measured electron concentration tends to saturate towards low temperature (77 K) and the C-V curves stay constant, even for temperatures down to 10 K. Consequently, the temperature dependence of the carrier concentration is explained by the action of a partially ionised compensating acceptor, just below the conduction band, which changes its occupation with temperature, in agreement with Massies et al. (1983).

Therefore, the remaining smaller and *temperature* dependent residual carrier deficit is proposed to be related to an acceptor-type trap concentration, which depends slightly on the Si-doping concentration at fixed growth temperature. This acceptor-like trap captures electrons especially at low-temperature and re-emits them partly at elevated temperature into the conduction band. Thus, the observed slightly increasing carrier concentration with increasing measurement temperature can be explained, which otherwise can not be understood in III/V-material, where the shallow donor level merges with the tail of the density of states distribution of the conduction band.

The energetic depth of the re-chargeable acceptor is estimated from carrier concentration modelling, as shown within the dashed-dotted lines in Fig. 2. Therefore, the standard neutrality condition for semiconductors (see Sze 1981) comprising Fermi-Dirac statistics for all dopant occupations was extended by the charge contributions (p, N_A^-) of an additional acceptor and solved numerically for various temperatures. An energetic depth of the shallow acceptor of around 70 meV below the conduction band resulted, to explain the observed carrier concentration change with temperature. The concentration of this shallow acceptor for

increasing Si-doping levels is well related to the concentration of the trap TDE 2, determined by DLTS.
Thus, it is assumed that carrier capture into the level TDE 2 is the origin for the observed moderate carrier concentration change with temperature. Further work has to be done to investigate the actual discrepancy between the trap activation energy determined by DLTS in the order of 400 meV compared to the results of our carrier concentration modelling, where only about 70 meV must be assumed. A possible explanation may be given by considering the temperature dependence of the capture cross section of TDE 2. Measurements on AlGaInAs with low gallium content show a quenching of the DLTS emission peak for the equivalent trap level under reduced filling pulse widths of less than 150 μs in contrast to the TDE 1 related emission peak (Schramm 1992). Thus, a quite large energetic shift of the TDE 2 trap position towards the conduction band seems likely. A similar mechanism is well known for the DX-center energetic position in AlGaAs, where the capture barrier is in the order of 250 meV (Mooney 1990).

5. CONCLUSION

The Si-doping characteristics in AlInAs were investigated under variation of the MBE growth temperature and chemical Si-concentration. The measured free electron concentration is reduced by an amount of 0.6 and $1.3*10^{17}$ cm^{-3} for Si-concentrations of 1.3 and $4.3*10^{17}$ cm^{-3}, respectively. Our investigations suggest the existence of two additional T_g- and Si-doping related acceptor-type compensation mechanisms. We developed a model to simulate the free carrier concentration, which includes besides the fully ionised Si donor a *T_g-induced deep acceptor* below midgap and a *(shallow) re-chargeable acceptor* near the conduction band edge, the latter of which may be related to the electron trap TDE 2 in AlInAs. This leads to a good agreement with the measured Hall data for all growth temperatures.

6. ACKNOWLEDGEMENT

We thank RTG Mikroanalyse GmbH, Berlin for providing SIMS measurements on the silicon contents in AlInAs. Partial funding of this work by Deutsche Bundespost TELEKOM under the OEIC project is greatfully acknowledged.

7. REFERENCES

Higuchi M., Ishikawa T., Imanishi K., and Kondo K. 1991 *J. Vac. Sci. Technol.* **B9**, pp 2802-4

Hoenow H., Bach H.-G., Böttcher J., Gueissaz F., Künzel H., Scheffer F., and Schramm C. 1992 *Proc. 4th Int. Conf. on InP and Rel. Mat. (IPRM)*, pp 136-9

Künzel H., Passenberg W., Böttcher J., and Heedt C. 1991 *Microelectronic Engineering* **15**, pp 569-72

Massies J., Rochette J.F., Etienne P., Delescluse P., Huber A.M., and Chevrier J. 1983 *J. Crystal Growth* **64**, pp 101-7

Mooney P. 1990 *J. Appl. Phys.* **67**, pp R1-26

Passenberg W., Bach H.-G., Böttcher J., and Künzel H. 1993 *J. Crystal Growth* **127**, pp 716-9

Schramm C. 1992 *Doctorate thesis (in german) TU Berlin (D83)*, pp 61-2

Sze S. M. 1981 *Physics of Semiconductor Devices* 2nd Ed. Wiley & Sons, pp 22-7

Behaviour of misfit dislocations in modulus-modulated layers of GaAs/$In_xGa_{1-x}As$/GaAs on Si

H.Katahama, K.Asai, Y.Shiba and K.Kamei
Advanced Technology Research Laboratories, Sumitomo Metal Industries, Ltd.,
1-8 Fuso-cho, Amagasaki, Hyogo 660, Japan

ABSTRACT: We investigate the effect of the insertion of a single $In_{0.1}Ga_{0.9}As$ layer into GaAs grown on Si on the dislocation reduction. The inserted layer thicker than 0.25μm reduces the threading dislocations. This is mainly caused by the sweeping-out effect due to the misfit stress. The misoriented growth is observed at the interfaces and the behavior of the misoriented growth differs at the two GaAs/InGaAs interfaces, which indicates that the lower interface plays an important role to reduce the threading dislocations.

1. INTRODUCTION

Heteroepitaxial growth of GaAs on Si substrates has been studied because it will lead to monolithic integration of GaAs- and Si-based devices. However, a large number of threading dislocations are induced by the difference in the lattice constants and the thermal expansion coefficients between GaAs and Si. To reduce the threading dislocations, insertion of layers into GaAs such as strained-layer supperlattices (SLSs) is very effective (Watanabe et al 1988) because the inserted layers can affect the motion of the threading dislocations. This dislocation reduction may be caused by two different effects; one is a sweeping-out effect which is mainly induced by the strains in the inserted layers. The other is a blocking effect due to the difference in the shear modulus (μ) between GaAs and the inserted layers. To clarify which effect is dominant to reduce the dislocations is very important to design more effective filtering layers. Tamura and Hashimoto (1992) have studied the dislocation reduction effects in modulus modulated structures with thin Si inserted layers, in which the blocking effect is dominant. In this case, the shear modulus of the Si layer is larger than that of GaAs. However, the modulus-modulation structure with smaller shear-modulus than GaAs has not been investigated well and InGaAs can be used for this modulus-modulated structure. If a single layer is inserted, we can clarify the contribution of the two effects to the dislocation reduction because of its simple structure. In this work, we have grown the modulus-modulated structures of GaAs/InGaAs/GaAs on Si and studied the behavior of dislocations at the interfaces.

2. EXPERIMENTAL

Heteroepitaxial growth of GaAs on (100) Si substrates (3° off towards [011] direction) was carried out by metalorganic chemical vapor deposition (MOCVD) method at 76Torr. Trimethylgallium (TMG), Trimethylindium (TMI) and AsH_3 were used as Ga, In and As sources, respectively. After an initial cleaning in H_2 gas at 1000°C, the first-step GaAs layers of 20nm were grown at 500°C and the second-step GaAs layers with a thick $In_xGa_{1-x}As$ layer were grown at 700°C. Figure 1 shows the structures of epitaxial layers with total thickness of 3μm, which consisted of three layers: Lower GaAs (GaAs1), inserted $In_xGa_{1-x}As$ with the thickness of h and upper GaAs (GaAs2). The In content of $In_xGa_{1-x}As$ was fixed to be 0.1 and, in this case, the modulus-modulation $(\mu_{GaAs} - \mu_{InGaAs}) / \mu_{GaAs}$ is about 4%. The position of inserted layers (d) was changed from 0 to 2μm. In addition, the thickness of $In_xGa_{1-x}As$ layers (h) was also changed from 0.05 to 0.5μm. The crystalline quality of these samples was characterized by a double-crystal X-ray diffraction method and transmission electron microscopy (TEM) at 400kV. The diffraction rocking curves were recorded in the vicinity of the (400) peak. The etch-pit density (EPD) was determined by Nomarski optical microscopy after etching in molten KOH at 350 °C.

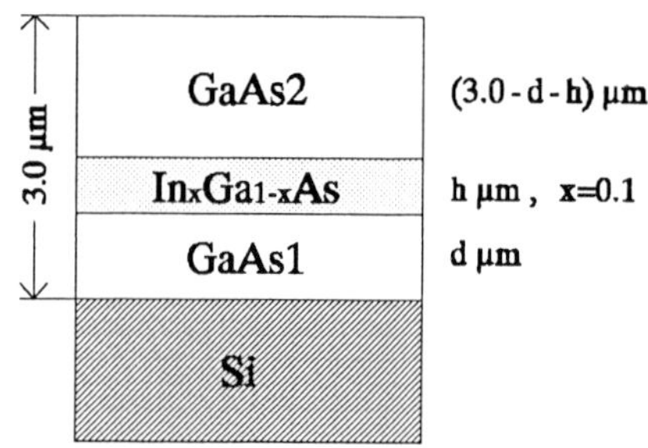

Fig.1 Structure of modulus-modulated layers of GaAs/ $In_{0.1}Ga_{0.9}As$/ GaAs with total thickness of 3μm on Si.

3. RESULTS AND DISCUSSION

3.1 Effect of single layer insertion

We have investigated the insertion-effect of a thick single $In_{0.1}Ga_{0.9}As$ layer on the dislocation reduction. In the X-ray diffraction patterns, three (400) peaks are separately observed from three layers: GaAs1, InGaAs and GaAs2, as shown in Fig.2. In these experiments, the thickness of the inserted $In_{0.1}Ga_{0.9}As$ layers is 0.5μm, which is thicker than the critical thickness. Therefore, the strains in the InGaAs layers were considerably relaxed and the diffraction peak from InGaAs can be observed. On the other hand, the

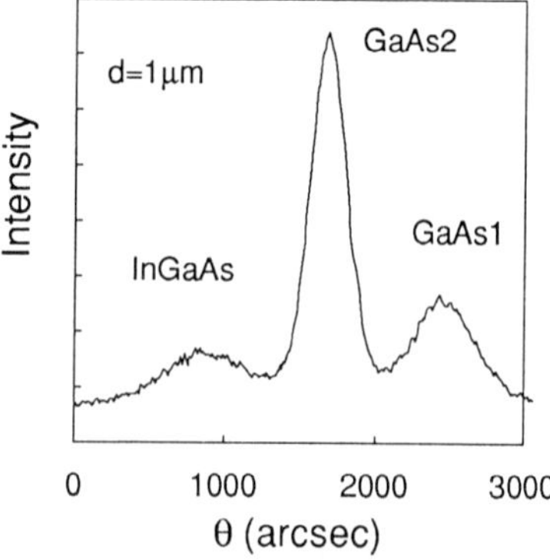

Fig.2 X-ray diffraction of GaAs/ InGaAs/ GaAs on Si (d=1μm, h=0.5μm).

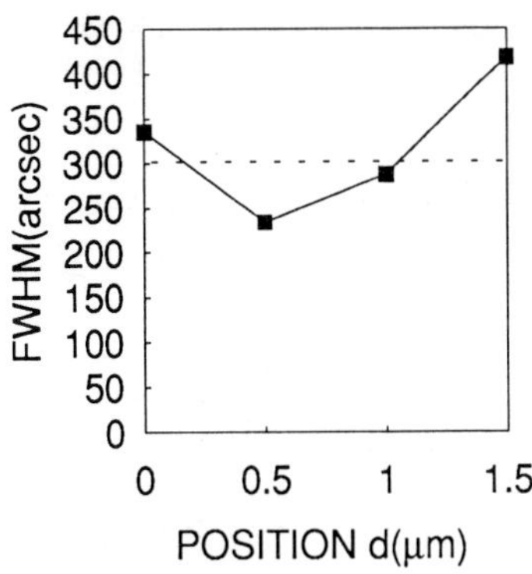

Fig.3 FWHM of X-ray diffraction from upper GaAs2 layers as a function of inserted position of InGaAs layers. Broken line represents FWHM without InGaAs layer.

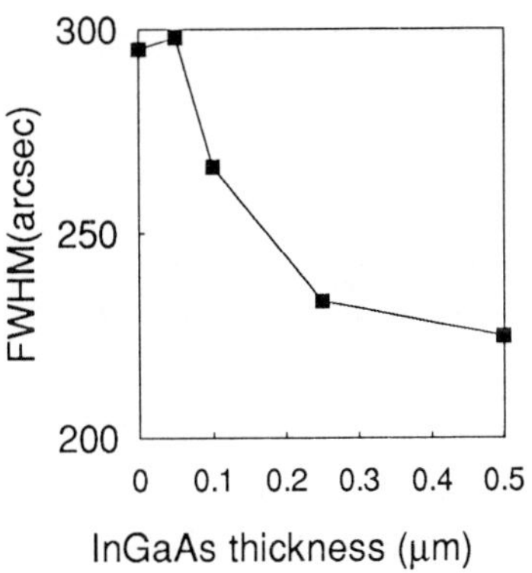

Fig.4 FWHM of X-ray diffraction from upper GaAs2 layers as a function of inserted InGaAs thickness.

separation of the two GaAs peaks is caused by the misorientation of the growth directions, as discussed later. Figure 3 shows the full width at half maximum (FWHM) of (400) X-ray diffraction peaks from GaAs2 layers, as a function of d. The X-ray FWHM in the samples without InGaAs layer was 295 arc sec. On the other hand, the minimum value of 230 arc sec has been obtained in the sample with the $In_{0.1}Ga_{0.9}As$ layer inserted at 0.5μm from the GaAs/Si interface. These results indicate that even the single layer of 0.5μm $In_{0.1}Ga_{0.9}As$ is effective to reduce the threading dislocations. In fact, the KOH etching reveals that the EPD is reduced from 4 $x10^7$ to 1~2 $x10^7 cm^{-2}$.

Figure 4 shows the dependence of the X-ray FWHM on the inserted-layer thickness from 0.05 to 0.5μm. The FWHM in the sample with 0.05μm $In_{0.1}Ga_{0.9}As$ layers is comparable to that without the inserted layers. According to the mechanical equilibrium model (Matthews and Blakeslee 1974), the critical thickness (hc) is about 0.08μm in the $In_{0.1}Ga_{0.9}As$/GaAs system. Therefore, this result indicates that the single layer below hc does not reduce the threading dislocations. With increasing h, the FWHM is gradually decreased and it is saturated in the samples thicker than 0.25 μm .

In general, the reduction of threading dislocations by the layer insertion is caused by the blocking and sweeping-out effects, which are shown schematically in Fig. 5. The blocking effect is caused by the difference in the shear modulus (μ) between InGaAs and GaAs. The dislocation energy in InGaAs is lower than that in GaAs, because it is proportional to μb^2, where b is Burgers vector; therefore, it is expected that the dislocations are confined in the InGaAs layers. In case of the modulus-modulated multilayer (MMML) with a period of h_M,

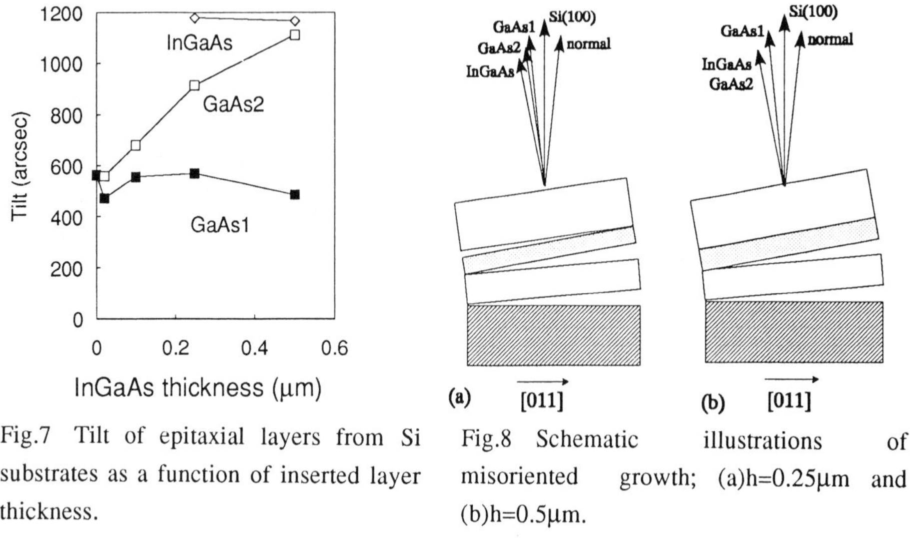

Fig.7 Tilt of epitaxial layers from Si substrates as a function of inserted layer thickness.

Fig.8 Schematic illustrations of misoriented growth; (a)h=0.25μm and (b)h=0.5μm.

will act as a new origin of the threading dislocations, the lower limit of the sweeping-out effect exits. To reduce the dislocations more effectively, the enhancement of the blocking effect is necessary by the multiple InGaAs layers. The insertion of six InGaAs layers with h=0.2μm has reduced the EPD to 3×10^6 cm^{-2}.

3.2 Misoriented growth

The misoriented growth has been reported in the mismatched heteroepitaxy on a vicinal substrates, including InGaAs/GaAs and GaAs/Si (Ghandhi and Ayers 1988), and it may relate to the formation of the misfit dislocations in the epitaxial layers (Ayers et al 1991). As shown in Fig.2, the three peaks have been observed in the X-ray diffraction patterns. This result indicates that the growth directions of the epitaxial layers are slightly misoriented. Since the shift of the peaks of the epitaxial layers from the (400) Si peak shows the sinusoidal variation with azimuthal angle and its amplitude corresponds to the tilted angle (Nozawa and Horikoshi 1993), we can determine the tilted angle of the epitaxial layers from the Si substrates by changing the incident X-ray directions.

The tilted angle from Si substrates is shown in Fig.7 in the samples with d=0.5μm, as a function of h. The two directions for the tilt have been reported (Matyi et al 1988); one is away from the surface normal, which is defined as positive, and the other is toward the surface normal (negative). The tilt observed in this work is away from the surface normal. This behavior is not consistent with the prediction of the preferential glide model (Ayers et al 1991) that the negative tilt is induced if the lattice constant of the epilayer is larger than that of the substrates. In this model, the preferential glide of the 60° dislocations nucleated at the surface results in the misorientation. The positive tilt of GaAs1 layers from Si

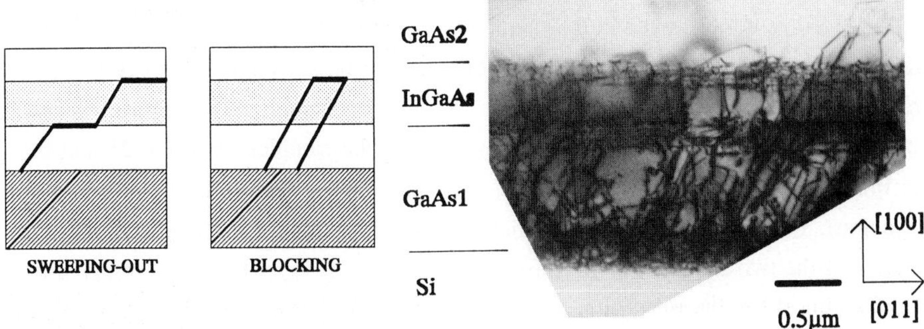

Fig.5 Dislocation reduction effects of inserted layers: Sweeping-out and blocking effects.

Fig.6 Cross sectional TEM image of GaAs/ InGaAs/ GaAs on Si (h=0.5μm, d=1μm).

the impeding stress, which blocks the dislocation motion, has been derived (Maeda et al 1993);

$$\tau_\mu = \frac{b}{4h_M}\Delta\mu\ \ln\left(\frac{4h_M}{b}\right). \tag{1}$$

Using this equation, the impeding stress is calculated to be 1.7×10^8 and 2.4×10^7 dyn/cm^2 for $h_M = 0.05$ and 0.5 μm, respectively. The impeding stress in case of the single layer insertion is considered to be smaller than that in the MMML. Next, we consider the sweeping-out effect due to the difference in the lattice constant between InGaAs and GaAs. The misfit stress in the coherently strained InGaAs layer is 4.1×10^9 dyn/cm^2. The mechanical equilibrium model predicts that the misfit strain has not relaxed completely even for h>hc; if h=0.5μm, the estimated residual stress is 8.4×10^8 dyn/cm^2. This value is larger than the impeding stress due to the blocking effect, which suggests that the dislocation reduction by the single layer insertion is mainly caused by the sweeping-out effect due to the misfit stress. During the relaxation process of the misfit stress, the misfit-dislocation network should be formed at the interfaces between InGaAs and GaAs. This network can enhance the probability of the annihilation and the coalescence of the threading dislocations; as a result, more reduction of the threading dislocations has been achieved in the sample with more relaxed InGaAs layers.

Figure 6 shows the TEM image in the $[01\bar{1}]$ cross section of the sample with h=0.5μm and d=1μm. The misfit dislocation network has been formed at both interfaces between InGaAs and GaAs. The running direction of the threading dislocations is changed at the InGaAs/GaAs1 interface. The dislocations running along the [100] direction in the InGaAs layers are observed. In addition, some dislocations have not affected by the interfaces and threaded into the GaAs2 layers. Since the dislocation network at GaAs2/InGaAs interface substrates is about 500 arc sec and does not depend on h. The X-ray diffraction peaks from InGaAs layers is not observed in the samples with h<0.1μm, because of their weak intensity. Taking it into consideration that the tilt of GaAs2 is as large as that of GaAs1, it seems that the misoriented growth does not occur at InGaAs/GaAs interfaces. Figure 8 shows the schematic illustrations of the misoriented growth in the samples with h=0.25 and 0.5 μm. The InGaAs layers reveals the further positive tilt from the GaAs1 layers in both samples. In contrast, the behavior of the misoriented growth at the InGaAs/GaAs2 interfaces is different in the two samples; the negative tilt of the GaAs2 layer has been observed in case of h=0.25μm, while the tilt of the GaAs2 layer from the InGaAs layer is very small in case of h=0.5μm. Considering that the dislocation-reduction effect is comparable in the two samples as shown in Fig. 4, the lower interface plays an important role to reduce the threading dislocations rather than the upper interfaces. These results suggest the formation mechanism of the dislocation networks; the dislocation network at the lower interface is originated by the threading dislocations in the GaAs1 layers, while the network at the upper interface is formed by the glide of the 60° dislocations nucleated at the growing surface.

4. SUMMARY

The reduction of the threading dislocations in the modulus-modulated structure of GaAs/InGaAs/GaAs grown on Si has been investigated. The insertion of the single $In_{0.1}Ga_{0.9}As$ layer thicker than 0.25μm improves the crystalline quality. This is mainly caused by the sweeping-out effect due to the misfit stress and the dislocation network is formed at the GaAs/InGaAs interfaces during the relaxation process. The misoriented growth is observed at the two GaAs/InGaAs interfaces. The different behavior of the misoriented growth indicates that the lower interface plays an important role to reduce the threading dislocations rather than the upper interfaces.

REFERENCE

Ayers J E, Ghandhi S K and Schowalter L J 1991 J. Cryst. Growth 113 430

Ghandhi S K and Ayers J E 1988 Appl. Phys. Lett. 53 1204

Maeda K, Yamashita Y, Fujita K, Fukatsu S, Suzuki K, Mera Y and Shiraki Y 1993 J. Cryst. Growth 127 451

Matthews J W and Blakeslee A E 1974 J. Cryst. Growth 27 118

Matyi R J, Lee J W and Schaake H F 1988 J. Electronic Materials 17 87

Nozawa K and Horikoshi Y 1993 Jpn. J. Appl. Phys. 32 626

Tamura M and Hashimoto A 1992 J. Electrochem. Soc. 139 865

Watanabe Y, Kadota Y, Okamoto H, Seki M and Omachi Y 1988 J. Cryst. Growth 93 459

Recognition of point defects and clusters and their distribution in semi-insulating GaAs

J.Vaitkus, V.Kažukauskas, R.Kiliulis, J.Storasta

Semiconductor Physics Department, Vilnius University

Saulėtekio al.9, 2054 Vilnius, Lithuania

ABSTRACT. Electric field dependence of thermally stimulated conductivity (TSC) was used to identify the point defects and clusters in si-GaAs. The transition of EL2 from normal to metastable state enabled to scan the electron and hole traps independently. The transient photo-Hall effect was used to determine the effective volume of conductivity channels of n- and p-type as well as the bipolar TSC state. The irregular behavior and modulation by low frequency oscillations of TSC have been observed. The effects are influenced by rearrangement of drift and recombination barriers.

1. INTRODUCTION

The influence of percolation effects on conductivity in si-GaAs was discussed in the works of Pistoulet et al (1983) and of Ferre and Farvacque (1990) and a model including separate electron and hole conduction channels was defined by Vaitkus et al (1992). As the si-GaAs substrate influence on properties of layers has been analyzed a more detailed investigation of carriers transport and defects in si-GaAs is still actual. In this work the previous model (Vaitkus et al 1992) has been developed and the attention to defect distribution and their parameters has been paid.

2. RESULTS AND DISCUSSION

Crystals grown by the high pressure LEC technique have been investigated. The average dislocation density as measured by etch-pits counting is about 10^4 cm^{-2}. The different etch-pits allow to propose the dimensions of impurity tubes around dislocations. The same thing was possible to evaluate from impurity distribution by ESXA mapping. If one paid attention on the main impurity of carbon the parameters of tube were found as follows: radius is equal aproximately to 300 nm and the cross section of "window" was about 1.5 μm. The distribution of carbon (and other defects) along the tube which was inhomogeneous and the concentration change was found up to 7-8 times.

The schematic picture of the sample which follows also from the results of Vaitkus et al work (1992) is given in Fig.1a,b. The p-type conductivity dominates in the tubes and an electron conductivity around them. If the barriers along the tubes are significant (a shadow area in the Fig.1b) the non-equilibrium holes are localized, i.e. the p-type conductivity is excluded. The

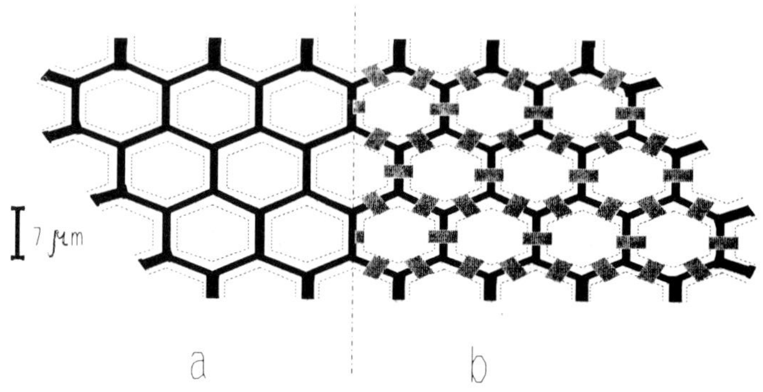

Fig. 1. The cellular structure of the dislocation distribution in the si-GaAs crystal. The gettering of residual impurities and native defects at dislocations induced local modifications of the conductivity: a) p-type is shown by the dashed lines surrounding areas; b) the inhomogeneity of impurity distribution is shown as a shadow area.

comparison of maximal experimental electron mobility and predicted theoretical value for this type of GaAs allows to calculate (Voronkov et al 1971) the volume of n-type material. For the sample the results of which are given in this paper the p-type channels fill the 3-5% of the sample. When the n-type conductivity is quenched the p-type conductivity in the tubes increases. Then the model of paralel conductors can be used (Look 1990) and the cross section of p-type conductivity tube can be estimated. If the TSHM in case of the quenched photoconductivity was measured the cross section of p-type conductivity tubes for the same sample is found appropriate equal to 20% of all the cross section of the sample. As it is easy to show all data support the predicted model. According to which, the p-type conductivity is defined by drift barriers due to inhomogeneous impurity distribution and the n-type conductivity is defined by barriers due to overlapping of space charge regions caused by the impurities around dislocations.

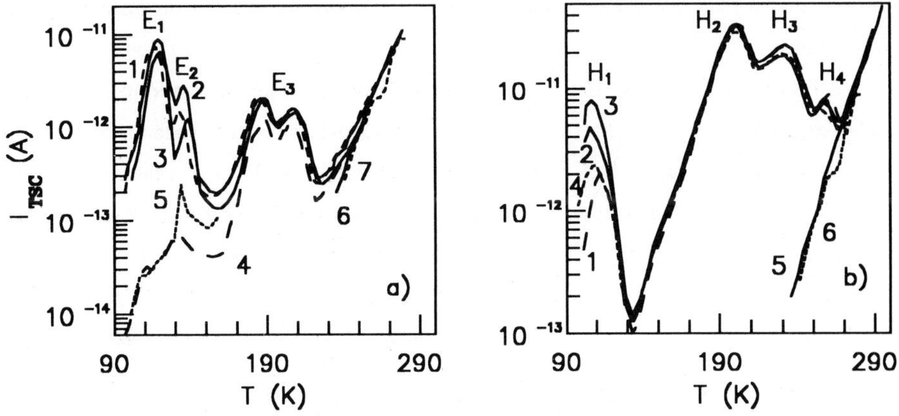

Fig. 2. TSC dependence (normalised) on electric field (V/cm) applied during the temperature scan: a) measured at normal EL2 state, 1-2.9, 2-29, 3-290, 4 and 5-820, 6 and 7-dark current corresponding 2.9 and 820V/cm; b) measured after the EL2 has been quenched, 1-2.9, 2-29, 3-290, 4-820, 5 and 6-dark current corresponding 2.9 and 820V/cm.

The indentification of traps and their distribution is possible from TSC spectra. TSC spectra measured at different voltages and normalized to the unit electric field in cases of normal and quenched photoconductivity (Fig.2). In the case of quenched photoconductivity with the p-type prevailing, the two peaks in TSC spectra labeled H_1 and H_4 are sensitive to electric field, i.e., they are cluster type defects surrounded by a barrier. From the dependence of the initial part of H_1 peak on the electric field the barrier height equal to approximately 70meV is found, and the defect activation energy is $\Delta E_{H1} = E_{HI} - E_V = 0.12$eV. The activation energies of H_2, H_3 and H_4 levels have been found to be equal to 0.40, 0.45, 0.50eV respectively. The H_2 and H_3 are associated with the point defects, because the small decrease of current at a higher electric field could be connected with an effective mobility change. In case of the unquenched sample (initial n-type conductivity) the influence of the electric field on TSC is different (Fig.2a). The E_1 level (E_C-E_1=0.19eV) and E_3 level (E_C-E_3 =0.34eV), the double structure of which caused by the thermal quenching of lifetime due to excitation of H_2 level (Vaitkus et al 1992) are independent on the electric field up to high field. It shows that E_1 and E_3 levels are associated with the point defects. Level E_C-E_2=0.29eV is influenced by the electric field and it shows the defect is a cluster-type. The significant decrease of E_1 peak and of TSC at higher temperature is possible to explain by the influence of hole injection from p-type channel to n-type channel. The decrease of the lifetime of carriers caused by the free holes is similar to thermal quenching effect.

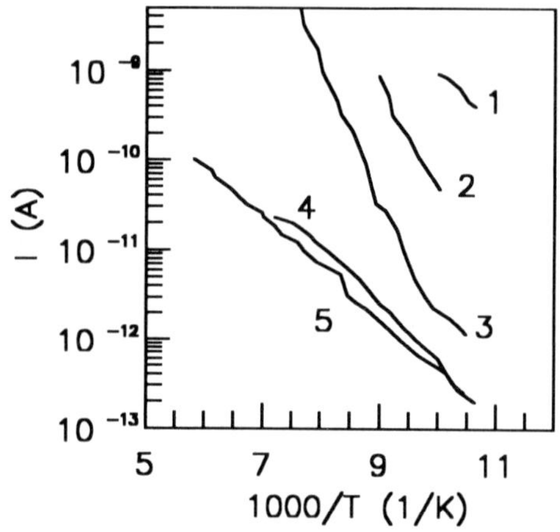

Fig. 3. Arrhenius plot of thermal trial TSC. Activation energies in eV: 1-0.11, 2.-0.20, 3-0.29, 4 and 5-0.12.

As the result of barriers dependent on the electric field and on the defect reconstruction the TSC instabilities have been seen at low temperature in the region of EL2 modification and E_2 TSC peak region. The instabilities have been observed also at high temperature if the H_4 level is excited. The instabilities are observed at higher temperature also, but their nature can be different. The TSC instabilities were observed also previously (Kaminska et al 1982), but details of this effect were not given. A chaotic behavior of instabilities and their ordering was reported in our previous paper (Vaitkus et al 1989) and the nature of instability was proposed. These results confirmed the model. There are more possibilities for the appearance of chaos, i.e., if the injection of carriers from n- to p-type regions is included (see the results below).

The parallel connected n-type and p-type channels sometimes gave unusual effects. Fig.3 presents the TSC results for another sample when temperature was increased until the TSC peak value and then the sample was cooled down. The next increase of temperature has started to release carriers from the next level. The procedure was repeated and the activation energies of peaks have been found: 0.12eV, 0.20eV, 0.29eV and the last one is 0.12eV again. This example shows, that a single trap H_1 (0.12eV) and the thermal quenching of lifetime causes the two peaks in TSC spectra in p-type conductivity part of the sample and at the same time between these peaks E_1 (0.20eV) and E_2 (0.29eV) traps cause TSC peaks in n-type channel. This result emphasizes the great influence of the both sign carriers.

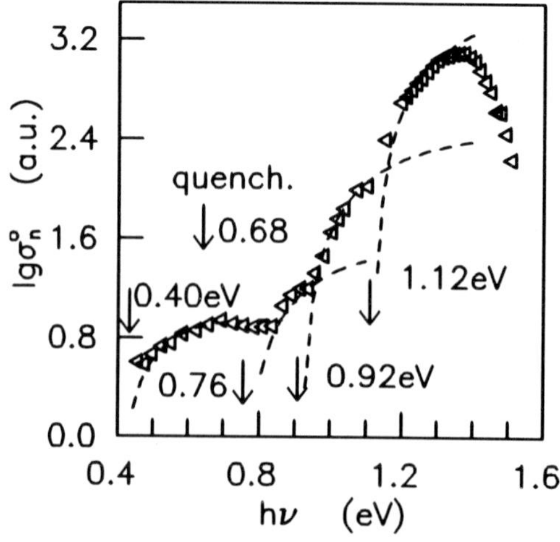

Fig. 4. Photoionization cross section spectrum.

The photo-Hall effect mobility dependence on temperature in thermo-

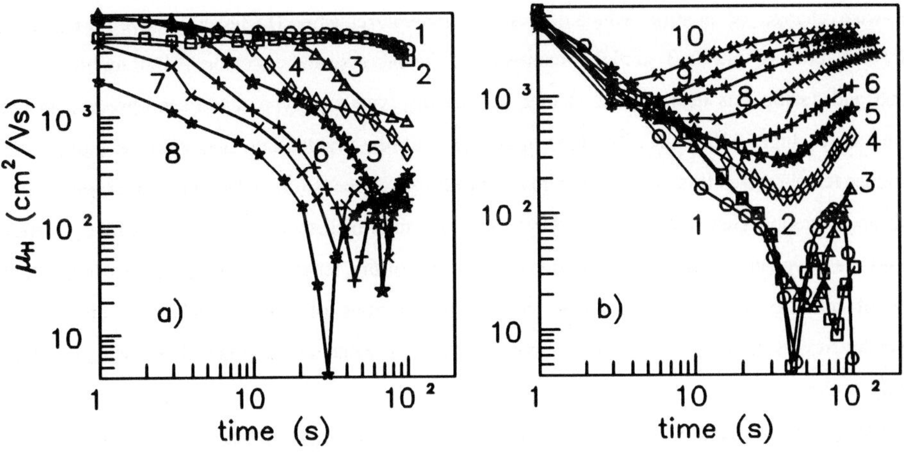

Fig. 5. Absolute values of Hall mobility transients measured after 1s of excitation with 1.08eV photons at temperatures (K): a) 1-162, 2-181, 3-193, 4-202, 5-211, 6-221, 7-227, 8-246; b) 1-257, 2-259, 3-261, 4-263, 5-267, 6-271, 7-276, 8-281, 9-283, 10-288.

stimulated regime and under constant illumination has different stages: the decrease at about 140K and at about 210K and the increase at higher temperature. The change of mobility in all periods depends differently on the excitation intensity. The details of this dependence will be described elsewhere but here it is important to pay attention to the fact of the influence of the electron traps on thermostimulated mobility and to possible relation of mobility quenching with antisite defects transformation (Weber et al 1983, Wang et al 1987). Looking at the details of high temperature TSC peaks and recombination processes in n-type and p-type channels, the two levels of EL2 family are available. The spectral dependence of photoionization cross section given in Fig.4 shows the two levels which could be attributed to EL2 type levels: E_a-E_V=0.76eV and E_b-E_V=0.92 eV that correspond to EL2a (E_C-0.83eV) and EL2b (E_C-0.76eV) (Wang et al 1987).

The trap filling influence on the temperature dependence of mobility allowed to investigate the transient behavior of mobility at different temperatures corresponding to characteristic regions of TSC dependence. Fig.5a illustrates the dependence of the full mobility transient after excitation with 1.15 μm wavelength light pulse and Fig.5b shows the mobility vs time when the change of it was of the main importance at different temperatures. The mobility transient dependence has a few features comparable with TSC spectra: 1) the step-wise change due to H_2 level emptying, and the value of the change of the product ΔSN- of carrier scattering cross section S to charged centers concentration N change equal to 4000cm^{-1} (it corresponds to the change of the charged centers concentration approximately 10^{15} -10^{16}cm^{-3} is rather available; 2) also the step-wise change of mobility which occurs at temperatures when H_3 level is active (in this case ΔSN was about 10^4 cm^{-1}); 3) at higher temperatures the decrease and restoring of Hall mobility was found and during this decrease the change of mobility type to p-type has been observed in the temperature range 215-260K. This type of change could be divided into two maxima: a dominant at higher temperatures (the time of maximum appearance depends on the temperature exponentially with an activation energy equal to 0.62eV) and an intermediate temperature maximum which causes the p-type conductivity. Due to the coincidence of the temperature range where the certain type of mobility change is dominant and the peculiarities of TSC and TSHM temperature dependencies are similar, it is possible to conclude that these maxima are interrelated with both EL2 family centers.

The presented results show the complicity of electrical properties of si-GaAs and also demonstrate the possibilities and sensitivity to sample parameters of not yet widely known methods of thermally stimulated Hall effect and transient photo-Hall effect. It has been found that point defects are dominant in samples but the main influence of the macroscopic n- and p-type conductivity channels on transport properties was shown.

ACKNOWLEDGEMENT

This work was supported , in part, by Foundation for Promotion of Material Science and Technology of Japan.

REFERENCES

Baraff GA and Schlüter M 1986 Phys. Rev. B **33** p 7346
Ferré D and Farvacque JL 1990 Rev. Phys. Appl. **25** pp 323-32
Kaminska M, Parsev JM, Lagowski J and Gatos HC 1982 Appl. Phys. Lett. **41** p 989
Look DC 1990 J. Electrochem. Soc. **137**pp 260-66
Pistoulet B, Girard P and Hamamdjan G 1983 J.Appl.Phys. **54** p 5176
Vaitkus J, Baubinas R, Kažukauskas V, Kiliulis R and Storasta J 1992 Int. Phys. Conf. Ser. N 129 Chapter 6 pp 549-54
Vaitkus J, Kiliulis R-P and Storasta J 1989 Soviet Phys. Collection (Allerton Press Inc. NY) **27** pp 514-16
Voronkov VV, Voronkova GI and Iglytsin MI 1971 Sov. Phys. Semicond. **4** pp 1949-52
Wang WL, Li SS and Lee DH 1987 J.Electrochem. Soc. **133** pp 196-99
Weber ER and Schneider J 1983 Physica **116B** p 398

Inst. Phys. Conf. Ser. No 136: Chapter 11
Paper presented at the Int. Symp. GaAs and Related Compounds, Freiburg, 1993

Non DX like deep donor states in AlGaAs

M L Fille*, U Willke**, D K Maude**, J M Sallese*°, M Rabary**, J C Portal** and P Gibart*

* Laboratoire de Physique du Solide et de l'Energie Solaire -CNRS, Av. Bernard Gregory, Parc de Sophia Antipolis, 06560 Valbonne (France)

** Service National des Champs Intenses-CNRS, Av. des Martyrs 38042 Grenoble and Institut National des Sciences Appliquées, CNRS, 31077 Toulouse cedex (France)

ABSTRACT: non DX like donor levels are reported in Sn, Si and Te-doped AlGaAs. The levels which are occupied only after illumination at low temperature to persistently photoionize the DX centres, are investigated using hydrostatic pressure. The levels are in thermal equilibrium with the Γ-conduction band even at low temperatures but are higher in energy than the DX centres. They are nevertheless capable of pinning the Fermi level after illumination and represent a previously unsuspected mechanism which limits the persistent carrier concentration.

1. INTRODUCTION

DX centers in III-V semiconductors have been extensively studied since the pioneering work of Lang (1977) in the late seventies. The main feature which characterizes the DX centre is the occurrence of carrier freeze-out and persistent photoconductivity (PPC) at low temperatures. Illuminating the sample with an infrared light at low temperature induces an increase in the free carrier concentration which is persistent as long as the sample is not heated above a critical temperature which depends on the chemical nature of the dopant. At low temperature, recapture on the ground state of the DX centre is prevented by the large thermal barrier to capture. A subject of considerable controversy was the charge state of this level. Using pseudopotential calculations, Chadi and Chang (1988) have shown that the DX is a negative-U, large lattice relaxation centre which is negatively charged. In this model, for group IV dopants, the donor atom in its ground state captures two electrons in a bond breaking reaction. However, Yamagushi (1986) (see also Ohno and Yamagushi 1991), using a similar pseudopotential calculation found that the DX center is associated with a single electron small

° Present address: Ecole Polytechnique de Lausanne, Inst. de Micro et Optoelectronique, DP-EPFL, 1015 Ecublens-Lausanne (Switzerland)

lattice relaxation A1 anti-bonding state of the donor. Most of the experimental work which has been performed up to now support the negative U model (Gibart 1990, Fockele 1990, Mosser 1991, vBardeleben 1992a,b).

In this paper, we use transport measurements under hydrostatic pressure in AlGaAs alloys to show the coexistence of at least two levels related to the same donor species. The ground state level is the DX centre characterized by the PPC effect at low temperature. The excited state is seen only after illumination at low temperature when the Fermi level is sufficiently high. It is in thermal equilibrium with the Γ-conduction band even at low temperature and it is able to pin the Fermi level after illumination.

2. DOPING LEVEL LIMITATION IN AlGaAs

One of the most important problem linked to the presence of the DX centres in the AlGaAs alloys is the doping limitation it induces. When growing samples by MOVPE for example, with the same partial pressure of dopant in the reactor, the carrier concentration measured at 300K is found to be nearly constant from $x=0$ to $x=0.22$. For $x>0.22$, it decreases sharply because the DX centres enter and trap electrons. This is because the value of the carrier concentration measured at 300K does not take into account the electrons trapped onto the DX centres. Up to now the total number of electrically active donor atoms in the sample was found using the "PPC property" of the DX levels. At low temperature, the sample is illuminated with an infrared LED and the carrier concentration measured then corresponds to the total concentration of electrically active dopant (N_D-N_A) in the sample.

Hydrostatic pressure is a very useful tool for studying direct band gap AlGaAs alloys ($x<0.37$) because an increase of 1% of the aluminium percentage has nearly the same effect on the band structure as increasing of hydrostatic pressure by 0.1 GPa. Hydrostatic pressure allows us to change the band structure while keeping the same doping concentration and avoids oxygen incorporation during the growth which occurs for high aluminium percentages. The position of the DX center has been evaluated from transport measurements in GaAs (Maude 1987, Sallese 1990) and in AlGaAs (Lavielle 1989, Sallese 1991, Basmaji 1987) by using the measured carrier density before and after illumination at 4.2K. For the samples studied, the same carrier density was always recovered after illumination independent of the applied pressure. That is why it could be considered as the total electrically active dopant concentration in the sample. In this paper, we show this is no longer the case for samples with a high carrier concentration and high aluminium percentage which exhibit a PPC quenching at high pressures.

3. EXPERIMENTAL PROCEDURE

The (Al,Ga)As samples used in this study, were grown by MOVPE in a vertical atmospheric pressure reactor on semi-insulating GaAs substrates. TMGa, TMAl, arsine, TMSn, silane and DETe were used as precursors. To perform reliable transport measurements, the presence of a 2-dimensional electron gas at the GaAs/AlGaAs interface must be avoided. Therefore, between the substrate and the active layer, two undoped buffer layers were grown, the first with an aluminium percentage of 70% and a second with the same aluminium composition as the active layer. All the growth runs were performed at 750°C and with a III/V ratio maintained at 100 in order to limit the incorporation of oxygen. All the layers have a thickness of about 3μm. The aluminium percentage was measured using double X-ray diffraction and checked using microprobe analysis. SIMS measurements were also performed in order to evaluate the metallurgical donor concentration present in the sample. For the calibration of the doping level in the AlGaAs alloys, the one done in GaAs was used because in GaAs, the DX centres are resonant with the Γ band and for doping levels less than 10^{19} cm^{-3} no PPC effect is observed at ambient pressure. Therefore the value of the carrier concentration measured at ambient pressure is an accurate measure of the achieved doping level. This assumed equivalence of the calibration for GaAs and AlGaAs was found to be correct up to an effective doping level of $2x10^{18}cm^{-3}$.

In order to allow a comparison between the three different dopants used (Si, Sn, Te), three samples were grown with identical aluminium percentage (21%) and doping level ($2x10^{18}$ cm^{-3}). Transport measurements were performed using both Hall and Shubnikov-de Haas effects to determine the carrier concentration. The sample was mounted in a liquid clamp pressure cell which was cooled to 4.2K for the measurements. The sample could be illuminated at low temperature and under pressure by means of an infrared light emitting diode (hν=1.4 eV).

4. RESULTS AND DISCUSSION

The carrier concentration measured in the dark and after illumination to saturation, for the three samples, is shown as a function of hydrostatic pressure in Figure 1. For the three samples, the dark carrier concentration falls rapidly as electrons trap-out onto the DX centres with increasing pressure. In previous transport measurements in n-type AlGaAs alloys, the same value of the free carrier concentration measured after illumination was always recovered independent of the applied pressure and was therefore considered to represent the concentration of electrically active dopant in the sample. For our samples this is no longer the case. For pressures above a critical value Pc (0GPa for Sn, 0.5GPa for the Si and 1GPa for the Te), we are unable to restore the ambient pressure illuminated carrier concentration. We interpret this behavior as Fermi level pinning by a level which is in thermal equilibrium with the Γ-conduction band minimum even at low temperature. The observed chemical shift between the three dopants (ie different values of Pc) suggests that the levels involved are highly localized. For the Sn doped sample, the illuminated carrier concentration falls immediately upon the application of the pressure indicating that the Fermi-level is already pinned after illumination even at ambient pressure. SIMS measurements which give a metallurgical Sn concentration of $2x10^{18}cm^{-3}$ confirms this hypothesis. For the Sn and Si-doped samples at high pressure we observe a complete "PPC quenching". Illuminating the sample does not induce any change in the population of the conduction band. For the Te-doped sample the pressure required for a complete "PPC quenching" is beyond the range of our pressure cell. The decrease of the illuminated carrier concentration has also been observed in GaAsP alloys doped with Te and Sn (Sallese 1992). This was interpreted as Fermi level pinning by an L-band effective mass state.

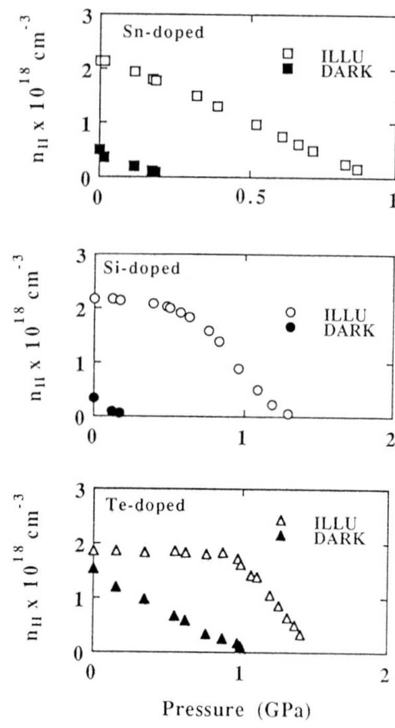

Fig. 1. Hall carrier concentration measured in the dark and after illumination at 4.2K versus pressure

Assuming that the level pins the Fermi level, we calculated its energy from the Fermi level computed using the value of the illuminated carrier concentration at 4.2K. The Fermi energy is calculated taking into account the non parabolicity and the pressure dependance of the effective mass in the Γ_{1C} minimum (three-band Kane model). The evolution of the level responsible for the "PPC quenching" with pressure is presented in Figure 2 for the three samples (the energy values and pressure coefficients are quoted relatively to the Γ conduction band). The band structure was taken from Guzzi and Staehli (1990).

In the Te-doped sample, the calculated energy of the level remains constant up to 1GPa and then starts to decrease with a slope of 108±14meV/GPa close to the pressure coefficient of the X band. It is therefore likely that the level could is an effective mass state tied to the X band. In the case of the Si-doped sample, the free carrier density starts to decrease for pressures above 0.5GPa. Then, two slopes can be observed when trying to fit the evolution of the Fermi level. The first one has a value, 47±8meV/GPa, close to the L-band pressure coefficient.

Therefore the level involved is most probably be the effective mass state tied to the L-band. The second slope which value is close to 110±10meV/GPa is again similar to the pressure coefficient of the X band. For the Sn-doped sample, the curve can also be fitted with two slopes which values are 56±6meV/GPa and 87±13meV/GPa. The first slope of the level energy is probably due to Fermi level pinning by the L-effective mass state because the value of its pressure coefficient is nearly equal to the one of the L-band.The second slope is identical to that reported by Wasileski (1986) and Stradling (1990) for the A1 antibonding state in GaAs. These authors performed FIR studies under high pressure in GaAs doped with Ge and Si and evidenced the existence of a highly localized A1 state by the occurrence of an anti-crossing with the shallow Γ(A1) level and a large chemical shift. This level had a pressure coefficient of 86 meV/GPa. Using EPR technics, von Barbeleben (1992) and Fockele (1989) have shown the existence in $Al_xGa_{1-x}As$:Sn (0,3<x<0,7) of a neutral A1 antibonding state of the simple substitutional donor. It was evidence by the observation of an hyperfine splitting in the EPR spectrum after illumination.

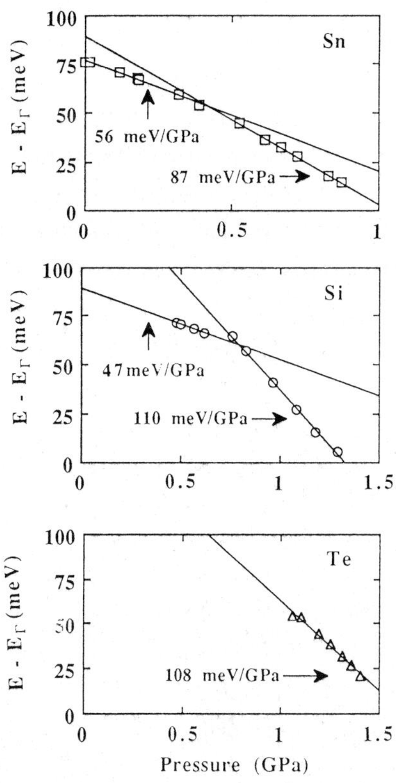

Fig. 2. Evolution of the level responsible for the Fermi level pinning as a function of pressure

It is important to note that DX is the ground state level which under normal circumstances (300K) controls the free electron population. It is only after illumination at low temperatures to persistently photoionise the DX centre that the higher lying excited states have the possibility to be occupied. Whether or not they are occupied depends on the Fermi level after illumination. There are two different types of excited states of the simple substitutional donor, "effective mass" states associated with the higher conduction band minima, and an A1 antibonding state. The level which is observed in transport and EPR measurements after illumination at low temperatures is the lowest of all the possible excited states and therefore depends on the Aluminium composition and applied pressure. The various excited states are observable in different regions of composition or pressure because each chemical species of dopant has a different chemical shift.

5. CONCLUSION

We have demonstrated for the first time using electrical transport measurements the occupation of excited states in Al GaAs upon illumination at low temperatures to photoionise the DX centres. For sufficiently high composition or hydrostatic pressure the excited states limit the free carrier concentration after illumination. The measured pressure coefficients of the excited states suggest two different types of levels are involved, "effective mass" like levels associated with a higher conduction band minima and the Al antibonding states. A conclusive identification requires further work including EPR measurements under pressure.

ACKNOWLEDGEMENTS:U Willke and DK Maude thank the EEC for financial support

REFERENCES

von Bardeleben H J, Buyanova I, Belyaev A and Sheinkman M 1992a Phys.Rev.B 45,11667

von Bardeleben H J, Sheinkhan M, Delarue C and Lannoo M 1992b Pro.Mat.Sci.Forum 787

Basmaji P, Portal JC, Aulombard R L and Gibart P 1987 Solid State Com. 63, 73

Bourgoin J C 1990 Physics of DX centres in GaAs alloys (Scientific Technological Publications)

Chadi N and Chang K J 1988 Phys. Rev. Lett. 61, 873

Dmochowski J E, Stradling R A, Wang P D, Holmes S N, Li M, McCombe B D and Weinstein B 1991 Semicond.Sci.Techno 6, 476

Fockele M, Spaeth J M and Gibart P 1990 Mat.Scien.Forum 65/66, 443

Gibart P, Williamson D L, Moser J and Basmaji P 1990 Phys.Rev.Lett. 65, 1144

Guzzi M and Staehli I L 1990 Physics of DX centres in GaAs alloys (Scientific Technological Publications)

Lang D V and Logan R A 1977 Phys. Rev. Lett 39, 635

Lavielle D, Sallese J M, Gouthiers B, Dmowski L, Basmaji P, Portal J C and Gibart P 1989 Int. Phys. Conf. Ser. 106, Chap. 5, Karuizawa, Japan

Lifshitz N, Jayaraman A, Logan R A and Card H C 1981 Phys.Rev.B 21, 670

Maude D K, Portal J C, Dmowski L, Foster T, Eaves L, Nathan M, Heiblum M, Harris J J and Beall R B 1987 Phys.Rev.Lett. 59, 815

Mooney P M 1990 J.Appl.Phys. 67, R1

Mosser V, Contreras S, Robert J L, Piotrzkowski R, Zawadzki W and Rochette J F 1991 Phys.Rev.Lett. 66, 1737

Ohno T and Yamagushi E 1991 Phys.Rev. B 44, 6527

Sallese J M, Lavielle D, Singleton J, Leycuras A, Grenet J C, Gibart P and Portal J C, 1990 Phys. Stat.Sol. 119, K14

Sallese J M, Ranz E, Leroux M, Portal J C, Gibart P and Selmi A. 1991 Semicond.Sci.Techno 6, 522

Sallese J M, Maude D K, Fille M L, Willke U, Gibart P and Portal J C 1992 Semicond.Sci.Techno 7, 1245

Wasilewski Z and Stradling R A 1986 Semicond.Sci.Techno. 1, 264

Wiklening W, Kaufmann U and Bauser E 1991 Semicond.Sci.Techno. 6, B84

Yamagushi E 1986 Japan.J.Appl.Phys 25 L643

Paper presented at the Int. Symp. GaAs and Related Compounds, Freiburg, 1993

Near infrared quasi-elastic light scattering spectroscopy of electronic excitations in III–V semiconductors

B H Bairamov
Department of Physics, Cavendish Laboratory, Cambridge, CB3 0HE, United Kingdom.
A F Ioffe Physico-Technical Institute, 194021, St Petersburg, Russia
V K Negoduyko, V A Voitenko, V V Toporov
A F Ioffe Physico-Technical Institute, 194021, St Petersburg, Russia
G Irmer, J Monecke
Fachbereich Physik der Bergakademie Freiberg, 9200, Freiberg, Germany

ABSTRACT: We present results of the first observation of the concentration dependencies of integrated intensities and line shapes of quasi-elastic scattering from the free electron gas in III-V semiconductors in the concentration range from ~ 10^8 up to ~ 10^{19} cm^{-3} under a condition of non-resonant excitation by using the 1064 nm line of a *cw* Nd^{3+}:*YAG* laser, when the observation of light scattering from energy- and momentum-density fluctuations is unambiguously demonstrated.

I. INTRODUCTION

One of the most important subjects of semiconductor physics remains the electron gas and inelastic light scattering by electronic excitations, which is of strong current interest and is a fast developing branch of optical spectroscopy. In particular, the papers concerning the electronic light scattering cover one of the most exciting chapters in the optics of microstructures and superlattices (Burstein et al 1991).

A common feature of all types of single particle electronic Raman scattering (RS) processes comes from conservation laws, which govern them and make the energy $\hbar\omega$, and the momentum $\hbar q$, transferred during the scattering process, small. Within the framework of the effective-mass approximation, the light scattering mechanisms in semiconductor quantum wells and superlattices are similar to those of the parent 3D systems (Pinczuk et al 1989).

There are all conditions for the observation of the majority of known single-particle excitations by the RS technique in III-V semiconductors. These excitations form the basic mechanisms of scattering which include charge-density fluctuations, energy-and momentum-density fluctuations as well as spin-density fluctuations .

Most of the recent inelastic electronic light scattering experiments (Pinczuk et al 1989) have been carried out in GaAs with excitation photon energies $\hbar\omega_i$ in the range 1.5-1.9 eV close to the resonance with the fundamental E_o or spin-orbit split-off $E_o + \Delta_o$ optical gaps, or with quantum well excitons providing large scattering enhancement required for sufficient sensitivity. On the other hand the simultaneous appearance of strong hot luminescence in such cases allows one to carry out measurements only in a limited electron concentration range and prevents temperature dependent measurements which, as we will show below, are essential for the observation of new features in the light scattering spectra from electronic excitations in semiconductors.

In this paper we examine the different scattering mechanisms from electronic excitations by using non-resonant near-infrared excitation of n-InP: $E_o + \Delta_o = 1.43 + 0.11 = 1.54$ eV (at T = 10K) and $\hbar\omega_i = 1.17$ eV. The absence of intense background luminescence in this case has provided us with an opportunity to carry out not only the conventional concentration dependent polarisation studies but also temperature dependent measurements of the integrated intensities and spectral linewidths of the quasi-elastic electronic scattering spectra. We find rather different effects of electronic excitations on polarised and depolarised scattering and undertake both theoretical and experimental comparison between intensities of scattering from energy- and spin-density fluctuations. We find these to be approximately equal at room temperature which removes any possible doubts concerning scattering from energy-density fluctuations, since scattering from spin-density fluctuations is well established.

2. EXPERIMENTAL SET-UP

Our polarised quasi-elastic light scattering measurements were performed by using the 1064.2 nm line of a *cw* Nd^{3+}:*YAG* laser and doped n-InP samples with carrier concentration from 10^8 up to 10^{19} cm^{-3} and the temperature range 27-300K. All measurements were conducted in right angle scattering geometry with the incident light always along $(11\bar{2})$ and scattered light along $(\bar{1}11)$ directions: for polarised spectra, hereafter labelled by $\bar{e}_i \| \bar{e}_s$ and $(11\bar{2})\{(\bar{1}10),(11\bar{2})\}(\bar{1}11)$ for depolarised spectra, labelled as $\bar{e}_i \perp \bar{e}_s$. The scattered light was analysed by double monochromator with a spectral resolution of 2.1 cm^{-1} and detected with a cooled photomultiplier tube with a photon-counting electronic system.

3. RESULTS AND DISCUSSION

The important aspect in the discussion of RS from charged excitations is the electrical screening of fluctuations. If the concentration of free carriers is low ($n \leq 10^{16}$ cm^{-3}) and the screening radius r_S is large enough to satisfy the condition $qr_S >> 1$, then every excitation created by the scattered light with the wave vector q is not screened. On the other hand, the quasi elastic light scattering spectra are recorded at room temperature when there are enough carriers to provide screening of the charged impurities. The condition of linear screening holds if $nr_s^3 >> 1$, which means that isolated ionised impurities are screened. Under this condition there still remain the long-range fluctuations in impurity potential with the mean square value $\gamma(r_s) = u(r_s)\sqrt{Nr_s^3}$, where $u(r) = e^2/\varepsilon r$ is the potential of an isolated impurity atom, and N is the total concentration of impurities. So single particle electronic excitations, created by light, are scattered by the long-range fluctuation potential and one can get the following expression for the light scattering cross section

$$\frac{d^2\sigma}{d\omega d\Omega} = n\left(\frac{e}{m^*c^2}\right)^2 (e_i.e_s)^2 \int \frac{dt}{2\pi}\cos(\omega t)\ \exp\left|-\frac{(qv_T t)^2}{4} - \frac{\rho_o q^2 t}{24m^{*2}}\right|, \tag{1}$$

where $v_T^2 = 2T/m^*$ and ρ_o is the root mean square fluctuation of the random force acting on the electron. The line shape of the spectrum differs slightly from a Gaussian and gives an approximately linear dependence of linewidth Γ on the concentration, in good agreement with the experimental observations (Figure 1).

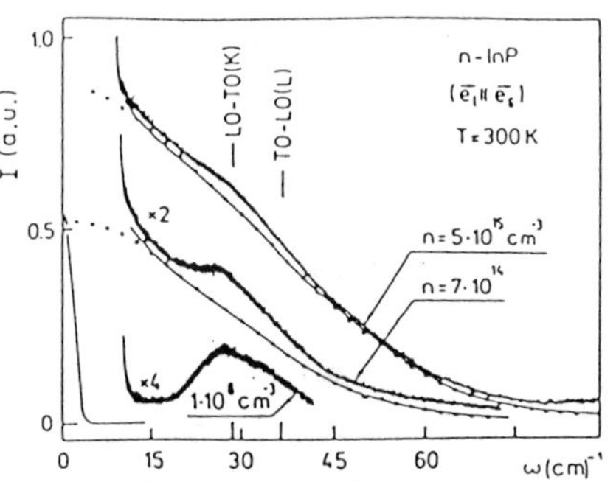

Fig.1 Raman spectra of n-type InP samples in the low-concentration range. The solid curves obtained after subtraction of the difference frequency combination two-phonon contribution, represent spectra for single-particle scattering from conduction electrons. The theoretical fit with a Gaussian approximation is shown by open circles with half-width Γ=35.0cm^{-1} for the sample with n = 5 x 10^{15} cm^{-3} and Γ = 24.9cm^{-1} for the sample with n = 7 x 10^{14} cm^{-1}. $e_i || e_s$, λ_i=1.06 μm, T = 300K.

With increasing concentration up to n = 3.0 x 10^{16} cm^{-3} when the condition of frequent collisions holds, (i.e. $q\ell << 1$, where ℓ is the electron mean free path) and due to the realisation of conditions for scattering from spin-density fluctuations $(\bar{e}_i \perp \bar{e}_s)$, the Gaussian type line shape transforms into a Lorentzian (Bairamov et al 1993). In the concentration range from 7.2 x 10^{16} up to 3.6 x 10^{17} cm^{-3}, collision controlled narrowing of the Lorentzian single particle spectra from 52 to 32 cm^{-1} is found experimentally. The narrowing of the linewidth with the concentration increase is a direct indication on the diffusion nature of the scattering mechanism.

Figures 2 (a and b) show typical Raman spectra in the polarised scattering configuration $(\bar{e}_i || \bar{e}_s)$ obtained on n-InP sample with n = 1.1 x 10^{18} cm^{-3} in the frequency range from 150 up to 450 cm^{-1} at different temperatures at 300K and 27. Figures 3 (a and b) show the same for the depolarised scattering configuration $(\bar{e}_i \perp \bar{e}_s)$. The scan is linear in wavelength and the positions of the lines are indicated in wave numbers.

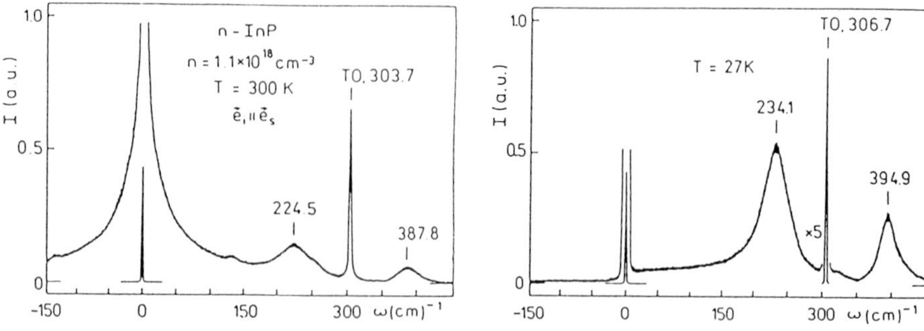

Fig. 2(a and b). Experimental Raman spectra of n-InP with n = 1.1×10^{18} cm^{-3} for polarised scattering configuration $e_i \parallel e_s$ at different temperatures which are indicated in the figures. The quasi-elastic Lorentzian contours discussed in the text are near the laser line.

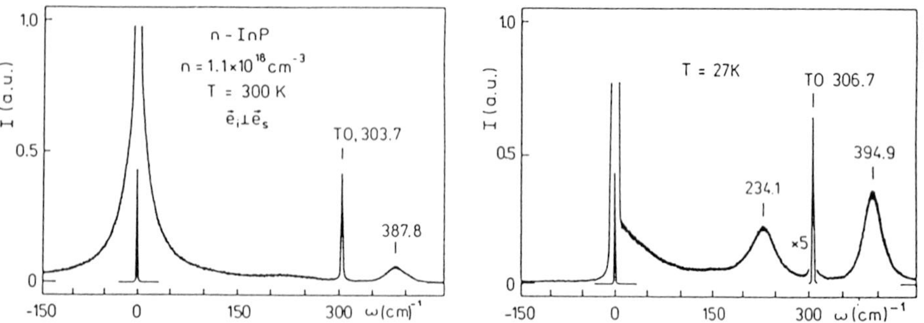

Fig. 3(a and b). Same as in Fig.2 but for the depolarised scattering configuration for crossed polarisations $e_i \perp e_s$. The quasi-elastic Lorentzian contours correspond to spin density fluctuations.

We unambiguously identify in all the spectra, sharp quasi-elastic parts of Raman scattering from the free electron gas, located within ± 150 cm^{-1} with the required characteristic finite intensity at frequencies close to zero. The Stokes and anti-Stokes components at near room temperatures are equal in intensity due to the response of the detector and spectrometer used.

In order to determine the integrated values of the intensities of the quasi-elastic electronic scattering we used uncoupled TO(Γ) phonon spectra for calibration, so that the spectra at different temperatures may be directly compared. Figure 4 compares the corresponding experimental integrated intensities. It is particularly striking, that a pronounced difference in the temperature dependencies of the intensity, as well as the linewidth, of quasi-elastic electronic scattering between the polarised and depolarised scattering configurations is readily observable.

To understand the data obtained it is necessary to briefly consider results of the developed hydrodynamic model of the electron gas (Bairamov et al 1993). In order to describe the temperature dependence of the cross-section in the case of parallel polarisations $e_i \parallel e_s$ one should take into account both contributions from energy and momentum density fluctuations

$$I_{\varepsilon,p} = \left(\frac{e^2}{mc^2}\right)^2 \left[\varsigma B_p(\omega_I)\right]^2 V \left(\frac{\delta n}{\delta \varsigma}\right)_T T \left[1 + \left(\frac{10T}{\varsigma}\right)^2\right], \tag{2}$$

where V is the crystal volume and ς is the chemical potential of electrons. One more contribution to the quasi-elastic scattering in the case of the depolarised scattering configuration is determined by spin-density fluctuations

$$I_\sigma = \frac{3}{2} V \left[\frac{e^2}{mc^2} B_\sigma(\omega_I)\right]^2 \left|e_I \times e_s^*\right|^2 n \frac{T}{\varsigma}\left(1 + \frac{\alpha}{2}\frac{T}{\varsigma}\right) \tag{3}$$

There is a small parabolic correction to usual the linear temperature dependence in this expression. Here $B_{p,\sigma}$ are band structure dependent coefficients. The ratio of the cross-sections in parallel and crossed polarisations of incident and scattered light is given by

$$\frac{I_\varepsilon}{I_\sigma} = \left(\frac{T}{\hbar\omega_I}\right)^2 D^2 \text{ where } D = \frac{\sum_j \frac{E_{gj}^2 + (\hbar\omega_1)^2}{\left[E_{gj}^2 - (\hbar\omega_1)^2\right]^2}\left(1 + \frac{m_\varepsilon^*}{m_j^*}\right)}{\frac{1}{E_g^2 - (\hbar\omega_I)^2} - \frac{1}{(E_g + \Delta)^2 - (\hbar\omega_I)^2}} \tag{4}$$

The estimation for n-InP and with $\hbar\omega_I = 1{,}17eV, E_g = 1{,}43eV, E_g + \Delta = 1.54eV$ gives

$$D = \frac{25.77}{0.78} = 53.47; \left(\frac{T}{\hbar\omega_I}\right)^2 = 3.88.10^{-4} \text{ and thus } \frac{I_\varepsilon}{I_\sigma} = 53.47^2 \times 3.88;10^{-4} = 1.11 \tag{5}$$

So equations (4) and (5) show that mechanisms of spin and energy density fluctuations give comparable intensities for the conditions of our experiments. Theoretical temperature dependencies of the integrated intensities $I_{\varepsilon,p}$ and I_σ from equations (2) and (3) are plotted in Figure 4 by solid and dot-dashed lines respectively, while filled and open circles represent the corresponding experimental points. Adjustable parameters α which is the temperature derivative $d\varsigma / dT /_{T=0}$ and $\alpha = 4, \varsigma = 99meV$. The dashed line represents theoretical curve for $I_{\varepsilon,p}$

$$I_{\varepsilon,p} = \left(\frac{e^2}{mc^2}\right)^2 \left[\xi B_p(\omega_i)\right]^2 V \left(\frac{\delta n}{\delta \varsigma}\right)_T T\left(1+\frac{100T}{\varsigma}\right), \tag{6}$$

which is obtained using temperature independent classical value of the electron heat capacity $C_\upsilon = \frac{3}{2} K_B n$, K_B being the Boltzmann constant. It gives a poor fit to the experimental points as compared with the solid line. The temperature dependence of C_υ is therefore significant in accordance with the rather large value of $\varsigma >> T$.

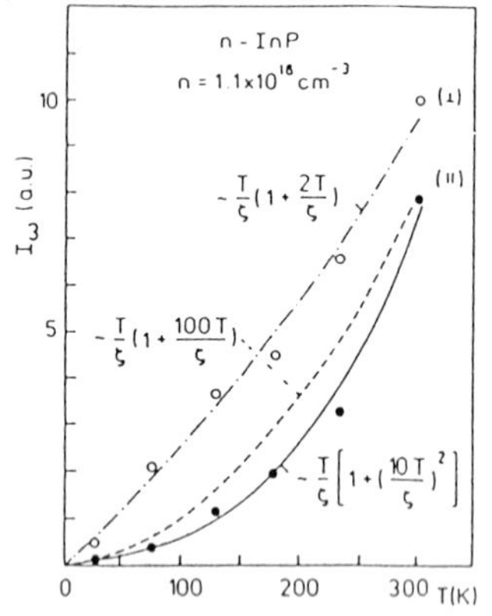

Fig.4. Temperature dependence of the integrated cross-section from the same sample as in Figures 1 and 2. Solid and dashed line-theory for energy-momentum density fluctuations corresponding to equations (2) and (6),filled circles are corresponding experimental points available from polarised scattering. Dash-dotted line-theory for spin density fluctuations (equation 3), open circles are corresponding experimental points available from depolarised scattering.

4. CONCLUSIONS

Finally the new approach to the problem and the observed results provide clear evidence for the existence of strongly temperature dependent free electron gas fluctuations in semiconductors with nonparabolic dispersion of energy bands. This proves unambiguously the observation of light scattering from energy- and momentum-density fluctuations and gives a sensitive measure of the magnitude of the different relaxation times of the electron gas.

REFERENCES

Bairamov B H, Voitenko V A, Ipatova I P and Toporov V V 1989 Proc. Int. Conf. held in connection with celebrations of the birth centenary of C V Raman and Diamond Jubilee of the Discovery of Raman Effect, Recent Trends in Raman Spectroscopy ed S S Banerjee and S S Sha (Singapore: World Scientific) pp 386-398

Bairamov B H, Voitenko V A, Ipatova I P 1993 Phys.Reports (in press)

Burstein E, Cardona M, Lockwood D J, Pinczuk A and Young J F 1991 In: "Light Scattering in Semiconductor Structures and Superlattices", by D J Lockwood and J F Young, Proc. of the NATO Symposium on Light Scattering in Superlattices and Microstructures, Quebec, Plenum Press, New York, pp 1-17

Pinczuk A and Abstreiter G 1989 In: "Light Scattering in Solids V" ed. by M Cardona and G Guntherodt, Topics Appl.Phys. V.66, Springer-Verlag, Berlin, pp 153-211

Paper presented at the Int. Symp. GaAs and Related Compounds, Freiburg, 1993

Fourier transform photoluminescence spectroscopy of *n*-type bulk InAs and InAs/AlSb single quantum wells

F. Fuchs, J. Schmitz, J.D. Ralston, and P. Koidl

Fraunhofer Institut für Angewandte Festkörperphysik, Tullastrasse 72, D-79108 Freiburg, Fed. Rep. of Germany

ABSTRACT: An optical study of n-type bulk InAs and epitaxially grown InAs-AlSb single quantum wells is presented. Fourier transform PLE spectra from a series of bulk samples with varying dopant concentration give evidence for the existence of a Fermi-edge singularity in the bulk semiconductor. In addition, for the first time the spatially indirect photoluminescence of type II InAs-AlSb single quantum wells is presented. The spectra in the mid infrared show the influence of the varying well widths on the confinement effects. The luminescence peaks in such quantum wells are observed ≈ 80 meV below the transition energy expected from selfconsistent calculations using a two-band envelope function model. This behaviour is tentatively explained by the presence of deep acceptor-like states in the interface region of the barrier material.

1. INTRODUCTION InAs is of great scientific and technological interest for several reasons. Due to the small electron mass, degeneracy of the electron gas is reached at low doping concentrations where the mobility is high. InAs is therefore well suited to study the influence of many-body-effects on the electronic and optical properties. The combination of InAs and AlSb in MBE-grown heterostructures is of great technological interest because of the large conduction band offsets and the high electron mobility of InAs. The possibility to fabricate long-wavelength detector structures which are compatible with III-V processing technology (Chow 1991) motivates such interest together with the promising transport properties of InAs/AlSb/GaSb polytype structures (Brown 1991). High-speed resonant interband tunnelling devices with large peak-to-valley ratios based on the above materials have also been demonstrated recently (see e.g. Yang 1990, Beresford 1990)).

InAs/AlSb heterostructures form type II superlattices (Fig 1). The band alignment leads to spatially indirect optical transitions, due to the separation of the confined electrons and holes. In addition, the low value of the band gap of InAs leads to degenerate doping at electron concentrations on the order of 10^{16} cm^{-3}, as well as strong non-parabolicity effects. Furthermore, in hetero-epitaxial growth the lattice constant of the underlying AlSb buffer leads to biaxial tensile strain of the InAs quantum wells (Yang 1992 and Yang 1993), which lowers both the band-gap energy and electron effective mass considerably. One of the most serious problems is the control of the electron concentration in the InAs quantum well

(Tuttle 1990). The sheet concentration is typically in the 10^{12} cm^{-2} range even in the absence of intentional doping. The influence of the cap layer, as well as the thickness of the top barrier, on the electron concentration have been both previously investigated (Chadi 1993, Ideshita 1992). However, very few publications concerning the optical properties of InAs-AlSb heterostructures are available (Brar 1993).

Using Fourier transform photoluminescence excitation (FTPLE) spectroscopy, we have performed an optical study of a series of n-type InAs bulk samples with doping concentrations between $2x10^{16}$ cm^{-3} and $3x10^{17}$ cm^{-3}. The PLE data obtained on the bulk samples are explained by the theory of optical interband transitions under degenerate conditions presented by Mahan 1967. Thus, direct evidence for the existence of the Fermi-edge singularity (FES) in a bulk semiconductor can be given.

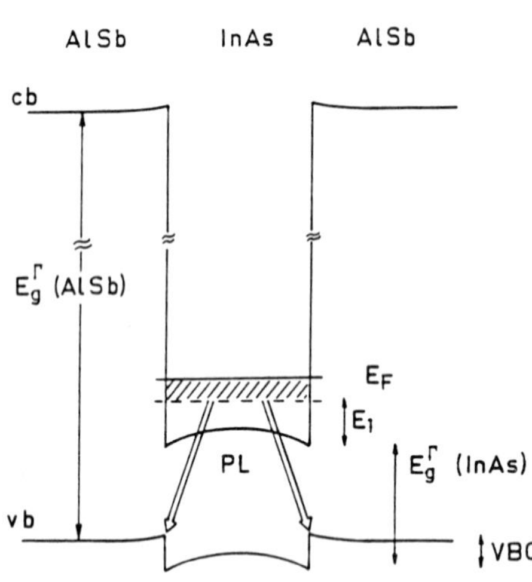

Fig. 1 Band alignment of InAs and AlSb. Spatially indirect luminescence is indicated.

Furthermore, we have investigated the photoluminescence of a series of InAs single quantum wells (SQW's) with varying well width embedded in AlSb barriers. The photoluminescence spectra of the SQW's show the expected quantum confinement with varying well width. Despite the spatial separation of electrons and holes it was possible to detect the mid-infrared photoluminescence (PL). The optical transition energies were found to be lower than those predicted by self-consistent envelope-function calculations which include the non-parabolicity effects. In these calculations, the shrinkage of the InAs band-gap to a value of 287 meV at the Γ-point was assumed. This value is lower than the bulk value of 412.5 meV because of the bi-axial tensile strain resulting from growth on an AlSb buffer layer (Yang 1993). The observed red shift of the PL is explained tentatively by recombination involving deep acceptor-like states in the AlSb barrier or at the Interface.

2. EXPERIMENTAL The QW samples were grown by solid-source MBE on (100) GaAs substrates. The present samples included a 1μm AlSb buffer layer for strain accommodation. On top of this buffer layer a superlattice of 10 periods 2.5 nm GaSb/2.5 nm AlSb was grown, followed by the AlSb/InAs/AlSb single quantum well. To prevent oxidation of the AlSb top barrier, the samples were capped with a 10 nm thick GaSb layer. The arsenic flux was controlled by a valved cracker effusion cell, which enabled us to reduce the arsenic

background pressure during growth by up to three orders of magnitude compared to conventional effusion cells. Growth of the barrier without a valved cell resulted in formation of ternary $AlSb_{1-x}As_x$, with an arsenic concentration in the range of x=10 %. The shutter sequence during growth of the InAs single quantum wells was such as to promote the formation of InSb-like interfaces (Tuttle 1990). For more details of the growth technique of the present samples see Schmitz 1993. The formation of InSb-like interfaces was confirmed by Raman spectroscopy (Wagner 1993). The mobilities of the confined electrons in the InAs channel were determined from measurement of Shubnikov de Haas oscillations (Schmitz 1993), and reach values up to $1.4x10^5$ cm^2/Vs at 4.2 K. It is known that such high values are only obtained with InSb like interfaces (Tuttle 1990). The AlAs like interface typically results in mobilities which are lower by more than one order of magnitude. A series of samples with the well thickness varying between 5 and 20 nm were grown. The four investigated InAs *bulk* samples showed electron concentrations ranging from $2x10^{16}$ cm^{-3} to $3x10^{17}$ cm^{-3} with 77 K mobilities of around 20,000 cm^2/Vs.

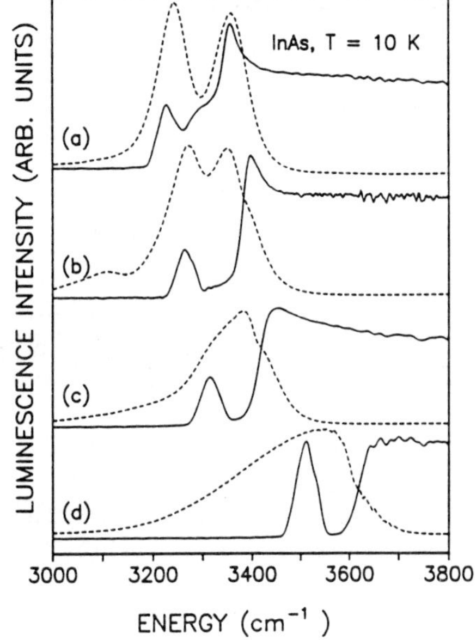

Fig. 2 PL (dashed) and PLE (solid lines) of InAs are shown. The increasing degeneracy can be directly observed by the blue shift.

The experiments were performed with a Fourier transform spectrometer (FTIR). For the PLE measurements on the bulk samples the FTIR spectrometer was utilised as the excitation source. Excitation of the PL was realised with interferometrically modulated glowbar radiation (For details see Fuchs 1993a). The photoluminescence experiments with the SQW samples were carried out using double modulation techniques as described by Fuchs 1989. For excitation of the PL the 514 nm radiation (2.41 eV) of an Argon ion laser was used.

3. FERMI-EDGE SINGULARITY. In Fig. 2 the photoluminescence (broken lines) and the FTPLE spectra (solid lines) of the four InAs bulk samples are shown. The isolated line on the low energy side of the PLE spectra is induced by stray-light (Fuchs 1993a). Due to the low effective electron mass of $0.0236m_0$, the degeneracy due to band filling can be observed at doping concentrations as low as 10^{16} cm^{-3}. The degeneracy leads to a pronounced Burstein-Moss shift, which is observed in both type of spectra. In addition, a resonance-like structure can be observed at the onset of the PLE spectra. Such an

enhancement under degenerate conditions is quite unexpected. At degenerate carrier concentrations the relaxation of electron-hole pairs into free or bound excitons is expected to be suppressed due to the screening of the Coulomb interaction. The position and width of this resonance depends on the electron concentration. These conditions are described by Mahan 1967, who treated the optical absorbance of direct band-to-band transitions under degenerate conditions. Excitonic states were shown to lead to a logarithmic singularity of the interband oscillator strength at the Fermi energy, even in the limit of a high-density electron gas. In contrast to the well known excitonic states under dielectric (undoped) conditions, the Fermi-edge singularity (FES) undergoes the same shift to higher energies as does the Fermi-level. In low dimensional structures the FES was first observed by Skolnick et al. 1987. The possibility to separate dopant atoms from the quantum well enables the formation of a high-density, high mobility electron gas, which is not possible in large gap semiconductors because of ionised impurity scattering. The present data demonstrate that similar conditions can be achieved in low gap semiconductors. In the present case the electron concentration is sufficient to allow the observation of degeneracy, while the electron mobility remains at values sufficiently high to make the observation of the FES possible.

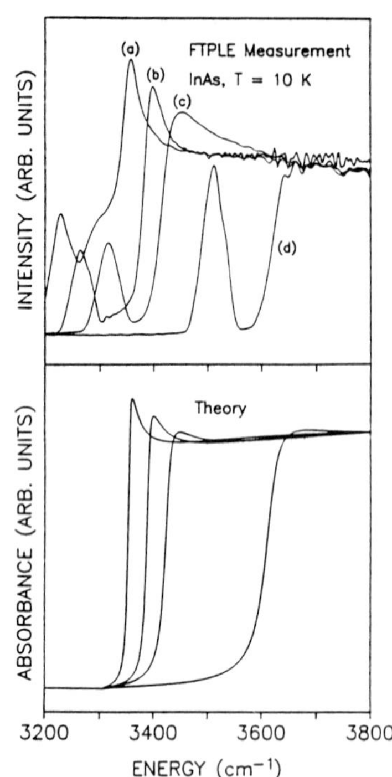

Fig. 3 Comparison of the FTPLE measurement and the calculated absorbance of bulk InAs (The low energy PLE line is induced by stray-light).

In Fig. 3 the PLE spectra and the absorbance according to the theory of Mahan are presented. The only free parameter used in the calculations is a broadening lifetime. In view of the very simple model we obtain a surprisingly good agreement between the calculated and measured data. Thus, these first PLE spectra on InAs prove the existence of the FES in a bulk semiconductor. For more details concerning these results see Fuchs 1993b.

4. LUMINESCENCE OF InAs SQW's. In Fig. 4 the photoluminescence spectra of a series of InAs-AlSb single quantum wells obtained by Fourier transform spectroscopy with 514 nm excitation radiation are shown. The cut-off energy of the measurement system, due to detection with an InSb photodiode operating at 77 K, is indicated with a bar. Detection of the

luminescence by excitation with the 1.06 μm or 1.32 μm line of a Nd:YAG laser was not possible with the present sensitivity. Thus, we have to assume that the dominant excitation mechanism is the creation of electron-hole pairs in the AlSb barrier material. The electrons will be trapped in the InAs quantum well. Recently, it has been shown by Tournie et al. 1992 with photoluminescence experiments in the near infrared (1.3 μm) on InAs single quantum wells embedded in the ternary compound $Ga_{0.47}In_{0.53}As$ lattice-matched on InP substrates that this trapping mechanism is indeed very efficient. In the present case the observed luminescence is due to recombination with the holes located at the interface region with the confined electrons (see Fig. 1).

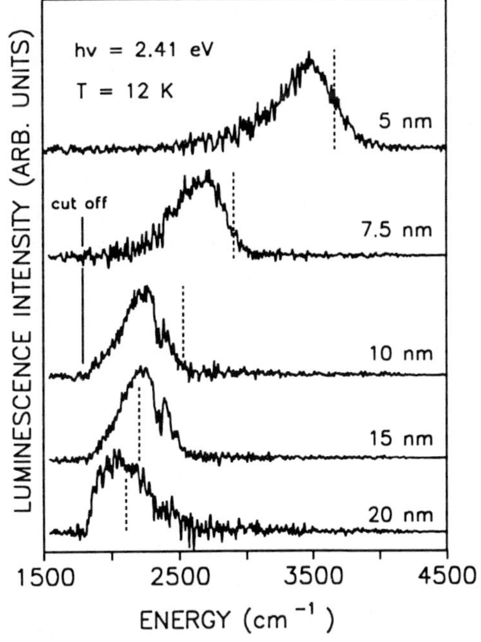

Fig. 4 Photoluminescence of a series of InAs-AlSb single quantum wells. The quantum well thickness is indicated. Dashed lines correspond to the calculated values for the low energy onset of the luminescence

A blue shift of the PL due to the enhanced electron confinement with decreasing well width is clearly observed. Self-consistent calculations of the transition energy using a two-band envelope-function model yield the transition energies indicated by broken bars in Fig. 4. The low-energy onset of the luminescence lines is consistently 80 meV lower than expected from the calculations. The confinement energy E_1 (see Fig. 1) varies between 10 meV for the 20 nm well and 100 meV for the 5 nm well. Note that the increase of the effective mass due to non-parabolicity is very well reproduced using a two-band model. Following Yang 1993 for the InAs well a value for E_g^Γ of 0.287 eV was used. This low value is a consequence of the bi-axial tensile strain of the system due to the 1000 nm AlSb buffer layer. Higher values, e.g. the bulk value of 412 meV, would result in a *blue shift*, in contrast to the observed red shift. An error in the effective mass or renormalisation effects due to penetration of the confined electrons into the barrier material can be excluded, such effects only result in a small error in the confinement energy. The red shift of 80 meV of the PL has several possible explanations. One possibility is the assumption of a valence band offset of about 180 meV. Following the argumentation of Kroemer 1983 we do not favour this explanation. Photoluminescence data are, in any case, not very well suited for determining valence-band offsets. Ideshita et al. 1992 found evidence for the presence of a deep acceptor in the ternary compound $Al_{0.5}Ga_{0.5}Sb$ with an activation energy of 145 ± 15 meV. Thus, it is reasonable to postulate the presence of a deep acceptor in the binary AlSb barrier or the interface.

4. SUMMARY An optical study on bulk InAs and InAs-AlSb single quantum wells has been performed. FTPLE experiments reveal for the first time the FES in a bulk semiconductor. The photoluminescence of the spatially indirect transitions of the SQW's exhibit the expected confinement effects. The line positions are about 80 meV lower in energy than those predicted by a two-band envelope-function calculation. This is explained by the presence of deep acceptor-like states in the AlSb barrier material.

5. ACKNOWLEDGEMENTS Helpful discussions with J. Wagner and support in performing the calculations by H. Schneider are gratefully acknowledged. We would like to thank G. Bihlmann and K. Schwarz for valuable technical assistance. This work was supported by the Bundesminister für Forschung und Technologie (BMFT) within the framework of the III-V Electronics program.

6. REFERENCES

Beresford R, Luo L F, Longenbach K F, and Wang W I 1990 Appl. Phys. Lett. **56**, 952
Brown E R, Söderström J R, Parker C D, Mahoney L J, Molvar K M, McGill T C 1991, Appl. Phys. Lett. **58**, 2291
Chadi D J 1993, Phys. Rev. B **47**, 13478
Chow D H, Miles R H, Nieh C W, McGill T C 1991, J. Crystal growth **111**, 683
Fuchs F, Lusson A, Wagner J, and Koidl P 1989, 7th Int. Conf. on Fourier Spectroscopy, Fairfax VA, SPIE Vol. 1145, p. 323-326
Fuchs F, Kheng K, Schwarz K, and Koidl P 1993a, Semicond. Sci. Technol. **8**, S75-S80
Fuchs F, Kheng K, Koidl P, and Schwarz K 1993b, Phys. Rev B, in press
Ideshita S, Furukawa A, Mochizuki Y, and Mizuta M 1992, Appl. Phys. Lett. **60**, 2549
Kroemer H 1983 Surface Science 132, p 543-576
Mahan G D 1967, Phys. Rev. **153**, 882
Schmitz J, Wagner J, Obloh H, Koidl P, Ralston J D 1993, J. of Electron. Mater. 1993, to be published
Skolnick M S, Rorison J M, Nash K J, Mowbray D J, Tapster P R, Bass S J, and Pitt A D 1987 Phys. Rev. Lett. **58**, 2130
Tournie E, Brandt O, Ploog K H 1992 Appl . Phys. **A56**, pp. 109-112
Tuttle G, Kroemer H, and English J H 1990 J. Appl. Phys. **67**, 3032
Wagner J, Schmitz, Maier M, Ralston J D, and Koidl P 1993, Proc. MSS-6, Solid State Electronics, to be published
Yang L, Chen J F, and Cho A Y 1990 J. Appl. Phys. **68**, 2997
Yang M J, Wagner R J, Shanabrook B V, Moore W J, Waterman J R, Twigg M E, and Fatemi M 1992 Appl. Phys. lett. **61**, 583
Yang M J, Lin-Chung P J, Wagner R J, Waterman J R, Moore W J, and Shanabrook B V 1993 Semicond. Sci. Technol. **8**, S129

Multiple excitonic features in low carbon content $Al_xGa_{1-x}As$

S.M. Olsthoorn[1], F.A.J.M. Driessen[1], D.M. Frigo[2], and L.J. Giling[1]

[1] *Department of Experimental Solid State Physics, RIM, Faculty of Science, University of Nijmegen, Toernooiveld, 6525 ED Nijmegen, The Netherlands*

[2] *Billiton Research B.V., P.O. Box 40, 6800 AA Arnhem, The Netherlands*

Abstract

Photoluminescence (PL) results are reported on $Al_xGa_{1-x}As$ containing very low carbon concentrations grown by metalorganic vapour-phase epitaxy. Multiple excitonic transitions are observed which cannot be attributed to variations in aluminum fractions. The localization energies of these multiple excitonic features, which are only observed in layers with aluminum fractions greater than 0.2, varies from 2 to 8 meV. This indicates that some of these transitions must be due to excitons bound to non-shallow centres, although their precise origin is still unknown.

1 INTRODUCTION

Photoluminescence (PL) spectroscopy is a powerful nondestructive technique to obtain information about both the intrinsic and extrinsic radiative transitions in $Al_xGa_{1-x}As$ alloys. In the PL spectra of nominally undoped direct bandgap $Al_xGa_{1-x}As$ ($x<0.4$) usually one bound-exciton peak and one acceptor related peak [(e, A^0),(D^0, A^0)] are detected (Stringfellow 1980, Pavesi 1992). The latter is accompanied by a phonon replica lying 36 meV lower in energy (Pavesi 1992).

For shallow impurities the localization energy of the exciton on the donor or acceptor is proportional to the ionization energy of that donor or acceptor (Haynes' rule (Haynes 1960)). Because acceptor and donor ionization energies are known in $Al_xGa_{1-x}As$ (Pavesi 1992), it is possible to calculate the separation between the different bound-exciton transitions as a function of Al-fraction.

Separate excitonic peaks (FX, (D^0, X), (D^+, X) and (A^0, X)) cannot be resolved in $Al_xGa_{1-x}As$ with $x>0.2$ owing to alloy broadening (Olsthoorn 1991), whereas for $x<0.2$ and in GaAs separate excitonic transitions are observed (Reynolds 1985).

By measuring the excitation density and temperature dependence and the linewidths of PL transitions their excitonic or acceptor-related nature can be determined. The intensity I of excitonic peaks depends superlinearly on the excitation density P ($I \propto P^k$, with $1 \leq k < 2$), whereas that of acceptor-related peaks depends sublinearly on P. With increasing temperature the intensity of excitonic peaks decreases much faster than acceptor peaks as a result of the much lower binding energies of excitons. Finally, the linewidth of an excitonic peak is typically 2 to 5 meV (Olsthoorn 1991, Schubert 1984), whereas that of an acceptor-related transition is around 15 meV (Schubert 1984).

In this paper we present PL and PLE spectra of high purity $Al_xGa_{1-x}As$ samples with x>0.2. Multiple exciton peaks are observed the nature of which cannot always be assigned as bound to a shallow impurity.

2 EXPERIMENTAL

The $Al_xGa_{1-x}As$ epilayers were grown by metalorganic vapour-phase epitaxy (MOVPE) on (100) 2° off towards (110) semi-insulating GaAs substrates. For the growth of sample 1 (reference sample) the precursors used were trimethylaluminum (TMAl), trimethylgallium (TMGa) and arsine. For samples 2, 3 and 4, mono- or bis-dimethylethylamine alane was used as the aluminum precursor (Olsthoorn 1992, Wilkie 1992). Hall measurements at 77 K were used to determine the electrical properties of the $Al_xGa_{1-x}As$ layers; all layers showed n-type conductivity. The Al-fractions x of the samples were determined from the PL spectra to be 0.22, 0.22, 0.28 and 0.24 for samples 1 to 4, respectively.

The PL and PLE experiments were performed in an optical flow cryostat with the samples in He exchange gas. Unless mentioned otherwise, the spectra were obtained at a temperature of 4.2 K. For the PL measurements optical excitation was provided by the 2.41 eV line from an Ar^+ laser with excitation densities ranging from 2.6×10^{-3} to 2.6 W/cm^2 at a spotsize of 3.8×10^{-2} cm^2. For the PLE measurements optical excitation was provided by a standing-wave dye laser with DCM as dye pumped by an Ar^+ laser. During these measurements the laser wavelength was scanned using a high-precision stepper motor connected to the birefringent filter of the dye laser. The etalons were removed from the laser cavity, so that the laser output was constant over the scanned wavelength range. The luminescence was dispersed by a 0.6 m double monochromator with 1200 lines/mm gratings and detected by a cooled photomultiplier tube with a GaAs photocathode.

3 RESULTS and DISCUSSION

Figure 1 shows PL spectra of several $Al_xGa_{1-x}As$ samples. The bottom spectrum (sample 1) shows a typical spectrum with one bound exciton peak and one acceptor-related peak. The energy separation of 20 meV implies that carbon is the acceptor involved (Stringfellow 1980). The other 3 spectra all show additional peaks. Rapid thermal quenching of the PL of these additional peaks was observed, as was a superlinear dependence on P, from which we conclude that they are excitonic. This is reinforced by their narrow linewidths (~5 meV, as compared with ~14 meV for the low energy peak).

X-ray diffraction measurements were performed on every sample. For all samples the x-ray peak originating from the $Al_xGa_{1-x}As$ epilayer had similar linewidth to that from the GaAs substrate (Bartels 1978). Furthermore, the broadening of the acceptor peak in the PL spectra was minimal (~14 meV for every sample, which is the theoretical limit for an ideal acceptor peak in $Al_{0.25}Ga_{0.75}As$ (Schubert 1984)). And finally, only one step-edge was observed with PLE. These three observations indicate that the aluminum fraction over the $Al_xGa_{1-x}As$ layer is constant, and therefore we conclude that fluctuations in composition are not the origin of the new PL emissions.

As mentioned in the introduction, with Haynes' rule and the known ionization energies it is possible to calculate the separation between the excitons bound to different shallow impurities (the broadening of these peaks is usually larger than their separation, so only one unresolved excitonic peak is observed (Olsthoorn 1991)). The separations between some of the exciton peaks observed in our spectra exceed these calculated values, and so all peaks cannot be attributed as bound to shallow impurities; those that do are indicated with an 'S' in figure 1.

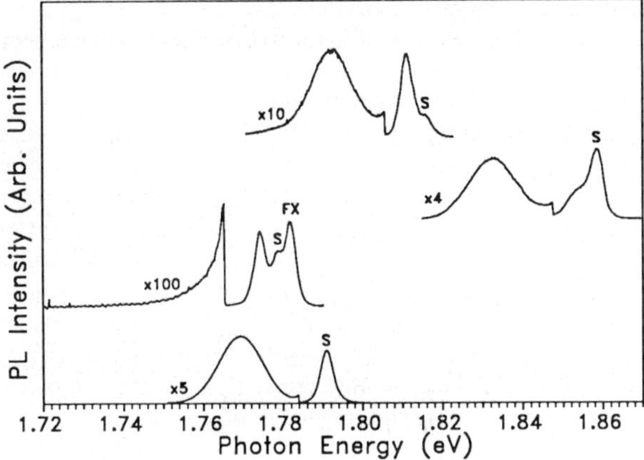

Figure 1: PL spectra of several $Al_xGa_{1-x}As$ samples.

For all samples the carbon concentration in the $Al_xGa_{1-x}As$ epilayers is very low. This was previously determined with PL (Olsthoorn 1992), and with SIMS and electrical measurements (Wilkie 1992). Because carbon usually predominates in PL spectra of unintentionally doped $Al_xGa_{1-x}As$, its presence at unusually low concentrations in our samples might be the reason these unprecedented emissions are observed.

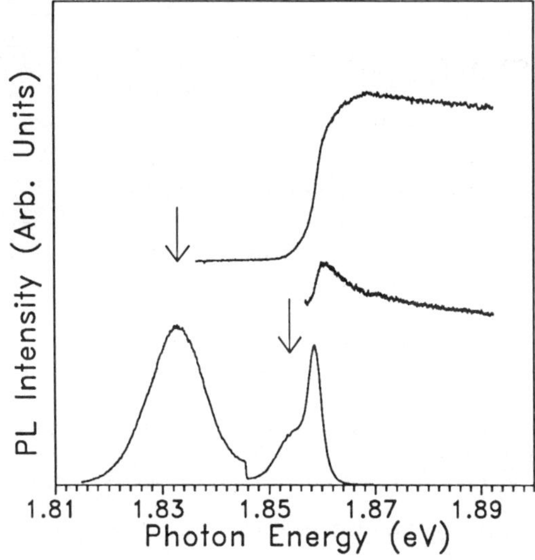

Figure 2: PLE and PL spectra of sample 3. The arrows indicate the detected energy.

Another remarkable result is that these multiple excitonic features are only observed in $Al_xGa_{1-x}As$ with $x>0.2$. In $Al_xGa_{1-x}As$ with $x<0.2$ (also grown with the new alane precursors) we only observed one unresolved exciton peak. The possibility that these new excitonic features are bound to DX centres (which become significant in $Al_xGa_{1-x}As$ with $x>0.2$) can not be proved by our measurements.

Finally, we performed PLE experiments on these $Al_xGa_{1-x}As$ samples. Figure 2 shows the PLE spectra of sample 3 along with its PL spectrum; the detection energies are indicated by arrows. The upper curve shows the PLE spectrum with the monochromator fixed on the acceptor-related peak. It shows one step-edge which is typical for a band-to-band transition (Schubert 1985). In the middle spectrum the monochromator is fixed on the lower energy exciton peak: A clear maximum appears which is indicative of excitonic absorption (Sturge 1962). From the energy difference between this free excitonic absorption and the maxima of the PL exciton emission peaks, we determined that their localization energies are 2 and 6 meV. The PLE measurements on the other samples resulted in localization energies of 2 and 6.7 meV for sample 4, and 0, 3 and 7.8 meV for sample 2. The latter result (on sample 2) shows that the exciton at highest energy in the PL spectrum is a free exciton. Along with the absence of acceptor-related peaks this indicates the extreme purity of this sample. Some of the localization energies are considerably larger than that of reference sample 1, for which the bound exciton peak had a localization energy of 2 meV.

4 CONCLUSIONS

PL measurements on low carbon content $Al_xGa_{1-x}As$ showed multiple exciton peaks in samples with Al-fractions higher than 0.2 which were not due to multiple Al-fractions. From the relatively high localization energies (up to 8 meV) as determined with PLE, we conclude that some of these excitons are not bound to shallow impurities but their origin is still unknown. They are probably observable because of the low carbon contamination.

Acknowledgement

The authors would like to thank Billiton Precursors B.V. the Netherlands for providing the samples, and A.M. Joel and M.W. Jones of EPI Cardiff and G.J. Bauhuis for performing x-ray diffraction measurements. This work was supported by the Nederlandse Organisatie Voor Energie en Milieu (NOVEM).

References

Bartels W J and Nijman W 1978, J. Cryst. Growth **44**, p.518.

Haynes J R 1960, Phys. Rev. Lett. **4**, p.361.

Olsthoorn S M, Driessen F A J M and Giling L J 1991, Appl. Phys. Lett. **58**, p.1274.

Olsthoorn S M, Driessen F A J M, Giling L J, Frigo D M, and Smit C J 1992, Appl. Phys. Lett. **60**, p.82.

Pavesi L 1992, in *Properties of Aluminium Gallium Arsenide* edited by S. Adachi, emis datareview series **No. 7**, INSPEC, the Institution of Electrical Engineers p.245-268.

Reynolds D C, Bajaj K K, Litton C W, Singh J, Yu P W, Henderson T, Pearah P, and Morkoç H 1985, J. Appl. Phys. **58**, p.1643.

Schubert E F, Göbel E O, Horikoshi Y, Ploog K and Queisser H J 1984, Phys. Rev. B **30**, p.813.

Schubert E F and Ploog K 1985, J. Phys. C **18**, p.4549.

Stringfellow G B and Linnebach R 1980, J. Appl. Phys. **51**, p.2212.

Sturge M D 1962, Phys. Rev. **127**, p.768.

Wilkie J H, Eyden G J M v., Frigo D M, Smit C J, Reuvers P J, Olsthoorn S M and Driessen F A J M 1992, Proc. of the 19th Intern. Symposium on GaAs and Related Compounds, Karuizawa Japan, Inst. of Phys. Conf. Ser. **129**, p.115.

Paper presented at the Int. Symp. GaAs and Related Compounds, Freiburg, 1993

Characterization of GaAs devices using the Franz–Keldysh effect

Randy A. Roush, David C. Stoudt, K. H. Schoenbach*, and J. S. Kenney*

Naval Surface Warfare Center, Dahlgren, Virginia 22448-5000 USA
*Physical Electronics Research Institute, Old Dominion University, Norfolk, Virginia 23529 USA

ABSTRACT

The Franz-Keldysh electro-absorption effect has been used to characterize diffusion profiles as well as electrical effects in GaAs. In the presence of strong electric fields, the absorption edge of GaAs shifts to longer wavelengths. Therefore, the transmission through the semiconductor (with electric field applied) can be monitored in order to determine the spatial distribution of the electric field. We have investigated two uses of this technique which involve (1) creating a p-n junction and observing the transverse band-edge light transmission in order to observe diffusion profiles, and (2) probing semiconductor devices to characterize electric field distributions.

1. INTRODUCTION

Characterization of semiconductors is primarily accomplished by thermal spectroscopy techniques such as Photo-Induced Transient Spectroscopy (PICTS)[1,2] and Deep Level Transient Spectroscopy (DLTS)[3], mass spectroscopy such as Secondary Ion Mass Spectroscopy (SIMS) and Glow Discharge Mass Spectrometry (GDMS), and various microscopy techniques. The latter two techniques involve the destruction of the sample by sputtering in order to measure the masses of the atoms. The thermal techniques are designed to provide information concerning the energy levels within the bandgap of semiconductors. This information includes the densities, cross-sections, activation energies, and in some cases the carrier type can be identified. The DLTS technique involves the use of a junction in order to measure the change in junction capacitance with temperature. The PICTS technique requires the use of a small voltage across the sample and an irradiation source, such as a laser with a photon energy greater than the bandgap which fills the deep traps with charges. Upon extinguishing the laser source, the trapped charges are thermally emitted, and this process is observed by measuring the current through the sample. The temporal characteristics of this current give the information desired. All of these techniques are designed to provide different information about the nature of impurities and defects in semiconductors. None of these methods are capable of providing accurate information concerning the diffusion profiles in the semiconductor over hundreds of microns in depth and large scale surfaces (cm^2).

A new technique to electrically characterize the diffusion profiles in semiconductors is absorption spectroscopy which relies on the Franz-Keldysh absorption shift[4]. This effect is observed in regions of the semiconductor where high electric fields are present. A strong electric field causes the bandgap of GaAs to effectively shift such that light with a previously sub-bandgap photon energy will be absorbed. One way to create these high field regions is to apply the voltage to a p-n junction. The majority of the voltage will appear across the depletion region, which is on the order of a few microns, and therefore high electric fields are formed. This idea may be extended to say that large area p-n junctions (1 cm^2) can be formed (through diffusion processes) which correspond to large area depletion regions. Observation of these depletion regions in the transverse direction via the Franz-Keldysh effect allows the diffusion profile (ie. p-n regions) to be examined.

The Franz-Keldysh effect is characterized by a shift in the absorption edge due to an applied electric field as follows[5]:

$$\Delta E = \frac{3}{2}(m^*)^{-\frac{1}{3}} q \hbar \xi^{\frac{2}{3}}$$

where ΔE is the change in the energy gap, m^* is the effective mass of electrons, q is the electronic charge, $\hbar$ is Planck's constant, and ξ is the electric field. This means that absorption is enhanced in a depletion region. This is due to the probability of finding an electron in the energy gap following the relation,

$$e^{-\frac{|E-E_g|}{\Delta E}}$$

where E_g is the gap energy, and E is the energy in the gap at any position. Therefore, this effect is sometimes called "photon-assisted tunneling".

2. EXPERIMENTS

In our experiments, a microscope equipped with a CCD camera was used to observe light (900 nm) transmitted through the GaAs substrate. The light source was a tungsten lamp with a 900 nm notch filter (± 1 nm) located under the substrate. In this way, only the 900 nm light is transmitted through the GaAs, into the microscope objective and CCD camera. The GaAs samples were fabricated by thermally diffusing copper into an isolated region of a silicon-doped GaAs wafer. Copper forms several acceptor levels in GaAs which compensate the shallow silicon donors. Thermally diffused copper has been shown to compensate shallow donors in GaAs and the solubility of copper is exponentially dependent on the temperature of the anneal. Therefore, by adjusting the annealing temperature, semi-insulating GaAs as well as strongly p-type (overcompensated) GaAs may be created using this process.[6] In this experiment the samples were intentionally over-compensated in order that the p-type regions were formed on the n-type wafer. Shown in figure 1 are the region around the n-type cathode without an applied voltage (a), and the same region with an applied voltage of 100 V (b). This figure shows that the diffusion profile in the sample is exposed by observing the Franz-Keldysh absorption shift. The sub-bandgap laser is used as a probe, and the dark areas of the semiconductor are identified as high-field regions. Therefore, the dark regions correspond to the p-n junctions. These figures show the nature of the non-uniform diffusion of copper into silicon-doped GaAs.

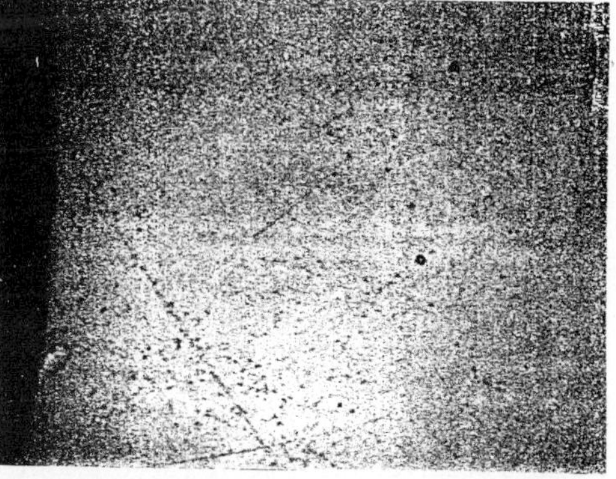

(a)

(b)

Figure 1. Image of a copper-doped GaAs sample without (a) and with an applied voltage of 100 V (b). Scale: 11 cm = 1 mm

The samples used in this study can be formed either by thermal diffusion of copper into n-type GaAs or by ion implantation of n-type dopants into a substrate which has been doped with a p-type impurity. In fact, we have studied samples with known implantation layers, and we have observed the Franz-Keldysh effect. This technique allows for the close control of the n-type region depth. Also, using a known p-n junction created by ion implantation, the electric field can be calibrated with a grey-scale such that impurity concentrations may be extracted from the pictures. We have used one n-type sample with a p-type implant region, and we have observed the expected absorption shift, however, the effect was such that the contrast was not high enough. Therefore, we will improve our image processing methods (using background subtraction, for example) in order to extract the required information.

(a)

(b)

Figure 2. Images of a GaAs-based photoconductive switch during device operation. Scale 1.7 cm = 1 mm.

The Franz-Keldysh effect has also been shown to be useful in the characterization of photoconductive switches[7]. These devices are used to deliver current to a load after excitation from a laser source. Of particular interest is the Bistable Optically Controlled semiconductor Switch (BOSS)[8], which is a device that relies on deep copper centers to allow closing and opening of the device. Nonlinear current transport and electric field distributions have been observed by probing this device with band-edge light during the application of high electric fields. Figure 2 shows images of a photoconductive GaAs device during operation. Moderate applied voltages give an image shown in 2a, and an increase in voltage causes the material to collapse into filamentary current flow (2b). The bright channel extending from anode to cathode in 2b is a current filament, which demonstrates the difference between recombination radiation from a dense electron-hole plasma (bright channel) and the absorption shift due to high local electric fields (dark areas).

3. CONCLUSIONS

To our knowledge, the Franz-Keldysh has not been used to characterize the diffusion profiles in semiconductors. The major advantages of this technique are depth profiling, sample size, and the ability to observe the diffusion profile either with or without microscopy. First, depth profiling is important when dealing with thick (>100 μm) substrates. Our technique allows us to see into the material because the sample is illuminated from the back, and this gives the capability to focus the image on any plane throughout the thickness of the sample (within the resolution of the microscope). In this way, the diffusion profile in the material can be observed for samples with thickness in the mm range. Our technique can also be used in conjunction with other microscopy techniques in order to see smaller scale diffusion profiles. Therefore, this technique may be applied to observe large scale (cm^2) or small scale (μm^2) effects.

In related work, it has been shown that characterization may be performed to determine the distributions of electric fields within small or large scale devices. This may become useful to micro-electronics designers who wish to observe these effects to provide experimental feedback to the design process. The results are roughly the experimental analog of solving the two dimensional Poisson's equation. Calibration of these results is presently being investigated in order to quantify the electric field with respect to the observed dark regions[7].

4. REFERENCES

1. Ch. Hurtes, M. Boulou, A. Mitonneau, Appl. Phys. Lett., **32**, 821, (1978).
2. L. Young, W. C. Tang, S. Dindo, and K. S. Lowe, J. Electrochem. Soc., **133**, 609, (1986).
3. D. V. Lang, J. Appl. Phys., **45**, 3023, (1974).
4. T. S. Moss, J. Appl. Phys., **32** (10), 2136, (1961).
5. J. I. Pankove, Optical Processes In Semiconductors, General Publishing Company,

Ltd., 1971, pg. 29.

6. R. A. Roush, D. C. Stoudt, M. S. Mazzola, Appl. Phys. Lett., **62** (21), 2670, (1993).
7. J. S. Kenney, K. H. Schoenbach, F. E. Peterkin, Ninth IEEE International Pulsed Power Conference, Albuquerque, New Mexico, 1993.
8. K. H. Schoenbach, V. K. Lakdawala, R. Germer, and S. T. Ko, J. Appl. Phys., **63**, 2460, 1988.

Inst. Phys. Conf. Ser. No 136: Chapter 11
Paper presented at the Int. Symp. GaAs and Related Compounds, Freiburg, 1993

Novel low-magnetic-field-dependent Hall-technique

H. Koser, O. Völlinger, and H. Brugger

Daimler-Benz AG, Forschungszentrum Ulm, D-89081 Ulm (Germany)

ABSTRACT: The measurement of the resistivity (ρ_{xx}) and the Hall coefficient (ρ_{xy}/B) over a magnetic field range $\Delta B = 0.5$ T allows a selective determination of the mobility (μ) and the carrier density (n_s) of individual layers in multi-layer conducting samples. The measured $\rho_{xx}(B)$ and $\rho_{xy}(B)$ values are numerically transformed into a conductivity S (or n_s) vs. continuous function of mobility plot. A selected number of measured data at different magnetic fields is used for the numerical calculation of $S(\mu)$, followed by a non-linear least squares fit on the measured $\rho_{xx}(B)$ and $\rho_{xy}(B)$ values. This novel technique is applied for the characterization of different modulation-doped field-effect-transistor layers. It allows a determination of the two-dimensional electron-gas transport properties (n_s and μ) with high precision even in the presence of dominating parallel conducting cap layers.

1. INTRODUCTION

The vertical layer sequence of novel semiconductor device structures (MODFETs, HBTs, etc.) consists of several hetero-epitaxial regions with different carrier mobilities, densities and polarities. The knowledge of the electronic properties of individual layers in the presence of parallel conducting regions is of basic interest for the optimization of the material quality. The routinely used Hall method and the more sophisticated Shubnikov-de Haas (SdH) technique do not allow a selective determination of mobility and carrier density in multi-layer conducting samples.

We report about a magnetic-field-dependent (B-dep.) Hall-technique which allows a selective determination of transport properties of single-layers with different mobility values in multi-layer conducting samples. The applicability of this novel method is demonstrated on modulation-doped field-effect-transistor (MODFET) structures. From the B-dep. Hall measurements of the resistivity (ρ_{xx}) and the Hall coefficient ($R_H = \rho_{xy}/B$) the transport properties of two-dimensional (2D) electron gas (2DEG) and hole gas (2DHG) systems are selectively determined in the presence of highly doped cap layers. Different MODFET samples on GaAs, InP and SiGe substrates are investigated with carrier densities between 5×10^{11} cm^{-2} and 3.2×10^{12} cm^{-2} and mobility values in the range between 7 500 cm^2/Vs and 140 000 cm^2/Vs measured at 80K.

2. SAMPLES AND EXPERIMENTAL SETUP

A typical layer sequence of investigated GaAs-MODFET structures is shown schematically in Fig. 1. On a semi-insulating GaAs substrate an undoped buffer layer is grown, followed by a quantum-well (QW) layer (channel), where the 2DEG is localized. In standard samples the QW consists of GaAs material. To enhance n_s a pseudomorphic (PM) InGaAs QW layer is used. The 2DEG is formed by a transfer of carriers from the doped and wide-band-gap AlGaAs supply layer which is separated by a spacer layer to reduce ionized impurity scattering. The transfer rate is improved by using a δ-doped or planar-doped layer. Finally, a thin highly doped GaAs cap is grown to allow the fabrication of low-resistance ohmic contacts.

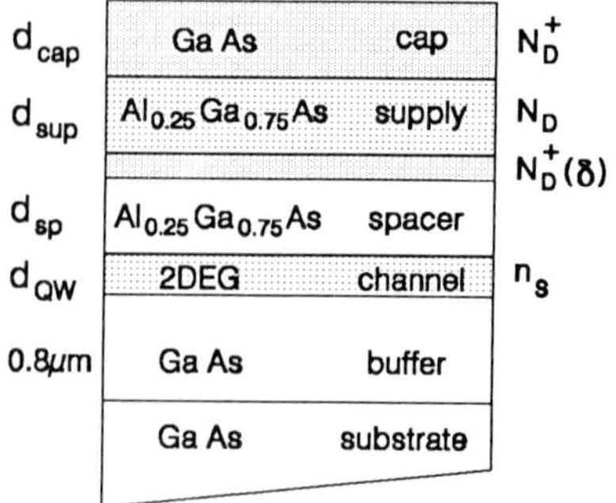

Fig. 1. Schematic layer sequence of a typical MODFET structure for device fabrication.

Additionally, MODFET structures on InP-material with an AlInAs supply layer and an InGaAs 2DEG channel layer (Dickmann 1992) and two types of pseudomorphic SiGe samples on Si-substrates are also investigated, which consist of an n-type SiGe (supply) / Si (2DEG) or p-type SiGe (supply) / Ge (2DHG) layer sequence (Schäffler 1993).

The 2D n_s- and μ-values of investigated samples at T = 80 K range from 5×10^{11} cm^{-2} up to 3.2×10^{12} cm^{-2} and 7 500 cm^2/Vs up to 140 000 cm^2/Vs, respectively. The bypass carrier density in the cap and supply layers range from below 10^{11} cm^{-2} (surface-depleted structures) up to above 10^{14} cm^{-2}. The cap layer mobility values are between 1 000 cm^2/Vs and 5 000 cm^2/Vs.

The magneto-transport measurements are performed at liquid nitrogen temperature with a standard Hall equipment, consisting of a Bruker B-E10 electro-magnet and a HP Data Acquisition/Control Unit, including current source and digital voltmeter. The bias current is set between 10 μA and 1 mA. Van der Pauw or Hall bar contact configurations are used. The magnetic field is varied between 0.04 T and 0.5 T. Circulating currents at the Hall contacts, as suggested by Syphers (1986), are neglected in the low-field-limit (B < 0.5 T). Thermoelectric effects, offset potentials and sample asymmetries are eliminated by forward and reverse data acquisition (Look 1989).

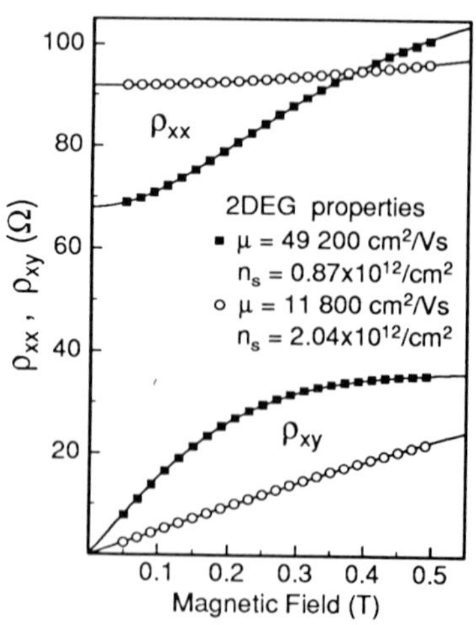

Fig. 2. Measured ρ_{xx} and ρ_{xy} on n^+-cap MODFET structures with different 2DEG n_s- and μ-values: GaAs/AlGaAs (filled squares) and GaAs/InGaAs/-AlGaAs (open circles). The solid lines are calculated results from B-dep. Hall.

In Fig. 2 the measured $\rho_{xx}(B)$ and $\rho_{xy}(B)$ magneto-resistances are shown for two different MODFET structures (standard and PM MODFET on GaAs) at T = 80 K over a magnetic field range of 0.5 T. The strong B-dependence is typical for a multi-layer conducting sample with layers of different mobilities. The cap layer carrier density and mobility values are 2.56×10^{13} cm^{-2} (2.83×10^{13} cm^{-2}) and 1720 cm^2/Vs (1730 cm^2/Vs) for PM (standard) MODFET samples. The 2DEG transport properties are shown in Fig. 2 (see inset).

ρ_{xx} increases parabolically in the low field limit ($\mu B < 1$) in contrast to the B-independent behaviour of a single-layer conducting sample. ρ_{xy} increases non-linearly with B, rather than with a constant slope like on a single-layer sample. Therefore a simple Hall measurement at one fixed magnetic field value yields only average numbers of n_s and μ and does not allow a selective determination of transport properties in the presence of parallel conducting layers.

3. NUMERICAL ANALYSIS

Based on the measured $\rho_{xx}(B)$ and $\rho_{xy}(B)$ data points the conductivity components $\sigma_{xx}(B)$ and $\sigma_{xy}(B)$ are calculated by tensor inversion. The conductivity is then described as a continuous function of mobility by a integral transformation from B-space into μ-space. Following the pioneering work of Beck and Anderson (1987) a continuous function $s(\mu)$ is derived such that the $\underline{\underline{\sigma}}(B)$ tensor can be expressed by

$$\sigma_{xx}(B) + i\sigma_{xy}(B) = \int_{-\infty}^{\infty} \frac{s(\mu)}{1 - i\mu B} \, d\mu \tag{1}$$

where $s(\mu)$ is a B-independent conductivity density function. In general there exist numerous possible solutions for $s(\mu)$ because of the finite number of experimental $\sigma_{xx}(B)$ and $\sigma_{xy}(B)$ data. To overcome this non-uniqueness problem an envelope function $S(\mu)$ over all possible $s(\mu)$ solutions is calculated. Details about the mathematical procedure are published by Beck and Anderson (1987). In Fig. 3 $S(\mu)$ is shown for a PM and a standard GaAs/AlGaAs MODFET layer sequence. The curves are obtained from the experimental data shown in Fig. 2. Two components appear whose abscissa is the mobility of each conducting layer and whose amplitude is the layer conductivity. The corresponding carrier densities are calculated via $n(\mu) = S(\mu)/e\mu$. The low-mobility peaks are related to the highly-doped cap layers. The high-mobility peaks are due to the 2DEG in the QW region. The 2DEG n_s-values obtained are 2.04×10^{12} cm^{-2} and 0.87×10^{12} cm^{-2} for the PM and standard MODFET layers, respectively. Independently, SdH measurements are performed on identical samples. The values obtained are 2.2×10^{12} cm^{-2} and 0.87×10^{12} cm^{-2} and are shown by open circles in Fig. 3. The data obtained from B-dep. Hall and SdH agree excellently.

There are several advantages of this novel Hall technique over conventional methods: No assumptions are necessary regarding the number of layers involved, and energy-dependent scattering mechanism, non-parabolicity effects and multiple-band conductivity are included. The method is limited to the classical region below the onset of quantum mechanical oscillations (SdH effect). The resolution and accuracy depends sensitively on the used $\rho_{xx}(B_k)$ and $\rho_{xy}(B_k)$ data points, where B_k are the K different B-values.

In practice the system is overdetermined, because the number of conducting layers is less than the number of measured data points. Additionally, there are data errors due to measurement accuracy and experimental limitations in B-field tuning. Beck and Anderson (1987) pointed out, that a maximum resolution is achieved if the field values are $B_k \propto 1/\mu$. Colvard et al. (1989) have shown that only an additional final fit of the $S(\mu)$ results to the data points $\rho_{xx}(B)$ and $R_H(B)$ yield to a 20% accuracy of the whole procedure; this was based on PM MODFET layers. But there were still problems on samples with highly conducting cap layers. Brugger and Völlinger (1993) have shown that the accuracy of the B-dep. Hall results is improved, if $K = M+1$ different data points are used for the calculation, and if the $\rho_{ij}(B_k)$-values are selected via their maximum relative sensitivity on the expected n_S- and μ-values of the M parallel conducting layers.

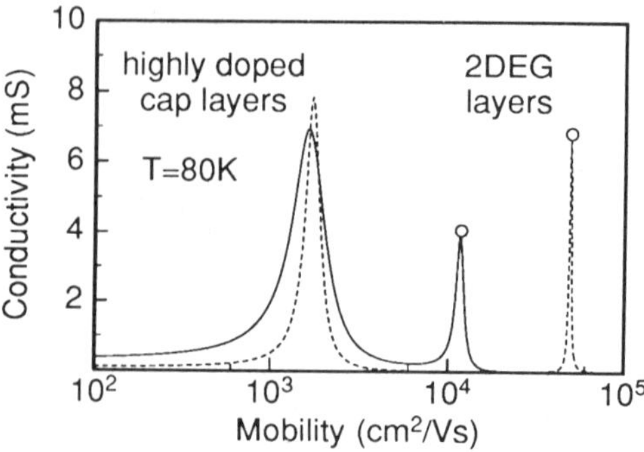

Fig. 3. Conductivity vs. continuous function of mobility plot for two MODFET structures with different 2DEG n_s- and μ-values: GaAs/AlGaAs (dashed line) and GaAs/InGaAs/AlGaAs (solid line).

A more sensitive procedure is the measurement of a whole set of data points over a field range of 0.5 T as shown for example in Fig. 2. Based on an optimized $\rho_{ij}(B_k)$ data selection for the $S(\mu)$ calculation an initial (n_j, μ_j) data set is determined which is used for the calculation of the resistivity tensor components by the formula

$$\rho_{xx}(B,n,\mu) \;+\; i\,\rho_{xy}(B,n,\mu) \;=\; \left(\sum_{k=1}^{M} \frac{e n_k \mu_k}{1 - i\mu_k B} \right)^{-1} \qquad (2)$$

The calculated values of equation (2) are fitted to the measured values $\rho_{xy}(B) = B \times R_H(B)$ and $\rho_{xx}(B)$. This is achieved by a minimization of the merit function

$$\chi^2(n,\mu) \;=\; \sum_{k=1}^{N} \left(\frac{(\rho_{xx})_k - \rho_{xx}(B_k,n,\mu)}{(\Delta\rho_{xx})_k} \right)^2 + \left(\frac{(\rho_{xy})_k - \rho_{xy}(B_k,n,\mu)}{(\Delta\rho_{xy})_k} \right)^2 \qquad (3)$$

which is solved iteratively using a modified algorithm for least-squares estimation of non-linear parameters (Marquardt 1962). The experimental uncertainties are considered by the relations

$$\frac{\Delta\rho_{xx}}{\rho_{xx}} = \sqrt{\left(\frac{\Delta U_x}{U_x}\right)^2 + \left(\frac{\partial\rho_{xx}}{\partial B}\frac{\Delta B}{\rho_{xx}}\right)^2} \quad , \quad \frac{\Delta\rho_{xy}}{\rho_{xy}} = \sqrt{\left(\frac{\Delta U_H}{U_H}\right)^2 + \left(\frac{\partial\rho_{xy}}{\partial B}\frac{\Delta B}{\rho_{xy}}\right)^2} \qquad (4)$$

Relative experimental errors considered in the calculation are 10^{-4} for the measured voltages and 5×10^{-3} for the magnetic field. Additionally, an absolute error for magnetic field tuning of 1 mT is assumed. This algorithm allows the 2D transport properties on MODFET structures with highly conducting cap layers to be determined with the highest ever reported resolution and accuracy.

4. RESULTS

The reliability of the B-dep. Hall method based on the above mentioned algorithm is demonstrated experimentally on a series of samples which are etched stepwise to reduce continuously the bypass conduction in the doped cap layers. A standard GaAs/AlGaAs MODFET is used with the layer sequence shown in Fig. 1 with $d_{cap} = 45$ nm, $d_{sup} = 48$ nm and $d_{sp} = 10$ nm. A series of selected curves at different etch depths is shown in Fig. 4. The initial total cap layer carrier density is 1.9×10^{13} cm^{-2} in comparison with the 2DEG $n_s = 5 \times 10^{11}$ cm^{-2}. The 2DEG signal appears at 118 000 cm^2/Vs. The bypass conductivity (resp. carrier density) of the cap layer decreases with increasing etch depth, whereas the 2DEG-peak remains unchanged as expected. In Fig. 5 the measured overall n_s- and μ-values obtained from a standard Hall experiment at $B = 0.1$ T and $B = 0.3$ T on the same sample are shown as a function of etch depth, together with 2DEG results from B-dep. Hall. It is clearly seen that the standard Hall results do not reflect the 2DEG transport properties. The higher the used magnetic field the larger the deviation of the measured n_s and μ results from the real 2DEG properties. The n_s and μ curves approach continuously the 2DEG values with increasing etch depth i.e. decreasing bypass contribution. The reliability of the B-dep. Hall method is demonstrated by the etch-depth-independent 2DEG n_s- and μ-values. The 2DEG density obtained agrees excellently with the $n_s = 5.3 \times 10^{11}$ cm^{-2} value from SdH, which is indicated in Fig. 5 by the solid line.

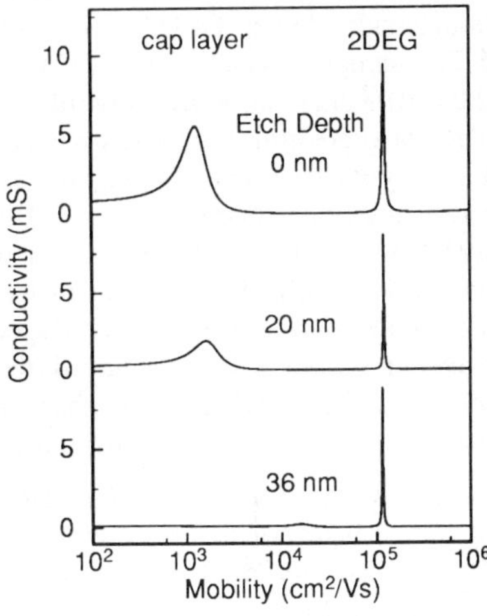

Fig. 4. S(μ) functions from stepwise etched GaAs/AlGaAs MODFET structures.

The applicability of this novel technique for a selective determination of transport properties in MODFET layers is demonstrated on varies heterostructures on GaAs, InP and SiGe substrates (Brugger 1993). Samples with highly conducting, weakly doped (surface depletion) and nominally undoped cap layers are used. The 2D carrier densities are determined independently by low-temperature SdH oscillations. In Fig. 6 a comparison of the 2D n_s-values from B-dep. Hall and SdH is shown. The 2DEG μ-values are in the range between 7 500 cm^2/Vs and 140 000 cm^2/Vs. The heterostructure Ge/SiGe consists of a 2DHG in the strained Ge layer. The other data points are from 2DEG samples. The results are obtained with the samples in the dark. The agreement between Hall and SdH results is excellent.

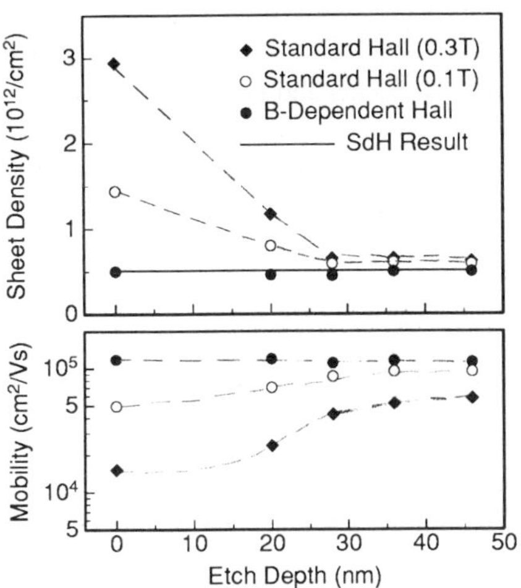

Fig. 5. Measured n_s- and μ-values of a GaAs/AlGaAs MODFET structure as a function of different etch depths.

The present accuracy and resolution of the B-dep. Hall method is demonstrated on the result of a PM double-side doped GaAs/InGaAs/AlGaAs MODFET sample which shows a 2DEG $n_s = 2.04\times10^{12}$ cm^{-2} at a mobility of 11 800 cm^2/Vs. The corresponding conductivity curve is shown in Fig. 3 by the solid line. The highly doped cap layers have a total carrier density of 2.56×10^{13} cm^{-2} at the mobility of 1 720 cm^2/Vs. A narrow 2DEG conductivity peak is observed, which can be well separated from the extremely doped cap layers even for such a small mobility difference of only 10 000 cm^2/Vs and a more than one order of magnitude higher cap layer carrier density.

Further work on multi-layer samples with different p- and n-type conducting regions are underway. The applicability of this novel method for room temperature characterization is investigated.

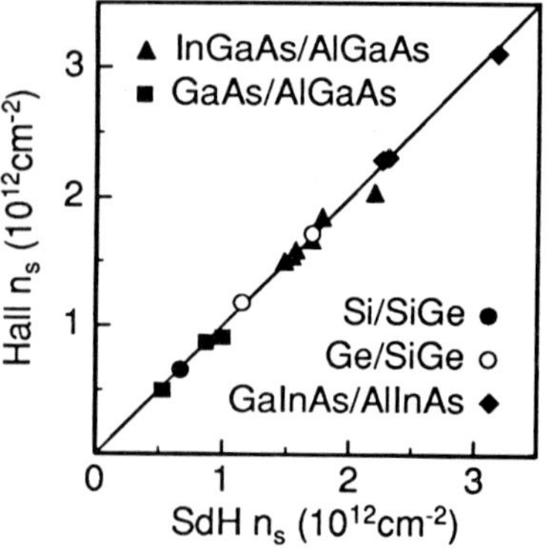

Fig. 6. 2D n_s-values from B-dep. Hall and SdH on different MODFET structures.

5. CONCLUSION

The presented B-dep. Hall-technique allows the transport properties of individual layers to be determined in the presence of highly conducting bypass regions of different mobilities. The applicability of this technique at liquid nitrogen temperature was demonstrated on different MODFET structures. The 2D carrier density and mobility values obtained agree excellently with SdH-data and Hall results from stepwise etched samples. The method is useful for routine characterization and requires only a low-cost 0.5 T equipment. The method is an alternative technique for time-consuming stripping Hall measurements.

ACKNOWLEDGEMENT

The authors thank C. Wölk, Th. Hackbarth, F. Schäffler, H. Nickel and H. Künzel for MBE-material, W. Limmer for SdH-data and U. Meiners for expert technical help. The work was partly supported by the Bundesministerium für Forschung und Technolgie under contract number 01 BM 120/7.

REFERENCES

Beck W A and Anderson J R 1987 J. Appl. Phys. 62 541

Beer A C 1963 Galvanomagnetic Effects in Semiconductors, Solid State Physics Suppl. 4 (New York: Academic Press)

Brugger H and Völlinger O 1993 (submitted for publication)

Colvard C Nouri N Ackley D and Lee H 1989 J. Electrochem. Soc. 136 3463

Dickmann J Dämbkes H Nickel H Lösch R Schlapp W Böttcher J Künzel H 1992 IEEE Microwave and Guided Wave Letters 2 472

Look D 1989 Electrical Characterization of GaAs Materials and Devices (Chichester: Wiley)

Marquardt D W 1963 J. Soc. Indust. Appl. Math. Vol.II, No 2

Schäffler F 1993 Proc. of Int. Conference on Modulated Semiconductor Structures MSS-6 Garmisch-Partenkirchen (Germany) Aug 23-27 (in press)

Syphers D A Martin K P and Higgins R J 1986 Appl. Phys. Lett. 49, 534

A novel *in-situ* characterization method of quantum structures by excitation power dependence of photoluminescence

Toshiya Saitoh, Hideki Hasegawa and Takayuki Sawada

Research Center for Interface Quantum Electronics and Department of Electrical Engineering, Hokkaido University, North 13, West 8, Sapporo 060, Japan

ABSTRACT: Excitation power dependence of photoluminescence from AlGaAs/GaAs quantum wells is analyzed both theoretically and experimentally. It is shown that the recombination processes at the sample surface and the quantum structure interfaces, which have not been considered previously, can cause marked non-linearity. Extending the recently proposed photoluminescence surface state spectroscopy, PLS^3, technique, state distributions at quantum well interfaces are determined in an in-situ, contactless and non-destructive fashion. Growth interruption is shown to produce high density of interface states and lower the PL efficiency.

1. INTRODUCTION

Intensity of photoluminescence (PL) from quantum wells and wires frequently exhibit non-linear dependence on the excitation intensity. Such non-linear dependence was previously interpreted in terms of intrinsic reduction of radiative recombination life time with excitation power (Komiya et al. 1984), increase of non-radiative recombination life time through defect saturation (Patel et al. 1991), non-linear D-A pair transition (Schmidt et al. 1992) etc. However, none of these interpretations has paid due consideration to the recombination processes that take place on the sample surface and the quantum structure interfaces.

On the other hand, in the traditional analysis of bulk photoluminescence (for example, Mettler 1977), the rate of surface recombination is described by a constant parameter called surface recombination velocity, S. However, our recent computer analysis has shown that S is not a characteristic constant of surface, but its magnitude depends in a complicated fashion on the excitation power and wavelength and the conduction type and the doping level of the semiconductor, etc (Saitoh et al. 1990). Briefly, such complicated behavior is primarily due to photo-induced variation of surface state occupancy. This analysis, in turn, has led to development of a new contactless, non-destructive and in-situ measurement method of surface state density (N_{ss}) distribution on semiconductor "free" surfaces called photoluminescence surface state spectroscopy (PLS^3) (Saitoh et al. 1991). In this method, non–linear dependence of PL intensity on excitation power is utilized to determine the N_{ss} distribution. PLS^3 technique has been successfully applied to processed free surfaces (Sawada et al. 1993) and MBE regrowth interfaces (Sawada et al. 1992).

In this paper, such a computer analysis is applied to the analyses of photo–luminescence from quantum wells, and it is shown that the recombination processes at the sample surface and the quantum structure interfaces cause marked non–linearity between PL intensity and excitation power. Based on this, the PLS^3 method is extended and applied for the in–situ assessment of AlGaAs/GaAs quantum wells prepared by MBE with and without growth interruption.

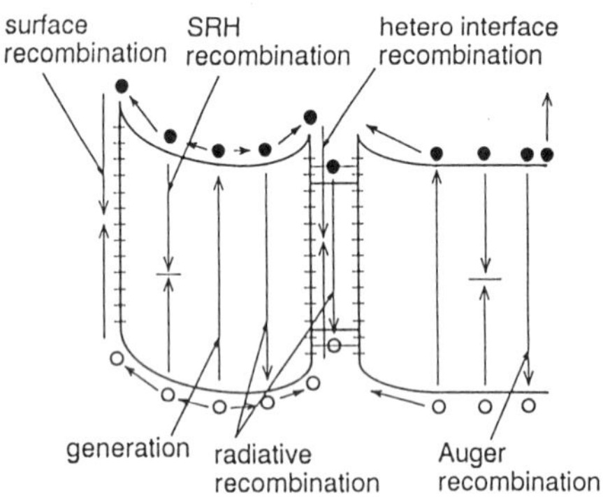

Fig.1 Schematic representation of physical situation of illuminated quantum structure.

2. METHOD OF ANALYSIS

The physical situation for PL measurement is schematically shown in Fig.1. To analyze such a complicated situation, one–dimensional Scharfetter–Gummel type vector–matrix simulation program has been developed. In the program, several hundreds of mesh points are appropriately chosen, and the continuity equations and Poisson's equation are satisfied at each mesh point. The recombination rates U at surface and hetero–interface are calculated by the following integral given by

$$U = \int_{E_v}^{E_c} \frac{\sigma_p \sigma_n v_{thp} v_{thn} (pn - n_i^2) N_{ss}(E)}{\sigma_p v_{thp}(p+p_1) + \sigma_n v_{thn}(n+n_1)} \, dE. \qquad (1)$$

Such rates are used as the boundary conditions. PL intensity from QW, I_{PL}, is assumed to be proportional to the pn product at QW.

In the analysis, the excitation power is expressed in terms of the photon flux density, ϕ and it is related to various recombination processes including bulk and QW radiative recombination process, surface recombination, recombination at hetero interfaces, bulk SRH and Auger recombination processes. To represent non–linear dependence of PL intensity, I_{PL}, on ϕ, the PL efficiency, I_{PL}/ϕ, is calculated as a function of ϕ. Thus, non–linearity exists between I_{PL} and ϕ, if efficiency is not constant, and depends on ϕ.

3. RESULTS AND DISCUSSION

3.1 Calculated Behavior of PL Efficiency in Quantum Well

In order to see the effects of surface and interface recombination processes on the PL intensity from QW, PLS[3] calculations were made using the model shown in Fig.2(a) and the possible various distributions of surface and interface states shown in Fig.2(b).

Figure 3 shows an example of the calculated excitation power dependence of PL efficiency from a sample having an ideal quantum well with no interface states. Presence of surface states at the sample surface with a U–shaped distribution shown in the inset was assumed. The calculation was done by changing the depth x_q of the quantum well from the surface. It is seen that the PL efficiency depends strongly on the excitation power indicating strong non–linearity even for ideal QW. This non–linearity is caused by surface recombination whose effective velocity changes with ϕ and changes the distribution of excess minority carriers within the sample. It is also seen that non–linearity depends strongly on the surface–QW separation, x_q. The behavior can be explained by the fact that most of the photo–generated carriers contributes to the surface recombination current rather than to the radiative recombination in the QW, if the QW lies near the surface.

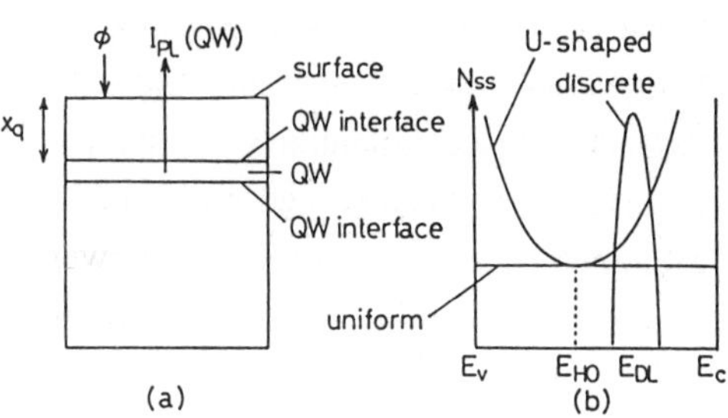

Fig.2 (a) Model of PLS[3] calculation and (b) the possible various distributions of surface and interface states.

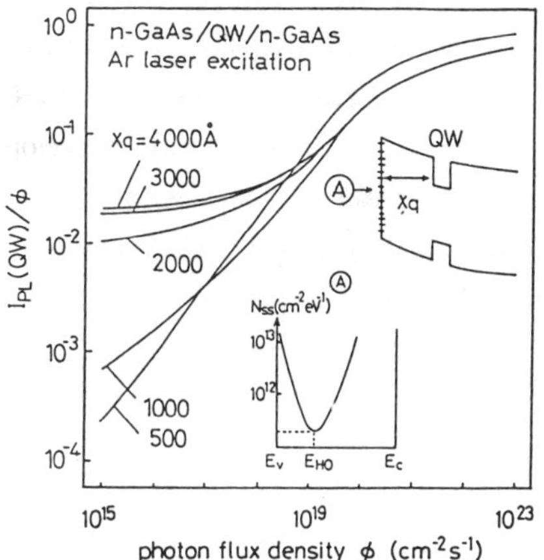

Fig.3 Calculated excitation power dependence of PL efficiency of a sample having an ideal heterointerface.

Interface states at the hetero–interfaces of the quantum well if they are introduced through the fabrication process should also affect the PL efficiency. Figure 4 shows an example of the calculated dependence of PL efficiency on the density N_{ssI} of the hetero–interface. A U–shaped surface state distribution and a uniform distribution of hetero–interface states were used in the calculation as shown in the inset. It is shown that the PL efficiency depends strongly on the N_{ssI}. This suggests that the determination of N_{ssI} can be made by computer fitting to measured PL data, if the surface state density distribution is separately determined using a reference sample which does not have QWs.

In order to see such a possibility, computer fitting was attempted to the reported non-linear PL data on a AlGaAs/GaAs/AlGaAs double hetero structure (V.Swaminathan et al. 1983). It was found that use of surface state density distribution usually present on chemically etched AlGaAs surface cannot at all explain the observed non-linearity, and it can only be explained by assuming presence of a uniformly distributed interface state continuum with $N_{ssI} = 10^{12}\ cm^{-2}eV^{-1}$ as shown in Fig.5.

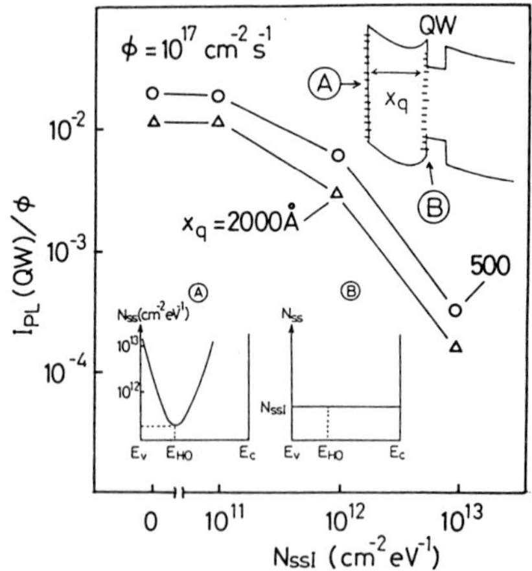

Fig.4 Calculated dependence of PL efficiency on the heterointerface state density, N_{ssI}.

3.2 Application to Experimentally Fabricated Quantum Wells

In order to apply the PLS[3] technique to the assessment of heterointerfaces of quantum wells, AlGaAs/GaAs QW structures shown in Fig.6 were fabricated by MBE. A UHV-based total system consisting of MBE, EB/FIB, CVD, XPS and PL chambers connected by a UHV transfer tunnel was used. In order to investigate the effect of growth interruption, five different samples with and without growth interruption at interfaces A and B in Fig.6 were prepared. They are the samples (1) without growth interruption, (2) with growth interruption at interface A at growth temperature (T_g = 580 °C) under As_4 flux for 1 hour, (3) with growth interruption at interface A below 100°C without As_4 flux for 1 hour, (4) with growth interruption at interface A at room temperature without As_4 flux for 3 and half hours, (5) with growth interruption at interface B at T_g under As_4 flux for 1 hour. These samples were transferred through a UHV transfer chamber to the UHV PL chamber after the growth without the exposure to air. For the measurement of the PL efficiency, an automatic PL measurement system was used where the excitation power was changed by using the

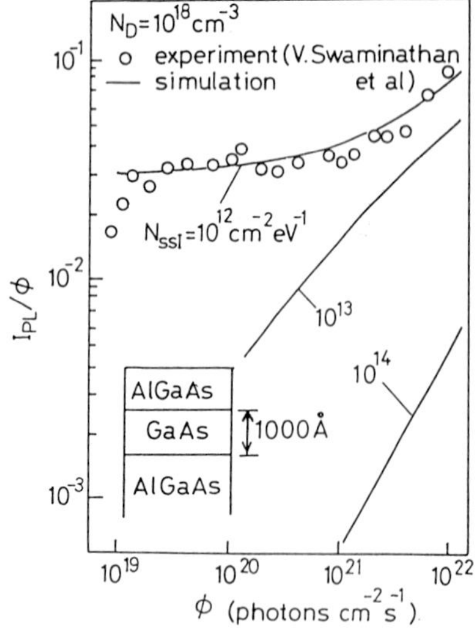

Fig.5 Computer fitting of the reported PL data.

computer controlled ND filter. PL measurements were done at room temperature using excitation by an Ar laser (514.5nm).

All the samples showed clear QW photoluminescence corresponding to $E_{1n}-E_{1hh}$ transition whose PL efficiency depended strongly on fabrication conditions and excitation power. Figure 7 shows the measured excitation power dependencies of PL efficiency. It is difficult to explain such non-linear behavior observed at room temperature by previous models. On the other hand, since generation of continuously distributed interface states has been detected separately by C–V (Ikeda et al. 1988) and PLS[3] (Sawada et al. 1992) after growth interruption at homo and hetero interfaces, it is highly likely that the present non-linearity comes from formation of interface states at quantum well interfaces.

For this, the surface state density distribution was determined by using a sample without QW with the ordinary PLS[3] technique and this was assumed to be the same for all the samples with QW.

It is seen that the interface state density is quite low in the sample without interruption being consistent with the C–V analysis (Tomozawa et al.

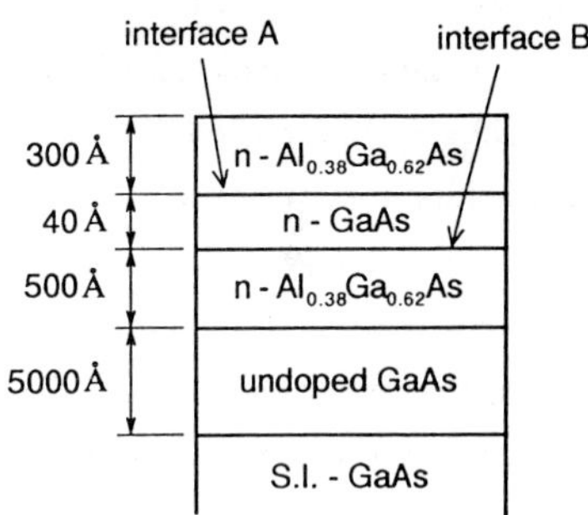

Fig.6 Fabricated AlGaAs/GaAs QW structure by MBE

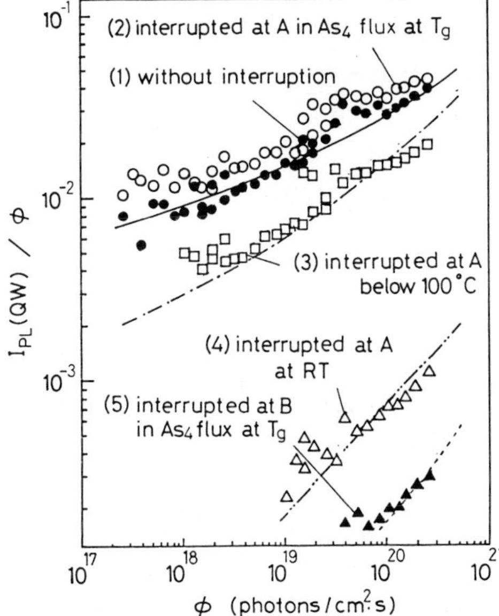

Fig.7 Experimental result of the excitation power dependence of PL efficiency of AlGaAs/GaAs QW structure.

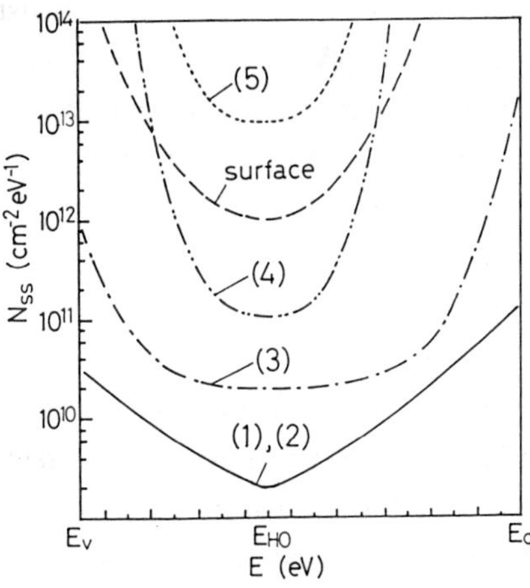

Fig.8 State density distributions at surface and at heterointerface determined by computer fitting.

1992). On the other hand, growth interruption has extremely large effects and can produce high density of interface states depending on the interruption conditions. All the hetero–interface state distributions are U–shaped continuum around the charge neutrality level E_{HO} being consistent with DIGS model (Hasegawa et al. 1986). It is also seen that the growth interruption at interface B seems to have a much larger effect than that at A. Thus, the growth interruption should be minimized in the fabrication process of quantum structures.

4. CONCLUSION

Excitation power dependence of photoluminescence from the AlGaAs/GaAs quantum wells was investigated both theoretically and experimentally. It is shown that the recombination processes at the sample surface and the quantum structure interfaces can cause pronounced non–linearity. Extending the PLS^3 technique to quantum structures, interface state distributions of quantum well interfaces with and without growth interruption were determined in an in–situ, contactless and non–destructive fashion. Growth interruption was found to introduce high density of interface states which are U–shaped in accordance with DIGS model.

REFERENCES

Komiya S, Yamaguchi A, Umebu I and Kotani T 1984 Jpn.J.Appl.Phys. **23** 308

Patel S, Kamata N, Kanoh E and Yamada K 1991 Jpn.J.Appl.Phys. **30** L914

Schmidt T, Lischka K and Zulehner W 1992 Phys.Rev.B **45** 8989

Mettler K 1977 Appl.Phys. **12** 75

Saitoh T and Hasegawa H 1990 Jpn.J.Appl.Phys. **29** L2296

Saitoh T, Iwadate H and Hasegawa H 1991 Jpn.J.Appl.Phys. **30** 3750

Sawada T, Numata K, Tohdoh S, Saitoh T and Hasegawa H 1993 Jpn.J.Appl.Phys. **32** 511

Sawada T, Tohdoh S, Saitoh T, Tomozawa H and Hasegawa H 1992 Proc. of 19th Int. Symp. on Gallium Arsenide and Related Compounds 387

Swaminathan V, Anthony P, Pawlik J and Tsang W 1983 J.Appl.Phys. **54** 2623

Ikeda E, Hasegawa H, Ohtsuka S and Ohono H 1988 Jpn.J.Appl.Phys. **27** 180

Tomozawa H, Numata K and Hasegawa H 1992 Appl.Surf.Sci **60/61** 721

Hasegawa H and Ohno H 1986 J.Vac.Sci.Technol. **B4** 1130

Paper presented at the Int. Symp. GaAs and Related Compounds, Freiburg, 1993

The spatial distribution of deep centers in semi-insulating GaAs measured by means of electron-beam induced current transient spectroscopy

T. Tessnow[1], K.H. Schoenbach[1], R.A. Roush[2], R.P. Brinkmann[3], L. Thomas[4], and R.K.F. Germer[5]

[1]) Physical Electronics Research Institute,
Old Dominion University, Norfolk, VA, 23529-0246, U.S.A.
[2]) Navel Surface Warfare Center, Dahlgren VA, U.S.A.
[3]) Siemens AG, Munich, Germany
[4]) Institut für Metallphysik, Technische Universität Berlin, Germany
[5]) Telecom Berlin Germany

ABSTRACT

The design of semi-insulating GaAs photoconductive switches requires a knowledge of the physical properties of deep centers and their spatial distribution. A method which allows us to obtain the deep level parameters for different depths is Electron Beam Induced Current Transient Spectroscopy (EBICTS). This diagnostic method can be applied to any wide band gap material. Copper doped GaAs was investigated with this method and strong variations of the impurity concentration were found in the surface layer.

1. INTRODUCTION

The most widely known methods used to measure the deep level impurities of semiconductors are DLTS (Deep Level Transient Spectroscopy) and PICTS (Photo Induced Current Transient Spectroscopy) (Schroder 1990). PICTS is suited for highly resistive semiconductors, where a laser is used to activate the traps. In this method the excitation depth is determined by the optical properties of the semiconductor. Generally the wavelength of the laser light is such that it either penetrates the entire sample or is absorbed in the surface layer. To achieve specified penetration depths, the wavelength of the laser light must be tunable about the bandedge of the semiconductor. A much more simple way to obtain the depth profile of deep centers is to use an electron beam instead of a light source to activate the deep centers. Except for the excitation source EBICTS uses the same concept as the PICTS. Since the penetration depth (range) of electrons depends on their energy (figure 1 and 2, Martinell 1973), it is possible to vary the electron range in a reproducible manner by regulating the e-beam voltage (electron energy). The difference of two measurements for electron ranges d_1 and d_2 provide the deep center density in the layer between d_1 and d_2. The deep center profile in semiconductors can therefore be obtained by varying the electron range.

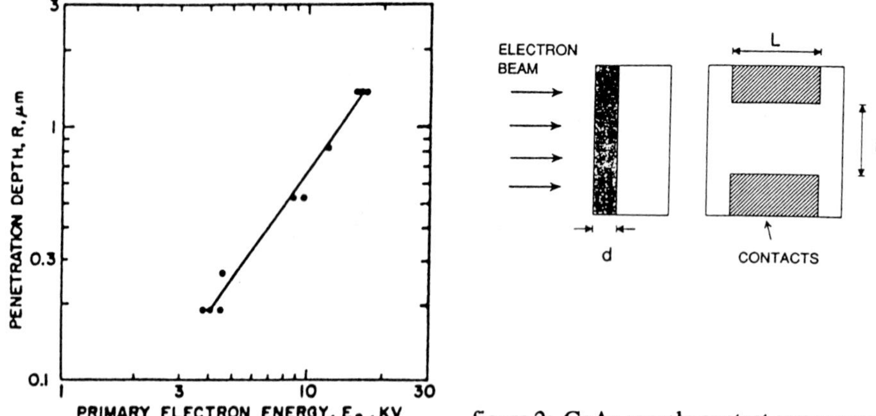

figure 2: GaAs sample,contact arrangement

figure 1: electron penetration depth in GaAs (Martinell 1973)

2. CONCEPT

The electron beam creates free charge carriers in the GaAs sample which generates a current, I_0 and fills the deep centers, for the duration of the e-beam pulse. After the termination of the pulse, an initial fast decay of the free carrier density (current) is caused by band-band recombination. Following this process the remaining charge carriers in the traps contribute to the current through thermal emission. The current carried by the emitted charges decays exponentially, and the emission time constant strongly depends on temperature since multi phonon processes are involved. Several traps will contribute to the current, therefore the total current, I(t), can be described by a sum of exponential functions:

$$I(t) = \sum_j i_j(t) = \sum_j K_j \exp(-\frac{t}{\tau_j}) \tag{1}$$

If I(t) is acquired by a digitizer, the constants K_j and τ_j can be found by a curve fit of I(t). By varying the temperature T and plotting the data on an Arrhenius plot, the energy level and the emission cross-section can be obtained. A knowledge of K_j and τ_j allows us to also calculate the trap density. This is valid only if all traps have been filled by the electron beam pulse (Schroder 1990).

3. EXPERIMENTAL SETUP

The electron beam is generated in a vacuum photodiode (figure 3). The sample, which also serves as the anode, is connected to ground potential. A negative voltage of up to 30 kV (DC) can be applied to the cathode. An excimer laser (193 nm) creates electrons at the cathode with a current density of about 1 A/cm^2. The pulse duration and shape of the electron current pulse is basically determined by the laser pulse. The laser intensity is chosen to be high enough to operate the e-beam in the space charge limited region in order to avoid fluctuations in the e-beam current. A voltage is applied to the sample and the current is measured upon excitation by the e-beam source. The signal is then digitized by an oscilloscope and loaded into a computer.

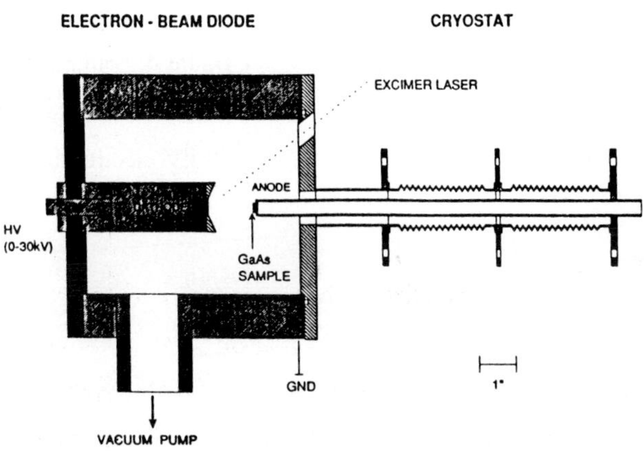

figure 3: schematic of experimental set up

In order to detect different levels, the sample temperature is varied by cooling (liquid nitrogen) or heating (electric heater). In order to insulate the sample thermally from the vacuum system, the sample holder is made of stainless steel which has very low thermal conductivity . A flange on one end of the cryostat allows for the insertion of an electric heater or the attachment of a reservoir for liquid nitrogen. The bellows of the cryostat are used to change the e-beam diode gap spacing and therefore the current density according to Child Langmuir's law.

4. EXPERIMENTAL RESULTS

The penetration depth can be changed by varying the electron energy. Measurements in this work involve two electron beam diode voltages, 14 kV and 20 kV, which corresponds to penetration depths of 1.2 μm and 2.0 μm, respectively (figure 1). Current transients have been recorded for a temperature range of 300 K to 450 K and examples are displayed in figure 4. The exponential currents decay faster with higher temperatures because the thermal emission processes are accelerated. The current at 20 kV electron beam voltage is higher than that for 14 kV due to the larger activated volume. Since more than one type of deep center contributes to the current, the current transients consist of a superposition of several exponentially decaying currents, which represent the emission from each deep level.

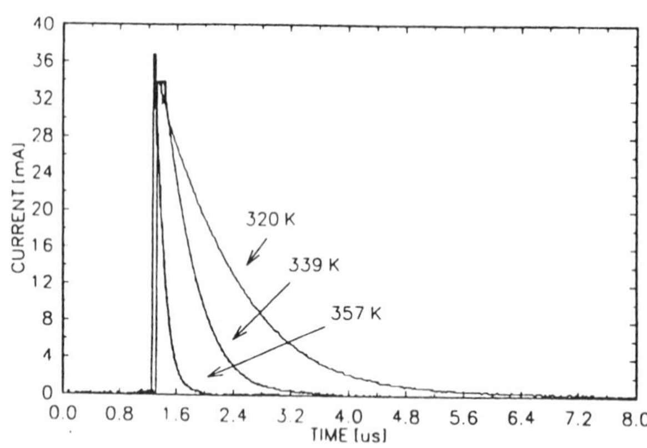

figure 4: sample currents at different temperatures for 20 kV electron beam voltages

5. EVALUATION METHOD

There are several methods to extract the decay constants and amplitudes of the current components from the current transients. The simplest one is the window method (Schroeder 1990). It is fast and simple but has a low energy resolution. Higher resolutions were accomplished by a curve fitting method (Lakdawala 1992) or an spectral analysis (Brinkmann 1993). Since the last one was not available at this time, the curve fitting method was used in this study.

The curve fitting method (Lakdawala 1992) uses a least square fit of a sum of exponentials (equation 1). The number of modes used in the fit is unknown, therefore the simulation of the experimental data starts with one mode and increases the number of modes until the fit is acceptable. Figure 5 and 6 show the Arrhenius plots for two different electron energies: 14 keV and 20 keV.

When calculating the densities, the result for 14 keV needed to be subtracted from the result for 20 keV in order to get an average over only the layer between 1.2 µm and 2.0 µm. The current amplitude of the 14 keV experiment was only 10 % of that for the 20 keV experiment. Therefore 10 % of the results for 20 keV were contributed by the first layer (0..1.2µm) and 90 % by the second layer (1.2 .. 2.0 µm). Surface recombination probably reduced the signal of the 14 keV experiment. Therefore absolute numbers for the densities could not be calculated, but the current signals contributed by each level can be compared with the total current, resulting in a relative concentration, where 100 % is the total density of all detected levels.

electron energy	depth	level	activation energy	relative concentrations
14 keV	0 .. 1.2 µm	Cu-A	(0.17 ± 0.02) eV	60 %
		Cu-C	(0.33 ± 0.04) eV	10 %
		EL-2	(0.62 ± 0.08) eV	30 %
20 keV	1.2 .. 2.0 µm	Cu-A	(0.19 ± 0.07) eV	< 1 %
		Cu-C	(0.36 ± 0.04) eV	66 %
		EL-2	(0.76 ± 0.08) eV	33 %

(notations for Cu-levels taken from Lang 1975 and Roush 1993)

In the first layer (0 .. 1.2 µm) the concentration of Cu-A centers was dominating, while in the second layer (1.2 .. 2.0 µm) Cu-A centers could not be detected anymore. The Cu-C signal was much stronger in the bulk than at the surface. More accurate measurements need to be performed in order to prove this assertion. The EL-2 level seems to be uniformly distributed in the sample, whereas Cu-A and Cu-C show a strong gradient.

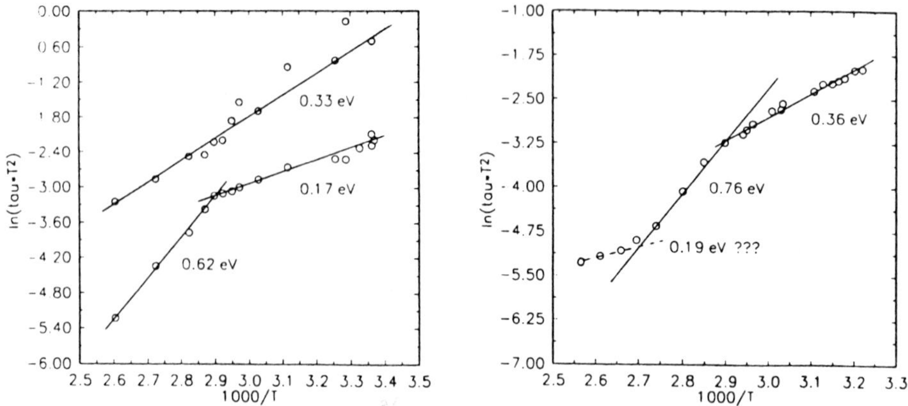

figure 5+6: Arrhenius plots for two penetration depths (1.2 μm, 2.0 μm)

6. SUMMARY

A new diagnostic method, the Electron Beam Induced Transient Spectroscopy (EBICTS) has been developed that allows us to determine the depth profile of deep centers and their properties such as activation energies, densities and cross-sections. The method was applied to the investigation of copper doped GaAs. It was found that the concentration of deep centers depends strongly on the depth. In a layer 0..1.2 μm, the Cu-A level was found to be the dominating center, but between 1.2 and 2.0 μm Cu-A was negligible and Cu-C was dominant. The obtained results indicate that the results of PICTS measurements, which rely on activation of a surface layer by using short wavelength activation radiation, might not be relevant for the bulk of the material.

REFERENCES

Brinkmann R.P., Tessnow T., Schoenbach K.H. and Roush R.A. 1993 9th IEEE Int. Pulsed Power Conf. PII-49 (page no. not known yet)

Lakdawala V.K., Panigrahi J., Thomas L.M. and Brinkmann R.P. 1992 SPIE Optically Activated Switching II, 1632-15

Lang D.V., Logan R.A., 1975, Journal of Electron. Materials, Vol 4, p 1053-1066

Martinell R.U. and Wang C.C., July 1973, J.Appl.Phys., Vol.44, No. 7

Roush R.A., Stoudt D., Mazzola M., Trans. on Electron Devices June 1993, Vol 40, No 6, p 1081-1086

Schroder,D.K. 1990 Semiconductor Material and Device Characterization, chapter 7 (New York: John Wiley)

Optical investigation of MBE overgrown InGaAs/GaAs wires

K.Pieger, Ch.Gréus, J.Straka, A.Forchel;

Technische Physik, Universität Würzburg, Am Hubland, D-97074 Würzburg, Germany;

ABSTRACT: We have prepared buried InGaAs/GaAs wires by MBE overgrowth of wet etched wires with widths down to 40 nm. The overgrown structures show a good surface morphology and good luminescence properties. In contrast to the monotonous decrease of the emission intensity with decreasing wire widths, observed for etched only structures, we observe an increase of the emission intensity in the overgrown wires. This effect is correlated with the passivation of surface recombination centres by the overgrowth and an efficient carrier capture into the wires after overgrowth.

1. INTRODUCTION

Novel optoelectronic devices for future applications based on low dimensional semiconductor structures are of increasing interest. Various technological processes have been investigated for patterning two dimensional heterostructures (Izrael et al. 1991). Most commonly these structures are defined by high resolution lithography and dry (Itoh et al. 1991), (Bickl et al. 1992) or wet (Schmidt et al. 1992), (Notomi et al. 1991) chemical etching. In these cases however, defects are often created near the open surfaces at the etched interface (Clausen et al. 1989). Especially ultra narrow semiconductor structures, which are interesting for e.g. quantum wire lasers, show a significant decrease of radiative quantum efficiency with decreasing lateral width. We show, that it is possible to reduce nonradiative surface recombination by MBE overgrowth.

2. PREPARATION OF QUANTUM WIRES

We have fabricated two-dimensional heterostructures, to be used for patterning, which consist of a MBE grown 5nm InGaAs SQW with an indium content of 10 to 18% cladded between a GaAs/AlGaAs barrier of 800nm and a GaAs top layer of 15 to 50nm thickness. Most samples were coated with 100nm negative resist SAL 601 and exposed with a high resolution electron beam of a scanning electron microscope. Fields of etch masks with wires between 1000 and 50nm lateral width were realised. Positive acting electron beam resist PMMA was used in order to achieve ultra narrow aluminium masks for wet chemical etching.

Two types of wires were investigated: In the first case, deep etched wires were prepared by a non selective etchant using H_2SO_4:H_2O_2:H_2O (1:8:1200). In this case the active InGaAs - layer is etched through and surface recombination is a very effective nonradiative recombination channel in particular for narrow wires. In the second case wires were created by selective top barrier removal. Here the top barrier material is removed by a selective etchant H_2O_2 buffered with NH_4OH (pH 7), GaAs-etch rate: 70 nm/min at 23°C, $In_{.18}Ga_{.82}As$-etch rate: 7 nm/min. Because of the increase of the energy band discontinuity in the unmasked areas (exchange of the barrier material from GaAs effectively to vacuum, i.e., the discontinuity is replaced by the electron affinity of about 5eV), we obtain a confinement of the carriers in the masked sections. One main advantage of this technique is the reduction of nonradiative recombination at the etched sidewalls compared to deep etched structures (Gréus et al. 1992). As shown on figure 1, we obtain lateral wire widths down to 30 nm, which are even smaller than the etch mask because of lateral undercut. The SAL resist mask is removed before overgrowth by use of dimethylformamide in ultrasonic bath.

Fig.1: Barrier modulated InGaAs/GaAs quantum wires. Lateral width: 30nm

3. PARAMETERS OF THE MBE OVERGROWTH PROCESS

In order to reduce the nonradiative recombination at the open sidewalls in the case of deep etched quantum wires and to passivate the surface states of the lateral barriers in the case of barrier modulation, we have overgrown the etched structures with a 100nm thick layer of GaAs (deep etched wires) or AlGaAs (barrier modulation). The samples were prepared for this second epitaxy by an organic cleaning step, followed by HCl (30%) and finally H_2SO_4 (96%) etching at 20°C, both stopped by rinsing in water.

We have systematically optimized the growth parameters of the second epitaxy including the in situ annealing conditions, the growth temperature, the III-V ratio and the growth rate. The surface morphology of the structures was evaluated by high resolution scanning electron microscopy. A small residual surface roughness is obtained for the following technological steps: a) Annealing of the sample for oxide desorption during one minute at 600°C under arsenic overpressure in the vacuum chamber. b) Growth, using a growth rate of $0.2 Ås^{-1}$, a V/III ratio of 25 up to 30 and a growth temperature of 600±10°C. Only fairly small irregularities occur close to the edges of the overgrown structures, which result from the etch induced roughness before overgrowth.

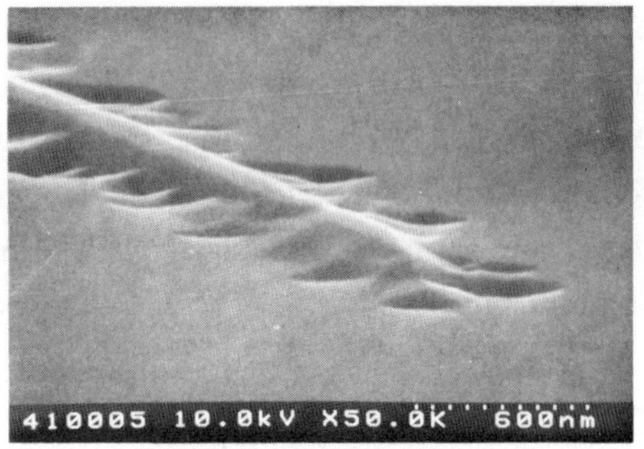

Fig.2: Overgrown deep etched wires of 200nm lateral width

We have studied the influence of the crystallographic orientation by examining <011> and <01-1> oriented wires. Our wet chemical etching procedure always generates <111> sidewall planes, which are either gallium or arsenic terminated. The arsenic terminated <111>B surfaces of the <011> oriented wires offer the best results after MBE overgrowth.

4. PHOTOLUMINESCENCE MEASUREMENTS

We have carried out photoluminescence measurements using an Ar^+-laser with an excitation wavelength of 514 nm and a multichannel analyzer as a detector. The open triangles in fig. 3

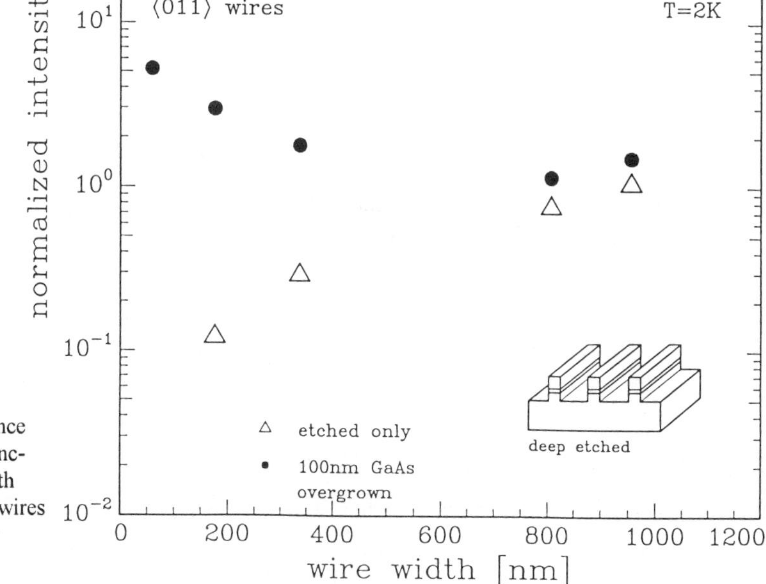

Fig.3: Photoluminescence intensity as a function of wire width for deep etched wires before and after overgrowth.

show the wire width dependence of the emission intensity, normalized by the respective area filling factors, of deep wet chemical etched wires before overgrowth. Fig. 4 shows the width dependence for etched only barrier modulated wires.

Deep etched wires with widths smaller than 400 nm show a significant decrease of the emission intensity, which is due to nonradiative recombination at the open sidewalls. In contrast, the normalized intensity of larger wires with widths in the micrometer range, which have a smaller surface to volume ratio is about one. The photoluminescence intensity of overgrown deep etched wires shows an enhancement of more than one order of magnitude for narrow wires due to the good interface quality and the efficient surface passivation (see filled circles in fig 3). The values of the normalized intensity of more than unity are due to carrier capture from the overgrown GaAs barrier.

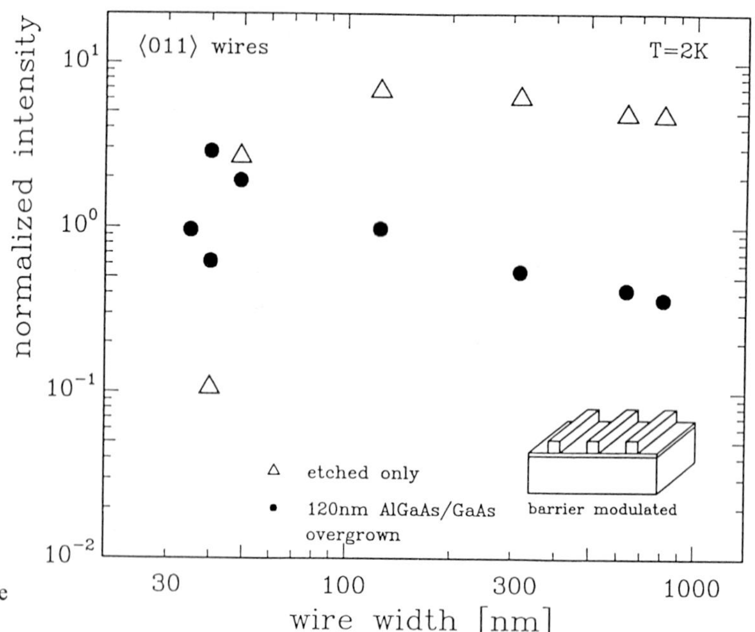

Fig.4: Dependence of photoluminescence intensity on the wire width for barrier modulated wires.

The emission intensity of barrier modulated wires (see open triangles in fig. 4), which have no open sidewalls, decreases considerably only for wire width below 150 nm. This decrease is due to the influence of the etched surface in the unmasked region. Above 150 nm width these wires show a normalized intensity which is more than unity because of carrier capture from the lateral barrier. For overgrown barrier modulated wires the lateral asymmetric InGaAs barrier is partly evaporated during annealing in the growth chamber and replaced by an AlGaAs barrier during the second epitaxy (aluminium content: 15%). The photoluminescence signal (filled dots in fig 4) shows an enhancement of the emission intensity of ultra narrow wires because of the absence of the vacuum interface in the unmasked section and reduced nonradiative recombinations. Large overgrown wires show a normalized intensity which is reduced by a factor of three because of the absorption of laser photons in the overgrown layer, which results in a less dominant barrier excitation and a reduced carrier capture from the lateral barrier.

For both types of structures, deep etched and barrier modulated, our optical studies show, that it is possible to reduce the influence of nonradiative recombination channels for narrow quantum wires.

5. CONCLUSION

We have shown the importance of MBE overgrowth for the quantum efficiency of wet chemical etched low dimensional semiconductor wires. For optimized regrowth conditions we obtain a mirror like surface morphology of the 100 nm to 120 nm thick overgrown layer. Low temperature photoluminescence measurements show an enhancement of the emission intensity by more than one order of magnitude after overgrowth. This improvement is due to the absence of open surfaces and the good semiconductor-semiconductor interface of the overgrown structures compared to the semiconductor-vacuum interface of etched only structures. Nonradiative surface recombination especially at process induced defects, a dominant recombination channel for etched only narrow wires, is significantly reduced.

We acknowledge the Deutsche Forschungsgemeinschaft for the financial support of this project.

REFERENCES

Izrael A, Marzin J Y, Sermage B, Birotheau L, Robein D, Azoulay R, Benchimol J L, Henry L, Thierry-Mieg V, Ladan F R, Taylor L, Jap. Jour. Appl. Phys. **30,** pp. 3256 (1991)

Itoh M, Honda T, Tsubaki K, Jap. Jour. Appl. Phys. **30,** pp. 2455 (1991)

Bickl Th, Jacobs B, Forchel A, Röntgen P, . Gyuro I, Speier P, Zielinski E, Inst. Phys. Conf. Ser. **129,** pp. 933 (1992)

Schmidt A, Forchel A, Straka J, Gyuro I, Speier P, Zielinski E, J. Vac. Sci. Technol. B **10,** pp. 2896 (1992)

Notomi M, Naganuma M, Nishida T, Tamamura T, Iwamura H, Nojima S,. Okamoto M, Appl. Phys. Lett. **58,** pp. 720 (1991)

Clausen Jr E M, Craighead H G, Worlock J M, Harbison J P, Schiavone L M, Florez L, Van der Gaag B, Appl. Phys. Lett. **55,** 1427 (1989)

Gréus Ch, Forchel A, Straka J, Pieger K, Emmerling M, Appl. Phys. Lett **61,** 1199 (1992)

Paper presented at the Int. Symp. GaAs and Related Compounds, Freiburg, 1993

High quality GaInAs/GaAs/GaInP laser structures grown by CBE using new organometallic precursors

Ph. MAUREL, J.C. GARCIA, J.P. HIRTZ

Laboratoire Central de Recherches, Thomson-CSF, Domaine de Corbeville, 91404 Orsay Cedex, FRANCE.

Abstract: High quality GaInAs/GaAs/GaInP laser structures have been grown by chemical beam epitaxy, using new organometallic precursors: tri-isopropylgallium, tertiarybutylarsine and tertiarybutylphosphine. Uncoated single diodes 100 µm wide and 300 µm long exhibited continuous wave output optical powers as high as 500 mW.

Recently, a great deal of attention has been paid to develop new organometallic precursors for replacement of standard group III organometallic sources and group V hydrides in growth techniques such as metalorganic chemical vapour deposition (MOCVD) or chemical beam epitaxy (CBE). Tri-isopropylgallium (TiPGa) has been demonstrated to reduce significantly the carbon incorporation in the CBE grown GaAs epilayers (Lane et al 1992). Parallelly, it was successfully used as Ga precursor for the growth of GaInP (Abernathy et al 1993). In combination with TBP, further reduction of C incorporation is found (Garcia et al 1993a). Tertiarybutylarsine (TBAs) and tertiarybutylphosphine (TBP) are now widely used as group V element sources in MOCVD for the fabrication of high quality optoelectronic and microwave devices (Miller et al 1990, Kim et al 1991, Ougazzaden et al 1991, Ogasara et al 1992). In contrast, only a few reports have been done on CBE grown material (Ritter et al 1990, Hincelin et al 1992, Garcia et al 1993b). We present in this paper significant advances in CBE growth and fabrication of 980 nm GaInAs/GaAs/GaInP laser structures, using TiPGa, TBAs and TBP as alternative precursors.

In our experiments, hydrogen sulfide and a solid source of beryllium are used as starting sources for n-type and p-type doping respectively in addition to TBAs, TBP, TiPGa and TMIn. The growth temperature of the laser structures, with an active layer consisting of three strained $Ga_{0.8}In_{0.2}As$ quantum wells of 80Å, with a GaAs optical cavity, was fixed to 545°C. After growth, a standard photolithographic technique has been used to define broad area laser diodes with widths varying between 40 and 100 µm for high power delivery. These diodes have recently proven to be of interest for pumping Er doped glasses in order to fabricate compact solid-state lasers (Laporta et al 1992). Additional details, concerning growth and technology of the structures are given in a previous paper (Maurel et al 1993).

First, structures grown with TiPGa were processed into 550 µm long diodes. They exhibit both low current density of 450 A/cm^2 and high quantum efficiency of 0.53 W/A. To the best of our knowledge, this is the first report, concerning laser diode fabrication from material using this newly developed precursor (Lane et al 1992). Parallelly, diodes with lengths varying between 300 µm and 1 mm have been processed from material grown with TBAs and TBP as group V element sources. Pulsed measurements exhibit state-of-the-art characteristics for the broad area lasers. Threshold current densities as low as 230 A/cm^2 are obtained for 1 mm long diodes (fig. 1a). Figure 1b shows the variation of the inverse of the quantum efficiency as a function of the cavity length. It is linear and yields an internal quantum efficiency as high as 70 %, together with internal losses as low as 8.5 cm^{-1}. For CW measurements, the diodes were mounted p side down on copper heatsink. With, as cleaved uncoated facets, output powers as high as 500 mW are delivered by 300 µm long, 100 µm wide single diodes (fig. 2).

The authors wish to thank Mr. D. Leguen for expert technical assistance and Dr. B. Groussin from Thomson CSF/TCS for the CW characterizations.

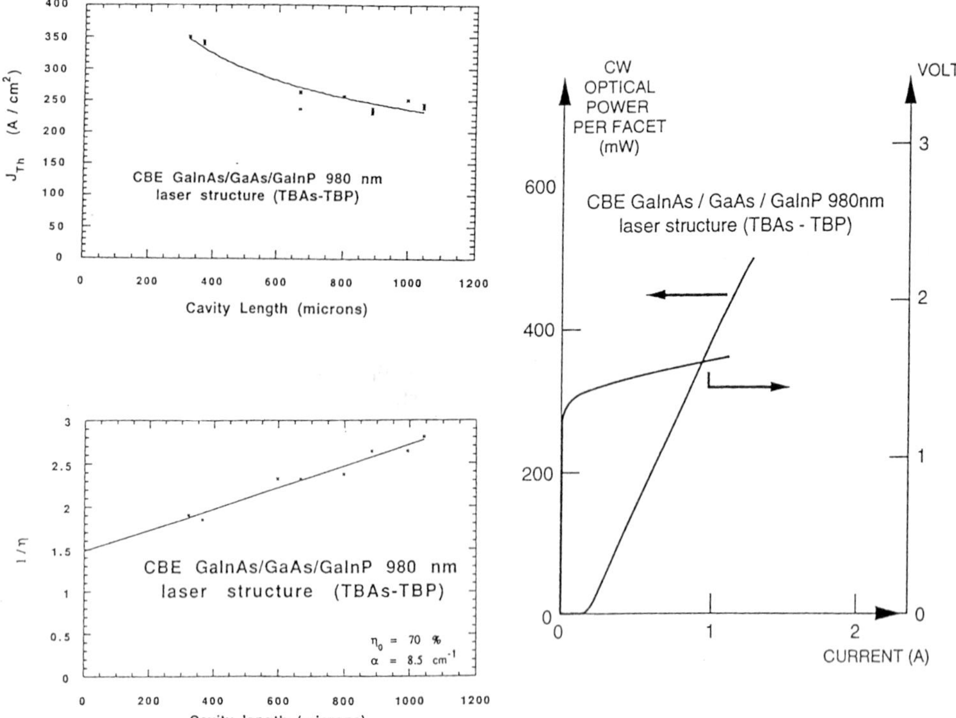

Figure 1a: Threshold current density versus cavity length (structure grown with TBAs and TBP).
Figure 1b: Inverse of quantum efficiency versus cavity length (structure grown with TBAs and TBP)

Figure 2: CW output power per facet for an uncoated single diode 300 µm long, 100 µm wide (structure grown with TBAs and TBP)

References.

Lane P A, Martin T, Freer R W, Calcott P D J, Whitehouse C R ,Jones A C, Rushworth S 1992 *Appl. Phys. Lett.* **61** 285

Abernathy C R, Wisk P W, Ren F, Pearton S J, Jones A C, Rushworth S A 1993 *J. Appl. Phys.* **73** 2283

Garcia J C, Regreny Ph, Delage S L, Blanck H, Hirtz J P 1993a *J. Cryst. Growth* **127** 255

Miller B I, Young M G, Oron M, Koren U, Kisker D 1990 *Appl. Phys. Lett.* **56** 1439

Kim T S, Bayraktaroglu B, Henderson T S, Plumton D L 1991 *Appl. Phys. Lett.* **58** 1997

Ougazzaden A, Mellet R, Gao Y, Kazmierski K, Rhein C, Mircea A 1991 *Electron. Lett.* **27** 1005

Ogasawara M, Sato K, Kondo Y 1992 *Appl. Phys. Lett.* **60** 1217

Ritter D, Panish M B, Hamm R A, Gershoni G, Brener I 1990 *Appl.Phys.Lett.* **56** 1448

Hincelin G, Zahouh M, Mellet R, Pougnet A M 1992 *J. Crystal Growth* **120** 119

Garcia J C, Maurel Ph, Hirtz J P 1993b *Electron. Lett.* **29** 432

Laporta P, Taccheo S, Svelto O 1992 *Electron. Lett.* **28** 172

Maurel Ph, Garcia J C, Hirtz J P, Vassilakis E, Baldy M, Parent A, Carrière C 1993 *Proc. of MRS Conf. San Francisco*

Paper presented at the Int. Symp. GaAs and Related Compounds, Freiburg, 1993

Wannier–Stark effect in $In_{0.53}Ga_{0.47}As/In_{0.40}Ga_{0.60}As$ superlattices

B Opitz, A Kohl, J Kováč*, S Brittner, F Grünberg and K Heime

Institut für Halbleitertechnik, RWTH Aachen, Templergraben 55, D-52056 Aachen, Germany

ABSTRACT: We report photocurrent spectroscopy at 77K on MOVPE grown PIN diodes containing $In_{0.53}Ga_{0.47}As/In_{0.40}Ga_{0.60}As$ strained layer shallow superlattices as a function of bias voltage. Pronounced excitonic features were observed, and clear evidence for the formation of Stark ladders was found for the first time in this material system. The absorption edge is close to 1.55μm.

1. INTRODUCTION

The Wannier-Stark effect (WSE) in superlattices (SL) (Bleuse et al. 1988) induces fast electrooptic modulation at low electric fields and is therefore well suited for a variety of optoelectronic devices. Goossen et al. (1991) observed the WSE in extremely shallow GaAs/AlGaAs SL. Compared to conventional short period SL shallow SL offer several advantages: The coupling between the wells is strong even in case of comparatively long periods resulting in a larger potential drop per period. A rapid sweepout of photoexcited carriers enables modulation at high optical intensities. The InGaAs / InGaAs system on InP offers the additional freedom of strain variation and is compatible to fiber transmission at λ=1.55μm (Osbourn 1983).

2. EXPERIMENTAL RESULTS AND DISCUSSION

MOVPE-grown shallow InGaAs/InGaAs SL were materialized by improvement of the layer quality. The growth is facilitated because switching of group V constituents is avoided. Imperfections due to interdiffusion and memory effects of group V material, which are a severe problem in most InP based systems, are therefore reduced.

Photocurrent spectra were taken from PIN diodes containing a 10-period SL with $In_{0.53}Ga_{0.47}As$ wells and $In_{0.40}Ga_{0.60}As$ barriers in their I-regions. Fig. 1 exemplarily shows such spectra for the case of 8nm wells and 7nm barriers. They clearly exhibit the evolution of numerous transitions. The excitonic peak (at 0.84eV in fig. 1) from the spatially direct e1hh1 0-transition first becomes more pronounced with increasing fields due to electronic localization. At larger fields the quantum confined Stark effect (QCSE) weakens the excitonic peak again. The transition energies, which have been read out of fig. 1, are plotted in fig. 2 together with simulation data from transfer matrix calculations. For the type-I-SL of electrons and heavy holes two Stark ladders (e1hh1 0, ±1 and e1hh2 0, +1, +2) based on the first and second heavy hole subband can be clearly identified. In the case of the type-II-

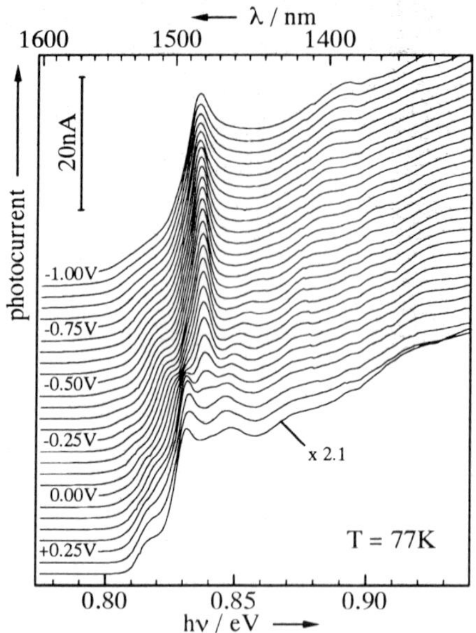

Fig. 1. 77K photocurrent spectra at different bias voltages. The spectra are shifted by an amount proportional to the applied bias. (8nm wells and 7nm barriers)

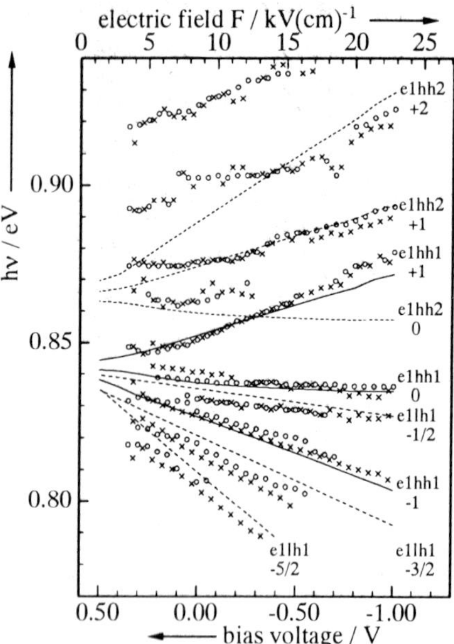

Fig. 2. Fan diagram. Transitions obtained from spectra by visual inspection (circles) or by digital data processing (crosses). Lines represent calculated transition energies.

SL of electrons and light holes the Stark ladder (e1lh1 -1/2, -3/2, -5/2) is less pronounced. In fact, the light holes are actually bound to the total SL rather than to the $In_{0.40}Ga_{0.60}As$ layers. However, calculations of the overlap of wavefunctions show some enhancement of transitions with the symmetry of a type-II-Stark-ladder. The identification of light and heavy hole related Stark ladders is based not only on their slopes, but also on polarisation-dependent photocurrent measurements.

3. CONCLUSION

Since experimental data matches well with theory, there is clear evidence for the formation of light and heavy hole Stark ladders (i. e. the WSE) in strained $In_{0.53}Ga_{0.47}As/In_{0.40}Ga_{0.60}As$ SL on InP. They exhibit strong excitonic features and the operation wavelengths are close to $\lambda=1.55\mu m$. In the light of these results the strained $In_{1-x}Ga_xAs/In_{1-y}Ga_yAs$ on InP appears to be a promising material for electrooptic modulators in fiberoptic transmission systems.

4. REFERENCES

Bleuse J, Bastard G and Voisin P 1988 Phys. Rev. Lett. 60 220
Goossen K W, Cunningham J E and Jan W Y 1991 Appl. Phys. Lett. 59 3622
Osbourn G C 1983 Phys. Rev. B 27 5126

The work was partially supported by Deutsche Forschungsgemeinschaft.
*permanent address: Slovak Technical University, Ilkovičova 3, 81219 Bratislava, Slovakia

Ultrafast optical nonlinearity in low-temperature grown InGaAs/InAlAs superlattices on InP and its application to MSM–PDs in the 1·55 μm wavelength region

R. Takahashi, Y. Kawamura, T. Kagawa, and H. Iwamura

NTT Opto-electronics Laboratories
3-1 Morinosato Wakamiya, Atsugi-shi, Kanagawa, 243-01 Japan

Semiconductor materials, especially those with multiple quantum well (MQW) structures, have extremely high optical nonlinearities, but have long recombination carrier lifetime of nanosecond region, making it difficult to use them for ultrafast optical devices. To overcome this slow photoresponse, several approaches have been tried, including such structural approaches as using tunneling bi-quantum well and staggered type-II quantum well structures, and approaches introducing non-radiative recombination centers by low-temperature (LT) growth (S.Gupta et al. 1992) and ion-implantation. Most of these reports concentrate on GaAs, and there have been no reports of materials with subpicosecond photoresponses in the 1.55 μm wavelength region.

We report, for the first time, subpicosecond recovery time of saturable absorption in Be-doped InGaAs/InAlAs MQWs on InP substrates grown by gas-source MBE at low substrate temperatures, and ultrafast electrical pulse generation of less than 3 ps by MSM-PDs in the 1.55 μm wavelength region. Our samples consisted of 100 periods of 9nm InGaAs wells and 7nm InAlAs barriers. The absorption spectra of the non-doped and Be-doped samples exhibited clear excitonic features, although broadening and decreasing of the exitonic absorption are observed with decreasing growth temperature and Be-doping (Fig.1).

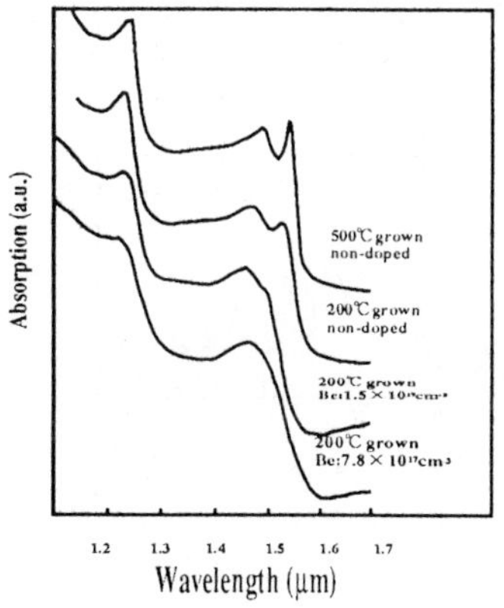

Fig. 1 Absorption Spectra

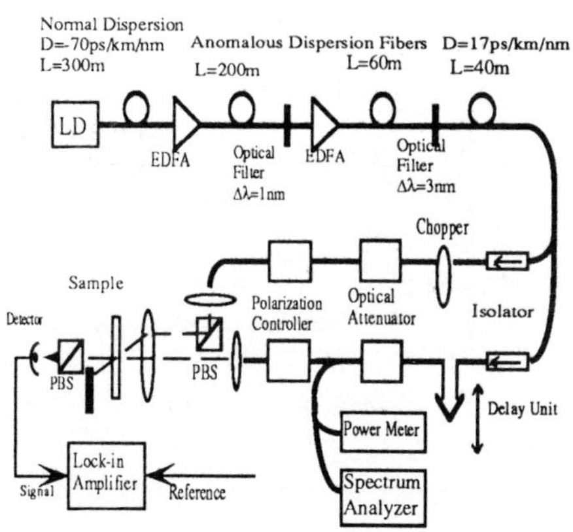

Fig. 2 Experimental set-up, pump-probe method

The carrier lifetime was investigated by a time-resolved pump-probe method (Fig.2). The semiconductor-laser-based optical source, consisting of a gain-switched DFB-LD and optical pulse compressors, generates extremely short optical pulses with 540 fs pulsewidth, 1.535 μm wavelength, and 100 MHz repetition rate. While the carrier lifetime can certainly be shortened even more by using LT-growth, it is only 160 ps in the 200℃-grown samples. Even in the samples grown at 500℃, Be-doping also shortens the carrier lifetime to 240 ps at a dose of $7.8\times10^{17}cm^{-3}$. On the contrary, the combination of LT-growth and Be-doping can successfully improve the photoresponse speed (Fig.3). As the Be-dose increases, the recovery time decreases and reaches the subpicosecond region. It is noted that, unlike GaAs, the sheet resistance of the LT-grown non-doped samples is very low even after annealing (at 500℃ for 60 min), meanwhile that of the 200℃-grown Be-doped sample ($7.8\times10^{17}cm^{-3}$) is reasonably high (350 kΩ). In addition to that, the electron Hall mobility is as high as 1000 V/cm^2sec.

In an application of this material, we investigated the MSM-PDs by using electro-optic sampling. The excitonic absorption peak of the MSM-PD was tuned to 1.54 μm by adjusting the well thickness to 10 nm. The MSM-PD with 5 mm gap interdigital electrodes generated the fastest electrical pulses of 2.7 ps pulsewidth (Fig.4).

In conclusion, the absorption recovery time in Be-doped LT-grown InGaAs/InAlAs MQWs was found to be extremely fast, that is, in the subpicosecond region, and the annealed samples exhibited high resistivity, high mobility, and exitonic features. Furthermore, an ultrafast MSM-PD in the 1.55 μm wavelength region was demonstrated using the material. In addition to MSM-PDs, the LT-grown and Be-doped MQWs have possible application in ultrafast optical devices that can operate above the 100 GHz bandwidth in the 1.55 μm wavelength region with very low energy.

[Reference] S. Gupta et al. IEEE J. Quantum Electron., vol. 28, p.2464, 1992.

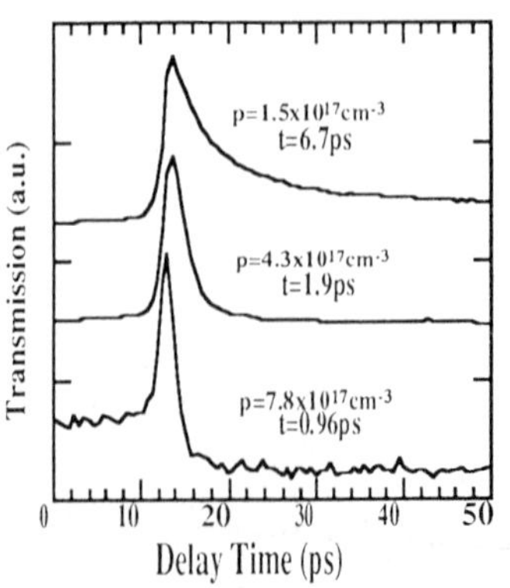

Fig. 3 Absorption recovery time
The samples grown at 200℃ and Be-doped

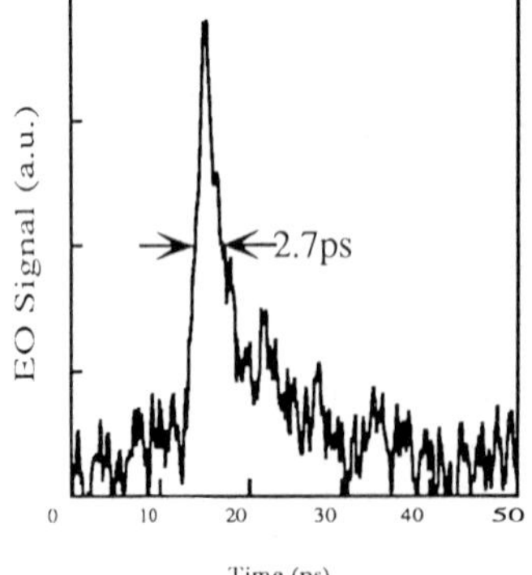

Fig. 4 Measured temporal response of MSM-PD to 1.535 μm wavelength

Paper presented at the Int. Symp. GaAs and Related Compounds, Freiburg, 1993

Photoluminescence studies of AlGaAs/GaAs single quantum wells grown on GaAs substrates cleaned by electron cyclotron resonance (ECR) hydrogen plasma

N. Kondo, Y. Nanishi, and M. Fujimoto

NTT Opto-electronics Laboratories, 3-1 Morinosato Wakamiya, Atsugi-Shi, Kanagawa Pref., 243-01 Japan

ABSTRACT: AlGaAs/GaAs single quantum wells are grown on GaAs substrates without thick buffer layers by using hydrogen ECR plasma cleaning. This process can be performed at a temperature as low as 500 °C, followed by MBE growth without breaking the vacuum. Clear contrast to conventional thermal cleaning, photoluminescence shows intense and narrow spectra. A flat and clean surface obtained by plasma cleaning improves the quality of grown layers. This technique is promising for device fabrication processes in which active layers are required to be grown closely on air-exposed surfaces.

1. INTRODUCTION

There are emerging interests in regrowth processes on patterned and/or masked substrates for quantum well, wire and box structure devices. Since these patterns should be finely structured, it is necessary to obtain high-quality regrown layers without thick buffer layers prior to second structure growth. These buffer layers, however, have been unavoidable in layers regrown on air-exposed surfaces by conventional thermal cleaning because of surface roughness and high impurity accumulations. There have been reports on *in situ* etch-regrowth for AlAs/AlGaAs structures (Choquette *et al.* 1992) and AlGaAs/GaAs structures (Kizuki *et al.* 1993). There was no mention, however, of regrown layers used as active layers.
Electron cyclotron resonance (ECR) hydrogen plasma has been effective in reducing impurity accumulations on AlGaAs surfaces at low temperatures. (Kondo *et al.* 1993) This paper describes the use of ECR plasma cleaning in growing device-quality layers on GaAs substrates without a buffer layer.

2. EXPERIMENTAL

ECR plasma cleaning was performed at 500 °C for 90 minutes with a hydrogen flow rate and microwave power of 40 sccm and 100 W. The wafers were then transported to a MBE growth chamber through a UHV tunnel. Single quantum well (SQW) structures were grown at 650 °C, both with and without a GaAs buffer layer (600 nm). The well widths for the GaAs layer sandwiched between two AlGaAs barrier layers (x=0.35, 17 nm) from bottom to top in the grown layer were 4.3, 2.2, and 1.1 nm, as shown in Fig. 1. "Without the buffer" means that the first layer grown on substrates is the AlGaAs barrier. The same structures were grown for comparison by conventional thermal cleaning at 650 °C for 10 minutes. A photoluminescence (PL) spectrum was studied with an Ar laser excitation of 6 W/cm^2 at 10 K.

3. RESULTS AND DISCUSSION

Figures 2 (a) and (b) show PL spectra of SQWs for plasma cleaning and thermal cleaning, respectively. For thermal cleaning, the resulting PL intensity corresponding to the narrowest well is very weak, even with the GaAs buffer layer. Without the buffer layer, the peaks shift toward longer wavelengths and the spectrum broadens, suggesting that interface roughness and/or impurity accumulations degrade the PL spectra. In contrast, ECR hydrogen plasma cleaning results in peaks that can be clearly observed, even without the buffer layer. These peaks exhibit almost no shift in position. In all cases, the full widths at half maximum (FWHM) values are smaller than those attained by thermal cleaning. This can be attributed to the flat and clean surface obtained by plasma cleaning, which results in better layer growth. Atomic force microscopy (AFM) studies also confirm that surface roughness for plasma cleaning is decreased to about 1/7 as compared with that for thermal cleaning.

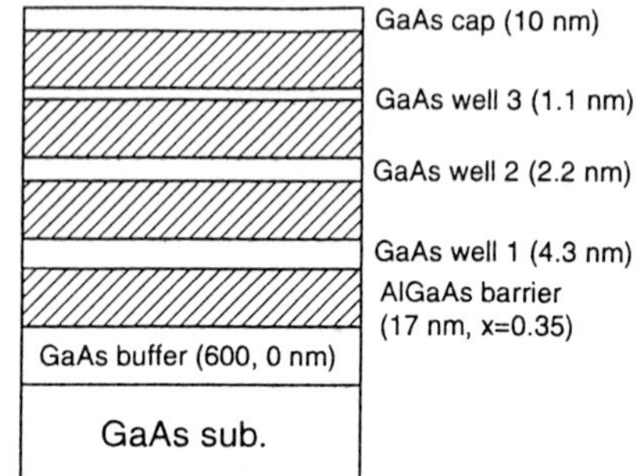

Fig. 1. A schematic diagram of the SQW structure.

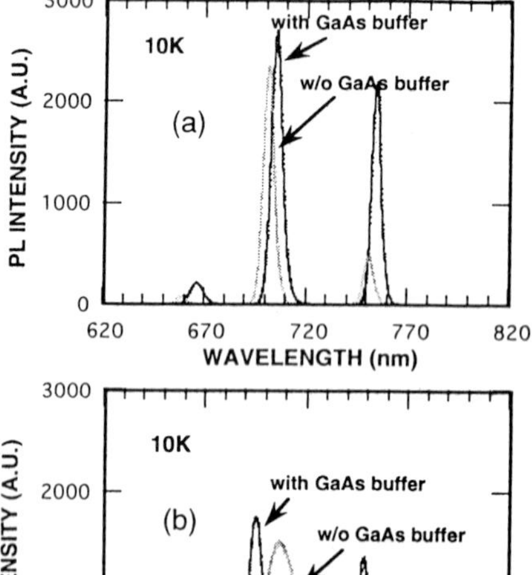

Fig. 2. 10 K PL spectra for AlGaAs/GaAs SQWs on GaAs substrates with and without a GaAs buffer layer prior to growth of SQW structures after (a) plasma cleaning and (b) thermal cleaning. The well widths of the SQWs are 1.1, 2.2, and 4.3 nm for the luminescent peaks, from left to right.

4. CONCLUSION

A flat and clean surface can be obtained by ECR hydrogen plasma cleaning, resulting in intense and narrow PL spectra for overgrown SQWs. Thus, this technique is a promising technology for regrowth processes. It can also be applied to device fabrication processes requiring a thin active layer to be grown closely on etched or air-exposed surfaces.

ACKNOWLEDGMENTS

The authors are grateful to Drs. Tomofumi Furuta and Mineharu Suzuki for their valuable discussions on PL and AFM analysis, respectively.

REFERENCES

Choquette K D, Hong M, Freund R S, Chu S N G, Mannaerts J P and Wetzel R C 1992 Appl. Phys. Lett. **60** 1738

Kizuki H, Hayafuji N, Fujii N, Kaneko N, Mizuguchi K, Murotani T and Mitsui S 1993 Inst. Phys. Conf. Ser. **129** (Bristol: IOP Publishing Ltd.) pp 603-608

Kondo N, Nanishi Y and Fujimoto M 1993 Inst. Phys. Conf. Ser. **129** (Bristol: IOP Publishing Ltd.) pp 585-590

Low voltage vertical-cavity, surface-emitting lasers (VCSELs) with low resistance C-doped GaAs/AlAs mirrors

R. Hey, A. Paraskevopoulos*, J. Sebastian**, B. Jenichen, M. Höricke, S. Westphal

Paul Drude Institut für Festkörperelektronik, Hausvogteiplatz 5-7, D-10117 Berlin.
* Heinrich Hertz Institut für Nachrichtentechnik GmbH, Berlin.
** Ferdinand Braun Institut für Höchstfrequenztechnik, Berlin.

ABSTRACT: Binary (GaAs/AlAs) Distributed Bragg Reflector (DBR) structures with very low resistivities ($1.7 \cdot 10^{-5}\ \Omega \cdot cm^2$) were grown by solid source Molecular Beam Epitaxy (MBE) using carbon as p-type dopant. VCSELs with C-doped mirrors were realized for the first time showing low threshold voltages (2.1 V). In addition, C-doping appears to stabilize the structure against relaxation by misfit dislocation formation.

INTRODUCTION

A low series resistance mainly of the p-type doped DBR is a prerequisite for the acceptable performance of a VCSEL. For a laser emission at λ = 980-1000 nm, the binary GaAs/AlAs combination provides the adequate choice, considering the difference in refractive index and the thermal conductivity. However, in this case, the potential barriers at the heterointerfaces, particularly in the p-type doped DBR, impede the current flow and lead to high series resistance.

EXPERIMENTAL RESULTS

Recently, we were able to show that very low resistivity values ($1 \cdot 10^{-5}\ \Omega \cdot cm^2$, 10 periods) are achievable in MBE-grown, Be-doped DBR structures (Paraskevopoulos 1993). For that, Beryllium outdiffusion from AlAs to GaAs had to be suppressed by choosing a lower growth temperature ($T_S \approx 400°C$). In this paper, we present equivalent resistance values with C-doped mirrors which were utilized, to our knowledge for the first time, in MBE-grown VCSEL structures. Carbon provides a promising dopant alternative because of its low diffusion coefficient. Contrary to Be-doping, the entire VCSEL structure was grown at a constant temperature (550°C).

The drastic reduction of the series resistance is mainly due to the implementation of δ-doped ($1 \cdot 10^{13}\ cm^{-2}$) $Ga_{0.5}Al_{0.5}As$ transition layers (4 nm thick) at the GaAs/AlAs interfaces. With this design no special adjustment of the growth rates or the doping level is needed, as it is the case in other types of graded layer structures (Lear 1993). The electrical characteristics of p-doped ($p = 4 \cdot 10^{18}\ cm^{-3}$), dry etched DBR mesas (∅10-70 μm, 20 periods) indicated (fig. 1) an exponential decrease of the resistivity with increasing current density, with best values approaching $1.5 \cdot 10^{-5}\ \Omega \cdot cm^2$. Index-guided VCSELs (∅20 μm) with C-doped mirrors showed threshold currents of 4-6 mA (cw operation) at a threshold voltage of 2.1 V (fig. 2), which is among the lowest values reported for VCSELs with binary mirrors.

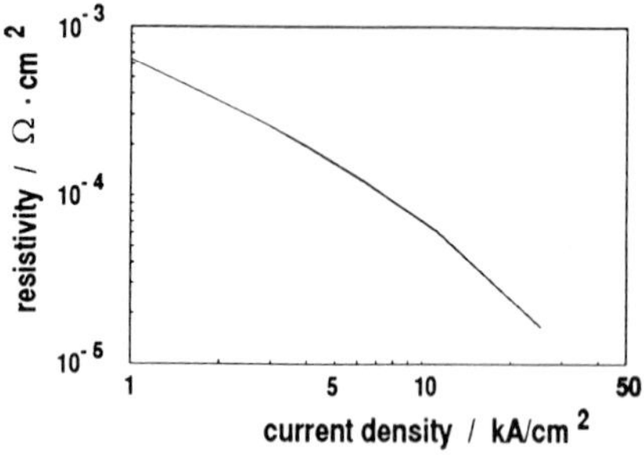

Fig. 1. Measured dependence of the resistivity of C-doped DBRs (20 periods) on the current density.

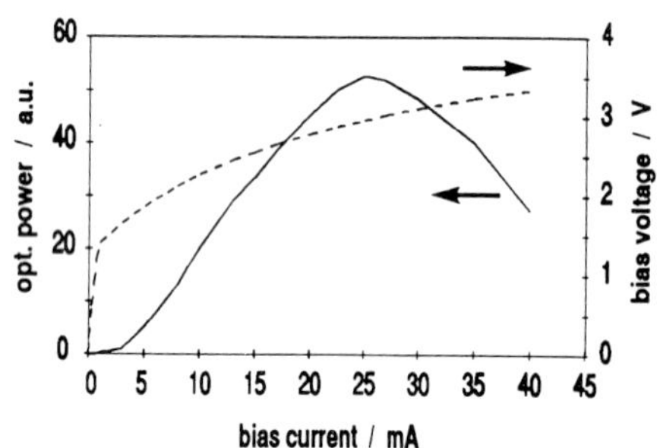

Fig. 2. L-I, V-I characteristics (cw-operation) of an index-guided VCSEL (∅20 μm) with C-doped DBR.

Depending on the amount of carbon incorporated on lattice sites, the lattice constants of GaAs and AlAs are reduced and, respectively, the critical thickness of the multilayer system is increased. As a result, relaxation due to misfit dislocation (MD) formation was not observed in C-doped 20 periods DBRs. On the contrary, numerous MDs were present in equivalent Be-doped mirrors, as the effect of lattice constant shrinkage is less pronounced. X-ray topography studies showed that the formation of MDs was also remarkably suppressed in VCSELs with C-doped mirrors. This demonstrates that C-doping not only reduces the series resistance of the VCSELs but also leads to an improved crystalline structure of these devices.

ACKNOWLEDGEMENT

This work was supported by the Bundesministerium für Forschung und Technologie, Bonn, within the Photonics program.

REFERENCES

Paraskevopoulos A et al. 1993, Proc. Conf. on Lasers and Electrooptics (CLEO '93), p80.

Lear K L et al. 1993, Proc. Conf. on Lasers and Electrooptics (CLEO '93), p80.

Paper presented at the Int. Symp. GaAs and Related Compounds, Freiburg, 1993

A single-wavelength all-optical GaAs/AlAs phase modulator: towards an optical transistor

G W Yoffe, J Brübach, F Karouta*, and J H Wolter

Eindhoven University of Technology, Physics Department, P.O. Box 513, 5600 MB Eindhoven, The Netherlands. *Electrical Engineering Department

ABSTRACT: We report a new single-wavelength all-optical transistor-like device, exploiting the polarisation-dependent absorption coefficient in quantum wells. The device is a GaAs/AlAs multiple-quantum-well hetero-*nipi* waveguide structure. A TE-polarised pump beam changes the refractive index through carrier-induced effects. A TM probe beam at the same wavelength, to which the material is transparent, undergoes a phase shift. We obtained a 180° phase shift in the probe beam with 4.2 mW of absorbed pump light. The effect was larger with a DC bias.

We present a new optical transistor-like device based on a GaAs/AlAs hetero-*nipi* waveguide. The phase of the transmitted light is controlled by a pump beam at the same wavelength which can be weaker than the transmitted beam. This effective gain means that the device could be used for all-optical signal processing. The device exploits the polarisation-dependent absorption coefficient in multiple-quantum-well (MQW) structures which arises from the different selection rules for electron-heavy-hole (e-hh) and electron-light-hole (e-lh) transitions (Weiner et al. 1985). TE light is absorbed by both transitions, whereas TM light is only absorbed by the e-lh transition. In MQW structures the e-lh transition is at shorter wavelength than the e-hh transition. There is therefore a wavelength range where TE light is absorbed while TM is not. TE light tuned near the e-hh exciton can be used to pump the structure, changing the refractive index. TM probe light at the same wavelength undergoes a phase shift, which can be converted to amplitude modulation by making an interferometer. "Hetero-*nipi*" structures (Döhler 1990), containing several undoped MQW regions separated by alternating *p*- and *n*-type doped layers, give very large optical non-linearities (Kost et al. 1988) so they are highly suitable. The main mechanism is electrorefraction through the quantum-confined Stark effect: photogenerated carriers are separated by the built-in field and have long lifetimes, producing a large photovoltage at a low pump intensity.

We have demonstrated the concept using a hetero-*nipi* waveguide, grown by molecular beam epitaxy, used earlier as an efficient electro-optic phase modulator (Yoffe et al. 1993). The waveguide core comprised two complete *n-i-p-i* periods, each *i*-region containing ten 75 Å GaAs quantum wells with 45 Å AlAs barriers. The interleaved doped layers comprised 500 Å of GaAs/AlAs superlattice, doped at $2 \times 10^{18}cm^{-3}$. The cladding layers were 1.5 μm of undoped $Al_{.45}Ga_{.55}As$. Waveguide ridges 8 μm wide were formed by wet etching. Selective *n*- and *p*-type contacts were made by etching a wider mesa through the *n-i-p-i* region and depositing AuSn and AuZn on the sloping sidewalls.

A waveguide 600 μm long was tested using light from a tunable CW Ti:sapphire laser. Phase shifts in the TM probe beam resulting from absorption of the TE pump beam (TE power 2 to 3 mW) were obtained for wavelengths from 828 to 838 nm with no electrical connections, as shown in Figure 1. Figure 2 shows the dependence on pump power at 832 nm. A 180° shift was obtained with 4.2 mW absorbed. The effect was larger with an applied reverse bias. In this case, the photocurrent causes a voltage drop across the contacts and doped layers, reducing the electric field. 2.8 mW of absorbed light caused a 180° phase shift with 0.5 V bias at 831 nm, 360° with 2 V bias at 836 nm.

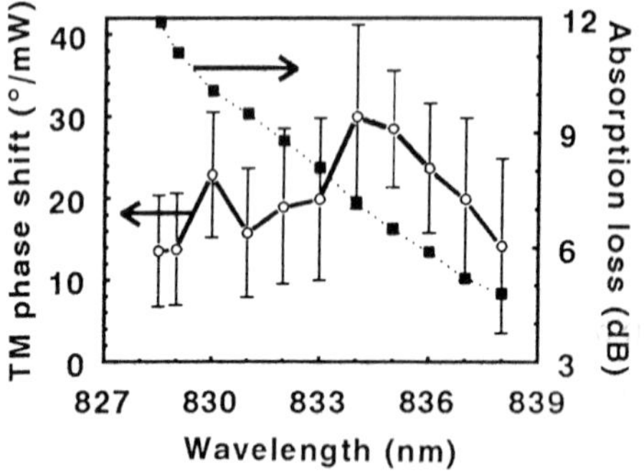

Figure 1. TM phase shift per mW of absorbed TE light, and TM absorption loss, as functions of wavelength.

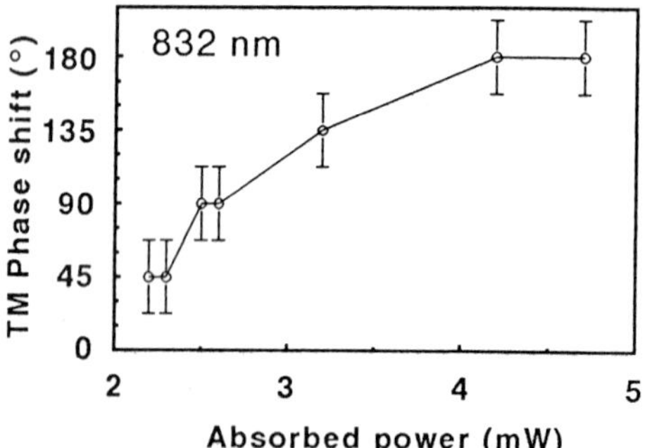

Figure 2. TM phase shift versus absorbed TE power at 832 nm.

It should be possible to reduce the required TE power. We (Yoffe et al. 1993) obtained a 180° phase shift electrically in 900 μm devices at less than 1 V applied bias. In GaAs *p-n* diodes (solar cells) and *n-i-p-i* structures without electrical connections (Döhler 1990), photovoltages of this order can be obtained with a pump intensity under 1 Wcm^{-2}. With waveguide geometry, an optimised device with thinner *i*-regions, giving a larger built-in field, should give a 180° phase shift with around 1 μW of absorbed TE light. With an applied bias, the sensitivity and speed are determined by the load resistance.

REFERENCES

Döhler G H 1990 Opt. and Quantum Electron. **22** S121
Kost A et al. 1988 Appl. Phys. Lett. **52** 637
Weiner J S et al. 1985 Appl. Phys. Lett. **47** 664
Yoffe G W et al. Appl. Phys. Lett. **63** to be published September 13 1993

Paper presented at the Int. Symp. GaAs and Related Compounds, Freiburg, 1993

GaAsSb grown by low pressure MOCVD using TEGa, tBAs and TMSb as precursors

Naozo Watanabe and Yasuo Iwamura

Department of Electrical Engineering, Kanagawa University
3-27-1, Rokkakubashi, Kanagawa-ku, Yokohama, 221 Japan

ABSTRACT: GaAsSb layer was grown by low pressure MOCVD at 420°C, using TEGa, tBAs and TMSb as precursors. Changing V/III ratio in the vapor from 12 to 5, a convex solid-vapor compositional relation (Sb content) changed and approached to a linear relation and the X-ray diffraction rocking curve of the layer in the immiscibility region becomes sharper.

We have recently reported InAsSb could be grown by low pressure MOCVD at 420°C using TMIn, tBAs and TMSb as precursors [1]. The relation of solid composition, x in $InAs_{1-x}Sb_x$, to that of vapor composition, y=[TMSb]/ ([TMSb]+[tBAs]), was different from the previously reported results obtained with arsine as an arsenic precursor. In addition to thermodynamics, kinetics may be necessary to explain the data.

In this paper we are reporting on GaAsSb growth using TEGa, tBAs and TMSb as precursors. Using TEGa, tBAs and TMSb as prcursors, GaAsSb was grown at a low temperature of 420°C by low pressure MOCVD at a reactor pressure of 1/10 atm. or 10,000 Pa. A reactor was horizontal with a cross section of 10x50 mm². A (100) oriented semi-insulating GaAs wafer was used as substrate throughout the experiment. Total hydrogen carrier gas flow rate was 480 sccm. A partial pressure of TEGa was 1 Pa calculated from the bubbler temperature and the carrier gas flow. V/III ratio in the reactor was 12 and 5. Under these growth conditions, the growth rate was about 0.4 μm/h and the layers were grown for 1.5 h.

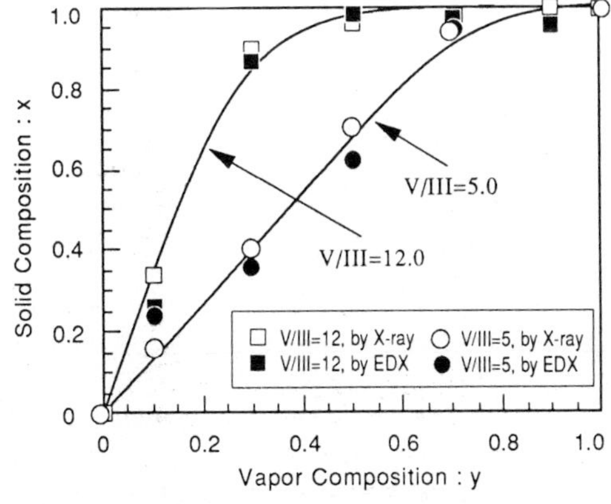

Fig. 1. Relation of the solid composition to the vapor composition.

Figure 1 is a relation of the solid

composition x in $GaAs_xSb_{1-x}$ to the vapor composition y = [tBAs]/([tBAs]+[TMSb]). The relation, when the V/III ratio is 12, is convex upward as it has been reported for the growth experiment with arsine[2]. The relation with V/III ratio of 5 approaches a linear relation. A concave upward relation may be expected with smaller V/III ratio.

Figure 2(a),(b) shows X-ray rocking curves of the grown layers. In Fig. 2(a), the curves of the layer grown with the V/III ratio of 12 were displayed. Strong double peak on the right end is due to the GaAs substrate. The curve with y=0, GaSb, is also a sharp double peak. When the composition goes into an immiscibility region, the layer with the corresponding composition is grown but the peak becomes lower and broader suggesting compositional inhomogeneity.

In Fig. 2(b), the curves for the layers grown with V/III of 5 were shown. Compared with Fig.2(a), the curves in the immiscibility region become slightly sharper. Smaller V/III ratio implies reduced surplus V species supply and the incorporation ratio of V species approaches to the supply ratio. It will also result in reduced composition diversity.

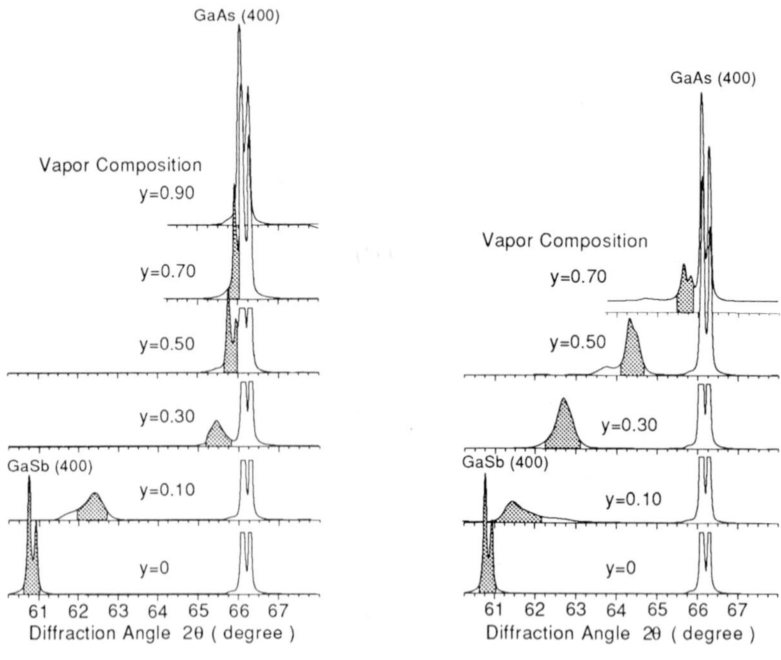

Fig. 2(a). Rocking curves for V/III=12.

Fig. 2(b). Rocking curves for V/III=5.

Supports of Nissin Electric Co. Ltd., Sony Corporation, Sumitomo Chemical Co. Ltd., Furukawa Co. Ltd. and Sumitomo Electric Industries Ltd. are acknowledged.

References

1) Y.Iwamura and Naozo Watanabe: Electronic Materials Conference 1993 (Santa Barbara) L8.

2) G.B.Stringfellow and M.J.Cherng: J.Cryst.Growth, 64 (1983) 413.

Pseudomorphic $Ga_{0.5}In_{0.5}P$ barriers grown by MOVPE for high drain breakdown and low leakage HFET on InP

W.Prost, C. Heedt, F. Scheffer A. Lindner, R. Reuter, F.-J. Tegude

Universität-GH-Duisburg, Sonderforschungsbereich SFB 254, Solid-State-Electronics Department, Kommandantenstr.60, D-47048 Duisburg, Germany

ABSTRACT: Highly strained $Ga_{0.5}In_{0.5}P$ spacer layer, is inserted in HFET on InP-substrate. The large band gap energy of $Ga_{0.5}In_{0.5}P$ results in improved gate leakage current, drain-conductance and drain-breakdown. At $V_{DS} = 10$ V a voltage gain of $V_u \geq 100$ is maintained for HFET with $L_g = 0.7$ µm. The output conductance obtained by $Ga_{0.5}In_{0.5}P$ spacer is all important for high microwave gain at high drain bias and high drain-current (cut-off frequency $f_{max} \geq 150$ GHz at $I_D = 500$ mA/mm, $V_{DS} = 5.5$ V).

InAlAs/InGaAs Heterostructure Field-Effect Transistor (HFET) suffer from low voltage gain, low turn-on drain-breakdown voltage and high gate leakage current. Besides more basic considerations like suitable doping concentration in the donor layer (Buchali et. al. 1992) and mesa side-wall leakage current (Bahl et. al. 1992) most work has been carried out on MBE grown samples (Bahl et. al. 1993, Heedt et. al. 1993). In this work we will show that highly strained $Ga_{0.5}In_{0.5}P$ spacer layers grown by MOVPE are very attractive due to the large band gap (1.9 eV at 300K) resulting in higher conduction- ($\Delta E_C = 0.77$ eV) and valence band ($\Delta E_V = 0.37$ eV) discontinuity (Tiwari and Frank, 1992). This provides a good 2DEG confinement and suppresses the contribution of holes to the gate leakage current (Buchali et al. 1992).

Growth started on s.i.(100) InP:Fe substrates with a 30 nm InAlAs buffer layer followed by a lattice matched 30 nm thick $In_{0.47}Ga_{0.53}As$ channel layer. On top a 2.5 nm thick highly strained $Ga_{0.5}In_{0.5}P$ spacer layer ($\Delta a/a = -3.6\%$) was grown. The switching sequence for the AsH_3/PH_3 exchange at the channel interface consists of three growth interruption steps with both group-V precursors are switched off during the second step. Next the InAlAs donor layer (20 nm, Si, $n = 1.3 \cdot 10^{18}$ cm^{-3}), an undoped InAlAs Schottky layer (20 nm) and a undoped InGaAs cap layer (10 nm) finished the strukture. Hall measurements exhibit electron mobility and sheet carrier density at T = 300 K (77K) of $\mu_{300K} = 11{,}300$ (47,400) cm^{-2}/Vs, and $n_{s,300K} = 2 \cdot 10^{12}$ ($2.3 \cdot 10^{12}$) cm^{-2} indicate the excellent quality of the MOVPE grown material. The HFET device technology is described elsewere (Heedt et. al. 1993).

The DC characteristic is plotted in fig. 1. A broad transconductance profile with $g_{m,max} = 360$ mS/mm is achieved and a remarkable breakdown voltage as high as $V_{DS,\,Br} = 8$ V at very high drain-current of $I_D = 400$ mA/mm can be observed.

Excess gate leakage current occurs for $V_{DS} \geq 5$ V where at negative gate bias leakage current is attributed to electron tunneling through the Schottky barrier and for $V_{GS} \geq -0.2$ V to band-band tunneling (cf. fig. 2a). Leakage due to holes reaching the gate-contact close to pinch-off bias is very effectively suppresed. In contrast to comparable devices with InAlAs spacer we found a decrease of output conductance with increasing gate bias (cf. fig. 2b).

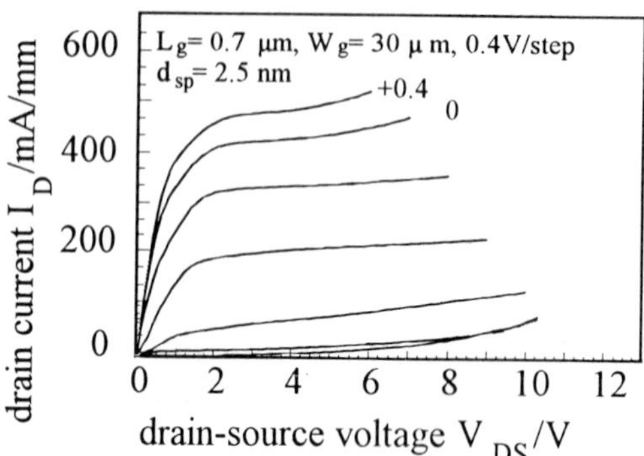

fig. 1: *Current voltage output characteristic of a HFET with $Ga_{0.5}In_{0.5}P$ spacer layer*

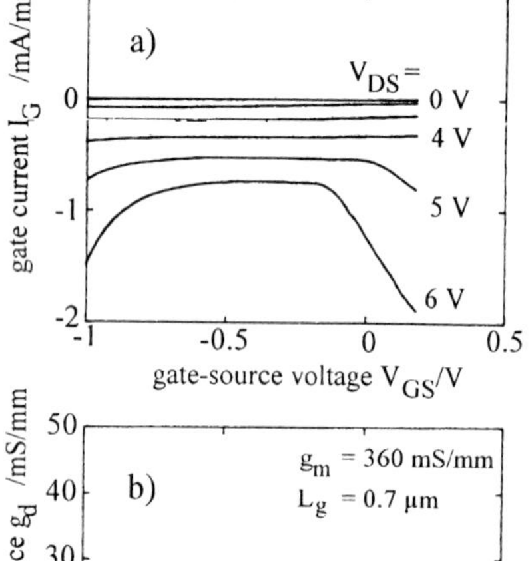

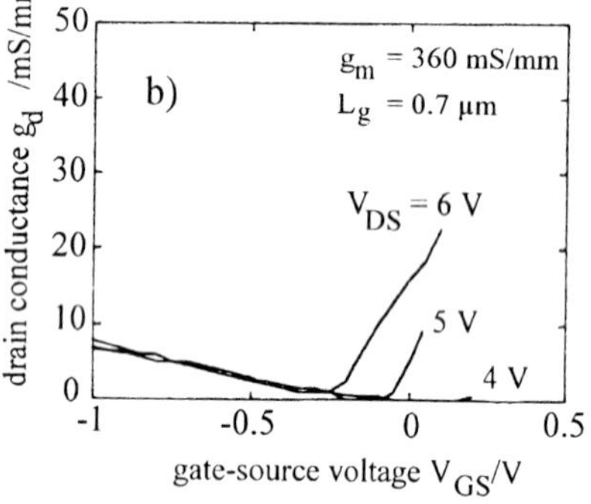

fig. 2: *DC-characteristics of HFET on InP-substrate with 2.5 nm $Ga_{0.5}In_{0.5}P$ spacer*

Even for devices with $Ga_{0.5}In_{0.5}P$ spacer a direct link between gate current and output conductance has been observed as can be seen in fig. 2a,b. The rapid increase of output conductance at forward bias (cf. fig.2b at $V_{DS} \geq 5$ V) is directly correlated to a negatively increasing gate leakage current. It is worth noting that there is no relationship between absolut values of gate leakage current and drain conductance. At $V_{DS} = 5.5$ V a cut-off frequency for maximum available gain of $f_C = 116$ GHz and for unilateral gain of $f_{max} \geq 150$ GHz is provided with a gate length of $L_g = 0.7$ µm due to the very low output conductance.

In summery we have introduced a highly strained $Ga_{0.5}In_{0.5}P$ spacer layer in a InAlAs/InGaAs HFET structure for the first time. The efficiency was proven by DC- as well as RF-measurements. The excellent breakdown behaviour ($V_{DS, Br} = 8$ V) at very high drain-current ($I_D = 400$ mA/mm) and the superior RF performance ($f_{max} \geq 150$ GHz, $V_{DS} = 5.5$ V) as well as the low gate leakage current give evidence of the advantages using strained $Ga_{0.5}In_{0.5}P$ spacer in HFET for high voltage applications.

References

Bahl S R, del Alamo J A 1992 IEEE Electron Device Letters, Vol. 13, No. 4, pp.195

Bahl S R, Bennett B R, del Alamo J A 1993 IEEE Electron Device Letters, Vol.. 14, No. 1, pp. 22

Buchali F, Heedt C, Prost W, Gyuro I, Meschede H, Tegude F J 1992 Microelectronic Engineering 19, pp.401-404

Heedt C, Buchali F, Prost W, Fritsche D, Nickel H, Tegude F J 1993 Inst. Phys. Conf. Ser. No.129 Chapter 12, pp. 941

Tiwari S, Frank D J 1992 Appl. Phys. Lett. 60 (5) pp. 630

Paper presented at the Int. Symp. GaAs and Related Compounds, Freiburg, 1993

GaAs layers grown on 100 mm diameter substrates in a liquid phase epitaxy centrifuge

M Konuma*, I Silier, E Czech, and E Bauser

Max-Planck-Institut für Festkörperforschung, Heisenbergstrasse 1, D-70569 Stuttgart, Germany

ABSTRACT: A liquid phase epitaxy centrifuge is used to grow GaAs epitaxial layers from gallium solution on 100 mm diameter GaAs substrates. Layer thicknesses in wide ranges from below 1 μm to more than 60 μm are obtained. Thickness uniformity is better than $\pm$ 15 % except for the peripheral 10 mm of the 100 mm diameter substrates. Hall effect measurements reveal carrier concentrations and mobilities typical of LPE grown GaAs.

1. INTRODUCTION

Liquid phase epitaxial (LPE) layers of GaAs and other III-V semiconductors serve well for optoelectronics, high speed electronics, and detectors for x-rays and particles. We used to grow GaAs and GaAlAs epitaxial layers on 2x2 cm^2 planar, non-planar, and patterned substrates (Bauser 1987). In order to increase productivity, assure reliability, and reduce costs, the size of the wafers must be increased. Wafers of 100 mm in diameter (so-called 4") are already used for industrial production of GaAs LSI. So far, attempts to grow GaAs LPE layers on 100 mm diameter substrates are, however, scarce. We have successfully grown Si and SiGe epitaxial layers on 100 mm diameter Si wafers in an LPE centrifuge (Konuma et al 1993, Hansson et al). In this paper we describe the LPE centrifugal technique for growing GaAs epitaxial layers on 100 mm GaAs substrates.

2. EXPERIMENTAL

The liquid phase epitaxy centrifuge for 100 mm diameter substrates is described elsewhere in detail (Konuma et al 1993). It consists of a magnetically suspended rotor and a graphite crucible attached to the lower end of the rotor shaft. The crucible has vertically stacked four chambers which are connected by channels as is schematically shown in Fig.1. Owing to centrifugal forces and gravity the solutions move from one chamber to the next.

Gallium arsenide epitaxial layers are grown on (100) GaAs substrates whose off-orientations is below 0.05°. The gallium solvent (7N) is saturated with semi-insulating

* On leave from Universität Erlangen-Nürnberg, Institut für Werkstoffwissenschaften, Cauerstr. 6, D-91058 Erlangen, Germany.

GaAs at 750 °C. The layers are grown within different temperature intervals in the temperature range between 750 °C and 200 °C, and at constant cooling rates of between 15 and 6 K/h. Growth is performed in hydrogen atmosphere at normal pressure.

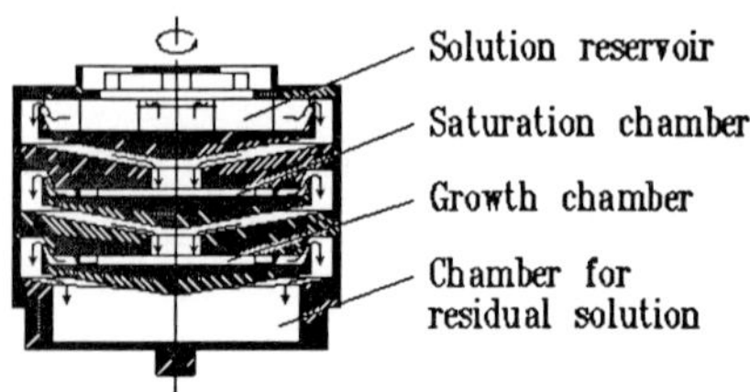

Fig.1 Graphite crucible for 100 mm diameter substrates. Longitudinal section.

3. RESULTS AND DISCUSSIONS

The tipping boats have been shown to be suitable for the growth of high purity GaAs layers on 4 cm^2 substrates. The electron mobility at 77 K in such layers had values up to $2.2x10^4$ $cm^2V^{-1}s^{-1}$. Carrier concentrations were as low as $2.7x10^{12}$ cm^{-3} (Silier et al). By using mesa patterned substrates extremely flat surfaces of GaAs layers have been obtained (Weishart et al).

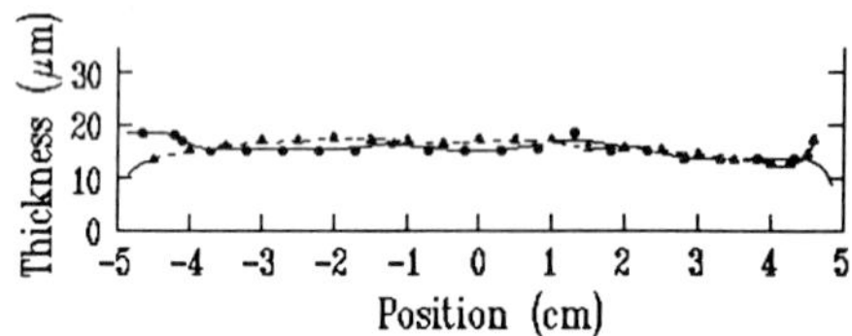

Fig.2 Thickness uniformity of the GaAs LPE layers grown on 100 mm diameter (100) GaAs substrates.

Using similar experimental conditions we grow GaAs layers on 100 mm GaAs substrates. The surfaces of these epitaxial layers show no meniscus lines in the center region of the substrates. We obtain layer thicknesses in wide ranges from below 1 μm to more then 60 μm. The thickness uniformity of about 16 μm thick layers is shown in Fig. 2. The layers are grown in the temperature interval between 750 to 730 °C and at a constant cooling rate of 6 K/h. Thickness uniformity is better than $\pm$ 15 % except for the peripheral 10 mm of the 100 mm diameter substrates. Reproducibility of the layer thickness in growth run to run is better than several %. Since the layers grow in the step flow mode in our experiments, surface morphology and layer thickness are strongly affected by dislocations in the substrates. The substrates used have dislocation densities of $< 10^5$ cm^{-2}. The edge growth may be caused by dislocations. However, by employing mesa structures in size of sub-square mm, extremely flat layer surface can be obtained. Hall effect measurements reveal carrier concentration and mobilities typical of LPE grown GaAs.

REFERENCES

Bauser E 1987 Thin Film Growth Techniques for Low Dimensional Structures ed Farrow R F C, Parkin S S P, Dobson P J, Neave J H, and Arnott A S (New York: Plenum)pp 171-194

Hansson P O, Czech E, Konuma M, and Bauser E to be published

Konuma M, Czech E, Silier I, and Bauser E 1993 Appl.Phys.Lett. **63** 205

Silier I, Subramanian S, and Bauser E to be published

Weishart H, Bauser E, and Konuma M to be published

Paper presented at the Int. Symp. GaAs and Related Compounds, Freiburg, 1993

Hot electron tunnelling photodetector

R.Redhammer, J.Kováč and Š.Németh

Dept. of Microelectronics, Faculty of Electrical Engineering, Slovak Technical University, SK-81219 Bratislava, Slovak Republic

ABSTRACT: A novel hot electron tunnelling photodetector based on the modulation of the height of a shielding potential barrier using the charge retaining effect is presented. The device concept and the first results of DC characterization are shown.

1. DEVICE CONCEPT

Both the charge retaining effect (Redhammer and Allsopp 1992) and effective mass filtering (Capasso et al 1985) were used for optical modulation of the current of hot electrones. The shielding AlGaAs barrier layer is placed asymmetrically in a GaAs intrinsic region between two n^+ GaAs contacts in a certain distance from the emitter. Under the applied voltage hot ballistic electrons are emitted from the emitter contact through an emitter spacer and they tunnel through the shielding barrier. Under illumination, the light generates electron - hole pairs mainly in the wider collector spacer and the electric field separates them. Electrons increase the total electron collector current, while holes, moving in opposite direction, are retained by the valence band edge potential barrier. The hole accumulation changes the potential profile across the structure and reduces the height of the potential barrier against the energy of hot tunnelling

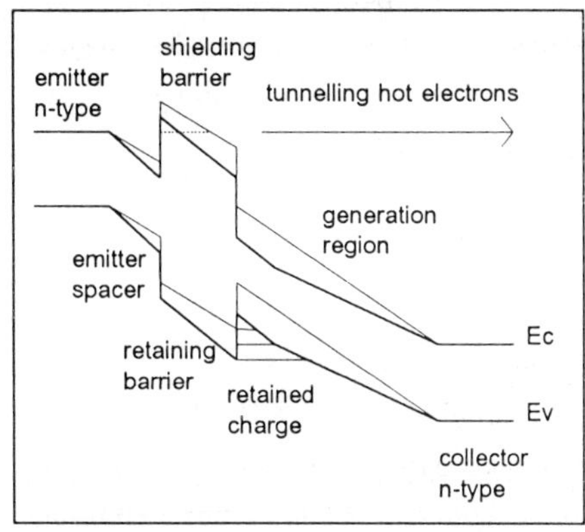

Fig. 1. Schematic band gap diagram of a hot electron tunnelling photodetector.

electrons. In this way the accumulation of retained hole charge modulates the tunnelling emitter current and amplifying is obtained. The electron charge retaining effect at the conduction band edge potential barrier (Redhammer and Allsopp 1992) is significantly smaller than the hole charge retaining effect due to different effective masses of holes and electrons.

2. EXPERIMENTAL RESULTS

A sample grown by an MBE has the potential barrier of AlGaAs 15 nm wide and the emitter intrinsic spacer 70 nm wide. The collector generation region is 1 μm wide. The I-V characteristics for different incident optical power intensities is plotted in Fig. 2. The light with the wavelenght of 850 nm and with the maximum optical power of 9 μW was used. The responsivity of this structure was up to 4 A/W, which corresponds to photogain of up to 10. Further structures, with two 5 nm thick shielding barriers separated by 5 nm GaAs layer, with low doped contact regions and with shorter emitter and collector regions (2 nm and 20 nm for structure B1 and 10 nm and 200 nm for structure B2), show the responsivity of up to 20 A/W, which corresponds to the photogain of approximately 50.

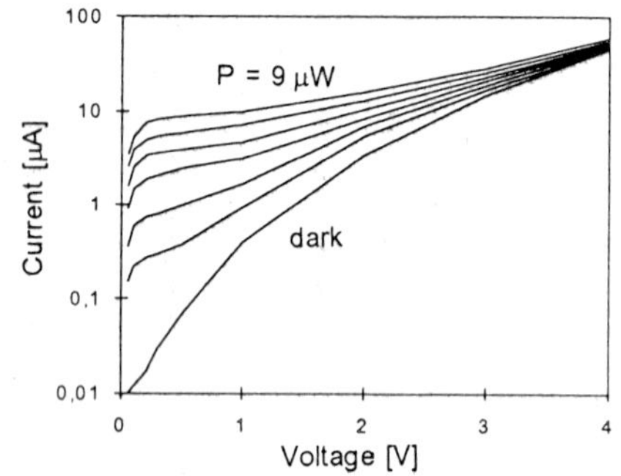

Fig. 2. I-V characteristics of the sample A for various incident optical power.

3. CONCLUSION

The first achieved results confirm that the charge retained by the potential barrier may have a significant influence on the tunnelling current and that it can be used for practical application. We believe that this type of structure can find a wide application, after some optimization, as a relatively fast amplifying photodetector.

REFERENCES

Redhammer R and Allsopp D W E 1992 *Proc. of ESSDERC '92*, Leuven Belgium pp 899-902; *Microelectronics Engineering* **19** 899.

Capasso F, Mohammed K, Cho A Y and Hull R 1985 *Appl. Phys. Lett.* **47** 420.

Keyword Index

Author Index